APES AND HUMAN EVOLUTION

Apes and Human Evolution

RUSSELL H. TUTTLE

Harvard University Press

Cambridge, Massachusetts, and London, England   2014

Library of Congress Cataloging-in-Publication Data

Tuttle, Russell H., 1939–
  Apes and human evolution / Russell H. Tuttle.
    pages   cm
  Includes bibliographical references and index.
  ISBN 978-0-674-07316-6 (alk. paper)
  1. Primates—Evolution.   2. Fossil hominids.   3. Human evolution.   I. Title.
  QL737.P9T86 2014

  599.93'8—dc23   2013014310

To two beautiful, brave women who have graced my life,
Nicole Irene Tuttle and Marlene Benjamin Tuttle,
and
my main man, Matt

Contents

1 Mongrel Models and Seductive Scenarios of
Human Evolution   **1**

*Theories of human evolution have been biased by folk beliefs about the meaning of
individual, sexual, and group differences in appearance and customs, and by world
events, like warfare, and personal experiences of theorists. As apes became subjects of
detailed study their behavior served variously to reinforce or refute notions of close
similarity between them and us.*

Part I: Terminology, Morphology, Genes, and Lots of Fossils

2 Apes in Space   **15**

*Although the old argument about which nonhuman primates are closer to humans
has settled on the apes, many puzzles remain regarding the extent to which we can
draw on them as models for specific aspects of our variable genomes, morphology, and
behavior. This chapter contains the vocabulary and features that are essential to
explore our phylogenic position vis-à-vis living apes.*

3 Apes in Time   **60**

*Rare major and many minor fossil discoveries underpin phylogenic models of primate
evolution over a ≥65–Ma span of geologic time. A variety of refined geochemical
and excavation methods allow ever more precise placement of specimens in time;
however, small samples of fragmentary specimens and very patchy spatiotemporal
representation usually limit their informational value.*

4 Taproot and Branches of Our Family Tree   **126**

*The past century of field research has produced a trove of fossil specimens indicating
that our Linnaean family, the Hominidae, contained a notable number and variety
of species that is difficult to organize into phylogenic lineages, only one of which
terminated in modern Homo sapiens.*

Part II: Positional and Subsistence Behaviors

5 Apes in Motion   189

Apes display highly diverse repertoires of locomotive behaviors that inspire models on the evolution of human anatomy. Chimpanzees, bonobos, and most gorillas engage in vertical climbing to move from ground to canopy, and orangutans and gibbons frequently climb up and down trunks and vines. Gorillas, chimpanzees, and bonobos knuckle-walk on the ground and on large horizontal branches. Gibbons, orangutans, and less frequently, chimpanzees and bonobos arm-swing (brachiate) beneath branches.

6 Several Ways to Achieve Erection   225

Modelers of the evolution of human obligate erect bipedal posture employed living apes to exemplify precedent behavior and anatomy upon which various selective forces acted to produce the human form, viz., brachiators and knuckle-walkers, closely resembling chimpanzees. Instead, a small-bodied vertical climbing, bipedal branch-running ape is a more likely model because similar lower limb mechanics are involved in vertical climbing and human bipedal walking.

7 Hungry and Sleepy Apes   261

The daily food quest and need for secure lodge trees affect the ranging patterns of all ape species. They prefer fruits; however, during fruit shortages they are remarkably versatile in accommodating to a wide variety of vegetal and animal foods. Great apes build arboreal nests at night. Gorillas, chimpanzees, and bonobos sometimes also rest in ground nests during the day. Like Old World monkeys, gibbons roost in trees on ischial callosities (sitting pads).

8 Hunting Apes and Mutualism   304

All apes ingest invertebrates inadvertently or via active searches. Chimpanzees and, to a lesser extent, bonobos are notable for capturing small and medium-sized mammals. Chimpanzees also engage in infanticidal cannibalism. The extent to which chimpanzee hunting is cooperative is arguable, as is the relation between food sharing and mating behavior in chimpanzees and bonobos.

Part III: Hands, Tools, Brains, and Cognition

9 Handy Apes   331

Unlike human hands, which are free of obligate locomotive functions, ape hands are evolutionary compromises serving various locomotive activities, grasping, and fine manipulation, including tool making and use. Their hands are highly sensitive, which allows them to feel fruits for ripeness and to engage in social grooming. Fossil hominid hands before 1.5 Ma retain features that would facilitate arboreal climbing.

10 Mental Apes   355

Scientists employ a diverse array of approaches and devices to explore the structure and functions of outsized human brains in comparison with those of apes. There is general consensus that apes possess self-awareness and think about situations and actions, albeit not comparable to or necessarily on the same basis as those underpinning human cognitive abilities.

Part IV: Sociality and Communication

11 Social, Antisocial, and Sexual Apes   397

Apes are a rich resource for theorists who model emergent hominid and early human social structures. Gibbons are attractive to those who believe that human monogamy is deeply rooted in the past. Orangutans are usually ignored because adults do not form stable groups. Gorillas and especially chimpanzees are favorites of persons who emphasize male dominance and aggression, while bonobos are preferred by modelers who view our apish ancestors living in quasi-egalitarian societies with notable female agency.

12 Communicative Apes   507

Laboratory studies revealed that great apes can learn to communicate with humans and one another via sign language and non-iconic symbols. Their natural vocalizations offer fewer clues to a system of communication that might have been used by early hominids. There is notable disagreement about when fully human language emerged and the sequence of events that culminated in it.

Part V: What Makes Us Human?

13 Language, Culture, Ideology, Spirituality, and Morality   569

Humans live in a symbolic niche: virtually everything we say, do, create, and make is consciously or unconsciously dependent upon symbols. Although many other animals probably think about proximate situations, humans have beliefs about phenomena and relationships. Humans have social and moral codes, while apes are probably amoral. Their survival and perhaps that of our species ultimately depends upon us.

Notes   603

References   691

Illustration Credits   1017

Index   1025

Note: Color Illustrations follow page 384.

Preface

In various respects anthropology has developed along paths leading into isolation. It no longer should hesitate to stake clear claims for all of primatology, now that the comforting old gap between the morphology of man and ape has practically disappeared with the latest fossil finds and that comparative psychology threatens our last and feeble definition of separation by which a science of man is supposed to begin only with creatures who can use *and* make implements.

ADOLPH H. SCHULTZ (1955. p. 56)

Introductory courses . . . are never introductions. They are always surveys, [in which] . . . nothing is allowed to be truly troublesome.

JONATHAN Z. SMITH (1990, p. 62)

Apes are increasingly part of popular culture in the Westernized world, Japan and other nations exposed to globalizing mass media, even as they and their habitats are rapidly vanishing from Earth. Although we have learned much about apes and ourselves since the millennium of Western exploration and expansion, many puzzles remain unresolved, and new ones undoubtedly are yet to be encountered, particularly regarding our evolutionary histories and common versus unique perceptual, cognitive, and behavioral traits.

Are we, along with bonobos, chimpanzees, gibbons, gorillas, and orangutans, apes? If so, is there any merit to a claim that we are the greatest apes? Are bonobos, chimpanzees, gibbons, gorillas, and orangutans sufficiently humanoid to merit any, many, or fully equal human rights? Are apes so special that they deserve privileged status among the organisms on Earth, particularly in relation to the goals of conservationists and the depredations of hunters, medical researchers, and other consumers? These are troublesome issues indeed, for which it would be hubristic to pretend to have definitive answers (Tuttle 2006a, b).

Nonetheless, questions of human evolution and our place among organisms are excellent challenges to one's ability to think critically and to assist students who wish to learn how to think critically instead of merely being titillated and told what to think. Accordingly, I hope, via this exploration of findings and ideas on the evolutionary biology and behavior of apes and humans, to provide a substantial factual base on topics for which there are actual data and to challenge interpretations of them and of other ideas for which there is little or no tangible evidence. My chief goal is to provoke thought and discussion in classes and among other readers who might want to explore aspects of the human career and condition and how knowledge of living apes and other animals might infuse insight into the exercise. I care much less to persuade others of my particular views, some of which have not been mainstream.

The volume is arranged in five parts, with the understanding that instructors and other readers might use it to inform specific topics. Nonetheless, I would hope that, regardless of specialty, readers will initially turn to Chapters 1, 2, and 13. One reviewer reasonably requested that I highlight what I believe to be the major contributions to primatology over the past fifty years, but I leave this to you, the readers, especially when you encounter inflated claims of specialists in the several subfields of primate evolutionary biology and human evolution in particular, and pundits who rely on their narratives.

I thank the many undergraduates, former graduate students, and staffers with whom I have been privileged to interact at the University of Chicago over the past forty-six years, and Andrew Rodriguez, Ian Kalman, Paige Davis, John Jewell, Kyle Wagner, Evan Scott, Timothy Murphy, Bart Longacre, Joshua Letzer, Zaid Alawi, Nelson Balbarin, Sandra Hagen, Fione Dukes, and Brandon Daniel for help in retrieving references and for technical assistance during the several decades over which the project gestated. Special thanks are due to Duane M. Rumbaugh, Karen B. Strier, Constantine Nakassis, Philip Blackwell, and Ray Jackendoff for helpful expert comments and criticisms on parts of the manuscript or its entirety. The Anthropology Department's Lichtstern Committee and Chair Judith Farquhar, along with Christa Modsceidler, Jiaxun Benjamin Wu, and other professional bibliographers and assistants in the Crerar and Regenstein Libraries, also contributed greatly to this project. The Marion and Adolph Lichtstern Fund, Department of Anthropology, the University of Chicago provided much-needed financial assistance. Photographs of fossils by Russell H. Tuttle are from a research trip to Eurasia and Africa in 1985 and 1986 funded by a Guggenheim Fellowship. Others are from research trips to Yerkes Primate Research Center in Orange Park, Florida, and Atlanta, Georgia; Osaka, Japan; and Peruvian Amazonia funded by the National Science Foundation, the National Institutes of Health, the Japan Society for the Promotion of Science, and the Marion and Adolph Lichtstern Fund.

The project was greatly enhanced by the artistic talents, computer skills, diplomacy, and professionalism of Alba Tomasulo y Garcia. Indeed, the book probably would not exist in its present form without her.

Others who have assisted generously with figure and table preparation and submission of permission requests include Leslie Aiello, David Alba, Peter Andrews, Freiderun Ankel-Simons, Emily Beech (British Museum of Natural History), David Begun, Lee Berger, Laura Watilo Blake, Christophe Boesch, Louis de Bonis, Michele Borzoni, Tom Bourdon, Jeanne Brewster (Nature Publishing Group), David Brill, Peter Brown, Michel Brunet, Perry Cartwright, Anne Ch'ien, David Chivers, Russell Ciochon, Elisabetta Cioppi, Dominique Cléré, Bonita De Klerke, Eric Delson, Nathaniel Dominy, Perry von Duijnhoven, John Fleagle, Elizabeth Garland, Melissa Gatter, Kristi Gomez, John Gurch, Michael Guven, Sandra Hagen, Terry Harrison, Adam Hirschberg, William Hopkins, Alain Houle, Jennifer Jones, Rick Jones, Takayoshi Kano, Sonia Katzen, Kenji Kawanaka, Bruce Latimer, Mary Leakey, Daniel Lieberman, David Liggett, Lu Quingwu, Henry de Lumley, Randy Marasigan, Robert D. Martin, John Mitani, Salvador Moyà-Solà, Peter Murray, Charles Musiba, Masato Nakatsukasa, Claudia Nebel, Melody Negron, Shauna Peffer, Martin Pickford, David Pilbeam, Penny Pivoriunis, Sue Savage-Rumbaugh, Jim Schultz, Eric Seiffert, Brigitte Senut, Becky Sigmon, Elwyn Simons, Fred Smith, Chris Stringer, Ian Tattersall, Javier Treuba, Caroline E. G. Tutin, Carel van Schaik, Linda Vigilant, Alan Walker, Tobias Wechsler, Tim White, Bernard Wood, Richard Wrangham, and Xu Qinghua. The text was copyedited by Vickie West.

Finally, but foremost, I sing the praises of the wonderful editorial, contract, permissions, and art departments at Harvard University Press, and especially Michael Fisher, Lauren Esdaile, Stephanie Vyce, Karen Peláez, and Eric Mulder, for their patient, professional interactions with a person who described himself as one with whom he would least like to deal on this massive, multifaceted project. They stuck with me in sickness and in health, a virtue that sustains some of the best human relationships.

APES AND HUMAN EVOLUTION

1

Mongrel Models and Seductive Scenarios of Human Evolution

The expectations of theory color perception to such a degree that new notions seldom arise from facts collected under the influence of old pictures of the world. New pictures must cast their influence before facts can be seen in different perspective.

NILES ELDREDGE and STEPHEN JAY GOULD (1972, p. 83)

As we have found out more about how bonobos behave in the wild, they have declined in favour as a model for our ancestor . . . This makes me wonder how much our use of models is influenced by how we would like our ancestor to have behaved—clearly the bonobo has fallen from grace because it shows what, for many, is behaviour that is socially unacceptable for a close relative of ours.

FRANCES WHITE (1992a, p. 47)

Human evolution is a topic about which much is known and about which a great deal more remains to be learned. Given that more people believe in extraterrestrial beings, ghosts, evil spirits, and angels than in evolution as the process whereby humans and other organisms developed on Earth, the pedagogic mission of evolutionary anthropology is formidable indeed.[1]

The United States lags behind thirty European nations, Cyprus, and Japan in the number of people who accept that we are natural products of evolution. In 2005, 40 percent of respondents accepted evolution, 39 percent rejected it, and 21 percent were uncertain.[2]

Since the mid-nineteenth-century Darwinian revolution, there has been a steady increase in the data and means to probe questions of where, when, how, and why our species—*Homo sapiens*—evolved, what our ancestral and collateral species might have looked like, and how they might have lived. In Darwin's day, comparative morphological and embryological comparisons of apes, monkeys, humans, and other animals and the geographic distribution of apes and indigenous peoples were the main sources of data for modeling human phylogeny. Fossils were few and fragmentary, were only sampled from Eurasia, and could not be reliably dated geologically.[3]

By the mid-twentieth century, more fossils were available, including African ones. Theorists aligned them toward culminant *Homo sapiens* and guessed where our species had achieved the human condition, albeit still with little control over geologic time. The second half of the twentieth century was characterized by an exponentially increased number and diversity of fossils, refined excavation methods and records, further developed geochemical dating techniques, more controlled employment of a vast quantity and variety of data from genetics, and extensive ecological and behavioral studies of primates and other animals as well as paleoenvironmental and behavioral modeling based on contextual information from archeological, faunal, floral, paleosol, and taphonomic studies.[4] In short, the multidisciplinary historical science of paleoanthropology was born, boomed, and continues to inform the human career and to excite the curious.

The primary resource for detailing the path of human evolution is fossil specimens. The trove from Africa and Eurasia indicates that for most of the human career more than one hominid species lived contemporaneously in the Old World. These underpin the bushy phylogenic models, within which it is impossible to connect a full chronological lineal series of species leading to *Homo sapiens* upon which experts can agree. At best, one can relay the nature of the specimens and species, where they were found, and when they lived. Questions of how they lived and why they might have died out or evolved into other species are addressed with mere scenarios, albeit scientifically informed ones, based on contextual information from the localities where they were collected.

Humans are distinguished among the approximately 400 extant species of primates by a constellation of morphological and behavioral characteristics, some of which only can be traced precisely through the fossil and archeological records. Obligate terrestrial bipedalism, precision-gripping hands, reduced teeth and jaws, and ballooned brains can be identified if fossils are complete enough in the skeletal regions under study. Archeological artifacts and features can indicate the presence of tool use and manufacture, control of fire, fabricated shelters, bodily ornamentation, mortuary practice, plastic and graphic arts, and other indications of cognitive skills and culture. There are no tangible empirical means to measure linguistic and mathematical capabilities from the fossils or archeological traces that predate graphic records, particularly in the early periods of the hominid career.

Reconstruction, Models, and Scenarios

> Current models of human behavioral evolution adopt a reductionist approach to its stages while relying upon a referent species for each. These models pretend to an empiricism that they do not exhibit and simply restate hominid-origin views of the 1960s.
>
> CRAIG B. STANFORD and JOHN S. ALLEN (1991, p. 61)

The term *reconstruction* appears often in evolutionary biological literature, particularly in regards to phylogenic schemes. Accordingly, so-called phylogeny reconstructions are based on various fossil, genetic, or comparative morphological analyses. However, it is misleading to imply that a holistic construction has been performed and represents a tangible biological reality. Instead, because we rarely have enough bits to conduct a reconstruction, we devise a *model:* a systematic description of an object or phenomenon that shares important characteristics with the object or phenomenon. Experts can reassemble fossil skulls and other bones and artifacts if enough pieces are recovered, but even in fossil restorations, a good deal of modeling may be required in lieu of the missing pieces.[5] Adding flesh and epidermal features relies on a few bony markers and a lot of imagination. Because fragmentary and warped fossils frustrate even the finest models, new discoveries keep paleoanthropology perpetually in ferment and flux.

Similarly, we can compare blood proteins, chromosomes, and mitochondrial and nuclear DNA sequences from extant taxa to construct models of genomic relationships, but in the absence of homologous materials from myriad extinct taxa, we cannot be certain that a particular model replicates the actual evolutionary histories of the survivors. Moreover, accurate microsatellite genotyping based on DNA from feces and shed hairs can be misleading if not conducted with sufficient sample amounts and via careful procedures to avoid contamination.[6]

In the speculative realms of behavioral and ecological modeling, it is often the case that one proposes *scenarios* instead of models, particularly in the sense that physical scientists employ the term model.[7] Indeed, paleoanthropologists are almost always reduced to writing scientifically informed stories because behavior and habitats are transitory, leaving only tantalizing traces.[8] Here, the key phrase is *scientifically informed*. Persuasive paleoanthropological scenarios differ from science fiction in that the former is bounded by facts. The extant apes and other animals, including all humans, provide us with a wealth of suggestive anatomical, behavioral, and ecological data that complement the patchy information derived from archeology and paleontology.[5]

Man the Hunter

> "Man the Hunter" is used to explain not only human biology but also human morality. The morals described, however, often reflect ancient beliefs and appear to be new ways of justifying old morality codes.
>
> ROBERT W. SUSSMAN (1999, p. 453)

Probably the oldest, and certainly the most enduring, scenario of human evolution portrays our ancient bipedal ancestors, particularly the ones with male genitalia, as weapon-wielding, meat-eating hunters.[9]

The views of Charles Darwin (1809–82) are restrained in comparison with the speculations by the advocates of killer ape scenarios, which flourished for several decades after the horrors of World War I (1914–19) and World War II (1939–45).[10] Darwin portrayed early man (his term) as having "sprung from some comparatively weak creature," who was not speedy and who lacked natural bodily defenses, namely, formidable canine teeth.[11] Consequently, this bipedal creature was stimulated to use his intellectual powers to make weapons for defense and hunting and to cooperate with "his fellow-men."[11]

In contrast to Darwin's puny survivalist, Henry Fairfield Osborn (1857–1935)—the pompous, affluent, influential, elitist, racist, classist, and sexist eugenicist founder of the American Museum of Natural History and *Natural History*—portrayed Dawn Man as a noble hunter.[12] Like another mammalian paragon, the horse, Dawn Man evolved "in the relatively high, invigorating uplands of a country such as central Asia."[13] Osborn's chief hard evidence for Dawn Man was the Piltdown fabrication *(Eoanthropus dawsoni),* which he ecstatically swallowed like a Eucharist wafer:

After attending on Sunday morning . . . a most memorable service in Westminster Abbey, a building which enshrines many of the great of all time, the writer repaired to the British Museum . . . to see the remains of the now thoroughly vindicated 'dawn man' of Great Britain . . . At the end of two hours . . . the writer was reminded of an opening prayer of college days, attributed to his professor of logic in Princeton University: "Paradoxical as it appears, O Lord, it is nevertheless true, etc." So the writer felt. Paradoxical as it appears to the comparative anatomists, the chinless Piltdown jaw, shaped exactly like that of a chimpanzee and with its relatively long, narrow teeth, does belong with the Piltdown skull . . .[14]

The Australian cum South African anatomist Raymond Dart (1893–1988), who described the holotype specimen of *Australopithecus africanus,* portrayed our ancient apelike ancestors as extremely violent beings. His two-legged killer ape is a fearsome, bloodthirsty, Grand Guignolesque creature, whose inventiveness was supposedly expressed by his employment of virtually every bone, tooth, and horn—the osteodontokeratic culture—from a wide variety of four-footed savanna prey to dispatch more of the same and members of his own species.[15] In the bone breccias of Makapansgaat, South Africa, Dart perceived the ungulate hemimandibles as saws and chopping, cleaving, and Samsonian hacking tools; long bones as clubs; horn cores as diggers and stabbing weapons; carnivore and warthog hemimandibles as slashing, ripping, or tearing weapons; antelope and zebra shoulder blades and hip bones as cleavers or splitting tools; and practically any skeletal bit as a projectile. Reading the entire inventory and their uses is enough to induce an empathetic headache—one easily cured by Dart's *Australopithecus,* conceived as a headhunter and a cannibal to boot.[16] The absence of caudal vertebrae indicated that they used the severed tails as signals and whips outside the cave.[15] (At least Dart imagined no kinkiness within the cave.)

In the 1960s and 1970s, as the Vietnam conflagration raged and the Cuban missile crisis menaced the Northern Hemisphere with nuclear Armageddon, Robert Ardrey (1908–80), Konrad Lorenz (1903–89), and other dramatists and writers of popular books envisioned Dart's killer ape in every man, not just the ancient cave dwellers.[10]

The hunting hypothesis was also promoted by leading physical anthropologists and human biologists of the day, preeminently Sherwood Washburn (1911–2000). Like Darwin, Washburn stressed the interrelatedness of tool use and bipedalism, which evolved coadaptively in male hunters, who needed weapons and the capacity for long-distance travel in pursuit of game and needed hands free of locomotor duties so that they could carry the meat back to camp.[17] Incidentally, females could carry their long-dependent young-sters as they gathered plant foods and small animals.[18] Washburn averred "our ancestors were striving creatures, full of rage, dominance and the will to live," but, unlike Dart and Ardrey, he believed that *Australopithecus* was predomi-nantly a vegetarian and could kill only the smallest animals.[19]

Washburn proposed that dramatic change occurred in the Early Pleisto-cene with the appearance of *Homo erectus,* a big game hunting, fire-making strider who had a clear-cut sexual division of labor, absence of estrus, an incest taboo, endogamy, reduced jaws and teeth, sizeable brains, enhanced mental capacities, handedness, manual dexterity, special cooperation among males, food sharing, nursing of the impaired, male pleasure in killing, and perhaps the beginnings of language.[20] They purportedly lived in small groups of nuclear families "ordered by positive affectionate habits and by the strength of personal dominance."[19]

In retrospect, it was naïve of men (and compliant women) to subscribe to heavily androcentric scenarios of hominid emergence and subsequent develop-ment. However, correctives were forthcoming in the second half of the twenti-eth century as fully credentialed female scientists entered the academy and other research institutions although some heavily gynocentric ideas might be viewed as too narrowly focused and intent on peripheralizing males.

Woman the Gatherer

> Of physiology from top to toe I sing, Not physiognomy alone
> nor brain alone is worthy for the Muse, I say the Form complete
> is worthier far. The Female equally with the Male I sing.

WALT WHITMAN (1982, p. 165)

As the smoke from napalm and burning draft cards and brassieres cleared, woman the gatherer strode forth to challenge the overwhelming importance of man the hunter in the human career.[21] Dubious from the outset though it was supported by Phillip Tobias, Dart's osteodontokeratic culture of *Aus-tralopithecus* was basically discredited by Charles K. Brain, whose careful taphonomic analyses supported Washburn's surmise that *Australopithecus* was more likely to have been the prey of carnivores than predators on large mammals.[22] Elizabeth Vrba's faunal analyses of several caves with australopithecine remains confirmed that they were probably the prey of carnivores; in fact, the Taung child might have been killed by an eagle.[23]

Concurrently, Richard Lee, Lorna Marshall, George Silberbauer, and Jiro Tanaka showed that vegetal products were a greater proportion of the diet of San people who lived in the inhospitable Kalahari Desert, where able men hunted with spotty success even the largest mammals via poisoned arrows, while women gathered daily provisions of smaller food items, water, and firewood.[24] The importance of food collected by women, including sizeable mammals, in the diets of other hunter-gatherers, particularly Australians of precolonial heritage, also enhanced the role of women in the minds of some observers and in new scenarios of evolutionary anthropologists.[25]

In a seminal essay, Sally Linton (b. Slocum) stripped the male transitional hominid of the supreme role in human development and stressed instead the centrality of females as basic provisioner and mother-infant units as the nexus of social groups.[21] She argued that the theory of man the hunter is unbalanced because it gives too much importance to aggression, which is only one factor of human life, and derives culture from killing.[26] Instead, "as the period of infant dependency began to lengthen, *the mothers would begin to increase the scope of their gathering to provide food for their still dependent infants.*"[27] Extension of the mother-infant bond would deepen and expand social relationships and initiate intragroup food sharing, which would lead to development of the family. She supported her argument with the facts that in contemporary human foraging societies and in nonhuman primate groups, females collect enough vegetal foods and small animals to support themselves and their dependent offspring.

Linton noted that among nonhuman primates, female choice of sexual partners for temporary consortships is common and that pair-bonding is rare; therefore, there is no reason to expect transitional hominid male hunters to hold females in long-term monogamous relationships that would guarantee their paternity.[28] Indeed, among *Homo sapiens* "changing sexual partners is frequent and common."[29] The early male hunters probably would initially share a kill with their mothers and siblings—that is, the individuals who had shared food with them—instead of with a wife or sexual partner.[29] Male aggressiveness and intermale cooperation were more likely to have served to protect the group than to motivate hunting.[30]

Linton partially disarmed man the hunter, claiming that many early tools—bones, sticks, and handaxes—could have been used to gather and to process plant foods.[28] Like Richard Lee, she further advanced that two of the earliest and most important cultural inventions were containers to hold the products of gathering and slings or nets with which to carry babies.[31] The cognitive demands of gathering, increasingly complex social-emotional bonds, and symbolic communication selected for increased brain size.[32]

While some anthropologists, notably Elaine Morgan, Nancy Tanner, and Adrienne Zihlman, promoted woman the gatherer over man the hunter as the best scenario of emergent hominid development, in *The Woman That Never Evolved* (1981) sociobiologist Sarah Hrdy discussed contemporary research on women and other female primates in order "to expand the concept of 'human nature' to include both sexes" and to correct male bias within evolutionary biology.[33] Hrdy effectively showed that female assertiveness and competition is the rule, not the exception among primates, though it is commonly expressed somewhat differently from that of conspecific males, as we shall see in Chapters 7, 8, 11, and 13.

Intersexual Sharers

> It should . . . be possible to extrapolate upward from ecological data on other mammals and suggest the biological attributes of the protohominids and to extrapolate downward from ethnological data on hunting and collecting peoples and suggest the minimal cultural attributes of the protohominids.
>
> GEORGE A. BARTHOLOMEW JR. and JOSEPH B. BIRDSELL
> (1953, p. 481)

In the midst of the feminist revolution, Glynn Issac (1937–85) employed rubbish from prehistoric archeological sites to stress the importance of intersexual food sharing and mutualism among early hominids.[34] He thought that meat sharing probably preceded the invention of carrying devices that would allow bulk collection of vegetal foods to be shared at a home base.[35]

Because one often cannot distinguish between scavenged and hunted carcasses at sites that evidence hominid collection, one need not assume that males were the sole providers of meat. Even when one has good reason to believe that humans killed large mammals, the sex of the hunters is indeterminate. Likewise, the melange of broken and cut-marked bones and cutting and pounding stone tools at early hominid sites cannot indicate whether females or males or both sexes made and used the tools.[36]

Isaac concluded that the archeological record of Plio-Pleistocene Africa is most compatible with scenarios "of human evolution that stress broadly based subsistence patterns rather than those involving intensive and voracious predation."[37] Nonetheless, he envisioned early development of a sexual division of labor with males procuring meat and marrow bones from sizeable animals and females gathering small creatures, eggs, and plant foods to share communally.[38] Selection for implemental, hunting and gathering skills and for advanced communication to mediate reciprocal social relationships, food sharing, and information about resources over a wide landscape led to enlargement and reorganization of the brain in Pleistocene hominids.[39]

In 1983, Isaac conceded to critics, particularly Lewis Binford, that an accumulation of smashed and cut-marked bones with stone artifacts does not definitively document intentional provisioning or meat sharing.[40] Accordingly, he diluted his food-sharing hypothesis to a central-place foraging hypothesis.[41]

Man the Provisioner

Linton was at a loss to explain what started the early hominids on a path of lengthened infant dependency—she used the term neoteny—and increased brain size, which would require increased parental care, transport, and provisioning.[28] Claude Owen Lovejoy developed an explanatory scenario that put males back on top.[42] Although he rejected the idea that hunting had played a major role in early hominid evolution, he imagined males to be essential provisioners of females and their young, who remained close to a home base. Bipedalism evolved so males could schlep foods home, where they shared them with their safely ensconced sole mates and progeny. According to Lovejoy, the *Father Knows Best* ideal of faithfully pair-bonded monogamous nuclear families was a fundamental feature of earliest hominid development. Needless to say, Lovejoy's scenario did not persuade feminists, and it appeared to be incongruent with the fact that the earliest recognized bipedal hominids at the time—*Australopithecus afarensis*—were characterized by notable dimorphism in body size.[43] Hill accepted a meat-for-sex hypothesis among early hominids who were sexually dimorphic, albeit with caveats that females simply copulated with the best providers and were not wholly provisioned by males.[44]

Two decades later, and based on much the same collection of fossils, Reno and colleagues used statistical methods to resurrect the possibility of man the monogamous provisioner.[45] They claimed that, like *Pan troglodytes* and *Homo sapiens*, *Australopithecus afarensis* is "non-dimorphic" in overall body mass.

Further, because, like *Homo sapiens, Australopithecus afarensis* lacks dramatic sexual dimorphism in canine teeth, monogamy is a feasible scenario for them. Contrarily, Gordon and coworkers demonstrated via a multivariate analysis of the femora, tibiae, humeri, and radii that *Australopithcus afarensis* was as dimorphic as *Gorilla* and *Pongo,* neither of which is monogamous.[46]

Return of the Hunter

In the final decade of the twentieth century—notorious for genocidal warfare in Europe and central and western Africa and for periodic mayhem and murder in U.S. and European cities, the Middle East, and south and southeast Asia, Indonesia, Latin America, and elsewhere globally—man the hunter reappeared in Wrangham and Peterson's nail-biting *Demonic Males* (1996) and Craig Stanford's meaty scenario in *The Hunting Apes* (1999).[47] Both were inspired by extensive observations of chimpanzees capturing monkeys and other mammals of modest size and apparently sharing the prey among individuals that were present at the kill site.

Wrangham and Peterson rooted human warfare, murder, cannibalism, infanticide, assault, rape, and intense intermale competition in >5 Ma (million-year-old) hunting, homicidal common ancestors of chimpanzees and humans.[48] Bonobos escaped the scourges of Mars and Cain by speciating from chimpanzees and developing interfemale bonds and mechanisms whereby females can modulate the influence of overly aggressive males. Ancestral women failed to bond with one another because they feared to lose "the investment and protection of the most desirable men" and thereby nurtured "the continued success of demonic males."[49] As Phyllis Dolhinow aptly quipped, "It sounds as though the authors really blame female human and chimpanzee bad judgement for the alleged sorry plight of their males."[50]

Stanford stressed the importance of meat as "money" instead of male bloodlust and violent intraspecific aggression in the evolution of human mentality and brain enlargement. He argued that "Man the Hunter was fatally flawed, first by its emphasis on the role of cognition in meat acquisition rather than meat sharing, and second by its unconscious ignorance of the role of females in the meat-control system."[51] Nonetheless, in Stanford's scenario, males remain in the limelight because they secure most of the meat, which requires notable motor skill and ingenuity, and they are predominant in determining who gets a share of it: meat's patriarchy.[52] Hominid females had to be politically adept and otherwise clever and seductive in order to obtain brain food (i.e., meat) from the alliances of males to share with their children. Consequently, they too advanced in intelligence along with the male hunters.

Zooarcheological research in the form of faunal analysis of cut-marks on bones of large mammals and other traces of hominid activity in East African Plio-Pleistocene sites are arguably evidence of hunting versus passive scavenging.[53] Of course, such traces do not indicate the sexes of the hunters, collectors, tool-makers, and tool-users. It is commonly assumed that male *Homo sapiens* were the hunting sex because globally human males are predominantly the subsistence hunters in their societies, but there are exceptions such as the Mossapoula Aka female net-hunters, who regularly bag a notable number of duikers and porcupines in the Central African Republic.[54]

People the Prejudiced

Clearly, it is impossible to contemplate the panorama of hominid evolution and the myriad diversity of human behavior totally free of our personal experiences and aspirations, local social context, and contemporary global events. Clearly, the fewer the facts that are available at a given time, especially in periods of personal, social, and political turmoil, the greater will be the chance that biases will affect scholarly and popular scientific writings.

We could lapse into cynicism and eschew the scientific enterprise that is portrayed in Donna Haraway's provocative, poignant *Primate Visions* (1989).[55] Conversely, we could view our current state of knowledge cautiously, as it may help illuminate proximate evolutionary puzzles, even as we counter the attempts to use it to support destructive popular folk beliefs and Machiavellian political adventures.[56] One need look no further than the millennia of female repression and the eugenics movements of the twentieth century to see how faulty and fabricated biology can cause needless, immeasurable, long-lasting human misery.[57]

Refined genomics has been key to solidly establishing the unity of humankind.[58] Accordingly, I believe that biological science and evolutionary anthropology in particular have much to offer regarding how we view our species, our fellow organisms, and the future of Earth. However, like virtually any human enterprise, they can be commandeered readily to support selfish acts and harmful beliefs that may undermine our social policies.

PART I

TERMINOLOGY, MORPHOLOGY, GENES, AND LOTS OF FOSSILS

2

Apes in Space

> Any discussion should start with a clear understanding as to
> what is to be discussed. A second prerequisite of useful discus-
> sion is the designation . . . of a vocabulary.
>
> GEORGE GAYLORD SIMPSON (1961, p. 1)

Linnaean Lineup

In the Linnaean system of classifying organisms, people and apes are common
members of the kingdom Animalia, phylum Chordata, class Mammalia, order
Primates, suborder (or infraorder) Anthropoidea, and superfamily Hominoidea. Stated colloquially, the apes and we are hominoid, anthropoid, primate,
mammalian, chordate animals. Accordingly, phrases like "primate and human
evolution" and "humans and animals" are redundant because humans are primates and animals, too. If one wishes to highlight people, the proper phrasings
would be "human and nonhuman primate evolution" and "humans and other
animals."

Similarly, one need not say anthropoid ape because the phrase is redundant. Ape suffices to designate individuals and specimens and various groupings of extant and fossil species. However, one can use anthropoid primates
when referring to the apes, monkeys, and people collectively because together they constitute the Anthropoidea. Simian(s), singe(s), and Affe(n) in-
formally refer to monkey(s), ape(s), or both in English, French, and German,
respectively.

15

Placing Our Order

For many decades, experts groped for a definition of the Primates, particularly one that would distinguish them from other orders of the Mammalia. Eventually, an uneasy consensus was reached whereby the Primates would be defined according to evolutionary trends that arrogantly culminated in *Homo sapiens,* instead of a suite of discrete features.[1]

This dogma held sway until Robert D. Martin enumerated a complex of features that served well to bound the order.[2] He accomplished this by scratching the clawed, and otherwise exceptional, tree shrews (family Tupaiidae, order Scandentia) and the extinct Plesiadapiformes, whose heyday was the Paleocene epoch.[3] This maneuver was not wholly acceptable to all experts.[4] Martin's Primates accommodate fossil species of modern aspect from the Eocene and younger epochs and extant modern forms. With several notable exceptions [indicated between brackets], they have the following characteristics:

- Typically arboreal inhabitants of tropical and subtropical forests. [Humans and a few species of monkeys are fundamentally terrestrial, and some lemurs, monkeys, and the African apes are arboreal only part time.]
- Extremities adapted to grip arboreal supports.
- Widely divergent hallux [except in *Homo*].
- Hand exhibiting at least some prehensile capacity.
- Most digits bearing flat nails instead of claws, and the hallux always bearing a nail.
- Ventral surfaces of the hand and foot (and sometimes also of the tail) having tactile pads with ridges (dermatoglyphics) that reduce slippage and enhance tactile sensitivity.
- Locomotion commonly, but not always, hind limb dominated.[5]
- Visual sense emphasized:
  - Eyes are relatively large, and the orbits sport either a postorbital bar or a more complete bony parietal separation between the orbit and the temporal fossa.
  - Frontal rotation of the eyes ensures binocular overlap of the visual fields.[6]
  - Some optic nerve fibers from each eye pass to the right and left cerebral hemispheres, providing a basis for stereoscopic vision.
- Olfactory system unspecialized or reduced.
- Brain enlarged and possessing special sulci (such as the Sylvian sulcus and the calcarine sulcus) on the outer surface of the cerebrum.

- Males with permanent precocial descent of the testes into a postpenial scrotum.
- Females lacking a urogenital sinus (that is, the urethra and vagina have more or less separate external openings).
- Long gestation periods, small litters (usually one), and precocial neonates.
- Slow reproduction relative to body size, slow fetal and postnatal growth, late sexual maturity, and long life spans.
- Maximum primate dental formula for upper and lower incisor, canine, premolar, and molar teeth.
- In half upper and lower jaws $\frac{2.1.3.3}{2.1.3.3}$ versus the basic mammalian dental formula of $\frac{3.1.4.3}{3.1.4.3}$.
- Cheek teeth relatively unspecialized.

A majority of the seventeen features—not any one or few of them—distinguishes the Primates. Other mammalian orders contain variable numbers of species that are arboreal (such as marsupials, squirrels, and civets), sport postorbital bars (tree shrews, cattle, antelopes, horses, and whales), are long-lived and have single births (elephants and whales), and have relatively unspecialized cheek teeth (pigs and peccaries). But none of them possess all or even a majority of the features that together characterize the Primates.

Martin's exclusion of the Plesiadapiformes from Primates has been challenged by paleontologists based on plesiadapiform skeletons with modern primate features of the auditory bulla and foot skeleton.[7] From molecular data, Janecka et al. inferred that Dermoptora (colugos or flying lemurs) and not Scandentia (tree-shrews) or Chiroptera (bats) are the closest living relatives of Primates, while Bloch et al. concluded that the Plesiadapiformes, which they consider to be ancestors of later primates, were not closely related to the Dermoptera or Scandentia.[8]

The first undoubted fossil Primates are from earliest Eocene deposits in Asia, North America, and Western Europe, which indicates that the order emerged ≥ 60 Ma.[9] Contrarily, a molecular clock of extant Mammalia and a model derived from the estimated duration of species preservation predicted that ca. 85 Ma the earliest primates were probably living in Gondwanaland—the combined Antarctica, South America, Africa, Madagascar, Australia, New Guinea, New Zealand, Arabia, and the Indian subcontinent (Figure 2.1).[10]

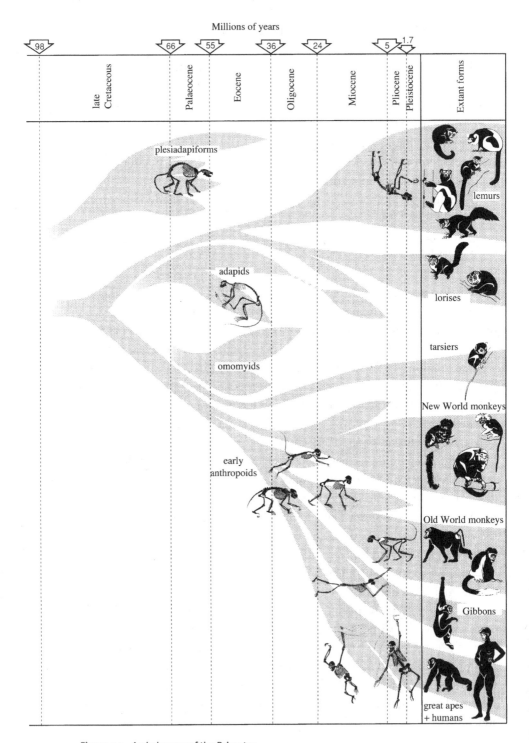

Figure 2.1. A phylogeny of the Primates.

Subordinal Discriminations

Traditionally, the Primates were divided into two suborders: the Prosimii and the Anthropoidea. The extant prosimians include the bizarre aye-aye and the tooth-combed lemurs, lorises, and galagos. The Anthropoidea encompasses all extant monkeys, apes, and humans and their fossil collaterals and intermediate forms.

Some authorities have argued that the most carnivorous of the nonhuman primates—the tarsiers (Tarsiidae)—have closer affinities with the anthropoid primates than with the prosimians. They have either placed them in the Anthropoidea or grouped them with the monkeys, apes, and us in the Haplorhini.[11] In the latter system, the aye-aye *(Daubentonia madagascariensis)* and the tooth-combed primates are classified as Strepsirhini. Other authorities have either rejected a classification of the Primates into Haplorhini and Strepsirhini or grouped tarsiers with the prosimians.[12]

The Anthropoidea share the following complex of features (Figure 2.2):[13]

- Relatively voluminous, rounded braincase and flatish face [or minimal projection of the nasal bones into the muzzle].
- Fusion of the frontal cranial elements to form a single frontal bone.
- Eyes close set and face directly forward.
- Bony postorbital closure.
- Relatively small corneas related to highly acute diurnal vision [except *Aotus*].[14]
- Fused mental symphysis.
- Integration of the trigonid and talonid of the molar teeth (Figure 2.3).
- Relatively small ears with limited mobility.
- Versatile facial muscles.
- Absence of a rhinarium (habitually moist nose tip).
- Fully developed hand with independent thumb [except some colobine and ateline monkeys that have mere nubbins].
- Flattened nails on all fingers and toes [except some orangutan halluces that are nailless and the clawed digits in marmosets, tamarins, and Goeldi's monkey].[15]

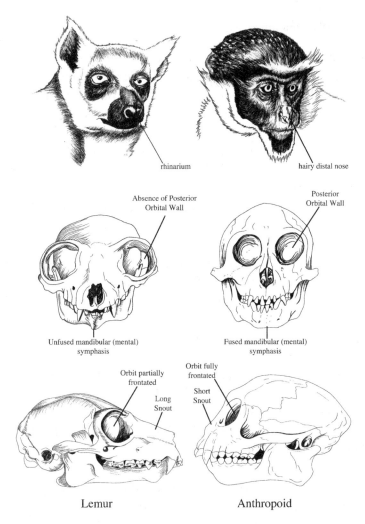

Figure 2.2. Prosimian (crowned lemur) and anthropoid (Diana monkey) heads and skulls.

Superfamily Circles

Extant Anthropoidea comprises three superfamilies: the Ceboidea, the Cercopithecoidea, and the Hominoidea. Each superfamily represents a major adaptive radiation on an extensive landmass. Extant Catarrhini comprises the Cercopithecoidea and the Hominoidea, which live in the Old World. Accordingly, a collective term for apes, people, and Eurasian and African monkeys is catarrhine primates. The Platyrrhini comprises monkeys of North

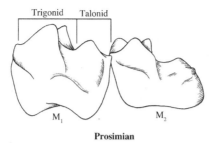

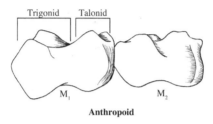

Figure 2.3. Fossil prosimian and anthropoid first and second lower molars. Note the lower talonids in the prosimian versus the subequal height with trigonids in the anthropoid.

and South America. Colloquially, they are platyrrhine, ceboid, or New World monkeys.

The extant ceboid monkeys are confined to the rain forests and adjacent wooded areas of southern Mexico, Central America, and South America. All species (>109) of New World monkeys are arboreal. Except for the night monkeys (*Aotus* spp.), they are diurnal foragers for fruits, nuts, leaves, resins, other vegetal parts and products, insects, and other arthropods. White-faced capuchins *(Cebus capucinus)* also catch and eat small mammals.[16]

The largest ceboids—*Ateles* (spider monkeys), *Lagothrix* (woolly monkeys), *Brachyteles* (muriquis) and *Alouatta* (howlers)—are uniquely equipped with powerful prehensile tails that sport extensive dermatoglyphics distally on their ventral surfaces. The atelid monkeys are particularly interesting to anthropologists because their trunks and shoulder blades have some features that resemble counterpart structures in apes, and some atelid societies have features that resemble those of chimpanzees and bonobos.[17]

The extant Cercopithecoidea are widespread in Africa and tropical and subtropical Asia. Except for *Homo sapiens,* they are the only other primate superfamily with natural residents in temperate regions, notably *Macaca fuscata*

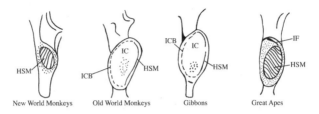

New World Monkeys   Old World Monkeys   Gibbons   Great Apes

Figure 2.4. In New World monkeys, great apes, and humans, the proximate attachment of the hamstring muscles (HSM) occupies much of the rounded surface of the ischial tuberosity. In Old World monkeys and gibbons, an ischial callosity (IC) attaches to a flat ischial tuberosity, and the HSM attach to its border (ICB). IF indicates proximal attachment of the ischiofemoralis muscle.

(Japanese macaques). Although most cercopithecoid species are extensively arboreal, some habitually move and forage on the ground. In the Ethiopian highlands, geladas *(Theropithecus gelada)* are almost as naturally earthbound as people are.

No cercopithecoid monkey has a prehensile tail. Further, unlike the ceboid monkeys, cercopithecoid monkeys have special sitting pads of cornified hairless skin—ischial callosities—attached firmly to their ischial tuberosities (Figure 2.4).[18]

The single family of extant cercopithecoid monkeys—the Cercopithecidae (approximately ninety-nine species)—is divided into two subfamilies: the Cercopithecinae (approximately fifty-seven species) and the Colobinae (approximately forty-two species).[19] A major difference between them is that whereas cercopithecine monkeys have variably developed cheek pouches lined with mucus membrane, in which food items can be stored briefly and perhaps predigested, colobine monkeys have complex, sacculated stomachs, in which quantities of young and mature leaves are digested by microbial fermentation (Figure 2.5).[20] All cercopithecid monkeys forage diurnally on variable proportions of fruit, young and mature leaves, flowers, petioles, other plant parts, and arthropods, and a few species, notably baboons (*Papio* spp.), catch and eat small mammals.

Both colobine and cercopithecine monkeys inhabit Africa and Asia. However, Asia is graced with the greater variety of colobine species (approximately thirty-one species), while Africa is home to the greater number of cercopithecine species (approximately thirty-eight species).[21]

The Hominoidea includes apes and humans, and their fossil collaterals and intermediate forms. They exhibit the following features in common (Figures 2.6, 2.7, and 2.8):

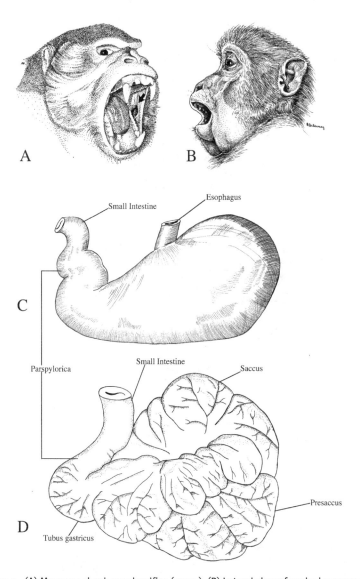

Figure 2.5. (A) Macaque cheek pouch orifice (arrow). (B) Lateral view of packed macaque cheek pouch. (C) Cercopithecine one-chambered stomach. (D) Colobine multichambered stomach.

- Dorsoventrally flattened chest.
- Recurved ribs.
- Long collar bones and mobile shoulders.
- Shoulder blades on the dorsal chest wall.

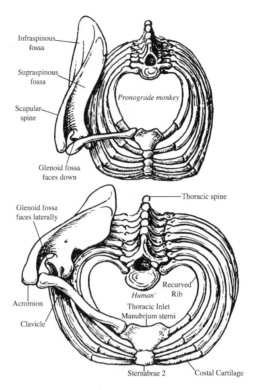

Figure 2.6. Pronograde monkey and human illustrating how the shape of the thorax and the length of clavicles position the scapulae on the sides of the thorax in pronograde monkeys and on the back of the thorax in humans (and apes). Consequently, the glenoid fossae in monkeys articulate atop the humeri, while hominoid glenoid fossae face laterally or cranially.

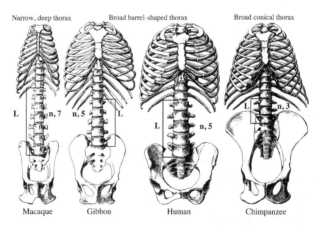

Figure 2.7. Skeletal torsos of macaque, gibbon, human, and chimpanzee rendered to the same length. Note the similarity in the thoracic shape and lumbar region (L) in the gibbon and human. (L_n is the number of lumbar vertebrae.)

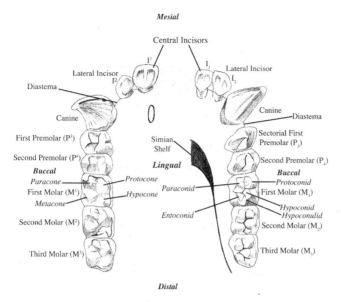

Figure 2.8. Occlusal views of gorilla permanent upper (left) and lower (right) teeth. Note the Y-5 pattern of grooves on M_1, M_2, and M_3 in which the stem of the Y is between the paraconid and entoconid and the arms of the Y embrace the hypoconid.

- Reduced lumbar region.
- Reduced olecranon on the ulna.
- Taillessness.
- Dental formula (in common with Cercopithecoidea) $\frac{2.1.2.3}{2.1.2.3}$.
- On the lower molars a Y-5 or other dryopithecine cusp pattern.
- Absence of cheek pouches.
- Large cranial capacity.
- Close apposition or attachment of viscera to the posterior abdominal wall and inferior surface of the diaphragm.
- Vermiform appendix.

Family Ties

Classically, the hominoid primates were divided into three families: the Hominidae, the Pongidae, and the Hylobatidae.[22] According to this scheme, only people (that is, humankind, or the outdated, sexist terms "man" and "mankind") and fossil creatures that resemble us are hominid primates.

Traditionally, the Pongidae or pongid apes included orangutans (*Pongo* spp.), chimpanzees *(Pan troglodytes),* bonobos *(Pan paniscus),* and gorillas (*Gorilla* spp.), known collectively as the great apes, and numerous fossil forms. They have in common a diploid chromosome number (symbolized *2n*) of 48 (versus *2n* 46 in *Homo sapiens*), the habit of nest building, and many anatomical, physiological, and biomolecular features.[23]

The Hylobatidae (hylobatid apes) encompass the gibbons or lesser apes. The term "lesser ape" refers only to the fact that they are much smaller than bonobos, chimpanzees, gorillas, and orangutans, not that they are of diminished importance in the animal kingdom.

Other schemes have been proposed that include all apes, or at least the chimpanzees, bonobos, and gorillas, in the Hominidae.[24] Morris Goodman sparked the trend by boldly recommending that the African apes join us in the Hominidae because two-dimensional starch-gel electrophoretic patterns of their blood proteins were more similar to those of humans than to orangutan and gibbon proteins. He left the orangutans hanging alone in the Pongidae, and the gibbons remained in the Hylobatidae.[25] Like other unorthodox suggestions that are based on novel scientific methods, Goodman's idea was vigorously attacked initially. But with the efflorescence of cladistics and after many more proteins and DNAs were studied, experts became increasingly comfortable with the idea of cofamilial status for bonobos, chimpanzees, gorillas, orangutans, gibbons, and humans, with the *Pan* species being the closest to the *Homo* species.[26]

Within an expanded Hominidae, we can segregate ourselves from the African apes by placing the two groups in separate subfamilies.[27] Accordingly, we are in the Homininae, while the African apes are nested in the Gorillinae or the Paninae.[28]

Another option has *Homo, Pan, Gorilla,* and *Pongo* in the Hominidae, with *Homo, Pan,* and *Gorilla* in the Homininae and *Pongo* segregated in the Ponginae.[29] Further, some biomolecular studies have indicated that there is a chimpanzee-bonobo-human clade to the exclusion of gorilla.[30]

Goodman's extremely cladistic scheme had all apes and people in the Hominidae, with *Pan* and *Homo* joined in a subtribe, the Hominina (Table 2.1).[31] Were one to adopt this classification, vernacularly classic great apes would be nonhylobatine, nonhuman hominids, and people would be nonchimpanzee, nonbonbo hominins. Goodman et al. further stressed the propinquity of bonobos, chimpanzees, and humans by declaring that we are only subgenerically distinct (Table 2.2).[32] Accordingly, chimpanzees are *Homo (Pan) troglo-*

dytes, bonobos are *Homo (Pan) paniscus,* and people are *Homo (Homo) sapiens.* Their scheme mirrors that of Diamond, who proposed that, as a third chimpanzee, *Homo sapiens* is congeneric with *Homo troglodytes* and *Homo paniscus.*[33]

One must wonder when someone will not only tree the bonobos and chimpanzees with people but also put them in the same bed. Considering the stubborn resistance to having African apes in the Hominidae, it is difficult to imagine ready acceptance of *Homo troglodytes* or *Pan sapiens* in our Linnaean vocabulary or heartfelt references to humans as the wise chimpanzee.[34]

Table 2.1.   Cladistic classification of extant apes and humans by Goodman (1992–96)

Superfamily **Hominoidea**	
Family **Hominidae**	
Subfamily Hylobatinae	*Hylobates:* gibbons
Subfamily **Homininae**	
Tribe Pongini	*Pongo:* orangutans
Tribe **Hominini**	
Subtribe Gorillina	*Gorilla:* gorillas
Subtribe **Hominina**	*Pan:* chimpanzees and bonobos
	Homo: humans

Table 2.2.   Cladistic classification of extant catarrhine primates by Goodman et al. (1998)

Infraorder Catarrhini
   Superfamily Cercopithecoidea
      Family Cercopithecidae: Old World monkeys
      Family Hominidae
         Subfamily Homininae
         Tribe Hylobatini
            Subtribe Hylobatina
               Symphalangus: siamang
               Hylobates: gibbons
         Tribe Hominini
            Subtribe Pongina
               Pongo: orangutans
            Subtribe Hominina
               Gorilla: gorillas
               Homo
                   Homo (Pan): chimpanzees, bonobos
                   Homo (Homo): humans

Orangutans also have champions for being closer relatives to humans than the molecular biological data would seem to support. Schwartz and Grehen cited many morphological features of *Pongo pygmaeus* and of fossil species to support their ideas.[35] Contrarily, shotgun reads of aligned whole genomes of human, chimpanzee, orangutan, gibbon, and macaque against a human reference genome indicate 97 percent average and median identity between orangutan and human versus 99 percent identity between chimpanzee and human. Interestingly, the gibbon genome also evidenced 97 percent similarity to the human genome.[36] Unfortunately, a gorilla genome was not included in the comparison.

Based on analyses of data from hominoid DNA hybridization experiments, Marks et al. challenged the existence of a special chimpanzee-human clade.[37] Furthermore, data from DNA sequences and karyology failed to resolve unequivocally the branching order of chimpanzees, gorillas, and humans.[38] For instance, in an analysis of >24,000 base pairs of sequences from fifty-three autosomal, intergenic, nonrepetitive DNA segments from gorillas, chimpanzees, and humans, only thirty-one segments appeared to support a *Homo-Pan* clade. Ten segments appeared to support a *Homo-Gorilla* clade, and twelve segments appeared to support a *Pan-Gorilla* clade.[39]

Indeed, we remain inadequately equipped to state how many genes, in what combinations, and through what interactions with the several environments that shape organisms throughout their careers might determine the distances among them after furcation of the human lineage, whether it be from

- a dichotomy of humans and chimpanzee/bonobos,
- a tritomy of African apes and humans, or
- a polytomy that also included extinct collateral lineages for which there is no genetic material.[40]

Here are the bald facts:

- We do not know how many genes mark levels of separation among apes and people.
- We cannot discretely recognize their phenotypic expressions.
- They probably are not of equal value to sort people from apes and apes from other apes.[41]

Accordingly, until the variation and developmental and functional biology of our genomes is much better understood, I recommend conservatism in attempts to sort out our bushy phylogeny and to resolve puzzles regarding the largely uncharted lineages of the extant apes.[42]

We do not even know how many genes overall and how much of the various kinds of genes characterize the genomes of *Homo sapiens* or of any species of apes.[43] Estimates of the number of genes in the human genome and presumably also in those of chimpanzees and other great apes have ranged between 20,488 and 150,000.[44] Accordingly, if humans and chimpanzees share 98.4 percent of their genes, between 328 and 2,400 of them could be different. However, if Britten is correct that the overall difference is 5 percent, then there are 1,024 to 7,500 different genes.[45] And if the human-chimpanzee difference in DNA has been underestimated "possibly by more than a factor of 2," the difference could be more than 2,050 to 15,000 genes.[46]

Nozawa et al. concluded that because different humans have different numbers (≤ 100) of genes asking about the exact number is meaningless.[47] Individual chimpanzees and other animals have a high prevalence of copy number variation of DNA segments, ranging from 103 to 106 nucleotides.[48] Moreover, an analysis of approximately 14,000 genes of chimpanzees and humans has indicated that the number of positively selected genes is substantially smaller in humans because of our smaller long-term effective population size. Accordingly, the anthropocentric view that a grand enhancement in Darwinian selection underpins human origins is refuted. Bakewell et al. concluded that the positively selected genes they identified support the association between human Mendelian diseases and past adaptations but not the widespread gene acceleration hypothesis of human origins.[49]

In view of the many ambiguities and uncertainties about how to interpret the molecular genetic data, for extant Hominoidea I employ a possibly gradistic four-family scheme, in which Hominidae is reserved for culturally dependent, ecologically specialized humans and their bipedal ancestors; Panidae comprises the African great apes (gorillas, chimpanzees, and bonobos); Pongidae includes the orangutans; and the gibbons aggregate in the Hylobatidae (Table 2.3).[50]

Final Assortment

Over the past 30,000 years, there has been only one species of humankind: *Homo sapiens*. Global samples of human mitochondrial and nuclear DNA are less variable than those from single populations of chimpanzees, probably because *Homo sapiens* is a recently evolved species that deployed rapidly from Africa to Eurasia, Oceania, and the Americas.[51] The genes underlying

Table 2.3. Extant taxa of the Hominoidea, specific and subspecific designations based on Brandon-Jones et al. (2004), Grubb et al. (2003), and Mootnick and Fan (2011)

Taxon	Common name
Hominoidea	hominoid primates; hominoids
Hominidae	hominid primates; hominids
Homininae	hominine primates; hominines
Homo sapiens	Humans: people
Pongidae	great apes
Paninae	chimpanzees
Pan troglodytes verus	Western chimpanzee
Pan troglodytes vellerosus	Nigeria chimpanzee
Pan troglodytes troglodytes	Central chimpanzee
Pan troglodytes schweinfurthii	Eastern chimpanzee
Pan paniscus	Bonobo; elya (Maniacky 2006)
Gorillinae	gorillas
Gorilla gorilla gorilla	Western gorilla
Gorilla gorilla diehli	Cross River gorilla
Gorilla beringei beringei	Mountain gorilla
Gorilla beringei graueri	Grauer's gorilla
Ponginae	orangutans
Pongo pygmaeus pygmaeus	Northwestern Bornean orangutan
Pongo pygmaeus morio	Northeastern Bornean orangutan
Pongo pygmaeus wurmbii	Southern Bornean orangutan
Pongo abelii	Sumatran orangutan
Hylobatidae	gibbons
Hylobates agilis	Agile gibbon (dark-handed gibbon)
Hylobates albibarbis	Bornean white-bearded gibbon

Hylobates klossii	Mentawai gibbon
Hylobates lar lar	Malayan white-handed gibbon
Hylobates lar carpenteri	Carpenter's white-handed gibbon
Hylobates lar entelloides	Central white-handed gibbon
Hylobates lar vestitus	Sumatran white-handed gibbon
Hylobates lar yunnanensis	Yunnan white-handed gibbon
Hylobates moloch moloch	West Java silvery gibbon
Hylobates moloch pongoalsoni	Central Java silvery gibbon
Hylobates muelleri muelleri	Müller's gray gibbon
Hylobates muelleri abbotti	Abbott's gray gibbon
Hylobates muelleri funereus	Northern gray gibbon
Hylobates pileatus	Pileated gibbon
Hoolock hoolock	Western hoolock gibbon
Hoolock leuconedys	Eastern hoolock gibbon
Nomascus concolor concolor	Western black gibbon
Nomascus concolor furvogaster	West Yunnan black gibbon
Nomascus concolor jingdongensis	Central Yunnan black gibbon
Nomascus concolor lu	Laotian black gibbon
Nomascus gabriellae	Buff-cheeked gibbon
Nomascus leucogenys	Northern white-cheeked gibbon
Nomascus nasutus	Eastern black gibbon
Nomascus hainanus	Hainan black gibbon
Nomascus siki	Southern white-cheeked gibbon
Symphalangus syndactylus syndactylus	Sumatran siamang
Symphalangus syndactylus continentis	Malayan siamang

phenotypic differences that were used to assign people to races vary much more between the presumed races than genes vary in general.[52] Further, in contrast with chimpanzees and bonobos, millennia of admixture among waves of humans across and between continents and islands has acted against the emergence of discrete human subspecies.[53] Consequently, *Homo sapiens* is devoid of races in the sense of zoological subspecies.

The specific classification of the extant apes is in flux as new information from comparative genomics expands the data sets for chimpanzees, gorillas, orangutans, and gibbons.[54] Therefore, I arbitrarily employ a scheme with six species of great apes: *Gorilla gorilla, Gorilla berengei, Pan troglodytes, Pan paniscus, Pongo pygmaeus,* and *Pongo abelii.* My scheme has approximately sixteen species of lesser apes: *Hylobates agilis, Hylobates albibarbis, Hylobates klossii, Hylobates lar, Hylobates moloch, Hylobates muelleri, Hylobates pileatus, Hoolock hoolock, Hoolock leuconedys; Nomascus concolor, Nomascus hainanus, Nomascus leucogenys, Nomascus gabriellae, Nomascus nasutus, Nomascus siki,* and *Symphalangus syndactylus* (Table 2.3).[55]

Anthropologists and human biologists have failed dismally in their attempts to subdivide *Homo sapiens* into subspecies (called races) largely because of the prehistoric and historical global admixture among the great variety of people who grace the planet and the disagreements on which features should be used to classify them.[56] Indeed, some argue that the exercise is not only futile but also unethical: past and current abuses of individuals and ethnic groups have been premised on narrowly conceived featural and folk racial classifications.[57]

Apes travel less widely than people do largely because we command a remarkable variety of transportation contraptions and beasts of burden, and we can ourselves carry heavy loads of food, water, firewood, and tools. Apes might wade, but deep water is an effective barrier to them because they do not swim. They also depend heavily on trees or other herbage for cover and sustenance. In brief, they are more ecologically restricted than we are, so they have experienced more subspecific variation (Table 2.3).

Some conservationists have stressed subspecies as the fundamental taxonomic category because were we to preserve only one or even a few isolated and/or small populations of a species, much genetic potential would be lost forever and the species could not recover from pandemics, poaching, fires, storms, commercial logging, and other catastrophic environmental assaults.[58] Geography alone should not be the sole criterion for subspecific classification. A subspecific scheme should have biological underpinnings, usually consisting of morphological features that are safely inferred to have genetic bases or actual genomic markers.

Each of the two species of gorillas contains two subspecies, and the single species of chimpanzees includes four subspecies. Bonobos and Sumatran orangutans are not divided into subspecies, but there might be three subspecies of Bornean orangutans (Table 2.3).[59]

Anecdotes about a gorilla-chimpanzee hybrid, the koolokamba, in Western Africa have not been substantiated empirically.[60] Suspicions that a long-captive male ape dubbed Oliver, which often walks bipedally, might be a gorilla-chimpanzee hybrid, a chimpanzee-bonobo hybrid, a human-chimpanzee hybrid, or a unique species have been dispelled by karyotypic and mitochondrial DNA analysis. Indeed, Oliver is probably a central chimpanzee, perhaps transported to the United States from Gabon.[61]

Several viable female and male hybrids were sired in France by a male bonobo with two female chimpanzees after each one had aborted a hybrid fetus.[62] Viable ape-human hybrids are unknown scientifically, albeit not for want of trying to propagate them.[63] Now there are compelling ethical and legal proscriptions against attempts to produce them.[64]

Making a Baby Ape-Man

In the 1960s, there were rumors that Russians were attempting to hybridize humans and chimpanzees, and one might assume such stories were merely Cold War propaganda.[64] However, when the Soviet archives opened in the 1990s as a part of perestroika, scholars found the records of Il'ya Ivanovich Ivanov, who attempted to mechanically inseminate two female chimpanzees, Babette and Sylvette, intravaginally with human sperm in Guinea, West Africa. He also inseminated Russian female "volunteers" in Sukhumi, Georgia, with the sperm of an orangutan named Tarzan. None of the experiments were successful.[65]

Ivanov succeeded in conducting his crude experiments, but Robert Yerkes failed with similar experiments on apes from Rosalià Abrue's colony in Cuba. She initially objected to natural insemination of her female chimpanzees by human males because the large human penis might cause them pain.[66] Other U.S., Dutch, German, and French scientists also encouraged or planned ape-human hybridization experiments for a variety of reasons apart from mere curiosity.[67] Some believed that success would support Darwinian evolution and counter the rising anti-Darwinist mood in the United States, Soviet Union, and other areas of the Northern Hemisphere.[68]

Racism raised its ugly, ignorant head also in the belief by some arrogant hybridization advocates that greater success might be achieved by using humans who were thought to be evolutionarily closer to the apes than they were.[68] Classism too might have been in play, given Yerkes's interest in bioengineering chimpanzees to become the new booboisie to replace troublesome servants, particularly Irish ones.[69]

Orthographic Order

Neophytes and experts alike should follow the long-established rules that are often ignored in English textbooks, mass media, and primary scientific publications:

- Generic nomina and those of all infrageneric ranks—subgenus, species and subspecies—are always italicized.
- Suprageneric nomina, such as subfamilies, families, superfamilies, suborders, orders, and classes, should not be italicized.
- Subgeneric nomina are located parenthetically between the generic and specific nomina.
- Within sentences, subgeneric, generic, and suprageneric nomina are always capitalized.
- Specific and subspecific nomina are never capitalized.
- Specific nomina never stand alone; for instance, our species is always written *Homo sapiens,* not simply *sapiens.*
- Linnaean binomials should not be used adjectivally. Accordingly, one should say "the evolution of *Homo sapiens*" instead of "*Homo sapiens* evolution" and "the browridges of *Gorilla*" instead of "*Gorilla* browridges."
- Like sheep, *Homo sapiens* (and other taxa than end in -*s*) are both singular and plural: *Homo sapien* is a no-no.
- Colloquial adjectival forms for suprageneric taxa are not capitalized except when they begin a sentence. For example, the adjectival form of Anthropoidea is anthropoid, of Hominidae is hominid, and of Australopithecinae is australopithecine.[70]

Hearing for the Little Guys

Gibbons are the most numerous and diversified of extant apes: sixteen species and twenty-seven subspecies.[71] Genomic studies indicate that the hylobatid lineage diverged from other Miocene anthropoids 18–20 Ma.[72] Extant gibbons are widely distributed in southeastern Asia, and they inhabit a variety of forests in Bangladesh, Assam (India), Myanmar, southern China, Hainan Island, Vietnam, Laos, Thailand, Kampuchea, Peninsular Malaysia, Sumatra, Java, Borneo, and the Mentawai Islands (Map 2.1).[73] During the Pleistocene epoch, siamang probably also lived on Java.[74] Hoolocks or concolor gibbons or both species were more widely distributed in China well into historical times.[75] Although gibbons once lived as far north as 31°N latitude, today their range is only to 26°N.[76]

The four genera of lesser apes are distinguished by different diploid numbers *(2n)* of chromosomes.[77] Most species are in *Hylobates* (seven species), which is characterized by $2n = 44$ chromosomes. None of them has laryngeal air sacs. *Nomascus* spp. (six species) have $2n = 52$ chromosomes. *Hoolock spp.* (two species) have $2n = 38$ chromosomes, and *Symphalangus syndactylus* has $2n = 50$ chromosomes. Male and female *Symphalangus syndactylus* have well-developed laryngeal sacs whereas those of *Hoolock hoolock* are small (Plate 2 bottom). In *Nomascus,* only males possess air sacs that are diminutive.[22]

Unlike the great apes and us, lesser apes have ischial callosities that they employ as sitting pads while foraging and sleeping (Figure 2.4). In possessing ischial callosities, the gibbons resemble the Old World monkeys. However, they develop postnatally in the Hylobatidae versus fetally in the Cercopithecidae.[78] They are the only extant apes that do not build nests.

Except for siamang, hylobatid apes have the densest coats among the catarrhine primates. Further, siamang, and, to a lesser extent, Mentawai gibbons are outstanding among lesser apes for high frequencies of webbing between the second and third toes.[78]

Hylobatids are easily the smallest of the living apes. The largest of them—the siamang—weigh between 9 and 13 kg. The smallest gibbons are about half as heavy as siamang.[22] Although males are commonly somewhat heavier than females, some adult females outweigh conspecific adult males. The lesser apes evidence minimal sexual dimorphism in body size, cranial features, and canine tooth dimensions, but they exhibit sexual dimorphism in some overall body dimensions, and several hylobatid species are dramatically dimorphic in pelage color or pattern or both.[79]

Hoolock hoolock
1. Garo Hills
2. Hollongapar Forest

Nomascus concolor
3. Mt. Wuliang

Nomascus hainanus
4. Hainan Island

Hylobates lar
5. Chiengmai
6. Khao Yai NP
7. Kuala Lompat (19)
8. Tanjong Triang
9. Ketambe Research Station,
   Gunung Leuser NP (22)(41)

Hylobates pileatus
10. Khao Ang Rue Nai

Hylobates agilis
11. Sungai Dal
12. Bukit Barisan Selatan NP (20)
13. Way Canguk (21)

Hylobates agilis x H. muelleri
14. Barito Ulu (18)

Hylobates klossi
15. Sirimuri, Siberut

Hylobates muelleri
16. West Kalimantan
17. Kutai NP (36)

Hylobates mulleri mulleri
18. Barito Ulu (14)

Symphalangus syndactylus
19. Kuala Lompat (7)
20. Bukit Barisan Selatan NP (12)
21. Way Canguk (13)
22. Ketambe Research Station,
    Gunung Leuser NP (9)(41)

Hylobates moloch
23. Ujung Kulon NP
24. Mt. Halimun Salak NP
25. Gunung Gede Pangrango NP
26. Pegunungan Pembarisan
27. Gunung Slamet
28. Pengunungan Dieng

Pongo pygmaeus pygmaeus
29. Sarawak
30. Gunung Palung NP
31. Tanjung Puting NP

Pongo pygmaeus morio
32. Sabah
33. Kinabatangan Wildlife Sanctuary
34. Ulu Segama
35. Sangkulirang Peninsula
36. Kutai NP (17)

Pongo pygmaeus wurmbii
37. Sabangau Ecosystem
38. Tuanan Field Station
39. Mawas Reserve

Pongo abelii
40. Suaq Balimbing, Gunung Leuser NP
41. Ketambe Research Station,
    Gunung Leuser NP (9)(22)

*NP=National Park

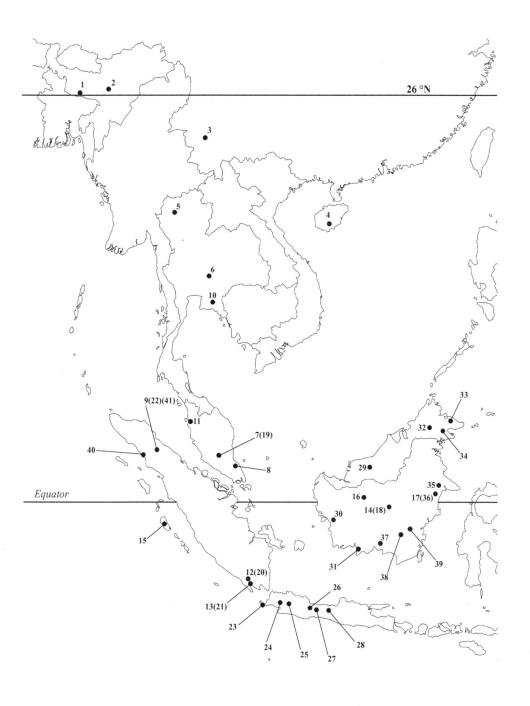

Map 2.1. Gibbon and orangutan study sites

There are four basic coat conditions among the Hylobatidae: monochromatism, polychromatism, asexual dichromatism, and sexual dichromatism (Plates 1 and 2). Siamang and Mentawai gibbons *(Hylobates klossii)* are monochromatically black: males and females have uniform coats of black hair.

The coats of *Hylobates muelleri* are polychromatic, ranging from shades of brown to gray. Both males and females can be any of several dark, light, or intermediate hues. Contrarily, *Hylobates agilis* of Sumatra, Borneo and peninsular Malaysia, and Thailand are asexually dichromatic. Adults of both sexes might be either dark or light. Further, lightly and darkly coated individuals are common within the same monogamous family. The widespread white-handed gibbons *(Hylobates lar)* are also asexually dichromatic (Plate 1).

Java silvery gibbons *(Hylobates moloch)* are moderately sexually dimorphic: the pelage of both sexes is grey, but females sport dark or black caps and chests (T. Geismann, personal communication) Six species of gibbons exhibit more dramatic sexual dichromatism. *Hoolock hoolock* and *Nomascus* spp. males are born and remain black, but the females are fawn colored at birth and turn black at six to nine months of age. As the females mature, their coats bleach back to fawn, except for a black patch on the crown of the head (Plate 1).

The sexual dichromatism of *Hylobates pileatus* is developmentally different from that in *Hoolock* and *Nomascus*. Young pileated gibbon males and females are buffy, with black spots on their crowns and chests. As they mature, the spots spread. The males become basically black, except for white markings on the head, hands, feet, and pubic region. Females only acquire a dark ventral patch that extends from throat to groin (Plate 1).

Hylobatid species have sexually distinctive vocalizations, and gibbon populations differ markedly in the kinds of sounds that they make, which sex makes them, how the sounds are combined, and which individuals display athletically and chase after calling.[22] Accordingly, these behavioral characteristics are valuable for sorting the different species and subspecies.[80] Hybridization of free-ranging and captive lesser apes provides excellent opportunities to study the genetic bases of their vocalizations and physical displays.[81]

Skeletally and fully fleshed, the lesser apes have a gracile, almost spidery appearance, though siamang are somewhat stockier than the smaller species.

Gibbon skulls are lightweight, small-faced, and relatively free of prominent crests, bumps and ridges (Figures 2.9 and 2.10). However, their capacious orbits are surrounded by bony rings that viewed frontally resemble

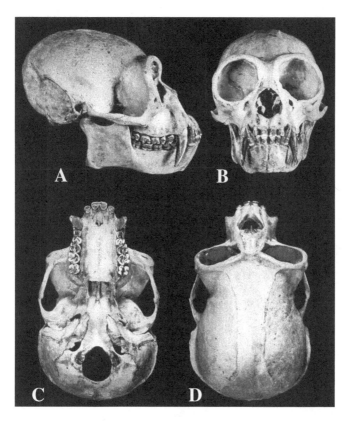

Figure 2.9. Skull of a male white-handed gibbon *(Hylobates lar)*. (A) Right lateral view. (B) Frontal view. (C) Basal view. (D) Dorsal view.

horn-rimmed spectacles. A sharp horizontal ridge rings the back of the cranium, beneath which the neck muscles attach. The forehead rises little above the orbits, which are separated medially by a wide interorbital pillar. The entire braincase is low, elongate, and constricted behind the orbits, more so in siamang than in other gibbons.[78] Gibbons lack frontal sinuses.[82]

Viewed from above, the hylobatid face, and especially the muzzle, projects conspicuously in front of the braincase. Hylobatid facial skeletons tend to be flexed dorsally relative to the cranial base, a condition known as airorynchy.[83] Siamang are more airorynchic than the smaller gibbons are (Figure 2.10).[84]

Hylobatid mandibles have low rami and slender bodies. The mandibular condyles are much closer to the plane of the occlusal surface of the cheek teeth in lesser apes than in great apes.

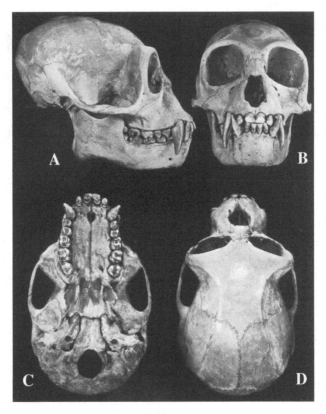

Figure 2.10. Skull of a male siamang *(Symphalangus syndactylus)*. (A) Right lateral view. (B) Frontal view. (C) Basal view. (D) Dorsal view.

The most striking aspect of the dentition in gibbons is their elongate, saber-like upper canine teeth, which they can employ with telling effect. They are honed to razor sharpness against the sectorial first lower premolars (P_3). The acute lower canines also project prominently above the occlusal surface of the lower incisors; they are about two-thirds the length of the upper canines. The unisex size of hylobatid canines is unique among nonhuman primates. This condition is particularly striking vis-à-vis the marked canine dimorphism of great apes.[85]

Gaps—diastemata—occur in the jaws of primates that have sizeable projecting canines. In apes, the maxillary diastemata, which might accommodate the lower canines, are generally between the lateral incisors and canine teeth. The lower diastemata, which might accommodate the upper canines, are between the mandibular canines and first premolar teeth.

Hylobatid upper incisors are heteromorphic.[86] Whereas the upper central (that is, first and medial) incisors have narrow bases and wide cutting edges, the upper lateral (second) incisors are relatively small and have rhomboid crowns. As in the other hominoid primates, the four vertically implanted lower incisors of gibbons are smaller than their upper incisors, and they do not vary as much morphologically; they are homomorphic.[87] Ungar reported that Sumatran white-handed gibbons seldom used their front teeth to eat fruits, insects, and spiders and used them less than sympatric monkeys and orangutans did when eating leaves, flowers, and stems.[88]

As Eve and Adam calamitously realized (Genesis 3:6), the upper incisors are particularly good for nipping through the pericarps of fruits. That the smaller lesser apes have relatively wider upper incisors than those of siamang might be related to the greater emphasis on fruits versus leaves in the diets of small gibbons.[89]

Gibbon upper premolars (P^3 and P^4, first and second, respectively) and second lower premolars (P_4) are bicuspid; that is, each has prominent buccal and lingual cusps. Contrarily, the sectorial P_3 is basically unicuspid. It has a high, sharply pointed cusp that slopes steeply mesially from the tip. Consequently, the lower premolars of hylobatid (and other extant) apes are said to be heteromorphic. In the lesser apes, the upper and lower second premolars evidence molarization.[85]

As is characteristic of all hominoid primates, gibbon upper molars are typically four cusped—protocone, paracone, metacone, and hypocone—and their lower molars are five cusped: protoconid, metaconid, hypoconid, entoconid, and hypoconulid. In both upper and lower molars, four main cusps rise from the corners of the quadrilateral enamel crown. On the lower molars, a fifth cusp, the hypoconulid, is located centrally or more commonly buccally behind the second row of main cusps. In this configuration, the grooves between the main cusps form the Y-5 (the *Dryopithecus*) pattern, in which the arms of a Y-shaped groove open buccally to embrace the hypoconid while the stem of the Y lies lingually between the metaconid and the entoconid (Figure 2.8).

In addition to the main cusps, cingula and supernumerary cusps appear variably on hylobatid cheek teeth.[86] Siamang have well-developed shearing and crushing mechanisms that are probably related to their folivory.[90] Kay and Nagatoshi described hylobatid molar enamel as thin; however, according to Shellis et al., relative to body mass among primates their molar enamel is of average thickness.[91]

In most hylobatid species, the M^3 shows signs of reduction. The M_3 is reduced in some species, but it is robust in others. When M_3 is reduced, the hypoconulid is absent or diminutive compared with its development in M_1 and M_2. With marked reduction or loss of the hypoconulid, the Y-5 pattern becomes a plus-4 pattern because the intersecting grooves separating the protoconid, metaconid, entoconid, and hypoconid form a plus sign [+]. The plus-4 pattern is rare in extant Hylobatidae.[85]

The postcranial axial skeleton of gibbons is the most humanoid among the apes.[92] Their appendicular skeleton is replete with slender, long bones. Both the forelimbs and the hind limbs are remarkably elongate in comparison with truncal length and body mass. Indeed, gibbons have the longest fore-limbs and second longest hind-limbs among the anthropoid primates.[93] However, siamang have relatively shorter hind-limbs than those of other gibbon species, which accounts for the higher intermembral index of siamang versus other gibbons.[94]

Les Singes Oranges

Orangutans (*Pongo* spp.) are the only extant Asian taxa of great apes. Seven genetic loci indicate that orangutans are more diverse than African apes and humans, and the Sumatran population is approximately twice as diverse as the Bornean population in most loci.[95] Genomic studies indicate that the pongid lineage diverged from other Miocene anthropoids 18–20 Ma.[96] During the Pleistocene epoch, *Pongo* was widely distributed from northern India through southern China, Thailand, northern Laos and Vietnam, Malaysia (Sarawak), and Indonesia, including Java.[97] Now orangutans are restricted to rapidly shrinking forest enclaves on Borneo, where they are sometimes sympatric with agile or Mueller's gibbons, and northern Sumatra, where they are sympatric with siamang and white-handed gibbons (Map 2.1).[98] Genomic data indicate that Bornean and Sumatran orangutans are 99.7 percent similar and that they have been separate for about 400 Ka.[96] After enduring centuries of human atrocities that cumulatively surpass the bloody fiction of the rue Morgue, the orangutans hang perilously on the brink of extinction after wholesale devastation of their habitats.[99]

Orangutans are extremely sexually dimorphic in body mass but less so in canine height. On average, males are twice as heavy as females; the mean masses of male and female samples are ca. 77 kg and 36 kg and ca. 86 kg and 38 kg, respectively.[100] Moreover, fully adult males sport striking epigamic features, including large cheek flanges and voluminous inflatable

Figure 2.11. Prime adult male Sumatran orangutan *(Pongo abelii)*. Note cheek pads (flanges), large deflated laryngeal sac, beard, and moustache.

laryngeal air sacs, with which they produce loud, leonine, long calls (Figure 2.11).[101]

The sagittal and nuchal crests are generally more developed on the skulls of males, which are, overall, more robust than the skulls of most female orangutans.[102] Indeed, sagittal crests are common in the larger males, but female orangutans rarely have them.[103] Cranial dimorphism is identifiable in infancy, and via different growth rates and durations it develops further among juveniles; it becomes more pronounced throughout their maturation due to additional dramatic transformations in the males.[104]

The pelages of male and female orangutans are polychromatic, ranging from yellowish and reddish orange through chestnut and maroon to chocolate brown. Like other catarrhine primates, orangutans exhibit piloerection when the sympathetic nervous system is appropriately stimulated. This makes the already bulky males look even bigger. Assuming that combatants aim at the surface of an opponent, an attacker might bite only hair and air instead of flesh and bone. The long hair on their backs and forelimbs further enhances the apparent size of displaying adult male Sumatran orangutans.[105]

The skin of orangutans varies from light to dark brown, probably depending, at least partly, on tanning. The faces of captives that have been deprived of sunlight can be quite pinkish.[22]

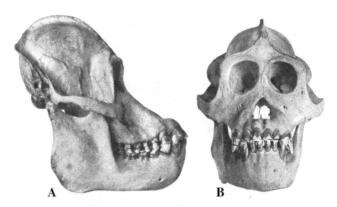

Figure 2.12. Skull of a male Bornean orangutan *(Pongo pygmaeus).* (A) Right lateral view. (B) Frontal view.

Skulls of female and male *Pongo* (Figure 2.12) are characterized by

- short, rounded brain cases,
- pronounced postorbital constriction,
- unimpressive browridges that do not join medially to form a supraorbital torus,
- absence of frontal sinuses,[106]
- receding foreheads that nevertheless rise above well-rounded, often ovoid orbits,
- a narrow interorbital pillar relative to biorbital breadth,
- dished faces, as viewed in profile,
- large, projecting muzzles,
- airorynchy, and
- massive mandibles.

The mandibular rami are tall, thereby placing the condyles high above the occlusal plane of the cheek teeth.

The heteromorphic anterior dentition of orangutans is especially conspicuous.[107] The procumbent upper central incisors are large and wide, somewhat dwarfing the neighboring lateral incisors. Ungar observed that Sumatran orangutans often employed their incisors to nip or crush hard, brittle fruits and to ingest arthropods, flowers, stems, and leaves.[88]

The lower incisors are homomorphic. Like the lower incisors of other extant Hominoidea, they articulate vertically in the mandibular alveoli. The lower and especially the upper canine teeth are stout. They project menac-ingly beyond the occlusal levels of the incisors and postcanine teeth. In keep-ing with the marked sexual dimorphism of the species, the conical canine teeth of male *Pongo* are notably larger than those of female *Pongo*. This di-morphism is more manifest in the upper than in the lower dentition.[108]

The upper premolars (P^3, P^4) and second lower premolar (P_4) are basically bicuspid, though P^4, and more rarely P_4, might also sport small supernumer-ary cusps that attest to their molarization.[109] The sectorial P_3 is dominated by a high, robust protoconid, but it too bears a small metaconid and, variably, other cusps. Cingula are reduced or absent from the postcanine teeth in ex-tant *Pongo*.[110]

Orangutan molars share the basic hominoid pattern of four main cusps on M^{1-3} and five main cusps on M_{1-3}. However, because the occlusal enamel on their molars and (to a lesser extent) premolars is crenulated, the identity of main versus supernumerary cusps is somewhat obscured.

The inherited pattern of enamel wrinkling distinguishes the postcanine teeth of *Pongo* from those of *Pan* and *Homo*.[111] Further, the molar enamel is thicker in *Pongo* than in *Pan*.[111] In this feature, *Pongo* resembles the Homini-dae and certain fossil Hominoidea, though Shellis et al. found in a small sample that relative to body mass the enamel in *Pongo* is only of average thickness or thin, depending on the type of tooth.[112] Vogel et al. associated the crenulations and thicker enamel of *Pongo pygmaeus* versus *Pan troglodytes* with field observations that *Pongo* consume more tough and hard fallback foods when fruits are scarce than do *Pan*.[113]

The bodies of *Pongo* must be counted among the most bizarre in the Mammalia.[114] Their squat, pot-bellied torsos anchor immense forelimbs and stubby hind limbs that nonetheless are noteworthy for long, powerful feet.[115] Orangutans are unique among primates for dramatic reduction of the hal-luces. Indeed, the hallucal nail, distal phalanx, and occasionally even the proximal phalanx are absent. Extreme hallucal reduction is more commonly expressed in females than in males.[15] All in all, they are extraordinary arbo-real foraging machines.

Die Afrikanischen Affen

Casual observers might initially have trouble distinguishing among isolated gorillas, chimpanzees, and bonobos. Whereas one can quickly learn to identify a fully adult male gorilla, it is more challenging to distinguish a large chim-panzee from an adult female or young male gorilla and especially between a chimpanzee and a bonobo. Genomic studies indicate that gorillas diverged from other anthropoids 6–8 Ma and that humans and the stem culminating in chimpanzees and bonobos diverged from other anthropoids 4.5–6.0 Ma.[72] Estimates of furcation of bonobos and chimpanzees vary from 0.9–2.5 Ma, which includes the period when the Congo River probably formed, namely, around 1.5 Ma.[116]

Because of the many physical similarities among bonobos, chimpanzees, and gorillas, especially their shared unique mode of quadrupedal positional behavior—knuckle-walking—they were tentatively placed by Tuttle and by Groves together in genus *Pan,* with two subgenera (*Pan* and *Gorilla*).[117] Accordingly, bonobos, chimpanzees, and gorillas would be *Pan (Pan) paniscus, Pan (Pan) troglodytes,* and *Pan (Gorilla) gorilla* and *Pan (Gorilla) beringei,* respectively. Tuttle and Groves no longer subscribe to this scheme.

Ecologically, chimpanzees are the most versatile of the extant apes, which might have characterized them for many centuries or millennia.[118] They inhabit a remarkable variety of primary and secondary forests, deciduous woodlands, riverine forests, and forest-savanna ecotonal regions, extending from the Gambia and Guinea-Bissau through West and northern Central Africa to western Uganda and northwestern Tanzania (Map 2.2).[119] Most long-term studies of free chimpanzees are situated in more open canopy locations instead of closed-canopy rain forests.[120] Two upper central incisors and an M1 from Middle Pleistocene deposits of the Kapthurin Formation, Tugen Hills, Central Kenya, indicate that >545 Ka, chimpanzees were penecontemporaneous with *Homo* sp.[121]

Bonobos are confined to the humid, swampy forests of central Democratic Republic of the Congo (Map 2.2).[122] They are also misleadingly called pygmy chimpanzees because of their presumed smaller overall body size. Although they are probably somewhat lighter than western subspecies of chimpanzees, their average body masses are remarkably similar to those of Tanzanian chimpanzees.[123] Further, because other features identify bonobos as a distinct species, terming them pygmy chimpanzees is rather like calling chimpanzees pygmy gorillas.[22]

Gorillas are the largest extant apes. They live in two widely separated areas of tropical Africa: western Africa and northeastern Central Africa and southwestern Uganda (Map 2.2).[124] Although some populations are sympatric with chimpanzees, others inhabit montane and secondary forests where chimpanzees are absent. Like chimpanzees and orangutans, gorillas exhibit greater genetic variation than do humans.[125]

Gorillas are among the most sexually dimorphic mammals. Adult males weigh between 132 and 218 kg, whereas adult females are between 68 and 98 kg. Like orangutans, average female gorillas are about half the size of average males.[126] Until about ten years of age, blackbacked males are difficult to distinguish from adult females. Thereafter, male secondary sexual characteristics, namely, their massive cranial superstructures, elongate palates and mandibles, large canine teeth, and a distinctive fibrofatty pad over the posterior part of the sagittal crest, develop rather rapidly (Figure 2.13).[127]

Like people, chimpanzees and bonobos are moderately sexually dimorphic in body size.[127] They are greatly outmatched by orangutans and gorillas in this characteristic. Adult male chimpanzees weigh around 60 kg and females weigh around 46 kg; male bonobos weigh around 45 kg and the females weigh around 33 kg.[100]

Chimpanzees have relatively large ears in comparison with those of other apes and most humans. Their pelage is polychromatically black or dark brown before graying as they mature. Youngsters have a shock of white hair—the perianal tuft—on the dorsal apex of the perineum. Like distinguished humans, some chimpanzees are predisposed to frontal baldness. Their skin varies from white and pinkish through bronze, coppery hues, and black. They tan readily.

From an early age, wild bonobos have black faces, ears, palms, and soles, although some captives that have been kept out of sunlight exhibit lighter pigmentation. Bonobos sport thick side-whiskers and long, fine, black body hair. Some adults retain an infantile white perianal tuft, and they usually escape frontal baldness. Bonobo ears are more like those of gorillas than like chimpanzee pinnae.

Except for fully mature males, gorillas have black or dark brown hair, sometimes with reddish tinges in some subspecies and individuals. Adult males are called silverbacks or silverbacked males because of the stunning patches of grayish-white hair over their lower backs.[128]

The skin of newborn gorillas is pinkish gray. Within the first year, it darkens to dark gray or black. Like chimpanzees and bonobos, baby gorillas have a white perianal tuft. It disappears sometime after the fourth year.[129] Gorilla ears, like those of orangutans and many humans, lie close to the head. Their faces are graced with squashed-tomato noses that are so variable that they distinguish individuals in the field.[130]

Gorilla and chimpanzee skulls are low, elongate, and rugged. The muzzle projects prominently forward, and the upper face is highlighted by a supraorbital torus that partly accommodates the frontal sinuses. The interorbital

Pan troglodytes verus
1. Mt. Assirik, Nikolo-Koba NP (Senegal)
2. Fongoli (Senegal)
3. Kanka Sili (Guinea)
4. Bossou & Nimba Mountains (Guinea)
5. Taï Forest (Côte d'Ivoire)

Pan troglodytes vellerosus
6. Gashaka Gumti NP (Nigeria)

Pan troglodytes troglodytes
7. Dja Biosphere Reserve (Cameroon) (36)
8. Campo Animal Reserve (Cameroon)
9. Okorobikó Mountains (Rio Muni)
10. Lopé Reserve (Gabon) (37)
11. Fernan Vaz (Gabon) (38)
12. Loango NP (Gabon) (39)
13. Goualougo, Noubalé-Ndoki NP (RC) (43)
14. Bai Hokou, Dzanga-Ndoki NP (CAR) (42)

Pan troglodytes schweinfurthii
15. Kahuzi-Biega NP (DRC) (45)
16. Gombe NP (Tanzania)
17. Kasoje, Mahale Mountains NP (Tanzania)
18. Filabanga & Kasakati Basin, Ugalla Tonguwe FR (Tanzania)
19. Rubondo Island (Tanzania)
20. Bwindi Impenetrable NP (Uganda) (50)
21. Kalinzu FR (Uganda)
22. Ngogo & Kanyawara, Kibale NP (Uganda)
23. Semliki (Uganda)
24. Budongo Forest (Uganda)
25. Tongo Forest (DRC)
26. Kibale Forest & Kyambura Game Reserve, Queen Elizabeth NP (Uganda)

Pan paniscus (DRC)
27. Eyengo, Lomako FR
28. Wamba
29. Lake Tumba
30. Lui Kotale, Salonga NP
31. Lilungu
32. Iyoko & Ikela, Yalosidi
33. Lukuru

Gorilla gorilla diehli
34. Cross River NP (Nigeria)

Gorilla gorilla gorilla
35. Kindia (Guinea)
36. Dja Biosphere Reserve (7)
37. Lopé Reserve (10)
38. Fernan Vas (11)
39. Loango NP (12)
40. Lossi Forest (RC)
41. Lokoué, Parc national d'Odzala (RC)
42. Mondika & Bai Hokou, Dzanga-Ndoki NP (14)
43. Mbeli Bai, Nouabalé-Ndoki National Park (13)

Gorilla beringei graueri
44. Maiko (DRC)
45. Kahuzi-Biega NP (15)
46. Itombwe (DRC)

Gorilla beringei beringei
47. Kabara, Parc National de Virunga (DRC)
48. Karisoke, Parc des Volcan (Rwanda)
49. Magahinga Gorilla NP (Uganda)
50. Bwindi Impenetrable NP (20)
51. Kigezi Gorilla Sanctuary (Uganda)

*CAR = Central African Republic
*DRC = Democratic Republic of the Congo
*FR = Forest Reserve
*NP = National Park
*RC = Republic of the Congo

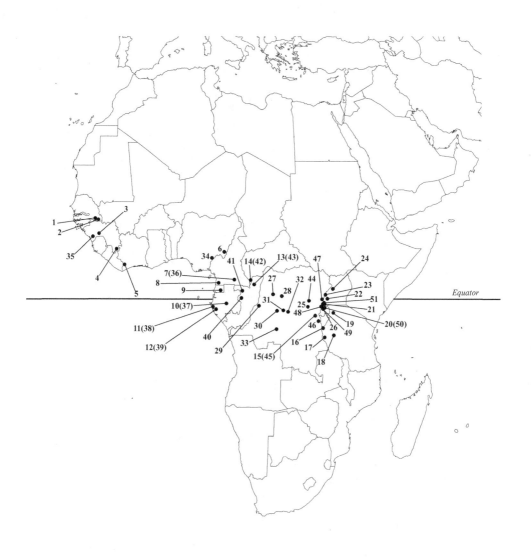

Map 2.2. African ape study sites

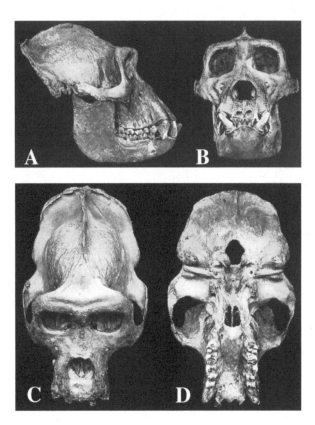

Figure 2.13. Skull of a male western gorilla *(Gorilla gorilla)*. (A) Right lateral view. (B) Frontal view. (C) Dorsal view. (D) Basal view.

pillar is relatively broad. They have marked postorbital constriction and flaring zygomatic processes.

Chimpanzee skulls are much less sexually dimorphic than those of gorillas and orangutans, though the difference appears quite early in development.[131] Sagittal crests are uncommon (19 percent of fifty-eight skulls) in male chimpanzees and virtually nonexistent in female chimpanzees (Figure 2.14). Male gorillas typically have large sagittal crests, which may sometimes occur in adult females. Nuchal crests occur regularly in both sexes of both species. They are especially well developed laterally, where they continue as supramastoid crests.[103]

The crania of male and female bonobos are gracile.[132] Their muzzles are less projecting, their foreheads are more prominent, and their skulls are

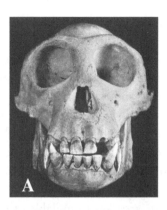

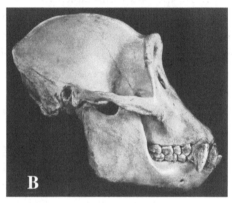

Figure 2.14. Skull of a male chimpanzee *(Pan troglodytes)*. (A) Frontal view. (B) Right lateral view.

rounder than those of chimpanzees and gorillas. Overall, the crania of adult bonobos are strikingly like those of juvenile chimpanzees. These and other features such as persistence of the perianal tuft indicate that bonobos might be neotenic.[133]

African ape mandibles have stout bodies and tall, wide rami; the condyles are high above the occlusal plane. The mandibles of gorillas are the most massive among the African apes.[134] The mandibles of *Pan paniscus* are distinctly shorter and less robust than those of *Pan troglodytes*.[135] In profile, the outer surface of the chin slopes downward and posteriorly. Sometimes there is a horizontal simian shelf anteroinferiorly where the two halves of the mandibular body converge at the mental symphysis. The simian shelf is a highly variable feature in the great apes.[136]

The African apes share a common dental pattern that is grossly similar to that of orangutans. It varies interspecifically and, to a lesser extent, dimorphically largely because of variations in body size between the sexes and among the species. Gorillas have the largest teeth among extant African apes. Bonobo teeth are generally smaller and less sexually dimorphic than chimpanzee teeth.[137] They contrast markedly with the condition in chimpanzees and gorillas, which have notable dimorphism in all teeth.[138]

The upper incisors of *Pan* and *Gorilla* are heteromorphic, with I^1 wider than I^2. The lower incisors are basically homomorphic, with I_1 typically somewhat smaller than I_2.[86]

The canines of African apes are formidable. They are largest and most sexually dimorphic in *Gorilla,* less so in *Pan troglodytes,* and least dramatic in *Pan paniscus.*[139] The upper premolars are bicuspid and homomorphic. Their lower premolars are heteromorphic. P_4 is basically bicuspid, though extra small cusps are common. And the sectorial P_3 is essentially unicuspid, with variably developed additional cusps and cingula.

Among apes, the molars of chimpanzees and bonobos are metrically and morphologically most similar to human molar teeth.[108] African ape upper molars are quadricuspid, and they frequently sport lingual cingula. The lower molars are usually pentacuspid, with the hypoconulid set buccally instead of centrally. The lower molars are commonly embellished with buccal cingula that are better developed in gorillas than in other great apes. In *Pan* and *Gorilla,* the Y-5 pattern of grooves is more characteristic than the plus-4 pattern, especially on M_1 (Figure 2.8).[138]

Chimpanzees, and especially gorillas, have relatively short broad torsos, elongate forelimbs, and short hind limbs (Figures 2.15 and 2.16). However, their forelimbs are not as elongate as those of orangutans. In the African apes, forelimb length and hind limb length scale negatively relative to body mass. This negative allometric relationship is stronger in the hind limb. Consequently, the intermembral index increases with increasing size from *Pan paniscus* and *Pan troglodytes* to *Gorilla* spp.[140]

Postcranially, bonobos are distinguished from chimpanzees by their narrower shoulders and slender, long-limbed physiques. They are less sexually dimorphic than chimpanzees in several postcranial features.[22]

Bonobo, chimpanzee, and gorilla genitalia distinguish them from humans and from one another. A bonobo clitoris is quite discernible, even during the monthly sexual swelling of her labia minora. The perineal swell reorients the clitoris so that it points ventrally between her thighs. Like a penis,

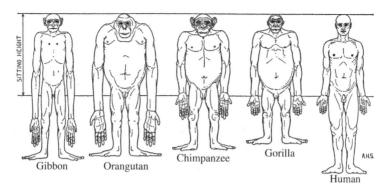

Figure 2.15. Length of upper and lower limbs of apes and human relative to sitting height.

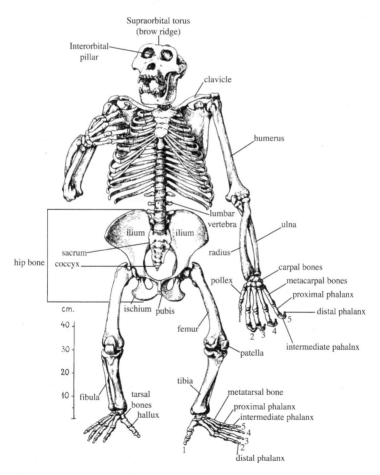

Figure 2.16. Skeleton of a western gorilla *(Gorilla gorilla)*.

it enlarges when the female is libidinous or otherwise excited. A chimpanzee clitoris is much less visible, especially during monthly perineal swells. As in orangutans, gibbons, and humans, female gorillas do not have prominent perineal swells with their ovulatory cycles.[141]

Like human males, male gorillas sport a well-developed glans penis that is apparent during penile erections. Contrarily, the chimpanzee and bonobo penes taper to a point. Upon erection a penis of *Pan* is conspicuous primarily because of its length and pinkness. Nevertheless, what chimpanzees and bonbos lack in penile size they recoup via their huge testicles.[142] Even gargantuan gorillas and overblown orangutans do not match them in this feature. Men can compete with them only through their fecund imaginations and CinemaScope.

The Peopled Earth

During the Late Pleistocene and Holocene epochs, humans achieved global distribution, fathomed the oceans, became airborne, and penetrated the heavens.[143] Now, all seven continents, the major archipelagos, and many lesser islands—ranging from paradisal to godforsaken—host human residents and visitors. Demographic expansions in Europe, southeastern Asia, and sub-Saharan Africa during the last 10,000 years were facilitated by the invention of agriculture, which ultimately led to a fivefold increase in population growth relative to more ancient expansions of hunter-gatherers, a trend that shows no sign of abatement despite wars, epidemics, and natural catastrophes.[144]

Genomic diversity of *Homo sapiens* is less than that of most populations of *Gorilla* spp., *Pan* spp, and *Pongo* spp. that have been tested; indeed, the pongid species are notably more diverse than all of *Homo sapiens*.[145] Nonetheless, there might have been a relatively recent acceleration of human adaptive evolution due to demographic growth and changes in human ecologies and cultures.[146]

Humans have much longer potential life spans than those of well-kept incarcerated apes, which rarely live more than 50 years. In addition to longevity, human life histories are characterized by an extended period of juvenile dependence, reproductive support by older postreproductive individuals, and male reproductive support via regularly sharing food with females and their offspring.[147] There are also notable differences between the incidence of cancer, ischemic heart disease, and neurodegeneration in humans (higher)

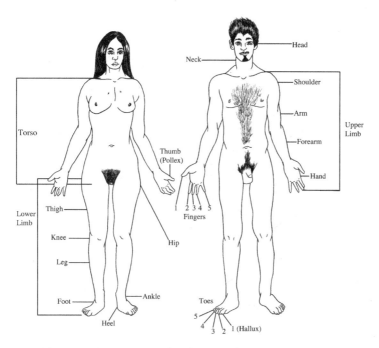

Figure 2.17. Major anatomical segments of the human body. Note epigamic features and moderate sexual dimorphism in overall body size.

versus apes, which might have genetic bases.[148] Although it is reasonable to suspect that susceptibility and resistance to diseases was important in the evolution and diversification of the Hominoidea, we need more longitudinal data on apes in their natural habitats and more specific information on the action and penetrance of genes that might be implicated as selective factors for life span and immunity versus susceptibility to disease vectors.

Homo sapiens exhibit only moderate sexual dimorphism in stature, weight, and many skeletal features, including the dentition. But humans are strikingly dimorphic in certain epigamic features. Postpubertal males typically have more hirsute faces and bodies; their pates commonly bald progressively with age. Human females are more glabrous, usually keep their head hair, and have special fat depots in their breasts, hips, and thighs that make them more curvaceous than the common man (Figure 2.17).[149]

Human skin pigmentation ranges from swine-pink to very dark, purplish brown.[150] Both light-skinned and dark-skinned people commonly tan, often to very dark hues. Gorilloid, gibbonoid, and bonoboid black is nonexistent

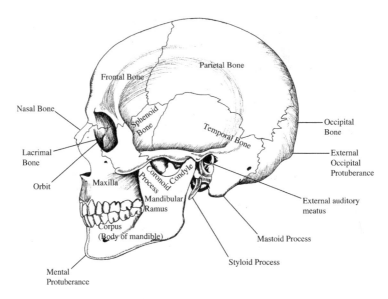

Figure 2.18. Skull of an adult male *Homo sapiens*.

among untanned, and perhaps even heavily tanned, human skins. Human hair pigmentation and texture vary widely also. Sometimes they match conditions in the apes, but in other instances humans transcend them. For example, the tightly coiled hair of various African, Asian, and Oceanic people are unmatched by apes. Similarly, the thin external membranous lips of apes do not rival the prominent lips of many people.

Human skulls (the crania) are dramatically different from the skulls of extant apes. In humans, the neurocranium is relatively large, and the splanchnocranium is quite abbreviated. The human cranium is high, rounded, and unadorned by superstructures. The human brow rises prominently above the orbits, and postorbital constriction is notably reduced in comparison with that of apes. The face is mainly tucked under the neurocranium; the palate and mandible are relatively short (Figure 2.18). Because the muzzle and upper face are massive in great apes, their faces are higher and more protruding than ours are. However, gibbons have proportionately smaller faces than ours.[136]

The human face is further distinguished from those of apes by the fact that our nasal bones project anteriorly from the facial plane, thereby forming an awning over the piriform aperture. Further, the external surface of the mandibular symphyseal region sports the mental protuberance (also known as the mental eminence) that distinguishes us among the primates.[151] Otherwise, human mandibles are rather middling. They fall far short of great ape robustness but do not approach hylobatid gracility.

Human palates are basically parabolic, which contrasts with the elongate U-configuration of great ape palates. The difference relates partly to the fact that in some dimensions the human dentition, especially the canines, is smaller than those of great apes. Hylobatid apes more closely match humans in palatal shape; among catarrhine primates, hylobatid mandibles are closest to the human configuration.[152]

Although human upper central incisors are larger than the lateral ones, they are usually homomorphic. The lower incisors are also homomorphic with I_1 somewhat smaller than I_2.[108]

Human canines are small, incisiform, and bluntly pointed. They do not project much, if at all, beyond the occlusal edges of the neighboring teeth.[153] In their low degree of canine sexual dimorphism, humans compare most closely with gibbons among the apes. However, some modern human populations exhibit slightly greater canine sexual dimorphism than that of gibbons.[96] Concordant with the puny upper canine, the human P_3 is nonsectorial and bicuspid. The other three premolars generally bear two principal cusps, but they are quite variable.[108] Diastemata are rare in *Homo sapiens*.[153]

Human molars exhibit the basic hominoid patterns of cusps and grooves, but there is a strong tendency in some populations for third molar reduction and even absence. Human lower molars are generally more bunodont than those of great apes, and the enamel is thick.[154]

The human body is easily distinguished from those of apes because of its unique proportions. We have the longest lower limbs relative to truncal height among anthropoid primates, and our upper limbs are also rather long (Figure 2.15). Our torsos are proportioned more like those of hylobatid apes than of great apes (Figure 2.7). Like *Pongo,* we depart dramatically from the other extant Hominoidea in pedal structure.[155] Our hands are quite special, too.[154]

While at first glance our bodies appear to be markedly different from those of apes, it is our elaborate cultural behavior and artifacts that seemingly set us furthest from them. This is all the more remarkable when we recall how closely related we are to the African apes genetically. Nonetheless, humans grow at rates slower than those of chimpanzees, which allows more time for increased cognitive development and lower body maintenance costs.[156]

The apparent chasm is reasonably linked to our ballooned brains, though we remain sadly deficient in knowledge about exactly how this fabulous fatty blob works. We should consider ourselves most fortunate to have faculties that allow us to investigate other natural phenomena and ourselves. Collectively and individually, we not only possess great stores of knowledge but also relentless curiosity about many worldly and spiritual puzzles that invite boundless exploration.

Prospective Human Evolution

During the twentieth century, when eugenics was ascendant in Western scientific and political circles, the grand bogey that powerful men hoped to exorcise from humanity via repressed breeding and extermination of persons who might have undesirable genomes, termed a genetic load, that reduces mean fitness of a population due to the presence of deleterious alleles. The misery and havoc caused by their biologically ill-informed and misguided sociopolitical efforts haunt most regions of the globe today, and they have caused many persons to distrust scientists and science. Tellingly, there is no hint that the human condition, biological or otherwise, was improved an iota by eugenic theories or practice.

Although human evolution is outside the realm of predictive sciences, we may expect certain changes if current global environmental conditions and trends persist. Genetic recombination will increase as people travel more, enjoy opportunities to reproduce with persons distant from their natal areas, contrive designer babies, and adopt children from other continents. The variety of human genomes will change also in response to euphenic recombinant DNA interventions, the spread of new viral and other microbial diseases, and the eradication of existing diseases.

As the U.S. mall diet spreads, so also will the mean body mass of its aficionados, and if nutritious diets from birth through adolescence become available to populations that were not so privileged in the past, there should be increases accordingly in stature, body mass, brain mass, and brain function.

Air, water, noise pollution, and the stress of living in overcrowded conditions might select for or against certain genomes, directly effecting individuals before birth or reproduction or inducing young people to migrate where they might breed with persons who are different from potential mates in their natal areas. Differential numbers of children in various societies, due to either parental choice or a governmental proscription will alter the representation of some genes in the phenotypes of the global population. Were cloning to become a reality for *Homo sapiens,* redundant genomes would be added to the mix.

Sadly, war—the grand facilitator of disease, malnutrition, forced pregnancy, and selective killing—shows no sign of passing out of fashion among humanity. Accordingly, there is the grim possibility that global warfare or a pandemic of rapidly evolving viruses or both might decimate *Homo sapiens* so that a few survivors would constitute a population bottleneck in a restricted area of the globe, thereby greatly reducing the variety of genomes in the species. The creatures that would deploy and evolve in such a much-altered landscape are perhaps best left for writers of science fiction to describe.

3

Apes in Time

Despite several recent, wonderful discoveries, our chest contains too few solidly dated fossils that are in a reasonably complete state. The rest of it is filled with hope and air.

RUSSELL H. TUTTLE (1988a, p. 392)

Rigorous and critical anatomical studies of fewer morphological characters, in the context of molecular phylogenies, is a more fruitful approach to integrating the strengths of morphological data with those of sequence data. This approach is preferable to compiling larger data matrices of increasingly ambiguous and problematic morphological characters.

ROBERT W. SCOTLAND, RICHARD G. OLMSTEAD, and JONATHAN R. BENNETT (2003, p. 539)

The Search for Fuzzy Adams and Eves

The paleoanthropological specialties are legion and sometimes tongue twisters. They include vertebrate and invertebrate paleontology, taphonomy, geology, geochronometry and geochronology, palynology, prehistoric archeology, behavioral primatology, the cultural anthropology of human hunters and gatherers, comparative and functional morphology, biomechanics, systematics, evolutionary theory, and molecular biology.[1] Paleoprimatologists are vertebrate paleontologists and other organismal biologists who specialize in the study of fossil apes, monkeys, and prosimians.

Historically, as a multidisciplinary team effort, paleoanthropology is most firmly rooted at Olduvai Gorge, Tanzania, where Louis and Mary Leakey attracted a diverse group of scientists to assist in their arduous quest for human origins.[2] F. Clark Howell, Richard Leakey, Yves Coppens, Donald Johanson, and many others refined the multidisciplinary approach and made it global practice.[3] Advanced geochemical dating methods are a special boon to paleoanthropological science.

The most accurate means to date fossil sites is by radiometric methods.[4] Potassium-argon (K/Ar), fission track, uranium-disequilibrium, and other radiometric techniques have been used to date some Oligocene-Pleistocene (34 Ma to 10 Ka) fossil anthropoid-bearing sites in Egypt and Eastern Africa. Many contemporary fossiliferous European, Asian, and southern African sites are less precisely dated becaue they lack the minerals that might provide geochemical dates. Instead, they are correlated with radiometrically dated sites elsewhere via faunal and sedimentological similarities: geochronology.[4] The indirect methods are fraught with problems, and they are less dependable than directly dating the fossils or the deposits in which they are found.[5] Radiocarbon dating is only useful to date bones, teeth, plants, and other organic materials in some Late Pleistocene archeological and paleontological sites.[4]

The Dating Game

By and large, the boundaries between various geological epochs and their subdivisions are drawn arbitrarily. We will employ the following scheme of Gradstein et al.[6]

Early Oligocene	34–28 Ma
Late Oligocene	28–23 Ma
Early Miocene	23–16 Ma
Middle Miocene	16–11 Ma
Late Miocene	11–5 Ma
Early Pliocene	5–3.6 Ma
Middle Pliocene	3.6–2.6 Ma
Late Pliocene	2.6–1.8 Ma
Early Pleistocene	1.8 Ma-780 Ka
Middle Pleistocene	780–126 Ka
Late Pleistocene	126–10 Ka
Recent (Holocene)	10-Ka-1800 AD
Anthropocene	1784–0 AD[7]

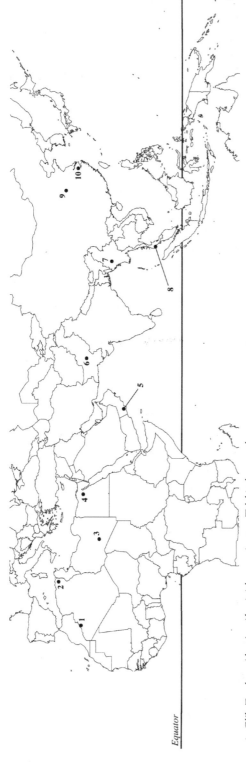

1. Glib Zegdou, Algeria; *Algeripithecus minutus, Talebia hammudae*
2. Bir el Ater locality, Algeria; *Biretia piveteaui*
3. Dur At-Talah, Libya; *Afrotarsius libycus, Biretia piveteaui, Talahpithecus parvus, Karanisia arenula*
4. Jebel Qatrani Formation, Fayum, Egypt; *Abuqatrania basiodontos, Catopithecus browni, Serapia eocaena, Proteopithecus sylviae, Arsinoea kallimos, Biretia fayumensis, Biretia megalopsis, Oligopithecus savagei, Qatrani wingi*
5. Taqah locality, Oman; *Oligopithecus savagei*
6. Bugti Hills, Balochistan, Pakistan; *Phileosimias kamali, Phileosimias brahuiorum*
7. Pondaung Formation, Myanmar; *Amphipithecus mogaungensis, Pondaungia cotteri, Pondaungia savagei, Myanmarpithecus yarshensis, Bahinia pondaungensis, Eosimias paukkaungensis, Afrasia djijidae*
8. Krabi Basin, Thailand; *Siamopithecus eocaenus*
9. Heti Formation, Shanxi Province, China; *Eosimias centennicus, Eosimias dawsonae, Phenacopithecus xueshii, Phenacopithecus krishtalkai*
10. Shanghuang, Jiangsu Province, China; *Eosimias sinensis*

Map 3.1. Eocene (54–34 Ma) fossil sites of Eurasia and Africa

The Neogene comprises the Miocene and Pliocene epochs, and the Paleogene comprises the Oligocene and preceding Eocene (54–34 Ma) and Paleocene (65–54 Ma) epochs.

Paleoecological evidence indicates that the Oligocene and Neogene habitats of anthropoid primates were forested to a greater or lesser extent, especially during the earlier time ranges. Africa was separated from Eurasia until ca. 18–22 Ma. Then the Afro-Arabian plate joined Eurasia, and faunal interchanges commenced between the two great landmasses.[8] Ancestors of Eurasian hominoids were probably among the creatures that left Africa for Eurasia during the Middle Miocene.[9]

Asian Beginnings or Afro-Arabian Origins?

Contenders for the earliest anthropoid primates have been found in Paleogene deposits in Myanmar, Thailand, China, Pakistan, Libya, Algeria, Egypt, Oman, and Morocco (Map 3.1).[10] In body size, they range from tiny to the modest mass of gibbons.[11] The search for the superlative "first primates" has been confounded by disagreement over the traits that securely identify fossil Anthropoidea, the variable possibilities for dating sites where suggestive fossils have been recovered, and the incompleteness and distortion of the specimens. Indeed, it is distinctly possible that the region where indisputable anthropoids primates emerged will never be found due to geological obliteration of the strata that contained their remains.

Based on an overview of the morphology and inferred paleoenvironmental contexts for available Eocene and Oligocene specimens, Kay et al. suggested that the ancestors of all Haplorhini were probably diurnal, insectivorous, leaping, and solitary.[12] The basal anthropoid ancestors were probably similar to them except that, based on sexual dimorphism of their canine teeth, they were more quadrupedal, more herbivorous, and probably exhibited greater social complexity, including group living and polygyny.[13]

The search for the beginnings of the Anthropoidea is complicated by the fact that the traits whereby we identify anthropoids seem to have developed via mosaic evolution. Moreover, unlike the Neogene period for which there is little chance of confusing an anthropoid craniodental specimen with a prosimian one, in the Paleogene the rare fragmentary specimens can be quite challenging even to the veteran expert. Because of differential preservation of dental enamel versus bone, identification of basal anthropoids often relies heavily on the structure of the molar teeth and temporal considerations.

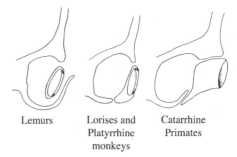

Lemurs   Lorises and   Catarrhine
             Platyrrhine   Primates
             monkeys

Figure 3.1. Three positions of the petrosal tympanic annulus: within the petrosal bulla (left), external to the bulla (center), and external to the bulla and extended into a tubular external auditory meatus (right).

The craniodental features that potentially qualify fossil specimens for membership in the Anthropoidea include (Figures 2.2, 2.3, and 3.1):

- fusion of the mandibular symphysis,
- early loss of the metopic suture to form a single frontal bone,
- reduction of the olfactory apparatus,
- frontation of the orbits,
- bony separation of the orbits from the temporal fossae,
- integration of the trigonid and talonid segments of the lower molars,
- bunodont molar cusps,
- vertical incisors, and
- tubular external auditory meatus.

Beard listed five Paleogene families as basal Anthropoidea—Afrotarsiidae, Amphipithecidae, Eosimiidae, Parapithecidae, and Protopithecidae—comprising thirty-one species in eighteen genera, plus *Algeripithecus minutus* and *Tabelia hammadae* in family *insertae sedis* (Table 3.1).[14] Simons et al. added the Arsinoeaidae to accommodate *Arsinoea kallimos* (Table 3.1).[15] Rasmussen supplemented the list with two late Eocene–Oligocene families— Oligopithecidae and Propliopithecidae—comprising nine species in six genera (Table 3.2).[16] Increasingly, they are considered to constitute the Propliopithecoidea, a superfamily that might be basal to the Catarrhini to the exclusion of the Platyrrhini.

MYANMAR.   Fragmentary gnathic, cranial, and dental specimens of *Amphipithecus mogaungensis* and *Pondaungia cotteri* are long-standing controver-

Table 3.1.   Basal (Paleogene) Anthropoidea

Afrotarsiidae		
Afrotarsius chatrathi	Early Oligocene	Egypt
Afrotarsius libycus	Late Middle Eocene	Libya
Amphipithecidae		
Amphipithecus mogaungensis	Late Middle Eocene	Myanmar
Bugtipithecus inexpectans	Early Oligocene	Pakistan
Pondaungia cotteri	Late Middle Eocene	Myanmar
Pondaungia savagei	Late Middle Eocene	Myanmar
Siamopithecus eocaenus	Late Eocene	Thailand
Arsinoeaidae		
Arsinoea kallimos	Latest Eocene	Egypt
Eosimiidae		
Afrasia djijidae	Late Middle Eocene	Myanmar
Bahinia pondaungensis	Late Middle Eocene	Myanmar
Eosimias centennicus	Late Middle Eocene	China
Eosimias dawsonae	Late Middle Eocene	China
Eosimias paukkaungensis	Late Middle Eocene	Myanmar
Eosimias sinensis	Middle Eocene	China
Phenacopithecus xueshiii	Late Middle Eocene	China
Phenacopithecus krishtalkai	Late Middle Eocene	China
Phileosimias brahuiorum	Early Oligocene	Pakistan
Phileosimias kamali	Early Oligocene	Pakistan
Parapithecidae		
Abuqatrania basiodontos	Late Eocene	Egypt
Apidium bowni	Earliest Oligocene	Egypt
Apidium moustafai	Earliest Oligocene	Egypt
Apidium phiomense	Early Oligocene	Egypt
Biretia piveteaui	Late Middle and Late Eocene	Libya, Algeria
Biretia fayumensis	Early or early Middle Eocene	Egypt
Biretia megalopsis	Early or early Middle Eocene	Egypt
Lokonepithecus mania	Late Oligocene	Kenya
Qatrania fleageli	Early Oligocene	Egypt
Qatrania wingi	Latest Eocene	Egypt
Parapithecus fraasi	Early Oligocene	Egypt
Parapithecus (= *Simonsius*) *grangeri*	Early Oligocene	Egypt
Proteopithecidae		
Proteopithecus sylviae	Latest Eocene	Egypt
Serapia eocaena	Latest Eocene	Egypt
Family incertae sedis		
Algeripithecus minutus	Middle-Late Eocene	Algeria
Altiatlasius koulchii	Late Paleocene	Morocco
Myanmarpithecus yarshensis	Late Middle Eocene	Myanmar
Tabelia hammadae	Late Early or Middle Eocene	Algeria

Table 3.2.   Early (Late Middle and Late Eocene–Oligocene) catarrhine Anthropoidea

Propliopithecoidea		
Oligopithecidae		
Talahpithecus parvus	Late Middle Eocene	Libya
Catopithecus browni	Latest Eocene	Egypt
Oligopithecus savagei	Early Oligocene	Egypt
Oligopithecus rogeri	Basal Oligocene	Oman
Propliopithecidae		
Moeropithecus markgrafi	Basal Oligocene	Oman, Egypt
Propliopithecus haeckeli	Early Oligocene	Egypt
Propliopithecus chirobates	Early Oligocene	Egypt
Propliopithecus ankeli	Early Oligocene	Egypt
Aegyptopithecus zeuxis	Early Oligocene	Egypt
Saadanioidea		
Saadaniidae		
Saadanius hijazensis	Late Oligocene	Saudi Arabia

sial contenders for the superlative first anthropoid.[17] They are from deposits in the late Middle Eocene Pondaung Formation of northern Myanmar.[18] Faunal correlation indicates that the gibbon-sized beasts lived ca. 41–37 Ma, which agrees with a fission-track age of ca. 37–38 Ma.[19] From molar teeth and post-cranial bones, one can estimate the body masses of *Amphipithecus* (4.9–7.0 kg), *Pondaungia savagei* (5.5–8.8 kg), and *P. cotteri* (3.9–5.8 kg).[20]

Their mandibular bodies are robust and deep, and their molars are bunodont and lack hypoconulids. Whereas *Pondaungia* has wrinkled molar enamel, that of *Amphipithecus* is smooth. *Pondaungia* might have been sexually dimorphic, as evidenced by different sizes of the jaws. The mandibles and molar teeth with weak shearing crests of *Amphipithecus* and *Pondaungia* indicate they subsisted on a low-fiber diet of fruits and seeds. The thick molar enamel in *Pondaungia* suggests that particularly hard seeds were in their diet. Large spatulate upper central incisors and robust projecting upper and lower canines would serve both genera well to husk fruits.[21]

Historically, the Pondaung specimens were variously assigned to the Prosimii, the Anthropoidea, the Pongidae, and the Condylarthra (early ungulates). The premolars and molars of both species sport mixtures of adapid prosimian and anthropoid features.[22] Accordingly, Ciochon and other researchers abandoned *Pondaungia* and *Amphipithecus* as Anthropoidea, considering them instead to be adapiform prosimians.[23] Their decision was based not only on craniodental evidence but also on traits of humeri, ulnae, and part of a calcaneus of one individual that they judged to be *Pondaungia* based on estimates of its body mass (4–9 kg). Contrarily, others were disposed to keep *Pondaungia* in the basal Anthropoidea.[24]

Takai et al. suggested that *Myanmarpithecus yarshensis,* a non-amphipithecid species from the Pondaung Formation, might be an early anthropoid.[25] The fragmentary upper and lower jaws and teeth of three individuals that were smaller (1.5–2.1 kg) than the Pondaung amphipithecids represent *Myanmarpithecus.*[26] Their low-crowned bunodont molars sport wrinkled enamel that suggests a frugivorous diet.[21]

A fourth species from the Pondaung Formation rekindled the idea that Myanmar might have been an important area of anthropoid emergence. The teeth and upper and lower jaw fragments of a small marmoset-sized primate (about 400 grams or 550–660 grams) represent *Bahinia pondaungensis.*[27] Like marmosets, its tooth structure indicates that it was probably insectivorous, and it also was probably arboreal.

The Pondaung Formation also yielded two tantalizing specimens of three additional species that might be basal anthropoids.[28] Gebo et al. assigned a calcaneus to a tiny unnamed species of Eosimiidae, with a body mass of 52–180 grams.[20] Takai, Sein, and coworkers provisionally named a new species *Eosimias paukkaungensis* based on its toothless left and right madibular corpora, except for the right M_3.[29] *Eosimias paukkaungensis* is larger (approximately 420 grams) than the other Pondaung Eosimiidae and two species of *Eosimias* in China.[20]

Pondaungia savagei is the most abundant species in the Pondaung Formation, followed closely by *Amphipithecus mogaungensis.* The next most abundant are *Pondaungia cotteri* and *Myanmarpithecus yarshensis.* Singular specimens represent *Bahinia* and *Eosimias.*[20]

THAILAND.   The geographical range of early anthropoids was expanded southeastward by the discovery of *Siamopithecus eocaenus* at three Late Eocene sites in southern Thailand.[30] Ducrocq concluded that the gnathic specimens evidence close affinities with *Pondaungia* in Myanmar, *Aegyptopithecus* in Egypt, and *Moeripithecus* in Egypt and Oman.[31] Accordingly, *Siamopithecus* potentially strengthens the argument for eastern Asia as an important region for early anthropoid evolution. *Siamopithecus* is represented by upper and lower jaws and teeth, from which it is estimated to have weighed 8–9 kg.

CHINA.   Beard et al. have argued that *Eosimias sinensis,* recovered from Middle Eocene (ca. 45 Ma) fissure-fillings at Shanghuang, in Jiangsu Province in southeastern China, is a basal anthropoid that lived before the diversification of the Parapithecidae, Oligopithecinae, Platyrrhini, and Catarrhini.[32]

They view *Eosimias* as a reasonable ancestor of *Hoanghonius stehlini*, a later, enigmatic species from the Late Eocene of China, and *Amphipithecus* and possibly *Pondaungia* of Myanmar. Subsequently, some researchers have placed *Hoanghonius* among the adapiform prosimians.[33]

Beard et al. estimated the body mass of *Eosimias sinensis* to be 67–137 grams.[32] Its dental formula is 2.1.3.3, and it has an unfused mental symphysis. Likewise, an isolated petrosal part of the temporal bone, provisionally assigned to *Eosimias*, indicates that the ear region had not developed the derived anthropoid morphology.[34] The Shanghuang fissure-fillings also contain adapiforms, omomyids, and an endemic tarsiid among its rich and varied mammalian fauna.

Eosimias centennicus from the Heti Formation in Shanxi Province of central China is a bit larger (91–179 grams) and is probably geochronologically younger than *Eosimias sinensis*.[35] It is roughly contemporaneous with the *Bahinia pondaungensis* of Myanmar (Table 3.1). The dentition is slightly more derived than that of *Eosimias sinensis*, the mandibular symphysis is unfused, and the lower dental formula is 2.1.3.3. The low-crowned, bluntly crested molars suggest that the diet of *Eosimias centennicus* comprised fruits and insects.[36]

Beard and Wang proposed three additional Eosimiidae—*Eosimias dawsonae*, *Phenacopithecus xueshii*, and *Phenacopithecus krishtalkai*—from the Heti Formation, which also contains sivaladapid and tarsiid species.[37] Eeosimiids are the most diverse and abundant primates at the site. The single specimen of *Eosimias dawsonae*—represented by a hemimandible with M_{2-3}—is the largest species in the genus.

Phenacopithecus xueshii and *Phenacopithecus krishtalkai* are larger than *Eosimias* spp. but smaller than *Bahinia pondaungensis*.[35] The hypodigm of *Phenacopithecus xueshii* comprises sixteen isolated teeth, including upper and lower premolars and molars and an I_2.

The single specimen of *Phenacopithecus krishtalkai* is a right maxillary fragment with P^4–M^2. Because the depth of the muzzle is greater in *Phenacopithecus krishtalkai* than in big-eyed, nocturnal *Tarsius* and Omomyidae, Beard and Wang inferred that its orbit was relatively small and that the Eosimiidae were diurnal.[37]

PAKISTAN.   Based on several dozen primarily dental specimens, Marivaux et al. has proposed three species of amphipithecid and eosimiid primates from the Early Oligocene locality of Paali Nala in the Bugti Hills of Balochistan, Pakistan. *Bugtipithecus inexpectans* is a small (approximately 350 grams) amphipithecid whose molar and premolar teeth have smooth versus crenulated enamel and more acute cusps than those of earlier, larger amphipithecid species from Myanmar.[38]

The Bugti eosimiids are *Phileosimias kamali* and the somewhat smaller *Phileosimias brahuiorum*. They are distinguished from Chinese eosimiids by cuspidal morphology and shapes of the molars and premolars. Their body sizes are estimated to be comparable to those of *Phenacopithecus* (about 250 grams).[38]

Phileosimias and *Bugtipithecus* and a variety of strepsirrrhine primates lived in the tropical Bugti habitat while primates were disappearing from the cooler, more northern latitudes of Eurasia and North America. Marivaux et al. concluded that *Phileosimias* and *Bugtipithecus* are stem Anthropoidea and that they support Asia as the ancestral homeland of the suborder.[38]

ALGERIA.   De Bonis et al. assigned a tiny primate M_1 from a Late Eocene locality in Eastern Algeria to the Anthropoidea.[39] Further, they placed the new species *Biretia piveteaui* in the Parapithecoidea and argued that it supports Africa as the early homeland of the Catarrhini. If the date is upheld, *Biretia piveteaui* might be a bit older than anthropoids from the Fayum, Egypt.

Based on three bunodont teeth, Godinot and Mahboubi reported an even earlier diminutive anthropoid, *Algeripithecus minutus,* from Glib Zegdou, Algeria.[40] It dates to the Middle-Late Eocene (ca. 42 Ma). The teeth of *Algeripithecus minutus* are smaller than the teeth of *Biretia piveteaui* and the small Fayum anthropoids *Catopithecus browni, Proteopithecus sylviae,* and *Qatrania wingi*. Like extant marmosets and small prosimians, *Algeripithecus* probably weighed only 150–300 grams.

The Glib Zegdou site also contained bunodont teeth of the somewhat larger potential anthropoid *Talebia hammadae*.[41] Godinot and Mahboubi linked *Algeripithecus* and *Talebia* more with the Propliopithecidae than the Oligopithecidae or Parapithecidae, suggesting that the two families diverged in the Early Eocene.[42] Contrarily, after more complete specimens of *Algeripihecus minutus* were found—especially a mandible indicating that it might have had a toothcomb—Tabuce et al. reasonably argued that it belongs with stem Strepsirrhini instead of Anthropoidea.[43]

LIBYA.   Jaeger et al. claimed that diminutive teeth from a Late Middle Eocene (38–39 Ma) site at Dur At-Talah in central Libya represent the oldest diverse assemblage of African anthropoids.[44] The three species—*Afrotarsius libycus, Biretia piveteaui,* and *Talahpithecus parvus*—represent three families: Afrotarsiidae, Parapithecidae, and Oligopithecidae, respectively. They also collected the teeth of a lorisiform prosimian: *Karanisia arenula*. The estimated adult body masses of the four taxa range between 120 and 470 grams. Jaeger et al. concluded: "Given the apparent absence of anthropoids in significantly older . . . Eocene African localities such as Glib Zegdou in western Algeria" one must consider the hypothesis "that multiple Asian anthropoid clades may have colonized Africa more or less synchronously during the Middle Eocene."[45]

EGYPT.   Indisputable early anthropoids are among the diverse species of fossil primates from Late Eocene–Early Oligocene deposits in the Fayum region of Egypt. The Jebel Qatrani Formation, in which many of them were located, is arguably dated radiometrically between 29.5 Ma and 41.2 Ma.[46] There are twenty-one species in fourteen genera of Anthropoidea in the collections (Tables 3.1 and 3.2).[47] All species for which adequate samples exist evidence sexual dimorphism, from which Fleagle et al. and Simons et al. inferred that they were polygynous.[48]

In ancient times, the Fayum was an open, lightly wooded area with swamps and gallery forests near the river. Away from the river, the region was probably seasonally arid. The Jebel Qatrani primates are from the heavily forested area.[49]

There are five species of dental apes in the Jebel Qatrani collections: *Aegyptopithecus zeuxis, Moeropithecus markgrafi, Propliopithecus ankeli, Propliopithecus chirobates,* and *Propliopithecus haeckeli.*[50] Numerous jaws and teeth and various other skeletal parts represent the *Aegyptopithecus zeuxis* and *Propliopithecus chirobates* (30 Ma).

Two or three mandibular specimens represent *Propliopithecus ankeli. Propliopithecus haeckeli* and *Moeropithecus markgrafi* are known only from holotype specimens. They are so primitive dentally, cranially, and postcranially that no family of extant Hominoidea—Hylobatidae, Panidae, Pongidae, or Hominidae—can be traced directly back to them.[51] Instead, they constitute a unique family, the Propliopithecidae.[52] Others found the Propliopithecidae distinctive enough to merit a separate superfamily, the Propliopithecoidea.[53] Dentally, they are similar to Miocene Hominoidea, Proconuloidea, Dendropithecoidea, and Pliopithecoidea. Therefore, irrespective of their superfamilial status, they accommodate reasonable candidates for ancestry to the Miocene Hominoidea. The propliopithecid dental formula is $\frac{2.1.2.3}{2.1.2.3}$.

Aegyptopithecus has a prominent, though basically anthropoid, snout with large canines and heteromorphic mandibular premolars. The cranium is ai-

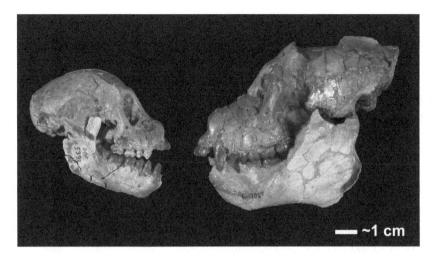

Figure 3.2. Female (left) and male (right) composite skulls of *Aegyptopithecus zeuxis*. Note striking sexual dimorphism.

rorynchic.[54] The face is hafted to the brain case such that it is low slung relative to the forebrain.[55] A nuchal crest is located high on the long, low neurocranium (Figure 3.2).[56]

Aegyptopithecus lacks the supraorbital torus. The prominent sagittal crest suggests robust masticatory muscles. Like modern Catarrhini, *Aegyptopithecus* has complete bony septa between the orbits and temporal fossae.[55] The orbits are oriented frontally and somewhat dorsally and are encircled by upraised rims. Expansion of the interorbital pillar gives the orbits a lima bean shape. The cheek region is robust and deep.

Aegyptopithecus lacks the bony tubular external auditory meatus of later Catarrhini.[56] In the Fayum anthropoids, as in ceboid monkeys, the ectotympanic bone is a simple annulus.[57]

The mandible of *Aegyptopithecus* is deep, and its symphyseal region sports a large superior transverse torus, a smaller inferior transverse torus, and prominent digastric impressions.[58] Their bunodont lower molars have medially placed hypoconulids; P_3 is sectorial.[59]

Relative to its estimated body size, the brain of *Aegyptopithecus* is more in the range of prosimian brains than of extant anthropoid brains. Indeed, the combination of large prognathic face and small neurocranium gives it a prosimian visage.[60] The postcranial bones assigned to *Aegyptopithecus* are more like those of ceboid monkeys than those of apes.[61]

Eight species of Fayum primates—*Abuqatrania basiodontos, Apidium bowni, Apidium moustafi, Apidium phiomense, Parapithecus fraasi, Parapithecus (=Simonsius) grangeri, Qatrania fleaglei,* and *Qatrania wingi*—along with *Biretia piveteaui* constitute the Parapithecidae, a separate anthropoid family that is uncertainly assigned to a superfamily or linked phylogenically with later primates. Like *Aegyptopithecus, Parapithecus grangeri* had a relatively small brain compared with those of extant Primates.[62]

Whereas Simons and Kay viewed the Parapithecidae as possible ancestors for the Cercopithecoidea, Cachel cited dental features that might ally them with the Hominoidea. She noted a five-cusp arrangement on the mandibular molars—the Fayum pattern—that harbingered that of Miocene Hominoidea. The paired protoconid and metaconid is followed by paired hypoconid and entoconid, with the hypoconulid centrally behind them.[63] Later paleoprimatologists placed the Parapithecidae in a unique catarrhine superfamily, the Parapithecoidea.[64]

Currently, parapithecid primates are not thought to be direct ancestors to cercopithecoid monkeys.[65] In several features they are more like platyrrhine than like catarrhine primates.[66] Indeed, Hoffstetter proposed that the platyrrhines originated in Africa from parapithecoid ancestors.[67] *Proteopithecus sylviae* or another proteopithecid taxon might be even better candidates.[68] Fleagle agreed that the ceboid monkeys and Fayum catarrhines probably have a common African ancestry.[69] How they might have they gotten from Africa to South America is a challenging mystery.

Like Miocene hominoids, the cercopithecoids might have evolved from propliopithecid primates.[70] However, first we need fossils from the immense span between the Jebel Qatrani Formation and the Early and Middle Miocene sites in Egypt, Libya, and eastern Africa that have yielded recognizable cercopithecoid monkeys (*Prohylobates* and *Victoriapithecus*), sometimes associated with dental apes.[71] The oldest cercopithecid specimen is a tooth (ca. 19 Ma) of *Victoriapithecus* from Napak V, Uganda, followed by *Prohylobates tandyi* (17–18-Ma) from Moghara, Egypt.[72]

An ischium of *Victoriapithecus* from the Middle Miocene deposits on Maboko Island, Kenya, shows that ischial callosities evolved before 14.7 Ma.[73] There is no evidence for ischial callosities on the Fayum hip bones, which are assigned to the Parapithecidae.[74]

The earliest Fayum primates (ca. 36 Ma) include *Abuqatrania basiodontos, Catopithecus browni, Serapia eocaena, Proteopithecus sylviae,* and perhaps *Afrotarsius chatrathi* and *Arsinoea kallimos,* followed by (33 Ma) *Oligopithecus savagei* and *Qatrani wingi.*[75] Gunnell and Miller viewed the Propliopithecidae as derived from *Catopithecus* (34 Ma) and the Parapithecidae from *Serapia.*[76]

Oligopithecus has been variously classified supragenerically. Although Simons claimed that the tiny specimens represented a hominoid, others have argued that they are adapid prosimians.[77] If *Oligopithecus* was adapid, it would expand the array of prosimian species from the Jebel Qatrani deposits that lived among the diverse anthropoid fauna in the Fayum.[78] In any case, the similarities between *Oligopithecus* and certain adapid forms suggest that they and other Anthropoidea are descended from the Adapidae.[79]

Harrison, Fleagle and Kay, and Rasmussen and Simons placed *Oligopithecus* closer to the Propliopithecidae than to other Anthropoidea and between the Platyrrhini and the Catarrhini.[80] Indeed, Rasmussen and Simons proposed a new propliopithecid subfamily, the Oligopithecinae, to accommodate *Oligopithecus savagei*, *Catopithecus browni*, and *Proteopithecus sylviae.*[81] The latter two species are from a low level in the Jebel Qatrani Formation that probably is Late Eocene.[82] Rasmussen placed only *Catopithecus browni* and *Oligopithecus savagei*, along with the Omani *Oligopithecus rogeri*, in the Oligopithecidae of the Propliopithecoidea (Table 3.2), a decision supported by dental and humeral features of *Catopithecus* vis-à-vis *Proteopithecus.*[83]

Based on analogies with the dentitions and body sizes of modern primates, we might infer that most of the Fayum primates were arboreal and basically frugivorous. Soft-fruit frugivory is suggested in species whose molars lack or show only modest cresting.[84] The larger species such as *Aegyptopithecus* might have supplemented their diets with leaves and harder objects.[85] Small species probably garnished meals with arthropods or resins.[86]

OMAN. Thomas et al. tentatively assigned an M_2 from a Late Eocene or Early Oligocene Taqah locality in the Sultantate of Oman to *Oligopithecus savagei.*[87] Thomas et al. assigned several other dental and mandibular specimens from Taqah to the propliopithecid species *Moeripithecus markgrafi.*[88] Gheerbrant et al. proposed a new species of Early Oligocene oligopithecine *Oligopithecus rogeri* based on around 120 isolated teeth from Taqah and Thaytiniti in the Sultanate of Oman.[89] The Omani fossils notably extend the geographic range of propliopithecoid primates southeastward of the Fayum.

MOROCCO. *Altiatlasius koulchii*, which was proposed on the basis of about ten isolated postcanine teeth, appeared to be the best candidate for a Late

Paleocene predecessor to stem Anthropoidea.[90] Others doubted the status of *Altiatlasius koulchii* as the first anthropoid and proffered instead that it is an omomyid tarsiiform or pleisiadapiform.[91]

If more complete 60-Ma specimens of *Altiatlasius koulchii* were to demonstrate that it represents basal Anthopoidea, the spotlight could shift from Asia back to Africa as a homeland of the earliest Anthropoidea.[92]

Eastern Africa Abounding in Apes

> We have little faith in *any* of the current phylogenetic analyses for Miocene hominoids, "analytically rigorous" or otherwise, nor do we believe that things will improve without some significant new material (especially from the tropical regions inhabited by extant apes.
>
> DAVID R. PILBEAM and NATHAN M. YOUNG (2001, p. 359)

After the Fayum propliopithecids, the next oldest prehominoid primates are the Late Oligocene–Miocene dental apes from eastern Africa (Table 3.3). Bracketed by K/Ar dates of 27.5 Ma and >24 Ma, *Kamoyapithecus hamiltoni* is the oldest among them.[93] According to K/Ar dating and faunal inferences, the other species variously lived between 23 Ma and 7 Ma.[94] They are concentrated areally at Rusinga Island, Maboko Island, Mfangano Island, Fort Ternan, and other localities in western and northern Kenya and the volcanoes of eastern Uganda. Kenya has the greatest trove of fossil catarrhines, but the total area within which they have been found is only about 30,000 km^2, which is 0.1 percent of the continental African land surface (Map 3.2).[95] Taphonomic studies of floral remains from a site in the Early Miocene Hiwegi Formation (ca. 17 Ma) on Rusinga Island in Kenya indicate that the area was covered by deciduous broadleaf woodland with a continuous canopy, trees, shrubs, lianas, and climbers.[96]

Much of eastern Africa was probably covered by lowland rain forest until 16 Ma. Thereafter, volcanic activity commenced, rifting occurred, and more open woodlands appeared. However, extensive savannas, like those of contemporary eastern Africa, probably did not develop there during the Miocene.[97]

SPLIT, LUMP, AND SPLIT AGAIN. Chiefly teeth and fragmentary gnathic specimens represent the African Miocene apes. There are precious few crania, and they are incomplete, sometimes distorted, and difficult to assign to species.[98] Consequently, as with earlier Anthropoidea, researchers

Table 3.3. Mostly Miocene catarrhine Anthropoidea (excluding Cercopithecoidea)

Proconsulidae	**Proconsuloidea**	
Proconsulinae		
Proconsul africanus	Early Miocene [ca. 19–20 Ma]	Kenya
Proconsul heseloni	Early Miocene [ca. 17–18.5 Ma]	Kenya
Proconsul nyanzae	Early Miocene [ca. 17–18.5 Ma]	Kenya
Proconsul major	Early Miocene [ca. 19–20 Ma]	Kenya
Ugandapithecus major	Early Miocene [ca. 19–20 Ma]	Uganda
Afropithecinae		
Afropithecus turkanensis	Early Miocene [ca. 17–18 Ma]	Kenya
Equatorius africanus[a]	Middle Miocene [14–15.5 Ma]	Kenya
Heliopithecus leakeyi	Early Middle Miocene	Saudi Arabia
Nacholapithecus kerioi	Middle Miocene [ca. 15 Ma]	Kenya
Nyanzapithecinae		
Mabokopithecus clarki	Middle Miocene [ca. 15–16 Ma]	Kenya
Nyanzapithecus harrisoni[b]	Middle Miocene [ca. 13–15 Ma]	Kenya
Nyanzapithecus pickfordi[b]	Middle Miocene [ca. 15–16 Ma]	Kenya
Nyanzapithecus vancouveringorum[b]	Early Miocene [ca. 17–18.5 Ma]	Kenya
Rangwapithecus gordoni	Early Miocene [ca. 19–20 Ma]	Kenya
Rukwapithecus fleageli	Late Oligocene [25 Ma]	Tanzania
Turkanapithecus kalakolensis	Early Miocene [ca. 16.6–17.7 Ma]	Kenya
Dendropithecidae	**Dendropithecoidea**	
Dendropithecus macinnesi	Early Miocene [ca. 17–20 Ma]	Kenya, Uganda
Micropithecus clarki	Early Miocene [ca. 19–20 Ma]	Kenya, Uganda

(continued)

Table 3.3. (continued)

Micropithecus leakeyorum	Middle Miocene [ca. 15–16 Ma]	Kenya
Simiolus enjiessi	Early Miocene [ca. 16.6–17.7 Ma]	Kenya
	Pliopithecoidea	
Dionysopithecidae		
Dionysopithecus orientalis	Latest Early-Middle Miocene [15–17 Ma]	Thailand
Dionysopithecus shuangouensis	Early Miocene [17–18 Ma]	China
Platodontopithecus jianghuaiensis	Early Miocene [17–18 Ma]	China
Pliopithecidae		
Pliopithecinae		
Pliopithecus vindobonensis	Middle Miocene [ca. 15–15.5 Ma]	Slovakia
Pliopithecus antiquus	Middle Miocene [ca. 15 Ma]	France, Germany, Spain, Switzerland, ?Poland
Pliopithecus piveteaui	Latest Early Middle Miocene [16–17 Ma]	France
Pliopithecus platyodon	Early Middle Miocene [15–17 Ma]	Austria, Switzerland
Pliopithecus zhanxiangi	Middle Miocene [ca. 15 Ma]	China
Pliopithecus canmatensis	Middle Miocene [11.7–11.6 Ma]	Spain
Crouzeliinae		
Anapithecus hernyaki	Late Miocene [10–11 Ma]	Hungary, Austria
Egarapithecus narcisoi	Late Miocene [9 Ma]	Spain
Laccopithecus robustus	Late Miocene [8 Ma]	China
Plesiopliopithecus auscitanensis	Middle Miocene [14.5 Ma]	France
Plesiopliopithecus lockeri	Middle Miocene [ca. 13? Ma]	Austria
Plesiopliopithecus priensis	Late Miocene [9.5 Ma]	France
Plesiopliopithecus rhodanica	Middle Miocene [11–13 Ma]	France
Damiao upper molar[c]	Latest Middle Miocene	Mongolia

Family incertae sedis		
Paidopithex rhenanus[d]	Late Miocene [10–11 Ma]	Germany
Lomorupithecus harrisoni	Early Miocene [19–20 Ma]	Uganda
Hominoidea		
Griphopithecidae		
Griphopithecus africanus[e]	Late Early Miocene–Middle Miocene [14–17 Ma]	Kenya
Griphopithecus alpani	Middle Miocene [13.5 Ma]	Turkey
Griphopithecus darwini	Middle Miocene [13.5–15 Ma]	Slovakia, Austria, ?Germany
Dryopithecidae		
Dryopithecus brancoi	Late Miocene [10 Ma]	Hungary, Germany, ?Georgia
Dryopithecus carinthiacus	Middle Miocene [11–12 Ma]	Austria
Dryopithecus crusafonti	Late Miocene [10.5 Ma]	Spain
Dryopithecus fontani	Middle Miocene [11–12 Ma]	France, Germany, Spain, ?Austria
Dryopithecus laietanus[f]	Late Miocene [9.5–10 Ma]	Spain
?*Dryopithecus uraduensis*	Late Miocene	China
Pierolapithecus catalaunicus	Middle Miocene [12.5–13 Ma]	Spain
Pongidae		
Ankarapithecus meteai	Late Miocene [10 Ma]	Turkey
Gigantopithecus blacki	Pleistocene	China, Vietnam
Gigantopithecus giganteus	Late Miocene [6.3 Ma]	India, Pakistan
Khoratpithecus chiangmuanensis	Middle Miocene [<10–13.5 Ma]	Thailand
Khoratpithecus piriyai	Late Miocene [7–9 Ma]	Thailand
Lufengpithecus hudienensis	Late Miocene [7–9 Ma]	China
Lufengpithecus keiyuanensis	Late Miocene [8–10 Ma]	China
Lufengpithecus lufengensis	Late Miocene [8–9 Ma]	China

(*continued*)

Table 3.3. (continued)

Ouranopithecus macedoniensis	Late Miocene [9 Ma]	Greece
Ouranopithecus turkae	Late Miocene [7.4–8.7 Ma]	Turkey
Sivapithecus indicus	Late Miocene [9.1 Ma]	Pakistan
Sivapithecus parvada	Late Miocene [10 Ma]	Pakistan
Sivapithecus sivalensis	Late Miocene [7.4–8.8 Ma]	India, Pakistan
Chororapithecus abyssinicus	Late Miocene [10–10.5 Ma]	Ethiopia
Nakalipithecus nakayamai	Late Miocene [9.8–9.9 Ma]	Kenya
Oreopithcidae		
Oreopithecus bambolii	Late Miocene [6–8.4 Ma]	Italy, Sardinia, ?Romania
Family incertae sedis		
Anoiapithecus brevirostris	Middle Miocene [11.9 Ma]	Spain
Morotopithecus bishopi[?]	Early [>20.6 Ma] or Middle [≈15 Ma] Miocene	Uganda
Kogolepithecus morotoensis	Late Early-basal Middle Miocene [17–17.5 Ma]	Uganda
Kenyapithecus wickeri	Middle Miocene [14 Ma]	Kenya
Kenyapithecus kizili	Middle Miocene [13.5 Ma]	Turkey
Otavipithecus namibiensis	Middle Miocene [13 ± 1 Ma]	Namibia
Samburupithecus kiptalami	Late Miocene [9.5 Ma]	Kenya

Orrorin tugenensis	Late Miocene [6 Ma]	Kenya
Ryskop hominoid	Middle Miocene	South Africa
Graecopithecus freybergi	Late Miocene [7–8 Ma]	Greece
Udabnopithecus garedziensis[h]	Late Miocene [8–8.5 Ma]	Georgia

Superfamily incertae sedis

Family incertae sedis

Ardipithecus kadabba	Late Miocene [5.2–5.7 Ma]	Ethiopia
?*Dionysopithecus* sp.	Early Miocene [16 Ma]	Pakistan
Kalepithecus songhorensis	Early Middle(?) Miocene	Kenya
Kamoyapithecus hamiltoni	Late Oligocene [23.9–27.8 Ma]	Kenya
Krishnapithecus posthumus	Late Miocene [7.4 Ma]	India, Mongolia
Limnopithecus evansi	Early Miocene [ca. 19–20 Ma]	Kenya, ?Uganda
Limnopithecus legetet	Early Miocene [ca. 19–22 Ma]	Kenya, Uganda

a. Others consider this species to be *Kenyapithecus africanus*.
b. *Nyanzapithecus* probably should be sunk into *Mabokopithecus*.
c. Zhang and Harrison (2008).
d. The Eppelsheim femur is the type specimen.
e. Begun (2002) would lump *Kenyapithecus africanus* and *Equatorius africanus* in this species.
f. Moyà-Solà, Köhler et al. 2009 reassign to *Hispanopithecus laietanus*.
g. Senut et al. (2000) and Gommery (2005) consider the specimens to be *Afropithecus turkanensis* and *Ugandapithecus major*.
h. Gabunia et al. (1999) consider the specimen to be *Dryopithecus garedziensis*.

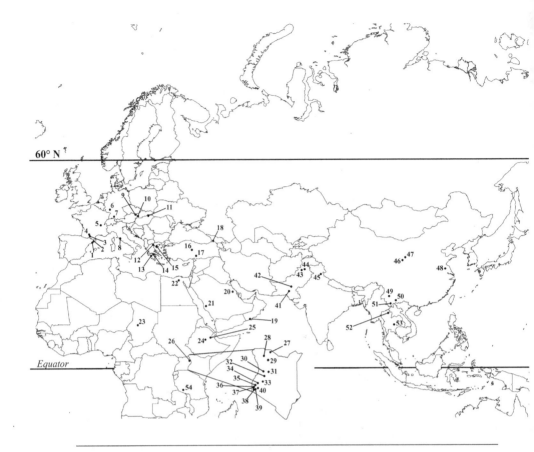

Map 3.2. Oligocene-Miocene (32–5 Ma) fossil sites

1. Barranc de Can Vila 1, Barcelona, Spain; *Pierolapithecus catalaunicus*
2. Vallès-Penedés Basin, Spain; *Pliopithecus, Anoiapithecus brevirostris*
3. St. Gaudens, France; *Dryopithecus fontani*
4. Sansan, France; *Pliopithecus*
5. Priay, France; *Pliopithecus priensis*
6. Eppelsheim, Germany; *Pliopithecus*
7. Engelwies, Germany; *Griphopithecus darwini*
8. Maremma District, Italy (Monte Bamboli; Bacinello); *Oreopithecus bambolii*
9. Klein Hadersdorf, Austria; *Austriacopithecus weinfurteri*
10. Neudorf an der March, Slovakia; *Pliopithecus vindobonensis*
11. Rudabànya, Hungary; *Dryopithecus brancoi, Anapithecus hernyai, Dryopithecus carinthiacus*
12. Ravin de la Pluie, Greece; *Ouranopithecus macedoniensis*
13. Xirochori, Greece; *Ouranopithecus macedoniensis*
14. Pyrgos, Greece; *Graecopithecus freybergi*
15. Nikiti 1, Greece; *Ouranopithecus macedoniensis*
16. Çandir, Turkey; *Griphopithecus alpani*
17. Pasalar, Turkey; *Griphopithecus alpani, Kenyapithecus kizili*
18. Udabno, Georgia; *Udabnopithecus garedziensis*

19. Dhofar Province, Oman (contains Taqah and Thaytiniti); *Oligopithecus savegei, Oligopithecus rogeri*
20. Dam Formation near Ad Dabtiyah, Saudi Arabia; *Heliopithecus leakeyi*
21. Shumaysi Formation, Saudi Arabia; *Saadanius hijazensis*
22. Jebel Qatrani Formation, Fayum, Egypt; *Aegyptopithecus zeuxis, Propliopithecus chirobates, Propliopithecus ankeli, Propliopithecus haeckeli, Moeropithecus markgrafi, Afrotarsius chatrathi, Apidium bowni, Apidium moustafai, Apidium phiomense, Qatrania fleageli, Parapithecus fraasi, Parapithecus grangeri*
23. Toros-Menalla locality, Djurab Desert, Chad; *Sahelanthropus tchadensis*
24. Middle Awash, Ethiopia; *Ardipithecus kadabba*
25. Beticha locality, Chorora Formation, Afar Rift, Ethiopia; *Chororapithecus abyssinicus*
26. Moroto I and II, Uganda; *Proconsul major, Morotopithecus bishopi, Afropithecus turkanensis*
27. Buluk, Kenya; *Afropithecus turkanensis*
28. Kalodirr, Kenya; *Afropithecus turkanensis, Turkanapithecus kalakolensis, Simiolus enjiessi*
29. Lokone, Turkana Basin, Kenya; *Lokonepithecus manai*
30. Nachola, Kenya; *Kenyapithecus africanus, Micropithecus, Xenopithecus, Nyanzapithecus harrisoni, Victoriapithecus, Nacholapithecus kerioi*
31. Namurungule, Samburu Hills, Kenya; *Samburupithecus kiptalami*
32. Nakali Formation, Kenya; *Nakalipithecus nakayamai*
33. Lukeino Formation, Tugen Hills, Kenya; *Orrorin tugenensis*
34. Kipsaramon, Tugen Hills, Kenya; *Equatorius africanus*
35. Songhor, Kenya; *Kalepithecus songhorensis, Proconsul major*
36. Fort Ternan, Kenya; *Ramapithecus wickeri, Kenyapithecus wickeri, Proconsul nyanzae*
37. Maboko Island, Kenya; *Simiolus leakeyorum, Kenyapithecus africanus, Mabokopithecus clarki Nyanzapithecus pickfordi, Victoriapithecus macinnesi*
38. Mfangano Islands, Kenya; *Nyanzapithecus vancouveringorum, Proconsul heseloni*
39. Rusinga Island, Kenya; *Nyanzapithecus vancouveringorum, Proconsul heseloni, Proconsul nyanzae*
40. Koru, Kenya; *Kalepithecus songhorensis, Proconsul major*
41. Manchar Formation, Sind, Pakistan; *Dionysopithecus*
42. Paali Nala, Pakistan; *Bugtipithecus inexpectans*
43. Kamlial Formation, Pakistan
44. Potwar Plateau, Pakistan; *Sivapithecus parvada, Sivapithecus indicus, Sivapithecus sivalensis, Gigantopithecus giganteus*
45. Haritalyangar, India; *Sivapithecus sivalensis; Gigantopithecus giganteus*
46. Gansu Province, China; *Dryopithecus wuduensis*
47. Tongxin County, Ningxia Hui Automonous Region, China; *Pliopithecus zhanziangi*
48. Xiacaowan Formation, Songlinzhuang, China; *Dionysopithecus shuangouensis, Platodontopithecus jianghuaiensis*
49. Lufeng County, Yunnan Province, China (contains Lufeng coalfieds & Shihuiba); *Lufengpithecus lufengensis, Laccopithecus robustus*
50. Xiaolongtan colliery, Keiyuan County, Yunnan Province, China; *Lufengpithecus keiyuanensis*
51. Xiaohe Formation, Yuanmou Basin, Yunnan Province, China; *Lufengpithecus hudienensis*
52. Phayao Provice, Thailand (contains Ban San Khlang & Ban Sa); *Dendropithecus orientalis, Khoratpithecus chiangmuanensis*
53. Nakhon Ratchasima Province, Thailand; *Khoratpithecus piriyai*
54. Nsungwe 2B locality, Rukwa Rift, Tanzania; *Rukwapithecus fleaglei*

primarily distinguish the many fossil species by morphological variations of their dentitions—such as the presence and distribution of cingula and cuspules and the shapes of primary cusps—maxillae and mandibles, especially the symphyseal region. Based on a cladistic analysis of extant hominoid molars, Hartman cautioned that dental measurements are unreliable indicators of phylogenic propinquity at low taxomonic levels.[99] For example, arguing against Uchida, Matsumura et al. proposed that dietary differences better explain the interspecific differences between chimpanzees and gorillas in M_1 morphology.[100] Consequently, relative overall body size, often based on estimates from molar size, has weighed heavily in all classificatory exercises on Miocene catarrhines.[101]

Some experts linked two species of African Miocene anthropoid species to the Hylobatidae. *Dendropithecus macinnesi* is supposed to show some suspensory features in its limb bones.[102] *Micropithecus clarki,* the smallest of the Miocene apes (Table 3.3), has a gibbonish face, but it evidences no clear connection with Oligocene Propliopithecoidea or with the Pliopithecoidea of Eurasia.[103]

Pickford noted that if *Micropithecus* or *Dendropithecus* incorporated bilophodonty into their molars, they would be viable ancestors for *Victoriapithecus,* a Middle Miocene monkey that is especially abundant on Maboko Island, Kenya.[104] If this scenario were correct, the Cercopithecoidea evolved in Africa from small-bodied dental apes during the Early Miocene.[105]

Lomorupithecus harrisoni is among the more tantalizing species of Pliopithecoidea, given that it is the only African Early Miocene member of the superfamily. In available morphological features, it groups with Eurasian species instead of later African ones.[106] The hypodigm comprises a fragmentary lower face, maxilla with left and right P^3-M^1, and a juvenile left mandibular fragment with M_1 and unerupted M_2 from Napak IX, a locality in Uganda that dates to ca. 10–20 Ma.[107]

Traditionally, the largest group of Eastern African Miocene apes—the Proconsuloidea—was generally and arguably linked to the living African great apes.[105] Before 1965, many species of Eurasian and African Miocene apes were named accordingly. Then, between 1965 and the mid-1970s, a schematic reduction in the number of fossil large ape taxa left only two genera: *Dryopithecus,* with three subgenera *(Dryopithecus, Proconsul,* and *Sivapithecus),* and *Gigantopithecus.*[108]

During the 1980s, there was a return to splitting, and the tendency to name and rename has continued unabated (Table 3.3).[105] For instance, Andrews reasserted the generic statuses of *Dryopithecus, Proconsul,* and *Sivapithecus,* and he further revised the taxonomy of African Miocene apes by proffering the following species, based largely on the relative sizes of their teeth: *Proconsul (Proconsul) africanus, Proconsul (Proconsul) nyanzae, Proconsul (Proconsul) major, Proconsul (Rangwapithecus) gordoni, Proconsul (Rangwapithecus) vancouveringi, Limnopithecus legetet, Micropithecus clarki,* and *Dendropithecus macinnesi.*[109]

Pickford listed no fewer than fifteen African Miocene anthropoid species binomially and noted eight more that awaited full and formal nomination. He also raised *Rangwapithecus* and the newly revived *Xenopithecus* from subgeneric to generic rank, resurrected *Kenyapithecus,* and accepted several new species, including *Mabokopithecus clarki, Micropithecus songhorensis,* and *Limnopithecus evansi.*[110]

Harrison changed *Proconsul (Rangwapithecus) vancouveringi* to *Nyanzapithecus vancouveringorum* and added a new species: *Nyanzapithecus pickfordi.*[111] He proposed that *Nyanzapithecus vancouveringorum* is a Middle Miocene ancestor for the Late Miocene European oddball *Oreopithecus bambolii.* He sustained *Rangwapithecus gordoni,* which might approach the Early Miocene ancestor of *Nyanzapithecus.*

Kunimatsu concluded that a novel species of *Nyanzapithecus* was probably present during the Middle Miocene at Nachola, Kenya, while *Nyanzapithecus vancouveringorum* might have been limited to Rusinga and Mfangano Islands, Kenya.[112]

Harrison also implemented a major taxonomic revision of the smaller Early Miocene catarrhine primates from western Kenya and Uganda.[113] He not only sustained *Dendropithecus macinnesi, Micropithecus clarki,* and *Limnopithecus legetet* but also recognized *Limnopithecus evansi* and a new species *Kalepithecus songhorensis* (formerly *Micropithecus songhorensis*), which is represented at Koru and Songhor.[114] Moreover, Harrison referred some 15-Ma specimens from Moboko Island to a new species, *Micropithecus leakeyorum.*[115] Benefit and McCrossin would sink it into *Simiolus,* as *S. leakeyorum.*[116] Harrison suggested that one of two small catarrhine primates at Fort Ternan is probably a novel species of *Simiolus.*[117]

Maboko is one of the richest paleoprimatological sites in eastern Africa. Researchers have recovered several thousand primate specimens and myriad remains of other mammals there. It is the type-site of *Kenyapithecus africanus, Mabokopithecus clarki, Nyanzapithecus pickfordi,* and *Victoriapithecus macinnesi.*[105]

Richard Leakey and coworkers recovered remarkable specimens, representing three distinct anthropoid species, at Kalodirr, an early Middle Miocene (16.8–17.8 Ma) locality west of Lake Turkana in northern Kenya. They named

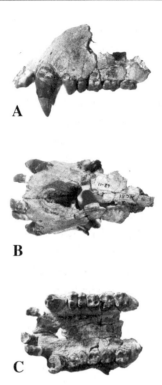

Figure 3.3. Ca. 17–20.6 Ma hominoid palate (UMP 62–11) from Moroto, Uganda, variously assigned to *Proconsul major, Dryopithecus major,* or *Morotopithecus bishopi.* (A) Left lateral view. (B) Superior view. (C) Occlusal view.

them *Afropithecus turkanensis, Turkanapithecus kalakolensis,* and *Simiolus enjiessi.*[118]

At the Moroto II site in Uganda, the 1963–64 expedition collected fifty-six additional pieces that Bishop ascribed to the same individuals of *Proconsul major* as the palatal and mandibular fragments that they had collected in 1961.[119] Allbrook and Bishop gave preliminary descriptions of the cranial remains.[120] In 1968, Walker and Rose reported remarkably African apelike vertebral remains, presumably of one individual, from Moroto II.[121]

The Ugandan remains of *Proconsul* constituted the primary empirical base for Pilbeam's doctoral thesis at Yale University. He was inclined to associate the palate, two mandibular fragments, and vertebral fragments from Moroto II as a single male of *Dryopithecus (Proconsul) major* (Figure 3.3). He concluded that all the large hominoid specimens from the three Napak sites belong to *Dryopithecus major.* He described the face as a scaled-down, long-snouted male gorilla with a gracile upper face. From the relatively low-crowned teeth, shallow palate, and other features, Pilbeam inferred that the Ugandan pongids were less well adapted than modern gorillas are to chewing tough vegetable matter.[122]

Pilbeam's reassessment of Kenyan specimens of *Proconsul* and *Sivapithecus* indicated close morphological affinities and conspecific status among *Dryopithecus major* from Koru and Songhor and the large Ugandan hominoids. He concluded that *Dryopithecus nyanzae* probably represents remnant populations of the ancestral stock that gave rise to *Dryopithecus major*.[122]

Gebo et al. collected additional postcranial specimens from Moroto I and II.[123] Although dated at 20.6 Ma, the specimens evidence more apelike features than any other Early or Middle Miocene anthropoid. Gebo et al. created a new species, *Morotopithecus bishopi,* for the entire collection of large anthropoid specimens from Moroto.[123] Unfortunately, they might have included a nonprimate scapular fragment in the hypodigm.[124] Further, Senut et al. noted that there are two large hominoids represented at Moroto: *Ugandapithecus major* (formerly *Proconsul major*) and *Afropithecus turkanensis,* of which *Morotopithecus* is a synonym.[125] In 1998–2003 at Moroto II, Pickford et al. recovered additional teeth of *Ugandapithecus* sp., isolated teeth of *Prohylobates macinnesi,* and a new species of small-bodied ape, *Kogolepithecus morotoensis*.[126] They also argued that the micromammalian fauna indicates that the site is 17–17.5 Ma instead of the older date suggested by Gebo et al.[123]

Nachola and Namurungule in the Samburu Hills south of Lake Turkana provided additional novel catarrhine specimens as a result of the efforts of the 1980 and subsequent Japan/Kenya Expeditions.[127]

Primate remains (*n* >200) are common components of the Nachola fauna.[128] Pickford proposed that the 15-Ma Nachola anthropoid collection comprised *Kenyapithecus* cf. *africanus, Micropithecus* sp., and perhaps *Xenopithecus* sp.[129] Contrarily, Kunimatsu and Sawada et al. noted that the Nachola catarrhines are *Nyanzapithecus harrisoni, Kenyapithecus* sp., and *Victoriapithecus*.[130]

Based on a partial skeleton from Kipsaramon, Tugen Hills, north-central Kenya, Ward et al. proposed a new species, *Equatorius africanus,* into which they sank presumed species of *Kenyapithecus* from the Aiteputh and Nachola Formations at Nachola and from the Muruyur Formation at Tugen Hills, and all specimens of *Kenyapithecus africanus* from the Maboko Formation of Maboko Island, Majiwa, Kaloma, Nyakach, and Ombo.[131]

Expanded samples of dentognathic and postcranial remains of the Nachola large proconsuloid also cast doubt on its being a species of *Kenyapithecus*.[132]

Accordingly, Ishida et al. named a new species: *Nacholapithecus kerioi*.[133] Postcranial remains have indicated that *Nacholapithecus kerioi* were tailless arboreal quadrupeds and climbers that engaged in little suspensory behavior.[134]

The Namurungule Formation contains one of the precious few Late Miocene sites in eastern Africa.[135] The single anthropoid specimen is a left maxilla with P³–M³ that Ishida and Pickford named *Samburupithecus kiptalami*.[136] It dates to ca. 9.5 Ma, which makes it the most complete hominoid specimen from early Late Miocene Eastern Africa. Pickford and Ishida concluded that dentally it is closest to early Hominidae, namely, *Australopithecus afarensis* and *A. anamensis,* which collectively they renamed *Praeanthropus africanus*.[137]

Fleagle placed *Dendropithecus, Limnopithecus, Micropithecus, Proconsul, Rangwapithecus, Simiolus, Kalepithecus,* and *Kamoyapithecus* together with two Chinese Miocene species—*Dionysopithecus and Platodontopithecus*—in the Proconsulidae.[138] He accepted *Nyanzapithecus* and *Mabokopithecus* as members of the Oreopithecidae and left *Afropithecus, Kenyapithecus, Turkanapithecus, Morotopithecus, Otavipithecus,* and *Samburupithcus* in the family *incertae sedis* of the Hominoidea.[139] Fleagle's suprageneric scheme resembles that of de Bonis, except that de Bonis placed *Rangwapithecus* in the Oreopithecidae and *Turkanapithecus* in the Proconsulidae and did not classify *Dendropithecus* familially.[140]

MISLEADING AND STILL MISSING LINKS.   Premature attempts to link specific African Miocene apes to extant African apes, such as *Proconsul africanus* to *Pan troglodytes* and *Proconsul major* to *Gorilla gorilla,* have been discredited.[141] Indeed, it is difficult even to derive a species of Eurasian Middle Miocene apes directly from the eastern African Miocene species, though it is generally thought that the latter must have evolved from African precursors.[142]

Pilbeam speculated that progressive adaptation to a tough vegetal diet on the heavily forested slopes of active volcanoes such as Moroto and Napak had transformed *Dryopithecus (Proconsul) major* into *Gorilla gorilla.* He believed that *Dryopithecus major* might well have been a knuckle-walker, though probably it was more active and less terrestrial than extant gorillas are.[122]

Pilbeam also concluded that *Dryopithecus (Proconsul) africanus,* though probably lacking knuckle-walking adaptations, was a likely ancestor to *Pan troglodytes*.[122] This would mean that the lineages leading to modern chimpanzees and gorillas were specifically separated ≥20 Ma.

Pilbeam retreated from the assignment of certain medium-sized African specimens to *Dryopithecus sivalensis* on the grounds that they were probably earlier than the Indian forms, though he thought they might represent ancestors of Eurasian *Dryopithecus*, especially *Dryopithecus (Sivapithecus) sivalensis*.[143]

The poor fossil record of apes particularly exacerbates the problem of linking Early and Middle Miocene apes to *Pan* and *Gorilla* during the Late Miocene and Pliocene of Africa.[144] There are few specimens to represent the overall morphology of Late Miocene African prehominid Hominoidea:

- ca. 9.5 Ma, maxillary fragment of *Samburupithecus kiptalami* from the Namurungule Formation in Samburu, northern Kenya.[145]
- ?5.2–5.8 Ma, isolated teeth that evidence an apish upper canine/P_3 honing complex, associated fragmentary upper limb bones, and a pedal phalanx of *Ardipithecus kadabba* from the Middle Awash in Ethiopia.[146]
- ca. 5.8 Ma, mostly postcranial bits of *Orrorin tugenensis* from the Lukeino Formation in Kenya.[147]
- ca. 7 Ma, relatively complete but distorted skull and several mandibular fragments and teeth of *Sahelanthropus tchadensis* from the Toros-Menalla locality in Chad, central Africa.[148]
- 10–10.5 Ma, teeth ($n = 9$) of *Chororapithecus abyssinicus* from the Beticha locality in the Chorora Formation of the Afar Rift in Ethiopia (Figure 3.4).[149]
- 9.8–9.9 Ma, mandible with three molars and eight isolated teeth of *Nakalipithcus nakayamai* from the Upper Member of the Nakali Formation in northern Kenya.[150]

The six species are too incompletely documented to provide links to extant apes, and *Ardipithecus kadabba*, *Orrorin tugenensis*, and *Sahelanthropus tchadensis* are arguably lineally closer to the Hominidae.[151] For example, if the femora of *Orrorin* are truly from a terrestrially bipedal hominoid, and were *Sahelanthropus* and even *Ardipithecus kadabba* verified to have been terrestrially bipedal via discoveries of associated postcranial skeletal remains, they could be basal Hominidae.[152]

The Samburu maxilla is gorilloid. However, unlike *Gorilla gorilla*, its postcanine teeth (P^3–M^3) are low crowned. Further, unlike modern African apes its molars have relatively thick enamel.[153]

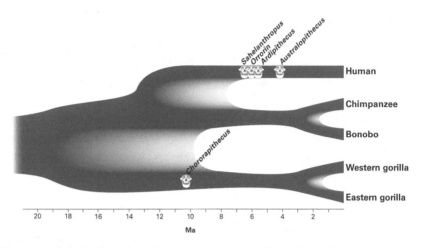

Figure 3.4. Phyletic scheme based on genetic parent age information on chimpanzees, mountain gorillas, and humans indicating the human–chimpanzee split at ≥7–8 Ma and Neandertal–modern human split at 400–800 Ka.

Chororapithecus abyssinicus is more compelling as an ancestor of *Gorilla.* The hypodygm comprises four upper and four lower molars or bits thereof and a lower canine tooth missing the tip and proximal root. The molar crowns are higher than those of *Samburupithecus* (but less than in *Gorilla*), and they have thick enamel. The cuspidal morphology indicates a capacity to grind and crush seeds and fibrous foods. Primates dominate the associated fauna, and the mosaic habitat bordering a river system included forest.[149]

Nakalipithcus nakayamai, represented by a right mandibular corpus with M_1–M_3 and eight isolated postcanine teeth, most closely resembles counterparts of *Ouranopithecus macedoniensis* in size and some dentognathic traits.[150] Stable isotopic and faunal analyses, particularly the presence of small noncercopithecoid catarrhine primates and a colobine monkey species, suggest a wooded or forested habitat for *Nakalipithecus.* Kunimatsu et al. have further suggested that *Nakalipithecus* might be close to the common ancestors of extant African great apes and humans.[150]

Early Pliocene bits are extremely sparse, minimally informative, and arguably more closely linked to Hominidae than to *Pan* or *Gorilla.*[154] There are no fossil specimens from the Pliocene to inform the evolution of *Pan* or *Gorilla.*

DIVERSITY THEN AND NOW.   All in all, the Miocene dendropithecoids, proconsuloids, and hominoids of Eastern Africa represent a remarkable adaptive radiation of forest, and perhaps woodland, primates that can be compared with the radiation of the Cercopithecoidea.[155] The four species of modern African apes are an impoverished faunal component in comparison with their Early and Middle Miocene cousins and cercopithecoid contemporaries.[115] Given that the number and diversity of extant primate species is highly positively correlated with the area of tropical forest and with mean annual rainfall, it is reasonable to ascribe the reduction of hominoid species through the Neogene to the aridification of major areas of Africa and Eurasia.[156]

The Miocene African catarrhines were diverse in size, morphology, and presumably habits. *Micropithecus clarki* was the size of capuchins (*Cebus* spp.). A Rusinga *Proconsul*—*P. heseloni*, represented by the subadult (female?) skeleton KNM-RU 2036, was as large (9–12 kg) and robust as a male *Colobus guereza*.[157] Another Rusinga *Proconsul*—*P. nyanzae*—was the size of female *Pan troglodytes;* consequently, it might have weighed as much as 40 kg, though Kelley and Pilbeam and Ruff et al. have argued that an average between 26 and 38 kg is more likely.[158] The Samburu hominoid was the size of a female gorilla.[159]

The 17.8-Ma *Proconsul heseloni* of Rusinga and Mfangano Islands, Kenya, is one of the best-known African Miocene species (Figure 3.5).[160] Although it had been classified traditionally as *Proconsul africanus,* Kelley and Pickford have argued that it is, in fact, *Proconsul nyanzae*.[161] Others have argued that two species of *Proconsul,* one of which is *P. nyanzae,* lived on Rusinga Island.[162] Based especially on size variation in tali and calcanei, Walker et al. reclassified the smaller *Proconsul* from Rusinga and Mfangano Islands as *Proconsul heseloni,* while continuing to recognize the larger species as *Proconsul nyanzae*.[163]

Mary Leakey found the most complete skull (KNM-RU 7290) of *Proconsul heseloni* at Rusinga locality R106 in the Hiwegi Formation.[164] Because of its gracility, we presume that it is from an adult female. It was grossly distorted taphonomically. However, guided by undeformed bits of a broadly contemporaneous subadult specimen (KNM-RU 2036) from Rusinga locality R114, its basic shape can be approximated.[165]

The cranium of female *Proconsul heseloni* is short, rounded, and remarkably devoid of bony superstructures such as brow ridges and the sagittal crest. Its nuchal crest is quite high; in fact, it lies higher than the upper borders of the orbits. Its face is orthognathous, and its orbits are large and rectangular. The zygomatic processes are modestly developed. Its nasal bones are long and narrow, though they lie in a wide interorbital pillar. The ectotympanic region is

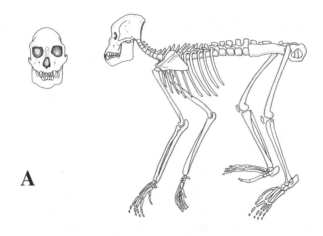

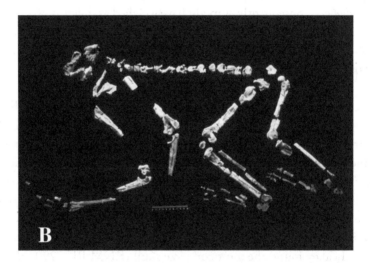

Figure 3.5. (A) Skeleton of 17.8-Ma *Proconsul heseloni*. (B) Ca. 15-Ma partial skeleton of *Nacholapithecus kerioi* from Nachola in northern Kenya.

tubular. Its estimated cranial capacity—167 cc—is arguably larger than those of modern monkeys with similar body size.[166]

The mandibular body of *Proconsul heseloni* is low and narrow, and its rami are relatively tall and thin. The coronoid process rises higher than the condylar process. Its mental symphysis is buttressed by a superior transverse torus, but it lacks an inferior transverse torus. The incisors are procumbent, and the canines are sizeable relative to the postcanine teeth. The P_3 is unicuspid, though it seems not to have functioned as a sectorial tooth. The M_2 is markedly reduced, but the M_3 has the largest surface area among the lower molars. The Y-5 pattern is manifest on M_1. The upper molars of *Proconsul* are characterized by lingual cingula, and the lower molars sport buccal cingula.[167]

Cranially, two species from Kalodirr are among the best-known Miocene apes. *Afropithecus turkanensis,* also from Buluk in northern Kenya, is a large snouty beast with a long, narrow palate, procumbent incisors, large projecting canines, and sectorial P_3. It sports a wide diastema between I^2 and the upper canine. The upper central incisors are large and broad, and the lateral incisors are relatively small and narrow mesiolaterally. The mandibular canines flare laterally. The upper molars increase in size from front to back: $M^1 < M^2 < M^3$. Similarly, the pattern in the lower molars is $M_1 < M_2 < M_3$, though the mandibular tooth rows are relatively short.[168] The dentition indicates a frugivorous diet.[169]

In the type specimen of *Afropithecus* (KNM-WK 16999), the long, narrow nasal bones lie in a wide interorbital pillar. The intact right orbit is higher than wide and is rounded perimetrically. Its zygomatic processes are well developed and sweep upward toward the temporal bones, indicating that the cranial vault was high relative to the face. The frontal bone slopes acutely backward from a shallow supraorbital notch that lies behind thin brow ridges. There is a median frontal sinus.[168]

The ramus (KNM-WT 17021) is oblique to the mandibular body. In a relatively complete anterior mandibular specimen of *Afropithecus* (KNM-WK 16840), the body is gracile and deep anteriorly; its depth decreases posteriorly. It lacks a superior transverse torus and has a poorly developed inferior transverse torus. This feature sets it apart from *Proconsul major,* which has a well-developed superior transverse torus and no inferior transverse torus.[170]

The colobus-sized skull of *Turkanapithecus kalakolensis* (KNM-WK 16950) is strikingly different from that of *Afropithecus.* It has a short snout, rather small orbits, a wide interorbital pillar, long, broad nasal bones, and relatively thick brow ridges. There is slight development of the supraorbital notch and a median frontal sinus. The zygomatic bone is deep below the orbit; the zygomatic process flares widely between it and the temporal squama.[171]

In *Turkanapithecus,* the gracile mandibular ramus is wide and low; the body is generally shallow but deepens at the symphysis. There is no superior or inferior transverse torus. The upper canine is large and $M^1 < M^2 < M^3$. The upper postcanine teeth sport supernumerary cuspules and relatively thin enamel.[171]

Although the jaws of several other Miocene African species—*Proconsul major, Proconsul nyanzae,* and *Rangwapithecus gordoni*—are represented fairly well, their neurocranial and basicranial morphologies are near blanks.[172] Further, except for *Proconsul* and *Nacholapithecus,* the postcranial skeletons of African Miocene hominoids are poorly known.

Thin dental enamel, spatulate upper central incisors, and little evidence of postcanine shearing features indicate that *Proconsul* spp. and *Limnopithecus* were soft-fruit frugivores. The more cuspidate shearing teeth of *Rangwapithecus gordoni* suggest that they were more folivorous.[169] The trove of dentognathic specimens from the Maboko and Aka Aiteputh Formations have allowed reliable inferences on probable vegetal components in the diets of *Kenyapithecus, Nacholapithecus, Victoriapithecus, Mabokopithecus,* and *Simiolus.*[105]

Kenyapithecus have thick enamel on broad molars that would serve well for crushing objects. Further, the markedly procumbent lower incisors, robust canines, strongly proclined mandibular symphyseal axis, and prominent inferior transverse torus of *Kenyapithecus* recall trait complexes of New World pitheciine monkeys *(Chiropotes, Pithecia, Cacajao).*[173] Accordingly, McCrossin and Benefit reasonably concluded that, like pitheciine monkeys, *Kenyapithecus* probably fed on hard fruits and nuts.[174] Dental microwear analyses by Palmer et al. on the posterior and anterior dentition of *Kenyapithecus africanus* confirmed that they were sclerocarp feeders that, like South American pitheciine primates, employed their procumbent lower incisors to bite open hard fruits and nuts.[175]

In Suriname, bearded sakis *(Chiropotes satanas)* and white-faced sakis *(Pithecia pithecia)* occupy hard immature seed-eating niches, with sakis feeding more frequently in the understory and lower canopy and bearded sakis in the taller trees of terra firme forests. *Chiropotes* specialize on Brazil nuts *(Bertholletia excelsa)* and also ingest at least six species of insects.[176] In Venezuela, on average over a twelve-month period, a white-faced saki diet comprised 88 percent fruit (63 percent of which was masticated seeds), 6 percent young leaves, 3 percent insects, and 2 percent flowers.[177]

In a mosaic of terra firme, chavascal and caatinga forests of northeastern Brazil, the diet of black-headed uacaris *(Cacajao melanocephalus)* consisted of 89 percent fruits, 5 percent flowers, 4 percent leaves, petioles, and bromeliads, and 2 percent arthropods. Boubli stated that the latter three food types are probably underestimated due to poor visibility. Seeds were the uacari staples year-round; they constituted 81 percent of fruit feeding records.[178] The most frequent uacari foods were unripe fruits with very hard green husks and no fleshy mesocarp.

Victoriapithecus probably also fed on hard fruits and nuts, as evidenced by dental morphology and dental microwear analyses.[179] However their diet might have been less specialized than that of *Kenyapithecus*. The craniodental studies of Benefit and McCrossin indicate that because *Victoriapithecus* is not specialized for leaf-eating, the niche of ancestral cercopithecoid monkeys was basically frugivorous.[180]

From dental microwear analyses, Palmer et al. concluded that on Maboko Island *Mabokopithecus pickfordi* and *Simiolus leakeyorum* were folivores and *Mabokopithecus clarki* was a mixed feeder.[179] Kunimatsu agreed that *Simiolus* was a folivore because of the sharp, prominent occlusal crests on their molar teeth, but he concluded that the relatively thick enamel, inflated cusps, and poorly developed crests on the molars of *Mabokopithecus* recommended them as hard-object feeders instead of folivores.[181]

It is unfortunate that we cannot reliably infer from their dentitions the extent to which *Kenyapithecus, Victoriapithecus, Mabokopithecus,* and *Simiolus* were faunivores. As seems to be true of most extant anthropoid primates, all of them might have eaten some arthropods via direct foraging or inadvertently as infestations of the plants that they ingested. The extent to which vertebrate prey was part of the diet of any of them is beyond our ability to guesstimate. Nonetheless, it is intriguing to consider that *Kenyapithecus* might have preyed upon *Victoriapithecus* and other smaller vertebrates, including *Simiolus* and *Mabokopithecus*.[173]

Predation on other mammals is rather widely distributed among anthropoid primates. Ducros and Ducros tabulated three species of apes, eleven species of Old World monkeys, and three species of New World monkeys that occasionally or commonly capture and consume other mammals, including primates.[182] Chimpanzees regularly hunt monkeys arboreally in the Taï National Park, Outambe-Kilimi National Park, Sierra Leone, Kibale Forest, Gombe National Park, and Mahale Mountains National Park.[183] Four female Sumatran orangutans killed and consumed eight slow lorises between 1989 and 1996, and an orangutan ate a baby gibbon.[184] Mandrills catch and eat small forest antelope in Cameroon.[185] Capuchins capture and eat arboreal squirrels, nesting young coatis, birds, and eggs in Latin American forests.[186] Wamba bonobos occasionally kill and eat flying squirrels, and Lomako bonobos sometimes kill and eat young duikers and squirrel-sized mammals.[187] In the Lilungu region of the Democratic Republic of the Congo, bonobos eat a variety of animals, including bees, ants, beetles, termites, earthworms, bats, and rodents.[188] Sabater Pí et al. also observed Lilungu bonobos handling, sometimes fatally, red-tailed guenons and Angolan black-and-white colobus, but they did not see them eat monkeys.[189] At Lui Kotale, bonobos caught and consumed monkeys, duikers, galagos, and rodents.[190]

During the dry season of 1969, in Amboseli National Park in southwestern Kenya, I observed baboons *(Papio cynocephalus)* catching vervets *(Cercopithecus aethiops)* that failed to escape in the fever trees *(Acacia xanthophloea)* near a waterhole.[173] Middle Miocene *Kenyapithecus* were more than twice the size of *Victoriapithecus;* both taxa seem to have been semiterrestrial, and *Victoriapithecus* were closely sympatric with *Kenyapithecus*.[191] Therefore, one might expect that in open woodland and floodplain areas, and even in some types of forest, like the Amboseli baboons preying on vervets, *Kenyapithecus* might have caught and eaten *Victoriapithecus*.[173]

Southern African and Arabian Rarities

Outside Kenya and Uganda, the record of African Miocene dental apes is pitiful. Indeed, there are only few specimens from Berg Aukus in the Otavai Mountains of northern Namibia to represent southern African Middle-Late Miocene anthropoids, and there are none from west and central Africa. Conroy et al. assigned the right mandibular corpus with P_4–M_3, a manual intermediate phalanx, a proximal ulna, and an axis to *Otavipithecus namibiensis*.[192] Pickford et al. added a frontal bone from Berg Auku to the hypodigm of *Otavipithecus namibiensis* and suggested that it resembles those of African apes and humans more closely than those of Eurasian great apes, which are more like orangutan frontal bones.[193]

One can look hopefully to Arabia for fossils that will elucidate the kinds of creatures that emigrated from Africa in the early Middle Miocene, thereby expanding the hominoid radiation into Eurasia. The good news is that there are Miocene deposits with vertebrate fossils in eastern Saudi Arabia; the bad news is that their yield of fossil hominoids has been meager.[194]

A lateral incisor from a 15–17-Ma locality in the Dam Formation at Asis is only provisionally assigned to the Primates. However, Andrews and Martin assigned four isolated teeth and a left maxilla with P^3–M^2 (M.35145) from continental equivalents of the Dam Formation near Ad Dabtiya, Saudi Arabia, to a new hominoid species: *Heliopithecus leakeyi*.[195] From associated fauna, *Heliopithecus* is estimated to be dated ca. 17 Ma, which is near the time when interchange between African and Eurasian mammals was facilitated by a landbridge. Andrews and Martin concluded that *Heliopithecus* has the closest affinities with the roughly contemporaneous *Afropithecus* of Kalodirr and Buluk in northern Kenya.[196] Like *Afropithecus, Heliopithecus* was probably frugivorous.[197] The habitat of *Heliopithecus* was tropical. The area contained fresh water ponds that were probably bordered by palms. Associated ruminants and rhinoceroses indicate that woodlands were nearby.[198]

Zalmout et al. recovered a partial skull and upper dentition of a medium-sized (15–20 kg) catarrhine—dubbed *Saadanius hijazensis*—from a late Oligocene (29–28 Ma) site in the Shumaysi Formation of western Saudia Arabia.[199] Fauna and palynological evidence indicate that *Saadanius* lived in a nonmarine, estuarian, back-mangrove habitat.[200] Like apes, hominids and catarrhine monkeys, it has a tubular tympanic annulus (ectotympanic bone). The face is hafted low on the neurocranium, the nasal bones are are large and narrow, and the muzzle projects prominently as in *Aegyptopithecus* and *Afropithecus*. *Saadanius* lacks frontal sinuses. Its temporal lines converge to form an anterior sagittal crest. The molar teeth are large and broad, and the palate is shallow. Because *Saadanius hijazensis* appears too unique to be in any existing catarrhine family or superfamily, Zalmout et al. created a new family and superfamily to accommodate it: Saadaniidae and Saadanioidea, respectively. They further inferred that the furcation of Hominoidea and Cercopithecoidea probably occurred sometime between 29 and 24 Ma.[199]

Eurasia's Earthy Assets

Eurasian Neogene apes are concentrated in Middle and Late Miocene deposits. They comprise fourteen species and five genera in two subfamilies of the Pliopithecidae, two species of Griphopithecidae, six or seven species of Dryopithecidae, thirteen species of Pongidae, one species of Oreopithecidae, three species of the Dionysopithecudae, and four species of uncertain familial status (Table 3.3).[201]

PLIOPITHECID PIONEERS.   *Pliopithecus* is the first fossil primate known to Western science. The type mandible was discovered at Sansan, France, in 1834. In 1820 at Eppelsheim, Germany, Schleichermacher found a femur that resembles Slovakian femora of *Pliopithecus* (Figure 3.6).[202]

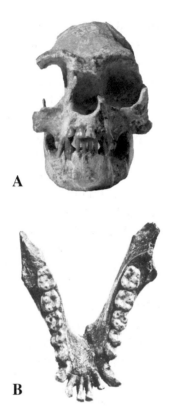

Figure 3.6. (A) Frontal view of skull of *Pliopithecus vindobonensis*. (B) Occlusal view of *Pliopithecus antiquus* from Sansan, France, the first fossil ape to be described in 1837.

Apparently, *Pliopithecus* was the earliest ape to reach Europe.[203] Deposits between 17 and 11 Ma in Spain, France, Germany, Switzerland, Austria, Slovakia, Poland, and Hungary yielded pliopithecid specimens.[204] They seldom occur directly associated with *Dryopithecus, Griphopithecus,* or other Hominoidea, perhaps because they had somewhat different ecological requirements.[205] Whereas deposits with *Pliopithecus* suggest swamp forest and forest/grassland mosaics as their dominant vegetation types, those of some *Dryopithecus* indicate fringing forest and open woodland or forest/grassland mosaics. At Rudabànya, Hungary, which had a subtropical forest habitat, *Dryopithecus brancoi* is associated with *Anapithecus hernyai,* a pliopithecid.[206] In Greece and Turkey, which have no pliopithecoid, Neogene hominoid specimens are correlated with sites where forest/grassland mosaics, single-species-dominated forests interspersed with grassy meadows, or parkland and grassland predominated.[207]

Nagatoshi lumped all of the European pliopithecid apes into *Pliopithcus antiquus*.[208] Previously, Ginsburg and Mein had recognized not only three species of *Pliopithecus*—*P. piveteaui, P. antiquus,* and *P. vindobonensis*—but also three additional genera of the Pliopithecidae: *Plesiopliopithecus lockeri, Anapithecus hernyaki,* and *Crouzelia,* which included *C. auscitanensis* and *C. rhodanica.*[209] Harrison et al. basically accepted the scheme of Ginsburg and Mein.[210] They recognized three European species of *Pliopithecus*—*P. antiquus, P. vindobonesis,* and *P. platyodon*—in the Pliopithecinae and ≥2 European species, including *Plesiopliopithecus lockeri* and *Anapithecus hernyaki,* in the Crouzeliinae of the Pliopithecidae. Welcomme et al. proposed a new species—*Pliopithecus (=Plesiopliopithecus) priensis*—for an 11-Ma mandibular fragment from Priay, France.[211] Moyà-Solà et al. added *Egarapithcus narcisoi,* based on 9-Ma gnathic specimens from Spain.[212] Harrison et al. also described an upper molar tooth from Cataluña that might be a ca. 12-Ma species of *Pliopithecus.*[213]

Three partial gibbon-sized skeletons of *Pliopithecus vindobonensis* from Slovakia best represent *Pliopithecus.*[214] A humerus of *Pliopithecus vindobonensis* sports an entepicondylar foramen that Simons viewed as a primitive trait because it occurs regularly in certain prosimian and platyrrhine species.[215] However, it also occurs as a variant in modern catarrhine primates, including humans.

The bony face of *Pliopithecus vindobonensis* is quite gibbonish. It has a relatively short snout, deep cheekbones, a broad interorbital pillar, and large, roundish orbits that are rimmed by spectacle-like ridges. The nasal bones are short and broad, and the supraorbital torus is prominent.[216]

Pliopithecus vindobonensis contrasts with *Hylobates* in having incomplete ossification of the external auditory tube and a more robust mandible, with high, vertical rami and variably developed symphyseal tori.[217]

Like *Hylobates, Pliopithecus* have thin molar enamel. But unlike gibbons, they have highly dimorphic upper and lower canine teeth. In *Pliopithecus,* I^1 is more spatulate than I^2, and the lower incisors are narrow, peg-like teeth. P_3 is sectorial, and P_4 is molariform. $M^1 < M^2 < M^3$ and $M_1 < M_2 < M_3$ or $M_1 < M_3 < M_2$. The molar enamel is commonly wrinkled. Cingula are variably developed on the postcanine teeth; M_3 sometimes lacks the hypoconulid.[218]

The dentitions of *Pliopithecus* species suggested different diets to several researchers. Nagatoshi inferred that they subsisted primarily by folivory on soft-leafed plants.[218] Andrews et al. agreed but added that perhaps the diet was supplemented by frugivory.[219] Further, they concluded that cruzeliines might have eaten even more leaves and other soft plant parts than *Pliopithecus* did. Based on complementary evidence from studies of shearing crest development and dental microwear, Ungar and Kay concluded that *Pliopithecus vindobonensis, P. platyodon,* and *Anapithecus hernyaki* were frugivorous, while Spanish *Pliopithecus* cf. *antiquus* was notably more folivorous.[220]

Nagatoshi concluded that *Pliopithecus* might have been driven to extinction by climatic cooling and concomitant diminution of their food sources.[208] As the climate cooled, species diversity might be expected to decrease and seasonality in leaf and fruit production to include spans when too little food was available to sustain viable pliopithecid populations.

The fossil record of small-bodied apes in South Asia is both impoverished—five isolated teeth from three localities—and portentous. A 16-Ma M^1 or M^2 from the Kamlial Formation of northern Pakistan might be one of the oldest securely dated catarrhine specimens in Eurasia. It most closely resembles eastern African *Micropithecus* and Chinese *Dionysopithecus.*[221]

Bernor et al. classified the Kamlial molar with three specimens (an upper canine, a P^4, and an M_1) from the 16–17-Ma Manchar Formation in Sind, Pakistan, as *Dionysopithecus* sp.[222] Further, they noted their remarkable resemblance to *Dionysopithecus shuangouensis* from the Xiacaowan Formation at Songlinzhuang, China. Contrarily, Harrison and Gu argued that the teeth are quite different from counterparts in Chinese *Dionysopithecus* and might in fact not belong to a species of Pliopithecoidea.[223]

The fifth South Asian small-bodied ape specimen is a single M^3, dubbed *Pliopithecus krishnaii,* from the Late Miocene (7.4 Ma) Quarry D at Haritalyangar, northern India.[224] Ginsburg and Mein placed it and another Miocene M^3 from southern Mongolia, *Krishnapithecus posthumus,* in the Hylobatidae.[225] Harrison et al. rejected *Krishnapithecus* because of the impoverished sample.[226]

Suteethorn et al. announced a new species—*Dendropithecus orientalis*—on the basis of an isolated M_1 from a Middle Miocene (ca. 15–17 Ma) locality at Ban San Khlang, northern Thailand.[227] Harrison and Gu reassigned it to *Dionysopithecus* because of its resemblance to *Dionysopithecus shuangouensis,* a judgment with which Kunimatsu et al. agreed.[228]

Dionysopithecus shuangouensis and *Platodontopithecus jianghuaiensis* from ca. 17–18-Ma localities in the Xiacaowan Formation in Sihong County, Jiangsu Province, People's Republic of China, are the earliest Catarrhini in Eurasia.[223] Subtropical deciduous and evergreen forest was the characteristic vegetation of Jiangsu Province (eastern China) during the Miocene.[229]

Li proposed *Dionysopithecus shuangouensis* on the basis of a left maxillary fragment with M^1–M^3 and two isolated molars.[230] Other Chinese scientists recovered additional isolated teeth of *Dionysopithecus shuangouensis* and isolated teeth, an edentulous mandibular fragment, a calcaneus, a pedal phalanx, and a pollical phalanx of a larger species: *Platodontopithecus jianghuaiensis*.[231] Harrison and Gu estimated the masses of *Dionysopithecus shuangouensis* and *Platodontopithecus jianghuaiensis* to be 5.5 kg and 14.9 kg, respectively.[223]

The earliest sample of the Pliopithecinae in China might be seven craniodental specimens of *Pliopithecus zhanxiangi* from the several localities in Tongxin County, Ningxia Hui Automonous Region. They are dated faunally at ca. 15 Ma.[232] Craniodentally, *Pliopithecus zhanxiangi* is larger than the European species of *Pliopithecus;* its dimensions approximate those of *Hoolock hoolock.* With an estimated body mass of 15 kg, *Pliopithecus zhanxiangi* is the heaviest species of *Pliopithecus.* Like other Eurasian Pliopithecidae and unlike modern Hylobatidae, *Pliopithecus zhanxiangi* exhibits notable dimorphism in canine size.[226] An isolated upper molar from Mongolia indicates that a somewhat bigger crouzeline pliopithecid might have been present during the latest Miocene in East Asia.[233]

The latest sample of Pliopithecidae in Asia is *Laccopithecus robustus,* from Late Miocene (6.2–6.9 Ma) deposits at Shihuiba, in Lufeng County, Yunnan Province of southwestern China. Nearly ninety specimens, including a crushed face (PA 860) with all upper teeth except the right I^2, an unspecified number of gnathic specimens, and a manual proximal phalanx, represent the species.[234] The estimated body mass is 12 kg. The bony face is remarkably gibbon-like in size and morphology, though it also resembles that of *Pliopithecus* in some features. Harrison, Bernor et al., and Fleagle placed it in the Pliopithecidae, and Harrison et al. further assigned it to the Crouzeliinae.[235] However, Tyler has suggested that *Laccopithecus* might be a derived sivamorph instead of a pliopithecid.[236]

Laccopithecus has a short, broad snout, a broad interorbital pillar, short, wide nasal bones, prominent rims circling the orbits, and large canines.[237] The small sample of *Laccopithecus* suggests that its canines and P_3 are more sexually dimorphic than those of *Hylobates.*[238]

GRIPHOPITHECID GROUNDBREAKERS.   Griphopithecidae might have
lived contemporaneously in Europe with earliest *Pliopithecus* spp., though
the evidence consists of a lone fragmentary M³ from Engelwies, Germany.[239]
The Griphopithecidae had thickly enameled teeth, which indicates a diet
that included tough foods.

Griphopithecus darwini, which might include the Engelwies tooth, was a
chimpanzee-sized hominoid that is modestly represented by a few teeth from
Slovakia and a damaged humerus and ulna from Austria. Excellent paleobotanical specimens at Engelwies, including fruit and leaves of evergreen broad-
leafed plants, indicate a warm, humid subtropical-tropical climate.[239] The robust forelimb bones suggest that in Austria *Griphopithecus darwini* was an
able climber that mostly moved atop branches while foraging in trees.[240]

The larger of two hominoid species at Pasalar, Turkey, with dentognathic
specimens from Çandir, Turkey, is *Griphopithecus alpani.*[241] Gnathic and
dental specimens much better represent the *Griphopithecus alpani* than *Griphopithecus darwini.*[242] Associated fauna and carbon isotopic data from tooth
enamel indicate that the Pasalar fauna, including *Griphopithecus,* foraged in
a treed area that was somewhat similar to seasonal deciduous woodlands of
tropical Africa.[243] Manual phalanges indicate that at Pasalar *Griphopithecus
alpani* engaged predominantly in pronograde quadrupedal locomotion in
trees but also could have sometimes engaged in terrestrial activities.[244] Dental microwear indicates that *Griphopithecus alpani* was chiefly frugivorous
and included unripe and other hard fruits and items in the diet.[245]

Kelley et al. described a smaller species—*Kenyapithecus kizili*—at Pasalar
based on a partial left maxilla with P³–M², a second maxillary fragment
with I², and seventy-one isolated teeth including upper and lower incisors,
canines, and first premolars.[246] A few of the Pasalar phalanges might also be
from *Kenyapithecus kizili.*[244]

DRYOPITHECID DEBUTS.   The Dryopithecidae lived in Europe between
12.5 and 9 Ma.[247] There were at least two, and perhaps as many as six species
in western Europe (see Table 3.3).[248]

The earliest European dryopithecid species is *Pierolapithecus catalaunicus*
from the Barranc de Can Vila 1 site in Barcelona, Spain, which is dated at
12.5–13 Ma.[249] The type and only specimen is an adult male partial skeleton
(IPS 21350) that includes the face, upper teeth, hand and foot bones, ribs,
clavicle, lumbar vertebrae, pelvic fragments, patella, and long bone shafts (Figure 3.7). Like extant great apes, the muzzle is more orthognathous than those

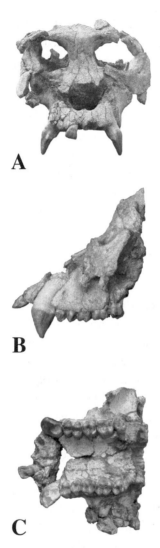

Figure 3.7. Partial cranium of 12.5–13.0-Ma *Pierolapithecus catalaunicus* from Catalonia, Spain. (A) Frontal view. (B) Left lateral view. (C) Occlusal view.

of Late Miocene Eurasian hominoid species, and the wrist, lower back, and markedly curved ribs resemble conditions in extant Hominoidea.[250]

Dryopithecus fontani is about the size of *Proconsul nyanzae*. The smaller species was variously named *Dryopithecus laietanus, Hispanopithecus laietanus,* and *Dryopithecus brancoi*. In overall size, they probably approximated *Proconsul africanus*. *Dryopithecus* spp. appear to have had quite dimorphic canines and molar enamel that, on average, is thicker than that of *Pongo* but thinner than that of Griphopithecidae.[251] Mein hinted that there might have been only one sexually dimorphic species of *Dryopithecus* in Europe.[252] But Begun advocated an adaptive radiation of frugivorous, forest-dwelling *Dryopithecus: D. fontani* in France, *D. brancoi* in central and eastern Europe, and *D. crusafonti* and *D. laietanus* in northeastern Spain.[253] Mottl, Andrews et al., and Kordos recognized a fifth taxon, *Dryopithecus carinthiacus,* from 12.5-Ma deposits in the western margin of the Carpathian Basin.[254]

The relatively thicker molar enamel of *Dryopithecus* would enable them to process tougher foods more frequently than *Pliopithecus* could.[255] This might account for their longer persistence in Europe than *Pliopithecus* and their common occurrence in habitats with sclerophyllous vegetation.[208] Dental microwear indicates that *Dryopithecus brancoi, D. crusafonti,* and *D. laietanus* were frugivores, and molar shearing crests support frugivory for *D. laietanus* and *D. fontani.*[256]

Dryopithecus brancoi is fairly well represented cranially. There are two relatively complete mandibular specimens, a 11.8-Ma lower face with partial maxillary dentition, perhaps one or two additional maxillary bits, and virtually no neurocranial specimens of *Dryopithecus fontani.*[257] The holotype mandible of *Dryopithecus fontani* is probably subadult and thus does not represent the adult population.[258] Pace Pilbeam and Simons, the few available upper-limb bone fragments of *Dryopithecus fontani* are largely uninformative as to the creature's habits.[259] *Dryopithecus carinthiacus* and *D. crusafonti* are represented by gnathic bits and isolated teeth.

Cranially, *Dryopithecus laietanus* (syn. *Hispanopithecus laietanus*) is revealed more completely via a young, presumedly male partial face and temporal bone (CLI-18000).[260] The interorbital pillar is wide, and there are small frontal sinuses. The brow resembles those of *Pongo* and *Sivapithecus* instead of African apes. But, unlike the faces of *Pongo* and *Sivapithecus sivalensis,* the face of *Dryopithecus laietanus* is not dished.[261] *Dryopithecus laietanus* is known postcranially from the informative partial skeleton (CLI-18000) (Figure 3.8).

Anatomical descriptions of *Dryopithecus* have been hampered because the larger apes from Rudabánya, Hungary, were uncertainly classified. Kretzoi named two species, *Bodvapithecus altipalatus* and *Rudapithecus hungaricus,* on the basis of numerous fragmentary gnathic and dental specimens, and provisionally assigned isolated postcranial bits to one or the other species, largely on the basis of their sizes.[262] Kay and Simons concluded that *Bodvap-*

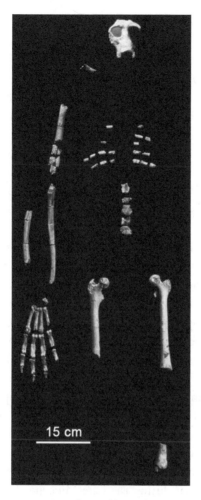

15 cm

Figure 3.8. Partial skeleton, 9.5–10 Ma, of *Dryopithecus laietanus* from Can Llobateres, Catalonia, Spain.

ithecus resembles *Sivapithecus indicus* and that *Rudapithecus* should be sunk into *Dryopithecus fontani*.[263] Nagatoshi provisionally placed *Rudapithecus* in *Dryopithecus brancoi* and *Bodvapithecus* in *D. fontani*.[218] Kelley and Pilbeam and Kordos concluded that the large hominoid remains from Rudabánya represent a single, highly dimorphic species: *Rudapithecus hungaricus*.[264] Begun placed the entire Rudabánya sample in *Dryopithecus brancoi*.[265] Andrews et al. agreed that there is only one dryopithecid species at Rudabánya, but they designated it *Dryopithecus carinthiacus*, the type of which is a fragmentary mandible from St. Stefan, Austria.[266] It is perhaps the oldest record of *Dryopithecus* in Europe.[267]

An aged, presumably female, fragmentary cranium (RUD-77) from Rudabánya indicates that *Dryopithecus brancoi* has a short muzzle, a wide interorbital pillar, D-shaped orbits, an endocranial volume of about 320 cc, and a body mass of about 25.5 kg.[268] It lacks true frontal sinuses, though the maxillary sinuses might have invaded the glabellar region, as they occasionally do in *Pongo*.[269] There is no supraorbital torus or supraorbital sulcus separating the superior orbital rim from the frontal squama that slopes upward and backward. The lateral orbital rim is quite robust.[270] The calvaria is low and narrow and lacks a sagittal crest; the palate is U-shaped.[271] A smaller, more complete young adult female cranium (RUD-200) has a small frontal sinus, broad interorbital region, feeble supraorbital torus, sizeable neurocranium (approximately 305 cc), thin enamel, and other features that extend diversity and add detail for *Dryopithecus brancoi*.[272]

The best-preserved mandibles of *Dryopithecus brancoi* appear to be from females. They have superior and inferior transverse tori. The rami are high and are rooted forward on the body, namely, between M_2 and M_3.[273]

Dryopithecus brancoi have narrow, stout upper central incisors and conical upper lateral incisors. The rather low-crowned lower incisors are also heteromorphic. The upper and lower canines are slender and tapered; the P_3 is sectorial. The incisors and canines of *Dryopithecus brancoi* are greatly reduced, compared with those of Siwalik *Sivapithecus*. Their molars have scant cingula and wrinkled enamel before occlusal wear.[273]

Based on a mandibular fragment with P_{3-4} and other damaged teeth, Xue and Delson claimed that during the Late Miocene, *Dryopithecus wuduensis* lived in Gansu Province, central China.[274] Associated fauna indicate that the habitat of *Dryopithecus wuduensis* was a woodland and grassland mosaic, with watercourses nearby.[275]

PONGIDS AND OTHER PONDERABLES.   Large apes with thick tooth enamel inhabited Europe and western Asia between 13.5 and 9 Ma.[276] Nagatoshi proposed four species of *Sivapithecus* in Europe and Turkey, but just as *Sivapithecus africanus* of Kenya later became *Kenyapithecus africanus* or *Equatorius africanus*, European and Turkish *Sivapithecus* spp. were variously renamed *Ankarapithecus, Griphopithecus, Graecopithecus,* and *Ouranopithecus*.[277] Accordingly, now nominally *Sivapithecus* is confined to South Asia.

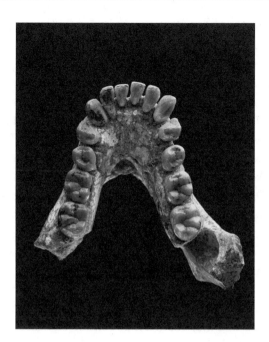

Figure 3.9. Occlusal view of mandible of 9-Ma *Ournaopithecus macedoniensis* from Ravin de Pluie in Greece.

The largest fossil ape in Europe is *Ouranopithecus macedoniensis* from several Late Miocene (9 Ma) localities in Greece.[278] The males might have weighed approximately 72 kg.[279] Several gnathic specimens and isolated teeth from Ravin de la Pluie and Nikiti 1 and a face and palate from Xirochori represent *Ouranopithecus macedoniensis* (Figure 3.9). Dental, mandibular, nasal, and subnasal structures and proportions set *Ouranopithecus* apart from *Sivapithecus* and *Pongo,* and align it more with *Australopithecus* and *Pan* according to de Bonis et al., Kelley and Pilbeam, and Koufos et al.[280] Accordingly, they retained the nomen *Ouranopithecus macedoniensis* instead of sinking them in *Sivapithecus.* Ravin de la Pluie and Xirochori were relatively open habitats, but giraffid and artiodactylan fauna indicate that Nikiti 1 was more forested.[281]

The poorly preserved Pyrgos mandible *(Graecopithecus freybergi)* might also belong in the hypodigm with *Ouranopithecus macedoniensis.* Faunal associations indicate that it is 8–9 Ma.[252] Because the nomen *Graecopithecus freybergi* has priority, Andrews et al. argued that all the Greek Late Miocene large hominoids should be called *Gaecopithecus freybergi* instead of *Ourano-pithecus macedoniensis.*[282]

Collectively, the Greek specimens indicate that, like *Gorilla* and *Pongo*, *Ouranopithecus macedoniensis* is a very dimorphic species. The presumed males were the size of female gorillas, and the females were considerably smaller.[283]

The face of *Ouranopithecus macedoniensis* is large, and the interorbital pillar is wide. The orbit is quadrangular, with the transverse diameter the longest. *Ouranopithecus macedoniensis* sports a sizeable supraorbital torus, and the upper face is vertical, unlike the concave lateral profiles of *Pongo* and *Sivapithecus.*[284] The mandibles of *Ouranopithecus macedoniensis* are robust and bear superior and inferior symphyseal tori. The gonial area indicates heavy development of the masseter muscles.[284]

The teeth of *Ouranopithecus macedoniensis* have very thick enamel.[169] The upper incisors are heteromorphic, with I^1 much larger than I^2. The lower incisors are relatively small and homomorphic. De Bonis and Koufos considered the upper canines of *Ouranopithecus macedoniensis* to be relatively low, but Kelley argued that they could not be considered reduced significantly in height.[285] P_3, particularly of presumed males, sports a honing facet for the upper canine. The molars are large, the cusps are low and bunodont, and the cingula are vestigial or absent from them; M^2 is larger than M^1 and M^3, and $M_1 < M_2 < M_3$. Molar shearing-crest development and dental microwear indicate that *Ouranopithecus macedoniensis* ate hard objects such as seeds, nuts, roots or tubers, and stripped stalks and perhaps bark with their incisors, which fits well with the model of an open habitat.[256]

In the Late Miocene (ca. 10 Ma), Turkey was also inhabited by two large hominoid species: *Ankarapithecus meteai* (ca. 10 Ma) and *Ouranopithecus turkae* (7.4–8.7 Ma). *Ankarapithecus meteai* is represented chiefly by an adult female face and palate with complete maxillary dentition and an associated mandible with complete dentition (AS95-500), a male lower face and palate with complete dentition (MTA 2125), and mandibular fragments with right I_2–C and left C–M_3 of a second male (MTA 2124).[286] Based on orbital dimensions, Andrews and Alpagut estimated that AS95-500 was slightly smaller (23–29 kg) than a female bonobo.[287]

There is distortion in areas of the face of AS95-500, but the mandible is not deformed. The orbits are square and separated by a relatively narrow interorbital pillar. The bony arches above the orbits are moderately developed, but the supraorbital sulcus is faint, and there is no true supraorbital torus. The frontal sinuses are extensive, but it is unclear whether they are extensions of maxillary air cells, as in *Pongo* and *Sivapithecus,* or ethmoidal air cells, as in *Pan* and *Gorilla.* The midface is tall, broad, and prognathic, and unlike *Pongo* and *Sivapithecus sivalensis,* the lateral profile is relatively straight instead of concave. Because the large zygomatic bones are attached to the maxilla at approximately right angles, the midface is flat, and the muzzle juts out from it. Like most other fossil apes, the palate of *Ankarapithecus* is shallow.[288]

Mandibles of *Ankarapithecus* are robust and sport well-defined superior and inferior transverse tori. The rami are tall, which positions the condylar and coronoid processes high above the occlusal plane of the cheek teeth.[288]

In *Ankarapithecus,* the upper incisors are highly heteromorphic and low crowned, while the lower incisors are homomorphic, narrow, and relatively high crowned. The upper canines are relatively small vis-à-vis the thickly enameled molars, and the lower canines are low crowned. P_3 is large and evinces wear from contact with the upper canine. The molar pattern is $M^1<M^2<M^3$ and $M_1<M_2\geq M_3$. The molars lack cingula, and the cusps are broad and flat.[289] All in all, the skull of *Ankarapithcus meteai,* like that of *Ouranopithecus macedoniensis,* bespeaks a powerful chewing machine adapted for incisal stripping and grinding hard and tough vegetation.[287]

Ouranopithcus turkae is gnathodenally the second largest hominoid from the Late Miocene of Asia; it is surpassed in size by *Gigantopithecus blacki.*[290] *Ouranopithcus turkae* is composed of a fragmentary maxilla with right C–M^2 and left I^1–M^3; a subadult mandible with right C–M_2 and left P_3–M_1; and an adult right mandibular fragment with P_3–M_3.[291] Associated fauna in the Çorayerler locality, especially the rodents, indicate a savanna-like environment, though suids and cervids suggest somewhat more closed woodland.[292] Very thick molar enamel and wear indicate that *Ouranopithcus turkae* probably ate tough and hard foods that required notable mastication.[291]

Udabnopithecus garedziensis is a small maxillary fragment with P^4 and M^1 from a Late Miocene locality at Udabno, Georgia.[293]

Sivapithecus lived in South Asia between 13 Ma and perhaps as late as 7 Ma.[294] Apart from *Sivapithecus sivalensis, S. indicus* and *S. parvada,* it is unclear what species the numerous specimens from northern India, Pakistan, and Nepal represent.[295] Kay concluded that both *Sivapithecus indicus* and *S. sivalensis* are sexually dimorphic to a low or moderate degree, at least in their dentitions.[296] However, Andrews concluded that in canine dimensions and body size, *Sivapithecus indicus* (and the large Greco-Turkish Hominoidea) were as sexually dimorphic as *Gorilla* and *Pongo.*[297] Needless to say, until the specific taxonomy of the South Asian hominoids is resolved, it is problematic to broach discussions of their sexual dimorphism and the behaviors that might have been bounded by it.

Kelly diagnosed a very large species—*Sivapithecus parvada*—on the basis of craniodental and postcranial specimens representing ≥4 individuals from the 9–10-Ma Sethi Nagri locality on the Potwar Plateau in northern Pakistan.[298] Analysis of associated fauna have indicated that the site was forested, perhaps with some areas where the canopy was not continuous.[299]

Badgely et al. detected two distinct upper and lower canine morphs in the hefty sample from Potwar, Pakistan.[300] They proposed that the larger species is *Sivapithecus indicus* and the smaller species should be designated *Ramapithecus punjabicus*. Because *Ramapithecus punjabicus* and Kay's poorly sampled *Sivapithecus simonsi* probably should be sunk into *Sivapithecus sivalensis*, we are left with four species of sizeable Hominoidea from the South Asian Miocene: *Sivapithecus parvada, S. indicus, S. sivalensis,* and *Gigantopithecus giganteus* (=*G. bilaspurensis* and *Indopithecus giganteus*), a truly monstrous Late Miocene (6.3 Ma) form that might have evolved from *Sivapithecus*.[301] All of them have low, thickly enameled, bunodont molars that lack cingula.[302] They are known predominantly from dental and gnathic, especially mandibular remains.

The 8.5–9.5-Ma deposits in the Potwar Plateau also yielded a spectacular bony face of *Sivapithecus sivalensis* (GSP 15000) and a smattering of postcranial bits of medium-sized and large hominoids that were initially assigned variously to *Sivapithecus* spp., *Ramapithecus,* and *Gigantopithecus*.[303]

The face, mandible, and dentition of an adult male *Sivapithecus sivalensis* is tellingly represented by GSP-15000, which unfortunately lacks the basicranium and all except a frontal segment of the neurocranium (Figure 3.10). The chimpanzee-sized individual was advanced in age, as indicated by his heavily worn teeth and fully fused sutures. As a final insult, evidenced by a suspicious puncture mark on his left cheekbone, he might have been a victim of predators or scavengers.[304]

The upper face of *Sivapithecus sivalensis* is remarkably like that of female *Pongo*. Its midface also bears some striking resemblances to *Pongo*, though here more differences can be noted between them such as the greater midfacial length in *Sivapithecus*. The jaws, particularly the mandible, depart notably from conditions in *Pongo*.[305]

Sivapithecus sivalensis has a modest superior orbital rim and no supraorbital sulcus; the frontal squama rises acutely behind the orbits. The lateral orbital rim is gracile. There is no true frontal sinus, though the maxillary sinuses

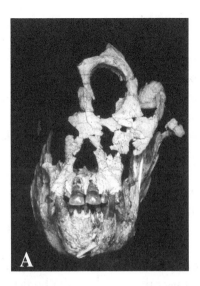

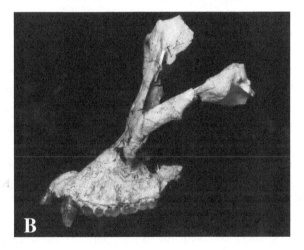

Figure 3.10. Partial cranium of 7.4–8.8-Ma *Sivapithecus sivalensis* [GSP-15000] from Potwar, Pakistan. (A) Frontal view. (B) Left lateral view. Note dished face profile in B.

might have invaded the region.[306] The orbit is ovoid and taller than wide. The interorbital pillar is narrow and thin walled; the nasal bones are long.[307]

Viewed in profile, the prognathic midface of *Sivapithecus sivalensis* is concave dorsally, like the dished face of *Pongo*. Its cheekbones are high, broad, and vertical, but the zygomatic arch is rather gracile. A combination of laterally flaring zygomatic arches and notable postorbital constriction indicates that *Sivapithecus sivalensis* had large anterior components of the temporalis muscle.[308]

The canine jugae and adjacent canine fossae are quite prominent in *Sivapithecus sivalensis*. Among extant apes, the canine fossa is most pronounced in *Pongo*.[308] Further, as in *Pongo*, the palate of *Sivapithecus sivalensis* appears to be airorynchic, and the upper incisors are very procumbent.[309] Ward and Kimbel demonstrated that *Sivapithecus, Ankarapithecus,* and *Pongo* share a unique subnasal morphology that is quite distinct from those of *Pan, Proconsul, Australopithecus,* and *Homo*.[310]

The mandibular body of *Sivapithecus sivalensis* is robust and deep, especially at the symphysis and anterior roots of the rami. The superior and inferior tori are strongly developed.[311] The lightly built rami are tall, wide, and vertically oriented. The condylar process is gracile; the coronoid process probably projected somewhat higher than the condylar process.[312] Digastric impressions are present near the antero-inferior border of the mandible, indicating that, unlike *Pongo pygmaeus, Sivapithecus sivalensis* had anterior bellies of the digastric muscles.[313]

In *Sivapithecus sivalensis,* the upper central incisors are sizeable and spatulate; the conical upper lateral incisors are much smaller. The narrow lower incisors have high crowns and are crowded together between the canines. Both the upper and lower canines are large and projecting. In their pristine state, they must have been fearful organs. P_3 is moderately sectorial and the upper premolars are molarized: $M^1 < M^2 = M^3$ and $M_1 < M_2 = M_3$.[314]

The postcranial skeletal bits of Potwar *Sivapithecus* are more similar to counterparts of *Proconsul* than to those of extant Pongidae.[315] Kappleman's analysis of bovids from the Chinji Formation (10.5–12.5 Ma) of Pakistan indicates that sympatric *Sivapithecus* were forest dwellers.[316] The largest Potwar *Sivapithecus—S. parvada*—probably ranged between the sizes of male *Pan troglodytes* and female *Gorilla*.[317] The other species of Potwar *Sivapitheus* were variously similar to large macaques, mandrills, and chimpanzees in body size.[318]

On the Potwar Plateau, the Late Miocene habitat of *Sivapithecus sivalensis* was a mosaic of forest, woodland, and grassland, as inferred from the diverse vertebrate faunal remains in the Dhok Pathan Formation. It is unclear whether the forests were deciduous or evergreen. An abundance of mammalian browsers with grazers suggests that a rich shrub layer was part of the Potwar floral community.[319] Based on isotopic evidence and dental microwear analyses of the fauna associated with Potwar *Sivapithecus,* Nelson concluded that they were as frugivorous as extant rain forest–dwelling apes. The vegetation mosaic and rainfall regime (five to six dry months) most resembled those of monsoonal forests in southern China.[320]

Although Kay reasonably inferred from cuspidal morphology and enamel thickness that Potwar *Sivapithecus* ate fruits with hard tough rinds, the microwear studies of Teaford and Walker, and Nelson indicated that, like *Pan troglodytes,* they ate softer upper canopy fruits, at least in the period shortly before their deaths.[321] Given that *Sivapithecus* probably experienced longer annual dry spells and concomitantly reduced availability of preferred soft fruits, they might have needed a sturdy mandible and dentition to process tough, hard, fallback foods during lean fruiting seasons.[322]

During earlier periods, Siwalik and southern Himalayan environments appear to have been wetter, and the forests were probably more extensive.[323] Beginning at 9.2 Ma, a period of transition to very open woodlands and more extensive grasslands began; by 7.4 Ma, they were predominant in the Potwar region.[324]

Two or three species of Pongidae lived in Yunnan Province of southwestern China during the Late Miocene: *Lufengpithecus lufengensis, L. keiyuanensis,* and *L. hudienensis.*[325] Although initially two hominoid primates that are larger than *Laccopithecus* were discerned in the rich 6.2–6.9-Ma faunal assemblage of the Lufeng coalfields near Shihuiba, detailed studies revealed that there is probably only one highly dimorphic pongid—*Lufengpithecus lufengensis*—preserved there.[326] Indeed, *Lufengpithecus* is more dimorphic dentally and probably also in body size than *Pongo pygmaeus* or perhaps any other living species of the Anthropoidea.[327] Based on the regression formula of Kay and Simons, Xu and Lu determined that females weighed 21–38 kg and males 39–60 kg.[328]

Lufengpithecus lufengensis is represented at Lufeng by >750 specimens, including five taphonomically damaged crania, ten mandibles, forty-seven more cranial and gnathic fragments, and 650 isolated teeth plus those articulated in twenty-nine bits of jaw, which include generous samples of both upper and lower incisors, canines, premolars and molars, and seven postcranial specimens: a scapula and clavicle of one individual, a radius, two manual proximal phalanges, a femur, and a hallucal metatarsal bone.[329]

In *Lufengpithecus lufengensis,* supraorbital ridging is modest overall and deficient medially so that there is no torus (Figure 3.11). Further, no sulcus separates the superior orbital rims from the frontal squama. The interorbital

Figure 3.11. Partial cranium (PA664, frontal view) and mandible (PA848 occlusal view) of 8–9 Ma *Lufengpithecus lufengensis* from Shihuiba, Lufeng, Yunnan, China.

pillar is wide; the nasal aperture is narrow; the orbits are ovoid, with the transverse diameter widest; and the palate is broad, short, and shallow.[330] The lateral orbital rims are more robust in *Lufengpithecus lufengensis* than in *Sivapithecus sivalensis.*[273]

The zygomatic processes are located more anteriorly on the maxillae in *Lufengpithecus lufengensis* than in extant Pongidae. Concomitantly, the snout is short and modestly prognathic.[331] Corresponding with robust canines in presumed male *Lufengpithecus lufengensis,* their canine fossae are very deep, and the canine jugae are prominent. The face is short, broad, and dished; their subnasal morphology is also similar to that of *Pongo.* A sagittal crest distinguishes one of the larger, presumedly male crania.[332] Like *Pongo, Sivapithecus, Dryopithecus brancoi,* and *Ouranopithecus macedoniensis, Lufengpithecus* has superiorly expanded maxillary sinuses, which might be one of several synapomorphic traits that relate them cladistically.[333] Conversely, Kelley and Etler, and Xu and Lu explicitly denied *Lufengpithecus* membership in the *Sivapithecus-Pongo* clade.[334]

Mandibles of *Lufengpithecus lufengensis* are characterized by a thick, deep symphysis in which the incisors are implanted vertically. Whereas the superior transverse torus is poorly developed, the inferior transverse torus is thick and prominent. The symphyseal surface sports digastric impressions. Like the symphysis, the mandibular body is deep and thick, but its depth decreases posteriorly. The rami are broad and presumably were tall; they rise perpendicular to the corpus.[335]

In most dental dimensions, *Lufengpithecus lufengensis* are even more dimorphic sexually than are gorillas and orangutans.[336] The central upper incisors are very broad, and their lateral counterparts are small. The lower incisors of females are homomorphic, and all are relatively small. The lower incisors of males are taller and more heteromorphic than those of females. Females have low canines with blunt tips; the males have large canines with long sharp cusps and posterior edges. In females, the P_3 is nonsectorial and bicuspid, though with a marked size difference between the two cusps. The P_3 of males is sectorial and usually unicuspid, and $M^1 < M^3 < M^2$ and $M_1 < M_3 < M_2$. The molars lack cingula and sport elaborate crenulations on the thick occlusal surfaces. They also differ from the molars of *Ankarapithecus* in having higher cusps.[337]

Overall the molars of *Lufengpithecus* are remarkably like counterparts in *Pongo,* but their skull morphologies are quite distinctive.[338] Molar wear patterns and cuspidal morphology indicate that *Lufengpithecus lufengensis* consumed relatively soft plant parts like leaves and berries.[339] Xu and Lu speculated that arboreal fruits, nuts, and seeds might also have been part of their diet.[340]

At Lufeng, *Lufengpithecus lufengensis* and *Laccopithecus robustus* inhabited mixed forest/grassland habitats near a swamp and lake, as determined by faunal, palynological, and stratigraphic analyses.[341] The Lufeng fauna are similar to those in the Siwalik Nagri and Dhok Pathan formations.[342] The climate at Lufeng during the Late Miocene was probably warm and humid, though the high incidence of linear enamel hypoplasia indicates nutritional stress due to seasonal reductions in food sources.[343]

A palate and mostly disarticulated teeth of three individuals from lignite deposits in the Xiaolongtan colliery of Keiyuan County, Yunnan Province, represent *Lufengpithecus keiyuanensis*. Some of the teeth have cingula and supernumerary cusps. The P_3 is sectorial and basically unicuspid, and $M_1 < M_2 < M_3$.[344]

The Xiaohe Formation in the Yuanmou Basin of Yunnan Province provided a tantalizing trove of Late Miocene pongid fossils that are provisionally assigned to *Lufengpithecus hudienensis*.[345] The hypodigm comprises a juvenile face with left and right c, dm^1–dm^2 and M^1 (YV 0999), eight maxillary and eleven mandibular fragments, one relatively complete mandible, and >1,500 teeth from four localities.[346]

Chaimanee et al. named two fossil species of large Hominoidea in northern Thailand: *Khoratpithecus chiangmuanensis* from the Middle Miocene Ban Sa locality in the Chiang Muan Basin, and *K. piriyai* from a Late Miocene sand pit in Nakorn Ratchasima Province (Khorat).[347] Floral remains indicate that both species lived in tropical areas, unlike the more temperate habitats of *Lufengpithecus* and seasonal open environments of *Sivapithecus*.[348]

Khoratpithecus chiangmuanensis comprises disarticulated upper and lower teeth of ≥2 individuals that notably resemble counterparts in *Lufengpithecus lufengensis* in evidencing marked sexual dimorphism and molars with thick enamel and crenulations. The major difference between the species is in details of the anterior dentition and premolars, and M$_3$>M$_2$.[349]

Khoratpithecus piriyai is a mandible (RIN 765) lacking the rami and anterior teeth, except the right canine. It is comparable in size to that of *Lufengpithecus,* but apparently it had a larger anterior dentition and M$_3$ and distinct canine structure. Although the premolar and molar teeth of *Khoratpithecus* closely resemble those of *Lufengpithecus* and harbinger *Pongo,* the absence of facets for the attachment of the anterior bellies of the digastric muscles on RIN 765 is a feature that *Khoratpithecus piriyai* shares uniquely with *Pongo* and might indicate similar decoupling of mandibular and hyoid movements.[350]

After a Middle Miocene heyday, Eurasian hominoid species declined in number and diversity during the Late Miocene. By the succeeding Pliocene epoch, hominids and monkeys largely replaced them over much of the vast area, though orangutans managed to evolve and persist in an ever-shrinking range and gibbons fared somewhat better until historic times.[351]

CERCOPITHECOIDS CATCH UP.   Fossil monkeys first appear in Europe ca. 11 Ma, followed by western Asia 7–9 Ma, and the Siwalik formations ca. 6.3 Ma.[352] Colobinae were present throughout the Late Miocene, but *Macaca* (Cercopithecinae) were latecomers.[282] Apparently, monkeys first occurred in China even later than in South Asia, namely, in the Late Pliocene or Early Pleistocene, and a colobine species had reached Japan by ca. 2.5 Ma.[353]

Treeing the Fossil and Extant Apes

The ultimate goal of paleoanthropologists and other evolutionary biologists is to discover the ancestral, descendant, and collateral relationships of fossil and living creatures and to articulate them into readily comprehensible schemes that are presented figuratively as dendrograms. For more than a century, these schemes took the form of phylogenic trees, shrubs, lattices, and bushes that depicted presumed evolutionary relationships, times of branching, and the extent of change since speciation.[354]

Cladistically inclined paleoanthropologists have deforested the field of florid graphics and proffered instead linear cladograms that simply show the branching points of successive sister groups based on presumed synapomorphic (that is, shared derived) features.

Phylogenetic classifications are based solely on the character information that is depicted by the phylogenic tree. Sister groups are classified together instead of apart. Phylogenetic systematists (cladists) treat fossils, regardless of their geochronological age, precisely like extant species.[355]

Cladists recognize only monophyletic groups as natural taxa. A monophyletic taxon includes all the descendents of a common ancestor and the ancestral species, if it is known. The monophyletic group is called a clade. Taxa that do not include all the descendents of an ancestral species are termed paraphyletic taxa.[355] Vis-à-vis molecular genetic data, the Panidae in Table 2.3 is a paraphyletic taxon because humans, chimpanzees, and bonobos might be more closely related to one another than any of them is to gorillas.

Phylogenetic systematists variously place the African apes, all great apes, and sometimes even the lesser apes with our fossil relatives and us in a more comprehensive, monophyletic Hominidae because we are presumed to have radiated from a common Miocene species. In contrast, evolutionary taxonomists segregate humans and their bipedal terrestrial fossil relatives in the Hominidae because of our remarkable adaptive zone, which is characterized by habitual terrestrial bipedalism, tool-assisted subsistence behavior, and symbol-based culture, language, beliefs, ideology, and moral codes.[356]

The Catarrhini had probably commenced by Early Oligocene times. Simons and Fleagle further posited that *Aegyptopithecus* and *Propliopithecus* document the presence of Hominoidea in the Jebel Qatrani fauna.[357] The evolutionary pathway between the Propliopithecidae and Hominoidea of modern aspect is utterly obscure because of the immense temporal and morphological gaps in the fossil record. However, Simons proposed a lineage extending from *Aegyptopithecus* to *Proconsul* to *Afropithecus*.[55] Leakey et al. documented similarities between the faces of *Aegyptopithecus* and *Afropithecus,* even though the former is the size of a small macaque and the latter is the size of a female chimpanzee.[358] They concluded that the complex of facial features had been retained over millions of years. However, because the face of *Victoriapithecus* is similar to those of *Aegyptopithecus* and *Afropithecus,* Benefit and McCrossin concluded that the shared facial traits are plesiomorphic catarrhine characters that indicate no particularly close phylogenic relationship among the three taxa.[359]

Fleagle favored the conservative view that the Early and Middle Miocene catarrhine radiation in eastern Africa might contain links to later Eurasian and African Hominoidea, including the remote ancestors of modern Hylobatidae, *Pongo, Pan, Gorilla,* and *Homo.*[360] Within his scheme, only the Hylobatidae might have arisen from a species of the Proconsulidae, while *Kenyapithecus* and its ilk are more likely links to the lineage of modern *Pongo, Pan,* and *Gorilla* (Figure 3.13B). De Bonis and Andrews and Martin accepted *Proconsul* as a stem hominoid, and Andrews and Martin noted that *Proconsul* marks the minimum time for hominoid origins at ≥22 Ma.[361] Rae also argued that the proconsulids are cladistic hominoids, based on their facial features.[362]

Harrison denied hominoid status to *Pliopithecus* and the Proconsulidae that, in his scheme, includes *Proconsul, Limnopithecus, Rangwapithecus, Dendropithecus, Kalepithecus,* and *Micropithecus.*[363] Instead, he conferred separate superfamilial status on them: Pliopithecoidea and Proconsuloidea (Table 3.3). Accordingly, Hominoidea are unknown during the Early and perhaps much of the Middle Miocene, and neither the Hominoidea nor the Cercopithecoidea can be linked clearly with the Proconsuloidea.

Moreover, the bonanza of additional craniodental and postcranial specimens of *Kenyapithecus africanus* that Benefit and McCrossin have collected since 1992 at Maboko Island in Kenya effectively eliminated *Kenyapithecus* as a candidate for basal ancestry to extant Hominoidea. McCrossin and Benefit persuasively argued that *Kenyapithecus* was a peculiar creature with a special anterior gnathodental complex adapted for feeding on hard nuts and seeds and postcranial features related to running on the ground, like the more terrestrial species of Old World monkeys.[364]

In 1987, de Bonis ventured one of the fullest scenarios on the suprageneric relationships and evolution of Miocene-Pleistocene Hominoidea.[365] Like Louis Leakey, he proffered a five-family scheme composed of Proconsulidae, Hylobatidae, Oreopithecidae, Pongidae (a clade with only *Pongo* and *Sivap-*

ithecus), and Hominidae (a clade that includes *Pan, Gorilla, Homo,* and five fossil genera).[366]

The clearest lineage, and exceptionally the one with which paleoanthropologists generally agreed at the time and that has been reaffirmed by Begun, linked *Sivapithecus* to *Pongo.*[367] Their synapomorphic traits include thick molar enamel, subnasal morphology, orbital and interorbital structure, and a dished face. Fifteen-million-year-old *Griphopithecus darwini* (formerly *Sivapithecus darwini*) might be a link between *Afropithecus* (17–18 Ma) and the Late Miocene species of *Sivapithecus,* which was proximately antecedent to Pleistocene Chinese and Southeast Asian *Pongo.*[368]

Because the teeth of *Dionysopithecus shuangouensis* resemble counterparts in *Micropithecus clarki,* Ciochon, Fleagle, Bernor et al., and Harrison et al. aligned *Dionysopithecus shuangouensis* with Early Miocene dental apes of eastern Africa.[369] Later, Harrison proposed that the Dionysopithedae were primitive ancestors of all other Pliopithecoidea 6–17 Ma.[370] In Europe, a pliopithecine species evolved into the crouzeliines, and one species of them returned to Asia eventually to beget *Laccopithecus.*[370] Because of similarities between the upper molars of *Bugtipithecus* and those of *Dionysopithecus shuangouensis,* Marivaux et al. raised the possibility that *Bugtipithecus* gave rise to the Pliopithecoidea via the Dionysopithecinae that Harrison and Yumin posited to be stem Pliopithecidae.[371]

In the phylogeny of de Bonis, *Oreopithecus bambolii* is a Late Miocene culminant European descendent of western Kenyan *Nyanzapithecus,* and *Dryopithecus* is a possible Middle Miocene ancestor of the extant African apes. He placed *Pan* and *Gorilla* with *Dryopithecus* in a hominid subfamily—the Dryopithecinae—because of similarities in their dentitions.[368] Contrarily, Moyà Solà and Köhler argued that because the supraorbital torus of *Dryopithecus laietanus* is not homologous with those of *Pan, Dryopithecus* is not part of the clade of *Pan/Australopithecus/Homo;* instead, based on derived features of the zygomatic bone, *Dryopithecus* is in the clade of *Pongo.*[372] Further, Moyà Solà and Köhler suggested that *Dryopithecus laietanus* and *Oreopithecus bambolii* were close phylogenetically.[373]

De Bonis and coworkers considered *Ouranopithecus macedoniensis* to be basal in their other hominid subfamily the Homininae, which also includes *Australopithecus, Gigantopithecus blacki,* and *Indopithecus* (=*Gigantopithecus giganteus*) (Figure 3.12).[374] De Bonis and Koufos considered *Kenyapithecus* to be the most likely Middle Miocene ancestor of *Ouranopithecus macedoniensis.*[375] The prime synapomorphy that unites the hominine clade of de Bonis is

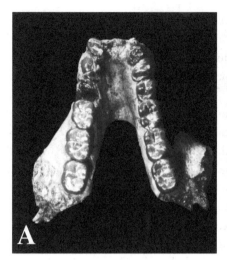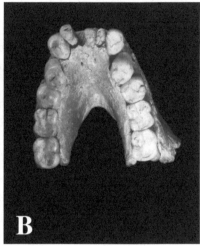

Figure 3.12. Occlusal view of mandibles. (A) *Gigantopithecus giganteus*, 6.3-Ma, from Haritalyangar, India. (B) Pleistocene *Gigantopithecus blacki* from southern China.

absence of a honing facet for the upper canine on the buccomesial aspect of P_3. The subnasal region also recommended *Ouranopithecus macedoniensis* as a Late Miocene predecessor to *Australopithcus.* De Bonis indicated that the furcation between the lineage of *Australopithecus* and *Homo* and that of *Gigantopithecus* occurred about 10 Ma. He showed the divergence between the Dryopithecinae and the Homininae at 15 Ma, which is discordant with most biomolecular dates for the furcation of recent African ape and human lineages.[368]

In the scenario of de Bonis, *Proconsul,* the Dryopithecinae, and the Oreopithecidae evolved in forests.[376] *Kenyapithecus, Sivapithecus,* and the Homininae evolved in more open environments (Figure 3.13A). Accordingly, *Pongo,* which is highly adapted to canopy life, ascended to its current state from ancestral denizens of woodlands or more open habitats instead of tropical rain forests.

Xu and Lu proffered a cladogram in which 6.2–6.9-Ma *Lufengpithecus* branches after the lineage of African apes and before the common ancestor of *Australopithecus* and *Homo;* the clade of *Sivapithecus-Pongo* furcated before that of the African apes (Figure 3.14C).[377]

Andrews et al. sketched a dendrogram that cautiously has more dashed lines than solid ones (Figure 3.13C).[219] Extant apes and humans are ultimately rooted with the Proconsulidae. In the Early Miocene, ancestors of the gibbon lineage furcate, followed by a tritomy of ancestral afropithecines, dryopithe-cines, and kenyapithecines. They proffered no specific root for their Ponginae (*Pongo* and *Sivapithecus* sensu lato) or Homininae (*Graeceopithcus,* including *Ouranopithecus macedoniensis, Homo* spp., and *Australopithecus* sensu lato); in-stead, they sprout from question marks. Nonetheless, Andrews et al. noted a tritomy of lineages leading to *Gorilla, Pan,* and *Homo* that are rooted in *Grae-copithecus* (=*Ouranopithecus*) ca. 10 Ma.[219]

Andrews and Bernor reviewed the morphology-based cladograms of An-drews, Begun et al., Harrison and Rook, and Schwartz and derived a con-sensus cladogram (Figure 3.14E):

- *Proconsul* is a stem-hominoid that precedes branching of *Hylobates.*
- Next, *Afropithecus, Griphopithecus,* and *Kenyapithecus* branch, with the latter two taxa as distinct clade.
- *Dryopithecus* and *Oreopithecus* branch next and constitute a clade, largely due to similarities in postcranial structures related to suspen-sory behavior.
- The next branching clade is composed of *Lufengpitheus, Graecopithecus, Ankarapithecus, Gigantopithecus, Sivapithecus,* and *Pongo* (dashed lines signal uncertainly as to whether *Pongo* is an offshoot of *Sivapithecus* or an independent branch).
- It is the sister clade of a partly terrestrial and less suspensorially adapted clade composed of *Gorilla, Pan,* and *Ardipithecus.*[378]

In the scheme of Andrews and Bernor, the Hominoidea began in the Early Miocene of Africa, but after the entrance into Eurasia of *Dionysopithecus* (16–18 Ma) and founding hominoids that resemble *Griphopithecus* (15–17 Ma), Eurasia was a major hotbed of hominoid evolution during the Mid-dle Miocene.[235]

Begun et al. also had cast their lot with Europe as the Miocene homeland of ancestral African great apes, humans, and orangutans.[379] Via cladistic analysis on a data set of 240 cranial and postcranial skeletal traits in eight Miocene Hominoidea sensu lato and *Hylobates, Pongo, Gorilla, Pan,* and *Australopithecus,* they generated a cladogram (Figure 3.14G):

- *Proconsul* and *Afropithecus* are excluded from stem Hominoidea, though only one additional step is required to place *Afropithecus* as a sister clade of *Kenyapithecus.*

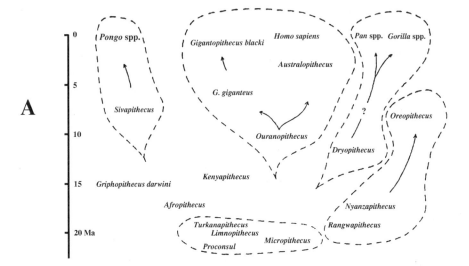

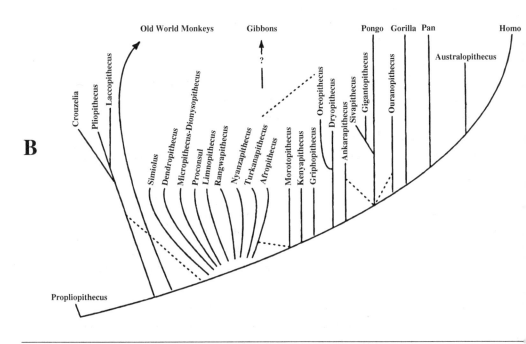

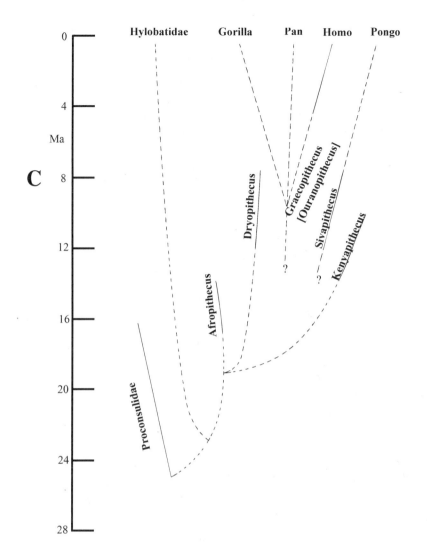

Figure 3.13. Sample of phyletic schemes based on morphology.

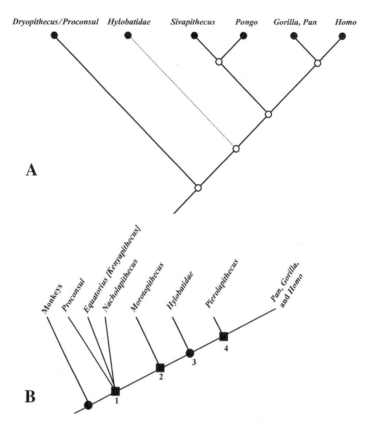

A

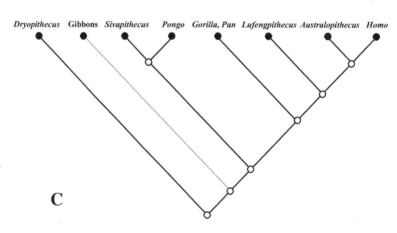

B

C

Dryopithecus   Gibbons   Sivapithecus   Pongo   Gorilla, Pan   Lufengpithecus   Australopithecus   Homo

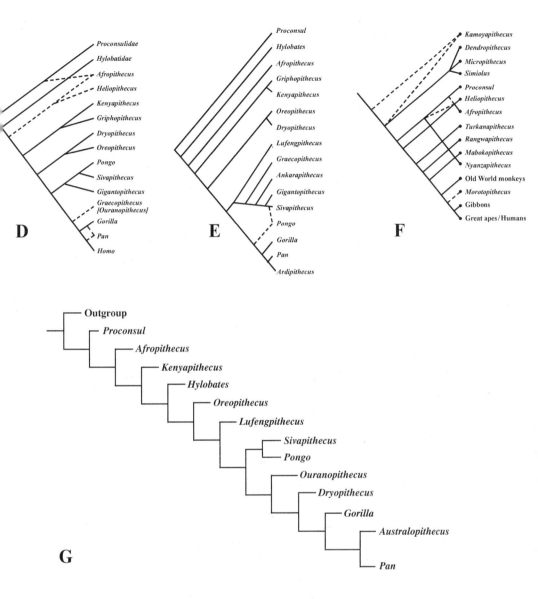

Figure 3.14. Sample of cladograms based on morphology, Ciochon 1983 (A); Moyà-Solà et al. 2004 (B); Xu and Lu 2008 (C); Harrison and Rook 1997 (D); Andrews and Bernor 2001 (E); Harrison 2002 (F); and Begun 2001 (G).

- *Kenyapithecus*, which is more primitive than *Hylobates,* is included among stem Hominoidea.
- *Oreopithecus* branches after *Hylobates* and is part of the large hominoid clade.
- *Pongo* and *Sivapithecus* constitute a sister clade, and only one additional step is needed for *Lufengpithecus* to join them.
- *Pan* is closer to *Homo* than either is to *Gorilla* (four additional steps are required for a Pan-Gorilla clade).
- An *Ouranopithecus-Australopithecus* clade with a *Pan-Gorilla* clade as the out-group requires an additional three steps.[379]

Although their analysis produced a most parsimonious cladogram, three others differ from it by only one additional step, and six others are also feasible, depending on further analyses of the evolutionary histories of particular traits and trait complexes.

Begun et al. concluded that the great ape and human clade originated from ancestors that were dentally like *Kenyapithecus* and postcranially like *Hylobates, Oreopithecus,* and *Dryopithecus,* which were adapted to arboreal suspensory behavior.[379] They suggested two biogeographic scenarios on the origins and deployments of the gibbons, Asian great apes, African great apes, and hominids. In both scenarios, *Griphopithecus* appeared to be important, though they had not included it in their maximum parsimony analysis owing to insufficient fossils.

In one scenario, during the Middle Miocene there were successive waves of African immigrants to Eurasia. Because the Engelwies molar—ascribed to *Griphopithecus darwini*—is dated at 15 Ma, *Dionysopithecus,* which is earlier, would have no place in hylobatid phylogeny. A second clade from Africa evolved into the Asian great ape clade; a third clade that is more closely related to African great apes and hominids deployed somewhat later to Eurasia.[379]

The emigration routes are elusive, but Moyà Solà and Köhler speculated that an anonymous African Middle Miocene hominoid species might have deployed to Europe via the Iberian Peninsula, giving rise to *Dryopithecus,* which remained in Europe.[380] Their descendent species, which are more like *Pongo,* namely, *Lufengpithecus* and *Sivapithecus,* evolved during later deployments to East Asia. Moyà-Solà et al. proposed that ca. 12.5–13 Ma *Pierolapithecus* is close to the last common ancestor of the great apes and humans after furcation of the hylobatid apes.[381]

Subsequently, Moyà-Solà et al. argued that *Anoiapithecus brevirostris*, a younger (11.9 Ma) Middle Miocene species from Abocador de Can Mata in Cataluña, Spain, is a Eurasian stem ancestor of the Hominidae and Pongidae.[382] The locality where the fragmentary facial skeleton and mandibular body with most of the dentition of the adult male type specimen was retrieved is the same area as that of *Pierolapithecus*—Els Hostalets de Pierola in the Vallès-Penedès Basin—which has a trove of fossiliferous localities spanning ca. 12.5–11.3 Ma.[383]

Alternatively, an ancestral stock of *Griphopithecus,* represented by specimens from Slovakia and Turkey, trifurcated to stock lineages of the gibbons, Asian great apes, and African great apes and humans. In response to the drier climate, a more terrestrially adapted branch of the third group returned to Africa ca. 9 Ma and continued to evolve into *Pan, Gorilla,* and the Hominidae (sensu Table 2.3).[379]

Begun et al. stated no preference for one of the two scenarios, though clearly they seemed to favor relatively poorly documented *Griphopithecus* spp. and Eurasia as primary in the evolution of extant African apes and humans.[379] Later, Begun seemed more convinced that the stocks leading to chimpanzees, bonobos, gorillas, and humans are derived from Eurasian taxa—*Dryopithecus, Graecopithecus,* or *Sivapithecus*—instead of autochthonous African forms.[384] The dearth of substantial collections of Late Miocene African hominoids could indicate that the scenario of Eurasian origins is correct. However, to date, productive Miocene hominoid sites are concentrated in a tiny fraction of the African landmass.

It is also possible that during the Late Miocene, the African apes and earliest hominids evolved in tropical forests and woodlands of central, west, or southern Africa. Accordingly, Late Miocene hominoid species with some cranial and dental features like those of European *Dryopithecus* and *Ouranopithecus* might have inhabited Africa, where they were basal to radiations of the Panidae and the Hominidae.

Presently, there is no reason to gnaw at Asian roots for the Hylobatidae, *Pongo,* and *Gigantopithecus.* The precise regions where they emerged are obscure, though *Khoratpithecus* renews hope that the ancestry of *Pongo* might be forthcoming. Based on cytochrome b gene sequences, Chatterjee estimated that the radiation of the Hylobatidae dates to ca. 10.5 Ma from ancestors in Yunnan, China, Vietnam, and Laos, which points toward *Laccopithecus* as possibly basal to extant gibbons, though its relatively young date (6.2–6.9 Ma) is problematic.[385]

4

Taproot and Branches of Our Family Tree

Numerous lines of molecular evidence now firmly place the human-ape divergence at 4–8 Ma . . . The molecular support for a human-chimpanzee clade is now overwhelming . . .

B. J. BRADLEY (2008, p. 337)

Our analysis shows that human-chimpanzee speciation occurred less than 6.3 million year ago and probably more recently, conflicting with some interpretations of ancient fossils. Most strikingly, chromosome X shows an extremely young generation time, close to the genome minimum along its entire length. These unexpected features would be explained if the human and chimpanzee lineages initially diverged, then later exchanged genes before separating permanently.

NICK PATTERSON, DANIEL J. RICHTER, SANTE GNERRE, ERIC S. LANDER, and DAVID REICH (2006, p. 1103)

Welcome to Ancestors Anonymous

Unless *Ouranopithecus* (including *Graecopithecus*) (Figure 3.9), *Griphopithcus,* or *Dryopithecus* is indeed the proximate ancestor of *Australopithecus,* we have precious few clues to the nature of Late Miocene and earliest Pliocene Hominidae. Indeed, there are only exiguous hominoid bits from the span between 11 Ma and 4 Ma in Africa.[1] In addition to *Samburupithecus kiptalami* (9.5 Ma), *Sahelanthropus tchadensis* (6–7 Ma), *Orrorin tugenensis* (5.8 Ma), *Ardipithecus kadabba* (?5.2–5.8 Ma), and *Ardipithecus ramidus* (4.4 Ma) (Figure 4.1), they comprise the following bits:

Figure 4.1. Sampler of Late Miocene–Middle Pliocene Hominoidea. (A) Left lateral and (B) occlusal views of left maxilla of 9.5-Ma *Samburupithecus kiptalami* from Samburu Hills, northern Kenya. (C) Cranium of 6–7 Ma *Sahelanthropus tschadensis,* from the Djurab Desert, Chad. (D) Right P³, C¹, C₁, P₃ of 5.6-Ma *Ardipithecus kadabba* from Asa Koma, Middle Awash Valley, Ethiopia, compared with chimpanzee. Note sectorial P₃ in both species. (E) Fragmentary skeleton of 5.8-Ma *Orrorin tugenensis* from Tugen Hills, Kenya. Note short neck and bulbous head of femora (a, b). (F) Partial skeleton of 4.4-Ma *Ardipithecus ramidus* from Asa Koma, Middle Awash Valley, Ethiopia.

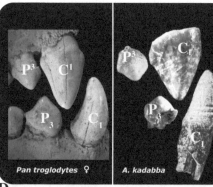

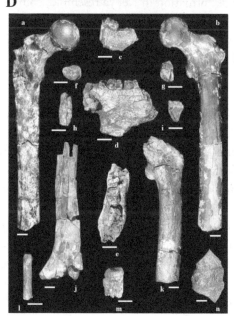

- Lukeino (KNM-LU 335) and Ngorora (KNM-BN 378) molars.
- Lothagam hemimandibular fragment (KNM-LT 329).
- Chemeron proximal humerus (KNM-BC 1745).
- Kanapoi distal humerus (KNM-KP 271).

Most of them have had both advocates and detractors, and none is proba-bly sufficient to document hominid status for the individual from which it came.[2]

Although Howell, Simons, Pilbeam, and Pickford placed the ca. 6-Ma chimpanzee-sized M_1 or M_2 from the Lukeino Formation in western Kenya in the Hominidae, McHenry and Corruccini found that it is most like those of *Pan troglodytes*.[3] Ungar et al. considered it to be most similar to a 4.1-Ma M_1 of *Australopithecus* from Allia Bay, East Lake Turkana, Kenya.[4] Pickford and Senut concluded that an apish upper molar and a central incisor that Senut et al. had assigned to *Orrorin tugenensis* from the Lukeino Forma-tion (Tugen Hills, Kenya) are probably from a second anonymous hominoid species.[5]

Eventually, the 11-Ma Ngorora M_2, which is at the transition between the Middle and the Late Miocene, might be assigned to *Kenyapithecus* or *Nachol-apithecus*. It resembles 15-Ma counterparts from the Nachola Formation, which is also in northern Kenya.[6] Pickford and Senut concluded that a 12.5-Ma M_2 or M_3 from the Ngorora Formation in Kenya resembles counterparts in *Pan troglodytes*.[7]

Fleagle designated the Samburu maxilla *Kenyapithecus*.[8] Neither he nor Ishida et al. claimed hominid status for *Samburupithecus kiptalami,* though the latter allow that it might be linked to the common ancestor of *Pan* and *Homo*.[9]

Several authors argued that the >5–7-Ma Lothagam mandibular speci-men from northern Kenya can be assigned to the Hominidae, but White concluded that it belongs to an indeterminate species of the Hominoidea.[10] Similarly, the early Pliocene (ca. 5 Ma) Chemeron humeral fragment from northern Kenya is hominid to Pickford et al. but an indeterminate hominoid to Senut.[11]

Most experts agree that the Kanapoi distal humerus is hominid, even hu-manoid.[12] It has been linked variously to *Australopithecus, Paranthropus,* and *Homo*.[13] Coppens placed it in *Homo habilis*.[14] The Kanapoi specimen is gen-erally dated at ca. 4 Ma, but as a surface find it could be younger.[15] M. G. Leakey et al. included it in the hypodigm of *Australopithecus anamensis,* which comprises several ca. 4-Ma mandibular and maxillary specimens, isolated teeth, and a partial tibial shaft from Allia Bay in northern Kenya.[16]

Kingston and Hill employed stable carbon isotopic analyses to infer the kinds of vegetation (C_3 and C_4 plants) that covered the landscape during the last 15.5 years in the Baringo Basin of western Kenya.[17] They concluded that if Hominidae originated in East Africa during the Late Miocene, they inhabited a heterogeneic mosaic of environments instead of adapting to an abrupt transition from rain forest to grassland and woodland biomes. Macho et al. inferred from microstructure and finite-element stress analyses that the thickly enameled teeth of *Australopithecus anamensis* were adapted for a diet of hard, tough items.[18] Sepulchre et al. argued that tectonic uplift was a major cause of the aridification and the expansion of grasslands in East Africa during the late Neogene, especially the past 8 Ma.[19]

Will the Real First Hominid Please Stand Up?

Historically, there were three main Miocene contenders for the coveted superlative "earliest known hominid." They are, in order of increasing temporal popularity in the twentieth century, *Oreopithecus, Gigantopithecus,* and *Ramapithecus.*

Beginning in 1954, Hürzeler championed *Oreopithecus bambolii* as an early member of the Hominidae (Figure 4.2).[20] He resisted advancing it as a direct Late Miocene ancestor of *Australopithecus* or *Homo* and later placed it in a new family, the Heterohominidae.[21] Viret, Kurth, and Heberer enthusiastically endorsed Hürzeler's initial proposal.[22] Woo and Robinson were also inclined to agree with his idea, but most other experts remained skeptical of its validity.[23] The fact that Hürzeler rooted the Hominidae in the Paleocene (ca. 60 Ma) undoubtedly fueled skepticism because many anatomists, anthropologists, and paleontologists who are acquainted with the detailed morphological similarities of apes and humans eschew extremely early origins for the Hominidae.[24]

The holotype of *Oreopithecus bambolii* is a juvenile mandible described in 1872 by Gervais, who placed it among the apes.[25] It came from Monte Bamboli in the Maremma District of Tuscany, Italy. Subsequently, additional craniodental and postcranial specimens were recovered from Monte Bamboli and five other Maremma sites.[26] Harrison listed 393 skeletal items, including twenty-two mandibles, twenty-seven crania, thirty-two isolated teeth, and representative bits from the vertebral column, ribs, forelimb, and hind limb,

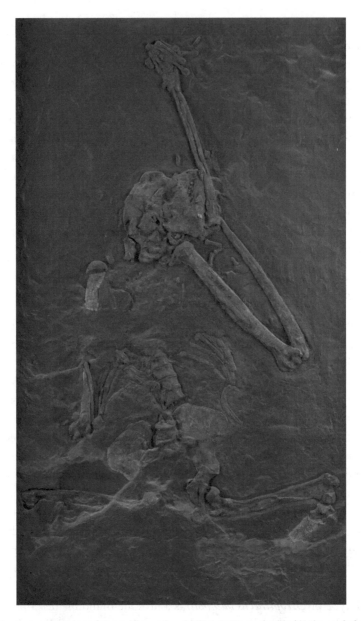

Figure 4.2. Skeleton of *Oreopithecus bambolii*, 6–8-Ma, embedded in the roof of a lignite mine in Baccinello, Italy.

including pectoral and pelvic girdles.[27] From a single potassium–argon (K/Ar) date and faunal correlations, Azzaroli et al. inferred that they are 8.5–11 Ma; later studies by Rook et al. placed them between 7 and 9 Ma.[28] The most spectacular specimen (IGF 11778) is a crushed adult male skeleton that was recovered from the roof of a coal mine at Baccinello.[29]

Although over fifty individuals are represented in the hypodigm of *Oreopithecus bambolii,* its place in anthropoid phylogeny and classification remains enigmatic and controversial. Paleontologists and anatomists have placed it variously with the Hylobatidae, the great ape/human clade, the Hominidae, or the Cercopithecidae. Making a queer monkey of it has been the more popular alternative to considering it an early hominid.[30]

Hürzeler listed nineteen cranial, dental, and postcranial traits that would link *Oreopithecus* with *Homo* and *Australopithecus* versus the Pongidae: short face, vertical incisors, small canines, reduced canine sexual dimorphism, absence of diastemata, homomorphic mandibular premolars, relative lengths of individual teeth in the tooth rows, anteriorly projecting nasal bones, anterior placement of the zygomatic bones (over P^4 or M^1), five lumbar vertebrae, and absence of a tail.[31]

The case for special cercopithecoid affinities of *Oreopithecus* rests chiefly on dental features: the bilateral alignment of four main molar cusps and apparent crests between them suggests incipient bilophodonty, and reduced hyopconulids.[32] Further, Szalay and Berzi's remodeling of the crushed skull from the Baccinello skeleton is much more monkey-like than Hürzeler's archaic humanoid model.[33] Both show sizeable circular orbits surrounded by prominent bony rims and a broad interorbital pillar. In the Szalay-Berzi model, the muzzle is more projecting, the nasal bones are less prominent, the nuchal muscular attachment is much higher on the occiput, the mandibular rami are taller, the angle of the mandible is expanded markedly, and there is a prominent sagittal crest and supraorbital torus.

Szalay and Berzi guessed that the cranial capacity of the Baccinello *Oreopithecus* is about 200 cc.[34] Straus and Schön's apish estimate of 276–529 cc is probably too high because they used Hürzeler's anthropomorphic model, in which he had inadvertently extended the braincase with bits of cervical vertebrae.[35] Given that the Baccinello beast, which Harrison inferred is a large male, weighed about 32 kg, a cranial capacity of 200 cc casts it as a peculiarly small-brained hominoid.[36] The estimated cranial capacity of a ca. 15-kg female *Oreopithecus* is only 128 ± 45 cc.[37]

Postcranial evidence seems to clinch hominoid status for *Oreopithecus,* though it by no means resolves whether its lineage emerged before the hylobatid furcation, broadly contemporaneously with it, or sometime between the hylobatid and the pongid furcations.[38] Its postcranial skeleton, especially the forelimb and thorax, is remarkably similar to those of extant apes.[39]

Some craniodental features of *Oreopithecus* represent the primitive catarrhine condition, but others, especially in the teeth and jaws, are autapomorphies related to heavily masticating a folivorous diet.[40] Both the postcranial adaptations for versatile climbing and the folivorous masticatory apparatus of *Oreopithecus bambolii* fit well with the fact that this markedly sexually dimorphic species probably inhabited warm temperate or subtropical mesophytic woodland, perhaps mixed with swampy areas.[41] Szalay and Langdon inferred from the foot structure that *Oreopithecus* was probably not a habitual biped particularly a terrestrial one.[42] Subsequent studies by Köhler and Moyà-Solà and Rook et al. indicate that *Oreopithecus* probably engaged in bipedal positional behaviors.[43]

Several renowned paleontologists, including Weidenreich, von Koenigswald, Broom, Robinson, Heberer, and Woo recommended *Gigantopithecus,* an Asian Late Miocene–Pleistocene hominoid, as a hominid (Figure 3.12). This suggestion has not enjoyed wide acceptance. Generally, *Gigantopithecus* is nested with the apes.[44] Regardless whether *Gigantopithecus* are the greatest apes or humongous hominids (264–545 kg), *Ouranopithecus* appears to be the best candidate for their Late Miocene ancestry.[45]

There are two species of *Gigantopithecus: G. blacki* of China and Vietnam and *G. giganteus* of South Asia.[46] Weidenreich concluded that three monstrous molars of *Gigantopithecus blacki* that von Koenigswald had procured from Chinese apothecaries were morphologically identical with other hominid teeth; they differed only in overall size.[47] This led him to propose that *Gigantopithecus* is the southern Chinese ancestor of *Pithecanthropus* (the Javanese *Homo erectus*) and possibly also *Sinanthropus* (Peking Man, the Chinese *Homo erectus*).

Because of the hypsodonty and occlusal wrinkling pattern, von Koenigswald seconded a hominid status for the molars of *Gigantopithecus.*[48] But he argued that they are too specialized to align with *Homo.* Instead, he sidelined *Gigantopithecus,* just as Hürzeler had done with *Oreopithecus.* After a mandible and fuller dentition of *Gigantopithecus blacki* were found, von Koenigswald became cautious about their hominid status.[49]

Broom et al. concluded that *Gigantopithecus blacki* is linked with *Australopithecus,* a robust member of which was basal to the human lineage.[50] Dart, Woo, and tentatively Heberer agreed that *Gigantopithecus blacki* was an Early Pleistocene Asian hominid cousin of the southern African *Australopithecus.*[51]

Beginning in the early 1970s, a second wave of support for *Gigantopithecus* as the ancestral hominid developed, partly because of skepticism about the hominid status of *Ramapithecus.*[52] The discovery of a >6-Ma mandible of *Gigantopithecus giganteus* in northern India checked the challenge that because some specimens of *Australopithecus* are probably older than *Gigantopithecus,* the latter were not ancestors to the former; that is to say, a descendent cannot be its own ancestor.[53]

Robinson further developed Broom's idea. He observed that *Gigantopithecus giganteus* is morphologically transitional between a pongid, like *Sivapithecus indicus,* and *Paranthropus.* Therefore, he erected the Paranthropinae, a new hominid subfamily to accommodate *Gigantopithecus* and *Paranthropus.* Because he viewed the latter as ancestors to later Hominidae, including *Homo, Gigantopithecus giganteus* is basal to the bifurcation that culminated in *G. blacki* in China and emergent *Homo* in Africa or elsewhere.[54]

The molars of *Gigantopithecus blacki* are characterized by thick, crenulated enamel, a shallow pulp cavity in the crown—taurodontism—and high, blunt cusps that bulge basally and are separated by deep, narrow furrows.[55] Indeed, von Koenigswald observed that the cusps appear to have been carved into solid blocks of enamel.[48] The lower molars sport the Y-5 pattern, and supernumerary cusps are abundant. Zhang noted that presumed Middle Pleistocene molars are significantly larger than those of the Early Pleistocene, but their anterior dentitions are basically alike.[56]

Contrarily, *Gigantopithecus giganteus* lack the high crowns, deep wrinkles, and elaboration of cusps that characterize the postcanine teeth of *G. blacki*[57]. Accordingly, they are well suited for direct ancestry to *G. blacki.* Further, insofar as they are known, the anterior dentition and mandible of *Gigantopithecus giganteus* represent conditions that evolved toward *G. blacki.*[57]

The degree of sexual dimorphism of *Gigantopithecus giganteus* is indeterminate because there are too few specimens. However, the three mandibles and >1,300 teeth of *Gigantopithecus blacki* evince notable size dimorphism, which is reasonably ascribed to sex.[58] The canines of *Gigantopithecus blacki* are robust but low in comparison with those of modern great apes. In *Gigantopithecus blacki,* diastemata are small or nil, and the P_3 is bicuspid but semisectorial.[59] In males, the upper canines sharpened against P_3.[60]

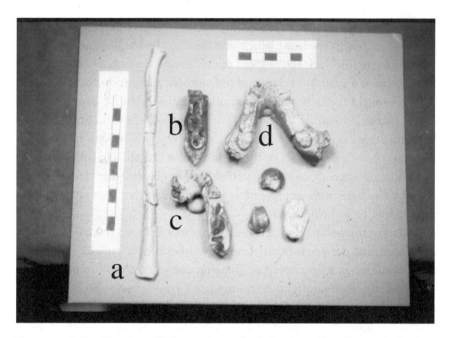

Figure 4.3. Radius (a) and mandibular specimens (b–d) of ca. 8-Ma *Sivapithecus sivalensis* (formerly *Ramapithecus punjabicus*) from Potwar, Pakistan.

Without postcranial bones, we cannot know the locomotor mode of *Gigantopithecus* or even how big they were. From molar dimensions and an extant ape equation, Conroy predicted a mass of 166 kg for *Gigantopithecus giganteus.*[61] Stature guesstimates for *Gigantopithecus blacki* range between 152 cm and 366 cm; its predicted mass is twice that of a male gorilla (between 264 kg and 436 kg) and reaches the weight class of the polar bear (a whopping 545 kg).[62]

What might these behemoths have eaten, and where did they rest and roam? Given the King Kong-like size estimates, we might respond, Whatever and wherever they wanted. Alas, comparisons with modern mammals lead us to infer, more soberly, that *Gigantopithecus* was a hand-to-mouth forest herbivore that ate quantities of grasses (probably including bamboo) and dicotyledonous fruits instead of being a weapon-wielding, open woodland or savanna carnivore.[63]

For two decades, Simons and Pilbeam vigorously promoted *Ramapithecus,* a Middle to Late Miocene hominoid from India, Pakistan, and Kenya, as the first hominid (Figure 4.3).[64] Although Lewis, who collected the type speci-

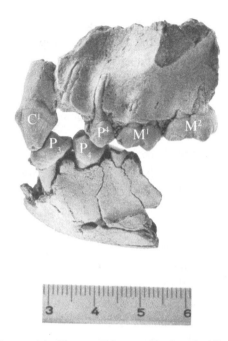

Figure 4.4. *Kenyapithecus wickeri* (composite), ca. 14-Ma, from Fort Ternan, Kenya.

men (YPM 13799), stopped short of publishing *Ramapithecus* in the Hominidae, Simons boldly took the plunge.[65]

Before 1975, South Asian *Ramapithecus punjabicus* was represented by fewer than a dozen fragmentary bits of upper and lower jaws; the available dentition was mostly postcanine, and no limb bone had been linked definitely with the craniodental remains that were assigned to *Ramapithecus*.[66]

One of the best specimens in Simons and Pilbeam's hypodigm of *Ramapithecus* came from Fort Ternan, a Middle Miocene locality in western Kenya.[67] Louis Leakey named it *Kenyapithecus wickeri*, but Simons promptly sank it with South Asian *Ramapithecus* (Figure 4.4).[68]

Although *Ramapithecus* seemed to show a mixture of hominid and pongid features, its champions stressed the hominid ones. Initially, the presumed parabolic dental arcade, reduced maxillary canines, vertically oriented incisors, and diminished diastemata appeared to confirm its hominid status. Like Darwin, who had linked reduced canines with tool use and bipedalism, Pilbeam and Simons and Prasad speculated that *Ramapithecus* might have been a biped that employed simple tools.[69]

The case for *Ramapithecus* did not emerge from carefully controlled comparisons of fossil and extant hominoid and other samples. Thus, it is no surprise that when such studies were conducted, particularly in light of the sketchy hypodigm of *Ramapithecus,* the distinctiveness of the taxon evaporated. Currently, most experts sink South Asian *Ramapithecus* in *Sivapithecus,* and *Kenyapithecus* has been reinstated as a distinct genus as originally proposed by Leakey and upheld by Aguirre.[70]

Several experts challenged the humanoid parabolic shape of the dental arcade that Lewis and Simons had claimed for *Ramapithecus.*[71] Indeed, the more complete specimens from Turkey and Pakistan indicate that the short, deep jaws are approximately V-shaped instead of parabolic. Further, the maxillary canines are neither incisiform nor reduced to hominid configuration; the incisors are procumbent, P_3 is sectorial, and the mandible sports a simian shelf. Molar and mandibular size and shape features, differential wear on the molars, and enamel thickness also fail to separate presumed *Ramapithecus* from *Sivapithecus.*[72] Faced with such overwhelming negative evidence, the champions of *Ramapithecus* retreated.[73]

Zhang et al. reprised *Ramapithecus* as the hominoid ancestor of *Homo* on the basis of a well-preserved juvenile maxillofacial cranial specimen with c, dm^1-dm^2, and M^1, three partial maxillae, and 109 isolated teeth from the lower Shagou Formation, Hudieliangzi, Yunnan Province, southwestern China. They dubbed the creature *Ramapithecus hudienensis* and suggested that it gave rise to *Homo orientalis,* which is represented by six isolated teeth from the upper Shagou Formation and basal Yuanmou Formation at Zupeng near Hudieliangzi. Associated fauna in the lower Shagou Formation indicate a Pliocene date (3–4 Ma) for *Ramapithecus hudienensis.*[74]

Ho reported that 250 hominoid teeth had also been recovered at Hudieliangzi, but he did not allocate them taxonomically. He seemed to favor affinities between *Lufengpithecus lufengensis* and *Ramapithecus hudienensis,* with *Ramapithecus hudienensis* having advanced features derived from *Lufengpithecus lufengensis,* which is 3–5 million years older. But the phylogenic placement and even the alpha taxonomy of the Hudieliangzi hominoid collection is hampered by juvenility of the most complete specimen from the locality.[75]

Xu and Lu claimed that although the upper limb bones of *Lufengpithecus lufengensis* indicate arboreal adaptation, the femur and hallucal metatarsal bespeak terrestriality, including bipedal locomotion.[76]

In brief, we probably have no clearly recognized ancestral Hominidae before perhaps *Ardipithecus ramidus* from the early Pliocene sites of Aramis, Middle

Awash, Ethiopia (4.4 Ma) and Duma, Gona Western Margin, Afar, Ethiopia (4.5–4.3 Ma) and more definitively *Australopithecus anamensis* (4.1–3.5 Ma) and *A. afarensis* (3.9–3 Ma).[77] As a hominid, *Ramapithecus* always was on shaky footing because the specimens were so few and fragmentary. Its credibility was weakened further by biomolecular evidence that indicated that the divergence of the African apes and our lineage probably occurred after its heyday.[78]

Our Selfish Genomes Keep Their Prize Secret

> Macroevolutionary patterns of diversification can be discerned with molecular studies only if detailed evolutionary interpretation of clade history is conducted with direct reference to fossil evidence.
>
> P.-H. FABRE, A. RODRIGUES, and E. J. P. DOUSERY
> (2009, p. 823)

The comparative morphological studies of Thomas Henry Huxley, Arthur Keith, William King Gregory, Dudley J. Morton, Sherwood L. Washburn and succeeding generations of anthropologists and human biologists, combined with pioneer serological discoveries, narrowed the realm of the closest extant nonhuman relatives of humans to the great apes.[79] Discovery and study of African Plio-Pleistocene homonid fossils that combined pongid and hominid features supported a pongoid ancestry of humans and steadily became a major resource for modeling hominid phylogeny and relationships with extant great apes.[2]

At that point, the revolution in molecular biology overran paleoanthropology.[2] It took nearly two decades for its influence to become manifest. Although Goodman documented that the electrophoretic patterns of blood proteins from chimpanzees and gorillas are remarkably similar to those of human and congealed the propinquity of the African apes and humans with immunological diffusion tests, Simons and Pilbeam proposed very early origins for the human and ape lineages, based on Fayum Oligocene and eastern African and South Asian Miocene fossils.[80]

Building on Goodman's pioneering efforts, Sarich and Wilson, devised a molecular evolutionary clock whereby serum albumins would tell when humans, apes, and other creatures had parted ways.[81] This sparked a war of words between advocates for supremacy of biomolecular data versus morphological (particularly paleontological) evidence for revealing major events in primate phylogeny.

Goodman never ticked with a molecular clock. He argued that the longer generation times of apes and humans perturbed its accuracy—a prophetic caveat in view of the recent studies of Langergraber et al. (Figure 3.4).[82] Further, most molecular clocks are not independent of the fossil record—they are calibrated with arguably dated fossils.[83]

It is now generally acknowledged that the evolutionary rates of molecules might vary during vertebrate evolution and that, like morphological features, various molecules and even different parts of the same molecule evolve at different rates.[84] Regardless, in the absence of an adequate fossil record, ingenious modelers have sustained the marketability of molecular clocks as the best available instruments for bounding hominoid evolutionary events.[85]

Sarich and Wilson ruffled some prominent fellow molecular anthropologists and paleoanthropologists by suggesting that we parted ways with the African apes at merely 5 Ma.[86] Nonetheless, the commentaries of persons who favor 5 Ma as the date of furcation(s) between us and the African apes are often taken to be authoritative despite the fact that a sampler of molecular clock dates, based on a variety of molecules and analytic methods, have revealed a notable range of possible dates between 2 Ma and 12 Ma for the divergence of the human and African ape lineages (Table 4.1).[87]

The younger dates, derived from a mitochondrial DNA (mtDNA) clock, would exclude not only *Ramapithecus* but also the Pliocene Awash, Hadar, and Laetoli hominids from our lineage. That is to say, *Australopithecus* would have branched from the lineage of *Homo* before the furcation of *Gorilla gorilla, Pan troglodytes,* and *Pan paniscus.* Hasegawa et al. speculated that mtDNA was transferred via hybridization between protohumans and bipedal protochimpanzees and that *Australopithecus afarensis* might be linked more closely to *Pan paniscus* than to *Homo.*[88] Although one might challenge that the mtDNA clock should be returned to the shop for adjustment or shelved, Patterson et al. entertained that the early genetic exchange might have occurred before the two lineages diverged permanently.[89]

The intriguing idea that viruses and other factors might mediate interspecific—horizontal or lateral—gene flow in Hominoidea has little currency among molecular anthropologists.[90]

In sum, the older molecular clock dates that extend into the latest Middle Miocene fully accommodate *Australopithecus* in our lineage and allow for even more apelike intermediates between them and earlier the Middle Miocene Hominoidea (Figure 4.5; Table 4.2).[91] However, if the appearance of

Table 4.1. A sampler of molecular clock divergence dates (Ma) for the chimpanzee, human, gorilla, orangutan, and gibbon lineages

Chimpanzee/Human	Gorilla/Human	Orangutan/Human	Gibbon/Human	Source
2.7 ± 0.6	3.7 ± 0.6	10.9 ± 1.2	13.3 ± 1.5	Hasegawa et al. 1985
4.9 ± 1.2	5.9 ± 1.2	11.9 ± 1.7	—	Hasegawa et al. 1987
5.0 ± 1.5	5.0 ± 1.5	10.0 ± 3.0	12.0 ± 3.0	Cronin 1983
5.5–7.7	7.7–11.0	12.2–17.0	16.4–23.0	Sibley and Ahlquist 1987
5.5 ± 0.2	6.7 ± 1.3	8.2 ± 0.8	14.6 ± 2.8	Kumar and Hedges 1998
5.9 ± 0.9	8.5 ± 1.1	—	—	Ueda et al. 1986
6.3 ± 1.2	6.6 ± 1.3	14.0 ± 2.8	—	Li et al. 1987
6.3–7.7	8.0–10.0	13.0–16.0	18.0–22.0	Sibley and Ahlquist 1984
7.4–9.5	7.7–9.9	15.0–19.3	—	Li and Tanimura 1987
9.2 ± 1.7	9.8 ± 2.1	16.0 ± 3.5	18.7 ± 2.2	Gingerich 1985
5.0–6.0	6.0–7.0	13.8 ± 0.8	—	Wildman et al. 2003
5.0–7.0	6.0–8.0	12.0–15.0	—	Glazko and Nei 2003
5.0–7.0	—	—	—	Kumar et al. 2005
4.5–6.0	6.0–8.0	12.0–16.0	18.0–20.0	Locke et al. 2011
6.0	10.0	—	—	Scally et al. 2012
8.5	10.0	—	—	Langergraber et al. 2012

Table 4.2. Pliocene–Holocene Hominidae

Hominidae

Paranthropinae

Paranthropus aethiopicus	2.2–2.6 Ma	Omo, Ethiopia; West Turkana, Kenya; Laetoli, Tanzania, Malawi
Paranthropus boisei	1.2–2.6 Ma	Olduvai Gorge and Peninj, Tanzania; East and West Turkana, Kenya; Konso, Ethiopia
Paranthropus robustus	1.2–1.8 Ma	Swartkrans, Kromdraai, Drimolen, Gondolin, and Cooper's D, South Africa

Australopithecinae

Australopithecus afarensis	2.9–3.6 Ma	Hadar, Maka, Behlodelie, Fejej, and Omo, Ethiopia; ?Laetoli, Tanzania; Baringo and West Turkana, Kenya; ?Sterkfontein, South Africa
Australopithecus africanus	2.3–3.0 Ma	Taung, Makapansgat, Sterkfontein, and Gladysvale, South Africa
Australopithecus anamensis	3.5–4.2 Ma	Kanapoi and Allia Bay, Kenya; Middle Awash, Ethiopia
Australopithecus bahrelghazali	3.0–3.5 Ma	Bahr el Ghazal, Chad
Australopithecus garhi	2.5 Ma	Middle Awash, Ethiopia
Australopithecus (=Homo) habilis	1.6–2.3? Ma	Olduvai Gorge, Tanzania; ?East Turkana and Chesowanja, Kenya; ?Sterkfontein, South Africa
Australopithecus sediba	1.78–1.95 Ma	Malapa, South Africa

Homininae

Homo antecessor	780 Ka–1.0 Ma	Gran Dolina, Atapuerca, Spain; Buia, Eritrea
Homo erectus	52? Ka–1.8 Ma	Africa, Asia, and ?Europe
Homo ergaster (=*Homo erectus*)	1.5–1.9 Ma	East and ?West Turkana, Kenya
? *Homo floresiensis*	?17.0–95.0 Ka	Flores Island, Indonesia
Homo georgicus	1.8 Ma	Georgia
Homo heidelbergensis	200–600 Ka	Europe, ?Ethiopia, ?Zambia, ?South Africa
Homo neanderthalensis	≥25≥250 Ka	Europe and western Asia
Homo rudolfensis	1.8–2.4 Ma	East Turkana, Kenya; Uraha Hill, Malawi, Ethiopia (?)
Homo sapiens	0–≥200 Ka	≥Global

Subfamily incertae sedis

Ardipithecus kadabba	5.6 Ma	Asa Koma, Middle Awash, Ethiopia
Ardipithecus ramidus	4.4 Ma	Aramis, Middle Awash, Ethiopia
Kenyanthropus platyops	3.5 Ma	West Turkana, Kenya
Orrorin tugenensis	<6.0 Ma	Tugen Hills, Kenya
Sahelanthropus tchadensis	6.0–7.0 Ma	Toros-Menalla, Chad

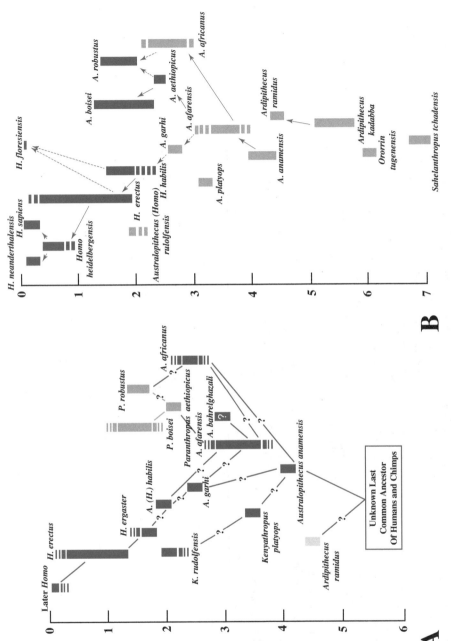

Figure 4.5. (A) 2001 and (B) 2009 phyletic schemes of Daniel Lieberman.

Homo is to be envisioned in the narrowest time frame, the event must have occurred by special creation via bonoboid bones, chimpoid chunks, and gorilloid gristle instead of deified dirt and an estrogenized rib.

Oooh, Look What I Found!

From the early 1960s onward, the pioneering work of the Leakey family at Olduvai Gorge in Tanzania, and later Richard Leakey's expeditions to East Turkana (East Rudolf and Koobi Fora) and West Turkana in northern Kenya, attracted wide public attention.[92] Subsequently, an American-French team in Hadar, Ethiopia, Mary Leakey's expeditions at Laetoli, Tanzania, and the Kimeu-Leakey-Walker group in West Turkana uncovered even more dramatic clues to our ancient heritage (Map 4.1).[93] The excitement and controversy that these paleontological treasures generated have remained unabated (Figures 4.6 and 4.7).[94]

This period of discovery and debate is outstanding in the history of science because it has been filmed and broadcast profusely to lay and scientific audiences. The attendant problems are that voiceless fossils may become mere props in popular media dramas, and inexpert viewers might receive the impression that more is known about our ancestry than is substantiated by the fossils and their shifting dates, including those derived via the best available geochemical methods.[95]

Some of the most securely dated, earliest Hominidae are from Laetoli, Tanzania, where three bipedal creatures left extensive trails of humanoid footprints in volcanic ash, which is a prime mineral for K/Ar analysis (Figure 4.8).[96] We do not know whether the skeletal bits of >25 individuals from Laetoli are the same species that made the prints, though they appear to be broadly from the same time period (3.5–3.8 Ma).[97]

Two fragmentary specimens (BEL-VP-1/1 and MAK-VP-1/1) from the Middle Awash Valley in Ethiopia, and a mandibular bit (KNM-TH 13150) from Tabarin in the Tugen Hills of Baringo, Kenya, are probably older than the Laetoli specimens. The Belohdelie adult cranial fragments and the Maka subadult left proximal femur were, respectively, below and above a tuff that is dated radiometrically at 3.8–4 Ma.[98] They are like counterparts in the Pliocene Hadar Hominidae.[99]

The Tabarin specimen is a partial right mandibular body with M_{1-2}. It is related to tuffs that are ca. 5-Ma according to K/Ar determinations. Faunal

FOSSIL SITES

ETHIOPIA
1. Woranso-Mille: *Au. anamensis, Au. afarensis*
2. Duma: *Au. anamensis, Au. afarensis*
3. Middle Awash Valley: *Ar. kadabba, Ar. ramidus*
   Au. anamensis, Au. afarensis, Au. garhi,
   H. erectus, H. sapiens
4. Omo: *P. aethiopicus*
5. Melka Kunturé: *H. erectus*
6. Konso: *P. boisei*
7. Bouri: *H. erectus*
8. Aramis: *Ar. ramidus*
9. Hadar Formation: *Au. afarensis*

SOUTH AFRICA
10. Kromdraai: *P. robustus*
11. Malpa: *Au. sediba*
12. Taung: *Au. africanus*
13. Saldanha: *H. erectus*
14. Klasies River Mouth: *H. sapiens*
15. Swartkrans: *P. robustus, H. erectus*
16. Sterkfontein: *Au. africanus, Au. habilis*
17. Drimolen: *P. robustus*
18. Makapansgat: *Au. africanus*

ZAMBIA
19. Kabwe: *H. heidelbergensis*

MOROCCO
20. Jebel Irhound: *H. sapiens*

ALGERIA
21. Tighenif: *H. erectus*

NAMIBIA
22. Berg Aukas; Otavipithecus namibiensis

KENYA
23. Nariokotome: *P. aethiopicus*
24. West Turkana: *P. aethiopicus, P. boisei, H. ergaster*
25. Allia Bay: *Au. anamensis*
26. East Turkana: *P. boisei, Au./H. habilis, H. ergaster, H. rudofensis*

TANZANIA
27. Laetoli: *P. aethiopicus, Au. afarensis, H. sapiens*
28. Peninj: *P. boisei*
29. Olduvai Gorge: *P. boisei, H. habilis, H. erectus*
30. Ndutu: *H. erectus*

MALAWI
31. Uraha Hill: *H. rudolfensis*

CHAD
32. Toros-Menalla: *Sahelanthropus tschadensis*
33. Bahr el Ghazal, Karo Toro: *Au. bahrelghazali*

Map 4.1. African Late Miocene–Holocene hominid fossil sites

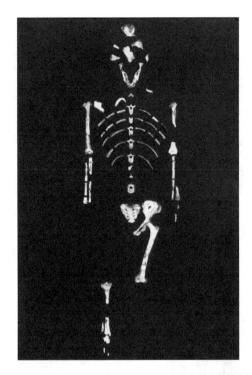

Figure 4.6. Partial skeleton of 3.2-Ma *Australopithecus afarensis* (A.L. 288–1) from Hadar, Ethiopia.

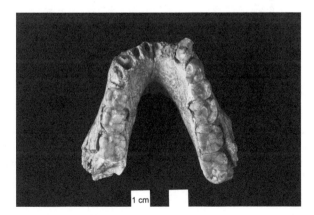

Figure 4.7. Occlusal view of 3.66-Ma mandible (Laetoli hominid 4), type of *Australopithecus afarensis,* from Laetoli in northern Tanzania.

Figure 4.8. Laetoli Site G hominid footprint trail. A small individual (G-1) made the left trail, and two individuals (G-2 and G-3) made the right trail. G-3 partially overprinted the tracks of G-2. A young hoofed mammal (ym), perhaps *Hipparion,* diagonally traversed the southern section of the trails and joined a mature hoofed mammal (mm).

analysis confirms that it is >4 Ma.[100] Ward and Hill concluded that it is quite similar to Pliocene hominid counterparts from Hadar and Laetoli.[101]

The mother lode of Pliocene hominid specimens is from Hadar, Ethiopia. The 316 fossilized bits represent between thirty-six and sixty-five individuals, the most famous of which is A.L. 288-1 (nicknamed Lucy) (Figure 4.6).[102] Their dating was variable.[103] For example, whereas Walter and Aronson radiometrically determined a span between 2.8 and 3.6 Ma, Schmitt and Nairn obtained a much narrower range of dates between 2.8 and 3.1 Ma for hominid sites in the Hadar Formation.[104] Via analysis of the Hadar megafauna, White et al. supported the dates of Walter and Aronson.[105]

Initially, on the basis of early collections, Taieb, Johanson, and Coppens thought that more than one hominid species might have inhabited the Hadar region contemporaneously during the Pliocene.[106] But only one species emerged after Johanson and White thoroughly studied the full craniodental samples from Hadar and Laetoli.[107] Johanson named it *Australopithecus afarensis*.[108] Johanson, White, and Coppens proposed that the following diagnostic features collectively distinguish *Australopithecus afarensis* from other species of the Hominidae:

- strong prognathism,
- rather procumbent incisors,
- long palate,
- nonparabolic upper tooth row,
- wide upper central incisors,
- large canines,
- P_3 with wear by the upper canine (i.e., partly sectorial),
- $M_3 > M_2 > M_1$ and $M^3 > M^2 > M^1$,
- heavy nuchal cresting,
- low cranial capacity, and
- transverse tori on the mandible.[109]

It was especially important that *Australopithecus afarensis* be distinguished from *Australopithecus africanus* because Tobias, Boaz, and Logan et al. challenged that the former might be sunk into the latter.[110] Accordingly, Johanson listed twenty-two craniodental features that characterize *Australopithecus afarensis* as more "primitive" than other known species of the Hominidae, including *A. africanus*.[111] They include:

- large anterior teeth relative to the postcanine dentition,
- a high frequency of diastemata between the mandibular canine and P_3,
- a high frequency of oval, unicuspid P_3,
- modest development of the symphyseal superior transverse torus,
- extensive pneumatization of the temporal squama, and
- a chimpanzee-like temporonuchal crest.

Based on fragmentary specimens from Hadar, *Australopithecus afarensis* also seems to be somewhat smaller-brained than later Hominidae, except perhaps *Australopithecus sediba* (Table 10.1).[112]

The abundant postcranial bones from Hadar did not figure in the formal diagnosis of *Australopithecus afarensis*.[113] Nevertheless, Johanson, White, and a host of other authors commonly refer to them—and to the makers of the Laetoli footprint trails—by this nomen.[114] McHenry noted that many postcranial bones of *Australopithecus afarensis* are quite similar to those of *Australopithecus africanus*.[115]

Johanson and White interpreted the notable variation in the hypodigm of *Australopithecus afarensis* to reflect marked sexual dimorphism in the species.[116] Contrarily, Coppens quickly defected from the single species camp and led a chorus of critics, who variously argued that there were more than one Pliocene hominid at Hadar and Laetoli.[117] Tellingly, there was notable disagreement among them about which taxa should join or replace *Australopithecus afarensis,* if, in fact, Johanson and White were incorrect.[2] The confusion is understandable, given that extant great apes and humans do not share a clear pattern of dental size variation that might provide secure models of dental size variation in a fossiliferous site that might contain more than one hominid species.[118]

No advocate of an alternative, dual-species hypothesis has presented arguments that are as thorough and compelling as those proffered by the defenders of *Australopithecus afarensis*.[119] Tobias and Boaz eventually accepted *Australopithecus afarensis* as a distinct species.[120]

Radiant *Australopithecus*

Because of the great variety of Plio-Pleistocene species that have been retrieved over the past seventy-five years and uncertainties about how they are to be arrayed, our family tree has given way to a tangled shrub blooming with question marks and tentative tendrils (Figure 4.5A).[121]

Three major adaptive radiations and probably some nonradiant lineages of Hominidae followed *Orrorin, Sahelanthropus,* and *Ardipithecus.* The earliest group extends from >4 Ma to ca. 2.3 Ma and is rooted in Kenyan and Ethiopian *Australopithecus anamensis,* followed by Ethiopian and perhaps Kenyan *Australopithecus afarensis,* and *Australopithecus bahrelghazali* in Chad.[122] The distinction between *Australopithcus anamensis* and *Australopithecus afarensis* is blurred by a collection of teeth and gnathic fragments from Woranso-Mille, Central Afar, Ethiopia. The specimens are intermediate between them in age (3.58–3.8 Ma) and morphology, which might mean that the two species are not distinct and minimally that they are in an ancestor-descendent relationship (Figure 4.9).[123]

Possible descendent species include *Australopithecus africanus* in South Africa, *A. habilis* in Tanzania and Kenya, and *A. garhi* in Ethiopia (Figures 4.10 and 4.11).[124] Except for *Australopithecus bahrelghazali,* all of them have locomotor skeletal bits that indicate bipedality, though other body parts attest to continued arboreal activity (Figure 4.12).[125] *Australopithecus* spp. also have great ape-sized brains and large molar teeth with thick enamel.[126]

Leakey, Spoor et al. described a new species, *Kenyanthropus platyops,* based on a cranium from 3.5-Ma Lomekwi deposits at West Turkana, Kenya.[127] However, White pointed out that taphonomic distortion is responsible for its appearance and reassigned it to *Australopithecus afarensis.*[128] Associated fauna indicated that Lomekwi was a mosaic habitat with woodland predominant and forest-edge species most prominent in the samples. The area might have been more vegetated and also wetter than contemporary localities in the Hadar Formation, Ethiopia.[128]

The initial radiation of *Australopithecus* probably occurred in forests or in mosaics of riverine forest, adjacent woodlands, and more open areas. African forests were being reduced by global climatic changes—cooling and drying—that expanded open woodlands and grasslands.[129] They provided novel habitats for new species and greater numbers of antelopes, pigs, monkeys, giraffes, elephants, and other animals for adventurous hominids to scavenge and perhaps to kill. Tubers, seeds, and grasses also might have become more common in the diets of the hominids that foraged in the open areas. Large cats, canids, and hyenas also flourished in the new environments.[130] They not only provided meat for alert scavengers but also posed a threat to hominids with whom they competed and upon whom they probably preyed.[131]

The radiation of robustly skulled *Paranthropus* partially overlapped the radiations of both *Australopithecus* and *Homo.*[132] Beginning with 2.6–2.2-Ma

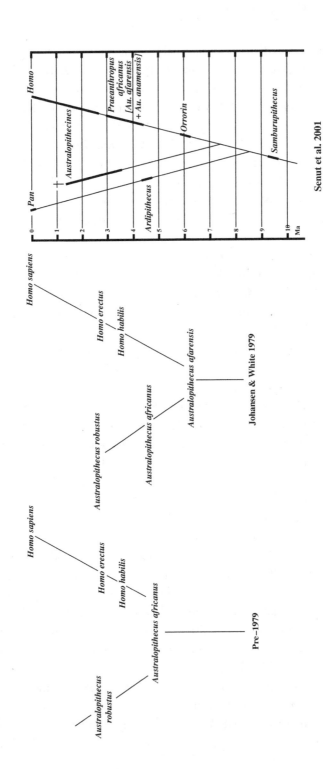

Figure 4.9. Sampler of phylogenetic schemes, >1979–2001.

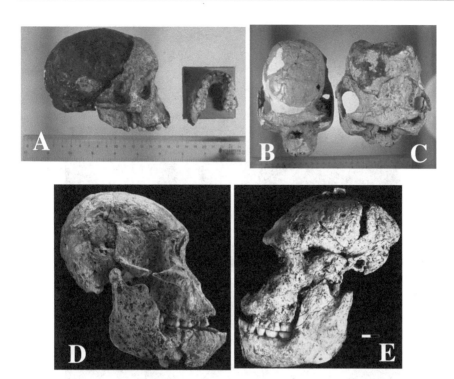

Figure 4.10. (A) Partial skull and endocast of *Australopithecus africanus* from Sterkfontein, South Africa (Taung 1, type) and top view of crania: (B) *Australopithecus africanus* (Sts 5) and (C) *Paranthropus robustus* (SK 48) from South Africa. Right and left lateral views of composite skulls from South Africa: (D) *Australopithecus africanus* (Sts 71 and Sts 36) and (E) *Paranthropus robustus* (SK46/SK23).

Paranthropus aethiopicus in Ethiopia and northern Kenya, the genus survived without issue until ca. 1 Ma as *P. boisei* in Kenya, Tanzania, and Ethiopia and until ca. 1.2 Ma as *P. robustus* in South Africa (Figure 4.13).[133] *Australopithecus afarensis* is the most likely ancestor of *Paranthropus*.[134] Like *Australopithecus, Paranthropus* sport ape-sized brains and even larger cheek teeth. Further, their upper and lower limbs are more like those of *Australopithecus* than like those of *Homo sapiens*.[126] Although they were probably terrestrially bipedal, their locomotion was different from that of *Homo,* and they retained features facilitating arboreal foraging, lodging, and escape from terrestrial predators, rivals, and pests.[135] Based on a study of nineteen facial or maxillary specimens and sixteen mandibular specimens of *Paranthropus robustus* from 1.5–2.0-Ma deposits at Swartkrans, Kromdraai, and Drimolen, South Africa, Lockwood et al. concluded that the species was markedly sexually

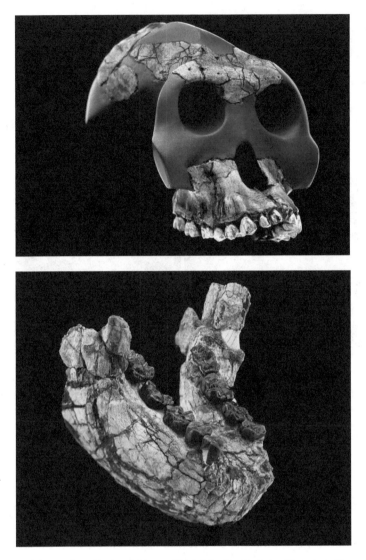

Figure 4.11. (A) Partial cranium and (B) mandible of 2.5-Ma *Australopithecus garhi.*

dimorphic. They further suggested that it took males longer to mature, and like modern primate species with prolonged male maturation and striking bodily dimorphism, they probably had a polygamous mating system.[136] Johanson and White proposed that *Australopithecus afarensis* gave rise to two hominid lineages ca. 2.5 Ma (Figure 4.9).[137] One population of *Australopithecus afarensis* evolved into *Australopithecus africanus,* represented by the classic

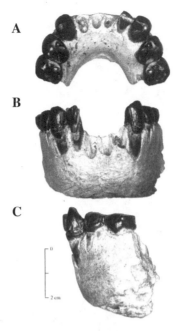

Figure 4.12. (A) Occlusal, (B) anterior, and (C) left lateral views of anterior mandible of 3–3.5 Ma *Australopithecus bahrelghazali* from Bahr el Ghazal, Koro Toro, Chad.

fossils from Taung, Makapansgat, and Sterkfontein, South Africa. They, in turn, gave rise to *Paranthropus robustus,* which in their scheme lived not only at Swartkrans and Kromdraai, South Africa, but also in Kenya, Tanzania, and Omo, Ethiopia.

Johanson and White's second population of *Australopithecus afarensis* hypothetically gave rise to *Homo habilis,* which in turn generated *Homo erectus* about 1.5 Ma. Later still, *Homo erectus* begot *Homo sapiens.*[138] There is solid evidence that *Homo erectus,* or a form like it, was contemporary with *Paranthropus boisei* at East Turkana, Kenya.[139]

Upon re-examining the data of Johanson and associates, several scientists concluded that *Australopithecus afarensis* has the firmest affinity with *Paranthropus.*[140] Accordingly, a less megadont variant of *Australopithecus africanus* probably would be a better candidate for proximate ancestry to *Homo.*[141] It should be sought in 2–3-Ma African deposits because *Australopithecus* was confined to Africa.[2]

In 2010, Berger et al. diagnosed another species of *Australopithecus*—*A. sediba*—from Malapa, South Africa, located about 15 km from Sterkfontein,

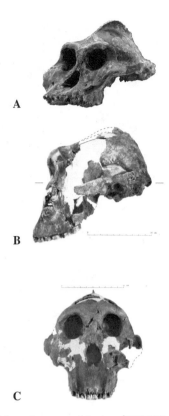

Figure 4.13. (A) Cranium of *Paranthropus aethiopicus* (KNM-WT 17000), 2.5-Ma, from West Turkana, Kenya. (B) Left lateral and (C) frontal views of 1.8-Ma cranium of *Paranthropus boisei* from Olduvai Gorge, Tanzania.

Kromdraai, and Swartkrans (Figure 4.14). The Malapa cave system yielded two fragmentary skeletons designated MH1 and MH2, which Dirks et al. concluded were deposited together in a debris flow that lithified soon thereafter between 1.95 and 1.78 Ma.[142]

The type specimen (MH1) is a juvenile male with second molars in occlusion, leading Berger et al. to infer that the 12- to 13-year-old individual had achieved 95 percent of the adult brain size and that any further growth in the skull and postcranial anatomy would not alter their diagnosis. MH2 comprises isolated maxillary teeth, a partial mandible, and a partial postcranial skeleton of an adult that Berger et al. consider to be female. If Berger et al. are correct that MH1 and MH2 are male and female, respectively, and

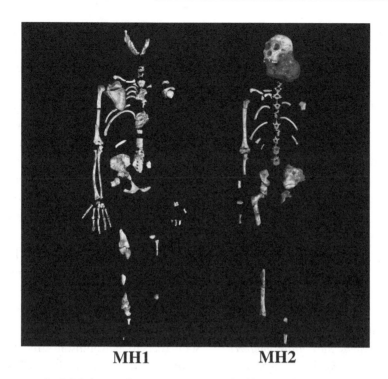

MH1   **MH2**

Figure 4.14. Partial skeletons of ca. 1.8–2.0-Ma *Australopithecus sediba* from Malapa, South Africa. MH1 young male, and MH2 adult female.

that MH1 had attained virtually adult size, *Australopithecus sediba* was modestly sexually dimorphic based on the teeth and postcranial skeletons.[143]

Among the several named species of *Australopithecus, A. sediba* compares most closely with *A. africanus,* but Berger et al. consider them to be somewhat more advanced craniodentally toward *Homo* than are specimens of *A. africanus.*[143] The cranial capacity of MH1 is 420 cc, which fits within the range of values for *A. africanus,* and the teeth are smaller than those of *A. africanus.* Accordingly, *Australopithecus sediba* might be a better candidate for ancestry to *Homo* than the other *Australopithecus* spp.

The 2.5-Ma Black Skull (KNM-WT 17000) and associated hominid specimens from Nariokotome, west of Lake Turkana in northern Kenya, evidence that *Paranthropus* were more ancient than we had thought.[144] Because they share anatomical features with precedent *Australopithecus afarensis,* it is reasonable to assume that they are part of the same lineage.[145]

Like the Miocene apes, specimens that are assignable to *Australopithecus* and *Paranthropus* have experienced extensive taxonomic shuffling, especially at generic and specific levels. Also like them, taxonomically valid, multiple species signify adaptive radiation in the early Hominidae.

Dart designated the first discovered specimen (Taung 1) *Australopithecus africanus* (Figure 4.10).[146] Thereafter, following customary practice of the day, Broom, Dart, and others bestowed different nomina upon the hominid samples from each southern African locality: Sterkfontein *(Plesianthropus transvaalensis)*, Makapansgat *(Australopithecus prometheus)*, Kromdraai *(Paranthropus robustus)*, and Swartkrans *(Paranthropus crassidens)*.[147] Additional names *(Telanthropus capensis, Zinjanthropus boisei, Paraustralopithecus aethiopicus)* were devised as distinctive specimens were discovered at the classic southern African sites and during intensive research at Olduvai Gorge and Omo (Figures 4.15 and 4.16).[148]

In 1954, Robinson lumped together some of these specimens, thereby reducing the number of southern African early hominid species to three: *Australopithecus africanus, Paranthropus robustus,* and *Telanthropus capensis.*[149] The nomenclature of Washburn and Patterson and Howell was even simpler: they sank *Paranthropus* into *Australopithecus.*[150] Soon thereafter, Robinson sank *Telanthropus* into *Homo,* and Tobias sank *Zinjanthropus* into *Australopithecus.*[151]

In 1978, Howell initiated a new era of splitting the early Hominidae.[152] Accordingly, some experts recognized five or more species of *Australopithecus,* including *A. afarensis, A. africanus, A. robustus, A. crassidens,* and *A. boisei.*[153] In due course, the Black Skull and other specimens from Nariokotome and Omo were designated *Australopithecus aethiopicus* or *Paranthropus aethiopicus,* as experts surmised that they are indeed specifically distinct from *Paranthropus boisei.*[154]

Clarke asserted that the Sterkfontein breccias yielded not only *Australopithecus africanus* and *Homo* but also *Paranthropus* and its ancestor, a second species of *Australopithecus.*[155]

Contrary to Simpson, Wood, and Simons, who recognized subgenera *(Paranthropus* and *Australopithecus)* in the genus *Australopithecus,* some contemporary authors reinstated *Paranthropus* as the generic nomen for the robustly skulled species.[156]

Before discovery of the Black Skull, paleoanthropologists generally assumed that *Australopithecus* predated and were ancestors to *Paranthropus* (Figure 4.13A). A notable exception was Robinson, who instead granted precedence to *Paranthropus.*[157] Ambiguities in dating the southern African early hominid sites permitted both views to persist.[158]

Figure 4.15. Pedestal (OH5) marking location of *Paranthropus boisei* (type) in Bed I, Oluvai Gorge, Tanzania.

In the Turkana Basin, species of *Paranthropus* span the period between 2.5 and 1.4 Ma.[159] In South Africa, the trustworthiest dates are for Makapansgat, which at ca. 3 Ma is probably the oldest of the southern African early hominid sites.[160]

Dated somewhere between 2 Ma and 800 Ka, Taung might have been the youngest, which induced Tobias to toy with the idea that the Taung child is actually a *Paranthropus*-like creature; however, the suggestion soon flagged.[161] Cooke argued persuasively that the Taung fauna is intermediate between those of Makapansgat and Swartkrans but is closer to the former.[162] Based on fuller collections, McKee inferred that the fauna associated with the Taung child indicates an age of 2.6–2.8 Ma.[163]

The presumed young geochronological age of the Taung child induced Verhaegen to consider *Australopithecus* an ancestor to bonobos or chimpanzees.[164] Further, impressed by select craniodental features and biomolecular dating, she suggested that *Paranthropus boisei* and some *Australopithecus afarensis* are close relatives of *Gorilla*. She did not discuss the hurdle that this scenario implies: the evolution of obligate knuckle-walkers from accomplished bipeds.

The skulls of *Paranthropus* are shaped differently from those of *Australopithecus*. Most particularly, the greater development of bony crests in many *Paranthropus* indicates a more massive configuration for their fleshed heads. Accordingly, it was generally assumed that overall they were much bigger beasts. For instance, Robinson guessed that although *Australopithecus africanus* was 122–135 cm tall and weighed 18–27 kg, *Paranthropus*, though not much taller (135–152 cm), weighed 68–91 kg. In each guesstimate, the lower figure is for females, and the higher figure is for males. Nevertheless, he conceded that male *Australopithecus* and female *Paranthropus* might have overlapped in size.[165]

Body size estimates, based on larger samples of postcranial specimens, have narrowed the presumed gap between *Paranthropus* and *Australopithecus*.[166] Now, at best the terms robust and gracile are meaningfully applied only to their skulls. Further, if the various species are indeed assignable to different genera or subgenera, then robust *Australopithecus* and gracile *Australopithecus* are unnecessary terms because they can be referred to as *Paranthropus* and *Australopithecus*, respectively.[167]

Although both *Australopithecus* and *Paranthropus* have large posterior teeth, *Paranthropus* have remarkably small canines, incisors, and P_3 relative to their more posterior teeth.[168] Further, their premolars are molariform, and the roots of their teeth are more robust than those of *Australopithecus africanus*.[169] *Paranthropus* has thicker molar enamel than that of *Australopithecus africanus*, which is comparable to conditions in *Homo sapiens*, early *Homo*, and *Sivapithecus*.[170]

Cranially, too, *Paranthropus* seem to have been powerful chewing machines.[171] As a rule, their calvariae are more heavily crested than those of *Aus-*

tralopithecus africanus; their zygomatic arches, to which the masseter muscles attach proximally, flare outward. Their neurocrania are remarkably constricted behind the orbits, where the anterior portions of the temporalis muscles attach proximally; and the sagittal crest, which provides purchase for the temporalis muscles, is forward atop the braincase. The frontal bone rises more prominently above the orbits in *Australopithecus,* giving them a more prominent forehead than *Paranthropus.* The more rugged calvariae of *Paranthropus* are extensively pneumatized, especially in the mastoid and neighboring regions.[172]

Paranthropus have especially high and broad vertical mandibular rami, to which the masseter muscles attach distally, and backward curving coronoid processes, where the temporalis muscles gain distal attachment. Their mandibular bodies are deep and thick.[173] In keeping with the reduced anterior dentition and emphasis on directing vertical forces onto the posterior dentition, the faces of some *Paranthropus* are less projecting than those of *Australopithecus.*[174] However, the two genera overlap in this feature; and the Black Skull, like *Australopithecus afarensis,* is very prognathic.[175] In both *Australopithecus* and *Paranthropus,* the nuchal area is lower on the occiput than it is in the Pongidae, which indicated that they were bipedal.[176] Their bipedality is confirmed by evidence from the postcranial skeleton.[177]

From Handy Man to Apocalyptic Primate

The inclination to split fossil Hominoidea and to view them as adaptive radiations instead of a series of broadly variable lineal relations has also stirred students of *Homo,* at least one species of which—*H. rudolfensis*—might have been afoot in eastern Africa at 2.4 Ma.[178] Wood concluded that there might be three hominid species, in addition to *Australopithecus,* at East Turkana.[179] Stringer suggested that there was a Plio-Pleistocene radiation in Africa that gave rise to at least three species of early *Homo.*[180]

By the end of the twentieth century, in addition to classic *Homo erectus* in Southeast Asia, China, and East Africa and *Homo neanderthalensis* in Europe and western Asia, *Homo ergaster, Homo heidelbergensis, Homo rhodesiensis,* and *Homo antecessor* had joined the pantheon with *Homo sapiens.* Tattersall reasonably argued the majority view that only one of the Middle to Late Pleistocene variants of *Homo* could have been an ancestor to *Homo sapiens.*[181] The nagging unresolved question is which one?

In 1964, Leakey, Tobias, and Napier sparked a controversy by assigning scattered battered bits of several Early Pleistocene Olduvai hominids to a new species: *Homo habilis,* the handy man (Figure 4.16).[182] Initially, several senior paleoanthropologists disclaimed the new taxon.[183] For instance, Robinson argued that the earlier specimens in the hypodigm are, in fact, *Australopithecus* and the more recent ones are *Homo erectus.*[184] But as additional and more complete contemporaneous specimens were recovered, especially at East Turkana, *Homo habilis,* like *Ramapithecus punjabicus,* achieved common usage. Subsequently, additional specimens from Olduvai, Tanzania, South Africa, and Omo, Ethiopia, tentatively joined the hypodigm.[185]

Homo habilis was distinguished from *Australopithecus africanus* chiefly by its somewhat larger endocranial volume and relatively minor metrical differences in the dentition (Figure 4.16).[186] In a monumental monograph focused on four Olduvai specimens (OH 7, OH 13, OH 16, and OH 24), Tobias reaffirmed the original diagnosis of *Homo habilis,* comprising the following key traits:

- Mean cranial capacity greater than that of *Australopithecus,* but smaller than that of *Homo erectus.*
- Muscular ridges on the cranium ranging from slight to strongly marked.
- Chin retreating, with slight or no development of a mental trigone.
- Jaws smaller than those of *Australopithecus* and within the range of *Homo erectus* and *Homo sapiens.*
- Incisors relatively large in relation to those of *Australopithecus* and *Homo erectus.*
- Canines large relative to premolars.
- Premolars narrower than those of *Australopithecus,* but within the range of *Homo erectus.*
- Molar dimensions between the lower range of *Australopithecus* and the upper range of *Homo erectus.*
- Buccolingual narrowing and mesiodistal elongation of all teeth, especially the lower premolars.
- Clavicle resembling that of *Homo sapiens.*
- Hand bones differing from those of *Homo sapiens* due to robustness, though the distal phalanges are broad and stout.
- Foot bones and the pedal model (as assembled by Napier) having a striking resemblance to counterparts in the human foot.[187]

Because of uncertainties in assigning most postcranial specimens from Olduvai and East Turkana specifically, it is difficult to employ them to resolve

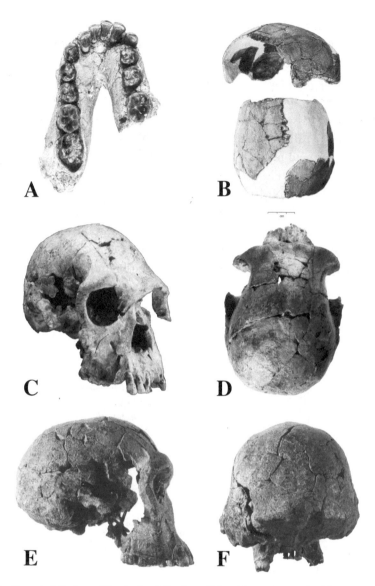

Figure 4.16. *Homo habilis* (type OH 7, top), 1.8 Ma, from Olduvai Gorge, Tanzania. (A) Occlusal view of mandible and anterolateral and superior views of calotte (B). *Homo habilis* (KNM-ER 1813), 1.9-Ma, from East Turkana, Kenya. Three-quarters facial view (C). Superior view (D). Right lateral view (E). Posterior view (F).

puzzles of hominid phylogeny.[188] Moreover, because the dental evidence is also ambiguous, students should focus on the basicranium, endocranial vault and facial shapes, and metrics.[189]

Among paleoanthropologists who accept its legitimacy, there is no consensus on which specimens belong in *Homo habilis*. From East Turkana localities in northern Kenya, KNM-ER 992, KNM-ER 1470, KNM-ER 1805, and KNM-ER 1813 have been especially problematic (Figure 4.17). Accordingly, there is a trend not only to retain *Homo habilis* but also to name additional species of Early Pleistocene *Homo* (Tables 4.2 and 4.3). Groves and Mazák proposed that KNM-ER 992 and 1805 represent a discrete species: *Homo ergaster*, the workman (Figure 4.17E).[190] Stringer entertained the possibility that KNM-ER 1813 and certain specimens from Olduvai Gorge (OH 13, OH 16) are also *Homo ergaster*.[191] Alexeev and Groves considered KNM-ER 1470 to represent another species: *Homo rudolfensis* (Figure 4.17 and 4.18).[192]

Table 4.3.   Tuttle preferred scheme [not employed consistently in the text]

Hominidae
    Australopithecinae
        Australopithecus africanus
        Australopithecus anamensis
        Australopithecus barelghazali
        Australopithecus garhi
        Australopithecus habilis
        Australopithecus sediba
    Homininae
        Homo antecessor
        Homo erectus
        Homo ergaster
        Homo georgicus
        Homo heidelbergensis
        Homo neanderthalensis
        Homo rudolfensis
        Homo sapiens
    Paranthropinae
        Paranthropus aethiopicus
        Paranthropus boisei
        Paranthropus robustus
    Subfamily *incertae sedis*
        Ardipithecus kadabba
        Ardipithecus ramidus
        Kenyanthropus platyops
        Orrorin tugenensis
        Sahelanthropus tchadensis

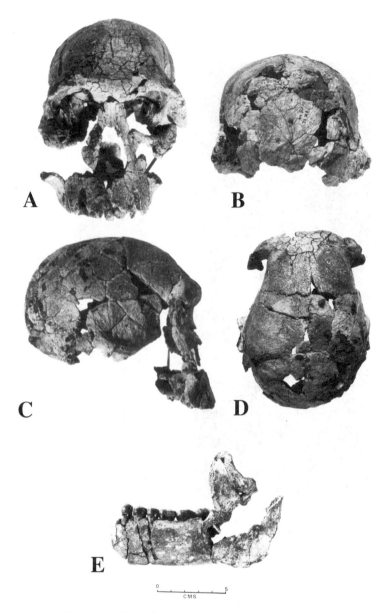

Figure 4.17. Cranium of *Homo rudolfensis* (KNM-ER 1470), 1.8-Ma. (A) Frontal view. (B) Posterior view. (C) Right lateral view. (D) Superior view. (E) Left lateral view of mandible of *Homo ergaster* (type, KNM-ER992) from East Turkana, Kenya.

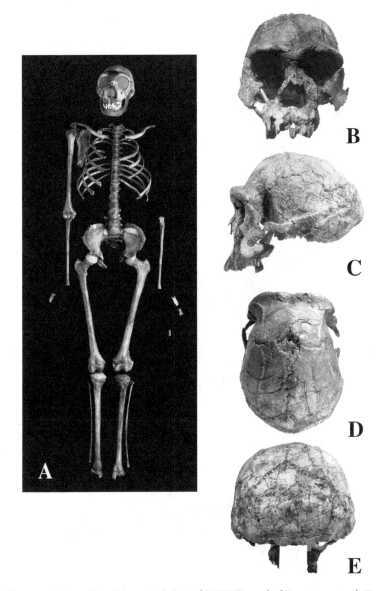

Figure 4.18. A 1.5-Ma adolescent skeleton (KNM-WT15000) of Homo ergaster (= Homo erectus) from Nariokotome, West Turkana, Kenya (A). Adult cranium of *Homo ergaster* (KNM-ER 3733), 1.8-Ma, from East Turkana, Kenya (B) Frontal view. (C) Left lateral view. (D) Superior view. (E) Posterior view.

Homo erectus was first discovered in Java, and its hypodigm was expanded greatly by excavations at Zhoukoudian, China (Map 4.2).[193] The species appears to be rooted in Africa (Figure 4.19). Whereas Olduvai hominid 9 (Chellean Man) establishes *Homo erectus* in eastern Africa at 1.2 Ma, few, if any, of the Javanese specimens predate 1 Ma. *Homo erectus* inhabited the cave at Zhoukoudian only 500–230 Ka.[194]

Swisher et al. obtained ^{40}Ar/^{39}Ar dates of 1.81 Ma and 1.66 Ma for volcanic units that are directly related to sites (Perning, Modjokerto, and Sangiran) in East Java where three specimens of *Homo erectus* had been collected decades before their geological sampling.[195] Further, Swisher et al. gave mean ages of 53–27 Ka for the strata presumed to have contained younger specimens (Ngandong and Sambungmachan) of *Homo erectus* in central Java, based on electron spin resonance and mass spectrometric U-series dating of bovid teeth from the sites.[196]

If the dates were accurate, Indonesia might boast the earliest *Homo erectus* and successive populations of the species would have been contemporaneous not only with species of Middle and Late Pleistocene *Homo* in Eurasia and Africa but also with *Homo sapiens* in Eurasia, Africa, and Australia. Further, if *Homo floresiensis* is a valid species, it too would have been contemporaneous with them (Table 4.4).[197]

The oldest dates raised the possibility that *Homo erectus* evolved in Asia, migrated to Africa before 1.2 Ma, and then perhaps gave rise to some other species of *Homo* or simply died out there. However, the trustworthiness of the dates is compromised because the deposits from which the mineral samples were taken have complex histories of deposition and perturbation.[198] For example, excavations in the Solo River Terrace of Ngandong cast serious doubt that Javan *Homo erectus* survived past the Middle Pleistocene: ^{40}Ar/^{39}Ar dating of pumices in the deposits indicate a mean age of 546 ± 12 Ka.[199] Moreover, the validity of the holotype of *Homo floresiensis* as a species versus a pathologically microcephalic *Homo sapiens* is moot (Figure 4.20).[200]

Acheulean handaxes and cleavers occur in at Attirampakkam, a ca. 1.5-Ma excavated site in South India. Because they were not associated hominid skeletal remains, it would be imprudent to ascribe them to *Homo erectus*.[201]

Hobbit Fever and Possible Cures

Homo floresiensis comprises the cranium and partial skeleton (LB1) of a short (1 m), small-brained (417 cc) adult female hominid and bits of several other

1. Forbes' Quarry, Rock of Gibraltar, Gibraltar: *Homo neanderthalensis*
2. El Sidrón Cave, Asturias, Spain: *Homo neanderthalensis*
3. Gran Dolina Cave; Sima le lost Huesos, Spain: *Homo sp.; Homo heidelbergensis*
4. Arago, Tautavel, France: *Homo heidelbergensis*
5. La Ferrassie rock shelter, Dordogne, France: *Homo neanderthalensis*
6. La Chapelle-aux-Saints, France: *Homo neanderthalensis*
7. Spy I and II, Jemeppe-sur-Sambre, Belgium: *Homo neanderthalensis*
8. Engis and Scladina Cave, Belgium: *Homo neanderthalensis*
9. Heidelberg, Germany: *Homo heidelbergensis*
10. Feldhofer Cave, Neander Valley, Germany: *Homo neanderthalensis*
11. Lehringen, Germany: *Homo neanderthalensis (?)*
12. Schöningen, Germany: *Homo neanderthalensis*
13. Vindija Cave, Croatia: *Homo neanderthalensis*
14. Petralona, Greece: *Homo heidelbergensis*
15. Kabara Cave, Mt. Carmel; Wadi Amud; Kafzeh, Israel: *Homo neanderthalensis*
16. Mezmaiskaya Cave, Russia: *Homo neanderthalensis*
17. Dmanisi, Georgia: *Homo erectus*
18. Shanidar Cave, Iraq: *Homo neanderthalensis*
19. Teshik Tash, Uzbekistan: *Homo neanderthalensis*
20. Okladnikov Cave; Denisova Cave, Siberia: *Homo neanderthalensis*
21. Narmada, India: *Homo erectus*
22. Attirampakkam, India: *Homo erectus*
23. Shagou Formation; Yuanmou Formation, China:
    Ramapithecus hudienensis; Homo orientalis
24. Locality 1, Zhoukoudian, China: *Homo erectus*
25. Sangiran; Ngandong; Sambungmacan; Trinil; Solo, Java, Indonesia:
    Homo erectus
26. Perning & Modjokerto, Java, Indonesia: *Homo erectus*
27. Liang Bua, Flores Island: *Homo floresiensis (?)*
28. Mata Menge, Flores Island: *Homo floresiensis (?)*

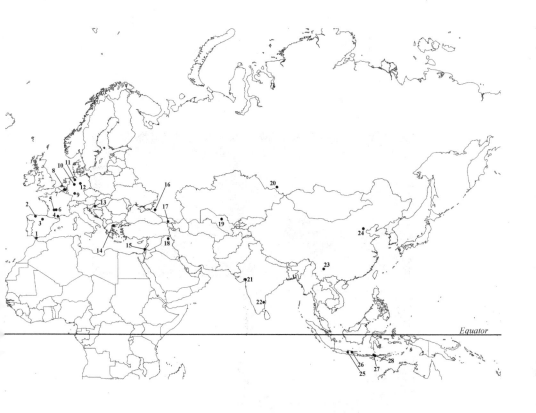

Map 4.2. Eurasian Late Middle Miocene–Holocene fossil sites

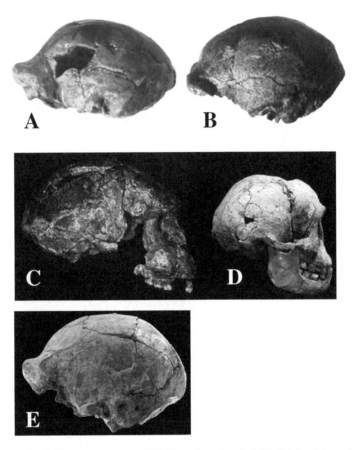

Figure 4.19. (A–C, E) *Homo erectus* and (D) *Homo floresiensis*. (A) Left lateral view of 1.2-Ma OH 9 calvaria from Olduvai Gorge, Tanzania. (B) Left lateral view of 500-Ka Zoukoudian calvaria from China. (C) Right lateral view of 550-Ka Sangiran 17 cranium, Sangiran, Java, Indonesia. (D) Right lateral view of 18-Ka LB1 skull from Flores Island, Indonesia. (E) Left lateral view of 1-Ma BOU-VP-2/66 calvaria from Daka, Bouri, Ethiopia.

individuals in the 95–13-Ka strata of Liang Bua, a cave on Flores Island, Indonesia (Figure 4.20).[202] At 18 Ka, LB1, dubbed "the Hobbit," is the youngest specimen assigned to *Homo floresiensis*.[203] Because crude stone tools like ones excavated in Liang Bua were recovered in Mata Menge deposits, dated ca. 840 Ka, in central Flores, and because of morphological similarities between *Homo erectus* sensu lato and LB1, it is reasonable to assume that *Homo floresiensis* evolved as an insular species from immigrant *Homo erectus,* if they cannot indeed be accommodated in the hypodigm of *Homo erectus*.[204]

Table 4.4. Geochronological spans and geographic distributions of the Hominidae and collateral species

ca. 7.0 Ma	*Sabelanthropus tchadensis*	Torros-Menalla, Chad
ca. 6.0 Ma	*Orrorin tugenensis*	Tugen Hills, Kenya
4.4 Ma	*Ardipithecus ramidus*	Aramis, Middle Awash, Ethiopia
5.6 Ma	*Ardipithecus kadabba*	Asa Koma, Middle Awash, Ethiopia
3.6–2.9 Ma	*Australopithecus afarensis*	Hadar, Maka, Behlodelie, Fejej, and Omo, Ethiopia; ?Laetoli, Tanzania; Baringo and
	[= *Praeanthropus africanus*]	West Turkana, Kenya; ?Sterkfontein, South Africa
3.5 Ma	*Kenyanthropus platyops*	West Turkana, Kenya
4.2–3.5 Ma	*Australopithecus anamensis*	Kanapoi and Allia Bay, Kenya; Middle Awash, Ethiopia
	[=*Praeanthropus africanus*]	
3.0–3.5 Ma	*Australopithecus barelghazali*	Bahr el Ghazal, Koro Toro, Chad
3.0–2.3 Ma	*Australopithecus africanus*	Taung, Makapansgat, Sterkfontein, and Gladysvale, South Africa
1.95–1.78 Ma	*Australopithecus sediba*	Malapa, South Africa
2.5 Ma	*Australopithecus garhi*	Middle Awash, Ethiopia
2.6–2.2 Ma	*Paranthropus aethiopicus*	Omo, Ethiopia; West Turkana, Kenya; Laetoli, Tanzania; Malawi
2.6–1.2 Ma	*Paranthropus boisei*	Olduvai Gorge and Peninj, Tanzania; East and West Turkana, Kenya; Konso, Ethiopia
1.8–1.2 Ma	*Paranthropus robustus*	Swartkrans, Kromdraai, Drimolen, Gondolin, and Cooper's D, South Africa
2.3?–1.6 Ma	*Homo habilis* (=*Australopithecus habilis*)	Olduvai Gorge, Tanzania; East Turkana and Chesowanja, Kenya; Omo and Hadar, Ethiopia; Sterkfontein, South Africa
2.4–1.8 Ma	*Homo rudolfensis*	East Turkana, Kenya; Uraha Hill, Malawi; Ethiopia (?)
1.8 Ma–52? Ka	*Homo erectus* [if *H. ergaster* included]	Africa, Asia, and Europe (?)
1.9–1.5 Ma	*Homo ergaster*	East and West Turkana (?), Kenya; ?Dmanisi, Georgia
1.8 Ma	*Homo georgicus*	Dmanisi, Georgia
1.0 Ma–780 Ka	*Homo antecessor*	Gran Dolina, Spain; Buia, Eritrea (?)
600–200 Ka	*Homo heidelbergensis*	Europe, Ethiopia (?) and South Africa (?)
≥ 250≥25 Ka	*Homo neanderthalensis*	Europe and western Asia
≥200–0 Ka	*Homo sapiens*	Earth globally
95–?17 Ka	?*Homo floresiensis*	Flores Island, Indonesia

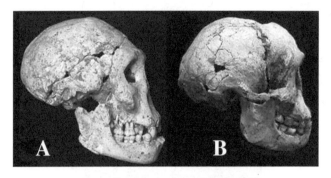

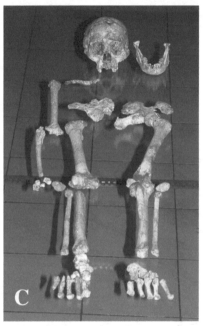

Figure 4.20. (A) Right lateral view of skull of 1.8-Ma *Homo georgicus* from Dmanisi, Republic of Georgia. (B) Skull of *Homo floresiensis* (LB1), 18-Ka, from Liang Bua, Flores Island, Indonesia. (C) Partial skeleton of LB1.

Alternatively, LB1 probably was an unfortunate *Homo sapiens* afflicted with one of about 500 syndromes that can include microcephaly.[205] A diverse cohort of paleoanthropologists, anatomists, medical scientists, and neuroscientists have raised numerous questions that should induce caution concerning the taxonomic status and phylogenic importance of LB1.[206]

How the hominids that left crude stone artifacts on Flores crossed the Pleistocene strait between the Sunda landmass and the islands of Wallacea, including Flores, is a major puzzler. The Pleistocene nonhomind fauna of Flores was impoverished relative to mainland faunas and included only species whose founders could have swum, flown, or landed serendipitously via natural rafts.[207]

The Liang Bua hominids were associated with bones of fish, frogs, snakes, birds, rodents, and bats, some of which probably represent natural accumulation, but others are charred, indicating that someone might have toasted them.[208] Early hominids also shared the tropical habitat of Flores with pygmy elephantids *(Stegadon),* giant tortoises (*Geochelone* sp.), Komodo dragons *(Varanus komodoensis),* and an even larger varanid lizard.[209]

Meanwhile, Back on Solid Ground

Dmanisi in the Republic of Georgia provided a firm base from which to argue that *Homo erectus* might have evolved in Asia, just beyond the modern boundary between eastern Europe and western Asia. The vertebrate-rich site is reliably dated at ca. 1.77 Ma by paleomagnetic, geochronological, faunal, and archeological data (Figure 4.20).[210]

In 1995, Gabunia and Vekua announced the 1991 discovery of a Plio-Pleistocene mandible of *Homo erectus* associated with Late Villafranchian fauna and stone tools similar to Oldowan artifacts. The basalt upon which the bone-bearing layer lay was dated at 1.8 Ma via K/Ar and at 2 Ma via laser-fusion $^{40}Ar/^{39}Ar$.[211] Bräuer and Schultz and Dean and Delson agreed with Gabunia and Vekua that the Dmanisi mandible has the closest affinities with African and East Asian *Homo erectus,* whereas Rosas and Bermúdez de Castro declined to assign it to an established hypodigm and designated it *Homo species indeterminata,* with affinities to *Homo ergaster.*[212] There followed recovery of three additional hominid cranial specimens and postcranial remains from the stratum that had contained the mandible.[213] Much of the argument over whether the Dmanisi hominids are *Homo erectus* or *H. ergaster* is due to the fact that some experts do not accept *H. ergaster* as anything more than a variant of *Homo erectus.* Moreover, Gabounia et al. assigned the Dmanisi specimens to a new species—*Homo georgicus*—that has been accepted by some authors.[214]

The mean endocranial volume of the Dmanisi skulls is ca. 650 cc, which is closer to that of *Homo habilis* (600 cc) than to those of *H. erectus* and *H. ergaster* (900 cc). The rounding of the back of the skull also recalls *Homo habilis* more than *H. erectus* or *H. ergaster.*[215] One adolescent was about 145 cm tall, and an adult was 166 cm tall, which is less than the stature of many specimens of *Homo erectus* and *Homo ergaster*. The estimated body masses of the three adults and the adolescent range from 40 to 50 kg.[216] In addition to the hominid specimens, Dmanisi Site 1 contained >3,000 identifiable vertebrate fossils that indicate at latest Pliocene–Early Pleistocene age, and >1,000 Oldowan lithic artifacts: chopping tools, rare choppers, a few scrapers, and many flakes. Indeed, Dmanisi Site 1 has the oldest Oldowan artifacts in Eurasia.[217] Carnivores are abundant in Dmanisi Site 1, and tooth pits on the bones are evidence carnivore damage, including on one of the hominid bones. Cutmarks on a few bones indicate that the hominids probably ate some meat, but how they acquired it has not been determined.[218]

The Dmanisi fauna also include ostrich, rhinoceros, deer, antelope, wolves, bears, horses, saber-toothed cats, hyenids, and abundant rodents such as voles, gerbils, and hamsters. Faunal and paleobotanical analyses have indicated that the Dmanisi area was a mosaic of open steppe and gallery forests.[211]

It appears that *Homo habilis* or *Homo ergaster* emigrated from Africa and might have given rise to Asian *Homo erectus* and one or more later species of European *Homo*. According to Gabunia et al., technological innovation might have been less important to their deployment than were more strictly biological and ecological factors, such as increased energy requirements due to larger body size and the quest for animal protein.[219] Accordingly, the quest for meat might have fueled the emergence of the first global primate.[220]

Although *Homo erectus* is a more bona fide species than either *H. habilis* or *H. ergaster,* there remains notable disagreement about which specimens constitute its hypodigm. In addition to the 1.77-Ma Dmanisi specimens, numerous specimens from Zhoukoudian (China), Trinil and Sangiran (Java), Tighenif (Algeria), Daka, Middle Awash (Ethiopia), and Olduvai Gorge (Tanzania) seem to lie securely in *Homo erectus*.[221] But disagreement has reigned regarding the presence of *Homo erectus* in Europe and whether the hominid fossils from Solo and Sambungmacan (Java), Narmada (India), Bodo, Melka Kunturé, and Bouri (Ethiopia), East and West Turkana (Kenya), Ndutu (Tanzania), Kabwe (Zambia), Saldanha and Swartkrans (South Africa), Dmanisi (Georgia), and sundry other Early and Middle Pleistocene Asian and African specimens are *Homo erectus*.[222]

Even relatively complete northern Kenyan specimens such as KNM-ER 3733, KNM-ER 3883, and KNM-WT 15000 raise problems for those who would assign them specifically. If they are indeed *Homo erectus,* the species dates at least to 1.6 Ma and therefore was contemporaneous with *Homo habilis* in eastern Africa.[223]

Based on the lengths of eight femora from East Turkana, Kenya, and Old-uvai, Tanzania, the mean adult stature of Early Pleistocene *Homo* spp. was 159.4 cm (range: 149–180 cm) and the mean body mass was 52 kg (range: 45–62 kg).[224] Ruff and Walker estimated the stature and body mass of the juvenile Turkana individual (KNM-WT 15000) to be >160 cm and 48 kg; had he attained full adult growth, the estimates were 185 cm and 68 kg, which would have him towering over hominid contemporaries and predecessors.[225] Later, Ruff revised his estimates to >157 cm and 50–53 kg.[226] Further, if H. Smith and Dean and Smith were correct that the juvenile Turkana individual's maturation, like that of other early *Homo* sp., *Australopithecus* spp., and *Paranthropus* spp., was closer to that of chimpanzees than to that of modern humans, more of his growth would have been completed at the time of death, and he would not have notably outstripped the stature of his contemporaries.[227]

The earliest evidence for a dental eruption pattern that might indicate a slower life history pattern like that of *Homo sapiens* is a ca. 1-Ma juvenile (5.3–6.6 years old) mandible of *Homo* sp. from Gran Dolina Cave in Sierra de Atapuerca, Spain.[228] The high wear rate evidenced by the mandibular incisors of eleven 400–500-Ka *Homo heidelbergensis* at Sima de los Huesos, Atapuerca, indicates that the life expectancy in the population might have been ≤50 years.[229]

Hard evidence for *Homo habilis* and *H. erectus* overlapping in time comes from Ileret in the Koobi Fora Formation at East Turkana, Kenya. A palate of the former species and a cranium of the latter species date from ca. 1.44 Ma and 1.55 Ma, respectively. Spoor et al. argued that the coexistence of the two species over nearly 500 Ka in eastern Africa makes *Homo habilis* an unlikely ancestor of *Homo erectus,* though they allowed that the transition might have occurred elsewhere followed by *Homo erectus* emigrating to the Turkana Basin.[230]

Homo erectus is distinguished from *Homo habilis* by its larger brain size. The long, low, flat cranium of *Homo erectus* is widest basally and narrows superiorly. Indeed, unlike the roundish calvariae of *Homo habilis* and *Homo sapiens,* the platycephalic calvariae of classic *Homo erectus* recall a fallen soufflé.[231] *Homo erectus,* especially as represented by East Asian specimens, have

- thick cranial vault bones relative to those of other *Homo* spp.,
- prominent brow ridges,
- a median sagittal keel on the frontal and parietal bones,
- postorbital constriction,

- a parietal angular torus,
- a transverse occipital torus,
- broad, variably prognathic faces,
- salient nasal bones, and
- robust mandibles with broad rami and no mental protuberance.

Homo erectus have notably reduced postcanine teeth by comparison with the dentition of *Australopithecus,* but they are larger than counterparts in *Homo sapiens.*[232]

For decades, *Homo erectus* had a secure place in the lineage of modern *Homo sapiens.*[233] Now, considering that it might have been contemporaneous with *Homo habilis* and that additional species of *Homo* might have inhabited the globe between 2 Ma and 200 Ka ago, there is room for doubt that *Homo erectus* sensu stricto, versus another species or variant of *Homo,* was an ancestor to late Middle–Late Pleistocene *Homo sapiens.*[234]

Although there is no agreement about which species of *Homo* gave rise to the younger species of *Homo,* the candidacy of *Homo ergaster*—or African *H. erectus*—gained popularity, particularly among experts who accepted *H. ergaster* as a species distinct from *Homo erectus.*[235] *Homo heidelbergensis* might have arisen from *H. ergaster, H. erectus,* or *H. antecessor,* and any or none of them could have been ancestors of the two latest species of *Homo: H. neanderthalensis* and *H. sapiens* (Figures 4.21 and 4.22).[236] Neandertalian populations, particularly as represented by specimens from western Europe, are generally thought not to be direct ancestors to modern humans (Figures 4.23 and 4.24).[237] Like Middle Pleistocene *Homo sapiens, H. neanderthalensis* were more robust obligate bipeds than Late Pleistocene and *Homo sapiens.*[238] Nonetheless, in addition to a suite of morphological differences, the dental maturation of *Homo neanderthalensis* was more rapid than that of Middle Pleistocene and later *Homo sapiens,* though dental maturation was slower in both species than in Pliocene and Early Pleistocene Hominidae.[239]

Mothers of Us All?

Africa is the probable cradle for the emergence of anatomically modern *Homo sapiens.*[240] Several fragmentary yet telling specimens from the Klasies River Mouth and Border Cave sites in South Africa are the primary evidence for an African origin of modern *Homo sapiens.*[241] Genetic data suggest that ca. 150 Ka

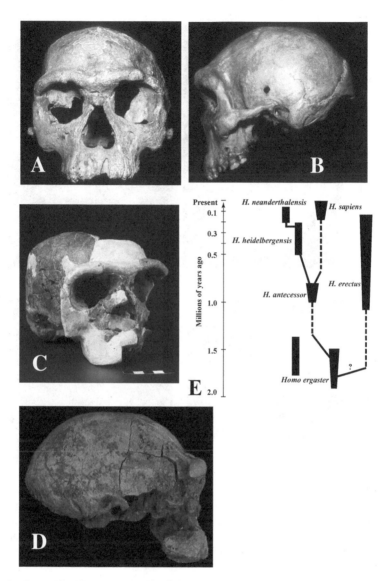

Figure 4.21. (A–C) African *Homo heidelbergensis* and (D) *Homo antecessor* (type). (A) Frontal view of 125–190-Ka cranium from Jebel Irhoud, Morocco. (B) Left lateral view of >125-Ka cranium from Kabwe, Zambia. (C) 400-Ka cranium from Ndutu, Tanzania. (D) Right lateral view of ca. 1.0-Ma cranium (type) from Buia, Eritrea. (E) Phyletic scheme of J. Bermúdez de Castro and J. Arsuaga.

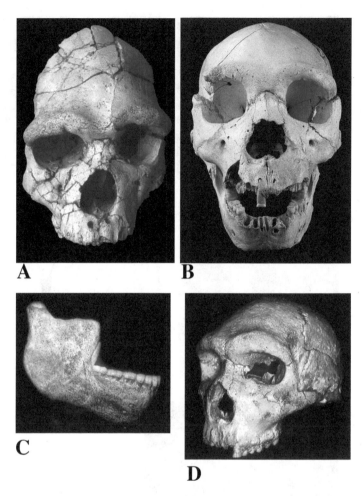

Figure 4.22. European *Homo heidelbergensis.* (A) Frontal view of 450-Ka cranium (Arago 21) from Arago, Tautavel, France. (B) Frontal view of 400–450-Ka skull (Atapuerca 5) from Sima de los Huesos, Sierra de Atapuerca, Spain. (C) Right lateral view of 600-Ka-Mauer mandible (type) from Heidelberg, Germany. (D) Three-quarters facial view of 200–300-Ka cranium from Petralona, Greece.

several African populations diverged into several branches and that polymorphisms present in extant populations might have introgressed via relatively recent interbreeding with the hominid forms that diverged from the ancestors of modern humans in the Lower-Middle Pleistocene.[242] A relatively small founder population from northeastern Africa emigrated to the Middle East

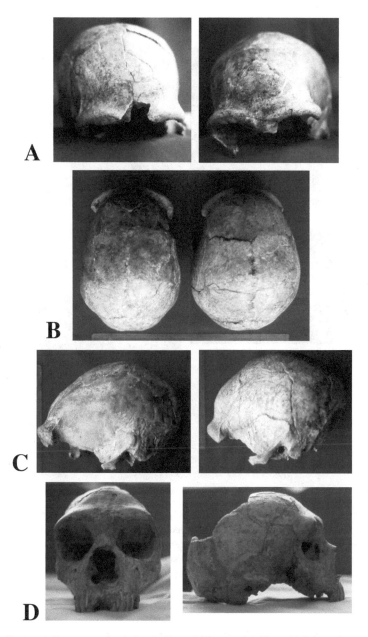

Figure 4.23. European *Homo neanderthalensis*: Frontal (A), superior (B), and left lateral views of Spy I (left) and Spy II (right) calvariae from Belgium. (D) Frontal and right lateral views of Gibralter I cranium.

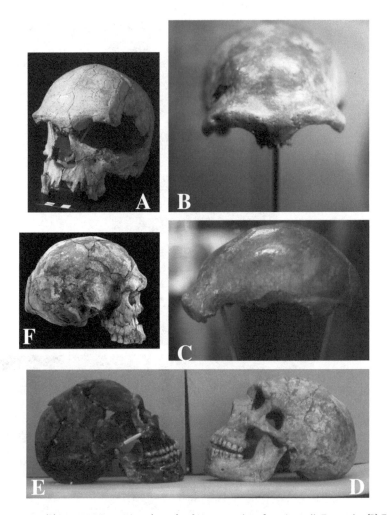

Figure 4.24. (A) Ca. 200-Ka cranium (LH 18), of *Homo sapiens* from Laetoli, Tanzania. (B) Top and (C) left lateral views of 40-Ka *Homo neanderthalensis* (type) calotte from Neander Valley, Germany. (D) Left lateral view of 50–60-Ka skull of *Homo neanderthalensis* from Wadi Amud, Israel. (E) Right lateral view of 80–120-Ka skull of *Homo sapiens* from Kafzeh, Israel. (F) Right lateral view of 160-Ka cranium of *Homo sapiens* from Bouri, Herto, Middle Awash, Ethiopia.

and Europe ca. 100 Ka, followed by the spread of *Homo sapiens* to Asia and Austro-Melanesia 30–50 Ka and to the Americas 15–30 Ka.[243]

Overall, the Klasies mandibles, teeth, and other cranial bits are indistinguishable from modern counterparts. They span a period between >60 Ka and ≤130 Ka. Similarly, the infant skeleton and adult cranial fragments and mandibles from Border Cave are clearly *Homo sapiens*.[244] Beaumont placed them between 90 Ka and 110 Ka.[245]

Other candidates for earliest African *Homo sapiens* are Laetoli Hominid 18 from the Ngaloba Beds at Laetoli, Tanzania; three crania from Herto Bouri, Middle Awash, Ethiopia; and a 160-Ka juvenile mandible and two adult crania from Jebel Irhoud, Morocco.[246]

The Ngaloba cranium was in situ with a Middle Stone Age industry. Manega employed amino acid and ^{40}Ar/^{39}Ar methods to establish a date of ca. 200 Ka for LH-18.[247] Its archaic features include a low, thickened vault and a flattened frontal contour. The cranial capacity is ca. 1,200 cc. LH-18 had little postorbital constriction, but the parietal vault is wider than the base. The incomplete face of LH-18 is modern in known features, and its maxilla sports canine fossae. LH-18 lacks a clearly defined supraorbital torus, though there is some toral development.

Clark et al. dated the crania of two adults and a child and associated artifacts from Herto at 160–154 Ka via the ^{40}Ar/^{39}Ar method.[248] Morphologically, the skulls are intermediate between fully modern *Homo sapiens* and older African specimens of *Homo*.[249] The 640 Herto artifacts sport Acheulean and Middle Stone Age features, and cutmarked and percussion-fractured hippopotamid and bovine bones indicate that the hominids butchered large mammals. Clark et al. interpreted the cutmarks on the skulls as evidence of defleshing as part of mortuary practice.[248]

Researchers have marshalled an impressive array of morphological and molecular anthropological evidence—including mtDNA, which singularly traces maternal inheritance, nuclear DNA, and a variety of proteins—to support the emergence of modern *Homo sapiens* in Africa between 100 Ka and 200 Ka.[250] Further, archaeological and geological evidence indicates that *Homo sapiens* emigrated relatively recently (ca. 42 Ka) from Africa to Asia and Europe, where they rapidly (≤16,000 years) replaced Neandertal and other archaic populations of *Homo,* probably with little or no admixture (Figure 4.25).[251] Concomitantly, modern human populations cannot be subdivided into subspecies or races, and we all have the same basic human potential.[252] The role of climatic changes such as megadroughts (135–75 Ka) followed by much wetter conditions after ca. 70 Ka in provoking or facilitating emigration from Africa to Eurasia remains to be determined.[253]

The single-origin theory provoked formidable critics, who instead favored multiple origins for modern *Homo sapiens* from archaic populations in several regions of the Eastern Hemisphere, including Africa. Their schemes contain

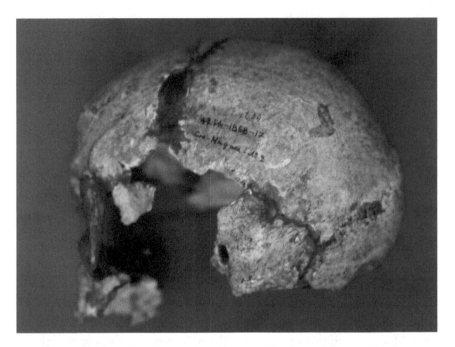

Figure 4.25. Left lateral view of ca. 30-Ka cranium (Cro-Magnon 2) of *Homo sapiens* from Cro-Magnon, France.

notable gene flow among early people, and continuity of selected morphological features between them and modern people in regions where the transitions occurred.[254]

Regardless of whether the African-mother hypothesis or the regional-transition theory is correct, the paleontological and genetic facts argue overwhelmingly that Africans made inaugural and enduring contributions to the human condition and that no extant population of *Homo sapiens* is biologically more or less human than the others.[255]

While a snowballing consensus favors Africa as the cradle of *Homo sapiens*, there actually are four basic models purporting to explain the evolution of *Homo sapiens* between ca. 200 and 30 Ka in Africa, Europe, continental Asia, and Australasia.[256] Their advocates severally use fossil hominid skeletal remains, genetic traits in global samples of modern people, and archeological and morphological indications of cognitive, linguistic, and technological capabilities in their arguments, but none of them provide definitive resolution.

Polar extremes are the multiregional evolution (regional continuity) model versus the African replacement (out-of-Africa) model.[257] The African hybrid-

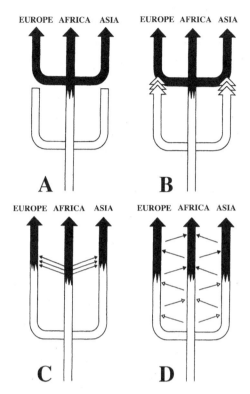

Figure 4.26. Models of modern human origin(s). (A) Recent African emergence with complete replacement of other *Homo* spp. in Eurasia. (B) Hybridization with and replacement by migrant African *Homo sapiens*. (C) Assimilation. (D) Multiregional evolution.

ization and replacement model and the assimilation model are intermediate between them (Figure 4.26). All except multiregionalists accept that *Homo sapiens* uniquely evolved recently (ca. 200–100 Ka) in Africa and then deployed to Eurasia and eventually to the Americas and Oceania.[256]

Replacementists argue that anatomically modern African emigrants replaced resident Eurasian and Australasian species of *Homo* with little or no hybridization.[258]

Hybridization-replacementists accept hybridization with some archaic indigenous populations, but view the effects to be of relatively minor significance.[259]

Assimilationists view continuity between archaic and anatomically modern hominids as having operated notably in some areas of Eurasia along with gene flow, admixture, and local environmental selective factors that would also produce morphological changes. Unity of the species was maintained by periodic admixture of *Homo sapiens* across wide areas.[260]

Multiregionalists reject the idea that *Homo sapiens* evolved solely in Africa. Instead, they advocate that discrete Middle Pleistocene archaic populations of *Homo* in Africa, Asia, and Europe evolved locally into Late Pleistocene *Homo sapiens*.[261] Throughout their tenures, both the archaic populations and descendent populations of *Homo sapiens* experienced some admixture with contemporaries from other areas.[262]

In 1997, Krings et al. announced the extraction of a short segment of mtDNA from the arm bone of the type specimen of *Homo neanderthalensis* from Feldhofer Cave in the Neander Valley, Germany.[263] With it they determined the entire sequence of hypervariable sequence 1 in the mtDNA control region. Comparison of the sample with a global reference sequence of *Homo sapiens* indicated that

- the Neandertal sequence is outside the variation of modern human mtDNAs,
- the age of the common ancestor of the Neandertal mtDNA and modern human mtDNAs is about four times greater than the common ancestor of modern human mtDNAs, and
- after deploying from Africa modern humans likely replaced Neandertals with little or no interbreeding.[263]

There followed refinements in retrieval and isolation of Neandertal mtDNA from Vindija Cave (Croatia), Mezmaiskaya Cave (Russian Federation), Engis and Scladina Cave (Belgium), La Chapelle-aux-Saints (France), and El Sidrón Cave (Spain), which strongly supported the specific status of *Homo neanderthalensis* and the out-of-Africa evolutionary scenario that indicated their virtual reproductive isolation from immigrant *Homo sapiens* in Europe and central Asia.[264]

Briggs et al. reported that they had reconstructed six complete mtDNA genomes from El Sidrón, Feldhofer, Vindija, and Mezmiaskaya Neandertal specimens.[265] Although they span ca. 30,000 years and the El Sidrón and Mezmiaskaya Caves are around 4,200 km (600 miles) apart, the mtDNA diversity in the overall sample is about one-third that of modern *Homo sapiens*. The research confirmed the monophyletic statuses of Neandertal and of modern human mtDNA. The most divergent among the six genomic samples is from Mezmiaskaya Cave, but it is geologically the oldest. Based on complete mtDNA genomes, the estimated mean date of the most recent common Neandertal mtDNA ancestor is 136,100 Ka before present. Further, it appears that the effective population of Neandertals at the time included <3,500 females. Partial mtDNA genomes from Teshik Tash, Uzbekistan, and Okladnikov Cave, in the Altai Mountains, Siberia, confirm the morphological evidence that Neandertals had deployed 2,000 km even farther east by 38–44 Ka.[266]

Krause et al. reported sequencing the complete mtDNA genome from a child's little fingertip bone, dated as 30–48 Ka. It is from Denisova Cave in the Altai Mountains of southern Siberia, about 100 km from Okladnikov Cave.[266] Despite its association with a Mousterian industry, the Denisova mtDNA specimen has nearly twice as many differences from modern human mtDNA as those of Neandertal mtDNAs, which indicated that the Denisova mtDNA lineage branched off well before the Neandertal and modern human lineages; the most recent common mtDNA shared by them is ca. 1 Ma. Accordingly, the authors suggested that there was temporal concurrence of Neandertals, *Homo sapiens,* and a third genetically distinct hominid living in the Altai region ca. 40 Ka.

The picture of relationships between *Homo neanderthalensis* and *Homo sapiens* was further confounded by the announcement of Green et al. that they had produced a draft sequence of a Neandertal nuclear DNA genome from three Vindija samples, probably representing two individuals.[267] In contrast with the mtDNA, the inferred nuclear DNA sequence indicated that there might have been some, albeit limited, interbreeding between Neandertals and European and Asian *Homo sapiens,* but probably not with the Africans. The most parsimonious scenario posits gene flow from Neandertals into *Homo sapiens,* but still allows that the reverse also might have occurred. There might have been positive selection in genomic regions of ancestral *Homo sapiens* that are involved in metabolism and in cognitive and skeletal development.

One of the more interesting findings indicates that like *Homo sapiens, Homo neanderthalensis* had the derived condition in the gene FOXP2, which might be involved, albeit modestly, in speech and human language capability.[268] One can expect future revisions of hypotheses and scenarios regarding Neandertal/modern human relations given the many challenges in extracting trustworthy DNA samples from long-buried bones that proffer diverse opportunities for degradation and contamination both taphonomically and in processing during excavation and laboratory manipulation.[269]

Compelling interpretation of the hominid fossil inventory is fraught with problems of patchy geographic and temporal sampling, small samples and poor preservation of specimens, insufficient dates for some specimens, limited

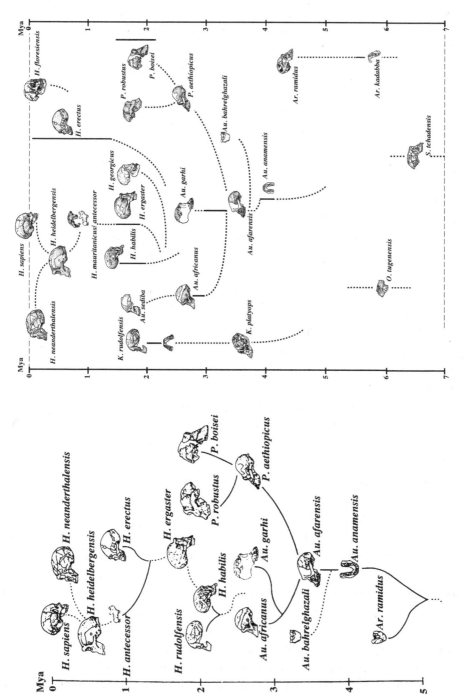

Figure 4.27. Tattersall's phyletic schemes: 2001 (left). 2011 (right). Symbols: *Ar, Ardipithecus; Au, Australopithecus; H, Homo; K, Kenyanthropus; O, Orrorin; P, Paranthropus; S, Sahelanthropus.*

choices of morphological traits, mostly from skulls and teeth, and ignorance of their genetic and developmental underpinnings. Behavioral inferences derived from archeological remnants and estimates of relative brain size and the vocal apparatus are particularly inadequate to reveal where and when modern human behavior evolved, even were we to agree on a definition of the human behavioral complex—human nature—and trace it into the brain.

Tempo Largo or Allegro Assai?

Genetic data from modern people, particularly mtDNA, arguably support substantially the African replacement model, but the profound limitations of available empirical data confound our efforts to discern whether distinctive features and lineages of the Hominidae developed gradually or during periods of relatively rapid change followed by stasis—punctuated equilibria.[270]

The occurrence of punctuated equilibria in the human career must be assessed on a case-by-case basis because there are claims for about twenty fossil hominid species in the past 6 million years. For example, it appears that *Homo neanderthalensis* is the terminal taxon of a series of ancestral species—*H. antecessor* and *H. heidelbergensis*—in Europe between ca. 700 Ka and 30 Ka years ago (Figure 4.27).[271] Whereas some commentators posit that *Homo sapiens* likewise evolved gradually through a series of species represented by specimens in Africa, others envision a dramatic shift in cognitive capacity and behavior that qualifies as a punctuational change in a small African population, followed by a period of biological stasis that we now enjoy.

PART II

POSITIONAL AND SUBSISTENCE BEHAVIORS

5

Apes in Motion

Are the knuckle-walkers or the brachiators, i.e. the long-armed gibbons, more nearly and essentially related to the human subject?

RICHARD OWEN (1859, p. 75)

The gibbon I saw walked without either putting his arms behind his head or holding them out backwards. All he did was to touch the ground with the outstretched fingers of his long arms now and then, just as one sees a man who carries a stick, but does not need one, touch the ground with it as he walks along.

THOMAS HENRY HUXLEY (1868, p. 180)

The positional behaviors—posture and locomotion—of apes and the morphologies that underpin them are of special interest to anthropologists because they might provide clues about the precursors and development of human adaptive complexes. Accordingly, we will explore the locomotor movements and postures of living apes and will sketch ideas about how their anatomies are related to them. This not only will introduce vocabulary and concepts that are necessary to comprehend models and scenarios on the evolution of human bipedalism but also will show how wild apes variously gain access to food and lodging sites and traverse their habitats to meet other apes for social interaction and reproduction while avoiding predators and other environmental dangers.[1]

189

Up, Up, and Away

To those who make a special study of the evolution of man,
the . . . gibbon is the most important of all surviving apes.

ARTHUR KEITH (1912b, p. 272)

Ricochetal arm-swinging is the most dramatic and probably also the most mechanically challenging component of hylobatid positional behavior.[2] It is unique to gibbons among the tetrapods. Consequently, it seems reasonable to interpret the anatomical peculiarities of hylobatid forelimbs as features of a derived (apomorphic) adaptive complex for it.

Gibbons ricochetally brachiate by synchronously flexing the elbow and retracting the shoulder joint of each forelimb alternately or of both forelimbs together. A gibbon can flex its forelimbs so powerfully that it is propelled upward and forward in free flight, with no hand contact until it lands on a target branch as far as 9 meters (30 feet) from the base branch.[3]

Richochetal arm-swinging is a particularly effective means for gibbons to project themselves instantaneously from beneath boughs to bipedal or crouched postures atop them. Gibbons sometimes employ ricochetal brachiation to traverse gaps in the canopy, such as during rapid bouts of locomotion. They probably use it even more often during acrobatic territorial displays and chases.[4] Although among wild apes the positional behavior of gibbons has been the most thoroughly and systematically studied, their brachiation has seldom been quantified discretely because it is often mixed with swinging beneath boughs, running bipedally on them, and leaping.

Traditionally, the elongated forelimbs of apes were viewed as the products of selection for brachiation, wherein subjects swing in a series of pendular arcs via alternate handholds beneath horizontal boughs and branches (Figure 5.1). But after field studies revealed that long-forelimbed great apes seldom brachiate, functional morphologists focused on other positional behaviors to explain the elongated forelimbs of all apes, including gibbons. Washburn and Ellefson proffered reaching while foraging and suspended feeding behavior as an explanation, and Washburn and other observers also have postulated climbing, particularly vertical climbing (Figure 5.2).[5]

Of course, because gibbons strategically employ their long forelimbs for all of these tasks, we might infer, minimally, that they are not impaired unduly in their performance by long forelimbs. Indeed, one can see distinct advantages to having long forelimbs for all these locomotor and foraging activities.

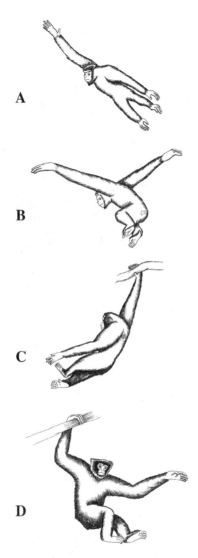

Figure 5.1. Male *Hoolock hoolock* brachiating. (A, B) In free flight. (C) Landing.
(D) Postlanding suspension.

For instance, long-limbed apes can brachiate over a distance with fewer handholds and perhaps faster than the short-limbed primates. And while sitting proximally atop sturdy branches apes with long reaches can forage relatively securely for choice fruits, buds, and new leaves that grow at the end of twigs.

Figure 5.2. Suspensory postures of male *Hylobates lar.*

Moreover, suspended feeding is advantageous because a pendant ape's weight draws dangling food toward it instead of pushing it away, as occurs when it reaches while seated atop the food-bearing branch. Further, suspended feeding increases the area that an ape can forage—from the upper segment of a sphere while it is seated to a full sphere encompassing the food that hangs from the base and nearby branches (Figure 5.3).[6]

Elongated forelimbs also enable apes to ascend trees with sizeable diameters, which unclawed sympatric mammals with shorter forelimbs cannot climb (Figure 5.4).[7] This is important because the choice sleeping and feeding sites are located in the crowns of tall emergent trees that rise so far above the main canopy that they can be reached only via their trunks or aerially.[8]

The elongated forearms and relatively short arms of gibbons enhance the leverage of their elbow flexors when they operate from suspended positions.[9] Hylobatid humeral heads have less medial torsion than those of other hominoids. Moreover, hylobatid apes possess a unique arrangement of muscles crossing the

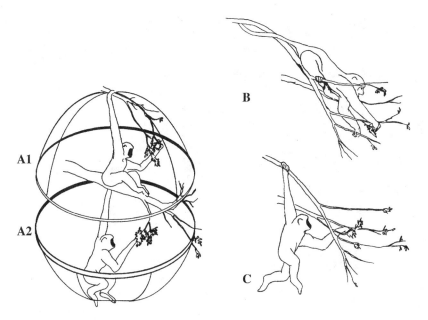

Figure 5.3. (A) 1. Both monkeys and gibbons can reach foods overhead while sitting on sturdy branches. 2. Long flexible upper limbs of gibbons also facilitate prolonged suspensory feeding. (B) A monkey's mass pushes terminal food away on thin flexible branches. (C) A suspended gibbon's mass bends the food toward it.

shoulder, elbow, wrist, and manual joints that might facilitate ricochetal arm-swinging, though electromyographic experiments on caged captives have failed to support this functional interpretation.[10] Electromyographic recordings from unrestrained gibbons engaged in vigorous naturalistic locomotor bouts very well might support the hypothesis that their forelimb muscle chains are an adaptive complex related to ricochetal arm-swinging.

In gibbons, orangutans, African apes, and certain New World monkeys— the spider and woolly monkeys, howlers, and muriquis—the glenoid cavity of the scapula, which articulates with the head of the humerus, is oriented cranially (Figure 5.5). This feature facilitates positional behaviors in which the forelimbs are employed overhead.[11] Further, in extant apes, the shoulder muscles, which raise the arm at the humeral joint, are well developed, and the clavicle, which struts the arm away from the chest wall for a wide range of movement, is long (Figure 2.6).[12]

Gibbon wrists contain special arrangements of the carpal bones that enhance rotation.[13] In both the proximal and the distal carpal rows, key elements

Figure 5.4. Male *Hylobates lar* climbing vertical trunks. Note abducted right (A) and left (B) thumbs gripping the trunks.

approximate ball-and-socket joints that proffer the greatest range of movement among all joint types, though ligaments can notably restrict them (Figure 5.6). This flexibility is particularly useful during versatile climbing in terminal branches, during unimanual suspended foraging, and during armswinging beneath boughs, which appears to entail more rotation of the body than ricochetal arm-swinging.[14]

Hylobatid hands and feet are remarkably versatile grasping organs. Deep clefts between the first and second digital rays augment their functional lengths, thereby increasing the range of diameters that they can grip securely (Figure 5.7). Combined with their elongated forelimbs, they allow gibbons to climb vines and vertical trunks ranging from small to large (Figure 5.4). Mobility of the hylobatid thumb is further enhanced by its ball-and-socket carpometacarpal joint—a feature that is unique to gibbons among primates.[15]

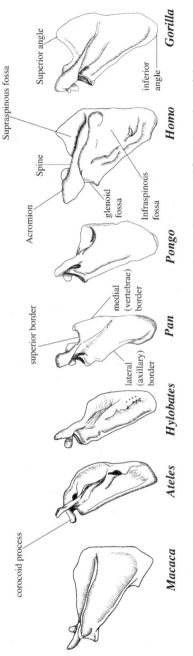

Figure 5.5. Dorsal view of scapulae in monkeys, apes, and humans. Note cranial orientation of the glenoid fossa in *Ateles*, *Hylobates*, *Pan*, *Pongo*, and (less so) *Gorilla*.

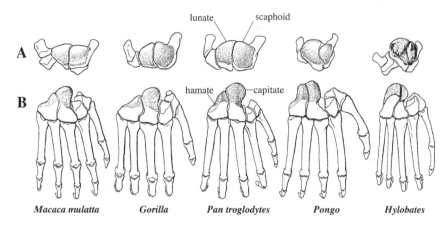

Figure 5.6. (A) Proximal carpal bones that articulate with the distal radius and (B) distal carpal bones that articulate with them. The more knoblike the articular areas (stippled) are, the greater potential for rotation and other actions

The proximal and intermediate phalanges in the second to fifth fingers and toes of hylobatid apes, like those of many other arboreal primates, are long and curved downward such that they conform to the convex surfaces of vines, branches, boughs, and trunks during prehension (Figure 5.7). Further, their interphalangeal joints have a flexion set that accentuates the overall curvature of the digits. Powerful flexor muscles attach to the distal and intermediate phalanges, especially in the fingers.[15]

Massive intrinsic and extrinsic muscles also service hylobatid thumbs and great toes. Notwithstanding an emphasis on power grips, which are requisite for their versatile, athletic positional behavior, gibbons have a well-developed manual precision of grip in fine manipulation. Because of their elongated fingers, they lack true precision grips, wherein the palmar pulp of the thumb tip can oppose the other fingertips. Nevertheless, their fine neurological control, flexible pollical carpometacarpal joints, and well-developed long digital flexor muscles of the pollical distal phalanx permit them to groom and to feed selectively on small suspended food objects with noteworthy finesse.

Gibbons have the second longest hind limbs compared with their truncal height among the anthropoid primates. In this index, only humans surpass them (Figure 2.15). In gibbons, like humans, the quadriceps femoris muscle, which extends the knee, is massive. However, their gluteal and calf muscles are more reminiscent of pongid than hominid counterparts.[16]

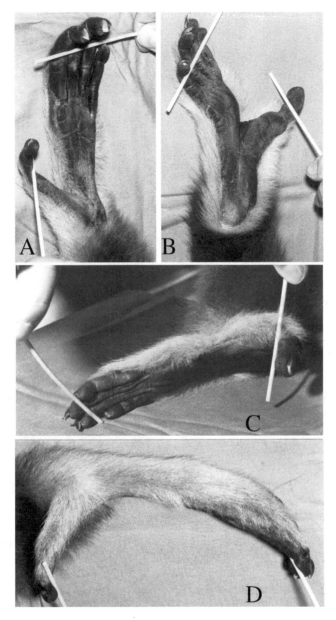

Figure 5.7. Left hand and right foot of *Hylobates lar*. (A) Palmar aspect of hand. (B) Plantar aspect of foot. (C) Ventromedial aspect of foot with hallux widely abducted. (D) Lateral aspect of pronated hand with pollex widely abducted.

Figure 5.8. Male white-handed gibbon *(Hylobates lar)* (A–C) walking and (D) running bipedally.

Powerful development of the quadriceps femoris muscle in the Hylobatidae is probably related to using the hind limb as a propulsive strut during vertical climbing. In addition, it facilitates bipedal running by resisting flexion of the knee in response to gravitational force, and by supplying propulsive force

(Figure 5.8).[17] The rather puny calf muscles—the triceps surae—of gibbons complement their poorly developed bony heels, to which they attach via the tendo calcaneus, the Achilles tendon.[18]

Unlike bipedal bonobos, chimpanzees, gorillas, and humans, gibbons do not initially contact the substrate with the heel at the outset of the stance phase.[19] Instead, they lead with the tip of the hallux, followed by midfoot/plantigrade contact.[20]

Climbing Any Which Way You Can

Orangutans are preeminently versatile climbers—a remarkable accomplishment in view of their massive bodies. Indeed, *Pongo* is the largest tetrapod that is committed fully to an arboreal career.[21]

The middle layer of Indonesian and East Malaysian forests offers the greatest continuity among supports for foraging, fleeing, and cruising orangutans. The tall emergent trees are isolates, which provide orangutans with no opportunity for horizontal travel. The low canopy is a broken highway that exposes peripatetics to predation by tigers, clouded leopards, and people in Sumatra, and clouded leopards and people in Borneo. In Borneo, adult male orangutans are more terrestrial than the females, perhaps because tigers are absent.[22]

Sugardjito and van Hooff reported that in the Gunung Leuser National Park, Sumatra, although both sexes fed opportunistically at all levels of the forest, the adult males tended to travel at lower levels than the females and youngsters.[23] Thorpe and Crompton argued that support diameter and type had the strongest association with Sumatran orangutan locomotion, so canopy height did not directly influence their locomotion. They speculated that there might be arboreal highways that individuals of all ages and both sexes employ. Parous females were more cautious than the adult males and immature individuals.[24]

In the middle canopy, unhurried orangutans move quadrupedally atop stout, horizontal, moderately inclined boughs and branches that they clasp with their powerful hands and feet.[25] When they reach areas where the branches are thin and there are short gaps between adjacent trees, they commonly hang below the current support and cautiously grab and test a new one before transferring the other hand and footholds to it. After such bridging transfers, orangutans hoist themselves atop available boughs and proceed quadrupedally through the core of the tree.

Gaps that cannot be negotiated by bridging transfer require orangutans to move vertically higher or lower in order to find more continuous supports.

Figure 5.9. Brachiating orangutan in free flight. Note overhand swing of forelimbs.

They seldom leap, brachiate, or drop over notable distances, except when flee-ing or as part of display behavior.[26] Orangutan brachiation differs markedly from that of gibbons in that it is nonricochetal.[27] Moreover, between hand-holds, orangutans swing the free forelimb overhand, whereas brachiating lesser apes swing it underhand (Figure 5.9). In the lower canopy, where trees are young and supple, hefty males will sway them until they can reach the next support.[28]

Like gibbons, orangutans employ a wide variety of feeding postures, rang-ing from seated to suspensory. Feeding orangutans commonly sit on firm branches and hold onto an overhead branch with one hand while gathering food with the free hand. Their long reach and great strength enable them to grab, draw in, and break off terminal branches that are inaccessible to many other nonflying mammals.[29]

More distally in a tree, an orangutan hangs out from or below a branch, holding it with the ipsilateral hand and foot, while foraging with the oppo-site hand and foot (Figure 5.10). They even hang headlong by their feet alone while manually collecting and opening fruits. However, three-point-suspension feeding postures are more common than two-point-pedal or unimanual postures.[30]

Powerful prehension and overall flexibility are at a premium for bulky beasts that climb and hang precariously in terminal branches, tens of meters above the forest floor. Accordingly, the digits and limb joints of *Pongo pyg-maeus* are especially adapted for secure grips and versatile postures.

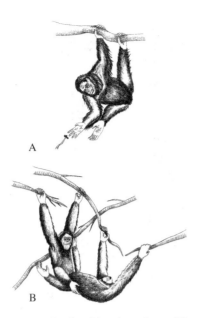

Figure 5.10. Orangutan suspensory feeding (A) and copulatory (B) postures.

Special proportions of the finger bones enable orangutans to grip slender twigs and vines with vice-like, double-locked grips.[31] Further, their elongated second to fifth manual and pedal digits are concavely curved ventrally due to bending in the proximal and intermediate phalangeal shafts and the fact that the joints between them are set in flexion (Figure 5.11). Both the long and short digital flexor muscles, which power the second to fifth digits, are massive. Indeed, the long digital flexor muscles are so fully committed to the second to fifth fingers and toes that they send no tendon or only vestigial ones to the pollex and hallux (Figure 5.12).[32] In *Pongo pygmaeus,* the concentration of extrinsic muscle power on the second to fifth digits has compromised the hallux more than the pollex, though according to Straus neither organ has become truly rudimentary.[33]

Having carried young orangutans, I can attest that their hallucal grip is strong enough to hurt and to bruise, even though they commonly lack a hallucal distal phalanx, nail, and long digital tendon to the hallux. Similar forceful prehension, powered by well-differentiated intrinsic hallucal muscles, is probably quite useful when wild orangutans climb vertically and walk quadrupedally atop boughs.[34]

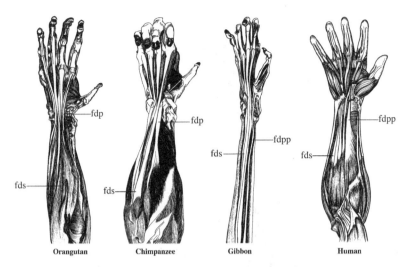

Figure 5.11. In orangutans, the radial deep digital flexor muscle (fdp) attaches to the distal phalanx of the index finger. Chimpanzees have the same condition, except some also have a vestigial tendon to the pollical distal phalanx. In gibbons and humans, the radial deep digital flexor muscle (fdpp) attaches entirely to the pollical distal phalanx. The flexor digitorum superficialis muscle (fds) lies atop fdp; its tendons of insertion to the intermediate phalanges are pierced by those of fdp en route to the distal phalanges.

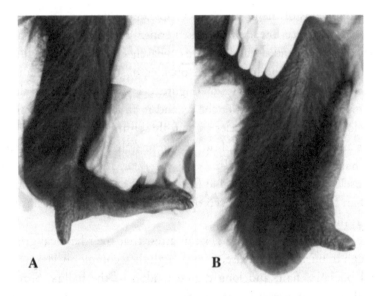

Figure 5.12. (A) Left foot of orangutan with toes maximally extended. Note curvature of toes II–V and brevity of the hallux. (B) When the foot is dorsiflexed, toes II–V nearly reach the knee because the leg is short and the lateral toes are long.

Like its hallux—but to a greater extent—the pollex of *Pongo* is powered by an elaboration of intrinsic muscles in lieu of an extrinsic flexor tendon.[34] However, unlike many halluces, orangutan thumbs contain a full complement of phalanges, and they bear nails. Tuttle and Rogers suggested that strong natural selection for fine pollical manipulation as part of selective feeding led to the orangutan pollex being less compromised than the hallux by demands for second-to-fifth digital prehension during versatile climbing.[35]

Orangutans on the move are strikingly loose limbed. Their shoulders, elbows, hips, knees, and wrists facilitate remarkable maneuverability, rotations about the limbs, and limb positions that could make all but the most accomplished contortionist cry with pain and envy. Orangutans can circumduct not only their shoulders but also their hip joints.[32] And, like gibbons though to a lesser extent, they have ball-and-socket mechanisms in their wrists (Figure 5.6).[36]

Among extant apes, orangutans are the least likely terrestrial bipeds. Their hind limbs are stubby, and their massive upper torsos and forelimbs make them top-heavy. Their long curved toes, short heels, and modest triceps surae muscles provide little support, and probably notable hindrance, to plantigrade propulsion by the feet (Figure 5.11).

Instead, the hind limbs of *Pongo pygmaeus* are adapted overwhelmingly for prehension and suspensory maneuvers.[37] In the leg, the long digital flexor muscles are developed outstandingly. In the thigh, the hamstring muscles are exceptionally large; one of them—the biceps femoris muscle—attaches quite low on the leg, which might be a special adaptation for powerful flexion of the knee to recover from hind limb suspended postures (Figure 5.13).[32]

Because of the modest development of their knee extensors and of the mechanism that plantarflexes their ankle joints, *Pongo* might depend heavily on their forelimbs to climb vertically; indeed, probably more so than gibbons and African apes, which have relatively robust hind limbs.[38]

The latissimus dorsi muscle of orangutans and other great apes probably plays an important role in hoisting owing to its robust attachment to the lateral half of the crest of the ilium (Figure 5.14).[39] This allows the lower body mass to be transferred directly to the retracting arm during hoisting actions.[40] The broad, elongated ilia and remarkably foreshortened lumbar region of the spine might also facilitate climbing by bringing the attachments of the latissimus dorsi muscle closer together, thereby increasing the power of their contraction.[32]

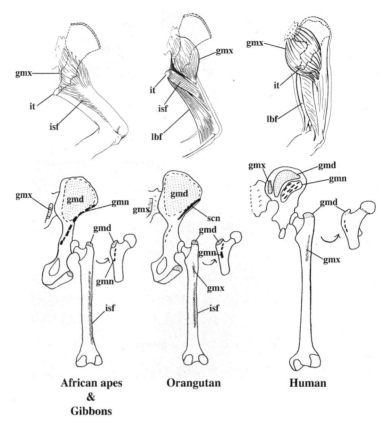

African apes & Gibbons

Orangutan

Human

Figure 5.13. Attachment patterns of hip and thigh muscles in apes and humans. *Symbols:* gluteus maximus muscle (gmx), gluteus medius muscle (gmd), gluteus minimus muscle (gmn), ischiofemoralis muscle (isf), scansarius muscle (scn), ischial tuberosity (it), long head of biceps femoris muscle (lbf).

Knuckling Down to Business

To traverse the forest, African apes habitually knuckle-walk on the ground; however, because many foods and preferred lodging sites of chimpanzees, bonobos, and some rain forest–dwelling gorillas are in trees, they are also arboreal frequently. They commonly climb tall, vertical trunks and adeptly forage in springy regions of the canopy. Among apes, only the mountain gorillas that inhabit montane forests where stout trees are few and far between are predominantly terrestrial.[41]

Accordingly, the knuckle-walking hands and plantigrade feet of African apes reflect adaptive compromises between terrestrial quadrupedal locomo-

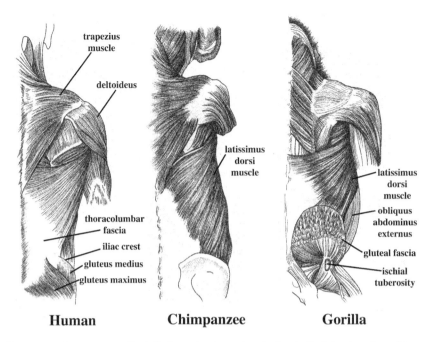

Human   **Chimpanzee**   **Gorilla**

Figure 5.14. In great apes, the latissimus dorsi muscle attaches to the lateral surface of the crest of the ilium. In humans, it has a modest attachment to the medial surface of the iliac crest.

tion and squatting and the demands of arboreal climbing and selective manual foraging and feeding.[42]

The term *knuckle-walking* derives from the fact that its employers flex the fingers so that the backs (dorsa) instead of the palmar surfaces contact the substrate.[43] Actually, the intermediate phalanges, not the knuckles (i.e., the finger joints), are the primary manual contacts during knuckle-walking (Figure 5.15). This unique form of dorsal digitigrady contrasts sharply with the palmigrade or ventral digitigrade postures of baboons and other terrestrially adapted monkeys.[44]

Knuckled postures allow long-fingered pongid hands, which are useful for arboreal climbing and foraging, to function effectively as terrestrial supports and propulsive organs. If the long fingers were extended during quadrupedal locomotion, they would have to be oriented laterally to avoid injury (Figure 5.16A). In lateral orientation, they probably would snag vegetation on the forest floor. For example, pinnipeds manage quite well on beaches that are relatively unobstructed, but their laterally extended pectoral flippers impede them in areas with notable ground cover. Accordingly, when walking on the

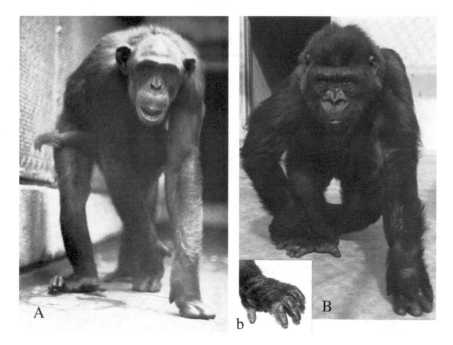

Figure 5.15. (A) Knuckle-walking adult female chimpanzee carrying infant ventrally. (B) Young female gorilla knuckle-walking. Inset *b* shows dermatoglyphic knuckle pads on right hand of B.

ground, orangutans commonly adopt a variety of fisted and modified palmigrade postures with their long fingers flexed, even though they are capable of fully palmigrade postures. And even on clear floors, palmigrade captives orient their hands laterally like pinniped flippers (Figure 5.16A).[45]

If we assume that the ancestors of *Pan* had long fingers as part of an arboreal adaptive complex, then a fisted hand posture appears to be a logical precursor to knuckle-walking.[46] Indeed, some incarcerated orangutans knuckle their hands while sliding on slippery cage floors, despite the absence of the morphological features that underpin true knuckle-walking in African apes (Figure 5.17).[47]

The preeminent, incontrovertible morphological feature that links obligate knuckle-walkers is epidermal: pads of friction skin on the dorsa of the second to fifth manual digits (Figure 5.15).[48] Osseoligamentous and myological features that predetermine and facilitate knuckle-walking are more variably expressed ontogenetically in and interspecifically among gorillas, chimpanzees, and bonobos.[49]

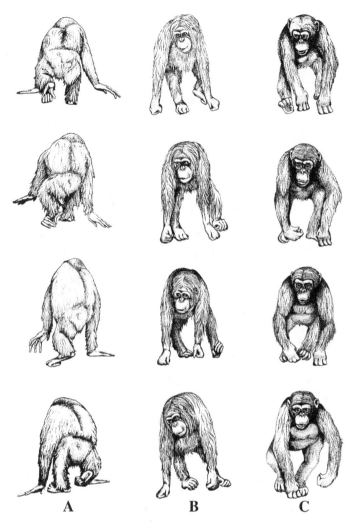

Figure 5.16. (A) Male orangutan moving rapidly away from viewer (note palmigrade hands). (B) Young orangutan fist-walking toward viewer. (C) Young chimpanzee knuckle-walking toward viewer.

Manipulation of the wrists of gorillas, chimpanzees, orangutans, and gibbons shows that the African apes are particularly limited in dorsiflexion (extension), adduction (ulnar deviation), and abduction (radial deviation). I documented special osseoligamentous mechanisms that limit carpal movements in gorillas and chimpanzees.[50] They include dorsal ridges on the distal radius

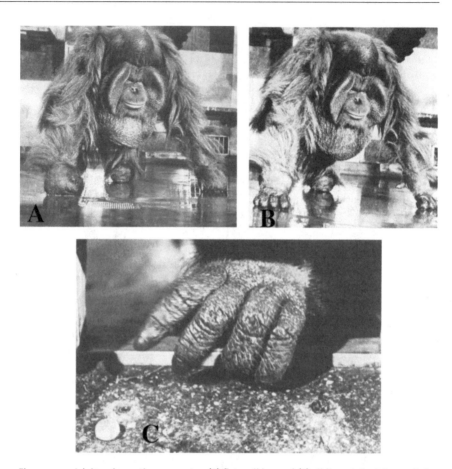

Figure 5.17. Adult male captive orangutan (A) fist-walking and (B) sliding on the intermediate phalanges of his fingers. (C) Note absence of knuckle pads and presence of hair on his fingers.

and scaphoid bone and ventro-ulnar inclination of the distal articular surface of the radius. Corruccini's extensive morphometric study on catarrhine wrist joints showed that chimpanzees and gorillas share a distinct morphological pattern related to knuckle-walking.[51] We need measurements of potential wrist movements and morphology in bonobos to test Doran's speculation that their wrists might differ from those of chimpanzees because bonobos more frequently use palmigrade postures to climb on boughs.[52] Chimpanzees and gorillas commonly knuckle-walk atop horizontal and modestly inclined boughs that are wide enough to accommodate their knuckled hands, but they grip branches that would make knuckle-walking precarious.

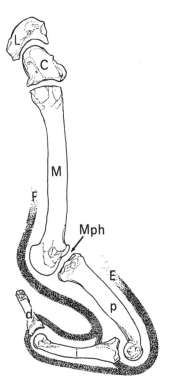

Figure 5.18. Exploded third manual digital ray and associated tendons and carpal bones. Symbols: L, lunate bone; C, capitate bone; M, 3rd metacarpal bone; p, proximal phalanx; i, intermediate phalanx; d, distal phalanx; F, long digital flexor tendons; E, long digital extensor tendon.

Metacarpal tori on the dorsodistal surfaces of weight-bearing metacarpal bones II-V indicate knuckle-walking, particularly if the articular surface of the metacarpal head extends dorsally, thereby facilitating overextension of the metacarpophalangeal joint (Figure 5.18 and 5.19).[53] Unfortunately for students who would trace the history of knuckle-walking, the metacarpal torus is variably expressed interdigitally and interspecifically in African apes. It develops ontogenetically, probably in response to the stresses of knuckle-walking.

Metacarpal tori might not be salient on all adult metacarpal bones II–V, particularly those of the second and fifth digits, due to disuse during knuckle-walking. Moreover, some adult *Pan paniscus* sport poorly developed metacarpal tori, even though, like *Pan troglodytes* and *Gorilla gorilla,* they commonly knuckle-walk.[54] We have no quantitative datum on overextension

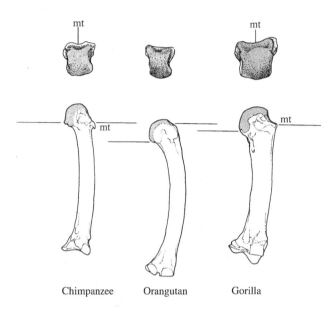

Chimpanzee   Orangutan   Gorilla

Figure 5.19. Extension of articular surface (stippled) onto the back of the metacarpal head with formation of a metatarsal torus (mt) in chimpanzee and gorilla but not in orangutan or human (not shown).

of metacarpophalangeal joints II–V in bonobos to compare with those for chimpanzees, gorillas, orangutans, and gibbons.[55]

The overextended metacarpophalangeal joints of a knuckled hand are particularly vulnerable to collapse downward during load bearing. The flexor digitorum superficialis muscle, which attaches distally to the intermediate phalanges of digits II–V, resists mechanical failure of the metacarpophalangeal joints. In chimpanzees, the intrinsic muscles of the palm—interossei and lumbricals—are relatively inactive during knuckled stances and do not seem to play a major role during knuckle-walking.[56]

In obligate knuckle-walkers, the distal and even the central and more proximal regions of many antebrachial flexor muscles, including the flexor digitorum superficialis, appear to be highly tendinous by comparison with counterparts in orangutans, which are not habitual knuckle-walkers; their muscular segments are characterized by relatively short fibers and pennate architecture.[57] Mammals that are highly adapted for economy of energy in running have long compliant tendons and short muscular fascicles. As forces stretch their tendons, they store energy that is released during elastic recoil.[58]

During knuckle-walking, tendons of the flexor digitorum superficialis muscle are loaded in tension (Figure 5.18). In chimpanzees and gorillas, the flexor digitorum superficialis muscle is relatively shortened by comparison with counterparts in other hominoid primates. Consequently, passive extension of the wrist and metacarpophalangeal joints II–V is accompanied by flexion of interphalangeal joints II–V in the African apes but not in Asian apes or humans.[59] Simultaneous extension of the wrist and metacarpophalangeal joints, which occurs at the beginning of support phase in knuckle-walking, loads the flexor digitorum superficialis muscle in tension. Electromyographic activity indicates an increase of the tensile load on the flexor digitorum superficialis tendons early in the support phase.[60]

Potential energy—represented by the strain of the superficial long digital flexor tendons when they are loaded with the mass of the subject and muscular stretch and contraction—is released as elastic recoil during carpal and metacarpophalangeal flexion, which occurs during the latter two-thirds of the support phase of knuckle-walking. Because some, perhaps most, of this potential energy is gained by placing the upper body mass on the load-bearing forearm instead of from muscular contraction, the elastic recoil of the flexor digitorum superficialis muscle contributes to mechanical energetic efficiency when gorillas and chimpanzees knuckle-walk.[60]

Although much of the propulsive force to move the body forward during knuckle-walking is probably provided by the hind limbs, the forelimbs must be coordinated with them if the overall quadrupedal locomotor pattern is to be effective.[61] The elastic recoil mechanism of the flexor digitorum superficialis muscle may provide energy-efficient propulsion to the hand, and perhaps to the entire forelimb, immediately before the swing phase.[60]

In sum, the adaptive complex for knuckle-walking in African ape hands includes the following traits:

• Friction-skin pads on digits II-V.
• Restricted wrist motions relative to those of Asian apes.
• Enlargement of the dorsal articular surfaces of metacarpal heads II–V and associated metacarpal tori.[62]
• Shortened tendons of the flexor digitorum superficialis muscle (Figure 5.12).

Except for the friction-skin pads, which are equally represented in all species of *Pan* and *Gorilla,* the knuckle-walking features are most complete and

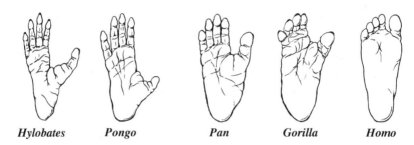

Hylobates   *Pongo*   *Pan*   *Gorilla*   *Homo*

Figure 5.20. Soles of apes and humans.

_____

evident in adult *Gorilla* and adult *Pan troglodytes*. Further detailed research is required to document their ontogeny and functions in gorillas and chimpanzees and the extent of their manifestations in bonobos.[63]

Traits uniquely related to knuckle-walking have not been identified unequivocally in other organs. In gorillas and chimpanzees, the deep olecranon fossa and its steep, flattened, lateral wall, the prominent lateral trochlear ridge, and the reduced olecranon of the ulna imply that the elbow joint is specially adapted to resisting stresses of knuckle-walking, particularly when the knuckled hand is in a coronal plane.[64] Still, Feldesman argued that no morphological trait that clearly distinguishes knuckle-walking African apes from orangutans has been identified on anthropoid distal humeri.[65] Moreover, a reduced olecranon and deep olecranon fossa also characterize orangutans, in which they facilitate versatile climbing and suspensory behavior instead of knuckle-walking.[32]

The plantigrade feet of African apes are quite distinct from those of Asian apes and monkeys (Figure 5.20). Indeed, revolutionary evolutionists have cited the resemblance of gorilla and chimpanzee feet to human feet as particularly compelling evidence for our close phylogenic affinity with them.[66]

Unlike the narrow-heeled catarrhine monkeys and gibbons, African apes regularly plant their long, wide heels fully on the ground while moving and standing quadrupedally and bipedally. As in human feet, their heels are robust and often neatly rounded. But unlike humans, African apes have a flexible midtarsal joint that allows the heel to be raised while the distal segment of the sole is still on the ground. Accordingly, they leave flat-footed impressions.

African ape feet are equipped with a powerful propulsive mechanism that can supplement the thrust of the extending hind limb during vertical climbing and plantigrade walking.[34] It consists of a relatively long power arm, including a prominent tuber calcanei, with hefty calf muscles that attach to it. Robust heels also support their bulk while they squat to feed and rest on the ground and on large boughs.[67]

The long, heavily muscled, divergent great toe in African ape feet not only provides powerful prehension for climbing but also may facilitate terrestrial locomotion by serving as a strut to distribute part of the load toward the lateral aspect of the foot (Figure 4.20).[68] Among chimpanzee pedal phalanges III–V, only the phalanges of the third toe are more curved than those of gorillas. Further, the pedal digital curvature is greater in young versus older chimpanzees, indicating that the younger chimpanzees climb in trees more than adults do.[69]

In several of their features, gorilla feet evince greater adaptation to terrestrial positional behavior—probably at the expense of versatile climbing—than the feet of bonobos and chimpanzees. Gorilla heels are more prominent and their nonhallucal toes are shorter and are usually more extensively webbed interdigitally, at least up to the proximal interphalangeal joints. Interdigital webbing provides a functional extension of the sole distally.[70]

Based on nineteen days of following a chimpanzee group in Kibale Forest, Pontzer and Wrangham estimated that chimpanzees expend ten times more energy traveling terrestrially than climbing tree trunks.[71] They concluded that chimpanzee postcranial anatomy reflects more selective pressure to minimize falls from trees than to facilitate maneuverability or to avoid predators, or that it simply reflects phylogenetic inertia.[72] The anatomical compromises related to climbing versus terrestrial travel surely allow them to restore energy most efficiently from arboreal sources of fruit, new leaves, and buds that are inaccessible without climbing, and to seek secondarily nutritious terrestrial fallback foods when the canopy is less fruitful. The energy gained by feeding arboreally must at least offset the cost of climbing the trees; on balance, terrestrial travel is probably more efficient than trying to negotiate arboreal highways, assuming that the trees in their habitats are contiguous enough to allow regular travel across the canopy.

Subterranean Secrets

Comparative observations and experiments on living apes, humans, and other animals are invaluable for devising hypotheses about the evolution of adaptive complexes such as bipedalism. But ultimately we must test these hypotheses with evidence from the fossil record, which for many reasons is a very challenging endeavor.

Nature seldom discloses its secrets fully and never easily. Even if bones alone could inform us definitively about muscular actions, positional behavioral repertoires, and phylogenic history, the brutal truth is that full documentation probably has not been preserved. Myriad telltale bits of evidence have been fragmented and pulverized beyond recognition by natural and anthropogenic processes. Moreover, for a variety of practical and political reasons, we may never salvage enough of the fossil record that is buried in Africa and Eurasia. Accordingly, we are far better prepared to discuss what we do not know from available evidence than to describe ancient apes in action and to argue dogmatically for specific transitional sequences that culminated in extant apes and people. One should keep these caveats in mind, especially when considering models on the positional behavior of Eocene–Oligocene basal and early catarrhine anthropoids, which are far removed temporally from the intact animate primates and other mammals that happen to sport bony features like those of fossil specimens.

Their tali and calcanei indicate that *Eosimias* walked quadrupedally atop branches and engaged in arboreal leaping.[73] Amphipithecidae are somewhat better represented postcranially than are Eosimiidae.[74] The humeri, proximal ulnae, and calcaneus of a one individual suggest that *Pondaungia savagei* or perhaps *Amphipithecus mogaungensis* was probably a slow, cautious, above-branch quadruped like the howlers and lorises.[75] The shoulder was capable of a wide range of movement, including reaching overhead, but the elbow was habitually flexed; the individual was not equipped for leaping or clinging to vertical substrates.[76] Further, the Pondaung primate might have moved on vertical branches, atop horizontal branches, and perhaps also below them. The tali of *Pondaungia* sp. optionally indicate that it was an arboreal quadruped that might have moved rapidly along branches and engaged in vertical postures or springing between branches.[77]

The Fayum Late Eocene–Oligocene quarries have been notably richer in postcranial elements, but a specific assignment in the absence of a clear association with craniodental remains has required that many of them be assigned specifically according to body mass, usually estimated from dental remains.[78] Accordingly, two distal humeri and a proximal femur from the Late Eocene Fayum Quarry L-41 were assigned to *Catopithecus browni* (body mass ca. 768 grams) because it is the most abundant species of the appropriate size in the locality.[79] A smaller proximal femur might be *Proteopithecus sylviae* (body mass ca. 542 g), and the acetabular portion of a hip bone also might be from *Proteopithecus sylviae* or perhaps *Arsinoea kallimos* (body mass ca. 355 g).[80]

One of the Fayum humeri might be from a nonanthropoid primate because it differs from one of two specimens that were excavated later in quarry L-41 near a proximal ulna and teeth of *Catopithecus*. Two additional humeri from L-41 probably represent *Proteopithecus*. Seiffert et al. concluded that *Proteopithcus sylviae* was an arboreal quadruped that engaged in rapid running and pronograde leaping and that *Catopithecus browni* probably moved quadrupedally more deliberately and leapt less than the smaller sympatric quadrupeds.[81]

The rather abundant, though isolated, postcranial bones that are assigned to *Aegyptopithecus, Propliopithecus,* and *Apidium* indicate that they were capable climbers and above-branch quadrupeds that lacked specializations for suspensory behavior or running on the ground.[82] Overall, the nearest postcranial analogue to propliopithecid primates is *Alouatta* (howling monkey), a platyrrhine versatile climber that sometimes leaps across gaps in the canopy and hangs beneath branches to feed via various combinations of pedal, caudal, and manual grips.[83]

The intimate apposition of the distal tibia and fibula and features of anklebones that are ascribed to *Apidium phiomense* imply that they might have been adapted for saltation, though not to a degree that would enable transatlantic island hopping to South America.[84] Fundamentally, *Apidium* is unique. If a modern analogue must be noted, neotropical squirrel monkeys *(Saimiri)* are recalled because they too run quadrupedally and leap about in forest canopies.[85] However, the forelimbs of *Apidium* indicate more robust climbing ability than that of *Saimiri,* and some of its pedal features hint a heritage of terrestrial habits.[86]

There is no direct evidence for catarrhine anthropoid postcranial evolution during the ≥10-million-year span that separates the Fayum anthropoids from the Early Miocene dental apes of Eastern Africa and Eurasia. The smaller species of Early and Middle Miocene Anthropoidea of Eastern Africa— *Dendropithecus* (8 kg), *Limnopithecus* (5 kg), *Micropithecus* (4 kg), *Simiolus* (5 kg), and *Kalepithecus* (6 kg)—are poorly or, at best, patchily represented postcranially.[87] Except for one lower thoracic vertebra, our knowledge of the postcranium is confined to the appendicular skeleton. The vertebra might belong to *Micropithecus*, the smallest of the Miocene apes.[88] It resembles those of suspensory anthropoids.[89]

Like neotropical spider monkeys *(Ateles), Dendropithecus* has long limbs.[90] Details of the humeri and hand bones imply that the smaller African Miocene anthropoids chiefly employed their forelimbs for arboreal quadrupedal locomotion and climbing, and perhaps some suspended foraging and bridging.[91]

Evidence from the hind limb, including a partial hip bone, not only complements this basically quadrupedal/adept-climbing model but also bolsters inferences of suspensory behavior among the smaller African Miocene apes.[89]

Postcranial features of Early and Middle Miocene *Proconsul, Afropithecus,* and *Turkanapithecus* appear to be logical precursors for those in later ape skeletons, though they still lack special features related to obligate suspended feeding and knuckle-walking.[92] The first carpometacarpal joint in *Proconsul* spp. and *Afropithecus turkanensis* is saddle shaped, like those of extant Pongidae and Hominidae, which suggests similar multiaxial movements that would facilitate fine manipulation.[93]

Unlike the Hylobatidae and extant Pongidae of comparable body size, *Proconsul* spp. were pronograde arboreal quadrupeds and rather conservative palmigrade climbers (Figure 3.5). Interspecifically, they differ primarily in size: *Proconsul heseloni,* 15 kg; *P. nyanzae,* 30 kg; and *P. major,* 50 kg.[87] They were not specially equipped for dramatic arboreal leaping and bridging behavior or for cursorial terrestrial locomotion.[94] They probably placed their hands palmigrade and their feet plantigrade during quadrupedal locomotion and stance.[95]

The torsos of *Proconsul* most closely resemble those of the Cercopithecoidea. Like Old World monkeys, they had relatively long, flexible, heavily muscled lower backs and narrow, dorsoventrally deep chests. Accordingly, their monkey-like scapulae lay on the sides of the thorax, with their humeral joints facing ventrally; the forelimbs moved chiefly in sagittal planes when *Proconsul* walked atop branches and if they ventured to the ground.[96]

Subequal limb proportions also confirm the basically quadrupedal, pronograde locomotor posture of *Proconsul.*[97] However, certain features of the hip, knee, elbow, wrist, fingers, tarsus, robust hallux, and other toes of *Proconsul* attest to a somewhat greater agility and security in arboreal niches than could be expected of similarly sized cercopithecoid monkeys.[98]

A sacrum of 18-Ma *Proconsul heseloni* implies no external tail or a greatly reduced one. Further, Nakatsukasa et al. showed that the vertebra that Harrison claimed were caudal vertebrae are actually lumbar vertebrae of *Proconsul heseloni.*[99] Accordingly, in at least some Miocene dental apes, dramatic caudal reduction occurred before the development of most other hominoid postcranial adaptive complexes. This condition is reminiscent of the dramatic caudal reduction in many species of macaques *(Macaca)* that inhabit a wide range of temperate, montane, tropical, and subtropical climates and occupy a spectrum of terrestrial and arboreal niches.[100] The remarkable adaptability of short-tailed and stump-tailed *Macaca* confounds attempts to model the specific selective complex that effected caudal reduction in *Proconsul*.

Compared with *Proconsul heseloni* and *P. nyanzae, Afropithecus turkanensis* (30 kg) from the early Middle Miocene site of Kalodirr, Kenya, is poorly represented postcranially. In known parts, its appendicular skeleton is closely similar in size, shape, and proportions to that of *Proconsul nyanzae*.[101] Its torso is unknown.

Interestingly, the 15–20.6-Ma vertebrae from Moroto, Uganda, are strikingly like those of extant great apes and dissimilar from those of *Proconsul nyanzae*.[102] Others have assigned them to *Proconsul major* and mentioned their possible future assignment to *Afropithecus* on the basis of excellent associated craniodental specimens.[103] However, Ward noted that if the Moroto vertebrae indeed belong to *Afropithecus,* either remarkable postcranial evolution occurred during the 2–4 million years that separate the Kalodirr and Moroto hominoids or exceptional locomotor diversity characterized species of *Afropithecus*.[96]

Of course, the Moroto vertebrae could belong to neither *Proconsul* nor *Afropithecus*. Indeed, Gebo et al. referred them to *Morotopithecus bishopi,* and Senut et al. dubbed them *Ugandapithecus major*.[104]

Appendicular bones and a second rib of an individual that weighed approximately 12 kg postcranially represent *Turkanapithecus kalakolensis,* a contemporary of *Afropithecus* at Kalodirr.[105] *Turkanapithecus* appears to be somewhat longer limbed than *Proconsul* but not as long-limbed as *Pliopithecus;* it was probably pronograde. Its thorax was probably narrow, like those of pronograde monkeys.[93] Rose inferred that, although it sports some unique features, the elbow complex of *Turkanapithecus* probably acted powerfully in flexion and extension, had the capacity to rotate extensively, and was most stable in full pronation.[89] Accordingly, *Turkanapithecus* was a slow quadruped atop stable boughs and branches but engaged in some acrobatic climbing, hoisting, and perhaps suspensory behavior in springy regions of trees. Palm and finger bones of *Turkanapithecus* are like those of *Proconsul,* so they too were probably palmigrade.[93] Based primarily on femoral evidence, Rose concluded that *Turkanapithecus* is most like *Alouatta* among the living anthropoid primates.[106] They might have been more versatile climbers than *Proconsul* was, though probably less suspensory than European *Pliopithecus* and the smaller Miocene apes of Eastern Africa and Asia.

During the past thirty years, there has been a steady increase in the number of postcranial specimens of East African Middle Miocene anthropoids.

We now have informative partial skeletons or representative parts of *Nacholapithecus kerioi, Kenyapithecus wickeri, Equatorius africanus, Mabokopithecus harrisoni, Nyanzapithecus pickfordi,* and *Victoriapithecus macinnesi,* which facilitates the estimates of body size and modeling their positional behavior.

Isolated skeletal elements suggest that *Nacholapithecus* were basically pronograde arboreal quadrupeds that engaged in some vertical climbing and clambering that involved wide extension of the elbow and powerful hoisting (Figure 3.5).[107] They were probably tailless.[108] The extent to which they were terrestrial is indeterminate. *Nacholapithecus* resemble *Proconsul* more than any other genus of Miocene Anthropoidea. They were the size of baboons, and they exhibit a similar degree of sexual dimorphism in body size: males weighed about 22 kg and females 11 kg.[109] Detailed study of an adult male partial skeleton, particularly the manual and pedal phalanges, has confirmed that *Nacholapithecus* were well adapted for arboreal climbing on vertical and horizontal substrates.[110] They exhibit derived features, in comparison with those of *Proconsul,* indicating that *Nacholapithecus* more frequently progressed on flexible branches in the canopy.

The proximal femur of Maboko *Equatorius africanus* is very similar to that of *Turkanapithecus kalakolensis,* but its proximal humerus implies cursorial pronograde locomotion on the ground like that of semiterrestrial and terrestrial cercopithecoid monkeys.[111] Based on isolated appendicular specimens, McCrossin and Benefit initially modeled Maboko *Equatorius africanus* as a macaque-like terrestrial quadruped that might have been palmigrade in trees and ventrally digitigrade during progression on the ground.[112] When they entered trees to lodge, escape terrestrial predators, and feed on fruits, they engaged in vertical climbing and versatile forelimb-dominated actions, including hoisting. Pedal specimens suggest a notable contribution of the hind limb while climbing vertically and plantigrade postures on the ground.[113]

Additional postcranial specimens, particularly a distal radius with a deep carpal articular surface and a metacarpal bone sporting a dorsodistal torus, have indicated to McCrossin et al. that Maboko *Equatorius africanus* was a knuckle-walker.[113] However, based on a comparison of the third metacarpal of *Equatorius africanus* with those from a broad sample of counterparts from catarrhine monkeys and apes, Patel et al. rejected the idea that *Equatorius* was a knuckle-walker.[114] They agreed that *Equatorius* probably engaged in more terrestrial behavior than earlier Miocene hominoids did; indeed, the 15-Ma species might be the earliest semiterrestrial hominoid.

Although a distal humerus of *Kenyapithecus wickeri* (12.8–14.4 Ma) from Fort Ternan, Kenya, exhibits a mix of hominoid and semiterrestrial monkey-like features, McCrossin and Benefit inferred that the individual might have been at least partially terrestrial.[115] Kappleman argued that *Kenyapithecus wickeri* inhabited a woodland habitat with areas of broken cover.[116] Rose concluded that *Kenyapithecus* is fundamentally a quadruped.[95] The upper canine teeth of *Kenyapithecus africanus* and *Kenyapithecus wickeri* indicate that both species were sexually dimorphic.[117]

Like *Kenyapithecus wickeri,* for *Mabokopithecus harrisoni* a single specimen—a proximal humerus from Bed 5 in the Maboko Formation on Maboko Island—represents its postcranial anatomy. From it, McCrossin inferred that, like *Cebus* and *Cacajao, Mabokopithecus harrisoni* was a capable arboreal climber with moderate mobility of the shoulder.[118] A 15-Ma proximal humerus, provisionally assigned to *Nyanzapithecus pickfordi* from Maboko Island, Kenya, is also basically that of an arboreal quadrupedal climber with limited mobility of the shoulder joint.[118]

Maboko *Victoriapithecus macinnesi* was a cursorial terrestrial or semiterrestrial quadruped.[119] They sported ischial callosities and a tail. They were much smaller than *Kenyapithecus africanus.* The estimated average body mass of *Victoriapithecus macinnesi* is 3 kg, with a range of 2–4 kg, which is somewhat smaller than *Cercopithecus aethiops.*[120] *Victoriapithecus* was probably sexually dimorphic, with the females weighing 2.4–3.1 kg and males 3.3–4.1 kg, which is coincident with notable sexual dimorphism in their upper and lower canine teeth.[121]

Although the forelimb bones of *Victoriapithecus* show many affinities with extant Cercopithecidea, especially the Cercopithecinae, they exhibit other features that are more like counterparts in the Ceboidea.[122] Because analyses of the humerus, radius, ulna, carpals, and phalanges have indicated that *Victoriapithecus* was as terrestrially adapted as some extant species of Cercopithecidae, Blue et al. speculated that stem Cercopithecoidea might have been terrestrial and digitigrade. Paleoenvironmental evidence indicates that *Victoriapithecus* favored a woodland habitat.[122]

In Europe, the early Middle Miocene postcranial bones of *Pliopithecus vindobonensis* indicate that it was basically an arboreal quadruped that could climb adeptly and perhaps sometimes foraged in suspended postures. But it had not developed the ricochetal arm-swinging complex, which characterizes modern Hylobatidae.[15] Fleagle estimated its body mass to be 7 kg, which is like that of its smaller African contemporaries.[123]

The forelimbs and hind limbs of *Pliopithecus* are not only subequal in length but are also relatively long. Accordingly, *Pliopithecus* variously resembles the sometimes suspensory neotropical howlers *(Alouatta)* and spider monkeys *(Ateles).*[90] However, overall the hind limb is closest to that of gibbons among the extant anthropoids.[124]

Features of the elbow and shoulder complexes of *Pliopithecus* suggest degrees of mobility that are associated with climbing and bridging in ceboid monkeys, but that were less versatile than the Atelinae, whose arboreal acrobatics are assisted by huge prehensile tails.[125] The sizeable olecranon of *Pliopithecus* implies regular quadrupedism atop boughs and larger branches. Its wrist is very monkey-like.[126]

The well-developed hallux and ventral curvature and other features of manual and pedal phalanges II–V in *Pliopithecus vindobonensis* imply that it was adapted for climbing in a variety of stable and flexible regions of trees.[15] Manual and pedal phalanges attributed to *Anapithecus* from the Late Miocene site of Rudabánya in Hungary indicate that they also engaged in versatile climbing and perhaps suspensory behavior. The Hungarian specimens are quite similar to their counterparts from Slovakia.[127]

The sternum of *Pliopithecus* is broad, which, like breadth of the lumbar vertebral bodies, implies that the thorax might have been wide transversely as in extant Hominoidea and Atelinae. Nevertheless, unlike modern hominoids and ateline monkeys, its lumbar region might not be reduced, and its relatively narrow iliac blade argues against a hominoid breadth of the thorax.[128] Although the lumbar vertebrae of *Pliopithecus* are more hominoid than those of *Proconsul,* the caudal reduction had not progressed to the condition in *Proconsul heseloni* and *Nacholapithecus kerioi.*[129]

Ankel devised an ingenious method to estimate caudal development in anthropoid primates based on the relative dimensions of the entrance and exit of the sacral canal, which houses the sizeable caudal nerves, arteries, and veins that serve large tails.[130] From its monkey-like sacrum, she initially inferred a long, nonprehensile tail for *Pliopithecus vindobonensis,* but later she communicated that it evidenced caudal reduction, possibly like that of *Cacajao,* the shortest-tailed ceboid monkey.[131]

Although there are revealing partial skeletons of *Pierolapithecus catalaunicus, Dryopithecus laietanus,* and *Oreopithecus bambolii,* otherwise the large apes of the European Middle and Late Miocene are meagerly represented postcranially. They have been variously assigned to *Dryopithecus* spp., *Austriacopithecus, Griphopithecus, Paidopithex, Sivapithecus,* and *Lufengpithecus.*

Although features of the trunk and wrist of late Middle Miocene (12.5–13-Ma) *Pierolapithecus catalaunicus* (body mass 30–35 kg) suggest orthograde posture, the short, straight manual phalanges and features of the metacarpophalangeal joints indicate that it was palmigrade and had not developed an adaptive complex for knuckle-walking or below-branch feeding and locomotion (Figure 3.7).[132] Others have argued that *Pierolapithecus* engaged in climbing and suspension, in addition to limited palmigrade locomotion.[133]

Inferred limb proportions of an early Late Miocene (9.6 Ma) partial skeleton of *Hispanopithecus laietanus* (= Drypopithecus laietanus CLI 18000) are intriguingly like those of *Pongo,* though the individual was probably not as versatile as *Pongo* in trees (Figure 3.8).[134] The individual weighed around 34 kg, a body mass quite comparable to that of *Pierolapithcus.*[135] The features that indicate an orthograde posture for the young male include lumbar vertebrae that are proportionally shorter than those of quadrupedal proconsulids and cercopithecoids and other traits that indicate stiffness of the lower back. The structure of the thoracic vertebrae indicates that the chest might have been broad, which together with the long clavicle would position the scapula on the back of the chest wall. The hands are very large, and the fingers are long. The femoral structure indicates notable hip mobility. Accordingly, like *Pongo, Dryopithecus laietanus* probably possessed many of the features that facilitate versatile climbing and suspensory postures to a greater degree than occur in *Pan* and *Gorilla.*[135] However, some features, such as the short metacarpals and longish lumbar vertebral bodies, testify that they had not fully developed modern pongid postcranial anatomy. The subtropical forest habitat is compatible with the positional behavior suggested by the morphology of Can Llobateres *Dryopithecus laietanus.*[136]

Begun considered various other skeletal bits to represent two distinct taxa: *Dryopithecus* and *Austriacopithecus.*[137] A humeral shaft from St. Gaudens, France, is *Dryopithecus fontani,* and certain forelimb and pedal specimens from Rudabánya, Hungary, are *Dryopithecus brancoi.*[138] Begun assigned a humeral shaft and an ulna from Klein Hadersdorf, Austria, to *Austriacopithecus weinfurteri.*[137] Whereas the humeri and ulnae of *Dryopithecus* are similar to those of extant hominoids, *Austriacopithecus* resembles earlier Miocene forms: *Kenyapithecus, Proconsul,* and *Turkanapithecus.* Accordingly, *Dryopithecus* might have engaged in notable suspensory behavior and vertical climbing in addition to quadrupedism, while *Austriacopithecus* was more restricted to above-branch quadrupedism.[139]

Hand and foot bones from Rudabánya also imply that *Dryopithecus bran-coi* engaged in forelimb and hind limb suspensory behavior.[140] The attribution of suspensory behavior for *Dryopithecus brancoi* and *Anapithecus hernyaki* fits comfortably with the paleoecological model of Rudabánya as a swamp forest (11.5–12.5 Ma).[141]

Among the Miocene primates, the fullest suite of apish postcranial features are present in *Oreopithecus bambolii* (Figure 4.2).[142] Indeed, were it not for its peculiar craniodental morphology, this 9-Ma postcranial ape would hang neatly among orangutans and extant African apes. Harrison argued that *Oreopithecus bambolii* provides the most clues to the kind of beast that gave rise to the postcranial morphologies of extant Pongidae and perhaps also the Hominidae.[143]

The recurved ribs and broad transversely oriented iliac blades provide evidence that *Oreopithecus* possessed a wide, dorsoventrally flattened chest. The lower back of *Oreopithecus bambolii* contains only five lumbar vertebrae, its sacrum consists of six fused vertebrae, and it probably·had no external tail.[144] The inferred shape of the thorax and long collar-bones imply that the shoulder blades lay on the back and that the broad, deeply concave glenoid cavities were oriented laterally as in apes and people, instead of ventrally as in quadrupedal monkeys.

The robust acromion of the scapula and prominent globular head and other proximal humeral features indicate that *Oreopithecus bambolii* was capable of versatile shoulder movements.[143] Likewise, details of the distal humerus, the proximal ulna—particularly the abbreviated olecranon and circular radial head—indicate full extensibility and extensive rotatory abilities of the elbow joint, which would facilitate versatile climbing and suspended posturing in *Oreopithecus bambolii*. The carpal bones imply notable degrees of adduction and abduction and powerful flexion of the wrist.[145] Moreover, it has longish forelimbs and relatively shorter hind limbs, which underscore the potential for clambering, vertical climbing, and pendant suspension.[146]

Details of the hip bone, proximal femur, knee joint, and ankle also indicate that *Oreopithecus bambolii* was an accomplished vertical climber and versatile arboreal forager. The hip joint probably allowed notable rotatory movements and full extension. The knee joint was probably also fully extensible, and the foot inverted and everted freely as required by a habitual arboreal climber.[147]

The nonpollical fingers and nonhallucal toes of *Oreopithecus bambolii* bear the hallmarks of powerful arboreal climbers that engage in suspensory activities: the proximal and intermediate phalanges are long, slender, and ventrally curved, and ventrally they bear prominent lateral crests, to which the retaining sheaths for robust, long digital flexor tendons were anchored. The opposable pollices were short, stout, and probably quite mobile; the halluces were also stout and were probably powerful prehensile organs.[147] Based on pedal morphology, Sarmiento inferred that gibbonish arboreal bipedalism might have been a common behavior of *Oreopithecus bambolii*.[142]

Like *Proconsul, Sivapithecus* lacks the distinctive postcranial traits of knuckle-walkers and acrobatic suspensory apes. However, some features indicate that, like *Dryopithecus*, they were well equipped for arboreal locomotion and foraging.[148] Moreover, postcranially *Sivapithecus* is more derived toward extant hominoid conditions than *Proconsul, Dendropithecus,* and other Eastern African Early and Middle Miocene Proconsuloidea and Dendropithecoidea.[149]

There is no postcranial axial bit of *Sivapithecus,* but Rose has inferred from the proximal humerus that its thorax was narrow, as in quadrupedal monkeys.[95] Unlike the humeral joint, the elbow region is much more ape-like, which suggests arboreal climbing in addition to quadrupedism. Carpal and manual phalanges of *Sivapithecus* are most like those of palmigrade, pronograde monkeys, though the ventral curvature of the phalanges recalls climbing and suspensory Hominoidea.[150]

Proximal femoral features indicate that *Sivapithecus* had mobile hips that would facilitate climbing and perhaps suspensory behavior.[87] The foot bones bespeak quadrupedism, and the robust hallux implies a powerful grasp of trunks, boughs, and branches.[95] Terrestrial quadrupedism also seems likely for *Sivapithecus parvada* and perhaps other large individuals.[151]

The upper limb specimens of *Lufengpithecus lufengensis* from the Late Miocene deposits at Shihuiba in Lufeng County, Yunnan Province, southern China, include associated left scapular and clavicular fragments from one individual and two manual proximal phalanges. The shoulder girdle is orangutan-like, and the phalanges, especially the notable longitudinal curvature, also suggest that *Lufengpithecus lufengensis* was firmly committed to arboreality.[152]

A finger bone of *Laccopithecus robustus* from the Lufeng site is very like that of a siamang.[153] Thus, it is reasonable to infer that *Laccopithecus* was highly adapted for arboreality, perhaps including suspensory behavior.[154] Pan estimated its body mass to be 10–12 kg, which is like modern siamang.[155]

In brief, the patchy fossil evidence of Oligocene and Miocene anthropoid postcranial anatomy is a very limited data set from which to proffer detailed comprehensive models of hominoid morphological and positional behavioral developments, particularly of ricochetal arm-swinging, pongid versatile climbing, knuckle-walking, and obligate bipedalism, which characterize extant Hylobatidae, *Pongo, Pan,* and *Gorilla,* and *Homo,* respectively.

The remains of *Oreopithecus bambolii* indicate that by 9 Ma at least one catarrhine primate had achieved many of the postcranial skeletal hallmarks of the Hominoidea. Interestingly, its lower back is not as reduced as those of extant great apes. Instead, it is more like those of extant Hylobatidae and the Hominidae.

Bits of other Eurasian Late and late Middle Miocene Anthropoidea suggest some development of postcranial features like those of modern apes, but no species is represented well enough to make overall comparisons with *Oreopithecus* and extant species. The better-known species of African Miocene dental apes, particularly of *Proconsul* and of *Nacholapihecus,* are either less derived toward hominoid conditions than *Oreopithecus* or are very poorly represented postcranially. In general, like the Oligocene anthropoids, they imply more basic quadrupedism and other arboreal behavior than occur in extant apes and that we infer for *Oreopithecus.* Regrettably, the Miocene roots of obligate human bipedalism are equally obscure.

6

Several Ways to Achieve Erection

> It is clear that the later the separation and the closer the
> relationship between man and the African apes, the more likely
> it is that our ancestors went through a stage of knuckle-walking.
>
> SHERWOOD L. WASHBURN (1968a, p. 14)

> The common ancestor of humans and chimpanzees was
> probably chimp-like a knuckle-walker with small thin-enameled
> cheek teeth. If correct, this scenario implies that known Miocene
> hominoids, most of which are postcranially archaic and have
> large, thickly enameled cheek teeth, throw little if any direct light
> on hominid origins.
>
> DAVID R. PILBEAM (1996, p. 155)

Bipedalism is not unique to humans, but our particular form of it is distinctive: while most other mammalian bipeds hop or waddle, we stride. Nonetheless, humans share remarkably similar locomotor neural circuits with other mammals, birds, and at least some reptiles.[1] We do not know for certain why it evolved. As mammals go, humans are not very fast runners: an Olympian four-minute miler runs only about 15 miles per hour, though Usian Bolt achieved a speed of 27.45 miles per hour (12.27 meters/second) in a record-setting 100-meter race.[2] A pedestrian walking regularly employs a speed of about 4.5 miles per hour and when walking as fast as possible achieves a speed of about 11 miles per hour.[3]

Because bipedalism frees the forelimbs from support functions, Darwin and later essayists linked it to tool use—especially to weapons.[4] This hit 'em-where-it-hurts hypothesis still has sympathizers, even after the debunking of the Piltdown forgery, and even though recent paleoanthropological evidence has uncoupled presumed links among bipedalism, stone tool use, and cerebral expansion.[5] The earliest uncontested evidence for bipedal Hominidae is the 3.66-Ma Laetoli G footprint trails, while the earliest solid evidence for tool use dates to 2.6 Ma, at Kada Gona, Hadar Ethiopia.[6] Hominid brain expansion probably began up to a half million years later.[7] If the claim that cutmarks on two bovid bones collected on the surface of 3.39-Ma deposits at Dikika, Ethiopia, are indeed that ancient, *Australopithecus afarensis* might be upgraded with the superlative: first stone-tool user.[8]

Nonetheless, other explanations on the emergence of bipedalism might be entertained. There is a marvelous variety of positional behavioral models to consider, including Keith's troglodytian brachiator, Morton's hylobatian brachiator, Washburn's knuckle-walker, and my hylobatian vertical climber, bipedal forager, branch runner, and leaper.[9]

Ups and Downs of an Ingenious Theory

Thomas Henry Huxley, the bulldog defender of Darwin's theory of evolution by natural selection, documented that humans and apes anatomically resemble one another more than any of them are like monkeys or prosimians.[10] Subsequently, Keith detailed further similarities among apes and humans on the basis of numerous dissections.[11] Further, inspired by his observations of wild gibbons in Thailand, Keith proposed that brachiation is responsible for the development of many postcranial features that humans share with apes. Keith's model has three successive stages: hylobatian (i.e., gibbon like), troglodytian (i.e., African ape like), and plantigrade (i.e., bipedal on the ground, like us).[12]

The upper limbs and torso figure prominently in arguments for a brachiating ape ancestry of the Hominidae.[13] The following features, most of which are diagnostic of the Hominoidea (Chapter 3), are cited to support brachiationist hypotheses:

- Long forelimbs relative to truncal length.
- Especially long forearms.
- Long hands, particularly the second-to-fifth fingers.
- Notable development of the thumb.
- Chests broader than deep.
- Broad single-unit sternum.
- Scapulae on the back of the thorax.

- Long clavicles.
- Laterally directed shoulders.
- Mobile shoulders and elbows.
- Large arm-raising and scapular muscles.
- Short trunk, particularly in the lumbar region.
- Reduced lower back muscles.
- Sacrum with more segments ($n = 4–8$) than cercopithecoid sacra ($n = 3$).
- No external tail.
- A muscular diaphragm in the pelvic outlet.
- A domed diaphragm between the thoracic and abdominal cavities.
- Visceral attachments to the diaphragm and posterior abdominal wall.[14]

Keith believed that the hylobatians possessed a full-fledged anatomical apparatus for brachiating and that the troglodytians were basically larger versions of them.[11] Eventually, some brachiating apes became too large for the trees. When they came to the ground, they adopted bipedal postures, and their lower limbs were transformed toward human morphology via selection for bipedal, plantigrade locomotion.

However, because ancestral protopongids continued to evolve in arboreal settings, they became top-heavy and thereby were committed to quadrupedalism both in trees and on the ground.[15] Consequently, Morton reasoned, emergent hominids must have been small-bodied creatures that lacked special postcranial features related to dramatic hylobatid brachiation. He imagined them to be gibbon sized but stockier than extant hylobatid apes, with forelimbs and hind limbs of nearly equal length and with well-developed, prehensile halluces and pollices.[16] They were tailless and probably lacked ischial callosities.

Morton's hylobatians were agile arboreal apes that sometimes brachiated in a chimpanzee manner but also, like modern gibbons, habitually moved bipedally in trees and on the ground. Accordingly, the protohominids that evolved from them were upstanding bipeds from the outset of terrestrial tenure; they did not pass through a phase of semierect quadrupedism on the ground.

The troglodytian brachiator theory, which was defended most eloquently by William King Gregory, was popular until about 1940, when Keith abandoned it.[17] Thereafter, it began to slip out of favor, partly because Rusinga *Proconsul* lacked many postcranial features that are associated with brachiation.[18] Nonetheless, Washburn carried on the brachiationist cause, with a bent toward very chimpanzee-like models.[19]

Data from long-term field studies threatened a coup de grace for troglody-tian brachiationist arguments. Although they confirmed that gibbons are brachiators par excellence, they showed that the African apes, particularly adult gorillas and chimpanzees, seldom locomote via brachiation. Instead, African apes travel between trees on terrestrial routes, and most orangutans traverse the forest canopy via versatile climbing.[20]

The Upwardly Mobile Hypothesis

Concurrent with revised notions about the naturalistic positional behavior of pongid apes, I discovered that special features of African ape hands could be explained as adaptations to their terrestrial knuckle-walking, instead of bra-chiation as per prior dogma.[21] Close study of orangutans, which are far more arboreal than African apes, showed that they lack the following manual fea-tures that characterize the knuckle-walking African apes:

- Foreshortened tendons of the flexor digitorum superficialis and flexor digitorum profundus muscles that flex the second to fifth digits.
- Metacarpal tori.
- Bony and ligamentous restrictions on wrist extension, adduction, and abduction.
- Pads of friction skin on the backs of the second to fifth digits (Figures 5.11, 5.15, 5.18, and 5.19).

Like orangutans and gibbons, human hands sport no trace of the knuckle-walking manual features. Still, looking toward the savanna and betting on the close genetic similarities between *Homo* and *Pan,* Washburn claimed that our ancestors were knuckle-walkers.[22] However, I looked to the trees and pro-posed instead that because all apes are climbers, climbing and other arboreal positional behaviors must have predisposed our ancestors for bipedalism.[23] Vertical climbing and bipedal foraging, bipedal running atop branches, and hind-limb-powered leaping across short gaps in the canopy were probably im-perative in the evolution of hominid bipedalism. Further, the protohominids probably were relatively small-bodied, about 10 to 15 kg.[24]

In brief, although hylobatian ancestors of the Hominidae probably did not closely resemble any of the diverse species of the Hylobatidae, they might have exhibited the postcranial and positional behavioral features of apes in the late pre-hylobatine phase of my model of hylobatid evolution, except that sometimes they walked bipedally on the ground instead of leaping and brachiating to traverse gaps in the canopy.[25]

The protopongids advanced more in the direction of general arboreal climbing and engaged in more suspensory behavior than our protohominid ancestors. Their forelimbs and chests enlarged, and their centers of mass were high in the torso. Consequently, when they came to the ground, they tipped forward and walked quadrupedally. Because their arboreally adapted fingers were long, they flexed them, thereby paving the way toward knuckle-walking adaptations in their forelimbs.[26]

Our stem ancestors were used to standing and running bipedally on branches and had lower centers of mass. Hence, they stood and moved bipedally on the ground, and, except during infancy, they rarely resorted to terrestrial quadrupedal locomotion.[27] During diurnal respites, they probably squatted or sat on the ground or on tree platforms.[28]

Other Upstanding Ideas

Additional explanations based on presumed behavioral and adaptive advantages of bipedalism include

- Carrying dependent young, tools, food, and other objects to a home base.[29]
- Terrestrial bipedal foraging.[30]
- Hand-assisted foraging, standing, and moving on small flexible branches.[31]
- Feeding while squatting on the ground.[32]
- Sitting and sleeping with the trunk erect.[33]
- Scavenging or hunting mammals.[34]
- Energetic considerations.[35]
- Thermoregulation.[36]
- Trekking after migratory herds.[37]
- Vigilance behavior, particularly peering over tall savanna grass.[38]
- Throwing projectile weapons.[39]
- Bipedal threat displays, mock fights, and appeasement.[40]
- Phallic display to attract breeding females or to intimidate other males.[41]
- Female display to attract mates.[42]
- Arboreal predation via stand-and-wait ambush.[43]
- Wading and other aquatic activities.[44,45]

Of course a combination of some, few, or many of these factors could be correct, though most of them would be impossible to trace via fossils. Carrying, scanning the environment, foraging, thermoregulating, or standing in an energetically efficient manner, walking, running, and trekking may be causally related to the innovation of hominid obligate bipedalism, but probably they were vital for the survival and further evolution of our genus at various points in the evolution from the first largely bipedal Hominidae to *Homo sapiens*.[46]

Mechanisms of Human and Ape Bipedal Stance and Locomotion

Because our pelvis is oriented like those of quadrupedal primates, human truncal erectness is chiefly a suprasacral accomplishment, focused in the lower back. It is evidenced by the anteriorly convex curvature of the lumbar vertebral column: lordosis (Figure 6.1).[47]

Illustrators often show the human bony pelvis—two hip bones, a sacrum, and a coccyx—unnaturally with the pubic symphysis as an anterior wall, the sacrum and coccyx as a posterior wall, and the ilia curving between them to complete a bottomless bony basin. In fact, like the pelves of pronograde quadrupeds, the human sacrum is the bony roof, and the pubis is part of the pelvic floor (Figure 6.2). One can simulate the natural bodily position of an articulated human pelvis by placing it against a vertical surface with the paired pubic tubercles and anterior superior iliac spines as contacts.

Transformations to the human pelvis and lower limbs entailed adjustments for energetically efficient bipedal posturing and leverage for rising, climbing, walking, and running solely on long, markedly extensible lower limbs that are hinged to a quadrupedal pelvis. The remarkable differences among human, pongid, and hylobatid pelvic limb skeletons (Table 6.1) reflect the difference between obligate use of the lower limbs for virtually all natural terrestrial locomotion and much of posturing versus a greater dependence on the forelimbs for positional behavior and negotiating arboreal habitats via heavy-duty, flexible hind limbs.

We stand with fully extended hip and knee joints, such that the thighbones are aligned with the leg bones to form continuous vertical columns. Healthy humans expend remarkably little muscular effort to retain a relaxed bipedal stance. Indeed, our sizeable gluteal, quadriceps femoris, and calf muscles are virtually silent while we stand quiescently. Instead, the human bipedal stance depends more upon the way the joints are constructed and upon strategically placed ligaments that hold the joints in position.[48]

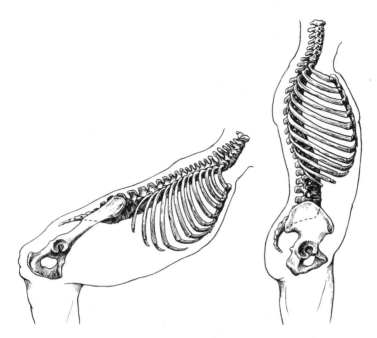

Figure 6.1. Like in apes and other quadrupeds, in humans the sacrum constitutes the roof of the pelvis. Thus, human obligate upright posture is largely presacral—that is, it is based on lumbar lordosis.

In a trim person standing fully erect, the line of gravity passes medianly through the body between the mastoid processes, closely in front of the shoulder joints, between or just behind the hip joints, and closely in front of the knee and ankle joints. The center of mass is located on this line at the level of the second sacral vertebra and posterior superior iliac spines (Figure 6.3).[49] While standing erect, one actually sways a bit, thereby eliciting activity in the soleus muscle, which is large and has more extensive proximal attachment in the legs of humans than in nonhuman primates.[50]

To walk, one simply tilts forward slightly and then keeps up with the center of mass, which is comparable to balancing a long stick vertically on the tip of one's forefinger. As the stick tips forward, one must move forward in order to keep it upright. The further the stick tips, the faster one must move to keep it from plummeting to earth.

The large muscle masses in human lower limbs power our locomotion, including climbing, and enable us to rise from squatting and seated postures (Figure 5.13). They include the gluteus maximus, gluteus medius, and gluteus

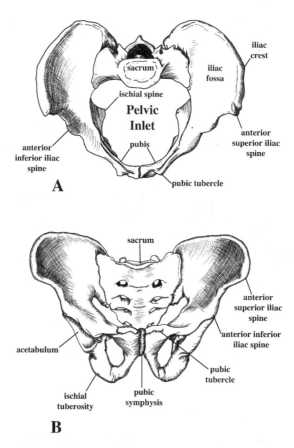

Figure 6.2. Human bony pelvis oriented (A) approximately as in bipedal individual and (B) as commonly portrayed in other books. Note that in anatomical position the sacrum is in the roof and the pubis is part of the pelvic floor.

minimus muscles at the hip joint; the quadriceps femoris muscle at the knee joint; and the triceps surae muscle at the ankle joint (Figure 6.4).

Superficially, our upper limbs are more like the forelimbs of apes than our lower limbs resemble their hind limbs.[51] This is the result of the special demands placed on the lower limbs of humans, who must keep the center of mass within a small area of support formed by the heels and toes. By contrast, great apes and other quadrupeds have a comparatively large area of support between their hands and feet, particularly relative to their natural quadrupedal standing height. One can illustrate this by sketching a quadrilateral around the feet of a standing person and comparing its area with one marked around the four cheiridia of a mounted medium-sized monkey skeleton or a taxidermic

Table 6.1. Contrasting features of the pelvic limb skeletons in apes and humans

Ape	Human
Elongate ilium	Short ilium
Moderately developed sacral area	Large sacral area
Long sacrum	Broad sacrum
Frontally oriented ilia	Medianly orientated, basinlike ilia
Great distance between sacral area and acetabulum	Sacral area close to acetabulum
Long ischium	Short ischium
Femoral head smaller than humeral head	Femoral head larger than humeral head
Short lower limb (except gibbons)	Long lower limb
Moderately developed ischial spine	Large ischial spine
Shallow sciatic notch	Deep sciatic notch
Unremarkable lateral patellar ridge	Well-developed lateral patellar ridge
Knee not aligned with hip	Knee more under trunk due to femoral condylar angle
Prehensile great toe	Adducted great toe
Long toes II–V	Short toes II–V
Mobile transverse tarsal joint	Rigid longitudinal arch

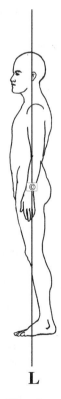

L

Figure 6.3. Approximate center of mass (©) and line of gravity (L) in *Homo sapiens*.

Fig-6.4. Dissection of a chimpanzee left leg with (A) medial and (B) lateral heads of the gastrocnemius muscle reflected to show the (C) soleus, (D) plantaris, and (E) popliteus muscles. A–C constitute the triceps surae muscle, which attaches to the calcaneus (heel bone) via the tendo calcaneus (Achilles tendon).

specimen. Although a human's center of mass is much higher than that of a monkey, their quadrilaterals of support are similar in area. As long as the center of mass stays within the perimeter of the supporting cheiridia, the creature should not topple over.

In fully quadrupedal monkeys, two-joint muscles are predominant in the hind limb (Figure 6.5A). They maintain flexure of the hip, knee, and ankle joints so that the limb acts effectively as a propulsive lever—analogous to paddling a canoe—wherein force is applied angularly on the lever. In order to move forward efficiently, the monkey must keep its back parallel with the ground or a branch via hip and knee joint flexures (Figure 6.5A).[52] During the final thrust, the hind limb acts as a propulsive strut.

In humans, one-joint muscles that extend the hip and knee joints are emphasized, which allows the lower limb to act as a propulsive strut—analogous

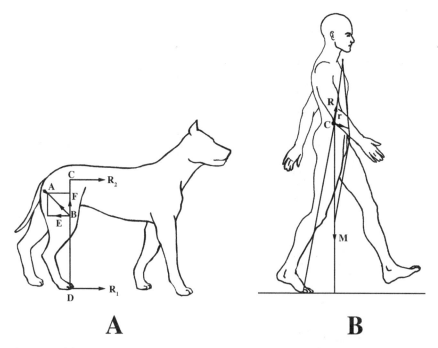

A   **B**

Figure 6.5. (A) Hind limb as a propulsive lever. Action of the hip extensor muscles in a quadruped's hind limb (*AB*) resolves as force *E* along *CD* to extend the hip and a force *F* along *AB*. Reaction of the ground acts against backward pressure of the foot (R_1), and the acetabulum acts against the backward pressure of the femur (R_2). Forces *E* and R_2 produce a force couple that extends the thigh at the hip joint. (B) Lower limb as a propulsive strut. When a human walks the resultant (*r*) of body mass (*M*) acting through the center of mass (*C*) and ground reaction force (*R*) propels the individual forward.

to punting a boat—wherein force is applied along the lever (Figure 6.5B).[52] A similar mechanism is likely useful for vertical climbing (Plate 4 right).[53] Apes are intermediate between monkeys and humans in the development of one-joint versus two-joint muscles of the hind limb.

When African apes walk and run bipedally, they retain notable flexure of the hip and knee joints, and the knee rarely, if ever, passes directly below or behind the hip joint. They also rather widely abduct and laterally rotate the thigh at the hip joint and shift their weight over the supporting foot in each step (Figure 6.6A).[54] Although the upper body sways from side to side when they run bipedally, the lateral motion of the center of mass is less in walking subadult chimpanzees than in human ambulators.[54] Tardieu documented that bipedal chimpanzees exhibit a tightrope walker's gait wherein they employ

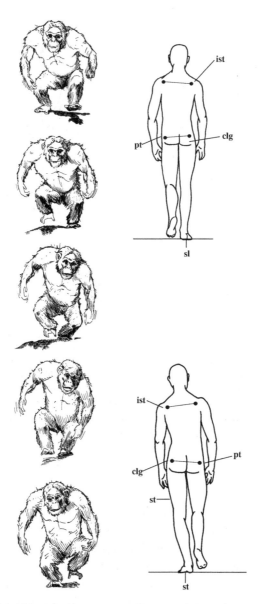

Figure 6.6. (A) Adult male chimpanzee running toward viewer (top to bottom). Note gross anterolateral shifts over alternate supporting feet. (B) Human pelvic tilt mechanism (pt). As one walks, the lesser gluteal muscles (clg) on the side of the supporting limb contract, thereby keeping the pelvis level or preventing depression to a point that lateral balance is lost. In some people, the ipsilateral shoulder (ist) tilts downward; in others, the shoulder remain level over the supporting lower limb (sl).

Figure 6.7. Female siamang walking rapidly.

large and irregular movements, particularly of their forelimbs, to maintain balance.[55] The chimpanzee alternately rotates its entire body about the hip of each supporting foot, which enhances the impression of unsteadiness vis-à-vis smooth human striding.[56]

Bipedal gibbons can exhibit fuller extension of their hind limbs and less truncal inclination than most African apes do during bipedalism on the ground, and some individuals move quite smoothly, albeit with notable lateral rotation and abduction of their thighs (Figures 5.8 and 6.7).[57] The energy loads on gibbon hind limbs are less when they run on the ground than when they run on branches or climb vertically.[58] Accordingly, an arboreal biped comes to the ground with an advantage toward adapting to terrestrial bipedalism.[59]

Orangutans sometimes extend their knees and overextend their thighs at the hip joints while standing and walking bipedally (Figure 6.8).[60] They throw their shoulders back and move rather stiffly like Frankensteinian cyborgs.

By contrast, people walk on nearly fully extended lower limbs; the knee passes behind the hip joint, and the upper body does not sway markedly from

Figure 6.8. Orangutan exhibiting capacity to fully extend its hip and knee joints.

side to side during alternate steps (Figures 6.5B and 6.6B). At most paces, human upper limbs alternately swing forward with an advancing contralateral lower limb. When walking very slowly, both upper limbs move pendularly forward with each step.[61] Although it is barely perceptible, we require a tad of flexion in the knee joint so that it is sufficiently mobile for us to walk smoothly. Accordingly, we are slightly shorter when we walk than when we stand. If you want to stand tall, stand still.

Unlike chimpanzees, humans synchronize the vertical and lateral oscillations of the center of mass, which lessens the amount of muscular effort required to walk smoothly for extended spans.[55] The economy of muscular activity during human bipedal walking has been demonstrated by electromyography. Individual muscles generally act momentarily to coordinate motion (Figure 6.9A).[62] They need not be active constantly in order to maintain the bipedal position against the force of gravity. In contrast, when apes move bipedally, most hind limb muscles are active for longer periods, probably to maintain the upright posture vis-à-vis gravitational forces (Figure 6.9B).[63]

Humans have a unique pelvic tilt mechanism that keeps us balanced over each supporting foot as we stride (Figure 6.6B). Unlike apes and other quadrupeds, our ilia curve forward so that the right and left iliac surfaces face one another. This places the lesser gluteal muscles—gluteus medius and gluteus minimus—lateral to the hip joint, where they act as abductors of the thigh at the hip joint. When we swing a foot forward, the lesser gluteal muscles of the contralateral hip contract, thereby preventing falling to the unsupported side. For example, when one steps forward with the left foot, the right lesser gluteal muscles contract. Although the pelvis of most people tilts downward somewhat on the unsupported side, the corrective action of the opposite lesser gluteal muscles limits the motion. Both females and males have a characteristic rocking motion of their hips while walking, which can be exaggerated or curtailed behaviorally and culturally but is an essential component of all regular human walking gaits.

Because the ilia of apes and other quadrupeds lie in the same plane, with the iliac fossae facing ventrally, the lesser gluteal muscles pass behind the hip joint and probably act as extensors of the thigh at the hip joint (Figures 2.7 and 5.13). The remarkable elongation of the ilium in great apes probably increases the power of the lesser gluteal muscles, which would serve them well when they climb vertically. Moreover, the great and the lesser apes have a sizeable extension of the gluteus maximus muscle—the ischiofemoralis muscle—extending from the ischial tuberosity far distally on the femur. It serves as a powerful extensor of the thigh at the hip joint. The modest remainder of the gluteus maximus muscle in apes might assist to abduct and to rotate the thigh at the hip joint laterally during vertical climbing.[64]

Humans have no ischiofemoralis muscle, and the gluteus maximus muscles are large and relatively compact blobs that attach near the hip joint (Figure 5.13). Accordingly, they can serve as powerful extensors of the thigh at the hip joint. They are more active when we climb steep gradients, rise from squatting postures, and run than they are during routine walking.[65] They can also be quite active during copulation, though this behavior is probably too variable and infrequent for them to be considered especially adapted for sexual intercourse.

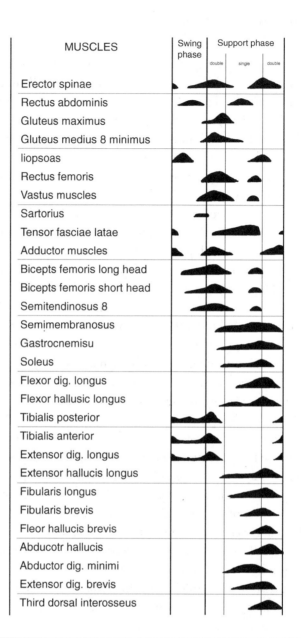

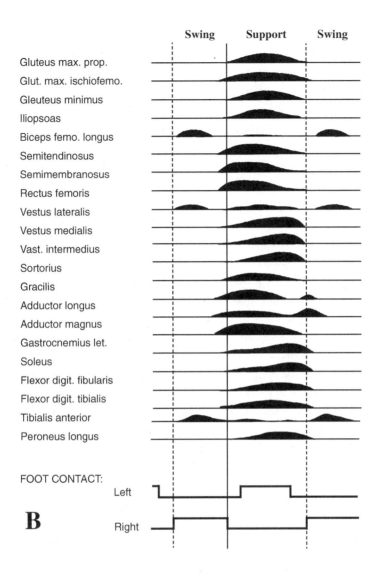

Figure 6.9. Muscle activity during support and swing phases of (A) bipedal human and (B) gibbon. Note phasic brief bursts in human muscles versus the sustained contraction of most muscles in gibbons during single support phase.

Our feet must bear our entire body weight; they are on the frontline against ground reaction forces when we stand, squat, and move. Their peculiar morphology reflects the special demands that are placed upon them.[66]

As we stand, our weight is evenly balanced between the feet. Within each foot, approximately half of each pedal load falls on the heel, and the other half is distributed among six contact points under the metatarsal heads. Paired sesamoid bones beneath the first metatarsal head and the heads of the lateral four metatarsals evenly share the load on the distal foot during our stance.

The axis of balance of the foot bisects the heel and sole longitudinally before passing between the second and third toes.[67] Squatting can shift notable weight from the pedal ball to the heel, thereby eccentrically loading it.[68]

When we walk, the pedal load shifts medially so that the leverage axis falls between the great and second toes.[69] After a heel strike, one's weight shifts quickly to the ball of the foot during the first third of the gait cycle.[70] Accordingly, human footprints are characterized by prominent heel and ball impressions (Figure 6.10C). Because the weight shifts medially during the stance phase, the great toe is the final contact of the foot before it swings clear of the ground. The hallucal pad is usually prominent in human footprints; commonly the lateral four toes leave fainter impressions. Because of the medial longitudinal arch, much of the central sole is unimpressive in human footprints. The lateral border of the sole might be evident but only lightly because the quick shift from the heel to the medial ball of the foot acts as a rigid lever.[71]

Running is the supreme mechanical challenge to the ball and toes, particularly the great toe, especially when we sprint quasi-digitigrade. Human top speeds are achieved more by applying greater support forces on the ground than by more rapidly repositioning the lower limbs in the air.[72]

The large human heel serves as a powerful lever for the triceps surae muscle, which plantarflexes the foot at the ankle joint during bipedal walking and running. But the robust heel probably also evolved to serve the postural functions of standing and squatting. African apes, particularly gorillas, also have robust heels, yet they rarely engage in bipedal locomotion. However, they squat for long periods of time to forage and to rest.[73] Squatting and bipedalism, versus sitting, increase the height that foragers can reach overhead, and squatters are better prepared for locomotion than they would be if they were sitting or reclining. Accordingly, calcanean robustness might have evolved in response to squatting, bipedal foraging, and short-distance bipedal travel before early hominids were fully adapted to obligate bipedalism and long-

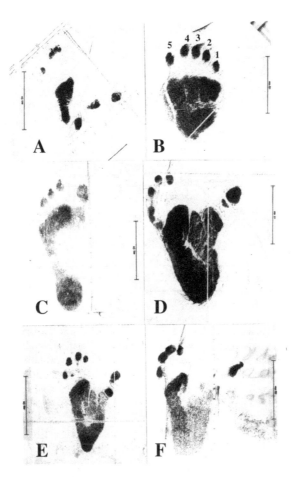

Figure 6.10. Pedal pressure points of left feet during bipedalism. (A) Lar gibbon. (B) Brown bear. (C) Human. (D) Gorilla. (E) Bonobo. (F) Chimpanzee. Note faint print of human 5th toe. In bears, the hallux is the shortest pedal digit.

distance travel. In this scenario, the toes might have retained arboreal features while the heel and ankle joint were more truly humanoid.[74] Elongate toes would be in jeopardy of dislocation and breakage during rapid running like that which characterizes *Homo sapiens*—aligned with the direction of travel—and perhaps also if the individual often rose onto the balls of the feet or toe tips to forage overhead.

Homo sapiens also appear to be adapted uniquely among primates for endurance running.[75] As we walk, we can conserve energy because our lower

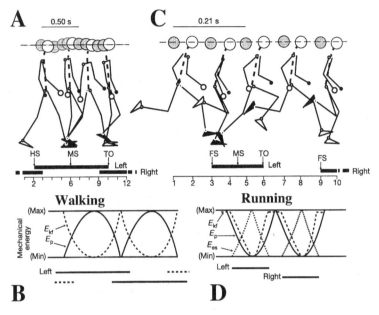

Figure 6.11. (A) When a human walks, the head and center of mass are lowest near toe-off (TO) and highest at midstance (MS) when the lower limb is relatively straight. (B) An inverted pendulum mechanism exchanges forward kinetic energy (E_{kf}) for gravitational potential energy (E_p) between heel strike (HS) and MS. (C) When a human runs, a mass-spring mechanism causes E_p and E_{kf} to be in phase. Lower limb tendons and ligaments convert decreases in E_p and E_{kf} to elastic strain energy during the first half of stance. It is released via recoil between MS and TO.

limbs, acting as inverted pendula, exchange forward kinetic energy for gravitational potential energy between heel strike and midstance when the limb is relatively straight (Figure 6.11A, B). The exchange is reversed between midstance and toe-off. When running, we rely on a mass spring mechanism, whereby tendons and ligaments partially convert gravitational potential energy and forward kinetic energy into elastic strain energy during the first half of stance phase and release it via recoil between midstrike and toe-off (Figure 6.11C, D).[76]

Human endurance runners can sustain speeds that are comparable to the preferred galloping speeds of cursorial quadrupeds, such as African hunting dogs *(Lycaon pictus)*, wolves *(Canis lupus)*, hyenas *(Crocuta crocuta)*, wildebeests *(Conochaetes* spp.*)*, and zebras and horses *(Equus* spp.*)*. Further, humans can employ a wide range of economical speeds while running continuously without changing their gait or metabolic penalty.[76] The bodily hairlessness,

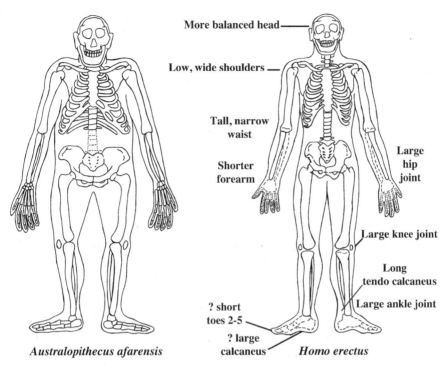

More balanced head

Low, wide shoulders

Tall, narrow waist

Shorter forearm

Large hip joint

Large knee joint

Long tendo calcaneus

Large ankle joint

? short toes 2-5

? large calcaneus

Australopithecus afarensis

Homo erectus

Figure 6.12. Features of 1.5-Ma *Homo erectus* that would underpin their potential for endurance running versus that of 3–4-Ma *Australopithecus afarensis*. Dashed lines and "?" indicate structures absent from the fossil record.

proliferation of eccrine sweat glands, ability to adjust breathing patterns without changing gait, and cranial venous sinus and emissary venous system of humans allow dissipation of metabolic heat and prevent the brain from overheating during physical exertion.[77] The structures and physiological features that allow humans to pursue slower game relentlessly until they can be dispatched probably developed during the emergence of *Homo* spp. (Figure 6.12).[78] Bramble and Lieberman also made the far-reaching suggestion that endurance running would allow scavenging *Homo* spp. to access carcasses before hyenas could get to them.[79]

Because humans lack a protective covering of body hair and sweat profusely during exertion and exposure to high temperatures, Newman hypothesized that emergent humans with similar characteristics probably inhabited well-watered tropical parklands and grasslands.[80] He further suggested that because of these specializations, the evolving species probably had a relatively limited geographic distribution until they could better protect themselves in a wider variety of niches via cultural innovations, including bodily coverings, containers to carry water, and control of fire.

Fossil Evidence for the Evolution of Bipedalism

Lovejoy et al. proposed that a 4.4-Ma partial skeleton (ARA-VP-6/500) assigned to *Ardipithecus ramidus* represents the last common ancestor of chimpanzees and hominids, albeit tilting it toward emergent Hominidae because of features indicating bipedalism and the absence of large projecting canine teeth.[81] The individual clearly lacked the manual features of knuckle-walkers, which supports my nontroglodytian model.[23] Claims that *Ardipithecus ramidus* did not engage in suspensory behavior and regular vertical climbing are on less secure footing, given that there is no substantive evidence from which to discern the anatomy of its pectoral girdles, and the humeri are not available for ARA-VP-6/500. Further the speculation that *Ardipithecus ramidus* had a long flexible lumbar region, unlike the relatively short, inflexible lower backs of extant Pongidae, is not supported (Figure 2.7). Indeed, there is no available description or illustration of lumbar vertebrae for the species.[81]

Purportedly *Ardipithecus ramidus* was a woodland dwelling, facultative terrestrial biped and a cautious climber/scrambler who subsisted predominantly on C_3 plants. They moved atop branches with palmigrade hands and plantigrade feet, gripping the substrate. However, one illustration has an upright *Ardipithecus ramidus* standing on a narrow branch with the second to fifth long toes extended along it as though on the ground; the long hallux limply draped over the side of the branch implies that the foot was basically a humanoid platform organ instead of a prehensile one.[82]

The 3.66-Ma footprint trails at Laetoli in Tanzania are the earliest definitive evidence for obligate hominid bipedalism. In all observable features of foot shape and walking pattern, the three creatures that made the trails are indistinguishable from modern habitually barefoot human beings walking at a leisurely pace (Figures 4.8 and 6.13).[83] Indeed, if the prints were undated or if they had come from a younger time period, they probably would be designated *Homo*.[84] That they were accomplished bipeds is beyond dispute because their regularly placed footprints ($n = 69$) extend over 27 meters of relatively open habitat with no hand impression anywhere along the trails.[85]

Although some commentators are convinced that the trackways were made by *Australopithecus afarensis* because the type specimen and other fragmentary

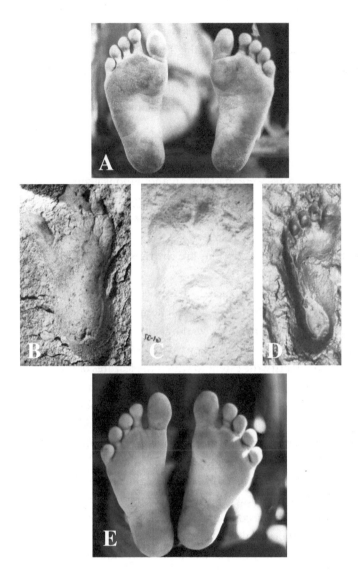

Figure 6.13. (A, E) Soles of short never-shod Amazonian people. Prints in mud made by a wild eastern chimpanzee (B) and (D) a never-shod Amazonia person. (C) A cast of a print of 3.66-Ma hominid G-1, Laetoli, Tanzania. Note the marked similarity among A, C, D, and E versus B due to the widely abducted hallux, the long, curved lateral toes, and the absence of a medial longitudinal arch in B.

gnathic and dental remains are from the same time span at Laetoli, none were at the footprint locality. Confounding the definitive assignment of the Laetoli G prints to *Australopithecus afarensis* is the fact that the foot bones assigned to the species from Hadar, Ethiopia, are unlikely to have come from feet that would make such humanoid prints.[86] Several optional conclusions are possible:

- There was a species of *Australopithecus* with virtually human feet.
- *Homo* was present at Laetoli around 3.66 Ma.
- The trackways were made by an anonymous genus for which we have no identified skeletal remains; accordingly, they should be designated Hominidae *genus* and *species indeterminata* or *Ichnanthropus bipes*.[87]

The first and the third options are the most likely, given that the trackways date to 3.66 Ma.

Although their feet appear to have been remarkably human, it would be foolhardy to conclude that the Laetoli hominids regularly stood for long spans, walked, and ran as *Homo sapiens* do, with virtually full extension of the lower limbs. We do not know what they were like beyond their soles, and researchers have yet to quantify the extent to which the features of human feet are particularly adapted to standing versus walking and running. Further, in order to determine that their lower limbs were notably humanoid, we need an extension of the Laetoli G trails that evidences longer strides and a sequence of pedal ball and toe prints with no heel impression—that is, evidence of sprinting.[88]

Laetoli hominid G foot indices that indicate that the pedal length versus pedal breadth fit comfortably within a global sample of human foot indices.[84] The hallux is aligned with the lateral four toes, and the interdigital gap between it and the second toe is quite human, particularly when compared to the nonpathological feet of persons who have never worn constraining footgear (Figure 6.13).[89] The lateral four toes are arrayed relative to the hallux and to each other as in a modern human foot, approximately 30 percent of total foot length, which is not significantly different from mean relative toe length of Peruvians of precolonial ancestry and Tanzanian Hadzabe.[89]

The Laetoli hominid G prints evidence a medial longitudinal arch. Apparently, the transfer of body weight during bipedal walking was humanoid: from robust heel strike, more lightly along the lateral sole, then more heavily medially across the ball of the foot so that the brunt of toe-off was borne by the hallux, which, unlike the lateral toes, regularly left prominent impressions in the substrate.[87]

The Laetoli hominid G footprint trails provide important information not only on the pedal morphology of their makers but also about the way they walked. Because their step lengths are short compared with the statures inferred from foot length, we are confident that the Laetoli hominids walked quite slowly in the moist volcanic ash.[90] Regrettably, we cannot measure their actual speeds or know how often they might have paused.[87]

Observable features of gait (foot angle, step length, stride length, stride width, relative stride width, foot lengths per stride) and inferred features of gait (relative stride, velocity, relative speed, cadence) based on studies of never-shod Peruvians, Tanzanian Hadzabe, and captive bipedal apes fail to place the Laetoli G printmakers unequivocally with either humans or apes.[91]

Moreover, inferences about the relative humanness of Laetoli G hominid gaits are confounded by our ignorance of their actual statures and bodily proportions. Because the morphology of their feet is so like that of *Homo sapiens,* Tuttle et al. used the regression of human stature on foot length to predict the statures of G-1 (122 cm) and G-3 (141 cm).[92] Yet it is unlikely that the Pliocene Laetoli hominids were built like modern humans. They might have had relatively shorter lower limbs (but perhaps relatively longer lumbar regions) than we do, and surely their neurocranial heights were lower than ours because of their smaller brains.[93]

The most striking difference between the gaits of G-1 and G-3, and by inference G-2, is their foot angles. Whereas G-1 out-toed and exhibits notable anisometry in its foot angles (mean: 17° left, 29° right), G-3 sports modest in-toeing and alignment of its feet with the line of progression and isometry in its foot angles (Figure 6.14). Indeed, G-1 is more anisometric in foot-angling than any person among sixty-nine never-shod Peruvians and fifty-four Tanzanian Hadzabe.[94]

The right lower limb of G-1 might have exhibited a pathological condition, likely a tibial torsion from an imperfectly healed fracture.[84] Although this could explain the slow pace of the three Laetoli G hominids, assuming that they were traveling together and adjusting their speeds to that of G-1, there are less dramatic explanations for their moderate progress. In fact, although G-2 must have followed G-3, we cannot discern whether G-1 preceded or followed them or whether any of them passed within sight of one another.[87]

The most accessible evidence against the pathology hypothesis is the absence of any indication that G-1 limped; that is to say, there is no anisometry in its step and stride lengths. In a modern clinical context, anisometry between steps

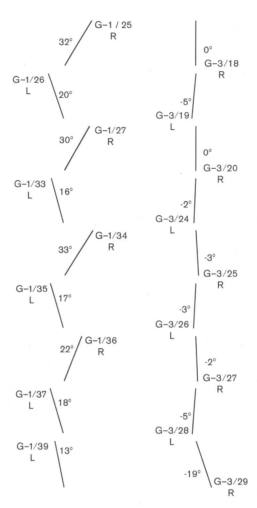

Figure 6.14. Left (L) and right (R) foot angles of Laetoli hominids G-1 and G-3. The two patterns are dramatically different (but within the range of *Homo sapiens*). Note the feet of G-1 are markedly out-toed, and those of G-3 are slightly in-toed. The right lower limb of G-1 might have had a tibial fracture or other pathology.

on the normal and painful sides, with the latter being longest, is only 1 cm.[95] This small difference might not show up in a series of prints as brief as the section of the trail that I studied.[96]

Alternatively to pathology, variations in or objects on the substrate might have induced G-1 to alter its gait. Although G-1 maintained its left foot at about the same angle over a series of eight steps, it did vary the angle of its acutely out-toed right foot. Concurrently, it seemed to slow down—indicated by decreasing step and stride lengths—and sometimes widened its stride width. Modern people employ all of these tactics to avoid slipping.[87]

It is unlikely that G-1 walked snugly side by side with G-2 or G-3. Variations in the gaits that are exhibited by the two trackways are sufficient to argue that G-1 was not intimately in contact with either G-2 or G-3. Because they were walking in an exotic substance, it is unremarkable that the two trails are quite close together and that one individual (G-3) attempted to walk in the tracks of a predecessor, G-2.[87]

The sexes and ages of the three Laetoli individuals cannot be inferred from available evidence. Accordingly, actual sexual dimorphism of the species is inaccessible. However, assuming that G-1 is an adult female and G-3 is an adult male, their inferred statures suggest that, like modern humans and chimpanzees, the species was moderately dimorphic (about 86 percent) in body size.[97] However, if Robbins's guesstimate of 175 cm for the stature of G-2 is correct, one could speculate that G-3 and G-2 are an adult male and an adult female, respectively, and that G-1 is a subadult.[98] Accordingly, the species remains moderately dimorphic sexually (approximately 82 to 87 percent). Optionally, if we assume that G-1 represents an adult female, G-2 an adult male, and G-3 a subadult male, the species was quite dimorphic sexually (approximately 70 percent), though falling short of conditions exhibited by extant orangutans and gorillas.[87]

The remarkable humanness of the Laetoli G footprints contrasts with the apelike features of the Hadar foot bones (Figure 6.15). Indeed, the longish, curved toes of the 3-Ma Hadar feet and features indicating absence of a medial longitudinal arch make them unlikely candidates for the species that made the Laetoli trails.[99] Therefore, I advise against ascribing the footprints to *Australopithecus afarensis*.

The lack of consensus about what kind of feet made the Laetoli G hominid footprint trails is fueled by the following factors:

- Conditions of the prints when they were originally excavated and weathering immediately thereafter were followed by reburial and further damage due to subsidence and growth of vegetation atop them.
- After re-excavation and before the second reburial, relatively few experts and commentators had access to them.
- Bias persists about what the feet of 3.66-Ma Laetoli hominids should look like, given that foot bones of broadly contemporaneous *Australopithecus*

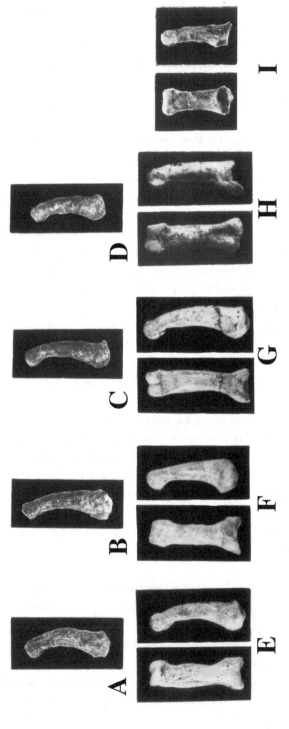

Figure 6.15. Hadar pedal phalanges attributed to *Australopithecus afarensis*. (A–D) Proximal phalanges II–V from one individual. (E–H) Isolated proximal phalanges. (I) Intermediate phalanx. Note the curvature of the shafts and flexion set of the distal ends of the proximal phalanges.

from Ethiopia and South Africa exhibit notable apelike and unique features.

• Students of the puzzle often focus on a few of the Laetoli print casts that seem to exhibit apelike versus humanoid features.

• Postcranial, and especially pedal remains of Laetoli Pliocene hominids that could be compared with those of *Australopithecus* from Ethiopia and South Africa have not been found.

• The relatively small individual that made the most distinct prints might have had a right lower limb pathology that affected its gait, leading to dramatically out-toed foot placement and hallucal drag marks, which Deloison, in particular, inferred to support the ape-foot concept.[100]

It is doubtful that agreement will be reached about specific features of the feet that made the Laetoli G trails. Accordingly, they are of limited value for modeling the suprapedal anatomy and fine details of how the individuals walked in 3.66 Ma. Because of a second flush of studies and commentaries, I reexamined the large set of casts that Mary Leakey provided for me to use when I prepared my first account of the Laetoli G hominid footprints.[101] There is no compelling reason to alter my earlier conclusions about the overall humanness of the feet that produced them. Further experiments to tease out clues to their gaits might be somewhat more productive than those that have been conducted to date.[102]

Whereas the feet of the Laetoli hominids were well adapted for obligate bipedalism, those of the broadly contemporaneous Hadar hominids retained telltale features related to climbing into trees.[103] The proximal and intermediate phalanges of the second to fifth toes are longish and curve downward like those of arboreal apes and monkeys. However, the structure of the metatarsophalangeal joints indicates the capacity to extend the toes dorsally, which would facilitate a plantigrade foot posture on the ground despite ventral curvature of the toes.[104] Further, Latimer et al. interpreted the ankle joint of the Hadar hominids to be fully capable of supporting sustained bipedal posture and locomotion.[105]

The hallucal bones are more robust than those of the lateral four toes, and there might have been contact between the base of the hallux and that of the second toe. Grooves on the backs of the distal fibulae suggest a well-developed peroneus longus muscle that facilitates a powerful hallucal grip in climbing apes and everts the foot and supports the longitudinal arch in humans.

The proximal articular surface of the hallucal metatarsal bone in *Australopithecus afarensis, A. africanus,* and *Paranthropus robustus* is more apelike than human, but that of *Homo habilis* (OH 8) is indistinguishable from that of *Homo sapiens*.[106] Curiously, a complete fourth metatarsal bone from Hadar, Ethoipia, of *Australopithecus afarensis* indicated to C. Ward et al. that it had a humanoid transverse arch, lacked a flexible midtarsal joint, and probably functioned within the foot like that of *Homo sapiens*.[107] Given Marchi's conclusion that second-to-fourth metatarsals are not useful to diagnose the specific locomotor habits of extant hominoids, the brazen declaration of Ward et al. that *Australopithecus afarensis* made the Laetoli site G hominid footprints is unwarranted.[108]

Other fossil hominids from Hadar, including A.L. 288-1 (nicknamed Lucy) sport additional features that might be linked to tree climbing, though they also manifest features that signal bipedalism.[109] The apish arboreal features include curved finger bones, a long pisiform bone, and laterally flared ilia.[103]

In 2000–2003, Alemseged et al. recovered the skull and associated scapulae, clavicles, cervical, thoracic, and upper two lumbar vertebrae, ribs, humeral fragments, manual metacarpals and phalanges, femoral and tibial fragments, patellae, and foot bones of an approximately 3-year-old *Australopithecus afarensis* from a 3.3-Ma locality (DIK-1) in the Sidi Hakoma Member of the Hadar Formation south of the Awash River. Many of the anatomical elements were articulated.[110] Alemseged et al. inferred that the leg and foot bones indicate bipedalism and that the long, curved manual phalanges, scapular shape, and wide olecranon fossa attest to climbing abilities.[110] Associated fauna indicate that the mesic deltaic habitat at DIK-1 was a mosaic of wooded areas and more open grasslands with a permanent water source nearby.[111]

Like the toe bones, the ventrally concave curvature allows the fingers to conform to the rounded surfaces of tree trunks, branches, and vines, and the marked lateral ridges (and inferred robust fibrous flexor sheaths) on the proximal and intermediate phalanges indicate strong grips by the long digital flexor muscles. The pisiform bone also can be interpreted as adapted to climbing because it resembles that of a chimpanzee instead of the puny pea present in humans. In the Laetoli hominids, the flexor carpi ulnaris muscle, in whose tendon of distal attachment the pisiform bone lies, probably acted as a powerful adductor and flexor of the wrist during vertical climbing and hauling activities.[103]

The pelvic structure of A.L. 288-1 comprises a relatively broad short sacrum that is more humanoid than like the great apes. The hip bone has a broad short ilium and clearly defined anterior iliac and ischial spines, femoral intertrochanteric lines, and other features that are markers of human bipedal stance (Figure 6.16). However, the orientation of the laterally flared iliac blades suggests that the gluteal muscles were more adapted for climbing

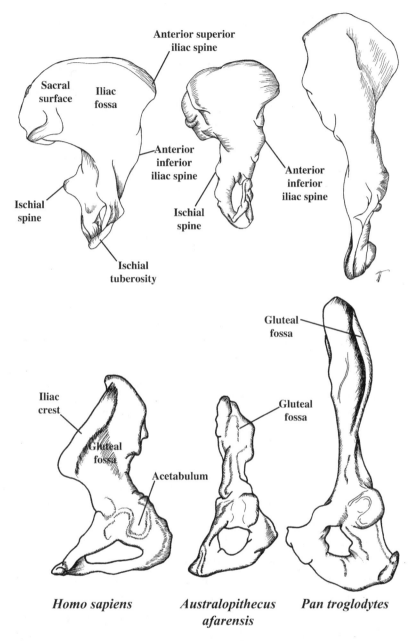

Figure 6.16. Hip bones of *Homo sapiens*, *Australopithecus afarensis* (A.L. 288–1), and *Pan troglodytes*. Note that the orientation of the iliac blade of *Au. afarensis* is more like that of *Pan troglodytes* than that of *Homo sapiens*.

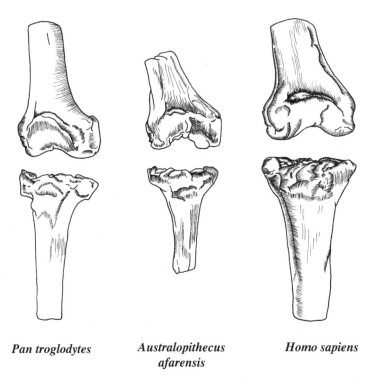

Pan troglodytes   **Australopithecus**   **Homo sapiens**
afarensis

Figure 6.17. Right femur and proximal tibia of *Pan troglodytes, Australopithecus afarensis* (AL 129–1) and *Homo sapiens*. Note that the femoral shaft is aligned with the tibial shaft in *Pan,* but it will sit at an acute angle on the tibial plateau in *Au. afarensis* and *Homo sapiens.*

vertical substrates and other activities that required powerful extension of the thigh at the hip joint than for humanoid bipedal striding and running.[112]

Distal femora and proximal tibiae testify to the capacity of the Hadar hominids to fully extend the knee joint (Figure 6.17).[103] Unlike the condition in apes, the Hadar hominid distal femur exhibits a condylar angle that positions the knee, leg, and foot narrowly beneath the broad pelvis. Hadar hominid knees were not totally like those of humans: they are characterized by a single attachment of the lateral meniscus on the tibial plateau instead of a dual attachment like those of *Homo,* which could have affected stability of the joint during some activities and facilitated it in others.[113]

As modeled by Schmid, the torso of A.L. 288-1 was remarkably pongid in having a funnel-shaped thorax, though the lumbar region probably comprised six vertebrae that, like the thoracic vertebrae, were pathological.[114]

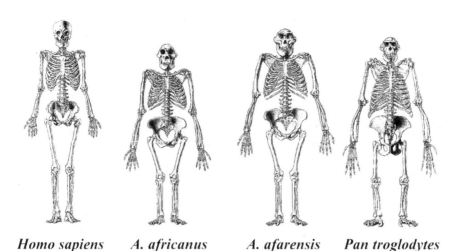

Homo sapiens   A. africanus   A. afarensis   Pan troglodytes

Figure 6.18. Artist's models of skeletal morphology of *Homo sapiens, Australopithecus africanus, Australopithecus afarensis*, and *Pan troglodytes*.

The scapular fragment of A.L. 288-1 is peculiar among hominids in having a relatively piriform instead of round glenoid cavity and sporting a projecting supraglenoid region that recalls conditions in cercopithecid monkeys. The glenoid cavity appears to have faced somewhat cranially at least as much as in humans, but probably not to the extent that characterizes the scapulae of the apes and atelin monkeys that engage in forelimb suspensory behaviors.

A scapular fragment assigned to *Australopithecus africanus* appears to have had the glenoid fossa oriented upward to a degree comparable to that of apes, particularly gibbons or orangutans, which is quite distinct from those of Pleistocene and recent hominids.[115] Indeed, hominid scapular evolution is very poorly documented before *Homo neanderthalensis* and *H. sapiens,* though 1.5-Ma remains of *Homo erectus* (or *Homo ergaster*) from West Turkana, Kenya, are very similar to those of *Homo sapiens.*

In other known parts, South African australopithecine postcranial anatomy does not differ markedly from that of *Australopithecus afarensis,* except in the relative proportions of the upper and lower limbs (Figure 6.18). *Australopithecus africanus* has shorter lower limbs relative to its upper limbs than the condition in *A. afarensis.*[116] Accordingly, although a younger species than *Australopithecus afarensis,* whose upper-to-lower limb proportions are closer to those of *Homo habilis, A. africanus* appears to have the more apelike relative

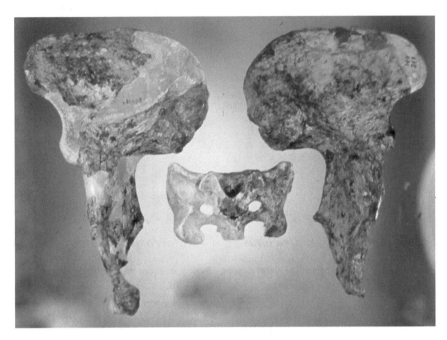

Figure 6.19. Hip bones and sacral fragment of *Australopithecus africanus* (Sts 14) from Sterkfontein, South Africa. Compare with Figure 6.16.

limb length.[117] Nonetheless, the lengths of the upper limb bones of *Australopithecus africanus* fall within the range of variation of *Homo sapiens* except for the brachial index, which indicates that they had either relatively longer forearms or shorter arms than those of *Homo sapiens*.[118]

The estimated humerofemoral index of *Australopithecus garhi* from 2.5-Ma deposits in Ethiopia is closer to that of modern *Homo* than the other species of *Australopithecus* are, but the brachial index (Radial length × 100/ Humeral length) is apelike.[119]

Models of the pelves of the Sterkfontein *Australopithecus africanus* (Sts 14) and *Australopithecus afarensis* (A.L. 288-1an, -1ao) are essentially similar regarding features related to posture and locomotion, but they have different shapes of the cavity that in females serves as a birth canal (Figures 6.16 and 6.19). Accordingly, either they were of different sexes or their fetuses might have birthed with different positions of the head.[120] Sexual dimorphism in lumbar vertebrae of *Homo sapiens* and *Australopithecus africanus* suggested to Whitcome et al. that over at least 2 Ma bipedal hominids had undergone strong selective pressure related to the forward shift of the center of mass in pregnant females.[121]

Postcranially, *A. sediba* is similar to other *Australopithecus* spp. in having small body size (30–33 kg), long upper limbs with large joint surfaces, cranially directed glenoid fossa, and primitive foot structure (Figure 4.14). The proximal and intermediate phalanges are long, robust, and curved toward the palm, and the intermediate phalanges attest to a strong attachment of the flexor digitorum superficialis muscle, all of which signal that *Australopithcus sediba* was a capable arboreal climber. However, the hip bone sports an enlarged area for attachment of the iliofemoral ligament, increased buttressing of the ilium, reduced distance between the sacroiliac and hip joints and between the ischial tuberosity and acetabulum, and strong femoral diaphyses. Berger et al. speculated that collectively features of the upper and lower limbs, thorax, and vertebral column indicate that, in addition to arboreal climbing, like *Homo erectus*, *Australopithecus sediba* engaged in more energetically efficient walking and running than that of other *Australopithecus* spp.[122]

By 1.8 Ma some hominids had evolved into obligately bipedal species.[123] Regrettably, postcranial bones that would provide clues to the important transition period between australopithecine and hominine bipedalism are rare indeed. Some of the Olduvai postcranial bones that are attributed to *Homo habilis* and *Homo erectus* and pelvic bones of Pleistocene *Homo* spp. from Eurasia, Israel, and Zambia are not fully like those of modern *Homo* sapiens.[124]

It is moot whether the seemingly minor variant features in the hip bones of Pleistocene *Homo* spp. signal a different pattern of bipedalism or of obstetric challenges or adaptation to local climates and more rugged lifeways.[125] The differences between the upper and lower limb bones of *Homo habilis* (as represented by OH 62) and *Homo erectus* (represented by KNM-WT 15000 and KNM-ER 1808) indicate more clearly that *Homo habilis* frequently engaged in arboreal behavior, though bipedal when on the ground, while 1.5-Ma *Homo erectus* were obligate terrestrial bipeds like *Homo sapiens*.[126] This adds to the case for moving specimens of *Homo habilis* to *Australopithecus* as *Australopithecus habilis*.[127]

A female pelvis of approximately 1 Ma from the Busidima Formation at Gona in the Afar Regional State of Ethiopia is sufficiently complete to challenge the inference of Bramble and Liberman that *Homo erectus* was adapted for endurance running.[76] Further, the Busidima habitat was semiarid and dominated by C_4 grasses and grazing herbivores.[128] The birth canal was capacious enough to accommodate passage of a fetus with a brain in the range of

Homo erectus, with discomfort like that which challenges female *Homo sapiens.*[128] Ruff's study of the Gona pelvis indicated that the specimen has acetabular and estimated femoral head dimensions that are closer to those of *Orrorin tugenensis, Australopithecus africanus,* and *Paranthropus robustus* than to those of *Homo* spp.[129] Further, his estimate of the individual's body mass—33.2 kg—is far outside the ranges of *Homo.* Accordingly, the Gona specimen is unlikely to belong to *Homo erectus,* which is simply too large to accommodate it without the species being more dimorphic sexually than any other hominid taxon.[129]

Although the lower limbs of the Dmanisi hominids are long and their foot bones indicate a capacity for long-distance bipedal locomotion, the upper limb and shoulder evince differences from *Homo sapiens.* Some of the features are reminiscent of conditions in the humeri and the model of the crushed pelvic girdle assigned to *Homo floresiensis.*[130]

7

Hungry and Sleepy Apes

Human life histories, as compared to those of other primates . . .
have at least four distinctive characteristics: an exceptionally long
lifespan, an extended period of juvenile dependence, support of
reproduction by older post-reproductive individuals, and male
support of reproduction through provisioning of females and
their offspring . . . Those four life history characteristics and
extreme intelligence are co-evolved responses to a dietary shift
toward high-quality, nutrient-dense, and difficult-to-acquire food
resources.

HILLARD KAPLAN, KIM HILL, JAND LANCASTER,
and MAGDELENA HURTADO (2000, p. 156)

Because of variation in individual groups over time . . . and in
different groups of the same species . . . more studies on each
species must be conducted before any conclusive statements
about apparent interspecific differences can be made.

DONNA ROBERTS LEIGHTON (1987, p. 135)

Like the plight of Earth's millions of homeless and displaced persons, the
daily quest for nutriment and safety from predation greatly determines the
daily rounds of forest apes. As heterotrophs, hominoids must forage for foods
that will provide them with sustaining energy for basic metabolic processes,
cell replacement, body growth, and, in females, gestating fetuses, lactation,
and transporting dependent youngsters.[1] On balance, daily foraging should
not often require more energy than is required to fuel the daily quest, and one
also must avoid predators, lengthy and potentially injurious altercations with
competitors, weak substrates, deep water, and places infested with parasites

and microbes. There are daily challenges in good times, spells of low food availability, and epidemics that can be lethal.

Data on the naturalistic diets of apes can be obtained in several ways:

- Observations of foraging subjects.
- Collection of leavings at foraging sites.
- Analysis of stomach contents from dead subjects.
- Analyses of urine and feces, including gross examination and DNA sequencing of undigested animal and plant fragments.[2]
- Stable isotopic analyses of bone, teeth, hair, and other tissues to indicate the kinds of plants and animals eaten.[3]

Much additional information can be provided via chemical analyses of food samples, which produce essential and quantitative data on nutrients, including water content, digestibility, fiber content, and potential toxins.[4] But plant parts vary individually, seasonally, and even over the course of a day in the availability of nutriments and the presence of self-protective toxins.[5] Also, the different parts of a plant vary in chemical content.[3] Although we have made notable advances in knowing why apes eat what they do, we fall short of understanding in detail how their feeding habits relate to their specific physiologies, health, and energy budgets.[6] But this is also somewhat true for the most thoroughly studied primate, *Homo sapiens,* particularly from an evolutionary perspective.[7]

In general, fruit pulp and edible seeds are sources of caloric energy via carbohydrates or lipids. Leaves, stems, flower buds, and seeds provide protein for growth. Insects contain protein, amino acids, and lipids.[8] Depending on the plant part, a notable amount of protein might not be available to an animal because it is bound to fiber.[9] All plant foods contain water, and some parts may be rich in vitamins and minerals. For example, relative to other forest fruits, on average, figs contain three times more calcium and have a calcium-to-phosphorus ratio sufficient to promote bone growth.[10] Figs provide fluid and quick energy in the form of carbohydrates, and they contain amino acids, protein, and lipids when infested by fig wasps and their parasites.[11]

Digestion begins in the mouth and continues in the stomach, small intestine, cecum, and colon, with each organ adapted to process specific components of the plants and animals that have been chewed and swallowed. Monosaccharide sugars (fructose and glucose) are readily absorbed by the small intestine without enzymatic action, but disaccharides (such as sucrose) are hydrolyzed into simple sugars in the small intestine. Lipids are emulsified by acids in the stomach and are further digested and absorbed in the small intestine.[12]

Amylase in the mouth and small intestine releases glucose units of polysaccharide starch in seeds and other structural plant parts for absorption in the small intestine. In the ceca and colons of anthropoids with simple stomachs, and in the first of the four chambers in the stomachs of colobine monkeys, microbial fermentation of the polysaccharide cellulose, hemicellulose, lignin, and pectin that form the walls of plant cells produces short-chain fatty acids that are absorbed into the bloodstream (Figure 2.5).[13]

Simple sugars and short-chain fatty acids in the bloodstream can serve as quick sources of metabolic energy or can be stored as glucose in the liver. Similarly, absorbed lipids can be used immediately or can be stored in the liver and many other fat depots throughout the body.[12]

Lambert has cautioned against the overly simple assumption that the major types of food items are eaten specifically for their macronutrient contents without the chemical testing to establish that they contain significant amounts of those nutrients.[14] She cited a feeding ecological study on a colobine *(Piliocolobus rufomitratus tephrosceles)* and a cercopithecine *(Cercopithecus ascanius)* in Kibale National Park, Uganda, that indicated that the leaves eaten by *Piliocolobus* contained more sugars than the fruits eaten by *Cercopithecus.*[15]

Another major challenge for primate socioecologists is to identify and to determine the effects of preferred versus staple and seasonal fallback foods on foraging behavior and group structure and dynamics.[16] Apes and other vertebrates adjust their daily and seasonal ranges, their feeding, and the traveling substrate and composition of their foraging parties according to the availability of their favorite and fallback foods.[17]

Generally, fallback foods are more abundant and of poorer nutritional quality than preferred foods. Accordingly, their use is negatively correlated with the availability of preferred foods.[18] One should further distinguish whether items are staple fallback foods—available and eaten year-round, with the potential to constitute 100 percent of a seasonal diet—versus filler fallback foods, which never constitute 100 percent of the diet.[18]

The first half-century of field studies on apes (1935–85) supported the following generalizations about the feeding habits of the apes, much of which holds true today: the Asian apes, chimpanzees, and bonobos are heavily frugivorous, with a special predilection for or periodic reliance on figs in the regions where they are available.[19] Chimpanzees and bonobos are somewhat more euryphagous: they eat a wider variety of foods than the Asian apes and are open to more variation than the Virunga mountain gorillas. In addition to a wide variety of fruits, arthropods, vertebrates, eggs, honey, and resins, chimpanzees eat many leafy, pithy, and fibrous foods, though not to the extent that some gorillas do.[20] In humans, fermentation of plant fiber by microbes in the colon produces short-chain fatty acids that provide metabolic energy.[21] One might reasonably expect that the same is true in apes and other mammals.[22]

Chimpanzees can also subsist chiefly on quantities of rather hard seeds and desiccated fruits—reminiscent of trail mix (gorp)—over notable periods of time. With the exception of Virunga mountain gorillas, the apes subsist on a remarkable variety of plant species and parts. Because of seasonal fluctuations in the availability of preferred plant species, their subsistence staples differ dramatically from month to month and season to season.[23] Chimpanzees and gorillas living near fields and plantations forage cultigens, thereby incurring the wrath of farmers.[24]

Mountain gorillas are heavily herbivorous, while western and Grauer's gorillas have a taste for fruit and will forage for them when they are accessible.[25] Indeed, small samples of captive *Gorilla gorilla* ($n = 6$) and *Pan troglodytes* ($n = 4$) preferred foods high in nonstarch sugars and with high sugar-to-fiber ratios and low in total dietary fiber, and they did not avoid foods containing tannins.[26] Further, tests have indicated that chimpanzees and gorillas have broadly similar senses of taste.[27]

Like the savanna-woodland chimpanzees at Fongoli in Senegal, some western gorillas seek insects, even ranging into the interior of savannas to forage at termite mounds.[28] Year-round, at Bai Hokou in Central African Republic, termites are the most commonly tallied food in the pooled diets of two monitored groups. Fecal samples of silverbacks contain proportionately more insect remains than the samples from females and juveniles. Displacements at termite mounds occurred more frequent than by chance. The gorillas did not employ tools; instead, they simply broke open the termite cells manually and licked out the prey. Feeding on termites was more common by the group living where the mounds were of higher density.[29]

Mountain gorillas have larger teeth, higher postcanine crowns with sharper cusps and transverse ridges, longer, more robust mandibles, and more developed jaw muscles and associated bony cranial morphology than western gorillas; Grauer's gorillas are intermediate between them. This might reflect the fact that mountain gorillas are more consistently herbivorous than the other two taxa, with Grauer's gorillas in montane forest being more herbivorous than the western gorillas.[30] Similarly, northeastern and southern Bornean orang-utan mandibles are more robust than those of Sumatran orangutans, which might reflect a greater reliance on hard and tough fallback foods by the Bornean subspecies.[31]

Some orangutans, gibbons, and gorillas actively prey on arthropods in addition to ingesting them inadvertently with plant foods.[32] Observations of them killing vertebrates and avian eggs for food are extremely rare.[33] Bonobo predation on insects and vertebrates is poorly documented, which precludes detailed comparisons with predatory behavior and preferences of other apes.[34]

Chimpanzees and gorillas eat soil (geophagy) and more rarely their feces (coprophagy).[35] Uehara suggested that ingestion of small bits of soil from termite mounds might inform Mahale chimpanzees about the reproductive state of the insects, thereby cueing them when to fish for them.[36] Bonobos, orangutans, and Sumatran white-handed gibbons also eat soil.[37]

Drinking behavior is relatively infrequent in free-ranging apes. The Asian apes, chimpanzees, and bonobos probably obtain most of their fluids from juicy fruits and other foods, though they also drink opportunistically from tree boles or, when on the ground, from puddles and streams. The less frugivorous gorillas meet their hydric needs mostly by ingesting lush herbage.[38]

Let's Eat, but What and Where?

Primates are unique among eutherian mammals in having color vision.[39] Insofar as we can be informed by current data, like humans, the apes and Old World monkeys are more or less equally endowed with trichromatic color vision that enables them to discern relatively tender red and yellow proteinaceous young leaves and flowers and colorful sugary and fatty fruit amid tougher mature green foliage (Plates 3, 4, 5, 6, and 7).[40]

Based on comparisons of human attention to animals versus vehicles, New et al. proposed that the human visual attention system includes a high-level category-specialized system that monitors animals in an ongoing manner.[41] Accordingly, we might have mechanisms that evolved to direct attention differentially to animals because they belong to an ancestrally important category, regardless of their current utility. Hominid interest in nonhominid animals and conspecifics could indicate fear of harm, or attention to possible food sources, or interest in social and sexual partners.[42] For example, like catarrhine monkeys and apes, humans also seem to pay particular attention to snakes and commonly react fearfully when they encounter them. Indeed,

Isbell argued that detection and avoidance of snakes better explains visual specialization, orbital convergence, and brain expansion in primates than does the development of visually guided reaching and grasping.[43]

Primates are probably endowed with somewhat different abilities to taste, smell, and palpate potential dietary items, though one cannot assume that the senses are controlled by the same alleles.[44] Likewise, although one might assume that primates hear with equal facility, this might not be the case.[45] Although the hands of apes might appear clumsy because of the proportions of their fingers, they are extremely sensitive and adept at manually collecting and preparing well-protected food items.[46]

Body mass, including that due to intraspecific sexual dimorphism, can affect diet in apes.[47] Clearly an adult male gorilla cannot enter some fruitful regions of the forest canopy that are readily accessible to smaller apes. Gross body mass, compromises in relative size among the abdominal organs involved in digestion and other functions, seed size, and the amount of oral processing necessary also affect the time needed for food to pass through the gut.[48] Jaw size, gape, and tooth size along with dental morphology can restrict or enhance the capacities of apes to access the nutritious inner regions of plant parts that are protected by hard and tough peripheral structures.[49] The capacity to tolerate toxins increases with body size, especially in primates with simple versus sacculated stomachs (Figure 2.5). Further, the energy demand per unit body mass is less the larger the individual. Accordingly, whereas the smaller gibbons best rely on sugary fruits and tender young leaves, the great apes can more readily fall back on tougher leaves, bark, unripe fruit, and stems.[50] For example, in Budongo Forest in Uganda, eastern chimpanzees selected foods according sugar content regardless of tannin level, though they spat out wadges of fig seeds high in tannins. They also appear to tolerate higher levels of tannin than do the three sympatric species of Cercopithecoidea.[51]

As researchers document the diets of increasing numbers of ape species and their populations in varied habitats via studies ranging from a few weeks to decades, it becomes more challenging to link general food categories to their overall adaptive complexes. They underscore the flexibility of ape feeding behavior, at least for the short term. Anthropogenic effects of humans hunting the apes and their potential competitors and other predators, in addition to logging, gardening, and collecting of firewood and other forest products for human use have historically and contemporaneously altered the habitats of most study sites directly or through removal of peripheral forest, which limits expansion of the forests and the emigration of their mobile inhabitants.[52]

Gibbons

Bartlett's compilation of dietary data on ten species and one hybrid population of gibbons at twenty localities indicates that fruits, including figs, constituted as much as 89 percent *(Hoolock hoolock)* or as little as 21 percent *(Nomascus concolor)* of their total diet during the span of a study. In these populations, the fig consumption varied from 45 percent *(Hylobates lar)* to zero *(Nomascus concolor).*[53] Marshall and Leighton noted that ripe fruit was predominant (93 percent of observations) in Bornean white-bearded gibbon feeding.[54] Leaves varied between 2 percent *(Hylobates klossii)* and 72 percent *(Nomascus concolor).* Flowers never surpassed leaves or fruits in the diets of any population, but insects constituted more of the total diet than leaves and flowers in three populations: *H. klossii* (26 percent), *H. lar* (24 percent), and *H. pileatus* (15 percent). Three species appear to be either more folivorous than frugivorous *(Nomascus concolor, N. leucogenys)* or about equally folivorous and frugivorous (West Malaysian *Symphalangus syndactylus),* though one population of Sumatran *Symphalangus* was as frugivorous as many populations of *Hylobates* (Plate 4).

Bartlett suggested that leaves and insects might be alternative sources of protein because individual gibbon diets generally have a negative correlation between them (Table 7.1).[55] Rainfall might account for the greatest portion of variance in hylobatid diets, as it appears to be the best predictor of folivory and frugivory.[56]

The diet of gibbons in the lowland dipterocarp forest at Barito Ulu in Central Kalimantan, Indonesia, was most strongly influenced by the availability of flowers, which provide relatively high levels of protein.[57] Flowers were also an important component of the siamang diet at Way Canguk, Sumatra. Indeed, flowers might constitute a fallback food, given that siamang ate them most when they fed least often on nonfig fruit.[58] Overall, siamang spent 12 percent of feeding time on flowers; in one month, eating flowers reached >40 percent of their feeding time. In addition to having to adjust to the spatiotemporal availability of preferred and fallback foods, gibbons must compete for them with other catarrhine primates, squirrels, and hornbills and other birds.[59]

Orangutans

Overall, female and male orangutans have different subsistence practices.[60] Depending on fruit abundance, orangutans employ one of two foraging strategies: sit-and-wait, whereby they conserve energy during low fruit availability,

Table 7.1. Gibbon feeding and ranging

Species	Study Site	Diet (%) Fruit (Fig)[a]	Leaves	Flowers	Arthropods	Range Use Day Range (km)	Home Range (ha)	Territory (ha, %)[b]	Source
Hoolock sp.	Hollongapar, Assam, India	67 (—)	32 (leaves and flowers)	—	1	—	—	18–30 (mean: 22)	Tilson 1979
	Lawachara, Bangladesh	89 (38)	6	5	0	1.2	35	32 (91)	Islam and Feeroz 1992
	Chunati, India	77 (44)	12	4	5	1.7	63	51 (81)	Ahsan 2001; Chivers 2001
		71 (30)	13	9	1	0.9	26	—	Ahsan 2001; Chivers 2001
Hylobates agilis	Sungai Dal, West Malaysia	58 (17)	39	3	1	1.3	29	22 (76)	Gittins 1982
H. klossii	Paitan, Siberut, Indonesia	72 (23)	2	0	25	1.5	32	21 (66)	Whitten 1984
H. lar	Tanjon Triang, West Malaysia	67 (—)	33	0	0	—	—	—	Ellefson 1974
	Kuala Lompat, West Malaysia	50 (22)	29	7	13	1.4	57	—	Raemaekers 1979
	Kuala Lompat, West Malaysia	64 (27)	31	1	5	1.8	53	28 (53)	MacKinnon and MacKinnon 1980
	Ketambe, Sumatra, Indonesia	71 (45)	4 (leaves, vine, and shoots)	1	24	—	—	—	Palombit 1997
	Mo Singto, Khao Yai, Thailand	66 (19)	24	1	9	1.2	23	19 (83)	Bartlett 1999, 2009a
H. moloch	Klong Sai, Khao Yai, Thailand	66 (20)	27	2	7	1.3	43	40 (93)	Suwanvecho 2003
H. moloch	Ujong Kulan, Java, Indonesia	61 (—)	38	1	0	1.4	27	16 (94)	Kappeler 1984a
	Gunung Halimun-Salak National Park, Java, Indonesia	62.5 (33.4)	24.7	11.8	—	1.18	37	—	Kim et al. 2011
H. muelleri	Kutai, Kalimantan, Indonesia	62 (24)	32	4	2	0.9	44	39 (89)	Leighton 1987

H. muelleri x agilis	Barito Ulu, Kalimantan, Indonesia	62 (17)	24	13	1	1.5	53	42 (79)	Bricknell 1999; McConkey et al. 2002
H. agilis albibarbis	Gunung Palung National Park, Kalimantan Barat, Indonesia	93 (—)	—	—	—	—	—	—	Marshall and Leighton 2006
H. pileatus	Khao Soi Dao, Thailand	71 (26)	13	0	15	—	—	—	Srikosamatara 1984
	Klong Sai, Khao Yai, Thailand	64 (21)	27	2	7	1.3	38	37 (96)	Suwanvecho 2003
Nomascus concolor	Mt. Wuliang, Yunnan, China	21 (7)	72	7	0	—	—	—	Lan 1993
	Mt. Wuliang, Yunnan, China	44 (—)	46	10	—	—	—	—	Fan et al. 2008
	Mt. Wuliang, Yunnan, China	25.5 (18.6)	46.5	9.1	—	Yes	129	29(22)	Fan et al. 2009; Fan and Jiang 2008b
	Ailao, Yunnan, China	24 (0)	54	6	14	1.3	87	—	Chen 1995
N. leucogenys	Meng La, Yunnan, China	39 (—)	53 (leaves and buds)	5	4	1.3	—	—	Hu et al. 1989, 1990
Symphalangus syndactylus	Ulu Sempan, West Malaysia	47 (41)	50	2	1	0.8	15	13 (87)	Chivers 1974
	Kuala Lompat, West Malaysia	32 (24)	58	9	2	0.9	35	26 (76)	Chivers 1974
		36 (22)	43	6	15	0.7	47	—	Raemaekers 1979
		45 (31)	44	4	8	0.6	28	18 (65)	MacKinnon and MacKinnon 1980
	Ketame, Sumatra, Indonesia	59 (42)	24	4	2	—	—	—	West 1981
	Ketambe, Sumatra, Indonesia	61 (43)	17	1	21	—	—	—	Palombit 1997
	Way Canguk, Sumatra	18.5 (32.5)–41.7 (14.8)	27.0–36.6	7.6–18.8	>1.0	—	19.5	—	Lappan 2009a

Source: Updated and based on Bartlett 2007.

a. Percentage of total diet.

b. Percentage of home range.

and search-and-find, wherein they move frequently and search for food.[61] Adult females spend more time feeding and less time resting per day than the males. Flangeless and flanged males spent about equal time moving, but flangeless males covered greater distances than the females per day and moved at higher speeds than adult females and flanged males (Table 7.2).[62]

Female orangutans forage more than the males on insects, but there is no significant difference among females or flanged and unflanged male orangutans in the tool extraction of insects. Flanged males consume harder, tougher foods than the flangless males and females, but there is no significant difference between the mechanical properties of the foods that females and flangless males eat.[62]

Orangutans feast on fruits when they are abundant and seek them when they are less readily available; during lean fruiting seasons, they also subsist on leaves, shoots, stems, vines, epiphytes, wood pith, seeds and flowers, sugary phloem, and the cambial layer of bark, which provides essential fatty acids.[63] For example, at Gunung Palung in Borneo, during a mast year—a period of heavy gregarious flowering and fruiting probably caused by an El Niño–Southern Oscillation—the diet of *Pongo pygmaeus* was 100 percent fruit. During a subsequent period of low fruit availability, they ate 37 percent bark, 25 percent leaves, 21 percent fruit, 10 percent pith, and 7 percent insects.[64]

Throughout their range, orangutans also opportunistically or more regularly add to their daily fare with honey, fungi, arthropods (mostly ants, bees, and termites), and mineral rich soil, which might absorb and neutralize plant

Table 7.2. Significant contrasts between adult female (F), flangeless male (FlM), and flanged male (FM) orangutans at Tuanan, Central Kalimantan (Borneo), and Suaq Balimbing, northwestern Sumatra

Variable	Simultaneous Follows
Resting time	M > F
Feeding time	F > M
Feeding insects	F > FlM > FM
Capturing slow loris	F > FlM and FM
Feeding fruit	FM > FlM
Feeding bout length	FM > F > FlM
Moving time	F > FM
Day journey length	FlM > F
Travel speed	FlM > F and FM

Source: Modified from van Schaik et al. 2009, p. 266.

toxins and digestion inhibitors (Table 7.3).[65] Further, the diets of Bornean orangutans vary within local populations in part because youngsters learn to adopt foods by watching their mothers, then later supplement their diets with low-level individual sampling.[66]

Based on a two-year study of Bornean orangutans in a dipterocarp lowland tropical rain forest, Leighton argued persuasively that his subjects preferred non-fig fruits over figs because of the high energy return from sugar.[67] Even protein was secondary in their quest for sugary fruits. Moreover, they foraged the mature fruit of some species for juicy pulp and other species for unripe seeds, but they rarely sought the same species for both items. Overall, in the orangutan diet, the non-fig fruit pulp and seeds contained more digestible carbohydrate than found in figs; the figs were more fibrous and contained higher levels of condensed tannin and phenolics.[67]

Like many human mall and fast-food customers who forage for sugar, salt, and lipids, orangutans have the potential to fatten dramatically, which serves them well in the forest but can predispose them to diabetes in captivity.[68] A large body size is probably advantageous to heavily frugivorous orangutans because it provides a greater capacity to store fat that can be used during lean fruiting and flowering periods.[69]

Knott monitored Bornean orangutan caloric intake and fat metabolism over a thirteen-month period in a primary tropical rain forest.[70] She calculated daily total caloric intake based on direct observation of feeding behavior and nutritional analyses of sample foods and determined the degree of metabolism of body fat deposits by measuring excreted byproducts—ketones—in the urine of focal subjects that she followed from dawn until they settled for the night. She estimated that during their highest fruit consumption, males consumed 8,422 kcal/day and females consumed 7,404 kcal/day, whereas their caloric intakes during lowest fruit consumption dropped to 3,824 kcal/day and 1,793 kcal/day, respectively. Their urine contained ketones only during the fruit-poor period.[64] Given that Knott estimated the daily energy requirements for adult males and females to be 3,344 kcal/day and 1,512 kcal/day, respectively, it is little wonder that they fatten when fruit is abundant and become slim when it is scarce.[64]

During the month of greatest fruit scarcity, males consumed more calories than the females did without spending more time feeding, probably because they were better able to access the lipid-rich seeds of large hard-shelled fruits (*Neesia*) that are protected by irritating hairs (Plate 3). Urine from females contained more ketones than that of males, indicating that they were more

Table 7.3. Orangutan feeding

Species/Study Location	Fruit (%)	Flowers	Leaves Mature	Young	Bark	Arthropods	Other	Source
Pongo abelii								
Ranun, Sumatra	84.7	—	← 10.2 →		4.5	0.6	—	MacKinnon 1974
Gunung Leuser Reserve, Sumatra	58.0	—	← 25.0 →		3.0	14.0	—	Rijksen 1978
Suaq, Gunung Leuser Reserve, Sumatra	68.0	—	← 18.0 →		1.0	12.0	1.0	Fox et al. 2004
Ketambe, Sumatra	67.5		← 16.5 →		2.6	8.8	4.8	Wich et al. 2006
Pongo pygmaeus								
Tuanan, Borneo	68.6	5.9	← 17.2 →		1.0	6.3	0.6	Morrogh-Bernard et al. 2002
Mentoko, Kutai National Park, Borneo	53.8	—	← 29.0 →		14.2	0.8	2.2	Rodman 1977
Tanjung Puting Reserve, Borneo	60.9	3.9	← 14.7 →		11.4	4.3	4.0	Galdikas 1988

Ulu Segama, Borneo	62.0	—	← 35.6 →	11.2	2.1	—	MacKinnon 1974
Tanjung Puting Reserve, Borneo	61.0	—	—	—	—	—	Hamilton and Galdikas 1994
Tanjung Puting Reserve, Borneo	61.0	—	—	—	—	—	Galdikas 1988
Gunung Palung National Park, Borneo	66.8	—	10,8	9.3	4.2	8.9	Knott 1998[1]
Gunung Palung National Park, Borneo	70.0	5.1	← 13.4 →	4.9	3.7	2.9	Morrogh-Bernard et al. 2002
Kinabatangan, Borneo	68.0	1.3	← 22.9 →	6.7	1.2	—	Morrogh-Bernard et al. 2002
Danum, Sabah, Malaysia	60.9	2.5	← 22.2 →	12.3	0.6	1.6	Kanamori et al. 2010

1. Based on estimates in Wich 2006.

stressed.[64] The total energy intake of Gunung Palung orangutans during three months of extremely low fruit availability was low, and energy from fermenting fiber was insufficient to balance their energy expenditure.[71] To the extent that dental pathologies reflect substandard diets, overall the 131 females fared less well than the 89 males in a population of western Bornean orangutans.[72]

In marked contrast with Bornean *Pongo pygmaeus, P. abelii* at Ketambe, Sumatra, experienced no prolonged negative energy budget from fluctuations in fruit availability.[73] Only one individual had ketones in her urine, probably because she was ill. She died ten days after the ketones appeared in her urine. Copulations and conceptions in the Ketambe population did not vary depending on fruit availability.[74] Figs (as fallback foods) and other fruits were the most important dietary items, on average constituting 67.5 percent of Ketambe orangutan foods, followed by nonreproductive plant parts (leaves, stems, shoots, and pith: 16.5 percent; cambium: 2.7 percent; other: 4.8 percent) and insects (8.8 percent).[75]

Although primarily frugivorous, *Pongo abelii* at Suaq Balimbing, Sumatra, regularly ate quantities of arboreal social insects and honey, and the females increased their feeding time on these foods during pregnancy and lactation.[76] Indeed, social insects were a staple instead of a fallback food of Suaq Balimbang orangutans.[77] Their average diet comprised 66.2 percent fruits, 15.5 percent nonreproductive plant parts, 13.4 percent insects, 1.1 percent cambium, and 3.8 percent other items. Like the diet of their Ketambe conspecifics, that of Suaq *Pongo abelii* always contained more than 50 percent fruit and never fell as low as those of Bornean *P. pygmaeus.*[75]

Overall Sumatran forests are more fruitful than Bornean forests. Further, on Sumatra large freestanding strangler figs are abundant, whereas on Borneo there are many more smaller fig species (Plate 3).[59] Indeed, Bornean orangutans feed on more plant families, genera, and species and eat more plant parts, including cambium and perhaps leaves, per species than their Sumantran congeners consume.[78] The more fruitful landscape of Sumatra might be due to it being a younger island with richer volcanic soils.[79]

Chimpanzees

While acknowledging that chimpanzee diets vary widely due to the diverse habitats that they occupy across Africa from Senegal to Uganda, Rwanda, Burundi, and Tanzania (Table 7.4), based on twenty-three field reports spanning much of the range of the four subspecies, Stumpf generalized the dietary composition of *Pan troglodytes* to be 64 percent fruit, 16 percent leaves, 7 percent terrestrial herbaceous vegetation, 4 percent bark and miscellaneous items, 4 percent animal prey, 3 percent seeds, and 2 percent flowers (Plates 5, 6 and 7).[80] They must compete for fruit and other desired plant parts with an extensive variety of monkeys, ruminants, rodents, and arthropods.[81] Nonetheless, by feeding selectively, chimpanzees and bonobos in markedly different habitats can obtain diets of similar nutritious quality and gross energy content (Table 7.5).[82]

Pruetz focused on the differences between the diets of populations in three open, seasonally relatively dry habitats versus eight more densely treed wetter habitats.[83] Annual fruit intake is lower at one of the dry sites—Semliki, Uganda—and to a lesser extent Mount Assirik, Senegal, but not at the savanna site of Fongoli, Senegal: 60.8 percent fruit, 4.3 percent leaves, 5.5 percent flowers and inflorescences, 23.9 percent invertebrates, 1.4 percent pith and stems, 2.3 percent cambium/bark, 0.4 percent vertebrates, and 1.6 percent other.[84] Fongoli chimpanzees ate unripe fruit (55 percent of sixteen bouts) more than ripe fruit (45 percent of bouts), and occasionally ate galagos *(Galago senegalensis)*.[85] Chimpanzees at the dry sites fed on fewer plant species (mean: 45) than did the chimpanzees in wetter habitats (mean: 136). At Fongoli, *Ficus* spp. are staples, and the chimpanzees might rely more on invertebrates than the chimpanzees elsewhere.[83] In the savanna woodland of Ugalla, Tonge East Forest Reserve, Tanzania, the chimpanzees sometimes excavated and ate underground storage organs of several plant species.[86] The area is one of the driest habitats where chimpanzees live.[87] Chimpanzees in a rain forest habitat at Bossou, Republic of Guinea, manually excavated cassava from human gardens, probably as fallback food, when wild and cultivated fruits were scarce.[88]

Like Bornean orangutans, neighboring chimpanzee groups vary in which plant and animal species they feed upon most frequently and intensively and whether their prey are rare or more commonly available, as illustrated by three adjacent groups in the Taï Forest in the Côte d'Ivoire.[89] Whereas some of the variation is probably due to the distribution, density, and nutritional and toxin contents of the prey species, it appears that some choices are simply individual and group preferences.[89]

In Kibale Forest, because of habitat heterogeneity (i.e., different plant communities), the diet of the Ngogo community is of higher overall quality

Table 7.4. Chimpanzee feeding

Subspecies/Study Location	Fruit (%)	Figs	Seeds	Flowers	THV	Leaves Mature	Leaves Young	Bark	Arthropods	Vertebrates	Other	Source
Pan troglodytes verus												
Fongoli, Senegal	62.5	—	—	11.0	3.0	16.0	—	2.5	5.0	—	—	Pruetz 2006
Bossou, Republic of Guinea	60.7	—	6.3	1.1	5.3	10.8	Yes	0.9	1.8	—	13.1	Yamakoshi 1998
Bossou, Republic of Guinea	52.0	—	6.5	4.5	13.4	17.9	Yes	1.6	—	—	4.0	Sugiyama and Koman 1992
Bossou, Republic of Guinea	49.2	—	3.4	6.3	—	36.1	Yes	Yes	Yes	Yes	5.9	Sugiyama and Koman 1987
Mount Assirik, Senegal, West Africa	57.0	—	10.0	10.0	3.0	10.0	Yes	7.0	Yes	Yes	3.0	McGrew et al. 1988
P. t. troglodytes												
Lopé Reserve, Gabon	69.2	—	7.7	6.4	0.05	9.9	Yes	—	5.3	—	1.0	Tutin et al. 1997
Ndoki Forest, Republic of the Congo	87.7	—	2.6	1.8	Yes	2.6	Yes	Yes	Yes	Yes	5.3	Kuroda et al. 1996
Lopé Reserve, Gabon	63.8	—	6.9	3.4	—	11.5	—	1.7	Yes (included in Other)	Yes (included in Other)	12.6	Tutin and Fernandez 1992; Tutin and Fernandez 1993
Ndoki-Nouabale Reserve, Republic of the Congo	80.4	—	5.9	—	Yes	Yes	Yes	—	Yes	—	13.7	Kuroda 1992
Lopé Reserve, Gabon	67.6	—	7.0	2.1	2.1	1.4	9.9	0.7	5.6	2.1	1.4	Tutin et al. 1991
Okorobikó Mountains, West Africa	44.8	—	7.7	—	<15.0	31.8	—	—	3.8	—	—	Sabater-Pi 1979

Gabon	62.6	—	—	—	← 33.3 →	—	4.1	—	—	Hladik 1977
Gabon	← 68.0 →		—	—	← 28.0 →	—	<5.0	Yes	0.05	Hladik 1973
P. t. schweinfurthii										
Rubondo Island, Tanzania	93.5	Yes	Yes	Yes	Yes	—	Yes	Yes	—	Moscovice et al. 2007
Montane Forest of Kahuzi-Biega, Democratic Republic of the Congo	99.0	92.0	Yes	Yes	Yes	—	4.0	9.0	5.6	Basabose 2002; Basabose and Yamagiwa 1997
Semliki, Uganda	39.0	15.0	3.0	9.0	30.0	3.0	—	—	—	Hunt and McGrew 2002
Budongo Forest Reserve, Uganda	64.5	Yes	Yes	Yes	19.7	Yes	—	—	15.8	Newton-Fisher 1999a
Mahale Mountains, Tanzania	64.5	0.4	—	14.8	5.8	—	13.8	—	—	Matsumoto-Oda and Hayashi 1999
Mahale Mountains, Tanzania	30.5	4.6	8.8	—	← 35.7 →	Yes	Yes	Yes	20.5	Sugiyama and Roman 1987; Nishida and Uehara 1983
Kibale National Park, Uganda	78.5	Yes	0.6	10.8	9.3	0.4	Yes	Yes	0.2	Wrangham et al. 1998
Kibale Forest, Uganda	78.0	1.0	4.2	—	4.0	1.5	Yes	Yes	2.2	Ghiglieri 1984
Mahale Mountains, Tanzania	27.0	9.0	8.0	8.0	37.0	11.0	—	—	—	Peters and O'Brien 1981
Gombe National Park, Tanzania	43.0	7.0	5.0	13.0	← 18.0 →	3.0	—	—	2.0	Wrangham 1977

Note: Yes = no percentage given separately from "other" combinations. THV = terrestrial herbaceous vegetation.

Table 7.5. Community and home range size of chimpanzees and bonobos

Country	Site	Source	Community Size	Range Size (km²)
Republic of Guinea	Bossou	Sugiyama 1994a	16–24	6
Republic of Guinea	Kanka Sili	Albrecht and Dunnett 1971	50	5
Côte d'Ivoire	Taï	Boesch and Boesch-Achermann 2000b	33–82	19.1–21.6
Côte d'Ivoire	Taï	Boesch 1997	29–82	13–26
Senegal	Mount Assirik	Baldwin et al. 1982	28	278–333
Republic of Guinea	Bossou	Sugiyama 1998, 1994, 2004	16–22	15–24
Senegal	Mount Assirik	McGrew et al. 1996, 2004; Tutin et al. 1983	<15	50
Tanzania	Gombe	Wrangham 1979a, 1979b; Goodall 1986; Wallis 1997	38–60	4–24
Tanzania	Mahale	Nishida et al. 2003; Hasegawa and Hiraiwa-Hasegawa 1983	45–101	7–14
Uganda	Kanyawara	Chapman and Wrangham 1993; Wrangham et al. 1996	44	16
Uganda	Ngogo	Watts 1998; Watts and Mitani 2001; Mitani, Watts, and Lwanga 2002; Mitani (unpublished data cited in Nishida et al. 2003)	>140	35
Uganda	Semliki	Hunt and McGrew 2002	>29	38.3
Uganda	Budongo	Newton-Fisher 1999a, 2003	32–56	7
Nigeria	Gashaka-Gumti	Sommer et al. 2004	>35	26
Democratic Republic of the Congo (DRC)	Kahuzi-Biega	Basabose 2002; Basabose and Yamagiwa 2002	22	10
DRC	Wamba	Kano 1992; Badrian and Malenky 1984	28	22 (*Pan paniscus*)
DRC	Lomako	Thompson-Handler 1990	33	14.7 (*Pan paniscus*)
DRC	Lukuru	Meyers Thompson 2002	?	? (*Pan paniscus*)

Source: Kormos et al. 2003; Stumpf 2007.

than that of the Kanyawara community that lives only 12 km away (Plates 6 and 7). Concomitantly, the density of the Ngogo community is three times that of the Kanyawara community, and its members rest significantly more, eat more ripe fruit, have a less diverse menu, and spend less time in the food patches than is characteristic of the Kanyawara chimpanzees.[90]

At Mahale National Park, from 2 years of age eastern chimpanzees markedly increased solicitation of food from their mothers; the mothers responded positively, but they never offered food to their youngsters. They also shared items that were difficult to process more often than those that were easily accessed. Competition between them was rare, and sharing diminished over time as proximity decreased between them or the mother became aggressive. Food transfer from a mother to her progeny virtually ceased by the time her offspring was 7 years old.[91]

Gorillas

Thorough studies on habituated Virunga mountain gorillas by Watts and Vedder confirmed incidental observations of Fossey and Harcourt that the gorillas spent the vast majority (>90 percent) of their feeding time ingesting nonreproductive parts of plants, often to the exclusion of fruits (Plates 8 and 9).[92] Later field researchers showed that western, Cross River, Grauer's, and even some mountain gorillas are more frugivorous (including seeds) than the Virunga mountain gorillas (Table 7.6).[93] Their data sets are chiefly based on analyses of fecal samples and leavings at field sites because of the difficulties of habituating the subjects; the relative brevity or primary foci of their studies did not include systematic dietary observations. Moreover, fecal samples do not accurately represent the items that were ingested; for example, nonreproductive plant parts can be underrepresented compared with seeds and other fruit parts.[94]

Over a six-year span, Rogers et al. and Williamson et al. recorded fruit remains in 98 percent of gorilla fecal samples and documented ninety-five species of fruit eaten by western gorillas in a lowland rain forest at Lopé, Gabon. Most of the fruits were succulent.[95] The gorillas also ate immature seeds in some of these fruits. In Gabon, gorilla feces contained the heads and mandibles of termites (*Cubitermes* sp.).[96]

Terrestrial herbaceous vegetation (THV) and aquatic plants attract foraging western gorillas even in regions with fruitful trees, vines, and shrubs. For example, surveys have indicated high densities of gorillas in the swamp forests

Table 7.6. Dietary and ranging patterns of gorillas

Subspecies/Study Location	Altitude (m)	Number of Food Species Eaten	Number of Fruit Species Eaten	Degree of Frugivory (%)	Day Journey Length (m)	Home Range Size (km²)
Mountain gorilla (*Gorilla beringei beringei*)						
Virungas–Karisoke	2,680–3,710	36	1	<1	570	3–15
Virungas–tourist group	2,500–2,800	42	2	<1	756	5
Bwindi–Ruhija	2,100–2,500	112	30	16–66	1,034	21–40
Bwindi–Buhoma	1,450–1,800	140	36	68–89	547–978	16–22
Grauer's gorilla (*G. b. graueri*)						
Kahuzi-Biega–Kahuzi	1,800–2,600	79	24	96	850	23–31
Kahuzi-Biega–Itebero	600–1,300	142	67	89	—	23–31
Cross River gorilla (*G. gorilla diehli*)						
Cross River-Afi	400–1,300	166	83	90	—	ca. 32
Western lowland gorilla (*G. g. gorilla*)						
Lopé, Gabon	100–700	134	95	98	1,105	22
Mondika, Central African Republic	<400	100	70	99	2,014	16
Bai Hokou, Central African Republic	460	138	77	99	1,580	11

of north-central People's Republic of the Congo, where they forage on palm hearts, hearts of *Raphia* sp., and petioles of *Lasiomorpha senegalensis* in ≤30 cm of water.[97] Mitani et al. and Blom et al. also attributed the high densities of western gorillas in Ndoki Forest, Republic of the Congo, and Dzanga-Nkoki National Park, Central African Republic, to the presence of aquatic plants, for which the chimpanzees seemed not to compete.[98] Forest elephants probably play a significant role in maintaining the stands of THV species that are foraged and used as nesting materials by gorillas. Accordingly, they can survive at high densities in primary forest when free of human predation.[99]

In tropical African landscapes with major elevational variance, the composition of forests changes from a tropical rain forest zone (≥1,100–1,300 meters), through transition forest zones (1,100–1,300 meters > 1,650–1,750 meters), to a montane rain forest zone (1,650–1,750 meters ≤ 2,300–3,400 meters).[100] The overall structure of forest components (tree height, layering of tree crowns, and presence of lianas and epiphytes) and arboreal physiognomy (buttressing, crown shape, and leaf size, shape, and thickness, location of flowers and fruit, and relative deciduousness) determine the potential foods that are available to the animals therein.[101] High-elevation African montane forests like those in which Watts and Vedder conducted their studies lack the many fruitful vines and trees that characterize the lower elevations, but they sustain a nutrient-rich herbaceous forest floor, including patches of bamboo in some areas.[102]

Grauer's gorillas in the montane forest of Kahuzi, Democratic Republic of the Congo (DRC), neatly illustrate the effects of elevation on gorilla diets. Whereas they share a fruit component with the diets of gorillas that live at lower elevations, they also feed heavily on leaves, bark, and pith like the mountain gorillas that also range in montane forests.[103] During a nine-year period, fruit constituted 19.7 percent of the 236 foods from 116 plant species that Kahuzi gorillas ate, while other plant parts constituted 70.2 percent. Nonetheless, their fondness for fruit is clear: 53.2 percent of Kahuzi gorilla fecal samples contained fruit remains.[103] They rarely ate earthworms or other fauna, so their protein source was floral. Yamagiwa et al. noted that they ate figs frequently when other fruits were also available, from which they concluded that figs are not a fallback food.[103] Like Virunga gorillas, seasonally some Kahuzi gorillas focused heavily on bamboo shoots (about 90 percent of their diet).[104]

Wild mountain gorilla groups in the Bwindi Impenetrable National Park preferred leaves with relatively low fiber and high protein, fat, and phenols, sugary fruit, and the pith of herbs. Foraging groups combined a variety of plant parts of select species in their ranges to satisfy their nutritional needs.[105] They also ate ants. Two groups with overlapping ranges consumed ants on 3.3 percent and 17.6 percent of days. In the latter group, as evidenced by fecal samples, adult females (13.2 percent) and juveniles (11.2. percent) ate significantly more ants than did the silverbacks (2.1 percent).[106]

Bonobos

Like chimpanzees, bonobos feed at all levels of the forest.[107] At Wamba, highly frugivorous bonobos collect the major sources of nutriment at heights of 24–40 meters, but in Iyoko, like some western gorillas, they regularly feed on aquatic and amphibious herbs and grasses.[108] Year-round, even during the fruiting season, Wamba and Lomako Forest bonobos also eat THV, which is a reliable, abundant protein source in large areas of secondary forest, forest gaps, and swamp forest (Table 7.7).[109] Malenky and Wrangham suggested that because Lomako bonobos eat THV during the season of fruit abundance, they use it as a source of protein, while Kibale chimpanzees, which eat less THV, probably use it as a fallback source of carbohydrates when fruit is scarce.[110] Ikela bonobos augment their vegetarian diet with earthworms, larvae, termites, ants, small mammals, honey, truffles, and aquatic plants, all of which are nutritious and tasty.[111]

Although African apes ingest most species of leaves and THV for their nutritional effects, field researchers have argued compellingly that, like sympatric humans, they also forage some species primarily to counter endoparasitic infestations.[112] After an initial report by Wrangham and Nishida that eastern chimpanzees at Mahale and Gombe in Tanzania swallowed unchewed young leaves of *Aspilia* sp. for pharmacological effects, and Huffman and Seifu reported that Mahale chimpanzees sucked the bitter, astringent juice from the pith in chewed stems of *Vernonia amygalina,* researchers in other areas reported similar behavior by chimpanzees and bonobos at sites extending from the Republic of Guinea to East Africa.[113] Tests have shown that the pith of *Vernonia amygalina* contains antiparasitic compounds.[114] Additionally, the extent to which the consumption of plant hormones (e.g., from *Vitex* spp.) might affect female reproductive physiology is an intriguing question that merits additional research.[115]

Table 7.7. Bonobo feeding

Study Sites in Democratic Republic of the Congo	Fruit	Figs	Seeds	Flowers	THV	Leaves Mature	Leaves Young	Arthropods	Vertebrates	Other	Source
Lomako Forest	54.9	Yes	7.1	4.4	6.2	←21.1→		Yes	Yes	—	Badrian and Malenky 1984
Lomako Forest	49.0	Yes	9.0	6.0	15.0	←21.0→		Yes	Yes	—	Badrian et al. 1981
Wamba Forest	←83.4/93.1[a]→		—	—	—	15.2/5.7	—	←1.5/1.1→		—	Kano 1992
Wamba Forest	93.1	Yes	—	—	—	15.8	Yes	←1.1→		Yes	Kano and Mulavwa 1984
Wamba Forest	—	—	—	—	33.0	—	—	—	—	—	Wrangham 1986
Yalosidi	73.5	—	—	2.9	38.2	5.9	—	—	—	2.9	Kano 1983
Lomako Forest	72.1	—	—	0.5	2.1	—	24.9	0.1	—	—	White 1992
Lilungu (Ikela)	—	—	—	—	—	—	—	Yes	Yes	Honey, truffles, and aquatic herbs	Bermejo et al. 1994

Can We All Get Along?

Fruitful as they are in the best of times and even during leaner periods compared with many other biomes, the tropical rain forests nonetheless pose major challenges to the great variety of animals that would subsist on their produce. Climate and seasonality affect habitat structure and plant phenology. The manner in which the latter are affected by the environmental factors plays a major role in the daily lives, life histories, and community ecology of forest, woodland, and savanna apes.[116]

The Asian and African forests that accommodate apes have markedly different characteristics. There are more plant families, genera, and species in Asian rain forests than in African rain forests.[117] Southeast Asian rain forests are especially rich in species of the Dipterocarpaceae, which, along with other large-seeded species, experience irregular supra-annual peaks in synchronous flowering and fruit production (masting) often followed by much lower yields during most years per decade. Possible results of this phenomenon are lower mammalian biomass and specific richness, including fewer species of frugivorous cercopithecine monkeys but more colobine species in Southeast Asian rain forests.[118] The densities of orangutan populations might also be lower in forests with relatively high densities of Dipterocarpaceae. However, when compared with the limestone karst forests, the lowland dipterocarp forest accommodated the highest density of orangutans in the Sangkulirang Peninsula of East Kalimantan, perhaps due to the low diversity of tree species in the limestone karst forests.[119]

Apes must compete with other arboreal frugivorous and folivorous primates, bats, squirrels, civets, colugos, birds, arthropods, and other non-mammalian and leaf, fruit, and seed predators.[120] Pigs, deer, antelopes, rodents, monkeys, and arthropodans can reduce the quantities of fallen fruits and terrestrial vegetation in biomes that they share with terrestrially foraging species of apes, although competition is greatly reduced in many cases because the feeding preferences and spatiotemporal ranging patterns of the various potential contenders vary.

Active competition can require a notable expenditure of energy and occasionally may lead to injury. Direct conflicts between different species of apes in areas where they are sympatric are rare or at least are seldom observed by field researchers. Bonobos and many species of gibbons are allopatric with regard to other ape species. But chimpanzee and gorilla populations are sympatric in many areas of tropical African rain forest, and in areas of Southeast Asia pairs of gibbon species are sympatric with one another. In Sumatra and Borneo, two and one hylobatid species, respectively, are sympatric with orangutans.[121]

Relations among Sympatric Gibbons and Orangutans

Gibbon species that are fruit-pulp specialists compete for small, sweet, colorful fruits with birds more than with sympatric monkeys and orangutans.[122] During a ten-month span, Sumatran *Hylobates lar* and *Pongo abelii* in the Gunung Leuser National Park shared only 32 of 134 (23 percent) fruit species in their diets. Based on fruit-feeding events, gibbons ate figs more frequently than the orangutans (about 50 percent versus 30 percent).[123] They also fed more often than the orangutans on small fruits and less often than the orangutans on medium and large fruits. Moreover, gibbons chose ripe fruits that had softer skins and were more often mildly (pH <4.25) or moderately (pH 4.25–5.00) acidic than the fruits preferred by the orangutans.[123] Although they visited all levels of the canopy, gibbons usually fed high in the main canopy or in emergents, and they rarely descended to the understory.[124] Orangutans were more versatile and variable, using all levels of the forest including occasional forays to the floor.[123] The Bornean white-handed gibbon population density is constrained by the availability of figs, which are a fallback food in their diet.[125]

It appears that the feeding and ranging patterns of some *Hylobates* spp. are so similar that they are ecological equivalents that eventually interbreed when they become sympatric.[126] *Symphalangus syndactylus* seems to be sufficiently different from *Hylobates* spp. to avoid hybridization where they are naturalistically sympatric, though in West Malaysia a male siamang traveled and called with a female white-handed gibbon for a brief span before she paired with a male white-handed gibbon.[127]

In Peninsular Malaysia, siamang are intolerant of lar and agile gibbons being with them in feeding trees, and male siamang will chase them out. At valuable fruit sources, smaller gibbons must wait their turn or seek another feeding site when siamang are present, but gibbons can prevail over formidable groups of macaques.[128] Their dominance may be why siamang can have much smaller ranges than those of smaller sympatric gibbons, though other factors, such as more selective feeding by the smaller gibbons, or intraspecific social factors, or both might also be involved.[129] Habitat factors, especially intersite variations in forest structure, phenology, elevation, and competitive fauna, probably are also operant.[130]

Pongo pygmaeus are so much larger and have sufficiently different diets from *Hylobates muelleri* that they can afford to be tolerant of them.[131] Sumatran orangutans and white-handed gibbons have much the same tolerant relationship as that of their Bornean counterparts, but siamang can be troublesome where the three species are conspecific. In northern Sumatra, Rijksen observed siamang near orangutans only in the vicinity of large strangler figs.[132] Siamang sometimes attacked young and adult female orangutans and could displace youngsters from a food site. Contrarily, a group of siamang that was occasionally harassed by another siamang group sometimes traveled for brief spans with a group of white-handed gibbons. Youngsters of the two species also allogroomed and played together.

Relations between Sympatric Chimpanzees and Gorillas

Reports of direct interactions of sympatric chimpanzees and gorillas are as rare as hen's teeth. It is difficult to discern whether this is because they seldom occur or because one or both species are insufficiently habituated to the presence of observers and other humans.[133] Therefore, I will summarize case studies and make minimal attempts to generalize to species or genera; most reports lack balance between the observations of contemporary activities of apes suspected of competing for the same foods.

There are rare isolated incidents of western gorillas and chimpanzees feeding peacefully together in fruit trees. For instance, in the Nouabalé-Ndoki Forest in the Republic of the Congo, Suzuki and Nishihara observed them feeding together in fig trees four times.[134]

In Bwindi Impenetrable National Park in Uganda, Stanford and Nkurunungi reported the only incident of contest competition between sympatric chimpanzees and gorillas.[135] A group of about thirteen mountain gorillas arrived and fed on fallen fruits beneath a tree in which nine eastern chimpanzees were feeding. Then two adults climbed within 5 meters of the chimpanzees and were met by two male chimpanzees, which approached them and intermittently displayed in the tree crown for about an hour. The chimpanzees fled immediately when others approached, and the gorillas ascended into the crown and fed on fruits therein.[135]

Conversely, another encounter between Bwindi mountain gorillas and eastern chimpanzees was pacific. A group of gorillas entered a fig tree in which around fifteen chimpanzees were feeding. The silverback sat and fed within 3 meters of two adult male chimpanzees, and all parties appeared to ignore one another.[135]

Jones and Sabater Pí provided the first report on the comparative ecology of African apes, based on a seventeen-month study primarily at three sites in Equatorial Guinea, only one of which had both western gorillas *(Gorilla gorilla gorilla)* and central chimpanzees *(Pan troglodytes troglodytes)*.[136] They rarely detected chimpanzees and gorillas near one another, which left the impression that there was little competition between them. Gorillas fed and rested in mature forest, regenerating forest, and montane forest (≥3,050 meters), but nested at night only in regenerating forest. Chimpanzees fed and rested in all three types of forest and nested at night in mature and montane forest, but not in regenerating forest.

During the dry season, both species increased their foraging forays on human plantations. However, during the wet season gorillas spent even more time in relatively open areas, where they fed on thickets of *Afromomum* or stands of *Musanga,* while the chimpanzees foraged chiefly in the upper strata of the forests. Adult gorillas were almost completely terrestrial foragers, whereas chimpanzees commonly foraged high (>20 meters) in trees. The differences in their daily activity patterns probably also kept the two species separated. Although they all left their nests shortly after dawn, chimpanzees usually moved to another area to feed, while gorillas foraged at their sleeping site. Chimpanzees tended to feed more intensively than the gorillas in the afternoon; however, whereas chimpanzees settled down rather promptly, gorillas fed and moved about in their sleeping sites.[136]

In Lopé, Gabon, there is notable overlap in the diets of central chimpanzees and western gorillas, especially in regard to the consumption of fruit. Lopé gorillas ate 77 percent of the food items and 82 percent of the fruits that chimpanzees ate.[137] Some of the fruits are fibrous instead of pulpy. Both species collected most of their food in trees: chimpanzees, 76 percent; gorillas, 69 percent. Tutin and Fernandez found that although Lopé gorillas ate more stems, pith, and leaves than the sympatric chimpanzees, the chimpanzees also used 59 percent of the same species. Moreover, Lopé chimpanzees ate 79 percent of the fruit species that the gorillas ate. Nonetheless, Tutin and Fernandez noted that there were few overt signs of interspecific competition.[137] When fruit was scarce, diets of the two species diverged most, with gorillas resorting to more THV, fibrous fruits, and bark.[138]

In a mature forest habitat of Loango National Park, Gabon, the dietary overlap between gorillas and chimpanzees appeared to be lower than that at the other sites where they are sympatric. Further, the paucity of terrestrial herbs did not increase the overlap between their diets. Both populations are frugivorous, albeit with gorillas eschewing species high in crude lipids and chimpanzees exploiting them heavily.[139] Loango chimpanzees consistently ate more fruit than the gorillas did, and gorillas regularly ate more leaves and bark.

Like the more herbivorous and folivorous diets of cercopithecid monkeys, those of gorillas versus chimpanzees are reflected in the topographies of their molar teeth. In addition to being larger, gorilla molars have higher cusps and longer shearing crests with which to shred leaves and THV.[140]

Pith, stems, and leaves of *Afromomum* and Marantaceae are abundant staple foods year-round at Lopé.[141] Young leaves of Marantaceae are high in protein, but the pith of one tested species of *Afromomum* was low in both protein and soluble sugars.[142] Gorillas will eat fruits, seeds, leaves, and bark containing high levels of condensed tannin and phenols.[143]

The relatively long gut retention time of great apes facilitates not only release of short-chain fatty acids by microbes in the cecum and colon but possibly also the detoxification of plant chemicals via the mucosa lining the intestines.[144] The relatively long retention time of chimpanzees probably affects forest composition and renewal because swallowed seeds are transported notable distances from the parent trees.[145] Moreover, passage through the gut probably increases the speed and probability of the seeds' germination.[146]

Adult gorillas have much longer gut retention times than chimpanzees (mean: gorilla, 50 hours; chimpanzee, 31.5 hours).[147] Although fundamentally frugivorous, like some sympatric monkeys and chimpanzees, Lopé gorillas are outstanding for "their ability to eat large fibrous fruit, mature leaves and stems, and to overcome high levels of . . . phenols and condensed tannins."[148]

Insect remains were about equally frequent in Lopé gorilla feces (30 percent) and chimpanzee feces (31 percent). Both species most frequently ate weaver ants; otherwise, their insect diets differed markedly. Lopé chimpanzees used tools to forage bees, honey and two large species of ants, while the gorillas manually captured three species of small ants.[149]

The average insect prey biomass intake per day by chimpanzees was twice that of the gorillas at Dja Biosphere Reserve in Cameroon. Neither species seemed to consume ants for specific nutrients, but the gorillas intake of the termites *Cubitermes* and *Toracotermes* met their estimated iron requirements; the chimpanzees intake of *Macrotermes* met their estimated manganese requirements and protein intake per day.[150] One needs much more detailed entomological data to explain great ape dietary choices of arthropodan species. For instance, at Campo Animal Reserve in Uganda, the chimpanzees employ sticks and probes to obtain *Macrotermes lillijeborgi*, though they could more easily dig manually into the nests of *M. vitrialatus*.[151]

A sixteen-month study in Nouabalé-Ndoki National Park, Republic of the Congo, confirmed that the diets of central chimpanzees and western gorillas are dominated by fruits and exhibit notable dietary overlap.[152] Ndoki gorillas ate 51 percent of the species that chimpanzees ate, and chimpanzees ate 59 percent of the species eaten by gorillas. Although Ndoki chimpanzees chewed and swallowed the seeds from pods of Caesalpiniaceae, they swallowed the pulpy fruit seeds whole. Ndoki gorillas frequently masticated seeds of pulpy fruit species; the chimpanzees did not masticate the seeds of pulpy fruit.[153] Ndoki gorillas also ate more leaves, tree bark, and THV than the chimpanzees did, and only the gorillas commonly ventured into swamp forest, where they fed on aquatic herbaceous vegetation. Ndoki chimpanzees were more insectivorous than the gorillas. A major difference between Ndoki gorillas and chimpanzees is that gorillas obtain more of their protein from seeds, leaves, and THV, while the chimpanzees rely on young leaves and insects as protein sources.[154]

Chemical analysis of sixty-three plant samples collected during periods of relatively low fruit availability from twenty families and thirty-five species at Bai Hokou in Dzanga-Ndoki National Park, Central African Republic, indicated that herbaceous stems and tree leaves were reliable sources of crude protein, whereas fruit samples contained variable amounts of soluble sugars (1.4 to 48.9 percent) and more tannins than were in the foliage.[155] If the fruits contained enough sugar, the gorillas were undeterred by moderate levels of tannin.

During an eight-year span in Kahuzi Forest, 38 percent of the foods eaten by gorillas and 64 percent of those eaten by sympatric chimpanzees were the same. Chimpanzee fecal samples contained more fruit species (mean: 2.7 spp.) than those of gorillas (mean: 0.8 spp.). Both gorillas and chimpanzees increased fruit in their diets when they were abundant, but only chimpanzees continued to eat the same number of fruits when fruit availability declined. Fig fruit was a preferred food of Kahuzi chimpanzees, and it might have served also as a staple fallback food. Honey, bees, and ants were filler fallback foods, whose collection was facilitated by tools. Readily available staple fallback food such bark enabled gorillas to remain in cohesive groups and to maintain the same home range despite fluctuations in fruit abundance, while filler fallback foods such as terrestrial herbs accommodated chimpanzees in small home ranges in the forest and animal foods allowed them to expand into arid areas.[156]

The separation might have been caused by one or more of the following factors: chimpanzees might have been avoiding contact with gorillas, and the gorillas ranged widely while chimpanzees repeatedly used a small area where fruits were available.[157] Moreover, they had different staple foods: gorillas ate leaves and bark, and chimpanzees ate figs.[158]

A yearlong study of sympatric eastern chimpanzees and mountain gorillas in the Bwindi Impenetrable National Park, Uganda, indicated that their dietary patterns and potential for competition over access to their preferred foods in common resembled those of other sympatric chimpanzees and gorillas.[159] In particular, like Grauer's and western gorillas, the Bwindi gorillas were notably frugivorous in comparison with Virunga mountain gorillas [160]

Each month, Bwindi gorillas and chimpanzees often ate the same plant species; however, Bwindi gorillas ate more plant parts ($n = 133$) of more species (≥ 96) than the chimpanzees ($n = 60$ plant parts of ≥ 32 species). Further, the gorillas foraged more species of fruit than the chimpanzees.[161]

The Bwindi chimpanzee diet comprised 64.6 percent fruit and 27.1 percent leaves and pith, supplemented with ants, bees, bee larvae, honey, and mammalian meat. The remains of ≥ 1 (mean: 2) fruit species were found in 98.4 percent of chimpanzee fecal samples. Figs were the most common fruits in Bwindi chimpanzee feces: 29 percent versus 14 percent for the next most common dietary fruit species.[161]

In contrast, the Bwindi gorilla diet comprised 24.6 percent fruit and 75.4 percent leaves and THV. More than one fruit species (mean: 2.2) were found in 47.2 percent of gorilla fecal samples. Particles of decayed wood were found in 19.4 percent of gorilla fecal samples, and they also ingested small stones.[161] Decaying wood provided 95 percent of their dietary sodium, though it represented only 3.9 percent wet weight of their food intake.[162]

Frequent overt feeding competition and interspecific aggression can be averted, even during lean fruiting seasons:

- Bwindi gorillas forage figs, the preferred fruit of chimpanzees, much less often.
- During lean fruiting seasons, gorillas rely more on fibrous filler fallback foods.
- Bwindi chimpanzees eat a variety of animals that gorillas do not eat.[163]

Food for the Ancestors

Given the problem of discerning the diets of organisms that we can observe in the wild, interpreting the results of experiments on captives, and testing for nutrients in natural foods, discovering the diets of extinct Primates would appear to be a hopeless endeavor. Nonetheless, we can glean a few salient clues upon which to construct scenarios on the ancient diets of our ancestors.[164] Apart from stimulating creative evolutionary stories, the exercise provokes us to reexamine and to modify our ideas about the costs and benefits of modern human diets and dietary fads.

Divergent hominid dietary energetics from great ape patterns probably played a key role in altering reproductive parameters leading to enhanced female fecundity. Exploitation of new foods and processing foods in new ways increased dietary quality. Cooked foods and increased consumption of meat and marrow, especially as weaning foods, would have enhanced female energetics, shortened the interbirth interval, and decreased juvenile nutritional dependence.[165] Further, the long view of hominid dietary history indicates that as our Plio-Pleistocene ancestors became increasingly cultural beings, their dietary practices developed via genomic-cultural adaptation.[166]

The anterior dentition of *Homo sapiens* and earlier species of *Homo* indicates the utility of slicing foods into smaller bits for further processing by the posterior teeth, which are shaped to crush and grind objects. Although having relatively small teeth in comparison with earlier hominids and many other mammals, thick enamel extends the longevity of human teeth.[167] Like the anterior dentition of apes and many other primates, those of *Australopithecus* spp. and *Homo* spp. suggest that frugivory, probably combined with variable amounts of opportunistic insectivory and eating young leaves, other soft plant parts including flowers, unripe seeds, and underground storage organs, manually collected tree exudates, and honey were part of early hominid diets.[168]

Findings from experimental studies of jaw robusticity, shape, and mechanics are helpful to model how similar features in Pliocene–Early Pleistocene Hominidae, particularly *Paranthropus* and *Australopithecus,* were adapted to masticate hard and tough foods.[169] Differences between the teeth, jaws, and cranial structures that accommodate the muscles of mastication indicate that *Paranthropus* was a specialized herbivore, whereas *Australopithecus* and early *Homo* spp. were more euryphagous, including some carnivory.[170]

Comparison of dental microwear in fossils with that in extant species can also provide approximations of physical properties of foods eaten by extinct hominoids, but researchers must be alert to taphonomic processes—sedimentary abrasion and acid erosion—that can be misleading.[171] Analyses of the microwear patterns on the anterior dentition of 3.66-Ma Laetoli hominids and

Hadar *Australopithecus afarensis* suggest that in addition to being frugivorous, they employed their incisors to strip leaves and grasses and the C/P$_3$ complex to puncture-crush food.[172] Concave erosion on the second molars of a specimen (OH-16) of *Homo habilis* indicates that this individual's diet included highly acidic plants.[173]

Microwear patterns on the posterior dentition of *Paranthropus boisei* suggest that they were as frugivorous as chimpanzees, orangutans, and mandrills and probably did not eat significant quantities of grass seed, leaves, or bone.[174] Stable isotopic analyses (C$_4$) support the hypothesis that, like *Australopithecus bahrelghazali,* their diet was predominantly grasses or sedges.[175] Indeed, their teeth contain higher levels of C$_4$ than those of South African *Paranthropus robustus.* Therefore, Cerling et al. concluded that the remarkable craniodental morphology of *Paranthropus boisei* was adapted to process large quantities of low-quality vegetation instead of hard objects.[176]

Comparisons of dental microwear in geographically diverse Neandertal samples and recent populations of human hunter-gatherers with documented diets indicate that meat predominated in Neandertal diets, albeit with plants also being important dietary components in some areas, particularly in mixed and wooded habitats.[177]

Unfortunately, even with palynological and other paleoecological data, it is impossible to model habitats with enough specificity to approximate a full menu of available food sources, let alone to determine what our potential ancestors ate as preferred and fallback foods, though there might be anatomical features indicating that some species were more likely to employ fallback strategies than others.[178] At best, such information can allow us to set boundaries on the range of possibilities. At sites where vegetal and animal food remains are preserved, we can begin to piece together some potential dietary components, but inferences about spatiotemporal usage range from elusive to phantasmal. Nonetheless, as it is reasonable to assume that during the earliest stages of stone tool-use hominids also had tools of vegetal materials that left no trace, one must allow that what paleoanthropologists find in the way of potential food remains are predominantly woeful indications of what the individuals actually ate over the course of their lives.

Stable isotopes in tooth enamel complement the gross morphological studies of teeth and jaws. For example, carbon isotope ratios (^{13}C/^{12}C) assist in discriminating between animals that ate tropical grasses, which use C$_4$ photosynthesis, and those that consumed plants that employ C$_3$ photosynthesis.[179] In East Africa, the C$_4$ grasses appeared in the Middle Miocene and expanded thereafter. During the Late Miocene and Pliocene, the diets of equids, rhinocerotids, bovids, hippopotamids, and suids consisted predominantly of C_4 plants; the elephantids and gomphotheriids transitioned from predominantly C_3 to C_4 herbivores; and the giraffids and deinotheriids mainly ate C_3 plants (trees, bushes, shrubs, and forbs).[180] Unfortunately, based on carbon isotopic ratios, we cannot determine whether an animal ate a particular kind of plant or animal that had eaten such plants.[181] Carbon isotopic ratios from a sample of ten *Australopithecus africanus* dated to ca. 2.5–3.0-Ma indicate that they were intensively engaged in the savanna foodweb via highly opportunistic and adaptable feeding habits.[182]

Strontium isotopic ratios ($^{87}Sr/^{86}Sr$) and strontium calcium ratios (Sr/Ca) from ca. 1.8-Ma skeletal specimens at Swartkrans, South Africa, have indicated that the diet of *Homo* sp. included more underground storage organs— bulbs, tubers, corms, and rhizomes—of geophytes than the diets of *Paranthropus robustus*.[183] Indeed, based on Sr/Ca and $^{13}C/^{12}C$ ratios, it appears that the diet of *Paranthropus robustus* might have been more euryphagous than indicated by their dental morphology and microwear.[184]

Archaeological data play an increasingly important role in models and scenarios on reduction of tooth size, particularly the models that stress increased technological versatility and innovation, cooking, or both.[185] For instance, Lucas hypothesized that during human evolution tool use had a major influence on the anterior teeth by reducing the input particle size, while cooking influenced the size of the posterior teeth by reducing the toughness or stress-strain gradient of foods.[186] The unique morphology of the human symphyseal structure, particularly the presence of a salient mental eminence, might well be related to reductions in human dentition and concomitant changes in the mechanical properties of the foods eaten by *Homo sapiens*.[187]

Long after the beginnings of crude stone-tool making in the Late Pliocene, the ancestors of *Homo* spp., including *Homo sapiens,* continued to subsist, sometimes quite successfully, via foraging for natural plants and animals. For instance, in Europe wild grains and other starchy plants appear to have complemented the meaty diets of Middle-Upper Paleolithic *Homo sapiens* about twenty millennia before the advent of farming.[188] In most areas, farming emerged slowly if at all via additions of horticultural produce from gardens to what had been basically foraging economies.[189]

The shift to heavy reliance on cultivated plants and domesticated animals did not begin until the end of the Late Pleistocene and early Holocene in at least seven major areas of the globe.[190] Further, in historic times and even today relatively small populations continue to subsist in forests and much more open areas via foraging, though many of them are in contact with neighbors who subsist on cultigens and domesticated animals.[191]

Viewed narrowly from a privileged status in modern society, it might seem that foragers became cereal farmers simply because crop cultivation was a better way to subsist; however, it is also likely that the social and demographic aspects of farming—its facilitation of population growth and military prowess, instead of its productivity of labor in contrast to foraging—might have been essential to its emergence and spread. Indeed, one might expect that the energetic output per unit of direct and indirect input of work was initially higher for early farmers than for many foragers. Further, there is paleoanthropological evidence that the health of populations declined during the Neolithic demographic transition from foraging to farming.[192]

Homo sapiens differ notably from other extant Primates in the extent to which adults regularly share food with weaned progeny, kin, and other group members. Because of food sharing, groups of hunter-gatherers can survive and reproduce in harsh habitats where juveniles cannot be self-sustaining.[193] For example, after the age of 5 years, Hadze children forage only about 50 percent of their caloric needs.[194]

The beginnings and diversification of extensive intragroup food sharing are obscure, but they might reasonably be expected of fossil hominid species whose maturation was slower than those of extant great apes and who left traces of social residence and feeding.[195]

To Sleep, Perchance to Die!

Like virtually all human denizens of Earth, safe daytime resting and nocturnal sleeping sites are vitally important for forest apes if they are to conserve energy, shelter from inclement weather, and avoid terrestrial, arboreal, or volant predators, parasites, and pests.[196] It is reasonable to expect that deprivation of sleep and daytime rest could predispose apes, like people, to illness and the risks related to daytime inattentiveness.[197]

Although all great ape species build tree platforms—nests—arboreally at night, some African apes rest diurnally on the ground, and in some areas the heavier gorillas nest nocturnally on the ground. Field studies recommend no secure generalizations about the determinants of lodge sites by apes, largely because there have been few nocturnal and crepuscular observations on factors that might disrupt the sleep or terminate the lives of apes nesting in forests and less densely treed areas. Pioneer and some later field primatologists were more inclined to use nests to indicate population distribution and density and group composition than to focus on why their subjects lodged where they did, particularly when the subjects were not habituated to their presence.[198] Notable exceptions have been the researchers who have proffered ideas about why their focal species lodge where they do.[199]

Commonly, primatologists emphasize the food quest—sometimes combined with social factors—as the major feature of the ranging behavior of apes. From this perspective, the apes simply lodge near the place where they ceased foraging, but this inference does not take several factors into account:

- Nocturnal predatory attempts by felids, humans, and snakes.
- Food preferences and activities of sympatric nocturnal foragers.
- Arboreal floral niches of ectoparasites and intermediate hosts of endoparasites that infest apes.
- Horizontal, vertical, and temporal fluctuations in temperature and moisture.
- Preemption of potential apes lodge sites by sympatric primates and other vertebrates.[38]

Gibbon Roosting

Like cercopithecoid monkeys, gibbons roost arboreally with the aid of ischial callosities. By contrast, great apes support their bulk by building nests, in which they recline.

Whitten and Tilson and Tenaza examined factors that potentially affect the choice of lodge trees by one species of ape, *Hylobates klossii*.[200] Like white-handed gibbons, Mentawai gibbons generally roosted centrally in their ranges. They preferred lianaless emergents, perhaps because local hunters climb vines in order to shoot primates with arrows. Pythons also might reach them via vines. At Whitten's study site, the Mentawai gibbons avoided lodge trees that hosted not only lianas but also epiphytes of a species *(Myrmecodia tuberosa)* that accommodates biting ants. Similarly, *Hylobates pileatus* at Khao Ang Rue Nai Wildlife Sanctuary in Thailand preferred tall (mean: 38.5 meters) emergent trees free of lianas, perhaps chiefly to avoid predators.[201]

More direct observations are needed to establish the empirical reality of statistical correlations between Mentawai lodge trees and the avoidance of biting ants. Whitten did not mention whether, in fact, the ants are active at night or simply keep gibbons out of their host trees before nightfall.[202]

Hylobatid species vary in the cohesiveness of family groups in lodge sites. Siamang and Mentawai gibbons are very cohesive roosters; each family occupies branches in a single tree or in closely juxtaposed trees. Other species such as white-handed, agile, and pileated gibbons have somewhat more dispersed roosting patterns, albeit in the centers of their ranges. All species roost at notable heights, and some seem to prefer emergents that project above the main canopy or trees on high ridges that might be good vantage areas from which to emit morning calls and to restrict access by human hunters and nocturnal predators such as clouded leopards and pythons.[203]

Great Ape Nesting

> Nesting behavior illustrates the appearance and phylogenetic development of constructivity, and, coincidentally, the transition from complete dependence on self-adjustment to increasing dependence on manipulation or modification of the environment as a method of behavior adaptation.
>
> R. M. YERKES and A. W. YERKES (1929, p. 564)

Great apes usually build a new nest or reline an old nest each night, but there is little evidence that they possess full-fledged nesting instincts.[204] They need safe new or refurbished nests each day because their foraging rounds take them to different localities. Moreover, their transience probably assists in confounding competitors and potentially also predators. There are reports of leopards chasing great apes but few reliable accounts of free-ranging apes being killed by lions, leopards, tigers, or other carnivores. The remains of apes in lion and leopard feces are inconclusive as to whether they were killed or scavenged by the cats. Moreover, some leopards appear to avoid chimpanzees, perhaps because of group defense or because they lack a taste for them.[205]

Frequent changes of bedding probably also reduce the parasitic load of apes and infestations by nocturnal bugs. During rare instances of nest reuse, apes usually refresh the lining with newly picked leafy sprigs. Reused nests and nest sites are commonly located near plentiful food supplies.[206]

Except for females with dependent youngsters or consorting couples, adult orangutans are quite solitary during the day, and they nest alone at night.[207] In Sumatra, the orangutans that were most likely candidates for predation by clouded leopards—the adolescents and females with infants—built nests higher in trees and farther from their last food tree of the day. Their lodge trees were also more cryptic and inaccessible than the nesting trees of adult males and solitary adult females, who nested in or close to their last feeding sites.[208]

At several sites in Indonesian Borneo, flanged male orangutans nested lower and more centrally in trees more frequently than the immature individuals, perhaps because the heavier males needed more secure supports in addition to being less vulnerable to predation by clouded leopards *(Neofelis diardi)*.[209] Adult males reused nests more frequently than other orangutans, and females with infants nested near the main stem under the tree crown or at the ends of branches covered by the crown of a neighboring tree that had overhanging branches to provide protection against rain, wind, and hypothermia.[210]

While being followed by humans in Gunung Palung National Park, adult male *Pongo pygmaeus* were significantly less vigilant, built nests that were closer to the last food tree of the day and lower in the canopy, and traveled less far than the females. Accordingly, they probably conserved energy.[211]

Sumatran orangutans cover their heads with large leaves or the ends of leafy branches during rain and sometimes roof their nests so well that they emerge dry after a downpour. Van Schaik linked this practice with the fact that they have less social contact with other orangutans to explain their low level of respiratory infections compared with other apes.[212]

Like orangutans, bonobos and chimpanzees prefer trees for night nesting (Plate 7). In some areas, they seem to select a particular species of trees for nesting. For example, at Yalosidi, though bonobos nested in 103 species of trees, nineteen species accommodated 81 percent of them.[107] At Lomako, bonobos nested in only twenty-six of fifty-two (50 percent) potentially available tree species in seventeen (of 48 total censused) forest plots that contained nests.[213] Like Sumatran orangutans, Wamba bonobos sometimes used leafy branches as rain covers.[214]

There were dramatic differences in the heights of bonobo nocturnal lodge sites at the several study localities.[215] In some regions, they nested relatively low in saplings; elsewhere, they lodged at medium or great heights and sometimes in the giant trees of the forest.[216]

Modal Lomako bonobo nests were located in trees of the middle canopy, 15–25 meters tall, with stem diameters >15 centimeters. More than half of the nests had no vegetation overhead. The height of the lowest branch, on average 10 meters above the ground, also seemed important for avoiding terrestrial predators, particularly leopards. Leaf size did not seem to be important in comparison with the availability of solid side-branches and flexible, densely leafed twigs.[213]

In Lomako Forest, bonobos commonly nested diurnally in trees bearing ripe fruit but not at night, unless the tree no longer bore fruit. Further, on average, the day nests were higher than those in nocturnal lodge trees.[213] Terrestrial nests occur occasionally in bonobo habitats, but they probably use them diurnally.

Like bonobos, which sometimes nest in groups of ≥20 individuals (mean: 9), chimpanzee nesting is generally much more communal than that of orangutans.[213] Chimpanzee foraging parties nest together, split into smaller nesting groups, or join neighboring foraging parties for the night.[217] There was a cluster of twenty-three chimpanzee nests at Ugalla, Tanzania.[218] A Gombe chimpanzee nesting party produced a cluster of seventeen nests, and another resulted in ten nests in one tree. Independent Gombe males tended to lodge ≥90 meters from females and juveniles that nested within a few meters of one another.[219]

In relatively open arid areas, especially ones inhabited by lions, chimpanzees seem to seek the tallest available trees with suitable branches for nesting.[220] But in primary forests they rarely lodge at the greatest heights.[217] In this respect, they contrast with the Malaysian siamang, who roost in tall emergents. Chimpanzees generally avoid fruiting trees as lodge sites, though eastern chimpanzees in the montane forest of Kahuzi-Biega National Park are a notable exception.[221] Occasionally nests occur on the ground in chimpanzee areas, but most of them probably are for daytime rests. However, Lawick-Goodall noted an adult male nesting beneath the palm tree in which his consort lodged.[219]

Generally, day nests are structurally simpler than night nests, but Nimba Mountains chimpanzee terrestrial nests, which accounted for 6.1 percent of all nests in a sizeable sample ($n = 994$), might have served as night nests. Fifty-two percent of the sixty-one day nests were "elaborate," and 48 percent of them were simple.[222] The elaborately constructed terrestrial nests were mostly at sites with groups of arboreal nests. A DNA analysis of shed hairs showed that males had occupied them. Low or nil predation pressure might account for the chimpanzees nesting on the ground, and they do not seem to have been affected by climatic conditions or a lack of appropriate lodge trees or relative abundance of TVH. Social factors also likely played a role.[222]

A comparison of the nesting behavior of chimpanzees in two similarly treed habitats in Senegal supported the hypothesis that chimpanzees nest higher in trees and closer together in a habitat with large predators than in one with no large predatory species.[223] In the latter area, during the dry season chimpanzees sometimes diurnally rested and fed in caves, probably to avoid the higher temperatures of the open grassland, woodland, and gallery forest.[224]

In Budongo Forest, male eastern chimpanzees commonly nested lower in trees than the females, perhaps because of their greater mass.[225] However, the argument based on different relative mass is weakened by the fact that, on average, both sexes nested at similar heights and higher in trees during the day than at night.

In Kalinzu Forest Reserve, Uganda, the nest group sizes and locations of eastern chimpanzees varied seasonally in response to where trees bore the most fruit.[226] Further, they had clear preferences for species in which they built nests. Although they nested in 43 of 111 tree species in Kalinzu Forest, >90 percent of them were in only fourteen species, and they preferred trees of medium size.[227]

Eastern chimpanzees in Bwindi Impenetrable National Park lodged at heights of 2 to 46 meters in 38 of 163 tree species, only four of which accounted for 72.1 percent of all lodge trees. There was no significant relationship between their choice of tree species and its abundance. Food tree species accommodated 93 percent of all chimpanzee nests, though not always while they were providing food items. Indeed, only one of the four species that chimpanzees used for lodging is an important food species, and the trees shared no obvious structural features. Although nearly all Bwindi chimpanzee nests were in trees, only 22 percent of sympatric mountain gorilla nests were arboreal.[228]

In Kahuzi-Biega National Park, larger foraging parties (evidenced by ≥5 contemporary nests) from a group of fourteen adult and eight immature eastern chimpanzees commonly nested in trees bearing their preferred ripe fruit, especially in the trees wherein the sympatric Grauer's gorillas also ate the fruit.[229] Because the gorillas, who outnumbered and outweighed the chimpanzees, frequently foraged in secondary forest, the chimpanzees rarely nested in the fruit-bearing trees there. When they did so, they were in larger numbers. They were especially attracted to the secondary forests when figs were available, which was reflected in larger chimpanzee nest groups.[229]

Figure 7.1. Soiled terrestrial gorilla ground nest.

Among apes, only some gorillas commonly lodge at night on the ground (Figure 7.1). For example, at the high-elevation (>2,200 to <4,200 meters) site of Kabara, where sturdy trees are scarce, 97 percent of 2,488 mountain gorilla nests were terrestrial.[230] But when sturdy trees are available and ground cover is sparse, some gorillas seem to prefer to nest arboreally.[231]

In contrast, a group of Bwindi mountain gorillas built 96.5 percent of 446 nests at forty-seven nest sites on the ground, sometimes after traveling through primary forest with tall trees. They used herbaceous plants, especially bracken fern *(Pteridium)*, herb/shrubs *(Mimulopsis)*, and vines *(Ipomea)* to build 99.7 percent of their nests in canopy gaps and secondary forest.[232] In this respect, they resemble the western gorillas in the Central African Republic and Gabon.[233] Nest reuse also varies among gorilla groups, but overall it is rare.[234]

Gorilla groups nest cohesively. The spatial distribution of terrestrial nests at a site seems to be random with regard to the age and sex of the lodgers. Thus, it appears that the silverback may not be placing his nest strategically in order to protect group members from nocturnal predatory attacks.[235]

Terrestrial nesting by Grauer's gorillas in the montane Kahuzi Forest—88 percent of nests over three years—did not correlate significantly with the seasonal changes in rainfall. The age of the individual influenced whether it nested on the ground or in trees, regardless of the microhabitat in which the group lodged. Immature members of the groups, who are more vulnerable during contested visits by lustful extragroup males and perhaps foraging leopards, most frequently nested arboreally and at greater heights (≤15 meters) than the silverbacks (≤4 meters) and other adults (≤9 meters).[236] After the focal silverback male of a group died, the other members, particularly the immature individuals, nested even less often on the ground until another silverback joined them. Moreover, the youngsters continued to eschew terrestrial nesting for a longer span than the other group members.[237]

In contrast with Kahuzi gorillas, western gorillas at Mondika in the Dzanga-Nkoki National Park built nests more frequently when the climate was wet and cool, and their nesting habits were independent of diet, ranging, and group size.[238] Of 3,725 nests, 79.1 percent were terrestrial, and 20.7 percent were arboreal. Many of the terrestrial beds (45 percent) were simply a flattened area, with no construction using surrounding vegetation. Like Kahuzi gorillas, at Mondika the silverback built arboreal nests less frequently (2 percent) than the other adults and youngsters (21 percent). With warmer temperatures, the Mondika gorillas slept more often on bare earth without bothering to construct a nest, and they increased nest construction on the ground (the silverback) and in trees (the other group members) as rainfall increased.

At Bai Hokou, there was a strong influence not only of rainfall but also of microhabitat on western gorilla nest construction.[239] Twenty-eight percent of 1,231 gorilla tree nests were in primary forest, where undergrowth was sparse, versus 8 percent in light gaps that had abundant herbaceous plants. Further, tree nests were more common during the wet season (19 percent) than the dry season (7 percent), and bare ground lodge sites were more common during the dry season (56 percent) than the wet season (45 percent).

Western gorillas in the Lopé Reserve preferred to nest on the ground (64 percent of 2,435 nests), especially in terrestrial herbaceous vegetation (40 percent), as opposed to arboreally (35 percent), even though trees were abundant in their habitat. Exclusively arboreal nests occurred at 8.3 percent of 373 Lopé gorilla nest sites; 10.2 percent of total nest sites were arboreal, except for one terrestrial nest attributable mostly to silverbacks.[240] When elephants were attracted to a localized food source that gorillas also ate, they nested more frequently in trees. They also nested arboreally more frequently during the rainy season for reasons that Tutin et al. could not determine precisely.[240] To some extent, foraging for arboreal fruits in localities with limited THV might have been a factor. In Cameroon, western gorillas nested in vegetation that limited visibility, especially in patches that were difficult to penetrate.[241]

Ground nests not only insulated the Lopé gorillas from damp ground and allowed them to have a better quality sleep but also might have reduced skin irritation and exposure to parasitic larva.[240] Nonetheless, Mondika gorillas slept on bare ground more often that the Lopé gorillas, even though they had higher parasitic loads of protozoans and helminths.[242]

The role played by parasites and diseases more broadly in hominoid evolution is a topic for which the empirical base is sparse but growing.[243] Studies on free-ranging extant apes have begun, but in most areas the epidemiology of apes must be studied comparatively with that of the humans and other anthropoid primates that live in or near and are active in the forests.[244]

When Did Our Ancestors Leave the Nest?

If we assume that our apish ancestors also nested arboreally at night, one is tempted to speculate on when in hominid history the practice was abandoned.[245] Obviously, once our ancestors quit the trees and spread into open country, tree lodging would cease. Given the uncertainties of sorting out the ecological and behavioral factors that influence why some African apes rest or lodge on the ground, it is quite problematic to construct persuasive scenarios on the role that secure, comfortable, and healthful lodge sites played in the evolution of earthbound hominids during the transition from obligate arboreal lodging to other rest sites in sparsely treed areas that could not accommodate even a modestly sized social group. Indeed, there is evidence that caves, rock shelters, and other structures took the place of arboreal platforms in a variety of biomes. But during earlier periods, continued reliance on arboreal nesting as a safeguard against nocturnal predation might have affected the size and composition of early hominid groups and their choice of habitats in which to range.

Archeological traces of human-made shelters occur, albeit rarely, in Middle Paleolithic Europe from 60 Ka, and become common by the Upper Paleolithic 40–10 Ka, particularly in regions with notable seasons of inclement weather. Body coverings derived from other animals or vegetal matter and control of fire would be boons not only to the hominids who spread into temperate and even colder climes but also during the night in subtropical and tropical regions where temperatures drop notably at night and after chilling rains. Cases for pre–Middle Pleistocene hominids regularly using fire and body coverings are arguable. Increasingly persuasive dates for the earliest use of fire by Hominidae range from ca. 1.6 Ma in East and South Africa to ca. 230 Ka in Eurasia.[246]

Archaeological evidence indicates that early hominid migrants to Europe were not habitual users of fire. Apparently, control of fire did not become part of European hominid behavioral repertoires until ca. 300–400 Ka, after which both Neandertal and Late Upper Paleolithic people employed it versatilely for cooking, warmth, and perhaps lighting. Further, Neandertalers also employed fire to produce hafting materials.[247]

The persistence of features regularly associated with arboreal primates—such as curved fingers and toes and a pelvic structure that would facilitate vertical climbing—in Pliocene *Australopithecus* spp. and Pliocene–Early Pleistocene *Homo* spp. indicates that they probably were still entering trees to forage, to escape terrestrial predators, and perhaps to rest during their travels and at night.[248] Such arboreal behavior should not be surprising to persons who have traveled globally to places where one can witness barefoot people scaling notable heights to obtain honey, coconuts, succulent fruits, and other arboreal delights.[249] Moreover, I recall being warned to always have an escape tree nearby as I tracked monkeys in a forest on Mount Meru, Tanzania, which also had resident buffalo, rhinoceroses, and elephants. From experiences in my youth, I was confident that even shod I had a fair chance of scaling many of the trees were it necessary.

8

Hunting Apes and Mutualism

Vegetarian—that's an old Indian word meaning "lousy hunter."

ANDY ROONEY

By routinely including animal protein in their diet, [humans]
were able to reap some nutritional advantages enjoyed by
carnivores, even though they have features of gut anatomy
and digestive kinetics of herbivores . . . Using meat to supply
essential amino acids and micronutrients frees space in the gut
for plant foods . . . Because these essential dietary requirements
are now being met by other means, evolving humans would
have been able to select plant foods primarily for energy
rather than relying on them for most or all nutritional requirements . . . This dietary strategy . . . would have permitted
ancestral humans to increase their body size without losing
mobility, agility, or sociality . . . [and] also have provided the
energy required for cerebral expansion.

KATHARIN MILTON (1999a, p. 11)

One of the most persistent theories about the evolution of hominid bipedalism links it with tool use for hunting and defense. According to the basic
man-the-hunter scenario, hominid males hunted prey with weapons and
shared meat with females, who serviced them sexually thereby producing
more hunters. Language developed so that they could discuss hunting tactics
and group movements and to keep intragroup rivalries over meat and mates
in check without having to use their weapons on one another. Although the
idea had wide appeal for many decades, to some observers it appeared to be

little more than a corporate male fantasy. Hence, as prime rib became less popular in the Yankee diet, red meat began to slip from primacy in models of hominid evolution. Field studies on apes helped both to dismantle and to reinforce aspects of the man-the-hunter scenario. Scenarios that emphasize scavenging, collecting vegetal foods, arthropods, and small vertebrates by both females and males, and bisexually sharing resources threatened the credence of hardcore man-the-hunter scenarios. Nonetheless, they remain the most attractive option to some theorists.

Scenarios of man-the-hunter were refueled by pioneer reports of meat eating and hunting among Gombe chimpanzees, followed by recognition that it is characteristic of many chimpanzee communities across Africa.[1] Concomitantly, as the United States flaunted its firepower in Vietnam, the aggressive male regained ascendency.

Chimpanzee Hunters and Meat-Eaters

> Cooperative hunting can evolve only when it increases individual feeding efficiency.
>
> CRAIG PACKER and LORE RUTTAN (1988, p. 161)

Chimpanzees, bonobos, baboons, and capuchins constitute the rare nonhuman primates that systematically hunt sizeable vertebrates, and bonobos and chimpanzees are the most inclined to share prey.[2] Goodall initially noted chimpanzees eating an infant bushpig, an infant bushbuck, and a small mammal, but no hunting. The first successful hunt that she witnessed was two adolescent males capturing an adult red colobus that they and several other chimpanzees consumed.[3] Nonhuman primates were the chief mammalian prey of chimpanzees at eleven of twelve research sites.[4]

In 1963, researchers set up a banana-provisioning area in the Gombe camp to facilitate filming and observations of their social behavior.[5] Sometimes when baboons also visited the camp, chimpanzees captured and ate the young ones.[6] Teleki summarized the first decade of data on the predatory behavior of Gombe chimpanzees with a special emphasis on baboon hunting during a one-year span.[7] The modes of acquisition, killing, division, and consumption of mammalian prey that Teleki described for Gombe chimpanzees are basically similar to those reported for chimpanzees at other localities in East, West, and Central Africa.

Like carnivoran predators, chimpanzees variably employ methods ranging from opportunistic lunges and grabs for unwary or poorly protected victims to stalking and chasing the prey. They might act singly or in larger groups. Individuals are more likely to act opportunistically, whereas sustained chases and stalks enlist more individuals. For instance, during an aggressive display, an adult male suddenly grabbed an infant baboon from its mother's lap and flailed it about like a branch. On another occasion, four males rose abruptly and moved quietly toward a small group of baboons; later one of them raised a juvenile baboon overhead and bashed its head on the ground.[8]

In many cases it is difficult to determine whether groups of individuals are actually hunting cooperatively.[9] Busse argued that Gombe chimpanzees did not.[10] Similarly, collective chimpanzee predation probably was not cooperative at Mahale, Tanzania.[11]

Contrarily, Boesch and Boesch concluded that in Taï National Park in Côte d'Ivoire the western chimpazees conducted some hunts in somewhat larger parties that seemed to have a specialized prey image and to search intentionally for adult prey. Taï chimpanzees might follow the same direction for long stints, and if they changed direction, they pant-hooted and drummed, presumably to maintain contact with others that were out of view. Like chimpanzees at other localities, they remained silent otherwise and followed one another closely single-file, stopping frequently to scan the canopy for potential prey.[12]

Tomasello et al. disputed the claims that Taï chimpanzees hunted cooperatively in complementary roles versus each individual being stationed where he might grab red colobus as they fled in various directions through the canopy or fell to the ground.[13] These researchers reasonably noted that using terms like driver, blocker, chaser, and ambusher conveyed a greater sense of cooperative activity based on a prior plan or agreement on a joint goal or assignment of roles than might be actually operant.[14] Nonetheless, given that captive chimpanzees quickly develop and remember which individual is the most effective partner with which to work to accomplish novel tasks that require cooperation, one should not be surprised that wild chimpanzees might possess or develop cooperative tactics and other reciprocal and affiliative dyadic and polyadic relationships as required to increase their hunting success.[15]

Thirteen adult males performed the vast majority of Gombe chimpanzee hunts.[7] Other researchers have concurred that males are the predominant hunters at Gombe, in the Taï Forest, and at Ngogo (Plate 10).[16] Although females and adolescents sometimes hunt successfully, there is no report of females killing baboons at Gombe.[17]

In Kasakati Basin, five adult and subadult female eastern chimpanzees vigorously participated with an adult male in chasing, capturing, and struggling over a red-tailed guenon.[18] Mahale female chimpanzees are also active hunters of birds and mammals.[19] They tend to grab young ungulates, while the males chase monkeys.[20] A 12-year-old Mahale female probed a tree hole with a stick that she had stripped of leafy side branches, thereby rousing a squirrel which she ate and shared with a 4.5-year-old companion. Functionally, the behavior was like that which Mahale females perform to obtain large ants from tree holes.[21]

At Fongoli in Senegal, Pruetz and Bertolani documented twenty-two instances of western chimpanzees searching for *Galago senegalensis* nesting in tree holes then attempting to jab them with branches that they had stripped and trimmed on the ends. Only one male and one female among the ten tool-bearing searchers were adults; the others comprised six adolescents, one juvenile, and one infant. Females were predominant (64 percent) among the hunters, and the only successful individual was an adolescent female. After jabbing into and widening the nest hole, she reached in and manually extracted the motionless prey.[22] Pruetz and Bertolani observed other instances of chimpanzees eating galagos and a vervet *(Cercopithecus aethiops),* but the researchers did not see how the prey had been obtained. As one might expect given the diminutive size of lesser galagos,[23] there was minimal sharing of the meat. In contrast, when pairs of Fongoli males captured young patas monkeys, they shared the prey.[24]

During a six-year span, a group of about twenty western chimpanzees at Bossou, Republic of Guinea, caught five tree pangolins *(Manis tricuspis)* and one wood-owl *(Strix woodfordii).*[25] Females, adolescents, and juveniles were the opportunistic captors, though adult males sometimes commandeered a carcass, albeit without the pandemonium that can characterize hunts by eastern chimpanzees. There was no begging or active sharing; individuals simply waited until a possessor discarded or dropped a carcass to feed on it.[26]

Boesch and Boesch maintained that Taï chimpanzees cooperated to hunt monkeys because the Taï rain forest canopy is much higher and otherwise more challenging than the woodland canopy at Gombe.[27] At Taï Forest, Gombe, and Kibale Forest, chimpanzee success in hunting colobus was positively correlated with the number of males in the hunting party (Plate 12).[28] Nonetheless, in Taï Forest, 13 percent of hunters were females, and over a two-year period they accounted for 18 percent of kills.[12] Adult females killed 10.7 percent of the 429 mammals that Gombe chimpanzees hunted between 1982 and 1991; adult and adolescent males were responsible for the other killings.[29] In Kibale Forest, during a thirty-four-month span, females in the adult Ngogo community killed only two colobus among 261 prey; adult (90 percent) and adolescent males (8 percent) were the predominant killers.[30]

Chimpanzees employ a variety of methods to dispatch their prey:

- Battering them against the ground or standing objects.
- Biting them on the nape of the neck or spine.
- Wringing their necks.
- Eating them alive.
- Pulling them apart while competing with others for the carcass.

Unlike large predatory cats and human killers, chimpanzees do not throttle their prey orally or manually. Further, with the exception of the Fongoli chimpanzees which probed and extracted galagos from tree holes with sticks, chimpanzees hunt and kill vertebrates without tools.[23]

Between March 1968 and August 1970, 48 percent of forty-four Gombe chimpanzee predatory attempts—mostly via seizures and chases of baboons—were successful.[7] Over a five-year period, Stanford recorded a 54.5 percent success rate for Gombe chimpanzees hunting red colobus. During a twelve-year span, Taï chimpanzees succeeded in 64 percent of red colobus hunts and 65 percent of all predatory attempts on seven primate species.[31] Ngogo chimpanzees achieved the highest success rate, 81.7 percent, for hunts on red colobus during thirty-four months.[32] There is good reason to infer that the decline in populations of red colobus and blue monkeys at Ngogo is due to predation by chimpanzees.

Intrasite and intersite fluctuations in chimpanzee hunting success could be due to variable numbers of skilled arboreal hunters, in addition to variable structure of forest canopies and the evasive and defensive actions of targeted monkeys.[33] It takes about twenty years for male chimpanzees to master progressively most aspects of hunting, especially skillful ambushes and blocking the escapes of colobus prey.[34]

Direct observations and fecal analyses have revealed that across tropical Africa wild chimpanzees consume individuals of ≥40 species of Mammalia, ≥22 of which are Primates (Table 8.1). Between 1960 and 1970, fifty chim-

Table 8.1. Mammalian prey of chimpanzees

Primates	Common Name
Cercopithecidae	
Cercocebus atys	Sooty mangabey
Cercopithecus ascanius	Red-tailed monkey
Cercopithecus aethiops	Vervet; Green monkey
Cercopithecus campbelli	Campbell's monkey
Cercopithecus diana	Diana monkey
Cercopithecus l'hoesti	L'Hoest's monkey
Cercopithecus mitis	Blue monkey
Cercopithecus mona	Mona monkey
Cercopithecus petaurista	Lesser spot-nosed monkey
Cercopithecus pogonias	Crowned monkey
Colobus guereza	Mantled guereza
Colobus polykomos	King colobus
Colobus satanus	Black colobus
Erythrocebus patas	Patas monkey
Lophoocebus albigena	Grey-cheeked mangabey
Papio anubis	Olive baboon
Papio cynocephalis	Yellow baboon
Procolobus badius	Red colobus
Procolobus pennantii	Pennant's red colobus
Procolobus verus	Olive colobus
Pongidae	
Pan troglodytes	Chimpanzee
Hominidae	
Homo sapiens	Human
Galagidae	
Galago senegalensis	Lesser galago
Otolemur crassicaudatus	Thick-tailed greater galago
Sciurocheirus alleni	Allen's galago
Lorisidae	
Perodicticus potto	Potto
Artiodactyla	
Bovidae	
Cephalophus callipygus	Peter's duiker
Cephalophus monticola	Blue duiker
Tragelaphus scriptus	Bushbuck
Neotragus moschatus	Suni antelope
Suidae	
Potamochoerus larvatus	Bushpig
Potamochoerus porcus	Red river hog
Pholidota	
Manidae	
Manis tricuspis	Tree pangolin

Table 8.1. (continued)

Rodentia
  Anomaluridae
    Anomalurus derbianus   Scaly-tailed flying squirrel

  Cricetidae
    Cricetomys emini   Giant pouched rat

  Sciuridae
    Funisciurus sp.   Squirrel
    Protoxerus stangeri   African giant squirrel

  Thryonomyidae
    Thryonomys swinderianus   Greater cane rat

Hyracoidea
  Procaviiidae
    Heterohyrax brucei   Yellow-spotted hyrax

Carnivora
  Viverridae
    Ichneumia albicauda   White-tailed mongoose
    Mungos mongo   Banded mongoose
    Viverra civetta   African civet

Insectivora
  Macroscelididae
    Rhynchocyon cirnei   Giant elephant shrew

Source: Modified from Wrangham and Riss 1990; Uehara 1997; and supplemented by data in Kuroda et al. 1996; Boesch and Boesch-Achermann 2000; Hamai et al. 1992; Nakamura 1997; Hashimoto et al. 2000; Newton-Fisher et al. 2002; Pruetz and Bertolani 2007; Watts and Mitani 2002; Pruetz and Marshack 2009; O'Malley 2010; Bogart et al. 2008.

panzees around the Gombe research camp killed and ate ≥95 mammals and attempted to capture thirty-seven others. Of the identified prey species, 65 percent were primates, and 35 percent were ungulates. Most primate prey were baboons and colobus. Ungulate prey was about equally divided between bushpigs and bushbuck. In the Gombe camp, baboons were the staple prey of the chimpanzees after the provisioned bananas attracted both species.[7]

Before Jane Goodall began her studies at Gombe, two documented adult male chimpanzee attacks on humans had occurred, presumably to snatch infants.[35] When searchers found the body of one human infant, it had been partially eaten, presumably by the kidnapper, although it is also possible that dogs or other scavengers had eaten it. In 2002 at Gombe National Park, the alpha-male of Kasekela community snatched a 14-month-old infant from a woman's back, killed it, and ate some of the body.[36]

Away from camp, Gombe chimpanzees most commonly preyed on red colobus. During 1973 and 1974, chimpanzees killed 8 percent to 13 percent of the Gombe red colobus population.[37] Others confirmed that red colobus are the most common prey of Gombe chimpanzees; annually they kill between 15 percent and 35 percent of the population.[38] About three-quarters of the victims are infants and juveniles.[29] Indeed, red colobus are the major mammalian prey of chimpanzees at all four sites where their hunting has been studied: Gombe, 58.8 percent in 1960–81 and 84.5 percent in 1990–95; Mahale, 56.3 percent in 1983–89 and 83.3 percent in 1990–95; Taï Forest, 77 percent; and Ngogo, 88.4 percent of mammalian prey.[39]

Predatory chimpanzees probably reduce population and group size and composition of other monkeys where they are sympatric. For example, in Kyambura Gorge, Uganda, groups of guerezas *(Colobus guereza)* were smaller and included fewer juveniles and infants in centers of chimpanzee activity than in groups where chimpanzees were less active. Further, guereza density was much lower—525 per km² versus 186 per km²—in areas of chimpanzee activity.[40]

Chimpanzees seldom kill baboons outside Gombe National Park; after food provisioning was stopped, baboons were rarely victims there.[41] Baboon killing was also very rare in Mahale Mountains National Park.[42] Hence, it seems reasonable to attribute past victimization of Gombe baboons to the banana bonanza.[43]

Division of the Prey

Often, all hell breaks loose when a chimpanzee catches a monkey in the company of other chimpanzees. This contrasts sharply with their silence when approaching prey. Unless the captor gets out of reach of other chimpanzees, they can snatch portions of the prey immediately after a catch. At that point, they calm down and beg, take bits from the possessor's mouth or hand, and glean dropped or abandoned bits of the carcass. Unlike Mahale chimpanzees, those at Gombe and Taï Forest usually did not engage in prolonged struggles for carcasses.[44]

Beggars get very close to the possessor and stare at its face or at the prey. The cadger might reach forward and touch the possessor's chin, mouth, or meat. Sometimes the beggar is rewarded with a choice morsel, a bit of skin or bone, or the masticated wadge of leaves that the possessor has chewed with bits of meat. Alternatively, the possessor may turn its back, move away, or express irritation by vocalizing, pushing, or striking at the beggar. At Gombe camp, 29 percent of solicitations were rewarded. Teleki never saw a vigorous fight over meat.[7]

Chimpanzees take between 1.5 and 9.0 hours (mean: 3.5 hours) to devour a monkey carcass. Unlike canids, chimpanzees do not bolt chunks of meat. Instead, they ingest small bits with leaves and spend relatively long periods sucking and chewing on the wadges. Gombe chimpanzees eventually spit out the wadges, but Taï chimpanzees swallow them.[45]

Although chimpanzee prey are not large on average, in Gombe camp eight individuals got bits of a carcass (range: 4–15 individuals). Sometimes only the possessor of the carcass ate the brain, but on other occasions more chimpanzees would obtain some of it.[46]

Extensive sharing of prey inspired suggestions that chimpanzee hunting is motivated by undefined social factors in addition to having some unmeasured nutritional importance.[47] Gombe chimpanzees often ate young baboons after stuffing themselves with bananas, so general hunger pangs probably are not the sole reason for their carnivory.[48] Researchers provisioning the site with bananas probably affected the intensity of meat-sharing by Gombe chimpanzees in the camp. Larger groups would visit when the bananas were plentiful.[45]

Factors that determine which chimpanzees will receive meat from a possessor are poorly understood for most populations.[49] Chimpanzees seem to have no vocal or gestural signal that is specific to sharing vertebrate versus vegetal foods. Moreover, laboratory experiments and field observations indicate that generally chimpanzees are selfish, in part due to their limited understanding and sharing of intentions, and in part due to their indifference to the welfare of unrelated group members, though they might be altruistic and empathetic toward some group members, especially kin or reciprocating partners.[50]

Food trials in an established captive group of twenty chimpanzees, including one adult male and eight mostly unrelated adult females demonstrated that grooming dramatically increased the probability that a groomee would share a bundle of leafy branches with a recent (≤2 hours) groomer. The effect of grooming was greatest in adult dyads that rarely groomed. Further, food possessors would resist approaches by individuals that had not groomed them.[51]

At Gombe, Taï, and Budongo Forest, access to meat did not conform neatly to a predictable social pattern, and it was not determined by overt competition.[52] For instance, possessors sometimes rewarded subadult and adult female beggars but refused higher-ranking males without incurring immediate penalties. Further, possessors could evade or hold onto their morsels vis-à-vis the strong-arm tactics of more dominant group members. In the only observation of insect-eating at Budongo, a dominant male broke off a piece of termite-infested mound and willingly shared it with a subordinate male, perhaps in order to receive support from him in the future.[53]

Conversely, at Mahale and Ngogo, dominant males often got the lion's share, regardless of their not being the original captors.[54] Ngogo data support the male-bonding hypothesis: the number of males per party was significantly higher when they hunt.[30]

Stanford et al. inferred that the number of estrous females influences the size of Gombe foraging parties, which are more inclined to hunt.[55] The male hunters share meat with these females, and sometimes they copulate. The Gombe females that were favored with meat produced more surviving infants than the other females did.[56] The dominant male in a Mahale chimpanzee group preferentially shared meat with his current sexual consorts, the mothers of his offspring, his own presumed mother, and very old females.[57]

Although Teleki stated that Gombe chimpanzees guarded bananas more selfishly than meat and that competition for bananas depended more upon brawn and social status, begging for bananas was nearly as successful (27 percent) as cadging for meat (29 percent).[58] Gombe chimpanzees rarely passed animal bits to tertiary individuals, except when mothers shared them with their dependent youngsters. One Gombe mother shared meat with her five progeny with decreasing frequency from the youngest to the eldest. Food plant parts, particularly ones that are difficult to open, are also most commonly shared between mothers and their subadult offspring. The greatest number of nonfamilial sharing events consisted of adult males giving provisioned bananas to presumably unrelated adult females. Adult Gombe males rarely begged for bananas.[7]

Teleki proposed that estrus affects the likelihood that begging females will acquire meat from adult males. Although he described no instance in which an estrous Gombe female obtrusively solicited and copulated with a male in exchange for meat, his figures are suggestive. The success rate of estrous beggars was 69 percent while that of anestrous females was 44 percent. Estrous females stayed closer to adult males and begged more persistently than anestrous ones did.[7]

Figure 8.1. Mahale chimpanzee eating a bushbuck calf *(Tragelaphus scriptus)*.

In contrast, Gilby's observations on the alpha male in the Gombe Kasakela community supported Wrangham's sharing under pressure hypothesis
that it is more energetically efficient to comply with a persistent beggar than to
defend a carcass.[59] Indeed, the alpha male shared in 100 percent of bouts with
persistent beggars versus 38 percent with low-intensity beggars. Moreover, he
did not preferentially share with estrous females (75 percent) versus anestrous
ones (71 percent).

At Mahale copulation also sometimes seemed to play a role in meat acquisition by females from males. Five of seven females that copulated with male
possessors received portions of the carcass between four and twenty minutes
afterward.[60] However, whether a female receives meat or not might be more
dependent upon her relationship with the male possessor than upon her estrous condition.[61]

At Mahale and Taï, males shared meat with male and female political allies
in their groups (Figure 8.1).[62] Further, Stanford argued that meat is social currency at Gombe: males used it as a tool of nepotism and social maneuvering.[63]
The best predictor of hunting was the presence of at least one estrous female

in the party. Hunting peaked in the season when the most females were in estrus. In 33 percent of successful hunts by parties with estrous females present, Stanford witnessed copulations in exchange for meat from males.[64]

Contrarily, based on a multivariate statistical analysis of data on Gombe chimpanzees over twenty-five years (1975–2001), Gilby et al. drew the following conclusions:

- The presence of sexually receptive females significantly decreased the probability of hunting; therefore, meat or sex instead of meat for sex seemed to prevail.
- Hunting party size did not seem to affect hunting by specific males, which casts doubt on the hypothesis that hunting facilitates male social bonding.
- Dietary quality was not associated with the probability of hunting or doing so successfully; that is, neither food shortages (nutrient shortfall) nor plentiful food (nutrient surplus) explained their hunting.
- Because per capita meat availability from prey decreased with adult male party size, hunting is probably not cooperative.
- Instead, simpler ecological factors appeared to apply, namely, visibility and prey mobility: hunts were more likely to occur and to be successful in open habitats—woodland and semideciduous forests—than in evergreen forests.[65]

The never-provisioned Ngogo chimpanzee community in Kibale National Park, Uganda, mostly hunt red colobus and make many more kills per hunt than the chimpanzees at other sites (Plate 10).[66] Ngogo is the largest known chimpanzee community and comprises the most males. They kill 6 percent to 12 percent of the red colobus population annually. Hunts are more likely to occur and are more successful where there are gaps in the canopy. Although the number of kills and the offtake of meat per hunt increase with the number of hunters, the per capita meat intake is independent of hunting party size. There is evidence of cooperation during some hunts. Relative dominance rank partly affects hunting success and meat intake among the Ngogo males. Ngogo chimpanzees hunted most frequently and intensively when fruit was most plentiful. Accordingly, it appears that hunting is more likely when energy needs can be met from other sources.[30]

The Ngogo data provide only limited support for the meat-for-sex hypothesis. Males achieved no greater share of matings with females during cycles when they shared meat than during cycles when they did not share meat with them.[67] Stronger support for the meat-for-sex hypothesis was provided by analysis of twenty-two months of data on Taï chimpanzees: "Although males were more likely to share meat with estrous than anestrous females given their proportional representation in hunting parties, the relationship between mating success and sharing meat remained significant after excluding from the analysis sharing episodes with estrous females"; accordingly, Gomes and Boesch concluded that "wild chimpanzees exchange meat for sex, and do so on a long-term basis."[68] The extent to which a twenty-two-month study is representative of other wild chimpanzees and even of Taï communities over a longer span remains to be demonstrated.

Chimpanzee Scavenging

Many species of the Carnivora that are notorious hunters also scavenge or pirate prey from conspecifics or other species. Among apes, chimpanzees are the only species reported to scavenge meat, albeit relatively rarely.[69] Ingesting uncooked carrion might be unwise because it is likely to be contaminated with bacteria that can cause gastrointestinal distress or worse.[70]

Goodall compiled twenty-seven incidents during about twenty-five years of Gombe chimpanzees eating bushbuck calves that they had or were suspected to have commandeered from baboons.[71] Ten (37 percent) of the probable pirates were female. Gombe chimpanzees even more rarely (ten reports) ate meat from carcasses on the ground in the absence of the killer.[72] Commonly, they ignored carcasses of mammals that they or their cohorts had not killed.[73]

Mahale chimpanzees eagerly devoured the meaty remains of adult or near-adult bushbucks (n = 2), blue duikers (n = 3), a red colobus, and a red-tailed monkey that had died of disease or perhaps leopard predation.[74]

Taï chimpanzees ate four primates that had been killed by crowned hawk-eagles *(Stephanoaetus coronatus)* and three times pirated red colobus from an eagle before the bird could kill its prey. Three times chimpanzees returned to feed on colobus that they had killed a day or two before.[75]

Two adult male Budongo chimpanzees ate part of a young blue monkey that an older blue monkey had been eating; during an eleven-year period, Ngogo chimpanzees scavenged two red duikers *(Cephalophus rubidus),* a red colobus, and a red-tailed monkey.[76] Like Tanzanian and Taï chimpanzees, those of Uganda commonly eschewed eating carrion that they had not killed.

Chimpanzee Cannibalism

Anthropologists argue endlessly about the incidence and symbolic meaning of cannibalism in human prehistory and more recent human societies, but there can be no doubt about its occurrence among chimpanzees (Figure 8.2).[77] Eastern chimpanzee cannibalism has been documented firsthand by scientists and professional field assistants in the Budongo Forest, Kibale National Park, Gombe National Park, and Mahale Mountains National Park.[78] Given their capacity for gentleness and protectiveness toward youngsters in certain situations, the gruesomeness of some instances is prima facie perplexing and horrific.[45]

The first observed instance was one of the more grotesque. In Budongo Forest, Suzuki encountered a familiar arboreal male holding a live, partly eaten, newborn chimpanzee.[79] Other males sat close to him, groomed him, and reached for the victim. The possessor and his groomer passed the body back and forth. The initial possessor ate from its lower body with wadges of leaves and allowed a few male beggars to touch the prey; after two hours, he descended to the ground with the baby in his mouth and disappeared into the bush, followed by the others. In 1995, Budongo chimpanzees were seen eating two more infants.[80]

Bygott recorded another incident in which five adult Gombe males encountered two adult females and vigorously attacked the eldest, who probably was alien to the area.[81] After momentarily losing track of them, Bygott found the males clustered in a tree where one of them held a live 1.5-year-old infant. Initially, the possessor held it by the hind limbs and battered its head against a branch; he bit flesh from its thighs, while another male wrenched off a foot and ate it. Over 1.5 hours, the possessor intermittently abused, toyed with, and ate bits of the carcass. After he had abandoned it, the second male dined on the body for 1.5 hours. After the initial possessor returned and flailed the carcass, more males snacked on it. However, they mainly ate other foods while holding the carcass. When they abandoned the carcass, six hours after initial capture, they had consumed one hand, the hind limbs, and the perineum. Bygott speculated that the infant was less appetizing as food and more a curiosity than the other mammalian prey of the Gombe chimpanzees.[81] These and other observations at Gombe were probably instances of *exocannibalism*—eating a member of an out-group of the same species.[82]

Beginning in 1975, Gombe researchers also documented instances of *endocannibalism*—eating conspecific members of the same social group.[83] For

Figure 8.2. Mahale chimpanzees eating an infant chimpanzee.

instance, an adult female attacked and killed the 3-week-old infant of a group member and shared the carcass with her adolescent daughter and infant son. About a year later, the daughter killed the recently born infant of the same female and shared it with her mother and brother. One month later, she killed another neonate in the group. Demographic records for the Gombe chimpanzee population indicated that over a three-year span only one infant survived longer than one month. The cannibalistic killings seemed to end, though Goodall intervened on one occasion to stop another attempted infanticide by the infanticidal daughter.[84] There have been too few incidents to draw generalizations about male versus female cannibalistic behavior at Gombe, though the contrast between females and males in which sex cannibalizes youngsters within the community versus infants from outside the community is striking.

Despite the fact that chimpanzees sometimes brutally attack and kill other adults, there have been no reports of chimpanzee cannibalism in which the victim was over 3 years old.[85] Further, attackers sometimes exhibit ambivalent behaviors toward the victims. For example, Mahale chimpanzees viciously attacked a female and her juvenile son, but they also groomed, embraced, and kissed their targets intermittently. The youngsters were 40 months old and 49 months old at the times of the attacks. Perhaps the youngsters' somewhat advanced age confused the muggers, but we will never know because human observers intervened to save the victims in both instances.[86]

One must wonder why chimpanzees do not cannibalize the older individuals that they kill during territorial altercations. They devour the bones, flesh, skin, and connective tissues of infant chimpanzees, monkeys, and young ungulates, and experiments have indicated that their powerful dental battery and jaw muscles can inflict damage to bovid and cervid ribs and long bones equivalent to that caused by Carnivora and perhaps some Pliocene and Early Pleistocene Hominidae.[87]

Between 1977 and 1990, Mahale researchers documented the intragroup killing of seven 1- to 10-month-old male infants of immigrant females that had been part of M group for zero to five years (Figure 8.2).[88] Six of the infants were cannibalized, and on six occasions the captors or the first-observed possessors were alpha or beta males in M group. The alpha male, Ntologi, killed ≥4 of the infants, and all the infants died while being eaten. In all cases, up to fourteen group members obtained bits of the prey via sharing or gleaning scraps. No mother ate bits of her own infant, and none of the mothers left the group. The mothers had mated more with immature and subadult males than with higher-ranking males, and none was known to have engaged in restrictive matings with high-ranking males.[89] Although infanticide occurred sporadically among Mahale chimpanzees between 1979 and 1990, it appeared to cease thereafter, until 1998, when the alpha and beta males of M-group launched a series of attacks on a 12-year-old natally born female apparently to seize her female infant.[90] They failed because of intervention by the infant's grandmother and her female associate.[91] Moreover, attacks on infants are not risk free even for sizeable groups of attackers. Six of nine Mahale M-group males incurred wounds when they attempted to capture an infant of a neighboring group. It is unknown whether the mother or other members of the infant's group caused the injuries or they occurred when the males fought over the carcass.[92]

Other Apes Eating What They Are

Unlike chimpanzees and humans, bonobos and orangutans seldom engage in cannibalism. Although bonobos appear to be free of infanticide, one day after a 2.5-year-old infant died, her mother and older sibling and other members of the community shared the corpse. Over a seven-hour span, during which possession of the corpse changed fourteen times, six adult females and two adult males ate bits of it.[93] Likewise, unrelated adult Sumatran orangutan mothers that had been reintroduced to the forest ate their decomposing babies several days after they had died.[94]

The only reported case of cannibalism in gorillas is based on arguable circumstantial evidence: fecal contents and the disappearance of an infant. Dian Fossey concluded that a mother and her young adult daughter had shared the first infant of a female in their group.[95] The male victim might have been the product of father/daughter incest. The silverback of Visoke group 5 probably sired the infant, and the older presumed cannibal was the dominant female in the group. Needless to say, this scenario does not take into account the possibility that the baby may have died at a site soiled with gorilla feces, thereby giving the impression that he had been eaten.

Bonobo Meat Eating and Sharing

Bonobos are the only other apes that offer opportunities to study the complex phenomena of hunting, meat eating, and food sharing. Because they live in a human war zone and are elusive, our knowledge of nonprovisioned bonobos is relatively rudimentary compared with chimpanzees. Insofar as their habits are known, bonobos appear to be more pacific than chimpanzees. In some localities, they kill and eat other vertebrates. The roster of their mammalian prey species ($n = 10$) is a mere one-fourth that of chimpanzees (Table 8.2). Their primate prey include Demidoff's dwarf galagos and cercopithecine monkeys.[96]

Lilungu bonobos caught and handled (sometimes fatally) two red-tailed guenons and an Angolan black-and-white colobus, but there is no evidence that they ate their captives.[97] An adolescent female Bossou chimpanzee exhibited similar behavior with a tree hyrax *(Dendrohyrax dorsalis)*; she groomed and carried it for fifteen hours and slept with it in her nest, but neither she nor other group members ate it.[98]

Although Lilungu bonobos eat a variety of animals, including bees, ants, beetles, termites, earthworms, bats, and rodents, there is no evidence that they eat monkeys or other sizeable mammalian prey.[99] A solitary male red colobus briefly groomed young Wamba bonobos in the presence of other bonobos. The bonobos touched the red colobus but did not reciprocate groom-

Table 8.2.   Mammalian prey of bonobos

Primates	Common Name
Galagidae	
Galagoides demidovii	Demidoff's dwarf galago
Cercopithecidae	
Cercopithecus ascanius	Red-tailed monkey
Cercopithecus pogonias wolfi	Congo Basin Wolf's monkey
?*Lophocebus aterrimus*	Black mangabey
Artiodactyla	
Bovidae	
Cephalophus dorsalis	Bay duiker
Cephalophus monticola	Blue duiker
Cephalophus nigrifons	Black-fronted duiker
Rodentia	
Anomaluridae	
Anomalurus fraseri	Scaly-tailed flying squirrel
Family *incertae sedis*	Rodent
Chiroptera	
Pteropodidae	
Eidolon helvum	Straw-colored fruit bat
Insectivora	
Family incertae sedis	Shrew

Source: Based on Ducros and Ducros 1992; Sabater Pi and Veà 1994; Uehara 1997; Hohmann and Fruth 2008; Surbeck and Hohmann 2008; Hirata et al. 2010.

ing.[100] Interactions between Wamba guenons and bonobos were also pacific, leading Ihobe to conclude that bonobos probably do not hunt sympatric primates.[101]

Conversely, Surbeck and Hohmann observed adult female LuiKotale bonobos capturing and eating three immature monkeys.[102] Apparently, Lui-Kotale bonobos consume meat as often as chimpanzees at some sites do, and the frequency of their meat eating might be higher that that of Lomako bonobos because prey species are more plentiful in Salonga National Park. They stated further that mammalian prey was always shared among adult members of the foraging party and included transfers of meat from adults to immature individuals.[103] Contrarily, Surbeck and Hohmann reported that only two adult females in a party of five bonobos received bits of a red-tailed monkey; only an adolescent male in a party of seven bonobos received a portion of a Wolf's monkey, and no cadger in a party of five bonobos was rewarded by the captor of another Wolf's monkey.[102]

[13]C and ^{15}N isotopes in hair samples from about twenty-one mature Salonga bonobos were homogeneous over a nine-month span and indicated no intersexual dietary difference in the group. Further, the isotopic data supported observations that the diet of Salonga bonobos is predominantly based on plant foods, with faunivory contributing marginal amounts of protein.[104]

Wamba bonobos that were provisioned with pineapple and sugarcane occasionally killed and ate flying squirrels. In two instances, there was unrewarded begging by an estrous female and other followers while a male ate a squirrel; there was also no sharing when a female ate a squirrel leg.[105]

At Lomako, nonprovisioned bonobos sometimes killed and ate young duikers and squirrel-sized mammals. Hohmann and Fruth saw two female beggars receive part of the prey of another female; a male beggar, three females, and three infants obtained meat from a female with a duiker during a 3.5-hour span.[106] There was no apparent agonistic behavior in the first instance and only displacements in the second. Three times during the duiker-eating episode, females engaged in frottage.[106] In a party of seven Lomako bonobos, a male shared duiker meat with a female seven times before she took possession of the carcass. Females without infants performed the most begging.[107]

Lone Lomako bonobos caught duikers twice at their resting sites.[108] While their prey was alive, they opened its belly and dipped fingers into the abdomen and licked them. Then they ate the intestines and stripped the flesh off the hind limbs. Later, the owner inverted the pelt and scraped flesh off the dermis with her incisors. Finally, the owner poked an index finger into the foramen magnum and dipped out brain tissue, then bit into the skull and ate the contents. Consumption averaged three hours. Females comprised 91 percent of carcass owners during twenty-three episodes. Fifteen females and eight males showed interest in the meat of an owner. Among eighty-seven episodes of meat eating at Lomako, all the females got some meat from an owner at least once, but no male did.[108]

Among apes, bonobos are most likely to share vegetal food across a wide range of individuals in addition to mother-infant pairs. Moreover, two rounds of tests on captives revealed that although pairs of bonobos (mean ages: 7.9 and 7.1 years) and chimpanzees (mean ages: 7.5 and 9.0 years) were equally likely to cooperate to retrieve food rewards, cofeeding bonobos were more socially tolerant than the cofeeding chimpanzees. Once they had obtained clumped pieces of fruit cooperatively, one chimpanzee monopolized 93 percent of it; the bonobos never took more than 63 percent of the pieces.[109]

Provisioned Wamba bonobos passed foods both ways between members of all age-sex classes, except from infants to other group members. They shared large, rare fruits mostly among adults, sometimes by passing them around begging clusters.[110] Most sharing occurred between females and much less between males.[111] One-fourth of adult sharing consisted of males giving food to females; the reverse was rare. Females commonly begged, while males seldom did. Estrous and anestrous females copulated with males before taking some of their provisioned sugar cane or pineapple. One young female placed a piece of her pineapple in a male's palm after they had copulated.[112]

Lomako bonobos shared several species of medium-sized and very large fruit, usually on the ground. Begging occurred at lower frequencies for fruit than for meat. Males shared fruit with females and a juvenile, but not with other males. Females without infants shared with males, juveniles, and a female with an infant. Two males fought for a large fruit as females approached. Overall, females begged more successfully than males did.[107]

In general agreement with White, Hohmann and Fruth summarized their observations on Lomako bonobos as follows:

- They share both plant and animal food.
- Shared food items are large and heavy.
- Food sharing generally occurred among individuals of different ages and sexes, but females were more often in possession of food and shared more often than did males.
- Females most frequently shared with infants, often with other females, and least often with males.
- Infants received and took food more often from females other than their mothers.
- Food transferred from males to females and among females at equivalent rates, but never from females to males or among males.[113]

The markedly different food-sharing behavior of bonobos and chimpanzees is reflected in their rapid endocrine shifts. Food competition tests on female, male, and bisexual dyads showed that males of both species experienced an anticipatory decrease in steroids when they shared food with a partner and an anticipatory increase when they were with a dominant partner that obtained more food. However, although bonobo males experienced shifts in cortisol, chimpanzee males experienced shifts in testosterone, perhaps due to differences in perception of the situation, namely, viewing the event either as a stressor or a dominance contest.[114]

Gibbon Meat Eating and Sharing

Fan and Jiang observed eleven attacks by black gibbons on giant flying squirrels, four of which were successful.[115] The nocturnal squirrels rest diurnally on tree branches or in tree hollows and park their infants in nests or tree hollows. Most members of a polygynous gibbon group attempted to capture them, but only one adult female was able to secure three infants and one subadult. She grabbed their long tails and threw them from the canopy. She variously allowed her dependent youngsters and an adult female to obtain some of the meat in three instances. Fan and Jiang also observed them eating eggs or nestling birds and a lizard.[115]

When and Why Primates Might Share Food

Prima facie, it might appear nonadaptive for an individual to give food without resistance to another individual—that is, to share—particularly when the object is relatively rare or not easily processed for ingestion.[116] In a phylogenetic analysis of food sharing in sixty-eight nonhuman primate species, including bonobos, chimpanzees, mountain gorillas, Western gorillas, Sumatran and Bornean orangutans, siamang, and white-handed and buff-cheeked gibbons, Jaeggi and Van Schaik concluded the following:

- Sharing with offspring is predicted by the relative difficulty of processing dietary items, as measured by the degree of extractive foraging; not by overall dietary quality.
- Interadult food sharing only evolved in species that already shared with offspring, regardless of diet.
- Intersexual sharing coevolved with the opportunity for female mate choice.
- Homosexual sharing coevolved with coalition formation.[116]

Jaeggi and Van Schaik further concluded that their analysis supports the hypothesis that among adults food sharing is traded for mating and coalition support. Accordingly, they predicted that food sharing should occur in any species with opportunities for partner choice. All nine of the ape species in their sample evidenced food sharing with offspring, but adult gorillas, white-handed gibbons, and siamang evidenced no food sharing even though some partner choice occurs among all of them (Chapter 11).

Man Meets Meat

Ever since stone tools were discovered associated with bones, sometimes including the bones of fossil hominids, archeologists have associated them with the consumption of meat from scavenged or hunted animals, giving minimal or no consideration of their possible uses in processing vegetal items for consumption or as foraging implements.[117] Experimental cutting and bashing fresh bones with stone implements allowed researchers to distinguish similar cut marks on fossil bones from the tooth marks left by other animals and to discern the sequence in which the bones at archaeological sites had been scarred by stone tools versus carnivoran teeth.[118] Other experiments indicated that different patterns of polish and wear on stone tools might distinguish those employed to deflesh hides, scrape bones, or disarticulate and fragment bones for brains and marrow from those potentially used to cut soft plants or make wooden implements for digging or stabbing.[119]

Given the predilection of apes and many other primates to acquire animal protein opportunistically or more deliberately, it is reasonable to expect that over the course of evolution the practice was characteristic of hominids also. The rub is that we have no unequivocal indicator of the kinds and degree of reliance on fauna as food in the Hominidae until very late in our history, and even then only in special cases before the Holocene. A major and persistent problem has been discerning the extent to which Plio-Pleistocene hominids were hunters or scavengers of sizeable mammals.[120] Associations of faunal remains with the tools of *Homo erectus* ca. 900 Ka at Olorgesailie, Kenya, and Middle-Late Pleistocene *Homo neanderthalensis* in Europe exemplify the problem of assessing hunting versus scavenging in pre-Upper Paleolithic hominid subsistence behavior.[121]

The remarkable accumulations of hefty Acheulean hand axes (mean mass: 1.59 kg), cores, scrapers, cleavers, and other stone artifacts often mixed with the bones of giant geladas (*Theropithecus oswaldi,* male mass ca. 65 kg), equids, elephants, hippopotami, rhinoceroses, suids, bovids, and other vertebrates have indicated circumstantially that *Homo erectus* might have butchered and eaten some of them.[122] Although Pat Shipman et al. presented a persuasive case for hominids butchering predominantly young geladas by smashing their bones and joints, the absence of cut marks is a major drawback to a definitive conclusion that Olorgesailie *Homo erectus* hunted or scavenged other animals.[123]

The case for meat eating (albeit supplemented with a variety of plant foods) by *Homo neanderthalensis* is clinched by isotopic analyses (^{15}N) and the cut marks, burns, and splintering of marrow bones that are clearly associated with Mousterian implements and sometimes hominid remains in Eurasia.[124] The tool kits of Neandertal people were more diverse than those of the Early Paleolithic people, but interpretive problems persist regarding their functions. For example, were some of them hafted? Were they used on animal carcasses or vegetal materials? Were some of them components of hunting weapons?

Several wooden shafts from the Lower Paleolithic (400 Ka) site of Schöningen, Germany, and younger sites in England and Germany suggest that select Mousterian lithics were used for woodworking and perhaps were hafted to them.[125] Whether any of the spears were employed as projectiles versus thrusting into prey at close quarters is unresolved. The ca. 126–118 Ka Lehringen yew wood lance was located between the ribs of a straight-tusked elephant *(Elephas antiquus),* and Schöningen spruce and silver fir shafts were associated with flint artifacts and a rich fauna comprising straight-tusked elephant, rhinoceros, red deer, bear, horse, numerous small mammals, birds, fish, and reptiles. Cut marks and fracturing were clear evidence of hominid butchering at Schöningen.[126]

Faunal analysis of ungulate remains from six stratified caves in Italy have indicated that Mousterian levels contained old-biased ungulate specimens like those of carnivoran hunter-scavengers, but, like Holocene *Homo sapiens,* Upper Paleolithic hunters appear to have selectively killed prime individuals.[127] Stiner's findings do not support the idea that cursory hunting was a common method of hominid hunters. Her faunal samples and those at other sites across Eurasia indicate that Neandertals also hunted and scavenged a notable variety of sizeable species and collected diverse small vertebrates and marine animals. She concluded that, unlike nonhuman predators, modern people are a specialized variety of ambush predator whose regular acquisition of prime-aged prey might be facilitated by technology, especially long-range weapons, cooperative ambush tactics, or both.

Diversification in the meat quest continued in the Upper Paleolithic, when clearer evidence of hafting and projectile weaponry appear, along with the use of new materials—bone, ivory, and antler—and fishing gear, traps, and other equipment. It would be interesting to know when poisons were discovered and first applied to projectile weapons and used for collecting fish and other aquatic fauna.

PART III

HANDS, TOOLS, BRAINS, AND COGNITION

9

Handy Apes

And the narrowest hinge in my hand puts to scorn all
machinery.

WALT WHITMAN (1982, p. 217)

Tool use is the external employment of an unattached or
manipulable attached environmental object to alter more
efficiently the form, position, or condition of another object,
another organism, or the user itself, when the user holds and
directly manipulates the tool during or prior to use and is
responsible for the proper and effective orientation of the tool.

ROBERT W. SHUMAKER, KRISTINA R. WALKUP, and
BENJAMIN B. BECK (2011, p. 5)

How many times per day do you employ tools? How long could you go without using a tool? Were you to try to function without tools, surely you would
soon realize that your efficiency and in many contexts your well-being have
been compromised. Accordingly, evolutionary anthropologists are especially
interested in both the tool behavior of nonhuman primates and other animals and the archeological traces of technology as they might inform the
sorts of mental and manipulatory capabilities upon which the development
of human intelligence and motor skills were developed.[1]

Neogene and Pleistocene fossil anthropoid forelimb bones can provide
important clues to the evolution of manual morphology in *Homo sapiens* and
Homo neanderthalensis, but we need genetic, behavioral, and artifactual information to craft scenarios of the selective factors that shaped our structurally unique hands and their fine motor control and tactile sensitivity.[2]

Archeological sites in Ethiopia evidence that 2.5-Ma *Australopithecus garhi* made and used stone artifacts to access meat and marrow.[3] The earliest lithic artifacts are from 2.5–2.6-Ma Kada Gona near the area where *Australopithecus garhi* lived.[4] Although McPherron et al. claimed that two bovid bone fragments from the ca. 3.4 Ma site of Dikika in the Sidi Hakoma Member of the Hadar Formation, Ethiopia, bore cutmarks indicating stone-assisted defleshing, Domínguez-Rodrigo et al. argued that the marks were produced by common taphonomic processes.[5] Accordingly, *Australopithcus afarensis* probably did not make and use stone tools. Of course, early hominids could have used artifacts of perishable vegetal and animal materials before the Late Pliocene.[6]

Pongid Tool Tutorials

> To conceive the idea of shaping a stone or stick for use in an imagined future eventuality is beyond the mental capacity of any known apes. Possession of a great capacity for this conceptual thinking, in contrast to the mainly perceptual thinking of apes and other primates, is generally regarded by comparative psychologists as distinctive of man. Systematic making of tools implies a marked capacity for conceptual thought.
>
> KENNETH P. OAKLEY (1959, p. 4)

The instrumental talents of captive chimpanzees weighed heavily in estimates of their relative intelligence.[7] During the first quarter of the twentieth century, Köhler noted that the stick was a multipurpose tool for captive chimpanzees.[8] They employed provisioned sticks for a wide range of functions: to swat down, probe for, and draw in foods; to dig up buried fruits; to brandish, stab, and throw as weapons; to use as prizing levers; and to pole vault for suspended foods. They placed sticks together and stacked boxes to reach desired objects. Further, they made tools by orally and manually tearing splinters from boards, peeling the ends of sticks, and trimming other objects. They employed straws to fish for the ants that crawled outside their enclosure and to dip fluids to drink, and they used twigs, rags, and bits of paper to wipe feces off their feet.[9]

Subsequent observations of incarcerated, experimental, and more freely ranging individuals continued to underscore the chimpanzee's status as a tool whiz and showed that bonobos, gorillas, and orangutans also are remarkable implemental apes.[10] Early experience with a variety of materials to solve recurring problems often led to creative solutions when later faced with novel problems, whereas deprivation of such opportunities could impair creativity.[11]

Orangutans readily solve instrumental problems of the sort that Köhler had used to challenge chimpanzees. Further, they are very clever and innovative when manipulating objects and making tools outside the venue of structured experiments.[12] For example, via notable human assistance a captive 5.5-year-old male orangutan, Abang, managed to use a tool to make a tool. He imitatively struck flakes from nodules of flint with a 1.36-kg hammer stone and used them to cut cords, thereby gaining access to food in a metal box.[13]

Abang's accomplishment encouraged an enterprising team of archeologists interested in the Paleolithic era and a cognitive psychologist to teach a male bonobo, Kanzi, to make stone tools that he could employ to cut cords and simulated tough animal skins to obtain otherwise inaccessible food rewards.[14]

After 9-year-old Kanzi had watched Nicholas Toth knap stone flakes and use them to access edible treats from a box by cutting a cord so that the box could be opened, it took only a few more human demonstrations for Kanzi to master the task with flakes provided by the experimenters.[15] He also learned relatively quickly by trial and error to select from sets of five quartz flakes the only one in each set that could cut the cord.

Next, the experimenters provided Kanzi with cobbles and stone nodules with which he had to make his own cutting tools. Sometimes while watching a knapping human Kanzi would bang rocks together by striking the one in his left hand with the other in his right hand (Plate 11 top). Optionally, he right-handedly hit a rock against another braced between his feet or against a stone anvil on the ground.[15]

With further practice, Kanzi became more successful at knapping by striking nearer the edge of a core, but his blows were too weak to produce effective flakes that were long enough to grip firmly and still leave an exposed cutting edge. Then, independently and insightfully, Kanzi struck on a new technique: right-handedly throwing the core against a hard surface (cobbles on the ground) to dislodge large flakes. Thereby, he could produce more force than with the handheld hard-hammer method.

Kanzi wisely preferred the toss-to-flake technique. Thereafter, the experimenters had to alter the context of future experiments by working in an area lacking hard objects against which he could throw his cores. From the initial period of experimentation and studies of Kanzi's flakes and cores, Toth et al. concluded that Kanzi's stone-flaking skill contrasted markedly with that of Oldowan hominids, who produced sharp-edged bifacial and polyfacial cores presumably by searching for acute angles from which to detach flakes efficiently.[15]

As a 12-year-old with about 120 hours of experience, Kanzi was a somewhat more proficient toolmaker and user, albeit predominantly with his toss-to-flake technique rather than the hand-held percussion method preferred by his human models.[16] Again, his flakes and cores fell short of Oldowan artifacts and those that proficient modern human knappers can produce. Although Kanzi was disinclined to retouch the dull edges of his flakes to sharpen them, he had learned that larger, heavier flakes were easier to hold and to use for cutting. Further, he astutely assessed the potential sharpness of flakes visually and perhaps orally.[16]

In the third experimental phase of the study, Kanzi was 20 years old and had been knapping for ten years. His 15-year-old half-sister, Panbanisha, had begun to knap four years earlier after observing Katherine Schick.[17] From cobbles selected and imported from Gona, Ethiopia, both bonobos produced large, usable cutting flakes that Toth et al. compared with the Oldowan flakes, cores, and debitage that had been left by 2.5-Ma Gona hominids and those that Toth and Schick had recently manufactured from Gona cobbles.[18] Both Toth and Schick had knapped for >20 years (they did not state their ages). After watching others and particularly their mother, Panbanisha, knap stone, two young males also began to flake stones.[17]

Toth et al. concluded that the bonobo lithic collection was sufficiently distinct from naturally broken stones to be discerned as artifactual if it were found in a prehistoric context, but that it fell short of resembling the Oldowan assemblages in Africa.[18] Although cores dominated both the bonobo collection and some Oldowan occurrences, it is reasonable to assume that the Oldowan hominids might have taken many flakes from the knapping spot to use elsewhere.[19] The bonobos produced flakes that attest to notable knapping skill, at least to the extent that they might be considered pre-Oldowan; in later experiments, they used them to open logs and dig for food rewards.[20] Ultimately, Toth et al. claimed that neither the bonobos nor the Gona hominids produced artifacts that collectively were as refined as their own handiwork.[18]

Tool Behavior of Free Apes

Great apes and gibbons variously drop, toss, shake, and otherwise trash vegetation as part of displays toward conspecifics, predators, and other alien nuisances.[21] Further, individuals in some wild populations of chimpanzees and orangutans use foraging implements, and some of them modify natural objects to render them functional for extracting food items. Wild gibbons, gorillas, and bonobos are undocumented as tool users during extractive foraging.[22]

If I Had a Hammer

Jane Goodall provided the first major field evidence for regular tool manufacture and use by wild apes.[23] When alate (winged) termites prepared for nuptial flights in the wet season, Gombe chimpanzees routinely inspected mounds during their daily rounds and fished for prey with slender twigs and vines, stems, and strips of bark fiber. They scratched into passages near the surface of a mound with an index finger, inserted a tool, withdrew it, and labially nibbled off clinging prey (Figure 9.1). Tools enabled the chimpanzees to obtain termites before the baboons and other predators who had to wait until they flew from the mounds.

No individual moved over ten yards from a feeding site in search of new tools.[24] However, occasionally they would collect tools before a termitarium came into view, within 100 yards of the tool source. Some individuals took several tools to a mound for sequential use. Gombe chimpanzees facultatively modified termiting tools: denuding stems of leaves, dividing wide grasses and bark fiber longitudinally, and nipping off frayed and bent ends.[24]

Gombe chimpanzees stood bipedally and levered sturdy sticks back and forth to enlarge entrances to subterranean bee nests and then manually dipped out honey. They also used slimmer sticks to extract ants from the subterranean and arboreal nests of two species of ants that attacked their intrusive probes.[25] They reused tools that were lying nearby or made new ones by breaking off a straight, tough, relatively pliable stick, defoliating it, and sometimes stripping off the bark. During ant-dipping bouts, they dentally trimmed frayed ends and replaced worn tools.[26]

Before inserting a stick, Gombe chimpanzees dug rapidly into subterranean driver ant *(Dorylus nigricans)* nests with their hands, thereby enlarging the entrances and inducing the inhabitants to swarm forth in its defense. To

Figure 9.1. Adult male Gombe *Pan troglodytes* fishing for termites.

avoid painful bites, the dippers usually stayed back from the hole and stood bipedally if they held the tool. Some individuals also perched on saplings, rocks, tree trunks, and fallen logs until the prey swarmed up the probes. They dashed forward, inserted a tool, moved away, and returned to retrieve it. Then, with the distal end of the tool near the mouth, a chimpanzee rapidly slid the other hand along the shaft, thereby sweeping the ants into the mouth. One quickly munched a bolus of ants, which might be the size of a hen's egg and contain as many as 300 ants.[27]

Gombe chimpanzees also used leaves as drinking sops and toilet wipes. They often inserted wads of masticated leaves into tree holes and then sucked out the steeped liquid instead of finger or hand-dipping water directly.[28]

In the seasonally arid Tongo Forest of the Democratic Republic of the Congo (DRC) where the fruits are small and low in water content, eastern chimpanzees not only manually dipped water from tree hollows and branch forks, but also employed wads of moss, often collected before arrival at a drinking tree, as sponges. They also chewed rotten wood that contained water after it had rained, and they went to considerable effort to access tubers for the water they contained. After chewing and sucking wood or tubers, they spat out the fibrous wadges.[29]

Chimpanzees learn tool behavior by watching other group members, especially their mothers.[30] They are not proficient termite collectors until after their third birthday. Ant-dipping takes longer (≥7 years) to develop, probably because the soldier ants deliver painful bites.[31] Two Taï Forest chimpanzee mothers actively taught their youngsters how to crack nuts on anvils with handheld stone hammers.[32] Further, controlled experiments with captives have demonstrated that novel technologies learned by an individual can spread to other members of a chimpanzee group.[33]

At Gombe, female chimpanzees exhibited more dietary tool use than the males; at Mahale, stools from known individuals in a large group revealed that females consumed more ants than the males.[34] The ants are difficult to obtain directly by hand or mouth because they nest inside the trunks and large boughs of standing and fallen trees. Mahale chimpanzees collect, trim, and insert probes into these nests (Figure 9.2). By slapping or bipedally pummeling the tree trunk at a lower level and shaking a tool in the passage, they attempt to move the ants toward the entrance to bite their probes.[35]

Mahale chimpanzees made tools from the vines in ant-infested trees or gathered other materials close by. They generally made tools from materials close to the ant holes before checking to see whether a site might be productive. Competition was often keen for anting sites. Individuals waited between thirty seconds and fifty-one minutes (average: sixteen minutes) for spots to be vacated. Occasionally, a gesturing, posturing, or vocalizing individual displaced a fisher.[35]

There is notable variation in the tool behavior of chimpanzees in Africa.[36] For example, there is no record of dietary stick-tool use in the well-studied Budongo chimpanzees, and some populations south of Gombe exhibit relatively little termite fishing.[37] Perhaps in relatively moist regions termites can be obtained simply by wrecking mounds instead of patiently probing them.[38] At some localities, individuals sequentially employ and sometimes reuse tool-composites of sticks and flexible plant parts to extract insects or water and stones to obtain nut kernels.[39] Although a western chimpanzee made and

Figure 9.2. Adult female Mahale *Pan troglodytes* fishing for arboreal ants *(Camponatus)*.

persistently employed a series of four wooden jabbing sticks and probes to reach a store of stingless bee honey in the Gambia, the central chimpanzees of the Congo Basin and Loango National Park, Gabon, have exhibited the greatest diversity of tool use, including up to five tools deployed sequentially to obtain honey.[40] To open beehives, the chimpanzees at Goualougo, Republic of the Congo, often pounded them intermittently with wooden clubs or hammers and then inspected them with probes, which they then used to dip out honey. Between bouts of pounding, they stored their percussive tools in the canopy or variously held them bodily while making dipping probes.[41] Goualougo chimpanzees also sequentially used two different types of tools to access army ants *(Dorylus):* a woody sapling to puncture the nest, and an herbal stem to dip into it.

Year-round at Gashaka, Nigeria, *Pan troglodytes vellerosus* variously employ sticks and dipping wands to capture ants and to collect honey from bee's nests. However, unlike *Pan troglodytes verus* and *Pan troglodytes schweinfurthii,* they neither hammer nuts nor fish for termites, respectively.[42]

Eastern chimpanzees apparently use sticks to harvest the honey and larvae from the ground nest of stingless bees in Kahuzi-Biega National Park, DRC, and central chimpanzees might have used stout sticks to extract decaying wood from a fallen tree.[43]

At several localities in West Africa, chimpanzees use wooden and stone hammers against rocky outcrops and exposed tree roots to open seedpods and five species of hard nuts (Figure 9.3).[44] At Bossou, Republic of Guinea, 3.5-year-olds began to use stones to crack oil-palm nuts on stone anvils. Infants as young as 6 months handled stones and nuts while observing their mothers and other individuals engaged in nut cracking.[45]

In the Taï Forest, Côte d'Ivoire, chimpanzee nut cracking might extend back 4,300 years.[46] Taï chimpanzees depend or at least rely heavily on one of five species of seasonally available nuts during four months of the year. All five species are high in calories, mostly from fats, and one species is very proteinaceous.[47] The chimpanzees collect the nuts from the trees or the ground, then carry handfuls and mouthfuls to anvils where they carefully position them one by one and crack them open with a hammer. Terrestrially, they commonly employ hammers that have been left near an anvil.

Alternatively, to harvest nuts arboreally, a chimpanzee carries a hammer up a tree, holds onto it while gathering nuts, and cracks them one by one on a horizontal bough.[48] Although some stone hammers are quite heavy (range: 0.9–41.7 kg), Taï chimpanzees sometimes carry them between nutting sites.[48] Clearly, arboreal nutting requires considerable coordination and some forethought.[49]

In two of three methods for harvesting nuts, adult female Taï chimpanzees are generally more proficient and persistent than adult males at opening coula nuts *(Coula edulis)* arboreally and panda nuts *(Panda oleosa)* terrestrially. Though cracking panda nuts requires less postural balance, notable skill and cognition are necessary, and the stone hammers might have to be transported over several hundred meters.[50] To avoid pulverizing the kernels, the cracker must precisely control the force of each blow Further, because the shells are so hard, cracking panda nuts is a more energetic, time-consuming job than cracking coula nuts. Nevertheless, the females are more adept than the males with the heavier hammers.[47]

Figure 9.3. Taï western chimpanzees cracking nuts. (A) Mother sharing kernels of *Panda oleosa* extracted with a large granite stone. (B) Taï male (<1-year-old) intently watches his mother crack a nut. (C) Taï infant begs her mother for a nut *(Coula edulis)*, she allows access to it, the mother places another nut on the anvil, and they share the kernel. (D) Nut species eaten by Taï chimpanzees: a. *Panda oleosa;* b. *Coula edulis;* c. *Detarium senegalensis;* d. *Parinari excelsa;* e. *Sacoglottis gabonensis.*

Unlike the learning processes for ant dipping and termite fishing, those for panda nut cracking and arboreal coula nutting extend into adolescence in both sexes and into adulthood in males. Indeed, most males never master either technique. Further, males crack coula nuts terrestrially when the nuts are dry and easier to open, whereas the females excel at cracking the fresh, tougher nuts.[51]

Males probably fail to master panda nut cracking and arboreal coula nutting because they

- monitor the activities of other group members,
- can see farther on the ground than in trees,
- tend to move away from nutting sites in response to others rather than continuing to feed on nuts, and
- would have to concentrate on keeping their balance and their grasp on nuts and hammer to crack coula nuts in a tree.[51]

This makes sense given that chimpanzee intermale relationships are complex (Chapter 11) and they must also be attuned to predators and other chimpanzee groups that might attack them or trespass their range.[52]

The more frequent and effective nut cracking by wild female chimpanzees and the fact that captive female chimpanzees and bonobos displayed a more extensive range of tool behaviors than males cast serious doubt on the silly sexist notion that males must have been the first hominoid technologists.[53] Such observations help to sideline the man-the-hunter hypothesis and to foreground woman-the-gatherer and sharing-people hypotheses.

Instances of tool-assisted hunting and lethal tool use by wild chimpanzees during predation on mammals are very rare, and they can be exaggerated in secondary scientific and lay publications.[54] After chimpanzees began to visit and trash the Gombe research camp, they took up rock throwing. Like captive apes, over time they refined and expanded their throwing. Initially, the rocks were mostly unaimed, and the chimpanzees were poor shots. However, as the frequency of aggression toward other chimpanzees, baboons, and humans increased in the banana-provisioning area, more males engaged in throwing, some youngsters and adult females also began to throw objects, and the percentage of hits versus misses improved.[55]

A Gombe adult male, already notorious for knocking kerosene cans about as part of displays, threw a sizeable rock that hit one of a group of bushpigs that he and several companions had surrounded. After the pigs fled, they captured and ate a piglet, but they used no tool to dispatch the prey.[56]

Although Pruetz and Bertolani presented a more compelling case for lethal tool use at Fongoli, Senegal, it is unclear whether the galago died from stab wounds or from the predator's grip.[57] Accordingly, it might be premature to draw too close comparisons between human use of hunting spears and the chimpanzees employing probes in what resembles in many respects tool-assisted predation on insects.

Oral Tool Users

Like chimpanzees, some orangutans employ leaves to wipe feces from their bodies and as vessels to hold food, obtain water, and enhance kiss-squeak calls.[58] They also employ leafy branches to protect themselves from bees and rain.[59]

Though not as extensive or as diverse as among *Pan troglodytes,* dietary tool use is part of the foraging regime in several populations of *Pongo abelii* inhabiting lowland swamp forests of northwestern Sumatra. Bornean orangutans engage in a lesser variety of tool use than that of Sumatran orangutans.[60]

At Suaq Balimbing in Gunung Leuser National Park, orangutans broke off and modified branches to extract insects, larvae, and honey from termite and apian nests in tree holes and to extract the well-protected, highly nutritious seeds and arils from cemengang *(Neesia malayana).* All tool manufacture and use was arboreal. Their toolkit contains different tools for preying on insects versus cemengang, and they sometimes used a series of tools in single bouts of food extraction.[61]

Their insect probes are longer and wider than their seed-extraction tools. Further, orangutans employed three different tools to prey on termites (widest and seldom stripped), ants (medium width and more often stripped), and stingless bees (thinnest and almost always stripped).[62] They usually broke off relatively straight live branches near target tree holes and variably hammered and poked with them to break into the nests. They then probed inside and scraped out insects, larvae, eggs, and honey. Intermittently, they might replace or trim initial or secondary tools. Unlike chimpanzees, the orangutans usually—79 percent of thirty-eight bouts and 83 percent of seventy-two cases—held tools in their mouths as they probed insect nests.[63]

Suaq Balimbing orangutans often used one denuded stick to extract seeds from many cemengang instead of changing tools between fruits. After picking a dehisced cemengang, one held it with one or both hands or in combination with or in lieu of a foot or two and inserted an orally held stick into the crack. By repeatedly scraping toward the apex, the irritating hairs that protect the arillate seeds are displaced; via lips or fingers, the individual can safely remove the seeds that were pushed near the top of the cracked fruit.[63]

Might Is All Right for Some Termites

Wild gorillas harvest termites and other insects by breaking into their mounds, rotten logs, and other objects where they reproduce.[64] Although they are undocumented using tools directly to access food items, at Mbeli Bai, Nouabalé-Ndoki National Park, Republic of the Congo, two adult female western gorillas employed detached branches in other contexts.[65] One individual used a walking stick to plumb the depth of a pool in which she waded bipedally. The other female detached a long, thick, leafless trunk of a dead shrub that she used as a stabilizer while she scooped food items from a pool with her free hand. She then laid the trunk across a patch of swampy ground and walked bipedally across her bridge.

Show-Offs

Observations of *Pan paniscus* in sanctuaries and other captive environments leave no doubt that they are just as capable of employing objects as tools as *Pan troglodytes,* though wild bonobos pale in comparison with many wild chimpanzee populations as tool-assisted foragers. Instead, much of their use of external objects that satisfies the definition of tool is in contexts of display.[66] They variously drop, drag, kick, slap, push over, throw, brandish, wave, and shake branches during play, aggression, and courtship.[67]

Tool Behavior of Monkeys

Captive monkeys are quite adept at manipulating objects in their enclosures and psychological test apparatuses.[68] Further, there are isolated reports of dietary tool use in some species of wild catarrhine monkeys:

- Hohmann reported that an adult female, a juvenile male, and a subadult male lion-tailed macaque *(Macaca silenus)* avoided the irritating protective hairs on 3–4-cm long chrysalides by rolling them in detached leaves before consuming them.[69]
- In southern Thailand, long-tailed macaques *(Macaca fascicularis)* unimanually and bimanually used stones to smash open rock oysters, gastropods, bivalves, and crabs after placing them on rocks.[70]
- In the Namib Desert, chacma baboons *(Papio ursinus)* removed fish from waterholes and covered them with piles of sand to subdue them before eating them.[71]
- In Nairobi National Park, Kenya, an adult male olive baboon *(Papio anubis)* used a twig sequentially to scrape, probe, and pry small tectonic stones from a loose clay soil matrix before ingesting them.[72]

Probing Capuchin Tool-Using Talents

Capuchins (*Cebus* spp.) are the chief monkey rivals of chimpanzees as tool whizzes, even though phylogenically they are farther removed than catarrhine monkeys from the apes and humans.[73] Captive and some wild capuchins engage in a wide variety of tool use in foraging and other activities (Plate 11 middle and bottom).[74]

Laboratory tests of the abilities of *Cebus apella* suggest that, in contrast with chimpanzees and bonobos, they did not understand the requirements of tool tasks presented to them and perhaps the physical forces, though they can discover solutions via trial-and-error manipulation of objects and with no evidence of imitating others.[75]

Capuchin Atlases Unshrugged

One of the most impressive tool-assisted feats of wild capuchins is exhibited by *Cebus libidinosus* at Fazenda Boa Vista, Piauí, Brazil.[76] The habitat is dry, open woodland with abundant palms.

Reminiscent of some western chimpanzees, the Boa Vista capuchins transport river pebbles to crack open four species of palm nuts on wooden or, more commonly, rock anvils. On average, the hammers weigh >1 kg, which is 25 percent to 40 percent of the body masses of adult female and male bearded capuchins (2.5–3.7 kg). The heaviest hammer weighed 2.53 kg.[77] The capuchins used heavier hammers to open the tougher and more internally partitioned nuts, the toughest of which (piassava: *Orbignya* sp.) are similar to the panda nuts *(Panda oleosa)* that wild western chimpanzees process with tools.[78] *Cebus libidinosus* at two sites in Caatinga, Brazil, also selectively employed stones weighing from <200 grams to >3 kg according to the hardness of the nut species.[79]

A monkey places a nut in a shallow concavity in a log or boulder, grasps a hammer bimanually, raises it, and forcefully strikes the nut several times. Individuals might remain seated or rise bipedally before strikes. They can achieve high momentum by rising quickly while concurrently raising the hammer via shoulder protraction and elbow extension, sometimes with the tail braced on a substrate, and then quickly flexing the hind limbs, retracting the shoulders, and flexing the elbows.[80]

The differences between the nut-cracking behavior of capuchins and chimpanzees are probably due to the gross difference in size and morphology of the participants. To generate sufficient force to crack nuts, capuchins grip a heavy hammer stone bimanually and stand bipedally, while chimpanzees usually sit or even recline and pound nuts unimanually with their greater forelimb strength. Further, chimpanzee hands are large enough to grip most hammers in one hand, leaving the other hand free to reposition the nut on the anvil. They also transport anvils to nutting sites, while capuchins carry nuts to the anvil sites.

Manus from Adaptive Processes

> Given the motivation and the cortical development necessary
> for tool making, great apes and most Old World monkeys could
> manufacture implements to fit their hands. The Oldowan pebble
> tools demonstrate a rudimentary level of craftsmanship and do
> not appear to have been specially fashioned to fit hands like
> those of *Homo sapiens.*
>
> RUSSELL H. TUTTLE (1967, p. 199)

> The appearance of stone tools does not necessarily reflect an
> abrupt dramatic shift in food *acquisition* behavior at all. What it
> does seem to show is a shift in food *processing* techniques and
> *discard* patterns.
>
> SALLY MCBREARTY and MARC MONIZ (1991, p. 76)

The human hand (manus) occupies a special place among the interests of evolutionary anthropologists because of its presumed uniqueness as a skillful manipulatory organ and expresser of emotions and even thoughts. In this arena, it is equaled or surpassed only by the brain and bipedal complex. As hominids evolved into obligate terrestrial bipeds, their manual morphology was freed from evolutionary compromises between manipulatory and locomotor functions and was further adapted morphologically and perhaps neurologically for manipulatory functions alone, including tool manufacture and use.

In *Homo sapiens,* the thumb (pollex) is relatively large and is specially adapted for fine manipulation and strong grasping in opposition to the other four fingers and the palm in a *power grip*, as when one holds the handle of a hammer (Figure 9.4).[81] The thumb is relatively large, and its saddle-shaped carpometacarpal joint facilitates medial rotation and pulp-to-pulp opposition with the tips of the second to fifth manual digits in *precision grips*.[82] Napier astutely observed that the hand postures that a person employs depend chiefly

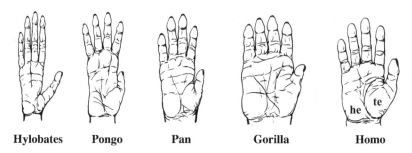

Hylobates   **Pongo**   **Pan**   **Gorilla**   **Homo**

Figure 9.4. Palmar view of ape and human right hands drawn to same length. Note short thumbs of the great apes and long thumbs of gibbon and human. Symbols: he, hypothenar eminence; te, thenar eminence.

on how she or he intends to use objects, and secondarily on the objects' shapes and sizes.[83] Indeed, Landsmeer suggested that the large thumb and the orientation of the fourth and fifth metacarpals set the stage for the perception of form and cerebral development.[84] The areas of the central nervous system that underpin cognition, motor control, and sensation are vitally involved with manual dexterity.[85]

Humans also have a unique muscle, the flexor pollicis longus, which powers the broad, stout distal phalanx of the thumb (Figure 5.11).[86] The distal phalanges of all the fingers are broad versus conical in apes. Abundant Meissner's corpuscles in the dermal papillae of the thumb and fingertips ensure a highly developed haptic sense.[87]

Pulp-to-pulp opposition of the thumb and index finger is easy for humans, but is much less commonly used by apes.[88] Some Old World monkeys and capuchins use it routinely.[89] Apes have short thumbs relative to lengths of the other fingers (Figure 9.4). They usually grasp small objects between the tip of the thumb and lateral side of the index finger instead of pulp-to-pulp (Figure 9.5). Consequently, although they have precision of grip due to their refined haptic sense and well-developed intrinsic hand muscles, they infrequently use pulp-to-pulp precision grips like those of humans.[90] Orangutans, chimpanzees, and gibbons are more disadvantaged in this regard than gorillas, whose digital proportions allow them to hold objects the size of grapes with a pulp-to-pulp thumb-index grip, albeit by acutely flexing the index finger toward the tip of the thumb.

Gibbon hands depart notably from the hands of great apes. The carpometacarpal joint approximates a highly mobile ball-and-socket configura-

Figure 9.5. *Pongo pygmaeus* and *Homo sapiens* gripping halves of a grain of rice. Whereas the human employs a tip-to-tip precision grip, the orangutan presses the pollical tip against the side of the index finger. The pollex in orangutans is quite strong and versatile due to short muscles in the thenar eminence, while the flexor pollicis longus muscle powers the strong human pinch.

tion, and the area that is bridged by the adductor pollicis muscle in great apes and humans is cleft, thereby freeing the entire digital ray for a wide range of grips on trunks, branches, and vines (Figures 5.4 and 5.7). Further, the tip of the long thumb is powered by a close approximation to a human flexor pollicis longus muscle, though its tendon is joined in the wrist by a translucent band of connective tissue to the flexor digitorum profundus muscle, which flexes the distal phalanges of the second to fifth fingers (Figure 5.11). This arrangement allows gibbons not only to grip sizeable substrates strongly between the thumb and other fingers but also to employ the thumb with precision to manipulate a wide range of objects.[91]

The short thumbs of great apes tend to move as a unit from the carpometa-carpal joint with little flexion of the distal phalanx, which at best receives a

filamentous tendon from the radial side of the indicial flexor digitorum profundus tendon (Figure 5.11). Indeed, orangutan thumbs usually lack any
connection to the flexor digitorum profundus muscle. In great apes, the large
radial portion of the deep digital flexor musculature, which in humans attaches to the distal phalanx of the thumb, instead attaches to the distal phalanx of the second finger, thereby notably increasing the power of their grasp
when climbing and engaging in suspensory behavior. Consequently, they
must rely almost exclusively on intrinsic hand muscles to power their thumbs.[91]
Because chimpanzee thumb muscles have shorter moment arms than those of
humans, their potential torques are less.[92]

In sum, the manual morphologies of African apes, orangutans, and gibbons
not only evidence adaptive compromises between manipulative functions and
locomotor functions—knuckle-walking and climbing, versatile climbing with
suspensory behavior, and vertical climbing and brachiating, respectively—but
also highlight how hominid hands, once freed from locomotor functions,
could be selectively adapted chiefly by manipulatory functions, particularly
tool manufacture and use.[93]

Limits of Fossils as Helpful Hands

Of the following twelve gross morphological features that facilitate a full spectrum of human precision and power grips, only half could be traced through
the fossil record were we blessed with a series of properly dated representative
fossil hands, and even fewer of them are unique to humans.[94]

- Broad, spatulate terminal phalanges that support relatively extensive areas of sensitive friction skin.
- A powerful, fully independent flexor pollicis longus muscle to the terminal pollical phalanx.
- Saddle-shaped (i.e., concavo-convex) second and fifth carpometacarpal
  joints and other features that increase the mobility of second and fifth
  carpometacarpal joints.[95]
- A special arrangement of the pollical carpometacarpal ligaments that
  guides the thumb into opposition with the other four fingers.[96]
- Configurations of the first metacarpal head and base of the pollical proximal phalanx and attachments of the abductor pollicis brevis muscle that
  medially rotate and abduct the phalangeal portion of the thumb at the

metacarpophalangeal joint so that its distal phalanx is opposed strategically to the other fingertips.[97]

- An extensor pollicis brevis muscle that attaches to the base of the proximal phalanx of the thumb.
- The orientation of metacarpophalangeal ligaments and asymmetry of the heads of the second and fifth metacarpal bones that induce rotation at the metacarpophalangeal joints during opposition with the thumb.[98]
- Proportionate lengths of the thumb and lateral four fingers that allow pulp-to-pulp contact of the fingertips.
- Notable misfit between the distal articular surface of the trapezium and the proximal articular surface of the pollical metacarpal bone that allows rotation, in addition to flexion, extension, abduction, and adduction of the thumb at the carpometacarpal joint.
- Extensive proximal attachments of the first dorsal interosseous muscle onto both the pollical and the second metacarpal bones, and distal attachments at the second metacarpophalangeal joint that strengthen and stabilize precision grips between the thumb and index finger.[99]
- Separate tendons of the flexor digitorum superficialis and flexor digitorum profundus muscles that facilitate independent flexion of the second to fifth fingers.
- Extensor expansions on the dorsa of the digits, to which intrinsic palm muscles attach, that along with intrinsic muscular attachments to the proximal phalanges concurrently extend interphalangeal joints and rotate the metacarpophalangeal joints.[100]

According to the patchy fossil evidence, the fully modern human hand evolved rather late, albeit on a substratum of Miocene and Pliocene catarrhine manual features. The hand bones of Proconsuloidea show that dental apes of the African Early Miocene had rather generalized hands; there is no hint of humanoid fine manipulation or knuckle-walking (Figure 9.6D). They were all-purpose climbing and quadrupedal palmigrade locomotor organs.[101] The same holds for African and Eurasian Middle and Late Miocene Pliopithecoidea and Hominoidea, though some of them evidence features that suggest increased suspensory behavior (Figure 9.6C).[102]

The 4.4-Ma hand bones of *Ardipithecus ramidus* lack salient features of living knuckle-walking apes.[103] They indicate a flexible wrist, long, curved second to fifth fingers, and a robust thumb, all features that would facilitate

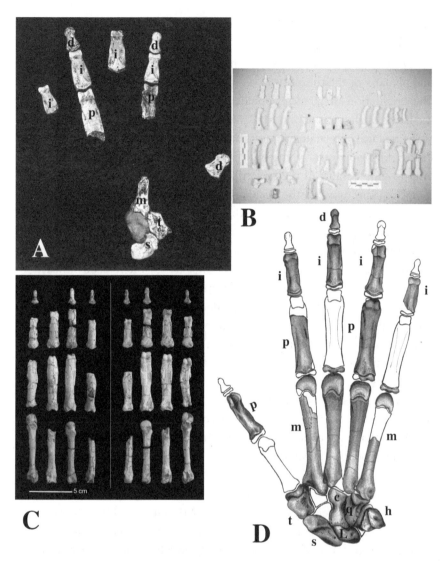

Figure 9.6. (A) Thirteen 1.8-Ma juvenile right-hand bones from Olduvai Gorge, Tanzania, attributed to *Homo habilis* (OH 7, type). (B) Hand bones of ca. 3.3-Ma *Australopithecus afarensis*. Note curvature of metacarpals and proximal and intermediate phalanges. (C) Dorsal (left) and palmar (right) views of right hand bones of 12.5–13-Ma *Dryopithecus laietanus* (= *Hispanopithecus laietanus*) [CL1-18800, type] (D) Palmar view of left hand of ca. 17.5–18-Ma. *Proconsul heseloni*. Symbols: c, capitate; d, distal phalanx; h, hamate; i, intermediate phalanx; L, lunate; p, proximal phalanx; m, metacarpal bone; q, triquetral; s, scaphoid; t, trapezium.

climbing and foraging in trees and a firm grip on branches, trunks, and vines. When they adopted quadrupedal postures on large boughs or the ground, they were probably palmigrade. Although it is conical as in apes instead of spatulate as in humans, the distal phalanx of the thumb indicates that *Ardipithecus ramidus* had a well-developed flexor pollicis longus muscle, which would facilitate firm power and precision grips. The mobile fifth carpometacarpal joint probably also facilitated firm grips between the fifth finger and the thumb.[103]

The ca. 3–3.6-Ma Hadar hypodigm of *Australopithecus afarensis* includes ≥50 hand bones.[104] The proximal and middle phalanges are curved and bear additional witness to climbing behavior.[105] There is no evidence for knuckle-walking. The thumb and wrist bones show a mixture of pongid, unique, and human features.[106] The long pisiform bone also indicates climbing by the Hadar hominids (Figure 9.6B).[107]

Two Pliocene–Early Pleistocene cave sites in South Africa—Sterkfontein and Swartkrans—contained hominoid hand bones that have variously been assigned to *Australopithecus, Paranthropus,* and *Homo* spp.[108] Based on a few isolated hominoid metacarpals and manual proximal phalanges, one middle manual phalanx, and a distal pollical phalanx from Sterkfontein, Ricklan concluded that, like the hands of *Australopithecus afarensis,* the highly dextrous hands of *Australopithecus africanus* were capable of powerful grips that would facilitate climbing, chopping, gouging, hammering, scraping, striking, and throwing, though their precision grips were probably less efficient or precise than those of *Homo sapiens.*[109] Green and Gordon concurred that the hands of Sterkfontein *Australopithecus africanus,* like those of *A. afarensis,* do not evidence the kinds of mechanical demands on the thumb experienced by more recent hominid stone toolmakers.[110]

The most complete specimen (StW 573) of *Australopithecus* from Southern Africa is from cave breccia in Member 2 at Sterkfontein.[111] Unfortunately, the date of the breccia is controversial. Whereas Clarke et al. insist on 3.3 Ma, others argued that StW 573 dates to ca. 2.2 Ma.[112] The hand of StW 573, particularly its curved manual phalanges, and possibly the foot skeleton bear hallmarks of arboreal behavior.[113] The individual was probably bipedal on the ground, given that its ilium is similar to those of other *Australopithecus,* and the upper and lower limbs were probably subequal in length.[114] The lateral four fingers of StW 573 are not notably elongate, and the thumb is relatively long and robust and sports a spatulate distal phalanx that would facilitate firm grips and object manipulation. The skeleton bears no feature that attests to knuckle-walking ancestors or practice.[114]

The ≥23 hominoid hand bones from the earliest hominid-bearing cave breccias at Swartkrans—ca. 1.8-Ma Member 1—are arguably attributable to *Paranthropus robustus,* which is the predominant species based on craniodental specimens, but there is a smattering of craniodental specimens of *Homo erectus* in Swartkrans Member 2, and possibly also in Members 1 and 3.[115] The collections from Members 1, 2, and 3 include first, third, and fourth metacarpals and proximal, intermediate, and distal manual phalanges.

Susman concluded that although the vast majority of the hand bones are probably from *Paranthropus robustus,* a pollical metacarpal from Swartkrans Member 1 might be from *Homo erectus.*[116] Trinkaus and Long contested Susman's result, arguing that the two bones are morphologically similar; indeed, both resemble counterparts in modern and late archaic *Homo sapiens.*[117] Their observation that the bones contrast chiefly in size did not persuade Susman that they are conspecific.[118]

Susman could reasonably conclude that *Paranthropus robustus* employed and probably made the bone digging tools and Developed Oldowan artifacts that occur in the Swartkrans Member 1–3 breccias.[119] Indeed, he claimed that *Paranthropus* had the same morphological potential as *Homo habilis* and *Homo sapiens* for refined precision grasping, tool using, and tool making.[116]

The distal pollical phalanx and the distal phalanges of the other fingers sport sizeable apical tufts, and the base of the pollical distal phalanx evidences attachment of a strong flexor pollicis longus tendon. The pollical metacarpal, other metacarpals, and proximal and intermediate phalanges are also morphologically and quantitatively more humanoid than apelike. The fingers are relatively short and straight. The thumb is long relative to the index finger, and the intrinsic pollical muscles are well developed.[120]

Based on the total, albeit sketchy, skeletal inventory of Swartkrans *Paranthropus robustus,* Susman concluded that the species was less arboreally and more bipedally adapted than *Australopithecus afarensis* and *Australopithecus africanus.*[116]

Like Swartkrans, Olduvai Bed I hominid hand bones and stone tools allow alternative scenarios regarding the species that produced the Oldowan pebble tools with which it was associated. At 1.8 Ma, *Paranthropus boisei* and *Homo habilis* were contemporaneous and perhaps even sympatric during the Late Pliocene–Early Pleistocene. Initially, Louis Leakey suggested that *Paranthropus* was responsible for the Oldowan artifacts, but after finding remains of a second hominid—*Homo habilis*—including thirteen juvenile hand bones and two adult manual phalanges at several other sites, it appeared more likely that *Homo habilis* was responsible for the Bed I artifacts.[121]

Robinson concluded that South African *Paranthropus* was more apish than *Australopithecus africanus* and that they probably climbed trees for food and shelter.[122] He cited the long ischium and somewhat apish foot of *Paranthropus robustus* to support this inference. He also implied that the 1.8-Ma hand bones from Olduvai Gorge might belong to *Paranthropus* instead of *Homo habilis.*

Whereas Napier emphasized the manipulatory potential of the Olduvai Bed I hand bones, I focused attention on the apelike features of the proximal and intermediate phalanges, which indicate strong development of the extrinsic digital flexors, particularly the flexor digitorum superficialis muscle (Figure 9.6A).[123] The bones needed to diagnose habitual knuckle-walking are absent, and others, such as the distal finger bones, are more hominid than apelike. Nevertheless, based on my commentary, Robinson suggested that *Paranthropus boisei* might have been a knuckle-walker.[122]

Subsequent studies have reinforced the inference that the hominid finger bones from Olduvai Gorge belong to *Homo habilis* and that the species sports apish features that are consistent with regular tree-climbing in addition to bipedalism.[124]

The hand of the type specimen of *Homo habilis* (OH 7) shows that its youthful possessor (10 to 12 years old) had a powerful grip that engaged the thumb.[125] The second to fifth fingers were well adapted for arboreal climbing; the fingertips and pollical carpometacarpal joint facilitated fine manipulation.[126]

In sum, on the basis of available manual specimens, *Australopithecus afarensis* has been widely judged to be less well adapted for humanoid tool use and manufacture than the Early Pleistocene hominids.[127] Contrarily, if the trapezium of Olduvai Hominid 7 *(Homo habilis)* is gorilloid, as judged by Tocheri et al., and its first distal phalanx is indeed hallucal instead of pollical, as intimated by Susman and Creel, its position as an implemental hominid could be more on a par with that of *Australopithecus.*[128] Nonetheless, like *Ardipithecus ramidus* and *Paranthropus* spp., they were all probably quite capable of gripping sticks and stones firmly for vigorous pounding and throwing.[129]

There is no published account of hand morphologies in *Homo erectus, Homo ergaster, Homo rudolfensis,* or *Homo georgicus* because there is insufficient material to model them. Neandertal hands are robust versions of our own, but they might not have been as dextrous as those of Upper Paleolithic and Holocene *Homo sapiens.*[130] Arboreal features are not manifest in other bodily regions of *Homo erectus, Homo ergaster,* and *Homo antecessor.* Accordingly, we can assume that hominids were fully adapted to life on the ground by 1.5 Ma and that their hands could have been developed notably toward the modern human condition. Nonetheless, one must avoid dogmatic statements on the topic, given that the fossil specimens from which to discern the skeletal morphology of their hands are practically nonexistent.

Linking refinement of tool technology from that exhibited by Oldowan industries to the more elegant hand axes and cleavers of Acheulean industries in coadaptation with manual evolution is reasonable but by no means definitive of actual evolutionary events.[131] Fortunately, manual specimens of archaic and prehistoric modern *Homo sapiens,* documenting hominid manual structure during the past 100,000 years, are substantial enough to support detailed descriptions of hands that facilitated the execution of the cave paintings and sculpture, fine beads, points, and hooks from bone, ivory, shell, and stone in the Upper Paleolithic and the subsequent technological and artistic developments.[129]

10

Mental Apes

> The evolutionary transition from ape to human cognition is to be characterized not as a loss of instincts, but as a gain in ways to learn.
>
> RAY JACKENDOFF (2007, p. 146)

> Humans have many cognitive skills not possessed by their nearest relatives.
>
> ESTHER HERRMANN, JOSEP CALL, MARIA VICTORIA HERNÀANDEZ-LLOREDA, BRIAN HARE, and MICHAEL TOMASELLO (2007, p. 1360)

Humans have big brains that they seldom use near full capacity. In gross size, the brains of elephants and whales surpass our brains, but relative to body mass human brains are the largest among placental mammals.[1] Many once presumably unique human qualities are attributable to our big brains: complex technology, insightful problem solving, self-awareness, theory of mind, language, culture, empathy, politics, deception, death awareness, love, and spirituality. Over the past half-century, bonobos, chimpanzees, gorillas, orangutans, and other animals have chipped away at this arrogant construct to the extent that few items have remained unqualified on the list. Nevertheless, there are notable indications of linkage between gross brain size and high-level cognition in extant Hominoidea because of "increased numbers of neural modules, increased size of the neocortex and of other higher neural processing areas, and increased neural connectivity."[2]

Experiments with children and observations on modern people who subsist by hunting and gathering suggest the existence of innate learning devices—modules—that enable infants to acquire the competencies on which adult mastery of the world depends: expectancies on how physical objects act, language, number and arithmetic, theory of mind, music, and spatial navigation.[3] When stimulated by information relevant to a problem for which the module is responsive, it can react rapidly, reflexively, and certainly to protect an infant from distracting information and entertaining false hypotheses.[4] Further, modular learning is distinct from learning by imitation.[5]

However one apportions factors of nature and nurture, one must admit that the brain is basic to virtually all human social and maintenance behaviors. Unfortunately, as difficult as it is to document and to explain the evolution of human bipedalism and tool behavior, such projects are truly empirical vis-à-vis elusive topics such as the origins of human language, non-material culture, and society. Few if any traces are left for paleoanthropologists to decipher. Accordingly, we resort to extant creatures for ideas, models, and scenarios. The exercise intrigues sociobiologists and laypersons, but it perturbs some sociocultural anthropologists and other social scientists and humanists. Cognitive neuroscientists and psycholinguists have joined physical anthropologists, archeologists, primatologists, and comparative psychologists as active contributors on human evolutionary projects.[6]

Brain Size, Structure, and Function

> There is no longer a good reason to consider encephalization as an index of some general functional capacity (intelligence) that is common to all mammals. We must face up to the fact that encephalization is largely uninterpretable in terms of cognitive and behavioral processes.
>
> TODD M. PREUSS (2001, p. 154)

To a large extent, brain size is determined by overall body size in primates.[7] Apes follow the general primate pattern, but humans have brains that are about three times larger than one would expect for primates of our size.[8] Samples of human and ape skulls indicate that human cranial capacities are 2.4 to 4.2 times larger than those of great apes and 10 to 17 times larger than those of gibbons (Table 10.1). Accordingly, modern humans are highly encephalized: the mass of the brain exceeds that expected for our body mass. Human brain

Table 10.1.   Endocranial volumes (cc) in Hominidae

Sahelanthropus tchadensis	320–380
Ardipithecus ramidus	300–350
Australopithecus afarensis	M: 445.8; R: 387–550 (N = 5)
Australopithecus africanus	M: 461.2; R: 400–560 (N = 9)
Australopithecus sediba	ca. 420 cc
Australopithecus garhi	M: 450; R: 450 (N = 1)
Paranthropus aethiopicus	M: 431.75; R: 400–490 (N = 4)
Paranthropus robustus	M: 493.3; R: 450–530 (N = 3)
Paranthropus boisei	M: 508.3; R: 475–545 (N = 6)
Homo floresiensis	ca. 380 (N = 1)
Homo habilis	M: 609.3; R: 509–687 (N = 6)
Homo georgicus	M: 715; R: 650–780 (N = 2)
Homo ergaster	M: 801 cc; R: 750–848 (N = 3)
Homo rudolfensis	M: 788.5; R: 752–825 (N = 2)
Homo erectus	M: 990.6; R: 727–1,390 (N = 25)
Homo antecessor	M: 1,218.3; R: 1,125–1,390 (N = 3)
Homo soloensis	M: 1,144.6; R: 1,013–1,251 (N = 7)
Homo heidelbergensis	M: 1,268; R: 1,165–1,740 (N = 12)
Homo neanderthalensis	M: 1,420.24; R: 1,172–1,740 (N = 29)
Homo sapiens (Middle-Late Pleistocene)	M: 1,456.95; R: 1,090–1,775 (N = 57)
Asian *Homo neanderthalensis*	1,537 ± 241 cc (n = 3)
European *Homo neanderthalensis*	1,510 ± 150 cc (n = 6)
Homo sapiens	1,330

Source: Partly based on Holloway et al. 2004.

mass is positively correlated with stature, body surface area, and body mass. At any given stature, body surface area, or body mass, male brains are about 100 grams larger than those of females.[9]

Although Passingham reported the cerebral neocortex and cerebellum to be especially large in humans compared with other primates, others have noted that the frontal lobes are not disproportionately enlarged in humans compared with those of great apes, and the human cerebellum is smaller than expected for an ape brain of human size.[10] The cerebellum and neocortex are associated with the motor coordination of fine movements and the integration of information from the several senses, respectively. In addition, the human cerebellum exhibits connectivity with the neocortical association areas involved in higher cognitive functioning and activation during a variety of cognitive tasks.[11]

Conversely, great apes and gibbons have cerebella that are relatively larger than predicted for their brain sizes.[12] The increase is particularly marked in hominoid lateral cerebella. MacLeod et al. suggested that the shift to a larger cerebellum—neocerebellum—took place in the common ancestors to the Hominoidea as they adapted to suspensory feeding behaviors and frugivory, which are facilitated by visual-spatial skills, planning complex movements, procedural learning, attention switching, and tactile discrimination of manipulated objects.[13] Cantalupo and Hopkins related volumetric differences and lateral asymmetry of the cerebellum to performance and hand preference in skilled actions such as tool use and aimed throwing.[14]

Although human frontal lobes might not be proportionally larger than those of great apes, they probably comprise more of the higher-order frontal and association cortex.[15] The ventral portion of the dentate nucleus, which connects the cerebellar and frontal association cortices involved with cognitive and language functions, is proportionally larger in humans than in great apes.[16] In humans, the most anterior part of the frontal lobes protects the execution of long-term mental plans from immediate environmental demands and generates new, perhaps more rewarding, behavioral or cognitive sequences, instead of controlling complex decision making and reasoning.[17]

The human cerebellum is also engaged in altering motor performances, but perhaps not in learning them initially.[18] Via positron emission tomography (PET), Stout et al. documented that while a veteran stone tool-maker knapped, there was significant activation of his superior parietal lobe, which is associated with complex spatial cognition integrating touch, vision, and proprioception, and the cerebellum, which integrates and controls motor activity.[19]

There appears to be a functional dichotomy along the anterior-posterior axis of the human cerebellum, with the anterior portion specialized for basic motor and somatosensory functions and the posterior portion more selectively involved in language, processing of spatial information, working memory, and executive function. Linguistic functions are lateralized to the right-posterior cerebellar hemisphere, and spatial processing is lateralized to the left posterior hemisphere.[20] The lateral asymmetry of chimpanzee posterior cerebella is also related to complex activities—especially tool use, aimed throwing, and handedness—that mirror human conditions.[15] Rilling summarized the distinctive features of hominoid brains as follows:

> Ape specializations include elaboration of the cerebellum (all apes) and frontal lobes (great apes only), and probably connectivity between

them. Human brain specializations include an overall larger proportion of neocortex, with disproportionate enlargement of prefrontal and temporal association cortices; an apparent increase in cerebellar connections with cerebral cortical association areas involved in cognition; and a probable augmentation of intracortical connectivity in prefrontal cortex.[21]

Much of the increase in human brain size is achieved postnatally. A newborn monkey has 60 percent and a newborn chimpanzee has 46 percent of its adult brain mass. But a human neonate has only 25 percent of its adult brain mass. Although the human brain continues to grow and develop into the late teens, most of its size is achieved by the fifth year.[22] One's first five years are an eventful period when major stages of language development occur: babbling, one-word utterances, and two-word combinations, all leading to sentences.[23]

The 10 percent difference between male and female gross brain size is a trait of early childhood, and appears to be due to increased cortical grey matter in boys (Table 10.1).[24] But by adulthood, females have the higher percentage of grey matter, while males have higher percentages of white matter and cerebrospinal fluid.[25] Grey matter is a major component of the central nervous system that is vital to muscle control, speech, sensory perception, memory, and emotions. It consists of neuronal cell bodies, dendrites and glial cells, and capillaries. White matter in the central nervous system consists chiefly of glial cells and myelinated (lipid-coated) axon tracts that conduct signals between the cerebral regions and between the cerebrum and the cerebellum and lower brain centers. The brain and spinal cord are bathed in cerebrospinal fluid, which protects the brain from traumatic injury, rids the central nervous system of metabolic waste, and prevents insufficient blood flow to the brain (ischemia).

In forty right-handed men, the percentage of grey matter was higher in the left hemisphere, the percentage of cerebrospinal fluid was higher in the right hemisphere, and there was hemispheric symmetry in the percentage of white matter. Forty right-handed women matched for age with the male sample (18–45 years) exhibited cerebral hemispheric symmetry in the values of the three components.[25] Consistently right-handed males had leftward gross morphological asymmetry in the motor cortex (a deeper central sulcus), whereas females evidenced hemispheric symmetry regardless of handedness.[26] Previously, Amunts et al. found that in right-handed males the deeper left central sulcus had microstructural complements, namely, leftward asymmetry in volume of neuropil—a tissue compartment of dendrites, axons, and synapses—in Brodmann's area 4, and increased intrasulcal surface of the precentral gyrus of the motor cortex.[27] Left-handers had a similar rightward microstructural asymmetry.

Allen et al. employed high-resolution magnetic resonance imaging (MRI) to measure the regional volumes of whole cerebral hemispheres, frontal, temporal, parietal, and occipital lobes, cingulate gyrus, cerebellum, and other structures in the brains of twenty-three men and twenty-three women. The right-handed subjects ranged from 22 to 49 years old.[28] Although the brains and various regions were significantly larger in men than in women, the proportional size of the regions relative to the total volume of the hemisphere was similar in the sexes. The only striking asymmetry was leftward, exhibited by the cingulate gyrus, which is involved with emotion formation and processing, learning, and memory as part of the limbic system.

Neuroimaging studies on small samples of human female and male subjects have indicated some sex-related differences in the neural correlates of general intelligence, cognitive performance on a variety of tasks, and appreciating beauty, but overall there was consistency in the outcomes.[29] Accordingly, there is sexual parity of performance, albeit based on different areas of the brain. In a relatively large sample of women ($n = 442$) and men ($n = 377$), there was no greater bilaterality of language representation in women than in men, from which Sommer et al. concluded that it is unlikely that differences in language lateralization underlie general sex differences in human cognitive performance.[30]

Tests of direct connections between Broca's and Wernicke's territories in a sample of twenty male and twenty female 18- to 22-year-old right-handed individuals indicated that half of them had extreme leftward lateralization, while only 17.5 percent of them had bilaterally symmetrical connections (Figure 10.1). The latter were better overall at remembering words using semantic association. Females were more likely to have the symmetrical pattern. Bilateral representation might ultimately be more advantageous than extreme lateralization for specific cognitive functions.[31]

Scanning with MRI revealed that ten male and ten female subjects had an extensive humor-response strategy.[32] Similar brain regions—the temporal-occipital junction and temporal pole (involved in semantic knowledge and juxtaposition) and inferior frontal gyrus (language processing)—were acti-

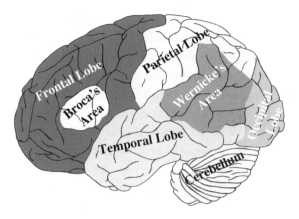

Figure 10.1. Broca's and Wernicke's areas.

vated as they rated seventy verbal and nonverbal achromatic cartoons as funny or not funny. However, the left frontal cortices of females were activated more than those of males, which suggested a greater degree of executive processing and language-based decoding. The "results indicate sex-specific differences in neural response to humor with implications for sex-based disparities in the integration of cognition and emotion."[33]

Cranial sutures close during one's early twenties. Thereafter, the main gross change in the brains of humans and chimpanzees is shrinkage. In European *Homo sapiens,* the mean mass of female brains is about 90 percent that of male brains until the eighth and ninth decades of life (94 percent): although the brains of both sexes shrink, those of males shrink more (Table 10.2).[34]

In humans, the heritability of intelligence changes over one's lifetime from about 30 percent at 3 years of age to as high as 91 percent at 65 years old and declining thereafter.[35] Statistically, heritability is the proportion of phenotypic variance attributed to genetic variance—in this instance, the extent to which individual genetic differences contribute to individual differences in observed behavior. Heritability is an abstract concept that says nothing about the specific genes that contribute to a trait. Indeed, as Posthuma et al., point out:

Most researchers . . . [are] in the dark about how many genetic variants are involved or how big their effects are. Genome-wide studies . . .

strongly suggest that there will be no genes with a large or moderate effect, and, therefore, studies aimed at securely identifying genetic contributions to cognitive differences will probably require very large samples, especially when a genome-wide approach is adapted. Statistically more powerful candidate gene studies have so far identified a handful of genes, but only a few of these have . . . shown replicated associations with intelligence. All of the identified genes have small effects, consistent with a polygenic view of the heritability of intelligence.[36]

The mean gross sizes of brains increase from the smaller gibbons through the siamang, bonobos, chimpanzees, and orangutans to the gorillas. Adult gibbon cranial capacities range from 70 to 152 cc. Siamang mean cranial capacities are about 25 percent larger than those of other gibbons. Like their body masses, there is low sexual dimorphism in the brain size of gibbons; for example, the mean cranial capacity of female lar gibbons is 97 percent of the male mean, and the siamang value is 98 percent (Table 10.1).

Great ape cranial capacities are 3.5 to 5.0 times larger than those of gibbons. Except among bonobos and perhaps Gombe chimpanzees, there is greater sexual dimorphism in brain sizes of great apes than in gibbons.[37] Bonobos are probably the least dimorphic apes in regard to cranial capacity: the female mean is 99 percent of the male mean (Table 10.2). On average, female chimpanzee cranial capacities are approximately 25 cc greater than those of bonobos, and the mean male chimpanzee value is approximately 55 cc greater than that of bonobos.

Chimpanzees are moderately sexually dimorphic in cranial capacity. The female means are 92.4 percent and 92.8 percent of the male means in large mixed samples of *Pan troglodytes* spp.; however, the population sample sizes varied. Whereas a small sample (eight males and eight females) of Gombe *Pan troglodytes schweinfurthii* indicated that there is no significant sexual dimorphism in cranial capacity, a larger sample (ten males, twenty-one females) of Taï Forest *Pan troglodytes verus* evinced 92 percent sexual dimorphism in cranial capacity (Table 10.2).[38]

As in body size, the brain sizes of gorillas and orangutans exhibit considerably greater sexual dimorphism than the brains of chimpanzees. Moreover, infant and juvenile male orangutans and gorillas have larger brains than the immature females. Nonetheless, the endocranial size dimorphism does not match the dimorphism in the body masses of gorillas and orangutans. The orangutan female mean is 81 percent of the male value, and the value for gorillas is 83 percent (Table 10.2).

Although female orangutan cranial capacities are smaller than those of chimpanzees, those of male orangutans are larger than those of chimpanzee males. Average cranial capacities of female gorillas are larger than those of average male orangutans and chimpanzees. Few adult male gorillas have brains that are as small as those of chimpanzees and orangutans. The largest great ape cranial capacity—752 cc—is from a male western gorilla. No other extant great ape skull had a cranial capacity near 700 cc (Table 10.2).[39]

Interestingly, two chimpanzees (Austin and Panzee) that achieved remarkable competence communicating via a computerized system of noniconic lexigrams had relatively large brains compared with those of predominantly wild conspecifics (Table 10.2). Indeed Austin's postmortem brain mass (529.2 grams) is far outside the ranges of the endocranial volumes in the samples of Schultz, Cramer, and Zihlman et al., and Panzee's brain (425 cc) extends the upper range of female chimpanzees. Austin's brain mass is also the largest among over 100 postmortem specimens at Yerkes National Primate Research Center.[40] Regular feeding of highly nutritious foods from infancy might explain their relatively sizeable brains, but the significance of brain size alone to account for their accomplishments is moot in the absence of more individuals with similar life histories and a broad range of brain sizes with controls for potential allometric effects.

Cerebral Cells and Connections

Neuroanatomical research pioneers noted nothing special about the cellular composition of large hominid brains, and particularly to what extent bigger cortices have proportionately more nerve cells and connections among them. For instance, increased cerebral volume was accompanied by a decreased density of nerve cells in the neocortical grey matter in a series of primates consisting of tarsiers, marmosets, Old World monkeys, chimpanzees, and humans.[41] The cortical nerve cells increased in size as cortical volume increased. As brain size increases, horizontal expansion of the cerebral cortex is not accompanied by a comparable increase in cortical thickness; instead, it becomes increasingly convoluted into gyri and sulci.[42]

A negative consequence of increased brain size is that interhemispheric transfer speed and its concomitant cognitive processing speed are decreased.

However, processing power can be maintained and increased if the components are concentrated in one hemisphere, which is probably at the root of hemispheric specialization.[43] Accordingly, one may expect that large brains, like those of great apes and humans, "will manifest a high degree of structural and functional hemispheric specialization."[44]

Cortical areas can be distinguished from one another by staining the cell bodies and myelinated fibers and by their connections and functional properties. Cortical neurons are biochemically variable across taxa. The basic structural-functional units of cortical organization are cell columns.[45] Neurons within a column are interconnected vertically, share extrinsic connectivity, and act as basic functional units for sensory, motor, and association areas that serve the highest cognitive functions.[46]

Rockel et al. reported that in all areas of the cortex, the columns have nearly the same number of cells—approximately 110—and that this number is nearly constant across species, albeit with the primate visual cortex purportedly constituting an exception in having approximately 270 cells per column.[47] Although it is widely cited by other authors, the columnar regional homogeneity reported by Rockel et al. is chiefly an artifact of their methods.[48] In fact, there is notable columnar variation among neocortical regions and taxa.[49]

Mammals differ widely in the number of cortical areas, with larger-brained mammals having more of them than the smaller-brained taxa.[50] Further, primates have primate-specific higher-order cortical territories that constitute a distinctive connectional system.[45] It is possible that some cortical and wider brain reorganization occurred between apehood and humanity, such as via the proliferation of dendritic interconnections. Suzanne Herculano-Houzel concluded:

> With 86 billion neurons and as many nonneuronal cells, the human brain is basically a scaled-up nonhuman primate brain in its cellular composition and metabolic cost, with a relatively enlarged cerebral cortex that does not have a relatively larger number of brain neurons yet is remarkable in its cognitive abilities and metabolism simply because of its extremely large number of neurons.[51]

A quantitative image analysis of the regions corresponding to the orofacial representation of the primary motor cortex (Brodmann's area 4) in embalmed macaque, baboon, orangutan, gorilla, chimpanzee, and human brains indicated that apes and humans have increased cortical thickness of layer III and lower cell volume densities compared with those of monkeys. Reduced cell volume density would allow more neuropil space for interconnections. Accordingly, differences in microstructure might relate to greater volitional fine motor control of facial expressions in the Hominoidea.[52]

Neurons in the ventral premotor cortices and inferior parietal lobes of macaques and humans fire not only in an individual performing an act but also in one observing or hearing the act, apparently indicating that there is action recognition and understanding based on motor simulation.[53] Some researchers have further proposed that the mirror-neuron mechanism might be involved in the evolution of speech and that mirror-matching mechanisms underpin the human capacity for empathy (sharing the emotions of another being, but knowing that they are not one's own), imitation, and understanding the intentions of others.[54] However, Lingnau et al. have challenged the existence of mirror neurons in humans, noting that although there was adaptation for motor acts that their subjects observed and then executed, there was no sign of adaptation for motor acts that were first executed and then observed.[55] For ethical reasons, precise single-cell recordings like those used to test mirror neurons in macaques have not been employed in humans; much of the evidence for mirror image systems in humans is from functional brain neuroimaging studies that show perception of action leading to activation of the brain areas that are also involved in the execution of action.[56]

Brainy Substrates in Pongids and People

Apes versatilely employ a variety of objects as tools and can learn to communicate via artifactual symbols according to simple rules. This has sparked questions about possible high-level neurological links between them and us, and the search is on for the specific areas of the ape central nervous system that underpin these capabilities. We are especially interested in homologous areas that make human speech and other symbolically mediated behaviors possible. Unfortunately, ape brains are only crudely charted by direct studies.[57] Ablation experiments to map ape cortices are forbidden for ethical reasons. Accordingly, until advanced neuroimaging techniques can be applied creatively to active apes, we must rely chiefly on interpolations between results from invasive experiments on unfortunate monkeys and clinical observations on humans.[58]

Nonetheless, contemporary neuroimaging studies, advanced histological techniques, and genomics have indicated that human brains are qualitatively distinct from those of chimpanzees, and we anticipate more detailed data once researchers can monitor the brains of alert, active apes telemetrically and resolve questions of genetic and environmental influences on brain development and functions.[59]

For example, two amino acids in the transcription factor FOXP2 (forkhead box P2), a gene implicated in Mendelian forms of human speech and language dysfunction, confer differential transcriptional regulation *in vitro*.[60] Although the structure of FOXP2 appears to have been highly conserved in the Mammalia, the human variant acquired two amino acids under positive selection after divergence from the common ancestors of *Pan* and *Homo*. Tests with tissue from adult chimpanzee and human brains, combined with information on FOXP2 in human fetal brains, have provided evidence of notable FOXP2 expression in the cerebellum and perisylvian cerebral cortex. Targets differentially enriched by FOXP2 are involved with cerebellar motor function, craniofacial formation, and cartilage and connective tissue formation. Konopka et al. inferred an important role for human FOXP2 in "establishing both the neural circuitry and physical structures needed for spoken language."[61] Prior researchers also concluded that FOXP2 is involved in the developmental process that culminates in speech and language, though the precise pathway remains to be discovered along with other genomic and cultural selective factors that might be involved.[62]

Surgical ablation studies on rhesus macaques indicate that particular areas of cerebral association cortex are related to specific modes of learning (Figure 10.2):

- Parietal association area: learning by touch.
- Superior temporal association area: learning by hearing.
- Inferior temporal association area: learning by sight.[9]

The prefrontal association area is involved with adaptability and the direction of actions that are appropriate to particular situations (Figure 10.2A).[63] Unlike ablations of the other three areas, removal of prefrontal cortex does not disrupt particular experimental tasks. But monkeys cannot adjust to a reversal (i.e., to a new task) if the prefrontal area is removed (Figure 10.2B).[9]

Humans who have lost similar cortical areas by trauma or disease confirm that the human neocortex is similar to that of monkeys. Passingham concluded

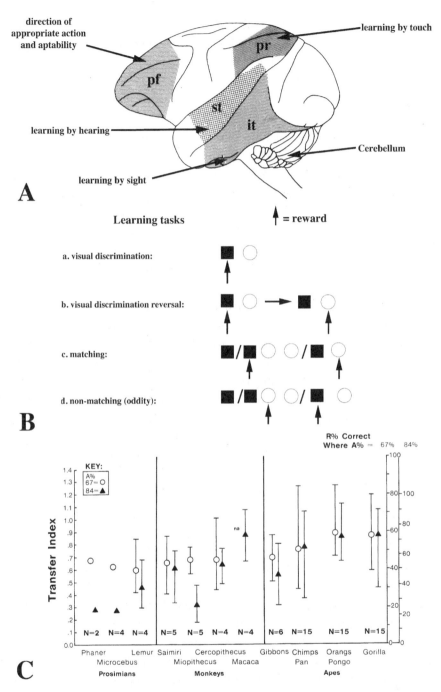

Figure 10.2. (A) Areas of the cerebral association cortex related to specific modes of learning and adaptability in *Macaca mulatta*. Symbols: it, inferior temporal; pf, prefrontal; pr, parietal; st, superior temporal. (B) Tests-a. Choose ■ and be rewarded; b. choose opposite (O) of ■ for reward; c. choose matching ■ (or matching O) for reward; d. choose non-matching ■ or O for reward. (C) Rumbaugh Transfer index.

that equal relative expansion of all association areas in the human brain indicates increases in all intellectual abilities.[9] Consequently, "we have a unique grasp of the world and how to act on it."[64]

Cerebral Asymmetry and Laterality

Behavioral asymmetries that reflect cerebral hemispheric specialization are relatively common in vertebrates, including fish, amphibians, reptiles, birds, and mammals.[65] Localization of linguistic functions, usually on the left side, and handedness indicate cerebral hemispheric dominance in human brains.[66] Nonetheless, the relationship between hemispheric dominance for speech and lateralized hand preference is not firm. There are exceptions, such as left-handers who are left-brain dominant instead of right-brain dominant for speech, and right-handers who may have a right-hemispheric dominance for speech.[67]

Human brains exhibit notable structural asymmetries during infancy as well as adulthood.[68] Two areas are particularly interesting because they are related to our capacity for language. An area of the parietal and temporal cortex containing Wernicke's area is commonly larger in the left cerebral hemisphere, and the frontal cortex, containing Broca's area, is also usually larger on the left side. Classically interpreted, Broca's area is basic to verbal articulation (production), and Wernicke's area is related to understanding speech (comprehension). The same areas are active in proficient bilingual persons regardless the language in use, and the left caudate nucleus monitors and controls it (Figure 10.1).[69]

The classic view of simple linguistic roles for Broca's area (left inferior frontal gyrus) and Wernicke's area is oversimplified.[70] Broadly speaking, Broca's area serves language production, and Wernicke's area underpins the production of speech and sign languages.[71] Apparently, Broca's area is involved in general grammatical processing instead of articulation per se.[72] Writing and manual signing can be disrupted as much as speech by an injury to Broca's area.[73] Moreover, the cerebellum and deeper areas of the brain are also implicated in disturbances of grammar and verbal fluency, though their specific interactions with Broca's cap are unclear.[74]

Wernicke's area appears to house the dictionary, which selects the words to be sent to Broca's area where they are assembled syntactically. Broca's area also may have more general involvement in recognizing meaningful action.[75]

The planum temporale, which comprises the higher-order auditory association cortex, is an area of the posterior superior temporal gyrus that constitutes a major part of Wernicke's area. In 100 human brains, it was predominantly larger—one-third longer—in the left cerebral hemisphere: 65 percent on the left, 11 percent on the right, and 24 percent with no bias.[76] High-resolution MRI of normal individuals supported the assumption that the symmetry and asymmetry of the planum temporale are related to functional lateralization.[77] For instance, musicians with perfect pitch have a pronounced leftward asymmetry of the planum temporale, even compared with conditions in nonmusicians and in musicians lacking perfect pitch.[78]

Certain occipital and frontal asymmetries in gross shape—petalias—can be related to handedness in humans. Humans have left occipital and right frontal petalias: the left occipital lobe projects farther posteriorly and is usually wider than the right one, and the right frontal pole projects farther anteriorly and is usually wider than the left one, particularly in right-handed people.[79] All patterns of asymmetry in gross cerebral shape in apes are different from those shared by all species of *Homo*.[80] There are hints of petalias in the endocasts of *Australopithecus* and *Paranthropus,* but the evidence is not sufficient to posit whether they reflect humanoid handedness and cerebral specializations (Figure 10.3).[81]

Analyses via MRI have shown rightward frontal lobe and leftward occipital lobe width asymmetries in the brains of bonobos, chimpanzees, gorillas, and orangutans but not in samples of catarrhine and platyrrhine monkeys.[82] The four taxa of great apes exhibited population-level leftward asymmetry in width and volume of a knob from the precentral gyrus, which controls hand and digital motion.[83] The knob was not apparent in samples of gibbons and monkeys. Hopkins et al. confirmed the population-level leftward asymmetry of the lateral sulcus (sylvian fissure) in a sample of great apes and noted a rightward asymmetry of part of the sylvian fissure in catarrhine monkeys but no sylvian fissure asymmetry in platyrrhine monkeys.[84] The lateral sulcus, which divides the parietal lobe from the temporal lobe, reflects the size of the planum temporale.

In a sample of fifty-three captive chimpanzees, MRI revealed significant inverse association in the asymmetry quotients $[(R - L)/[(R + L) \times 0.5]$ of the anterior and posterior cerebellar regions, indicating that the organ was torqued at the individual level, but not at the population level.[85] Further handedness for tool use, but not for simple reaching or digitally probing peanut butter

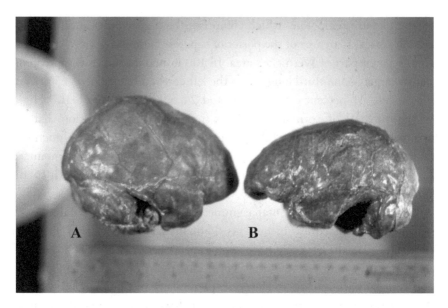

Figure 10.3. Natural endocasts: (A) Right lateral view of Sts 60 *(Australopithecus)*. (B) Left lateral view of SK 1585 *(Paranthropus)*.

from a pipe (the tube task), was associated with variation in cerebellar asymmetries. In contrast with chimpanzees, a sample of eleven capuchins exhibited cerebellar asymmetries that were significantly associated with handedness during the tube task. Humans have population-level rightward anterior/leftward posterior cerebellar torque, which suggests a direct association with neocortical asymmetry.[86]

An extensive review of research on handedness in nonhuman primates and observations of chimpanzees in Gombe National Park indicated that only chimpanzees evidence signs of population-level bias toward right-handedness, and then only in captivity and incompletely.[87] McGrew and Marchant concluded that lateralization of hand function at the specific level is not documented for any nonhuman primate taxon, task, or context.[88] Therefore, nonhuman primates are unreliable models upon which to construct theories on the evolution of human handedness.[89]

Contrarily, Lonsdorf and Hopkins documented right-handedness during termite fishing among seventeen members of the Kasekela chimpanzee community in Gombe National Park, and Humle and Matsuzawa concluded that chimpanzees at Bossou, Republic of Guinea, demonstrated a tendency for population-level right-handedness during ant dipping and manual extraction of crushed oil-palm hearts.[90] Moreover, Lonsdorf and Hopkins cited a host of findings supporting handedness in captive chimpanzees, bonobos, gorillas, orangutans, and gibbons during bimanual feeding, other coordinated bimanual actions, fine manipulation, simple reaching, throwing, and manual gestures.[91] Prior and later studies further support the case for handedness in wild and captive apes.[92] For example, right-handedness was predominant in a sample of ninety-seven captive chimpanzees of both sexes (fifty-eight female, thirty-nine male), ranging from 6 to 50 years old, when they gestured toward other chimpanzees and humans.[93] When handling fruits bimanually and during fine manipulations, Mahale male ($n = 15$) and female ($n = 24$) chimpanzees differed significantly: most females were right-handed, but males were predominantly left-handed.[94] Contrarily, when hand-dipping water from tree holes, forty-nine siamang in Bukit Barisan Selatan National Park of Sumatra, Indonesia, evidenced a population-level left-hand preference and no difference between female and male hand preference for dipping.[95]

A sample of forty-eight captive orangutans exhibited population-level left-handedness, and a small sample of orangutans ($n = 8$) in Gunung Palung National Park, West Kalimantan, Indonesia, showed evidence of hemispheric specialization in limb use, though their hands appeared not to be under strong lateralized hemispheric control. The difference between population-level right-handedness in chimpanzees, bonobos, and gorillas versus left-handedness in orangutans might be related to the positional behavioral versatility of orangutans in trees.[96]

Posture and variable contextual factors undoubtedly influence which hand apes might use for specific tasks. For instance, in samples of eight orangutans, forty chimpanzees, and twelve gorillas, there was significant dextral bias when the subjects reached for food while bipedal, but none when they reached while tripedal.[97] Eleven captive bonobos also were significantly more right-handed when standing bipedally, but five other captive bonobos were increasingly left-handed as they rose from seated to bipedal postures to reach for food.[98]

Posture significantly biased dextral throwing by thirty-six captive chimpanzees. Both males and females tended to throw objects with remarkable accuracy via the right hand regardless of standing bipedally or tripedally, but males were even more frequently right-handed when they threw objects from a bipedal posture.[99] Similarly, although disposed to reach with the left hand when tripedal, rhesus monkeys employed the right hand significantly more often when reaching while bipedal.[100]

Forty captive chimpanzees evidenced significant left-hand bias when descending onto their knuckles from platforms about 1 meter above the floor. Older individuals were more lateralized than younger ones, and males were more lateralized than females.[101] A similar bias for leading with the left forelimb probably existed in sixty-four wild chimpanzees as indicated by a greater cross-sectional area of the humeral shaft.[102]

Stone flakes produced by Early and Middle Pleistocene hominids indicate that the right hand was the preferred working hand for knapping flakes. Via experiments, Toth showed that between 1.4 and 1.9 Ma some hominids had become right-handed when knapping flakes from cores.[103] However, his inference that their brains might have been profoundly lateralized for different functions and that the feature is genetically determined is perhaps reading too much from the rocks of ages past.

Brains of catarrhine monkeys also sport cerebral asymmetries.[104] Some of these asymmetries are located in areas that are related to speech in humans. For instance, lesions in an area of the brain of *Macaca fascicularis* that is comparable to Wernicke's area impaired the ability of the monkey to distinguish between the human utterances /a/ and /u/. But it could still distinguish between a pure tone and a burst of noise and visually between the numerals 4 and 6.[105] Further, selective unilateral and bilateral removal of the left and right superior temporal gyri, including the auditory cortices, in *Macaca fuscata* confirmed that at least some monkeys have laterality in the receipt of salient communicative features of their vocal signals.[106]

If discernment of specific communicative features is located in the left temporal regions of monkey brains, then we should not be surprised to find that laterality for comprehension of communicative utterances is a basic catarrhine feature. Accordingly, it probably is not unique to humans. Even more compellingly, bonobos, chimpanzees, gorillas, and orangutans sport gross asymmetries in the planum temporale, with the left larger than the right, though the opposite was maintained in a sample of eight bonobos.[107]

While humans exhibit robust asymmetry in cortical cell columns of the planum temporale, chimpanzee and rhesus monkey brains lack them.[108] Because Hopkins et al. could not identify a key landmark—Herschl's gyrus—in their samples of gibbons, macaques, baboons, mangabeys, capuchins, and squirrel monkeys, they could not assess the degree of symmetry in their plana temporale.[109]

Brodmann's area 44, which is in the same region of the frontal lobe as Broca's area, is larger in the left cerebral hemispheres of bonobos, chimpanzees, and gorillas.[110] Further, the left inferior frontal gyrus and other cortical and subcortical areas were active when three chimpanzees directed gestures and vocalizations toward humans in response to food incentives outside their cages.[111] Nonetheless, before drawing too close comparisons between cerebral structure and functions in humans and great apes, one must heed the caveats of Sherwood et al.: "gross morphologic patterns do not offer substantive landmarks for the measurement of Brodmann's area 44 in great apes. Whether or not Broca's area homologue of great apes exhibits humanlike asymmetry can only be resolved through further analyses of microstructural components."[112]

Chimpanzees and orangutans, like humans, have rightward asymmetry of an area of the inferior parietal lobule: the planum parietale.[113] Only four of eight chimpanzees exhibited rightward asymmetry of the planum parietale, whereas the others were leftward asymmetric or symmetric.[114] The planum parietale is adjacent to the planum temporale, but in humans they appear to be independent of one another. In a sample of 106 human right-handers and 35 left-handers, balanced between males and females, there was strong rightward asymmetry of the planum parietale in right-handed men and left-handed women.[115] Given that the planum parietale is implicated in dyslexia and other communication disorders, its possible role in great apes might hold some clues to the evolution of human language.[116] It would be most interesting to know how the planum parietale of Matata (b. 1970), a bonobo that failed to employ a lexigramic system, compares with those of Kanzi (b. 1980) and Panbanisha (1985–2012), the bonobos that concurrently learned to do so readily.[117]

Increased temperature of the left tympanic membrane and a decrease in the right one indicated leftward asymmetry of cognitive functions in chimpanzees while they worked on matching-to-sample and visual-spatial discrimination tasks.[118] When they engaged in a motor task, there was no lateralized change in cerebral blood flow, as reflected by temperature of the right and left tympanic membranes.

A variety of experiments and methods indicate the humans, chimpanzees, and rhesus monkeys have rightward cerebral asymmetry related to emotions. For example, the left hemiface is more involved in both positive and negative facial expressions of emotional state, which suggests a right cerebral hemispheric dominance for the facial expression of emotions.[119] However, in adult humans the situation might be more complicated, with the right frontal region more active during negative/withdrawal emotions and the left frontal region more involved in positive/approach emotions.[120] The right prefrontal region controls the suppression of emotional memories, at least in humans who are not psychiatrically impaired.[121] In humans, the medial frontal cortex might have a special role in second-order representations needed to communicate acts when one has to represent someone else's representation of one's own mental state. The amygdala attaches emotional value to faces, enabling one to recognize expressions of fear and trustworthiness, and the posterior superior temporal sulcus predicts the end point of complex trajectories created when agents act upon the world.[122]

Functional magnetic resonance imaging (fMRI) studies on twenty-six healthy adult human subjects confirmed that the anterior temporal lobes play a central role for representing abstract conceptual knowledge and concepts denoted by composite expressions, and that the right temporal lobe is important for social cognition.[123]

Mystery of the Ballooned Brain

> But when the brain sizes of the different evolutionary phases of man are compared with each other, it becomes evident that we are still far from understanding what these differences in size really mean and whether the most suggestive interpretation of an increase in size is tantamount to an augmentation of reasoning power and cultural progress is correct.
>
> FRANZ WEIDENREICH (1946, p. 93)

In tracking brain evolution through time, we are hostage to serendipitous or more deliberate findings of fossil crania that are complete enough to measure endocranial capacity or even more rarely a natural fossil endocast.[124] Unfortunately, in sizeable mammals the brain does not fill the cranial cavity snugly, so measuring its volume overestimates brain size. Further, sulcal impressions are less common on the endocranial vaults of sizeable mammals, including great apes and humans, than on those of smaller ones, such as gibbons and monkeys (Figure 10.3).[125] Because fossil hominid specimens limit detailed consideration of which cerebral areas expanded relative to others, behaviors such as handedness and especially language remain elusive to paleoanthropologists.

Hominid brain mass has evidenced a threefold increase in size over the past 3 million years (Table 10.1).[126] Brain size doubled between *Australopithecus* (3.0 Ma) and *Homo erectus* (1.5 Ma). Further, there are indications of nonallometric (i.e., not related to body growth) widening of the temporoparietal areas in the brains of *Homo* spp., which is particularly evident in *Homo neander-thalensis*. Nonallometric widening of the anterior cranial fossa, which accommodates the frontal lobes, is also indicated for *Homo sapiens* and *Homo neanderthalensis* versus earlier Hominidae, including *Homo erectus*.[127] Stone tool technology was rather simple during the heydays of *Australopithecus, Paranthropus,* and *Homo habilis,* but it is problematic to weight the development and refinement of artifacts against other factors such as the development of language and other mental processing as selecting for cranial expansion in *Homo*.

Stone tool technology became more elaborate during the time of *Homo erectus* and truly burgeoned during the Late Pleistocene (125–10 Ka). Control of fire may have played a minor selective role in brain enlargement if it was not practiced until ca. 500 Ka, but its role might have been more significant if hominids used it regularly beginning in the Early Pleistocene.[128] Weighting the selective roles that avoidance of predation, subsistence foraging, and cultural and social behavioral factors might have played in hominid brain expansion depends on our ability to identify when our distant ancestors developed the several intelligences that depend on symboling to encode and express them and to organize their social, economic, and political lives.[129]

Like the organs of digestion—the stomach, liver, pancreas, and intestines—the brain is metabolically a very expensive organ that requires more fuel than the other organs. Further, "among important changes in both humans and chimpanzees, but to a greater extent humans, are the up-regulated expression profiles of aerobic energy metabolism genes and neuronal function-related genes, suggesting that increased neuronal activity required increased supplies of energy."[130]

Because humans have relatively small guts and very large brains that are not correlated significantly with an elevated high basal metabolic rate, we require high-quality diets.[131] Demands on pregnant females are even more severe given that the human placenta is a highly efficient extension of a voracious parasitic fetus.[132] Accordingly, we require foods that are calorically rich and readily digested. The more the foods are macerated and finely fragmented, the quicker they can be digested. Because of our puny teeth and relatively weak jaws, tools and cooking are vital to human development and survival.[133] The challenge is to weight these factors and place them geochronologically among other possible selective factors in scenarios of human brain expansion and cognitive enhancement.

Table 10.2. Adult hominoid cranial capacities or brain masses measured in cubic centimeters (cc) or weighed in grams (g), respectively

Species	Sex	No	Mean cc or g	Range cc or g	Source	Dimorphism (F × 100/M)
Hylobates klossii		10	87	78–103	Schultz 1933	—
Hylobates lar carpenteri	M	95	104	89–125	Schultz 1965	97.1%
	F	86	101	82–116	Schultz 1933; Tobias 1971	
Hylobates concolor		69	101	82–136	Tobias 1971	—
Hylobates syndactylus	M	23	126	100–150		97.6
	F	17	123	105–152		
Pan paniscus	M	6	356	334–381	Schultz 1969	92.4
	F	5	329	275–358		
	M	29	352	140	Cramer 1977	98.9
	F	30	348	160		
P-Suke*	M	1	371	—	W. D. Hopkins (personal communication)	
Nathan*	M	1	393	—		
Tamuli*	F	1	359	—		
Pan troglodytes	M	57	381	292–454	Schultz 1969	92.4
	F	59	352	282–418		
	M	33	404	140	Cramer 1977	92.8
	F	34	375	260		
P.t. verus	M	10	379	345–415	Zihlman et al. 2008	92.2
	F	21	350	300–395		

	Sex	N	Mean	Range	Reference	
P. t. schweinturthii	M	8	380	326–420	Zihlman et al. 2008	99.9
	F	8	379	337–406		
Austin	M	1	529	—	W. D. Hopkins (personal communication)	
Panzee	F	1	425	—		
Lana	F	1	398	—		
Pongo pygmaeus	M	57	416	334–502	Schultz 1965	81.2
	F	52	338	276–425		
Pan gorilla	M	72	535	412–752	Schultz 1965	82.8
	F	43	443	350–523		
Homo sapiens Swiss	M	70	1,463	1,250–1,685	Schultz 1965	91.7
	F	40	1,314	1,215–1,510		
Danes	M	724	1,440	—	Pakkenberg and Voigt 1964	89.0
19–25 years*	F	302	1,282	—		
Hungarians						
20–40 years*	M	80	1,386	1,050–1,670	Tóth 1965	90.4
	F	120	1,253	900–1,750		
40–60 years	M	405	1,375	1,000–1,900		90.2
	F	383	1,241	800–1,775		
60–80 years	M	571	1,335	1,000–1,700		90.5
	F	462	1,208	900–1,700		
80–100 years	M	54	1,254	1,000–1,500		94.0
	F	57	1,179	1,000–1,400		

Pioneering Explorations of Ape Mentality

> Squeezed into the framework of an experiment, mental activity
> is like a free bird, accustomed to unhampered flights and
> unlimited spaces, that has been caught and put into a cage. Like
> a live bird, the mental activity bumps into the walls of the
> experiment, unwilling to be confined by them, struggles to get
> out, and breaks the nets of theoretical expectations and the
> experimenter's plans. A contemplative observer who grants total
> freedom of expression to this kaleidoscopical change of mental
> conditions will be more likely to watch its development, follow
> it wherever it goes, catch it, and describe it more easily.
>
> NADEZHDA N. LADYGINA-KOHTS (2002, p. 9)

Systematic observational and experimental studies on captive great apes were
begun by Wolfgang Köhler at an Anthropoid Research Station in Tenerife,
Canary Islands (1913–17), Nadezhda Nikolaevna Ladygina-Kohts at the
Museum Darwinianum of Moscow (1913–16), and Robert M. Yerkes at the
Yale University primate facility (1925–30) in New Haven, Connecticut.
Other participants included several generations of comparative psychologists
at the Yale Laboratories of Comparative Psychobiology (1930–35) and Yale
Laboratories of Primate Biology in Orange Park, Florida (1935), which was
renamed Yerkes Laboratories of Primate Biology in 1942 and Yerkes National
Primate Research Center of Emory University in Atlanta, Georgia, beginning
in 1962.[134]

Köhler and other comparative psychologists noted that apes excel in problems that require the use of tools. Further, they were especially interested in
whether apes exhibited insight during problem solving.[135] Köhler set the criterion for insight as "the appearance of a complete solution with reference to the
whole lay-out of the field."[136] For example, a subject might manipulate experimental objects and use them inappropriately for a span. Then the subject may
stop and remain quiet or may engage in unrelated activities. Abruptly, the
subject returns to the situation and solves the problem straightaway. In this
situation, the subject appears to have thought about the task before its actual
resolution—the subject employed ideational processes.[137]

Experience with similar situations and objects enhance the problem-solving
abilities of apes, and play is especially important for learning the properties of
objects and for the maturation of motor and social skills.[138] Great apes are
more responsive to objects than are many other mammals, which is reminis-

cent of human curiosity and playfulness. It appears that, as in our increase of brain size, we have simply extended trends that were present in our apish ancestors. The more we must learn, the longer we must be nurtured and protected. Our sexual development is delayed, and we live to nurture the next generation or two. What gave the initial kick to this complex life-history pattern is unknown.

Because the studies of Köhler, Kohts, and Yerkes gave the appearance that chimpanzees provided the most promising exemplar of an ape mind, hominoid psychologists focused studies on them for decades after the 1930s. Unlike current students of great apes, Yerkes was keen to have a colony of chimpanzees in order to determine how their behavior might be altered, thereby serving as prototypes for human biological engineering.[139] Yerkes and Yerkes further speculated that "indeed were it capable of speech and amenable to domestication, this remarkable primate might quickly come into competition with low-grade manual labor in human industry."[140] Although they were not as diverse and imaginative as the techniques employed by many torturers, the more extreme experiments used to test and modify chimpanzees would strike most modern behavioral scientists and lay observers as extremely cruel and predictably unproductive.

Albeit after rather extensive discrimination-reversal training, chimpanzees are very capable performers on learning sets: "learning-how-to-learn as a function of cumulative positive transfer across as series of problems in which a common rule always defines the correct response."[141] Often on the first trial, chimpanzees grasp all the information that is necessary for problem solution and succeed with a "win-stay; lose-shift" strategy (Figure 10.2B).[142] The most widely publicized triumph of operant conditioning methods was the chimpanzee-manned *Mercury* space capsules. Astronauts Ham and Enos received electric shocks for wrong choices and banana pellets or sips of juice or water for correct ones.[143]

The chief function of the comparative method is to establish the generality of phenomena; however, before the late 1960s, many comparative psychologists assumed that chimpanzees were the most intelligent nonhuman primates, though they had not tested enough anthropoid species and devised sufficient protocols and methods to allow definitive scientific statements about how they ranked mentally.[144]

A twenty-first-century meta-analysis of published studies has indicated that great apes are more intelligent than other nonhuman primate species, from which van Schaik et al. concluded "primate cognition is distinguished by some generalized capacity rather than a collection of narrow, problem- or domain-specific abilities, supporting the view that great apes constitute a homogeneous group that outranks other primates in cognitive performance."[145]

Ape Intelligence Tests

Transfer Index

Beginning in the late 1960s, Rumbaugh notably advanced the study of comparative primate learning and intelligence by developing the transfer index and applying it to statistically significant small samples of apes, monkeys, and lemurs, initially at the San Diego Zoo and then at the Yerkes National Primate Research Center, which had complemented their large colony of chimpanzees with gorillas, orangutans, and gibbons (Figure 10.2C).[146] Before Rumbaugh's ambitious project, only a smattering of psychological tests had been conducted on great apes other than chimpanzees, and comparative psychologists had neglected siamang and most other hylobatid species.

The transfer index compares the complex learning skills of primates while reducing the potential effects of intertaxonal, ontogenetic, and morphological differences on the motivation, perception, and motor abilities of the subjects (Figure 10.2C).[147] Transfer indices measure the capacity for reversal after holding constant the amount learned during the prereversal acquisition period in a series of two-choice discrimination problems. One computes the transfer index by dividing the percentage correct in the second through tenth postreversal trials (R) by the percentage correct during acquisition training (A). Generally, one sets A at 67 percent and 84 percent. All primates so tested had rich histories of discrimination training. The transfer index values were based on blocks (sets) of over ten problems at each A percent level of training (Figure 10.2C).

Great apes performed significantly better than gibbons, Old World monkeys, and lemurs on 67 percent and 84 percent schedules. Gorillas and orangutans were significantly superior to chimpanzees on the 67 percent prereversal schedule. There was no statistically significant difference among the three great apes on the 84 percent schedule.[148] Hence, Rumbaugh rightly challenged the myth of the mental supremacy of chimpanzees among apes.[149] The transfer index indicated that relatively bright and dull gorillas employed the same processes to earn their scores and only differed in competence.[150]

Via the transfer index, Davenport et al. tested six chimpanzees that had endured the first two years of life in restricted, impoverished laboratory environments in comparison with eight wild-born chimpanzees that lived in social groups.[151] (Two other chimpanzees could not be tested: they had been so sorely traumatized by their impoverished rearing histories that they could not adapt to the transfer index testing situation.) Both sets of subjects had shared the same cages and test experiences after 3 to 4 years of age when they were tested. As one might expect, the cohort that experienced extreme early environmental privation exhibited inferior cognitive skills.

Direct comparisons of the ability of spider monkeys, capuchins, long-tailed macaques, bonobos, chimpanzees, gorillas, and orangutans to remember object locations, to track object displacements, and to obtain rewards that were out of reach failed to support an overall clearcut distinction in cognitive skills between the monkeys and the apes because of substantial specific variation across tasks. The only dichotomy might be between bonobos and chimpanzees versus gorillas, orangutans, and the monkeys that were tested.[152]

Abstraction

A variety of discrimination-reversal tasks indicated qualitative differences in the learning processes of cercopithecine monkeys, gibbons, and gorillas. Whereas the monkeys learned primarily via stimulus-response processes, gorillas learned abstractively. Gibbon performance was better than that of monkeys but fell short of the gorilla performance. Indeed, gibbons showed little evidence for abstractive learning.[153]

Retests showed that gorillas not only learn by abstractive strategies but also can retain an abstract concept for at least 2.5 years. Further, their long-term memory was not contingent upon the specific stimuli with which they were challenged.[154] Orangutans also performed well on a complex discrimination-reversal task in which they strategically employed a win-stay: lose-shift hypothesis, thereby showing that they too can adapt to a variety of learning situations.[155]

Extensive studies substantiated abstract reasoning and novel problem-solving skills by chimpanzees. For example, via plastic symbols on a magnetized board, a 16-year-old female language-trained chimpanzee, Sarah (b. 1962), completed analogies by choosing the correct alternative (B′) in forced-choice problems with the design: A:A′ same B:?. Moreover, she correctly noted analogy or absences of analogy by choosing symbols for same or different in the problems: A:A′ ? B:B′ and A:A′ ? B:C, respectively.[156] Chimpanzees untrained to communicate via the plastic symbols failed the tests for analogical reasoning.[157]

Juvenile chimpanzees can draw transitive inferences about relative amounts of food. For example, they can reckon that D > B in a transitive series wherein E > D > C > B > A. The mechanisms chimpanzees employ to make transitive inferences about amounts of food and to draw analogies are obscure. The two types of processes are so different that it is probably inadvisable to lump them together under the general term reasoning.[158]

Whereas Sarah assembled a realistic face from the picture of a blank-faced chimpanzee and cutouts of its eyes, nose, and mouth, seven other chimpanzees failed the test. Her success was probably predicated on memory of chimpanzee faces; that is to say, she used imaginal instead of abstract representation.[159] The failure of chimpanzees to draw, paint, or sculpt representational figures suggests that they are unable to analyze complex objects into their parts and to understand the relations among them.[160]

After Sarah viewed people in predicaments on videotapes that stopped short of solution, she correctly chose from among photographs of the correct versus incorrect means to resolve seven of the eight situations.[161] It is unlikely that she had encountered some of the situations, though she was acquainted with the objects and their functions. Thus, she seemed to recognize the nature of certain problems and to infer solutions for them, that is, to deal with them abstractly. Alternatively, Sarah might have solved the video puzzles by match-to-sample methods instead of abstract reasoning.[162] For instance, if the solution to the actor's video dilemma was to attach a hose to a spigot and one of her choices among the 35-mm photographs included a picture of a hose attached to a spigot, it is understandable that Sarah chose it over other objects absent from the videotape.

Conservation and Calculation

Sarah and a 4-year-old chimpanzee passed Piaget's classic tests in which matched quantities of fluid are placed in differently shaped containers and solids are deformed without loss of mass.[163] She succeeded via inference instead of mere perceptual acuity; however, the nature of her inferences is obscure.[164] When challenged with a back-and-forth foraging game, twelve juvenile and adolescent chimpanzees appeared to determine the inferences that another chimpanzee was likely to make and adjusted their competitive tactics accordingly.[165]

Although Sarah initially failed the pretests in which she was to discern whether the number of buttons on two trays were the same or different, she later passed, perhaps via analogical reasoning, the match-to-sample tests for number (1, 2, 3, 4) and proportionality (1/4, 1/2, 3/4, 1), despite the fact that the referents and alternatives had dissimilar shapes, colors, masses, areas, lengths, and other features. Four juvenile chimpanzees failed the tests.[166]

Studies with other subjects and different methods confirmed that chimpanzees and orangutans can discriminate more from fewer objects, especially when the maximum number per sample is ≤9.[167] Confusion occurs most commonly when the alternative cardinal numbers are adjacent ones above 3. Chimpanzees seem to have more difficulty with the concept of *some* than with *none, one,* or *all;* however, their confusion might also be attributed to the schedules for rewarding subjects rather than their conceptual abilities.[168]

In sum, if apes and perhaps monkeys have latent mathematical abilities, they are either well masked or very rudimentary by comparison with human measures based on arbitrary symbols.[169] Little more might be expected of other apes, though it would be prudent to test significant samples of them.[170] Nonetheless, study with two 7-year-old male rhesus macaques indicated that they could make ordinal judgments of numerical symbols: 0–9.[171] Further, two *Macaca mulatta* compared favorably with the performances of eleven Duke University students in tests wherein they were to arrange pairs of arrays of 1–30 items on a computer display from least to most as quickly as possible. Accordingly, Cantlon and Brannon claimed, "The qualitative and quantitative similarity in their performance provides the strongest evidence to date of a single nonverbal, evolutionary primitive mechanism for representing and comparing numerical values" that is shared by humans and other animals.[172] Indeed, some Brazilian Amazonian people share with other animals "a language-independent representation of number, with limited, scale-invariant precision."[173]

Studies on prelinguistic human children and speechless nonhuman catarrhine primates appear to clinch that knowledge of number can be independent of language.[174] Additional evidence for independence of human mathematical performance from language was provided by Varley et al., who found that "despite severe grammatical impairment and some difficulty processing phonological and orthographic number words, all basic computational procedures were intact" in three men with large left-hemispheric presylvian lesions.[175] Accordingly, it is reasonable to conclude that language and mathematics are functionally and neuroanatomically independent in *Homo sapiens*.[176]

Needless to say, foraging apes, who need to conserve energy, must find the ability to assess relative amounts of food before expending the effort to climb up and down trees and to move about in canopies composed of networks of compliant branches. It would also be useful to avoid wasteful competition with other groups, particularly unfamiliar ones. The ability to judge the difference between proximate low cardinal numbers is surely less important, perhaps even inconsequential, to them given that they lack systems of barter, let alone cash economies.

Interestingly, an inability to control the impulse to have the larger amount of desirable food treats can prevent chimpanzees from choosing the lower amount of food to avoid the larger amount being given to another chimpanzee. Yet they readily choose the smaller symbolic numeral once they have learned that this will earn them the larger amount of the desired treat, while the lesser amount will go to another chimpanzee.[177]

Self-distraction serves to curb impulsivity of chimpanzees in some test situations as it does in young children.[178] Unlike chimpanzees, two orangutans that were administered a test like that of Boysen's chimpanzees were able to inhibit their first response to choose the larger number of grapes in order to receive the smaller of two amounts.[179]

Cognitive Mapping

Cognitive mapping is a vital practical skill for free-ranging apes who must make a living and raise their young in vicissitudinous natural habitats.[180] Chimpanzees and bonobos readily learn the locations, directions, distances, relative quantities, and quality of desired and fearful objects. They often approach them by economical routes of their own design and can remember them over notable spans.[181] Chimpanzees groups subdivide in practical ways to harvest differently sized caches of preferred food. Although they achieved greatest success after directly viewing desirable objects, young chimpanzees performed significantly above chance after seeing the locations of incentives on a small black-and-white television monitor.[182] Chimpanzees also evinced planning as they employed a joystick to move an icon through the two-dimensional space of alley mazes on a computer screen.[183]

Plate 1 (top left) Sexual dichromatism in *Nomascus leucogenys* (male left and female right). (top right) Sexual dichromatism in *Nomascus leucogenys* (male left) and *Hoolock hoolock* (female right). (C) Female (left and upper right) and male (lower right) *Hylobates pileatus*.

Plate 2 (top) Asexual dichromatism in *Hylobates lar* (male left and female right). (bottom) Monochromatism in *Symphalangus syndactylus*. Note the inflated laryngeal sacs.

Plate 3 (top) *Pongo abelii* feeding on *Ficus racemosa* at Ketambe, Sumatra. (bottom left) Unripe fruit of *Neesia* sp. (cemengang). (bottom right) Ripe fruit of *Neesia* exposing seeds protected by sharp whiskers.

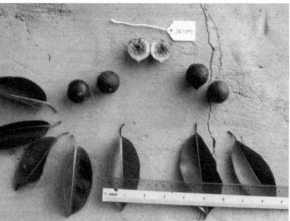

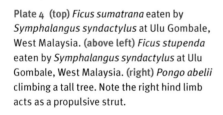

Plate 4 (top) *Ficus sumatrana* eaten by *Symphalangus syndactylus* at Ulu Gombale, West Malaysia. (above left) *Ficus stupenda* eaten by *Symphalangus syndactylus* at Ulu Gombale, West Malaysia. (right) *Pongo abelii* climbing a tall tree. Note the right hind limb acts as a propulsive strut.

Plate 5 (top left and top right) *Pan troglodytes schweinfurthii* palpating a fig *(Ficus sansibarica)* in Kibale National Park, Uganda. (middle left) *Caloncoba welwitschii* in Ituri Forest, Democratic Republic of the Congo. (middle right) *Cola gigantea* in Semliki National Park, Uganda. (bottom left) *Uvariopsis congolana* and (bottom right) *Aframomum sanguineum* in Ituri Forest.

Plate 6 **(top left)** Adult female *Pan troglodytes schweinfurthii* (Lia) of the Kanyawara community in Kibale National Park, Uganda, visually examining and smelling a fruit of *Ficus sur*, which she bit open and subsequently ate. **(top right)** Adult *Pan troglodytes schweinfurthii* at Kanyawara selecting via palpation a ripe fruit of *Ficus sur*. **(middle left)** Subadult male eating ripe fruit of *Ficus sur*, a species that all Kanyawara chimpanzees, including breastfeeding infants, eat only when red. **(middle right)** *Ficus sur* in various stages of ripeness from green to yellow to orange to red. Kanyawara chimpanzees occasionally eat orange but not yellow or green fruit of *Ficus sur*. **(bottom)** Adult Kanyawara female feeding on *Ficus sur* with her 7-month-old daughter that also ate part of a fig.

Plate 7 **(top left)** Adult Ngogo male *Pan troglodytes schweinfurthii* (Mulligan) eats *Uvariopsis congensis* in Kibale Forest, Uganda. **(top middle)** Infant Ngogo male (Fleck) forages for *Chrysophyllum albidum*. **(top right)** Female Ngogo chimpanzee (Anderson) and her infant son (Nelson) eat *Ficus mucoso*. **(middle left)** Adult Ngogo male chimpanzee (Berg) eats a large fruit of *Treculia africana*. **(middle right)** Adult Ngogo male in nest. **(bottom left)** Adult Ngogo chimpanzee male sleeping in arboreal nest. **(bottom right)** *Uvariopsis congensis*, a favorite food of Ngogo and Budongo chimpanzees.

Plate 8 (top left) *Gorilla beringei* eating *Galium ruwenzoriensis* on Mt. Visoke, Rwanda.
(top right) *Gorilla beringei* eating flowers of *Galium ruwenzoriensis* on Mt. Visoke, Rwanda.
(bottom left) *Gorilla beringei* eating giant senacio roots *(Senacio johnstonii)* on Mt. Visoke,
Rwanda. (bottom right) *Gorilla beringei* eating *Peucedanum linderi* (celery) on Mt. Visoke,
Rwanda.

Plate 9 (top) *Gorilla beringei* eating *Arundinaria alpina* (bamboo shoot) on Mt. Visoke, Rwanda. (bottom left) *Gorilla beringei* eating shelf fungus on Mt. Visoke, Rwanda. (bottom right) *Gorilla beringei* eating a *Carduus nyassanus* (thistle stalk) on Mt. Visoke, Rwanda.

Plate 10 (top left) Ngogo male *Pan troglodytes schweinfurthii* (Pincer) with *Procolobus badius* (red colobus). (top right) Ngogo male *Pan troglodytes schweinfurthii* (Lofty) with *Procolobus badius*. (middle left) Ngogo male *Pan troglodytes schweinfurthii* (Dolphy) with left forelimb of *Procolobus badius*. (middle right) Ngogo male *Pan troglodytes schweinfurthii* (Cash) with *Procolobus badius*. (bottom) Ngogo male *Pan troglodytes schweinfurthii* (Bartok) with *Cercopithecus ascanius* (red-tailed monkey).

Plate 11 (top) Captive male *Pan paniscus* (Kanzi) using a hammer stone in his right hand to strike flakes from a flint nodule. (middle and bottom) Capuchins *(Cebus capucinus)* attempting to open coconuts at La Vallée des Singes sanctuary, Romagne, France.

Plate 12 (bottom) Ngogo males begin to align for a patrol. (middle) The line tightens. (top) The line further tightens as the patrol enters tall grass.

Plate 13 **(top left)** Young Karisoke *Gorilla beringei* close to the silverback of Karisoke group 5 on Mt. Visoke, Rwanda. **(top right)** The silverback *Gorilla beringei* of Karisoke group 5 is playing with his 8-year-old son on Mt. Visoke, Rwanda. An adult female is to the silverback's left, and an infant is on his right. **(bottom left)** Karisoke silverback *Gorilla beringei* ascending a steep slope. Note the visibility of his dorsal patch and the lush ground vegetation. **(bottom right)** Group of *Pan paniscus* at Wamba, Democratic Republic of the Congo.

Plate 14 (top left and right and middle left and right) Ngogo male chimpanzee dyadic allogrooming. (bottom left and right) Ngogo male chimpanzee triadic allogrooming.

Plate 15 (top) Captive adult male *Pan paniscus* grooming the left eye of an estrous adult female in attendance with two other estrous females at La Vallée des Singes Sanctuary in Romagne, France. (bottom) Captive adult male *Pan paniscus* grooming the right ear of an estrous adult female at La Vallée des Singes Sanctuary in Romagne, France.

Plate 16 (top left) Gay nuclear family.
(top right) Lesbian nuclear family.
(above left) Polygynous nuclear family.
(above right) Polyandrous nuclear family.
(bottom) Trispecific nuclear family.

Nonetheless, though chimpanzees are adept at mapping familiar tracts from direct views and television screens, some fail to orient themselves toward incentives in a distant room on the basis of simple small-scale models (i.e., symbolic maps) like those comprehensible to 5-year-old humans.[184]

In experiments with German, Dutch, and Khosian children and adults, and 8- to 29-year-old bonobos, chimpanzees, orangutans, and gorillas, human spatial cognition, including relational thinking, systematically varied with language and culture, but there was a clear inherited bias for one spatial strategy in the great apes.[185] Concordantly, whereas Westerners attend more to focal objects, East Asians attend more to contextual information.[186]

Cross-Modal Skills

The ability to generalize representations of phenomena that are received via one of the five sensory modalities—visual, auditory, haptic, gustatory, and olfactory—so that they are recognizable from stimuli in another modality surely enhances the learning, adaptability, and survival of apes and other vertebrates. Concomitantly, "primates are motivated and partially prewired to look at, touch, and handle objects, and to put together all this information in cross-sensory representations."[187]

Before 1970, some scientists maintained that nonhumans lacked cross-modal representational abilities because they are mediated by language. Early experiments with monkeys seemed to support this view.[188] Then via innovative match-to-sample procedures, Davenport and Rogers demonstrated that chimpanzees and orangutans can abstract and exchange information from visual and haptic characteristics of familiar and novel objects.[189]

Apes trained to communicate with humans and one another via signs and artifactual symbols also evidence cross-modal transfer. For instance, some of Premack's language-trained chimpanzees exhibited intermodal transfer between gustatory and visual senses and auditory and visual senses.[190] The same is true of modified American Sign Language (ASL) signing of *Pan troglodytes* and *Gorilla gorilla* and of the *Pan troglodytes* and *Pan paniscus* that employ noniconic symbolic lexigrams.[191]

Monkeys—*Cebus apella, Macaca mulatta,* and *Macaca fascicularis*—also can match or recognize objects bidirectionally across haptic and visual modalities, and there is no apparent indication that chimpanzees are markedly superior to the monkeys in this skill.[192] Accordingly, verbalization probably should not be regarded as a prerequisite for cross-recognition and problem solving, though cross-modal capabilities are probably important enhancers, if not actual prerequisites, for human language.[193]

Self-Recognition and Self-Awareness

Among many arrogant conceits in which humans indulge, one of the greatest is the claim that only we are aware of ourselves, have individual identities, and think about what we are doing, have done, and will do. Cartesian dualists argued that consciousness and its concomitant mind are premised on human language.[194] According to this dogma, thoughts are impossible in the absence of human language.[195]

Scientists from diverse disciplines have vigorously challenged these entrenched beliefs. For instance, the renowned experimental zoologist Griffin argued that we must search for awareness and mind not only in great apes, but also in much more distant species such as the social insects.[196] He resolved that the complex versatile and limited symbolic behavior of ASL-signing chimpanzees and dancing honeybees were especially compelling evidence that nonhuman animals have mental experiences and conscious intentions.[197] He reasonably proposed that awareness—the experience of interrelated mental images—confers an adaptive advantage to the animals that have it because it enables them to react appropriately and versatilely with physical, biological, and social events and signals in their environments.

Griffin's argumentation was weakened by the suggestion that experimenters don chimpanzee disguises, dab themselves with chimpanzee pheromones, and conduct participant observations among wild groups.[198] Even if they tolerated the curious aliens, I doubt that chimpanzees would be fooled and that anyone could do what chimpanzees do in the forest without rapid deterioration of their costumes and health.[199] Nonetheless, as he drew more fully on the research with great apes that has preceded, overlapped, and followed his 1976 argument for animal consciousness, thinking, and intentionality, Griffin proved much more correct than the Cartesians about the minds of apes.[200]

Gallup et al. provided solid experimental evidence for self-recognition in chimpanzees and orangutans via mirror-image stimulation (MIS).[201] They applied spots of paint to areas on the heads of anesthetized subjects where they could not see them directly or via peripheral vision and that one could not feel on the skin. Once awake and presented with mirrors, the subjects that had prior exposure to mirrors digitally explored the marked areas and sometimes smelled or orally contacted the exploratory fingers.

Apes with no prior experience with mirrors initially react socially to their images. But after several days, some young and adult chimpanzees begin to use mirrors to explore parts of their bodies that are normally out of view or difficult to position for direct views.[202] Further, mirrors assisted them in inspecting food wadges in their mouths, cleaning between their teeth, picking their noses, blowing bubbles, and making faces. Such behavior led Gallup to conclude that they must recognize themselves in a mirror and that they might have a self-concept.[203] Working with larger samples, Lin et al. noted self-recognition in 2.5-year-olds, while Povinelli et al. and de Veer et al. concluded that self-recognition generally emerges at 4.5 to 8 years of age, declines at the population level in adulthood, and in group settings occurs within minutes of a subject's exposure to mirrors.[204] Contrarily, Swartz and Evans found that among eleven mostly wild-born chimpanzees between 4 and 19 years of age, only a 16-year-old female appeared to pass Gallup's mark test. The subjects were living in stable social groups in the Gabonese colony when tested.[205]

Gallup noted that in chimpanzees self-recognition must develop through exposure to reflecting surfaces and that it is influenced by early experience.[206] It appears that awareness of others, particularly those with whom the individual has social relations, is an important feature in the development of self-identity.[207] Exposure to a model of novel tool use influenced its acquisition in a cohort of 3- and 4-year-old chimpanzees but not in 2-year-olds. Bard et al. concluded that, as in older chimpanzees and humans, there is a link between the emergence of imitation, self-recognition, and a comprehension of cause-effect relationships in the tool tasks with which they tested their subjects.[208]

Mirrors can serve as social facilitators in diverse creatures, including humans. For example, chickens lay more eggs and students can perform better on tests in a mirrored room. Healthy ≥18-month-old humans correctly identify their own reflections in mirrors.[209] But fish, most birds, monkeys, and other mammals do not evidence self-recognition, and some react to their reflections as if they are viewing other individuals of their species, commonly in the form of aggressive or sexual displays.[210] Possible exceptions include magpies *(Pica pica)*, bottlenose dolphins *(Turciops truncates)*, orcas *(Orcinus orca)*, false killer whales *(Pseudorca crassidens)*, elephants *(Elephas maximus)* and pigs *(Sus scrofa)*.[211] Persons who were blind at birth and later had their sight established also initially responded to mirrors as though they represented real space and may fail to recognize themselves. Some psychotics, drunks, and otherwise drugged persons may not recognize themselves in mirrors.[206]

A male chimpanzee that had spontaneously learned to recognize himself in the steel cover over an electrical outlet in his cage groomed his heel while watching it projected on a monitor within one minute of his first exposure to a television system.[212] Moreover, two adult male chimpanzee paternal half-brothers—Sherman (b. 1973) and Austin (1974–96)—could track the images of their hands in a mirror and on closed-circuit television pictures in order to locate an object on the exterior wall of the enclosure.[213] They quickly adjusted to rotations and reversals of images on the monitor and distinguished live performance from playbacks on a second screen.[214]

Initially, gorillas and gibbons failed to evidence self-recognition in mirrors via Gallup's dye test.[215] While the hylobatid failure seemed reasonable, considering their other resemblances to monkeys, the results for gorillas were puzzling.[216] Later studies indicated possible self-recognition in gorillas and self-directed behavior before mirrors by bonobos.[217] Hyatt tested nine gibbons with the dye test after they had been exposed to mirrors. Although two subjects seemed to use the mirror to view parts of their bodies not visible to them by other means, none of the nine passed the dye test.[218]

We do not know whether self-recognition in mirrors is evidence of self-awareness of a human sort or whether lack of evidence for mirror-image recognition proves an absence of self-consciousness.[219] Gallup's experiments did not persuade all skeptics that chimpanzees and orangutans have a self-concept or more particularly that the level of their self-consciousness is established vis-à-vis the human condition.[220] For instance, Ristau and Robbins noted that the MIS data "do not require an interpretation that the chimpanzee has an awareness of self as a mind; an awareness of self as a body will suffice."[221]

Intentionality and Theory of Mind

> There is very little evidence that nonhuman primates of any species understand others as psychological beings with intentions and other psychological states that mediate their behavioral interactions with the world—as human children begin to do sometime during their second year of life.
>
> ELISABETTA VISALBERGHI and MICHAEL TOMASELLO
> (1998, p. 189)

Around 9 months old, human infants begin to evince new behaviors indicating that they understand others as intentional agents partially by analogy to themselves (i.e., by simulation). They monitor the attention of others to outside entities, including attention to them, leading to an understanding of "me" as an object in the world.[222] Contrary to claims by prior researchers, Tomasello noted, "Nonhuman primates . . . do not understand others as intentional agents because they cannot make the link between self and other in the same way as typically developing children."[223]

Premack and Woodruff conducted tests with the chimpanzee Sarah to discern whether she could attribute mental states to actors on tapes; that is, did Sarah have a theory of mind?[224] They decided that she did, though other psychologists were unconvinced.[225]

Sarah chose constructive solutions to problems for a trainer that she liked, but selected predominantly hurtful or otherwise negative outcomes for a person that she disliked. She failed tests in which she was to indicate which photograph another chimpanzee would choose in order to solve a familiar problem that confronted a familiar human in a videotape. Accordingly, it appears that the attribution of attribution is beyond the capability of at least one intentional ape.[226] Young chimpanzees also demonstrated intentional behavior by directing a well-regarded trainer to food rewards, misdirecting an untrustworthy one, and suppressing the urge to look toward a baited container after fooling a bogus villain.[227]

There are similar instances of other captive and free chimpanzees withholding responses to preferred objects while dominant chimpanzees or humans are watching them and misdirecting others before sneaking back to an incentive.[228] Further, they act against conpecifics that have served them badly and cooperate with those that have not.[229] Mitchell also documented instances of deception and hiding in a group of four captive western gorillas.[230] Tactical deception and other manifestations of Machiavellian intelligence are attributes not only of humans and apes but also of nonhominoid primates.[231]

Corballis distinguished between zero-order theory of mind—comprising the mental processes of thinking, knowing, perceiving, or feeling—which is probably common to many species, from first-order theory of mind, which is recursive—that is, thinking, knowing, perceiving, or feeling what others are thinking, knowing, perceiving, or feeling.[232] He proposed that recursion is a principal characteristic that distinguishes *Homo sapiens* from all other extant species. Humans are capable of several more than two levels of recursion as they try to discern what others are thinking. For example, in Goldman's play *The Lion in Winter* (Act I, Scene III), during a conversation about machinations of King Henry II regarding his sons Prince John and Prince Richard,

Prince Geoffrey says to Queen Eleanor, "I know. You know I know. I know you know I know. We know Henry knows, and Henry knows we know it. We're a knowledgeable family."[233]

Emergents and Rational Behaviorism

> What we call 'concepts' and 'communication' are both products of semiotic activity . . . We cannot devise a framework for calibrating gradable varieties of conceptual and communicative ability without attending to the structure of semiotic relationships which give rise to them . . . Our formulation of evolutionary hypotheses must be guided by our answers to the following question: Is the relationship between the two types of superimposed organization—biological and semiotic—itself systematic across species?
>
> ASIF AGHA (1997, p. 212)

Rumbaugh and Washburn concluded that because they exhibit emergent behaviors like humans, apes and other animals think.[234] They defined emergent behaviors as "abilities to acquire concepts, to learn insightfully, to learn complex skills and behaviors via observation, to make and use tools, to learn the basic dimensions of language, and in many other ways to manifest advanced intelligence."[235] Further, they advanced "rational behaviorism" as a new field to encompass all behaviors: respondent, operant, emergent, and instinctual.[236] Probably the most generative process among them for cognitive advance, behavioral flexibility and fitness is the instinct to learn.[237]

Rational behaviorism stresses the inclusion of cognition (rational: exercising reason) and the fact that behavior yields the only data available to comparative psychologists (behaviorism). This view endows other animals with agency and allows that they synthesize information, regardless of how it was learned; that is, to think thereby producing totally new behavior patterns and abilities in addition to variants on prior behaviors.[238]

Emergent behaviors have no clear history of past reinforcement; instead they appear unexpectedly and are consistent with a logical or adaptive function appropriate to both new and familiar situations. They are not subject to stimulus control, and their acquisition can rarely be charted. They reflect the organism's experience with classes of experiences instead of constrained training.[239]

For example, an 11-year-old female language-trained chimpanzee called Panzee (aka Panpanzee, b. 1985) observed Charles Menzel hiding a kiwi fruit in the woods outside her fenced yard.[240] When she was later indoors, Panzee recruited another researcher, who was naïve about the presence of the kiwi, to retrieve it for her. First, Panzee presented her rump to the researcher to get her attention, then she pointed to the lexigram for kiwi fruit until the person acknowledged its meaning. Panzee then moved to the tunnel leading to her outdoor area, covered her eyes with one hand to indicate "hide," and gestured toward the tunnel door with her other forelimb. When the researcher verbally asked if Panzee wanted her to go outdoors, she ran through the tunnel into the yard. Panzee vocalized and gestured with an extended forelimb toward the area with the kiwi, then more exactly with her finger while gazing at the precise spot where it was hidden. After a brief period of honest searching as the level of excitement in Panzee's vocalizations varied according to how close she was to the hidden fruit, the researcher found it. Beckoning the researcher to follow, Panzee hurried back indoors, where she received the kiwi.[241]

Many other trials followed, in which Menzel varied nonfood and more than twenty foods and their locations in the woods. Even after imposed delays of up to sixteen hours, Panzee continued to secure them by whatever means she could devise to get the attention of the naïve surrogate searchers who could read lexigrams.[240]

Emergent behaviors encompass not only instances of insight such as chimpanzee solutions to instrumental problems but also "numerical reasoning, mathematical frameworks, compositions and language" and great ape comprehension of spoken words, "which requires the ability to understand semantic definitions and to decode the syntax of novel sentences of request."[242]

Contrary to traditional behaviorists who described behaviors as responses to positive, negative, or absence of reinforcing stimuli, Rumbaugh et al. cogently argued, following Hebb, that the brains of species "are uniquely designed to perceive and to relate stimulus events that are contiguous, salient and relevant to adaptation."[243]

As learning events occur and cell assemblies strengthen their synaptic connections with related cell assemblies, learning can become a larger amalgam and the basis for cognition.[244] "The combination of cell assemblies, or the combination of associations between events, could then produce a higher cognitive function such as a concept, thought, or emergent behavior."[245]

Refined technology promises to reveal neurophysiological mechanisms whereby emergent behaviors alter and are encoded in the brain. For instance, via noninvasive MRI and voxel-based morphometry, Quallo et al. documented changes in brain structure—grey and white matter—in *Macaca fuscata* as they learned to use a rake to obtain food rewards.[246]

Herrmann et al. contested the view that humans simply have more general intelligence than other primates.[247] Instead, they concluded from a extensive inventory of cognitive tests administered to chimpanzees, orangutans, and 2.5-year-old preliterate children that whereas the children and chimpanzees had very similar cognitive skills for dealing with the physical world, the children had more sophisticated cognitive skills than the apes for dealing with the social world, namely, social learning and cognition, communication, and theory of mind as indicated by gaze following and intentions.[248]

Premack concurred that although there are small similarities (but not equivalences) between nonhuman animal and human cognitive abilities, the dissimilarities are large, based on examination of teaching, short-term memory, causal reasoning, planning, deception, transitive inference, theory of mind, and language.[249] Further, contrary to Cartesian dualism, he concluded that there is no disparity between brain and mind. Apparently, the sometimes rather small neuroanatomical and genomic differences between ape brains and those of humans can indeed be functionally significant.

Social Cognition

> The crucial difference between human cognition and that of other species is the ability to participate with others in collaborative activities with shared goals and intentions: shared intentionality.
>
> MICHAEL TOMASELLO, MALINDA CARPENTER,
> JOSEP CALL, TANYA BEHNE, and HENRIKE MOLL
> (2005, p. 675)

The ability to understand and to engage in social interactions in the context of one's culture and social institutions—social cognition—is essential to a person's quality of life and even survival.[250] The skill develops over many years via observation, imitation, emulation, and tutelage. Because apes and other social mammals exhibit some social behaviors and learning processes that resemble those of humans, one might reasonably accept that humans are not singularly endowed with social cognition any more than one should deny that apes have some level of meaningful social communication even though they lack humanoid speech. The challenge for evolutionary scientists is to refine knowledge of the neural underpinnings for social cognition and to determine the extent to which they are homologous in apes and humans.

Indeed, it is likely that human and perhaps other mammalian social cognition is a specialized system of the mind and a core domain of knowledge.[251] A major difference between the socialization of humans and that of other social mammals is surely the amount of conscious teaching by adults and other children, much of which depends on explicit verbal instruction, metaphors, similes, recursive narratives, and other modes that rely heavily on symbolism that is not available to nonhumans.[252]

Other ways that humans learn, especially in the early years, are like those of other mammals. This is particularly true regarding observation of social interactions and outcomes that have no explicit explanation but nonetheless register in an individual's mind and can be accessed later along with other models to assist decision making in social situations that are somewhat similar but also have novel components. The following factors might constitute cognitive specialization for social interaction in humans:

- An unlimited number of understandable social situations.
- A combinatorial rule system in the mind of the social participant.
- A rule system that is only partly available to consciousness.
- A rule system acquired by a child with imperfect evidence and only partly taught.
- Learning that requires inner unlearned resources, perhaps only partly specific to social cognition.
- Inner resources determined by the genome interacting with processes of biological development.[253]

The major challenge here is to sort out the extent to which the patterns of social intercourse are hardwired because of learning regimens with or without genomic factors in humans and apes—a tall order, indeed. Given the intraspecific variability and versatility of ape and human social interactions in response to demographic and ecological fluctuations, one might best view the genomic component as related more to a general capacity or disposition to be affiliative than to attributing specific social patterns to determinative genomic factors.

James K. Rilling and colleagues documented differences in the neural circuitry of bonobos and chimpanzees that relate to social cognition. Specifically, bonobos have more in brain regions involved in perceiving stress in self and others and a larger pathway implicated in top-down control of aggressive impulses and bottom-up biases against harming others. They concluded "this system not only supports increased empathic sensitivity in bonobos but also behaviors like sex and play that serve to dissipate tension, thereby limiting distress and anxiety to levels conducive to prosocial behavior."[254]

Theories on the selective factors that have shaped the high intelligence of apes and humans are also in flux.[255] Few would deny that hominoid cognitive skills were honed by selective forces in the physical environment: learning and recalling the availability of seasonal and fallback food resources, using tools to extract and process some foods, and navigating arboreal substrates and avoiding predators. However, a social intelligence hypothesis—that intense social life led to improved social cognition and greater general intelligence—has predominated since primatologists began to report on the variety, flexibility, and puzzling complexity of great apes' social interactions in the field and in sizeable captive social groups.

Van Schaik et al. argued that the classic social intelligence hypothesis cannot explain great ape distinctiveness.[256] Instead, they posited a revision that incorporates life-history variation as a prominent selective pressure arising from social life:

> If having large bodies has major ecological consequences, it also has dramatic social ones: it increases the costs of sociality, which leads to flexible grouping patterns through increased vulnerability to competition, and it substitutes vulnerability to predators for vulnerability to hostile conspecifics. These two consequences lead to a cascade of further social consequences including increased social leverage for subordinate individuals and cooperation among non-relatives. Possession of large brains in large bodies thus indirectly set the stage for uniquely elaborate cognitive solutions to non-unique social problems.[256]

Recently, Miller et al. documented that neocortical myelination is developmentally protracted in humans compared with chimpanzees. Among chimpanzees, adult-like levels are achieved around the time of sexual maturity; in humans, there is slower myelination during childhood with a delayed period of maturation that extends beyond late adolescence.[257] Perhaps they have located a key developmental process that underpins the further complexity of human social cognition and behavior.

PART IV

SOCIALITY AND COMMUNICATION

11

Social, Antisocial, and Sexual Apes

> While it is now obvious that social relationships must be
> described in terms of social interactions and, in turn, social
> structures in terms of social relationships, the influence of
> higher levels on lower, social structure on relationships and
> relationships on interactions, is not so clear.
>
> ALEXANDER H. HARCOURT and KELLY J. STEWART
> (1983, p. 307)

> The study of conflict as negotiation starts with the careful
> documentation of how and when these tools are being used.
> This is a formidable task indeed as it requires analysis of
> behavioral exchanges in their entirety over long intervals. But
> this may be exactly how the minds of social animals keep track
> of behavior: not in terms of mere frequencies and durations, but
> in terms of how the behavior functions within the relationship,
> and how it fits or violates expectations about one another.
>
> FRANZ B. M. DE WAAL (1996, p. 170)

Systematic socioecological studies of apes began in the 1930s, when Robert Yerkes inspired Clarence Ray Carpenter to inaugurate an intensive study of *Hylobates lar* in Thailand.[1] Although Carpenter's project was successful, his informative monograph was singular in the following quarter century. Yerkes also encouraged field studies of western chimpanzees and mountain gorillas that are minimally informative, especially regarding social behavior and structures.[2] Fortunately, during the past fifty years scientists have collected extensive information about naturalistic social behavior in orangutans, chimpanzees, bonobos, and gorillas, and many more populations of gibbons.

397

Solitary, but Sometimes Social, in Indonesian Forests

Unlike chimpanzees, bonobos, gorillas, and gibbons, one rarely encounters adult male and female orangutans together in most Indonesian forests. The most common social units of *Pongo* spp. are

- individual females with dependent youngsters,
- lone adult and subadult males, and
- various small groupings of adolescents or singletons.

Mother/young units and older youngsters sometimes mingle temporarily, and sizeable aggregations occur rarely at large fruit sources.[3]

Even in close quarters, captive family groups live together as compatibly as other primates for which more cohesive bisexual groups are the norm. Captive orangutans can be exceptionally affectionate, playful, and tolerant of one another.[4] Further, wild orangutans vary in the extent to which they form larger groupings. Hence, a major question for theorists is, Why are orangutans so solitary? Or more correctly phrased, Given their capacity in some forests to be sociable and to travel together for relatively brief periods between fruit sources, why aren't they more gregarious or group living?[5]

The Long Path to Parenthood

The life history of *Pongo abelii* and *Pongo pygmaeus* is the slowest among the great apes.[6] Further, they are unusual among mammals in expressing bimaturism. Mature male orangutans take two distinct forms: some have established territories, emit long calls, and sport prominent cheek flanges, long mantle hair, and large laryngeal sacs; the less established males lack these characteristics.[7] The fully mature morphotype appears to develop in response to sex steroids and luteinizing hormones, which have lower concentrations in males that have not developed the striking epigamic features.[8]

Orangutan males are sexually mature and active by about 8 years of age, but development of their epigamic features can be delayed markedly. Indeed, some Ketambe males that were not offspring of resident flanged males remained flangeless for up to twenty years after sexual maturity.[9] Flanged males are highly intolerant of other flanged males but tolerant toward flangeless ones. Flangeless males are generally tolerant toward other males and are more gregarious than flanged males.[10]

At Ketambe, *Pongo abelii* first give birth when they are 13- to 18-year-olds (mean: 15.5 years), and at Tanjung Puting, *Pongo pygmaeus* first give birth at 15 to 17 years of age (mean: 15.7 years). The mean interbirth intervals are 9.3 and about 8.2 years at two Sumatran sites and 7.7 and 7.0 years at two Bornean sites.[11] Wich et al. estimated the longevity of wild males and females to be ≥58 years and ≥53 years, respectively, with no evidence of menopause.[6]

Based on cross-sectional data from orangutans in Sabah, East Malaysia, Horr's chronicle of ontogenetic distancing of young Bornean orangutans from the natal unit is in general agreement with the accounts on Sumatran and Bornean orangutans at other localities.[12] In the first year, an infant depends upon its mother for nourishment, transport, and protection in the forest. It clings to her and sleeps in her nest. Occasionally she might play with it or groom it briefly. She introduces it to solid foods by allowing it to sample her meals, including bits from her mouth.

During the second and third years, an infant learns to build nests and is increasingly adventurous while climbing independently in the canopy. It might play briefly with an older sibling, often with little body contact; usually mothers rebuff attempts to touch her infant. Often by the time a youngster is independent enough to play vigorously—at 2.5 to 3.0 years old—no siblings are available. When a mother/young unit encounters other orangutans, the latter usually do not handle her infant.[13]

Juveniles are weaned and move on their own except when assistance is needed to traverse gaps in the canopy; then one might use its mother's body as a bridge. Juveniles begin to sleep in their own nests, especially after repeated rebuffs from their nesting mother, which sometimes induces tantrums. Until the fifth or sixth years, youngsters remain in the vicinity of their mothers. Juveniles eventually lag farther behind, engage in distinct activities at their own pace, and are seldom groomed. Older juvenile males seem to strike out on their own more readily than do females, who tend to remain near their mothers.[14] It is not known whether this is related to stronger negative behavior from the mother toward male offspring.[15]

In Sabah, East Malaysia, one Bornean orangutan mother offered food to her female juvenile offspring and shared food sources with her.[13] Conversely, four mothers in the Mawas Reserve, Kalimantan, only shared food with their offspring when they initiated the event. The bits of food the youngsters took from their mothers were those that were otherwise inaccessible to them and seemed not to depend on their quality. The sharing did not peak during weaning; indeed, mothers tolerated their offspring taking food from them for a span after weaning. The practice might have had more to do with young-sters obtaining information about foods that require complex feeding skills than with weaning per se.[16]

Distance Makes Good Neighbors

Independently roaming older juveniles did not engage in long social play se-quences even though they had increased opportunities to meet other young orangutans. One juvenile male seemed to be upset by the attentions of a ju-venile female and her mother. Given the general absence of adult male role models, the development of male behavior might be under more direct gene-tic control than that of females.[13]

Nine individuals in four mother/young units at Kutai, East Kalimantan, had extensively overlapping home ranges, and an independently roaming ju-venile female sometimes remained within the boundaries of her natal unit range.[17] The ranges of the two adult males did not overlap, but instead ex-tended widely to include parts of two or three female ranges. The adult female in one unit might have been the offspring of the other. Because two of the mother/young units used the same paths and fruit trees where their ranges overlapped, they met, fed, rested, and moved together for brief periods. A male and another mother/young unit shared an area but used different resources within it.

During a span of fifteen months, secondary groupings of six primary units accounted for only 1.65 percent of the total observation time at Kutai:

- Eight were temporary associations of a male with one or two mother/young units.
- Seven were chance encounters at food sources.
- Five comprised the two mother/young units that might have been re-lated matrilineally.
- One was clearly sexually motivated.[17]

As with the orangutans at Kutai, at Ulu Segama, Sabah, the average size of the Bornean orangutan social units was 1.8 individuals. Upon meeting, adult females ignored one another, while their youngsters might play. Sometimes independent juveniles joined other units to play for a few hours. Grooming was rare and was mostly confined to mothers grooming their infants.[18]

Initially, MacKinnon concluded that 160 nomadic Ulu Segama orangutans wandered over extensive areas.[19] Later he modified his surmise: "although some animals might have been nomadic, many of the Bornean orangutans occupied definite home ranges. The ranges of adult males were large, certainly several square kilometers, but two of the resident adult females might have had rather smaller ranges. Ranges of both sexes showed considerable overlap but there was some evidence that males defended range boundaries against others of their sex."[20]

The Ulu Segama orangutans were near an area of intensive logging and also might have been affected by recent tectonic events in the region. Further, during sixteen months in Borneo, MacKinnon was unable to follow individuals continuously enough to plot actual ranges. Nevertheless, his observation that units were more clumped sometimes stands as a caveat against overgeneralizing about orangutan ranging behavior on the basis of other studies. Apparently, although loosely banded, Ulu Segama orangutan social units maintained contact and coordinated their movements through the area in response to male long calls.[21]

Observations of fifty-eight Tanjung Puting orangutans basically support the inferences of Horr and Rodman concerning Bornean orangutan ranging patterns, particularly those of females.[22] Whereas Tanjung Puting females ranged over 5–6 km², adult males ranged over ≥12 km². Stable female ranges overlapped extensively.[23] Independent juveniles stayed within their natal unit ranges.[24] A male disappeared for a few years but returned and followed prior pathways.[24]

At Tanjung Puting, the social units were in contact much more often—19 percent of the observation hours—than the Kutai units.[24] Demographic and ecological factors in the regions might have been responsible, but the details are not available.[25] Galdikas considered Tanjung Puting orangutans to be more similar socially to Sumatran orangutans than to Sabah and Kutai populations.[26] She ultimately concluded that orangutans are only semisolitary.[27]

The largest temporary groups ($n = 2$) at Tanjung Puting included nine individuals, none of which were adult males.[28] Galdikas observed dyadic contacts between adult males only six times; half of them culminated in combat, and a fourth fomented a chase.[29] Subadult males avoided adult males.[30]

Adult Tanjung Puting females exhibited more variable relations than the males did. Some females ignored one another, but others were associative or reacted aggressively.[24] Adult females appeared to remember one another even though they had not met for many months.[31] Mutual reactions of specific pairs were consistent during successive encounters, and some mother/young units seemed to travel together for more than a week after meeting.[32] The longest associations between two mature females spanned only three days and two nights.[24] Peer contacts of youngsters in groups appeared to be "as important, if not more so, than direct genealogical ties in determining female associations in adulthood."[33]

Adult female orangutans in the Gunung Palung National Park of Indonesian Borneo actively avoided one another, though their home ranges overlapped notably (mean: 67.8 percent). They employ this foraging tactic mutually to reduce the likelihood of encountering food patches that were being or had been depleted by another feeding orangutan.[34] During focal follows of six adult females, the overall encounter rate was 5.7 percent: 33.3 percent for mother-daughter dyads, 5.2 percent for sisters, and 1.4 percent for nonrelatives. Several females never encountered one another. Accordingly, Knott et al. inferred that because scramble competition imposes a cost they engage in active defense of their somewhat overlapping core areas—small areas in which they spend half their time—and passive range exclusion.[35] Scramble competition occurs when needed resources are distributed more or less evenly over an area in amounts that are inadequate for the needs of all its occupants.[36] It more commonly affects relationships within and between groups of a species than those among species.[37]

Although they are rare, aside from incidents of mothers encouraging their juvenile and adolescent daughters to travel independently, agonistic encounters have revealed that there are clear dominance relationships between some of the Gunung Palung females. Accordingly, they also engage in contest competition. Residents won 89 percent of the fights that occurred exclusively in their core areas; however, there is no evidence that residents were more likely to attack females within their own core area.[35]

Ketambe *Pongo abelii* basically fit the definition of scramble competition.[38] Sugardjito et al. inferred that the seasonal high availability of fruit allows the orangutans to congregate with minimal competition, but when fruit becomes scarce, they avoid one another to escape competition and interference in obtaining food. However, Utami et al. documented that contest competition determined access to large fig trees and other large fruit patches at Ketambe.[39] Displacements between females were predominantly unidirectional, indicating a linear dominance hierarchy among them. One adult male displaced all other adult males, and they were in turn dominant to all subadult males. Adult males were more tolerant of adult females. They did not react to them entering a fig tree, and females did not hesitate to enter a tree in which a male was feeding. Relations between adult females and subadult males were more variable; sometimes the former displaced the latter and vice versa.

Both *Pongo abelii* and *Pongo pygmaeus* exemplify individual-based fission-fusion sociality—that is, group size and composition change facultatively throughout the year with different activities and situations. However, the Sumatran species, at least at Ketambe and Suaq Balimbing, appeared to be more gregarious than Bornean congeners, probably because Sumatran habitats, especially the Suaq Balimbing swamp forest, are more productive than Bornean forests.[40] Nonetheless, Singleton and van Schaik described the social organization of Suaq orangutans as a loose community comprising one or more female clusters with extensively overlapping home ranges and a more widely ranging adult male with which they all preferred to mate. The dominant male's range was smaller than those of other sexually mature males. Whereas females tend to be philopatric and might be related to one another, subordinate sexually mature males avoided clusters of adult females. There was a skewed sex ratio favoring females due to higher male mortality from intermale mating competition, emigration, or a combination of both factors.[41]

A comparison of female orangutan home ranges at nine Bornean and three Sumatran sites (Table 11.1) revealed wide variation—40 to >850 hectares (ha)—in which Singleton et al. discerned two main patterns: one linked to habitat heterogeneity and perhaps population density and an independent one linked to taxonomic affiliation.[42] At Suaq Balimbing, Gunung Palung, and Tanjung Puting, the female home ranges are >500 hectares to facilitate access to the top dietary plant species; elsewhere, such as in most dryland forests, the density of food species is much higher. There also appears to be a gradient of increasing female home range size related to the degree of reliance on nonfruit fallback foods (namely, leaves and bark) by *Pongo pygmaeus* in Sabah and East Kalimantan versus fruits by *Pongo abelii* on Sumatra.

Flanged male ranges in the same twelve areas are ≥3 times larger than those of resident females (60 to >2,500 hectares), and dominant flanged males can occupy somewhat smaller home ranges than those of less dominant males (Table 11.1).[43]

Table 11.1.　Estimates of home range size of female and male orangutans

Study Site	Study Area Size (ha)	Home Range (ha)	
		Female	Flanged Male
Lokan SaEK	390 up to 2,070	65	520
Mentoko-1 SaEK	300	40–60	60–120
Mentoko-2 SaEK	300	>150	>Females
Mentoko-3 SaEK	Unclear	>150	500–700
Kinabatangan SaEK	600	180	>225
Tuanan CWK	500	250–300	>Females
Sabangau CWK	900	250–330	>560
Tanjung Puting CWK	3,500	350–600	>Females
Gunung Palung CWK	2,100	600	>650
Ketambe-1 S	150	150–200	>Females
Ketambe-2 S	350	300–400	>Females
Suaq Balimbing S	500 up to >200	>850	>2,500

Source: Modified from Singleton et al. 2009 and Utami Atmoko, Singleton et al. 2009.

Abbreviations: CWK, Central and West Kalimantan; ha, hectares; S, Sumatra; SaEK, Sabah and East Kalimantan (Borneo).

Te Boekhorst et al. tested two hypotheses—sexual attraction and food attraction—to explain why nonresidents enter areas of the forest that are occupied by longer-term residents.[44] Their analysis is based on twelve years of data from forty-three identified residents and nonresidents at Ketambe. The predominant view has orangutan society comprising one dominant resident adult male, adult females, and subadults and adolescents of both sexes (and presumably juveniles and infants). Females occupy overlapping home ranges containing their core areas within a larger home range of the principal breeding prime adult male. Alternatively, orangutan society is a community in which subgroups of larger traveling bands, centered around an adult male, move in the same general direction and share a range rather that occupying distinct parts of it. Nonresidents are transient in both scenarios.[44]

The Ketambe study indicated that although there were nonresidents in the area, there is no apparent transient sex, and no positive correlation between the densities of nonresident males and resident females capable of conceiving. Accordingly, the sexual-attraction hypothesis was not supported, but it cannot be refuted because the results are not based on correlations with receptive females. The food-attraction hypothesis has more support; orangutan density and the presence of nonresidents of both sexes were higher when fruit was available.[44]

In Kutai National Park of East Kalimantan and Gunung Palung Nature Reserve of West Kalimantan, individual *Pongo pygmaeus* joined others primarily at food sources and to mate.[45] Kutai orangutans encountered one another more frequently and associated for longer spans than did Gunung Palung individuals. Gunung Palung males avoided conspecifics, whereas Kutai adult and subadult males were more associative. Likewise, adult females and subadult males associated less with others than their Kutai counterparts did.

Passive associations comprised feeding in a tree without interacting or traveling past one another without interacting overtly. Social associations included chasing and fighting, mating, traveling, and foraging with one another, and changing travel direction and otherwise increasing distance from one another immediately upon meeting.[45]

Kutai orangutans traveled and foraged together more frequently than their Gunung Palung counterparts did. Social interactions were frequent after encounters at food sources. Although agonistic interactions between subadult males occurred at Gunung Pulang, the observers saw no mating. In contrast, at Kutai there were matings and aggressive interactions involving subadult males. Because Gunung Palung females were not sexually active due to having small, clinging offspring, while the Kutai females were accompanied by older independent juveniles, John Mitani et al. concluded that female reproductive status accounts for the difference between the two sites.[45]

Bornean orangutans are probably most social before they mature. At Tanjung Puting, adolescent females and subadult males were with non-natal units 76 percent and 41 percent of the time, respectively.[46] Only adolescent females occasionally groomed individuals outside their natal units, such as one another or, more frequently, adult males.[31]

In Sumatra, observer encounters with small social units were more frequent than contacts in Borneo. Because of nearly complete overlap of Ranun subunits, it appeared that there was a group range instead of numerous discrete subunit ranges.[47] Presumedly, the subunits were coordinated by male long calls. Once four adult males, one subadult male, four females, and five youngsters fed in the same tree. The high frequency of males accompanying females with infants suggested that they were complete family units instead of temporary consortships or casual groupings.[47]

Perhaps siamang attacks on young orangutans and the presence of large felid predators, which are absent from Borneo, encourage the greater social cohesiveness of Sumatran orangutans.[48] However, MacKinnon was incorrect about the presence of leopards *(Panthera pardus)* on Sumatra, though clouded leopards *(Neofilis nebulosa)* that hunt arboreally are there.[49] He also suggested that Sumatran orangutans represent a stage between group-living ancestral orangutans and less socialized Bornean orangutans.[47]

At Ketambe also, the ranges of Sumatran orangutan male and female social units partly or entirely overlapped.[50] Males chiefly roamed more widely than females. The average group size was 1.5 individuals. Loners or mother/young units comprised 46 percent of observed orangutans. The majority (54 percent) were in larger social groups (17 percent) or temporary arboreal feeding associations (37 percent).[50] Most social groups—that is, individuals moving together in a coordinated manner—were composed of subadults (8- to 13-year-olds or 15-year-olds) and especially adolescents (5- to 8-year-olds). There was little direct social contact between adults. Adolescents of both sexes were in social groups during 33 percent of the observation time versus adult males (5 percent) and adult females (6 percent).[50]

Despite relatively high population density, there was no tendency for Ketambe orangutans to band together into larger social groups or to coordinate their movements through the area with reference to long calls by high-ranking males. Rijksen suggested that the wider interindividual social contacts that they had during immaturity probably provided the basis for stable relationships and relative dominance status during their chance encounters and temporary feeding associations in adulthood.[50] During immaturity, the dominance relations that order adult social intercourse might be established.[50] The provisioning station at Ketambe probably also affected population density and the frequency of encounters among individuals and social units.

Dyadic interactions between members of different age/sex classes of Sumatran orangutans are similar to those of Bornean orangutans. Grooming was rare. Play occurred more commonly than grooming, wherein immature individuals were principal participants. Females invited males to play with them more often than the reverse, and young males played together more frequently and roughly than the young females.[51]

Footnote for the Kinsey Report

Although creative orangutans in captivity commonly engage in it, solitary sex is probably no more satisfying to orangutans than it is to most adult humans.[52] Indeed, mounting evidence indicates that libidinous female and male orangutans devote considerable effort to the pursuit of sexual partners.[53]

Sexual behavior between females and between males is quite rare. At Suaq Balimbing in Sumatra, a subadult male appeared to copulate anogenitally several times with a somewhat smaller subadult male, whose vocalizations were similar to those emitted by female recipients of forced copulation. At Ketambe, a 9-year-old male fondled the genitalia of a 12-year-old and inserted a finger into his anus. The 12-year-old briefly mouthed the anogenital region of the other male then thrust his phallus between the spread thighs of the reclining younger male. The interaction occurred in silence.[54]

Immediately before the centennial of Darwin's *The Descent of Man and Selection in Relation to Sex,* Horr, Rodman, and MacKinnon amassed a trove of data on the socioecology of Bornean orangutans.[55] At Harvard University, where Horr and Rodman studied, Trivers, Wilson, and others synthesized sociobiological theory.[56] They were particularly astute in viewing the social organization and extreme sexual dimorphism of orangutans as a consequence of intrasexual competition for resources.[57] Later researchers, especially Galdikas, filled in important details, and Wrangham employed the approach to model other ape societies.[58]

Because adult orangutans are large animals that can consume most of the fruit in a particular tree, they forage in small units. Horr further argued that they have no serious natural predators to induce them to seek group protection.[59] Unencumbered by youngsters, males can roam more widely for food and have a better chance to breed with estrous females. Movements of adult females, even ones with clinging infants, might be limited somewhat by juvenile tagalongs that have trouble negotiating sections of the arboreal highway.[60] If bulk alone were a prime limitation to movement, adult males might be expected to have smaller ranges than those of females.

Sociobiological theory posits that the sex with the greater parental investment in offspring will be a limiting resource for the opposite sex. "Male orang-utans manifestly invest nothing in their offspring but the energy of sperm production . . ."[61] Patchy distribution of fruit is the main limiting resource for females. Accordingly, dispersed females are a limiting resource for males. Because the birth interval is rather long and estrus lasts only five to six days per month, males must travel widely and advertise their locations in order to contact a receptive estrous female. Concealed ovulation due to absence of perineal swelling during periovulatory periods gives females greater opportunity to choose high-quality mates—the males in prime condition.[62] The large body size and elaborate epigamic features—cheek flanges and throat sacs—of adult males probably function primarily for intrasexual competition (Figure 2.11). Thus, orangutans are prime exemplars of Darwinian sexual selection.[63]

At Kutai, a resident male orangutan seemed to subsist on less preferred food than the females.[64] Indeed, he moved away from them when fruit was scarce in their common area, which might represent altruistic behavior. Alternatively, males might seek the best source of food during times of shortage. Their knowledge of a more extensive area might enable them to feed at particular trees longer than females do.[65] In lean fruiting periods, they would seek richer sources of less preferred foods that their powerful jaws are able to process. It would be interesting to know what happens when one or more females come into estrus during a low fruiting period, assuming that estrus and food supply are not coupled.[66]

Were a male to try to keep up with a female, he would have to travel arboreally more often between feeding bouts because females change feeding locations more often than males. And if she were to follow him faithfully, she would sacrifice a degree of selectivity among food trees and modest food sources because of his formidable bulk and greater appetite. Rodman observed that when a male settled down for a big breakfast his temporary consort moved off, presumably to sample a more varied menu; after insemination, she had little to gain and much to lose by remaining with him.[65]

Who's Crying Rape?

At Segama, seven of eight incidents of apparent sexual behavior that MacKinnon termed "rape sessions" entailed aggressive males chasing and assaulting screaming, resistant females.[67] Youngsters of the victims also screamed in addition to biting, striking, and pulling the attacker's hair. Males hunched over the females irrespective of their postures and seemed not to achieve vaginal intromission. The most dramatic battle between the sexes began in a treetop and ended on the ground. The eighth copulation was between youngsters.[68] Needless to say, in the absence of more details about the motivation of male orangutans and potential muddles with ideas and motivations of rape in humans, it would be better to term the behavior in orangutans sexual coercion or forced copulation.[69]

Because females are not intentionally wounded, Knott argued, "Orangutan coercive sexual behavior is direct coercion and is not used as an indirect means to influence or control future female sexual behavior, as occurs in species such as chimpanzees and humans."[70] Further, female orangutans can afford to be solitary foragers because aboreality lowers their vulnerability to felid predation and there is no evidence that male orangutans are infanticidal.[71]

Hook-Ups and Speed Dating among Swingers

Bornean and Sumatran orangutans apparently form stable consortships.[48] At Segama, half the sexually mature males consorted with females in an anthropogenically disturbed forest, and 23 percent did so in a less disturbed one. At Ranun, half the adult and subadult males were involved in consortships. In both localities, more subadult than fully adult males were with females: subadults 75 percent at Ranun and 41 percent at Segama; adults 54 percent at Ranun and 22 percent at Segama.

One might have other interpretations of MacKinnon's figures on and few descriptions of intersexual interactions of orangutans.[72] For instance, subadult males might have been tagalongs that were intent upon contacting females regardless of their sexual receptivity. Indeed, three forced copulations punctuated one two-day association. In a second example, a subadult male tried to inspect an adolescent female's genitalia but did not attempt copulation.[66]

MacKinnon reported that consorting orangutans usually nested in the same tree at night, and twice Sumatran males groomed their companions during the day.[72] That both sexes initiated foreplay is more persuasive evidence for consortship. His further claim for long-term relationships between adult males and females, to the extent that bisexual family units exist, is not supported by longitudinal studies on known individuals.[73]

After confirming Rodman's observation that full-bodied males exert drag on the foraging females that accompany them, MacKinnon hypothesized that males must adopt different reproductive tactics as they grow from spermatic lightweights to full adults.[74]

MacKinnon's model of male sexual strategy differs in some features from that of Horr and Rodman. Because he saw more subadult than adult males with adult females and never saw old adult males copulating—though they called and performed aggressive displays—MacKinnon concluded: "sexual vigor and potency decline earlier in the male's lifetime than does agonistic vigor."[75] Accordingly, a subadult male either establishes long consortships with cooperative females or attempts forced copulation with resistent ones. He provided only circumstantial evidence for the former, which would have the best chance for reproductive success, especially if the female remained faithful when she came into estrus.

MacKinnon's claim that "adult males also sometimes indulge in rape" is unsupported.[76] Prime males mainly attract females and defend their priorities via long calls, but as they decline from prime, they often adopt the role of guardian over their progeny with which intruders might compete for food.[68]

The MacKinnon model is not supported fully by Galdikas's long-term studies of identifiable individuals.[77] She concluded that despite more frequent association with females and fumbling attempts to copulate, subadult males are virtually excluded from reproduction because females prefer large-flanged adults that fight for sexual prerogatives. Although, unlike MacKinnon, Galdikas noted vaginal intromission during some forced copulations, none of the victims became pregnant thereby.[78] Subadult males hastily retreat from female company, often on the ground, when a mature male approaches. Solitary subadults simply stay or move out of the way of closely passing adult males.[79]

Few male-female encounters at Tanjung Puting culminated in copulation. Adults of opposite sex generally seemed to ignore and to avoid one another. Whereas subadult males sometimes orally or digitally explored the perinea of passing females, adult males generally did not. Estrous females did not overtly avoid the males that they encountered. Pregnant females refrained from copulating with the subadult interlopers that followed them. One new mother went out of her way to avoid a subadult male.

Similarly to females as a class preferring fully mature males as sexual partners versus the subadult males that commonly associated with them, adult males appeared to be more interested in receptive adult versus adolescent females.[80] Adolescent females were particularly attracted to large males.[81] They eagerly approached, groomed, and orally and manually contacted the genitalia of would-be mates. One urinated on a potential partner. Adult females rarely engaged in such behavior, but they willingly submitted to the advances of mature consorts.

Tanjung Puting orangutans formed consortships and traveled together for three to eight days, during which they mated several times.[80] Copulations were generally conducted ventroventrally while suspended from overhead branches (Figure 5.10B), though they also mated in nests. A male usually mouthed or handled the female's genitalia before copulation and variably uttered a long call or part of one during the breeding bout. Sexually successful consortships were characteristic of adult pairs, but persistent adolescent fe-males sometimes also established them with adult males.

Subadult Tanjung Puting males were responsible for 95 percent of noncon-sort copulations, 86 percent (nineteen of twenty-two) of which were forced.[82] Further, they were most persistent. They spent 83 percent of their social time with adult females.[83] Besides this tactic possibly being optimal in the overall reproductive strategy of an individual, he or she might exhibit behavior based on personal idiosyncrasies or past social experiences.[84]

Unlike Galdikas, Rijksen rejected the Horr-Rodman model in favor of MacKinnon's suggestion that extreme arboreality and solitariness of recent *Pongo* is the result of interactions with *Homo*.[85] Purportedly, when faced with food competition and predation by *Homo sapiens* and perhaps *Homo erectus,* terrestrial, banded *Pongo* took to the trees and dispersed in the canopy where they are more elusive prey.

Nonetheless, Rijksen's data on the sexual behavior and his interpretations of the reproductive tactics of female and male Ketambe orangutans closely agree with those of Horr, Rodman, and Galdikas.[51] He also emphasized female choice and intermale competition for reproductive access. Adult Ketambe females that were most likely to be receptive—namely, those that were childless or with juveniles—were more often with adult males than with subadult males. The latter more frequently accompanied female-infant dy-ads, in which the mother would not have menstrual cycles due to lactational amenorrhea.[51]

Ketambe females initiated four consortships with adult males, and sub-adult males initiated four consortships. Researchers observed none of the eight couples *in copulo.* Guided by long calls, estrous females actively sought the highest-ranking male in order to enter a consortship.[51] This male appeared to be interested in a female only after she had presented herself for inspection. She terminated the relationship, presumably after insemination.

Subadult Katambe males were chiefly limited to sexual coercion, which Rijksen speculated was related as much to dominance as to sexual behavior. Some females submitted to forced copulation more readily than others, per-haps reflecting their social status. Some bouts included vaginal penetration and left evidence of insemination (ejaculate on the vulva).[51]

Although Rijksen clearly identified wild versus rehabilitants, his general in-ferences appear to be influenced by both classes.[51] For example, he stated that consorting Ketambe orangutans traveled together for days, sometimes even months, but provided no quantification that one can compare with other studies on wild subjects. Indeed, the longer Ketambe pairings—three and seven months—were between a wild subadult male and a rehabilitant female.[51]

Subadult male rehabilitants were involved in more forced copulations ($n = 22$) than the wild males ($n = 5$), and their behavior was more bizarre. They forced copulation with every new female at the feeding station. When two couples met in the forest, each subadult male promptly copulated forcefully with the female that was with the other male. Such behavior among rehabilitants that had experienced unorthodox, human-dominated childhoods is a questionable base from which to conclude that sexual coercion in orangutans is largely an expression of dominance behavior.[51]

Schürmann confirmed that Ketambe females prefer to mate with an alpha male.[86] He chronicled the sexual history of a young female that initially accepted subadult males but later progressed to higher-ranking mates. As an adolescent, she was keenly interested in an alpha male, but it took nearly five years for her to become his consort. He gradually showed interest in her—for example, allowing her to take bits of fruit from his hand and mouth. When she presented her genitalia close to his face, he sometimes sniffed. Further, she masturbated in front of him and handled his penis. For several months he ignored or gently rebuffed her advances, but finally he consummated the union. She manually assisted penile intromission and performed the pelvic thrusts while he remained inert.

Several months after the first copulation, they established consortships spanning about sixteen and twenty-one days, during which times he chased away males that approached them. Copulations occurred midway in the consortships, never at the beginning or near termination.[87] The female usually initiated mating, even during the most sexually active period of their consortships. She sometimes changed position from the common ventroventral one to lateroventral and dorsoventral ones and intermittently manipulated and mouthed his penis. Ultimately, she birthed.[87]

The alpha male evidenced reproductive advantage over the four subadult males that had pursued, sexually coerced, and peacefully copulated with the young female before her consortships with him. They appeared to be more active sexually than the adult male was, but, in fact, during a consortship he mated twenty-five times, which is more often than any of the subadult males had scored during one year.[87] Individual mating bouts lasted longer when the female initiated them, which she never did with subadults. Accordingly, Schürmann and van Hooff hypothesized "the extended subadult phase represents a submissive strategy, allowing subadult males to remain in the home range of adult males but with minimal reproductive success."[88]

If the dominant male uttered a long call, his consort presented to him, he sniffed her genitalia, and he mounted her and quickly climaxed. Once she crashed a tree limb to which he responded with a long call and quick mating bout. Schürmann thought that she might have acted tactically to arouse him sexually.[87] Males respond to loud noises nearby with long calls and evidence sexual arousal. A calling male may exhibit penile erection, precopulatory ejaculation, and solitary nonfrictional ejaculation.[87] However, the fact that the female had pushed the snag after a rebuff suggests an alternative interpretation: if she were venting frustration and the snag fell by chance, one need not ascribe to her as much intention as Schürmann did.[87]

Who's Your Daddy?

The mating behavior of Kutai male orangutans led Mitani to question the certainty of current ideas about the relative reproductive success of dominant, subordinate, and subadult males.[89] He forthrightly discussed the limitations of his substantial database and concluded that further research is essential to clarify the issue. The absence of genomic paternity data was particularly problematic. Although the alpha male regularly supplanted other males that were with females, and sometimes copulated with his partners, he probably did not sire the only infant born during Mitani's tenure at Kutai.[89]

Copulation occurred during all daylight hours. Most matings were arboreal although two were terrestrial. Only one female initiated copulation; she positioned her perineum in front of a male's face, and he mated with her. Ventroventral postures were most frequent, but occasionally they employed dorsoventral and lateroventral ones. Males sometimes inspected a female's genitalia orally or manually before positioning her for intromission. Ancillary sex play was not apparent.[89]

At Kutai, both adult and subadult males forced copulations, comprising lengthy struggles, including males slapping, biting, and grabbing, while females kiss-squeaked, grumphed, whimpered, cried, squealed, and grunted. Forced copulations lasted significantly longer than pacific couplings did. Males restrained struggling females by grasping their limbs and/or torso. Resistant females were almost equally successful in escaping subadult (8 percent) and adult (7 percent) males.[89]

Whether an individual engaged in forced or complaisant copulation seemed to be related to male rank and size. Most copulation between subadult males and adult females was forced. The alpha male employed sexual coercion once, but did not force eight other copulations. Four subordinate adult males forced twelve copulations and engaged in seven that appeared to be consensual. A poorly pouched and flanged lightweight among them accounted for three-fourths of subordinate male forced copulations. Although an initial copulation sometimes appeared to be forced by the male, subsequent ones were peaceful. In other instances of serial copulation, all were forced. During one association, the alpha male copulated pacifically six times.[89]

Like Galdikas, Mitani noted that subadult males are most persistent in their associations with fertile females.[90] Even though Kutai subadult males performed the majority of copulations, none resulted in conception. Consequently, Mitani agreed that subadult males might be missing ovulations or may have low sperm counts due to overindulgence.[89]

Armed with endocrinological data on the periovulatory periods of a small sample of Gunung Tanjung orangutans, Knott illuminated orangutan reproductive behavior, especially regarding occurrences of forced copulation and female choice.[91] Like humans, orangutan females lack perineal swells; consequently, there is no blatant visible indication of ovulatory status. Consequently, females can have notable control over when and with which male to mate. Gunung Tanjung females copulated almost exclusively with prime flanged males near or during a periovulatory period, and all such copulations were cooperative. As a female's fecundity waned, flangeless males were more successful, though they often met with resistance. Females were also less resistant to copulation during pregnancy.[92]

During periods of high fruit abundance at Suaq Balimbing, nearly all adult male-adult female consortships occurred, and up to six subadult males simultaneously increased attempts to copulate with individual females whose infants were weaned.[93] Forty-four percent of 207 copulations were between adult females and the subadult males that initiated 99 percent of them. Females resisted 36 percent of their attempts and did not resist in 28 percent of them. The status of resistance was indeterminate for the remaining 36 percent of cases.

Females could avoid mating with flangeless males by joining or staying near flanged males, sometimes without mating with them.[93] Nonetheless, at Suaq Balimbing, subadult males succeeded in 71 percent of attempts to copulate with adult females in consort with adult males. Resistance and compliance were not significantly different for consorting and unaccompanied females.

Resisting flangeless males can be successful, and whether mating occurs or not, the female is unlikely to be hurt. Occasional cooperative and modestly resisted copulations with flangeless and flanged males during pregnancy and nonperiovulatory periods might be an anti-infanticide tactic to confuse male paternity, though there is no documented case of infanticide among orangutans.[94]

Mean pairwise relatedness tests for seventeen adult orangutans indicated that the females might be less related than the males, and that both sexes disperse at Ketambe.[9] Paternity tests of fecal microsatellites from eleven offspring born over a fifteen-year span in a sample of thirty known Ketambe orangutans indicated that three flangeless males had sired six of them.[9] Accordingly, a female choice model in which they have an exclusive preference for flanged males is not supported, though female choice might be operant if some of them choose to mate with flangeless males. Of course, some or all of the successful copulations by flangeless males might have been coerced.

In view of the Ketambe data, Utami et al. proposed the alternative tactics hypothesis.[95] The flanged and flangeless conditions represent parallel alternative tactics with roughly equal returns in the equilibrium condition. Flanged males employ a sit-call-and-wait tactic, whereby they rely on estrous females to come when they call, while the flangeless males use a search-and-find tactic, whereby they encounter females that might accept their attempts to mate or be coerced to copulate.

Microsatellite markers from twenty-three adult Bornean orangutan fecal DNA samples indicated that resident females were as related to one another as resident males were in the Lower Kinabatangan Wildlife Sanctuary of Sabah, East Malaysia.[96] Moreover, residents of both sexes are more related to one another than they are to nonresident males and females. Like Ketambe male *Pongo abelii,* both flangeless and flanged male *Pongo pygmaeus* sired infants at Kinabatangan. Contrarily, microsatellite markers from sixteen adult *Pongo pygmaeus* at the Sabangau peat-swamp in Borneo showed significantly higher relatedness among females than among the males, which suggests that females are philopatric and males disperse.[97]

Overall, during the past quarter century there have been notable advances in data and ideas toward understanding orangutan reproductive ecology and behavior, assisted particularly by endocrinological, paternity, and nutritional analyses on a greater number of population samples and observations on known individuals of wild *Pongo pygmaeus* and *Pongo abelli* over notable time spans. Nonetheless, much more data and refinement of methods and hypotheses are needed.[98]

Definitive models on how the several factors interact to produce observed social interactions, social relationships, and overall social structure are inadequate because the data sets from virtually all studies, however elegant, constitute small samples; and most, perhaps all, are from anthropogenically disturbed habitats. The emerging picture is one of remarkable flexibility, resourcefulness, and adaptability by orangutans, surely facilitated by keen perception and cognitive ability and versatile bodies that allow them to negotiate complex arboreal habitats, to find and to subsist on a notable variety of foods, and to adjust physiologically and hormonally to wide fluctuations in food availability (females) and the presence of powerful rivals (males). In brief, *the* orangutan is not just elusive—it does not exist. Likewise, because of pervasive phenotypic, behavioral and genetic variation *the* chimpanzee, *the* bonobo, *the* gorilla, *the* gibbon, and *the* human do not exist.

Politicos and Female Choice Among Chimpanzees

Decipherment of wild chimpanzee social relationships and structure have progressed slowly because their fission-fusion social structure varies across populations.

Based on a 2.5-month study in the Republic of Guinea, Henry Nissen rejected as gratuitous the designation "family" for chimpanzee groups because it merely reflected the anthropocentric beliefs of the local people and gullible Europeans. He observed twenty-five groups of between four and fourteen subjects (mean: 8.5) that sometimes contained several adult males and females. In the end, he confessed that he had no sense of their social cohesiveness.[99]

However, after a thirty-year hiatus, field studies in western Tanzania, Uganda, and eastern Democratic Republic of the Congo (DRC) fueled expectance that the secrets of chimpanzee society would be revealed.[100]

Congenial Parties and Community Discord

Based on observations of individuals in a papaw and banana plantation in the DRC, Kortlandt dismissed the idea that chimpanzees live in closed harem groups.[101] Instead, he noted two kinds of aggregation when they traveled: bisexual groups comprised chiefly of adult males with childless females and a minority of females with youngsters; and nursery groups consisting mostly of females and youngsters and sometimes one or two adult males. The predominantly adult bisexual groups generally were composed of >20 members, and nursery groups usually included up to 15 individuals. The former were the more wide-ranging and demonstrative, and individuals freely joined other groups. Fission and fusion were manifest in their foraging rounds, and there was no particular order during group travel.[101]

While tracking elusive chimpanzees in the intimidating Budongo Forest, Vernon and Frances Reynolds confronted worse problems than those of Nissen in the Republic of Guinea. As a result, they failed to document the social organization of *Pan troglodytes*. Nonetheless, they surmised that chimpanzees do not live in closed social groups: fusions and fissions and scattered individual arrivals and departures, particularly at food sources, typified Budongo chimpanzee aggregations.[102]

The Reynoldses termed the loosely organized, unstable congeries "bands." Larger bands ($n > 15$) occurred during major fruit seasons than when fruit was scarce ($n = 3–4$, and singletons also). They recognized

- bisexual adult bands that occasionally included adolescents but no younger individuals,
- all-male bands,
- mother bands that occasionally included females unaccompanied by youngsters, and
- mixed bands that were amalgams of mother and all-male bands.[103]

Like Kortlandt, the Reynoldses noted that mothers generally traveled less widely than the childless adults did. Although they observed occasional instances of aggression, they inferred that Budongo chimpanzees lacked male and female linear hierarchies and permanent group leaders.[103] Prophetically, they entertained the idea that chimpanzees possess a highly developed social organization that can be sustained without continuous visual contact among its members. The chimpanzees literally drummed this into their heads (Chapter 12), but it took a while longer for the message to be broadcast.[104]

Goodall perpetuated the image of chimpanzee society as loose and fluid, with the only stable unit being a mother with her dependent youngsters; otherwise, there was no identifiable larger community.[105] During a two-year span before inaugurating intensive provisioning, she observed that 91 percent of 498 Gombe groups contained nine or fewer individuals, including 64 singletons. The largest group comprised twenty-three individuals, and there were three groups of twenty individuals. Thirty percent of them were mixed units, 24 percent were mother units, 18 percent were bisexual adult bands, 10 percent were unisexual units, and 18 percent were solitary males or less frequently solitary females.[106]

Provisioning has allowed *National Geographic* photographers, filmmakers, Goodall, and other primatologists to detail the interrelationships of known individuals and the ontogeny of social behaviors that would have been acquired much more slowly, if at all, from shy peripatetics. Once individuals were habituated, they could be followed from camp to see how the behaviors were expressed in the forest. How much provisioning perturbed their social relationships in degree or in kind could not be discriminated readily. We will never know how typical the individuals that Goodall observed are of non-provisioned Gombe chimpanzees.[107]

Itani and Suzuki provided the first solid evidence for a more complex social organization among *Pan troglodytes* after studying forty-three eastern chimpanzees traveling over a sparsely treed area between two patches of riverine forest at Filabanga, Tanzania.[108] Congeries of females and youngsters preceded and followed seven adult males in the center of the procession. They dispersed in smaller groups in the forest. Additional mixed groups of twenty-one, thirty-one, and thirty-two individuals at other localities also progressed with a clump of adult males in the middle; small, all-male groups also were common at Filabanga.[109]

Because Filabanga-43 was the only group of chimpanzees in the region, Itani and Suzuki concluded that they constituted a discrete group and inferred that, universally, chimpanzees live in groups of twenty to fifty individuals, comprising four to six adult males, around ten adult females, and their youngsters.[109] In the nascent Itani-Suzuki model, mother/infant was the only cohesive unit lower than the large group. Given that young adult males were absent or rare in Filabanga groups, they further speculated that chimpanzees might leave their natal group and eventually join a neighboring group. Conversely, Kano, who succeeded Itani and Suzuki at Filabanga, doubted that young males are driven from their natal groups.[110]

In 1985, Nishida began long-term studies of eastern chimpanzees at Kasoje in the Mahale Mountains, Tanzania, where he recognized that what had appeared *prima facie* to be loosely associating, variably constituted small bands of foraging individuals were in fact members of a larger cohesive unit for which he coined the term "unit-group."[111] Contemporary researchers accepted Nishida's insightful model but preferred to use the term "community" instead of unit-group.[112]

At least four unit-groups resided in the Kasoje area; Nishida et al. focused on two of them.[113] Like Goodall, they provisioned members of K-group with sugar cane and bananas, but not in the quantities that characterized the Gombe banana bonanza.[114]

Kasoje unit-groups contained between 20 and 106 members generally divided into temporary subgroups of variable composition, stability, and size (mean: 8; range: 1–28). Larger subgroups formed at the end of the dry season when food was most abundant. Adult males tended to stay together more consistently than did females. Nishida estimated that male cohesiveness was twice that of females.[115] Forty-six percent of grooming occurred between males, 39 percent between males and females, and only 10 percent between females. Estrous females were most commonly in sizeable, mixed subgroups with 11 or more individuals.[114]

Chimpanzees surpass other primates in their frequency of mutual and polyadic grooming.[116] In Mahale M-group, males groomed longer than any other age-sex class in small clusters of two to four individuals; females groomed as long as males, in clusters of five or more individuals. Grooming clusters ranged between two and twenty-three individuals, with nearly 70 percent of grooming time in polyadic clusters of three or more individuals. Simultaneous grooming and being groomed or being groomed by multiple individuals accounted for about 20 percent of individual grooming.[116] The variability and extent of mutual and polyadic grooming evidence the high cognitive ability of the chimpanzee social brain to sustain their fission-fusion society.

Adult males constitute the stable core of unit-groups, while females, especially estrous nulliparae, commonly emigrate to another unit-group.[117] Ultimately, most young females reside outside their natal unit-groups. Over an eight-year span, twenty-three of twenty-nine original members of K-group remained in it. Two males probably died, four subadult females disappeared, presumedly to join another unit-group, and six new ones immigrated into it. Of the nine adult females in K-group, three appeared to be attached strongly to the core males, but the other six were less so.[118]

Kasoje unit-groups had well-delineated boundaries that they would not cross, though large sections of their home ranges overlapped. The larger M-group, which ranged south of K-group, was dominant to them, as evidenced by displacement northward when M-group moved seasonally into the southern part of their range.[119] M-group used around 50 percent of K-group's home range, while K-group used only about 25 to 33 percent of M-group's home range.[118]

Estrous adult females transferred between the unit-groups most often during two short annual migratory periods that constituted 20 percent to 30 percent of the yearly round. Estrous adolescent and subadult females more commonly transferred during sedentary phases when the unit-groups were farther apart.[119]

Female immigrants were gradually accommodated in unit-groups, but males of different unit-groups were agonistic toward one another. For instance, when a lone old male of M-group appeared at the feeding ground, three K-group males attacked him, and he fled with a bitten thigh.[120] In contrast, when in the common section of the two ranges, K-group usually retreated rapidly and silently upon hearing calls from M-group. Retreats appeared to be premised on which subgroup included fewer adult males.

At Gombe, regardless of their location within small groups, dominant individuals generally regulated the group's movements, unless they moved rapidly, in which case the leader would be at the forefront. Leaders were not apparent in large, mixed groups and male parties.[121] Rank was indicated by priority of access to favored foods, right of way on the paths, and freedom from attack by other chimpanzees. Healthy adult males were dominant over all adult females, and both classes were dominant over youngsters.[121] In contrast, at Kasoje some females were dominant to two adult males.[122]

Gombe males lacked a strictly linear hierarchy. There was an alpha, followed by six others whose ranks varied according to which other males were present. Four males were subordinate to the first seven; relative rank among them also varied with the presence of specific males. Some individuals tended to assist one another if conspecifics or baboons threatened one of them.[123] Similar patterns of intermale alliance characterize other eastern and incarcerated chimpanzee groups.[124]

Although interfemale alliances were rare, Gombe females also showed a relatively clear dominance ordering. Their assertiveness seemed to depend on the presence of an adolescent son and perhaps their estrous cycle; perineal tumescence seemed to be coupled with assertiveness. The status of youngsters depended upon the presence and status of their mothers.[125]

Long-term ties persist between some chimpanzee females and their offspring.[121] These ties probably develop during the relatively long period of infant and juvenile dependency. Weaning is not completed until youngsters are between 4.5 and 7.0 years old, and individuals might continue to travel with their mothers and siblings until adolescence or early adulthood.[126]

Whereas infant Gombe males might mount and hunch their mothers, older sons rarely did so.[121] At Bossou, Republic of Guinea, the male youngsters also hunched their mothers.[127] The bond between mother and offspring is reinforced not only by tender loving care, including fiddling with the infant's genitalia, but also gentle play which decreases in frequency and mutual grooming which intensifies as the offspring matures.[121]

Allogrooming occupied a notable portion of Gombe chimpanzee leisure time, commonly among presumably unrelated adults, siblings, and parents. Although this was partly an artifact of provisioning, it is probable that in general wild chimpanzees are more inclined to allogroom than Asian apes.[128] Because both social play and grooming entail intimate physical contact and mutualism, they probably contribute to the somewhat greater cohesiveness of chimpanzee versus orangutan society.[129]

Gombe grooming clusters comprised up to 10 individuals, and allogrooming sessions that lasted up to 2.5 hours.[121] Most allogrooming occurred in pairs up to an hour. Adult males were groomed more than they allogroomed others, whereas adolescent males generally groomed more often than they were groomed. Adolescent females usually allogroomed as much as they were groomed, and youngsters received more grooming than they performed, largely from their mothers. Females with juvenile or older offspring spent more time grooming than being groomed. Indeed, they were the second most common class of groomers. Estrous females were groomed more than anestrous ones.[121]

In striking contrast with orangutans, high-ranking Gombe males spent notable spans grooming one another. They were the most common groomers and groomees on the provisioning ground. An alpha male (Mike) and a recently deposed alpha male (Goliath) appeared to prefer one another as grooming partners.[121] Nuclear DNA analysis indicated that Gombe Kasekela males significantly shared more alleles per locus than the females did; indeed, they were related on the order of half-brothers.[130]

During an intensive yearlong study on the grooming behavior of eleven adult males that visited the Gombe banana ground, the average grooming session lasted one hour, with three hours the longest span.[131] Seven elders groomed one another more than four younger males did. High-status males, as measured by higher frequencies of displaying, eliciting pant-grunts, and especially supplanting others, were most often engaged in grooming bouts. Although he was groomed a lot, Mike reciprocated for long periods and often was first to groom the males that presented to him. Other high-ranking males usually groomed subordinates less than they were groomed, but most bouts were characterized by some degree of mutualism. Both lower- and higher-ranking partners initiated bouts.

Simpson insightfully surmised that grooming facilitates social cohesion among male chimpanzees.[131] No specific dyad occurred >60 percent of the time that males were at the feeding ground. Nevertheless, individuals clearly showed consistent preferences for some males over others during the year. In brief, although Gombe male society appeared to be relatively open and fluid, it was not haphazard.

Like Nishida and Kawanaka, Goodall and coworkers documented instances of female transfer and territorial behavior in chimpanzees.[132] Goodall et al. speculated that male boundary patrols might serve to recruit young females into the resident community.[133] Over a seven-year span, adult Gombe males did not change communities.[134] Instead, they engaged in group boundary patrols during which they were unusually quiet, cohesive, and systematic while investigating the habitat. If they encountered aliens, they often called, displayed, and chased them. When they caught a transgressor, they attacked him viciously. Several individuals were mortally wounded during such encounters.

At Gombe, the alpha male seemed less inclined to patrol and attack aliens than his subordinates were. In addition to the risk of being injured during encounters with males of other communities, patrolling males incur energetic cost: for example, patrolling Ngogo males covered longer distances and spent more time traveling and less time feeding than when not on patrol (Plate 12).[135]

Long-term observations at Kasoje documented the near extinction of K-group due to expansions of the range of M-group and loss of males from K-group, some of which might have been killed by M-group males.[136] By 1979, K-group had been reduced to a one-male group from which cycling females moved to M-group, while the male attempted to herd them within their reduced range.[137] In 1982, the last adult male and a subadult male disappeared, leaving only two adult females, an infant, and an adolescent male in remnant K-group, while M-group comprised eleven adult males, thirty-nine adult females, nineteen adolescents, fourteen juveniles, and twenty-two infants.[138] Further, twelve adult females with three adolescents and five juveniles visited M-group, whose range extended over 14 km^2.[139]

Although the reports of conspecific killings have been predominantly among eastern chimpanzees, there is also strong evidence for similar occurrence among western chimpanzees. The fresh corpse of an adult male combined with vocalizations and movements of chimpanzees the night before and fecal DNA analysis argued strongly that he had met his fate at the hands, feet, and jaws of chimpanzees from another community in the Loango National Park in Gabon. Chimpanzees in the area are neither habituated nor provisioned. The victim was not cannibalized.[140]

Wrangham interpreted the reports of unit-group transfers of Kasoje females actually to represent core males of a neighboring unit-group having moved into the ranges of more or less resident females instead of the females having immigrated to a new unit-group.[141] In this scenario, only male chimpanzees were bonded in cohesive communities, while females generally existed in the same asocial state ascribed to orangutans by most observers at the time. Thereby, like *Pongo* spp., *Pan troglodytes,* especially mothers with young, avoided scramble and contest feeding competitions via foraging alone.[142]

Uehara acknowledged that K-group and M-group males were discrete in their associations and ranging patterns.[143] Some females resided within the boundary of K-group or M-group and associated exclusively with its males. Others ranged in the area or areas of one or both unit-groups and interacted alternately with males of both groups where the ranges of the unit-group males overlapped. But diachronic observations revealed that with maturity they settled in one community or the other. Consequently, Uehara dismissed Wrangham's scenario in favor of a bisexual community for *Pan troglodytes.*[143]

Kawanaka initially provided anecdotal evidence to counter Wrangham's scenario based an encounter between four adult females and a juvenile of K-group and two adult females and two young adult males of M-group characterized by tension—manifest by profuse diarrhea—but no explicit antagonism.[144] Indeed, some individuals greeted and groomed one another despite apparent agitation. Kawanaka concluded that fighting was avoided because fully adult males were absent from both groups.[144] Moreover, the behavior of the females during the meeting suggested that they identified with their own unit-groups, counterpart members of which would not induce so much diarrhea.

A more thoroughgoing proximity-matrix analysis of data from a yearlong study of K-group and M-group showed that except for three cycling females, one of which was accompanied by her juvenile male son, all females remained exclusively with one or the other group. The female/juvenile dyad ranged farther than other individuals, including adult males. The others did not remain in core areas within the ranges of unit-group males but instead roamed widely within the entire range of at least one unit-group.[145]

Itani noted that Wrangham's scenario failed to accommodate the large, bisexual Filabanga-43 group and a later, presumably seasonal, migration from the area for a span of ≥7 months.[136] Ghiglieri also favored the concept of a bisexual community versus Wrangham's scenario, based on Kibale chimpanzee travel companionship and allogrooming. He inferred that although males exhibited the greatest affinity among themselves, females were more social toward one another than toward the males. Moreover, they associated with males in ways that are not directly relatable to mating.[146]

Though not contesting Wrangham's scenario, Halperin noted that five anestrous Gombe females with dependents were homophilicly social.[147] Whereas mothers with single dependents associated with other females in nursery groups, those with two dependents were inclined to be separate from them. Perhaps the social needs of the latter females were satisfied by their offspring, whereas the females with singletons sought the company of other adult females.[147]

Pusey initially agreed with Wrangham but abandoned his scenario in favor of the bisexual unit-group. She noted that Gombe females sometimes participated with unit-group males on border patrols, and they did not always accept immigrant females peacefully into their community. Further, males of a unit-group attacked females of another unit-group if they were anestrous.[134]

Studies in a wide range of habitats outside Tanzania support the hypothesis that androcentric unit-groups and female transfer between unit-groups are universally characteristic of wild chimpanzees.[148] Territorial behavior also appears to be widespread among chimpanzees.[149] Although Sugiyama reported that Budongo chimpanzees from neighboring regional populations mixed freely and maintained friendly relationships, later studies showed them to be hostile toward neighbors.[150] At Mount Assirik in Senegal, there was only one unit-group of twenty-five to thirty western chimpanzees, so there was no opportunity to observe intercommunity responses and female transfers. Interestingly, rare male subgroups comprised only a pair of adults or an adult and an adolescent.[151] Border patrols are unnecessary for isolated populations.

At Mount Assirik, mixed bands were most common; on average, those containing estrous females were significantly larger (mean: 10; range: 3–22) than those without estrous females (mean: 7; range: 2–20). Among Taï western chimpanzees, a party size increased in the presence of estrous females, and they appeared to sacrifice efficient foraging in order to obtain the social and/or reproductive benefits of associating with estrous females.[152]

In the Ngogo community of eastern chimpanzees in Kibale National Park in Uganda, food abundance and the presence of estrous females were positively correlated with party size, and the two factors had independent effects.[153] Similarly, based on data from four different localities of eastern chimpanzees in Uganda, Republic of the Congo, and the DRC, Hashimoto et al. hypothesized that monthly fruit availability affects monthly party size when it is low enough to limit party size during a major part of the year, but party size does not increase with monthly fruit abundance.[154] Instead, social factors, particularly the presence of estrous females, appear to be more important than fruit among the factors that can increase a chimpanzee party size.[155]

Male chimpanzees in the Budongo Sonso community evidenced a preference for association partners that tended to form small parties, and the strength of dyadic male associations accounted for a significant portion of variance in party size.[156] Party size had no or a negative relationship with habitatwide measures of food abundance.[157] Data from Sonso support the hypothesis that the associations are the result of tactical decisions and refute the predictions of hypotheses that associations are related to passive attraction to fruit trees or other locational features.

Sonso party size increased with the number of estrous females present, but the number of males present was similar regardless of the number of cycling females (one to four) in the party. Newton-Fisher reasonably suggested that the cognitive demands of tactical decisions concerning with which individuals to associate might explain the high intelligence of chimpanzees.[156] The findings also appear to support the social brain hypothesis—that brain complexity evolved in response to selection for maintaining social relationships and negotiating social interactions, especially in primate societies that engage in fission-fusion.[158]

Whether the large Mount Assirik unit-group range (228 km²) is a consequence of food availability, predator avoidance, or the absence of competing territorial unit-groups is unclear; however, chimpanzee foraging and travel in large, mixed parties might reduce the risk of predation by lions, leopards, African wild dogs, and spotted hyenas.[159]

Likewise, leopard predation probably affected party composition among Taï chimpanzees, in which the sexes are more cohesive than congeners at eastern African sites. The vast majority (89 percent) of the twenty-seven females had infants that could use the protection of males. The females without infants, the juveniles, and the subadults also could not hold their own against a leopard.[160]

An eighty-member focal Taï community included seven adult males and twenty-seven adult females, giving an adult sex ratio of 1:3.86. Only one-third of Taï parties were unisexual. Both males and females associated with members of the same sex far less often than they did with the other sex: adult females were with adult males 82 percent of the time, and males were with females 74 percent of the time.[161] Christophe Boesch and Hedwige Boesch-Achermann concluded that, as in bonobos, whereas community size explains most of the differences in party size, the sex ratio in the community affects the frequency of mixed parties.[162]

Eastern chimpanzees in the Ngogo community of Kibale Forest were as gregarious as Taï chimpanzees and bonobos, though the group size, composition, and adult male:female ratio (1:1.62–1.72) were dramatically different from those of the Taï community. The Ngogo community included around 145 individuals: 25–26 adult males, ≥42 adult females, 6–14 adolescent females; 12–28 adolescent males, 18–22 juveniles, and 26–28 infants.[163] Ngogo males formed two distinct subgroups, thereby developing social bonds above dyadic pairs. Further, the males that were closely related genetically through the maternal line appeared not to affiliate or cooperate with one another.[164] Likewise, at Kanyawara there is no significant relationship between genetic relatedness and the strong male social bonds evidenced by association, proximity, grooming, meat sharing, or boundary patrols.[165]

Like other chimpanzee communities, Ngogo males were more gregarious than the females, but females that spent around 64 percent of time with other females formed associative cliques.[166] Interfemale aggression was very rare. There was no significant correlation between fruit availability and party size and numbers of anestrous females per party that might indicate food competition. Indeed, they seemed to avoid feeding competition by preferentially associating with a small subset of other females, perhaps because they are members of such a large community.[163]

These observations confirm Michael Ghiglieri's earlier report that Kibale females have stronger affiliations than one might expect according to an exclusive androcentric model.[167] Indeed, he observed that Ngogo females "frequently ranged together, almost as if they were members of a female community."[168] Because, like Sugiyama, Ghiglieri had limited spans of observation, his comment only served as a caveat to universally applying Nishida's model for chimpanzee society.

Informed critics dismissed the early model that virtually excluded female chimpanzees from the broader social framework.[169] Later, based on data from Kibale National Park, Wrangham acknowledged that, like orangutans, female chimpanzees are distributed in core areas in order to maximize access to preferred food for themselves and their youngsters, while the males range more widely to gain access to sexually receptive females.[170] The revised model is supported by data on some eastern chimpanzees.

Gregariousness of females (and males) probably varies in response to local resource distribution and abundance and threats by predators, chimpanzees of neighboring communities, human observers, and anthropogenic hazards.[171] Indeed, sociability can increase notably among captive chimpanzee females, sometimes to the extent that they appear to bond and coalesce to defend one another against male aggression.[172] That free females can be effective against male aggression is well illustrated by the combined action of two females to avert an intragroup infanticide of a female newborn by two high-ranking males in Mahale M-group. The successful defenders were the grandmother of the neonate and her adult female friend.[173]

Analyses of six years of Kanyawara data indicated a relationship between subgroup size and density and distribution of chimpanzee food sources; however, food density and travel costs appeared to determine subgroup size of males more than that of females. For females, the benefits of being in a subgroup did not exceed the costs, even when ecological conditions seemed to minimize subgroup foraging costs.[174] Quality foods are essential to the reproductive success of females; estrogen levels in cycling and noncycling Kanyawara females increased during seasonal peaks in the consumption of drupe fruits. Further, when average fruit consumption remained high over several months, females conceived more quickly.[175]

In the Kanyawara community of Kibale Forest, where females have stable core areas that vary in the density of preferred foods, those in the richer areas had higher levels of fitness, as evidenced by their elevated ovarian hormone production, shorter birth intervals, and higher infant survivorship than their neighbors in poorer areas.[176]

Kanyawara females—mostly mothers—aggressively defended their core areas, especially against incursions by immigrant females.[177] Indeed, inter-female aggression increased fourfold with the arrival of immigrant females, peaked when there were multiple immigrant arrivals, and sometimes exhibited an additional feature of coalitional aggression, which is unusual among chimpanzees. Kanyawara females whose core areas were high-quality foraging resources were dominant over those in poorer areas, even though the latter could be older than the former.[177]

At Kanyawara, female urinary cortisol levels increased with age, were elevated in immigrants, and were higher in estrous females due to aggressive sexual coercion by males. Cycling, anestrous females had relatively low levels of urinary cortisol, but low-dominance females had elevated urinary cortisol levels, indicating stress, especially during lactation in the months of low fruit consumption.[178] In the presence of males, lactating mothers increased vigilance and tried to avoid close proximity to the males, even to the extent of being less gregarious than the females without dependents.[179] High-ranking males, even in a stable hierarchy, also exhibit elevated urinary cortisol, probably due to the metabolic costs of their aggressive behavior.[180]

When eleven adult males experienced spans of suboptimal access to energy from sugary and fatty ripe fruits, their levels of urinary testosterone were lower than those of eleven well-fed captives.[181] Highly ranking Kanyawara males were more aggressive than low-ranking males, and they produced higher levels of urinary testosterone.[182] In chimpanzees, as in birds, variations in male testosterone seem to be more closely associated with aggression in reproductive contexts than with changes in reproductive physiology—known as the challenge hypothesis.[183] Interestingly, the males' testosterone levels increased in the presence of parous females with maximally tumescent perineums but not in the presence of nulliparae in the same condition. They copulated at similar rates with nulliparae and parous females, but they did not mate-guard or increase aggression in the presence of the former cohort, which supports the idea that adult males prefer older females.[182]

In Taï Forest, a female's dominance rank correlated positively with reproductive success. Based on greeting behavior directed at dominant individuals by subordinate ones, Taï females appeared to have linear dominance. Relative dominance rank was related to winning contests over food and was not related to the age of the contestants.[184] Females of two Taï communities were highly gregarious, and they did not monopolize core areas. Instead, like Taï males, they ranged over their entire territory. High-ranking females were more gregarious than low-ranking females only when food was scarce, and they traveled more than low-ranking females did, by which the latter might have conserved energy.[185]

At Gombe, although females exhibited a high degree of core-area site fidelity, they sometimes had to adjust to changes in the male-defended territorial boundaries. Females with core areas in richer areas had higher reproductive success than others, and adult females that moved from one community to another produced the fewest surviving offspring.[186] High-ranking females had narrower dietary breadth and higher quality food than the subordinate females that foraged less efficiently in lower quality habitats.[187]

Overt competition among females is usually less dramatic than that of males, but females also act violently, even murderously, to maintain the integrity of the core areas upon which their reproductive success depends. Pusey et al. documented severe female aggression on an immigrant female and an attack on a female and her successive newborn infants in areas that overlapped their core areas.[188] They also inferred that a female in the Gombe Mitumba community had killed an infant of another mother in the community. Kasekela males killed the 3.5-year-old male child of a 17-year-old primipara that lived in the center of the community core area. The incident was puzzling because previously at Gombe females had perpetrated the intracommunity infanticides on much younger infants of both sexes.[189] Perhaps to protect it from the females that had been aggressive toward her when she birthed twins and a singleton, another Kasekela female took her primiparous granddaughter's newborn and kept it until it died five months later. Her daughter resumed cycling and birthed again within one year; hence, the interbirth interval was notably shorter than that typical (5.5 years) for Gombe chimpanzees.[190]

A major difference from the orangutan pattern is that male chimpanzees travel in groups and act together to prevent incursions by extracommunity males, which conserves food resources in their area for females, their youngsters, and themselves. Williams et al. concluded that Gombe males proximately defend a feeding territory for their females and protect them from sexual harassment, though a large range might eventually attract more females.[191] At Gombe, the number of males was not correlated with community range size.

Rank ordering among males reduces the frequency of intragroup competition over the limited resources.[192] Dominance interactions are less frequent among female chimpanzees, but the Kibale Kanyawara data indicate that the intensity of selection on interfemale competition might be similar between males and females.[176]

Gombe males had longer active periods than those of anestrous females, and estrous females had the longest active periods. High- and middle-ranking males had shorter active periods than low-ranking males. Rank appeared not to influence the duration of the activity period in anestrous females, perhaps because they live less socially than males. Accordingly, the active period probably does not reflect any differences in female competitive abilities as it does in males.[193]

Chimpanzees are exceptional among social mammals, and primates in particular, in females being the dispersing sex.[194] Although chimpanzee females apparently are not averse to mating in the natal group, they are predominantly the individuals to emigrate to other groups, despite the notable risks of aggression by resident females and the initial or longer-term limited access to the best resources in the new residence because of low social status. The proximate causes for emigration are elusive and might vary among individuals and their natal social and ecological contexts. A ten-year study of twenty-two nulliparae—fourteen natal members of and eight immigrants to the Kanyawara community—indicated that dispersal is not strictly predicted by the physiological events of puberty or by community stress. Instead, subtle variations in her developmental trajectory play a role in a female's decision to migrate.[195] Females experiencing conditions of high fruit consumption emigrated soon after menarche, and others emigrated during periods of increasing fruit consumption; however, Stumpf et al. were careful not to overemphasize dietary quality per se versus possible correlated variables such as the changing size or composition of social parties or increased fecundity in the presence of other females.[196]

A principal components analysis of a sizeable group, comprising seventeen adult females, five adult males, and four to ten juveniles and infants at Chester Zoo in the United Kingdom indicated that three of nine variables were key components in relationship quality among chimpanzees.[197] Kin had more valuable, compatible, and secure relationships than those of non-kin; bisexual dyads were the most secure; and female dyads were more compatible than either male dyads or bisexual dyads. The relationships of individuals of similar age were more secure and valuable than ones in which there was a wide age differential. The relationships of long-term residents were more compatible and valuable than those of shorter-term members, and the latter were less secure. The nine variables in the principal components analysis are grooming, grooming symmetry, consistency of affiliation, proximity, tolerance to approaches, support, counterintervention, aggression, and successful begging.[197] The extent to which the results of the study might be approximated by free chimpanzees is a notable challenge to field researchers.

Postpandemonium Peacemaking

As attested by observations of bite wounds *in vivo* and on skeletons, chimpanzees can severely injure one another.[198] Although they sometimes engage in brutal attacks on strangers and members of their own communities, to restore relative calm in the group most intragroup altercations are followed by one or both of the following interactions: postconflict reconciliation in the form of an affiliative reunion between opponents, or consolation in the form of postconflict affiliative interaction directed to the recipient of aggression by a third individual.[199]

The role of consolation by third parties is uncertain. For example, tests on some captives have cast doubt on the assumption that it alleviates stress in the consoled individual, but others have supported the stress-alleviation hypothesis.[200] When reconciliation fails to occur, consolation might serve to manage conflict for victims of aggression and consolers.[201] Two social groups housed at Yerkes National Primate Research Center Field Station in Atlanta, Georgia, evidenced postconflict affiliation by bystanders toward aggressors, but apparently not by aggressors toward bystanders. Because appeasement occurred more often in the absence of reconciliation than after its occurrence Romero and de Waal suggested that appeasement might act as an alternative to reconciliation when reconciliation fails to occur.[202]

In African apes, especially those in fission-fusion societies—namely, chimpanzees and bonobos—reconciliation likely serves to undo the damage that aggression has inflicted on valuable social relationships between aggressors and victims, but consolation by third parties might not serve the same function.[203] Indeed, consolation might be a means to develop relationships, including alliances against aggressors. During reconciliation and consolation, chimpanzees engage in open mouth-to-mouth contact (kisses), embraces, touching, grooming, and hand extending, and they emit a submissive vocalization. Mouth-to-mouth contact appears to be more characteristic of reconciliation, while embraces characterize consolation.[204]

In Taï Forest, eighteen chimpanzees employed reconciliation to resolve conflicts among high-value partners and when approaching a former opponent with whom they were unlikely to incur further aggression.[205] Frequent reconciliation might actually benefit former opponents, in addition to creating calm in the group. Among same-sex partners with highly cooperative relationships, friendly interactions were up to 8 times less frequent when no reconciliation occurred.[206]

Consolation appeared to substitute for reconciliation when Taï chimpanzee opponents were of low value or when approaching the former opponent would be too risky, as when further aggression seemed likely. They renewed aggression after undecided conflicts and when the losers were unexpected. After long conflicts, they redirected their aggression toward others, perhaps because conciliatory behaviors were likely to fail. After brief conflicts, Taï chimpanzees usually resumed regular activities, especially if unlimited resources were available.[205]

Reconciliation occurred much less frequently in the Sonso community of Budongo Forest than in captive groups, and consolatory actions and explicit postconflict gestures appeared to be absent among them.[207]

Male Ngogo chimpanzees evidenced reciprocity and support via interchange of grooming given and support received, and grooming received and support given; this occurred independent of reciprocity in grooming and support, and of correlations of support and grooming with dominance rank (Plate 14). They cooperated in contests with third parties mostly when the risk was low—such as when both individuals outranked the opponent. Watts concluded that reciprocity and interchange between Taï males might be important not only to maintain intermale social bonds but also to attain and to sustain high dominance.[208] The Ngogo males were inclined to engage in peaceful postconflict interactions (PPCI) with the opponents that had sometimes formed coalitions with them, especially if they had exchanged coalitionary support repeatedly, more so than with individuals that had not joined them in coalitions. Allies groomed more than nonallies as part of PPCI; indeed, the latter made only brief physical contact.[209] Among Ngogo males, PPCI is probably mutualism (an action in which both partners stand to benefit) rather than reciprocal altruism (direction of beneficial action to one that is likely to benefit one's own future fitness).[210] As in Budongo chimpanzees, PPCI might reduce aggression-induced stress and reduce the risk of continued aggression.[211]

Initially, Silk argued that in primates peaceful postconflict contact among former opponents might merely serve "to signal the actor's intention to stop fighting and behave peacefully," but later she allowed that chimpanzees probably also use these signals "to enhance the quality of long-term relationships and to manipulate social relationships for political purposes."[212]

Captive chimpanzees responded to inequity of rewards based on the individual receiving the reward instead of the presence of the reward alone. Like people, intergroup variation in chimpanzee responses could be caused by group size, social closeness reflecting the amount of time the group has been together and group-specific traditions.[213]

Sex in Bananaland, Away from Food Fights

Studies at Gombe National Park generated intriguing hypotheses about the reproductive strategies and sexual tactics of female and male chimpanzees.[214] Early reports evoked an image of Gombe females as highly promiscuous and sexually voracious.[215] Nonetheless, given that mean ages of first reproduction in females at five sites ranged between 13.0 and 15.4 years and the mean interbirth intervals were between 5.1 and 6.2 years, one would expect them to have tactics to secure fit mating partners.[216] Further, if individual female reproductive histories are equivalent to or less successful than those of captives, negative birth outcomes will increase with maternal age instead of the number of infants she bears (i.e., parity).[217]

Estrous females in large groups were normally mounted in rapid succession by up to eight males, especially during periods of general social excitement. For instance, during perineal swells, one of the latter attracted retinues of six to ten mature and four or five adolescent males. She copulated with young males only twice versus fifty-two times with mature males.[121] Resultant postcopulatory sperm competition might constitute cryptic female choice that benefits them by ensuring fertilization and genetic compatibility with specific genotypes with highly viable genes.[218]

Of course, the quality of semen and spermatozoa is unknown to either copulatory partner, and quantity of ejaculate might be less important than the quality of spermatozoa. Masturbatory semen samples at five-hour intervals from six chimpanzees indicated that ejaculate volume increased from 2.6 ± 0.7 ml to 4.7 ± 0.6 ml, whereas the total sperm count decreased from $1,278 \pm 872 \times 10^6$ to $587 \pm 329 \times 10^6$ spermatozoa between the first to the sixth ejaculations.[219] The increase in ejaculate volume was due to an increase in the volume of coagulum, which congeals quickly and might impede spermatozoa of other males. Male chimpanzees contrast starkly with men, in whom the ejaculate volume decreases as the interval between ejaculations decreases.[220]

Four hypotheses to explain high frequencies of copulation in chimpanzees, bonobos, and other mammals merit consideration, especially insofar as they might benefit female fitness.

- Best male. Frequent copulation increases competition by males, thereby assisting females to identify partners with high fitness potential.
- Social passport. Mating partners might support females in agonistic situations, protect and otherwise care for their infants, and provide access to food and nest sites.
- Paternity confusion. If males are ignorant of paternity, they might prevent infanticide of a female's offspring.
- Optimal period for producing next offspring. Females may evince more proceptive behavior toward copulation.[221]

Adult males generally accepted the gang approach of other mature males, but randy adolescents were subjected to frequent interruptions by male elders and had to be sexually discreet in their presence. They seldom joined the mating frenzies that were triggered by group reunions and banana booms. Instead, most of their mating occurred peripherally during pacific moments. Given that the paternity testing of twins born to a captive female indicated that they were sired by different males, consortship is probably the best and perhaps the only assurance that a particular male is the sire of a female's offspring.[222]

Male mating behavior is almost fully developed during the first two years of infancy, but they do not become socially mature for at least 9 years.[223] Consequently, males have a long wait for the opportunity to fulfill their breeding potential.

At Gombe and Kibale Kanyawara, young females were less popular sexual partners than the older females.[224] As adolescent Kanyawara females reduced their association with their mothers, they increased their time with potential sexual partners. Before emigrating, they usually copulated with natal males when they had the first maximal perineal swell, and they often continued to do so until emigration, sometimes over a year later. A female that did not emigrate had the highest copulation rates and mated with two maternal brothers among others. Females initiated the majority of copulations, and natal females were as likely as immigrants to do so. Accordingly, Stumpf et al. concluded that inbreeding avoidance was not a major cause of female dispersal.[225] Genetic consequences of breeding within the group might have been minimized by the fact that the young females were not fecund during the relatively brief span of mating with natal males.

Despite their preference for older females as mating partners, Kanyawara males protected immigrant females from attacks by resident females. Indeed, they intervened more often in conflicts between residents and immigrants than between residents. Unsurprisingly, immigrant females associated more with males than resident females did, both during estrous and when anestrous.[177]

Gombe males sometimes appeared to force females, including anestrous ones, to accompany them for spans up to four days.[226] But in other instances, individual estrous females seemed to travel voluntarily with a male, perhaps to avoid serial gang sex. Moreover, Gombe females were not always passive objects of male attentions; females initiated 37 (17 percent) of 213 copulations and attempted copulations.[121] Moreover, in the Kalinzi Forest, Uganda, females initiated 37 percent of copulations.[227]

Over a sixteen-month span, nine of fourteen Gombe females were impregnated during restrictive associations with single males versus serial copulation with several opportunists.[228] There were two restrictive mating patterns: short-term possessiveness by high-ranking males that prevented estrous females from mating with other males, and temporary consortships, in which breeding pairs actively avoided the company of other chimpanzees. The former was male dominated, whereas the latter required female cooperation, wherein she could exercise choice. The brevity of possessive mating periods suggests that female choice could operate then also. Pairs usually remained together in possessive associations for up to 1 day, whereas consortships had a mean duration of 9.5 days (range: 3 hours to 28 days; median: 7 days). The majority (85 percent) of 209 unsuccessful courtship sequences occurred because females avoided or did not respond to the males. Only 4 percent failed because of male unresponsiveness.[229]

Female chimpanzees should choose mates strategically because they have limited reproductive potential and a high level of parental care. They are in estrus 4.2 to 6.4 percent of their adulthood.[227] Examination of a nineteen-year accumulation of data indicated that the theoretical lifetime reproductive potential of Gombe females was five or six offspring; in fact, the median is three births per female, and only two of them are expected to survive to reproductive age.[230] The banana bonanza might have reduced the birth rate in Gombe National Park, and in retrospect infant mortality might have been high because of infanticide.[231]

The lifetime reproductive potential of Gombe chimpanzee females approximates that of Gainj women in highland New Guinea, whose total fertility was 4.3 live births per woman although for reasons very different from that of the chimpanzees.[232] The women, who used no contraception, experienced menarche and married relatively late, had a long interval between marriage and first birth, had high probability of widowhood at later reproductive ages, and had low fecundity and prolonged lactational amenorrhea. Wood et al. concluded that the most important factor in birth spacing among Gainj women was prolonged lactational amenorrhea.[232]

At Gombe, consortships accounted for seven of nine (78 percent) conceptions that occurred during restrictive matings, while only 2 percent of 1,137 copulations involved consorting pairs. The vast majority of copulations occurred opportunistically, and the remaining 25 percent were with possessive, high-ranking males.[233] Mitochondrial DNA analyses confirmed a positive relationship between male rank and reproductive success but also indicated that consortships and opportunistic matings are effective male tactics.[234]

Males establish consortships by remaining close to an estrous female until there is an opportunity to lead her silently in a different direction when the foraging group begins to travel. The female might choose not to be chosen simply by not following the suitor or by responding to the calls of other males. Rejected males make no attempt to prevent a female from staying with the group.[229]

Consorting estrous females receive a higher frequency of grooming (mean: 14.8 percent of waking hours) than the average in larger mixed groups.[229] They probably also profit from decreased competition for food while staying with only one other adult.[235]

Male aggression toward fecund females might be a counterstrategy to female attempts to confuse paternity, thereby reducing risks of male intracommunity infanticide.[236] Data from the Kanyawara community in Kibale National Park cast doubt on the extent to which female chimpanzees have free choice of mates. Indeed, Kanyawara females initiated periovulatory copulations with the males that were most aggressive toward them throughout their monthly cycles, and even during their estrus periods despite high rates of copulation.[237]

The aggression of Kanyawara males toward females fits the criteria set by Smuts to qualify as sexual coercion: it intensified in reproductive contexts and correlated with increased mating activity, and the females would have been better off not experiencing it.[238] Indeed, Gombe males sometimes attacked females in an attempt to guard mates from copulating with others, and at Mahale such actions included severe bites to the perineal swell.[239] Nonetheless, the abuse might assist them in testing the quality of potential mates.[240]

While copulating with preferred high-ranking males, Budongo chimpanzees appear to suppress their copulation calls if high-ranking females are nearby to avoid being attacked by them. Further, endocrinological tests have indicated that calling is not related to their fertile periods or likelihood of conception. Accordingly, Townsend et al. concluded that copulation calling might be a female tactic to advertise receptivity to high-ranking males, to confuse paternity, and to secure future support from powerful mating partners.[241] They also cast doubt on the idea that copulation calls are a sexually selected trait that enables females to advertise their receptivity, thereby inciting intermale competition from which they might secure partners that will produce high-quality offspring and practice mate guarding to protect them from infanticide.[242] Moreover, voluntary suppression of copulation calls indicates that chimpanzees can control the vocal apparatus, at least in some contexts.

Findings from the Taï and Kanyawara communities indicate that western and eastern chimpanzee males can detect subtle changes in perineal swells within and between cycles that might indicate potential for fertilization.[243] Such an ability might act as a counteradaptation to female strategies to obscure paternity.[244] As evidenced by the Budongo Sonso and Kibale Kanyawara communities, copulation rates increase when females are most likely to ovulate—during the last week of perineal tumescence.[245] Were females to have estrous synchrony when in proximity with one another, they might have freer choice among mating partners; at least in Mahale M-group, this potential tactic was absent.[246] Indeed, in Taï Forest, when an increasing number of females were concurrently receptive, the alpha male's paternity decreased relative to that of other males.[247]

Watts noted that pairs or trios of top-ranking males in the Kibale Ngogo community commonly engaged in coalitionary mate guarding, wherein they were cooperatively aggressive to prevent periovulatory females from mating with other males, though they tolerated matings of their coalitionary partners. The male coalitions occurred in large mating parties, perhaps because they contained too many males for a single male to maintain exclusive access to the estrous females. Coalitionary duos occurred when the number of males present made it unlikely that either partner could get ≥50 percent of total copulations on his own, and coalitionary trios formed when the possibility of success for a singleton dropped below 33 percent.[248]

Female dispersal from the natal group varies widely among chimpanzee communities, and only one western chimpanzee community exhibited not only female but also male adolescent emigration.[249] The factors that determine female choice are elusive. Perhaps the extended period of adolescent sterility during which females copulate promiscuously provides clues about which males might be the best partners when they are physiologically equipped to conceive.[229] What induces some young females to engage in sexual experimentation outside their natal community where they will know much less about the behavior of the males?[250]

Estrous females prefer males that share food, groom, and spend time with them.[251] Indeed, food sharing, association, and proximity might reinforce the relationship between unrelated males and females apart from female reproductive state.[252]

Only an alpha male appears to gain reproductively from the position that he obtains via energetic displays, coalitions, and occasional fights or coercion. Other males are reduced to

- opportunistic copulation, which apparently is not the most reliable means for impregnation,
- breeding with young nulliparae that commonly are less fertile than mature females, and
- leading prime breeders away from the group, where copulation might lead to conception.

Except for the alpha male, consortship is probably the best reproductive tactic for adult males and females.[253] The major disadvantage to consorting couples is that they sometimes venture to the periphery of the community range and beyond where they could be attacked an extracommunity border patrol. Their quietness, especially regarding pant-hooting, might serve not only to isolate them from their own community but also to avoid detection by others. Another potential risk for a male engaged in a prolonged consortship is that other males might challenge him and he could lose rank when he reenters the group. The alpha male would seem to have the best of both worlds. He enjoys reproductive prerogatives and sustains his status while subordinates deflect potential extracommunity rivals. Inbreeding is generally avoided, but one high-ranking Gombe male mated with his mother, and she birthed his infant.[234]

Studies in the Mahale Mountains indicated that Tutin's model was at least broadly applicable to other communities of Tanzanian chimpanzees. Nishida noted female choice; indeed, he reported that females initiated almost all sexual interactions and frequently the males were unresponsive.[114] Though rare, instances of overt intermale sexual competition were dramatic. But during the initial twenty-two months of study, Nishida witnessed only fourteen copulations between members of K-group and three between other group members.[114]

During a subsequent yearlong study of K-group, a male in the alpha position exhibited possessive behavior wherein he maintained proprietary copulatory rights with estrous females.[254] The two subordinate adult males of K-group engaged in consortships but apparently most of the time their partners were anestrous. After the alpha had been bloodied and usurped by a coalition of the beta- and gamma-males, his sexual privilege was forfeit. His successor became the sexual powerhouse of K-group. The contests that led to the alpha male's fall always occurred when at least one estrous female was present.[254]

A twenty-eight-month study in the Mahale Mountains confirmed that chimpanzees engage severally in opportunistic, possessive, and consort mating. Hasegawa and Hiraiwa-Hasegawa focused on M-group, comprising twelve adult males, forty adult females, and fifty-four youngsters.[255] They also observed K-group, whose potential spermatic contingent had dwindled to a postprime male and a young adult male.

The majority (92 percent) of 660 copulations in M-group were opportunistic, and only 7 percent occurred during possessive relationships. Hasegawa and Hiraiwa-Hasegawa observed only a single consortship.[255] But their data are biased because the consorting couples were away from the feeding station and could not be located in the bush. Opportunistic mating occurred most frequently in large groups, when the alpha male could not monopolize all estrous females in M-group. Nonetheless, he was involved in 75 percent of twenty possessive matings, all with older females. Two other high-ranking males and a middle-ranking male consummated the remaining four possessive matings. Whereas most of the alpha male's possessive bouts lasted ≥2 days (range: 1–7 days), none of the three subordinate males engaged in possessive couplings longer than one day. Further, the subordinates managed possessiveness only when the alpha male was absent or was engaged with another estrous female.[255]

Immature Mahale males usually copulated with young nulliparae and sometimes with adult females that were unlikely to be ovulating at the time.

Mature females were more restrictive in their mating as their perineal swells peaked. Whereas 7 percent (thirty-five of fifty) of matings involving mature females were restrictive, only 1 of 115 copulations by younger females occurred in a possessive relationship. Young females copulated with young males more than older females did.[255]

A decade later, the data set from a study on six of the twenty-seven adult females in Mahale M-group comprised only four possessive matings versus 169 opportunistic ones. The sample subjects ranged from about 13 to 39 years old. Female choice was expressed by acceptance or rejection of male approaches to mate, by maintenance of proximity, and by female grooming the male. As in other groups, females copulated more promiscuously, including with adolescent males, during the early phase of estrus; they copulated with fewer partners, especially adult males, in the later phase of estrus. When fertilization was more likely to occur, they copulated repeatedly with high-ranking males and increased the frequency that they groomed them.[256]

The disintegration of K-group allowed Hasegawa and Hiraiwa-Hasegawa to compare the mating patterns of female transferees with those of resident females.[255] They noted that resident females generally mated more restrictively than the newcomers. Further, the alpha male restricted his possessive mating to resident females of M-group. He differed from an alpha male of K-group that most frequently mated with transferees.[118] Whether this variation reflects personal preferences by the two alpha males or was due to behavior by residents of M-group and K-group toward transferees cannot be discerned from available reports. Other high-ranking males of M-group also seemed to prefer resident females as mating partners.[255]

Mitochondrial DNA analyses indicated that two alpha males had sired five of ten infants in M-group, and other males had low reproductive success. Among the ten offspring, 15.6 percent were paternal half-siblings. Although falling short of average relatedness of half-siblings, average relatedness was significantly higher among adult males than among adult females, which might be due to the fact that one alpha male had long tenure and there was an old male in the group. Female immigrants from various groups maintained low relatedness within M-group.[257]

Facial recognition by captive chimpanzees of unfamiliar individuals and kin in photographs indicated that, in addition to recognizing specific individuals, they could significantly match mothers to sons but not mothers to daughters. Parr and de Waal suggested that this capacity might facilitate inbreeding avoidance because a migrating female should probably not settle in a community in which many males look like her mother.[258] Alternatively, or in addition, phenotypic matching might assist recognition of subsets of related males that might be allies.

In contrast with Mahale M-group, mitochondrial DNA analysis showed that average relatedness within each of three contiguous communities of Taï western chimpanzees and within the Budongo Sonso community of eastern chimpanzees was low, rarely significantly higher than average relatedness of females within a community or of males across communities. Accordingly, kin relationships at Taï might not be the primary cause of interactions among larger numbers of individuals in a community.[259] Contrary to the preliminary findings of Gagneux et al., more refined methodology and reanalysis of mitochondrial microsatellite data challenged the inference that extra-group paternity commonly occurred in Taï chimpanzees.[260]

Fourteen years of observational data indicated that Taï males might recognize their progeny. Although they did not interact preferentially with the mothers of their offspring, they spent significantly more time playing with them. Infants groomed and played more with their maternal siblings; however, with a choice between unrelated, similarly aged partners, the kids interacted with kith instead of kin. Although Taï males did not associate preferentially with the females that birthed their joint offspring, they might benefit because Taï males are generally less aggressive toward females that have birthed. Lehmann et al. concluded that despite the fact that paternal care does not play an obvious role in chimpanzee survival, kin recognition exists in aspects of the life of adult males and youngsters.[261]

The dominant males in three Taï communities sired 50 percent of the forty-eight offspring that were tested via mitochondrial DNA paternity analysis. Further, in a span of fourteen years, during which the Taï North community declined from around eighty to seventy-seven members to twenty-five members, the alpha male's reproductive success decreased from 67 percent when he had few competitors to 38 percent when he had four or more competitors.[262]

Baby Food

During much of our existence, human beings have engaged in infanticide, selective termination, and prevention of pregnancy, in addition to experiencing stressful conditions that induce abortion. Nonetheless, claims for and descriptions of infanticide by wild nonhuman primates shocked some scientists and lay readers, leading them to deem it a social pathology. Others began to search for evolutionary explanations.[263] Unlike the langur species *(Semnopithecus entellus)* that sparked and periodically reignited controversy on the occurrence and causation of primate infanticide, chimpanzees initially added a confounding factor—cannibalism—that was absent among langurs.

After four decades of arguments back and forth, the predominant opinion among evolutionary biologists and some anthropologists favored the sexual selection hypothesis that infanticide is a tactic of males to ensure that they sired newborn members of their group.[264] However, to test the sexually selected infanticide hypothesis thoroughly, one needs a substantive database including the following evidence:

- One can identify individuals in the infanticidal events based on prior study of focal and neighboring groups.
- Males are not killing their own offspring.
- Females become sexually receptive sooner than if they had not lost their infants (i.e., there is a reduction in birth intervals).
- The infanticidal male mated, preferably exclusively, with mothers in the group when ovulation occurred.
- Offspring of the male produced succeeding surviving generations of offspring, thereby evidencing his fitness. Accordingly, the behavior would indeed be part of an evolutionarily stable strategy—that is, it is fixed in the population, and only natural selection is sufficient to prevent alternative (mutant) strategies from invading successfully.[265]

This is a tall order, indeed one virtually impossible to fill. Noninvasive techniques to test for paternity and other kin relationships are a boon, but observations are never complete enough spatiotemporally to provide the behavioral data. Infant mortality, abortions due to harassment and other causes, anthropogenic perturbations of the focal study populations, political events, or conflicting personal demands on observers too often cause seasonal and longer disruptions in the continuity of observers and data sets.

Some studies seem to confirm predictions of the sexually selected infanticide hypothesis in chimpanzees, but other explanations such as the range-expansion hypothesis seem equally probable for others.[266] Confounds like female killers, female victims, and a male cannibalizing infants that he might have sired offset the development of an evolutionary stable strategy that is based solely on the actual behavior of males.[267] Finally, demographic factors in chimpanzee populations must be taken into account, particularly patterns of mortality. For example, over a thirty-five-year period in Mahale National Park, 50 percent of M-group died before weaning, and two samples of females gave birth to a mean of 3.9 and 2.7 offspring but weaned only 1.4 and 2.0 of them, respectively. The major cause of death in the population was disease (48 percent), followed by old age (24 percent) and intraspecific aggression, including infanticide. Anthropozoonotic disease epidemics might have been involved, as occurred in other African ape populations.[268] It is unfortunate that genetic pedigree data are not available for Mahale M-group over the thirty-five-year period so that one could sort out the long-term effects of infanticide vis-à-vis the fitness of the perpetrators.

Saturnalia and Sharing in Bonoboland

> Among hominoids, only humans and bonobos have been verified to make peace with different groups.
>
> TAKAYOSHI KANO, LINGOMO-BONGOLI, GEN'ICHI IDANI, and CHIE HASHIMOTO (1996, p. 69)

For nearly a half-century after bonobos were identified as a distinct species, the nature and adaptive meaning of their social organization and sexual behavior remained sketchy. Field observations at several localities were interrupted periodically by warfare and political turmoil in the DRC, and bonobos are the prey of poachers in some areas.[269]

Early studies near Lake Tumba were unproductive because researchers had brief or no direct contact with bonobos. Indigenous hunters' reports indicated that they often occurred in groups of fifteen to forty individuals, but sometimes traveled in pairs or alone.[270] Conversely, based on occasional sightings during a two-year period, Horn reported that Tumba bonobos lived in small groups, the largest of which comprised an adult bisexual pair with two youngsters. He also saw a lone male and a group of three, comprising an adult bisexual pair with a youngster. He commented that they probably formed larger groups occasionally.[271]

In 1973, Nishida and Kano surveyed the Salonga National Park in central DRC.[272] Based on large clusters of nests and fourteen encounters with bonobos, Kano posited that *Pan paniscus* live in groups larger than those of *Pan*

Figure 11.1. Female bonobos engaged in frottage. Note bottom participant is eating.

troglodytes. Though he saw up to 22 bonobos in a group, he guesstimated that groups comprised ten to ninety individuals.[273]

Alignment of Venus and Mars

Establishment of a base for long-term studies near Wamba village in northern DRC inaugurated an era of major adances in knowledge of free-ranging bonobo social and sexual behavior.[274]

Like *Pan troglodytes,* Wamba *Pan paniscus* had unit-groups of 50 to 120 individuals that usually foraged in smaller subunits (mean: 17 individuals; range: 2–54 individuals; $n = 147$ subunits).[275] Lone individuals were rare ($n = 10$). Small foraging subunits ($n = 10$ individuals) were usually quiet, nonvocal, pacific toward one another, and sexually abstemious. They became excited and vocalized in response to larger parties; these were clamorous, demonstrative, and given to sexual frenzies that were more dramatic than reunions of Gombe chimpanzees. However, excited bonobos tended to be less aggressive than Gombe chimpanzees.[276]

Nearly 75 percent of 163 subunits were mixed bands according to Reynolds and Reynolds.[103] Kuroda suspected that many of an additional 10 percent of groups of uncertain composition were probably also mixed bands. Both small and large groups were bisexual. The smallest mixed band comprised a male, an anestrous female, and an infant.[276] Other kinds of subunits were rare at Wamba: only four all-adult bands of three to thirteen individuals, three of which included two or three estrous females and four to six males. The only copulation Kuroda witnessed occurred in the largest adult band.[276]

Individual identification and observations of Wamba bonobo terrestrial behavior were facilitated via provisioning some members of a unit-group with sugar cane. Kuroda's successor, Kano, confirmed his conclusions about bonobo sociality.[277] Eighty-five percent of 156 encounters were with mixed bands. Adult bands ($n = 2$), mother bands ($n = 1$), and loners were rare. Estrous females were absent from adult and mother bands. The lone individuals were males, one of which was sometimes with mixed groups.[278] In a later study on provisioned bonobos, Kano reported that out of 180 sightings, 96 percent were of mixed bands; the others were male bands ($n = 2$) or lone males ($n = 6$).[275]

Affinity between adult males and females and among females was apparent from allogrooming patterns: 54 percent of eighty-four adult bouts were between males and females; 34 percent were interfemale, and 12 percent were intermale. On average, partners switched roles three or four times; some alternated up to 10 times per session. The highest number of grooming sessions ($n = 102$) involved mothers grooming their infants and young juveniles nonreciprocally.[278]

High-ranking adult females received the most allogrooming in each of four European captive groups ranging in size from six to eight individuals and including two to three adult females and one to two adult males. Franz included subadult males and females and a juvenile, but excluded infants from her analysis.[279] Grooming competition is positively correlated with dominance rank; the highest-ranking females performed the most displacements over allogrooming. However, within adult female dyads, allogrooming was not clearly associated with rank. Overall, the captives significantly directed the most allogrooming to the face (Plate 15). High-ranking individuals and females directed most grooming to the face and shoulders of their partners, while subordinates and males tended to avoid the face and focused instead on the back and anogenital region of partners.[279]

Propinquity while feeding, food-sharing, and frottage—consensual interindividual genital rubbing (Figure 11.1)—also suggested strong affiliations among bonobo females. Whereas Wamba females usually aggregated peacefully in large food trees, males bickered and dispersed more widely; ultimately, a choice resource would contain only one or a few males with many females.[278]

Observations of two Wamba unit-groups indicated that dominance relationships among males were greatly affected by bonds between mothers and sons. Unlike chimpanzees, Wamba males rarely united in aggressive actions, but males of the E-2 group engaged in agonistic and affinitive interactions more often than those of the E-1 group. Whereas E-1 males were spatially divided into several clusters, E-2 males were more cohesive (Plate 13). Ihobe suggested that male bonobos might achieve coexistence by decreasing inter-male intragroup and intergroup competition.[280]

Based on observations over a longer period of time and the advantage of more comprehensive individual identification, Furuichi agreed that the relative dominance of a female among females and her presence in a Wamba group affected the status of her son such that he could have high dominance only if he was the son of a high-ranking female.[281] Consequently, he could remain in the center of mixed parties and have greater access to females, which would advantage the fitness of both. If her status among females changed, his would also.[281]

This observation might explain the situation at the Apenheul Primate Park, Apeldoorn, Netherlands, where a bonobo group exhibited fairly non-linear hierarchy, albeit with an alpha male that engaged in significantly more copulations than the other two males.[282] Because the males and five females were taken from different Congolese localities, they were probably not related. The absence of strong mother-son relationships could have contributed to lack of a clear hierarchy among them.

At Wamba, about 61 percent of interadult food-sharing was interfemale. Behavior during begging and sharing was predominantly friendly, though not totally free of tension.[278] They probably alleviated tension via frottage, which occurred most frequently during the first five minutes after arrival at a feeding locality.[283] Thereafter, it decreased except when begging clusters formed. When a female joined a group, she serially solicited or was solicited by others to embrace ventroventrally whereupon they rapidly and rhythmically rubbed their pudenda together sidewise for spans of a few seconds to >60 seconds.[284]

In the Wamba region, there were six unit-groups of bonobos, one of the largest of which—the E[langa]-group—was provisioned.[285] Kano identified all sixty-three of the group's mature and immature members. Except for two adolescent females that moved between unit-groups, individuals associated peacefully with members of a single unit-group. E-group was composed of fifteen adult males (>13 years old), sixteen adult females (>13 years old), five adolescent males (8 to 13 years old), eight adolescent females (8 to 13 years old), nine juveniles (2 to 7 years old), and ten infants (<2 years old).[286]

In 1976, E-group appeared to have northern and southern subunits that periodically fused and fissioned until 1984, when they ceased to fuse and thereafter constituted two groups: E-1 and E-2. In the initial early period of reunions, relationships were replete with affinitive behaviors, but in the latter period relationships were increasingly agonistic.[286]

The ranges of neighboring unit-groups overlapped, especially where food was plentiful.[275] When bands from different unit-groups came into audible range, they avoided one another; this reaction contrasted with parties of the same unit-group that generally merged noisily within one to ten minutes of mutual awareness. Direct agonistic engagements were usually averted. Vocal exchanges probably allowed band members to determine whether another band belonged to their unit-group and, if not, whether the rivals might be supplanted versus avoided.

Idani provided a more varied account of unit-group encounters.[287] He frequently observed members of different unit-groups intermingling at provisioning sites and in the forest, during which time they engaged in frottage, copulation, and peering. Also transfers of nulliparae occurred, and aggressive interactions were rare. He concluded that bonobos might have a regional society above the level of unit-groups.[287]

The average size of mixed bands at the Wamba sugar cane field was nineteen ($r = 5$–37; $n = 172$), which is not much larger than bands of nonprovisioned bonobos.[275] Both sexes were nearly equally represented in mixed bands, except those that had fewer than ten members. Virtually all (98 percent) of them included at least one estrous female. Indeed, around 25 percent of members in most mixed bands were estrous females.

Kano recognized four bisexual subunits of E-group, whose members tended to forage and to travel together in particular sections of the community range; but Kitamura, whose study partially overlapped that of Kano, identified only two primary subunits in E-group.[288] E-group females generally ranged less widely than some of the males. Accordingly, Kitamura proposed that female clusters are the focal points of the bisexual groups.[288] Based on kinship inferred from facial likeness, Kano added that in addition to the coteries of females, cliques of males and mothers with sons might serve as cores of various subunits in bonobo communities.[289]

Subsequent observations on Wamba E-group indicated that bonobo unit-groups tend to form one large mixed aggregation comprising most of its members, including all females regardless of their estrous state.[290] Older females, which generally have high rank, occupy the center of the group, where their adolescent and adult sons are located more often than the sons of low-ranking or deceased females. Older juvenile and early adolescent females visit neighboring unit-groups and finally settle in one of them.

The immigrants seek social interaction with senior females, probably to secure high status among them and perhaps to be available as breeding partners with their adult sons. A young immigrant female would select an older resident female that she would follow frequently, groom, and engage in frottage. The immigrant female was the more common initiator of the affinitive behaviors that facilitated her integration into the unit-group. Resident males approach and mate with immigrant females, which appears to cement their relationships with them.[291]

Furuichi has speculated that cohesiveness of the multimale/multifemale bonobo unit-group and the prominent mother-offspring subunits, which are unique among apes, might be "a feasible model of the basic society from which human society evolved."[292]

Like Gombe, at Wamba provisioning increased the frequency of intragroup agonistic behavior, which was most frequent between males.[293] Nonetheless, unlike chimpanzees, bonobo males merely engaged in displays toward low-ranging males and supplanted adolescents from feeding sites; they rarely launched a severe physical attack.[294] Some individuals were excluded from the cane field, but multiparae boldly approached the feeding spots of high-ranking adult males and supplanted them with no overt aggression. It would be interesting to know whether they were closely related to the supplanted males or had powerful sons in the vicinity or both.

In a secondary forest at Yalosidi, near the southeastern limit of the range of *Pan paniscus,* the society is basically like that at Wamba. Mixed bands constituted 70 percent of foraging groups (mean: 8.5; range: 3–21 individuals). The ratio of females to males was 1.6.[295] There also were lone males ($n = 5$), a pair of males, two bands of subadult males, and two mothers with youngsters. Like Wamba males, Yalosidi males usually fed in separate trees, while females were gregarious and gathered near a few individual males; for instance, occasionally two males were in a fruit tree with females and youngsters, while others dispersed nearby.[296]

Intragroup relations were generally peaceful. During 107 allogrooming sessions at Yalosidi, bisexual dyads were most common, followed by female pairs. During the average six- to seven-minute bout, partners changed roles several times. The three longest reciprocal allogrooming bouts (forty-five to seventy-seven minutes) were between males and females; there was no adult male grooming dyad.[296] Juveniles often played while the adults fed. One juvenile entered the day nest of a male and stayed with him for an hour, during which they groomed one another reciprocally. Interfemale frottage occurred three times.[296]

At the northernmost locality where bonobo sociality has been studied—Lomako Forest—observers eschewed provisioning.[297] Initially, the Badrians sighted mostly (80 percent) small groups of two to four individuals and singletons.[298] Small groups were bisexual, like those in the Lake Tumba area.[299] Subsequently, researchers found that, socially and sexually, Lomako bonobos were remarkably like those at Wamba, except they foraged in smaller groups.[300]

Two unit-groups, with overlapping ranges, lived at Lomako: B[akumba]-group had around fifty members, and E[yengo]group was incompletely censused. Vocalizations facilitated mutual avoidance (Chapter 12).[300] The modal group of Lomako bonobos was two to five individuals ($n = 268$; mean: 7.6; range: 1–50). Bands of six to ten individuals were second most common. Indeed, 75 percent of encounters were with loners and bands of up to 10 individuals. Whereas larger groups vocalized and displayed toward other bonobos and humans, small bands were relatively quiet in the forest.[300]

Like Wamba, in Lomako Forest bisexual groups were the norm. They tended to contain approximately equal numbers of adult females and males, but as group size increased, females outnumbered males. Sixty-eight percent of 191 completely censused bands were mixed, and 8 percent were bisexual adult bands. Most (60 percent) adult bands consisted of adult pairs, seven of which included an estrous female. Five bisexual adult bands had two adult males with one to three females, many of which were estrous.[300]

Seven of nine Lomako male bands were dyadic, and the other two had three members each. Female and mother bands were also infrequent (5.2 percent of encounters) and small ($n = 2$–5 members). Males were alone on twenty-four occasions, and twelve adult females, five of which had dependent youngsters, were apart from other adults. Some lone bonobos were probably laggards from silently retreating bands. Vigilant males were especially inclined to lag.[300]

Although the composition of Lomako bands and other groupings varied according to fruit availability, some adults evidenced notable cohesiveness.[301] For example, a band of two males, one female, an infant, and a juvenile were together for more than two weeks.[302] As at other localities, reunions of large bands were clamorous and involved displays and sexual behavior more often than occurred during encounters between small bands at food trees.[300]

Allogrooming usually occurred during rest periods and was more frequent in groups of ten to twenty individuals than in smaller ones.[300] Half of seventy-two sessions involved mother/youngster dyads. Males rarely (1 percent of sessions) groomed youngsters. Adult males and females formed the next most common grooming dyads (25 percent). The former served as groomers more often than the latter did. Copulation was a common component of bisexual allogrooming sessions. Although the longest session (125 minutes) involved a male dyad, females (17 percent of sessions) groomed one another more often than the males did (7 percent of sessions).[300] Females and youngsters acquired bits of a young bushbuck from males that had captured it, but the possessors did not share it voluntarily.[303] Females copulated with the meat-possessing males and indulged in interfemale frottage while consuming meat.[300]

Data from Yalosidi adds to the data sets from other sites to underscore that the core of bonobo society consists of strongly bonded females and the males that associate with them.[304] Clearly, bonobo society is unique among ape societies and should not be considered a mere variant of chimpanzee society or an amalgam of chimpanzee and gorilla social patterns.[300]

Via a two-year study at Lomako, comprising observations and statistical tests with data on interactions, party (the term of Van Elsacker et al.) composition and proximity of nearest neighbors, White and Burgman further confirmed that, in contrast with androcentric chimpanzees society, bonobo society is characterized by strong affiliation among females and between males and females, but not among males.[305] Party composition varied with party size. On average, females outnumbered males, but the proportion of males increased as party size increased. Cohesion between males and females appeared to increase with party size; in contrast with relatively common parties of females without infants, White observed no all-male bands.[306] Because bonobos feed more frequently in larger trees than the eastern chimpanzees do, bonobo females can afford to be in larger parties and associate with other females, whereas the higher cost of feeding competition constrains sociality among female chimpanzees.[307]

Lomako males were dominant to females in dyadic interactions, but they deferred to female coalitions in nondyadic interactions. Female feeding priority in small food patches (but not large ones) with male dyadic social dominance implied that male deference during feeding could not be excluded as a factor in interpretations of bonobo female dominance.[308]

A study of six captive bonobo groups in Europe indicated that although females occupy the highest-ranking positions in the dominance hierarchy, in each group mature males held rather high ranks and were able to dominate at least one mature female. Because females could not consistently evoke submission from all males in all contexts, Stevens et al. concluded that bonobo hierarchies are based on nonexclusive female dominance.[309] The hierarchies of all six groups were highly linear, but they were steeper among the males. On average, the males were more despotic than the females, but females also exhibited despotic relations with males and other females. Accordingly, they suggested that captive bonobos are semidespotic instead of egalitarian.[309] Jaeggi et al. further noted that captive chimpanzees shared food more frequently, more tolerantly, and more actively than captive bonobos. Chimpanzees shared reciprocally, but bonobo transfers were mostly unidirectional. Overall, the bonobos had a steeper, more linear dominance hierarchy, indicating greater despotism, and they groomed less reciprocally than the chimpanzees did.[310]

Observations of three adult female and three adult male bonobos at Plackendael Wild Animal Park indicated that coalitions functioned to maintain existing ranks, to acquire ranks, to reduce tension, and to test or to strengthen bonds. Mutual support during polyadic agonistic interactions probably maintained the power of the two highest-ranking females over the males, but dominance relationships among the females were based on individual attributes—physical strength or aggressiveness—instead of coalitions.[311]

Field observations of bonobos at LuiKotale, Salonga National Park, DRC, tempered the prior generalizations about maternal influence on the social status and mating success of their sons.[312] Focal individuals in the thirty-three to thirty-five member study group comprised five adult and four adolescent males and eleven adult and ≥5 adolescent females. The community had a stable linear dominance hierarchy. Mothers and sons had high association rates, and the presence of a mother enhanced a male's proximity to and mating success with estrous females. High-ranking males copulated more with females in advanced stages of perineal swelling, and low- and mid-ranking males tended to have greater copulatory success when their mothers were present. Indeed, the highest-ranking male in a party performed 40.8 percent of copulations with maximally tumescent females in parties with no mother present versus 25 percent with all mothers present. The nine LuiKotale males did not form coalitions; instead, mothers supported their sons in agonistic contests with other males. Only six of the nine males in the LuiKotale community had mothers present. Nevertheless, the third- and fourth-ranking males in the steep linear male hierarchy had no mother present.[312]

Comparison of tropical forest-dwelling Kibale Kanyawara chimpanzees with Lomako bonobos further underscored interspecific differences in their social relationships and organization. Lomako bonobos had smaller nearest neighbor distances than those of Kanyawara chimpanzees, and the latter had a more restricted range of nearest neighbor distances. In particular, chimpanzees generally avoided situations of very close proximity. Male chimpanzees frequently had another male as nearest neighbor, but male bonobos rarely did. Conversely, chimpanzee females tended to move apart, but in bonobos males did the moving.[313]

When feeding, Lomako bonobos maintained distances of around 5 meters in a large range of tree sizes, whereas chimpanzees increased nearest neighbor distances in large trees. Aggregation increased while they rested, but they were more dispersed during travel. Female frottage occurred, usually at the beginning of feeding bouts, at plentiful food sources but not where food was limited.[314]

To maintain proximity with females, Lomoko males appeared to be attracted to parties that were based on larger cores of regularly associating females, thereby increasing the proportion of males in them. Food patch size and large patches with abundant food attracted the larger parties.[315]

Although Lomako and Wamba bonobos exhibit many behaviors and overall social structure in common, the differences are noteworthy in relative importance of affiliation between males and females versus affiliation among females. The average party size at Wamba ($n = 16.9$ individuals) was greater than at Lomako ($n = 6.2$ individuals), and at Lomako the male-female associations were more significant than the female-female associations only in large parties, White suggested that the differences in bonobo social organization might reflect party sizes at both sites.[316]

The challenge then was to discern the factors that affected party size and composition. White reasonably suggested that differences between the two habitats, the history of provisioning, and human predation most likely accounted for the variances in social organization at Lomako and Wamba. Wamba bonobos were provisioned with sugar cane and pineapple; Lomako bonobos were not. Provisioning sites could be comparable to sizeable natural food patches that accommodate large parties. Lomako bonobos compete for forest resources with humans, who also kill them, which would directly reduce party size; the Wamba bonobos were not hunted.

Habitat features eluded the theorists who might employ them to explain variations in party size not only between bonobo unit-groups but also between bonobos and chimpanzees. The Badrians and Wrangham hypothesized that year-round abundance and wide distribution of terrestrial herbaceous vegetation (THV) at Wamba reduced feeding competition such that female bonobos could consistently feed together when arboreal fruits were not abundant.[317] On average, THV constituted 33 percent of Wamba bonobo monthly food intake, but only 7 percent of monthly food intake by Gombe chimpanzees. Nonetheless, the idea was short-lived. At Lomako, athough THV was ubiquitous, the bonobos were quite picky about the species and plant parts that they would eat—the highly proteinaceous larger stems of *Haumania liebrechtsiana,* whose occurrence was patchy.[318] Further, based on a larger data set on average party size in chimpanzees (range: 2.6–10.1 individuals) and bonobos (range: 5.6–16.9 individuals) Chapman et al. realized that intraspecific variance could be as large as interspecific variance between study sites.[319] Consequently, they proposed, "that discussions of average party size help little in understanding the differences in bonobo and chimpanzee social organization and behavior."[320] Instead, White suggested that the close tie between female bonding and ecological parameters in bonobos, together with the ecological differences between them and chimpanzees, indicate that social differences between them might be related to greater variability in feeding competition among chimpanzees versus more stable levels in bonobos.[321]

Analyses of nuclear and mitochondrial DNA from feces and behavioral observations on individually identified members of the Eyengo community further clarified social relationships and structure among Lomako bonobos.[322] Although copulation occurred opportunistically and promiscuously both within and outside the community, the two most dominant Eyengo males had the highest paternity success, and resident males sired most Eyengo infants. High association indices were most consistent for mother-son dyads and between half-siblings, and there was no indication of mother-son mating. Accordingly, affinity between male and female bonobos is not always based on sexual attraction.[323] The genetic study also confirmed a pattern of male philopatry and the dispersal, indeed the frequent migration, of young females. Most adult and subadult Eyengo males had a matching Eyengo mother, but females did not. Gerloff et al. agreed with Furuichi et al. that the high degree of sociality and cooperation among resident female bonobos is more likely attributable to mutualism or reciprocity than to close genetic ties.[324]

Although Hohmann et al. confirmed the spatial relations among the resident females, they emphasized the stronger bonds between some males and females.[325] Indeed, while most interfemale associations lasted only one field season, the long-term associations were predominantly male-female dyads that involved both close kin and unrelated individuals. Nonetheless, allogrooming was related more to spatial association than to kinship. During all four field seasons, the largest proportion of close associations was between males and females (>50 percent), followed by female dyads (25 to 41 percent) and male dyads (4 to 19 percent). Close associations between unrelated individuals were short-term, opportunistic, and less frequent than expected, and grooming bouts were longer and more reciprocal between close associates than between random associates.

Hohmann et al. also underscored the previous suggestions that males might benefit from bonding with females by increased reproductive success via rank acquisition, whereas females might benefit from inclusive fitness, reduced food competition, and perhaps protection against male intruders.[325] The portrait of bonobos as less severely aggressive than chimpanzees also requires further study, given that whenever bisexual parties from the Eyengo community encountered two alien males nesting or feeding nearby, the males and females would attack them, with the males being the more severe aggressors.[325]

During the period of instability induced by the alien males, the alpha male had the highest circulating testosterone levels among seven males. The average testosterone levels of five resident Eyengo males reflected their relative dominance status, though there is no correlation between testosterone and dominance rank across all males; the testosterone levels of the two immigrant males were higher than those of the three lowest-ranking Eyengo males. Marshall and Hohmann concluded that it is unlikely that the relationship between dominance and testosterone is as direct in bonobos as it is in chimpanzees; dominance rank in bonobos reflects not only a male's competitive ability but also the rank of his mother and perhaps his relationships with other females in the community.[326]

Bonobo Sex Made Simple

At Wamba, estrus was the usual state of postpubertal females, and adolescent females were receptive and attractive to males during most of the menstrual cycle. Parous females resumed cycling within a year after delivery even though they were nursing infants. Approximately 25 percent of observed copulations involved females with youngsters clinging dorsally or ventrally.[327]

Because individual mating histories were not available, Kano's narrative might give the impression that promiscuity prevailed. For example, when two parties met, males copulated with both newly arrived females and females that had been accompanying them.[328]

Forty percent of 121 copulations occurred during feeding periods, 35 percent during travel, and 26 percent during rest periods. There was evidence for restrictive mating—that is, females that copulated at least twice a day with a single male—but females also copulated with more than one male per day.[278] Nearly two-thirds of adult copulations were dorsoventral, 35 percent were ventroventral, and 3 percent commenced canine and climaxed missionary. Seventeen copulations paired adults with youngsters. Immature females ($n = 2$) were not penetrated. Immature males ($n = 15$) achieved intromission with adult partners. The relationships of the couplers were unknown, so the state of incest cannot be assessed.

Pregnant females and mothers with neonates were inactive sexually, but the females with ≥3-year-old infants copulated as frequently as the females without infants.[329] Furuichi noted that copulation was mainly restricted to the phase of maximal perineal swelling.[330] The average interbirth interval was significantly shorter in Wamba bonobos (4.8 years) than in Mahale chimpanzees (6 years) but not Bossou chimpanzees (4.6–5.1 years).[331] Infant mortality was lower in Wamba bonobos than in wild chimpanzees, probably because of the greater abundance of fruit and herbaceous foods, female priority of access to food patches, and the absence of infanticide.[332] Moreover, bonobo mothers were more tolerant than chimpanzees about the attempts of their infants to suckle, and their infants were less likely to interfere during copulation.[333]

Data from 3.5 months of observations at Wamba feeding sites indicated that the dominance rank of a male bonobo affects his chances of mating. Copulation frequency was weakly and copulation rate was positively correlated with male rank, though males could increase rank within groups smaller than the unit-group by associating with individuals that were not highly dominant. In small parties, the dominant male had clear mating priority. Three males whose mothers were alive in E-group were dominant, so they copulated more frequently and at higher ranks than three others whose mothers probably were not with E-group. Adolescent males were subordinate to them but had higher mating frequencies.[334]

Adult male copulation rates at Wamba equaled those of Mahale and Gombe adult male chimpanzees, though adolescent male chimpanzees copulated more frequently than adolescent male bonbos.[335] Copulation rates of adult female bonobos and chimpanzees were approximately equal throughout the swelling cycle, but they were much higher in bonobos throughout the interbirth interval. Unlike Wamba adolescent females, those of Mahale M-group rarely copulated with adult males. Considering that adult male bonobos solicited young females, Takahata et al. concluded that they were particularly attractive to them.[335]

In wild bonobos, sexual coercion of females is subtle, if it exists at all. At Wamba, males initiated most copulation attempts by approaching or engaging in courtship behavior—hand raising, body swaying, or phallic display—when they were over 5 meters from females. The females approached or exhibited courtship behavior only when males solicited them within 5 meters. Most copulation involved females with perineal swells because they accepted male solicitations more readily than when their perinei were not swollen. Nonetheless, one-third of male solicitations to copulate involved females lacking swollen perinei, and half of those were consummated.[336] Males initiated 96 percent of copulations in bonobos versus 63 percent in chimpanzees.[337] Furuichi and Hashimoto suggested that the attractiveness and receptivity of females in the nonswollen perineal phase might have evolved to mollify intermale sexual competition and provide higher social status for females among bonobos.[338] They further suggested that, via higher proceptivity and copulation rates, female chimpanzees might benefit more from a high frequency of copulation than the female bonobos would.[339]

Data are not available to assess the relationships between male dominance rank, mating, and reproductive success in forest bonobos, but the genetic relationships among three sexually mature males in a group of ten captives at Twycross Zoo, Leicestershire, United Kingdom, indicated that the dominant male did not sire the most offspring. Indeed, lower-ranking males sired both infants that were conceived during the study. Further, although the top male had almost exclusive mating access to one of the females when her perineum was swollen maximally, the lowest-ranking male sired her infant. Because he mated with her when her perineum was not swollen maximally, Marvan et al. suggested that sperm competition was operant, ovulation is decoupled from the phase of maximal perineal swelling in bonobos, or both factors might be operant, thereby allowing greater female choice of mates.[340]

Observations from an eighteen-month study in Lomako Forest complemented information on bonobo sexual behavior at Wamba.[341] Because the Lomako bonobos were not habituated to observers, individual mating histories are not available. Most mating occurred during the morning feeding peak (07:00–09:00 AM).

Most (84 percent) copulations were performed by adult males, followed by infant (8 percent), juvenile (7 percent), and adolescent (1 percent) males. Whereas young males mated ventroventrally (83 percent of twelve cases), usually with their mothers, adult males rarely (14 percent of fifty-eight copulations) mated ventroventrally. Seven of the eight adult ventroventral couplings were with the mothers of dependent youngsters.[341]

Lomako females with partially swollen perinei copulated most frequently (60 percent of fifty-two cases), and those lacking swells did not copulate. The second most common female copulators had early swells (27 percent); fully tumescent individuals were the least common copulaters. Mothers engaged in 44 percent ($n = 32$) of seventy-three copulations.[341]

Early observations allowed researchers to entertain the idea that, like chimpanzees, bonobos might engage in temporary consortships.[342] The following observations recommended the hypothesis. Researchers regularly encountered bisexual adult pairs in the forest. The fact that the female partner commonly had dependent youngsters did not argue against consortship because often bonobo mothers are estrous and willing to copulate while their youngsters cling to their bodies. The apparent relatively low frequency of copulation by Lomako females in the peak of estrus might have been due to their absence from observed groups in order to engage in consortships. Although copulation is common at feeding sites, that which counts most could be occurring away from the promiscuous crowd, which includes not only libidinous males but also females engaged in frottage. Recall that it was only after observers could follow known individuals that the importance of chimpanzee consortship was discerned at Gombe. At Wamba, only opportunistic copulations seemed to be initiated by males.[335]

Thompson-Handler et al. not only confirmed the universality of interfemale frottage among bonobo females but also observed several instances of genital contact between males.[341] Interfemale frottage ($n = 25$) was fivefold more frequent than intermale genital contact ($n = 5$). Most (64 percent) frottage at Lomako occurred during feeding bouts, but it also occurred during travel reunions (8 percent) and play between infants (12 percent).[343] The latter indicates that interfemale frottage, like the heterosexual behavior of male bonobos, begins during infancy. Mothers do it with their daughters, and clinging infants were commonly sandwiched between females engaged in frottage. Immature individuals were one or both partners in 36 percent of female sessions of frottage; infants accounted for somewhat more than half of them.

Mothers and other adult females with partially swelled perinei were the most frequent participants in interfemale frottage. They constituted one or both partners in 60 percent of bouts. Fully tumescent females engaged in 20 percent of bouts, and those with early swells were involved in 24 percent of bouts.[343] At Wamba also, frottage occurred mostly between females with submaximal swellings. Further, exclusivity of partners in frottage was absent.[335]

Interfemale frottage probably serves multiple functions in bonobo society.[344] It appears to stimulate heterosexual copulation in a group by piquing the interest of males.[341] On several occasions, it was interspersed with copulation. Moreover, it might alleviate intragroup tension.[283] For instance, interfemale frottage and heterosexual copulation occurred during a meat-eating episode at Lomako.

Systematic observations over two months on captive bonobos during the formation of a group from diverse backgrounds comprising four adult males, four adult females, one with a female infant, and a juvenile female at Apenheul Primate Park support the sex-for-social-bonding hypothesis.[345] From the outset of assembly in an exhibit cage, there was a high level of sexual behavior ($n = 1,589$ occurrences)—interfemale and much less intermale frottage, copulation, manual-genital and oral-genital contact—versus only seventy occurrences of aggressive displaying, chasing, displacement, hitting, grabbing, or biting. Aggression occurred particularly between individuals that had been housed together before assembly. Most sexual couplings were between unfamiliar individuals at feeding time, which indicates that they also employed sex to alleviate tension. Males engaged predominantly in heterosexual acts, while females more often engaged in frottage than copulation.[345]

Data from six field seasons at Lomako strongly support the inference that interfemale frottage serves multiple functions. Among five possibilities derived from studies of interfemale mounting among other mammals, birds, and insects—reconciliation, tension regulation, expression of social status, social bonding, and mate attraction—the first three were operant among Lomako bonobos. The reconciliation hypothesis is supported by the fact that rates of postconflict frottage exceeded preconflict rates. The tension-relief hypothesis is supported by high rates of frottage when food could be monopolized and tension was high, though it was significantly more frequent between bystanders and owners of fruit than among bystanders that were being denied access to it. Rank-related asymmetries in initiation and perfor-

Figure 11.2. Male bonobos touching rumps.

mance of frottage support the social status hypothesis: although low-ranking females solicited frottage more often than high-ranking females did, the latter were more often the mounters than the ones being mounted.[346]

Intermale contacts that might be glossed "homosexual" by casual human observers are very rare in the forest. During eight years of observation at Lomako, Fruth and Hohmann noted twenty-two mounts with pelvic thrusts, rarely involving intromission, one ventrodorsal anogenital copulation, two instances of frottage with lateral movements resembling interfemale frottage, and two rump-to-rump contacts (Figure 11.2).[347] Reports of intermale fellatio and masturbation are more common in captivity than in the wild, perhaps because some individuals were juveniles.[348]

In conclusion, bonobos do not merit the epithet pansexual apes. Further, the variety of acts that have been labeled homosexual hardly matches that achieved by gay, lesbian, and heterosexual human lovers. Many interactions that appear to be sexually motivated are in fact greetings, modulations of social relations, and reaffirmations of group cohesion, much as the butt-pats, hugs, and kisses among politicians and athletes are essentially asexual, at least on the playing field.[349]

Rewards and Risks in Playing King of the Mountain

The first scientific accounts on the social grouping and behavior of free-ranging gorillas in natural habitats focused primarily on *Gorilla beringei*.[350]

Mountain gorillas continued to be foci of longer-term studies in Rwanda and Uganda, and Grauer's gorillas *(Gorilla gorilla graueri)* became subjects in DRC during the next fifty years.[351] Concurrently, research teams inaugurated long-term studies of western and Cross River gorillas (*Gorilla gorilla gorilla* and *Gorilla gorilla diehli*) at sites in West and West Central Africa.[352]

Western Challenges

As of 2003, about 80 percent of the world's gorillas and most of its chimpanzees lived in Gabon and the Republic of the Congo.[353] Between 1983 and 2000, half the apes in Gabon fell victim to commercial hunters due the rapid expansion of mechanized logging. The spread of Ebola hemorrhagic fever, killing gorillas for meat and infants, shifting agriculture and logging activities, and oil prospecting also disrupt and often reduce populations in Central and West African forests.[354] Further, most gorilla and chimpanzee populations in West Africa live precariously in forest isolates, and some populations suffered heavily when they were prey of local hunters and foreign collectors who killed them for museum specimens and captured them for zoo exhibits and research.[355]

An early census indicated that in Gabon the western gorilla groups averaged four weaned nesters (range: 1–19 individuals).[356] These figures are probably lower than actual group sizes. After Tutin et al. obtained actual numbers for several groups, they found that only 34 percent of 137 nest sites of one group actually corresponded with the number of weaned individuals in the group.[357] Indeed, during a 6.5-year focal study, Tutin et al. identified eight groups ranging from four to sixteen individuals (median: 10) in the Lopé Reserve.[358] Six of the groups had one silverback, and the largest groups (*n* = 15 and 16) each accommodated two fully mature males that might have been father and son. Lone silverbacks also roamed the forest. There was notable overlap between the ranges of three focal groups. Despite being more frugivorous than Virunga mountain and Kahuzi Grauer's gorillas, Lopé gorilla group composition and median size were similar: nine, twelve, and ten individuals, respectively. However, because fruit sources are widely dispersed, their range sizes and daily travel distances are probably greater than those of Virunga gorillas, as they proved to be at Mondika in the Dzanga-Ndoki National Park, Central African Republic.[359]

During eleven years at Lopé, Tutin et al. documented only twenty-two encounters between groups and sixteen between a lone male and a group, ≥3 of which left evidence of fighting.[360] An old silverback appears to have been killed during one of them. Nonetheless, some groups seemed to tolerate the proximity of others, and rare intergroup agonism was generally over food sources instead of to acquire females. Unlike chimpanzees, there is no evidence that Lopé gorillas engaged in fission and fusion of subgroups.

Based on direct observations and estimates from nests in a lightly hunted area and an anthropogenically undisturbed area, Mitani inferred that western gorillas in the Ndoki Forest of the Republic of the Congo forage in small groups of fewer than three individuals as part of a fission-fusion society like that of chimpanzees.[361] Contrarily, based on complete counts for five groups, Nishihara determined that Ndoki gorillas lived in groups of six to ten individuals (mean: 7.8), each group comprising one silverback, one or more adult females, immature individuals of various ages, and perhaps one or more blackback males.[362] Subsequent studies at Mbeli Bai in Nouabalé-Ndoki National Park support the integrity of western gorilla groups.[363] Further, eight monitored western gorilla groups in the Lossi Forest, near Ndoki in DRC ranged from seven to thirty-two individuals (mean: 17); there were also four solitary individuals in the area.[364] Based on nest counts, the Cross River gorilla groups in Nigeria and Cameroon fall within the ranges of other gorillas.[365]

When fruit was unusually scarce at Bai Hokou in Central African Republic, a group comprising thirteen weaned individuals, including two silverbacks occasionally (34 of 182 days) foraged and slept apart for a few days. Each of the two subgroups had the same individuals, including one of the silverbacks. On four occasions, they nested 800 to 1,000 meters apart before rejoining to form a single group. Unlike among chimpanzees, there were no all-female/offspring foraging or nesting parties.[366]

Studies of western gorillas at other sites in the Republic of the Congo and adjacent areas of the Central African Republic repeatedly underscored that although both young females and males commonly migrate from their natal groups to form new groups or to join an established group, the basic pattern of at least one silverback and usually multiple adult females and their youngsters constitute western gorilla groups, which do not engage in chimpanzee-like fission/fusion.[367]

A compilation of life-history patterns and social structures of *Gorilla gorilla* at three sites in the Republic of the Congo and one site in the Central African Republic indicated that they are broadly similar to those of eastern gorillas (*Gorilla beringei* and *Gorilla graueri*). Therefore, Robbins et al. concluded that "the ecological variability across gorilla habitats likely explains the flexibility in the social system of gorillas, but we need more information on the social relationships and ecology of western gorillas to elucidate the causes for the similarities and differences between western and eastern gorillas on the levels of individual, social group, and population dynamics."[368]

Observations at an 18-hectare sodium-rich swampy clearing in Parc National d'Odzala, Republic of the Congo, exemplify the adaptive advantages to gorillas of social stability, absence of human competition and poaching, and high-quality food sources. The mean group size of thirty-one stable focal groups and 163 other visiting groups was 11.2 (range: 2–22) and 10.7 (range: 2–22) individuals, respectively. No group had more than one silverback. Ninety percent of the solitary individuals were silverbacks; there was one solitary adult female. The area accommodated a remarkably high birth rate and low mortality rate for infants and juveniles.[369]

Analysis of lineages of mitochondrial and Y-chromosome markers from fecal samples indicated that over a 6,000-km^2-area male Odzala and Lossi gorillas belong to a single population, but the female lineages are more restricted locally. Douadi et al. concluded that wider dispersal is easier for males because most blackbacks experience a period of solitariness before becoming long-term members of groups in which they can breed.[370] Analyses of the DNA from fecal and hair samples revealed that twelve of fourteen silverbacks near Mondika Research Station in Central African Republic were related to one or more silverbacks in the approximately 50-km^2 area.[371] Their close relatedness might be responsible for the relatively pacific relationships between the groups.[372]

Eastern Enigmas and Potential Contrasts

The brevity of initial studies on Ugandan mountain gorillas and the ephemeral nature of encounters between them and observers greatly limited data on intragroup and intergroup social dynamics. During an eight-month study in the Kigezi Gorilla Sanctuary of southwestern Uganda, Donisthorpe sighted several groups that contained between three and twelve individuals.[373] Osborn, who had participated in a failed provisioning scheme before Donisthorpe's arrival, also observed small groups in the region: two with ≥3 members, one with four members, and one with four or five members.[374] Donisthorpe concluded that large groups sometimes divided into subgroups with irregular compositions that later reunited.[373]

Their successors, Kawai and Mizuhara, encountered thirty-nine gorillas in four groups ($n = 5$, 7, 9, and 18) at Kigezi, and censused two groups ($n = 17$ and 13) of mountain gorillas in the Bwindi Forest Reserve of Uganda.[375] The largest group contained a silverback, a blackbacked male, four adult females, six youngsters, and five members whose ages and sex could not be identified.

Contrary to Donisthorpe, Kawai and Mizuhara stressed the cohesiveness of mountain gorilla polygamous families that serve as the cores of larger groups.[375] They noted that each group had a silverback leader and protector. They further inferred that some silverbacks had particularly strong bonds with individual adult females in their groups. During the study, a secondary young silverback became peripheral to the group with which he had been associated. Although the ranges of adjacent groups overlapped, they tended to avoid one another.

Twenty years later, Harcourt noted that the mean and median size of four Bwindi gorilla groups was nine individuals (range: 6–11).[376] His census was premised on nests and the sizes of fecal boluses therein. An average group included a silverback, three to four adult females, and four to five youngsters. He also found spoor of three lone males.

In 1997, there were around 300 mountain gorillas in the Bwindi Impenetrable National Park, Uganda. The population increased to about 340 individuals by 2006. Based on three censuses, the mean group size was ten or eleven individuals (range: 2–28 individuals). By 2006, researchers had identified eleven solitary silverbacks and thirty groups, five of which were habituated to humans. Most groups, ranging from three to twenty-one individuals, had one silverback; four groups (range: 6–21 individuals) had two silverbacks; and three groups (range: 17–28 individuals) had three silverbacks. They did not distinguish between adult females and blackbacks, which build nests of similar size and are difficult to distinguish at first sight.[377] By 2009, there were probably around 380 Virunga gorillas and 302 Bwindi gorillas for a total of ≥682 mountain gorillas.[378]

Schaller was the first scientist to habituate mountain gorillas to a visible human observer.[379] Six of eleven groups at Kabara, Parc National des Virungas in eastern DRC tolerated his presence for spans of one to seven hours, but he rarely followed them on complete daily rounds. Dense foliage greatly limited his ability to monitor their individual and social behaviors.[379] The initial survey indicated ≥350 mountain gorillas inhabited the Virunga Massif; later Schaller revised the estimate to 400 to 500 individuals.[380]

The mean group size ($n = 10$; mean: 17; range: 5–27) of the Kabara mountain gorillas was more than twice the average size ($n = 6$; mean: 7–8; range: 2–15) of the Ugandan mountain gorilla groups. Seven of ten relatively stable, cohesive Kabara groups had one silverback; one group included two silverbacks; and two groups contained four silverbacks each when Schaller first censused them. Eight groups also had ≥1 blackbacks. Nevertheless, in all except one Kabara group, adult females outnumbered spermatic males. Accordingly, Schaller concluded that the groups were stable, except for the births and deaths of infants and the arrivals and departures of males. Fission into subgroups was rare; subunits generally remained near one another and rejoined soon after separation.[381]

By 1971 to 1973 the Virunga gorilla population had declined to 260–290 individuals, especially in DRC, as a result of human hunting; the same was the case in Rwanda and Uganda because of large-scale anthropogenic habitat destruction. By 1976 to 1978, the population had stabilized at about 252 to 285 individuals, comprising twenty-eight groups and six solitary males. Further, group size had increased in some areas (mean: 8.8; range: 3–21 individuals). One group had three silverbacks, ten contained two silverbacks each, and seventeen had one mature male each.[382] Based on a census in 1981, 242 to 266 gorillas were living in the Virungas.[383] Although the population of western gorillas dwarfs that of mountain gorillas, analysis of variation at multiple microsatellite loci indicates that there is only a slight reduction in heterozygosity in the latter versus the former.[384]

In Kabara groups, there were ≥4 silverbacks, only one of which appeared to be past his prime, and three blackbacks that were predominantly solitary. In Kabara group IV, only the alpha silverback was present yearlong; another three silverbacks visited and departed from the group. Only two of six groups were popular with the loners. Apparently visitors were welcome if they elicited no overt aggression from other males in the group. If the alpha male stared threateningly at an approaching silverback, he subsequently avoided the group.[381]

Schaller concluded that mountain gorillas are not territorial, despite the charges and glaring by dominant males during some group encounters and the uprooted tufts of hair where groups had met. Members of different groups usually ignored one another and occasionally mingled peacefully for several minutes before separating. Nevertheless, focal groups tolerated some neighboring groups more than others.[385]

Schaller emphasized the attractiveness of the silverback as much his general pacificity.[379] He is the hub of his group. Females and youngsters stay near him, and the males are peripheral. All members appeared to be attuned to the silverback's activities; he led them through daily rounds of feeding, travel, resting, and nesting. When extra silverbacks were with a group, its members usually reacted only to the alpha silverback as leader.[386] During rapid progressions, he headed the group; when danger threatened, he dropped back as the rearguard. Once a startled silverback grabbed a juvenile and carried it 10 to 13 yards away from Schaller.[381]

The alpha silverback supplanted all other group members for choice sitting and feeding sites and on the narrow trails. Multimale groups had linear hierarcies of silverbacks, and blackbacks and females were dominant over youngsters. Females seemed to lack stable hierarchies, and their relations with blackbacks were variably dominant or subordinate.[379]

Cohesiveness of the Kabara groups seemed not to be maintained via allogrooming. Grooming was rare and was never reciprocal. Schaller concluded that adult allogrooming was chiefly utilitarian—to remove irritants from otherwise inaccessible regions of the groomee.[379] In 89 of 134 (66 percent) bouts, females groomed youngsters. Three times a silverback groomed infants, and once a juvenile groomed him. The blackback never groomed, and he was groomed only once each by a female and a juvenile. Females rarely groomed each other (3.7 percent of bouts). Juveniles were the second-most frequent groomers (25 percent), and they focused on infants and other juveniles. Juveniles also groomed females nine times.[379]

Youngsters played more than they groomed others. They were predominant participants (97 percent) in ninety-six play bouts, and they often played alone (43 percent of observations). Social play bouts were relatively brief (up to 15 minutes) and never led to quarreling or injury, even though the partners were sometimes grossly different in size. Based on cross-sectional data, Schaller concluded that infant mountain gorillas remain physically close to their mothers for about three years.[379] They nest with and are carried by them, especially when danger threatens. Nutritional weaning begins during the infant's first year.

There is a trove of detailed information on mountain gorilla socioecology from long-term yearlong studies in the Parc National des Volcans, particularly on the Karisoke population, groups of which roamed the slopes of Mount Visoke and neighboring montane areas in Rwanda. Via sustained habituation, researchers established conditions for observing gorillas more intimately than Schaller could in DRC. Unfortunately, political turmoil prevented Fossey from continuing work on Mount Mikeno where Schaller had been based.[387] Further, human encroachments and depredations periodically disrupted gorilla studies in Rwanda and drastically reduced the populations of mountain and Grauer's gorillas after 1960.[388]

Tragic events accelerated new group formation among the Karisoke gorillas, so it is difficult to discern which aspects of the process are characteristic of gorilla social dynamics in the absence of destructive agencies. For example, shooters commonly target silverbacks. When there is no other silverback in the natal group, females disperse to other groups, thereby exposing their up to 3-year-old youngsters to infanticide.[389] We can only hope that some groups will be allowed to stabilize and to develop naturally over several generations in large tracts undisturbed by poachers, military traffic, tourists, or other anthropogenic agents in Rwanda, DRC, and Uganda. It is sobering to hear that despite decades of research "we still do not know what the appropriate population structure or interunit relationship is for gorillas."[390]

The 2010 census results offer hope for the survival of some populations of *Gorilla beringei* on the Virunga Massif: the total count is 480 individuals in thirty-six groups plus fourteen solitary silverbacks. The increase of 26.3 percent over a seven-year span (2003–10) occurred despite humans killing ≥9 individuals.[391] A high percentage of the Virunga gorillas are habituated to close observation by researchers, patrol guards, and tourists, which might be a mixed blessing. Nonetheless, the observed groups have grown faster than the overall population, chiefly due to a lower mortality rate versus a higher birth rate.[392] The Bwindi population has a lower growth rate, possibly because, being more frugivorous than Virunga gorillas, they have a slower life history or they are closer to the carrying capacity of their habitat. Like Virunga gorillas, Bwindi observed groups have a higher birthrate than the general population, probably due to veterinary care and protection from poachers.[393]

Although their temporary presence might protect the gorillas, human observers may expose the gorillas to pathogens, and unwary habituated gorillas are probably more vulnerable to the humans that would kill them. Nonetheless, three groups habituated for research by the Karisoke Research Center staff comprised forty-eight, twenty-two, and twenty-four individuals with three, six, and four silverbacks, respectively, and a habituated group that is frequently visited by tourists had thirty-nine members, including three silverbacks. In 2007, during intermingling of one of the largest research gorilla groups and a large gorilla group for tourist viewing, a subadult female transferred into the latter, and four females, one of which had an infant, transferred to the research group. Meanwhile, all the dominant and subordinate males of the groups engaged in conflict with no serious injury.[394]

Early censuses of the Karisoke population documented nine gorilla groups, averaging thirteen members (range: 5–19 individuals). Each cohesive group had a dominant silverback with blackbacks, females, youngsters, and in some instances additional silverbacks.[395] In Karisoke groups 4 and 5, each of which included a silverback and a blackback, the number of adult females ($n = 2$–4 in group 4; $n = 3$–6 in group 5) and youngsters ($n = 4$–5 in group 4; $n = 4$–8 in group 5) varied during a twenty-month span due to female transfers, births, deaths, and maturation.[396] By 1991, approximately 40 percent of Karisoke social units were multimale.[397]

There also was an unusual unit (group 8) that initially comprised five spermatic males, headed by a dominant silverback and an aged female that died shortly after Fossey encountered them.[398] Later, a silverback from group 8 became solitary; the others took females from a bisexual group. Group 8 experienced further additions, including a birth, and departures as it developed into a small bisexual group more like other groups.[399]

The ranges of Karisoke groups 4, 5, 8, and 9 overlapped notably. When mutually aware of another group, they exchanged vocalizations and sometimes displayed. They also commonly approached one another and occasionally intermingled.[400] Like Schaller, Fossey emphasized tranquility in gorilla life as attested by the near absence of aggressive behavior, though she noted that the gamma silverback left group 4 when the alpha silverback died and was replaced by the beta silverback.[395] Further, the new leader led his group away from group 8, which then began to chivy group 9. Aggression increased among the silverbacks in group 8 and between groups 8 and 9.[398]

Lone silverbacks ranged over large sections of the ranges of their natal groups.[401] They tended to overuse the core sections of their ranges, perhaps as defense against encroachments by other gorillas.[400] Conversely, such behavior might make the area unattractive to females.[16]

During a thirteen-week span, the daily rounds of two lone Karisoke silverbacks were very similar in area to those of their natal groups. Interactions were predominantly agonistic between a lone silverback from group 8 and another lone silverback and neighboring groups, including group 8, which is not surprising because lone silverbacks established new groups by taking young females from other groups.[402]

Harcourt and Fossey agreed with Schaller that mountain gorillas are not territorial in the sense that they defend specific tracts of habitat.[403] Overlap between group ranges and those of lone males is too extensive for them to fit the classic definition of territoriality: active defense of a section of habitat for one's exclusive use versus overlap with conspecific social units.

Harcourt confirmed that the alpha silverback is truly the focal member of mountain gorilla groups.[404] The proximity of individuals, particularly adult females, to one another occcurs because they are attracted to him instead of interfemale mutuality. Females with dependent offspring are especially inclined to stay near him.[405] In each Karisoke group, some females allogroomed with their silverbacks. All subadult females and the youngest adult female of group 4 solicited and were groomed occasionally by the silverback. They did not reciprocate. The two adult females of group 5 that spent the least time near the silverback groomed him most often. He only groomed one of them reciprocally. Unlike Schaller, Harcourt concluded that adult grooming facilitates affinitive interactions between the silverback and the females, the latter initiating most of the interactions. Often the grooming females had been having the least stable relationship with a silverback.[405]

Although most females were in proximity with one another because they were near the silverback, related females and those that had been together since childhood exhibited more affinity than unrelated, immigrant females did. Young females also spent greater than average spans with peers in Krisoke groups 4 and 5. Rarely, estrous and less commonly anestrous (and pregnant) females engaged in frottage.[406]

Nulliparous females were quite interested in the neonates of other females and tended to remain near them. Females with infants rarely touched one another; related females and those that had been together since childhood had more frequent body contact, usually initiated by the youngest of a pair. Daughters groomed their mothers much more frequently than the mothers reciprocated. There was no obvious dominance hierarchy among females in group 4 or group 5 even though some females tended to avoid aggressive approaches by others, which the former individuals never displaced.[407]

Contrarily, Watts documented weak linear hierarchies among the females in two Karisoke groups based on nonaggressive approach-retreat interactions, but they do not have clear agonistic dominance hierarchies.[408] Aggression usually erupted mildly in the form of pig-grunting. Louder, more demonstrative displays and injurious attacks were rare. Group females seemed to accept newcomers and were generally tolerant of one another, perhaps because of a silverback's presence and his acceptance of female transferees. Silverbacks repressed interfemale agonism and rowdiness among youngsters.[409] Nearly all male aggression toward females was moderate, and females typically reconciled with them.[410] It has been impossible to decide to what extent, if any, male display is essentially courtship behavior, indicating male vitality and ability to protect females and their young, leading females to choose them as mates or a coercive tactic to ensure female submission and mating with them.[411]

Because affiliative and agonistic interactions in dyads of female mountain gorillas vary rather widely in frequency and intensity, it is difficult to generalize about the population, though interactions between female kin are generally less aggressive and they tolerate closer proximity and grooming more than non-kin do.[412] Further, wounds are less frequent when female relatives fight. Watts also noted, "Presumed paternal half-sisters generally show levels of proximity and rates of affiliative and agonistic interaction intermediate between those of maternal relatives and nonrelatives."[413] Indeed, adult females with no adult relative in the group groomed immature individuals and with adult males, whereas adult female relatives groomed reciprocally and sometimes were the chief grooming partners.[414]

Based on displacements during feeding, particularly on fruit, Bwindi gorillas significantly exhibited a linear dominance hierarchy among males and females. A female hierarchy was not supported statistically due to the small sample size; however, lower-ranking females seldom if ever displaced the higher-ranking ones. Aggression, though mild, like that of Karosike gorillas, was more frequent at fruit sources than when they fed on other foods.[415]

Relationships between the silverbacks and blackbacks in Karisoke groups 4 and 5 were dramatically different. In group 4, they rarely interacted; in group 5, they spent notable time together, and the blackback even groomed the silverback. The blackback initiated most of their intimate interactions. Although the group 5 males had had a history of close association since the birth of the younger male, and they might even have been father and son, no such affinity existed between the males of group 4.[416] The silverbacks of both groups freely encroached on blackback feeding spaces, but the reverse was rare.[417]

An alpha silverback is highly attractive to youngsters. Indeed, whether in one-male or multimale groups with up to 9 adult males, youngsters are in the proximity of and interact most with the alpha silverback.[418] They freely romp on him during adult rest periods. Older youngsters preferred the silverback to their mothers when the group moved and fed. Four youngsters, ranging from 2.5 to 3.8 years old, survived the death or abandonment of their mothers because the silverback continued to parent them, including sharing his night nest, without which they probably would have succumbed to the low montane temperatures.[419] Additionally, if a mature male is near mothers of weaning infants, he has a good chance to mate with them when their estrus cycles resume.[420]

Intragroup agonistic interactions indicated that gorillas variably form alliances and have the means to prevent and to alleviate aggressive acts. Related Karisoke females commonly supported one another and maintained reciprocal alliances while living together.[421] Most nonrelated females rarely supported one another, though some developed alliances. Females with affinitive relationships mostly supported their partners against individuals that often engaged in dyadic aggression. Karisoke females had mostly undecided agonistic interactions. They often retaliated against aggressors and engaged in sequences of dyadic aggressive exchanges, including bidirectionality.[422] Nonetheless, the effectiveness of female coalitions was notably limited by male intervention.

Mature males competed to control female aggression. In a sizeable Karisoke group, all females interacted affiliatively and mated with a young silverback. Long-term resident females also interacted with an old nonbreeding silverback that was their relative, but recent immigrants spent little time near him. The older silverback intervened agonistically to support his relatives, while the younger silverback tended to ameliorate the lot of immigrant females in competition with residents.[423]

The fact that high-ranking males curtail the aggression of their subordinates toward females might make them attractive to females. In multimale groups, the dynamic can be complicated by formation of male coalitions against other males in the unit.[424] Male interveners were most effective (83 percent of interventions) when they approached females often while vocalizing, displaying, grabbing or hitting them, or otherwise contacting the opponents physically. Vocalization alone seldom sufficed. Males intervened in 80.7 percent of bisexual conflicts and 75 percent of intermale altercations. Female interveners screamed and sometimes lunged at individual opponents, but they were less likely than the males to physically contact them.[424]

In addition to intervening aggressively, Virunga gorillas sometimes interrupted the agonistic interactions of others by nonaggressively interposing themselves between the contestants. Sicotte described forty interpositions during mostly low-intensity conflicts, usually signaled by cough grunts, between the silverbacks in two bimale groups.[425] Escalated conflicts included mutually grabbing one another, screaming, and biting, occasionally leading to wounds. Interpositions occurred in 18 percent of conflicts in the larger group ($n = 25$) and 10 percent of conflicts in the smaller group ($n = 12$). Interposing immature and female individuals ran between contestants, the former emitting a scream that was unique to interposition; the females usually emitted appeasement mumbles toward one of the contestants. An individual might throw himself at the feet of a strutting male, and youngsters would cling to his chest and be carried a short distance. Even mothers carrying small infants sometime joined interpositions. Regardless of the dominance status of the silverbacks, initiators of aggression tended to be targets of interposers, and sometimes the interposition seemed not to be directed at a specific contestant. In the larger, but not the smaller group, interposition extended the interval between intermale conflicts. Sicotte suggested that interpositions might facilitate male coexistence by reducing the rate of intermale aggression.[425]

Unlike chimpanzees, Karisoke gorillas exhibit no postconflict reconciliation between adult females, between mature males, or between immature individuals. However, after male aggression toward a female, she can reconcile herself with him, which is prudent given that he is an important social partner and protector while she is with his group. Watts suggested that females might not need to reconcile with one another because relationships between coresident relatives are resilient, while those between nonrelatives are mostly neutral or antagonistic.[426] Further, consolation directed by a third party to the target of aggression appears to be absent in gorillas.[427]

Two independent studies on three captive groups demonstrated that, like mountain gorillas, reconciliation occurs among western gorillas, but unlike mountain gorillas they also exhibited consolation.[428] The difference might be due to the fact that captives have less freedom to get away from agonistic individuals and the stressful disruptions that arise from them. Indeed, in the Apenheul group, composed of one silverback, five adult females, two adolescents, five juveniles, and three infants, the levels of consolation were higher in the absence of reconciliation than in its presence, leading Cordoni et al. to suggest "consolation may function as an alternative mechanism in stress reduction of the victim."[429]

Harcourt, Fossey, Sicotte, and Watts shattered the initial image presented by Schaller and Fossey that the silverback is a gentle giant, highly tolerant of neighbors.[430] Violence was common during the formation of new Karisoke groups.[431] Accordingly, Harcourt concluded that aggression is the main tactic of resident males against males that try to associate with his females.[432] Fossey noted that female transfers between established groups or from a group to a lone silverback almost always entailed overt aggression between the dominant male in the female's current group and the seductive silverback.[433] The presence of estrous females and copulations within groups seemed to attract visitors and to set off displays and fights between the males of different social units.[434] Apart from the brevity of his contacts with Kabara gorillas, perhaps Schaller had observed only stable groups that had well-established intergroup dominance relationships.

After interunit encounters, males sometimes sported wounds on their heads and shoulders, presumably from their antagonists' canines, but the brunt of several contests was borne even more heavily by infants in the resident groups, ≥7 of which were bitten to death by impetuous young silverbacks in pursuit of mates.[435] Indeed, during a fifteen-year span, 38 percent ($n = 13$) of Karisoke infant deaths are attributable to infanticide, and another 23 percent of infant deaths also might have been infanticidal, though not necessarily all perpetrated by silverbacks.[433]

Watts tallied six additional infanticides, one attempt, and three suspected infanticides in Karisoke gorillas.[436] The instances occurred predominantly when a mother was unaccompanied by a mature male, usually because he had died or she and the infant were separated from him. Three infanticides occurred during group encounters with another heterosexual unit or a lone male. Once a male infant that had immigrated with his mother survived recurrent attacks by members of the resident group. Initially, both individuals received more aggression than the long-term members did, but after the female mated with the males and her youngster was weaned, aggression toward her decreased. The youngster was not so fortunate. Instead, aggression toward him increased and stopped only when he became a blackback in the group.[437] Watts concluded that field observations support the sexual-selection hypothesis for infanticide and strongly support the argument that intersexual mutualism and intraspecific aggression have been central factors in gorilla social evolution.[438] A model based on novel modifications of a gas molecule equation supported the anti-infanticide hypothesis regarding female association with a male, albeit without negating the antipredation hypothesis that large males protect females and their youngsters more effectively than they can themselves.[439]

Yamagiwa et al. noted that rapid changes in the density of Grauer's, western, and mountain gorilla social units and their relations after drastic environmental changes caused by human disturbance might increase the possibility of infanticide.[440] After thirty years of monitoring two groups of habituated Grauer's gorillas with no indication of infanticide, researchers documented three cases in 2003.[441] During an intergroup encounter with fierce fighting, the young silverback in one group snatched an infant from its mother and bit it to death. The infants of two females that transferred into and birthed in the other group were also bitten to death by the resident silverback. He did not kill an infant born later in the group, probably because he had sired it. These events occurred after 1998–99 when large-scale poaching of gorillas occurred in Kahuzi-Biega National Park.[441]

No direct observation of infanticide is available for western gorillas, but Stokes et al. suspected three instances at Mbeli Bai in the Nouabalé-Ndoke National Park, where they observed foraging groups from an observation platform.[442] Two victims were infants (aged 5 and 28 months) of transferring females, and a recent immigrant female that had probably been impregnated by a male in a prior group had birthed the third. The somewhat older (31 and 58 months) suckling offspring of two other immigrant females were not harmed in their new group. Both of the latter mother-offspring pairs joined the new group with males from the former group.[442]

As with the mountain gorillas, loss of habitat, hunters, pandemic and local diseases, and stress, possibly introduced by hunters, researchers, tourists, and other visitors, also threaten western, Cross River, and Grauer's gorillas with extinction and frustrate the attempts by scientists to study them continuously over several generations. Contact with apes also might lead to transmission of pandemic diseases to human populations. Release of sanctuary apes is also risky to resident apes.[443]

Cohesiveness of gorilla groups might be based on strong affinities between alpha silverbacks and youngsters. Females are bound to him through their young children. Childless young females have looser ties and thus are disposed to emigrate. Alien silverbacks can break male-female bonds by killing infants of parous females.[444] This scenario leaves unexplained how the father-daughter bond loosens, if, in fact, she was bonded to him as a child.

Founding and Finding a Group for Safe Sex

During early adulthood, young adult male and female mountain, western, and presumably also Grauer's gorillas usually emigrate from their natal groups.[445] Female Virunga gorillas begin to sexually mature around 6 years old, but they appear to be sterile for about two years thereafter; they first give birth at a median age of 10 years.[446] Males live alone or temporarily with a few other males for several years until they can acquire fertile females. Black-back transfers to another group are extremely rare. Females may transfer to neighboring groups and beyond, but commonly they visit their natal group before settling down to produce infants with a silverback in the new reproductive unit. Sometimes a son remains with his natal group and inherits it upon his father's demise. He might fail to keep the recently beheaded group together in the face of forays by silverback competitors and the wanderlust of its females.[447] Transfer did not delay reproduction in nulliparous and most parous Virunga females, but parous females might experience longer inter-birth intervals if they continued to transfer to other groups. Dispersal might be more likely for a given female when successful reproduction is delayed due to infertility, miscarriage, or infant death.[393]

At Mbeli Bai, new group formation was exclusively via acquisition of fe-males by solitary males.[442] Female immigration rates were negatively related to group size and the number of females per group, and emigration rates were positively related to group size only, resulting in larger groups losing females and smaller groups gaining them. Females not only emigrated from their natal groups but also sometimes transferred secondarily to another group, especially within six months of losing an infant or upon the death of the group's silverback. After the deaths of two silverbacks, a pair of females from each group transferred to other groups. Two of four females that mi-grated with infants lost them, perhaps because of infanticide, and some females left without their juvenile progeny.[442] Affiliative behavior was rare, perhaps because the gorillas were preoccupied with feeding. Female agonis-tic interactions varied between groups: most were undecided, and supplan-tations were infrequent. Agonistic behavior between silverbacks and females was higher and more consistent between groups.[448]

Based on 12.5 years of data, Breuer et al. suggested that (compare: Stokes et al.) larger group size might be preferred by Mbeli Bai females, which clus-ter around males to enhance offspring survival.[449] Males with larger groups of females had lower offspring mortality with no apparent reduction in fe-male fertility or tenure length. Accordingly, female feeding competition or risk of male takeovers and infanticide apparently did not limit the number of females in a group.

Immature western gorillas of Mbeli Bai are weaned later than their coun-terparts among mountain gorillas, and they appear to mature more slowly, from which Breuer et al. inferred that, like other frugivorous species, they support a risk-aversion hypothesis, whereby longer suckling enables them grow and survive during times when fruit is in short supply.[450]

DNA genotyping of most of the Bwindi mountain gorilla population indicated it is genetically and geographically structured by nonrandom movement by females. Guschanski et al. concluded that dispersing females prefer to join groups that range within an area having vegetation similar to that of their natal group.[451] Conversely, Bwindi males evidenced no association of geographic and genetic distance, suggesting that they disperse widely enough to eliminate close linkage to geographic factors.

Genotypic analysis based on fecal DNA from ninety-two Karisoke mountain gorillas indicated that the dominant or second-ranking male sired forty-eight infants born into four multimale groups. Dominant males sired most (85 percent) of the infants, but unrelated second-ranking males sired 15 percent of them. Accordingly, Bradley et al. concluded that mountain gorilla groups do not approximate family groups, but instead are more similar to chimpanzee groups in being long-term assemblies of related and unrelated individuals.[452]

The dynamics of male departure are clearer than those of spontaneous female emigration. Indeed, Harcourt commented that the emigration of female gorillas was always slightly surprising because it occurred so abruptly.[432] The dominant silverback prevents subordinate males in his group from breeding with the estrous primiparous and multiparous females in his group. He is more tolerant of their copulations with immature, nulliparous, and pregnant females, none of which are likely to be impregnated by him.[453] Consequently, if the leader remains robust, young spermatic males must leave the group and somehow obtain their own females to become full-fledged members of the breeding population. They might have to fight silverbacks to secure females for themselves. If a young silverback kills a female's infant, she will likely join his budding breeding unit.[433]

At least four tactics are available to a male gorilla that would head his own group:

- Mature in and remain as a follower until he can head the group of a senescent or deceased silverback that may be his father.[454]
- Depose the silverback of an established group.
- Emigrate and attract transferring females, perhaps accelerating the process by raiding established groups.

- During group fission, be accompanied by other members of the multi-male group.

Regardless which tactic they adopt, males are around 15 years old before they are fully grown and possibly become the dominant silverback of a heterosexual unit, though one Karisoke male procreated as an 11-year-old.[455] Based on thirty years of behavioral and demographic data on Virunga gorillas, a model predicted an average lifetime reproductive success of 3.2 for the philopatric males (followers) that remained in their natal groups versus 1.6 for the males that emigrated from the group, which indicates that the first tactic is advantageous. Regardless whether males remained or emigrated, those that became dominant in a group had similar values for lifetime reproductive success (4.5 and 4.6, respectively).[456]

Young Karisoke silverbacks (12 to 15 years old) most commonly employed the third tactic, even though males that leave their natal groups are less successful reproductively than those that remain in the unit and inherit its females.[457] Apart from incorporating females into his new unit, a male and his females have few opportunities to copulate outside the unit; however, there are a few observations of young adults briefly copulating during groups intermingling.[458]

Exemplary of the fourth type of new group formation: after fission, the subordinate silverback in a multimale Bwindi gorilla group established a new group with two females of unknown paternity from the natal group. Eight youngsters sired by the dominant silverback and their mothers remained in the natal group.[459] At Bwindi, frugivory was associated with an increased frequency of group encounters, but competition for fruit per se seemed not to increase aggression. Instead, competition was associated more with mate defense and acquisition. Encounters with solitary males tended to elicit more avoidance and aggression, less tolerance, and more herding than encounters with groups did.[460]

Although initially dominant silverbacks might try avoidance, hoots, or displays to keep raiders at bay, they commonly fight ferociously to hold their mates.[461] At Karisoke, 79 percent (fifteen of nineteen) of encounters between strange males led to violent threat displays. During eight of sixteen encounters in which Harcourt et al. could determine whether or not a fight had occurred, battles were manifest.[462]

Except during intergroup encounters, Karisoke silverbacks generally tend not to herd females aggressively or otherwise to direct overt efforts toward them to keep them from emigrating once they have chosen to leave.[463] The most commonly herded females are proceptive and lack dependent offspring. Herding is more likely in newly formed groups with significantly more males than those that are longer established with fewer males.[464] Dominant males benefit reproductively from maintaining the integrity of their units. For example, despite various vicissitudes, the dominant males of two Karisoke focal groups produced four to six and eight to ten surviving offspring between 1967 and 1977, and one dominant silverback had sired ≥19 offspring.[465]

In thirteen Karisoke groups, on average, females were menarcheal at 7 years of age (range: 6.4–8.6 years).[434] Menstruation and estrus are very difficult to detail in wild gorillas. Nulliparous females exhibit small labial swells, but parous females sport no visible perineal clue to their estrous condition. Adolescent females are receptive for three to five days, and their monthly cycles might be irregular. On the basis of observed copulations, regularly (twenty-eight-day) cycling females appear to be receptive for only one to three days per month. Several pregnant females also were estrous, though infrequently.[466] Karisoke females experience variable spans between menarche and first pregnancy. Indeed, emigration, which generally occurs shortly after menarche, might delay conception. The average age when nine transferees first conceived was ten years. The ages when females became primiparae varied between 8.7 and 12.75 years.[467]

Infant (up to 3 years old) mortality was high, accounting for 34 percent of deaths in the Karisoke study population. The median interval between surviving births was about four years. Because suckling induces lactational amenorrhea, loss of an infant shortened the birth intervals of the females that had lost them; indeed, they usually conceived during two to four menstrual cycles of postpartum estrus.[468]

Analysis of data from sixty-six Virunga females over thirty-seven years indicated that primiparous mothers had 50 percent higher offspring mortality and 20 percent shorter interbirth intervals than the mothers that had birthed a second time. Generally, reproductive success was relatively low for primiparous females; it improved as they matured and declined thereafter. They evidenced no extended postreproductive span before death.[469]

Like males, for young females emigration is the rule, to which there are exceptions. Transfer is probably not ecologically costly to mountain gorilla females because of the notable overlap of unit ranges and the lack of site fidelity by males and their groups. The highest priority for females is to be with quality males.[470] Over a thirteen-year span, twenty-six Karisoke females transferred between groups forty-three times.[434] Most of them were of low status vis-à-vis the dominant silverback.

The dominance of Karisoke females largely depended on the order of their acquisition by the dominant silverback. Over a thirteen-year span, only two of fifteen adult females transferred out of the groups in which they had conceived infants with the dominant silverback. Both had occupied relatively low status. Their solicitations to copulate had been rebuffed; perhaps because they had nearly 4-year-old youngsters that continued to nurse occasionally. Their status improved notably after emigration because they were the first and secondly acquired females of their respective new silverbacks. One of them promptly produced an infant with her new mate.[434]

Males that were born into a group and remained with them after becoming silverbacks might breed incestuously with a few young females that also were born and remained in the group. Over a fifteen-year span at Karisoke, one infant was conceived via a brother-sister coupling and half-sibs procreated ≥3 infants.[433]

Unlike males, females transfer directly to potential mates and do not experience prolonged periods of singleness. They generally passed from their natal or latest groups to new units very quickly; one female remained alone for 1.5 days.[432] Peaceful and violent transfers usually occurred when social units were close together and interacting. Transferring females spent between three days and forty-one months (median: 3.5 months) in alien units before settling down.[432]

Transferring females appeared to prefer new groups and lone silverbacks versus established groups.[471] Recall that the intragroup status of females depends directly upon the order in which they become mates of the dominant silverback: thus, the first mate has highest status, the second is beta, and so forth.[472] This might partially explain why 50 to 60 percent of Virunga females reside in one-male groups despite the higher risk of infanticide. Multimale groups did not have higher female immigration rates than those of one-male groups, and transfer destinations were not biased toward multimale groups.[473]

There are several possible reasons for female choice among available silverbacks. Harcourt emphasized quality of a male's range and especially a female's success in rearing offspring with him as reasons for her staying in a new unit.[432] If he cannot protect the group against infanticidal males, she will leave him for one with recently demonstrated prowess, even the murderer of her child.[474] Primiparous Karisoke females were three times more likely to lose infants than the multiparous females; males killed infants more often than other agents did.[472] With data from Virunga gorillas, Robbins et al. tested and found tentative support for the theory "that a male in good condition at the end of the period of parental investment is expected to outreproduce a sister in similar condition, while she is expected to outreproduce him if both are in poor condition."[475] Indeed, gorilla mothers in good condition, as indicated by their dominance rank, had longer interbirth intervals after birthing sons, while subordinate mothers had longer birth intervals after birthing daughters.

Ten of eighteen Karisoke females transferred in pairs from natal units to the same new group either simultaneously or within a few months of one another. Fossey speculated that such pairs are more likely to be integrated into new groups and they would be more successful reproductively because the reduced tendency to roam would limit their exposure to infanticidal males.[433] Further, a single bisexual pair of gorillas seemed not to be a viable breeding unit.[476]

Gentleness and Sex among Giants

Females are attracted to the silverbacks that not only can copulate upon command and protect their infants but probably also take an active part in parenting. Youngsters stayed near the silverback while their mothers moved off to feed. Indeed, during travel-feed periods, some youngsters spent as much time near the silverback as near their mothers. Even more than tolerating the presence and capers of 3- and 4-year-olds whose mothers had new infants, the silverback sometimes groomed and cuddled them.[477]

Most diurnal copulations occurred during travel-feed periods when group members were well separated.[434] Female mountain gorillas are proceptive for one to four days of each estrous cycle and at irregular intervals while they are pregnant.[478] Males, most of which were sterile adolescents (8 to 13 years old), initiated significantly more copulations, mostly with adolescent females, than they did with pregnant or fertile nonpregnant females, with which the adult males mated primarily.[479] Both dominant and subordinate males initiated and received mating harassment. It was usually mildly aggressive and terminated the copulations of subordinate males; however, about 25 percent of copulations involving an alpha male incurred harassment, often directed toward his female partner.[480]

Modes of male copulatory initiation include approaching a female, touching her, or gazing at her and emitting a train grunt.[481] Based on one month of observations of a Karisoke one-male group and two estrous females, Nadler concluded that the silverback had initiated most of the copulations via a tight-lipped, stiff-limbed stance or direct approach to the female; however, the display occurred after a female had moved within 5 meters and stared at him. Then he stared at her, and she backed into him.[482]

Sicotte noted that most copulation occurred without male display.[483] Males displayed about equally toward anestrous and estrous females. When display was associated with copulation, it usually followed the act. Often the female appeared to appease the male via approaching, establishing contact, and grumbling. Females reacted significantly more often in this manner after a male displayed than after any other type of interaction, including approaches and aggression. Males also neighed in order to negotiate proximity with females, more often in multimale than in one-male groups. When a female moved away from a male, he might neigh, after which she either stopped or he followed her with his gaze. Males neighed more toward cycling females than toward pregnant and lactating ones.[484] Contrarily, Robbins found that all males except the lowest-ranking male in one of the same groups exhibited higher rates of aggression toward estrous than toward anestrous individuals.[485] The group had two additional mature males during Robbins's study, which probably increased competition among them ($n = 4$).

In multimale groups, females might have preferred mating partners. Forced copulations were quite rare. Some females in two Karisoke groups solicited only one male, but most estrous females in multimale groups mated with more than one male, perhaps because they could not avoid some solicitors.[486] Estrous females, who initiate most couplings, slowly and hesitantly approached the alpha silverback. Commonly, he seemed oblivious to their over-the-shoulder glances and brief manual contacts with his body.[487] However, after he decided to copulate, she moved her rump toward him, and he drew her onto his lap, often while holding her about the waist. They adjusted their genitalia to achieve intromission and thrusted for 1.5 minutes on average. They maintained copulatory positions for spans of 15 seconds to 19.75 minutes (median: 96–80 seconds).[488] Thus, like the silverback's erection, copulatory bouts were remarkably brief.[489] The female usually broke the embrace and moved away first.[434]

When 2 or more females were estrous simultaneously, males copulated at significantly higher rates than when only one female was estrous, and silverbacks were more likely to refuse solicitations. However, copulation rates of the females were lower than when only one of them was in estrus. Watts reasonably speculated that, on average, females probably receive fewer sperm per copulation on days when they shared a silverback's sexual service, but it does not prevent conception.[490] Solicitations by pregnant females were refused more often than those of females that were not pregnant, and males rarely tried to copulate with females that were not estrous. Nonetheless, females solicited more copulation when pregnant (77 percent) than when cycling (60 percent).[485]

At Mondika, Central African Republic, pregnant females varied in timing and frequency of mating and used postconceptive mating conditionally, synchronizing copulations to occur when other females mated and refraining from mating when others did not.[491] Doran-Sheehy et al. concluded that among western gorillas postconceptive mating is a form of female competition instead of functioning to obtain immediate benefits from the male or to confuse paternity. Because males prefer high-ranking females and initiate copulations with them regardless whether they are pregnant or cycling, a female can reinforce her own status while perhaps delaying conceptions in her rivals.[491]

The precopulatory behavior of mature gorillas lacks faciogenital and manual-genital contact. Usually both partners thrust and vocalize, but the silverback normally sustains these activities longer than his mate does.[492] Unlike chimpanzees, the copulations of silverbacks with choice mates very rarely received interference from other group members.[493] Older alpha silverbacks tolerated copulation between their daughters and mature or immature males, but not between subordinate males and parous females that had been the silverback's mates or with recent immigrants. Nonetheless, subordinates could copulate any willing female when not in proximity to the dominant male.[490]

Usually only one female in a Karisoke group is estrous on a single day.[492] Only 5 percent of 580 copulations occurred on days when ≥1 female in a group was estrous. One day, a silverback copulated with an elder female five times and with a young female once during a four-hour span. In this and other instances of multiple copulations, the silverback's vigor flagged with each bout.[434]

Sexual experimentation begins early in Karisoke mountain gorillas. Fossey tabulated 240 mountings that involved youngsters.[434] The youngest mounter and mountee were 2.5 and 0.5 years old, respectively. In a sample of thirty-nine mountings that involved youngsters, the youngest mounter was 3.1 years old, and the youngest mountee was 1.8 years old. In about one-third of the mountings, both partners were immature, with the elder one virtually always on top. The remaining episodes involved adults, especially a black-back, mounting youngsters. Copulating couples employed both ventroventral and dorsoventral positions. Unlike chimpanzees, young gorillas did not mount their mothers, and unlike silverbacks, immature male gorillas attended to the pudenda of estrous females, probably in response to olfactory cues.[434]

There were ten instances of interfemale mounting by Karisoke mountain gorillas.[493] Unlike bonobo frottage, dorsoventral mounting ($n = 6$) was more common than ventroventral frottage ($n = 3$). Usually at least one partner uttered copulatory vocalizations. One female would approach another, then they would embrace, engage in frottage, and part. Unlike male bonobos, the silverback was not sexually aroused by their behavior. Instead, he attacked one pair engaged in frottage, the estrous one of which he had bred an hour before. Because estrous females were the active participants and had copulated with a silverback shortly before switching to a female, Harcourt et al. concluded that female mountain gorilla frottage is sexually motivated.[462] Conversely, they inferred that three cases of males mounting males were agonistic rather than sexual acts.

Abstinence Makes the Fond Grow Harder

As an alternative to ranging singly, some mountain and western gorilla males reside in all-male groups that can provide a long-term stable social environment despite turnover in membership.[494] All-male groups appear to be absent from the Kahuzi population of Grauer's gorillas.[495] In the Virungas, they might have formed initially because the anthropogenic disturbances in the 1970s and 1980s had reduced the available ranges and group sizes in the population.[496] Although group sizes have since increased, especially in multimale groups guarded by humans, all-male groups have persisted, perhaps because some males lose long-term in the mating game with other males.

At Lokoué, Republic of the Congo, five groups, ranging from two to fifteen individuals, predominantly consisted of blackbacks, subadult males, and juveniles the sex of which was indeterminate; four of them each contained one silverback. Although there were frequent changes in membership due to male migration, all-male groups probably play an important role in the life history of some migrating males.[497]

Lacking the opportunity to mate with females, some males engage in sexual behavior that researchers freely refer to as homosexual. In one respect, the activity might be more reasonably considered truly sexual versus social, in contrast to the view that female bonobo frottage is chiefly a social behavior. Male gorillas mirror the mating pattern of male-female pairs by engaging in courtship behavior that culminates with copulation and mate guarding.[498] Six unrelated Karisoke males—two silverbacks, two blackbacks, and two subadults—formed a highly cohesive group that was characterized by frequent intermale sexual activity. The silverbacks fought violently when their partners avoided them or ignored their courtship. They exhibited no submissive or reassurance behavior. Aggression was always unidirectional from elder and dominant males to younger and subordinate males, and supportive actions were the reverse.[496] Individuals in an all-male group of Karisoke gorillas "stayed closer together, affiliated more, exhibited more homosexual behavior, and were more aggressive toward each other than males in heterosexual groups."[499]

Pongid Sexual Selection Writ Large and Small

The trove of data on the social structures and mating behaviors of great apes has facilitated informed evolutionary interpretations of their epigamic features and genitalia.

Although male gorillas are twice the bulk of female gorillas, they have a relatively tiny penis and testes.[500] Further, whether tumescent or flaccid, their penis remains inconspicuous. Parous female gorillas exhibit no readily visible clues to their sexual condition, though nulliparous females sport small labial swells at peak estrogen levels.[501] A male gorilla's somatic massiveness is probably essential for him to establish and to maintain a harem of fertile females vis-à-vis competition with other silverbacks.[502] However, once a group has been established, neither he nor his females need to flash gaudy genitalia in order to realize their breeding potential. Females are estrous for very brief periods and remain close to the alpha silverback. They need only approach him seductively, and he needs only to close the embrace for them to procreate. The male's small testicles and concomitant modest sperm count is sufficient to impregnate females, particularly considering the infrequency of estrus and absence of intragroup male competition during estrus periods in gorilla groups.[503]

A female might be cued to try more fertile fields if the male repeatedly fails to respond sexually or if he actively rebuffs her overtures. Frottage might suffice for the short term, but eventually estrous females would be expected to emigrate, especially if the silverback is ignoring their initial invitations to copulate.

Like silverbacks, adult male orangutans are twice the size of conspecific adult females and have relatively diminutive testes, but perhaps a somewhat longer penis than that of gorillas; male oranguans also will fight fiercely with their rivals.[504] Unlike silverbacks, at least in some areas, flanged males compete by territorially commanding the range in which resident females forage instead of bonding with them to form stable cohesive breeding units. Further, though an orangutan's pink penis is hidden when flaccid, it is readily visible upon erection.[503] Still, a male's cheek pads are probably a better clue to his sexual capacity (his testosterone titer) than his penis.

Like gorillas, female orangutans sport no visible sign of estrus. The male might attract receptive females by vocal advertisements and grotesque epigamic features. When he is the only show in town or has somehow demonstrated himself to be the best available male, he will breed with all, or at least most, of the females in his territory or range. The postural versatility and patience of both sexes facilitate accommodation of a modest penis in unswollen female genitalia.

The longer period of estrus in orangutans versus that of gorillas allows time for prospective mates to find one another and to consort productively. Further, the sexual vigor of prime male orangutans seems to be greater than that of mountain gorillas, as attested by a Sumatran male seen consorting productively with a young female.[86]

Chimpanzees and bonobos stand in marked contrast with gorillas and orangutans regarding many features that are the focus of Darwinian sexual selection. Male *Pan spp.* are not inordinately larger than female *Pan spp.*, and the sexes can be confused facially.[503] The styloid, erect penis of chimpanzees and bonobos is relatively long and conspicuous, especially if flipped, and adult male chimpanzees and bonobos also have relatively titanic testicles. Chimpanzees have astronomical sperm counts, which is probably also true of bonobos given their large testes.[505] This is coupled with the greater intragroup competition for mates in chimpanzee and bonobo societies. Estrous females compete to attract the most robust males, and the latter compete with one another to inseminate the females, particularly those that have shown their fertility. In an arena where promiscuous sex is common, a male must quickly flood the field with sperm to have a chance to impregnate a female that might copulate straightway with other males.[506]

Although an alpha male chimpanzee can keep other males in his purview from mating, he also needs to be sexually vigorous and speedy because he might have more than one estrous female to service over a short period of time; he cannot dally with one of them lest the others cheat by going into consortship or behind a bush with another male. The rather long estrous periods of the female chimpanzees and bonobos must further challenge the sexual energies of the males. Of course, it might give a female chimpanzee time to attract and hold a choice mate in consortship.

Gibbon Monogamy?

Monogamous family groups constitute a basic social unit of hylobatid species; however, extra-pair mating and social polygamy occur in some populations and species, which calls into question whether monogamy accurately characterizes their social structure and mating pattern and sharply distinguishes them from the great apes.[507] Like many great ape populations, gibbons have suffered notable loss of habitat, which might explain the polygamous groups and the hybridization of some species.[508]

Strictly defined, monogamy is a mating pattern in which an individual "reproduces sexually with only one partner of the opposite sex."[509] Unlike birds, in which monogamy is common, among mammals, and especially primate species, it is an infrequent mating pattern.[510]

Brockelman and Srikosamatara attempted to explain the adaptive significance of hylobatid monogamy vis-à-vis a wealth of field data, albeit in several currencies, that was available.[511] They notably complemented the work of other theoretical biologists who puzzled over apparent obligate monogamy in the Hylobatidae. Perhaps most importantly, they acknowledged the limitations of their model, proffered several caveats, and thereby laid the groundwork for future fieldwork, including further experimentation.

In most, if not all, hylobatid species, the behavior of females is the key to understanding their territoriality and obligate monogamy.[512] Because gibbons are selective feeders on high-quality foods that are not generally abundant, females stake claims to sections of the forest that are large enough to support themselves and their progeny yet small enough to defend against other females. Males, which are similarly limited by available resources, cannot command larger territories with more than one female/young unit, so they become obligately mated monogamously, repel other males, and often assist to eject other intruders.

Long-term female acceptance of a male is most easily understood in *Symphalangus syndactylus,* wherein males sometimes carry infants and otherwise assist in their care and development and are effective defenders of group territorial integrity. Predator defense is another possibility, especially among Mentawai gibbons.[513] Brockelman and Srikosamatara did not explicitly address the role of lodge trees as an ecological factor selecting for gibbon social behavior; however, they proposed that a low degree of interspecific competition and niche overlap, which certainly could include the choice sleeping sites, are probably major factors selecting for their territoriality.[514]

Adult mated pairs and their immature offspring occupy small, stable areas of the tropical forest. They exclude conspecifics from their territories by calling and chasing intruders. Although physical contacts between females or between males are uncommon during altercations, they can be lethal.[515] Accordingly, gibbons fit the classic criterion for a territorial species as it was originally applied to birds—the active defense of a particular area where they reside.[516]

Like the diets of various hylobatid species, the sizes of the territories and ranging patterns of gibbons also vary. For instance, Malayan siamang occupy territories (20 hectares) that are less than half the size of conspecific white-handed gibbon territories.[517]

Thai white-handed gibbons become active shortly after dawn and first light. An adult or juvenile male might sing a solo before first light. After an initial bout of feeding and foraging, the group engages in a bout of calling with the female, male and juveniles emitting distinct vocalizations and swinging about in the branches. The male, female, or both give a series of whoops, then the female emits a great-call, after the climax of which the male joins with a coda. They forage away from the sleeping site, which is commonly centrally located in the territory. At the periphery, they might encounter one or more neighboring groups. If so, the adult males of the groups might engage in prolonged bouts of calling, display swinging, and chasing, commonly lasting more than an hour. Concurrently, females and youngsters might continue to forage and feed in the contested area or deeper within their respective territories.[518]

After an encounter, they forage away from the boundary region until midday, when adults rest and groom while youngsters play. After this brief respite (approximately ten minutes), they resume feeding. During the remainder of the early afternoon, the group sporadically forages, feeds, and pauses for brief bouts of rest, grooming, and play. By around 15:15 to 15:45 PM, they begin to settle into separate sleeping trees. Infants sleep with their mothers.[518]

In comparison with chimpanzees, bonobos, and especially orangutans, white-handed gibbon groups are remarkably cohesive, though group members do not travel, forage, and sleep as closely to one another as the siamang do. The mechanisms for white-handed gibbon cohesiveness are obscure. They call frequently, but grooming is infrequent compared with the grooming frequencies in many Old World monkeys. Adults groom during only 3 percent (15 minutes) of their waking time. A mated male and female groom each other about equally, or the male grooms the female more than she grooms him.[519] Youngsters are groomed far more than they groom adults. Often the mother grooms her infant after they have retired in the afternoon.[518]

Overt sexual activity may play a very minor role in maintaining hylobatid pair bonds. Mated pairs likely engage in sexual behavior for short periods about once every two years, for a mean birth spacing of 2.0 to 2.5 years.[519] In Khao Yai National Park, Thailand, the mean birth interval of white-handed gibbons is 41 ± 9.1 months.[520] Khao Yai females exhibited small perineal swellings that might indicate the probability of ovulation, albeit without allowing males to pinpoint the exact time.[521]

Within a white-handed gibbon group, dominance might be related more to age than to sex. Indeed, dominance per se might not exist in many mated pairs. Females commonly initiate group movements, ascents into sleeping trees, and other activities. However, intergroup dominance probably depends on a paired male's prowess in calling and displaying during territorial contests.[519]

White-handed gibbon groups range from two to six or seven individuals, the largest comprising a mated adult pair, an infant (or rarely twins), two juveniles, and a subadult. Generally no other fully adult conspecific animal lives with a white-handed gibbon group, and peaceful temporary aggregations are extremely rare.[522] Nonetheless, researchers found during a two-year study of *Hylobates lar* at two sites in Khao Yai National Park that, although social monogamy characterized 85 percent of groups, 9 percent ($n = 3$) of groups were socially polyandrous, and 3 percent ($n = 2$) comprised one male and two resident females. Of the sixteen groups that Reichard and Barelli monitored for ≥ 1.5 years, 25 percent experienced periods of social polyandry, and 13 percent exhibited social polygyny at some time.[520] Of thirteen groups monitored for over 5 years, 85 percent experienced social change of one adult: two old females probably died, and in three cases males immigrated into a social group and either displaced the resident male or associated polyandrously.[520] A subadult male displaced a Khao Yai resident adult male and was later joined by his two juvenile siblings after their mother died.[523] Female Sumatran white-handed gibbons also sometimes changes mates.[524]

At Kuala Lompat in peninsular Malaysia, a male white-handed gibbon and a male siamang jointly frequented fig-laden trees. They often sang together in the morning and traveled together during part of the day. One day, twenty-five minutes after they had called, a female gibbon appeared. The male gibbon attacked and briefly grappled with her and then followed her away from the siamang. The next morning, the bisexual pair sang together, with female climactic territorial great-calls. They fed and traveled together for three hours. Then a group of three gibbons viciously attacked them and left with the female.[525]

Both subjects were alone the next day. The male called, but the female began to travel and sing with the lone siamang. Three days after their separation, the two white-handed gibbons were reunited and resumed territorial morning song bouts. During the next attack by the three gibbons, the new pair was victorious. The female was an active combatant, which is unusual among white-handed gibbons. The pair became firmly established in the area and produced an infant thirty-two months after pairing.[526]

The MacKinnons concluded that lone male calls attract prospective mates and that they are tolerated by neighboring groups; but when lone females duet with lone males, other groups are threatened territorially and will attack them.[525] Their speculation that hylobatid territories are more permanent than the families that occupy them is less defensible.[15]

An eighteen-year longitudinal study of Khao Yai *Hylobates lar* indicated that the manner of new group formation and relationships between adult pairs and subadult group members have been overgeneralized. Only two of seven cases of dispersal involved pairings of subadults that dispersed into a new territory. The remaining five subadults—five males and one female—remained in their groups for two years after maturing and then moved only one or two territories from their natal groups. At least two of the males replaced adults in a neighboring group, one by forcing out the resident male, after which two juveniles that were probably his brothers joined him. A young juvenile that was probably fathered by the displaced male remained with them. They lived harmoniously with the youngsters freely playing together and the adults grooming them regardless of parentage.[527]

In seven pair-living and six multimale white-handed gibbon groups at Khao Yai, though females and males were codominant overall, females exhibited greater leadership in coordinating group activities. They consistently headed group travel, followed by a constant order of the other group members; the females also arrived first at feeding sites and had priority access to resources. Cycling females led movements more often than the pregnant and lactating females.[528]

The social behavior, activity budgets, territory sizes, and ranging strategies of lowland dark-handed (agile) gibbons at Sungai Dal in peninsular Malaysia are basically similar to those of white-handed gibbons.[529] Group sizes of seven monogamous families ranged from three to six individuals. Allogrooming and play were very rare in the focal family of four that Gittins followed from roost to roost until he lost contact with them.

Adults of the focal group were codominant. When they moved single file, either one was likely to lead. The family moved and slept as a unit. But during the day the male and juvenile were over 30 meters away from the female (and infant) 12 percent of the time and over 10 meters away from them for more than 50 percent of the time.[530] They spent 5 percent of their waking period calling. The focal male devoted 13 percent of his activity period to territorial behavior.[529] The chief protagonists in disputes were adult males (76 percent of observations) though immature males (15 percent) and adult females (10 percent) sometimes also participated. The focal family averaged 0.69 territorial disputes per day. There was no dispute on 47 percent of days, one dispute on 40 percent of days, and two disputes on 11 percent of days. Once there were four disputes on the same day.[531]

Males seemed unable to space out disputes. They occurred randomly in response to chance meetings of neighbors near the boundary of their territories. The exact location of disputes seemed to depend upon the availability of suitable trees from which the males could display. Disputes were not correlated with a time of day or particular seasons. They lasted between 1 and 147 minutes; the median duration was 28.5 minutes.[531] Initially, the combatants sat about 15 meters apart in the upper canopy while other group members rested 10 to 20 meters behind them. If the dispute lasted more than 10 minutes, they moved further away from the males and resumed foraging.[531]

The males stared, faced away, moved about their stations, lunged toward one another, and vocalized. Then they hung with all four limbs outstretched to expose maximum surface area to the opponent. After 15 to 20 minutes of displaying, they vigorously chased one another back and forth over the boundary while avoiding deep penetration into the other's territory. No physical contact occurred during chases, and no males bore facial scars from bites. Intermale dominance status was not apparent. Unlike white-handed gibbon contests at Tanjong Triang, at Sungai Dal all dark-handed gibbon contests were two-sided. Gittins concluded that the extensive display behavior of dark-handed gibbons enable them to assess their relative fitness conservatively and to break off contests before injuries occur.[531]

Malayan siamang groups comprise an adult pair and up to 4 immature individuals.[532] They are notable among hylobatid apes for cohesiveness and paternal care of weaned infants and juveniles.[533] They feed, rest, groom, and sleep in the same tree or in closely juxtaposed trees that are usually connected by climbers. A siamang is rarely more than 30 meters from its group.[534] One notable exception was a female that birthed while her group was out of sight.[535]

Siamang travel single file as a close-knit unit. In changing from one activity to another, there are only brief intervals between first and last commencements by members of the group. On average, siamang began daily activities within 14 minutes and retired within 19 minutes of each other.[534] The entire group engaged in the same activity 73 percent of the time; synchronism decreased steadily from dawn until early afternoon. Adults stayed closest to one another in food trees, with the infant nearby and older immatures farther away. The subadults closed the distance during rest bouts and travel. Chivers explained the marked cohesion of siamang as the product of social development in the young vis-à-vis patterns of parenting by the adults.[536]

Initially, a siamang mother is responsible for her infant. But when it is 8 to 10 months old, the male carries it during the day. Thereby, the youngster acquires many of its feeding habits from close association with the male. At night, the infant sleeps with its mother. Generally, by the time she has a new infant, its nearest sibling is a juvenile that can feed itself and negotiate all but the most challenging sections of their arboreal highways. At night, the juvenile sleeps with the male. A subadult siamang becomes intimately acquainted with group youngsters when it approaches the male for grooming, and then it might play briefly with his charge.[534]

Siamang allogroom during day rests and when they settle for the night. On average, one study group allogroomed 15 percent of their waking period.[534] The adults and subadult groomed during the day, and the adults groomed their young sleeping partners in the evening as they settled in the lodge tree. During rest periods, youngsters commonly played or rested while older individuals allogroomed. The male reciprocally groomed the subadult male even though there was tension between them during feeding bouts. Overall, the male groomed much more than he was groomed.[537]

Conversely, there was parity in bisexual allogrooming in three pairs of Sumatran siamang at the Ketambe Research Station in eastern Indonesia. Like conspecific white-handed gibbons, male siamang were more responsible than females were for close proximity. There were different mechanisms for its maintenance—substantial paternal care and less intragroup food competition in siamang compared with white-handed gibbons.[538]

Although male Malayan siamang were the foci of many group activities, females often led them around the day range. Females were generally the first to descend from and to enter a lodge tree, and to shift from one activity to another. Chivers summarized the siamang group's behavior as a compromise between following the female and remaining near the male.[534]

Female siamang are sexually receptive for several months every two to three years, after their current infant is fully weaned. Breeding in Malayan siamang is periodic, probably being timed so that pregnancy coincides with spans of increased food sources in the habitat.[539] They copulate ventroventrally and dorsoventrally.[540] Grooming frequency between the male and female correlates positively with frequency of copulation.[537]

Female and male subadults received the same peripheralization and grooming from the adult male and avoidance by the adult female.[541] Subadult departures commonly coincided with the births of infants. Nonetheless, siamang adults are generally more tolerant of subadult group members than are white-handed gibbons. At Kuala Lompat in western Malaysia, the male was the most and the female was the next most aggressive toward the subadult. The male was four times more aggressive than she was. The male juvenile was the second-most common target of intragroup aggression. The male attacked the juvenile more as he grew older. Male aggression also increased toward the female during the mating period.[534]

Mitochondrial DNA from eighteen adults in seven siamang groups at Way Canguk Research Station in southern Sumatra indicate that males more frequently immigrated to neighboring groups, but females dispersed farther: 50 percent of adult males versus only 16.7 percent of females in five contiguous groups shared a haplotype with a member of an adjacent group. In three multimale groups, each with two males and one female, they all had different haplotypes. In a fourth multimale group, the haplotype of one of the males was identical with that of the group female.[542]

At Kuala Lompat, a subadult male left his family after intensified episodes of aggression toward him by the adult male. He took up residence in an area of the forest that was unoccupied by siamang, and a young female joined him. Six months later, an older female periodically joined them. Whereas the male initially was most attentive to the young female, he ultimately shifted his interest to the older one. He played exclusively with the young female and copulated with the older female. The neighboring group from which he had been peripheralized increased their singing as the new group formed but later lapsed back to previous levels. No fight occurred between the groups. Eventually the old female disappeared.[543]

Siamang exhibit lower levels of territorial behavior than those of white-handed gibbons. Unlike white-handed gibbons, which call daily, and agile (dark-handed) gibbons, which call twice per day, siamang call once in three days. Overt intergroup conflicts are infrequent. When siamang groups encounter one another, males display and chase back and forth over the territorial border while females and youngsters hide. Otherwise, they maintain their relatively small territories by group morning song bouts and by regularly ranging near the boundaries.[544]

Because of low-level indigenous hunting pressure and the pristine state of a forest in central Siberut Island, Indonesia, researchers were able to document the processes of group formation, fragmentation, and territorial behavior by *Hylobates klossii* in a habitat that was free of logging and roads.[544] The basic social unit of Mentawai gibbons is a monogamous adult pair with zero to four youngsters.[545] The mean group size in central Siberut varied between 3.5 individuals ($n = 16$) and 4.2 ± 1.4 individuals ($n = 12$).[546] In addition to monogamous groups at Sirimuri, Tenaza observed fragmentary social units: two unmated (perhaps widowed) females, one accompanied by a subadult female and the other by a juvenile male; two unmated males that each held a territory; a courting pair; and four single peripatetic males that lived in a forest that was peripheral to the hill forest that was dissected into gibbon territories.[547] Resident females chased trespassing females, but males tolerated them, and males chased trespassing males while females were more tolerant of them. Females led group travel through their ranges, and the males followed and guarded them against predators.[548]

Although Tenaza's initial study indicated that Mentawai gibbons occupy very small areas (5 to 8 hectares), later research revealed that the home ranges at Sirimuri (15 and 20 hectares) are comparable with those of other hylobatid species.[549] Further, Mentawai gibbon groups at Paitan had home ranges between 31 and 35 hectares.[550] A greater density of gibbons in the Sirimuri area might have resulted in the smaller ranges there, as evidenced by the higher levels of territorial behavior at Sirimuri than at Paitan.[550]

Because the boundaries of gibbon territories commonly lie along ridge tops and rarely, if ever, occur along valley bottoms, the shapes of their ranges might be determined partly by topography.[550] Moreover, like the MacKinnons, Whitten speculated that gibbon territories and home ranges might be more permanent than the groups that inhabit them.[525] After a hunter shot the male of a focal study group, the female and juvenile dispersed, and another group of three gibbons entered the home range from beyond the study area.[550]

Tilson provided the fullest account of group formation in Mentawai gibbons, based on a twenty-one-month study in which he focused on social change, especially the transition from unmated to mated status.[551] During a thirty-month period, five *Hylobates klossii* disappeared from the Sirimuri area. Although one senile female might have died naturally, hunters probably killed the others.

Adults in Mentawai groups are increasingly aggressive toward like-sexed youngsters from the age of 4 years until they exclude them from the group, generally by 8 years old.[549] Females are not as markedly peripheralized as the males, probably because the latter must establish territories whence they can attract mates via songs.[551] Over the span of Tilson's study, three subadult males and one subadult female were excluded from family groups, and four others (three males and one female) were peripheralized.[551]

Although they aggressively repel maturing family members, adult Mentawai gibbons also help them to form partnerships. Adults that presumably were related to one partner in three new pairs facilitated successful pairings in two of them.[551] After the adult residents of a territory disappeared, two peripheral males from adjacent groups entered the area and competed for it with calls and chases. Then a third subadult male arrived with his family and joined the contest. There followed a period wherein the family returned to their own territory, leaving the male behind; then he rejoined them, then the entire group returned to the area and left him behind, and so on. Eventually, a solitary female joined him. After a period of breaks and reunions, they were bonded and produced an infant thirteen months later.[551]

In the second case, a lone male established himself in a space adjacent to a group that contained a peripheralized subadult female. The group expanded their range toward him. He began to follow them and even entered their home territory without being attacked by the resident male. After he embraced the squealing subadult female, they began to spend more time together. Eventually she joined him and directed her song displays toward her home group. Her natal group returned to their original territory, the new pair mated, and henceforth they successfully defended the boundary between the two territories.[551]

At Sirimuri, the third successful new group formed when a solitary male established a territory and attracted a subadult female to it from an adjacent group. They produced an infant sixteen months after the courtship began. The female returned to her home group several times before settling in the new territory.[549]

Another mechanism for new group formation and territorial establishment is suggested by a foiled attempt by Mentawai gibbons to assist new group formation. A family accompanied their peripheralized subadult male member into a neighboring territory where the males fought with the resident male and displaced the group. But their dominance was temporary. When the subadult was left to defend the area, the usurped group returned, and, after several encounters between the groups, took it back. The subadult male disappeared.[549]

A subadult male replaced the male in his family after the latter had disappeared. Three solitary males visited the area and commenced countersinging. When one of them moved on the female, she fought vigorously in the manner of male-male confrontations. The invader then chased the subadult male but failed to usurp him. Five days later, after six song bouts, the couple copulated twice. They continued to sing and mate and produced an infant 15 months after the adult male had disappeared.[551]

Another subadult male that had been excluded from his group for thirteen months rejoined them after the male disappeared. He embraced the youngsters and was tolerated by the female. They did not mate before the study ended. Widows lacking subadult family males with which to mate can hold their territories for some time against invading males. But whether they can do so indefinitely and whether they eventually accept extrafamilial males is unknown.[551]

Tenaza and Tilson concluded that there is high probability that Mentawai gibbons in neighboring territories are close relatives.[552] A major sociobiological advantage of this arrangement is that the far-carrying sirening and alarm trills of males warn neighboring groups and dependent relatives about the presence of aboriginal hunters. Their speculation that this also might hold true for other hylobatid species is not supported with data.

Tilson's assumption that subadult males that replaced the males in their families were the sons of the females is reasonable but unsupported. Recall that the younger siblings of a male usurper joined the new family of Thai *Hylobates lar*.[553] One of them might later replace him, giving short-term observers the erroneous impression that the couple exemplified mother-son inbreeding.

Tenaza might have overgeneralized the role of same-sex aggression in the peripheralization of subadults in hylobatid species.[548] For example, apparently among siamang the male is responsible for peripheralizing subadults of both sexes.[541] Clearly, further long-term studies on group formation in all species are needed before we can generalize about hylobatid monogamy and other social relations.

Western hoolock gibbons *(Hoolock hoolock)* live in monogamous family groups, though there are indications of polygamy.[554] Further, an all-male group comprising five adults and two juveniles and solitary females and males occurred in patches of forest in the West Garo Hills of northeast India.[555] In a three-member Garo group, an adult male took a crying male neonate from a subadult male, bit and otherwise abused it, and dropped it to the ground. He had been moving back and forth between his group and a lone female before taking the infant, which its mother had willingly given to the subadult male.[556]

In the Hollongapar Forest, Assam, India, the mean size of twenty-four hoolock groups was 3.2 ± 0.8 individuals. Families contained between zero and four youngsters. Two lone young males and a single subadult female roamed across several family territories. The lone males were quiet, and the female moved toward the distant calls of gibbon pairs. In one family, the male displayed toward the subadult when he entered their food tree; the female ignored him. Tilson estimated birth intervals to be between two and three years.[557] During the winter months, each family regularly sunbathed in a few sparsely leafed, emergent trees that were centrally located in their territories.[557] Perhaps the absence of intergroup encounters was because their visibility indicated to neighbors that the area was occupied.

Nomascus concolor in central Yunnan, China, and *Nomascus hainanus* on Hainan Island, China, indicate that *Nomascus* might be the most socially and perhaps reproductively polygynous genus in the Hylobatidae.[558] The mean group size of seven complete groups of *Nomascus concolor* at three sites in the Wuliang Game Reserve was 6.57 individuals (range: 3–10 individuals). One group comprised an adult female, a subadult, and an infant; the other six groups comprised an adult male, one to four adult females, zero to one subadult, and one to five juveniles, but no infant, perhaps due to low visibility.[559]

Another study, some of which was based on vocalizations instead of direct observations, indicated that Wuliang gibbons lived in smaller groups than those noted by Haimoff et al. and challenged the claim that they are commonly reproductively polygynous.[560] Nonetheless, via direct observations, Fan et al. determined that all five groups in a population on Mount Wuliang in Dazhaizi comprised one adult male, two adult females, and two to five subadults, juveniles, and infants for total mean group sizes of 6.2 and 6.4 individuals over a two-year span.[561] Total group size ranged from five to eight individuals. Three of the focal groups were stably polygynous and generally harmonious over a six-year span.[562]

The adult females in a group exhibited little overt aggression toward one another. In a group of seven, the adult females groomed one another and one another's infants, but were spatially separate while foraging and never slept in the same tree. Once a pair of resident females interrupted the duet of a lone female with their male, and then they chased her. Ten days later, she left the territory. Subadult females and males dispersed from their natal territories. Subadult males were often spatially distant from their groups while moving and especially from the adult male while foraging. They sang solos in the natal territory before dispersing, but never in the same tree with an adult male.[563]

Over a six-year span, two of the four groups in the remaining population of twenty-one *Nomscus hainanus* were stably polygynous, and the others were monogamous. Group sizes ranged from four to seven individuals with a mean of 5.25. Liu et al. observed no polygynous mating but inferred it from the age structure of the groups.[564] Adult males led group movements and other activities and exhibited paternal care of the young. The groups were cohesive and harmonious while foraging, feeding, playing, and resting.

The current state of research with free-ranging gibbons is inadequate to resolve questions about the extent of polygamous reproductive mating and long-term pair bonding in the Hylobatidae, let alone to weigh any potential causal factors in specific cases:

- spatiotemporal distribution of food, lodge trees, and other resources,
- the expanse of habitat available for dispersion and new group formation,
- the number of available mates, especially in small populations,
- cues and freedom of partner choice,

- paternal care, and
- potential for infanticide.

Genomic data, in addition to more numerous observations of intragroup and extragroup copulations, are essential to identify parentage and the nature of kin relations in groups, including those that are part of long-term studies with well-habituated subjects.

As with orangutans and gorillas, there appears to have been no single pan-specific mechanism of hylobatid dispersal or social structure for the individuals that matured in groups that were subject of long-term studies. Anthropogenic factors, especially deforestation, and in some populations a history of human predation by indigenous people and researchers collecting specimens surely have constrained the options of many gibbons for dispersal and new group formation. Nonetheless, one can reasonably conclude that regardless of causes, gibbon sociality is much more flexible than was suggested by the mid- to late-twentieth-century researchers.[565]

But What about Us?

> Food sharing in simple societies is both an index of cooperation and a key symbol of what it is to be human. More generally, an ethic of communalism and equal access to resource production is highly developed.
>
> BRUCE M. KNAUFT (1991, p. 393)

Elusive Ancestral Social Patterns

> The extent to which modern hunter-gatherer patterns represent ancestral patterns is a complex question. For example, most of our sample societies were censused after the elimination of warfare, and many had been geographically displaced or lived in environments that had been substantially depleted of large game. All modern hunter-gatherers use projectile weapons that were not available to our distant ancestors.
>
> KIM R. HILL, ROBERT S. WALKER, MIRAN BOZICEVIC, JAMES EDER, THOMAS HEADLAND, BARRY HEWLETT, A. MAGDALENA HURTADO, FRANK MARLOWE, POLLY WIESSNER, and BRIAN WOOD (2011, p. 1288)

Theorists who aim to devise scenarios of Plio-Pleistocene and early Holocene hominid mating patterns and societies are limited to skeletal features, especially those related to sexual dimorphism, that might assist in narrowing down possible analogous patterns among the notable variety of basic reproductive and social patterns exhibited by apes and other nonhuman mammals. The estimated gross body mass and degree of canine teeth dimorphism are the most common indicators.[566] With some modern hylobatid species in mind, minimal or no skeletal dimorphism might indicate monogamous mating and pair-bonded bisexual couples living with their progeny, both sexes of which as adults must form new pairings. Larsen's interpretation of a small sample of femoral heads that indicated monomorphy, combined with similar indications for the canine teeth of Hadar *Australopithecus afarensis,* tilted in favor of a monogamous social structure, perhaps with pair-bonded bisexual couples.[567] However, given that some gibbons engage in extra-pair matings and polygamous groupings despite sporting no sexual domorphism in canine teeth size, body mass, or other features is reason to doubt the validity of the monogamous scenarios for ancestral Hominidae.

When bimodal patterns emerge in analyses of morphological features, theorists turn to great apes patterns, usually those of African apes, and posit scenarios of polygamy or more specifically polygyny because it is assumed that the larger morph is male and the smaller morph is female in fossil hominid samples.[568] But different traits can recommend conflicting scenarios. For instance, whereas a high degree of sexual dimorphism in body mass of 3.0–3.9-Ma *Australopithecus afarensis* might indicate polygyny, the low-moderate degree of canine dimorphism might suggest monogamy. A third scenario emerged from an ingenious study of fourteen mandibular molars that came from eight individuals who probably died contemporaneously at Hadar Site 333/333w in Ethiopia. Discrete dental trait analysis of four molar cuspids and groove patterns in four males and four females (based on molar dimensions) led McCrossin and Reyes to conclude that *Australopithecus afarensis* were neither monogamous nor polygynous.[569] Instead, because the inferred females shared only one trait, while the males shared four traits, the social organization of *Australopithecus afarensis* at Hadar Site 333/333w seemed most similar to that of *Pan troglodytes*—male philopatry in a multimale-multifemale group.

Strontium isotopic readings ($^{87}Sr/^{86}Sr$ ratios) from the dental enamel of Sterkfontein *Australopithecus africanus* and Swartkrans *Paranthropus robustus* have suggested that smaller individuals assumed to be females in both species were more likely to have dispersed from their natal groups than the larger individuals (assumed to be males). A higher proportion (about 50 percent) of small individuals ($n = 10$) had nonlocal strontium isotopic compositions than found in the larger individuals (11 percent, $n = 9$). The larger individuals either had smaller home ranges than those of the small individuals or preferred dolomitic landscapes. Copeland et al. speculated that the dispersal patterns of the Plio-Pleistocene hominids that roamed over the geologically diverse substrates in the Sterkfontein Valley might have been more like those of chimpanzees and some historic human groups that are male philopatric than like the dispersal pattern of gorillas, in which both males and females leave their natal groups.[570] Nevertheless, they wisely concluded that it is unlikely that there is a modern analogue for the social structure of Transvaal Pliocene-Early Pleistocene Hominidae.

Testicles mutely testify that regardless whether some precedent hominid species had a multimale-multifemale breeding system like that of *Pan troglodytes,* at some point the ancestors of *Homo sapiens* began to evolve more toward polygynous or monogamous patterns. If the hypothesis is correct that sperm competition is important in the association of testicular size specifically with breeding system, humans do not fit the chimpanzee pattern. Human testes are not only notably smaller than those of chimpanzees, but also they have a lesser volume of seminiferous tubules versus interstitial tissue and produce fewer sperm.[571] Indeed, the ratio of tubules to connective tissue in human testes is like those of gibbons and other generally nonmultimale breeding catarrhines and is markedly unlike those of chimpanzees and other catarrhines in which males engage in intense sperm competition.[462]

Inferences on social organization in Middle and Late Pleistocene Hominidae based on tooth and body size are particularly unreliable in the absence of intentional burial of individuals who appear to have died closely in time. Although at Locality 1 Zhoukoudian, China, there were remains of about 51 individuals, including craniodental remains of ten adults and seven infant and juvenile females, and fifteen adult and ten infant or juvenile males of *Homo erectus.* The presence of nonhominid bite marks on 67 percent of the hominid bones and controversy surrounding controlled use of fire precludes clarity on how the bones got there, let alone regarding the foraging behavior, ranging patterns, and social organization of the hominid visitors to the cave.[572]

At least some Neandertals buried their dead in caves and rock shelters.[573] Although many of the grave sites contained single individuals, some contained several individuals. For example, the La Ferrassie rock shelter in Dordogne, France, contained an adult male, an adult female, two children (about 10 years old and 3 years old), a neonate, and two fetuses. Several levels in Shanidar Cave, Iraq, contained five adult males, one adult female, a young adult, and two infants, who appear to have been deliberately interred.[574] Because very few bodies were accompanied by unarguable configurations of objects that might indicate ritual or spiritual beliefs, some paleoanthropologists argue that Neandertal burials are not comparable to funerals or even intentional burials.[575] Accordingly, there are doubts that Neandertals had achieved full humanity despite their large brains and impressive tool kits.[576]

Indeed, there is insufficient data to discern specific social relationships of Neandertals other than to assume that they were probably affiliative, with groups of adult males, females, and young living together in groups large enough to reproduce and survive at low population levels for many millennia in a wide variety of temperate, Mediterranean, and cold environments. High adult mortality probably caused major demographic and social instability not only among Neandertals but also among Middle and Upper Paleolithic *Homo sapiens* in Eurasia and northeastern Africa.[577]

Similar interpretive limitations are true for anatomically modern people that were contemporary with Neandertals in the Late Pleistocene and early Holocene. Moreover, in the absence of written records by truthful, knowledgeable living informants, archeologists depend heavily on ethnographic reports, grave goods, and architectural configurations on the landscape to devise scenarios on the structure of past human societies.[578] This is no easy task in view of the challenges that greeted social and cultural anthropologists who tried to decipher the highly complex social relationships and structures among people who were less technologically outfitted than they were.

Globally, between 2.6 Ma and 10 Ka hominids relied on gathering and later variably with some hunting. An analysis of coresidence patterns in thirty-two extant foraging societies has indicated that a mean band included twenty-eight adults in which most residents were genetically unrelated; individuals of either sex might disperse or remain in their natal group; and adult brothers and sisters often coresided. Hill et al. concluded that such patterns produced large social interaction networks of unrelated adults, which might assist in explaining why humans evolved capacities for social learning that led to symbolic language and cumulative culture.[579] They further suggested that extensive cooperation in hunter-gatherer bands could not be explained by inclusive fitness—the effect of an allele or genotype not only on an individual but also on related individuals that possess it (i.e., kin selection).[580] Hamilton et al. demonstrated that the area required by an average individual hunter-gatherer decreases with increasing population size because social networks of material and information exchange introduce an economy of size.[581]

The economic environment often determines whether fair or selfish people dominate the equilibrium.[582] Because of remarkable individual heterogeneity, interaction between altruists and selfish individuals is vital to human cooperation in hunter-gatherer and other extant human societies. A minority of genetically related or unrelated altruists can force a majority of selfish persons to cooperate, and a few egoists can induce a large number of altruists to defect. Individuals are much more likely to cooperate in two-party interactions if future interactions are likely than when larger groups are involved. Language facilitates reputation building and loss. Broadcasting the trustworthiness of individuals allows one to avoid interacting with cheaters and to develop congenial relations with altruists.[583] Human language further provides the means to devise and to disseminate codes of behavior and sanctions that enforce them.

Kith, Kin, and Are You Kidding?

> The special quality of kinship . . . is "mutuality of being":
> kinfolk are persons who participate intrinsically in each other's
> existence; they are members of one another.
>
> MARSHALL SAHLINS (2013, p. *ix*)

Globally, there is large-scale variation in kinship systems, which might have deep roots, as evidenced by genetic, linguistic, and archeological findings.[584] With sufficient samples of DNA, geneticists can demonstrate paternity, maternity, and with lesser reliability siblingship and more remote biological kin in apes and other mammals. However, the extent to which genomics affects their social relationships, mating behavior, and dispersal patterns is elusive and open to speculation.[585] Nonetheless, we can be reasonably certain that although some of them evidence preferential affiliation with and support of genetic relatives, they lack humanoid concepts of kinship, and their perceptions of kin are probably different from those of humans. We learn via verbal teaching who is to be regarded as mother, father, and many other kinship relationships, and we also learn how we should behave among them according to societal norms and sanctions.[586]

In humans, unlike apes, much of the information is verbally transmitted and usually encompasses a wide spectrum of living and deceased persons, some of whom lack genetic or marital linkage but nevertheless are considered to be one's kin. In other words, individuals who might be considered kith in one society can be considered kin in another and vice versa.[587] Sports teams, military units, firefighters, police squads, gangs, religious groups, gay, lesbian, and transgendered couples, and communities and other affiliatively bonded people commonly employ kinship terms to refer to one another (Plate 16). Even pets can be sincerely referenced as kin: I commonly refer to our daughter's cat as our grandcat and our son's dog as our granddog, though I never referred to pets this way before our children became independent.[588]

Sires and Dads: Instinct or Invention?

One outcome of the expansion of behavioral flexibility from prosimians to humans is apparent in the spectra of social organization and sexual behavior of males and the variety of their involvement in the lives of their progeny.[589] Fatherhood entails much more than siring young. Although all men who inseminate women are sires, their participation in nurturing, socializing, and educating their progeny varies from none to an intimate lifetime commitment. The latter are dads, and the behavior of men at the other end of the spectrum is reminiscent of nonhuman males. Moreover, men who partner women with children that they did not sire often take on the role of father and are accepted as such by her children and the community, and gay men can love their adopted or sired children as much or more deeply than many heterosexual sires.

Healthy human males experience sexual urges year-round from their preteenage years to their twilight years. Low percentages of them are exclusively homosexual or both homosexual and sexually active with women.[590] When females are not accessible, many basically heterosexual males resort to onanism, homosexuality, pedophilia, incest, or bestiality despite religious and societal proscriptions against such behavior.[591] Attempts to link human homosexual activities phylogenically to the displays and play behavior of apes have failed to take account of the remarkable variety of sexual activities in which men engage, their reasons for doing so, and the benefits they might receive therefrom.[592] Interestingly, no field observer has demonstrated that, given free choice, male great apes in all-male or mixed groups that have engaged in sexual acts with other males would not mate exclusively with females were they free to do so.

Some men are content to sire children while having little or no long-term commitment to copulatory partners or progeny. Others enter long-term partnerships but covertly copulate opportunistically with women other than their spouses. The reported mean paternity discrepancy varied from <0.08

percent to 30 percent of children in seventeen sizeable samples from the Americas, Europe, and New Zealand: however, other studies have indicated values at approximately 2.0 percent.[593] In societies that sanction polygyny, wealthy, powerful men marry several women, have concubines, and engage in sex with other women, girls, men, and boys.[594] In brief, there is much in the behavior and fantasies of men and women to cast doubt on *Homo sapiens* being biologically predisposed toward monogamy.

Mothers, Grandmothers, and Others

> Functionally viewed, an act of fellow service may be more or less unselfish, but it is doubtful that any act, even in man, is purely altruistic.
>
> ROBERT M. YERKES (1939, p. 111)

Although it is hardly trivial, the mythic Old Testament punishment meted out to womankind—pain during childbirth—because of one female's fruity faux pas pales in comparison with the cumulative stresses women face while trying to ensure the survival and well-being of their children.[595] In many societies, sires are not reliable or available helpers, even if they remain with the family as providers and protectors. Consequently, women welcome daily assistance from their mothers, older children, and other female kin and friends and, less commonly, male relatives and friends. Fortunately, infants and young children have faces and vocalizations that attract older individuals, who want to hold, feed, and entertain them. The plight of fecund women, particularly those with no means of contraception or power to avoid copulation, can be especially difficult.[596] Polyandry, commonly comprising a woman and brothers who are unrelated to her by birth, is rare; in the human societies that sanction polyandry, it is not the norm.[597]

Grandmothers can be a boon to families blessed with them, women who have survived in reasonably good health long past their own reproductive years. Of course, they must live with the reproductive unit or at least nearby. When a woman lives with a mate away from her own mother, his mother might be a dedicated grandmother to her children. Grandmothers and grandfathers have a trove of knowledge and experience, much of which takes years for younger individuals to acquire by observation, direct instruction, and practice under the eye of an expert. They are also the source of narratives that inform younger generations about places and other physical resources, past events, and social relationships that might be important for group well-being and survival. They may also contribute significantly to the larder by collecting surplus food items, fuel, and other resources to share with group members, particularly their children and grandchildren.

Although the onset of postreproductive periods in female great apes, particularly captive ones (35 to 40 years old) is close to those of women (40 to 41 years old), apes commonly die much earlier.[598] Indeed, whereas ≥50 percent a woman's life span might be postreproductive, that of a captive female gorilla is about 25 percent of her life span.[599] Reproductive decline is gradual in human males and in both sexes in other animals, but it is relatively abrupt in human females.[600] Williams proposed that by being free of pregnancy and birthing, a 45- to 50-year-old woman could focus her energy on care of her living children instead of concurrently producing new ones: the stopping early hypothesis.[601]

The grandmother hypothesis emphasizes the evolutionary biological role of older postreproductive women in provisioning descendent kin with food plants that are difficult to acquire, such as digging up deeply buried underground storage organs and carrying heavy loads back to camp. Proponents of the grandmother hypothesis argue that because of uncertainty about paternity men might often focus on mating competition with other men instead of paternal activities, leaving the nutritional welfare of children to their mothers. Further, the meat and other foods that male hunters acquire are irregularly available and are customarily shared widely with many members of the band instead of exclusively with close kin: the showoff hypothesis. Consequently, nursing mothers would have difficulty provisioning both themselves and their weaned children without the consistent assistance of grandmothers.[602]

While not rejecting the grandmother hypothesis outright, Kaplan et al. pointed out that it underplays the importance of a long juvenile period to facilitate learning complex skills and to develop a large brain, which underpin high productivity later in life.[603] Further, they reinstate males as important providers for the group. For example, male Hadze hunter-gatherers produce more food than the postreproductive Hadze women, who had inspired the grandmother hypothesis. They correctly challenge that there is no evidence to support the hypothesis that postreproductive women have been the primary breadwinners in most societies in human history; indeed, they can be net consumers instead of net producers in resource-poor societies.[604] It takes notable bisexual production and sharing along with the oft-neglected role of children, who snack while accompanying older foragers, to sustain the fitness of hunter-gatherer bands in many habitats.

What's Love Got to Do with It?

> And though I have the gift of prophecy, and understand all mysteries, if I have not love, I am nothing.
>
> 1 Corinthians 13:2

> Love is more talked about than surrendered to.
>
> CHARLES WRIGHT (1997, p. 73)

What's love got to do with it? As a particularly blessed septuagenarian who loves, has been loved, and is loved, though often pushing his luck, my prompt response is "a great deal!"[605] Both men and women have the potential to feel and to behave tenderly toward their mates, children, and others, and sincere signals of love are important for the development of children and in many other human relationships. The extent to which such behavior is homologous with indications of affection in apes and other nonhuman beings is moot.

De Waal, a leading champion of similarities between apes and humans, eloquently declared:

> Human social organization is characterized by a combination of (1) male bonding, (2) female bonding, and (3) nuclear families. We share the first with chimpanzees, the second with bonobos, and the third is ours alone. It is no accident that people fall in love, are sexually jealous, know shame, seek privacy, look for father-figures in addition to mother-figures, and value stable partnerships. The intimate male-female relationship implied by all of this, which zoologists have dubbed a "pair bond," is bred into our bones. I believe that this is what sets us apart from the apes more than anything else.[606]

There is much with which to agree here, particularly when we move from naïve metaphorical genomics—"bred into our bones"—to general proclivities for the development of love as an emergent quality of human beings which might have precursors in other vertebrates.[606]

Metaphor is not an evolutionary mechanism. The employment of metaphors, such as "selfish gene" and "mitochondrial Eve," in scientific writing generally masks igororance of the actual, far more complex processes that might have been or are operant.

Mothers in many species exhibit tenderness toward their infants, at least until and sometimes beyond weaning. But, uniquely, human bonds not only can persist but also can intensify over a lifetime, regardless of negative behavior by either party in the relationship. These bonds also can expand to include fathers and other kin and kith.[607]

The extent to which human couples bond sufficiently to forego sexual activity with others is highly variable. Women and men can love their primary partners deeply yet still engage sexually with others whom they also love either secretly or with the knowledge or even acceptance of their initial mates. They also love many others asexually. Fundamentally, love is an emergent quality of individual humans that comprises hormonal, experiential, social, and cultural factors, providing an endless variety of personalities, sexual practices, and bonding relationships, only one of which is the monogamous nuclear family in which the parental couple is unfailingly faithful to one another sexually.[608] Indeed, love facilitates the formation and maintenance of multiple pair bonds, both sexually and asexually, which underpin a wide spectrum of marriage practices, including polygamy. It also facilitates less formal close human relationships that transcend age and gender, beyond the biological links between parents and children.

12

Communicative Apes

Ape Vocalizations, Facial Expressions, and Gesticulations

> Although we know a great deal about the communicative
> potential of chimps and gorillas, we understand almost
> nothing about whether animals make use of this potential (or
> the selective pressures that might cause them to do so) under
> natural conditions. It seems clear that research on captive apes
> poses a challenge to those who study primates in the field
> because it illustrates just how little we know about "semantics"
> of primate communication under natural conditions.
>
> PETER MARLER (1977, p. 505)

The remarkable ability of apes to achieve competence in artifactual languages reenergized the imperative to learn what their natural communicative signals mean to conspecific recipients. Have capacities for symbolic communication in ape minds simply laid dormant in their natural, autonomous lives, only to emerge under human-imposed learning regimes?[1] Or did they develop in response to selective factors for other cognitive purposes only to be recruited for artifactual language in laboratory settings? Playback experiments in the wild would seem to be an informative way to complement catalogues of gesture-calls based on observations, recordings, and films of apes that were sometimes included in long-range and short-term studies focused chiefly on their foraging behavior, ranging patterns, and social organizations.

Playback experiments are a valuable tool in studies of monkeys and apes.[2] In the late nineteenth century, Garner, a schoolteacher, businessman, and autodidactic evolutionist, conducted playback experiments with monkeys in a zoological garden near the Smithsonian Institution in Washington D.C. via an Edison cylinder phonograph.[3] In 1892, he ensconced himself in a protective cage at Fernan Vaz, Gabon, to document free-ranging chimpanzees and gorillas speaking in the forest; however, he recorded nothing phonographically.[4] It was not until the 1970s that field researchers productively employed feedback experiments with natural populations of apes, namely, gibbons, orangutans, and chimpanzees.

Gibbonese

> I cannot doubt that language owes its origin to the imitation
> and modification of various natural sounds, the voices of other
> animals, and man's own instinctive cries, aided by signs and
> gestures . . . Some early progenitor of man probably first used
> his voice in producing true musical cadences, that is in singing,
> as do some of the gibbon-apes at the present day.
>
> CHARLES DARWIN (1871, p. 463)

Carpenter was the first scientist to attempt a systematic study of ape vocalizations in the field.[5] He was candid about the limitations of his recording methods and his crude verbal descriptions of white-handed gibbon vocalizations. He urged that especially equipped laboratory studies be conducted to complement field studies. Whereas the functions of high-volume intergroup calls could only be discerned in the wild, subtler intragroup vocalizations need be recorded in a laboratory. He listed nine types of vocalizations and their probable functions, but acknowledged that verbal descriptions fail to convey their richness and complexity and sometimes border on absurdity.[5] Later studies indicated that although gibbon songs are species-specific, they have a capacity for contextual variation.[6]

A wide range of stimuli evoked particular vocal responses among Carpenter's white-handed gibbons, and there was considerable variation in individual responses to them. Nonetheless, because specific vocalizations persisted in captive assemblages of mixed species, Carpenter suspected that there are notable genetic factors underpinning hylobatid vocalizations.[5] His suspicion was reinforced regarding the female great-calls of at least some species.[7]

The loud choral morning song bouts of Carpenter's white-handed gibbon groups served to maintain the integrity of their territories in lieu of daily combat. Morning songs peaked between 07:30 AM and 08:30 AM during the period when groups were traveling the most. Less dramatic chirps, squeals, chatters, clucks, and facial expressions seemed to facilitate group cohesion and harmony and to coordinate group progression through the canopy.[5]

Ellefson augmented Carpenter's glossary of white-handed gibbon signals, most of which were not dramatic vocalizations, facial expressions, and gestures produced by both sexes. Ellefson's study of Malayan white-handed gibbons securely established sexual divocalism in white-handed gibbon territorial calls.[8] The distinction between male and female vocalizations appears very early in childhood.[9] The distinctive morning song bout consists of an elaborate female great-call with a duetted male hoot series coda. A great-call initiates vigorous swinging by adults and chases by males if another group is in view.[9] Contesting males conflict-hoo and perform acrobatic brachiating displays to dissipate agonistic tensions. Thereby, they often avoid actual fighting, particularly if they are away from preferred food sources.[10]

Baldwin and Teleki further expanded the list of white-handed gibbon signals based on six young, free-ranging captives on Hall's Island, Bermuda. Their descriptions of nine types of vocalizations, twelve facial expressions, and nineteen expressive postures, gestures, and movements correspond well with those of Carpenter and Ellefson and those of Chivers for wild West Malaysian siamang and other species of gibbons.[11] Waller and colleagues systematized the identification of eighteen independent facial movements in hylobatids and noted that they are similar to those of other hominoids.[12]

In a wild Tai population of central white-handed gibbons, unmated subadult males in natal territories sang solos along with mated adult males over spans that ranged from a few minutes to less than four hours around sunrise from lodge trees or elsewhere later in the morning. Subadults called more frequently and, on average, for longer periods. In addition, males uttered distinctive calls during territorial chasing matches.[13]

Females called solo less often than males did.[13] A female's single great-call induces duets in which she delivers complete arias that culminate in great-calls to which the male appends a coda. The great-calls were individualistic. Young females sometimes accompanied the great-calls of adults.[13] Most dueting (mean duration: 12 minutes) occurred during midmorning, peaking around 08:30 AM, from anywhere in a group's territory.

In Khao Yai National Park, Thailand, all eight focal groups in playback experiments approached stimulus duets in the centers of their territories, but only two of eight groups approached the sources of the duets on the borders of their ranges. Further, they called more often (four of eight trials) in response to stimulus duets on the border than at the core of their territories (one of five trials).[13]

Adult females led approaches to playbacks of conspecific female solos in the centers of their ranges significantly more often (seven of eight trials) than they led approaches to male solos (none of eight trials) or duets (one of seven tests) at a core. Adult males were generally among the leaders of approaches to stimulus duets (all eight tests) and to male solos (seven of eight trials). They were among the leaders toward only two of eight group approaches to female solos. Only centered playbacks of female solos consistently induced calling, including duets in which a target female was especially vigorous vocally.[13]

Field observations indicated that there is strong intrasexual competition in white-handed gibbons. Although resident males might tolerate alien females, a resident female would evict them. Likewise, a resident male's antagonism toward other males prompts him to chase off alien males and their cohorts.[13]

Playback experiments in Kao Yao National Park also indicated that white-handed gibbons use songs to announce the presence of felid predators and to a lesser extent reticulated pythons, but not crested serpent eagles. In predator contexts versus more common songs, females introduced significantly more hoos, the overall song duration was longer, the female great-call was delayed, males replied earlier to their partner's great-call, and there were fewer leaning *wa* notes and more *hoo* notes, and the song invariably contained sharp *wow* notes.

Neighboring groups appeared to differentiate predator-induced calls from regular duet songs, as indicated by the fact that they responded in kind. Because neighboring groups frequently changed composition and had close relatives of the current group living in them, Clarke et al. concluded that alarm calling is intended to alert kin and current group members to the presence of predators.[14]

Playbacks of solos by female pileated gibbons *(Hylobates pileatus)* in the center of white-handed gibbon ranges induced male-led approaches in 62 percent of trials. A target group called during only one of eight tests, indicating that female white-handed gibbons probably cannot identify the sex of pileated gibbon callers.[13] Because male white-handed gibbons approach them and female white-handed females might not react aggressively to them, hybridization could occur where the two species meet. Indeed, bigamous male gibbons were present in Khao Yai National Park. One of a male's mates was always a conspecific, and the second female was another species or a hybrid.[15]

The morning calls of agile gibbons *(Hylobates agilis)* are similar to, yet readily distinguishable from, those of closely related white-handed gibbons. The agile female's great-calls consist of shorter notes, are more ornamented, and have more sustained decrescendos. The short phrases of males are quite different, and his coda sometimes overlaps the female's finale.[16] Whereas an entire agile gibbon family might join the coda, only the subadult male white-handed gibbons sing it.[17] Sumatran agile gibbon vocalizations are like those of Bornean and Peninsular Malaysian conspecifics.[18]

Male agile gibbons sing solos for long spans, and females rarely sing alone.[19] Males begin morning song bouts with a hoot series, which the females later augment with great-calls. Gittins speculated that the male's singing advertises their territory, and the female's signals to potential rivals that they are already mated. Young males are discouraged from joining adult songs, but young females duet great-calls with their mothers.

The female parts of peninsular Malaysian agile gibbon duets are highly individualistic, as revealed by statistical analyses of field recordings on eight subjects.[20] The same is true for two populations each of Sumatran and Bornean agile gibbons.[21] This might allow them to identify one another while singing contemporaneously, though errorless recognition might be confounded because great-calls from the same individual can vary notably.[21]

Like white-handed gibbons, agile gibbon males conflict-hoo, move about on branches, chase, and hang stiffly in front of opponents, while the females and young stay peripheral to the fracas. In addition to hoos, agile males emit low whistling sounds during display bouts.[22] In West Kalimantan, acoustic analyses and call playbacks indicate that male agile gibbons employ a series of graded signals for intrasexual territorial defense, not to attract mates. There were few differences between the performances of unmated and mated males, and the reproductive status of singing males did not affect the responsiveness of other gibbons to their songs.[23]

In duets of *Hylobates pileatus,* male and female contributions overlap instead of alternating as in bisexual duets of other *Hylotates* spp. However, like white-handed gibbons, pileated gibbon males end the song with a solo coda.[24] A female great-call begins as a series of hoots that rises steeply in pitch and accelerates into a long bubbly trill.[25] A male calls "oh-ah" several times on notes nearly an octave apart and often follows with a brief, low, bubbly trill.

Female great-calls of Müller's Bornean gibbons *(Hylobates muelleri)* are like those of pileated gibbons.[26] According to Marler and Tenaza, instead of singing with a female, a male simply adds a brief coda to her finale; however, the Marshalls noted that instead of duetting, males solo before dawn, and females call later in the morning.[24] Contrarily, Haimoff and Mitani noted that although a female dominates the duet, a male participates complexly early and late in the song bout. They confirmed that males also sing solos.[27]

At Kutai Game Reserve, East Kalimantan, mated males usually soloed for ≤1 hour before dawn about once per week, and solitary males sang more sporadically. The female morning solos were less frequent and consisted of their part of a duet.[28] Pairs of Müller's gibbons usually duetted high in the canopy on dry mornings.[29] Approximately 30 percent of 113 duetting bouts involved interactive singing among neighbors, who tended to sing near the boundaries of their ranges.[30] Mitani could distinguish the different Kutai groups by individualistic temporal patterns and acoustic features of their morning song bouts.[31]

Mitani conducted playback experiments in order to discern the possible functions of song bouts.[31] In >1,500 hours, the focal group came within 100 meters of another group only seven times, all of which elicited duetting. The focal group countersang five times, engaged in one chasing match, and avoided contact with the other group by unilateral or mutual withdrawal six times.[30] According to the location of recordings in their range of around 40 hectares, Kutai gibbons reacted to playbacks of duets by some or all of the following: countersinging, display swinging, uttering alarm hoos, and approaching the stimulus. They appeared to not distinguish recordings of their own duets from those of other groups.[31] They reacted most dramatically to calls that were central in their range, not only by engaging in all of their usual behaviors but also by remaining within the playback area for extended periods and increasing the duration of their song bouts after the disturbance.[30] Playbacks of duets near the boundary of their range consistently elicited only countersinging from the residents, though they might also approach or move away from the stimulus thereafter. Duets within a neighboring range rarely stimulated countersinging or approaches by the target group.[30]

The female of a Kutai couple commenced the duet, usually alarm hooed, and led the group movement toward or away from duet playbacks in their territory. Playbacks of female solos elicited the same response by the target group. However, after playbacks of male solos, the male of the target group silently led them toward the stimulus. Mitani concluded that adults of each sex defend the group range against like-sex intruders, albeit with backup from her or his mate.[31]

Most morning song bouts of Javan silvery gibbons *(Hylobates moloch)* consist of female great-calls that end rallentando.[32] Male hoots occurred between female great-calls early in the bout. Indeed, like Mentawai gibbons *(Hylobates klossii),* Javan gibbon males and females do not perform duets.[33]

Field studies and recorded vocalizations show that West Javan silvery gibbons have four types of loud vocal bouts.[34] As in Müller's gibbons, adult females are the lead singers in territorial maintenance and defense. In fact, male West Javan silvery gibbons rarely sing. During an eighteen-month span, apparently five mated males did not sing, though an unmated male sang once, perhaps to attract a mate.[34]

Javan gibbon great-calls are quite individualistic, so much so that that it probably allows females to recognize one another.[35] When a female commenced a song bout in her territory, females of neighboring territories often began to sing immediately, sometimes producing a chorus. Subadult females commonly joined them in song bouts.

A soloist ascends to the crown of one of several emergent trees in her territory. When she sings more than once per day, she uses a new song tree for each encore. During the cadenza of great-calls, she brachiates rapidly through the crown of the song tree. Sometimes this sends dead branches crashing to the forest floor.[34] Songs of alien females inside or at the periphery of a resident female's territory prompt her to climb the nearest song tree and respond in kind. In three cases, the intruders became silent and fled with the resident males in silent pursuit of the trespassing pairs. Playbacks of alien great-calls also elicited song bouts by the resident female.[34]

In four of ten encounters between Javan gibbon groups, females countersang and parted with no overt aggression between them.[34] In five episodes, upon mutual awareness, all members of both groups screamed loudly. The males brachiated rapidly toward one another and faced off while the cacophony continued. Occasionally, while younger group members fell silent, females uttered great-calls that seemed to prompt vigorous bouts of chasing over all stories of the forest between males that had been sitting opposite one another. Other group members stayed back from the fracas and screamed. The contests usually ended abruptly after a vigorous chase when the silent males wandered off instead of sitting to face one another again. In the shortest conflict, only males screamed and chased while other group members remained silent. All group members freely uttered loud staccato screams and moved about agitatedly during harassing call bouts directed at a leopard.[34]

Mentawai gibbons *(Hylobates klossii)* have ten or eleven types of vocalizations.[36] Two of them—sirening and alarm trills—function as long-distance alarm calls. Males and females emit them when they detect humans on the forest floor. Their morning song bouts are quite distinct from those of siamang and instead are closer to those of other gibbons.[37]

Echoing Charles Darwin, the Marshalls proclaimed the Mentawai gibbon great-call to be "probably the finest music uttered by a land mammal."[38] Males sometimes sang from lodge trees during the four hours before sunrise.[39] At the onset of the fruiting season, some male singing shifted to postdawn. Females always sang during the four hours after dawn and after departure from lodge trees. Their mates remained silent or softly whistled during breaks in their arias.

Generally males respond to singing by other males, and females react to other females. Because of overlaps, the bouts usually developed into unisexual neighborhood choruses. The countersinging males appeared relaxed, and they tended to take turns while countersinging females engaged in vigorous locomotor displays and sang simultaneously.[40] The close proximity of adult males seemed to suppress the vocal efforts of younger males. Male songs are individualistic enough that Tenaza could identify the singers.[41]

Tenaza concluded that male singing functions to maintain exclusive use of their emergent lodge trees, which are scarce due to the high density of gibbons.[41] Starting with a single piping note, a Mentawai male progressed through whoos and whoops to a cadenza of complicated trilling phrases. The length and elaborations of his songs, particularly by comparison with those of females, led Whitten to infer that mere territorial advertisement is not adequate to explain Mentawai male song bouts.[42] They might also provide information about the fitness of potential rivals.

The louder songs and athletic displays of females, which are executed along their territorial boundaries, indicate readiness to chase out trespassers. The lack of bisexual duetting might be a response to human predation.[43] The quiet males, which stay below the calling females, can serve as lookouts while their mates defend the territorial border. Chorusing might reduce the risk of individual victimization from terrestrial predators such as humans.[41]

The morning song bouts of siamang *(Symphalangus syndactylus)* are well documented at several localities in peninsular Malaysia and in numerous laboratory pairs.[44] Detailed sonographic analyses have revealed no distinction between the morning songs of wild Sumatran and Peninsular Malaysian siamang.[45] Inflation of the laryngeal sac produces a boom that siamang pairs use to coordinate and synchronize their duets.[46] The resonant effect of the inflated sacs produces louder calls than those of white-handed gibbons.[47]

Typically, duets begin when stationary adults are close to one another in an emergent tree. They exchange single barks that escalate to variable sets of multiunit barks, after which the male moves away and barks rapidly. Then, as the female continues to execute bark series, the male periodically delivers bitonal and ululatory screams. During the former, he moves back toward the female. After an ululatory scream he moves vigorously about the tree while screaming loudly. Concurrently, a barking female also moves athletically. Maturing family members might chorus with the adult pair, but newly mated pairs seem to require notable spans to develop full-fledged, well-synchronized morning calls.[48]

Unlike white-handed gibbons, male siamang are mute during territorial chases, branch shaking, and staring matches.[49] However silent a siamang might be, they still mean business. Playback calls played centrally in a siamang territory elicited violent visual displays, repeated returns to the transgressed sites, and stepped up patrols of the area. Chivers and MacKinnon concluded that siamang territorial calling is largely innate, albeit with no supportive developmental or genetic datum.[50]

Without enumerating siamang signals, Chivers proposed that their intragroup repertoire is smaller and less often employed than that of white-handed gibbons (Chapter 11).[49] Further, siamang usually sing later in the morning than sympatric white-handed and agile gibbons do.[51] Perhaps siamang closeness and lifelong bonding obviate the need for a larger repertoire of conspicuous signals.

Nonetheless, fourteen captive siamang employed thirty-one different signals in social communication: twelve tactile gestures, four facial expressions, eight visual gestures, and seven actions. They employed tactile gestures and facial expressions most frequently. The repertoire increased from birth until 6 years old, after which it declined. They flexibly employed most signals in ≥3 social contexts, often in combination with other signals, and adjusted them for recipients; for instance, they employed visual signals most often when the recipient was looking at them.[52]

Like siamang, hoolock gibbons (Hoolock hoolock) sing elaborate bisexual duets and do not engage in unisexual choruses or countersinging bouts.[53]

Further, like the duets of *Hylobates pileatus,* the male and female arias overlap. Youngsters of both sexes might join the song. A male begins with a short series of high notes and repeats them until a female joins in. Then both execute accelerated lines of alternating high and low notes. Like siamang, hoolock neighbors tend to wait until one pair has finished a song before beginning theirs. Hence, the duets are audible sequentially through the forest.[54] Sonograms of captive hoolock copulatory sounds indicate that, like humans and baboons, hoolock female vocalizations begin earlier and are more complex than those of the grunting male. Both are elaborations of heavy breathing, but their functions remain obscure.[55]

Vocalizations of adult female and male central Yunnan black gibbons *(Nomascus concolor)* are audible 2 to 3 km from the source. Their songs appear to be male-dominated and controlled. The group male would always begin bouts by emitting single multimodulated notes that were weakly modulated and relatively brief. Females in a family group emitted great-calls synchronously at various times throughout a song bout, most often after the male had produced a multimodulated figure. They sang together while sitting. Males did not solo.[56]

There are specific differences between songs of male northern white-cheeked gibbons *(Nomascus leucogenys),* yellow-cheeked crested gibbons *(Nomascus gabriellae),* and Hainan black crested gibbons *(Nomascus hainanus).*[57] For instance, although the topography of segments is the same in white-cheeked and yellow-cheeked crested gibbons, the frequencies of emission and the duration of components in their segments differ. Differences between the songs of female white-cheeked and yellow-cheeked crested gibbons are less marked than those of males.[58] Further, there is notable individuality in the songs of male black crested gibbons *(Nomascus concolor).* At two localities in China, eight males evidenced that the songs of near neighbors were more individual than those of individuals between sites, suggesting that immediate neighbors can adjust their vocalizations to enhance individual identification.[59]

Calls of captive *Nomascus* sp. are high pitched.[60] The climax of the female great-call is higher pitched than those of other gibbons.[61] Males begin a duet with staccato *eek* phrases that lead to a variable number of inflected shrieks. A female enters with low growls that develop into a whine. Then she soars in pitch, intensity, and speed, abruptly breaks off, and finishes with a descending series of chirps. A male responds with a coda that is more elaborate than his prelude.[62] The utterance of great-calls appears rather suddenly in captive pubescent males.[63]

Wild Laotian white-cheeked gibbons *(Nomascus leucogenys)* duet just before sunrise and sometimes in the late afternoon. Duetting is more extensive on clear days than on cloudy days.[57] Experiments with different groupings of subjects have indicated that the great-calls of male, female, and young *Nomascus* chiefly serve to reinforce their social bonds.[64] In the wild, they probably also advertise territorial claims.[65]

Despite Darwin's conjecture that, like the musical cadences of gibbon calls, human vocality is probably rooted in singing as part of the courting behavior of our progenitors, the reports I have cited indicate that any model on the evolution of hominid communication based on gibbon song should be much more complex—probably not worth the attempt to devise it. The most dramatic gibbon calls seem to relate as much or more to territorial identity and defense than to wooing ready, willing, and able mating partners. Before the invasion of gibbon habitats by humans with projectile weapons, they were probably remarkably safe from predation in the high canopy of tropical forests. If our weaponless hominid predecessors were predominantly terrestrial but, like the gibbons, ranged daily in small territories and had relatively modest body mass and small group size, it would have been imprudent for them to be noisy.

Although researchers have noted a variety of quieter vocalizations, particularly in captive gibbons, an analysis of their role in mediating intrasexual and intersexual social relationships is needed. The task of implementing studies on the semantics and pragmatics of quiet vocalizations in forest gibbons is a formidable task because of their distance from the ground, which severely restricts regular systematic monitoring. Intervening foliage and their black faces proffer to crested gibbons additional difficulties for discerning the accompanying and vocally silent facial expressions.

Orangutanese

In comparison with the problems faced by students of wild gibbon vocalizations, facial expressions, and gesticulations, those of field researchers who would decipher the semantics and pragmatics of orangutan signals are compounded by the solitary lives and dispersal of their subjects.

Pioneer field researchers concurred with the supposition of Robert and Ada Yerkes that orangutans are probably the least vocal of the great apes.[66] Nonetheless, MacKinnon catalogued sixteen vocalizations of free and captive subjects.[67] He derived most of the information from free-ranging Bornean individuals. Rijksen described seventeen sounds that are uttered by wild, captive, and rehabilitant Sumatran orangutans.[68] Maple compiled a useful comparative table of their listings, in which only twelve are the same.[69] Based on analyses of recordings of wild Bornean, rehabilitant, and captive orangutans, Hardus et al. listed thirty-two call types, eleven of which are not in the lists of MacKinnon or Rijksen.[70] In addition, Wich et al. reported that two captive female orangutans uniquely emit brief piano-mezzo piano monotonal whistles both spontaneously and in response to human models, and wild orangutans place a hand or leaves in front of the mouth, which alters vocalizations.[71] Among great apes, only bonobos ($n = 36$) exceed orangutans in number of call types.[72] Chimpanzees have twenty-nine, and mountain gorillas have sixteen.[73]

Short-distance orangutan calls notably outnumber the middle-distance and long-distance ones. Youngsters have eleven call types, four of which are unique to them. Nulliparous females have twelve calls, and females with youngsters emit fourteen calls; collectively four calls are unique to them. Nonflanged males emit the most call types ($n = 17$), none of which is unique to them. Flanged males emit fifteen calls, one of which is unique to them.[74]

The most intensively studied vocalization of *Pongo* is the dramatic long call that only prime adult males emit.[75] Galdikas described the long call as "the most impressive and intimidating sound to be heard in the Kalimantan forest."[76] The voluminously inflated laryngeal sacs contract heavily and irregularly during the call, which begins as a low soft grumble with vibrato, increases to a leonine roar, and terminates decrescendo with soft grumbles and sighs. Primate long calls have relatively low sound frequencies to minimize their attenuation. Moreover, there is a significant negative correlation between long call acoustic frequencies and home range size among primate species.[77]

During the roar, the caller protrudes and parts his lips and extends his neck. His lips are closed during the grumbling finale.[78] The cheek flanges might act as parabolic reflectors to locate callers.[79] Orangutan long calls are audible 1 to 2 km from the caller.[80] In temperate and tropical forests, the transmission distance is maximal when the calls are loud, low in frequency, and emitted above the forest floor, but not in the canopy.[81]

Pongid long calls occur both day and night in Bornean and Sumatran forests.[82] For instance, at Tanjung Puting, Kalimantan Tengah, Borneo, during all-day follows, the calling frequency of males peaked between 09:00 and 11:00 AM, but the calls of distant subjects were distributed widely during the day and night with 31 percent of them between 05:00 and 08:00 AM. Afternoon calls occurred most often between 16:00 and 17:00 PM, and 29 percent of distant long calls were between 17:00 PM and 05:00 AM.[83] Focal males uttered long calls 907 times over a four-year span, with a mean of 1.5 per day. As preludes to or during 3 percent of long calls, they caused loud noises by pushing over snags.[83]

At Tanjung Puting, dominant males called more frequently than less dominant ones did; one dominant male called on average four times per day.[83] At Ketambe, Sumatra, the rate of calling by five males varied widely. The best predictor of calling was a male's response to the long calls of others: the stronger his tendency to approach them, the more he called.[84]

Rodman heard long calls on average twice per month in the Kutai Game Reserve of East Kalimantan.[85] There were only two adult males in the study area, one of which did not call, perhaps because he was past his prime. During a sixteen-month follow-up study, the Kutai resident population comprised an adult male, two subadult males, and three females with youngsters. In addition, six adult males, more than two subadult males, and a solitary adult female visited the area. Mitani observed and tested their reactions to playback long calls.[86]

Most (82 percent) of the dominant male's 157 long calls appeared to be spontaneous: eight followed loud branch or tree falls, six immediately preceded 67 percent of nine copulations, six overlapped playback calls of another male, one occurred as he chased another adult male, and three immediately followed similar chases. He called significantly more often on the days after he had been with females than on days before and during bisexual associations and when he was alone.[87] Subordinate male and adult female Kutai orangutans typically oriented toward the resident male's long calls ($n = 223$) when they were <400 m away and generally ignored those farther away. Two adult males kiss-squeaked after two of eleven calls. Other individuals remained quiet. Adult males usually increased the distance between themselves and the calling site. Adult females did not approach the caller, even when they were with subadult males who had engaged them in forced copulations who probably would have to leave were the females to approach the caller.[86]

Mitani's playback experiments supported his naturalistic observations that responses to long calls vary as a function of male dominance relationships.[87] Whereas the resident adult tended to countercall and approach the stimulus, subordinate adults and subadults moved away. Further, they tended to retreat quickly and not call for a while. A few subordinate targets expressed agitation by bubbling, kiss-squeaking, or shaking branches, but their displays were unimpressive. The three females tended to eschew the taped calls just as they seemed to shun live recitals.[88]

Researchers suggested several functions for long calls, all of which relate broadly to the reproductive behavior of the species. MacKinnon initially proposed that the long calls maintain spacing among adult males and do not serve to attract females.[89] Although Segama females typically showed no reaction to a call and occasionally even hid or moved away on hearing it, adult males responded by calling themselves, approaching, or perhaps chasing or displaying toward the caller. But sometimes a male receiver remained unusually quiet, and once a female seemed to hurry toward a caller.[67] Calls were commonly uttered spontaneously and immediately after loud percussive noises such as falling trees and breaking branches. Overcrowding due to deforestation might have caused an increased frequency of long calls in one part of the Segama area.[90]

It pays to advertise.[91] Many field researchers agree that long calls serve not only to space adult males in the forest but also to attract sexually receptive (and in some areas nonreceptive) females toward the caller.[92] Mitani had cast doubt on the female attraction hypothesis, but he acknowledged that parity might influence whether a female will pursue a distantly calling male. The three Kutai females in his study were parous and perhaps anestrous.[86]

Variants in long calls might serve to identify callers to other orangutans.[93] Humans could distinguish the calls of four Tanjung Puting males at ≤400 m.[83] Acoustic analyses support the likelihood that individual male Bornean and Sumatran orangutans are identifiable by their long calls.[94] Moreover, aspects of call structures can demarcate populations and perhaps subspecies and species of orangutans.[95]

Galdikas noted that only 6.7 percent of calls by focal males followed sudden sounds such as falling snags and trees or animal noises.[83] They were much more responsive to nearby crashes than to distant ones. The remaining 93.3 percent of the calls by focal males were spontaneous, though some of them might have been stimulated by human presence. Ten percent of spontaneous calls began while the subjects were on the ground, and about one-third of them were completed in trees. Eleven percent of spontaneous calls came from night nests, and another 10 percent occurred while they were in or entering food trees.

Two particular contexts commonly elicited long calls.[83] When two males were visible to one another, invariably one or both of them called. They also called on 31.5 percent of occasions when they met other orangutans that they had not seen for a while. By observing the reactions of known males to the individualistic long calls of others, Galdikas descerned a dominance hierarchy among four males at Tanjung Puting.[83] Less dominant males often fled terrestrially or arboreally upon hearing long calls of more dominant males, the foremost of which never moved away from a caller. Indeed, the more dominant males frequently crashed snags, called, and moved toward less dominant callers. At Ketambe, Sumatra, a male challenger that had fought and defeated an older male resident with eighteen-years' tenure emitted long calls during their interactions, but the old male did not.[96]

Copulation is another context of frequent long calls or parts thereof. Tanjung Puting males called or uttered the grumbling segment of long calls during twenty-six of thirty copulations. Galdikas mentioned no vocal peculiarity that might broadcast that the male had climaxed.[83] Females uttered low, hoarse grunts and groans while resisting copulation but later quieted down when apparently accepting the male's efforts.[97] During copulation, females sometimes emitted mating cries that are essentially like fear screams except that the former are more rhythmic and continuous.[98] Rijksen termed them mating squeaks.[99] Like humans, some Sumatran female orangutans habitually vocalized while others remained silent during copulation.

Because parturition is rarely witnessed, there are minimal data on the extent to which it is signaled vocally. A rehabilitant primiparous Bornean female whimpered and cried tearfully as she gave birth in an open cage.[100] Two births in tree nests suggest that parturition is basically nonvocal in the forest, where it might attract predators.[101]

Youngsters and, to a lesser extent, adults utter a variety of squeaks, grunts, moans, barks, and screams during playful and fearful situations.[102] The noisy tantrums of youngsters are especially audible in the forest.[103] Frightened rehabilitants and captives sometimes grind their teeth audibly.[68]

Alarmed and annoyed adults utter several peculiar sounds that might be accompanied by branch breaking and other physical displays. In Sarawak, East Malaysia, three Bornean orangutans that were presumably annoyed by George Schaller interchangeably produced kissing sounds by sucking air through their puckered lips and occasionally kissing the backs of their flexed fingers loudly, made gluck-gluck-gluck noises by gulping several times with the lips pursed, and emitted bitonal burps from a low note to a high note.[104] MacKinnon and Rijksen confirmed the first two sounds, and Rijksen termed the gulping sound grumph.[105] MacKinnon noted that Sumatran orangutans kissed the backs of their hands and their knuckles. Neither observer described bitonal burps, though they might be counterparts of MacKinnon's gorkums. Kiss sounds were the most common and grumphs the second most common vocalizations of his subjects, probably because they are reactions to humans.[67]

Very annoyed females and subadult males uttered lork calls and displayed violently. Lorks probably also advertise an adult female much as the long call serves adult males.[106] Kisses, grumphs, gorkums, and lorks seem to indicate increasing levels of annoyance. Instead of making kissing sounds, some subadults blow raspberries. Occasionally during intimidation displays the adult males produced ahoor calls by gasping sharply and then uttering an explosive grunt.[67]

Males use their bulk and other epigamic features to good advantage during intimidation displays. The inflated laryngeal sacs and piloerection on their shoulders and arms make them look even larger than usual.[107] Further, they adopt a variety of extended suspensory and rigid bipedal and quadrupedal postures and move their bodies in ways that sport their inflated shapes. The long hair on the backs and forelimbs of Sumatran males is quite impressive as they sway back and forth.[107]

Although snag crashing seems to be more common among Tanjung Puting orangutans, others commonly wave, shake, break, and drop branches at annoyances.[83] Snag crashing might have developed from branch shaking. Some individuals elaborate on branch shaking by executing dramatic dives and lunges wherein they release some grips and dangle headfirst below footholds or sidelong from an ipsilateral hand and foot.[108]

Branch-busting orangutans commonly stared and exhibited tense-mouth facial expressions wherein the lips were tightly pressed together.[109] Threatening individuals also gape and yawn widely thereby displaying their teeth.[110]

The teeth are exposed when fearful orangutans grimace and playful youngsters draw back the corners of their rather widely open mouths.[111] Orangutans sometimes protrude their prehensile lips in contexts wherein they seem to be mildly fearful, such as when trying to appease or approach another individual. This might accompany whimpers and moans, especially in youngsters.[107] There are many similarities between orangutan and chimpanzee facial expressions and their apparent meanings.[107]

Figure 12.1. A 7-year-old captive-born female orangutan (Ubar) requests an object.

Some Bornean and Sumatran orangutans housed socially in three European zoos have repertoires of gestures that appear to convey intentional meanings. They employed forty gestures predictably to achieve one of six social goals: to initiate contact, grooming, or play, to request objects, to share objects, to instigate locomotion, to cause a partner to move back, or to stop an action. Further, they consistently used twenty-nine of the gestures with a single meaning.[112]

During interactions with humans, some captive and rehabilitant orangutans, like captive chimpanzees, bonobos, gorillas, and gibbons, indicate that they want food or other objects by reaching forward with an outstretched forelimb and open hand and occasionally an extended index finger. The gestures strongly resemble human referential pointing (Figure 12.1).[113]

In the current state of research on wild orangutan communication, I demur from speculation on the semiotic roles of most of the catalogued vocalizations, facial expressions, and gesticulations in the context of their dispersed social pattern. Further, as with gibbons, extreme arboreality combined with the infrequency and ephemerality of postadolescent wild orangutan social interactions do not recommend a role for their reperatoire in the evolution of humanoid language. Nonetheless, their ability to meet and to mate productively in the wild and for captives to live harmoniously in larger bisexual social groups speaks well for their notable intelligence, sensitivity, awareness of others, and possession of a flexible effective system of communication, probably on a par with that of African great apes.

Chimpanzese

> The chimpanzee sound system is almost entirely made up of vocalic sounds and a few types of closants (consonant-like sounds resulting from a close approximation or closing and opening of the upper vocal tract with or without voicing . . .).
>
> HAROLD R. BAUER (1986, p. 332)

Chimpanzees are perhaps the noisiest and most demonstrative of apes. They could hold their own in the British Parliament. Despite experience with clamoring captives, Nissen was astounded by the cacophony of chimpanzees in a Guinean forest.[114] The din that they produce must make many a predator salivate or perhaps hesitate as there are few documented instances of nonhuman predation on boisterous chimpanzees.[115] Reynolds speculated that their outsized ears might be related to their vociferousness.[116]

Chimpanzees have a diverse repertoire of graded vocal signals that are variably combined with subtle and striking facial expressions, bodily postures, and athletic displays, during which they brutalize their surroundings. In quieter times, head orientation and eye-gaze cues are probably operant.[117]

Largely through the efforts of Marler and Goodall, the communication of provisioned Gombe chimpanzees was thoroughly described and to a lesser extent interpreted functionally.[118] On the basis of sound spectrographic and film analyses, Marler reduced the number of classes of chimpanzee vocalizations from twenty-four to thirteen.[119] He was hampered in this exercise because many calls grade into one another and sometimes only what seemed like parts of series were uttered.

Pant-hoots were the most frequent chimpanzee vocalization.[120] Males uttered them most often, perhaps to mark status.[121] In both sexes, these pant-hoots are individually distinctive to human and presumably also chimpanzee receivers.[122] Discrete populations differ in the acoustic measures of male pant-hoots, perhaps due to learning.[123] In Kibale National Park, males emitted calls that lack the buildup component that characterizes pant-hoots of fellow eastern chimpanzees in Gombe and Mahale National Parks.[124]

Although Mahale male pant-hooters do not emit similar call types repeatedly during successive choruses, some accommodate each other vocally via active alteration of their calls during choruses, which probably maintains and strengthens bonds among them. Mitani hypothesized that instead of calling to identify their locations to other members of the unit-group, they pant-hoot only to maintain contact with particular individuals.[125] Low-ranking individuals did not match the calls of an alpha male; however, the calls of higher-ranking, dependent individuals of one of the dyads seemed to converge on those produced by the other dyadic member.[126]

The vocal similarity between individual males in Gombe and Mahale populations is positively correlated with the amount of time they spent together.[127] When they called together, they matched one another's pant-hoots, though the resultant choruses cannot compete with the beauty of a cappella chants by the Benedictine monks of St. Michael's. Instead, they are reminiscent of group hooting by the fans of Western sports teams.

Males in three contiguous communities and one distant community of western chimpanzees in Taï Forest had different pant-hoots by which researchers could identify them. Moreover, pairwise comparisons of individuals via nuclear DNA analysis have indicated that the acoustic distances among communities were probably not due to genetic factors. Accordingly, chimpanzees might modify their pant-hoots to distinguish their community from others, which supports the vocal learning hypothesis.[128]

The learning hypothesis is further supported by a study of pant-hoots in male chimpanzees at two zoological parks. Although the hooters were assembled from different sources, the calls in each group converged structurally as a consequence of vocal learning. Nonetheless, the intragroup variation in call structure of the captives was similar to that of wild eastern chimpanzees in Kibale National Park.[129]

Pant-hoots are usually voiced on exhalation and inhalation. Complete male pant-hoots increase steadily in volume to a roaring or shrieking climax at the end of a crescendo. Female pant-hoots lack such dramatic climaxes and tend to begin with deeper pitches.[130] Thus, the sex of a caller should be apparent to listeners even if they do not know him or her personally. Chimpanzee pant-hoot choruses can be audible over more than 3 km.[116]

Pant-hoots probably serve to announce the locations of individuals so that friends can meet and foes can be discreet.[131] Pant-hoots occur in various contexts. Chimpanzees seem to listen carefully to distant pant-hoots and then answer them. They also pant-hoot spontaneously while feeding, when abed at night, when meeting other chimpanzees, while subdividing into smaller foraging parties, when eating mammalian prey, and when otherwise aroused.[132] Pant-hoot variants might broadcast information specific to social contexts such as an advertisement by males to attract females and male allies, or they may be related to particular ecological contexts such as announcing arrival at fruiting trees.[133]

Playbacks of pant-hoots by a male chimpanzee that was not a member of the Kanyawara community in Kibale Forest indicated that the nature of responses by community males depended on the number of adult males in the party that heard them.[134] In contrast with parties with ≥3 males, which chorused loudly and approached the speaker together, those with fewer than three males usually remained silent, approached the speaker less often, and traveled slower when they did so. Higher- and lower-ranking males responded similarly, and the location of the experimental intruder did not affect their responses.

Three adjacent communities of Taï Forest chimpanzees responded differently to playbacks of pant-hoots by group members, neighboring individuals, and strangers.[135] Like Kanyawara chimpanzees, Taï male party size affected their responses to pant-hoots of strangers but not to those of neighbors except that they approached the calling site of a neighbor more closely. Recorded pant-hoots of group members elicited pant-hoots from other community members, but those of strangers and neighbors induced screams. Overall, males responded more strongly than females did.

Eastern chimpanzees in the Budongo Forest were more likely to pant-hoot at abundant food sources and to incorporate a decrescendo in the call upon arrival there. Their pant-hoots were different while traveling terrestrially in small parties before merging with other members of their community. Accordingly, Notman and Rendall concurred with other researchers that pant hoots serve a general social function; "the subtle structural differences observed in some contexts reflect different effects on vocal production introduced by the variable arousal and motivation, physical demands, or vocal effort associated with calling in those contexts."[136]

In addition to pant-hooting, highly excited male chimpanzees rhythmically slap, stomp on, and pound resonant tree buttresses, trunks, and logs, thereby producing sounds that are audible several kilometers away.[137] Drumming is usually accompanied by vocalizations and sometimes also by other sorts of assaults on vegetation.

In the Republic of Guinea, chimpanzees drummed frequently during the day, quite rarely at dusk, and never at night.[114] In Equatorial Guinea also, chimpanzees drummed frequently, with 93 percent of sessions occurring during the first six daylight hours and few during the rest of the day or first hour after dawn.[138] The peak of drumming was between 07:30 and 09:00 AM.

Over a nine-month span, Budongo chimpanzees drummed and called continuously for several hours eight times; however, most drumming lasted only a few minutes. One day, eighteen outbursts of drumming occurred between 07:30 AM and 16:30 PM. The peak was between 13:00 and 13:30 PM, and 83 percent of them preceded 13:30 PM.[139]

In Gombe, male fellow travelers took turns drumming on specific tree buttresses along trails.[140] Ironwood trees were favorite substrates of drumming for Budongo chimpanzees.[141] The absence of nocturnal drumming (in contrast with calling) might be related to the fact that percussionists would have to leave their nests and venture near the ground.[142]

Six adult male western chimpanzees in Taï Forest produced individualistic drumming patterns, usually accompanied by pant-hoots, which differed in the number of beats per bout, the mean duration of the interbeat intervals, the mean number of beats per bout, and the mean bout duration. Collectively, their drumming patterns differed from those of male eastern chimpanzees in Kibale Forest. Further, Kibale chimpanzees drummed more often without pant-hooting.[143]

The remaining twelve chimpanzee vocalizations—pant-grunt, laughter, squeak, scream, whimper, bark, waa bark, rough grunt, pant, grunt, cough, and wraaa—appear to be directed at nearby listeners, though screams can carry notable distances. Pant-grunts of Mahale males are more individually variable and thus less individually distinctive than pant-hoots, as might be expected of short-distance signals wherein sender and receivers can see one another.[144] In addition, captive chimpanzees employ novel vocal signals—raspberry and extended grunt—to attract the attention of humans, which indicates they possess a generative capacity for communication in varied contexts.[145]

Great apes appear to be sensitive to the attentional state of humans and use appropriate signals to gain a person's attention.[146] Young chimpanzees might be selectively attracted to beings that make direct eye contact with them, but juvenile chimpanzees often engaged in gestures and vocalizations toward a person holding food rewards regardless whether the person's eyes were open or closed and looking away or looking inquisitively directly at them.[147] Further, some young captive chimpanzees were unable to use information about direction of human gaze or visual access to information; they simply chose a person facing them versus one facing away.[148] James Reaux et al. concluded that even by the age of 9 years they had no understanding of visual perception as an internal state of attention.[149]

Like human children, apes develop referential pointing spontaneously.[150] Whereas apes usually point to request things from others, human infants also point declaratively to show and to share attention upon things. Both probably share a common cognitive complexity at the level of understanding behaviors as connected to targets through attention, but they differ in where the ultimate goal for pointing lies. Apes point to make people do things with targets, whereas human infants make people look at and emotionally react about targets.[151]

Like orangutans and humans, young chimpanzees throw noisy tantrums when frustrated, such as during weaning.[152] Adults, including deposed alpha males, also exhibit tantrums.[153]

During agonistic interactions among Budongo chimpanzees, victims alter the acoustic structure of screams relative to the intensity of aggression, which apparently cues others about the severity of the attack. In severe assaults, they might exaggerate screams above the actual level of aggression if one or more spectators match or outrank the attacker. Accordingly, Slocombe et al. hypothesized that chimpanzees possess triadic awareness—understanding third-party relationships—that influences their vocal production.[154] Further, such control of vocalizations would indicate that some wild chimpanzees have more control over their vocal apparatus than indicated by attempts to provoke it in captives.

Often during copulation Gombe and Budongo females squeaked and grinned.[155] After copulation, Gombe females either ran off screaming or remained quite calm. Acoustic analysis of copulation calls from six Budongo females indicated no difference in calls of specific individuals based on fertility status from hormonal profiles. However, acoustic structure reliably encoded identity of the caller. Therefore, "in chimpanzees, the use and morphology of copulation calls have jointly been shaped by the selective advantage of concealing fertility."[156]

Several Gombe males invariably panted at the culmination of copulation and sometimes slowly smacked their lips before ejaculating.[146] Both sexes pant quietly during grooming sessions, peaceful reunions, and mutual greetings.[157]

Frolicking chimpanzees, especially tickled youngsters, emit rapid staccato panting sounds that vaguely recall human laughter.[158]

Chimpanzees threaten nuisances mildly with coughs and assertively with barks. Male western chimpanzees in Taï Forest emitted two graded, context-specific bark types during hunts and encounters with snakes, and context-specific signal combinations with different call types or drum tattoos. They also barked when traveling, upon hearing chimpanzees from a neighboring community or a different party of their own community, and observing or receiving aggression from a community member.[159]

Subordinates pant-grunt as distance is closed between more dominant individuals and themselves. Pant-grunts have been invaluable for determining the relative statuses of individual Gombe chimpanzees. They are species-specific; that is to say, unlike other chimpanzee signals, they were directed only at other chimpanzees and not toward the baboons that also interacted with them in Gombe camp.[160]

Rough grunts occur when chimpanzees approach and eat preferred foods. Rough grunts and other vocalizations are probably not learned; Viki, though isolated from other chimpanzees, uttered them.[161] Similarly, a female chimpanzee that grew up in a human home performed many of the sounds and facial expressions of Gombe chimpanzees despite the fact that she had no opportunity to learn them from fellow chimpanzees. She also produced some unique vocalizations.[162]

Adult females and adult and adolescent males utter wraaa calls in response to human intruders, predators, and other beasts that alarm them. Chimpanzees sometimes also slap or pedally thump the ground or a tree sharply. The percussions and wraaa calls not only threaten the stimulator but also warn companions of potential danger.[163] Wraaa calls were especially conspicuous among the reactions of Gombe chimpanzees after the death of an adult male that fell from a tree.[164]

Actions of twenty-three mimetic muscles provide *Pan troglodytes* with a great variety of subtle and more dramatic facial gestures in a highly graded facial communication system that is qualitatively different from those reported

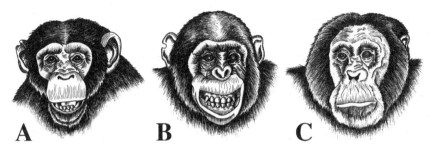

Figure 12.2. Chimpanzee facial expressions. (A) Relaxed. (B) Frightened/anxious.
(C) Angry/tense.

for other nonhuman primates (Figure 12.2). Moreover, there are few differences between the facial musculature of chimpanzees and humans.[165]

Intensive studies on the visual signals, and especially postures and facial expressions, of captive chimpanzees complement the available field observations, providing a solid ethogram for the species.[166] Some of the visual signals accompany particular vocalizations, including some that might be necessary for the production of certain sounds.[167]

Like their vocalizations, chimpanzee visual signals tend to combine with and to grade into one another.[168] Marler and Tenaza classified chimpanzee facial expressions into six categories: tense-mouth face, open-mouth threat face, grin, pout face, play face, and lip-smacking face. They omitted facial configurations that are associated with the physical production of vocalizations but have little apparent valence as visual signals.[157]

Before chasing or attacking a subordinate individual and before copulation, a male chimpanzee might glare fixedly at his target and exhibit a tense-mouth face in which his lips are tightly compressed (Figure 12.2C). His upper lip might bulge as though it holds a pocket of air. In less tense situations, particularly if the would-be intimidator is not unduly afraid, he stares at the target with his mouth open a variable amount and his eyebrows depressed. Observations on six male and five female chimpanzees at Edinburgh Zoo indicated that yawning is associated with a change in activity levels, but only modified yawns might relate to increased arousal.[169]

A frightened or squelched chimpanzee grins toward conspecifics with its mouth closed or open slightly (Figure 12.2B). If contacts are nonaggressive, chimpanzees grin with their mouths open widely and eyebrows elevated. Open-mouth grins also occur when subordinates threaten dominant group members and other species of which they are fearful. The open-mouth grins of chimpanzees are exploited in the human entertainment industry to make it appear that the individuals are happy or otherwise pleased, when in fact, they are probably fearful of the trainer or the context in which they find themselves. Humans would be wise to discern the difference between smiles and grins, especially when encountering strangers or dealing with solicitors and politicians.

Whimpering and mildly fearful, frustrated, and curious youngsters and adults protrude their lips, forming pout faces. Contentedly playing individuals allow their mandibles and lower lips to droop thereby producing play faces. Groomers stare closely at their tiny targets and slowly move their jaws in a lip-smacking face. A high-ranking female Budongo chimpanzee uttered acoustically unique loud whimpers as she persistently nagged a high-ranking male to share his fruit even though there were other fruits readily available to her.[170]

Hell knows no fury like a riled chimpanzee. If his glaring, tense-mouth face does not quell a disturbance or keep rivals apart, a dominant male might rise with the hair on his arms and shoulders erect and charge quadrupedally. If he catches an offender, he might grab its hair and stomp bipedally on its back. If the victim is small, it might receive one or more body slams or be dragged by a limb or hair. The attacker commonly mouths, and less frequently bites, an unfortunate. Handfuls of hair might mark the scene. Milder attacks, particularly those of females, might involve slapping and scratching.[171] Disturbed Gombe chimpanzees administered noteworthy knocks to humans and competitive baboons.[172]

Displays ranging from tenuous to athletic frequently occur instead of attacks. A disturbed individual might bark and jerk its head up and back by extending the neck. It might also rapidly fling a forelimb palmward or lunge and make hitting movements with the dorsum of the hand facing the nuisance. Downward swats and lunges might precede slapping bouts and follow attacks when the vanquished has been emboldened by a reassuring third party. Threatening individuals vigorously jerk and shake attached branches and wave free objects or hurl missiles toward disturbers.[146] Males might bluff-over subordinates by leaping or swinging a hind limb over the crouching target.[153] They also take to the trees where they dash about and brachiate.[155]

Among the most striking displays of chimpanzees are bipedal runs and sway walks (swaggers), especially when they are piloerected and waving their forelimbs and perhaps branches overhead. Males perform sway walks more commonly than females. Youngsters amusingly mimic them.[173]

Victims caught in attacks have little recourse but to crouch, scream, and hope for the best, especially when they are outweighed or outnumbered. But in a pinch, females and subordinate males will fight back and can score telling bites on overbearing males and other adversaries.[153] Nonetheless, most altercations are avoided or resolved via a rich repertoire of submissive, appeasement, and reassuring postures and gestures.

Flight can carry one away from a fight or fright. Occasionally, one can hide, or more commonly move near the protection of others, or move away from the source of disturbance. A threatened individual might also avoid attack by crouching close to the ground, usually with its rump toward the aggressor. After attacks, victims commonly crouch and might back toward their tormentors: presenting. Estrous females also adopt presentation postures before copulation. Various grins, screams, and squeaks accompany presentations, depending upon the intensity of the interaction. Subordinates might present to dominant individuals as they pass near them or are approached by them. Low-key presentations might be straight-faced.[174] Gombe females presented most often to males. Males frequently received presentations and seldom executed them. Youngsters also presented much more often than they received presentations.[146]

When approaching dominant individuals face first, especially after a display, subordinates might bow quadrupedally, flexing their elbows more than their knees. They also might bob one or more times when facing higher-ranking chimpanzees.[146]

A bowing or crouching chimpanzee might press its lips or teeth to the face or body, usually the groin, of a dominant individual. If distance is not closed, it might reach toward the higher-ranking individual or touch some part of its body, usually the head, back, or rump. The submissive one might also mount and lightly embrace its attacker. It might execute a few pelvic thrusts and gingerly touch or hold the scrotum of the dominant male with its foot.[146]

A dominant individual signals reconciliation by touching the extended hand or proffered buttocks or by kissing or patting some part of the suppliant. Passing subordinates might also be touched briefly with a hand or foot. Males sometimes thoroughly examine the genitalia of passing females. The dominant individual might mount or embrace a subordinate in manners similar to the reciprocals previously described. Occasionally, dominant chimpanzees present to subordinates and might receive scrotal manipulations from them.[146]

Figure 12.3. Reassurance behavior between high-ranking adult male eastern chimpanzees (Figan and Evered).

Reassurance gestures are sought and given not only between attackers and victims but also among other group members that are disturbed by intragroup altercations or extragroup scares (Figure 12.3). Many of them occur when subgroups merge and individuals greet one another.[175]

Libidinous males might stand bipedally and rock sideways from foot to foot, shake branches, and outdo Tarzan with arboreal swings and leaps.[146] Some flash their erect attenuated pink penis toward prospective mates by sitting with thighs widely abducted and the pelvis thrust forward.[176] They might also flip their phallus up and down. A male might have to grunt to get a female to regard his penis and then beckon her to approach with his upper limb.[177] If receptive to his advances, she crouches over his lap, and the event is quickly consummated.[153]

Libidinous adolescent females sometimes pursue males, and after wearing them out from copulation, attempt to stimulate them by fondling their flaccid penes. If the effort fails, they sometimes throw noisy tantrums.[178]

In a group of nineteen captive chimpanzees, individuals chiefly used visually based gestures when other individuals were already attending versus using tactile gestures, which they employed regardless of whether the recipient was attending to them. Moreover, to gain attention of a target individual, they might move in front of the recipient, vocalize, or do both before gesturing further.[179]

Menzel devised a group-oriented approach to chimpanzee communication that produced intriguing results that underscore the need for innovative projects to test the natural communication of apes as social phenomena.[180] He concluded that greater emphasis should be placed on "what they are communicating and how they do it" instead of their capacities for communication.[181] The behavior of groups should also be studied instead of merely compiling lists of individual vocalizations, facial expressions, and postures or trying to induce subjects to emulate features of human language.

Menzel conducted experiments over a span of six years with five female and four male juvenile wild-born chimpanzees in an enclosed 1-acre field in southeastern Louisiana. Initially the field was treed, but the subjects destroyed much of the natural vegetation.[182] Six of them lived together in the field for one year before testing, and they formed a cohesive and relatively compatible social group.

While the rest of the group was caged out of view, Menzel showed one individual—the leader—a cache of food or a fearful object somewhere in the field.[183] Five of the subjects served alternately as leaders; the others were too distressed by separation from the group to be a leader.[184] After the leader had rejoined the group, they were released together into the field. If food was the incentive, the group followed the leader and sometimes even ran ahead of him. The followers seemed to orient more to that thing out there than to the leader. If the incentive was a fearful object, the group moved cautiously.

In trials wherein one leader had seen hidden fruit that was a preferred food, and another had seen vegetables, the entire group tended to follow the fruitful leader. This indicated that there had been pooling of information in the holding cage. Further, the exchanges of information were quite subtle. Usually there was no conspicuous gesture, facial expression, or vocalization except that leaders sometimes emitted single alarm calls when they first sighted novel frightful incentives. In 88 percent of sixty test trials, the group followed the leader to caches of food instead of detouring to retrieve single foods that were clearly visible in the field.[183] When each of two agreeable subjects had seen a different quantity of hidden food, they often traveled together for the larger amount and then moved to the smaller cache. They generally ignored blank spots. In almost 80 percent of trials when the leaders split, the majority of the group followed the one that had seen the largest cache of food.[184]

How the chimpanzees communicated information in the holding cage or momentarily after release is a mystery. Indeed, the leader often wrestled with others or just sat by the door until they were freed. Isolation of the leader before the group was released did not affect their performance even though it was stressful for her or him. Menzel suggested that the "visual orientation and the locomotor postures and movements of the 'animal as a whole' contain sufficient information to account for the bulk of the communication about hidden objects."[185] One is reminded of Clever Hans, a horse more perceptive and responsive to subtle inadvertent movements of his owner than were human observers, who appeared not to perceive them. Audiences believed that Clever Hans could count until his owner stayed out of view whereupon the horse failed to stop tapping his hoof beyond the numeral on a test card.

It is also possible that Menzel's subjects obtained clues via vocal signals. Based on observations that individuals produced acoustically distinct types of rough grunts in feeding contexts, Slocombe and Zuberbühler employed playbacks to see whether a focal captive subject could use such signals produced by other group members to guide his search for food. Indeed, he did, which supports the conclusion that chimpanzees "can produce and understand functionally referential calls as part of their natural communication."[186]

Menzel noted that his subjects succeeded so well because they were well acquainted with one another.[183] The longer a pair had lived together before the test, the more closely interrelated were their movements, though the effects decreased markedly during adolescence when each of them became more independent.[184] No less should be expected of relatively stable wild populations of chimpanzees and perhaps other social mammals. Needless to say, the more subtly a social group communicates during routine foraging, nesting, and other maintenance activities, the less detectable they might be to predators and opportunistic competitors. A major challenge is to devise and conduct more informative observations and experiments on natural social groups.[187]

Close observation of habituated provisioned and nonprovisioned chimpanzees complemented by observations and experiments on sizable samples of captives has provided a fuller database from which one can begin to craft scenarios on the communication systems and social patterns of emergent hominids. Whereas most quiet communication of orangutans is performed intimately by dyads and gibbons interact closely with relatively few individuals during their lives, chimpanzees are challenged to a remarkable variety of dyadic and communal interactions with individuals of all ages and both sexes because of their fission-fusion social pattern. Individual and community well-being and survival depend on their ability to distribute themselves over the landscape in order to forage economically, often while keeping contact with other subgroups, and to quickly assess others and adjust socially when subgroups meet before dispersing again, often as part of a subgroup with a different composition. Needless to say, such a social pattern coupled with an even more sophisticated system of communication would serve our progenitors very well indeed, especially were they to pool and share collected foods, daily experiences, and knowledge of proximate changes in their habitat.

Gorillese

During the post–World War II renaissance of primate field studies, gorilla acoustic communication rivaled that of chimpanzees as a research topic. Fossey notably augmented Schaller's observations on vocalizations of mountain gorillas; however, except for the chest-beating display, the other gestures, facial expressions, and communicative postures of gorillas were not as fully documented as those of chimpanzees, especially photographically.[188] In the swamp forests of the Republic of the Congo, western silverbacked gorillas perform dramatic agonistic splash displays in addition to chest-beating, perhaps to intimidate potential competitors that aim to acquire females.[189]

Schaller listed twenty-two vocalizations of wild mountain and captive gorillas.[190] He heard four of them only once and seven others fewer than ten times each. Fossey lumped some of the calls that intergrade, thereby reducing Schaller's list to fifteen specific vocalizations, three of which she had not confirmed at Karisoke.[191] Based on 2,225 hours of contact with wild groups and observations on two juveniles in her camp, she described sixteen vocalizations in seven categories.[191]

Marler noted remarkable similarity between the vocal repertoires of gorillas and chimpanzees.[192] He matched all thirteen chimpanzee vocalizations with calls from gorillas, leaving roars, growls, whines, and whinnies as peculiar to gorillas among African apes. He was most confident of similarities between pant-hoots and hoot series, laughter and chuckles, screams, rough grunts and belches, and pants and copulatory pants; he was less confident of similarities between squeaks and whimpers and gorilla cries, waa barks, wraaas, and wraaghs, and grunts, coughs, and pig grunts. He was least confident in comparisons between pant-grunts and pant series and several barks.[126]

Radically different field conditions have frustrated quantitative comparisons between the calls of nonprovisioned gorillas in the process of habituation and the heavily provisioned, well-habituated Gombe chimpanzees.[193] Nevertheless, Marler reasonably related the outstanding vociferousness of silverbacks, including their singular roars, to their unique position as guardians of relatively close-knit groups versus the more flexible and fluid social groupings of chimpanzees.

Harcourt et al. agreed that the loud vocalizations of mountain gorilla silverbacks are meant to intimidate predators.[194] Further, they underscored the role of quieter vocalization—close-calls—in maintaining cohesion of gorilla groups versus the more fluid nature of chimpanzee communities. Indeed, on average, adult members of two Karisoke groups uttered close-calls at a median rate per individual of eight times per hour.[195] Like students of chimpanzee and orangutan communication, Harcourt and Stewart concluded that gorilla "calls . . . are means of negotiating social interactions, i.e. are means of assessment and manipulation of others' ability and intention (potential future behavior) by provision of information about one's own ability and intentions."[196] They are not mere contact calls to keep from losing one another in dense vegetation.[197]

In two groups of mountain gorillas, females consistently exchanged calls at a higher rate with adult males than with other group members. Among adults, use of close-calls correlated with one's dominance rank and that of the recipient, especially within the cohort of dominant individuals. Non-dominant adults uttered more double grunts in the presence of subordinates than when they were near dominant individuals, and they exchanged calls with adult males at a higher rate than with other members of the group. Kin were more likely to call in the presence of kin and to exchange close-calls with kin than with non-kin. Feeding individuals were more likely to cease feeding soon after being approached by a vocalizing adult male than were he to approach them silently.[198]

Aggressive gorillas roar, growl, and utter pant series.[191] Schaller described high intensity roars as "probably among the most explosive sounds in nature."[199] They stopped him cold on the trail and routed his trackers. Fossey usually persevered. Silverbacks roar to threaten humans, other gorilla intruders, and buffalo instead of immediately fleeing from them.

Silverbacks and rarely blackbacks and females utter low doglike growls when approaching individuals that annoy them. The latter usually retreat and sometimes growl back. Females, and less commonly silverbacks and juveniles, utter low whisperish pant series from the perimeter during disputes and skirmishes among other group members.[200]

Mildly alarmed and curious adults emit tritonal question barks in which the middle pitch is higher than the neighboring pitches. They usually spark other gorillas to seek the source of disturbing sounds. Hiccup barks, consisting of staccato disyllabic bursts of sound, occur under much the same conditions as question barks, such as when a human observer moves suddenly or out of view.[200]

Gorillas often fall silent and slip away as quietly as possible when they are extremely afraid, as when they detect approaching poachers or fresh aliens.[201] During intragroup altercations, individuals of all ages and both sexes are inclined to scream shrilly and repeatedly. Screaming spreads contagiously through the group during fearful quarrels and contests. Silverbacks, particularly the less dominant ones, are the main screamers.

Females might scream during and after copulation.[201] Copulating silverbacks utter rapid series of staccato hoots that progressively merge into a long howl.[201] Fossey heard copulatory pants during four out of eight copulations.[191] Harcourt et al. heard copulating mountain gorilla females whimpering rapidly and males panting and occasionally whimpering.[202] Hess noted copulatory sounds in captive western gorillas. His descriptions are not sufficiently detailed for comparisons with wild mountain gorillas, except that female western gorillas did not include screamers.[203]

Wraaghs are loud, explosive, monosyllabic alarm barks that are pitched between roars and screams. They are uttered by adults, mainly silverbacks, and rarely by juveniles, when startled by sudden, close contacts (e.g., with buffalo) in dense foliage, or upon hearing loud noises such as thunderclaps, branches breaking, or rockslides. Fossey noted individualistic wraaghs by the silverbacks with which she was intimately acquainted.[191] Following wraaghs, groups usually retreat from the source of a disturbance.[191]

Distressed infants emit a variety of screeches, sobs, wails, and other noises that Fossey treated generically as cries. She mentioned that occasionally they build to a crescendo of tantrumlike shrieks. Apparently, weanlings shriek, but Fossey's list of causes does not include maternal rejection of nursing attempts.[191]

Youngsters and females that fear injury or abandonment utter prolonged puppylike whines.[190] Fossey also witnessed whining by a silverback and a blackback but did not state their circumstances.[191]

Wrestling, chasing, and otherwise playful infants chuckle, particularly when engaged with one another or with tickling humans. A pair of black-backs chuckled during a low level tickling session.[201]

Gorillas emit series of pig grunts more often when moving from areas than when stationary. Silverbacks are the principal grunters, though the vocalization can spread contagiously to involve other adults. Silverbacks grunt while orchestrating group movements and during lustful pursuit of females. Others might grunt during quarrels over access to food, and a silverback grunts to quell them. A mother grunts at her wayward infant and it returns to her.[201]

Moving and stationary gorillas also emit disyllabic hoot-barks. Although silverbacks hoot barked the most, all categories of gorillas were inclined to hoot when mildly alarmed. Hoot-bark alarms induced group members to cluster and to climb for a better view. Fossey commented, "Of all vocalizations, the hoot bark was the most effective in clustering group members."[204]

To initiate group movements, a silverback might hoot-bark and posture. Other adults might answer with a chorus of pig grunts that he joins.[191] Contented gorillas produce deep, prolonged, soft belching sounds. This too is contagious in a stationary group as they feed, assemble for night nesting, rouse from sunning themselves, and play and groom. Eructating observers can induce response belches from habituated gorillas.[200]

There is considerable variability in the belch vocalizations. During prolonged spans of belching, they produced whinnylike sounds.[191] Isolated horsey whinnies were uttered twice by a blackback and more often by a sick silverback that produced no other vocalization. He whinnied in contexts wherein healthy silverbacks would hoot-bark or wraagh. Because his autopsy revealed pneumonia and pleurisy, one might suspect that the sound was caused by the pathologic condition of his respiratory tract.[191]

Hoot series occur during intergroup contacts and encounters between groups and single silverbacks.[191] Aside from silverbacks, other individuals rarely emit them. The displayer begins quietly with a series of low-pitched hoos that steadily rise in pitch until they are strained and sustained, rather like a dog's whine. A hoot series might end in a harsh growl, a lapse into silence, ground thumping, assaults on vegetation, running noisily through the foliage, or, most spectacularly, a chest-beating display.[201]

Tattooing the thorax versus a tree buttress carries the obvious advantage of always having a resonant chamber readily at hand. Whether there are particular communicative differences between the two actions or whether chest-beating merely developed in areas where appropriate drumming trees were insufficient for early gorillas is unknown.[205]

Although chest-beating was noted by Du Chaillu and was mentioned regularly in later accounts on gorillas, the complete display and its circumstances were not detailed until a century later.[206] Schaller observed that full-fledged chest-beating displays of silverbacks include nine more or less distinct acts.[190]

First is a hoot series, during which the displayer might pluck a leaf or sprig and place it between his lips. Just before the climax of the hoot series, he rises bipedally and often tears off and flings a handful of vegetation into the air. The hoots merge into a growl as he rapidly tattoos his inflated chest. The laryngeal sacs, which commonly extend into the pectoral region, resonate percussions that can be heard up to 1.6 km away. While slapping his chest with slightly cupped hands, the displayer might kick sideways with one limb and tear foliage with his foot. Then he runs a short distance sideways, often tearing and slapping at vegetation en route. He might also assault foliage at the end of his run. As a finale, he sharply smacks the ground with one or both palms. The entire display takes thirty seconds and often much less.

Silverbacks might delete the hoot series or hoot only once before proceeding to a climax. Indeed, Schaller noted that only 10 percent of chest-beating displays included a notable hoot series.[190] Hooting false-starts are also common and would-be show-offs seem to be easily distracted during the overture, which signals other group members to clear the stage.

Females and youngsters also strike their chests, but their slapping sounds lack the resonance of silverback thoracic percussions. Further, they do not perform complete displays. Sometimes group members rise and beat their chests as the silverback reaches his climax.[190] Often a silverback performs truncated displays and variably combines some of the middle acts in the nine-part sequence.

Schaller concluded that chest-beating is basically innate.[190] The following situations can induce chest-beating by mountain gorillas: the presence of humans, other gorilla groups, or lone males; hooting, chest-beating, and thumping by distant gorillas; displays by other group members; and during play. Further, some might simply be spontaneous.[207]

In Equatorial Guinea, most chest-beating was audible in the morning and after the gorillas had nested for the night. Causality could not be established, except in cases of human intrusion.[144] Nocturnal chest-beating was either accompanied by vocalizations or was solo.

The full chest-beating display is unique to gorillas. Although excited chimpanzees hoot, stand, and run bipedally, assault and throw vegetation, and thump the ground, they rarely beat their chests.[172] No one has described chimpanzees performing a complete silverback display or including a thoracic percussion section after a chest-beating male's cadenza.

Schaller sketched the facial expressions of his subjects under eight emotional states, ranging from placidness to fear.[190] He commented that many of them resemble those of captive chimpanzees as described by Foley and the Yerkes.[208] Indeed, we can recognize in Schaller's descriptions crude counterparts of the chimpanzee tense-mouth face, open-mouth threat face, grin, pout face, and play face. A sample of thirteen 1- to 6-year-old captive gorillas exhibited six auditory, eleven tactile, and sixteen visual gestures, some of which were group specific, leading Pika et al. to conclude that although ontogenetic ritualization was the main learning process, a form of social learning might be involved in acquisition of special gestures.[209]

Gentry et al. registered 102 gesture types in social groups of captive and wild western gorillas, noting that they pay specific attention to the attentional state of an audience. Although six gestures appeared to be traditions within single social groups, "overall concordance in repertoires was almost as high between as within social groups." In reference to the work of Pika et al., Gentry et al. found no support for ontogenetic ritualization being the chief means whereby gorillas acquire gestures because they used gestures derived from species-specific displays as intentionally and almost as flexibly as gestures whose form is consistent with learning by ritualization.[210]

Theorists who would model the emergence of human language should not ignore gorillas if they elect to emphasize ancestral beings living in relatively moderately sized stable groups that daily foraged, traveled, rested, and lodged together, particularly if indeed their close-calls might be as semantically and pragmatically sophisticated as the students of mountain gorillas have suggested.[211] Weighing against a model based predominantly on *Gorilla* spp. is the fact that their social pattern appears to be heavily influenced by extreme morphological sexual dimorphism, which is probably not a feature of early and recent species of *Homo*.

Bonoboese

Accounts from the field and reports based on captives have evidenced that bonobos have some unique postures, gestures, and vocalizations.[212] For instance, an alarmed arboreally fleeing female bonobo uttered eerie, high-pitched screams similar to the calls of gulls. MacKinnon concluded that bonobos are generally quiet and appear to have no long-range contact call or territorial signal.[213] Although the Badrians agreed that Lomako bonobos are quieter than Gombe chimpanzees, they noted short calling bouts.[214]

Like captive bonobos, wild ones emit two types of hoots: low and high.[215] Lomako bonobos directed low-hoots toward human intruders; when hearing vocalizations from members of a distant community, they emitted low-hoots accompanied by agonistic displays of stomping on the ground or drumming on trees. Their high-hoots coincided with the high-hoots of bonobos in other parties (48 percent), loud calls of mangabeys (16 percent), and less specific noises (3 percent), and sometimes they occurred spontaneously (33 percent). It appeared that high-hooting functioned to close the distance between and to increase cohesion among different bonobo foraging parties. Further, they seemed able to adjust the spectral parameters of high-hoot distant calls according to corresponding calls of other bonobos.[216]

Kano portrayed bonobos at other Democratic Republic of the Congo localities as much noisier than those at Lomako, stating that their incessant screaming made them easy to locate.[217] He agreed that chimpanzees are much noisier and that bonobos have a distinct high-pitched, metallic alarm call that both males and females utter. The higher pitch of bonobo vocalizations is one of the more dramatic differences between them and chimpanzees and gorillas.[218] The difference cannot be explained simply on the basis of gross body mass because bonobos and chimpanzees do not differ much in body mass. Instead, bonobos probably have smaller vocal tracts than those of chimpanzees.[219] Specifically, their vocal cords are probably shorter and thinner than those of chimpanzees.

Although captives, including those separated from their mothers at birth, exhibit most of the communicative repertoire typical of wild *Pan paniscus,* some also develop structurally unique vocalizations, which indicates a capacity for flexibility in their vocal communication.[220] Five-year-old Kanzi employed at least four structurally unique vocalizations—"Uhh," "Ennn," "Ii-angh," "WHAI"—that were absent from the vocal repertoires of five adult bonobos, including his father and birth and foster mothers at Yerkes National Primate Research Center (RPRC), and from the repertoires of adult and juvenile bonobos at the San Diego Zoo.[215] Nonetheless, he shared the majority ($n = 10$) of vocal types with them.

Kanzi most often uttered "Ii-angh" in response to human verbal questions, such as "Do you want a peanut?," and "WHAI" in response to queries or comments, such as "Do you want to hide?" "Unn" was also a response to human questions, such as "I was going to put some Kool-Aid in your bowl. Do you want some?" The most frequently recorded of the four unique vocalizations, "Ennn," served as a vocal request, often accompanied by a gesture.[221]

Several small captive bonobo groups emitted five acoustically distinct calls, usually as part of long, complex sequences during interactions with food. The composition of the call sequences related to the kind of foods and the food preferences of the caller, from which Zanna Clay and Klaus Zuberbühler concluded that bonobo call sequences convey meaningful information to other group members; they are not mere indications of being generally excited about the presence of food.[222]

Spectrographic analysis of videotapes of adult Kanzi's vocalizations while interacting with humans showed that he modulates his vocal output of peeps on both temporal and spectral levels when signifying banana, grape, juice, or yes. That is to say, his utterances varied systematically according to the semantic context in which he emitted them.[223] Turn-taking between a researcher's statements and Kanzi's vocal utterances, lexigamic statements, or both also notably facilitates communicative exchanges between them.[224]

Based on a nine-month study of a Wamba bonobo group, mostly at a feeding station, Mori identified fourteen vocalizations and forty-seven nonvocal behavior patterns in the communicative repertoire of *Pan paniscus*.[225] She compared them with data from a six-month study on the provisioned Mahale chimpanzee group and a one-month study on two groups of Grauer's gorillas in the Parc National du Kahuzi-Biega, Democratic Republic of the Congo.[226] She noted that many behavior patterns are similar between bonobos and chimpanzees. But she also noted that some behaviors such as interfemale frottage were specific to bonobos versus chimpanzees. Unlike displaying chimpanzees, terrestrially charging Wamba bonobos seldom ran bipedally.[226] However, at Lilungu and in captivity displaying bonobos sometimes drag branches or food bipdally for short distances.[227]

Bonobo vocal repertoires are also different from those of other African apes. For instance, they hiss in tense situations. Mori's brief descriptions of other vocalizations by *Pan paniscus* also indicate that they are distinctive.[225] Although she specifically mentioned that bonobo pant-hoots, threat calls, and wraahs are similar to those of eastern chimpanzees, they depart from them in some acoustic features.[226]

At Lilungu, three communities of bonobos communicated via fifteen vocal units—peep yelp, whistle (scream-whistle and whistle-whine), pout moan, bark, soft bark, hiccup, yelp, peep, scream, composed bark, low hooting, panting laugh, grunt, muffled bark, and croak—and nineteen sequences of iterated units or combinations of units.[228] When members of different communities detected one another, there was great agitation among the first to detect the other, followed by very long vocal exchanges of hooting and composed barks.

Lilungu bonobos were also quite vocal when they encountered trees containing quantities of popular food, which precipitated chases, begging, greeting, and appeasement displays. The vocalizations comprised soft barks, whistles, grunts, hooting, screams, bark-screams, hiccups, and barks. As they settled down they continued to vocalize with peep yelps, hiccups, soft barks, grunts, peeps, and soft mixed series. Individuals that located new food trees emitted barks and whistles that attracted other members of the foraging party.[228]

Bermejo and Omedes proposed that bonobo vocalizations while traveling, feeding in dense terrestrial herbaceous vegetation, or spreading out to search for earthworms function to maintain group contact.[228] Like the close-calls of mountain gorillas, they might also (or instead) function to regulate social relationships and behaviors beyond merely staying with the group.[198]

Juvenile play was punctuated with a variety of vocalizations that intensified and changed when playful interactions graded toward fighting. Infants and juveniles screamed when visually isolated from community members. When prevented from suckling, weaning youngsters initially softly peep yelp and peep then scream as they become more agitated.[228]

Copulating females commonly screamed during eye contact with the male. Females also screamed when they solicited frottage, which usually proceeded in silence, though one or two units of soft screams sometimes occurred. Screams also characterized intragroup fights and agonistic interactions comprising chasing, charging, attacking, fleeing, and crouching.[228]

At nesting sites, Lilungu bonobos frequently hooted and occasionally scream-barked, mainly at the beginning and end of group movements and as they rose in the morning. Individuals occasionally hooted during the night. If alarmed by the proximity of other bonobo communities, monkeys, humans, or a male display, they barked and uttered a variety of other sounds. Some individuals also vocalized as they huddled during a rain storm.[228]

Bermejo and Omedes concluded that all vocalizations of bonobos are far more structured than those of chimpanzees.[229] Their greater complexity underpins a greater potential to express different meanings or differences among individuals. They further stressed that whereas bonobo acoustic communication is always related to facial expressions, gestures, and tactile communication, that of chimpanzees is much less so.

Some captive bonobos exhibit more numerous gentle tactile gestures (55.8 percent of all gestures) than chimpanzees (34.6 percent).[230] Both species employ brachiomanual gestures more flexibly than their facial expressions and vocalizations across contexts. Unlike captive chimpanzees, captive bonobos engage in multimodal communication. Combinations of bonobo gestures and facial/vocal signals elicit more responses (87 percent) from recipients in all contexts than unimodal signals do (around 67 percent). Pollick and de Waal mused, "Could the relative scarcity of multimodal signaling in bonobos relate to a more deliberate combination of gestures with other forms of communication, perhaps in an attempt to add critical information to the message instead of merely simplifying it?"[231]

Furuichi coined the term *peering* for a unique behavior in which a bonobo moves within half a meter from the face of another individual and gazes intently at her or his mouth.[232] The facial expressions of both individuals convey calmness, but they generally avoid eye contact.[233] There are several functional interpretations of bonobo peering. Kano, who first noted the behavior at Wamba, associated peering with begging gestures and food sharing.[234] Later reports indicated that in the absence of overt begging gestures, peering rarely resulted in food transfer, especially between older animals.[235]

In a sample of 230 peering episodes, Wamba adult females of two unit-groups were most frequently paired with individuals of all age-sex cohorts, but males seldom participated in peering. Younger individuals were the most frequent peerers. Idani noted four outcomes of peering: most commonly the peerer left; or the peeree left; or both peerer and peeree remained with no further interaction; or they engaged in another interaction.[236] Dominant individuals tolerated subsequent actions directed toward them by subordinate younger peerers. Agonistic interactions were rare after peering interactions. Instead, grooming, female frottage, copulation, and food-sharing were more common follow-ups to peering. Accordingly, Idani concluded that peering served to initiate affinitive interactions with the peeree.[236]

In 617 observations of peering among three 8- to 14-year-old male and seven 7- to 20-year-old female captives, only fifteen instances of food exchange occurred. As at Wamba, the younger San Diego bonobos performed the most peering, and they primarily directed it at older females. Moreover, in a given dyad, the peerer was more likely to follow the peeree, but not necessarily to groom her. Although males were infrequent peerers, one was more likely to peer and to be a recipient of peering if he frequently groomed and infrequently aggressed toward a female. Johnson et al. suggested that peering might signal acknowledgement of female status.[233]

Six adult bonobos at Planckendael Zoo exhibited a social dominance hierarchy in which two of three females ranked higher than all three males and the third-ranking female was dominant to the two lowest-ranking males. Peering occurred almost strictly unidirectionally and seldom followed immediately after aggression, but it was equivalent to fleeing from aggression as a marker of dominance. Further, it was a better marker of the hierarchy than aggression, yielding, or teeth-baring.[237]

A subsequent study of peering in five bonobo groups in Europe indicated that the linearity in peering relationships was significantly high, but unidirectionality was low. Further, there was little correspondence between peering order and agonistic dominance rank. Individuals in four of the groups peered significantly more often at high-ranking individuals, particularly if the latter might monopolize resources, but when resources were distributed more evenly, high-ranking individuals might peer down the hierarchy.[238]

Like captive chimpanzees and bonobos that spontaneously develop pointing or readily learn it from humans, a wild bonobo at Ikela exhibited indexical referential pointing accompanied by vocalizations to draw the attention of other group members to poorly camouflaged human observers, and young chimpanzees at a nut-cracking site at Bossou, Republic of Guinea, pointed indexically to a stone once and to nuts nine times.[239] At Ngogo in Kibale National Park, male chimpanzees employ audible exaggerated directed scratches to areas of their bodies that their partners subsequently groom.[240]

Bonobos exhibit a wide variety of manual, pedal, facial, and whole body gesticulations in communication with one another and with humans. Some captives appear to exhibit voluntary control over their facial musculature during solitary games that include making unusual faces in situations that seem to lack emotional engagement.[241] Although not all gesticulations expressed by captives have been documented in wild bonobos, the majority of them, like vocalizations, have been observed, albeit in lower frequencies.[215] The generational resilience of bonobo vocal and gesticulatory systems in captives is remarkable; that is to say, despite being initially raised apart from their mothers in a nursery or living among humans for long periods of time, they retain a solid heritage of bonoboese.[242]

The remarkable variety of copulatory positions that bonobos exhibit sparks comparisons with the *Kama Sutra,* and like human mating, it probably also suggests a flexible system of communication.[243] Savage-Rumbaugh et al. described spontaneous forelimb gestures that orchestrated copulatory positions between captive bonobos.[244] They included positioning movements in which one individual touched another, thereby indicating how it was to posture its limbs, and various iconic hand motions that showed its partner what to do. Bonobos commonly engage in ventroventral copulatory positions, including sustained eye contact that clearly distinguishes them from chimpanzees and gorillas. One might wonder whether complex forms of sexual coordination were factors in the development of language during hominid evolution and, if so, when it might have played a role.

Based on research to date, bonobos might rival chimpanzees in proffering intriguing possibilities for modelers of language development during the course of human evolution, particularly were one to emphasize gestures and close-calling, though aspects of their greeting behavior (namely, female frottage) might cause some modelers to demur. Needless to say much more detailed understanding of the extent to which their natural communication system contains not only indexical and iconic referential signals but also some basic syntactic combinations that might allow them to comment on spatiotemporally distant situations, events, and objects in addition to those perceptible proximately.[245]

Speak or Say Uncle

After the earlier attempt by Furness to teach a young orangutan to say "papa" and "cup" and a young chimpanzee to say "mama," Winthrop Kellogg and Keith and Catherine Hayes fostered infant female chimpanzees in order to document the extent to which their bourgeois homes could humanize them.[246] At the time, no greater psychological triumph might be imagined than for them to say "mama" and "papa" in a prelude to more complex linguistic accomplishments.[247]

The results were pitiful. Viki (1946–54) lived with the childless Hayeses for around 6.5 years, and Gua (b. ca. 1930) lasted with the Kelloggs for only nine months.[248] The Kelloggs intended to raise Gua alongside their infant son Donald to facilitate her humanoid development, but when Donald seemed to be aping the ape more than Gua behaved like him, they abandoned the effort.

Gua did not even babble, let alone recite nursery rhymes, and Viki produced only four hoarse, breathy utterances—glossed *mama, papa, cup, up*—after much human tuition, including molding her lips.[249] These induced vocalizations would decay if her trainers did not practice with her regularly. Gua's vocalizations were predominantly emotional responses to provocations that were readily apparent. Viki was also silent between emotional calls and training sessions.

Laidler induced an infant male orangutan, Cody, to utter *kuh, fuh, puh,* and *thuh* for drinks in a mug, solid food and food in a pan, contact and comfort, and the continuance of brushing, respectively.[250] He did not attempt to teach Cody any English words, though he considered Cody's utterances to be referential words for each of the objects or actions. To get Cody to utter the desired sounds, Laidler employed even more extreme manipulations of Cody's body than the lip molding the Hayeses applied to Viki; for example, he blocked his nares and placed him in a headlock to induce him to utter *gruh.*

Clearly, apes do not easily imitate human spoken words even when they are immersed in human society and monotonously tutored and tormented. Other approaches had to be sought.

Ape Command of Artifactual Languages

Ameslan for Apes

Beginning in 1966, several teams of comparative psychologists implemented projects to test the suggestion of Yerkes that chimpanzees might "be taught to use their fingers somewhat as does the deaf and dumb person, and thus helped to acquire a simple, nonvocal, 'sign language.'"[251]

The Gardners obtained a 1-year-old female chimpanzee—Washoe (1965–2007)—and began to raise her in a middle-class Yankee social environment.[252] She slept alone at night in a fully furnished and provisioned house trailer and daily played and otherwise interacted with human adults who used minimal verbal utterances in her presence. Washoe also frequented a spacious treed yard equipped with a swing and other playground apparatus.[253]

The Gardners and numerous human playmate-companions, some of whom were deaf, trained and tested Washoe in American Sign Language (ASL or Ameslan) with the ultimate aim of engaging her in two-way conversations. ASL is a mixture of iconic and arbitrary signs that are analogous to spoken English words, and it is notable for the absence of finger spelling. ASL is supposed to contain units—cheremes—that are counterparts of phonemes in human speech formed by hand-sign positions, configurations, and motions.[254] Combinations of ASL signs translate very roughly and with difficulty into English sentences. ASL signed constructions are commonly quite telegraphic.[255]

Washoe's trainers encouraged her to name and to ask for objects and to ask questions about them with ASL. She acquired some of the signs by imitating her tutors, but others were shaped and molded by them. Indeed, molding was the most effective method to train her.[256] When necessary, teachers adjusted signs to Washoe's anatomy and postural preferences, which led commentators to dub her performance pidgin sign language.[257]

After mastering a sign, glossed "come-gimme," Washoe added "more," up," and "sweet" during the first seven months of training. By the time she was 5 years old, she had mastered 132 signs and understood several hundred more that were directed toward her by humans.[258] Her performance and those of her pongid successors pale in comparison with that of an average human, who beginning in postnatal year 2 learns around 60,000 words at a rate of eight to ten words a day between birth and adulthood.[259]

Washoe used signs for classes of referents: for instance, "dog" for individuals of many breeds, sizes, or colors, pictures of dogs, and barking. She could innovatively apply signs learned in one context to new situations. For example, she signed "open" not only for doors, rooms, and cupboards but also for containers and a water faucet.[260]

At about 21 months of age, Washoe began to combine signs, mostly in pairs, to make simple constructions that the Gardners compared with the early sentences of verbal children. For instance, this and that were in initial position. She could also respond to *Wh*-questions: who, what, where, how many, whose. Her signs were comprehensible to people who could not see her referents.[261]

Baby Washoe manually babbled and signed to herself while leafing through magazines and settling in for the night. She even corrected herself. Further,

Gardner and Gardner reported that she signed "quiet" to herself while creeping to a forbidden part of the yard and "hurry" while dashing to a potty chair.[262] To some observers, Washoe reestablished the expectance that the communication gap between apes and people would be narrowed notably if not actually bridged.[263]

In a follow-up to Project Washoe, the Gardners implemented a second project with four newly born chimpanzees—Moja (1972–2002), Pili, Tatu (b. 1975), and Dar (b. 1976)—to determine the development of two-way communication in chimpanzees compared with that of human infants and to learn whether they would communicate among themselves with ASL. Moja and Pili began to sign at 3 months old.[264] They commanded fifty signs each at about the age when human infants have fifty words (ca. 20 months). Like human infant vocabularies, nominals were predominant in those of the chimpanzees (about 50 percent). They mastered negative signs by 15 months of age.

In due course Washoe, Moja, Pili, Tatu, and Dar not only signed to each other, to themselves, and to strangers but also to cats, dogs, toys, tools, and trees. The development of their signing was ostensibly like that of signing and speaking human children.[265]

The Gardners schooled 3-year-old Moja. She sat at a desk and drew with chalk, crayons, and pens and named her creations.[266] For instance, Moja signed her first chalk markings "bird" and made circular squiggles with an orange pen when instructed to draw a berry there. Having done so, she signed "berry" when asked via ASL "What that?"

The interpretation that she actually had a berry and a bird in mind is suspect given that her squiggles are questionable representations of the objects glossed by her trainers, and we have no way to assess whether she created humanoid abstract art. Indeed, no ape has drawn, painted, sculpted, or molded true or even near representations of complex three-dimensional objects or geometric figures.

In 1970, Fouts took 5-year-old Washoe to Norman, Oklahoma, where he continued studies with her and initiated a new study with other young chimpanzees (Figure 12.4).[267] He also received subjects from the Gardners. In 1980, Fouts established the Chimpanzee and Human Communication Institute in Ellensburg, Washington, with Washoe, Moja, Tatu, and Dar.[268] Like the Gardners, he aimed to determine the range of individual variation for the acquisition of ASL by chimpanzees and to establish intraspecific communication in the colony. He hoped that the chimpanzees would ac-

Figure 12.4. Captive chimpanzees communicating via American Sign Language (ASL). (A) A 3-year-old chimpanzee (right) signs "drink" to a 6-year-old chimpanzee. (B) The 6-year-old (left) signs "come" to the 3-year-old.

quire signs from one another. Several of the subjects had lived in human homes, where, without the distraction of natural chimpanzee communication, they had learned ASL more quickly.

After human training, chiefly via molding, two young males (Booee, b. ca. 1964, and Bruno, b. 1968) learned to sign to one another about food, mutual comfort, and play, particularly tickling. A female, Lucy, and Washoe combined signs presumably to describe novel objects for which they had not been given the ASL signs. For example, Lucy designated oranges, grapefruits, lemons, and limes "smell fruits," and signed "cry hurt food" after tasting a radish and "drink fruit" and "candy drink" for watermelon. Washoe presumedly signed "water bird" for swans, "dirty monkey" for a feisty macaque, and "dirty Roger" for her trainer after he refused her requests. "Dirty" was her sign for feces and soiled objects.[269]

An alternative explanation of the gloss water bird would seem reasonable. Lucy had been taken down to a lake and asked "What that?" The referent could have been ambiguous to her, so she signed both "water" for the lake and "bird" for the swan instead of naming the swan "water bird." Fouts's interpretations of "dirty monkey" and "dirty Roger" are more difficult to challenge so simply.

Before Fouts learned the ASL sign for leash, Lucy invented an iconic sign to indicate that she wished to go outside on her leash.[270] His younger subjects combined signs much earlier than Washoe had, probably because they were taught ASL earlier than she had been.[271]

A 3-year-old male chimpanzee (Ally, b. 1969) learned to transfer object names in spoken English and their ASL signs in order to refer to their physical referents. Chimpanzees readily learn to understand verbal object names and simple commands. First, Ally learned the English names for ten objects. Then, without reference to the objects, he learned ASL signs for the words. Later, when trainers showed the objects to Ally and signed "what that," Ally could sign correctly for the ten referents. Thus, Ally exhibited notable cross-modal transfer and comprehension of spoken English in addition to mnemonic and cognitive skills.[272]

Fouts bred 14-year-old Washoe and Ally in order to study the transmission of signing from mother to child. While pregnant, Washoe vomited a meal of yogurt and raisins. When asked "what that?" she signed "berry lotion." As her pregnancy advanced, she was asked "What in your stomach?"; she repeatedly signed "baby."[273]

Because Washoe's infant died two months after birth, Fouts acquired a 10-month-old male, Loulis (b. 1978), for her to mentor and later other crossfostered chimpanzees (Moja, Tatu, and Dar) to communicate with him, Washoe, and one another in ASL. As a 2.5-year-old, Loulis had a vocabulary of seventeen signs, only one of which was used by humans in his presence. Concurrently, Washoe's ASL vocabulary was 180 signs. Loulis manually babbled before producing full-fledged signs such as "tickle" and "drink."[274]

People used only seven signs—mostly *Wh*-signs—in Loulis's presence. Washoe seemed to understand spoken English, though there was no formal, controlled testing to confirm this inference. Nonetheless, her human companions assumed they need not sign in order to communicate with Washoe when she was with Loulis. Thus, Loulis apparently acquired most of his signs by modeling after Washoe and Moja. Further, Washoe actively taught Loulis to sign. For instance, she molded his hand into the configuration for "food" and touched his mouth with it. He used the sign. She placed a chair in front of him and signed "chair"—glossed "sit down"—five times. He did not use the sign.

Fouts et al. concluded that signing and various examples of object manipulation show that chimpanzees acquire at least some behaviors via cultural transmission.[273] Washoe molding Loulis's hands to make signs also indicates that a chimpanzee can directly teach another chimpanzee, just as humans had taught her many signs.[267]

During high-arousal interactions, Washoe, Loulis, Moja, Tatu, and Dar emphatically modulated their signs by signing more vigorously, enlarging a sign's movement, prolonging or reiterating it, or by employing a bimanual version of a sign that was usually unimanual.[274]

Following fast upon the successes achieved by chimpanzees with ASL, other comparative psychologists ventured to explore the abilities of gorillas and orangutans to communicate via Ameslan.[275]

In 1972, Patterson initiated an ASL project with a 1-year-old, captive-born female western gorilla, Koko (b. 1971), at the San Francisco Zoo before moving her and her Washoesque house trailer to Stanford University and ultimately to Woodside, California, where her subject could enjoy walks in the woods and other outdoor experiences.[276] In 1977, Patterson purchased Koko from the San Francisco Zoo with the assistance of San Fransciso Mayor George Moscone, who was assassinated with Harvey Milk by Dan White in 1978. In 1976, Patterson also purchased two baby western gorillas from an illicit animal dealer in Vienna. The female died within a month, but the male, renamed Michael (1973–2000), survived and lived with Koko until his death. Patterson hoped that Michael and Koko would mate and enjoy a communicative family life.[277] They both learned ASL but did not procreate, which probably should not be attributed to a lack of communication.

Patterson and Gordon have claimed that Koko's ASL vocabulary comprises >1,000 signs and that she understands spoken English well enough to respond via Ameslan.[278] They further related that Koko was learning the letters of the English alphabet, can read some printed words, including "Koko," uses a computer, and scored 85–95 on a Stanford-Binet Intelligence Test, which is lower borderline normal or just below normal for people (100 ± 16), depending on corrections for age, nervousness during the test, and other factors. They did not include these data for Koko.

Koko's presumptive accomplishments continue with completing nursery rhymes, arguing, understanding pig Latin, directing insults at offenders, lying to avoid personal consequences of unacceptable behavior, anticipating the responses of others, indulging in imaginary solitary and social play, creating representational drawings and paintings, relating memories and discussing past events in her life, and showing command of time-related words such as before, after, later, and yesterday, responding to jokes via vocalization, gesticulations and ASL, showing empathy for others in pictures, and exhibiting death-awareness (i.e., she can discuss where she will go when she dies).[279] On the program *Nova*, Patterson even said that Michael used ASL to relate to her the horror of his bloody capture in Africa.[280] Koko's supposed nipple fetish led to sexual harassment lawsuits against Patterson by three assistants after allegedly being pressured to bare their breasts for Koko. The suits were settled out of court.[281]

Patterson and her assistants simultaneously spoke to Koko and Michael while signing. They quickly learned to sign but never spoke back. The first signs that Koko mastered were for food. During the first ten years of instruction, Koko acquired 876 signs, and she continued to add signs at a reduced rate thereafter. Michael's instruction began when he was 3.5 years old, and he learned ≥400 signs.[282]

Gary Shapiro trained two female orangutans—3.5-year-old Princess and 10- to 12-year-old Rinnie—in Ameslan at the Tanjung Puting National Park, Indonesia. Both were rehabilitants. Princess had spent brief periods in home-rearing and captive conditions.[283] Rinnie was born in captivity and was later released to live in the forest.[283] They established that orangutans share with African apes a capacity to acquire ASL signs. Moreover, the specific early Ameslan vocabularies of *Pongo pygmaeus* overlap notably with those of *Pan troglodytes* and *Gorilla gorilla*.[283]

During nineteen months of ASL training via molding, Princess learned thirty-seven of eighty-five different signs, 59 percent of which are nominals, at an average rate of 1.8 signs per month. Shapiro concluded that the relatively rapid rate of sign acquisition by Princess versus Washoe and Koko was

primarily because she was more mature than they were at the outset of training.[283] Further, he reinforced her with food and social rewards—patting and
verbal praise—when she signed correctly.

During twenty-two months, via molding free-ranging Rinnie learned
thirty-four Ameslan signs that mostly refer to objects, particularly edibles,
while sitting at the river's edge or on a feeding platform in the forest. Shapiro
tested her vocabulary by showing her a referent and asking in Ameslan and
Indonesian "What is this?" He also asked her questions, such as "What do
you want?" Or "How?" He reinforced her correct responses with food, drink,
scratching, grooming, or other contact.[284]

In 1978, Miles obtained Chantek (b. 1977), a 9-month-old captive-born
orangutan, from Yerkes NPRC in Atlanta, Georgia.[285] Over the next nine
years, she raised and trained him under Washoesque conditions in a five-
room house trailer at the University of Tennessee in Chattanooga. Although
he did not have privileges at the university gym, his fenced courtyard contained several trees between which ropes were hung, and it also included a
large jungle gym, a hammock, barrels, and a picnic table. In 1986, Chantek
returned to Yerkes NPRC and later moved to a special treed exhibit at the
Atlanta Zoo, where he could climb and swing.

Miles and a small group of assistants administered ASL training when
Chantek seemed to be most attentive and socially motivated. He was not
drilled or held to an acquisition schedule. They also spoke as they signed to
him. After one month of training, Chantek signed "food-eat" and "drink."
He mastered fifty-six signs by the time he was 28 months old. In due course,
he could use around 140 ASL signs, 127 of which met the acquisition criterion of spontaneity and appropriate usage on half the days per month.[286]
During the second month of training, Chantek spontaneously combined
signs into sequences like "come food-eat." After three years of training, he
produced three- and four-word sequences such as "key milk drink open" that
Miles glossed: Open the trailer door so that I can enter for a drink of milk.
The mean length of Chantek's utterances is 1.9 signs, with the longest consisting of five signs.[287]

Miles noted that Chantek signed more slowly and deliberately than Ally
and other chimpanzees of the Oklahoma project did.[288] Further, the rate at
which he acquired new signs was higher than those of the ASL chimpanzees.
Accordingly, she concluded that the more articulate signing and the distinctive insightful cognitive style of orangutans might indicate their superior
linguistic and cognitive abilities relative to those of other apes—perhaps

reflecting the common bias that my kid is smarter than your kid. Ultimately, Miles concluded that Chantek's skills were high in memory-related operations, combinatorial tasks, semantic relations, deception, iconicity, and reference; medium for imitation, symbolic complexity, displacement, and productivity; and low in elaboration of new information, symbolic play, modulation, and rule-following behaviors.[289]

Between 1973 and 1977, Terrace conducted ASL studies with a male chimpanzee that he dubbed Nim Chimpsky (1973–2000).[290] By 1977, Nim had acquired 125 ASL signs via human molding and guidance. Nim was 2 months old when he was sent from the Institute of Primate Studies (IPS) in Norman, Oklahoma, to live in a hectic New York townhome with a family of Terrace's friends in order to develop in an environment that would make him highly socialized and maximize his chance to develop a linguistic concept of self.[291] There was frequent turnover of his fifty-seven playmate-teachers and shifts of study location and living quarters before he was returned to IPS.[292]

Unlike previous workers, Terrace routinely videotaped Nim's signing sessions. Analysis of them revealed that his teacher's prior signs often prompted his signs. Commonly he merely repeated the instructor's signs, perhaps with additions like "Nim," "me," "you," "hug," and "eat." Terrace detected similar inadvertent prompting on films of Washoe and Koko and their trainers, which recalls the Clever Hans phenomenon wherein a performing horse reacted to subtle cues of his owner instead of truly counting or otherwise comprehending human symbols.[293]

Analysis of strings of signs produced by Nim revealed that they could be viewed simply as unstructured combinations of signs, in which each one is separately appropriate to the situation at hand. Although the mean of Nim's multisign utterances ($n > 19,000$) is 1.6, he produced strings up to sixteen signs long in order to emphasize what he wanted. The longer constructions were riddled with redundancy as shown by his most sustained utterance: "give orange me give eat orange me eat orange give me eat orange give me you." While reminiscent of poetry by e.e. cummings, the two really are worlds apart.[294] As the mean length of human utterances increases, their complexity also progressively increases.[295]

How Tweet It Isn't!

In 1966, David and Ann Premack inaugurated a study of two-way communication with a 5-year-old female chimpanzee, Sarah, and four other young subjects that were much less impressive students. They adhered to strict laboratory procedures instead of having freer, homier social interactions like those that characterized the great ape Ameslan projects.[296] Sarah had close contact with humans until she literally defaced a trainer and had to be kept locked up.[297]

The Premacks did not aim for Sarah to imitate human language.[298] Instead, they were interested in her capacity to employ a relatively small vocabulary productively and syntactically. The artifactual words were metal-backed pieces of plastic in various colors and random (noniconic) shapes. Sarah affixed them to a magnetized board in vertical series. Within six years, Sarah had a vocabulary of 130 units that she could arrange in series with up to eight units.

Ultimately, the Premacks concluded that while apes recognize equivalence between their world and the plastic symbols, they could not acquire even the weak grammatical system of young human children.[299] They agreed with Terrace and other critics that mere word order is not equivalent to syntax. Sarah's strings of plastic signs were constructions, not sentences.[300] Her plastic constructions were simple one-to-one correspondences between words and the actual items to which they refer. By contrast, human sentences are infinitely variable, often recursive, and can convey much more abstract concepts.

Savage-Rumbaugh et al. challenged that Sarah's successful choices of signs glossed as "name of," negation, and so on were often due to conditional match-to-sample strategies—that is, Sarah's tasks were not communicative.[301] Like Ameslan signs and lexigrams, the plastic signs might not be symbolically equivalent to human words.[302] They simply showed that she could solve "complex conceptual tasks, but the relation between Sarah's performance on such tasks and language as utilized by human speakers remained vague."[303] The Premacks were not persuaded by their arguments, and continued to reason that, like human words, the plastic signs represent concepts in chimpanzees.[304]

Sarah learned to designate "sameness" and "difference" and mastered the concept "name-of," which greatly facilitated her acquisition of new names. She seemed to learn interrogatives—who, what, why—the conditional if-then, and prepositional concepts like "on," "under," and "to the side of." Sarah could also describe features of objects in the absence of the objects when presented with their plastic signs. For instance, when shown the plastic sign for apple, she chose plastic signs for red versus green, round versus square, and having a stem. According to the Premacks, she seemed to exhibit meta-linguistic phenomena, using symbols to discuss other symbols.[305]

Apes joined the computer age in 1972 when Rumbaugh and a multidisci-plinary team began training and conducting experiments on language learn-ing and communication with the infant female chimpanzee Lana (b. 1970) at Yerkes NPRC in Atlanta, Georgia.[306] They particularly wanted to establish whether apes have the capacity for linguistic productivity, whether they could create and understand new messages. They devised a series of artificial lexigrams and created a new language, Yerkish, based on lexigrams and sim-ple grammatical rules. The Yerkish grammar was designed after correlational grammar, in which there is no distinction between syntax and semantics. The lexigrams were color coded according to seven gross semantic categories: au-tonomous actors, spatial objects and concepts, ingestibles, body parts, states and conditions, activities, and prepositions, determiners, and particles. Two additional colors signaled sentential modifiers that had to be placed in the initial position of messages (Figure 12.5).

After 9 months of testing with lever pressing and matching-to-sample tasks via hand-drawn noniconic geometric lexigrams, Lana entered the lan-guage training area. It had a keyboard for her, another for the experimenter, and panels on which they could read what was typed. A computer recorded all communications automatically. Lana could request a variety of foods, toys, contact with trainers, music, movies, a romp outside, or a view of the real world through a window. She began with please and had to end with a period. She could eliminate mistakes at any point in her messages by press-ing the period key. Periodically, the trainers shuffled the lexigrams on Lana's keyboard. Low backlighting indicated that a key could be activated, and no backlighting meant that the key was inactive. When she pressed active keys, they lit up more brightly. Lana generally had fifty keys back-lit out of a total of seventy-five keys on the board.[307]

Lana readily mastered the system and used it to obtain what she wanted and needed in order to subsist. The speed with which she typed her requests is quite impressive. Within thirty months of training and practice, Lana had mastered 75 of the 125 lexigrams in the program.[308] Twenty-five years later, she chose the correct lexigrams for five of seven objects and colors for which she had not seen the lexigrams during over twenty years.[309]

Lana spontaneously learned to read displays that were projected above the keyboard and to hit the period key when she erred while typing instead of completing the series that would not be rewarded. Once she had grasped the

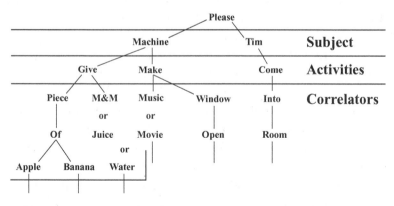

Figure 12.5. (A) A captive chimpanzee (Lana) operates a computer-controlled language training apparatus. She pulls an overhead bar (*B* to activate the system). As she presses keys (*Y*) with her right index finger, they are projected sequentially at *P* above the keyboard. The first projection is glossed "please," and the request must be completed with a lexigramic question mark (*?*). The requested food or drink dispensed at *F*. (B) Yerkish grammar.

basic mechanism for using names, she rapidly learned new ones and began to request the names of objects for which she had no lexigram. She combined available lexigrams in novel ways in order to communicate about things for which she had not been provided lexigrams. She used the lexigram for "this" to refer to specific incentives.[310]

Lana extended use of the lexigram glossed "no" from simple negatives to protests about malfunctioning food dispensers and problems with trainers. And she began to compose novel sequences of lexigrams in addition to using the standard chains of words that characterized her training period. Eventually, she initiated a two-way communicative exchange with a trainer, who was outside her chamber, about his Coca-Cola by typing "?Lana drink Coke out-of room."[308] Subsequently, other complex Yerkish exchanges occurred when Lana wanted something exceptional or when a practical problem arose within the system.[311]

Following the Lana project, Savage-Rumbaugh, who had worked on the Oklahoma projects, and Rumbaugh inaugurated an elegant study of symbolic communication between chimpanzees via the computerized lexigraphic apparatus.[312] Savage-Rumbaugh and assistants induced two arduously language-trained, socially coreared male chimpanzees—Lana's paternal half-brothers; 3.5-year-old Austin and 4.5-year-old Sherman—to communicate with one another about and to share food. When Sherman and Austin had learned to communicate with one another via the keyboard regarding the identity of one of eleven possible foods or drinks in a container, the researchers tested them to see whether they could convey the information by natural gestures and vocalizations. They failed to do so.[313]

When they made a mistake on the keyboard, it was another food or drink that they named instead of inedible objects that also were represented by lexigrams. At the culmination of the experimental series, the researchers provided one chimpanzee with an array of foods that could be seen by his partner with access to the keyboard. The possessor could see projected requests of the solicitor and usually read them accurately and handed correct tidbits through a small window in the transparent plastic partition that kept them physically apart.[313]

Savage-Rumbaugh et al. then tested the capacities of the chimpanzees to categorize objects as food or tools and to comprehend the representational function of lexigraphic symbols.[314] After learning the lexigrams for food and tools, Sherman and Austin were able to categorize photographs and lexigrams for specific foods and tools correctly.

Savage-Rumbaugh et al. cogently argued that Washoe, Sarah, Lana, Nim, and other chimpanzees that used symbols in ways that emulate human usage did not comprehend their representational functions.[314] Contrarily, Sherman and Austin "had discovered that the relationship that a lexigram has to an object *is a function of* the relationship it has to other lexigrams, not just a function of the correlated appearance of both lexigram and object. This is the essence of a symbolic relationship."[315]

Whereas Sherman and Austin could immediately label novel objects as food or tools, Lana could not. Sherman's only error was labeling a sponge "food," but he ate part of it. Lana could sort actual foods from tools, but she could not sort the lexigrams for specific foods and tools per the lexigrams for food and tool; that is to say, she could not encode the perceived functional relationships between the foods and tools symbolically. Thus, because of the training regimes employed Lana, Washoe, and their peers appeared to have learned contextually appropriate usages for their lexigrams or ASL signs, but they really did not know what they represented. They were probably problem-solving instead of using language with the intent of conveying meaning. By contrast, Sherman and Austin showed that chimpanzees could comprehend the meaning of lexigrams and presumably other sorts of symbols if appropriate training procedures were used.[314]

Sherman and Austin also learned to use the lexigramic system to work cooperatively for food that had to be obtained with tools (Figure 12.6). They were alternately placed in rooms containing manually inaccessible food and a kit with seven kinds of tools. The possessor of food or drink had to request an appropriate tool from the hardware chimpanzee. He would then share his reward with the tool supplier.[313] Accordingly, Savage-Rumbaugh et al. concluded that the Sherman-Austin study evidenced:

- Great apes can comprehend symbols, but production does not lead spontaneously to comprehension.
- In order to function representationally, the symbols must become decontextualized and freed for use in novel situations.
- Great apes can communicate with one another via symbols if they develop joint attention in an environment that requires mutual cooperation.
- Great apes can make informative statements regarding their intended future actions.
- Referential comprehension and usage are prerequisites to the development of syntactic competence.[316]

Figure 12.6. Via lexigramic communication chimpanzees engage in cooperation, tool-assisted foraging, and food-sharing. (A) Austin requests "stick," and (B) Sherman passes it to Austin. (C) Austin probes food from a tube and (D) shares it with Sherman.

Ape Language Trials and Advances

As investigators of two-way communication with apes realized the elusiveness of syntax, they turned to problems of semantics and vocalization, just as linguists had shifted from a heavy concentration on syntax to questions of semantics and phonology in human speech.[317] But it is much easier to demonstrate that great apes know how to employ the artifactual system than to discern what they actually know. Clearly great apes can label objects with arbitrary and iconic symbols and remember a remarkable number of artifactual signs. However, skeptics argued cogently that they had not exhibited comprehension of meaning such that English glosses that investigators apply are unequivocal equivalents of human words. Thus, although great apes can name and symbolize, we still do not know precisely what they might actually know or that their use of symbols is homologous with our own symbolic capabilities. A productive direction for future studies would involve more emphasis on symbolic communication in naturalistic social relations instead of focusing on food and bourgeois children's toys and activities.[318] Moreover, major advances in telemetric monitoring to map brain activity are needed to note the extent to which similar patterns occur in ape and human brains in a wide variety of contexts. Decoding possible genetic underpinnings would also be required to argue persuasively for neurological homologues.

Despite the impressive accomplishments of Sherman and Austin, a host of critics cast clouds of doubt over ape language studies.[319] Ape language researchers also commented critically on the relative merits of the projects of others that fell into two camps based chiefly on methodology: those who fostered their subjects (the Gardners, Fouts, Terrace, Miles, and Patterson) and those who elected more traditional psychology laboratory settings (the Premacks, Rumbaugh, and Savage-Rumbaugh).[320]

Because critics continued to challenge the results of the Sherman-Austin study, Savage-Rumbaugh et al. reiterated that their subjects have shown:

- Great apes can acquire symbols without being cued and without imitating experimenters.
- Great apes can comprehend symbols and produce them.
- Symbols can serve as internal representations of foods and other objects.
- The initial capacities arose through training and rewards, but new skills constantly emerged that went beyond training and required no reward to maintain.
- Great apes can employ symbols to communicate specific referential information to one another that they could not transfer by nonverbal means.[321]

Undeterred by skeptics and critics, Savage-Rumbaugh et al. proceeded to advance the field notably via lexigram studies with bonobos: Kanzi, his half-sisters Mulika (1983–86) and Panbanisha, and Panbanisha's son Nyota (b. 1998). Kanzi, Panbanisha, and Nyota reside with four lexigram-naïve bonobos at the Bonobo Hope Sanctuary that was founded in 2005 at the Great Ape Trust of Iowa to accommodate bonobos from the Language Research Center at Georgia State University (LRCGSU), Atlanta.[322] Highly productive studies continue at LRCGSU with a lexigram-proficient chimpanzee, Panzee.[323] She was coreared with Panbanisha and like her and other lexigram-proficient bonobos understands human speech (Chapter 10).[324]

During a four-year span, beginning when he was 6 months old, Kanzi learned lexigrams by observing unsuccessful attempts to teach his foster mother, Matata (b. 1970), to use the keyboard. He accomplished this while gamboling about the room and otherwise giving little indication to the investigators working with Matata that he was attending to his mother's lessons. When he was 5 years old, they noted that he also understood spoken English, a skill that had evaded lexigram-proficient Sherman and Austin.[325] Like Sherman, Austin, and short-lived Mulika, he had constant companionship of other apes and a stable cohort of human companions as he developed facility with lexigrams and comprehension of spoken English. Kanzi's comprehension of spoken English and production of lexigramic requests, responses, questions, and comments occurred within a social environment characterized by his human companions' respect for his agency as a curious, observant being.

Rumbaugh et al. concluded "it was through Kanzi's access to the patterned experiences afforded by the *logic structure* of his environments (e.g., the speech of the experimenters and their use of word-lexigrams on a keyboard that structured his mother's instructional sessions) that he perceptually *discerned* and *learned* the relationships between symbols and events that provided for him the basic processes and competencies with language."[326]

As an 8-year-old, Kanzi evidenced comprehension of spoken English (74 percent correct) comparable to that of Alia, a 2-year-old human child (65 percent correct). He readily responded to novel verbal instructions such as "Put the melon in the potty" and "Go outdoors and get the pine needles." His performance fell off dramatically (33 percent success) when asked to give two different objects collectively to an animate receiver or to perform two different actions with different objects. The researchers tested Kanzi and Alia with more than seven types of sentence, one of which (type 5C) had end phrase recursion. Kanzi's score (77 percent correct) on type 5C requests (Go get object X that's in location Y) was higher than Alia's score (52 percent correct).[327]

Both subjects evidenced comprehension exceeding their productive abilities, more in Kanzi's case because he could not speak and Alia could.[328] Whereas during six months of testing Alia's mean length of utterance (MLU) increased from 1.91 to 3.19 words, during nine months of testing Kanzi's MLU (1.15 lexigrams) did not change. Neither subject produced sentences with embedded phrases or used phrasal modifiers such as "that's," though they appeared to comprehend such structures. Kanzi could form two-symbol combinations that had order and construct simple ordering rules.[329]

An analysis of Kanzi's responses to forty-two verbal requests convinced Benson, Greaves et al. that because his "interpretations can plausibly be said to be based on symbolic relations between indices, there is evidence that a bonobo brain can process human symbolic language . . . and thus manifests some degree of human-like consciousness."[330] Further, a study with lexigram-proficient Panbanisha indicated that in ordinary conversation she competently engaged in co-constructing conversational turns, using shared knowledge and repetitions to achieve compliance with a request. Accordingly, Pedersen and Fields suggested that she possessed knowledge about sociolinguistic interactions that transcends pure informational content of words.[331] Conversely, at this point in time, it seems like a leap to equate use of artifactual language with consciousness in the bonobos. Computers can process some amount of human language, but this does not mean they have consciousness like that of the humans who program them.

To determine whether the differences between Lana, Sherman, and Austin versus Kanzi and Mulika were due to methodological or specific differences—being a chimpanzee versus being a bonobo—Savage-Rumbaugh coreared Panzee and Panbanisha during the first five years of their lives. Instead of learning lexigrams by standard shaping protocols of working for contingent rewards, they developed the skill by observational learning in a social environment that did not require them to employ lexigrams. Nonetheless, like Kanzi, they began to use lexigrams to communicate with their laboratory and field companions during daily routines indoors and sojourns in a fifty-acre woodland at LRCGSU.[332] Accordingly, it is reasonable to conclude that methodological, not specific, differences account for performance of the two lexigram-competent cohorts.[333]

PART V

WHAT MAKES US HUMAN?

13

Language, Culture, Ideology, Spirituality, and Morality

> In ordinary conversation, the typical number of different consonants and vowels that we produce per second is at least an order of magnitude greater than the unit output rate of any other behavior, or "output complex," either our own or that of any other living form.
>
> PETER F. MACNEILAGE (2008, p. 3)

Talk about Human Language

> Deacon and Donald are correct in seeing symbol use as the most fundamental factor in language evolution.
>
> RAY JACKENDOFF (1999, p. 273)

The emergence of symbolic thought and activity, particularly language, in *Homo sapiens* and its underpinnings in the brain constitute one of the grand mysteries that await empirical resolution by scientists from many disciplines.[1] Effective communication between individuals fundamentally requires a means of production by senders and comprehension by receivers. Physically, speech is produced via air propelled from the lungs by the diaphragm and abdominal muscles flowing upward through the trachea and across the laryngeal vocal cords. Laryngeal and pharyngeal muscles, soft palate (velum palatinum), tongue, and lips interrupt the airflow, thereby shaping the vocal emissions into sounds and articulations that are used to emit linguistic signals.[2] The analogue nature of such sounds is heard as discrete units—phonemes—that go into building larger linguistic units. Unfortunately, less is known about the morphology and functioning of nonhuman primate vocal tracts during call production.[3]

Comprehension occurs in the brain based on the communication of received sounds mediated by our auditory apparatus, accompanying facial expressions and gestures when the sender is in view, and various contextual factors.[4] Mentally healthy human communicators generally possess the following capabilities:

- To engage in joint attention,
- To share a common code,
- To recognize the objects and phenomena to which a signaler refers, or to calibrate social activity based on referent-focal joint attention,
- To surmise one another's intentions,
- To surmise a speaker's thoughts and level of knowledge and beliefs about the topic,
- To assess the validity and importance of messages exchanged, and
- To decide whether and how to respond to them.

From birth onward, humans must exponentially encode masses of information through listening, observation, imitation, emulation, and instruction.[5] Comprehension develops before children can speak, and most of the basics of human language are acquired without formal training.[6] By 18 months of age, children employ phonological knowledge of their native languages to guide interpretation of salient phonetic variation.[7]

Verbal and sign languages are uniquely digital. The information that they convey is composed of words and morphemes compounded from meaningless arbitrary sounds (phonemes) or hand postures and positions (signs).[8] While language partakes of iconic, indexical, and the symbolic signs, it is lexical symbolism and double articulation (also duality of patterning)—the combination of intrinsically meaningless phonemes to produce discrete meaningful words and morphemes—that makes human speech.[9]

The words of human languages are conventionally patterned to form sentences (syntax), often with embedded phrases (recursion), which make it possible to convey information and to share ideas and beliefs, among other things.[10] One of the great challenges facing scientists who would claim that great apes and other nonhuman beings are capable of replicating aspects of human language is to demonstrate that they can employ artifactual symbols syntactically and especially to communicate via sentences that contain nested recursive patterns.[11] More basically, it is the capacity to denote and to predicate about states of affairs not tethered to the here and now or communication that seems to be unique to human symbolism.

Gesture-calls, such as cries, frowns, whimpers, smiles, laughs, chuckles, giggles, and guffaws, in response to being hurt, annoyed, or pleased are analogue (graded) nonverbal signals, some of which are probably homologous with counterparts in apes.[12] The hand, head, and other body movements (gesticulations) that accompany human speech are also analogue signals. Human speech is further characterized by prosody—pitch, volume, timbre, and rhythm—that, like gesticulations, reveals more about a speaker's emotional state and intentions than about propositional meanings.[13] Most languages of the world are tonal—in which pitch or the pitch contour distinguishes the meanings of words that are otherwise the same phonologically.[14]

In both humans and great apes, social communication is an intrinsically creative process that unfolds as partners continuously adjust their behaviors to one another: coregulation. That is to say, our social communication is more than a linear transfer of information from a sender to a receiver who decodes the signal for its information content.[15] Nevertheless, while human and nonhuman communication systems share much of the same semiotic properties—of indexical and iconic types—research on the evolution of human language, which is uniquely symbolic, predominantly focuses on evidence from fossils, ape communication, and archaeology that might point to the emergence of symbolism as such.

Symbols and Symboling Beings

The modern field of semiotics—the study of signs and symbols as elements of communication—is often traced through two independent genealogies. The Swiss linguist Ferdinand de Saussure inaugurated the field he called "semiology," "the science that studies the life of signs within society," wherein his structuralist linguistics was a part. By contrast, the American philosopher Charles Sanders Peirce was concerned with a more general science of signs—semiotics—whose capacious reach included much more than language or the conventional social systems that Saussure's semiology would cover. Also a scientist, Peirce's semiotic was a more general metaphysics that provided a conceptual language to discuss issues from mathematics and logic to the general laws of physics and a phenomenology of mind. Peirce is perhaps most famous for his discussion of the three ways, or grounds, by which a sign-vehicle can be said to be articulated to its object: iconically, indexically, or symbolically.[16]

Iconic signs resemble the things for which they stand. Thus, for example, early Egyptian hieroglyphs and realistic paintings and photographs are iconic signs because there is a similarity in form between sign and referent. Indexical signs connect to their objects by virtue of time-space contiguity—that is, contextually or by causal relation. Like the index finger *(digitus indicis),* they "point to" their referents. Symbols are signs that are related to their object by virtue of habit, association, or rule. Symbols are otherwise arbitrary signs that signify through convention. It is this latter class of signs that Saussure was particularly concerned to explicate in his *Cours de linguistique générale.* Symbolic signs take on meaning and persist via customary usage, meaning that they gain legitimacy in the communicative system of beings that employ them. Note that these distinctions are not mutually exclusive: a sign may, and often does, partake in all three modalities. A Yankee flag, for example, has symbolic, indexical, and iconic aspects. However, symbolic signs associated with human language often, but again not always, lack iconic and indexical aspects.[17] Indeed, it is this symbolic ground that seems to set human communication apart from much of animal communication.

Asif Agha used this Peircean framework to proffer a hierarchy of types of sign-concept relationships. The size and erected plumage of a bird iconically signal to other birds that it is "strong" (the bigger the stronger), albeit probably lacking many aspects of what that gloss might mean to a person fluent in English. Likewise, size, piloerection, and epidermal adornments are iconic of robustness in many other animals.[18]

The waggle dance of bees comprises indexical icons: the angle of wag iconically represents the angular distance between the nectar source and the sun, while the duration of the wag iconically represents the distance of the nectar source from the hive.[18] Taken together, this dance is also indexical for it also provides directions or points to the location of nectar (calculated from the polar coordinates created by the dance itself).

Collaboration among female members of catarrhine monkey matrilines, such as in vervets *(Cercopithecus aethiops),* are indexical of kinship ties. Although vervets likely lack a symbolic concept of kinship, they appear to recognize maternal kin indexically on the basis of close association and interaction with a common mother and female siblings during development.[18]

In primates and other animals, learning sets are based on formative spans of co-occurrence of events leading to similar indexical cause-effect relationships based on event-memory. Classic conditioning of stimulus-response associations in laboratory animals establishes memory-based indexical regularities: for example, pressing levers or touch screens or traversing labyrinths results in food rewards.[18]

Agha labeled three alarm calls of vervets as indexical symbols.[18] The monkeys emit distinct vocalizations when they notice potential aerial, reptilian, and large carnivoran predators with which some members of the group might have had experience.[19] The calls are not the equivalent of human words because they lack grammatico-semantic properties. They simply indicate that there is a potential danger in an area of the habitat; at most, they pick out a predator as co-present to the event of utterance, not in any other way.[18] They do not lead to communication about the referent outside the immediate context of its presence, as humans can do with words such as "Martial eagle," "leopard," or "snake," in good part because the vervet calls have near-zero syntax. In this way, they are like the English deictics "this," "that," "here," there," "now," or "then," which also denote location in space or time in ways that are irreducibly contextual in nature. Unlike the vervet calls, however, such indexical elements of human language can be discussed via other words syntactically to achieve additional possibilities in reference, such as talking about nonpresent referents or even hypothetical or imaginary ones.[18]

In brief, it is difficult to imagine hominid symbolic language developing *de novo* without reliance on iconic and indexical signs. The grand mystery is to determine how and when iconic and indexical signs were combined into syntactic patterns to which arbitrary signs—that is, symbols—could be incorporated to create ever more complex communicative forms and customary expressions. Although the evolutionary history of the preeminently symboling species *Homo sapiens* might never be fully elaborated, there is much merit to continued exploration into the communication systems and minds of nonhuman primates, cetaceans, birds, and other brainy species.

Fossil Facts and Fancies

Prominent among tasks that confound evolutionary anthropologists, psychologists, linguists, and neouroscientists is to chart the evolution of speech over time and to relate it to other modes of communication such as facial expressions and other bodily gestures and the gesture-calls of nonhuman primates and modern people.[20] Patchiness of the fossil record combined with the fact that the human vocal mechanism, which is composed of mostly soft tissues, decomposes quickly postmortem forces researchers to resort to other sources of information and to creative imagination in order to construct plausible models and scenarios of the evolution of speech in the Hominidae.

Some paleoanthropologists have inferred speech from the endocranial features of fossil hominid specimens (such as *Homo habilis*), but others have denied speech to many fossil hominids on the basis of basicranial and mandibular morphology.[21] Still others have claimed that human speech was not developed fully until shortly before the Upper Paleolithic period (40–100 Ka)—perhaps as late as 50,000 years ago—in early anatomically modern *Homo sapiens* because of the simplicity of the tool kits and plastic arts before then.[22] Contrarily, to the extent that brain size can be inferred from fossils and related to the capacity for speech in *Homo,* one might expect *Homo neanderthalensis* to match or even to exceed early *Homo sapiens* as social communicators.[23]

The hyoid bone, which anchors many of the muscles involved in vocalization and swallowing, is fragile, variably shaped, and ambiguously diagnostic of speech.[24] There are two noteworthy fossil specimens: the hyoid bone of a 3.3-Ma juvenile *Australopithecus afarensis* from Dikika, Ethiopia, that Zeresenay Alemseged et al. compared favorably to those of extant African apes, and a 60-Ka hyoid bone of *Homo neanderthalensis* from Kabara Cave, Mount Carmel, Israel, that Baruch Arensburg et al. described as almost identical to those of *Homo sapiens*.[25]

Models based on relative flexion of the cranial base in extant primates, its ontogenetic development in modern humans, and various other basicranial and oropharyngeal details are suggestive but insufficiently convincing, except perhaps to delineate the kinds of sounds that might have been produced by hominoids that were otherwise neurologically wired and cognitively equipped to employ them in a language that linguists would accept as human.[26] For example, Kay et al. offered hope that relative metrics on the hypoglossal canal, through which the hypoglossal nerve passes en route to control movements of the tongue, might illuminate the evolution of human speech.[27] Unfortunately, later studies negated the diagnostic value of the

Figure 13.1. Three stages in the cleaning and assembly of the OH 24 *(Homo habilis)* cranium.

hypoglossal canal, and two senior proponents of the idea joined its refutation.[28]

Tobias claimed that he could detect well-developed Wernicke's areas and Broca's areas on the endocasts of KNM-ER 1470 and OH 24, both of which he classified as *Homo habilis.*[29] Recall that in *Homo sapiens* both areas are associated with human language (Chapter 10). The crushed distorted crania were assembled from many bits, and experts could not restore them to their original states (Figure 13.1).[30] Holloway noted that, almost without exception, the interior surface of the neurocranium is poorly preserved in early hominids from Olduvai Gorge and East Turkana, so no cerebral surface detail is present, especially in regions that might evidence Wernicke's area.[31] Nonetheless, he reported the presence of left Broca's caps in two specimens of *Homo rudolfensis* (KNM-ER 1470 and KNM-ER 3732). Analysis of three specimens of African *Homo erectus* (KNM-ER 3733, KNM-ER 3883, and KNM-WT 15000) from East and West Turkana implied Broca's caps that are more discrete on the right side.[31]

Tobias also noted only a hint of Broca's area and no trace of Wernicke's area in *Australopithecus.*[32] In contrast, Holloway speculated that the type specimen of *Australopithecus africanus*—the Taung child—and an adult specimen of *Australopithecus afarensis* (A.L. 162–28) provide evidence that Wernicke's area had emerged by 3 Ma, but he noted no evidence for Broca's area on their endocasts.[31]

Because speech areas are not readily apparent on the surfaces of fresh human brains, such claims are remarkable. For example, macroscopic landmarks such as the sulci do not accurately reflect the underlying cytoarchitecture of Brodmann's areas 44 and 45.[33]

Among Pliocene–Pleistocene hominids, the mean values of cranial capacity are larger for *Homo habilis, Homo rudolfensis,* and *Homo georgicus* than those of all species of *Australopithecus, Paranthropus,* and extant apes, but notably less than the mean cranial capacities of *Homo erectus* and especially of later Pleistocene and recent Hominidae (Table 10.1). A mere increase in brain size, particularly of <250 cc, and changes in the gross shape of the brain are insufficient to claim or to deny verbal abilities for Early Pleistocene *Homo.* If cerebral expansion was functionally important in them, it might imply general increases in cognitive abilities, manipulative skill, sociality, or other factors important to survival in the Early Pleistocene rather than rudimentary articulate speech. Indeed, prominent Broca's caps occur on some chimpanzee endocasts, yet claims to the contrary not withstanding, no ape has uttered a word despite laborious attempts to get them to speak.[34]

From Primal Language to Language in Culture

Speculation on the origin and evolution of human language, particularly glottogony, has evinced remarkable time depth and myriad variety; including so many silly just-so stories that by 1866 the chief nineteenth-century authority on the study of language, La Société de Linguistique de Paris, imposed a ban on the topic. Although the subject returned to scientific discourse in the mid-1970s, it is still impossible to devise a scenario, let alone a formal model, for the origin and evolution of the modern human communication system upon which linguists, psychologists, and evolutionary anthropologists would agree.[35]

Whereas some theorists advocated the initial predominance of gestural systems, others emphasized vocal systems or allowed some combination of vocalism and gestures as basal to evolutionary developments culminating in the modern human system.[36] The various theories on hominid language origins fall into three basic categories—gestural protolanguage, lexical protolanguage, and musical protolanguage—none of which in isolation suffices as a phylogenic precursor of fluent syntactic recursive vocal and signed propositional human language. Advocates of gestural protolanguage suggested that hominid language began in a manual modality and that syntax and semantics preceded speech. Those favoring lexical protolanguage proposed that language started with isolated meaningful spoken words (speech and semantics) with syntax developing last. Proponents of musical protolanguage proposed that speech began as complex learned vocalizations that were more like song than speech, with semantics developing later.[37]

Hands to Mouth or Vice Versa?

Hewes merits kudos for reviving the attempts to formulate scientifically informed grand scenarios and discussion on the beginnings and subsequent phylogeny of human propositional language, but a modern reading of his thesis leads one to sense quaintness instead of resolution to the puzzle.[38] He aimed "to make explicit the relationship between the emergence of human language and the rise of tool-making and tool-using" based on the hypothesis that hominid protolanguage was gestural rather than vocal.[39] His theory is a consilience of a few hard facts and many more arguable or discredited interpretations: hominoid cerebral lateralization and handedness, ca. 2 million years of hominid cerebral expansion, Plio-Pleistocene potential and actual tools, chimpanzee tool-use and manufacture, chimpanzee use of artifactual gestures and limited vocal control and cross-modal abilities, the functional equivalence of modern sign languages to speech, and models of vocal mechanisms in fossil hominids (all of which are updated in prior chapters or sections of this chapter).

In brief, the neural substrates for manual function and lateralization and language comprehension and production in both human sign and vocal languages are far more complex than those posited in the 1960s and 1970s. Dart's osteodontokeratic tool kit has been returned to the nonartifactual boneyard.[40] Like humans, great apes have multimodal links among auditory, visual, haptic, gustatory, and olfactory senses. Great apes have much richer natural repertoires of vocal and gestural signals than were known to Hewes, and some individuals learn to control them, albeit not to human equivalence. Paleoneurological inferences from Plio-Pleistocene endocasts and models of fossil hominid vocal tracts are poor indicators of speech and gesture in species of Hominidae before *Homo sapiens*.[41]

A major stumper for Hewes and other advocates of gestural protolanguage involves the transition from primally gestural to predominantly spoken language.[42] Further, like Wallace, Hewes was aware that, unlike vocal language, a purely gestural language is limited because the hands must be free, there must be sufficient light, and communicators must be in view of one another.[43] Needless to say, strictly gestural communication would be notably restricted during tool use and in foliated areas and uneven landscapes, even during daylight hours. Nonetheless, Hewes might have been correct to suggest that because great ape gestures are more volitional than their vocal signals and they attend to the attentional states of others, the last common ancestors of hominids and great apes might have shared a potential to develop intentional semanticity, which is vital to human language.[44]

The discovery by Rizzolatti et al. of mirror neurons for manual action in macaques and probable mirror neuron systems in humans (Chapter 10) rejuvenated theories that gestural protolanguage preceded human speech.[45] The initial model of Arbib and Rizzolatti comprised four sequential temporal stages, which Arbib continued to revise, culminating in seven stages:

1. Manual grasping, which characterizes all Primates.
2. A mirror system for grasping shared with *Macaca*.
3. Simple imitation shared with *Pan troglodytes* but not *Macaca*.
4. Complex imitation that is beyond the capacity of *Pan troglodytes*.
5. Innovation of protosign, which underpins an open repertoire.
6. Innovation of protospeech (vocal control) by extending volitional neocortical motor control of the hands and face to the neighboring motor and premotor cortex for musculature of the tongue and larynx.
7. Human language.[46]

Stamenov challenged that the mirror neuron system does not perform the same way in monkeys and humans if one assumes a causal role for it in the origin of language.[47] Hurford added that mirror neurons fail to provide new insight into the meaning of signs, which Newmeyer declared to be "what linguists of all persuasions agree is the task of any linguistic theory, namely to relate sound and meanings (perhaps 'expressions' would be a more appropriate term than 'sounds'), so as not to exclude signed languages."[48]

The priority of gestural protolanguage was challenged by the discovery of audiovisual mirror neurons that respond both while monkeys perform hand movements and while they listen to sounds of similar actions.[49] Like manual mirror neurons, macaque audiovisual mirror neurons are located in area F5 and are adjacent to the motor cortex (area F1). Further, magnetic resonance imaging (MRI) studies indicated that manual and oral actions are separate in humans, with oral actions paralleling the mouth area in the primary motor cortex.[50] Given that mirror neurons are multimodal and code for oral and manual action in macaques and humans, hominid protolanguage was as likely to have been vocal or both gestural and vocal as it was to have been strictly gestural. Further, a major challenge for advocates of gestural and vocal protolanguage is to explain how a basically iconic, holistic, meaningful system might develop duality of patterning, which is a key feature of human communication.[51] Recall also that mirror neurons have not been documented in humans via single-cell recordings (Chapter 10), so their role if any in the evolution of the evolution of language is vague at best.

From Straight Talk to Syntax and Metaphor

> We do not know . . . what particular neurobiological organization gives rise to the possibility of grammar. It appears likely, however, that this question will remain biologically intractable as long as we are unable to answer the question of how grammar, as a distinctively human endowment, relates to nondistinctive endowments—such as instinct, event memory, lexical meaning, and the significance of pragmatic patterning—in the semiotic behavior of humans. Meanwhile, a better understanding of how such endowments interrelate in the semiotic activity of other animals must remain a parallel goal.
>
> ASIF AGHA (1997, pp. 212–213)

Advocates for a lexical protolanguage posited that the hominid communication system began with a large learned lexicon of meaningful words but no complex syntax, which was a culminant development in the evolution of human language.[52] Fitch concluded that while their hypotheses account for some components of our faculty of language in the broad sense, they leave unexplained vocal imitation and phonology.[51] He agreed with Bickerton and others in who have recognized that beings inhabiting a rich perceptual/cognitive world predated humans and therefore any form of human language.[53] Accordingly, in order for signals and meanings to give rise to protolanguage, they must first have been linked by concepts of specific objects or phenomena.

Bickerton's hypothetical protolanguage comprised vocal learning and expression via the auditory/vocal modality (signal), lexical items (individual form-meaning mappings: semantics), and motivation/drive to share information *(Mitteilungsbedürfis)*.[54] It lacked modern syntax composed of grammatical items (function words and inflectional morphemes), phrase structure, obligatory expression of argument structure, readily identifiable null elements, and varied word orders for varied semantic and pragmatic functions. Moreover, the transition to syntax was abrupt, perhaps as a result of a neural reorganization that interconnected cerebral regions for reasoning processes, concepts, and the lexicon. As the possible clues to a stage of lexical protolanguage lacking syntax, he proffered child language, pidgin languages, pidgin/creole transitions, utterances of some artifactual-language-trained apes, and the language of a horrifically abused child, all of which are problematic.[55]

Jackendoff concurred with Bickerton that all stages of language evolution were auditory-vocal and that rich conceptual structures and a basic symbolic capacity to match sounds with arbitrary referents were present before language; but unlike Bickerton, who envisioned a sudden transformation, Jackendoff proffered a detailed, multistep process by which *Homo sapiens* could have incrementally developed a full modern syntax from a basic lexical protolanguage.[56]

Assuming that "their 'fossils' are present in the grammar of modern language itself," Jackendoff employed insights from studies of child language, late second-language acquisition, aphasia, pidgin languages, and ape artifactual language skills to formulate a model (Figure 13.2).[57] The first step was to move beyond call systems like those of nonhuman primates that are situation specific (e.g., the alarm calls of vervets) to single symbol utterances like those of young children where use can be non-situation-specific. For instance, a baby might utter "doggy" to draw attention to a dog, to ask where the dog might be, to call it to him, or to comment on something that resembles a dog.

Once symbols could be used in a non-situation-specific fashion, a period of pre-protolanguage ensued, wherein the system could expand to the use of a large, open, unlimited class of phonological symbols (possibly syllables before phonemes) and their combination to convey basic semantic relations. At this point in Jackendoff's model, hominids had a protolanguage out of which developed a hierarchical phrase structure comprising systems that encode abstract semantic relations and grammatical categories, from which modern syntactic human language emerged via the sequential additive combination of three components, each of which dependent on the preceding one:

- Symbols that explicitly encode abstract semantic relationships (behind, in, over, the, a/an, some, if, because, should, and so forth).
- A system of grammatical relationships—phrase structure—to convey semantic relations.
- A system of inflections to convey semantic relationships—the modification of words to express tense, grammatical mood, grammatical voice, aspect, person, number, gender, and case.[58]

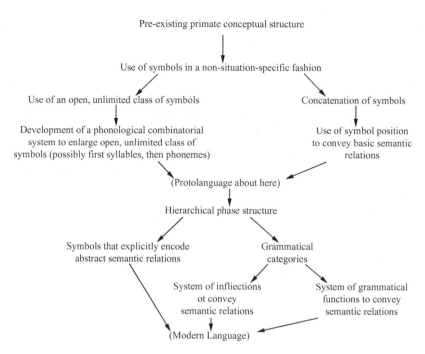

Figure 13.2. A linguist's model for the evolution of modern human language.

Let There Be Music

Vocal phonology is a powerful generative system in its own right. Obeying its own rules, it can generate a vast repertoire of acoustically distinctive vocal signals by combining a small set of sonic primitives (phonemes) that have no meaning in themselves to form meaningful morphemes (e.g., -ed) and words: duality of patterning. The core generative process in phonology is the hierarchical formation of larger units—syllables, phonological words, and phonological phrases—based on the discrete categorically interpreted phonemes. In some languages, the system limits the number and form of combinations and permutations; for example, one might allow only syllables that start with single consonants and end in single vowels.

Each language and musical tradition is culturally transmitted to children, who must learn its distinctive phonological system as a whole. Like language, music is a culturally transmitted human universal that entails a generative system constituted by tonality, timbre, and rhythm, composed of hierarchically combined small sets of notes or tonal nuclei to form syllabic sequences in nonlyrical and lyrical song and acoustic events in instrumental music. In each culture, generative systems of music are learned very early—like phonology.[59]

In addition to syntax, major unsolved puzzles pertaining to the evolution of speech and modern sign languages include how hominids developed a phonological system, how it might derive from a protomusical system, and whether vocal music and language developed together.[60] Some theorists believe that the vocal signaling systems of apes and other animals, including the more resonant and melodic among them, minimally inform the evolution of human language or music or both, while others consider nonhuman vocalizations in their scenarios.[61]

Far from Final Words on Human Language

> The capacity to generate a limitless number of meaningful expressions from a finite set of elements differentiates human language from other animal communication systems.
>
> TECUMSEH FITCH and MICHAEL HAUSER (2004, p. 377)

The potential relevance of the naturalistic vocalizations of primates, birds, whales, dolphins, seals, and perhaps other less well-studied species to inform scenarios on the evolution of hominid protomusic and protolanguage is persuasively discussed and endorsed by Fitch, albeit without resolution of the classic problems.[62] Although Fitch pointed out the problems with precedent ideas, he refreshingly avoided dogmatic championing of his own views or belittling those of his predecessors.[61]

Beginning with Darwin's model of musical or prosodic protolanguage and Jespersen's idea of holistic protolanguage and extensions, combined with tweaks by others, Fitch proposed a four-stage model for the evolution of human language.[63]

First, phonology developed via complex vocal learning during an initial period of songlike communication that lacked propositional meaning, perhaps driven by sexual selection or kin selection or a combination of both factors.[64] As Fitch opined, "Thus, the sharpest distinction between humans and chimpanzees—vocal imitation—arose first, along with simple 'phonological' aspects of syntax (sequencing, hierarchy, and phrase structure)."[65]

Second, perhaps driven by kin selection, whole, complex phonological signals (phrases or songs) and whole semantic complexes—activities, repeated events, rituals, and individuals—were linked by simple association, which served to influence other adults or young offspring. Accordingly, at this stage, musical protolanguage was not a mode for unlimited expression of thought but instead was simply "a manipulative, emotionally rounded vocal communication system."[66]

Third, there followed a phase wherein linked wholes were gradually disassembled and individual lexical items coalesced from them without further genetic changes. Preexisting conceptual primitives were the foundation for the semantic components of protowords and complex syntax. Mapping onto phonological components was arbitrary, was driven by chance associations, and gradually was regularized by cultural transmission. Moreover, "the mismatch between language as a successful vehicle of thought, and its frequently depauperate use in social intercourse, is intelligible by this model, due to the dual origin of the phonological and semantic components of language."[66]

Finally, as the language of a hominid community grew more analytic, strong selective pressure for rapid analytic learning by children and sharing truthful information among close relatives resulted in language that was mostly composed of anatomic meaning units (morphemes or words). Fitch further speculated that the analytic urge was genetically fixed in the final stage of human language evolution.[66]

Fitch commented that by his model some syntactic components—such as hierarchy, concatenation, and linearization—evolved early during the musical protolanguage stage, and the assembly of large, complex semantic structures that map compositionally onto such basic phonological structures occurred later.[67] Variations in inflection and conjugation, different restrictions on order, and agreement and many other factors would not be biologically evolved. Instead, "they would represent various culturally discovered solutions to the ill-defined problem of mapping high-dimensional conceptual complexes onto simple hierarchical phonological representations, and out to the sensory-motor interface of speech and sign."[68]

Most modelers and commentators seem to allow that *Homo erectus* sensu lato had language above that of extant apes, with the language and musical abilities of Australopithecinae and Paranthropinae and ≥2-Ma Homininae being equivalent to or somewhat greater than those of apes. Before accepting or denying any of the scenarios regarding the language capacities and functions of emergent and later Hominidae, we need more research on the naturalistic communication of apes and monkeys, especially on the nature and function of the close calls, which are difficult to record in the field.

My bias at this point is that emergent group-living hominids had or relatively quickly developed a fairly rich communication system comprising a mixture of vocal, facial, and bodily gestures that facilitated their social, mating, and subsistence systems. Eventually, facial gestures, body language, and musical features were subordinated to speech.[69] Nonetheless, they continued to have information content that astute recipients of speech could detect and by which they might assess the affective state, sincerity, and truthfulness of the sender. Music and dance continued to evolve as somewhat autonomous musical and bodily kinesthetic intelligences *sensu* Gardner that, like speech, sign language, and culture, were greatly facilitated by imperfectly understood symbolic processes.[70]

In addition to refinement of the vocal tract and its innervation, there were probably adjustments in thoracic motor control of the abdominal muscles that facilitate the breath control underpinning modern human speech and song.[71] Perhaps they developed from or concomitantly with the remarkable ability of modern humans to adjust our breathing rates over a range of locomotive speeds.[72] Further, as anyone who has studied singing or elocution with professional coaches might tell you, the mechanism of breath control involved is different from those employed during regular daily routines (including ordinary conversation) or when pregnant.

The Quest for Culture

> The meanings of culture and cultural transmission have been broadened over the past two decades, inspired by writings of evolutionary biologists who see evidence of culture in a variety of social activities in all kinds of animal species . . . Unfortunately, this shift in meaning has led to some unproductive debates in which different researchers use the same words to talk about different things.
>
> MICHAEL TOMASELLO (1994, p. 301)

A confluence of information from extensive field and laboratory observations with chimpanzees, comparative genomic studies of humans and apes, and cladistic analyses of extant and fossil hominoid morphology has persuaded an impressive cohort of behavioral and evolutionary scientists to declare that chimpanzees are cultural beings and people are simply other culturally endowed apes.[73] Correlatively, presenting people as merely privileged apes implies that chimpanzees are more humanoid than might be the case. If this is so, it is obscene to treat them as though they have fewer than human rights.[74]

One of the more remarkable examples of terminological abuse in the effort to assimilate chimpanzees with humans is that of McGrew, who referred to *Pan troglodytes* and *Homo sapiens* as "two sibling species of hominids."[75] Canonically, sibling species are "morphologically similar or identical natural populations that are reproductively isolated."[76] I doubt that McGrew would be mistaken for the Gombe chimpanzee icon David Greybeard; and, thus far *Homo sapiens* and *Pan troglodytes* are reproductively isolated species. The problem of culture is much more complicated, particularly for evolutionary scientists who would model the human career and condition.

While my heart is wholly with the plight of chimpanzees and against the ongoing holocaust our species has arrogantly unleashed on them and the other sentient beings with which we must share the Earth, I sense that opportunities to understand the minds and behavioral capacities of chimpanzees and other vertebrates like the cetaceans will be missed if too many behavioral scientists accept the recent declarations of chimpanzee culture and the coming of age of cultural primatology without the evidence that they truly are cultural beings.[77] To date, little more has been demonstrated than that great apes have relatively trivial local or demic behavioral practices that are variably influenced somehow by group conspecifics such as variant grooming postures, vocal signals, food handling techniques, and tool behavior (Figure 13.3).[78]

No one has shown that chimpanzees in nature have pervasive shared symbolically mediated ideas, beliefs, and values, the *sine qua non* of culture as understood by most students of culture.[79] Indeed, one rarely encounters mention, let alone detailed discussion, of symbols in the arguments for naturalistic chimpanzee culture. Instead, there is an emphasis on how behavioral traditions are learned and passed on socially and that the variations in or tangible products of chimpanzee behavior are not a consequence of physical environmental (i.e., ecological influences) or genetic transmission.[80] The actual nature of culture itself, and especially the mechanism(s) by which meanings are encoded for chimpanzees, is absent in most discussions.

While I agree that language—a cultural category rooted in linguistic intelligence—need not be invoked as the criterion for culture in other animals, the challenge remains to discern behaviors that are influenced by symbolically encoded meanings in wild chimpanzees and other nonhuman beings and to identify neurological substrates underpinning them that are homologous with ours. Combined with our growing understanding of human cultural

Figure 13.3. Hand-clasp allogrooming by mature male Ngogo chimpanzees.

cognition, particularly from studies of developmental and cultural psychologists, we might have a fuller base from which to model the evolution of human cultural capacities and to appreciate actual similarities with and unique features of chimpanzee minds.[81]

Boesch and Tomasello have argued that "culture is not monolithic but a set of processes."[82] Because there is "little agreement among anthropologists on precisely what is meant by the term 'culture' as it is applied to human social groups," they thoughtfully explored, albeit with a soupçon of jargon, how a concept of culture could be devised to embrace the chimpanzee case. There had been and continues to be confusion between product and process, with the assumptive former taken to indicate presence of the latter. Process should be preeminent in the exercise. Foremost, chimpanzees and other animals should be revealed as symbolizing beings before labeling their demic practices cultural.[79] That is to say, field researchers should look for evidence of cultural processes—patterns of behavior acquired and transmitted by symbols—in context before they construct elaborate explanatory schemata to fill in the cultural backstory projected from material objects and actions. It would be helpful if researchers more thoroughly digested the perspectives that emerged from the half-century of intensive anthropological research, particularly from the American school, on the nature of culture since

Alfred Kroeber and Clyde Kluckhohn first compiled the classic catalog of definitions.[83]

Although Kroeber and Kluckhohn confessed that "we have no full theory of culture," they articulated the central idea that was formulated by most social scientists approximately as follows:

> Culture consists of patterns, explicit and implicit, of and for behavior acquired and transmitted by symbols, constituting the distinctive achievement of human groups, including their embodiments in artifacts; the essential core of culture consists of traditional (i.e. historically derived and selected) ideas and especially their attached values; culture systems may, on the one hand, be considered as products of action, on the other as conditioning elements of further action.[84]

As a 1960s four-field-trained anthropologist and primatologist who has been embedded in an anthropology department for about fifty years, I find it ironic that given a half-century of contestation and even abandonment of "culture" by many anthropologists primatologists, most of whom originate from purely biological disciplines, would claim that nonhuman primates exhibit culture.[85] Further, I doubt that there is much disagreement among anthropologists and sociologists that symbols and symbolically mediated ideas, values, and beliefs are key to a concept of culture, however difficult it might be to explicate the precise psychological, neurophysiological, and social processes that underpin them, to discern them from behavior, narratives, or texts, and to employ them to demarcate cultures as discrete entities.[86] Accordingly, "symbolic culture" is redundant, and human culture with no symbolism is an oxymoron.[87]

Boesch and Tomasello, who attempted "to bridge the gap between the views of culture typical in . . . biology and psychology and to find common ground between them," concluded that "there seems to be enough common ground concerning processes of culture and cultural evolution that investigators from many different disciplines can begin to make their voices heard in a way that results in an accumulation of modifications to the concept of culture that will facilitate everyone's empirical work."[88] Until the advocates for primate culture engage the scholarly corpus from decades of research on anthropological concepts of culture and its attendant difficulties as produced by a notable roster of sociocultural anthropologists, I doubt that much empirical progress can be made toward discerning naturalistic humanoid cultural capacities in chimpanzees and other nonhuman beings, and especially the phylogeny of human cultural capacities.[89] While I do not deny the importance of understanding how animals, including people, learn and transmit behavior to conspecifics spatiotemporally, I believe we need greater focus on whether (and if so, how) symbolic mediation might be involved in naturalistic behaviors of other beings and whether they are homologous with human processes. Then, we might begin to construct refined models on hominid behavioral evolution over the past 5 million years.

It is unfortunate that, beginning with Tylor, definitions of culture restricted the phenomenon to *Homo sapiens* and that many sociocultural anthropologists believe that to search for counterparts in other animals is futile. This should not dissuade others from searching for symbolically mediated, shared systems of meaning among chimpanzees and other animals.[90] As we saw in Chapter 12, captive chimpanzees and bonobos can employ artifactual symbols to communicate with other beings.[91] Nonetheless, I expect that instead of coming of age, cultural primatology will come a cropper unless fresh, broader, interdisciplinary approaches are developed and applied in the search for naturalistic symbol use in nonhuman beings. Until then, perhaps the quest for nonhuman primate culture should be termed "primate culturology" instead of cultural primatology. It will be interesting to see whether primatologists can revive the nominal use of "culture" among a predominance of sociocultural anthropologists who have abandoned the concept.

Nature and Development of the Symbolic Niche

> Nearly 40% of potential terrestrial primary productivity is used directly, co-opted, or foregone because of human activities.
>
> PETER M. VITOUSEK, PAUL R. EHRLICH, ANNE H. EHRLICH, and PAMELA A. MATSON (1986, p. 368)

> The consequences are predictable: contraction of geographic ranges, reduction of population sizes, and increased probability of extinction for most wild species; expansion of ranges and increased populations of the few species that benefit from human activity; and loss of biological diversity at all scales from local to global.
>
> JAMES H. BROWN and BRIAN A. MAURER (1989, p. 1149)

The shared ideas, values, and beliefs of human beings are collective cognitive emergents that can facilitate broader subsistence and fill more social, political, economic, and spiritual needs than might be achieved by an overwhelming drive for individual reproductive fitness within a group, which is commonly arguably posited for naturalistic populations of apes and other nonhuman beings.[92] Once the neuroendocrinological capacity for symbols was established, the Gardnerian visual-spatial, bodily kinesthetic, interpersonal, intrapersonal, linguistic, musical, and logical-mathematical intelligences could be recruited to increase group cohesion and to expand intergroup communication, mediation of differences between individuals and conspecific groups, trade, division of labor, and prescriptive sharing of food and innovations beyond proximate genetic kin (if even recognized beyond mother/child) and building upon them. Language, music, dance, art, technology, logic, mathematics, spatial orientation, self-knowledge, and knowledge of others are all mediated by symbols, in many instances quite powerfully and often subconsciously.[70]

Accordingly, neurologically intact mature human beings possess enhanced, perhaps even unique abilities to control basic endocrinologically initiated communicative, sexual, agonistic, maternal, and empathetic urges via learned symbolically mediated codes and norms of behavior that enable small groups and larger societies to function efficiently in a remarkable variety of physical and social environments. Because reactions to stimuli from other beings and physical and imagined phenomena include proximate physiological factors, one cannot exclude the role of biological elements in culminant collective symbolically mediated cultural emergents such as death-awareness, morality, altruism, spirituality, and myriad political, economic, religious, and the scientific ideologies and practices that encompass them. Like apes and other beings, human youngsters learn by observation, social facilitation, rewards, and punishment how to function in their families, peer groups, and adult societies. However, tuition is a major additional factor in human socialization, one greatly facilitated by symbolically mediated language, music, dance, games, sports, and other group activities.

As noted by Chase, several factors are true for a large set of Holocene humans, whether hunter-gatherers, agriculturalists, or members of urbanized industrial societies:

- Culture is based on socially created codes.
- Socially created coding provides motivation for an individual's behavior.
- An all-inclusive system of emergent coding pervades and absorbs into itself almost all other coding and almost everything else perceived or thought by humans.[93]

A major puzzle for evolutionary scientists has been to determine how, when, and why the capacity to symbolize (and to create the socially created coding based upon it) occurred, spread, and diversified among the various Gardnerian intelligences. Via auditory and visual symbols, individuals and small groups can manipulate elements of cultural systems for better or worse in subgroups of their own societies or in response to those of other societies, often with wonderfully productive or horrifically destructive outcomes.

There are clear adaptive advantages to individuals in societies that have common values, share resources gathered or produced according to individual skills, protect one another, and can rely on others to assist in child care and tuition instead of each individual having to provide entirely for herself, himself, and dependent progeny. At some point, communication system(s) became vital for coordinating subsistence activities and resolving disputes within and between groups.

As noted in earlier, crania, endocasts, and hyoid bones are inadequate to mark the beginnings or later development of a communicative system based on symbols. If exposed to appropriate learning environments and training, captive apes exhibit command of arbitrary artifactual symbols with which to communicate with experimenters at the level of 2- to 3-year-old humans; however, we are remarkably ignorant as to whether wild apes employ any of their natural vocal signals and gestures symbolically for the well-being of the group (as opposed to their positions within it) and toward sufficient access by all members to sustaining food sources and lodge sites. Nonetheless, some African apes, both wild and captive, provide tantalizing hints that altruism, sympathy (feeling for another being and wishing to increase its welfare), empathy (Chapter 10), consolation, reconciliation, and cooperation might occur in beings with brains one-third the mean size of ours.[94]

Some archaeologists reasonably infer that artifactual beads, pendants, and especially objects of plastic and graphic art probably demonstrated symbolism in the Middle Pleistocene and assuredly in the Late Pleistocene perhaps as early as 77 Ka; however, the extent to which they reflect modern human language and social coding is arguable.[95] Tracing whether earlier hominid species had entered a symbolic niche is elusive to the point that attempted detailed scenarios easily tip into the realm of science fiction. Interpretations of social behavior and social structure of Pliocene–Middle Pleistocene hominid species based on fossil bone accumulations of nonhuman vertebrates and hominid sexual dimorphism, as indicated by fragmentary and even reasonably complete specimens, are also untrustworthy due to the stochastic nature of the evidence. The tendency of some paleoanthropologists, and especially science journalists, to provide dramatic single storylines instead of optional interpretations (including admission that there are insufficient data to support specific scenarios) can distract uncritical readers from assessment of data and inferences. Further, they sometimes unproductively shift focus to rivalries and personalities of researchers.[96]

That said, I suspect that capacity to symbolize began to develop in hominids long before tangible indications of it might be expected to appear in the archaeological record. Recall that like *Homo sapiens*, earlier Hominidae were not endowed morphologically to defend themselves against formidable felid, canid, hynaenid, and crocodilian predators with which they were sympatric.[97] Until effective manual, projectile, and poisoned weapons were invented and fire could be controlled, it would also have been unwise to engage frequently in attacks on neighboring conspecific groups. Factors favoring symbolic capabilities probably co-evolved with enhanced sharing: joint attention to tasks that were further facilitated by cooperation (including those that consoled and calmed others), and protection of the group from external threats. With repeated positive reinforcement, sharing would have led to mutually beneficial, predictable, trusting relationships in dyads and beyond. Although initially the core group for such relationships might have been close cohabiting genetic kin, rapid replacement by trusted others from within the group and immigrants would have been advantageous for group integrity, given the likelihood of notable mortality from accidents, pathogens, and predation that strikes individuals of all ages and physical conditions. Freedom from deterministic genetic restrictions would allow a much richer realm of possibilities for group cohesion, expansion of knowledge, and survival if the vital members could be pacifically replaced.

Death as a Fact of Life

> Death's still the secret of life, the garden reminds us, Or vice versa. It's complicated.
>
> CHARLES WRIGHT (1997, p. 73)

In their once widely used textbook, Dobzhansky et al. declared, "There is no indication that individuals of any species other than man know that they will inevitably die. In man, any individual past childhood who is not a low-grade mental defective is aware of the inevitability of death."[98] Great apes and many other mammals appear to recognize change from living to lifeless, and they sometimes exhibit behavior that one might compare to human states of sadness: lingering near a corpse or carrying a dead infant even when in advanced decomposition.[99] Nonetheless, one can agree with Dobzhansky et al. that there is no indication that they are aware, let alone worry, that they too will die.[100]

Death-awareness is closely linked with self-awareness beyond the simple recognition of oneself in a mirror or exhibiting a distinctive personality.[101] I doubt that apes possess the wherewithal to envision specific future life-history events and to contemplate their deaths and what might occur beyond that event, let alone consciously or subconsciously to recognize a postfrost garden, woods, or meadow as metaphor of inevitable death or its blossoming in spring as a sign of possible existence after death.

The simplistic statement of Dobzhansky et al. on human awareness of the inevitability of death is not so easily accepted given the wide spectrum of ideas and beliefs about what might happen after corporeal destruction. Indeed, many people believe that they will survive as an entity in a peaceful or painful place with or without corporeal reconstitution, or perhaps will be reborn within another corporeal being. Still others believe they may occasionally make ethereal appearances to console or to terrify select warm-blooded individuals. Past and current events attest that a belief in life after corporeal death is a powerful incentive for constructive and destructive behaviors and the ideologies related to them.

On the Warpath

> We are apes of nature, cursed over six million years or more
> with a rare inheritance, a Dovstoyevskyan demon.
>
> RICHARD W. WRANGHAM (1995b, p. 7)

It is tempting to view the violent acts of some chimpanzees against conspecifics as analogous to human homicide, infanticide, genocide, or other cruelties toward vulnerable individuals and groups. Some consider this evidence of a special close biological linkage between chimpanzees and humans, perhaps with roots as far back as the Early Pliocene, and they support such scenarios by placing *demonic males* predominant in hominid phylogeny.[102] Some speak loosely of the striking similarities between chimpanzee intercommunity aggression and primitive warfare among humans, namely, "the common occurrence of large male coalitions, systematic control of territory boundaries, and lethal attacks on isolated individuals from neighboring groups."[103]

Apart from the silly departure from any standard definition of demonic (belonging to an evil spirit, wicked, fiendish, or having a supernatural power or genius), there is no paleoanthropological support for the notion that our stem ancestors regularly engaged in intragroup and intergroup killing, infanticide, cannibalism, or female bashing. Indeed, given the lack of fearsome teeth and other morphological or technological means to kill one another, the arguable probability is that they lived consistently in groups with sizeable cohorts of adult males under an omnipresent possibility that they might be confronted by carnivoran predators, which were more formidable in number and variety than modern ones. It is equally or more likely that homosexual and bisexual cooperation and intergroup tolerance or avoidance were predominantly in force.[104]

In presumptive compliance with cost-benefit theory, Manson and Wrangham initially diagnosed the "demonic male" syndrome: "comparative data on nonhuman primates and cross-cultural study of foraging [human] societies suggests that attacks are lethal because where there is sufficient imbalance of power their cost is trivial, that these attacks are a male and not a female activity because males are the philopatric sex, and that it is resources of reproductive interest to males that determine the causes of intergroup aggression."[105] Unlike the fuzziness in *Demonic Males* about the extent to which the syndrome might be congenital, they and others eschewed inferences that chimpanzee and human males are hardwired to be killers and that the causes of their aggression might be homologous.[106] Consequently, there is no merit to claims that analogous lethal behavior supports an especially close phylogenic link between chimpanzees and humans extending into the Pliocene. The feature might simply be a capacity for facultative aggression that is shared widely in the Mammalia, and expression of it in *Pan troglodytes* and *Homo sapiens* is the result of chance similarity. Moreover, during chimpanzee and human evolution, different or similar ecological or social phenomena might have selected for male and female lethal behavior.

Although Boehm acknowledged the similarities between patterns of East African chimpanzee and patrilineal-patrilocal human male aggression, he noted that there are major differences between *Homo sapiens* and *Pan troglodytes:*

- Humans develop feuding systems between groups, but chimpanzee revenge behavior remains at the individual level.
- Patrilineal humans relatively frequently engage in intensive warfare between two large groups, but East African chimpanzees do not.
- Human communities often manage intensive external conflicts by making external alliances that balance power, and by making peace treaties, but chimpanzees do not.
- Humans are unique in being able to motivate individual males to engage in mass attacks in situations in which the lethally armed males of two entire communities face one another. Human warriors are moved to engage in mass combat by a combination of ideology and negative sanctioning of cowards, two features of macrocoalition competition that chimpanzees lack.[107]

The discussions of Manson and Wrangham's hypothesis and contemporary and later articles and commentaries have indicated that although their inferences from chimpanzee studies were a compelling explanation of their behavior, they had oversimplified the residence patterns and social behavior of human foraging societies and had not taken into account the powerful effect that food sharing, equitable division of labor with female agency, child care and tuition, language, and other cultural behavior would have played in premodern hominid evolution, particularly to mitigate intragroup female and male aggression and frequency of attacks on other groups who appear to be vulnerable.[108] Because there were fewer hominids in the Pliocene and Pleistocene, group migration away from agonistic groups would seem to have been an available option. Moreover, developing mutually beneficial relationships with them probably would be a better option for one or both parties, instead of having to plan and execute attacks in lieu of foraging and sustaining bonds within one's own group.

Further, globally, the residence practices and patterns of human foraging societies vary notably to include matrilocality and bilocality, in addition to patrilocality, all of which can facilitate many cooperative intergroup activities. Regardless of residence type, mixtures of kin and non-kin encourage exchange of information and tangible resources. Chance congenial encounters and planned intergroup rituals and ceremonies allow females and males to meet, mate, and perhaps establish more permanent and extensive bonds. Some chimpanzees also exhibit bilocal residency: in 1995, nearly half of the fully adult females born into the main Gombe research group resided there; in the Republic of Guinea, the natal females that survived into adulthood remained in the semi-isolated Bossou group over twenty-one years.[109]

Reflecting further on issues raised in and by Manson and Wrangham, Wrangham proposed an imbalance-of-power hypothesis whereby coalitional killing of adults in neighboring groups of chimpanzees and wolves "implies that selection can favor components of intergroup aggression important to human warfare, including lethal raiding."[110] Presumably, in the unneighborly quest for dominance two conditions are necessary and sufficient to account for coalitionary killing: a state of intergroup hostility and sufficient imbalances of power between groups such that coalitions of males can attack outgroup individuals with impunity in order to enhance their status among ingroup males.[111]

Arguing against hominid homicidal tendencies being an evolutionary development rooted in killer apes that initially hunted and devoured nonhuman mammals, Wrangham and Peterson entertained the idea that hominid and chimpanzee homicidal behaviors may have eased the way for them to become hunters.[112]

Bonobos have presented a major challenge to advocates of a common ape and human origin for the demonic male syndrome because bonobos are genomically closer to chimpanzees than either species of *Pan* is to *Homo sapiens,* yet they appeared from early field studies not to hunt monkeys, "perhaps as a consequence of their low interest in intraspecific killing."[113] Fuller data sets on bonobo behavior indicate that, like chimpanzees, they kill and eat other mammals, including monkeys, and occasionally engage in cannibalism (Chapter 8), and they can be quite aggressive toward one another (Chapter 11). One must also be mindful of the likelihood that all species employed in such comparisons have differentiated from one another, perhaps quite notably, over the past 6 million years.

Cognitive, social, and technological developments greatly extended the difference between interindividual and intergroup altercations among chimpanzees compared with those among humans. Human technological advances and perturbations have also indirectly affected the frequency and form of chimpanzee aggression.[114]

Humans engage in mass killings both in and out of view of their victims, and armed women can act as aggressors and protectors on a par with men.

Human encroachment has greatly limited the available habitat where chimpanzees can move, thereby potentially increasing intragroup and intergroup competition, while often humans can quit inhospitable environments. Moreover, only humans appear to be able to contrive belief systems that dehumanize others and devise social codes and ideologies that rationalize their exploitation and destruction. As Lieutenant Cable saliently sang in *South Pacific:*

> You've got to be taught before it's too late,
> before you are six or seven or eight,
> to hate all the people your relatives hate;
> you've got to be carefully taught![115]

What Is More Real: God or Race?

One of the most puzzling human cognitive capacities is the ability individually and collectively to believe in intangible as well as tangible phenomena. God and race *prima facie* appear to be very different because God requires belief in something wholly intangible whereas race is purportedly based on tangible scientific evidence. Both generate a wide spectrum of belief systems and social codes.

I believe that God is an ever-increasing collective emergent of the love of all beings past, present, and future, but this cannot be proven by available scientific methods of experimentation or controlled comparison. In contrast, the belief in race, in the sense of biological subspecies of *Homo sapiens,* lacks a tangible basis; indeed, it has been proven unsupportable genomically, behaviorally, and phenotypically.[116]

Individuals and political groups have manipulated both God and race for nefarious purposes, but actions rooted in the human capacity to affiliate with non-kin, to cooperate, and especially to unite in love and respect for the agency of others has given rise to a variety of constructive social codes that facilitate intragroup and extensive intergroup harmony and mitigate disruptive personal and social behavior.

Whereas scientists possess the means to eliminate belief in human races, they lack the means to eradicate belief in God, and frankly they are probably wasting time and treasure on the exercise. Clearly, religion, particularly when it is little more than agenda-driven politicized spirituality, has no place in science classes, though it might greatly enhance world tolerance and peace were comparative religion classes to be part of school curricula, preferably when youngsters are beginning to be aware that people have different beliefs and that "race" is insignificant once friendships are formed.

Like humans, group-living nonhuman primates are predominantly pacific during their daily activities, and those that live in stable social groups evince frequent acts of affiliation, mutual tolerance, and cooperation.[117] Perhaps such behavior was the basis from which humans developed moral codes that emphasize cooperation and consideration of the needs of others. Nonetheless, given the current state of research, it is premature to consider that nonhuman primates have anything like the symbolically mediated values and moral beliefs of *Homo sapiens*. Consequently, one should view nonhumans as amoral beings, and the acts of mature, cognitively intact humans can be assessed as moral or immoral within and across social groups, albeit with great difficulty and often without clear resolution.

Great Ape Debates

> And God said, Let us make man in our image, after our
> likeness, and let them have dominion over the fish of the sea,
> and over the fowl of the air, and over the cattle, and over all the
> earth, and over every creeping thing that creepeth upon the
> earth.
>
> Genesis 1:26

> Because humans view apes as mentally limited, some current
> captive environments may appear idyllic while offering only an
> illusion of appropriate care derived from a simplistic view of
> what apes are rather than what they might be. This perception
> of apes determines their handling, which determines their
> mental development, which perpetuates the prevailing percep-
> tion. Only breaking this cycle will allow the current perception
> of apes to change.
>
> SUE SAVAGE-RUMBAUGH, KANZI WAMBA, and PANBANISHA
> WAMBA (2007, p. 7)

A major ethical dilemma that faces the world community concerns the humane treatment of captive nonhuman primates and indeed whether any primate species should be confined or killed as part of medical experiments or other scientific studies. It is even arguable whether they should be incarcerated for educational and recreational purposes.[118] Minimally, everyone should agree that all extant species of free-ranging primates must be protected with their habitats in vast regions of the globe.

Dominion over aquatic, aerial, and nonhuman terrestrial animals requires stewardship, not thoughtless, ravenous exploitation and eradication. As the only extant beings capable of moral and immoral acts, *Homo sapiens* should conscientiously ensure the survival and well-being of most nonhuman amoral organisms with which we share Earth. Abuse and disrespect for apes and other animals has many dehumanizing ramifications for human behavior. It is time to acknowledge the major advances in behavioral and social sciences that demonstrate the remarkable cognitive abilities of apes. Although they might not be moral creatures, our treatment of them will surely reflect badly on our own morality until we change global public policy. A first step is universally to declare the captives whose habitats are shrinking to be endangered species.

The time is long past when people should cease

- referring to apes as monkeys. (Listen up, Stephen Colbert, Jon Stewart, David Letterman, and other highly public figures with large audiences!)
- displaying chimpanzees, orangutans, and other nonhuman animals in costumes.
- keeping them as pets.
- causing them mental and physical distress by using them as animal models for human disease while accommodating them in solitary confinement or otherwise minimally permitting them a regular social life with conspecifics.
- depriving them of natural lives and agency, which are essential for such highly sensitive, comprehending, social beings.

I have witnessed the degradation and torment of apes, monkeys, dogs, and other animals in laboratories and have contributed to the discomfort of several apes in the course of pioneering functional morphological research (Chapters 5 and 6).[119] I had few regrets when the projects ended. Although our test procedures are commonly performed with human volunteers, the pongid subjects did not volunteer. Moreover, in order to equip the subjects for observation, technicians frightened them with the restraints required by the rules of the research center. The face masks, goggles, heavy gloves, and full-body suits now worn to protect handlers and subjects from disease transmission strike terror in incarcerated nonhuman animals.

Among the organismal biological sciences, primatology is most strategically positioned to showcase the values of tropical environmental preservation and comprehensive conservation. Few organisms can rival primates as poster creatures for conservation, and they inspire myriad works by humanists and artists.[120] Primatologists hail from many primary disciplines: anatomy, anthropology, bioengineering, clinical medicine, cognitive science, communication science, conservation, demography, developmental biology, ecology, education, endocrinology, ethology, evolutionary biology, gerontology, husbandry, immunology, microbiology, molecular genetics, neuroscience, nutritional science, paleontology, parasitology, pathology, pharmacology, physiology, population genetics, psychiatry, psychobiology, reproductive biology, space biology, systematics, toxicology, veterinary medicine, and virology. In the twenty-first century, our challenge is to work together for a truly global primatology in which leadership is concentrated among scientists in the countries that are graced with natural populations of nonhuman primates.

Sixteen years ago, Russell Mittermeier noted that "wild populations of nonhuman primates are in trouble in all of the 92 countries where they occur. Some of the most serious problems are in those nations richest in primates, among them Brazil, Perú, Colombia, Madagascar, Indonesia, China and Vietnam."[121] For the vast majority of species, subspecies, and populations, the situation has worsened.[122] Clearly, the herculean efforts of environmental scientists, conservationists, national governments, and international funding agencies must be redoubled and rethought.[123] It is equally vital that primate behavioral ecologists, conservation biologists, cultural anthropologists, and wildlife managers increase mutual communication and work closely together with indigenous communities who share habitats with nonhuman primates.[124]

The nonhuman primate cause probably would be best served if scientists from the host countries—those that have indigenous populations of nonhuman primates—were more visible participants in governing the international scientific organizations. Moreover, they should be senior authors of more papers in the best peer-reviewed journals and serve more often on the editorial boards of primatological and other scientific journals and book series. National governments should respect and directly seek the counsel of their own scientists, who must be supported as equals by peers internationally. Surely local scientists, who understand the languages and customs of their compatriots, are best equipped to work with indigenous educators to develop awareness, respect, and love for tropical rain forests, other primate habitats, and their inhabitants.

Sadly, over the past millennium, Earth has become the Planet of People, where far too many individuals and societies behave like the omnipotent beings in Pierre Boulle's *Planet of the Apes*.[125] The foremost corrective is to deal effectively with human hubris, poverty, greed, political corruption, and other all-too-human failings and find a way to conserve biodiversity and natural ecological communities without privileging some nonhuman beings *vis-à-vis* others and while upholding the dignity and rights of humans who live most proximately to them.[126]

Accordingly, I am uneasy about aspects of the Great Ape Project, the Great Ape Survival Plan, and Great Ape World Heritage Species Project, which would privilege a few—albeit marvelous—species over others in efforts to save them from mistreatment and extermination. They privilege researchers whose careers depend upon access to localities with great apes. Accordingly, dubbing the great apes flagship species may be close to an unwelcome truth: in colonial times, a flagship carried elite opportunists to and precious commodities from the lands and peoples that they exploited.

We must hope that World Heritage–specific status will protect great apes better than some World Heritage Sites like Laetoli and Olduvai Gorge in Tanzania have fared under a similar designation, and that indigenous expert scientists can freely determine who conducts research in their nations by what they contribute to the conservation efforts at their work sites.[127] The argument that great apes will be umbrella species, thus protecting not only them but also their habitats and fellow denizens, is an improvement over the flagship moniker, but many more taxa of primates and other animals that do not live with the great apes should be targeted for special protection. Indeed, among the World's Top 25 Most Endangered Primates 2010–12, only two are great apes, which would be protected as World Heritage species; the other twenty-three of the imperiled taxa are less protected because they do not live with them.[128]

I agree with the ethicists, philosophers, and fellow scientists who believe that our conservative efforts should be based on broader moral and ethical considerations that do not depend on presumed biological closeness to humans. Further, to argue that African and Indonesian leaders, who are of precolonial heritage, should save great apes because they too are apes displays a remarkable ignorance of history and political naivety.[129] Moreover, the enlistment of analogies with injustices that ignorant people visit upon slaves, intellectually and physically challenged persons, women, children (who initially are probably amoral), gender extenders, and other groups of people whom they perceive to be different from themselves is unsettling in ways that would deter me from supporting aspects of the declaration on great apes were I not already intimately acquainted with them. Such comparisons neither promote the equal dignity and equal rights of all human beings nor lead to a full appreciation of the adaptive complexes and novel capabilities of apes.[130]

Figure 13.4. Left: Orphaned bonobo at Lola Ya Bonobo Sanctuary, Democratic Republic of the Congo.

Notes

1. King 2007, p. 220.
2. Miller et al. 2006.
3. Tuttle 2006c.
4. Tuttle 2006c; Kingston 2007; Magill et al. 2013a,b; Tipple 2013.
5. Tuttle 2001.
6. Morin et al. 2001.
7. *contra* Tooby and DeVore 1987.
8. Kingston 2007; Wells and Stock 2007.
9. Cartmill 1993, 1997.
10. Dart 1949, 1953, 1957; Lorenz 1963; Ardrey 1961, 1966, 1970, 1976; Morris 1967; Tiger 1969; Tiger and Fox 1971.
11. Darwin 1871, p. 444.
12. Rainger 1991.
13. Osborn 1927; Osborn 1929, p. 6.
14. Dawson and Woodward 1913; Osborn 1921, pp. 581–582.
15. Dart 1957.
16. Dart 1959.
17. Washburn 1960.
18. Washburn and Lancaster 1968.
19. Washburn 1960 p. 63.
20. Washburn and Lancaster 1968, p. 300.
21. Linton 1971; Slocum 1975.
22. Tobias 1967a; Brain 1968a, 1968b, 1969, 1981; Washburn 1957.
23. Vrba 1975; Berger and Clarke 1995, 1996; Sanders et al. 2003; McGraw et al. 2006.
24. Lee 1968, 1969, 1979; Marshall 1976; Silberbauer 1981; Tanaka 1980.
25. Fox 1952; Peterson 1978; Estioko-Griffin and Griffin 1981, 1985; Barbarosa 1985; M. J. Goodman, Czelusniak, and Beeber 1985; Boehm 1999; Dahlberg 1981; Meggitt 1962; Gould 1969; O'Connell et al. 1999.

26. Slocum 1975, p. 39.

27. Slocum 1975, pp. 43–45 (emphasis in original).

28. Linton 1971.

29. Slocum 1975, p. 44.

30. Slocum 1975, p. 47.

31. Lee 1979, pp. 489–494; Slocum 1975, p. 46.

32. Slocum 1975, pp. 46–47.

33. Morgan 1972; Tanner and Zihlman 1976; Tanner 1981, 1987; Zihlman 1978; Zihlman and Tanner 1978; Hrdy 1981, p. 189.

34. Isaac 1971, 1976, 1978, 1981; Isaac and Crader 1981.

35. Isaac 1971.

36. Etkin 1954.

37. Isaac 1971 p. 294.

38. Isaac 1978.

39. Isaac 1978, 1981.

40. Binford 1981.

41. Isaac 1983b, p. 16; Marlowe 2005.

42. Lovejoy 1981, 2009a, 2009b.

43. Fedigan 1986, Haraway 1989, Hrdy and Bennett 1981, Leibowitz 1983, McBrearty and Moniz 1991; Zihlman and Lowenstein 1983; Zihlman 1987; Tuttle 1981a.

44. Hill 1982.

45. Reno et al. 2003; Larsen 2003.

46. Gordon et al. 2008.

47. Hamburg 2008; Wrangham and Peterson 1996; Stanford 1999.

48. Wrangham and Peterson 1996, p. 219.

49. Wrangham and Peterson 1996, pp. 241–243.

50. Dolhinow 1999, p. 446.

51. Stanford 1999, p. 11.

52. Stanford 1999, p. 199.

53. Domínguez-Rodrigo and Pickering 2003.

54. Winterhalder 1996; Noss and Hewlett 2001.

55. Haraway 1989.

56. Strum and Fedigan 2000.

57. Kevles 1995; Marks 1995a; Gould 1996; Graves 2001.

58. Ingman et al. 2000; Feldman et al. 2003; Marks 2002; Koenig et al. 2008.

2. APES IN SPACE

1. Le Gros Clark 1971; Cartmill 1972, 1974a, 1982a; Gingerich 1984a.

2. Martin 1986, pp. 16–17, 1993, 2012.

3. Martin 2006, 2012.

4. Gingerich 1986a.

5. Demes et al. 1994; Hanna et al. 2006.

6. Heesy 2004.

7. Bloch et al. 2007; Sargis et al. 2007; Silox et al. 2007.

8. Janecka et al. 2007; Meredith et al. 2011; Bloch et al. 2007; O'Leary et al. 2013.

9. Ravosa and Dagosto 2007; Ni et al. 2005; Smith et al. 2006; Martin 2012.

10. Janke et al. 1994; Soligo et al. 2007; Martin 2012.

11. Beard et al. 1991; Cartmill 1980, 1982b; Cartmill and Kay 1978; Goodman et al. 1978; Groves 2001; Hershkovitz 1974; Hill 1955; Hoffstetter 1988; Kay et al. 1997; Koop et al. 1989; Luckett 1975, 1980; Martin 1990; Szalay 1975a; Ross 2000; Xing et al. 2007.

12. Aiello 1986; Gingerich 1981; Simons 1972; Simons and Rasmussen 1989; Gingerich 1984a; Rosenberger 1986; Fleagle 1988.

13. Le Gros Clark 1971; Fleagle 1988.

14. Kirk 2004.

15. Tuttle and Rogers 1966.

16. Rose et al. 2003.

17. Erikson 1963; Symington 1990; Chapman et al. 1995; Strier 2000, p. 7, 2003.

18. Miller 1945; Rose 1974a.

19. Grubb et al. 2003; Brandon-Jones et al. 2004; Davenport et al. 2006.

20. Murray 1975; Bauchop 1978.

21. Brandon-Jones et al. 2004: Grubb et al. 2003; Davenport et al. 2006.

22. Tuttle 1986.

23. Bianchi et al. 1985; de Grouchy 1987; Dugoujon and Hazout 1987; Khudr et al. 1973; Kluge 1983; Mai 1983; Marks 1983a, 2005; Martin 1990; Schultz 1936; Stanyon and Chiarelli 1982; Tuttle 1975a, 1986; Gibbs et al. 2002.

24. Delson and Andrews 1975; Groves 1986; Harrisson 1987; Andrews et al. 1996; Begun 1999.

25. Goodman 1962a, 1962b.

26. Goodman 1992, 1996; Goodman et al. 1998; Tuttle 1986, 1988a; Marks 2005.

27. Ciochon 1983.

28. Andrews and Cronin 1982; Gantt 1983.

29. Goodman 1982, 1986; Goodman, Olson et al. 1982; Goodman et al. 1983, Goodman, Czelusniak, and Beeber 1985; Hasegawa and Yano 1984; Andrews 1985; Groves 1986; Miyamoto et al. 1988.

30. Adachi and Hasegawa 1995; Andrews 1986a, 1987; Andrews and Martin 1987a; Arnason et al. 1996; Baba et al. 1982; Baldini et al. 1991; Barriel 1997; Barriel and Darlu 1990; Britten 1989; Caccone and Powell 1989; Clemente et al. 1990; Daiger et al. 1987; Felsenstein 1987; Fitch 1986; Goldman et al. 1987; Gonzalez et al. 1990; Goodman 1992; Goodman et al. 1971, 1982; Goodman, Olson et al. 1982; Goodman et al. 1983, 1989; Hasegawa et al. 1985; Holmquist et al. 1988a; Horai et al. 1992, 1993; Hoyer et al. 1972; Kim and Takenaka 1996; Lanave et al. 1986; Lucotte et al. 1990; Lucotte and Hazout 1986; Maeda et al. 1988; Makova and Li 2002; Miyamoto et al. 1987, 1988; Mohammad-Ali et al. 1995; Nei et al. 1985; Patterson et al. 1993; Ruvolo and Pilbeam 1986; Ruvolo and Smith 1986; Ruvolo et al. 1991; Ruvolo et al. 1994; Ruvolo 1994; Scally et al. 2012; Saitou and Nei 1986; Sakoyama et al. 1987; Sibley and Ahlquist 1984, 1987; Sibley et al. 1990; Slightom et al. 1985; Spuhler 1988; Takahata and Satta 1997; Ueda et al. 1986, 1989; Weiss 1987; Wildman et al. 2002: Wilson et al. 1984; Xing et al. 2006; Yang 1996; Yunis and Prakash 1982.

31. Goodman 1992.

32. Goodman et al. 1998; Page and Goodman 2001; Wildman et al. 2003.

33. Diamond 1984, 1988a, 1988b, 1992.

34. Diamond 1984, 1988a, 1988b; Marks 1988, 2005.

35. Schwartz 1984a, 1984b, 1988, 2005, 2007; Grehen 2006.

36. Lodke et al. 2011.

37. Marks et al. 1988; Sarich et al. 1989.

38. Bailey 1993; Bruce and Ayala 1978; Carroll 2003; Chen and Li 2001; Corruccini 1992; Ely et al. 1992; Saitou 1991; Smouse and Li 1987; Hasegawa and Kishino 1989, 1991; Hasegawa et al. 1989; Holmes et al. 1989; Kishino and Hasegawa 1989; Leeflang et al. 1992; Marshall 1991; Miyamoto and Goodman 1990; Fowler et al. 1989; Dangel et al. 1995; Kawamura and Ueda 1992; Meyer et al. 1995; Minghetti and Dugaiczyk 1993; Pellicchiari et al. 1982, 1990; Rapacz et al. 1991; Rasheed et al. 1991; Rumpler and Dutrillaux 1990; Ruano et al. 1992; Stanyon 1989.

39. Chen and Li 2000; Vigilant and Bradley 2004.

40. Robinson et al. 2008; Fowler and Schreiber 2008; Li and Saunders 2005; Pennisi 2005; Culotta 2005a; Dennis 2005; Bailey et al. 1992; Bauer and Schreiber 1995; Begun 1994; Borowik 1995; Bradley 2008; Diamond 1988; Gonzalez et al. 1991; Goodman et al. 1998, 1994; Hasegawa 1990, 1991, 1992; Horai et al. 1995; Janczewski et al. 1990; Kawamura et al. 1991; Laursen et al. 1992; Lockwood et al. 2004; Patterson et al. 2006; Pilbeam 1996; Ruvolo 1996; Salem et al. 2003; Shoshani et al. 1996; Suzuki et al. 1994; Takahata et al. 1995; Uddin et al. 2004; Williams and Godman 1989; Wood and Harrison 2011; Xing 2007; Marks 1992, 1993, 1995b; Bruce and Ayala 1979; Andrews 1992; Chaline et al. 1996; Deinard and Kidd 1999; Deinard et al. 1998; Djian and Green 1989; Kawamura et al. 1992; Livak et al. 1995; Samollow et al. 1996; Rogers 1993, 1994; Rogers and Comuzzie 1995; Corruccini 1994; Tuttle 2006a.

41. Culotta 2005a.

42. Tuttle 2006a; Asthana et al. 2007; Bailey and Eichler 2006; Bazin et al. 2006; Berezikov et al. 2006; Booth and Neufer 2005; Calarco et al. 2007; Carroll 2003; Cheng et al. 2005; Clark 1999; Clark et al. 2003; Cohen 2007a, 2007b; Crawford et al. 2005; Davidson and Levine 2005; Davidson and Erwin 2006; Dennis 2005; Ebersberger et al. 2002; Eichler et al. 2001; Enard, Khaitovich et al. 2002; Eyer-Walker 2006; Fujiyama et al. 2002; Gagneux and Varki 2001; Hamdi et al. 1999; Harris et al. 2007; Hauser 2005a; Hayakawa et al. 2005; Johnson et al. 2001; Macdonald and Long 2005; Martin et al. 2005; Medina 2005; Naylor and Brown 1998; Olson and Varki 2003, 2004; O'Rourke 2003; Pennisi 2005, 2006a, 2007a, 2007b; Perry et al. 2006; Pollard et al. 2006; Popesco et al. 2006; Schmidt et al. 2005; Seringhaus and Gerstein 2008; Sikela 2006; Uddin et al. 2008; Varki and Altheide 2005; Watanabe et al. 2004; West-Eberhard 2005; Wilson 2005; Winckler et al. 2005; Wise et al. 1997; Zhang et al. 2007.

43. Tuttle 2006a; Tay et al. 2009; Shen et al. 2011.

44. Culotta 2005b; Venter et al. 2001; Claverie 2001; Cohen 1997; Fields et al. 1994; Hattori et al. 2000; O'Brien et al. 1999; Reeves 2000; Pennisi 2005, 2007d.

45. Britten 2002.

46. Britten et al. 2003, p. 4664.

47. Nozawa et al. 2007.

48. Zhang 2007.

49. Bakewell et al. 2007, p. 7489.

50. Cheng et al. 2005; Mikkelsen 2005; Campbell 1985, xiv–xv; Cela-Conde 1998; Marks 1984, 2005; Martin 1990; Tuttle 1986; Fleagle 1988.

51. Arnason et al. 1966; Crouau-Roy et al. 1996; D'Andrade and Morin 1996; Gagneux 2002; Gagneux et al. 1999; Hacia et al. 1999; Vigilant et al. 1991.

52. Feldman et al. 2003.

53. Becquist et al. 2007.

54. Arnason 1998; Clifford et al. 2003; Garner and Ryder 1996; Goldberg and Ruvolo 1997b; Gonder et al. 1997; Hacia 2001; Hall et al. 1996; Hayashi et al. 1995; Hofreiter et al. 2003; Janczewski et al. 1990; Jensen-Seaman et al. 2003; Jolly et al. 1995; Kaessmann et al. 1999; Morin et al. 1994; Muir et al. 1998; Ruvolo et al. 1994; Ryder 2003; Ryder and Chemnick 1993; Saltonstall et al. 1998; Uchida 1996, 1998b, 1998c; Xu and Arnason 1996; Gonder et al. 2011.

55. Groves 2001; Grubb et al. 2003; Brandon-Jones et al. 2004; Tuttle 2003a; Brandon-Jones et al. 2004; Mootnick and Groves 2005; Geissmann 2007; Thinh et al. 2010; Mootnick and Fan 2011.

56. Washburn 1963a; Livingstone 1962a; Dobzhansky et al. 1977, pp. 138–146; Littlefield et al. 1982; Nei and Roychoudhury 1982; Feldman et al. 2003.

57. Lewontin 1972; Futuyma 1979, p. 205; Brace 1982; Lieberman and Jackson 1995; Marks 2002; Lieberman et al. 2003; Gilroy 2000; Graves 2001; Palmié 2008.

58. Brandon-Jones et al. 2004; Grubb et al. 2003; Mittermeier 2002.

59. Goossens et al. 2009.

60. Tuttle 1986; Shea and Coolidge 1988.

61. Ely et al. 1998.

62. Vervaecke and Van Elsacker 1992; Vervaecke et al. 2004.

63. Rossiianov 2002; Wynne 2008a.

64. Tuttle 1986, p. 16.

65. Rosiianov 2002; Wynne 2008a.

66. Wynne 2008a.

67. Rossianov 2002; Remington 1971; Moens 1908; Rohleder 1918.

68. Rossianov 2002.

69. Tuttle 1986, p. 18.

70. Simpson 1961; Blackwelder 1967.

71. Geissmann 2007; Brandon-Jones et al. 2004; Zihlman, Mootnick, and Underwood 2011.

72. Locke et al. 2011; Scally et al. 2012.

73. Tuttle 1986; Marshall and Sugardjito 1986; Ma et al. 1988; Hamard et al. 2010.

74. Hooijer 1960.

75. Tuttle 1986; Fooden et al. 1987.

76. Gu 1986, 1989.

77. Chiarelli 1972; Liu et al. 1987; Prouty et al. 1983; van Tuinen and Ledbetter 1983.

78. Schultz 1933, 1937; 1973; Miller 1945.

79. Smith and Jungers 1997; Thorén et al. 2006; Schultz 1973; Oxnard 1983b.

80. Haimoff et al. 1982, 1984; Marshall and Sugardjito 1986; Mitani 1987; Heller et al. 2010.

81. Brockelman and Gittins 1984; Brockelman and Schilling 1984; Geissmann 1984; Geissmann 1995; Shafer et al. 1984; Wolkin and Myers 1980; Tenaza 1985; Marshall and Sugardjito 1986.

82. Schultz 1936.

83. Delattre and Fenart 1956.

84. Shea 1985.

85. Frisch 1973.

86. Swindler 2002.

87. Frisch 1973; Swindler 2002.

88. Ungar 1994.

89. Kay and Hylander 1978; Curtin and Chivers 1978.

90. Kay and Hylander 1978.

91. Kay 1981; Nagatoshi 1984, 1990; Shellis et al. 1998.

92. Tuttle 1975a.

93. Schultz 1956, 1973; Erikson 1963; Drapeau and Ward 2007.

94. Schultz 1973; Jungers 1985; Jungers and Cole 1992; Zihlman, Mootnick, and Underwood 2011.

95. Steiper 2006.

96. Locke et al. 2011.

97. Hoang et al. 1979; Ciochon and Olsen 1986; Röhrer-Ertl 1988; Schwartz et al. 1994; Bacon and Long 2001; Rijksen and Meijaard 1999; Caldecott and McConkey 2005.

98. McConkey 2005a, 2005b; Tuttle 1986.

99. Poe 1975; Maple 1980; Van Schaik 2004.

100. Thorén et al. 2006; Markham and Groves 1990.

101. Kuze et al. 2005.

102. Winkler et al. 1988.

103. Schultz 1969a.

104. Masterson and Leutenegger 1990, 1992; Hens 2005.

105. MacKinnon 1974; Rikjsen 1978.

106. Blaney 1986; Rossie 2005.

107. Swindler and Olshan 1988.

108. Swindler 1976.

109. Hooijer 1948; Swindler 2002.

110. Frisch 1965; Swindler 2002.

111. Gantt 1983; Swindler and Olshan 1988.

112. Gantt 1986; Nagatoshi 1984, 1990; L. Martin 1986; G. Schwartz 2000.

113. Vogel et al. 2008.

114. Tuttle and Cortright 1988; Zihlman, McFarland, and Underwood 2011.

115. Drapeau and Ward 2007; Tuttle 1970; Tuttle and Cortright 1988.

116. Kaessmann et al. 1999; Gagneaux et al. 1999; Stone et al. 2002; Yu et al. 2003; Beadle 1981.

117. Tuttle 1967, 1969a, 1986; Groves 1970b.

118. Goldberg 1998; Goldberg and Ruvolo 1997b.

119. Tuttle 1986; Inskip; 2005; Sommer and Ross 2011.

120. Russak and McGrew 2008.

121. McBrearty and Jablonski 2005.

122. Kortlandt 1995; Lacambra et al. 2005.

123. Tuttle 1986; Uehara and Nishida 1987; Morbeck and Zihlman 1989.

124. Ferriss 2005; Ferriss et al. 2005.

125. Vigilant and Bradley 2001.

126. Tuttle and Cortright 1988; Thorén et al. 2006.

127. Leigh 1992.

128. Breuer et al. 2007.

129. Fossey 1979.

130. Schaller 1963; Fossey 1983.

131. O'Higgins et al. 1990; O'Higgins and Dryden 1993; Bromage 1992.

132. Shea et al. 1993.

133. Tuttle 1975a; Shea 1983, 1984, 1986.

134. Taylor 2006.

135. Cramer 1977; Taylor 2009.

136. Schultz 1963a.

137. Godefroit 1990.

138. Johanson 1974; Kinzey 1984; Swindler 1976.

139. Almquist 1974; Johanson 1974; Godefroit 1990.

140. Shea 1981, 1984; Jungers and Susman 1984.

141. Zinner et al. 2004.

142. Short 1979; Dixson 2009.

143. Goebel et al. 2008; Dugoujon et al. 2004.

144. Gignoux et al. 2011.

145. Burckhardt et al. 1999; Gagnaux et al. 1999; Garner and Ryder 1996; Goldberg and Ruvolo 1997b; Hacia 2001; Kaessmann et al. 1999, 2001; Wise et al. 1997.

146. Hawks et al. 2007; Kittles and Weiss 2003; Shriver and Kittles 2004.

147. Kaplan et al. 2000.

148. Finch 2010.

149. Montagna 1982, 1985; Pond 1987a.

150. Buettner-Janusch 1973; Parra 2007.

151. Daegling 2005.

152. Frisch 1973; Kinzey 1970.

153. Le Gros Clark 1971.

154. Swindler 1976; Shelli et al. 1998.

155. Tuttle 1988b, 1974, 1975, 1992; Tuttle et al. 1998.

156. Walker et al. 2006.

3. APES IN TIME

1. Hill 2000; Brown 2000; Thomas 2000.

2. L. Leakey 1937, 1965, 1974; M. Leakey 1984; Cole 1975; Morell 1995.

3. Coppens et al. 1976; Johnston 1982; Kalb 2001; Leakey and Harris 1987; Leakey and Leakey 1978; Rapp and Vondra 1981; Tuttle 1988a.

4. Brown 2000.

5. McKee et al. 1995.

6. Gradstein et al. 2004.

7. Crutzen 2002; Zalasiewicz et al. 2008; Kolbert 2011.

8. Adams et al. 1983; Bernor 1983; Rögl 1999; Thomas 1985; Whybrow 1984, 1987.

9. Bernor 1983; Thomas 1985.

10. Beard et al. 1994; Beard 2002; Dagosto 2002; Rasmussen 2002; Ross and Kay 2004; Senut 1988a; Seiffert et al. 2004.

11. Covert 2004, p. 141; Simons 1992; Gebo 1993a; Fleagle and Kay 1987; Rasmussen and Simons 1992; Egi, Takai et al. 2004; Egi, Tun et al. 2004.

12. Kay et al. 1997.

13. Plavcan 2004a; Seiffert et al. 2004; Kay et al. 1997; Simons et al. 1999.

14. Beard 2002.

15. Simons et al. 2001.

16. Rasmussen 2002.

17. Ciochon et al. 1985; Ba Maw et al. 1979; Simons 1971; Szalay 1970; Conroy and Bown 1976; Takai et al. 2002.

18. Aung 2004; Maung et al. 2005.

19. Tsubamoto et al. 2002.

20. Egi, Tun et al. 2004.

21. Kay and Simons 2005.

22. Ciochon et al. 1985.

23. Ciochon 2001; Ciochon and Gunnell 2004; Gunnell et al. 2002; Seiffert et al. 2004.

24. Chaimanee et al. 2000a; Beard 2002; Thein 2004; Takai and Shigehara 2004.

25. Takai et al. 2001; Takai and Shigehara 2004.

26. Egi, Tun et al. 2004.

27. Jaeger et al. 1999.

28. Gebo et al. 2002; Chaimanee et al. 2012.

29. Takai, Sein et al. 2005.

30. Chiamanee et al. 1997, 2000b; Chaimanee 2004.

31. Ducrocq 1998, 1999.

32. Beard et al. 1994.

33. Gebo 2002; Gebo et al. 1999; Qi and Beard 1998.

34. MacPhee et al. 1995.

35. Beard et al. 1996; Beard and Wang 2004.

36. Beard et al. 1996.

37. Beard and Wang 2004.

38. Marivaux et al. 2005.

39. De Bonis et al. 1988.

40. Godinot and Mahboubi 1992, 1994.

41. Godinot 1994.

42. Godinot and Mahboubi 1992; Godinot 1994.

43. Tabuce et al. 2009.

44. Jaeger et al. 2010.

45. Jaeger et al. 2010, p. 1097.

46. Fleagle et al. 1986a, 1986b; Kappelman 1992; Rasmussen and Simons 1992; Rasmussen et al. 1992; Van Couvering and Harris 1991; Seiffert et al. 2003; Seiffert 2006.

47. Simons and Rasmussen 1991; Simons 1992; Gebo 1993a.

48. Fleagle et al. 1980; Simons et al. 1999; Simons et al. 2007.

49. Bown et al. 1982; Kay and Van Couvering 1988; Olson and Rasmussen 1986; Rasmussen and Simons 1992.

50. Fleagle and Kay 1983; Simons et al. 1987.

51. Simons 1965 (pace).

52. Simons et al. 1987.

53. Andrews 1985; Harrison 1987; de Bonis 1987b.

54. Shea 1985, 1988.

55. Simons 1987.

56. Simons 1972, 1987; Fleagle and Kay 1983.

57. Cartmill et al. 1981.

58. Simons 1987; Harrison 1987.

59. Cachel 1975; Simons 1987.

60. Harrison 1987.

61. Simons 1972; Fleagle and Kay 1983, 1987.

62. Bush et al. 2004.

63. Simons 1972; Kay 1977; Cachel 1975.

64. Szalay and Delson 1979; Simons and Rasmussen 1991; Fleagle 1999.

65. Kay and Simons 1983a; Kay 1988; Simons 2004.

66. Fleagle and Kay 1987.

67. Hoffstetter 1972, 1980.

68. Miller and Simons 1997, Simons and Seiffert 1999; Seiffert et al. 2004.

69. Fleagle 1986a, 1999.

70. Delson 1975a; Fleagle and Kay 1983, 1987; Fleagle 1986b.

71. Benefit 1993; Benefit and McCrossin 1989, 1993a, 1993b; Delson 1975b, 1979a; Harrison 1989a, 1989b; Leakey 1985; McCrossin and Benefit 1994; Miller 1999; Pickford 1986b; Pickford and Senut 1988; Pilbeam and Walker 1968; Simons 1972; Tattersall et al. 1988; Walker and Leakey 1984.

72. Miller 1999; Benefit and McCrossin 2002.

73. McCrossin and Benefit 1992; Feibel and Brown 1991.

74. Fleagle and Simons 1979.

75. Seiffert 2006.

76. Gunnell and Miller 2001.

77. Simons 1972; Simons et al. 1987; Simons and Rasmussen 1991; Szalay 1970; Gingerich 1980.

78. Simons and Bown 1985; Simons et al. 1987; Simons 1997.

79. Rasmussen 1986; Rasmussen and Simons 1988.

80. Harrison 1987; Fleagle and Kay 1987; Rasmussen and Simons 1988.

81. Rasmussen and Simons 1992; Simons 1989a, 1990.

82. Simons 1989a, 1990; Rasmussen and Simons 1992.

83. Rasmussen 2002; Seiffert et al. 2000.

84. Kay and Simons 1980.

85. Teaford et al. 1996.

86. Kay and Simons 1980; Cachel 1983; Simons and Kay 1983; Rasmussen and Simons 1992; Kirk and Simons 2001.

87. Thomas et al. 1988, 1989; Rasmussen and Simons 1992; Gheerbrant et al. 1995; Seiffert 2006.

88. Thomas, Sen et al. 1991; Thomas, Roger et al. 1991.

89. Gheerbrant et al. 1995.

90. Sigé et al. 1990; Godinot 1994; Seiffert et al. 2004; Jaeger and Marivaux 2005; Seiffert et al. 2005.

91. Sigé et al. 1990; Hooker et al. 1999.

92. Jaeger and Marivaux 2005; Seiffert et al. 2005.

93. M. G. Leakey, Ungar, and Walker 1995; Stevens et al. 2013.

94. Pickford 1986b; Matsuda et al. 1986.

95. Hill and Ward 1988.

96. Collinson et al. 2009.

97. Andrews and van Couvering 1975; Andrews and Evans 1979; Andrews and Walker 1976; Evans et al. 1981; Hill et al. 1991; Kappelman 1991; Nakaya 1989; Pickford 1983; Shipman et al. 1981; Shipman 1986a.

98. Andrews et al. 1981.

99. Pilbrow 2006; Hartmam 1988, 1989.

100. Uchida 1998a; Matsumura et al. 1992.

101. Simons and Pilbeam 1965; Andrews 1978; Harrison 1981; Teaford et al. 1988; Tuttle 2006c.

102. Andrews and Simons 1977.

103. Fleagle 1984, 1999.

104. Pickford 1982; Harrison 1986a, 1989a; Senut 1987; Benefit 1993.

105. Tuttle 2006c.

106. Rossie and MacLatchy 2006.

107. Senut et al. 2000.

108. Simons and Pilbeam 1965.

109. Andrews 1978.

110. Pickford 1986b; Madden 1980.

111. Harrison 1986a, 1989b.

112. Kunimatsu 1992a, 1992b.

113. Harrison 1988.

114. Harrison 1981.

115. Harrison 1989b.

116. Benefit 1991; McCrossin 1992.

117. Harrison 1992.

118. Leakey and Leakey 1986a, 1986b, 1987; R. E. Leakey et al. 1988a, 1988b; Boschetto et al. 1992.

119. Bishop 1964.

120. Allbrook and Bishop 1963.

121. Walker and Rose 1968.

122. Pilbeam 1969b.

123. Gebo et al. 1997.

124. Benefit 1999a; Pickford et al. 1999; Senut et al. 1999.

125. Senut et al. 2000; Gommery 2005.

126. Pickford et al. 2003.

127. Ishida 1984; Nakatsukasa et al. 2005.

128. Sawada et al. 2005.

129. Pickford 1986b; Sawada et al. 1998.

130. Kunimatsu 1997; Sawada et al. 1998.

131. Ward et al. 1999; Hill et al. 1991; Pickford 1998; Feibel and Brown 1991; Pickford 1986a.

132. Kunimatsu et al. 1999; Nakatsukasa et al. 1999.

133. Ishida et al. 1999.

134. Rose et al. 1996; Nakatsukasa et al. 1996, 1998; Nakatsukasa, Tsujikawa et al. 2003; Nakatsukasa, Kunimatsu et al. 2003.

135. Sawada et al. 1998, 2005.

136. Ishida and Pickford 1997.

137. Pickford and Ishida 1998.

138. Fleagle 1999, p. 456.

139. Fleagle 1999, p. 456.

140. de Bonis 1987a.

141. Pilbeam 1968, 1969b; Tuttle 1975b, 2006c.

142. Andrews 1978; McHenry et al. 1980.

143. Pilbeam 1969b; Simons and Pilbeam 1965.

144. Tuttle 1988a; Pickford and Senut 2005; Suwa et al. 2007.

145. Pickford 1986b.

146. Haile-Selassie et al. 2003; 2009.

147. Senut et al. 2001; Pickford and Senut 2001.

148. Brunet et al. 2002, 2005; Guy et al. 2005; Vignaud et al. 2002; Zollikofer et al. 2005; Lebatard et al. 2008.

149. Suwa et al. 2007.

150. Kunimatsu et al. 2007.

151. Brunet et al. 2002, 2005; Wolpoff et al. 2006.

152. Senut et al. 2001; Pickford et al. 2002; Cela-Conde and Ayala 2003; Richmond and Jungers 2008; Gibbons 2008; Brunet et al. 2002, 2005; Zollikofer et al. 2005; Wolpoff et al. 2006.

153. Ishida 1984; Ishida et al. 1984.

154. Tuttle 1988a.

155. Andrews and Van Couvering 1975.

156. Reed and Fleagle 1995; Sepulchre et al. 2006.

157. Walker and Pickford 1983; Conroy 1987; Ruff et al. 1989; Walker et al. 1993.

158. Walker and Pickford 1983; Kelley and Pilbeam 1986b; Ruff et al. 1989; Ruff 1990.

159. Ishida 1984.

160. Drake et al. 1988; Walker, Teaford, and Leakey 1986; Walker and Teaford 1989; Walker et al. 1993.

161. Le Gros Clark and Leakey 1950, 1951; Walker 1992; Walker and Shipman 2005; Kelley 1986; Kelley and Pilbeam 1986b; Pickford 1986f.

162. Bosler 1981; Ruff et al. 1989; Cameron 1991, 1992.

163. Walker et al. 1993; Teaford et al. 1993.

164. Le Gros Clark and Leakey 1951.

165. Pickford 1986a, 1986b; Walker , Teaford, and Leakey 1986; Davis and Napier 1963; Walker et al. 1983.

166. Walker et al. 1983; Davis and Napier 1963; Szalay and Delson 1979; Leutenegger 1984; Smith and Walker 1984.

167. Andrews 1978.

168. Leakey and Leakey 1986a; R. E. Leakey et al. 1988b.

169. Andrews and Martin 1992.

170. Leakey and Leakey 1986a; R. E. Leakey et. al. 1988b; Andrews 1978; Pilbeam 1969b.

171. Leakey and Leakey 1986b; R. E. Leakey et al. 1988a.

172. Pilbeam 1969b; Szalay and Delson 1979; Ward and Pilbeam 1983.

173. Tuttle 2006d.

174. McCrossin and Benefit 1994, 1997.

175. Palmer et al. 1999; Tuttle 2006d.

176. van Roosmalen et al. 1988.

177. Norconk and Conklin-Brittain 2003.

178. Boubli 1999.

179. Palmer et al. 1998.

180. Benefit and McCrossin 1997; Benefit 1999b; McCrossin and Benefit 1994, 1997.

181. Kunimatsu 1997.

182. Ducros and Ducros 1992, p. 247.

183. Boesch and Boesch 1989; Alp and Kitchener 1993; Mitani and Watts 1999; Stanford 1998; Uehara 1997; Uehara and Ihobe 1998.

184. Utami and Van Hooff 1997; Sugarjito and Nurhuda 1981.

185. Kudo and Mitani 1985.

186. Rose 1997; Rose et al. 2003.

187. Ihobe 1992a; Hohmann and Fruth 1993; White 1994.

188. Sabater Pí and Veà 1994.

189. Sabater Pí et al. 1993.

190. Hohmann and Fruth 2008; Surbeck and Hohmann 2008.

191. McCrossin and Benefit 1994; McCrossin et al. 1998; Harrison 1989a; Zambon et al. 1999; Rose et al. 1996; Blue et al. 2006.

192. Conroy et al. 1992a, 1992b, 1993a, 1993b, 1996.

193. Pickford et al. 1997.

194. Thomas et al. 1982, 1985; Whybrow 1984, 1987; Whybrow and Bassiouni 1986; Zalmout et al. 2010.

195. Andrews and Martin 1987b.

196. Leakey and Leakey 1986a; Leakey and Walker 1985b; McDougall and Watkins 1985; Kelley and Pilbeam 1986a.

197. Andrews 1992.

198. Whybrow 1984, 1987.

199. Zalmout et al. 2010.

200. Srivastava and Binda 1991.

201. Begun 2002a; Begun 2002b; Kelley 2002.

202. Ginsburg 1986; Pohlig 1895; McHenry and Corruccini 1976; Begun 1989, 1992b; Köhler et al. 2002; Simons 1972.

203. Rögl 1999; Andrews and Bernor 1999.

204. Alba et al. 2010.

205. Nagatoshi 1987; Andrews et al. 1996.

206. Andrews and Bernor 1999.

207. Nagatoshi 1987; Alpagut et al. 1990a.

208. Nagatoshi 1987.

209. Ginsburg and Mein 1980; Ginsburg 1986.

210. Harrison et al. 1991; Ginsburg and Mein 1980.

211. Welcomme et al. 1991.

212. Moyà-Solà et al. 2001.

213. Harrison et al. 2002.

214. Zapfe 1960.

215. Simons 1972.

216. Zapfe 1960; Szalay and Delson 1979.

217. Zapfe 1960; Szalay and Delson 1979; Fleagle 1984.

218. Nagatoshi 1984.

219. Andrews et al. 1996.

220. Ungar and Kay 1995; Ungar 1996a, 1998.

221. Barry et al. 1985, 1986.

222. Bernor et al. 1988.

223. Harrison and Gu 1999.

224. Chopra and Kaul 1979; Chopra 1983; Barry 1986; Johnson et al. 1983.

225. Ginsburg and Mein 1980; Schlosser 1924.

226. Harrison et al. 1991.

227. Suteethorn et al. 1990.

228. Harrison and Gu 1999; Kunimatsu et al. 2005.

229. Wang 1984.

230. Li 1978.

231. Lei 1985; Gu and Lin 1983.

232. Qiu and Guan 1986; Bernor et al. 1988; Harrison et al. 1991.

233. Zhang and Harrison 2008.

234. Flynn and Qi 1982; Qiu et al. 1985; Wu and Pan 1985; Wu and Wang 1987; Pan 1988; Meldrum and Pan 1988.

235. Harrison 1987; Bernor et al. 1988; Fleagle 1988.

236. Tyler 1991.

237. Wu and Pan 1985.

238. Wu and Wang 1987; Pan et al. 1989.

239. Heizmann and Begun 2001.

240. Begun 1992a.

241. Kelley 2008; Kelley et al. 2008; Ersoy et al. 2008.

242. Andrews 1995; Kelley 2002.

243. Andrews 1990; Viranta and Andrews 1995a; Quade et al. 1995.

244. Esroy et al. 2008.

245. King et al. 1999; King 2001.

246. Kelley et al. 2008.

247. Mein 1986; Nagatoshi 1987; Pickford 1986c; Thenius 1982; Rögl and Daxner-Höck 1996; Andrews and Bernor 1999; Kordos 2000.

248. Nagatoshi 1987; Harrison 1991a.

249. Moyà-Solà et al. 2004, 2005; Culotta 2004.

250. Moyà-Solà et al. 2004.

251. Nagatoshi 1984; Andrews et al. 1996.

252. Mein 1986.

253. Begun et al. 1990.

254. Mottl 1957; Andrews et al. 1996; Kordos 2000.

255. Nagatoshi 1990.

256. Ungar 1996a, 1998; Ungar and Kay 1995.

257. Moyà-Solà, Köhler, et al. 2009.

258. Simons and Meinel 1983.

259. Pilbeam and Simons 1971; Morbeck 1983.

260. Moyà Solà and Köhler 1993, 1995.

261. Köhler et al. 2001.

262. Kretzoi 1975.

263. Kay and Simons 1983b.

264. Kelley and Pilbeam 1986b; Kordos 1987, 1988, 1991.

265. Begun 1992b.

266. Andrews et al. 1996.

267. Agusti et al. 2001.

268. Kordos 1987, 1988; Kordos and Begun 1998; Wolpoff 1980a.

269. Blaney 1986.

270. Kelley and Pilbeam 1986b; Kordos 1987, 1988; Ward and Brown 1986.

271. Kordos 1987.

272. Kordos 2000; Kordos and Begun 2001.

273. Kelley and Pilbeam 1986b.

274. Xue and Delson 1988, 1989.

275. Xue and Delson 1989.

276. Pickford 1986c; Nagatoshi 1984; Mein 1986.

277. Nagatoshi 1984; Leakey 1962; Ward et al. 1999.

278. Sen et al. 2000.

279. de Bonis and Koufos 1994.

280. de Bonis 1987a, 1987b, 1987c; de Bonis et al. 1981, 1986, 1990b; de Bonis and Melentis 1980, 1984, 1987; Kelley and Pilbeam 1986b; Koufos et al. 1991; Koufos 1993, 1995.

281. de Bonis and Koufos 2001; de Bonis et al. 1992; Kostopoulos et al. 1996; Kostopoulos and Koufos 1996.

282. Andrews et al. 1996.

283. de Bonis and Melentis 1980; Kelley and Pilbeam 1986b; Koufos 1995.

284. de Bonis and Koufos 2001.

285. de Bonis and Koufos 2001; Kelley 2001.

286. Ozansoy 1957, 1965; Andrews and Tekkaya 1980; Alpagut et al. 1996; Begun and Güleç 1998; Andrews and Alpagut 2001.

287. Andrews and Alpagut 2001.

288. Alpagut et al. 1996; Begun and Güleç 1998; Andrews and Alpagut 2001; Andrews and Takkaya 1980; Andrews et al. 1996.

289. Alpagut et al. 1996; Begun and Güleç 1998; Andrews and Alpagut 2001; Andrews and Takkaya 1980; Andrews and Martin 1992; Begun 2002b.

290. Güleç et al. 2007; Wang et al. 2007.

291. Güleç et al. 2007.

292. Ünay et al. 2007; Güleç et al. 2007.

293. Maschenko 2005.

294. Barry 1986; Barry et al. 1990; Flynn et al. 1990, 1995; Johnson et al. 1983; Kappelman et al. 1991; Kelly 2001.

295. Badgley et al. 1984; Barry 1986; Munthe et al. 1983.

296. Kay 1982a, 1982b; Kay and Simons 1983b.

297. Andrews 1983.

298. Kelly 1988, 2002; Flynn et al. 1990; Barry 2002.

299. Scott et al. 1999; Nelson 2003.

300. Badgely et al. 1984.

301. Kay 1982b; Andrews 1978; Johnson et al. 1983.

302. Andrews and Martin 1992; Kelly 2002.

303. Badgely et al. 1984; Pilbeam 1982; Pilbeam et al. 1977, 1980; Preuss 1982; Rose 1984a; Ward and Brown 1986.

304. Pilbeam 1982; Preuss 1982.

305. Brown and Ward 1988; Kelley and Pilbeam 1986b; Pilbeam 1982; Preuss 1982; Ward and Brown 1986.

306. Rossie 2005.

307. Preuss 1982; Kelley and Pilbeam 1986b; Ward and Brown 1986.

308. Ward and Brown 1986.

309. Preuss 1982; Pilbeam 1982.

310. Ward and Kimbel 1983; Ward and Brown 1986.   ·

311. Kelley and Pilbeam 1986b; Ward and Brown 1986.

312. Preuss 1982; Kelley and Pilbeam 1986b; Ward and Brown 1986.

313. Ward and Brown 1986; Brown and Ward 1988.

314. Preuss 1982; Kelley and Pilbeam 1986b; Ward and Brown 1986.

315. Rose 1993.

316. Kappleman 1991.

317. Kelley 1988.

318. Pilbeam et al. 1980.

319. Badgley and Behrensmeyer 1980; Badgley et al. 1984; Barry et al. 1982; Pilbeam 1982; Pilbeam et al. 1979.

320. Nelson 2003.

321. Kay 1981; Teaford and Walker 1984; Nelson 2003.

322. Kelley and Pilbeam 1986b; Nelson 2003.

323. Flynn et al. 1990; Mathur 1984; Prasad 1975; Sharma 1984; Tattersall 1969a, 1969b; Vishnu-Mittre 1984; West 1984.

324. Barry et al. 2002; Badgley et al. 2008.

325. Harrison 2006; Harrison et al. 2002; Kelley 2002; Liu and Zheng 2005a.

326. Xu and Lu 1979; Lu et al. 1981; Wu and Xu 1985; Wu et al. 1983, 1984; Wu and Oxnard 1983a, 1983b; Etler 1984; Martin 1991; Xu and Lu 2008; Wu 1987; Wu et al. 1986; Kelley and Etler 1989; Kelley and Pilbeam 1986b; Schwartz 1990.

327. Kelley and Xu 1991; Kelley 1993.

328. Kay and Simons 1980; Xu and Lu 2008.

329. Wu et al. 1986; Wu and Oxnard 1983a, 1983b; Kelley and Xu 1991; Kelley 1993; Qi 1993; Xu and Lu 2008.

330. Wu 1987; Wu et al. 1986; Lu et al. 1988; Kordos 1988; Schwartz 1990.

331. Wu et al. 1986.

332. Wu 1987; Wu et al. 1983; Wu and Xu 1985.

333. Schwartz 1990.

334. Kelley and Etler 1989; Xu and Lu 2008.

335. Wu et al. 1984; Wu and Xu 1985; Wu et al. 1986.

336. Wu and Wang 1987.

337. Wu 1987; Wu and Xu 1985; Wu et al. 1986; Qi 1993.

338. Wu et al. 1983, 1985; Kelley 2002.

339. Liu and Zheng 2005b; Liu et al. 2002.

340. Xu and Lu 2008.

341. Qi 1985, 1993; Qiu et al. 1985; Wang 1984; Wu and Xu 1985.

342. Qiu et al. 1985; Pan 1988.

343. Wang 1984; Wu and Xu 1985; Zhao 2004.

344. Woo 1957, 1958, 1980; Wu and Wang 1985; Harrison 2006; Harrison et al. 2002.

345. Kelley 2002.

346. Zhang et al. 1987; Ho 1988; Harrison et al. 2002; Kelley and Gao 2012.

347. Chaimanee et al. 2003, 2004.

348. Chaimanee et al. 2004.

349. Chaimanee et al. 2003.

350. Chaimanee et al. 2004; Wall et al. 1994.

351. Jablonski 2002; Bacon and Long 2001.

352. Andrews, Harrison et al. 1996; Ardito and Mottura 1987; Barry 1987; Barry and Flynn 1990; Barry et al. 1990; de Bonis et al. 1990a; Brunet et al. 1982; Delson 1975b; Gentili et al. 1998; Heintz et al. 1981; Hohenegger and Zapfe 1990; Jolly 1967; Simons 1970; Whybrow et al. 1990; Pickford 1998; Pan et al. 2004.

353. Pan and Jablonski 1987; Jablonski 1990, 1992, 1993; Jablonski and Gu 1991; Jablonski and Pan 1988; Jablonski 2002; Iwamoto et al. 2005.

354. Kennedy 1960; Wiley 1981.

355. Wiley 1981.

356. Simpson 1961, 1963, 1975; Tuttle 1986, 2006a; Marks 2005.

357. Simons 1987; Fleagle 1988.

358. Leakey et al. 1991.

359. Benefit and McCrossin 1991, 1993a.

360. Fleagle 1988.

361. De Bonis 1987a; Andrews and Martin 1987a; Andrews and Martin 1987b.

362. Rae 1999.

363. Harrison 1987, 1988.

364. McCrossin and Benefit 1993, 1994.

365. De Bonis 1987a, 1987b.

366. Leakey 1963; de Bonis 1987a, 1987b.

367. Begun 2005; Moyà Solà and Köhler 1993.

368. De Bonis 1987a.

369. Ciochon 1983; Fleagle 1988; Bernor et al. 1988; Harrison et al. 1991.

370. Harrison 2005.

371. Marivaux et al. 2005; Harrison and Yumin 1999.

372. Moyà Solà and Köhler 1993.

373. Moyà Solà and Köhler 1997.

374. De Bonis 1987a, 1987b, 1987c; de Bonis et al. 1990b, 1991; de Bonis and Koufos 1993; Koufos 1993.

375. de Bonis and Koufos 1993.

376. de Bonis 1987b.

377. Xu and Lu 2008.

378. Andrews and Bernor 1999; Andrews 1992a; Begun et al. 1997; Cameron 1997; Harrison and Rook 1997; Schwartz 1990.

379. Begun et al. 1997.

380. Moyà Solà and Köhler 1993.

381. Moyà-Solà et al. 2004.

382. Moyà-Solà, Alba, et al. 2009.

383. Moyà-Solà, Alba, et al. 2009; Alba et al. 2006.

384. Begun 2001; Andrews 1992a; Ward and Kimbel 1983.

385. Chatterjee 2006.

4. TAPROOT AND BRANCHES OF OUR FAMILY TREE

1. Hill and Ward 1988.

2. Tuttle 1988a.

3. Howell 1978; Simons 1981; Pilbeam 1979; Pickford 1975, 1986b; McHenry and Corruccini 1980; Corruccini and McHenry 1980.

4. Ungar et al. 1994; Coffing et al. 1994.

5. Pickford and Senut 2005; Senut et al. 2001.

6. Bishop and Chapman 1970; Hill et al. 1985; Ishida et al. 1984; Pickford et al. 1984; Tauxe et al. 1985; Tuttle 1988a.

7. Pickford and Senut 2005.

8. Fleagle 1988.

9. Ishida 1984; Ishida et al. 1984.

10. Simons 1978, 1981; Pilbeam 1979; McHenry and Corruccini 1980; Coppens 1983a, 1983b; Kramer 1986; Hill et al. 1992; White 1986.

11. Pickford et al. 1983; Senut 1983a.

12. Howells 1977; Oxnard 1975b.

13. Patterson and Howells 1967; Feldesman 1982; Senut 1979, 1980.

14. Coppens 1982a.

15. Boaz 1983; Tuttle 1988a.

16. M. G. Leakey, Feibel et al. 1995.

17. Kingston and Hill 1994.

18. Macho et al. 2005.

19. Sepulchre et al. 2006.

20. Hürzeler 1954, 1956, 1958, 1960.

21. Hürzeler 1968, 1978.

22. Viret 1955; Kurth 1956, 1958; Heberer 1956, 1959.

23. Woo 1962; Robinson 1963.

24. Hürzeler 1968.

25. Gervais 1872.

26. Rook 1993.

27. Harrison 1991b.

28. Azzaroli et al. 1986; Rook et al. 1999, 2000.

29. Hürzeler 1968; Schultz 1960; Straus 1963.

30. Delson 1986a; Delson and Rosenberger 1984; Harrison 1986b; Hürzeler 1968; Sarmiento 1987.

31. Hürzeler 1968; Alba et al. 2001.

32. Szalay and Delson 1979, Delson 1979b, 1987.

33. Szalay and Berzi 1973; Hürzeler 1960.

34. Szalay and Berzi 1973.

35. Straus and Schön 1960.

36. Harrison 1989c; Jungers 1987.

37. Harrison 1989c.

38. Sarmiento 1987; Harrison 1986b, 1991b; Rose 1988.

39. Köhler and Moyà-Solà 1997; Rook et al. 1999.

40. Ungar and Kay 1995, Ungar 1996a, 1998.

41. Azzaroli et al. 1986; Sarmiento 1987; Harrison 1986b, 1991b; Harrison and Harrison 1989; Jungers 1987.

42. Szalay and Langdon 1986.

43. Köhler and Moyà-Solà 1997; Rook et al. 1999.

44. Pilbeam 1970b; Simons 1972; Corruccini 1975a, 1975b; Pilbeam et al. 1977; Szalay and Delson 1979; Ciochon 1983; Fleagle 1988.

45. Gelvin 1980; de Bonis 1987a.

46. Hoang et al. 1979; Ciochon and Olsen 1986; Ciochon 1991.

47. Weidenreich 1945, 1946; von Koenigswald 1935.

48. von Koenigswald 1952.

49. von Koenigswald 1957, 1958.

50. Broom 1941; Broom and Schepers 1946; Broom and Robinson 1952.

51. Dart 1960; Woo 1962; Heberer 1959.

52. Robinson 1972a; Robinson and Stuedel 1973; Eckhardt 1973, 1974, 1975; Frayer 1973; Gelvin 1980; de Bonis 1987a, 1987b.

53. Simons and Chopra 1969.

54. Robinson 1972a.

55. Woo 1962.

56. Zhang 1982.

57. Simons and Ettel 1970; Simons 1972; Zhang 1982; Miller et al. 2008.

58. Woo 1962, 1980; Zhang 1985a; Krupinski and Rajchel 1985.

59. Woo 1962; Zhang 1985a.

60. Zhang 1985b.

61. Conroy 1987.

62. Pei 1957; Garn and Lewis 1958; Weidenreich 1946, p. 61; Simons and Ettel 1970; Krantz 1987; Ciochon 1991.

63. White 1975; Ciochon 1991; Dart 1960; Simons and Ettel 1970; von Koenigswald 1981.

64. Simons 1961, 1963, 1964a, 1964b, 1967, 1968a, 1968b, 1969, 1972, 1977; Pilbeam 1966, 1967, 1968, 1969a, 1969b, 1970a, 1972; Pilbeam and Simons 1965; Simons and Pilbeam 1965, 1972, 1978.

65. Lewis 1934; Simons 1961, 1964a.

66. Conroy and Pilbeam 1975; Khatri 1975; Lipson and Pilbeam 1982.

67. Simons and Pilbeam 1965.

68. L. S. B. Leakey 1962, 1967; Simons 1964a.

69. Darwin 1871; Pilbeam and Simons 1965; Pilbeam 1966; Prasad 1969, 1982.

70. Shea et al. 1993; Pickford 1982, 1985a, 1985b, 1986d, 1986e; L. S. B. Leakey 1962, 1963, 1967; Aguirre 1972, 1975.

71. Genet-Varcin 1969; Vogel 1975; Walker and Andrews 1973; Andrews and Walker 1976; Greenfield 1978; Lewis 1934; Simons 1961.

72. Yulish 1970; Greenfield 1974, 1975, 1978, 1979, 1980, 1983; Frayer 1976, 1978; Wolpoff 1982.

73. Cartmill et al. 1986; Lewin 1987; Pilbeam 1978, 1979, 1980, 1983, 1984, 1985, 1986; Pilbeam and Jacobs 1978.

74. Zhang, Lin et al. 1987; Zhang, Jiang, and Lin 1987.

75. Ho 1988.

76. Xu and Lu 2008, pp. 221–222.

77. White et al. 1994, 1995; Renne et al. 1999; Semaw et al. 2005; Haile-Selassie et al. 2010.

78. Andrews 1982; Oxnard 1983a.

79. Huxley 1863; Keith 1923; Gregory 1927a, 1927b, 1927c, 1927d, 1928, 1929, 1930, 1934; Morton 1926; Washburn 1950; Tuttle 1974; Friedenthal 1900; Nuttal 1904; Uhlen-huth 1904.

80. Goodman 1962a, 1962b, 1963a, 1963b; Simons 1965; Pilbeam 1969b, 1970a; Simons and Pilbeam 1978.

81. Sarich and Wilson 1967a, 1967b, 1973; Sarich 1968, 1971; Wilson and Sarich 1969; Wilson et al. 1977; Zuckerkandl and Pauling 1962, 1965; Zuckerkandl 1978, 1987.

82. Goodman 1981; Langergraber et al. 2012.

83. Avise and Aquadro 1982; Goodman et al. 1983; Jacobs and Pilbeam 1980; Tuttle 1988a; Glazko and Nei 2003; Fukami-Kobayashi et al. 2005; Steiper and Young 2008; Langergraber et al. 2012.

84. Avise 1986; Avise and Aquadro 1982; Ayala 1986; Beeber et al. 1986; Beneviste 1985; Britten 1986; Brown et al. 1979, 1982; Cann et al. 1984; Carlson et al. 1978; Chang and Slightom 1984; Cherry et al. 1978; Corruccini et al. 1979; Deininger and Daniels 1986; Dickerson 1971; Dover 1987; Ferris, Brown et al. 1981; Ferris et al. 1981; Fitch 1976; Fitch and Langley 1976; Gillespie 1986; Gillespie et al. 1982; Gingerich 1984b, 1986b; Goodman 1982; Goodman et al. 1972; Goodman and Cronin 1982; Harris et al. 1986; Hayasaka et al. 1988; Holmquist et al. 1988a, 1988b; Jukes 1980; King and Wilson 1975; Kohne 1970, 1975; Kohne et al. 1972; Koop et al. 1986; Kortlandt 1972; Laird et al. 1969; Li and Tan-imura 1987; Li et al. 1987; Liebhaber and Begley 1983; Lovejoy et al. 1972; Margoliash 1963; Margoliash and Fitch 1968; Marks 1983b, 1986; Nei 1986; Novacek and Norell 1982; O'Neil and Doolittle 1973; Radinsky 1978; Ramharack and Deeley 1987; Read and Lestrel 1970; Sarich and Cronin 1976; Sibley and Ahlquist 1987; Smith et al. 1987; Stone 1987; Uzzell and Pilbeam 1971; Vawter and Brown 1986; Weber et al. 1986; Wilson et al. 1974a, 1974b, 1975; Wu and Li 1985; Wu and Maeda 1987.

85. Hasegawa et al. 1987, 1988; Kimura 1983, 1987; Nei and Tajima 1985; Ochman and Wilson 1987; Preparata and Saccone 1987; Kumar et al. 2005.

86. Sarich and Wilson 1967a; Buettner-Janusch 1974; Romero-Herrera et al. 1979; Pilbeam 1969b; Simons 1981.

87. Andrews 1986a; Kumar et al. 2005; Seth and Seth 1986; Yang 2002; Cronin 1983; Gingerich 1984b, 1985; Hasegawa et al. 1985, 1987; Li et al. 1987; Li and Tanimura 1987; Sibley and Ahlquist 1984, 1987; Ueda et al. 1986.

88. Hasegawa et al. 1985.

89. Weiss 1987; Patterson et al. 2006.

90. Syvanen 1984, 1987; Mayer et al. 1988; Weiss 1987; Spuhler 1988.

91. Gingerich 1985.

92. Cole 1975; L. S. B. Leakey 1960b; M. D. Leakey 1975; R. E. F. Leakey 1970, 1973a, 1981; Leakey and Lewin 1977, 1978; Payne 1965, 1966.

93. Coppens 1982a; Day 1985; Hay and M. D. Leakey 1982; Herbert 1983; Hinrichsen 1978; Johanson 1976; Johanson and Edey 1981; Jones 1987; M. D. Leakey 1987a, 1987b; M. D. Leakey et al. 1976; M. D. Leakey and Hay 1979; R. E. F. Leakey 1980; R. E. F. Leakey and Walker 1985a; R. E. F. Leakey and Lewin 1992; Robbins 1987; Shipman 1986b; Tuttle 1987; White 1977, 1980a, 1980b, 1981, 1982, 1984; Willis 1989.

94. Lewin 1987; Reader 1988; Rensberger 1984; Tuttle 1988a, 2002; Weaver 1985.

95. Tuttle 1988a, 2003b; Clark 1988.

96. M. D. Leakey 1979, 1980, 1981, 1987b; M. D. Leakey and Hay 1979; Hay 1978, 1981, 1987; Hay and Leakey 1982.

97. Drake and Curtis 1987; M. D. Leakey 1987b; M. D. Leakey et al. 1976; White 1977, 1980a, 1981.

98. Hall et al. 1984; Williams et al. 1986.

99. Asfaw 1987; Clark et al. 1984; White 1984.

100. Hill 1985; Hill et al. 1986; Ferguson 1986a; Hill and Ward 1988.

101. Ward and Hill 1987.

102. Johanson 1980; Johanson and Coppens 1976, 1980; Johanson and Taieb 1976, 1978; Johanson, Taieb et al. 1978; Johanson et al. 1978, 1982; White and Johanson 1989.

103. Aronson and Taieb 1981; Aronson et al. 1977, 1980, 1983; Boaz 1988; Boaz et al. 1982; Brown 1982, 1983; Cooke 1978, 1983a, 1983b, 1985; Coppens et al. 1980; Gentry 1981; Schmitt et al. 1980; Taieb 1982; Taieb and Tiercelin 1980; Taieb et al. 1974, 1976, 1978; Tiercelin 1986.

104. Walter and Aronson 1982; Schmitt and Nairn 1984.

105. White, Moore, and Suwa 1984.

106. Taieb, Johanson, and Coppens 1975; Johanson and Coppens 1980; Johanson and Taieb 1976, 1978; Johanson 1977.

107. Johanson and White 1979.

108. Hinrichsen 1978; Logan et al. 1983.

109. Johanson, White and Coppens 1978.

110. Tobias 1978a, 1978b, 1978c, 1980a, 1980b, 1981a, 1981b; Boaz 1979, 1982, 1983; Logan et al. 1983.

111. Johanson 1985.

112. Boaz 1988; Falk 1985; Holloway 1983a, 1983b; Kimbel et al. 1982, 1984; Tobias 1987.

113. White et al. 1981.

114. Tuttle 1988a, 1996, 2008.

115. McHenry 1986.

116. Johanson and White 1979; White 1985.

117. Coppens 1977, 1980, 1981, 1982a, 1982b, 1983a, 1983b, 1983c, 1983d; Ferguson 1983, 1984, 1986b, 1992a, 1992b; M. D. Leakey 1981; R. E. F. Leakey 1981; Olson 1981, 1985a; Schmid 1983; Senut 1983b, 1986; Senut and Tardieu 1985; Tardieu 1982, 1983a, 1983b, 1986; Zihlman 1985.

118. Scott and Lockwood 2004.

119. Blumenberg and Lloyd 1983; Johanson 1985; Kimbel 1984; Kimbel and Rak 1985; Kimbel et al. 1984, 1985, 1986; Leonard 1991; Leonard and Hegmon 1987; Leuttenegger and Shell 1987; McHenry 1991b; Rak 1983; White 1982, 1985; White et al. 1981.

120. Tobias 1987; Boaz 1988.

121. Tuttle 1988a; Leiberman 2001.

122. M. G. Leakey et al. 1995; Andrews 1995b; Ward et al. 2001; White et al. 2006; Kimbel et al. 2006; Johanson and Edgar 1996; White 2003; Brunet et al. 1995, 1996; Lebatard et al. 2008.

123. Haile-Selassie et al. 2010; Senut and Pickford 2000.

124. Robinson 1963, 1972a; Berger and Hilton-Barber 2001; Tobias 1991; Wood and Richmond 2000; Asfaw et al. 1999.

125. Oxnard 1975a; Senut and Tardieu 1985; Stern and Susman 1983; Susman et al. 1984; Tuttle 1981b, 1988a, 2001.

126. McHenry and Coffing 2000.

127. Leakey, Spoor et al. 2001.

128. White 2003.

129. Vrba et al. 1995.

130. Werdelin et al. 2010.

131. Tuttle 2001; Hart and Sussman 2009.

132. Strait et al. 1997.

133. Arambourg et al. 1967, 1968; Walker, Leakey et al. 1986; Tobias 1967b; R. E. F. Leakey et al. 1978; Wood 1991; Suwa et al. 1997; Wood and Lieberman 2001; Wood and Constantino 2007; Grine 1988a; Keyser 2000; de Ruiter et al. 2009.

134. Rak et al. 2007.

135. Tuttle 2001.

136. Lockwood et al. 2007.

137. Johanson and White 1979; Johanson 1985.

138. Johanson and White 1979.

139. Leakey and Walker 1976; R. E. F. Leakey 1980; Walker and Leakey 1978; Walker 1981a; Feibel et al. 1989.

140. Falk 1986, 1988, 1992; Falk and Conroy 1983; Holloway 1983a; Olson 1985a, 1985b; Wood 1985a, 1985b, 1988; Wood and Chamberlain 1986.

141. Boaz 1988; McHenry 1984, 1985, 1994b; McHenry and Skelton 1985; Skelton et al. 1986; Tobias 1988; Wolpoff 1983a; Yaroch and Vitzthum 1984.

142. Dirks et al. 2010.

143. Berger et al. 2010; Berger 2013; Irish et al. 2013; de Ruiter et al. 2013.

144. Leakey and Walker 1988; Walker, R. E. Leakey et al. 1986; Walker and Leakey 1988.

145. Boaz 1988; Delson 1986b; Kimbel et al. 1988; Tuttle 1988a; Wood 1988.

146. Dart 1925.

147. Broom 1936a, 1936b, 1938a, 1938b, 1949a, 1949b, 1950; Dart 1948; Broom and Robinson 1952; Broom and Schepers 1946.

148. Broom and Robinson 1949a, 1950; L. S. B. Leakey 1959; Arambourg and Coppens 1967, 1968.

149. Robinson 1954.

150. Washburn and Patterson 1951; Howell 1959.

151. Robinson 1961; Tobias 1967b.

152. Howell 1978.

153. Grine 1984, 1985a.

154. Kimbel et al. 1988; Johanson 1989; Grine 1993a.

155. Clarke 1990.

156. Simpson 1945; Wood 1988; Simons 1989b; Clarke 1988; Dean 1986, 1987, 1988; Grine 1988, 1993a; Grine and Martin 1988; Jungers 1988; Jungers and Grine 1986; Kay and Grine 1988b; McCollum et al. 1993; Susman 1988a, 1988b; Tuttle 2006a.

157. Robinson 1961, 1963, 1967, 1968a, 1972a.

158. Brock et al. 1977; Butzer 1971, 1974, 1984; Cooke 1983b; Delson 1988; McFadden 1980; McFadden et al. 1979; Maguire 1985; Partridge 1973, 1978, 1982, 1985, 1986; Vogel 1980, 1985; Vrba 1975, 1985; Wilkinson 1985.

159. Brown and Feibel 1988.

160. Rightmire 1984a; McKee et al. 1995.

161. Butzer 1974; Partridge 1985; Tobias 1973, 1974, 1978d; Grine 1982, 1985a, 1985b; Olson 1985b; Rak 1983, 1985; Tuttle 1988a; Falk and Clarke 2007.

162. Cooke 1990.

163. McKee 1993; McKee et al. 1995.

164. Verhaegen 1990.

165. Robinson 1972a.

166. Grausz et al. 1988; Jungers 1988; McHenry 1988, 1991a.

167. Blackwelder 1967, p. 227.

168. Dean and Benyon 1991; White et al. 2000.

169. Robinson 1956, 1968b; Grine 1988b; Kay 1985; Kay and Grine 1988; Suwa 1988; Wood and Stack 1980.

170. Grine and Martin 1988; Beynon and Wood 1986.

171. Strait et al. 2009.

172. Robinson 1963, 1968a; Tobias 1967b; Howell 1978; Brown et al. 1993.

173. Hylander 1988.

174. Robinson 1963, 1968b; Tobias 1967b.

175. Rak 1988; Kimbel and White 1988.

176. Le Gros Clark 1950, 1955; Ashton and Zuckerman 1951; Robinson 1958, 1963.

177. Le Gros Clark 1967; Day 1985; Lovejoy et al. 1973; McHenry 1975; Robinson 1972a; Washburn 1950.

178. Hill et al. 1992; Wood 1992a.

179. Wood 1985b.

180. Stringer 1986.

181. Tattersall 1986.

182. Leakey, Tobias, and Napier 1964.

183. Robinson 1965, 1966, 1967, 1972b; Howell 1965; Le Gros Clark 1967; Campbell 1973.

184. Robinson 1965, 1966.

185. Blumenschine et al. 2003; Le Gros Clark 1978; Clarke and Howell 1972; Hughes and Tobias 1977; Tobias 1978a, 1991, 2003; Howell and Coppens 1976; Boaz and Howell 1977.

186. L. S. B. Leakey et al. 1964; Tobias 1965.

187. Tobias 1991.

188. Day 1976a, 1976b; Howell 1978.

189. Wood 1985b; Stringer 1986.

190. Groves and Mazák 1975.

191. Stringer 1986.

192. Alexeev 1986; Groves 1989.

193. DuBois 1894; Black 1927, 1934; Weidenreich 1935, 1937a.

194. de Vos 1985; Leinders et al. 1985; Matsu'ura 1986; Pope 1983, 1988; Semah 1984.

195. Swisher et al. 1994.

196. Swisher et al. 1996.

197. Brown et al. 2004; Morwood et al. 2004, 2005.

198. Langbroek and Roebroeks 2000; de Vos and Sondaar 1994; Sémah et al. 1977.

199. Indriati et al. 2010.

200. Tuttle and Mirsky 2007; Martin 2007; Henneberg 2007; Jungers and Kaifu 2011.

201. Pappu et al. 2011.

202. Brown et al. 2004; Morwood et al. 2005; Lahr and Foley 2004; Falk et al. 2005, 2009; Tocheri et al. 2007; Jungers et al. 2009; Brown 2012.

203. Morwood et al. 2005.

204. Moore et al. 2008; Culotta 2009; Brumm et al. 2006; Morwood et al. 1998, 2004, 2005; Brown et al. 2004; Argue et al. 2006; Moore et al. 2009; Westaway et al. 2009; Culotta 2009; Weston and Lister 2009; Lieberman 2009; Kaifu et al. 2009.

205. Hunter 2006; Martin et al. 2006a, 2006b; Martin 2007; Tuttle and Mirsky 2007; Vannucci et al. 2011.

206. Bowdler 2007; Bromham and Cardillo 2007; Conroy and Smith 2007; Culotta 2007; Eckhardt 2007; Jacob et al. 2006; Henneberg 2007, 2008; Henneberg and Thorne 2004; Hershkovits et al. 2007; Martin 2007; Martin et al. 2006a, 2006b; Obendorf et al. 2009; Richards 2006; Thorne and Henneberg 2007; Tuttle and Mirsky 2007; Weber et al. 2005; Zeitoun et al. 2007; Zollikofer and Ponce de Léon 2007.

207. Morwood et al. 1998, Lahr and Foley 2004.

208. Morwood et al. 2004.

209. Morwood et al. 1998, 2004, 2005; Brown et al. 2004.

210. Vekua et al. 2002; Lordkipanidze et al. 2007.

211. Gabunia and Vekua 1995; Gabunia et al. 2000a.

212. Bräuer and Schultz 1996; Dean and Delson 1995; Rosas and Bermúdez de Castro 1998.

213. Gabunia et al. 2000b; Vekua et al. 2002; Lordkipanidze et al. 2005, 2007; Fischman 2005.

214. Gabounia et al. 2002; Sawyer et al. 2007, p. 154; de Lumley et al. 2006.

215. Fischman 2005; Vekua et al. 2002.

216. Lordkipanidze et al. 2007; Lieberman 2007; Gibbons 2007a.

217. Gabunia et al. 2001.

218. Fischman 2005.

219. Gabunia et al. 2000b.

220. Shipman and Walker 1989; Shipman 2000; Balter and Gibbons 2000; Fischman 2005.

221. Gilbert and Asfaw 2009.

222. Aguirre et al. 1980; Aguirre and de Lumley 1977; Aguirre and Rosas 1985; Andrews 1984; Asfaw 1983; Asfaw et al. 2002; Bartstra et al. 1988; Bilsborough 1983, 1984; Bräuer 1984a; Chavaillon et al. 1974; Le Gros Clark 1978; Clarke 1985a; Conroy 1980; Conroy et al. 1978; Cook et al. 1982; Coon 1962; Cybulski 1981; Day 1986a; de Bonis 1986; de Bonis and Melentis 1991; Gibert et al. 1983; Gilbert et al. 2003; Hemmer 1972; Howell 1960, 1978, 1981; Howells 1980, 1981; Hublin 1985; Jacob 1975, 1981; Jaeger 1975, 1981; Jelinek 1978, 1980; Kennedy 1991; Kennedy et al. 1991; von Koenigswald 1975; Kraatz 1985; Kramer 1993; Laitman 1985a; Livingstone et al. 1961; de Lumley 1981; Maier and Nkini 1984, 1985; Mania and Vlcek 1981; Mallegni et al. 2003; Manzi et al. 2003; Protsch 1981; Rightmire 1979, 1980, 1981, 1983, 1984a, 1984b, 1984c, 1985, 1986, 1988, 1990; Santa Luca 1980; Sartono 1975; Sonakia 1985; Stringer 1984a, 1984b, 1985; Stringer et al. 1979; Thoma 1981; Tobias 1966; Vlcek 1978, 1980, 1983; Walker and Leakey 1978; Wolpoff 1977, 1980a, 1980b, 1985; Wolpoff and Nkini 1985; Wolpoff et al. 1984; Wood 1984, 1992b; Wu 1984; Wu and Dong 1985; Wüst 1951.

223. Bilsborough and Wood 1986; Brown et al. 1985.

224. Ruff and Walker 1993, table 11.15; Dean and Smith 2009.

225. Ruff and Walker 1993.

226. Ruff 2007.

227. H. Smith 1994; Dean and Smith 2009; Jungers and Kaifu 2011.

228. Bermúdez de Castro et al. 2010.

229. Bermúdez de Castro et al. 2003.

230. Spoor et al. 2007b; Gibbons 2007b.

231. Howells 1973.

232. Le Gros Clark 1978; Howell 1978; Rightmire 1988; Weidenreich 1936, 1937b, 1943.

233. Cronin et al. 1981.

234. Andrews 1984, 1986b; Bilsborough and Wood 1986; Eldredge and Tattersall 1982; L. S. B. Leakey 1963; Wood 1984.

235. Gabunia et al. 2000b; Walker and Leakey 1993; Tattersall and Schwartz 2000.

236. Tattersall 2000; Conroy et al. 1978; Rightmire 1990; Abbate et al. 1998; Arsuaga et al. 1997; Carbonell et al. 2005; Carretero et al. 1997; Ghinassi et al. 2009; Rightmire 2008; Tattersall and Swartz 2008; Wagner et al. 2010.

237. Bräuer 2008; Holliday 2000; Krings et al. 1997; Rak et al. 2002; Hublin 2009.

238. Ruff et al. 1993, 1994; Trinkaus et al. 1991, 1994; Weaver 2009; Walker et al. 2011.

239. Smith, Toussaint et al. 2007, 2010; compare Gautelli-Steinberg et al. 2005; Bayle et al. 2010.

240. Tuttle 1988a; Simons 1989b; Steward and Stringer 2012.

241. Singer and Wymer 1982; Bräuer 1984a; Rightmire 1984a, 1988, 2008; Stringer and Andrews 1988; Rightmire and Deacon 1991.

242. Campbell and Tishkoff 2008; Rightmire 2009; Hammer et al. 2011; Gibbons 2011. Schlebusch et al. 2012.

243. Campbell and Tishkoff 2008; Tishkoff et al. 2009; Demeter et al. 2012; Trinkaus 2003; Richards and Trinkaus 2009; Hoffecker 2009.

244. de Villiers 1973, 1976.

245. Beaumont 1980.

246. Day et al. 1980; Magori and Day 1983a, 1983b; Rightmire 1984c; White et al. 2003; Smith, Tafforeau et al. 2007.

247. Manega 1993.

248. Clark et al. 2003.

249. White et al. 2003.

250. Gunz et al. 2009; Campbell and Tishkoff 2008; Cann 1988; Cann et al. 1987; Cavalli-Sforza et al. 1986, 1988; Fagundes et al. 2007; Horai et al. 1986; Nei 1978; Nei and Roychoudhury 1982; Ruvolo et al. 1993; Stoneking et al. 1986; Stoneking and Cann 1989; Stoneking 1994; Vigilant et al. 1991; Wainscoat et al. 1986; Wilson and Cann 1992; Yu et al. 2002; Weaver and Roseman 2008; Henn, Gignoux et al. 2011; Henn, Busamante et al. 2011; Hublin and Klein 2011.

251. Adler et al. 2008; Balter 2011; Brockley et al. 2008; Roebroeks 2008; Slimak et al. 2011; Pearson 2004; Pinhasi et al. 2011.

252. Lewontin 1972; Goodman 2005; Edgar and Hunley 2009; Relethford 2009; Long et al. 2009; Hunley et al. 2009; Gravlee 2009; Koenig et al. 2008.

253. Scholz et al. 2007.

254. Wolpoff 1985, 1992; Wolpoff et al. 1984, 1988; Spuhler 1988; Thorne and Wolpoff 1992.

255. Templeton 1993.

256. Aiello 1993.

257. Coon 1962.

258. Howells 1993; Stringer 1989a, 1989b, 1990a, 1990b, 1992; Stringer and Andrews 1988; Pennisi 2007c; Hudjashov 2007; Thangaraj et al. 2005; Macaulay et al. 2005.

259. Bräuer 1982, 1984a, 1984b, 1984c, 1989, 1992; Trinkaus 2007; Liu 2010.

260. Smith 1982, 1984, 1985, 1992; Smith and Paquette 1989; Smith et al. 1989; Smith and Trinkaus 1991; Martinón-Torres et al. 2007.

261. Aigner 1976, 1978; Aigner and Laughlin 1973; Freedman and Lofgren 1979a, 1979b; Thorne 1971a, 1971b, 1976, 1980; Thorne and Macumber 1972; Thorne and Wolpoff 1992; Woo and Wu 1982.

262. Weidenreich 1943, 1946; Wolpoff 1989a, 1989b; Wolpoff et al. 1984, 1988.

263. Krings et al. 1997.

264. Serre et al. 2004; Lowe et al. 2012; Eriksson and Manica 2012; Hublin 2012.

265. Briggs et al. 2009.

266. Krause, Orlando et al. 2007; Pennisi 2013.

267. Green et al. 2010.

268. Krause, LaLueza-Fox et al. 2007; Burbano et al. 2010; Preuss 2012.

269. Höss et al. 1996; Green et al. 2010.

270. Eldridge and Gould 1972; Gould 2002.

271. Schmitz et al. 2002.

5. APES IN MOTION

1. Tsukahara 1993; Fay et al 1995.

2. Tuttle 1972; Hollihn and Jungers 1984; Swartz 1989, 1993; Swartz et al. 1989; Takahashi 1990; Usherwood et al. 2003.

3. Carpenter 1940; Tuttle 1972.

4. Ellefson 1968; Tuttle 1986; Sati and Alfred 2003; Michilsens et al. 2011.

5. Washburn 1963b, 1973a; Ellefson 1968, 1974; Kortlandt 1968; Fleagle 1976.

6. Grand 1972.

7. Tuttle 1972; Cartmill and Milton 1977; Jungers 1985.

8. Kortlandt 1968; Tuttle 1986.

9. Keith 1926; Tuttle 1972.

10. Tuttle 1972; Jungers and Stern 1980; Swartz 1993.

11. Erikson 1963.

12. Ashton and Oxnard 1963.

13. Jenkins and Fleagle 1975; Jenkins 1981.

14. Avis 1962.

15. Tuttle 1972.

16. Cortright 1983; Sigmon 1975.

17. Cortright 1983.

18. Schultz 1963b; Tuttle 1972.

19. Morton 1924, 1935; Tuttle 1972; Tuttle et al. 1992; Ishida et al. 1976, 1978, 1985; Gebo 1992, 1993b; Schmitt and Larson 1995.

20. Vereecke et al. 2005.

21. Tuttle 1986; Cant 1987; Thorpe and Crompton 2006, 2009; Manduell et al. 2011.

22. Tuttle 1986; Sugardjito and Cant 1994.

23. Sugardjito and van Hooff 1986; Sugardjito 1988.

24. Thorpe and Crompton 2005.

25. Cant 1987, 1992; Thorpe and Crompton 2006.

26. Thorpe and Crompton 2009.

27. Cant 1987.

28. Tuttle 1986; Sugardjito 1988; Sugardjito and Cant 1994.

29. MacKinnon, 1974a; Tuttle 1986, p. 42.

30. MacKinnon 1974; Rijksen 1978; Thorpe and Crompton 2006; Myatt and Thorpe 2011.

31. Napier 1960.

32. Tuttle and Cortright 1988.

33. Straus 1942.

34. Tuttle 1970.

35. Tuttle and Rogers 1966.

36. Jenkins and Fleagle 1975.

37. Tuttle and Cortright 1988; Zihlman, McFarland, and Underwood 2011.

38. Cortright 1983; Tuttle and Cortright 1988.

39. Tuttle and Basmajian 1977.

40. Tuttle 1975a.

41. Tuttle and Watts 1985.

42. Tuttle 1970; Tuttle et al. 1994, 1999.

43. Tuttle, 1975b.

44. Tuttle 1969c; Rose 1973; Whitehead 1993.

45. Tuttle 1967, 1969a, 1969b, 1970, 1975b.

46. Tuttle 1967, 1969a, 1969b; Marzke 1971.

47. Tuttle and Beck 1972; Tuttle 1975b; Susman 1974; Susman and Tuttle 1976.

48. Ellis and Montagna 1962; Montagna 1965.

49. Inouye 1992, 1994; Kivell and Schmitt 2009; Shea and Inouye 1993; Susman and Creel 1979.

50. Tuttle 1967, 1969a, 1969b, 1970.

51. Corruccini 1978.

52. Doran 1993.

53. Tuttle 1965, 1967; Preuschoft 1973b.

54. Doran 1992, 1993; Inouye, 1990, 1991, 1992; Susman 1979a, 1984; Susman et al. 1980.

55. Tuttle 1969a, 1969c, 1970, 1972.

56. Susman and Stern 1980.

57. Tuttle 1967, 1969a, 1969b, 1970.

58. Alexander 1991.

59. Tuttle 1967, 1969a, 1969b.

60. Tuttle et al. 1994.

61. Kimura 1992; Demes et al. 1994.

62. Lazenby et al. 2011.

63. Tuttle et al. 1999; Doran 1997a; Lazenby et al. 2011.

64. Senut 1986; McHenry 1976; Senut 1989a; Fischer 1906; Knussmann 1967; Martin 1934; Tuttle and Basmajian 1974; Tuttle et al. 1983.

65. Feldesman 1982.

66. Huxley 1863; Haeckel 1866, 1868, 1874; Darwin 1871; Gregory 1916; Morton 1922, 1924a; Keith 1923; Smith 1927.

67. Gregory 1916; Tuttle et al. 1998.

68. Tuttle 1970, 1985.

69. Congdon et al. 2012.

70. Schultz 1927; Midlo 1934; Tuttle 1970.

71. Pontzer and Wrangham 2004.

72. Jurmain 1997; Goodall 1986; Carter et al. 2008.

73. Gebo et al. 2000; 2001; Gebo and Dagasto 2004.

74. Ciochon et al. 2001; Ciochon and Gunnell 2004; Gunnell et al. 2002.

75. Kay and Simons 2005; Gunnell et al. 2002; Egi et al. 2004.

76. Kay and Simons 2005.

77. Dagosto et al. 2010; Marivaux et al. 2003; Kay and Simons 2005; Marivaux et al. 2010.

78. Gebo et al. 1994; Seiffert et al. 2000; Simons et al. 2001.

79. Gebo et al. 1994.

80. Kirk and Simons 2001.

81. Seiffert et al. 2000.

82. Conroy, 1976; Fleagle and Simons 1978, 1982a, 1982b; Fleagle 1983, 1988; Gebo 1986, 1993a; Gebo and Simons 1987; Ankel-Simons et al. 1998.

83. Fleagle et al. 1975; Schön Ybarra and Conroy 1978; Gebo 1993a.

84. Fleagle and Simons 1983; Fleagle and Kay 1987.

85. Fleagle and Simons 1995.

86. Gebo 1993a.

87. Harrison 1993.

88. Leakey and Walker 1985a; Rose et al. 1992.

89. Rose 1993.

90. Jungers 1984.

91. Senut 1988b; Rose et al. 1992; Rose 1993, 1994.

92. Beard et al. 1986; Corruccini et al. 1975, 1976; Gebo et al. 1988; Jouffroy et al. 1991; Kelley and Pilbeam 1986b; Leakey et al. 1988a, 1988b; McHenry and Corruccini 1983; Napier and Davis 1959; Rose 1983, 1993; Senut 1989b; Walker and Pickford 1983; Walker and Teaford 1989; Ward 1993; Ward et al. 1993a.

93. Rose 1992.

94. Rose 1993; Senut 1988b; Walker and Teaford 1989; Ward 1993; Ward et al. 1993a, 1995.

95. Rose 1994.

96. Ward 1993.

97. Corruccini et al. 1976; Ruff et al. 1989; Walker and Pickford 1983; Ward et al. 1993a.

98. Bacon 1994; Beard et al. 1986; Begun et al. 1994; Jenkins and Fleagle 1975; Langdon 1986; Lewis 1989; MacLatchy and Bossert 1996; McHenry and Corruccini 1975, 1983; Morbeck 1975, 1976; Napier and Davis 1959; Preuschoft 1973a; Rose 1993; Ward 1993; Walker and Pickford 1983; Ward et al. 1993b.

99. Ward et al. 1991; Nakatsukasa et al. 2004; Harrison 1998.

100. Fooden 1969, 1990; Schultz 1961; Wilson 1972.

101. Harrison 1993; Leakey et al. 1988b; Rose 1993, 1994; Ward 1993.

102. Walker and Rose 1968; Ward 1993; Sanders and Bodenbender 1994.

103. Allboook and Bishop 1963; Bishop 1964; Simons and Pilbeam 1965; Pilbeam 1969b; Szalay and Delson; Leakey et al. 1988b.

104. Gebo et al. 1997; Senut et al. 2000; Gomery 2005.

105. Leakey et al. 1988a; Harrison 1993.

106. Rose 1993, 1994.

107. Rose et al. 1996; Nakatsukasa et al. 1996, 1998.

108. Nakatsukasa, Tsujikawa et al. 2003.

109. Rose et al. 1996.

110. Nakatsukasa et al. 1998; Nakatsukasa, Kunimatsu et al. 2003.

111. Ward et al. 1999; McCrossin and Benefit 1994.

112. McCrossin and Benefit 1994.

113. McCrossin et al. 1998; McCrossin 1999.

114. Patel et al. 2009.

115. Dalrymple 1979; Shipman et al. 1981; Pickford 1986b; McCrossin and Benefit 1994.

116. Kappelman 1991.

117. McCrossin and Benefit 1997.

118. McCrossin 1992.

119. Blue et al. 2006; Harrison 1989a; McCrossin et al. 1998; McCrossin and Benefit 1994; Benefit 1999a.

120. Harrison 1989a; Zambon et al. 1999.

121. Zambon et al. 1999; Benefit 1993.

122. Blue et al. 2006.

123. Fleagle 1988.

124. Tuttle 1972; Simons and Fleagle 1973.

125. Zapfe, 1960; Simons and Fleagle 1973; Rose 1988, 1993, 1994; Ciochon and Corruccini 1977.

126. Corruccini et al. 1975.

127. Begun 1988, 1992b, 1993.

128. Zapfe 1960; Rose 1993, 1994.

129. Simons and Fleagle 1973; Ward et al. 1991; Nakatsukasa, Tsujikawa et al. 2003.

130. Ankel 1962, 1965, 1972.

131. Ankel 1965; Simons and Fleagle 1973.

132. Moyà-Solà et al. 2004, 2005.

133. Begun and Ward 2005.

134. Agustí et al. 1996; Köhler et al. 2001.

135. Moyà Solà and Köhler 1996.

136. Agustí et al. 1996.

137. Begun 1992b.

138. Begun 1992b; Begun and Kordos 1993.

139. Begun 1992b; Rose 1994.

140. Morebeck, 1983; Begun, 1988, 1992b, 1993; Kivell and Begun 2009.

141. Begun 1993; Begun and Kordos 1999.

142. Sarmiento 1987.

143. Harrison 1991b.

144. Schultz 1960; Straus 1963; Hürzeler 1968; Harrison 1986b, 1991b.

145. Sarmiento 1987, 1988; Harrison, 1991b.

146. Straus 1963; Harrison 1986b, 1991b; Jungers 1987; Cant 1987; Kortlandt 1968; Cartmill 1974b; Jungers 1977; Sarmiento 1987; Rose 1993; Harrison 1991b.

147. Sarmiento 1987; Harrison 1991b; Rose 1993.

148. Badgley et al. 1984; Flynn et al. 1990; Kelley and Pilbeam 1986b; Langdon 1986; Pilbeam et al. 1990; Rose, 1983, 1984a, 1987, 1989, 1993, 1994; Senut 1989a.

149. Rose 1989.

150. Rose 1986, 1993, 1994.

151. Spoor et al. 1991; Andrews 1983; Senut 1989a, 1991.

152. Wu et al. 1986.

153. Meldrum and Pan 1988.

154. Qi 1993.

155. Pan 1988.

6. SEVERAL WAYS TO ACHIEVE ERECTION

1. Inman et al. 1981; Dominici et al. 2011; Grillner 2011.

2. Cartmill 1983; Graubner et al. 2009.

3. Lau 2001.

4. Darwin 1871; Washburn and Avis 1958; Washburn 1960.

5. Weiner et al. 1953; Spencer 1990a, 1990b.

6. Tuttle 1987, 1990a; Lockley et al. 2008: Deino 2011; Corvinus and Roche 1980; Harris 1983; Howell et al. 1987; Roche and Tiercelin 1980; Semaw et al. 1997; Heinzelin et al. 1999.

7. Tobias 1971, 1987, 1991; Falk 1992.

8. McPherron et al. 2010; Braun 2010.

9. Keith 1912a, 1912b, 1923; Morton 1924a; Washburn 1967; Tuttle 1975a.

10. Huxley 1863; Darwin 1859.

11. Keith 1903, 1912a, 1923.

12. Keith 1923.

13. Corruccini 1975c.

14. Tuttle 1969a, 1974.

15. Morton 1922, 1924a, 1924b, 1926; Tuttle 1975a.

16. Morton 1924a.

17. Gregory 1927a, 1927b, 1927c, 1927d, 1928, 1929, 1930, 1934.

18. Le Gros Clark 1940; Jones 1940; Schultz 1950; Straus 1940, 1942, 1949; Napier and Davis 1959.

19. Washburn 1950, 1951, 1959, 1960, 1963b; Washburn and Avis 1958.

20. Tuttle 1977, 1986.

21. Tuttle 1967, 1969a, 1969b, 1970, 1975b.

22. Washburn 1967, 1968a, 1968b, 1972, 1973a, 1973b.

23. Tuttle 1969a, 1974, 1975a.

24. Tuttle 1974, 1975a, 1981b, 1994.

25. Tuttle 1972, pp. 201–202; Zihlman, Mootnick, and Underwood 2011.

26. Tuttle 1969a, 1975a.

27. Tuttle 1975a.

28. Tuttle 1981b, 1992, 1994; Tuttle et al. 1998.

29. Do Amaral 1989, 1996; Bartholemew and Birdsell 1953; Hewes 1961, 1964; Rumbaugh 1965; Isaac 1978, 1981; Lovejoy 1981, 1988; Parker 1987; Tanner 1981; DeSilva 2011.

30. DuBrul 1962; Hunt 1994; Jolly and Plog 1986; Leutenegger 1987; Rose 1974b, 1976, 1984, 1991; Sigmon 1971; Wrangham 1980.

31. Thorpe et al. 2007; O'Higgins and Elton 2007; Crompton et al. 2008.

32. Jolly 1970; Kingdon 2003.

33. Straus 1962.

34. Read-Martin and Read 1975; Szalay 1975b; Livingstone 1962b; Geist 1978; Merker 1984.

35. Rodman and McHenry 1980; Carrier 1984; McHenry 1994b; Leonard and Robertson 1995; Tuttle 1994a; Bramble and Lieberman 2004; Sockol et al. 2007.

36. Wheeler 1984, 1991a, 1991b, 1992, 1993, 1994a, 1994b.

37. Sinclair et al. 1986.

38. Dart 1959; Robinson 1963; Ravey 1978.

39. Fifer 1987; Knüsel 1992.

40. Wescott 1967; Kortlandt 1980; Jablonski and Chaplin 1992, 1993.

41. Tanner 1981; Sheets-Johnson 1989; Guthrie 1970; Mautz et al. 2012.

42. Montgomery 1988.

43. Eickhoff 1988.

44. Westenhöfer 1942; Hardy 1960, 1977; La Lumière 1981; Morgan 1972, 1990; Ellis 1993; Pond 1987b; Bender et al. 1997.

45. Isaac 1983a; Day 1986b; Rose 1991; Tuttle et al. 1991.

46. Tuttle 1991; Chaplin et al. 1994.

47. Schultz 1936.

48. Carlsöö 1972; MacConnail and Basmajian 1969; Basmajian and De Luca 1985.

49. Basmajian and De Luca 1985.

50. Carlsöö 1972.

51. Washburn 1968a.

52. Haxton 1947.

53. Washburn 1950.

54. Tuttle 1970.

55. Tardieu 1991, 1992a, 1992b; Tardieu et al. 1993.

56. Tardieu et al. 1993.

57. Prost 1967; Tuttle 1972; Yamazaki 1985, 1990; Schmid and Piaget 1994.

58. Yamazaki and Ishida 1984.

59. Tuttle 1994a.

60. Tuttle and Cortright 1988.

61. Webb et al. 1994.

62. Carlsöö 1972; Basmajian and De Luca 1985.

63. Ishida et al., 1978; Tuttle and Basmajian 1975; Tuttle, Basmajian, and Ishida 1975, 1978, 1979; Tuttle, Buxhoeveden, and Cortright 1979.

64. Sigmon 1975; Tuttle et al. 1975, 1978; Ishida et al. 1978.

65. Basmajian and De Luca 1985; Lieberman et al. 2006.

66. Carrier et al. 1994.

67. Morton 1935.

68. Tuttle, Hallgrímsson, and Stein 1998.

69. Morton 1935; Tuttle 2008.

70. Nicol and Paul 1988.

71. Tuttle et al. 1992; Tuttle, Hallgrímsson, and Stein 1998.

72. Weyland et al. 2000.

73. Schultz 1963b; Tuttle and Watts 1985.

74. Tuttle, Hallgrímsson, and Stein 1998; DeSilva 2009; Balter 2009a.

75. Bramble and Lieberman 2004; Zimmer 2004; Carrier et al. 2011.

76. Bramble and Lieberman 2004.

77. Montagna 1982, 1985; Bramble and Carrier 1983; Carrier 1984; Falk 1990; Tuttle 1994a; Bramble and Lieberman 2004.

78. Carrier 1984; Tuttle 1994a; Bramble and Lieberman 2004.

79. Bramble and Liebeman 2004; Zimmer 2004.

80. Newman 2002.

81. Lovejoy 2009a, 2009b; Lovejoy, Latimer et al., 2009a, 2009b; Lovejoy, Simpson et al. 2009a, 2009b; Gibbons 2009a; Lovejoy, Suwa, Simpson et al. 2009a, 2009b; Lovejoy, Latimer et al. 2009a, 2009b; Lovejoy, Suwa, Spurlock et al. 2009a, 2009b; Suwa et al. 2009a, 2009b; Senut 2012.

82. Lovejoy, Simpson et al. 2009; Lovejoy, Latimer et al. 2009a, 2009b; White et al. 2009; Gibbons 2009a.

83. Capecchi 1984; Clarke 1979; Day and Wickens 1980; Leakey and Hay 1979; Robbins 1987; Tuttle 1985, 1987, 1988, 1990a, 1994, 1996; Tuttle, Webb, and Baksh 1991; Tuttle et al. 1990, 1992; Tuttle, Webb, and Tuttle 1991; White 1980b; White and Suwa 1987; Raichlen et al. 2010; Bower 2010.

84. Tuttle 1987.

85. Bonnefille and Riollet 1987; Harris 1985, 1987; Hay 1987; Leakey 1987a; Leakey and Hay 1979.

86. Tuttle 1987, 1981a, 1981b.

87. Tuttle 1996; Kramer 2012.

88. Tuttle 1994a.

89. Tuttle, Webb, and Baksh 1991; Tuttle 1996; Musiba et al. 1997.

90. Alexander 1984; Charteris et al. 1981, 1982; Tuttle et al. 1990, 1992; Tuttle, Webb, and Tuttle 1991.

91. Tuttle et al. 1992; Musiba et al. 1997.

92. Tuttle et al. 1990; Tuttle 1996.

93. Tuttle et al. 1990; Tuttle 1996; Hartwig-Scherer and Martin 1991; Jungers 1982; Jungers and Stern 1983; Wolpoff 1983b, 1983c.

94. Tuttle et al. 1990; Tuttle 1996; Musiba et al. 1997.

95. Murray et al. 1971.

96. Tuttle 1987, 1996.

97. Tuttle 1987, 1988.

98. Robbins 1987.

99. Tuttle 1981b; Harcourt-Smith and Hilton 2005.

100. Tuttle 1987, 1988a, 1996, 2004; Clarke 1979, 1985a, 1985b, 1999, 2003; Clarke and Tobias 1995; Day 1985; Deloison 1991, 1992, 1997, 2004; Harcourt-Smith and Hilton 2005; Kullmer et al. 2003; Latimer 1991; Meldrum 2004; Schmid 2004; Berillon 2004; Senut 2003, 2005; White 1980b; White and Suwa 1987; Latimer 1991; Stern and Susman 1983; Susman et al. 1984; Wong 2005; Crompton et al 2011; Deloison 2004; Tuttle 2008.

101. Berillon 2004; Clarke 2003; Deloison 2004; Harcourt-Smith and Hilton 2005; Kullmer et al. 2003; Meldrum 2004; Schmid 2004; Tuttle 1987.

102. Tuttle 2008.

103. Tuttle 1981b.

104. Tuttle et al. 1998; Griffin and Richmond 2010.

105. Latimer et al. 1987.

106. Proctor 2010; Proctor et al. 2008.

107. Ward et al. 2011.

108. Marchi 2010.

109. Tuttle 1981b; McHenry 1986; Lovejoy et al. 2002; Haile-Selassie et al. 2010.

110. Alemseged et al. 2006; Green and Alemseged 2012; Larson 2012.

111. Wynn et al. 2006.

112. Tuttle 1981b; Schmid 1983; Häusler and Schmid 1995.

113. Tardieu 1999.

114. Schmid 1983; Cook et al. 1983.

115. Oxnard 1968; Broom et al. 1950.

116. McHenry 1986.

117. McHenry and Berger 1998a, 1998b.

118. Drapeau and Ward 2007.

119. Asfaw et al. 1999; Richmond et al. 2002.

120. Berge and Goularas 2010; Häusler and Schmidt 1995.

121. Whitcome et al. 2007.

122. Berger et al. 2010; Berger 2013; Churchill et al. 2013; DeSilva et al. 2013; Schmid et al. 2013; Williams et al. 2013.

123. Lordkipanidze et al. 2007; Lieberman 2007; Bennett et al. 2009; Crompton and Pataky 2009; Susman and Brain 1988.

124. Trinkaus 1983a, 1984; Rak and Arensburg 1987.

125. Trinkaus 1984; Rosenberg and Trevathan 1995; Weaver and Hublin 2009; Gibbons 2009b; Franciscus 2009; Trinkaus 1983b; Simpson et al. 2008.

126. Ruff 2009.

127. Wood and Collard 1999; Tuttle 2006a.

128. Simpson et al. 2008; Dunsworth et al. 2012.

129. Ruff 2010.

130. Lordkipanidze et al. 2007; Lieberman 2007; Larson et al. 2007.

7. HUNGRY AND SLEEPY APES

1. van Schaik et al. 2009; Jasienska et al. 2006; Leonard et al. 2007.

2. Mayes 2006; Knott 1997.

3. Ortmann et al. 2006; Blumenthal 2012; Fahy et al. 2013.

4. Waterman et al. 1980, 1983; Waterman and Kool 1994.

5. Houle et al. 2007; Perica 2001; Nishida et al. 1983.

6. Tuttle 1986; Knott 2005; Conklin-Brittain et al. 2006.

7. Ungar and Teaford 2002; Ungar 2007a; Whitney and Rolfes 2002; Wrangham 2009; Gremillion 2011; Ulijaszek et al. 2012.

8. Lambert 1998; Strier 2000, p. 172; O'Malley and Power 2012.

9. Rothman, Chapan et al. 2008.

10. O'Brien et al. 1998.

11. Tuttle 1986; Anstett et al. 1997; Machado et al. 2005.

12. Lambert 1998.

13. Lambert 1998; Kay and Davies 1994; Waterman and Kool 1994.

14. Lambert 2007a.

15. Danish et al. 2006.

16. Marshall and Wrangham 2007; Harrison and Marshall 2011.

17. N'guessan et al. 2009.

18. Marshall and Wrangham 2007.

19. Basabose 2002a; Conklin and Wrangham 1994; Curtin and Chivers 1978; Tweheyo and Babweteera 2007; Yamagiwa et al. 2005; White 1992b; Djojosudharmo and van Schaik 1992; Tuttle 1986.

20. Tuttle 1986, 1990b; Bermejo et al. 1994; Hohmann and Fruth 1996; McGrew 1996; McGrew et al. 2007; Ihobe 1992b.

21. Wrangham et al. 1991; Bourquin et al. 1993; Topping and Clifton 2001.

22. Conklin-Brittain et al. 2006.

23. Tuttle 1990b; Knott 2005.

24. Hockings et al. 2009; Hall et al. 1998.

25. Doran and McNeilage 1998; Voysey et al. 1999a, 1999b; Yamagiwa et al. 2005; Robbins 2007; Takenoshita and Yamagiwa 2008; Takenoshita et al. 2008.

26. Remis 2002, 2006.

27. Remis 2006.

28. Bogart and Preutz 2008, 2011; Carroll 1986, 1988, 1990.

29. Cipolletta et al. 2007.

30. Groves 1970b, 1970c; Uchida 1998b; Taylor 2009; Yamagiwa and Basabose 2006.

31. Taylor 2009.

32. Tuttle 1986; Carroll 1988, 1990; Watts 1989b; Yamagiwa et al. 1991; Deblauwe et al. 2003; Deblauwe and Janssens 2008.

33. Tuttle 1986; Utami and van Hoof 1997; Knott 1999; van Schaik 2004; van Schaik et al. 2009; Bartlett 2009a; Fan and Jiang 2009.

34. McGrew et al. 2007.

35. Goodall 1963a; Hladik 1977; Nishida and Uehara 1983; Aufreiter et al. 2001; Mahaney et al. 1999, 2005; Mahaney, Hancock et al. 1996, 1997; Tweheyo et al. 2006; Schaller 1963; Mahaney 1993; Mahaney et al. 1990, 1995; Tuttle 1986.

36. Uehara 1982.

37. Kano and Mulavwa 1984; Horr 1972; Mahaney, Stambolic et al. 1996; MacKinnon 1974.

38. Tuttle 1986.

39. Jacobs 1993, 2002, 2005.

40. Nei et al. 1997; Surridge et al. 2003; Lucas et al. 1998; Dominy and Lucas 2001, 2004; Dominy et al. 2003; McKey et al. 1981; Davies et al. 1988; Matsuno et al. 2006; Allen 1879; Mollon 1989; Regan et al. 2001; Sumner and Mollon 2000a, 2000b; Dominy et al. 2006; Osario and Vorobyev 1996.

41. New et al. 2007.

42. Öhman 2007.

43. Isbell 2006, 2009.

44. Hladik and Simmen 1996; Simmen and Hladik 1998; Hladik et al. 2002; Hellekant et al. 1996, 1997, 1998; Dominy 2004; Hoffman et al. 1997, 2004; Scheibert et al. 2009; Miller 2009; Wooding et al. 2006.

45. Coleman 2009; Coleman and Ross 2004; Heffner 2004; Dominy et al. 2001, 2004; Scally et al. 2012; Quam et al. 2013.

46. Byrne 1994, 2001; Byrne and Byrne 1993.

47. Lambert 1998; Hemingway and Bynum 2005; Plavcan et al. 2005.

48. Milton 1984, 1987, 1993, 1999b, 2000; Milton and Demment 1988; Lambert 1998, 2007b.

49. Smith 1984; Lucas et al. 1994; Hylander 2013.

50. Caldicott and Kapos 2005; Chancellor 2012.

51. Reynolds et al. 1998.

52. White and Tutin 2001; O'Brien and Kinnaird 2011.

53. Bartlett 2007.

54. Marshall and Leighton 2006.

55. Bartlett 2007, p. 278; 2009.

56. Elder 2009.

57. McConkey et al. 2003.

58. Lappan 2009a.

59. Marshall, Ancrenez et al. 2009.

60. Galdikas and Teleki 1981.

61. Morrogh-Bernard et al. 2009.

62. van Schaik et al. 2009.

63. Knott 2005; van Schaik 2004; Heller et al. 2002.

64. Knott 1998; Vogel et al. 2012.

65. Tuttle 1986, pp. 63–69; Galdikas 1988; Leighton 1993; Hamilton and Galdikas 1994; Knott 1998, 2005; Rodman 1988; Sugardjito et al. 1987; Ungar 1995; Delgado and van Schaik 2000; Rijksen and Meijaard 1999; van Schaik 2004; Fox et al. 1999, 2004; Caldecott and McConkey 2005; Wich et al. 2006a, 2006b; Russon et al. 2009.

66. Jaeggi et al. 2010; Bastian et al. 2010; Gustafsson et al. 2011; van de Waal et al. 2013; de Waal 2013.

67. Leighton 1993.

68. Gresl et al. 2000.

69. Wheatley 1982.

70. Knott 1998, 1999.

71. Conklin-Brittain and Knott 2006.

72. Stoner 1995.

73. Wich et al. 2006a, 2006b.

74. Wich et al. 2006a.

75. Wich et al. 2006b.

76. Fox et al. 2004; van Schaik 2009.

77. Fox et al. 2004.

78. Russon et al. 2009.

79. MacKinnon et al. 1996.

80. Stumpf 2007.

81. Gautier-Hion 1990.

82. Hohmann et al. 2010.

83. Pruetz 2006; Bertolani and Pruetz 2011.

84. Hunt and McGrew 2002; Pruetz 2006; Bogart and Preutz 2011.

85. Pruetz and Bertolani 2007.

86. Hernandez-Aguilar et al. 2007.

87. Moore 1994.

88. Hockings et al. 2010.

89. Boesch et al. 2006.

90. Potts et al. 2011.

91. Nishida and Turner 1996.

92. Watts 1984, 1985a, 1985b, 1987, 1988, 1990a, 1996, 1998a, 1998b, 1998c; Vedder 1984; Fossey and Harcourt 1977.

93. Bermejo 2004; Blake and Fay 1997; Calvert 1985; Fay 1989; Cipolletta 2003, 2004; Cousins and Huffman 2002; Doran et al. 2002; Doran-Sheehy et al. 2004; Goldsmith 1999a; Kuroda, Nishihara et al. 1996; Magliocca and Gautier-Hion 2002; Masi et al. 2009; Moutsabouté et al. 1994; Nishihara 1992, 1995; Popovich et al. 1997; Poulsen et al. 2001; Remis 1997a, 1997b; Remis et al. 2001; M. E. Rogers et al. 1988, 1990, 1992, 1994, 1998, 2004; Sabater Pí 1977; Tutin and Fernandez 1985, 1993a, 1993b, 1994; Tutin, Fernandez et al. 1991; Tutin, Williamson et al. 1991; Tutin, White et al. 1994; Tutin, Ham et al. 1997; Tutin, Fernandez et al. 1991; Tutin, Williamson et al. 1991; Voysey et al. 1999a, 1999b; Williamson et al. 1988, 1990; Oates et al. 2003; Casimir 1975; Yamagiwa and Basabose 2006a, 2006b, 2009; Yamagiwa , Mwanza, Spangenberg et al. 1992: Yamagiwa, Mwanza, Yumoto et al. 1992, 1994, Yamagiwa, Kaleme et al. 1996, Yamagiwa, Maruhashi et al. 1996; Yamagiwa, Basabose et al. 2003, 2005; Yamagiwa and Mwanza 1994; Mwanza et al. 1992; Yumoto et al. 1994; Sarmiento et al. 1996; McNeilage 2001; Goldsmith 2003; Stanford and Nkurunguni 2003; Robbins and McNeilage 2003; Nkurunguni et al. 2004; Ganas et al. 2004, 2008; Ganas and Robbins 2005; Rothman, Dierenfeld et al. 2006, 2008; Rothman, Pell et al. 2006; Harcourt and Stewart 2007; Stanford 2008.

94. Tutin and Fernandez 1985; Doran et al. 2002.

95. Rogers, Abernethy et al. 2004; Williamson et al. 1990.

96. Tutin and Fernandez 1983.

97. Fay et al. 1989; Blake et al. 1995.

98. Mitani et al. 1993; Blom et al. 2001.

99. Fay 1989.

100. Richards 1952, pp. 357–363.

101. Whitmore 1990.

102. Watts 1984, 1996, 1998c; Vedder 1984; Plumptre 1996; McNeilage 2001; Yamagiwa et al. 2005.

103. Yamagiwa et al. 2005.

104. Casimir and Butenandt 1973.

105. Ganas et al. 2008, 2009.

106. Ganas and Robbins 2004.

107. Kano 1983.

108. Kano and Mulavwa 1984; Kano 1992, p. 122; Idani et al. 1994; Uehara 1990.

109. Idani et al. 1994; Melenky and Stiles 1991; Melenky and Wrangham 1994a; Melenky et al. 1994; White and Wrangham 1988; White 1989a, 1992b, 1996a, 1996b, 1998, 2007; Hashimoto et al. 1998.

110. Malenky and Wrangham 1994b.

111. Bermejo et al. 1994.

112. Huffman et al. 1993; Wrangham 1995a; Huffman 1997; Huffman and Wrangham 1994; Huffman et al. 1998; Huffman 2001, 2007; McRae 1994; Dominy 2012; Lee-Thorp et al. 2012.

113. Wrangham and Nishida 1983; Huffman and Seifu 1989; Huffman 1995, 2001, 2007.

114. Jisaka et al. 1992; Koshimizu et al. 1994; Ohigashi et al. 1994; Huffman, Koshimizu, and Ohigahi 1996; Huffman et al. 1996.

115. Emery Thompson et al. 2008.

116. Whitmore 1990; van Schaik and Brockman 2005; van Schaik and Pfannes 2005; van Schaik et al. 2005.

117. Whitmore 1990.

118. Whitmore 1990; van Schaik and Pfannes 2005.

119. Marshall, Cannon, and Leighton 2009; Marshall et al. 2007.

120. Tweheyo and Obua 2001.

121. Tuttle 1986; Caldecott and Miles 2005.

122. Leighton and Leighton 1983; Chivers 2005.

123. Ungar 1995.

124. Ungar 1996b.

125. Marshall and Leighton 2006.

126. Gittins and Raemekers 1980; Raemakers 1984; Brockelman and Gittins 1984; Chivers 2005.

127. MacKinnon and Mackinnon 1977; Tuttle 1986, pp. 245–246.

128. Tuttle 1986, p. 70; Whitington 1992.

129. MacKinnon 1977; Chivers 1973, 1974, 1977a, 1977b; MacKinnon and MacKinnon 1977, 1978, 1980; Raemaekers 1978a, 1978b.

130. Chivers 1974; Caldecott 1980; Palombit 1997.

131. MacKinnon 1977; Rodman 1978.

132. Rijksen 1978.

133. Stanford 2008, pp. 92–95.

134. Suzuki and Nishihara 1992.

135. Stanford and Nkurunungi 2003.

136. Jones and Sabater Pí 1971.

137. Tutin and Fernandez 1993a.

138. Tutin et al. 1997.

139. Head et al. 2011.

140. Kay 1975, 1978; Hartman 1988; Ungar 2004.

141. Williamson et al. 1990.

142. Rogers et al. 1988.

143. Rogers et al. 1990.

144. Milton and Demment 1988; Lambert 1998; Caton 1999; Caton et al. 1999; Remis 2000; Remis and Dierenfeld 2004.

145. Lambert 1999; Tweheyo and Lye 2003.

146. Wrangham et al. 1994.

147. Remis 2000.

148. Rogers et al. 1990, p. 326.

149. Tutin and Fernandez 1985, 1992; Tutin, Fernandez et al. 1991.

150. Deblauwe and Janssens 2008.

151. Muroyama 1991.

152. Kuroda, Nishihara et al. 1996.

153. Kuroda 1992; Kuroda, Nishihara et al. 1996; Nishihara 1992, 1995.

154. Kuroda, Nishihara et al. 1996; Kuroda 1992.

155. Remis 2003.

156. Yamagiwa and Basebose 2009.

157. Yamagiwa, Maruhashi et al. 1996; Yamagiwa, Kaleme, and Yumoto 2008; Yamagiwa and Basabose 2006a, 2006b.

158. Yamagiwa and Basabose 2006a, 2006b.

159. Stanford and Nkurunungi 2003; Stanford 2008.

160. Stanford and Nkurunungi 2003; Goldsmith 2003; Stanford 2008.

161. Stanford and Nkurunungi 2003.

162. Rothman, Van Soest, and Pell 2006.

163. Stanford and Nkurunungi 2003; Stanford 2006.

164. Ungar 2007b.

165. Knott 2011.

166. Arjamaa and Vuorisalo 2010; Burger et al. 2007; Perry et al. 2007; Walker 2007; Wrangham and Conklin-Brittain 2003.

167. Teaford 2007.

168. Teaford and Ungar 2000; McGrew 2001; Laden and Wrangham 2005; Dominy et al. 2008.

169. Hylander 1988; Daegling and Grine 2007.

170. Robinson 1963; du Brul 1977; Kay 1985; Ryan and Johanson 1989; Grine et al. 1990; Teaford and Ungar 2002, 2004; Ungar and Scott 2009; van der Merwe 2008.

171. Wallace 1975; Grine 1981, 1986, Grine and Kay 1988; Ungar and Grine 1991; Ungar 2004, 2007c; Scott et al. 2005; Teaford 2007; King et al. 1999a, 1999b.

172. Puech and Albertini 1984; Puech et al. 1986; Ryan and Johanson 1989.

173. Puech 1984.

174. Walker 1981b; Jolly 1970; Szalay 1975b.

175. Cerling et al. 2011; Lee-Thorp 2011; Dominy 2012; Lee-Thorp et al. 2012.

176. Cerling et al. 2011.

177. El Zaatari et al. 2011; Richards and Trinkaus 2009.

178. Tuttle 2006d, Peters 2007; Reed and Rector 2007; Sept 1990, 1992a, 1994, 2001, 2007; Lambert 2007.

179. Denier and Epstein 1978; Lee-Thorp and Sponheimer 2006; Sponheimer and Lee-Thorp 2007.

180. Uno et al. 2011.

181. Sponheimer and Lee-Thorp 1999; Sponheimer et al. 2006, 2007; Kellner and Schoeninger 2007; Schoeninger 2007.

182. van der Merwe et al. 2003.

183. Sillen et al. 1995.

184. Sillen 1992; Lee-Thorp et al. 1994; Sponheimer et al. 2005; Lee-Thorp 2002.

185. Dahlberg 1963; Brace 1963, 1995; Brace et al. 1987; Wrangham et al. 1999; Wrangham and Conklin-Brittain 2003; Wrangham 2007, 2009; Brace 1962, 1967; Calcagno and Gibson 1991; Liu et al. 2013..

186. Lucas 2007, p. 33.

187. Gröning 2011.

188. Revedin et al. 2010.

189. Kennett and Winterhalder 2006; Denham et al. 2003; Denham and White 2007.

190. Larsen 2002, p. 19; Smith 1998.

191. Turnbull 1965; Lee and Devore 1968; Marlowe 2005, 2010; Kelly 1995; Gould 1969; Lee and DeVore 1976; Lee and Daly 1999.

192. Bowles 2011, p. 4760; Bocquet-Appel 2009; Gage and White 2009; Lambert 2009.

193. Hawkes et al. 1997.

194. Blurton-Jones et al. 1989.

195. Peccei 1995; Walker and Leakey 1993; Swisher et al. 1994; McHenry 1994a; Klein 1955; Trinkaus and Tompkins 1990.

196. Tuttle 1986, p. 126; Anderson 1998; Davidson et al. 2003.

197. McNamara et al. 2009.

198. Kano 1984; Williamson and Usongo 1996; Furuichi et al. 1997; Gonzalez-Kirchner 1997; Blom et al. 2001; Fenton et al. 2003; Matthews and Mathews 2004; Poulsen and Clark 2004; Kouakou et al. 2009.

199. A. J. Whitten 1982; T. Whitten 1982; Tilson and Tanaza 1982; Fruth 1995; Balwin et al. 1981, 1982; Koops et al. 2007; Bronslow et al. 2001; Furuichi and Hashimoto 2004a; Furuichi et al. 2001a, 2001b, 2001c; Basabose and Yamagiwa 2002; Yamagiwa 2001; Tutin et al. 1995; Remis 1993; Mehlman and Doran 2002.

200. A. J. Whitten 1982; T. Whitten 1982; Tilson and Tanaza 1982.

201. Phoonjampa et al. 2010.

202. A. J. Whitten 1982.

203. Tuttle 1986, pp. 126–128; Fan and Jiang 2008a; Reichard 1998.

204. Tuttle 1986, pp. 143–146; McGrew 2004; Videan 2006.

205. Watson 1999–2000; Fay et al. 1995; Rahm 1967; Schaller 1963, pp. 302–304; Boesch 1991a; Boesch and Boesch-Achermann 2000; Tsukahara 1993; Jenny and Zuberbühler 2005; Zuberbühler and Jenny 2003; Zuberbühler 2007; Robbins et al. 2004; D'Amour, et al. 2006; Rijksen 1978; van Schaik 2004, p. 73; Mehlman and Doran 2002.

206. Iwata and Ando 2007.

207. Tuttle 1986, pp. 128–132; Prasetyo et al. 2009.

208. Sugargito 1983.

209. Rayadin and Saitoh 2009; Setiawan et al. 1996.

210. Rayadin and Saitoh 2009.

211. Setiawan et al. 1996.

212. Van Schaik 2004.

213. Fruth 1995.

214. Kano 1982b.

215. Tuttle 1986, pp. 137–138.

216. Horn 1980; Kano 1983; MacKinnon 1976; Badrian and Badrian 1977; Fruth 1995; Fruth and Hohmann 1994.

217. Tuttle 1986, pp. 132–137.

218. Ogawa et al. 2007.

219. Lawick-Goodall 1968a.

220. Baldwin et al. 1981; Sept 1992b; Ogawa et al. 2007.

221. Tuttle 1986, pp. 132–137; Basabose and Yamagiwa 2002; Basabose 2002b.

222. Brownlow et al. 2001; Hirata et al. 1998; Koops et al. 2007.

223. Pruetz et al. 2008.

224. Pruetz 2001, 2007.

225. Brownlow et al. 2001; Reynolds 2005.

226. Furuchi et al. 2001a; Furuchi and Hashimoto 2004.

227. Furuchi et al. 2001a, 2001b.

228. Stanford and O'Malley 2008.

229. Basabose and Yamagiwa 2002.

230. Schaller 1963, p. 373.

231. Goodall 1978; Furuichi et al. 1997; Brugiere and Sakom 2001.

232. Rothman, Pell et al. 2006.

233. Remis 1993; Tutin et al. 1995.

234. Rothman, Pell et al. 2006; Iwata and Ando 2007.

235. Tuttle 1986, pp. 138–142.

236. Yamagiwa 2001.

237. Yamagiwa 2001; Yamagiwa and Kahekwa 2001.

238. Mehlman and Doran 2002.

239. Remis 1993.

240. Tutin et al. 1995.

241. Dupain et al. 2004.

242. Mehlman and Doran 2002; Lilly et al. 2002.

243. Huffman and Chapman 2009; Heeney et al. 2006; Nunn and Altizer 2006.

244. Fossey 1983; Goodall 1986; McGrew et al. 1989; Ashford et al. 1990, 1996, 2000; Landsoud-Soukate et al. 1995; Wallis and Lee 1999; Mudakikwa et al. 2001; Lilly et al. 2002; Muehlenbein 2005; Rothman, Bowman et al. 2006; Goldsmith et al. 2006; Isabirye-Basuta and Lwanga 2007; Leroy et al. 2004; Peeters 2004; Rouquet et al. 2005; Sharp et al. 2004; Vogel 2003, 2006. 2007; Wolfe et al. 2004; Lonsdorf et al. 2006; Köndgen et al. 2008; Bakuza and Nkwenguilila 2009.

245. Tuttle 1981b; Sabater Pí et al. 1997.

246. Gowlett et al. 1981; Clark and Harris 1985; Brain 1993a, 1993b; Brain and Sillent 1988; Rowlett 1990; Bellomo 1991, 1993, 1994; Bellomo and Kean 1997; Black 1931–32; Binford and Ho 1985; Binford and Stone 1986; James 1989, 1996; Benditt 1989; Pécsi 1990; Horváth 1990; Dobosi 1990; Osmond 1990; Wu and Lin 1983; Weiner et al. 1998, 2000; Wuethrich 1998; Patel 1995; Goren-Inbar et al. 2000, 2004; Goldberg et al. 2001; Boaz et al. 2004; Alperson-Afil and Goren-Inbar 2006; Alperson-Afil et al. 2009; Balter 2009c; Stiner et al. 2009; Berna et al. 2012.

247. Roebroeks and Villa 2011.

248. Tuttle 1981b.

249. Venkataraman et al. 2013.

8. HUNTING APES AND MUTUALISM

1. Goodall 1963b; Lawick-Goodall, 1968a; Teleki 1973a, 1973b; Alp, 1993; Alp and Kitchener 1993; Anderson et al. 1983; Boesch 1978; Busse 1977; Bygott 1972; Ghiglieri 1984; Itani 1982; Kawabe 1966; Kawanaka 1981, 1982b; McGrew 1983; Newton-Fisher et al. 2002; Nishida et al. 1979; Norikoshi 1982, 1983; Hashimoto et al. 2000; Pruetz and Bertolani 2007; Reynolds 2005; Roach 2008; Sugiyama 1981; Suzuki 1971, 1975; Takahata et al. 1984; Wrangham 1974a.

2. Rose 1997; Surbeck and Hohmann 2008.

3. Goodal 1963b; Lawick-Goodall 1968a.

4. Ihobe 1997a.

5. Wrangham 1974b.

6. Lawick-Goodall 1968a.

7. Teleki 1973a.

8. Lawick-Goodall 1967, 1968a, 1971.

9. Muller and Mitani 2005.

10. Busse 1978.

11. Takahata et al. 1984.

12. Boesch and Boesch 1989.

13. Tomasello et al. 2005; Tomasello 2008, pp. 173–185.

14. Boesch 2002; Tomasello 2008, p. 174.

15. Melis 2006.

16. Goodall 1986; Wrangham and Riss 1990; Stanford et al. 1994a, 1994b; Gilby et al. 2006; Boesch and Boesch 1989; Mitani and Watts 1999; Watts and Mitani 2002.

17. Goodall 1986, pp. 304–312.

18. Kawabe 1966.

19. Nishida et al. 1979; Fujimoto and Shimada 2008.

20. Takahata et al. 1984; Uehara et al. 1992.

21. Huffman and Kalunde 1993.

22. Pruetz and Bertolani 2007; Roach 2008.

23. Pruetz and Bertolani 2007.

24. Pruetz and Marshack 2009.

25. Sugiyama 1981, 1989.

26. Sugiyama 1989.

27. Boesch and Boesch 1989, 1994c.

28. Boesch and Boesch 1989; Stanford et al. 1994a.

29. Stanford et al. 1994a.

30. Watts and Mitani 2002.

31. Stanford 1998; Boesch and Boesch-Aschermann 2000, p. 160.

32. Watts and Mitani 2002; Mitani and Watts 1999, 2001.

33. Teelen 2007.

34. Boesch and Boesch 2000, pp. 182–185.

35. Goodall 1965, 1968, 1986, pp. 282–283; Thomas 1961.

36. Kamenya 2000.

37. Busse 1977.

38. Wrangham and Riss 1990; Stanford et al. 1994a, 1994b; Stanford 1995a, 1995b, 1996, 1998.

39. Goodall 1986, p. 269; Stanford 1998a; Uehara 1992; Hosaka et al. 2001; Boesch 1994a, 1994b; Boesch and Boesch 1989; Watts and Mitani 2002.

40. Krüger et al. 1998.

41. Busse 1978; Goodall 1986, p. 269.

42. Nakamura 1997.

43. Teleki 1973a; McGrew 1983; Tuttle 1986; Wrangham 1974a; Wrangham and Riss 1990.

44. Takahata et al. 1994; Teleki 1973a; Goodall 1986; Stanford 1998a; Boesch 1994b; Boesch and Boesch-Achermann 2000.

45. Tuttle 1986, p. 118.

46. Teleki 1973a; Goodall 1986, pp. 296–298.

47. Teleki 1973a, 1973b; Kortlandt, Beck, and Vogel in Tuttle, 1975c, pp. 301–304.

48. Wrangham 1974b; Teleki 1973a.

49. Feistner and McGrew 1989.

50. Thomasello et al. 2005; Silk et al. 2005; Jensen et al. 2006.

51. de Waal 1997.

52. Goodall 1986; Boesch and Boesch-Achermann 2000; Reynolds 2005; Fahy et al. 2013.

53. Newton-Fisher 1999b.

54. Nishida et al. 1992; Watts and Mitani 2002.

55. Stanford et al. 1994a, 1994b; Stanford 1995, 1996, 1998.

56. Goodall 1986.

57. Nishida et al. 1992.

58. Teleki 1973a; McGrew 1975.

59. Gilby 2001; Wrangham 1997.

60. Takahata et al. 1984.

61. Kawanaka 1982b.

62. Nishida et al. 1992; Boesch 1994b; Boesch and Boesch-Achermann 2000.

63. Stanford 1995, 1998.

64. Stanford 1995, 1998, p. 204.

65. Gilby et al. 2006.

66. Watts and Mitani 2002; Mitani and Watts 1999, 2001.

67. Mitani and Watts 2001.

68. Gomes and Boesch 2009, p. 1.

69. Goodall 1986; Muller et al. 1995; Boesch-Achermann 2000; Watts 2008.

70. Ragir et al. 2000.

71. Goodall 1986, p. 292.

72. Goodall 1986, pp. 293–295.

73. Muller et al. 1995.

74. Hasegawa et al. 1983; Nishida 1994.

75. Boesch and Boesch-Achermann 2000, p. 170.

76. Reynolds 2005, p. 74; Watts 2008.

77. Arens 1979; T.D.White 1992; Gibbons 1997; Turner and Turner 1999; Petrinovich 2000; Pennisi 2003; Travis-Henikoff 2008.

78. Newton-Fisher 1999c; Goodall 1986; Pusey et al. 2008; Hamai et al. 1992; Nishida et al. 1979; Nishida and Kawanaka 1985; Nishida 2008; Watts and Mitani 2000; Watts et al. 2002; Sherrow and Amsler 2007.

79. Suzuki 1971.

80. Newton-Fisher 1999c.

81. Bygott 1972.

82. Goodall 1986, pp. 283–285; Tuttle 1986, pp. 122–123.

83. Goodall 1979; Tuttle 1986, p. 123.

84. Goodall 1979, p. 619.

85. Goodall 1986, pp. 503–519; Nishida et al. 1985; Wrangham and Peterson 1996; Nishida 1996; Fawcett and Muhumuza 2000; Wilson et al. 2004; Watts et al. 2006.

86. Nishida and Hiraiwa-Hasegawa 1985.

87. Pickering 1997.

88. Nishida and Kawanaka 1985; Hamai et al. 1992.

89. Hamai et al. 1992.

90. Nishida 2008.

91. Sakamaki et al. 2001.

92. Kutsukake and Matsusaka 2002.

93. Fowler and Hohmann 2010.

94. Dellatore et al. 2009.

95. Fossey 1981, 1983, 1984.

96. Hohmann and Fruth 2008; Surbeck and Hohhann 2008.

97. Sabater Pí et al. 1993.

98. Hirata et al. 2001.

99. Pi and Veà 1994.

100. Ihobe 1990.

101. Ihobe 1997b.

102. Surbeck and Hohmann 2008.

103. Hohmann and Fruth 2008.

104. Oelze et al. 2011.

105. Ihobe 1992b.

106. Hohmann and Fruth 1993.

107. White 1994.

108. Fruth and Hohmann 2002.

109. Hare et al. 2007; Melis et al. 2006.

110. Kano 1980; Kuroda 1980.

111. Kano 1980.

112. Kuroda 1984.

113. Hohmann and Fruth 1996.

114. Wobber et al. 2010, p. 12457.

115. Fan and Jiang 2009.

116. Jaeggi and Van Schaik 2011.

117. L. Leakey 1960a; Washburn and Lancaster 1968; M. D. Leakey 1971; Isaac 1971; Isaac and Isaac 1997; Brain 1981, 1993c; Ragir 2000; Ragir et al. 2000; Pickering 2001; Shipman 2002; Schoeninger et al. 2001; Shea 2007; Bunn 2007; Blumenschine and Pobiner 2007.

118. Bunn, 1981, 2007; Bunn and Kroll 1986; Bunn et al. 1997; Potts and Shipman 1981; de Heinzelin et al. 1999; Pickering and Domínguez-Rodrigo 2006; Blumenschine and Probiner 2007; Probiner et al. 2008; Shea 2007; Braun et al. 2010; Steele 2010.

119. Keeley 1977; Keeley and Newcomer 1977; Keeley and Toth 1981.

120. Schaller and Lowther 1969; Binford 1977, 1981, 1983, 1985; Binford and Ho, 1985; Binford and Stone 1986; Blumenschine 1991; Stiner 1991a; Tappen 2001; Vandenburgh 2001; Boaz et al. 2004.

121. Leakey 1952; Isaac 1977; Shipman et al. 1981; Potts 1989; Potts et al. 1999, 2004; Sikes et al. 1999; Stiner 1990, 1991b, 1993, 1994; Suzuki and Takai 1970; Trinkaus 1983a; Trinkaus and Shipman 1993; Stringer and Gamble 1993; Mellars 1996; Shreeve 1995.

122. Jolly 1972; Leakey and Leakey 1973; Isaac 1977; Shipman et al. 1981.

123. Shipman et al. 1981; M. G. Leakey 1977; Potts et al. 1999.

124. Sponheimer and Lee-Thorp 2007; Henry et al. 2011; Stiner 1994, Mellars 1996; Akazawa et al. 1998; Speth and Tchernov 2001.

125. Thieme 1997; Thieme and Veil 1985; Jacob-Friesen 1959; Oakley et al. 1977; Mellars 1996.

126. Thieme 1997.

127. Stiner 1994, 2001.

9. HANDY APES

1. Tuttle, 1986; Harlacker 2006; Gibson and Ingold 1993; Wynn 1979, 1981; McPherron 2000; McPherron 2013; Caruana et al. 2013.

2. Webb and Fabiny 2009.

3. Heinzelin et al. 1999; Asfaw et al. 1999.

4. Semaw 2006; Semaw et al. 2003.

5. McPherron et al. 2010; Domínguez-Rodrigo et al. 2010, 2011.

6. Panger et al. 2002.

7. Tuttle 1986, pp. 148–151; Call 2013; Boesch 2013; Byrne et al. 2013.

8. Köhler 1925.

9. Köhler 1925; Tuttle 1986, p. 149.

10. Tuttle 1986, pp. 149–151; Menzel 1973a; Candland 1987; Sumita et al. 1985; Greenfield et al. 2000; Paquette 1992, 1994; Call and Tomasello 1994a; Gold 2002; Mulcahy and Call 2006; Van Elsacker and Walraven 1994; Nakamichi 1998, 1999; Mitchell 1999; Russon and Galdikas 1995; Russon 1999; Boysen et al. 1999; Parker et al. 1999; Takeshita and van Hooff 1996, 2001; Goustard 1986; Limongelli et al. 1995; Visalberghi et al. 1995; Visalberghi and Tomasello 1998; Brent 1995; Povinelli et al. 2001; Paquette 1992; Balter 2009b; Rijksen 1978; van Schaik 2004, p. 117; Herrmann et al. 2008; van Casteren et al. 2012; Meulman and van Schaik 2013.

11. Rumbaugh et al. 1972.

12. Lethmate 1977, 1979, 1982.

13. Wright 1978.

14. Wright 1972; Toth et al. 1993, 2006; Savage-Rumbaugh 1994; Schick et al. 1999; Savage-Rumbaugh and Lewin 2004; Savage-Rumbaugh et al. 2007.

15. Toth et al. 1993.

16. Schick et al. 1999.

17. Savage-Rumbaugh and Fields 2006.

18. Toth et al. 2006.

19. Wynn and McGrew 1989.

20. Toth et al. 2006; Ono 1995; Roffman et al. 2012.

21. Tuttle, p. 151; Wittiger and Sunderland-Groves 2007; van Schaik 2004, pp. 117–118.

22. Kano 1992; Ingmasson 1996; Takeshita and Waldraven 1996; van Schaik et al. 1999.

23. Goodall 1963a, 1963b, 1986; Lawick-Goodall 1968a; Tuttle 1986, pp. 152–156.

24. Goodall 1964.

25. Lawick-Goodall 1970.

26. McGrew 1974.

27. McGrew 1974; Tuttle 1986, pp. 154–155.

28. Lawick-Goodall 1965, 1970.

29. Lanjouw 2002.

30. Lawick-Goodall 1973a; 1975; Boesch and Boesch-Achermann 1991.

31. McGrew 1977; Lawick-Goodall 1975; Tuttle 1986, pp. 155–156.

32. Boesch 1993, pp. 176–177.

33. Whiten et al 2005.

34. McGrew 1979; Lonsdorf 2004; Lonsdorf et al. 2005; Uehara 1984.

35. Nishida and Hiraiwa 1982; Tuttle 1986, p. 158.

36. Kortlandt and Holzhaus 1987; Nishida and Nakamura 1993; Sugiyama 1993, 1994, 1995; Fay and Carroll 1994; Yamakoshi and Sugiyama 1995; Yamakoshi 1998; Matsuzawa and Yamakoshi 1996; McGrew 1992b, 2007; McGrew et al. 1997; Boesch 1995; Boesch and Boesch 1990, 1993; Tonooka et al. 1994; Alp 1997; Whiten et al. 1999, 2001; Humle and Matsuzawa 2002; Sherrow 2005; Hicks et al. 2005; Deblauwe 2006; Deblauwe et al. 2006; Hernandez-Auilar et al. 2007; Schöning et al. 2008; Yamamoto et al. 2008; Gruber et al. 2009.

37. Reynolds 2005.

38. Tuttle 1986.

39. Suzuki et al. 1995; Sugiyama 1997; Carvalho et al. 2009.

40. Brewer and McGrew 1990; Sanz and Morgan 2007, 2009; Sanz et al. 2007; Boesch et al. 2009.

41. Sanz and Morgan 2009.

42. Fowler et al. 2011.

43. Yamagiwa et al. 1988; Nishimura et al. 2003.

44. Tuttle 1986, pp. 160–165; Whitesides 1985; Sept and Brooks 1994.

45. Inoue-Nakamura and Matsuzawa 1997; Matsuzawa 1994.

46. Boesch and Boesch 1981, 1982, 1984a, 1984b, 1984c; Boesch and Boesch-Achermann 2000; Mercader et al. 2007.

47. Boesch and Boesch 1982, 1984a.

48. Boesch and Boesch 1982.

49. Boesch and Boesch 1981.

50. Mercader et al. 2002.

51. Boesch and Boesch 1981, 1984b.

52. Tuttle 1986, p. 165.

53. Tuttle 1986, p. 165; Nishida 1973; McGrew 1979; Lawick-Goodall 1970; Huffman and Kalunde 1993; Gruber et al. 2010.

54. Lawick-Goodall 1970; Tuttle 1986, p. 165.

55. Sarringhaus et al. 2005.

56. Plooij 1978a; Tuttle 1986, p. 165.

57. Pruetz and Bertolani 2007.

58. MacKinnon 1974; Rogers and Kaplan 1994; Knott 1999; Peters 2001.

59. Rijksen 1978.

60. van Schaik 2004; van Schaik et al. 1996, 1999, 2003a, 2003b; van Schaik and Knott 2001; Fox et al. 1999; Meulman and van Schaik 2013.

61. Byrne 1995, pp. 187–188; Fox et al. 1999.

62. Fox et al. 1999.

63. Fox et al. 1999; van Schaik et al. 1999.

64. Carroll 1986, 1988, 1990; Cipolletta et al. 2007.

65. Breuer et al. 2005.

66. Beck 1980.

67. Shumaker et al. 2011.

68. Beck 1975, 1980.

69. Hohmann 1988.

70. Malaivjitnond et al. 2007.

71. Hamilton and Tilson 1985.

72. Oyen 1979.

73. McGrew 1993.

74. Anderson 1990; Anderson and Henneman 1994; Boinski 1988; Boinski et al. 2001, 2003; Canale et al. 2009; Dindo et al. 2008; Jalles-Filho 1995; Fernandes 1991; Fragaszy and Adams-Curtis 1991; Fragaszy et al. 2004; Fragaszy, Izar et al. 2004a, pp. 173–201; Langguth and Alonso 1997; Mannu and Ottoni 2009; McGrew and Marchant 1997a; Moura and Lee 2004; Ottoni and Mannu 2001; Panger 1998; Perry and Manson 2003; Phillips 1998; Schrauf et al. 2008; Urbani 1999; Visalberghi and Néel 2003; Waga et al. 2006; Westergaard 1994, 1995; Westergaard and Fragaszy 1985, 1987; Westergaard et al. 1995, 1996, 1997; Westergaard, Lundquist et al. 1998; Westergaard and Soumi 1993, 1994a, 1994b, 1994c, 1994d, 1994e; 1995a, 1995b, 1995c, 1995d, 1995e, 1995f.

75. Visalberghi et al. 1995; Visalberghi and Fragaszy 2000; Visalberghi and Fragaszy 2013; Limongelli et al. 1995; Visalberghi 1990, 1993a, 1993b; Visalberghi and Trinca 1989; Fragaszy and Visalberghi 1989; Visalberghi and Fragaszy 1990; Visalberghi and Limongelli 1994, 1996; Visalberghi and Tomasello 1998; Vauclair and Anderson 1994.

76. Fragaszy, Izar et al. 2004b; Visalberghi et al. 2007, 2008, 2009; Liu et al. 2009.

77. Visalberghi et al. 2007.

78. Visalberghi et al. 2008.

79. Ferreira et al. 2010.

80. Liu 2009; Fragaszy, Izar et al. 2004b.

81. Marzke 1992; Marzke et al. 1992, 1998.

82. Napier 1956, 1961a, 1961b, 1965, 1966, 1980, pp. 67–83, 1993, pp. 55–71; Marzke 1983, 1986, 1997; Marzke and Shackley 1986.

83. Napier 1980, pp. 77–79; 1993, pp. 65–66.

84. Landsmeer 1986, 1987.

85. Darian-Smith et al. 1996.

86. Smith 1995, 2000; Shrewsbury et al. 2003.

87. Hoffman et al. 2004.

88. Tuttle 1970, 1972; Christel 1993; Marzke and Wullstein 1996; Jones-Engel and Bard 1996; Christel et al. 1998.

89. Napier 1961b; Bishop 1964; Fragaszy et al. 1989.

90. Tuttle 1969b, 1970, 1972.

91. Tuttle 1969b, 1972.

92. Marzke et al. 1999.

93. Tuttle 1969b; Landsmeer 1984.

94. Tuttle 1992; Susman 1998.

95. Lewis 1977.

96. Napier 1955.

97. Napier 1952.

98. Landsmeer 1955; Lewis 1977.

99. Masquelet et al. 1986.

100. Landsmeer and Long 1965; Long et al. 1970.

101. Napier and Davis 1959; Beard et al. 1986; Begun et al. 1994; Rose 1993, 1994; Nakatsukasa, Kunimatsu et al. 2003, 2007; Ishida et al. 2004; Walker, Teaford, and Leakey 1986; Walker and Shipman 2005.

102. Zapfe 1970; Tuttle 1972; Moyà-Solà and Köhler 1996; Moyà-Solà et al. 1999, 2004, 2005; Almécija et al. 2007; Lovejoy 2007; Begun 1988, 1993; Begun and Ward 2005; Morbeck 1983; Sarmiento 1987, 1988; Harrison 1991b; Meldrum and Pan 1988; Wu et al. 1986; Almécija et al. 2012.

103. Lovejoy, Simpson et al. 2009; Lovejoy, Latimer et al. 2009a, 2009b.

104. Bush et al. 1982.

105. Tuttle 1981b; Susman et al. 1984.

106. Marzke 1983, 1986, 1997; Marzke and Marzke 1987; Marzke and Shackley 1986; Marzke et al. 1994; McHenry 1983; Smith 1995; Tocheri et al. 2003.

107. Tuttle 1981b.

108. Clarke 1998, 1999, 2000a, 2000b; Pickering et al. 2004; Berger et al. 2002; Walker et al. 2006; Broom and Robinson 1949b; Robinson 1972a; Napier 1959; Susman 1988a, 1988b; 1989, 1991, 1993; Rightmire 1972; Trinkaus and Long 1990.

109. Ricklan 1987.

110. Green and Gordon 2008.

111. Clarke 1998, 1999, 2000b.

112. Clarke 2000a; Pickering et al. 2004; Berger et al. 2002; Walker et al. 2006.

113. Clarke and Tobias 1995.

114. Clarke 2000b.

115. Grine 1988b, 1993b.

116. Susman 1988a, 1988b, 1989, 1993.

117. Trinkaus and Long 1990.

118. Susman 1991.

119. Susman 1988b; Brain and Shipman 1993; Clark 1993.

120. Susman 1988a.

121. Leakey 1959, 1961; Leakey et al. 1964.

122. Robinson 1972.

123. Napier 1962a, 1962b; Tuttle 1967.

124. Tuttle 1981b; Susman and Stern 1982; Stern and Susman 1983; Susman et al. 1984, 1985; Senut and Tardieu 1985; McHenry 1986.

125. Tuttle 1967, Trinkaus 1989a.

126. Tuttle 1967, 1992; Susman 1979b; Susman and Creel 1979; Susman and Stern 1979, 1982.

127. Bush 1982; Johanson et al. 1982; Lewis 1989; Marzke 1983, 1986, 1997; Marzke and Marzke 1987; Marzke and Shackley 1986; Shrewsbury and Johnson 1983; Tuttle 1981; Susman 1994.

128. Leakey et al. 1964; Napier 1962b; Tocheri et al. 2003; Susman and Creel 1979.

129. Tuttle 1992.

130. Musgrave 1969, 1971, 1973, 1977; Trinkaus 1983a, 1986; 1989b; Niewoehner et al. 1997.

131. McBrearty et al. 1996; McBrearty and Brooks 2000; Bar-Yosef and Kuhn 1999; Tyron and McBrearty 2002; Beyene et al. 2013.

10. MENTAL APES

1. Eisenberg 1981; Martin 1990.
2. van Schaik and Deaner 2003; Gibson et al. 2001, p. 79; Rumbaugh et al. 1996; Beran et al. 1999; Falk 2001; Gilissen 2001; Krubitzer and Kahn 2004.
3. Premack and Premack 2003, pp. 6, 17–36; Barrett 2012.
4. Premack and Premack 2003, p. 18.
5. Premack and Premack 2003, p. 19; Chomsky 1965, 1980; Fodor 1983; Sperber 1994.
6. Bekoff et al. 2002; Dehaene et al. 2005; Fragaszy and Perry 2003; Gibson and Ingold 1993; Gómez 2004; Hauser 2000a; Hirschfeld and Gelman 1994; Maestripieri 2003; Matsuzawa 2001; Matsuzawa et al. 2006; Parker et al. 1990, 2000; Russon et al. 1996; Savage-Rumbaugh et al. 1998; Tomasello 1999a; Tomasello and Call 1997; de Waal and Tyack 2003; Washburn 2007a.
7. Jerison 1973; Harvey et al. 1987; Martin 1990.
8. Passingham 1982.
9. Ho et al. 1980; Akney 1992; Falk et al. 1999; Falk 2001.
10. Passingham 1979, 1982; Semendeferi and Damasio 2000; Holloway 2002.
11. Allen et al. 2005; Baillieux et al. 2008; Manni and Petrosini 2004; Ramnani et al. 2006; Schmahmann and Pandya 1997; Stoodley and Schmahmann 2009; Beaton and Mariën 2010; Murdoch 2010.
12. Rilling and Insel 2003; MacLeod et al. 1998; Cantalupo and Hopkins 2010.
13. MacLeod et al. 2003; Coffman et al. 2011.
14. Cantalupo and Hopkins 2010.
15. Preuss 2000b, 2004; Semendeferi et al. 2001.
16. Matano 2001.
17. Koechlin and Hyafil 2007.
18. Seidler et al. 2002; Hazeltine and Ivry 2002.
19. Stout et al. 2000; Stout and Chaminade 2012.
20. Manni and Petrosini 2004; Stoodley and Schmahmann 2009; Murdoch 2010.
21. Rilling 2006, p. 65.
22. Shaw et al. 2006; Passingham 2006; Miller 2006.
23. Sakai 2005.
24. Reiss et al. 1996.
25. Gur et al. 1999.
26. Amunts et al. 2000.
27. Amunts et al. 1996.
28. Allen et al. 2002.
29. Haier et al. 2005; Bell et al. 2006; Boghi et al. 2006; Cela-Conda et al. 2009.
30. Sommer et al. 2004.
31. Catani et al. 2007; Bishop 2013.
32. Aziz et al. 1995.
33. Aziz et al. 2005, p. 16496.
34. Tóth 1965; Raz et al. 1998; Cantalupo et al. 2008.
35. Posthuma et al. 2009, p. 99.

36. Posthuma et al. 2009, p. 112.

37. Zihlman et al. 2008.

38. Schultz 1969; Cramer 1977; Zihlman et al. 2008.

39. Schultz 1962; Tuttle 1986, pp. 172–174.

40. Schultz 1969; Cramer 1977; Zihlman et al. 2008; William Hopkins, personal communication, May 15, 2011.

41. Shariff 1953.

42. Chenn 2002; Chenn and Welsh 2002.

43. Gillesen 2001; Aboitiz et al. 1992; Ringo et al. 1994; Anderson 1999.

44. Gilissen 2001, p. 187.

45. Preuss 2001.

46. Rakic 2008; Kass 2012.

47. Rockel et al. 1980; Preuss 2001.

48. Rockel et al. 1980.

49. Preuss 2001; Herculano-Housel et al. 2008; Rakic 2008.

50. Kaas 1987; Preuss and Kaas 1999.

51. Holloway 1968, 1976, 1979; Preuss 2004, 2007a, 2007b; Tuttle 1986, p. 175; Herculano-Houzel 2012.

52. Sherwood et al. 2004.

53. Rizzolatti, Fadiga et al. 1996a, b; Gallese et al. 1996; Kohler et al. 2002; Gallese 2003; Rizzolatti and Craighero 2004; Rizzolatti and Buccino 2005; Fadiga et al. 1995; Nelissen et al. 2005; Rizzolatti and Sinigaglia 2006.

54. Gallese et al. 1996, 2004: Gallese and Goldman 1998; Rizzolatti and Craighero 2004; Gallese 2003; Gazzola et al. 2006; Iacoboni et al. 1999, 2005; Ferrari and Fogassi 2012; Iacoboni 2012; Singer and Hein 2012; Preston et al. 2002, 2003; Platek et al. 2005; Xu et al. 2009; Mathur et al. 2010.

55. Lingnau et al. 2009.

56. Brass and Rüschemeyer 2010, p. 140.

57. Preuss 2000a, b; 2001; 2004.

58. Press 2004.

59. Rilling and Insel 1999a, 1999b; Rilling and Seligman 2002; Rilling et al. 2007, 2008; Allman et al. 2005; Cáceres et al. 2003, 2007; Enard, Khaitovich et al. 2002; Enard, Przewoski et al. 2002; Evans et al. 2005; Hamer 2002; Mekel-Bobrov et al. 2005; Nimchnsky et al. 1999; Oldham et al. 2006; Preuss et al. 2004; Preuss 2009; Raghanti et al. 2008a, b; Varki and Altheide 2005.

60. Konopka et al. 2009.

61. Konopka et al. 2009, p. 217.

62. Lai et al. 2001; Enard et al. 2002; Pinker 2001; Fisher and Ridley 2013.

63. Rougier et al. 2005; Bault et al. 2011.

64. Passingham 1982, p. 119.

65. Bradshaw 1991; Bradshaw and Rogers 1993; Ward and Hopkins 1993; Hellige 1993; Fagot and Vauclair 1988, 1991, 1993; MacNeilage 1992, 1998c; Elliot and Roy 1996; Rogers 1995; Rogers and Andrew 2002; Hook 2004; Moorman et al. 2012.

66. Geschwind and Galaburda 1984.

67. LeMay and Culebras 1972.

68. Sun et al. 2005; LeMay 1976.

69. Crinion et al. 2006.

70. Sakai 2005.

71. Hopkins et al. 2003.

72. Grodzinsky and Santi 2008.

73. Corina et al. 1992; Corina and McBurney 2001; Grossi et al. 1996.

74. Baillieux et al. 2008.

75. Willems and Hagoort 2007, 2009; Hagoort 2005; Hagoort et al. 2009; Grewe et al. 2005; Gentilucci et al. 2006; Caplan 2006; Fadiga and Craighero 2006; Skipper et al. 2007.

76. Geschwind and Levitsky 1968; Gilissen 2001.

77. Steinmetz and Galaburda 1991; Habib et al. 1995; Gilissen 2001.

78. Schlaug et al. 1995.

79. Gilissen 2001; Holloway 2009.

80. Holloway and Coste-Lareymondie 1982.

81. Holloway et al. 2004.

82. Hopkins and Marino 2000.

83. Hopkins and Pilcher 2001; Hopkins and Cantalupo 2004.

84. Hopkins et al. 2000; Yeni-Komshian and Benson 1976.

85. Cantalupo et al. 2008.

86. Phillips and Hopkins 2007; Cantalupo et al. 2008; Snyder et al. 1995; Szabó et al. 2003.

87. McGrew and Marchant 1997b, 1996; Marchant and McGrew 1996.

88. McGrew and Marchant 1973; 1997b; Marchant and McGrew 1991.

89. McGrew and Marchant 1997b; Palmer 2002.

90. Lonsdorf and Hopkins 2005; Humle and Matsuzawa 2009.

91. Lonsdorf and Hopkins 2005.

92. Olson et al. 1990; Annett and Annett 1991; Morris et al. 1993; Bard et al. 1999; Boesch 1991b; Byrne 1999; Byrne and Byrne 1991; Byrne and Corp 2003; Colell et al. 1995a, b; Corp and Byrne 2004; Christel 1994; Fagot and Vauclair 1988; Finch 1941; Fletcher 2006; Hopkins 1993, 1994, 1995, 1996, 1998; Hopkins et al. 1989, 2004; Hopkins and Bard 1993; Hopkins and Bennett 1994; Hopkins and de Waal 1995; Hopkins and Fernández-Carriba 2000; Hopkins and Pearson 2000; Hopkins and Wesley 2002; Hopkins, Cantalupo et al. 2002; Hopkins and Cantero 2003; Hopkins and Rabinowitz 1997; Hopkins and Leavens 1998; Hopkins, Stoinski et al. 2003; Hopkins and Russell 2004; Hopkins, Russell, Cantalupo et al. 2005; Hopkins, Russell, Freeman et al. 2005; Hopkins, Russell, Hook et al. 2005; Hopkins, Russell, Hostetter et al. 2005; Hopkins, Russell, Remkus et al. 2007; Hopkins, Phillips et al. 2011; Llorente et al. 2009; McGrew and Marchant 1992; McGrew et al. 1999; Morange 1994; Nishida and Hiraiwa 1982; Olson et al. 1990; Rogers and Kaplan 1996; Shafer 1993, 1997; Stafford et al. 1990; Steiner 1990; Tonooka and Matsuzawa 1995; Vauclair and Fagot 1993; Ward and Hopkins 1993; Llorente et al. 2011; Chapelain et al. 2011; Morino 2011; Bogart et al. 2012.

93. Meguerditchian et al. 2010.

94. Byrne and Corp 2003.

95. Morino 2011.

96. Peters and Rogers 2008; Hopkins et al. 2011.

97. Hopkins 1993; Olson et al. 1990.

98. Hopkins, Bennett et al. 1993; De Vleeschouwer et al. 1995.

99. Hopkins, Bard et al. 1993.

100. Westergaard, Kuhn, and Soumi 1998.

101. Hopkins 2008.

102. Sarringhaus, Stock et al. 2005.

103. Toth 1985; Schick and Toth 1993.

104. Ghazanfar and Hauser 1999; Falk et al. 1990.

105. Dewson 1978.

106. Heffner and Heffner 1984.

107. Hopkins et al. 1998; Gannon et al. 1998; Pilcher et al. 2001; Gilissen 2001; Cantalupo et al. 2003; Hopkins et al. 2009.

108. Buxhoeveden et al. 2001.

109. Hopkins et al. 1998.

110. Cantalupo and Hopkins 2001.

111. Taglialatela et al. 2008.

112. Sherwood et al. 2003, p. 276; Romanski 2012.

113. Gannon et al. 2005; Jäncke et al. 1994; Steinmetz et al. 1990.

114. Gilissen 2001.

115. Jäncke et al. 1994.

116. Heiervang et al. 2000; Gannon et al. 2005.

117. Savage-Rumbaugh et al. 1986.

118. Hopkins and Fowler 1998.

119. Campbell 1982; Borod et al. 1997, 1998; Morris and Hopkins 1993; Fernández-Caribba et al. 2002; Hopkins and Fernández-Caribba 2002; Hauser 1993a; Vermeire et al. 1998.

120. Davidson 1992.

121. Depue et al. 2007.

122. Frith 2007.

123. Zahn et al. 2007; Bozeat et al. 2000; Garrard and Carroll 2006; Jefferies and Lambon Ralph 2006; McClelland and Rogers 2003; Rogers, Lambon, Ralph, et al. 2004; Spitsnya et al. 2006; Bozeat et al. 2000; Garrard and Carroll 2006; Jefferies and Lambon Ralph 2006; McClelland and Rogers 2003; Spitsnya et al. 2006; Liu et al. 2004.

124. Murrill and Wallace 1971; Conroy et al. 1998.

125. Hirschler 1942; Radinsky 1972; Tuttle 1986, p. 174; Holloway 2009.

126. McHenry 1994; Preuss 2008.

127. Bruner and Holloway 2010; Wu et al. 2011.

128. Wrangham 2009; Wrangham et al. 1999; Pennisi 1999.

129. Parker and Gibson 1977, 1979; Clutton-Brock and Harvey 1980; Milton 1981; Jolly 1966; Humphrey 1976; Alexander 1979; Beck 1982; Kummer 1982; King 1986; Pennisi 2006b; Dunbar 2003a; Dunbar and Shultz 2007; Gardner 1983, 1993.

130. Uddin et al. 2004, p. 2957.

131. Martin 1981, 1983; Aiello and Wheeler 1995; Fish and Lockwood 2003.

132. Martin 1981, 1983; Parker 1990; Gibbons 1998.

133. Brace 1962; Wrangham 2009.

134. Köhler 1959; Ladygina-Kohts 2002; Dewsbury 2006; Dukelow 1995.

135. Köhler 1925.

136. Köhler 1959, pp. 169–170.

137. Yerkes and Yerkes 1929, p. 575.

138. Martin and Caro 1985; Bekoff and Byers 1998; Gomez 2004; Pellegrini and Smith 2005.

139. Yerkes 1943, pp. 289–301.

140. Yerkes and Yerkes 1929, p. 376.

141. Harlow 1949; Rumbaugh et al. 2007, p. 978.

142. Schusterman 1962.

143. Tuttle 1986, pp. 181–182; Wolfe 1979, pp. 154–156; Cassidy and Davy 2003.

144. Tuttle 1986, p. 182–183.

145. van Schaik, Preuschoft, and Watts 2004, p. 190.

146. Rumbaugh 1969, 1970, 1971a, 1971b, 1974, 1975, 1993; Rumbaugh and Steinmetz 1971; Rumbaugh and Gill 1973; Rumbaugh and Pate 1984a, 1984b; Davenport et al. 1973; Essock and Rumbaugh 1978; Gill and Rumbaugh 1974a; Meador et al. 1987; Pate 2007; Tuttle 1986, p. 183–185.

147. Rumbaugh 1969.

148. Rumbaugh 1974, 1975; Rumbaugh and Gill 1973.

149. Rumbaugh 1971a; Tuttle 1986, p. 184.

150. Gill and Rumbaugh 1974a.

151. Davenport et al. 1973.

152. Amici et al. 2010.

153. Rumbaugh 1971b; 1985.

154. Patterson and Tzeng 1979.

155. Essock-Vitale 1978; Tuttle 1986, p. 186.

156. Gillan et al. 1981.

157. Premack and Premack 1983, p. 128.

158. Gillan 1981; Tuttle 1986, pp. 186–187.

159. Premack 1975; Premack and Premack 1983, p. 105.

160. Premack and Premack 1983; Tuttle 1986, p. 187.

161. Premack and Woodruff 1978a; Tuttle 1986, p. 187–190.

162. Savage-Rumbaugh 1983, p. 302.

163. Muncer 1983a; Piaget 1973.

164. Woodruff et al. 1978.

165. Schmelz et al. 2011.

166. Woodruff and Premack 1981; Premack and Premack 1983.

167. Ferster and Hammer 1966; Hayes and Nissen 1971; Dooley and Gill 1977a, 1977b; Brown et al. 1978; Muncer 1983a; Matsuzawa 1985; Matsuzawa et al. 1986; Rumbaugh et al. 1987, 1988, 1989; Rumbaugh 1990; Rumbaugh and Savage-Rumbaugh 1990; Rumbaugh and Washburn 1993; Boysen 1988, 1996, 1997; Boysen and Berntson 1989, 1990, 1995;

Boysen and Capaldi 1993; Boysen, et al. 1996, 1999; Biro and Matsuzawa 1999, 2001a, 2001b; Gadagkar 1995; Beran 2001, 2004a, 2004b, 2008, 2009a, b; Beran and Beran 2004; Beran and Evans 2006; Beran et al. 1998; 2007, 2009; Beran and Rumbaugh 2001; Beran et al. 1998; Call 2000; Shumaker et al. 2001.

168. Brown et al. 1978; King and Fobes 1982, p. 350; Tuttle 1986, pp. 190–191.

169. Hauser et al. 1996; Cantlon 2012.

170. Tuttle 1986, p. 191.

171. Washburn and Rumbaugh 1991.

172. Cantlon and Brannon 2006, p. 401.

173. Gelman and Gallistel 2004; Gordon 2004; Pica et al. 2004; Dehaene et al. 2008.

174. Carey 1998. 2001a, 2001b; Butterworth 1999; Dehaene et al. 1999; Dehaene 2001.

175. Varley et al. 2005, p. 3519.

176. Brannon 2005.

177. Boysen and Berntson 1989; Boysen and Capaldi 1993; Boysen et al. 1996, 2001; Beran 2001, 2004a, 2008, 2009; Beran and Beran 2004; Beran and Evans 2006; Beran et al. 2009; Beran and Rumbaugh 2001.

178. Evans and Beran 2007; Mischel et al. 1989.

179. Shumaker et al. 2001.

180. Tolman 1948; Tuttle 1986, p. 191; Boesch and Boesch 1984c; Tomasello and Call 1997, pp. 27–38.

181. Menzel 1973b, 1974, 1978, 1979; Menzel and Halperin 1975; Menzel et al. 1978, 2002; Menzel 1997, 2005; Premack and Premack 1983.

182. Menzel 1978; Tuttle 1986, p. 191.

183. Fragaszy et al. 2002.

184. Premack and Premack 1983, pp. 102–105; compare Boysen and Kuhlmeier 2002; Kuhlmeier and Boysen 2001, 2002; Kuhlmeier et al. 1999.

185. Huan et al. 2006.

186. Chua et al. 2005.

187. Gómez 2004, p. 171.

188. Tuttle 1986, p. 191.

189. Davenport and Rogers 1970; Davenport 1976, 1977; Davenport et al. 1973, 1975; Rogers and Davenport 1975.

190. Premack 1976a, b.

191. Linden 1974; Tanner et al. 2006; Savage-Rumbaugh et al. 1988.

192. Bolster 1978: Cowley and Weiskrantz 1975; Elliott 1977; Ettlinger and Garcha 1980; Jarvis and Ettlinger 1977, 1978.

193. von Wright 1970; Tuttle 1986, p. 192.

194. Descartes 1996.

195. Tuttle 1986, pp. 192–193.

196. Griffin 1976, 1978, 1981, 1983, 1984a, 1984b, 2001.

197. Griffin 1976, pp. 103–105.

198. Griffin 1976, p. 95; Griffin 1981, p. 157.

199. Tuttle 1986, p. 193.

200. Griffin 1984a, 1985, 2001a, 2001b; Griffin and Speck 2004; Gould 2004.

201. Gallup 1968, 1970, 1975, 1977a, 1977b, 1979, 1982, 1994; Gallup and McClure 1971; Gallup et al. 1971, 1977, 2002; Hill et al. 1970; Suarez and Gallup 1981; van den Bos 1999; Tuttle 1986, pp. 193–194.

202. Gallup et al. 2002.

203. Gallup 1970.

204. Lin et al. 1992; Povinelli et al. 1993; de Veer et al. 2003.

205. Swartz and Evans 1991.

206. Gallup 1975; Tuttle 1986, p. 194.

207. Gallup et al. 1971; Hill et al. 1970; Custance and Bard 1994.

208. Bard et al. 1995.

209. Piaget 1952; Amsterdam 1972.

210. Gallup 1966, 1968, 1975; Tuttle 1986, p. 193; Anderson 1993; Thompson and Contie 1994; Mitchell 2002; de Waal et al. 2005; Roma et al. 2007; Anderson and Gallup 2011.

211. Prior et al. 2008; Martin and Psarakos 1994, Marino et al. 1994; Reiss and Marino 2001; Delfour and Marten 2001; Plotnik et al. 2006; Broom et al. 2009.

212. Menzel et al. 1978.

213. Menzel et al. 1985.

214. Tuttle 1986, p. 194.

215. Shillito et al. 1999.

216. Tuttle 1986, pp. 194–195.

217. Patterson and Gordon 1993, 2001; Patterson and Cohn 1994; Allen and Schwartz 2008; compare Povinelli 1994; Posada and Colell 2007; Hyatt and Hopkins 1994; Westergaard and Hyatt 1994; Inoue-Nakamura 2001.

218. Hyatt 1988.

219. Ristau 1983; Epstein et al. 1981; Hart and Karmel 1996; Schilhab 2004; Fox 1982; de Veer and van den Bos 1999.

220. Swartz and Evans 1991; Heyes 1991; Mitchell 1992, 1993a, 1993b, 1997.

221. Ristau and Robbins 1982, p. 218; Tuttle 1986, p. 196.

222. Tomasello 1999b, pp. 73–74.

223. Tomasello 1999b, p. 74.

224. Premack and Woodruff 1978; Tuttle 1986, p. 187.

225. Heyes 1998; Tomasello 1999b.

226. Premack and Premack 1983, p. 67.

227. Premack and Premack 1983, pp. 51–57.

228. Lawick-Goodall 1971; Menzel 1973b, 1974, 1979; Menzel and Halperin 1975; Savage-Rumbaugh and McDonald 1988; Whiten and Byrne 1988; Whiten 1998; Hirata 2006.

229. Jensen et al. 2007; Silk 2007.

230. Mitchell 1991.

231. de Waal 1982; Byrne and Whiten 1988; Byrne 1996; Whiten and Byrne 1997; Maestripieri 2007.

232. Corballis 2007, p. 244.

233. Goldman 1981, p. 29.

234. Rumbaugh and Washburn 2003.

235. Rumbaugh and Washburn 2003, p. 252.

236. Rumbaugh 1997, p. 200; Washburn 2007b; Rumbaugh 2002; Rumbaugh and Washburn 2003, p. 253.

237. Marler 1991.

238. Rumbaugh et al. 2007, 2008; Rumbaugh, Washburn, and Hillix 1996; Naour 2009, pp. 91–115; Greenberg et al. 2007; Marr 2007.

239. Rumbaugh et al. 2007, p. 979.

240. Menzel 1999.

241. Rumbaugh and Washburn 2003, pp. 5–6.

242. Rumbaugh et al. 2007, p. 980; Savage-Rumbaugh et al. 1993; Hillix and Rumbaugh 2004; Köhler 1925.

243. Thorndike 1898; Rumbaugh et al. 2007, p. 973; Hebb 1949.

244. Hillix 2009, p. 105.

245. Rumbuagh et al. 2007, p. 988; Chang et al. 2013.

246. Quallo et al. 2009.

247. Herrmann et al. 2007.

248. Meltzoff et al. 2009; Call and Tomasello 2003; Povinelli et al. 1996, 2000; Bering and Povinelli 2003.

249. Premack 2007.

250. Jackendoff 2007.

251. Jackendoff 2007; Spelke 2003; Sallet et al. 2011; Miller 2011; Mascaro and Csibra 2012; Seyfarth and Cheney 2013.

252. Gergely and Csibra 2006; Tomasello 2009, p. 219.

253. Jackendoff 2007, p. 150.

254. Rilling et al. 2012, p. 369.

255. Holekamp 2007.

256. Van Schaik, Preuschoft, et al. 2004, p. 202.

257. Miller et al. 2012.

11. SOCIAL, ANTISOCIAL, AND SEXUAL APES

1. Carpenter 1940.

2. Nissen 1931; Bingham 1932.

3. Carpenter 1938; Schaller 1961; Harrisson 1962; Yoshiba 1964; Davenport 1967; de Silva 1971; MacKinnon 1971, 1974a, 1974b, 1979; Cohen 1975; Horr 1972, 1975, 1977; Rodman 1973, 1977, 1979, 1984; Mitani 1985b, 1985d; Galdikas 1978, 1979, 1981, 1983, 1984, 1985a, 1985b, 1985c; Rijksen 1975, 1978; Schürmann 1981, 1982; Tuttle 1986; Rodman and Mitani 1987; Rodman 1988; van Hooff 1995; van Schaik 1999; 2004; Knott and Kahlenberg 2007, Knott et al. 2008; Mitra Setia et al. 2009.

4. Edwards and Snowdon 1980; Edwards 1982; Maple 1980; Becker and Hicks 1984.

5. van Schaik 2004, pp. 69–70.

6. Wich et al. 2004.

7. Utami and Mitra Setia 1995; Utami et al. 2002; Utami, Singleton, et al. 2009; Utami Atmoto and van Hooff 2004.

8. Kingsley 1982, 1988; Maggioncalda et al. 1999, 2000, 2002.

9. Utami et al. 2002; Pradhan et al. 2012.

10. Utami Atmoto and van Hooff 2004.

11. Wich et al. 2004; Wich, de Vries, et al. 2009; Singleton and van Schaik 2002; van Noordwijk and van Schaik 2005; van Noordwijk et al. 2009; Galdikas and Wood 1990; Knott 2001.

12. Horr 1977; Noordwijk et al. 2009.

13. Horr 1977.

14. Horr 1975.

15. Tuttle 1986.

16. Jaeggi et al. 2008.

17. Rodman 1973.

18. MacKinnon 1971, 1974a.

19. MacKinnon 1971.

20. MacKinnon 1974a, p. 16.

21. MacKinnon 1974a, p. 17; 1974b, p. 210.

22. Galdikas 1978, 1981, 1984, 1985a, 1985b, 1985c.

23. Galdikas 1978, 1979.

24. Galdikas 1984.

25. Galdikas 1979, 1985b.

26. Rijksen 1975.

27. Galdikas 1985b.

28. Galdikas 1984, 1985b.

29. Galdikas 1978, 1985a, 1985b.

30. Galdikas 1985c.

31. Galdikas 1978.

32. Galdikas 1982c.

33. Galdikas 1978, p. 294.

34. Knott et al. 2008, p. 991.

35. Knott et al. 2008.

36. Nicholson 1954; Thornhill and Alcock 1983.

37. Nicholson 1954.

38. Sugardjito et al. 1987.

39. Utami et al. 1997.

40. Van Schaik 1999, 2004; Singleton and van Schaik 2002; Fox 2002; Mitra Setia et al. 2009.

41. Singleton and van Schaik 2002.

42. Singleton et al. 2009, p. 211.

43. Utami Atmoko, Mitra Setia, et al. 2009, p. 227.

44. Te Boekhorst et al. 1990.

45. Mitani et al. 1991.

46. Galdikas 1978; 1985b, 1985c.

47. MacKinnon 1974a.

48. MacKinnon 1974a.

49. 1978; Meijaard 2004; van Schaik and Griffiths 1996.

50. Rijksen 1975, 1978.

51. Rijksen 1978.

52. MacKinnon 1974a; Maple 1980; Nadler 1977; Rijksen 1978.

53. Tuttle 1986, p. 260.

54. Fox 2001.

55. Darwin 1871.

56. Horr 1972; Rodman 1973; Trivers 1972; Wilson 1975.

57. Horr 1972; Rodman 1973.

58. Galdikas 1978; Wrangham 1979a, 1979b; Tuttle 1986, p. 260.

59. Horr 1972, p. 49.

60. Horr 1975.

61. Trivers 1972; Williams 1975; Rodman 1973, p. 202.

62. Burley 1979.

63. Rodman 1973, 1977; Horr 1975.

64. Rodman 1977, 1979.

65. Rodman 1979.

66. Tuttle 1986, p. 261.

67. MacKinnon 1971, p. 176.

68. MacKinnon 1979.

69. Estep and Bruce 1981; Knott 2009.

70. Knott 2009, p. 82.

71. Wich et al. 1999.

72. MacKinnon 1974a, pp. 56–57.

73. MacKinnon 1974b, p. 209.

74. MacKinnon 1979, p. 264; Rodman 1973, 1979.

75. MacKinnon 1979, p. 268.

76. MacKinnon 1979, p. 270.

77. Galdikas 1979, 1983, 1985c.

78. Galdikas 1979.

79. Galdikas 1979, 1981b, 1984, 1985a.

80. Galdikas 1979, 1981.

81. Galdikas 1981.

82. Galdikas 1981, 1984.

83. Galdikas 1985c.

84. Galdikas 1979, 1981, 1985c.

85. Rijksen 1978; MacKinnon 1974a.

86. Schürmann 1981, 1982.

87. Schürmann 1982.

88. Schürmann and van Hooff 1981.

89. Mitani 1985d.

90. Galdikas 1985a; Mitani 1985d.

91. Knott 2009.

92. Knott et al. 2007.

93. Fox 2002.

94. Stumpf et al. 2008; Knott 2009.

95. Utami et al. 2002; Utami, Atmoto, and van Hooff 2004, p. 204.

96. Goossens et al. 2006.

97. Morrogh-Bernard et al. 2011.

98. Knott et al. 2009.

99. Nissen 1931.

100. Azuma and Toyoshima 1961–62; Izawa and Itani 1966; Itani and Suzuki 1967; Suzuki 1969; Nishida 1968, 1970, 1979, 1983a, 1983b, 2012; Izawa 1970; Nishida and Kawanaka 1972; Kawanaka and Nishida 1975; Uehara 1981; Uehara and Nyundo 1983; Kawanaka 1982a; Hasegawa and Hiraiwa-Hasegawa 1983; Hiraiwa-Hasegawa et al. 1984; Nishida and Hiraiwa-Hasegawa 1985, 1987; Nishida et al. 1985; Goodall 1965, 1983, 1986; Goodall et al. 1979; van Lawick-Goodall 1967, 1968a, 1971, 1973a, 1973b, 1975; Reynolds 1965; Reynolds and Reynolds 1965; Kortlandt 1962; Tuttle 1986, pp. 266–279; Stumpf 2007; Nakamura and Nishida 2012a; Watts 2012; Wilson 2012.

101. Kortlandt 1962.

102. Reynolds and Reynolds 1965; Reynolds 1965.

103. Reynolds and Reynolds 1965.

104. Tuttle 1986, pp. 230, 268.

105. Goodall 1965; van Lawick-Goodall 1967, 1968a, 1968b.

106. Goodall 1965; van Lawick-Goodall 1968a.

107. Reynolds 1975; van Lawick-Goodall 1975; Goodall 1983; Wrangham 1974b; Wrangham and Smuts 1980; Riss and Busse 1977; Ghiglieri 1984a, 1984b; Tuttle 1986, p. 268; Power 1991.

108. Itani and Suzuki 1967; Suzuki 1969.

109. Itani and Suzuki 1967.

110. Kano 1971.

111. Nishida 1968; Tuttle 1986, p. 271; Mitani, Watts, and Muller 2002, Mitani et al. 2006.

112. Van Lawick-Goodall 1973b, 1975; Goodall 1983.

113. McGrew and Collins 1985; Kawanaka and Nishida 1975.

114. Nishida 1968.

115. Nishida 1979.

116. Spruijt et al. 1992; Nakamura 2003. Kanngiesser et al. 2011.

117. Nishida 1968, 1979.

118. Nishida 1979.

119. Nishida and Kawanaka 1972.

120. Kawanaka and Nishida 1975.

121. Van Lawick-Goodall 1968a.

122. Nishida 1970.

123. Van Lawick-Goodall 1968a; Hamburg 1971; Riss and Goodall 1979; Bygott 1979.

124. Nishida1983a; Nishida and Hiraiwa-Hasegawa 1985; Inaba 2009; de Waal 1978, 1982, 1984; de Waal and van Roosmalen 1979.

125. Van Lawick-Goodall 1968a, 1975.

126. Van Lawick-Goodall 1973a, 1975.

127. Sugiyama and Koman 1979.

128. Van Lawick-Goodall 1968a, p. 263.

129. Bekoff and Allen 1998; Palagi 2006.

130. Morin et al. 1994.

131. Simpson 1973.

132. Nishida and Kawanaka 1972; Nishida 1968; Goodall 1986; Goodall et al. 1979; Bygott 1979; Halperin 1979; Pusey 1979, 1980; Teleki et al. 1976.

133. Goodall et al. 1979.

134. Pusey 1979.

135. Amsler 2010.

136. Itani 1980.

137. Nishida and Hiraiwa-Hasegawa 1984.

138. Nishida et al. 1985.

139. Hiraiwa-Hasegawa et al. 1984.

140. Boesch et al. 2007.

141. Wrangham 1979b.

142. Wrangham 1979b, 2000; Wrangham and Smuts 1980.

143. Uehara 1981.

144. Kawanaka 1982a.

145. Kawanaka 1984.

146. Ghiglieri 1984a.

147. Halperin 1979.

148. Sugiyama 1968, 1969, 1973, 1981, 1984: Sugiyama and Koman 1979; Suzuki 1975; Ghiglieri 1984a, 1984b; Yamagiwa, Mwanza, Spangenberg, et al. 1992; Basabose 2005; Watts 1998d, 2002, 2006; Watts and Mitani 2001; Mitani et al. 2000; Mitani, Watts, and Muller 2002; Wakefield 2008; Pepper et al. 1999; Wrangham et al. 1992; Kahlenberg et al. 2008; McGrew et al. 1982; Tutin et al. 1983; Baldwin et al. 1982; Reynolds 2005; Newton-Fisher 2002; Boesch 2009; Boesch and Boesch-Achermann 2000.

149. Manson and Wrangham 1991; Stanford 1998b; Wrangham 1999; Muller and Wrangham 2001; Muller 2002; Muller and Mitani 2005; Wilson et al. 2001; Wilson and Wrangham 2003; Wrangham and Wilson 2004; Wrangham et al. 2006; Boesch and Boesch-Achermann 2000; Boesch et al. 2008; Watts et al. 2006; Sherrow and Amsler 2007.

150. Sugiyama 1968, 1973a, 1973b; Reynolds 2005.

151. Tutin et al. 1983.

152. Anderson et al. 2002.

153. Mitani, Watts, and Lawanga 2002.

154. Hashimoto et al. 2003.

155. Hashimoto et al. 2001, 2003; Furuichi et al. 2001c.

156. Newton-Fisher 1999d.

157. Newton-Fisher et al. 2000.

158. Brothers 1990, 2002; Barton and Dunbar 1997; Dunbar 1998; Frith 2007; Mitchell and Heatherton 2009.

159. Tutin et al. 1983.

160. Boesch and Boesch-Achermann 2000; Boesch 1991a.

161. Boesch and Boesch-Achermann 2000; Boesch 1996.

162. Boesch and Boesch-Achermann 2000, pp. 96–97.

163. Wakefield 2008.

164. Mitani 2006.

165. Mitani et al. 2000.

166. Pepper et al. 1999.

167. Ghiglieri 1984a, 1984b.

168. Ghiglieri 1984a, p. 182.

169. Ghiglieri 1984a; Itani 1980; Kawanaka 1982a, 1984; Pusey 1980; Uehara 1981; Goodall 1983; Nishida et al. 1985; Wrangham 1979b.

170. Wrangham et al. 1992; Chapman and Wrangham 1993; Wrangham and Smuts 1980; Williams et al. 2002.

171. Wrangham 1986; Chapman et al. 1993; Doran 1997b; Sakura 1994; Mitani, Watts, and Lawanga 2002.

172. Baker and Smuts 1994; de Waal 1994; Boehm 1994.

173. Sakamaki et al. 2001.

174. Chapman et al. 1995.

175. Emery Thompson and Wrangham 2008a.

176. Thompson et al. 2007.

177. Kahlenberg et al. 2008.

178. Emery Thompson et al. 2010; Kahlenberg 2008.

179. Otari and Gilchrist 2005.

180. Muller and Wrangham 2004b.

181. Muller and Wrangham 2005.

182. Muller and Wrangham 2004a.

183. Muller and Wrangham 2005; Wingfield et al. 1987, 1990, 1997, 2000.

184. Wittig and Boesch 2003a.

185. Riedel et al. 2011.

186. Pusey et al. 1997; Williams et al. 2002.

187. Murray et al. 2006; Murray, Mane, and Pusey 2007.

188. Pusey et al. 2008.

189. Murray, Wroblewski, and Pusey 2007.

190. Wroblewski 2008.

191. Williams et al. 2004.

192. Wrangham 1979b.

193. Lodwick et al. 2004.

194. Pusey and Packer 1987.

195. Stumpf et al. 2009, p. 652.

196. Stumpf et al. 2009, pp. 249–650.

197. Fraser, Schino, and Aureli 2008.

198. Jurmain 1997; Carter et al. 2008.

199. De Waal and van Roosmalen 1979; de Waal 1982, 1989b; 2000; de Waal and Aureli 1996; Cords and Aureli 1996, 2000; Aureli et al. 2002; van Schaik and Aureli 2000; Fraser

and Aureli 2008; Fuentes et al. 2002; Kutsukake and Castles 2004; Preuschoft et al. 2002; Baker and Smuts 1994; Romero and de Waal 2011.

200. Fraser, Schino, and Aureli 2008; Koski and Sterck 2007.

201. Palagi et al. 2006; Fraser, Stahl, and Aureli 2008.

202. Romero and de Waal 2011.

203. De Waal 1993, 1996.

204. De Waal and van Roosmalen 1979.

205. Wittig and Boesch 2003b.

206. Wittig and Boesch 2005.

207. Arnold and Whiten 2001.

208. Watts 2002.

209. Watts 2006.

210. Watts 2002; Trivers 1971.

211. Arnold and Whiten 2001; Watts 2006.

212. Silk 1996, p. 39; Silk 1998, p. 361, 2002.

213. Brosnan et al. 2005.

214. Tutin 1975, 1979a, 1979b, 1980; McGinnis 1979; Tutin and McGinnis 1981.

215. Van Lawick-Goodall 1968a, 1973a, 1975.

216. Wich et al. 2004.

217. Roof et al. 2005; Littleton 2005.

218. Birkhead and Pizzari 2002; Pradhan et al. 2006; Dewsbury 1982.

219. Marson et al. 1988, 1989.

220. McLeod and Gold 1952; Rui et al. 1984.

221. Short 1979; Dixson 1998; Hrdy 1981; Whitten 1987; Wright 1990; Boesch and Boesch-Aschermann 2000; Hrdy 1977; van Schaik 2000; van Schaik, Pradhan, and van Noordwijk 2004; Furuichi and Hashimoto 2000; Hashimoto and Furuichi 2006; Wrangham 2002.

222. Ely et al. 2006.

223. Van Lawick-Goodall 1975.

224. Goodall 1968a; Muller et al. 2006.

225. Stumpf et al. 2009.

226. Van Lawick-Goodall 1968a; McGinnis 1979.

227. Hashimoto and Hashimoto 2002.

228. Tutin 1975, 1979a, 1979b, 1980; Tutin and McGinnis 1981.

229. Tutin and McGinnis 1981.

230. Tutin 1980.

231. Teleki et al. 1976; Goodall 1977, 1986.

232. Wood et al. 1985.

233. Tutin 1979a.

234. Constable et al. 2001.

235. Wrangham 1977.

236. van Schaik et al. 2000; Noordwijk and van Schaik 2000; van Schaik, Pradhan, and van Noordwijk 2004; Hrdy 1979, 1981; Wolff and Macdonald 2004; Muller et al. 2007, 2011; Pieta 2008; Stumpf et al. 2008.

237. Muller et al. 2007, 2011.

238. Smuts and Smuts 1993.

239. Goodall 1986, p. 452; Matsumoto-Oda 1998.

240. Muller et al. 2007.

241. Townsend et al. 2008; Townsend and Zuberbühler 2009.

242. Pradhan et al. 2006.

243. Deschner et al. 2003, 2004; Emery and Whitten 2003.

244. Emery Thompson and Wrangham 2008b.

245. Emery Thompson 2005.

246. Matsumoto-Oda and Kasuya 2005.

247. Boesch et al. 2006.

248. Watts 1998d.

249. Mitani, Watts, and Muller 2002; Sugiyama 1999.

250. Pusey 1980.

251. Tutin 1975, 1979a.

252. Slocombe and Newton-Fisher 2005.

253. Tutin 1979a.

254. Nishida 1983a.

255. Hasegawa and Hiraiwa-Hasegawa 1983.

256. Matsumoto-Oda 1999.

257. Inoue et al. 2008.

258. Parr and de Waal 1999.

259. Vigilant et al. 2001; Lukas et al. 2005.

260. Gagneux et al. 1979, 1999; Vigilant et al. 2001; Constable et al. 2001; Gagneux et al. 2001.

261. Lehmann et al. 2006.

262. Boesch et al. 2006.

263. Rees 2009.

264. van Schaik and Dunbar 1990; van Schaik and Kappeler 1997; Van Schaik and Janson 2000.

265. Maynard Smith and Price 1973; Sussman et al. 1994.

266. Hamai et al. 1992; Watts et al. 2002; Sherrow and Amsler 2007; Goodall 1977, 1986; Newton-Fisher 1999d.

267. Goodall 1986; Townsend et al. 2007; Takahata 1985; Sakamaki et al. 2001.

268. Wallis and Lee 1999.

269. Mubalamata 1984; Kano et al. 1996; Dupain et al. 2000; Dupain and Elsacker 2001a, 2001b; Van Krunkelsven et al. 2000; Van Krunkelsven 2001; Reinartz and Inogwabini 2001; Omasombo et al. 2005; Raffaele 2006; Balongelwa 2008; Hashimoto and Furuichi 2001; Furuichi and Thompson 2008; Thompson 2001; Thompson et al. 2008; Idani et al. 2008; Hart et al. 2008; Inogwabini et al. 2008; Tuttle 1986, pp. 279–280.

270. Nishida 1972.

271. Horn 1980.

272. Nishida and Kano 1973.

273. Kano 1979.

274. Kuroda 1979, 1980, 1984; Kano 1980, 1982a, 1996; Kano and Mulavwa 1984; Kitamura 1983; Mori 1983, 1984; Furuichi 1987, 1989, 1997; Idani 1990, 1991; Ihobe 1992a, 1992b, 1997; Furuichi and Ihobe 1994; Furuichi et al. 1998, 2008; Furuichi and Hashimoto 2002, 2004b; Takahata et al. 1996; Hashimoto and Furuichi 2006; Hashimoto et al. 2008; Mulavwa et al. 2008; Furuichi et al. 2012.

275. Kano 1982a; Kuroda 1979.

276. Kuroda 1979.

277. Kano 1980, 1982a.

278. Kano 1980.

279. Franz 1999.

280. Ihobe 1992a.

281. Furuichi 1997, 2011.

282. Paoli et al. 2006.

283. Kuroda 1980; Kano 1980.

284. Kuroda 1980; Kano 1980; Mori 1984.

285. Kano and Mulavwa 1984; Idani 1990; Kano et al. 1996.

286. Kano 1982a.

287. Idani 1990.

288. Kano 1982a; Kitamura 1983.

289. Kano 1983b; Kano and Mulavwa 1984.

290. Furuichi 1987, 1989.

291. Idani 1991.

292. Furuichi 1989, p. 173.

293. Kano and Mulavwa 1984; Mori 1984; Kuroda 1984.

294. Furuichi 1994.

295. Kano 1983, 1984b.

296. Kano 1983.

297. Badrian and Badrian 1977, 1980, 1984a, 1984b; Badrian and Malenky 1984; MacKinnon 1976; Thompson-Handler et al. 1984; White 1988, 1989b, 1992b, 1992c, 1996a, 1996b; White and Burgman 1990; White and Lanjouw 1992; White and Chapman 1994; Chapman et al. 1994; Fruth and Hohmann 1993; Hohmann and Fruth 2000; Hohmann et al. 1999; Gerloff et al. 1999; Marshall and Hohmann 2005.

298. Badrian and Badrian 1977.

299. MacKinnon 1976.

300. Badrian and Badrian 1984b.

301. Badrian and Malenky 1984; Badrian and Badrian 1984b.

302. Badrian and Badrian 1980; Badrian et al. 1981.

303. Badrian and Malenky 1984.

304. Kitamura 1983; Badrian and Badrian 1984b.

305. Van Elsacker et al. 1995; White 1988, 1989b; White and Burgman 1990.

306. White 1988.

307. Wrangham 1979a, 1979b; White and Wrangham 1988.

308. White and Wood 2007.

309. Stevens et al. 2007.

310. Jaeggi et al. 2010.

311. Vervaecke et al. 2000b.

312. Surbeck et al. 2011.

313. White and Chapman 1994.

314. White and Lanjouw 1992.

315. White 1989a.

316. White 1992c.

317. Badrian and Badrian 1984b; Wrangham 1986.

318. Melenky and Stiles 1991.

319. Chapman at al. 1994.

320. Chapman et al. 1994, p. 53.

321. White 1996b.

322. Gerloff et al. 1999; Hohmann and Fruth 1999.

323. Hohmann and Fruth 1999.

324. Gerloff et al. 1999; Furuichi 1989, 1997; Hohmann and Fruth 1996; Parish 1996; Parish and de Waal 2000.

325. Hohmann et al. 1999.

326. Marshall and Hohmann 2005, p. 90.

327. Kano 1980, 1982a.

328. Kano 1982a.

329. Furuichi 1987; Takahata et al. 1996.

330. Furuichi 1987.

331. Takahata et al. 1996; Furuichi et al. 1998; Nishida et al. 1990; Sugiyama 1994b.

332. Furuichi et al 2008; Mulavwa et al. 2008.

333. Furuichi et al. 1998.

334. Kano 1996.

335. Takahata et al. 1996.

336. Furuichi and Hashimoto 2004b.

337. Hashimoto and Furuichi 2006.

338. Furuichi and Hashimoto 2004b; Hashimoto and Furuichi 2006.

339. Furuichi and Hashimoto 2002.

340. Marvan et al. 2006.

341. Thompson-Handler et al. 1984.

342. Badrian and Badrian 1984b, p. 332; Tuttle 1986, p. 288.

343. Thompson-Handler et al. 1984, table II, p. 356.

344. Kano 1980, 1989, 1992; Thompson-Handler et al. 1984; de Waal 1987, 1990, 1995; Parish 1994, 1996; Mason et al. 1997; Hohmann and Fruth 2000.

345. Gold 2001.

346. Hohmann and Fruth 2000; Fruth and Hohmann 2006.

347. Fruth and Hohmann 2006.

348. de Waal 1988.

349. de Waal 1987, 1988, 1989a.

350. Bingham 1932; Donisthorpe 1958; Kawai and Mizuhara 1959–60; Osborn 1963; Emlen 1962; Emlen and Schaller 1960a, 1960b; Schaller 1963, 1965; Schaller and Emlen 1963.

351. Fossey 1970, 1971, 1972, 1974, 1979, 1981, 1982, 1983; Fossey and Harcourt 1977; Caro 1976; Elliott 1976; Harcourt 1977, 1978a, 1978b, 1979a, 1979b, 1979c, 1979d, 1981; Hall et al. 1998; Harcourt and Curry-Lindahl 1978; Harcourt and Fossey 1981; Harcourt, Fossey, and Sabater Pí 1981; Harcourt, Fossey et al., 1980; Harcourt and Groom 1972; Harcourt and Stewart 1977, 1978, 1981, 1983, 1984, 2007; Harcourt, Stewart, and Fossey 1976, 1981, 1987, 1989; Stewart 1977, 2001; Vedder 1984; Watts 1984, 1985a, 1985b, 1987, 1988, 1989a, 1989b, 1990a, 1990b, 1991, 1992, 1994a, 1994b, 1994c, 1995a, 1995b, 1996, 1997, 2000a, 2000b, 2001, 2003; Weber and Vedder 1983; Yamagiwa 1983, 1986, 1987a, 1987b, 1992, 1999, 2001, 2004; Yamagiwa and Goodall 1992; Yamagiwa and Kahekwa 2001; Yamagiwa, Kaleme, et al. 1996; Yamagiwa, Maruhashi, et al. 1996; Yamagiwa, Mwanza, et al. 1993; Yamagiwa, Yumoto, et al. 1993; Yamagiwa, Mwanza, et al. 1992; Yamagiwa, et al. 1999, 2009, 2012; Yamagiwa, Basabose, et al. 2003; Yamagiwa, Kahekwa, et al. 2003; Achoka 1993; Sholley 1991; Sicotte 1993,1995, 2001, 2002; M. Robbins 1995, 1996, 1997, 1999, 2003, 2008, 2009; Robbins and Robbins 2004, 2005; Robbins and Sawyer 2007; M. Robbins et al. 2009a; A. Robbins et al. 2006, 2007, 2009a, 2009b; McNeilage et al. 2001, 2006a; Hall, White et al. 1998; Hall, Saltonstall et al. 1998; Omari et al. 1999; Nkurunungi and Stanford 2006; Stanford 2008.

352. Sarmiento and Oates 2000; Jones and Sabater Pí 1971; Carroll 1988, 1990; Fay 1989; Tutin 1996; Tutin et al. 1992; Mitani 1992; Remis 1993, 1997a, 1997b; Remis et al. 2004; Nishihara 1994; Yamagiwa et al. 1995; Goldsmith 1999b; Parnell 2002; C. Cipolletta 2003; Stokes 2004; Stokes et al. 2003; Gatti et al. 2003, 2004; Bradley et al. 2004; Blom et al. 2004; Robbins et al. 2004; Douadi et al. 2007; Breuer 2008; Breuer et al. 2010; Doran-Sheehy et al. 2009; Arandjelovic et al. 2010; Bergl and Vigilant 2007; Bergl et al. 2008; De Vere et al. 2011; Oates et al. 2007; Sunderland-Groves et al. 2009; Wittiger and Sunderland-Groves 2007.

353. Tutin 2001; Walsh et al. 2003; Fay et al. 1992; Bermejo 1999; Magliocca et al. 1999.

354. Tutin and Fernandez 1984; Oko 1992; Blom et al. 1992; Kano and Asato 1994; Harcourt 1996; Olejniczak 2001; Huijbregts et al. 2003; Blom 1997; Blom et al. 2004; Garcia and Mba 1997; Bowen-Jones and Pendry 1999; Robbins et al. 2004; Bermejo et al. 2006; Devos et al. 2008; Rabanal et al. 2010.

355. March 1957; Sabater Pí 1980–81; Carroll 1988; Harris et al. 1987; Harcourt et al. 1989; Tutin 2001; Walsh et al. 2003; Oates et al. 2007; Bergl et al. 2008.

356. Tutin and Feranadez 1984.

357. Tutin et al. 1975, 1996.

358. Tutin et al. 1992.

359. Tutin et al. 1992; Doran-Sheehy 2004.

360. Tutin 1996.

361. Mitani 1992.

362. Nishihara 1994.

363. Parnell 2002; Stokes 2004; Stokes et al. 2003; Breuer et al. 2010.

364. Bermejo 1999.

365. Wittiger and Sunderland-Groves 2007; Sunderland-Groves et al. 2009; De Vere et al. 2011.

366. Remis 1997b.

367. Magliocca et al. 1999; Douadi et al. 2007; Bradley et al. 2004; Doran-Sheehy 2004, 2009; Carroll 1988; Fay 1989; Cipolletta 2003; Robbins et al. 2004; Arandjelovic et al. 2010.

368. Robbins et al. 2004, p. 145.

369. Magliocca et al. 1999.

370. Douadi et al. 2007.

371. Bradley et al. 2004.

372. Doran-Sheehy et al. 2004, 2009.

373. Donisthorpe 1958.

374. Osborn 1963.

375. Kawai and Mizuhara 1959–60.

376. Harcourt 1980–81.

377. McNeilage et al. 2006b.

378. Gray et al. 2009; Guschaski et al. 2009.

379. Schaller 1963, 1965.

380. Schaller 1960, 1963.

381. Schaller 1965, pp. 339, 347, 348.

382. Weber and Vedder 1983.

383. Aveling and Harcourt 1984.

384. Lukas et al. 2004.

385. Schaller 1963, pp. 111–120, 1965.

386. Schaller 1963, p. 238.

387. Fossey 1970, 1983.

388. Harcourt and Groom 1972; Harcourt 1977, 1986; Harcourt and Curry-Lindahl 1978; Harcourt, Fossey, and Sabater Pí 1981; Harcourt et al. 1983; Fossey 1981, 1983; Oates 1986; Butynski et al. 1990; Stewart et al. 2001; Steklis and Gerald-Steklis 2001; Kalpers et al. 2003; Mwanza et al. 1988; Mankoto et al. 1994; Hall, Saltonstall et al. 1998; Hall, White et al. 1998; Yamagiwa 2003.

389. Watts 1989b; Kalpers et al. 2003.

390. Yamagiwa et al. 2009, p. 301.

391. International Gorilla Conservation Programme (IGCP) 2010.

392. McNeilage 1996; Robbins and Robbins 2004; Robbins et al. 2011.

393. Robbins et al. 2009a.

394. Kalpers et al. 2003; Vecellio 2007.

395. Fossey 1970, 1971.

396. Harcourt 1979a; 1978a.

397. Robbins 1999.

398. Fossey 1971, p. 582.

399. Elliott 1976; Caro 1976; Fossey and Harcourt 1977; Fossey 1981, 1983.

400. Fossey 1974.

401. Fossey 1974; Caro 1976; Elliott 1976.

402. Caro 1976; Fossey 1974.

403. Harcourt 1978a, 1978b, 1979a, 1979b, 1979c, 1979d; Fossey and Harcourt 1977.

404. Harcourt 1978a, 1979a, 1979b, 1979c, 1979d.

405. Harcourt 1979c.

406. Harcourt 1979b; Stewart 1977.

407. Harcourt 1979b.

408. Watts 1984a, 1985b, 1994b.

409. Harcourt 1979b; Fossey 1979; Watts 1997.

410. Harcourt 1979b; Watts 1992, 1995a, 1977.

411. Robbins 2009.

412. Harcourt and Stewart 1987, 1989; Watts 1994b, 1994c, 1995a, 1995b, 1997, 2001, 2003.

413. Watts 1994b, 2001, p. 219.

414. Watts 2001.

415. Robbins 2008.

416. Harcourt 1979d.

417. Harcourt 1978a, 1979a, 1979d.

418. Stewart 2001; Rosenbaum et al. 2011.

419. Harcourt 1978a, 1979a; Fossey 1979, 1983.

420. Harcourt and Stewart 2007; Smuts 1999.

421. Harcourt and Stewart 1987, 1989; Watts 1997.

422. Watts 1994b, 1995a.

423. Watts 1992.

424. Watts 1997.

425. Sicotte 1995.

426. Watts 1995a, 1995b.

427. Watts 1995b; Watts et al. 2000.

428. Cordoni et al. 2006; Mallavarapu et al. 2006.

429. Cordoni et al. 2006, p. 1382.

430. Harcourt 1978b; Fossey 1983, 1984; Sicotte 1993, 2000; Watts 1994a, 1996, 2003; Schaller 1963; Fossey 1970, 1971.

431. Harcourt 1978b; Fossey 1981, 1983, 1984.

432. Harcourt 1978b.

433. Fossey 1984.

434. Fossey 1982.

435. Fossey 1981, 1983, 1984; Harcourt et al. 1981.

436. Watts 1989b.

437. Sicotte 2000.

438. Watts 1989b, p. 1.

439. Waser 1982; Harcourt and Greenberg 2001.

440. Yamagiwa et al. 2009, p. 29.

441. Yamagiwa and Kahekwa 2004.

442. Stokes et al. 2003.

443. Butynski and Kalina 1993; Inogwabini et al. 2000; Graczyk, Mudakikwa et al. 2001; Graczyk, DaSilva et al. 2001; Kalema-Zikusoka et al. 2002; Binyeri et al. 2002; Tuttle 2003; Harcourt 2003; Plumptre et al. 2003; Sarmiento 2003; Oates et al. 2003; Weber 1993; Bowen-Jones and Pendry 1999; Ammann 2001; Wilkie 2001; Woodford et al. 2002;

Williamson and Feistner 2003; Klailova et al. 2010; Sharp et al. 2013; Schaumburg et al. 2012.

444. Yamagiwa 1983.

445. Stokes et al. 2003; Parnell 2002.

446. Harcourt et al. 1980, 1981; Watts 1991; Czekala and Sicotte 2000.

447. Harcourt et al. 1976, 1980, 1981; Harcourt and Stewart 1977, 1978, 1981; Harcourt 1978b, 1979a, 1979b, 1979c, 1979d, 1981; Fossey 1982, 1983, 1984; Yamagiwa 1983.

448. Stokes 2004.

449. Breuer et al. 2010; Stokes et al. 2003.

450. Taylor 1997; Breuer et al. 2009.

451. Guschanski et al. 2008.

452. Bradley et al. 2005.

453. Harcourt et al. 1980, 1981; Harcourt and Stewart 1978b, 1981.

454. Harcourt and Stewart 1981; Watts 2000a; van Noordwijk and van Schaik 2004.

455. Watts 1991; Bradley et al. 2005.

456. Robbins and Robbins 2005.

457. Watts 1990b; Harcourt et al. 1976; Robbins 1995; Watts 2000a, 2000b, 2003; Sicotte 2001.

458. Czekala and Sicotte 2000; Sicotte 2001.

459. Nsubuga et al. 2008.

460. Robbins and Sawyer 2007.

461. Yamagiwa 1986.

462. Harcourt et al. 1981.

463. Watts 1991.

464. Sicotte 1993, 1994.

465. Harcourt et al., 1981.

466. Fossey 1982; Harcourt and Stewart 1978b.

467. Fossey 1982; Watts 1991.

468. Stewart 1988; Watts 1991.

469. Robbins et al. 2006.

470. Watts 1990a, 1994a.

471. Harcourt 1978b, p. 405.

472. Fossey 1982, 1984.

473. Robbins et al. 2009b.

474. Harcourt 1978b; Fossey 1984.

475. Robbins et al. 2007; Trivers and Willard 1973, p. 90.

476. Harcourt et al. 1976.

477. Harcourt 1978a, 1979c; Fossey 1979.

478. Harcourt et al. 1980, 1981; Fossey 1982; Watts 1990b.

479. Watts and Pusey 1993.

480. Robbins 1999, 2003.

481. Fossey 1972; Watts 1991.

482. Nadler 1989.

483. Sicotte 2002.

484. Sicotte 2001; Harcourt et al. 1993; Watts 1994b.

485. Robbins 2003.

486. Sicotte 2001; Watts 1991; Robbins 1999.

487. Fossey 1982; Watts 1990b, 1991.

488. Harcourt et al. 1980; Watts 1991.

489. Short 1979, 1980, 1981; Harcourt et al. 1980, 1981.

490. Watts 1990b.

491. Doran-Sheehy et al. 2009.

492. Harcourt et al. 1980.

493. Harcourt and Stewart 1978b.

494. Yamagiwa 1986, 1987a, 1987b; Yamagiwa 2006; Robbins 1996; Gatti et al. 2003, 2004; Watts 2003.

495. Yamagiwa and Kahekwa 2001.

496. Yamagiwa 1987a.

497. Gatti et al. 2004.

498. Yamagiwa 1992, 2006.

499. Robbins 1996, p. 942.

500. Dixson 2009.

501. Watts 1991; Czekala and Sicotte 2000.

502. Short 1979, 1980; Caillaud et al. 2008.

503. Short 1979.

504. Short 1979; Dahl 1988; Dixson 2009.

505. Short 1979; Dixson 2009; Gould et al. 1993; Harcourt, Harvey et al. 1981.

506. Short 1979; Harcourt 1981.

507. Kleiman 1981; Leighton 1987; Sommer and Reichard 2000; Fuentes 2000, 2007; Savini et al. 2009; Fan and Jiang 2010; Reichard and Barelli 2008; Reichard et al. 2012; Barelli et al. 2008; Srikosamatara and Brockelman 1987; Brockelman et al. 1998; Palombit 1994a; Palombit 1994b; Bleisch and Chen 1991; Fan et al. 2006, 2009; Haimoff 1987; Haimoff et al. 1986; Liu et al. 1989; Jiang et al. 1999; Malone and Fuentes 2009; Lappan 2007, 2008, 2009b; Bartlett 2009b; Zhou et al. 2008.

508. Muzaffar et al. 2007; Jiang et al. 2006.

509. Wickler and Seibt 1983, p. 46.

510. Crook and Gartlan 1966; Crook and Goss-Custard 1971; Eisenberg et al. 1972; Kleiman 1977; Emlen and Oring 1977; Clutton-Brock and Harvey 1977; Wrangham 1979a; Wittenberger 1979; Wittenberger and Tilson 1980; Rutberg 1983; Sussman and Kinzey 1984.

511. Brockelman and Srikosamatara 1984; Brockelman 1984.

512. Brockelman and Srikosamatara 1984; Brockelman 1984; Mitani 1984.

513. Chivers 1972, 1974; Palombit 1994a, 1994b; Lappan 2008, 2009b; Tuttle 1986, p. 255; Reichard 2003a.

514. Brockelman and Srikosamatara 1984; Tuttle 1986, p. 255.

515. Ahsan 1995; Palombit 1993.

516. Howard 1920.

517. Chivers 1973, 1974.

518. Ellefson 1968; Bartlett 2009a, 2009b; Palombit 1996.

519. Ellefson 1968.

520. Reichard and Barelli 2008.

521. Barelli et al. 2007.

522. Carpenter 1940; Ellefson 1968.

523. Treesucon and Raemaekers 1984.

524. Palombit 1995.

525. MacKinnon and MacKinnon 1977.

526. MacKinnon and MacKinnon 1977; Chivers and Raemaekers 1980; Gittins and Raemaekers 1980.

527. Brockelman et al. 1998.

528. Barelli et al. 2008.

529. Gittins 1980; Gittins and Raemaekers 1980.

530. Gittins and Raemaekers 1980.

531. Gittins 1980.

532. Carpenter 1940; McClure 1964; Kawabe 1970; Koyama 1971; Chivers 1971a, 1971b, 1972, 1973, 1974, 1975, 1976, 1978; Papaioannou 1973; Rijksen 1978; Fox 1972.

533. Chivers, 1971a, 1971b, 1972, 1974, 1975, 1976, 1977a.

534. Chivers 1974.

535. Chivers and Chivers 1975.

536. Chivers 1972.

537. Chivers 1976.

538. Palombit 1996.

539. Chivers 1978.

540. Chivers 1974; Koyama 1971; Aldrich-Blake and Chivers 1973.

541. Chivers 1974, Fox 1972.

542. Lappan 2007.

543. Aldrich-Blake and Chivers 1973; Chivers 1974; Chivers and Raemaekers 1980.

544. Tenaza 1975, 1976; Tilson 1981; Tenaza and Tilson 1977; Tilson and Tenaza 1982; A.J. Whitten 1982.

545. Tenaza and Hamilton 1971; Tenaza 1975, 1976; Tilson 1981; A.J Whitten 1982, 1984.

546. Tilson and Tenaza 1982; Tilson 1981.

547. Tenaza 1975, 1976.

548. Tenaza 1975.

549. Tenaza 1975; Tilson 1981.

550. Whitten 1965.

551. Tilson 1981.

552. Tenaza and Tilson 1977.

553. Treesucon and Raemaekers 1984.

554. McCann 1933; Tilson 1979; Mukherjee et al. 1988, 1991–92; Islam and Feeroz 1992; Dam 2006; Siddiqi 1986; Choudhury 1990, 1991.

555. Mukherjee et al. 1991–1992; Alfred and Sati 1991.

556. Alfred and Sati 1991.

557. Tilson 1979.

558. Haimoff 1987; Haimoff et al. 1986; Bleisch and Chen 1991; Jiang et al 1999; Fan et al. 2006; Fan, Xiao et al. 2009; Fan and Jiang 2010; Liu et al. 1989; Zhou et al. 2008.

559. Haimoff 1986, 1987.

560. Haimoff 1986, 1987; Bleisch and Chen 1991.

561. Fan et al. 2006, 2009; Fan and Jiang 2010.

562. Fan and Jiang 2010.

563. Fan et al. 2006.

564. Liu et al. 1989.

565. Brockelman et al. 1998; Sommer and Reichard 2000; Reichard 2003b.

566. Plavcan 2004b.

567. Larsen 2003; Reno et al. 2003.

568. McHenry 1994b.

569. McCrossin and Reyes 2010.

570. Copeland et al. 2011; Schoeninger 2011.

571. Schultz 1938; Dixson 2009.

572. Boaz et al. 2004; Binford and Ho 1985; Binford and Stone 1986; Goldberg et al. 2001; Weiner et al. 1998, 2000.

573. Bordes and Lafille 1962; Suzuki and Takai 1970; Lévêque and Vandermeersch 1980; Bar-Yosef et al. 1986; Mellars 1996, pp. 375–381.

574. Heim 1968, 1970, 1974; 1976, 1982a, 1982b, 1984; Solecki 1971; Trinkaus 1983.

575. Gargett 1989; Stringer and Gamble 1993; Mellars 1996; Sandgathe et al. 2011; Balter 2012.

576. Trinkaus and Shipman 1993; Klein 1995; Hublin et al. 2012.

577. Trinkaus 2011.

578. Gamble 2008; Gowlett 2008.

579. Hill et al. 2011; Ragir 1985.

580. W. D. Hamilton 1964a, 1964b; Trivers 1971.

581. M. J. Hamilton et al. 2007, p. 4765.

582. Fehr and Schmidt 1999.

583. Fehr and Fischbacher 2003.

584. Jones 2003, 2010; Levinson 2012; Kemp and Regier 2012.

585. Morin and Goldberg 2004; Rendall 2004; Cheney and Seyfarth 2004; Chapais and Berman 2004.

586. Gouzoules and Gouzoules 1987; Korstjens 2008; Lehmann 2008.

587. Morgan 1871; Murdock 1960; Eggan 1975; Schneider 1984, 1986; Hughes 1988; Carsten 2000.

588. Shir-Vertesh 2012.

589. Vasey 1995; Gettler et al. 2011; Gray 2011.

590. Whitam 1983; Cardoso 2004.

591. Kinsey et al. 1948; Hunter 2007; Scacco 1975, 1982; Wooden and Parker 1982; Feierman 1990.

592. de Waal 1997; Werner 2006.

593. Bellis et at. 2005; Anderson 2006; Peritz and Rust 1972; Simmons et al. 2004; Voracek et al. 2008.

594. Borgerhoff Mulder and Caro 1983; Dickemann 1979; Hartung 1982; Low 1988.

595. Cashdan 1996.

596. Hrdy 1999a, 1999b, 2009.

597. Prince Peter of Greece and Denmark 1955; Leach 1955; Goldstein 1971; Hiatt 1980.

598. Bronikowski et al. 2011.

599. Leridon 2004; Atsalis and Videan 2009a, 2009b; Atsalis and Margulis 2006, 2008; Caro et al. 1995; de Velde et al. 1988.

600. Peccei 1995, 200.

601. Williams 1957.

602. Hawkes 1991, 1993; Hawkes et al. 1989, 1991, 1997, 1998.

603. Kaplan et al. 2000.

604. Kaplan et al. 2000, pp. 179–180; Strassmann 2011.

605. Tuttle 2006b, 2009.

606. de Waal 2005, pp. 113–114.

607. Bernstein 1991.

608. Alvergne et al. 2010; Gangestad and Thornhill 2004; Lancaster and Lancaster 1983.

12. COMMUNICATIVE APES

1. Marler 1985, p. 505; Owren and Rendall 2001.

2. Cheney and Seyfarth 1980, 1982a, 1982b, 1990; Gouzoules et al. 1984; Marler 1985; Zuberbühler et al. 1999.

3. Garner 1890.

4. Radick 2008; Garner 1896.

5. Carpenter 1940.

6. Haraway and Maples 1998.

7. Brockelman and Schilling 1984; Geissman 1984; Shafer et al. 1984; Marler and Tenaza 1977; Tenaza 1985.

8. Ellefson 1974, pp. 127–132.

9. Ellefson 1968, p. 193.

10. Ellefson 1968, 1974.

11. Baldwin and Teleki 1976; Carpenter 1940; Ellefson 1974; Chivers 1972, 1974.

12. Waller et al. 2012.

13. Raemaekers and Raemaekers 1985.

14. Clarke et al. 2006.

15. Raemaekers and Raemaekers 1985; Brockelman and Srikosamatara 1984.

16. Marshall et al. 1972.

17. Marshall 1981.

18. Heller et al. 2010; Marshall et al. 1972.

19. Gittins 1978, 1984.

20. Haimoff and Gittins 1985.

21. Heller et al. 2010.

22. Gittins 1980.

23. Mitani 1988.

24. Marshall and Marshall 1976; Marler and Tenaza 1977.

25. Brockelman 1975.

26. Marshall and Marshall 1976.

27. Haimoff 1985; Mitani 1984, 1985c.

28. Mitani 1984.

29. Mitani 1985c.

30. Mitani 1985a.

31. Mitani 1984, 1985a.

32. Marshall and Marshall 1976.

33. Geissmann 2002.

34. Kappeler 1984b.

35. Dallmann and Geismann 2009; Kappeler 1984b.

36. Tenaza and Tilson 1977; Marler and Tenaza 1977.

37. Chasen and Kloss 1927; Tenaza and Hamilton 1971.

38. Marshall and Marshall 1976, p. 237.

39. Tenaza 1975.

40. Tenaza 1975, 1976.

41. Tenaza 1976.

42. Whitten 1984.

43. Keith et al. 2009.

44. Chivers 1971a, 1972, 1974, 1976; Chivers and MacKinnon 1977; Chivers et al. 1975; Kawabe 1970; McClure 1964; Papaioannou 1973; Lamprecht 1970; Tembrock 1974; Haimoff 1981, 1983, 1984; Haraway et al. 1981.

45. Haimoff 1983.

46. Haimoff 1981, 1983, 1984.

47. Chivers et al. 1975; Wich and Nunn 2002, p. 478; Hewitt et al. 2002.

48. Chivers 1972, 1974, 1976; Haimoff 1981.

49. Chivers 1976.

50. Chivers and MacKinnon 1977.

51. Gittins and Raemaekers 1980, p. 75.

52. Liebal et al. 2004.

53. Marshall and Marshall 1976; Marler and Tenaza 1977; Tilson 1979.

54. Tilson 1979.

55. Hamilton and Arrowood 1978.

56. Haimoff 1987; Fan et al. 2009.

57. Goustard 1979.

58. Goustard 1965, 1976.

59. Sun et al. 2011.

60. Marshall et al. 1972.

61. Marshall and Marshall 1976.

62. Marshall et al. 1972; Marshall and Marshall 1976; Goustard 1979.

63. Demars et al. 1978.

64. Goustard 1979, 1982, 1984.

65. Schilling 1984.

66. Yerkes and Yerkes 1929; Davenport 1967; Horr 1975; MacKinnon 1971; Galdikas 1983.

67. MacKinnon 1974a.

68. Rijksen 1978.

69. Maple 1980, pp. 80–81.

70. Hardus et al. 2009, p. 52.

71. Wich, Swartz, et al. 2009; Lameira et al. 2012.

72. Bermejo and Omedes 1999.

73. Goodall 1986; Fossey 1972; Hardus et al. 2009, p. 53.

74. Hardus et al. 2009.

75. MacKinnon 1971, 1974, 1979; Horr 1972, 1975, 1977; Rodman 1973; Galdikas 1978, 1979, 1982a, 1983, 1985a, 1995; Galdikas and Insley 1988; Rijksen 1978; Mitani 1985d; Utami and Mitra Setia 1995; van Hooff 1995; van Schaik 1999; Delgado and van Schaik 2000; Wich and Nunn 2002; Mitra Setia and van Schaik 2007; Ross and Geissmann 2007; Delgado 2007; Delgado et al. 2009; Lumier and Wich 2008; Utami Amoko et al. 2009; Hardus et al 2009; Spillmann et al. 2010.

76. Galdikas 1979, p. 212.

77. Mitani and Stuht 1998.

78. Hofer 1972; MacKinnon 1971.

79. Short 1981.

80. Galdikas 1982a; Wich and Nunn 2002; Hardus et al. 2009; Spillmann et al. 2010.

81. Marten and Marler 1977; Marten et al. 1977; Galdikas and Insley 1988.

82. Horr 1975; MacKinnon 1971, 1974; Rijksen 1978.

83. Galdikas 1983.

84. Mitra Setia and van Schaik 2007.

85. Rodman 1973.

86. Mitani 1985b, 1985d.

87. Mitani 1985d.

88. Mitani 1985d; Tuttle 1986, p. 224.

89. MacKinnon 1971, p. 170.

90. MacKinnon 1971, 1974, 1979.

91. Tuttle 1986, p. 225.

92. Horr 1972, 1975; Rodman 1973; Galdikas 1979, 1983; Rijksen 1978; Mitra Setia and van Schaik 2007; Hardus et al. 2009; Utami Atmoko, Singleton et al. 2009; and Mitra Setia et al. 2009.

93. Horr 1977; Rijksen 1978.

94. Delgado 2007; Lameira and Wich 2008; Spillmann et al. 2010.

95. Delgao 2007; Ross and Geissmann 2007; Delgado et al. 2009.

96. Utami and Mitra Setia 1995.

97. Galdikas 1979, pp. 208–209.

98. MacKinnon 1971, 1974a.

99. Rijksen 1978, p. 230.

100. De Silva 1971, 1972.

101. Galdikas 1982c; Tuttle 1986, p. 227.

102. MacKinnon 1974b; Niemitz and Kok 1976; Rijksen 1978; Maple 1980.

103. Horr 1977; Rijksen 1978.

104. Schaller 1961.

105. MacKinnon 1974a; Rijksen 1978.

106. Rijksen 1978, p. 233.

107. MacKinnon 1974a; Rijksen 1978; Caeiro et al. 2013.

108. Davenport 1967; MacKinnon 1974; Rijksen 1978.

109. Rijksen 1978, fig. 107.

110. MacKinnon 1974, p. 61.

111. Rijksen 1978, pp. 220–222.

112. Cartmill and Byrne 2010.

113. Woodruff and Premack 1979; Premack 1984; Savage-Rumbaugh 1986; Call and To-masello 1994b; Tomasello and Call 1997, pp. 262–263, 268; Miles 1993; Zimmermann et al. 2009; Povinelli et al. 1992, 1996, 2003; Gómez 1996, 2005; Cartmill and Byrne 2007; Leavens 2004; Leavens and Hopkins 1998, 1999, 2005, 2007; Leavens et al. 1996, 2004, 2009; Leavens, Russell, and Hopkins 2005; Krause and Fouts 1997; Patterson 1978b, 1978c; Patterson and Linden 1981; Segerdahl et al. 2005.

114. Nissen 1931.

115. Tuttle 1986, p. 229.

116. Reynolds 1964.

117. Bethell et al. 2007.

118. Goodall 1963b, 1965; Lawick-Goodall 1967, 1968a, 1968b, 1971; Marler 1965, 1969, 1976; Marler and Hobbett 1975; Marler and Tenaza 1977.

119. Lawick-Goodall 1968a, 1968b; Marler 1976; Marler and Tenaza 1977.

120. Marler 1976.

121. Clark and Wrangham 1994; Kajikawa and Hasegawa 2000.

122. Lawick-Goodall 1968a; Marler and Hobbett 1975; Bauer and Philip 1983; Mitani et al. 1996.

123. Mitani et al. 1992.

124. Mitani et al. 1999.

125. Mitani 1994.

126. Mitani and Gros-Louis 1998.

127. Mitani and Brandt 1994.

128. Crockford et al. 2004; Janik and Slater 1997.

129. Marshall et al. 1999.

130. Marler and Hobbett 1975.

131. Tuttle 1986, p. 229; Mitani and Nishida 1993.

132. Reynolds and Reynolds 1965; Izawa and Itani 1966; Lawick-Goodall 1968a, 1968b; Marler 1976; Teleki 1973a; Ghiglieri 1984a, 1984b.

133. Clark 1993; Clark and Wrangham 1993; Ghiglieri 1984b.

134. Wilson et al. 2001, 2002.

135. Herbinger et al. 2009.

136. Notman and Rendall 2005, p. 187; Mitani and Nishida 1993; Mitani and Brandt 1994; Clark and Wrangham 1994.

137. Nissen 1931; Kortlandt 1962; Reynolds 1965; Reynolds and Reynolds 1965; Goodall 1963b; Lawick-Goodall 1968a, 1968b, 1971; Izawa and Itani 1966; Jones and Sabater Pí 1971; Sugiyama 1973; Nishida 1979; Mori 1982; Ghiglieri 1984b; Arcadi et al. 1998.

138. Jones and Sabater Pí 1971.

139. Reynolds and Reynolds 1965.

140. Lawick-Goodall 1968a, 1968b.

141. Reynolds 1965.

142. Tuttle 1986, p. 230.

143. Arcadi et al. 1998.

144. Mitani et al. 1996.

145. Hopkins et al. 2007.

146. Poss et al. 2006; Hyatt and Hopkins 1998; Hostetter et al. 2001; Leavens, Russell, and Hopkins 2005; Call and Tomasello 2007; Cartmill and Byrne 2007.

147. Povinelli and Eddy 1996; Theall and Povinelli 1999.

148. Povinelli et al. 1996.

149. Reaux et al. 1999.

150. Leavens, Hopkins, and Bard 2005.

151. Gómez et al. 1993; Gómez 2007, p. 731.

152. Tomilin and Yerkes 1935; Andrew 1962; Lawick-Goodall 1968a, 1968b; Clark 1977; Nicolson 1977.

153. de Waal 1982.

154. Slocombe and Zuberbühler 2007, 2011; Slocombe et al. 2009, 2010.

155. Sugiyama 1969.

156. Towsend et al. 2011, p. 914.

157. Marler and Tenaza 1977.

158. Reynolds and Reynolds 1965; Lawick-Goodall 1968a, 1968b; Marler 1976; van Hooff 1981.

159. Crockford and Boesch 2003.

160. Bygott 1979.

161. Marler and Tenaza 1977; Hayes 1951.

162. Temerlin 1975.

163. Nissen 1931; Reynolds and Reynolds 1965; Izawa and Itani 1966; Lawick-Goodall 1968a, 1968b; Marler and Tenaza 1977.

164. Teleki 1973c.

165. Burrows et al. 2006.

166. Ladygina-Kots 1935; van Hooff 1962, 1963, 1967, 1971, 1972, 1973, 1976; Reynolds and Luscombe 1969a, 1969b, 1969c, 1976; Chevalier-Skolnikoff 1973; Berdecio and Nash 1981; de Waal 1982; de Waal and van Hooff 1981; Tomasello et al. 1985; Lawick-Goodall 1968a, 1968b; Reynolds and Reynolds 1965; Marler 1976; Marler and Tenaza 1977; Plooij 1978a, 1979.

167. Andrew 1963a, 1963b, 1963c, 1964, 1965; van Hooff 1981.

168. Bauer 1986; Crockford and Boesch 2005.

169. Vick and Paukner 2010.

170. Slocombe and Newton-Fisher 2005.

171. Tuttle 1986, p. 232.

172. Lawick-Goodall 1968a.

173. Lawick-Goodall 1968a, 1968b; van Hooff 1973.

174. Tuttle 1986, p. 233.

175. Lawick-Goodall 1968a, 1968b; de Waal 1982.

176. Tutin and McGrew 1973; Tutin and McGinnis 1981; de Waal 1982.

177. Tuttle 1986, p. 234.

178. de Waal 1982, p. 161; 2007, p. 156.

179. Liebal et al. 2004; Pika et al. 2007.

180. Menzel 1971a, 1971b, 1973, 1974, 1978, 1979; Menzel and Halperin 1975.

181. Menzel 1979, p. 360.

182. Menzel 1971b, 1974.

183. Menzel 1971b.

184. Menzel 1973b.

185. Menzel 1973b, p. 218.

186. Slocombe and Zuberbühler 2005.

187. Tuttle 1986, pp. 240–241.

188. Fossey 1972; Schaller 1963, 1965a.

189. Parnell and Buchanan-Smith 2001.

190. Schaller 1963.

191. Fossey 1972.

192. Marler 1976; Marler and Tenaza 1977.

193. Fossey 1972; Marler 1976.

194. Harcourt et al. 1993; Harcourt and Stewart 1996, 2001; Stewart and Harcourt 1994.

195. Harcourt et al. 1993.

196. Harcourt and Stewart 2001, p. 259.

197. Harcourt et al. 1986; Seyfarth et al. 1994.

198. Harcourt and Stewart 1996; Morell 2011.

199. Schaller 1963, p. 218.

200. Fossey 1972, 1983.

201. Fossey 1972; Schaller 1963.

202. Harcourt et al. 1981.

203. Hess 1973.

204. Fossey 1972, p. 48.

205. Tuttle 1986, p. 240.

206. Du Chaillu 1861, p. 98; Emlen 1962; Schaller 1963, 1965, 1970.

207. Schaller 1963; Fossey 1983; Carpenter 1937.

208. Foley 1935; Yerkes and Yerkes 1929.

209. Pika et al. 2003.

210. Gentry et al. 2009, p. 527.

211. Harcourt 1993; Harcourt and Stewart 1996.

212. Mori 1983, 1984; Badrian and Badrian 1984; Kano 1980; Kano and Mulavwa 1984; Kuroda 1980; Harcourt et al. 1993; Mitani and Gros-Louis 1995; Mitani 1996; Veá and

Sabater Pí 1995; Bermejo and Omedes 1999; Savage-Rumbaugh 1984b; Savage-Rumbaugh et al. 1978; Rumbaugh et al. 1978; T. Patterson 1979; de Waal 1988; Pika et al. 2005; Pollick and de Waal 2007; Pollick et al. 2008; Clay and Zubrbühler 2009, 2011; Clay et al. 2012.

213. MacKinnon 1976.

214. Badrian and Badrian 1977.

215. de Waal 1988.

216. Hohmann and Fruth 1994, 1995.

217. Kano 1979.

218. Yerkes and Learned 1925; T. Patterson 1979; de Waal 1988; Mitani 1996.

219. Mitani and Gros-Louis 1995; Mitani 1996.

220. de Waal 1988; Pollick et al. 2008; Owren et al. 2011.

221. Hopkins and Savage-Rumbaugh 1991.

222. Clay and Zuberbühler 2009, 2011; Clay et al. 2012.

223. Taglialatela et al. 2003.

224. Benson, Fries et al. 2002; Taglialatela et al. 2003, 2004.

225. Mori 1984.

226. Mori 1983.

227. Bermejo and Olmedes 1999; de Waal 1988.

228. Bermejo and Omedes 1999.

229. Bermejo and Omedes 1999, p. 355.

230. Pollick et al. 2008.

231. Pollick and de Waal 2007, p. 8188.

232. Furuichi 1989.

233. Johnson et al. 1999.

234. Kano 1980.

235. Kuroda 1984; Furuichi 1989; Idani 1991, 1995.

236. Idani 1975.

237. Vervaecke et al. 2000a.

238. Stevens et al. 2005.

239. Veá and Sabater Pí 1998; Inoue-Nakamura and Matsuzawa 1997.

240. Pika and Mitani 2009.

241. de Waal 1989a; Pollick and de Waal 2007.

242. de Waal 1998; Segerdahl et al. 2005.

243. Vatsyayana 1982.

244. Savage-Rumbaugh et al. 1978; Rumbaugh et al. 1978.

245. Agha 1997.

246. Furness 1916; Kellogg and Kellogg 1933; Hayes and Hayes 1951, 1954; Hayes 1951.

247. Tuttle 1986, p. 198.

248. Kellogg 1969.

249. Kellogg and Kellogg 1933; Hayes and Hayes 1951, 1954; Hayes 1951.

250. Laidler 1978, 1980.

251. Yerkes 1925, p. 180.

252. Gardner and Gardner 1969.

253. Fouts and Fouts 1999.

254. Stokoe 1960, 1970.

255. Tuttle 1986, p. 199.

256. Fouts 1972.

257. Terrace 1981; Ristau and Robbins 1982.

258. Gardner and Gardner 1969, 1971, 1972, 1974.

259. McMurray 2007; Bloom 1973.

260. Gardner and Gardner 1969, 1971.

261. Gardner and Gardner 1971, 1975; Van Cantfort and Rimpau 1982; Van Cantfort et al. 1989.

262. Gardner and Gardner 1974.

263. Linden 1974, 1981; Tuttle 1986, p. 200.

264. Gardner and Gardner 1975, 1978, 1980, 1984, 1988, 1989, 1998.

265. Van Cantfort and Rimpau 1982; Van Cantfort et al. 1989; Gardner and Gardner 1989, 1998.

266. Gardner and Gardner 1978.

267. Gardner and Gardner 1989.

268. Fouts and Mills 1997.

269. Fouts 1974, 1975a.

270. Fouts 1975a.

271. Fouts 1975b.

272. Fouts 1975a, 1975b, 1990; Fouts et al. 1976; Fouts and Couch 1976; Allen and Fortin 2013.

273. Fouts et al. 1982.

274. Cianelli and Fouts 1998.

275. Patterson and Linden 1981; Patterson 1978a, 1978b, 1978c, 1980, Patterson and Cohn 1990; Patterson and Gordon 1993, 2001; Bonvillian and Patterson 1999; Miles 1978, 1983, 1990, 1993, 1999; Shapiro 1982; Shapiro and Galdikas 1999.

276. Patterson and Linden 1981.

277. Patterson and Linden 1981; Tuttle 1986, p. 204; Anonymous 2000.

278. Patterson and Gordon 2001.

279. Patterson 1978a, 1978b, 1980, 1986; Patterson and Linden 1981; Patterson and Gordon 2001; Tuttle 1986, p. 204.

280. Tuttle 1986, p. 204.

281. Yollin 2005a, 2005b, 2005c.

282. Patterson and Cohn 1990; Patterson 2001.

283. Shapiro 1982.

284. Shapiro and Galdikas 1999.

285. Miles 1983.

286. Miles 1990.

287. Miles 1983, 1990.

288. Miles 1983, 1977, 1978.

289. Miles 1990, p. 537.

290. Terrace 1979a, 1979b, 1981; Hess 2008.

291. Terrace and Bever 1976.

292. Terrace 1979a, pp. 256–261; Marsh 2011.

293. Terrace 1979a, 1982; Pfungst 1911.

294. Cummings 1972, p. 614; Yang 2013.

295. Terrace et al. 1979a.

296. Mounin 1976.

297. Premack 1970, 1971a, 1971b, 1971c, 1976a, 1976b, 1978; Premack and Hines 1976; Premack and Premack 1972, 1975, 1983, 2003; Premack and Schwartz 1966.

298. Premack and Premack 1983, pp. 115–116.

299. Premack and Premack 1983, 2003.

300. Terrace 1979.

301. Savage-Rumbaugh et al. 1980; Savage-Rumbaugh and Rumbaugh 1978.

302. Savage-Rumbaugh and Rumbaugh 1978.

303. Savage-Rumbaugh et al. 1993, p. 70.

304. Premack and Premack 2003.

305. Premack 1970, 1976a.

306. Rumbaugh 1974, 1977a.

307. Rumbaugh 1974, 1977a, 1977b, 1977c, 1978; Rumbaugh and Gill 1975, 1976a, 1976b, 1977; Rumbaugh, Gill, Brown et al. 1973; Rumbaugh, Gill, and von Glasersfeld 1973, 1974; Rumbaugh, Gill, von Glasersfeld et al. 1975; Rumbaugh, von Glasersfeld, Gill, et al. 1975; Rumbaugh, von Glasersfeld, Warner et al. 1973, 1974; Rumbaugh et al. 1978; Essock et al. 1977; Gill 1977, 1978; Gill and Rumbaugh 1974a, 1974b; von Glasersfeld 1976; von Glasersfeld et al. 1973; Pate and Rumbaugh 1983.

308. Rumbaugh and Gill 1975.

309. Beran et al. 2000; Beran 2004.

310. Rumbaugh and Gill 1976a; Rumbaugh and Beran 2005.

311. Rumbaugh 1977b.

312. Savage-Rumbaugh and Rumbaugh 1978, 1980, 1982; Savage-Rumbaugh 1979, 1981, 1984a, 1986; Savage-Rumbaugh, Rumbaugh, and Boysen 1978a, 1978b, 1980; Savage-Rumbaugh et al. 1980, 1983; Savage-Rumbaugh and Sevcik 1984; Rumbaugh et al. 1982; Greenfield and Savage-Rumbaugh 1984; Sevcik and Savage-Rumbaugh 1994; Brakke and Savage-Rumbaugh 1995, 1996.

313. Savage-Rumbaugh, Rumbaugh, and Boysen 1978b.

314. Savage-Rumbaugh et al. 1980.

315. Deacon 1997, p. 86.

316. Savage-Rumbaugh et al. 1993.

317. Jackendoff 2007.

318. Tuttle 1986, p. 212.

319. Bronowski and Bellugi 1970; Katz 1976; Limber 1977; Mistler-Lachman and Lachman 1974; Terrace 1979a, 1979b, 1981, 1982; Terrace et al. 1977, 1979; Seidenberg and Petitto 1979, 1987; Sebeok and Umiker-Sebeok 1980; Sebeok and Rosenthal 1981; Epstein et al. 1980; Brown 1981; Muncer 1983b; Thompson and Church 1980; Umiker-Sebeok and Sebeok 1981, 1982; Ristau and Robbins 1982; Ristau 1996; Michael 1984; Sanders 1985; Wynne 2008b; Cohen 2010.

320. Hixson 1998.

321. Savage-Rumbaugh et al. 2009, p. 27.

322. Savage-Rumbaugh 1984b, 1987, 1988, 1990, 1991; Savage-Rumbaugh et al. 1983, 1985, 1986, 1990, 1992, 1993, 1998, 2000, 2001, 2005, 2006, 2009; Savage-Rumbaugh and Rumbaugh 1993; Savage-Rumbaugh and Lewin 2004; Savage-Rumbaugh and Fields 2000; Hillix and Rumbaugh 2004; Greenfield and Savage-Rumbaugh 1990, 1993; Lyn and Savage-Rumbaugh 2000; Menzel et al. 2002; Rumbaugh et al. 1991; Rumbaugh, Savage-Rumbaugh, and Washburn 1996.

323. Menzel 1999, 2005, Menzel et al. 2002.

324. Savage-Rumbaugh et al. 1998, p. 136.

325. Savage-Rumbaugh et al. 1986.

326. Rumbaugh et al. 1996, p. 119 (emphasis in original).

327. Savage-Rumbaugh et al. 1993; Bates 1996, p. 239.

328. Burling 2005, pp. 12–13.

329. Greenfield and Savage-Rumbaugh 1990; Savage-Rumbaugh et al. 1993; Rumbaugh er al. 1994.

330. Benson, Greaves et al. 2002, p. 55.

331. Pedersen and Fields 2009, p. 22.

332. Savage-Rumbaugh 1991; Lyn et al. 2011.

333. Savage-Rumbaugh et al. 1989; Rumbaugh et al. 2003.

13. LANGUAGE, CULTURE, IDEOLOGY, SPIRITUALITY, AND MORALITY

1. Tattersall 2001; Deacon 1997.

2. Crelin 1987; Perkins and Kent 1986; Greenfield 1968.

3. Fitch and Hauser 1995.

4. McNeill 1992; Burling 2005; MacNeilage 2008.

5. Tomasello 1996; Boesch and Tomasello 1998.

6. Jusczyk and Hohne 1997; Bales 1993; Mahmoudzadeh et al. 2013; Oller et al. 2013.

7. Dietrich et al. 2007; Bergelson and Swingley 2012.

8. Burling 2005; Emmorey 1999, 2002.

9. Martinet 1960; Hockett 1960, 1966; Fortuny 2010; Buchler 1956, pp. 98–119; Agha 1997; Deacon 1997.

10. Goldin-Meadow 2005; Sandler et al. 2005.

11. Tuttle 1986, p. 212; Pinker 1997, p. 124; Kako 1999a, 1999b; Shanker et al. 1999; Pepperberg 1999; Hauser et al. 2002; Pinker and Jackendoff 2005; Fitch et al. 2005; Jackendoff and Pinker 2005; Kinsella 2009.

12. Darwin 1872; Burling 2005.

13. Burling 2005.

14. Fromkin 1978.

15. King and Shanker 2003.

16. Saussure 2011, p. 16; Buchler 1955.

17. Deacon 1997.

18. Agha 1997.

19. Struhsaker 1967; Cheney and Seyfarth 1982a, 1990.

20. Burling 1993, 2005; Jack et al. 2012.

21. Tobias 1987; Wilkins and Wakefield 1995; Lieberman 2007.

22. Klein 1995, 2009; Aiello 1996a, 1996b.

23. Bruner and Holloway 2010.

24. Papadopoulos et al. 1989; Laitman et al. 1990; Frayer 1993; Frayer and Nicolay 2000; Lieberman 1993, 1994a, 1994b; Arensburg 1994; Stringer and Gamble 1993, p. 90.

25. Alemseged et al. 2006; Bar-Yosef et al. 1992; Arensburg et al. 1989.

26. Lieberman and Crelin 1971; Lieberman 1975, 1976, 1984, 1991, 1994; Burr 1976; Leiberman et al. 1969; Laitman 1983, 1985b; Laitman and Heimbuch 1982; Laitman et al. 1978, 1979, 1982, 1992; Tuttle 2006b; DuBrul 1958, 1976; Falk 1975; Duchin 1990; Houghton 1993; Schepartz 1993; Fitch 2000, 2002, 2009b; Nishimura 2005, 2006; Nishimura et al. 2003, 2008; Wind 1976; Enfield 2010.

27. Kay et al. 1998.

28. DeGusta et al. 1999; Jungers et al. 2003.

29. Tobias 1987.

30. Leakey 1973b; Walker 1981, pp. 202–203; Tobias 1991, pp. 53–54 and plates 1–15.

31. Holloway 1983b.

32. Tobias 1987, 1996.

33. Tuttle 1992; Amunts et al. 1999.

34. Tuttle 1986, pp. 198–199, 203.

35. Christiansen and Kirby 2003, p. 2; Liberman and Whalen 2002.

36. Condillac 1746; Rousseau 1755; Tylor 1868, 1874, 1877, p. 36; Morgan 1877; Romanes 1889; Wundt 1911; Paget 1930, 1944; Jóhannesson 1949, 1950; Diamond 1959, p. 265; Hewes 1973a, 1973b, 1976, 1983; Hockett 1978, pp. 299–300; Kimura 1979, pp. 201–203; Parker and Gibson 1979, p. 373; Stokoe 1974, 1980, p. 383; Rizzoletti and Arbib 1998; Armstrong et al. 1994, 1995; Armstrong 1999, p. 32, 2008; Armstrong and Wilcox 2007; Corballis 1991, pp. 156–159, 1998, 1999, 2002, p. 184, 2003a, 2003b; Gentilucci and Corballis 2006; Knight 2000; Arbib 2002, 2003, 2005a, 2005b; Arbib et al. 1996; 2008; Pika et al. 2005; Tomasello 2002, 2003, 2008, p. 328; Remedios et al. 2009; Darwin 1871; Wallace 1881, 1895; Carini 1971; Hockett and Ascher 1964; Steklis and Raleigh 1979: Steklis 1985; Bradshaw 1988; Bickerton 1990, 2003, p. 81, 2009, p. 59; Gibson 1991; Dunbar 1993, 1996, p. 115, 2003b, 2009; Hauser 1993b; Deacon 1997, p. 362; Goldin-Meadow and McNeill 1999: Wray 1998, 2000, 2002; King 2003, 2004; Falk 2004; Burling 2005, p. 123; MacNeilage and Davis 2005; Jackendoff and Lerdahl 2006; Jackendoff 2007; MacNeilage 1998a, 1998b, 2008, pp. 201, 287; Emmorey 2005; McNeill et al. 2005; Pinker and Bloom 1990; Pinker 1994, 2003: Hurford 2003, 2007, 2012; Hauser and Fitch 2003; Studdert-Kennedy and Goldstein 2003; Lieberman 2003; Fitch 2005a, 2005b, 2005c, 2006, 2010; Jackendoff 1999, 2002, 2007; Steels 2009; Mithen 2009; Cross and Woodruff 2009; Odling-Smee and Laland 2009; Woll 2009.

37. Fitch 2010, p. 9.

38. Hewes 1973a, 1973b, 1983.

39. Hewes 1973a, p. 101.

40. Dart 1957; Vrba 1975; Brain 1981.

41. Fitch 2009b, 2010.

42. Hewes 1983.

43. Wallace 1895; Hewes 1973a, 1973b.

44. Tomasello and Call 2007; Fitch 2010; Hewes 1973b.

45. Arbib 2002, 2003, 2004, 2005a, 2005b; Arbib and Rizzolatti 1996; Arbib et al. 2008; Rizzolatti and Arbib 1998; Rizzolatti and Buccino 2005; Rizzolatti and Craighero 2004; Rizzolatti and Sinigaglia 2006; Rizzolatti et al. 1996a, 1996b, 2001; Fadiga et al. 2002; Ferrari and Fogassi 2012; Fogassi and Ferrari 2007; Fogassi et al. 2005; Brass and Rüschemeyer 2010; Iacoboni 2012.

46. Rizzolatti 1996; Rizzolatti and Arbib 1998; Arbib 2002, 2003, 2005a, 2005b, 2012; Arbib et al. 2008; Fitch 2010.

47. Stamenov 2002, pp. 269–270.

48. Hurford 2004, p. 297; Newmeyer 1991, p. 6.

49. Kohler et al. 2002; Keysers et al. 2003.

50. Gazzola et al. 2006.

51. Fitch 2010.

52. Lieberman 1984; Bickerton 1990, 1993, 1998, 2000a, 2000b, 2003, 2007, 2009; Givón 1995, p. 435: Jackendoff 2002, 2007.

53. Fitch 2010; Bickerton 1990, 1993, 1998, 2000a, 2000b, 2003, 2007, 2009; Pinker and Bloom 1990; Newmeyer 1991; Hauser et al. 2002; Cheney and Seyfarth 2007; Hurford 2007.

54. Bickerton 2003; Fitch 2010, p. 410.

55. Curtiss 1977; Rymer 1994; Fitch 2010, pp. 404–41.

56. Jackendoff 1999, 2002; Fitch 2010, p. 410.

57. Jackendoff 2002, p. 236.

58. Jackendoff 2002.

59. Fitch 2006, 2010, pp. 466–468; Lerdahl and Jackendoff 1983; Trainor and Trehub 1992; Trehub 2000, 2001, 2003; Tramo 2001; Trainor et al. 2002; Trehub and Hannon 2006; Zatorre and Salimpoor 2013.

60. Darwin 1871; Jespersen 1922; Livingstone 1973; Lerdahl and Jackendoff 1983; Aiello and Dunbar 1993; Richman 1993; Brown et al. 2000; Brown 2000; Falk 2000; Jerison 2000; Mithen 2005; Hurford 2007, 2012; Fitch 2006, 2010; Patel 2008.

61. Reynolds 1968; Burling 1993, 2005; Weiss 1974; Marler 2000; Slater 2000; Bickerton 2000; Darwin 1871; Hockett and Ascher 1964; Merker 2000; Miller 2000; Ujhelyli 2000; Szathmáry 2001; Hauser and McDermott 2003; Patel 2008; Fitch 2010.

62. Bickerton 2000a; Brown 2000; Brown et al. 2000; Fitch 2006; Geissmann 2000; Gray et al. 2001; Hauser 2000b; Mâche 2000; Marler 2000; Merker 2000; Miller 2000; Payne 2000; Slater 2000; Ujhelyi 2000; Whaling 2000.

63. Darwin 1871; Jespersen 1922; Mithen 2005; Brown 2000; Wray 1998, 2000, 2002; Kirby 1999, 2000; Fitch 2010.

64. Darwin 1871; Disssanayake 2000.

65. Fitch 2004, 2009a, 2010, p. 503.

66. Fitch 2010, p. 504.

67. Fitch 2010, pp. 504–505.

68. Fitch 2010, p. 505.

69. Goldin-Meadow and McNeill 1999.

70. Gardner 1983, 2006.

71. MacLarnon and Hewitt 2004.

72. Bramble and Carrier 1983; Tuttle 1994a.

73. Whiten et al. 1999; Goodman et al. 1998; Begun 1999; de Waal and Bonnie 2009; McGrew 2009; van Schaik 2009; Whiten 2009.

74. Cavalieri and Singer 1993.

75. McGrew 1998b, p. 607.

76. Mayr 1963, p. 34.

77. Rendell and Whitehead 2001; de Waal 1999, p. 635; Wrangham et al. 1994; Whitehead 2009; Sargeant and Mann 2009.

78. van Lawick-Goodall 1973a; McGrew 1992a, 1998a, 1998c, 2004; McGrew et al. 2003; Whiten et al. 1999, 2001, 2003, 2007; van Schaik, Ancrenaz, et al. 2003, van Schaik, Fox, and Fechtman 2003; van Schaik, van Noordwijk, and Wich 2006; Nakamura and Nishida 2006; Whiten and van Schaik 2007; van Schaik et al. 1996, 1999, 2001, Fox et al. 1999; de Waal and Seres 1997; Bonnie and de Waal 2006; Byrne 2007; Marshall-Pescini and Whiten 2008; Boesch 2003; Boesch et al. 1994; Joulian 1994, 1995, 1996; Nakamura and Nishida 2012; Nishida et al. 1999; Foley and Mirazón-Lahr 2003; Tennie et al. 2008; Sawyer and Robbins 2009; Gruber et al. 2009; O'Hara and Lee 2006; Laland and Hoppitt 2003.

79. Tuttle 1986, 2001, 2006a; Hill 2009; Perry 2006, 2009; Galef 2009; Tomasello 2009; Rapchan 2011.

80. Boesch and Tomasello 1998; Laland and Galef 2009; Laland et al. 2009; Strier 2003.

81. Tomasello 1999a, Cole 1996.

82. Boesch and Tomasello 1998, p. 591.

83. Geertz 1973; Handler 2004; Harris 1999; Kuper 1999; Sahlins 1976; Schneider 1968; Shore 1996; Taylor 1948; Wagner 1975; White 1959; White and Dillingham 1973; Kroeber and Kluckhohn 1952.

84. Kroeber and Kluckhohn 1952, p. 357.

85. Brightman 1995; Abu-Lughod 1991; Bordieu 1977; Clifford 1986, 1988; Rosaldo 1989.

86. Cowgill 1993; Hodder 1982; Ortner 1984; Singer 1980; Watson 1995; Wolf 1984; Turner 1967, 1969; Yengoyan 1986; Marcus and Fischer 1986; Appadurai 1988, 1990; Turner 2000.

87. Compare: van Schaik et al. 2003.

88. Boesch and Tomasello 1998, p. 610.

89. Borofsky et al. 2001; Bourdieu 1977; Geertz 1973; Handler 2004; Harris 1999; Sahlins 1976, 1981; Schneider 1968; Shore 1999; Cronk 1999, pp. 132–133; White and Dillingham 1973; Ingold 2001; Brightman 1995.

90. Tylor 1874.

91. Savage-Rumbaugh et al. 1989, 1993; Menzel 1999; Menzel et al. 2002.

92. McElreath et al. 2003; Fehr and Fischbacher 2004; Gil-White 2001; Chase 2006; Sterelny 2009; Hill 2009; Strum 2012.

93. Chase 2006, p. 35; compare Radcliffe-Brown 1922; Durkheim 1938, 1965; Geertz 1973; Rappaport 1999.

94. de Waal 1989b, 2009; Rifkin 2009; Keltner 2009; Campbell et al. 2009; Campbell and de Waal 2011; Keltner et al. 2010; Sussman and Cloninger 2011; Silk and House 2011; Cheney 2011; Singer and Hein 2012.

95. Marshack 1972, 1990; Foster 1990; Mithen 1996; Klein 1999; Chase 2006; Sadier et al. 2012; Henshilwood et al. 2011; White et al. 2012.

96. Tuttle 2002.

97. Werdelin and Peigné 2010.

98. Dobzhansky et al. 1977, p. 454.

99. Goodall 1986, p. 203; Bekoff 2007; Bekoff and Pierce 2009; Teleki 1973c; Stewart et al. 2012.

100. Dobzhansky et al. 1977, p. 454.

101. Weiss et al. 2011.

102. Wrangham 1995, 1999; Wrangham and Peterson 1996.

103. Boesch et al. 2008, p. 519; Wrangham 1999; Ghiglieri 1987, 1989; Foley 1989; Foley and Lee 1989.

104. Marks 2002; Sussman 1999, 2004; Hart and Sussman 2005; Gavrilets 2012; Dijker 2011.

105. Manson and Wrangham 1991, p. 369.

106. Manson and Wrangham 1991, pp. 369–370; Wrangham 1999; Wilson and Wrangham 2003; Wrangham and Wilson 2004.

107. Boehm 1992, pp. 169–170; Mathew and Boyd 2011.

108. Knauft 1991; Rodseth et al. 1991; Boehm 1992; Burbank 1992; Sussman 1997; Kelly 2000, 2005.

109. Pusey et al. 1997; Sugiyama 1999.

110. Wrangham 1999, p. 1.

111. Wrangham 1999, p. 1; Wilson and Wrangham 2003; Wrangham and Wilson 2004.

112. Dart 1953; Washburn and Lancaster 1968; Wrangham and Peterson 1996; Wrangham 1999.

113. Wrangham 1999, p. 25.

114. Wrangham 1974b; Reynolds 1975; Nishida 1979; Power 1991.

115. Rogers and Hammerstein II 1949.

116. Koenig et al. 2008.

117. Sussman 2000; Fry 2007; Sussman et al. 2005; de Waal 1996b, 2013b; Roca and Helbing 2011.

118. Blum 1994; Roush 1997; Cavalieri 1996; Cavalieri and Singer 1993; Fouts and Mills 1997; Glatston 1996; Tuttle 1994b, 1998, 2006b; Radick 2000; Wolfe 2003, 2004; Schroepfer et al. 2011; Panksepp 2011; Stanford 2012.

119. Tuttle 2011.

120. Balog 1993; Boulle 1963a; Cook 1997; Peterson and Goodall 1993; Tuttle 1998.

121. Wallis 1997, p. xi; Rowe 1996, p. v; Brandon-Jones et al. 2004; Grubb et al. 2004.

122. Marsh 2003.

123. Struhsaker 1997; Caldecott and Miles 2005.

124. Carneiro da Cunha 1986, Bullock and Hodder 1997; Janzen 1997; Sarrazin and Barbault 1996; Sarukhán 1997; Sponsel 1997; Strier 1997; Stanford 2012.

125. Boulle 1963b.
126. Tuttle 1998; 2007.
127. Tuttle 1998, 2002.
128. Mittermeier et al. 2011.
129. Tuttle 1998, 2006.
130. Cavalieri and Singer 1993; Cavalieri 1996; Tuttle 1994b, 2007.

References

Abbate, E., Bianelli, A., Azzaroli, A., Benvenuti, M., Tesfamariam, B., Bruni, P., Cipriani, N., Clarke, R. J., Ficcarelli, G., Macchiarelli, R., et al. 1998. A one-million-year-old *Homo* cranium from the Danakil (Afar) Depression in Eritrea. *Nature* 393:458–460.

Aboitiz, F., Scheiel, A. B., Fisher, R. S., and Zaidel, E. 1992. Individual differences in brain asymmetries and fiber composition in the human corpus callosum. *Brain Research* 598:154–161.

Abu-Lughod, L. 1991. Writing against culture. In R. Fox, ed., *Recapturing Anthropology: Working in the Present,* 137–162. Santa Fe: School of American Research Press.

Achoka, I. 1993. *Home Range, Group Size and Group Composition of Mountain Gorillas (Gorilla gorilla beringei) in Bwindi Impenetrable National Park, South-Western Uganda.* MSc thesis, Makerere University.

Adachi, J., and Hasegawa, M. 1995. Improved dating of the human/chimpanzee separation in the mitochondrial DNA tree: heterogeneity among amino acid sites. *Journal of Molecular Evolution* 40:622–628.

Adams, C. G., Gentry, A. W., and Whybrow, P. J. 1983. Dating the terminal Tethyan event. *Utrecht Micropaleontological Bulletins* 30:273–298.

Adler, D. S., Bar-Yosef, O., Belfer-Cohen, A., Tshabraishvili, N., Boaretto, E., Mercier, N., Valladas, H., and Rink, W. J. 2008. During the demise: Neandertal extinction and the establishment of modern humans in the southern Caucasus. *Journal of Human Evolution* 55:817–833.

Agha, A. 1997. "Concept" and "communication" in evolutionary terms. *Semiotica* 116:189–216.

Aguirre, E. 1972. Les rapports phylétiques de *Ramapithecus* et *Kenyapithecus* et l'origine des Hominidés. *L'Anthropologie (Paris)* 76:501–523.

———. 1975. *Kenyapithecus* and *Ramapithecus*. In R. H. Tuttle, ed., *Paleoanthropology, Morphology and Paleoecology,* 99–104. The Hague: Mouton.

———. 1977. Aguirre, E., and de Lumley, M.-A. Fossil men from Atapuerca, Spain: their bearing on human evolution in the Middle Pleistocene. *Journal of Human Evolution* 6:681–688.

Aguirre, E., de Lumley, M.-A., Basabe, J. M., and Botella, M. 1980. Affinities of the mandibles from Atapuerca and L'Arago, and some East African fossil hominids. In R. E. Leakey and B. A. Ogot, eds., *Proceedings of the 8th Panafrican Congress of Prehistory and Quaternary Studies Nairobi, 5 to 10 September 1977,* 171–173. Nairobi: International Louis Leakey Memorial Institute for African Prehistory.

Aguirre, E., and Rosas, A. 1985. Fossil man from Cueva Mayor, Ibeas, Spain: new findings and taxonomic discussion. In E. Delson, ed., *Ancestors: The Hard Evidence,* 319–328. New York: Alan R. Liss.

Agustí, J., Cabera, L., and Garcés, M. 2001. Chronology and zoogeography of the Miocene hominoid record in Europe. In L. de Bonis, G. D. Koufos, and P. Andrews, eds., *Hominoid Evolution and Climatic Change in Europe,* Vol. 2: *Phylogeny of the Neogene Hominoid Primates of Eurasia,* 2–18. Cambridge: Cambridge University Press.

Agustí, J., Köhler, M., Moyà-Solà, S., Cabera, L., Garcés, M., and Parés, J. M. 1996. Can Llobateres: the pattern and timing of the Vallesian hominoid radiation reconsidered. *Journal of Human Evolution* 31:143–155.

Ahsan, F. 1995. Fighting between two females for a male in the hoolock gibbon. *International Journal of Primatology* 16:731–737.

———. 2001. Socio-ecology of the hoolock gibbon *(Hylobates hoolock)* in two forests of Bangladesh. In *The Apes: Challenges for the 21st Century, Conference Proceedings, May 10–12, 2000,* 286–299. Brookfield, IL: Brookfield Zoo.

Aiello, L. C. 1986. The relationships of the Tarsiiformes: a review of the case for the Haplorhini. In B. Wood, L. Martin, and P. Andrews, eds., *Major Topics in Primate and Human Evolution,* 47–65. Cambridge: Cambridge University Press.

———. 1993. The fossil evidence for modern human origins in Africa: a revised view. *American Anthropologist* 95:73–96.

———. 1996a. Terrestriality, bipedalism and the origin of language. *Proceedings of the British Academy* 88:269–289.

———. 1996b. Hominine preadaptations for language and cognition. In P. Mellars and K. Gibson, eds., *Modelling the Early Hominid Mind,* 89–99. Cambridge, UK: McDonald Institute for Archaeological Research.

Aiello, L. C., and Dunbar, R. I. M. 1993. Neocortex size, group size, and the evolution of language. *Current Anthropology* 34:184–193.

Aiello, L. C., and Wheeler, P. 1995. The expensive-tissue hypothesis. *Current Anthropology* 36:199–221.

Aigner, J. S. 1976. Chinese Pleistocene cultural and hominid remains: a consideration of their significance in reconstructing the pattern of human bio-cultural development. In A. K. Ghosh, ed., *Le Paléolithique Inférieur et Moyen en Inde, en Asie Centrale, en Chine et dans le Sut-Est Asiatique,* 65–90. Paris: UISPP Colloque VII, Centre National de la Recherche Scientifique.

———. 1978. Important archaeological remains from North China. In F. Ikawa-Smith, ed., *Early Palaeolithic in South and East Asia,* 163–232. The Hague: Mouton.

Aigner, J. S., and Laughlin, W. S. 1973. The dating of Lantian Man and his significance for analyzing trends in human evolution. *American Journal of Physical Anthropology* 39:97–110.

Akazawa, T., Aoki, K., and Bar-Yosef, O. 1998. *Neandertals and Modern Humans in Western Asia*. New York: Plenum Press.

Akney, C. D. 1992. Sex differences in relative brain size: the mismeasure of women, too? *Intelligence* 16:329–336.

Alba, D. M., Moyà-Solà, S., Casanovas-Vilar, I., Galindo, J., Robles, J. M., Rotgers, C., Furiò, M., Angelone, C., Köhler, M., Garcés, M. et al. 2006. Los vertebrados fósiles del Abocador de Can Mata (els Hostalets de Pierola, Anoia, Cataluña), una sucesión de localidades del Aragoniense superior (MN6 y MN7+8) de la cuenca del Vallès-Penedes, Campañas 2002–2003, 2994 y 2005. *Estudios Geoógicos* 62:295–312.

Alba, D. M., Moyà-Solà, S., and Köhler, M. 2001. Canine reduction in the Miocene hominoid *Oreopithecus bambolii:* behavioural and evolutionary implications. *Journal of Human Evolution* 40:1–16.

Alba, D. M., Moyà-Solà, S., Malgosa, A., Casanovas-Vilar, I., Robles, J. M., Almécija, S., Galindo, J., Rotgers, C., and Mengual, J. V. B. 2010. A new species of *Pliopithecus* Gervais, 1949 (Primates: Pliopithecidae) from the Middle Miocene (MN8) of Abocador de Can Mata (els Hostalets de Pierola, Catalonia, Spain). *American Journal of Physical Anthropology* 141:52–75.

Aldrich-Blake, F. P. G., and Chivers, D. J. 1973. On the genesis of a group of siamang. *American Journal of Physical Anthropology* 38:631–636.

Alemseged, Z., Spoor, F. Kimbel, W. H., Bobe, R., Geraads, D., Reed, D., and Wynn, J. G. 2006. A juvenile early hominin skeleton from Dikika, Ethiopia. *Nature* 443:296–301.

Alexander, R. D. 1979. *Darwinism and Human Affairs*. Seattle: University of Washington Press.

Alexander, R. M. 1984. Stride length and speed for adults, children, and fossil hominids. *American Journal of Physical Anthropology* 63:23–27.

———. 1991. Elastic mechanisms in primate locomotion. *Zeitschrift für Morphologie und Anthropologie* 78:315–320.

Alexeev, V. P. 1986. *Origin of the Human Race*. Moscow: Progress Publishers.

Alfred, J. R. B., and Sati, J. P. 1990. Survey and census of the hoolock gibbon in West Garo Hills, Northeast India. *Primates* 31:299–306.

———. 1991. On the first record of infanticide in the hoolock gibbon—*Hylobates hoolock* in the wild. *Records of the Zoological Society of India* 89:319–321.

Allbrook, D., and Bishop, W. W. 1963. New fossil hominoid material from Uganda. *Nature* 197:1187–1190.

Allen, G. 1879. *The Colour-Sense: Its Origin and Development*. London: Trübner.

Allen, G., McColl, R., Barnard, H., Ringe, W. K., Fleckstein, J., and Cullum, C. M. 2005. Magnetic resonance imaging of cerebellar-prefrontal and cerebellar-parietal functional connectivity. *NeuroImage* 28:39–48.

Allen, J. S., Damasio, H., and Grabowski, T. J. 2002. Normal neuroanatomical variation in the human brain: an MRI-volumetric study. *American Journal of Physical Anthropology* 118:341–358.

Allen, M., and Schwartz, B. L. 2008. Mirror self-recognition in a gorilla *(Gorilla gorilla gorilla)*. *Electronic Journal of Integrative Biology* 5:19–24.

Allen, N. A. 2008. Tetradic theory and the origin of human kinship systems. In N. J. Allen, H. Callan, R. Dunbar, and W. James, eds., *Early Human Kinship: From Sex to Social Reproduction*, 96–1152. Malden, MA: Blackwell.

Allen, T. A., and Fortin, N. J. 2013. The evolution of episodic memory. *Proceedings of the National Academy of Sciences of the United States of America* 110 (Suppl. 2): 10379–10386.

Allman, J. M., Watson, K. K., Tetreault, N. A., and Hakeem, A. Y. 2005. Intuition and autism: a possible role for Von Economo neurons. *Trends in Cognitive Sciences* 9:367–373.

Almécija, S., Alba, D. M., and Moyà-Solà, S. 2012. The thumb of Miocene apes: new insights from Castell de Barberà (Catalonia, Spain). *American Journal of Physical Anthropology* 148:436–450.

Almécija, S., Alba, D. M., Moyà-Solà, S., and Köhler, M. 2007. Orang-like manual adaptations in the fossil hominoid *Hispanopithecus laietanus:* first steps towards great ape suspensory behaviours. *Proceedings of the Royal Society B* 274:2375–2384.

Almquist, A. J. 1974. Sexual differences in the anterior dentition in African primates. *American Journal of Physical Anthropology* 40:359–367.

Alp, R. 1993. Meat eating and ant tipping by wild chimpanzees in Sierra Leone. *Primates* 34:463–468.

———. 1997. "Stepping-sticks" and "seat-sticks": new types of tools used by wild chimpanzees *(Pan troglodytes)* in Sierra Leone. *American Journal of Primatology* 41:45–52.

Alp, R., and Kitchener, A. C. 1993. Carnivory in wild chimpanzees, *Pan troglodytes verus,* in Sierra Leone. *Mammalia* 57:273–274.

Alpagut, B., Andrews, P., Fortelius, M., Kappelman, J., Temizsoy, I., Çelebi, H., and Lindsay, W. 1996. A new specimen of *Ankarapithecus meteai* from the Sinap Formation of central Anatolia. *Nature* 382:349–351.

Alpagut, B., Andrews, P., and Martin, L. 1990a. Miocene paleoecology of Pasalar, Turkey. In E. H. Lindsay, V. Fahlbusch, and P. Mein, eds., *European Neogene Mammal Chronology,* 557–571. New York: Plenum Press.

———. 1990b. New hominoid specimens from the Middle Miocene site at Pasalar, Turkey. *Journal of Human Evolution* 19:397–422.

Alperson-Afil, N., and Goren-Inbar, N. 2006. Out of Africa and into Eurasia with controlled use of fire: evidence from Gesher Benot Ya'aqov, Israel. *Archaeology, Ethnology and Anthropology of Eurasia* 28:63–78.

Alperson-Afil, N., Sharon, G., Kislev, M. Melamed, Y., Zohar, I., Ashkenazi, S., Rabonivich, R., Biton, R., Werker, E., Hartman, G., et al. 2009. Spatial organization of hominin activities at Gesher Benot Ya'aqov, Israel. *Science* 326:1677–1680.

Alvergne, A., Jokela, M., and Lummaa, V. 2010. Personality and reproductive success in a high-fertility human population. *Proceedings of the National Academy of Sciences of the United States of America* 107:11745–11750.

Amici, F., Aureli, F., and Call, J. 2010. Monkeys and apes: are their cognitive skills really so different? *American Journal of Physical Anthropology* 143:188–197.

Ammann, K. 2001. Bushmeat hunting and the great apes. In B. B. Beck, T. S. Stoinski, M. Hutchins, T. L. Maple, B. Norton, A. Rowan, E. F. Stevens, and A. Arluke, eds., *Great Apes and Humans: The Ethics of Coexistence,* 71–85. Washington, DC: Smithsonian Institution Press.

Amsler, S. J. 2010. Energetic costs of territorial boundary patrols by wild chimpanzees. *American Journal of Primatology* 72:93–103.

Amsterdam, B. 1972. Mirror self-image reactions before age two. *Developmental Psychobiology* 5:297–305.

Amunts, K., Jäncke, L., Mohlberg, H., Steinmetz, H., and Zilles, K. 2000. Interhemispheric asymmetry of the human motor cortex related to handedness and gender. *Neuropsychologia* 38:304–312.

Amunts, K., Schlaug, G., Schleicher, A., Steinmetz, H., Dabringhaus, A., Roland, P. E., and Zilles, K. 1996. Asymmetry in the human motor cortex and handedness. *Neuroimage* 4:216–222.

Amunts, K., Schleicher, A., Bürgel, U., Mohberg, H., Uylings, H. B. M., and Zilles, K. 1999. Broca's region revisited: cytoarchitecture and intersubject variability. *Journal of Comparative Neurology* 412:319–341.

Anderson, B. 1999. Commentary Ringo, Doty, Demeter and Simard, *Cerebral Cortex* (1994) 4:331–343: a proof of the need for the spatial clustering in interneuronal connections to enhance cortical computation. *Cerebral Cortex* 9:2–3.

Anderson, D. P., Nordheim, E. V., Boesch, C., and Moermond, T. C. 2002. Factors influencing fission-fusion grouping in chimpanzees in the Taï National Park, Côte d'Ivoire. In C. Boesch, G. Hohmann, and L. F. Marchant, eds., *Behavioural Diversity in Chimpanzees and Bonobos,* 90–101. Cambridge: Cambridge University Press.

Anderson, J. R. 1990. Use of objects as hammers to open nuts by capuchin monkeys *(Cebus apella)*. *Folia Primatologica* 54:138–145.

———. 1993. To see ourselves as others see us: a response to Mitchell. *New Ideas in Psychology* 11:339–346.

———. 1998. Sleep, sleeping sites, and sleep-related activities: awakening to their significance. *American Journal of Primatology* 46:63–75.

Anderson, J. R., and Gallup, G. G., Jr. 2011. Do rhesus monkeys recognize themselves in mirrors? *American Journal of Primatology* 73:603–606.

Anderson, J. R., and Henneman, M. C. 1994. Solutions to a tool-use problem in a pair of *Cebus apella*. *Mammalia* 58:351–361.

Anderson, J. R., Williamson, E. A., and Carter, J. 1983. Chimpanzees of Sapo Forest, Liberia: density, nests, tools and meat-eating. *Primates* 24:594–601.

Anderson, K. G. 2006. How well does paternity confidence match actual paternity? Evidence from worldwide paternity rates. *Current Anthropology* 47:513–520.

Andrew, R. J. 1962. The situations that evoke vocalization in primates. *Annals of the New York Academy of Science* 102:296–315.

———. 1963a. Evolution of facial expression. *Science* 142:1034–1041.

———. 1963b. The origin and evolution of the calls and facial expressions of the primates. *Behaviour* 20:1–109.

———. 1963c. Trends apparent in the evolution of vocalization in the Old World monkeys and apes. *Symposia of the Zoological Society of London* 10:89–101.

———. 1964. The displays of the Primates. In J. Buettner-Janusch, ed., *Evolutionary and Genetic Biology of Primates,* Vol. 2, 227–309. New York: Academic Press.

———. 1965. The origins of facial expressions. *Scientific American* 213:88–94.

Andrews, P. 1978. A revision of the Miocene Hominoidea of East Africa. *Bulletin of the British Museum (Natural History), Geology Series* 30:85–224a.

———. 1982. Hominoid evolution. *Nature* 295:185–186.

———. 1983. The natural history of *Sivapithecus*. In R. L. Ciochon and R. S. Corruccini, eds., *New Interpretations of Ape and Human Ancestry*, 441–463. New York: Plenum Press.

———. 1984. An alternative interpretation of the characters used to define *Homo erectus*. *Courier Forschungsinstitut Senckenberg* 69:167–175.

———. 1985. Family group systematics and evolution among catarrhine primates. In E. Delson, ed., *Ancestors: the Hard Evidence*, 14–22. New York: Alan R. Liss.

———. 1986a. Molecular evidence for catarrhine evolution. In B. Wood, L. Martin, and P. Andrews, eds., *Major Topics in Primate and Human Evolution*, 107–129. Cambridge: Cambridge University Press.

———. 1986b. Fossil evidence on human origins and dispersal. *Cold Spring Harbor Symposia on Quantitative Biology* 51:419–428.

———. 1987. Aspects of hominoid phylogeny. In C. Patterson, ed., *Molecules and Morphology in Evolution: Conflict or Compromise*, 21–53. Cambridge: Cambridge University Press.

———. 1990. Palaeoecology of the Miocene fauna from Pasalar, Turkey. *Journal of Human Evolution* 19:569–582.

———. 1992a. Evolution and the environment in the Hominoidea. *Nature* 360:641–646.

———. 1992b. Community evolution in forest habitats. *Journal of Human Evolution* 19:423–438.

———. 1995a. Time resolution of the Miocene fauna from Pasalar. *Journal of Human Evolution* 28:343–358.

———. 1995b. Ecological apes and ancestors. *Nature* 376:555–556.

Andrews, P., and Alpagut, B. 2001. Functional morphology of *Ankarapithecus meteai*. In L. de Bonis, G. D. Koufos, and P. Andrews, eds., *Hominoid Evolution and Climatic Change in Europe*, Vol. 2: *Phylogeny of the Neogene Hominoid Primates of Eurasia*, 213–230. Cambridge: Cambridge University Press.

Andrews, P., Begun, D. R., and Zylstra, M. 1996. Interrelationships between functional morphology and paleoenvironments in Miocene hominoids. In D. R. Begun, C. V. Ward, and M. D. Rose, eds., *Function Phylogeny, and Fossils*, 29–58. New York: Plenum Press.

Andrews, P., and Bernor, R. L. 1999. Vicariance biogeography and paleoecology of Eurasian Miocene hominoid primates. In J. Agusti, L. Rook, and P. Andrews, eds., *Hominoid Evolution and Climatic Change in Europe*, Vol. 1: *The Evolution of Neogene Terrestrial Ecosystems in Europe*, 454–487. Cambridge: Cambridge University Press.

Andrews, P., and Cronin, J. E. 1982. The relationships of *Sivapithecus* and *Ramapithecus* and the evolution of the orang-utan. *Nature* 297:541–546.

Andrews, P., and Evans, E. N. 1979. The environment of *Ramapithecus* in Africa. *Paleobiology* 5:22–30.

Andrews, P., Harrison, Delson, E., Bernor, R. L., and Martin, L. 1996. Distribution and biochronology of European and Southwest Asian Miocene catarrhines. In R. L. Bernor, V. Fahlbusch, and H.-W. Mittmann, eds., *The Evolution of Western Eurasian Neogene Mammal Faunas,* 169–207. New York: Columbia University Press.

Andrews, P., Harrison, T., Martin, L., and Pickford, M. 1981. Hominoid primates from a new Miocene locality named Meswa Bridge in Kenya. *Journal of Human Evolution* 10:123–128.

Andrews, P., and Martin, L. 1987a. Cladistic relationships of extant and fossil hominoids. *Journal of Human Evolution* 16:101–118.

———. 1987b. The phyletic position of the Ad Dabtiyah hominoid. *Bulletin of the British Museum of Natural History (Geology)* 41:383–393.

———. 1992. Hominoid dietary evolution. *Philosophical Transactions of the Royal Society, London B* 334:199–209.

Andrews, P., and Simons, E. L. 1977. A new African Miocene gibbon-like genus, *Dendropithecus* (Hominoidea, Primates) with distinctive postcranial adaptations: its significance to origin of Hylobatidae. *Folia Primatologica* 28:161–169.

Andrews, P., and Tekkaya, I. 1980. A revision of the Turkish Miocene hominoid *Sivapithecus meteai. Palaeontology* 23:85–95.

Andrews, P., and van Couvering, J. A. H. 1975. Palaeoenvironments in the East African Miocene. In F. S. Szalay, ed., *Approaches to Primate Paleobiology, Contributions to Primatology* 5:62–103.

Andrews, P., and Walker, A. 1976. The primate and other fauna from Fort Ternan, Kenya. In G. L. Isaac and E. R. McCown, eds., *Human Origins: Louis Leakey and the East African Evidence,* 279–304. Menlo Park, CA: W. A. Benjamin.

Ankel, F. 1962. Vergleichende Untersuchungen über die Skelettmorphologie des greifschwanzes südamerikanischer Affen (Platyrrhina). *Zeitschrift für Morphologie und Ökologie der Tiere* 52:131–170.

———. 1965. Der Canalis Sacralis als Indikator für die Länge der Caudalregion der Primaten. *Folia Primatologica* 3:263–276.

———. 1972. Vertebral morphology of fossil and extant primates. In R. Tuttle, ed., *Functional and Evolutionary Biology of Primates,* 223–240. Chicago: Aldine.

Ankel-Simons, F., Fleagle, J. G., and Chatrath, P. S. 1998. Femoral anatomy of *Aegyptopithecus zeuxis,* and Early Oligocene anthropoid. *American Journal of Physical Anthropology* 106:413–424.

Annett, M., and Annett, J. 1991. Handedness for eating in gorillas. *Cortex* 27:269–275.

Anonymous 2000. "Sign language" gorilla dies at 27. *Chicago Tribune,* section 1, p. 10.

Anstett, M. C., Hossaert-McKey, M., and Kjellberg, F. 1997. Figs and fig pollinators: evolutionary conflicts in a coevolved mutualism. *Trends in Ecology and Evolution* 12:94–99.

Appadurai, A. 1988. Putting hierarchy in its place. *Current Anthropology* 3:36–47.

———. 1990. Disjuncture and difference in the global cultural economy. *Public Culture* 2:1–24.

Arambourg, C., and Coppens, Y. 1967. Sur la découverte dans le Pléistocène inférieur de la vallée de l'Omo (Ethiopie) d'une mandibule d'Australopithécien. *Comptes rendus hebdomadaires des séances de l'Académie des Sciences, Paris, série D* 265:589–590.

———. 1968. Decouverte d'un australopithecien nouveau dans les gisements de l'Omo (Ethiopie). *South African Journal of Science* 64:58–59.

Arandjelovic, M., Head, J., Kühl, H., Boesch, C., Robbins, M. M., Maisels, F., and Vigilant, L. 2010. Effective non-invasive genetic monitoring of multiple wild western gorilla groups. *Biological Conservation* 143:1780–1791.

Arbib, M. A. 2002. The mirror system, imitation, and the evolution of language. In K. Dautenhhn and D. L. Nehaniv, eds., *Imitation in Animals and Artifacts,* 229–280. Cambridge, MA: MIT Press.

———. 2003. The evolving mirror system: a neural basis for language readiness. In M. H. Christiansen and S. Kirby, eds., *Language Evolution,* 182–200. Oxford: Oxford University Press.

———. 2004. How far is language beyond our grasp: a response to Hurford. In D. K. Oller and U. Griebel, eds., *Evolution of Communication Systems: A Comparative Approach,* 315–321. Cambridge, MA: MIT Press.

———. 2005a. From monkey-like action recognition to human language: an evolutionary framework for neurolinguistics. *Behavioral and Brain Sciences* 28:105–167.

———. 2005b. Interweaving protosign and protospeech: further developments beyond the mirror. *Interaction Studies* 6:145–171.

———. 2012. *How the Brain Got Language: The Mirror System Hypothesis.* New York: Oxford University Press.

Arbib, M. A., Liebel, K., and Pika, S. 2008. Primate vocalization, gesture, and the evolution of human language. *Current Anthropology* 49:1053–1076.

Arbib, M. A., and Rizzolatti, G. 1996. Neural expectations: a possible evolutionary path from manual skills to language. *Communication and Cognition* 29:393–424.

Arcadi, A. C., Robert, D., and Boesch, C. 1998. Buttress drumming by wild chimpanzees: temporal patterning, phase integration into loud calls, and preliminary evidence for individual distinctiveness. *Primates* 39:505–518.

Ardito, G., and Mottura, A. 1987. An overview of the geographic and chronologic distribution of West European cercopithecoids. *Human Evolution* 2:29–45.

Ardrey, R. 1961. *African Genesis: A Personal Investigation into the Animal Origins and Nature of Man.* New York: Atheneum.

———. 1966. *The Territorial Imperative: A Personal Inquiry into the Animal Origins of Property and Nations.* New York: Atheneum.

———. 1970. *The Social Contract: A Personal Inquiry into the Evolutionary Sources of Order and Disorder.* New York: Atheneum.

———. 1976. *The Hunting Hypothesis: A Personal Conclusion Concerning the Evolutionary Nature of Man.* New York: Atheneum.

Arens, W. 1979. *The Man-Eating Myth: Anthropology and Anthropophagy.* New York: Oxford University Press.

Arensburg, B. 1994. Middle Paleolithic speech capability: a response to Dr. Lieberman. *American Journal of Physical Anthropology* 94:279–280.

Arensburg, B., Tillier, A. M., Vandermeersch, B., Duday, H., Schepartz, L. A., and Rak, Y. 1989. A Middle Paleolithic human hyoid bone. *Nature* 338:758–760.

Argue, D., Donlon, D., Groves, C., and Wright, R. 2006. *Homo floresiensis:* microcephalic, pygmoid, *Australopithecus,* or *Homo? Journal of Human Evolution* 51:360–374.

Arjamaa, O., and Vuorisalo, T. 2010. Gene-culture coevolution and human diet. *American Scientist* 98:140–147.

Armstrong, D. F. 1999. *Original Signs: Gesture, Sign, and the Source of Language*. Washington, DC: Gallaudet University Press.

———. 2008. The gestural theory of language origins. *Sign Language Studies* 8:289–314.

Armstrong, D. F., Stokoe, W. C., and Wilcox, S. E. 1994. Signs of the origin of syntax. *Current Anthropology* 35:349–368.

———. 1995. *Gesture and the Nature of Language*. Cambridge: Cambridge University Press.

Armstrong, D. F., and Wilcox, S. E. 2007. *The Gestural Origin of Language*. Oxford: Oxford University Press.

Arnason, U. 1998. Response. *Journal of Molecular Evolution* 46:379–381.

Arnason, U., Gullberg, A., Janke, A., and Xu, X. 1996. Pattern and timing of evolutionary divergences among hominoids based on analyses of complete mtDNAs. *Journal of Molecular Evolution* 43:650–661.

Arnason, U., Xu, X., and Gullberg, A. 1996. Comparison between the complete mitochondrial DNA sequences of *Homo* and the common chimpanzee based on nonchimeric sequences. *Journal of Molecular Evolution* 42:145–152.

Arnold, K., and Whiten, A. 2001. Post-conflict behavior of wild chimpanzees *(Pan troglodytes schweinfurthii)* in the Budongo Forest, Uganda. *Behaviour* 138:649–690.

Aronson, J. L., Schmitt, T. J., Walter, R. C., Taieb, M., Tiercelin, J. J., Johanson, D. C., Naeser, C. W., and Nairn, A. E. M. 1977. New geochronologic and palaeomagnetic data for the hominid-bearing Hadar Formation of Ethiopia. *Nature* 267:323–327.

Aronson, J. L., and Taieb, M. 1981. Geology and paleogeography of the Hadar hominid site, Ethiopia. In G. Rapp Jr. and C. F. Vondra, eds., *Hominid Sites: their Geologic Settings*, 165–195. Boulder: Westview Press.

Aronson, J. L., Walter, R. C., and Taieb, M. 1983. Correlation of Tulu Bor Tuff at Koobi Fora with the Sidi Hakoma Tuff at Hadar. *Nature* 306:209–210.

Aronson, J. L., Walter, R. C., Taieb, M., and Naeser, C. W. 1980. New geochronological information for the Hadar Formation and the adjacent central Afar, Ethiopia. In R. E. Leakey and B. A. Ogot, eds., *Proceedings of the 8th Panafrican Congress of Prehistory and Quaternary Studies Nairobi, 5 to 10 September 1977*, 47–52. Nairobi: International Louis Leakey Memorial Institute for African Prehistory.

Arsuaga, J. L., Martínez, I., Garcia, A., and Lorenzo, C. 1997. The Sima de los Huesos crania (Sierra de Atapuerca, Spain): a comparative study. *Journal of Human Evolution* 33:219–281.

Asfaw, B. 1983. A new hominid parietal from Bodo, Middle Awash Valley, Ethiopia. *American Journal of Physical Anthropology* 61:67–371.

———. 1987. The Belohdelie frontal: new evidence of early hominid cranial morphology from the Afar of Ethiopia. *Journal of Human Evolution* 16:611–624.

Asfaw, B., Gilbert, W. H., Beyene, Y., Hart, W. K., Renne, P. R., WoldeGabriel, G., Vrba, E. S., and White, T. D. 2002. Remains of *Homo erectus* from Bouri, Middle Awash, Ethiopia. *Nature* 416:317–320.

Asfaw, B., White, T., Lovejoy, O., Latimer, B., Simpson, S., and Suwa, G. 1999. *Australopithecus garhi:* a new species of early hominid from Ethiopia. *Science* 284:629–635.

Ashford, R. W., Lawson, H., Butynski, T. M., and Reid, G. D. F. 1996. Patterns of intestinal parasitism in the mountain gorilla *Gorilla gorilla* in the Bwindi Impenetrable Forest, Uganda. *Journal of Zoology* 239:507–514.

Ashford, R. W., Reid, G. D. F., and Butynski, T. M. 1990. The intestinal faunas of man and mountain gorillas in a shared habitat. *Annals of Tropical Medicine and Parasitology* 84:337–340.

Ashford, R. W., Reid, G. D. F., and Wrangham, R. W. 2000. Intestinal parasites of the chimpanzee *Pan troglodytes* in Kibale Forest, Uganda. *Annals of Tropical Medicine and Parasitology* 94:173–179.

Ashton, E. H., and Oxnard, C. E. 1963. The musculature of the primate shoulder. *Transactions of the Zoological Society of London* 29:553–650.

Ashton, E. H., and Zuckerman, S. 1951. Some cranial indices of *Plesianthropus* and other primates. *American Journal of Physical Anthropology* 9:283–296.

Asthana, S., Noble, W. S., Kryukov, G., Grant, C. E., Sunyaev, S., and Stamatoyannopoulos, J. A. 2007. Widely distributed noncoding purifying selection in the human genome. *Proceedings of the National Academy of Sciences of the United States of America* 104:12410–12415.

Atsalis, S., and Margulis, S. W. 2006. Sexual and hormonal cycles in geriatric *Gorilla gorilla gorilla. International Journal of Primatology* 27:1663–1687.

———. 2008. Perimenopause and menopause documenting life changes in aging female gorillas. *Interdisciplinary Topics in Gerontology* 36:119–146.

Atsalis, S., and Videan, E. 2009a. Reproductive aging in captive and wild common chimpanzees: factors influencing the rate of follicular depletion. *American Journal of Primatology* 71:271–282.

———. 2009b. Functional versus operational menopause: reply to Herndon & Lacreuse. *American Journal of Primatology* 71:893–894.

Aufreiter, S., Mahaney, W. C., Milner, M. W., Huffman, M. A., Hancock, R. G. V., Wink, M., and Reich, M. 2001. Mineralogical and chemical interactions of soils eaten by chimpanzees of the Mahale Mountains and Gombe Stream National Parks, Tanzania. *Journal of Chemcal Ecology* 17:285–311.

Aung, A. K. 2004. The primate-bearing Pondaung Formation in the Upland Area, Northwest of Central Myanmar. In C. F. Ross and R. F. Kay, eds., *Anthropoid Origins,* 205–217. New York: Kluwer Academic/Plenum Press.

Aureli, F., Cords, M., and van Schaik, C. P. 2002. Conflict resolution following aggression in gregarious animals: a predictive framework. *Animal Behaviour* 64:125–143.

Aveling, C., and Harcourt, A. H. 1984. A census of the Virunga gorillas. *Oryx* 18:8–13.

Avis, V. 1962. Brachiation: the crucial issue for man's ancestry. *Southwest Journal of Anthropology* 18:119–148.

Avise, J. C. 1986. Mitochondrial DNA and the evolutionary genetics of higher animals. *Philosophical Transactions of the Royal Society, London, B* 312:325–342.

Avise, J. C., and Aquadro, C. F. 1982. A comparative summary of genetic distances in the vertebrates. *Evolutionary Biology* 15:151–185.

Ayala, F. J. 1986. On the virtues and pitfalls of the molecular evolutionary clock. *Journal of Heredity* 77:226–235.

Aziz, E., Mobbs, D., Jo, B., Menon, V., and Reiss, A. L. 2005. Sex differences in brain activation elicited by humor. *Proceedings of the National Academy of Sciences of the United States of America* 102:16496–16501.

Azuma, S., and Toyoshima, A. 1961–62. Progress report of the survey of chimpanzees in their natural habitat, Kabogo Point Area, Tanganyika. *Primates* 3:61–70.

Azzaroli, A., Boccaletti, M., Delson, E., Moratti, G., and Torre, D. 1986. Chronological and paleogeographical background to the study of *Oreopithecus bambolii*. *Journal of Human Evolution* 15:533–540.

Baba, M. L., Darga, L. L., and Goodman, M. 1982. Recent advances in molecular evolution of the primates. In A. B. Chiarelli and R. S. Corruccini, eds., *Advanced Views in Primate Biology*, 6–27. Berlin: Springer.

Bacon, A.-M. 1994. Interprétation fonctionnelle des proportions de la trochlée fémorale en relation avec l'aptitude à la rotation axiale du genou chez les Primates Simiiformes actuels. Comparaison avec *Proconsul, Australopithecus* et *Homo*. *Annales de Paléontologie (Invert.-Vert.)* 80:194–210.

Bacon, A.-M., and Long, V. T. 2001. The first discovery of a complete skeleton of a fossil orang-utan in a cave of the Hoa Binh Province, Vietnam. *Journal of Human Evolution* 41:227–241.

Badgley, C., Barry, J. C., Morgan, M. E., Nelson, S. V., Behrensmeyer, A. K., Cerling, T. E., and Pilbeam, D. 2008. Ecological changes in Miocene mammalian record show impact of prolonged climatic forcing. *Proceedings of the National Academy of Sciences of the United States of America* 105:12145–12149.

Badgley, C., and Behrensmeyer, A. K. 1980. Paleoecology of Middle Siwalik sediments and faunas, northern Pakistan. *Palaeogeography, Palaeoclimatology, Palaeoecology* 30:133–155.

Badgley, C., Kelley, J., Pilbeam, D., and Ward, S. 1984. The paleobiology of South Asian Miocene hominoids. In J. R. Lukacs, ed., *The People of South Asia: The Biological Anthropology of India, Pakistan, and Nepal*, 3–28. New York: Plenum Press.

Badrian, A., and Badrian, N. 1977. Pygmy chimpanzees. *Oryx* 13:463–468.

———. 1980. The other chimpanzee. *Animal Kingdom* 83:8–14.

———. 1984a. The bonobo branch of the family tree. *Animal Kingdom* 87:39–45.

———. 1984b. Social organization of *Pan paniscus* in the Lomako Forest, Zaire. In R. L. Susman, ed., *The Pygmy Chimpanzee: Evolutionary Biology and Behavior*, 325–346. New York: Plenum Press.

Badrian, N., Badrian, A., and Susman, R. L. 1981. Preliminary observations on the feeding behavior of *Pan paniscus* in the Lomako Forest of Central Zaire. *Primates* 22:173–181.

Badrian, N., and Malenky, R. K. 1984. Feeding ecology of *Pan paniscus* in the Lomako Forest, Zaire. In R. L. Susman, ed., *The Pygmy Chimpanzee*, 275–299. New York: Plenum Press.

Bailey, J. A., and Eichler, E. E. 2006. Primate segmental duplications: crucibles of evolution, diversity and disease. *Nature Reviews Genetics* 7:552–564.

Bailey, W. J. 1993. Hominoid trichotomy: a molecular overview. *Evolutionary Anthropology* 2:100–108.

Bailey, W. J., Hayasaka, K., Skinner, C. G., Kehoe, S., Sieu, L. C., Slightom, J. L., and Goodman, M. 1992. Reexamination of the African hominoid trichotomy with additional sequences from primate ß-globin gene cluster. *Molecular Phylogenetics and Evolution* 1:97–135.

Baillieux, H., De Smet, H. J., Paquier, P. F., De Deyn, P. P., and Mariën, P. 2008. Cerebeller neruocognition: insights into the bottom of the brain. *Clinical Neurology and Neurosurgery* 110:763–773.

Baker, K. C., and Smuts, B. B. 1994. Social relationships of female chimpanzees. In *Chimpanzee Cultures,* R. W. Wrangham, W. C. Mc Grew, F. B. M. de Waal, and P. G. Heltne, eds., 227–242. Cambridge, MA: Harvard Univerity Press.

Bakewell, M. A., Shi, P., and Zhang, J. 2007. More genes underwent positive selection in chimpanzee evolution than in human evolution. *Proceedings of the National Academy of Sciences of the United States of America* 104:7489–7494.

Bakuza, J. S., and Nkwenguilila, G. 2009. Variation over time in parasite prevalence among free-ranging chimpanzees at Gombe National Park, Tanzania. *International Journal of Primatology* 30:43–53.

Baldini, A., Miller, D. A., Miller, O. J., Ryder, O. A., and Mitchell, A. R. 1991. A chimpanzee-derived chromosome-specific alpha satellite DNA sequence conserved between chimpanzee and human. *Chromosoma* 100:156–161.

Baldwin, L. A., and Teleki, G. 1976. Patterns of gibbon behavior on Hall's Island, Bermuda. In D. M. Rumbaugh, ed., *Gibbon and Siamang,* Vol. 4, 21–105. Basel: Karger.

Baldwin, P. J., McGrew, W. C., and Tutin, C. E. G. 1982. Wide-ranging chimpanzees at Mt. Assirik, Senegal. *International Journal of Primatology* 3:367–385.

Baldwin, P. J., Sabater Pí, J. McGrew, W. C., and Tutin, C. E. G. 1981. Comparisons of nests made by different populations of chimpanzees *(Pan troglodytes). Primates* 22:474–486.

Balog, J. 1993. *Anima.* Boulder, CO: Arts Alternative Press.

Balongelwa, C. W. 2008. Foreward to conservation study section. In T. Furuichi and J. Thompson, eds., *The Bonobos: Behavior, Ecology, and Conservation,* 219–226. New York: Springer.

Balter, M. 2009a. Our ancestors were no swingers. *ScienceNOW Daily News,* April 13.

———. 2009b. Arrest that chimp! *ScienceNOW Daily News,* March 9.

———. 2009c. The origins of tidiness. *ScienceNOW Daily News,* December 18.

———. 2011. Did Neandertals linger in Russia's far north? *Science* 332:778.

———. 2012. Did Neandertals truly bury their dead? *Science* 337:1443–1444.

———. 2013. Archaeologists say the "Anthropocene" is here—but it began long ago. *Science* 340:261–262.

Balter, M., and Gibbons, A. 2000. A glimpse of humans' first journey out of Africa. *Science* 288:948–950.

Ba Maw, Ciochon, R. L., and Savage, D. E. 1979. Late Eocene of Burma yields earliest anthropoid primate, *Pongaungia cotteri. Nature* 282:65–67.

Barbarosa, A. 1985. The ethnology of the Agta of Lamika, Peñablanca, Cagayan. In P. B. Griffin and A. Estioko-Griffin, eds., *The Agta of Northeastern Luzon: Recent Studies,* 12–17. Cebu City, Philippines: San Carlos Publications.

Bard, K. A., Fragaszy, D., and Visalberghi, E. 1995. Acquisition and comprehension of a tool-using behavior by young chimpanzees *(Pan troglodytes):* effects of age and modeling. *International Journal of Comparative Psychology* 8:47–68.

Bard, K. A., Hopkins, W. D., and Fort, C. L. 1990. Lateral bias in infant chimpanzees *(Pan troglodytes). Journal of Comparative Psychology* 104:309–321.

Barelli, C., Boesch, C., Heistermann, M., and Reichard, U. H. 2008. Female white-handed gibbons *(Hylobates lar)* lead group movements and have priority of access to food resources. *Behaviour* 145:965–981.

Barelli, C., Heistermann, M., Boesch, C., and Reichard, U. H. 2007. Sexual swellings in wild white-handed gibbon females *(Hylobates lar)* indicate the probability of ovulation. *Hormones and Behavior* 51:221–230.

————. 2008. Mating patterns and sexual swellings in pair-living and multimale groups of wild white-handed gibbons, *Hylobates lar. Animal Behaviour* 75:991–1001.

Barrett, H. C. 2012. A hierarchical model of the evolution of human brain specializations. *Proceedings of the National Academy of Sciences of the United States of America* 109:10733–10740.

Barriel, V. 1997. *Pan paniscus* and hominoid phylogeny: morphological data, molecular data and "total evidence." *Folia Primatologica* 68:50–56.

Barriel, V., and Darlu, P. 1990. Approche moléculaire de la Phylogénie des Hominoidea l'exemple de la pseudo êta-globine. *Bulletin et Mémoires de la Société d'Anthropologie de Paris* 2:3–24.

Barry, J. C. 1986. A review of the chronology of Siwalik hominoids. In J. G. Else and P. C. Lee, eds., *Primate Evolution,* 93–106. Cambridge: Cambridge University Press.

————. 1987. The history and chronology of Siwalik cercopithecids. *Human Evolution* 2:47–58.

Barry, J. C., and Flynn, J. 1990. Key biostratigraphic events in the Siwalik sequence. In E. H. Lindsay, Fahlbusch, V., and Mein, P., eds., *European Neogene Mammal Chronology,* 557–571. New York: Plenum Press.

Barry, J. C., Flynn, J., and Pilbeam, D. R. 1990. Faunal diversity and turnover in a Miocene terrestrial sequence. In R. M. Ross and W. D. Allmon, eds., *Causes of Evolution: A Paleontological Perspective,* 381–421. Chicago: University of Chicago Press.

Barry, J. C., Jacobs, L. L., and Kelley, J. 1986. An Early Miocene catarrhine from Pakistan with comments on the dispersal of catarrhines into Eurasia. *Journal of Human Evolution* 15:501–508.

Barry, J. C., Johnson, N. M., Raza, S. M., and Jacobs, L. L. 1985. Neogene mammalian faunal change in southern Asia: correlations with climatic, tectonic, and eustatic events. *Geology* 13:637–640.

Barry, J. C., Lindsay, E. H., and Jacobs, L. L. 1982. A biostratigraphic zonation of the Middle and Upper Siwaliks of the Potwar Plateau of northern Pakistan. *Palaeogeography, Palaeoclimatology, Palaeoecology* 37:95–130.

Barry, J. C., Morgan, M., Flynn, L. J., Pilbeam, D., Behrensmeyer, A. K., Raza, S. M., Khan, I. A., Badgely, C., Hics, J., and Kelley, J. 2002. Faunal and environmental change in the Late Miocene siwaliks of Northern Pakistan. *Paleobiology* 28 (sp3):1–71.

Bartholomew, G. A., and Birdsell, J. B. 1953. Ecology and the protohominids. *American Anthropologist* 55:481–498.

Bartlett, T. Q. 1999. *Feeding and Ranging Behavior of White-handed Gibbons (Hylobates lar) in Khao Yai National Park, Thailand.* PhD thesis, Washington University, St. Louis.

———. 2007. The Hylobatidae. In C. J. Campbell, A. Fuentes, K. C. MacKinnon, M. Panger, and S. K. Bearder, eds., *Primates in Perspective,* 274–289. Oxford: Oxford University Press.

———. 2009a. *The Gibbons of Khao Yai: Seasonal Variation in Behavior and Ecology.* Upper Saddle River, NJ: Pearson/Prentice Hall.

———. 2009b. Seasonal home range use and defendability in white-handed gibbons *(Hylobates lar)* in Khao Yai National Park, Thailand. In S. Lappan and D. J. Whittaker, eds., *The Gibbons: New Perspectives on Small Ape Socioecology and Population Biology,* 265–275. New York: Springer.

Barton, R. A., and Dunbar, R. I. M. 1997. Evolution of the social brain. In A. Whiten and R. W. Byrne, eds., *Machiavellian Intelligence II: Extensions and Evaluations,* 240–263. Cambridge: Cambridge University Press.

Bartstra, G.-J., Soegondho, S., and van der Wijk, A. 1988. Ngandong man: age and artifacts. *Journal of Human Evolution* 17:325–337.

Bar-Yosef, O., and Kuhn, S. L. 1999. The big deal about blades: laminar technologies and human evolution. *American Anthropologist* 101:322–338.

Bar-Yosef, O., Vandermeersch, B., Arensburg, B., Goldberg, P., Laville, H., Meignen, L., Rak, Y., Tchernov, E., and Tilier, A.-M. 1986. New data on the origin of modern man in the Levant. *Current Anthropology* 27:63–64.

Basabose, A. K. 2002a. Diet composition of chimpanzees inhabiting the montane forest of Kahuzi, Democratic Republic of Congo. *American Journal of Primatology* 58:1–21.

———. 2002b. Factors affecting nest site choice in chimpanzees at Tshibati, Kahuzi-Biega National Park: influence of sympatric gorillas. *International Journal of Primatology* 23:263–282.

———. 2005. Ranging pattern of chimpanzees in a montane forest of Kahuzi, Democratic Republic of Congo. *International Journal of Primatology* 26:33–54.

Basabose, A. K., and Yamagiwa, J. 1997. Predation on mammals by chimpanzees in the montane forest of Kahuzi, Zaire. *Primates* 38:45–55.

Basmajian, J. V., and De Luca, C. J. 1985. *Muscles Alive.* 5th ed. Baltimore: Williams & Wilkins.

Bastian, M. L., Zweifel, N., Vogel, E. R., Wich, S. A., and van Schaik, C. P. 2010. Diet traditions in wild orangutans. *American Journal of Physical Anthropology* 143:175–187.

Bates, E. 1993. Comprehension and production in early language development. In *Language Comprehension in Ape and Child,* E. S. Savage-Rumbaugh, J. Murphy, R. A. Sevcik, K. E. Brakke, S. L. Williams, and D. M. Rumbaugh, eds. *Monographs of the Society for Research in Child Development* 58:222–242.

Bauchop, T. 1978. Digestion of leaves in vertebrate arboreal folivores. In G. G. Montgomery, ed., *The Ecology of Arboreal Folivores,* 193–204. Washington, DC: Smithsonian Institution Press.

Bauer, H. R. 1986. A comparative study of common chimpanzee and human infant sounds. In D. M. Taub and F. A. King, eds., *Current Perspectives in Primate Social Dynamics,* 327–345. New York: Van Nostrand Reinhold.

Bauer, H. R., and Philip, M. M. 1983. Facial and vocal individual recognition in the common chimpanzee. *Psychological Record* 33:161–170.

Bauer, K., and Schreiber, A. 1995. Tricky relatives: consecutive dichotomous speciations of gorilla, chimpanzee and hominids testified by immunological determinants. *Naturwissenschaften* 82:517–520.

Bault, N., Joffily, M., Rustichini, A., and Coricelli, G. 2011. Medial prefrontal cortex and striatum mediate the influence of social comparison on the decision process. *Proceedings of the National Academy of Science of the United States of America* 108:16044–16049.

Bayle, P., Macciarelli, R., Trinkaus, E., Duarte, C., Mazurier, A., and Zilhão, J. 2010. Dental maturational sequence and dental tissue proportions in the early Upper Paleolithic child from Abrigo do Lagar Velho, Portugal. *Proceedings of the National Academy of Sciences of the United States of America* 107:1338–1342.

Bazin, E., Glémin, S., and Galtier, N. 2006. Population size does not influence mitochondrial genetic diversity in animals. *Science* 312:570—572.

Beadle, L. C. 1981. *The Inland Waters of Tropical Africa: An Introduction to Tropical Limnology.* London: Longman.

Beard, K. C. 2002. Basal anthropoids. In W. C. Hartwig, ed., *The Primate Fossil Record,* 133–149. Cambridge: Cambridge University Press.

Beard, K. C., Krishtalka, L, and Stucky, R. K. 1991. First skulls of the Early Eocene primate *Shoshonius cooperi* and the anthropoid-tarsier dichotomy. *Nature* 349:64–67.

Beard, K. C., Qi, T., Dawson, M. R., Wang, B., and Li, C. 1994. A diverse new primate fauna from middle Eocene fissure-fillings in southeastern China. *Nature* 368:604–609.

Beard, K. C., Teaford, M. F., and Walker, A. 1986. New wrist bones of *Proconsul africanus* and *P. nyanzae* from Rusinga Island, Kenya. *Folia Primatologica* 47:97–118.

Beard, K. C., Tong, Y. Dawson, M. R., Wang, J., and Huang, X. 1996. Earliest complete dentition of an anthropoid primate from the late Middle Eocene of Shanxi Province, China. *Science* 272:82–85.

Beard, K. C., and Wang, J. 2004. The eosimiid primates (Anthropoidea) of the Heti Formation, Yuanqu Basin, Shanxi and Henan Provinces, People's Republic of China. *Journal of Human Evolution* 46:401–432.

Beaton, A., and Mariën, P. 2010. Language, cognition and the cerebellum: grappling with an enigma. *Cortex* 46:811–820.

Beaumont, P. B. 1980. On the age of Border Cave hominids 1–5. *Palaeontologia Africana* 23:21–33.

Beck, B. B. 1975. Primate tool behavior. In R. H. Tuttle, ed., *Socioecology and Psychology of Primates,* 413–447. The Hague: Mouton.

Becker, C., and Hick, U. 1980. *Animal Tool Behavior: The Use and Manufacture of Tools by Animals.* New York: Garland STPM Press.

———. 1982. Chimpocentrism: bias in cognitive ethology. *Journal of Human Evolution* 11:3–17.

———. 1984. "FamilienzusammenfUrung" als soziale Beschiiftigungstherapie und Aktivitiitssteigerung bei sieber Orang-Utans *(Pongo p. pygmaeus)* im KaIner Zoo. *Zeitschrift des Kainer Zoo* 27:43–57.

Becquet, C., Patternson, N., Stone, A. C., Przeworski, M., and Reich, D. 2007. Genetic structure of chimpanzee populations. *PLoS Genetics* 3:0617–0626.

Beeber, J. E., Czelusniak, J., and Goodman, M. 1986. Systematic position and evolution of Primates within Eutheria: amino acid and nucleotide sequence findings. In D. M. Taub and F. A. King, eds., *Current Perspectives in Primate Biology,* 89–106. New York: Van Nostrand Reinhold.

Begun, D. R. 1988. Catarrhine phalanges from the Late Miocene (Vallesian) of Rudabánya, Hungary. *Journal of Human Evolution* 17:413–438.

———. 1989. A large pliopithecine molar from Germany and some notes on the Pliopithecinae. *Folia Primatologica* 52:156–166.

———. 1992a. Phyletic diversity and locomotion in primitive European hominids. *American Journal of Physical Anthropology* 87:311–340.

———. 1992b. *Dryopithecus crusafonti* sp. nov., a new Miocene hominoid species from Can Ponsic (northeastern Spain). *American Journal of Physical Anthropology* 87:291–309.

———. 1993. New catarrhine phalanges from Rudabánya (northeastern Hungary) and the problem of parallelism and convergence in hominoid postcranial morphology. *Journal of Human Evolution* 24:373–402.

———. 1994. Relations among great apes and humans: new interpretations based on the fossil great ape *Dryopithecus. Yearbook of Physical Anthropology* 37:11–63.

———. 1999. Hominid family values: morphological and molecular data on relations among the great apes and humans. In S. T. Parker, R. W. Mitchell, and H. L. Miles, eds., *The Mentalities of Gorillas and Orangutans,* 3–42. Cambridge: Cambridge University Press.

———. 2000a. The Pliopithecoidea. In W. C. Hartwig, ed., *The Primate Fossil Record,* 221–240. Cambridge: Cambridge University Press.

———. 2000b. European hominoids. In W. C. Hartwig, ed., *The Primate Fossil Record,* 339–368. Cambridge: Cambridge University Press.

———. 2001. African and Eurasian Miocene hominoids and the origins of the Hominidae. In L. de Bonis, G. D. Koufos, and P. Andrews, eds., *Hominoid Evolution and Climate Change in Europe,* Vol. 2: *Phylogeny of the Neogene Hominoid Primates of Eurasia,* 231–253. Cambridge: Cambridge University Press.

———. 2005. *Sivapithecus* is east and *Dryopithecus* is west, and never the twain shall meet. *Anthropological Science* 113:53–64.

Begun, D. R., and Güleç, E 1998. Restoration of the type and palate of *Ankarapithecus meteai:* taxonomic and phylogenetic implications. *American Journal of Physical Anthropology* 105:279–314.

Begun, D. R., and Kordos, L. 1993. Revision of *Dryopithecus brancoi* Schlosser, 1991 based on fossil hominoid material from Rudabánya. *Journal of Human Evolution* 25:271–285.

————. 1999. Femora of *Anapithecus* from Rudabánya. *American Journal of Physical Anthropology* 28 (Suppl.): 173.

Begun, D. R., Moyá-Sola, S., and Kohler, M. 1990. New Miocene hominoid specimens from Can Llobateres (Vallès Penedès, Spain) and their geological and paleoecological context. *Journal of Human Evolution* 19:255–268.

Begun, D. R., Teaford, M. F., and Walker, A. 1994. Comparative and functional anatomy of *Proconsul* phalanges from the Kaswanga primate site, Rusinga Island, Kenya. *Journal of Human Evolution* 26:80–165.

Begun, D. R., and Ward, C. V. 2005. Comment on *"Pierolapithecus catalaunicus,* a new Middle Miocene great ape from Spain." *Science* 308:203.

Begun, D. R., Ward, C. V., and Rose, M. D. 1997. Events in hominoid evolution. In D. R. Begun, C. V. Ward, and M. D. Rose, eds., *Function, Phylogeny, and Fossils,* 389–415. New York: Plenum Press.

Bekoff, M. 2007. *The Emotional Lives of Animals.* Novato, CA: New World Library.

Bekoff, M., and Allen, C. 1998. Intentional communication and social play: how and why animals negotiate and agree to play. In M. Bekoff and J. A. Byers, eds., *Animal Play: Evolutionary, Comparative, and Ecological Perspectives,* 97–114. Cambridge: Cambridge University Press.

Bekoff, M., Allen, C., and Burghardt, G. M. 2002. *The Cognitive Animal.* Cambridge, MA: MIT Press.

Bekoff, M., and Byers, J. A. 1998. *Animal Play: Evolutionary, Comparative, and Ecological Perspectives.* Cambridge: Cambridge University Press.

Bekoff, M., and Pierce, J. 2009. *Wild Justice: The Moral Lives of Animals.* Chicago: University of Chicago Press.

Bell, E. C., Willson, M. C., Wilman, A. H., Dave, S., and Silverstone, P. H. 2006. Males and females differ in brain activation during cognitive tasks. *NeuroImage* 30:529–538.

Bellis, M. A., Hughes, K., Hughes, S., and Ashton, J. R. 2005. Measuring paternal discrepancy and its public health consequences. *Journal of Epidemiology and Community Health* 59:749–754.

Bellomo, R. V. 1991. Identifying traces of natural and humanly-controlled fire in the archaeological record: the role of actualistic studies. *Archaeology in Montana* 32:75–93.

————. 1993. A methodological approach for identifying archaeological evidence of fire resulting from human activities. *Journal of Archaeological Science* 20:525–553.

————. 1994. Methods of determining early hominid behavioral activities associated with the controlled use of fire at FxJj 20 Main, Koobi Fora, Kenya. *Journal of Human Evolution* 27:173–195.

Bellomo, R. V., and Kean, W. F. 1997. Evidence for hominid-controlled-fire at the FxJj site complex, Karari Escarpment, eds. In G. L. Isaac and B. Isaac, eds., *Koobi Fora Research Project,* Vol. 5: *Plio-Pleistocene Archaeology,* 224–233. Oxford: Clarendon Press.

Bender, R., Verhaegen, M., and Oser, N. 1997. Der Erwerb menschlicher Bipedie aus der Sicht der Aquatic Ape Theory. *Anthropologischer Anzeiger* 55:1–14.

Benditt, J. 1989. Cold water on the fire. *Scientific American* 260:21–22.

Benefit, B. R. 1991. The taxonomic status of Maboko small apes. *American Journal of Physical Anthropology* 12 (Suppl.): 50–51.

———. 1993. The permanent dentition and phylogenetic position of *Victoriapithecus* from Maboko Island, Kenya. *Journal of Human Evolution* 25:83–172.

———. 1999a. *Victoriapithecus:* the key to Old World monkey and catarrhine origins. *Evolutionary Anthropology* 7:155–174.

———. 1999b. The dentition of *Kenyapithecus africanus*. In H. Ishida, ed., *Abstracts of the International Symposium "Evolution of Middle-to-Late Miocene Hominoids in Africa," July 11–13, 1999,* Takaragaike, Kyoto, Japan.

Benefit, B. R., and McCrossin, M. L. 1989. New primate fossils from the Middle Miocene of Maboko Island, Kenya. *Journal of Human Evolution* 18:493–497.

———. 1991. Ancestral facial morphology of Old World higher primates. *Proceedings of the National Academy of Sciences of the United States of America* 88:5267–5271.

———. 1993a. Facial anatomy of *Victoriapithecus* and its relevance to the ancestral cranial morphology of Old World monkeys and apes. *American Journal of Physical Anthropology* 92:329–370.

———. 1993b. The lacrimal fossa of Cercopithecoidea, with special reference to cladistic analysis of Old World monkey relationships. *Folia Primatologica* 60:133–145.

———. 1997. Earliest known Old World monkey skull. *Nature* 388:368–371.

———. 2002. The Victoriapithecidae, Cercopithecoidea. In W. C. Hartwig, ed., *The Primate Fossil Record,* 241–253. Cambridge: Cambridge University Press.

Beneviste, R. E. 1985. The contributions of retroviruses to the study of mammalian evolution. In R. J. MacIntyre, ed., *Molecular Evolutionary Genetics,* 359–417. New York: Plenum Press.

Bennett, M. R., Harris, J. W. K., Richmond, B. G., Braun, D. R., Mbua, E., Kiura, P., Olago, D., Kibunjia, M., Omoubo, C., Behrensmeyer, A. K., et al. 2009. Early hominin foot morphology based on 1.5-million-year-old footprints from Ileret, Kenya. *Science* 323:1197–1201.

Benson, J., Fries, P., Greaves, W., Iwamoto, K., Savage-Rumbaugh, S., and Taglialatela, J. 2002. Confrontation and support in bonobo-human discourse. *Functions of Language* 9:1–38.

Benson, J., Greaves, W., O'Donnell, M., and Taglialatela, J. 2002. Evidence for symbolic language processing in a bonobo *(Pan paniscus)*. *Journal of Consciousness Studies* 9:33–56.

Beran, D. R. Australopithecine butchers. *Nature* 466:828.

Beran, M. J. 2001. Summation and numerousness judgements of sequentially presented sets of items by chimpanzees *(Pan troglodytes)*. *Journal of Comparative Psychology* 115:181–191.

———. 2004a. Chimpanzees *(Pan troglodytes)* respond to nonvisible sets after on-by-one addition and removal of items. *Journal of Comparative Psychology* 118:25–36.

———. 2004b. Long-term retention of the differential values of Arabic numerals by chimpanzees *(Pan troglodytes)*. *Animal Cognition* 7:86–92.

———. 2008. The evolutionary and developmental foundations of mathematics. *PLoS Biology* 6, no. 2: e19.

———. 2009. 3 questions I am often asked. *Eye on Psi Chi,* Fall, 15–19.

Beran, M. J., and Beran, M. M. 2004. Chimpanzees remember the results of one-by-one addition of food items to sets over extended time periods. *Psychological Science* 15:94–99.

Beran, M. J., and Evans, T. A. 2006. Maintenance of delay of gratification by four chimpanzees *(Pan troglodytes):* the effects of delayed reward visibility, experimenter presence, and extended delay intervals. *Behavioural Processes* 73:315–324.

Beran, M. J., Evans, T. A., and Harris, E. H. 2009. When in doubt, chimpanzees rely on estimates of past reward amounts. *Proceedings of the Royal Society B* 276:309–314.

Beran, M. J., Gibson, K. R., and Rumbaugh, D. M. 1999. Predicting hominid intelligence from brain size. In M. C. Corballis and E. G. Lea, eds. *The Descent of Mind: Psychological Perspectives on Hominid Evolution,* 88–97. New York: Oxford University Press.

Beran, M. J., Gulledge, J. P., and Washburn, D. A. 2007. Animals count: what's next? contributions from the Language Research Center to nonhuman primate cognition research. In D. A. Washburn, ed., *Primate Perspectives on Behavior and Cognition,* 161–173. Washington, DC: American Psychological Association.

Beran, M. J., Pate, J. E., Richardson, W. K., Rumbaugh, D. M. 2000. A chimpanzee's *(Pan troglodytes)* long-term retention of lexigrams. *Animal Learning and Behavior* 28:201–207.

Beran, M. J., and Rumbaugh D. M. 2001. "Constructive" enumeration by chimpanzees *(Pan troglodytes)* on a computerized task. *Animal Cognition* 4:81–89.

Beran, M. J., Rumbaugh D. M., and Savage-Rumbaugh, E. S. 1998. Chimpanzee *(Pan troglodytes)* counting in a computerized testing paradigm. *Psychological Record* 48:3–19.

Berdecio, S., and Nash, L. T. 1981. Facial, gestural and postural expressive movement in young, captive chimpanzees *(Pan troglodytes). Arizona State University Anthropological Research Papers,* No. 26.

Berezikov, E., Thuemmler, F., van Laake, L. W., Kondova, I., Bontrop, R., Cuppen, E., and Plasterk, R. H. A. 2006. Diversity of microRNAs in human and chimpanzee brain. *Nature Genetics* 38:1375–1377.

Berge, C., and Goularas, D. 2010. A new reconstruction of Sts 14 *(Australopithecus africanus)* from computed tomography and three-dimensional modeling techniques. *Journal of Human Evolution* 58:262–272.

Bergelson, E., and Swingley, D. 2012. At 6–9 months, human infants know the meanings of many common nouns. *Proceedings of the National Academy of Sciences of the United States of America* 109:3253–3258.

Berger, L. R. 2013. The mosaic nature of *Australopithecus sediba. Science* 340:163.

Berger, L. R., and Clarke, R. J. 1995. Eagle involvement in accumulation of the Taung child fauna. *Journal of Human Evolution* 29:275–299.

———. 1996. The load of the Taung child. *Nature* 379:778–779.

Berger, L. R., de Ruiter, D. J., Churchill, S. E., Schmid, P., Carlson, K. J., Dirks, P. H. G. M., and Kibii, J. M. 2010. *Australopithecus sediba:* a new species of *Homo*-like australopith from South Africa. *Science* 328:195–204.

Berger, L. R., and Hilton-Barber, B. 2001. *In the Footsteps of Eve.* Washington, DC: National Geographic/Adventure Press.

Berger, L. R., Lacruz, R., and de Ruiter, D. J. 2002. Brief communication: revised age estimates of *Australopithecus*-bearing deposits at Sterkfontein, South Africa. *American Journal of Physical Anthropology* 119:192–197.

Bergl, R. A., Bradley, B. J., Nsubuga, A., and Vigilant, L. 2008. Effects of habitat fragmentation, population size and demographic history on genetic diversity: the Cross River gorilla in comparative context. *American Journal of Primatology* 70:848–859.

Bergl, R. A., and Vigilant, L. 2007. Genetic analysis reveals population structure and recent migration within a highly fragmented range of the Cross River gorilla *(Gorilla gorilla diehli)*. *Molecular Ecology* 16:501–516.

Berillon, G. 2004. In what manner did they walk on two legs? An architectural perspective for the functional diagnostics of the early hominid foot. In D. J. Meldrum and C. E. Hilton, eds., *From Biped to Strider: The Emergence of Modern Human Walking, Running, and Resource Transport*, 85–100. New York: Kluwer Academic/Plenum Press.

Bering, J. M., and Povinelli, D. J. 2003. Comparing Cognitive Development. In D. Maestripieri, ed., *Primate Psychology*, 205–233. Cambridge, MA: Harvard University Press.

Bermejo, M. 1999. Status and conservation of primates in Odzala National Park, Republic of the Congo. *Oryx* 33:323–331.

———. 2004. Home-range use and intergroup encounters in western gorillas *(Gorilla g. gorilla)* at Lossi Forest, North Congo. *American Journal of Primatology* 64:223–232.

Bermejo, M., Illera, G., and Sabater Pí, J. 1994. Animals and mushrooms consumed by bonobos *(Pan paniscus):* new records from Lilungu (Ikela), Zaire. *International Journal of Primatology* 15:879–898.

Bermejo, M., and Omedes, A. 1999. Preliminary vocal repertoire and vocal communication of wild bonobos *(Pan paniscus)* at Lilungu (Democratic Republic of Congo). *Folia Primatologica* 70:328–357.

Bermejo, M., Rodríguez-Teijeiro, J. D., Illera, G., Barroso, A., Vilà, C, and Walsh, P. D. 2006. Ebola outbreak killed 5000 gorillas. *Science* 314:1564.

Bermúdez de Castro, J. M., Martinón-Torres, M., Prado, L., Gómez-Robles, A., Rosell, J., López-Polín, L., Arsuaga, J. L., and Carbonell, E. 2010. New immature hominin fossil from European Lower Pleistocene shows the earliest evidence of a modern human dental development. *Proceedings of the National Academy of Sciences of the United States of America* 100:11739–11744.

Bermúdez de Castro, J. M., Martinón-Torres, M., Sarmiento, S., Lozano, M., Arsuaga, J. L., and Carbonell, E. 2003. Rates of anterior tooth wear in Middle Pleistocene hominins form Sima de los Huesos (Sierra de Atapuerca, Spain). *Proceedings of the National Academy of Sciences of the United States of America* 107:11992–11996.

Berna, F., Goldberg, P., Horwitz, L. K., Brink, J., Holt, S., Bamford, M, and Chazan, M. 2012. Microstratigraphic evidence of in situ fire in the Acheulean strata of Wonderwerk Cave, Northern Cape Province, South Africa. *Proceedings of the National Academy of Sciences of the United States of America* 109:7593–7594.

Bernor, R. L. 1983. Geochronology and zoogeographic relationships of Miocene Hominoidea. In R. L. Ciochon and R. S. Corruccini, eds., *New Interpretations of Ape and Human Ancestry*, 21–64. New York: Plenum Press.

Bernor, R. L., Flynn, L. J., Harrison, T., Hussain, S. T., and Kelley, J. 1988. *Dionysopithecus* from southern Pakistan and the biochronology and biogeography of early Eurasian catarrhines. *Journal of Human Evolution* 17:39–358.

Bernstein, I. S. 1991. The correlation between kinship and behavior in non-human primates. In *Kin Recognition*, P. G. Hepper, ed. Cambridge: Cambridge University Press.

Bertolani, P., and Pruetz, J. D. 2011. Seed reingestion in savannah chimpanzees *(Pan troglodytes verus)* at Fongoli, Senegal. *International Journal of Primatology* 32:1123–1132.

Bethell, E. J., Vick, S.-J., and Bard, K. A. 2007. Measurement of eye-gaze in chimpanzees *(Pan troglodytes)*. *American Journal of Physical Anthropology* 69:562–575.

Beyene, Y., Katoh, S., WoldeGabriel, G., Hart, W. K., Uto, K., Sudo, M., Kondo, M., Hyodo, M., Renne, P. T., Suwa, G., and Asfaw, B. 2013. The characteristics and chronology of the earliest Acheulean at Konso, Ethiopia. *Proceedings of the National Academy of Sciences of the United States of America* 110:1584–1591.

Beynon, A. D., and Wood, B. A. 1986. Variations in enamel thickness and structure in East African hominids. *American Journal of Physical Anthropology* 70:177–193.

Bianchi, N. O., Bianchi, M. S., Cleaver, J. E., and Wolff, S. 1985. The pattern of restriction enzyme-induced banding in the chromosomes of chimpanzee, gorilla, and orangutan and its evolutionary significance. *Journal of Molecular Evolution* 22:323–333.

Bickerton, D. 1990. *Language and Species*. Chicago: University of Chicago Press.

———. 1993. Putting cognitive carts before linguistic horses. *Behavioral and Brain Sciences* 16:749–750.

———. 1998. Catastrophic evolution: the case for a single step from protolanguage to full human language. In J. R. Hurfod, M. G. Studdert-Kennedy, and C. Knight, eds., *Approaches to the Evolution of Language: Social and Cognitive Bases,* 341–358. Cambridge: Cambridge University Press.

———. 2000a. Can biomusicology learn from language evolution studies? In N. L. Wallin, B. Merker and S. Brown, eds., *The Origins of Music,* 153–163. Cambridge, MA: MIT Press.

———. 2000b. How protolanguage became language. In C. Knight, M. Studdert-Kennedy, and J. R. Hurford, eds., *The Evolutionary Emergence of Language: Social Function and the Origins of Linguistic Form,* 264–284. Cambridge: Cambridge University Press.

———. 2003. Symbol and structure: a comprehensive framework for language evolution. In M. H. Christiansen, and S. Kirby, eds., *Language Evolution,* 77–93. Oxford: Oxford University Press.

———. 2007. Language evolution: a brief guide for linguists. *Lingua* 117:510–526.

———. 2009. *Adam's Tongue: How Humans Made Language, How Language Made Humans.* New York: Hill and Wang.

Bilsborough, A. 1983. The pattern of evolution within the genus *Homo*. *Progress in Anatomy* 3:143–164.

———. 1984. Multivariate analysis and cranial diversity in Plio-Pleistocene hominids. In G. N. van Vark and W. W. Howells, eds., *Multivariate Statistical Methods in Physical Anthropology,* 351–375. Dordrecht: D. Reidel.

Bilsborough, A., and Wood, B. A. 1986. The nature, origin and fate of *Homo erectus*. In B. Wood, L. Martin, and P. Andrews, eds., *Major Topics in Primate and Human Evolution,* 295–316. Cambridge: Cambridge University Press.

Binford, L. R. 1977. Olorgesailie deserves more than the usual book review. *Journal of Anthropological Research* 33:493–502.

———. 1981. *Bones: Ancient Men and Modern Myths.* New York: Academic Press.

———. 1983. *In Pursuit of the Past: Decoding the Archaeological Record.* London: Thames and Hudson.

———. 1985. Human ancestors: changing views of their behavior. *Journal of Anthropological Archaeology* 4:292–327.

Binford, L. R., and Ho, C. K. 1985. Taphonomy at a distance: Zhoukoudian, "the cave home of Beijing Man"? *Current Anthropology* 26:413–442.

Binford, L. R., and Stone, N. M. 1986. Zhoukoudian: a closer look. *Current Anthropology* 427:453–475.

Bingham, H. C. 1932. Gorillas in a native habitat. *Carnegie Institute of Washington Publications* 426:1–66.

Binyeri, D. K., Hibukabake, D. M., and Kiyengo, C. S. 2002. The Mikeno gorillas. *Gorilla Journal* 25:5–7.

Birkhead, T. R., and Pizzari, T. 2002. Postcopulatory sexual selection. *Nature Reviews Genetics* 3:262–273.

Biro, D., and Matsuzawa, T. 1999. Numerical ordering in a chimpanzee *(Pan troglodytes):* planning, executing, and monitoring. *Journal of Comparative Psychology* 113:178–185.

———. 2001a. Use of numerical symbols by the chimpanzee *(Pan troglodytes):* cardinals, ordinals, and the introduction of zero. *Animal Cognition* 4:193–199.

———. 2001b. Chimpanzee numerical competence: cardinal and ordinal skills. In T. Matsuzawa, ed., *Primate Origins of Human Cognition and Behavior,* 199–225. Tokyo: Springer.

Bishop, A. 1964. Use of the hand in lower primates. In J. Buettner-Janusch, ed., *Evolutionary and Genetic Biology of Primates,* 133–225. New York: Academic Press.

Bishop, D. V. M. 2013. Cerebral asymmetry and language development: cause, correlate, or consequence. *Science* 340:1302.

Bishop, M. J., and Friday, A. E. 1986. Molecular sequences and hominoid phylogeny. In B. Wood, L. Martin, and P. Andrews, eds., *Major Topics in Primate and Human Evolution,* 150–156. Cambridge: Cambridge University Press.

Bishop, W. W. 1964. More fossil primates and other Miocene mammals from north-east Uganda. *Nature* 203:1327–1331.

Bishop, W. W., and Chapman, G. R. 1970. Early Pliocene sediments and fossils from the northern Kenya Rift Valley. *Nature* 226:914–918.

Black, D. 1927. On a lower molar hominid tooth from the Chou Kou Tien deposit. *Palaeontologia Sinica,* n.s., series D, 7:1–28.

———. 1931–32. Evidences of the use of fire by *Sinanthropus. Bulletin of the Geological Society of China* 11:107–108.

———. 1934. On the discovery, morphology, and environment of *Sinanthropus pekinensis. Philosophical Transactions of the Royal Society of London,* series B 223:57–120.

Blackwelder, R. E. 1967. *Taxonomy.* New York: Wiley.

Blake, S., and Fay, M. J. 1997. Seed production by *Gilbertiodendron dewevrei* in the Nouabalé-Ndoki National Park, Congo, and its implications for large mammals. *Journal of Tropical Ecology* 14:885–891.

Blake, S., Rogers, E., Fay, M., Ngangoué, M., and Ebéke, G. 1995. Swamp gorillas in northern Congo. *African Journal of Ecology* 33:285–290.

Blaney, S. P. A. 1986. An allometric study of the frontal sinus in *Gorilla, Pan,* and *Pongo. Folia Primatologica* 47:81–96.

Bleisch, W. V., and Chen, N. 1991. Ecology and behavior of wild black-crested gibbons *(Hylobates concolor)* in China with a reconsideration of evidence for polygyny. *Primates* 32:539–548.

Bloch, J. I., Silcox, M. T., Boyer, D. M., and Sargis, E. J. 2007. New Paleocene skeletons and the relationship of plediadapiformes to crown-clade primates. *Proceedings of the National Academy of Sciences of the United States of America* 104:1159–1164.

Blockley, S. P. E., Ramsey, C. B., and Higham, T. F. G. 2008. The Middle and Upper Paleolithic transition: dating, stratigraphy, and isochronous markers *Journal of Human Evolution* 55:764–771.

Blom, A. 1998. A critical analysis of three approaches to tropical forest conservation based on experiences in the Sanhga Region. *Yale F&ES Bulletin* 102:208–215.

Blom, A., Alers, M. P. T., Feistner, A. T. C., Barnes, R. F. W., and Barnes, K. L. 1996. Primates in Gabon—current status and distribution. *Oryx* 26:223–234.

Blom, A., Almasi, A., Heitkönig, I. M. A., Kpanou, J.-B., and Prins, H. H. T. 2001. A survey of the apes in the Dzanga-Ndoki National Park, Central African Republic: a comparison of estimating gorilla *(Gorilla gorilla gorilla)* and chimpanzee *(Pan troglodytes)* nest group density. *African Journal of Ecology* 39:98–105.

Blom, A., Cipolletta, C., Brunsting, M. H., and Prins, H. H. T. 2004. Behavioral responses of gorillas to habituation in the Dzanga-Ndoki National Park Central African Republic. *International Journal of Primatology* 25:179–196.

Blom, A., Yamindou, J., and Prins, H. H. T. 2004. Status of the protected area of the Central African Republic. *Biological Conservation* 118:479–487.

Bloom, L. *One Word at a Time: The Use of Single Word Utterances before Syntax.* The Hague: Mouton.

Blue, K. T., McCrossin, M. L., and Benefit, B. R. 2006. Terrestriality in a Middle Miocene context: *Victoriapithecus* from Maboko, Kenya. In H. Ishida, R. H. Tuttle, M. Pickford, M. Nakatsukasa, and N. Ogihara, eds., *Human Origins and Environmental Backgrounds,* 45–58. New York: Springer.

Blum, D. 1994. *The Monkey Wars.* New York: Oxford University Press.

Blumenberg, B., and Lloyd, A. T. 1983. *Australopithecus* and the origin of the genus *Homo:* aspects of biometry and systematics with accompanying catalog of tooth metric data. *BioSystems* 16:127–167.

Blumenschine, R. J. 1991. Prey size and age models of prehistoric hominid scavenging: test cases from the Serengeti. In M. C. Stiner, ed., *Human Predators and Prey Mortality,* 121–147. Boulder CO: Westview Press.

Blumenschine, R. J., Peters, C. R., Masao, F. T., Clarke, R. J., Deino, A. L., Hay, R. L., Swisher, C. C., et al. 2003. Late Pliocene *Homo* and hominid land use from western Olduvai Gorge, Tanzania. *Science* 299:1217–1221.

Blumenschine, R. J., and Pobiner, B. L. 2007. Zooarchaeology and the ecology of Oldowan hominin carnivory. In P. S. Ungar, ed., *Evolution of the Human Diet,* 167–190. Oxford: Oxford University Press.

Blumenthal, S. A., Chritz, K. L., Rothman, J. M., and Cerling, T. E. 2012. Detecting intraanual dietary viariability in wild mountain gorillas by staple isotope analysis of feces. *Proceedings of the National Academy of Sciences of the United States of America* 109:21277–21282.

Blurton Jones, N. G., Hawkes, K., and O'Connell, J. F. 1989. Modelling and measuring costs of children in two foraging societies. In V. Standen and R. A. Foley, eds., *Comparative Socioecology: The Behavioural Ecology of Humans and Other Mammals,* 367–390. Oxford: Blackwell Scientific.

Boaz, N. T. 1979. Hominid evolution in eastern Africa during the Pliocene and early Pleistocene. *Annual Review of Anthropology* 8:71–85.

———. 1982. American research on australopithecines and early *Homo,* 1925–1980. In F. Spencer, ed., *A History of American Physical Anthropology, 1930–1980,* 239–260. New York: Academic Press.

———. 1983. Morphological trends and phylogenetic relationships from Middle Miocene hominoids to Late Pliocene hominids. In R. L. Ciochon and R. S. Corruccini, eds., *New Interpretations of Ape and Human Ancestry,* 705–720. New York: Plenum Press.

———. 1988. Status of *Australopithecus afarensis. Yearbook of Physical Anthropology* 31:85–113.

Boaz, N. T., Ciochon, R. L., Xu, Q., and Liu, J. 2004. Mapping and taphonomic analysis of the *Homo erectus* loci at Locality 1 Zhoukoudian, China. *Journal of Human Evolution* 46:519–549.

Boaz, N. T., and Howell, F. C. 1977. A gracile hominid cranium from Upper Member G of the Shungura Formation, Ethiopia. *American Journal of Physical Anthropology* 46:93–108.

Boaz, N. T., Howell, F. C., and McCrossin, M. L. 1982. Faunal age of the Usno, Shungura B and Hadar Formations, Ethiopia. *Nature* 300:633–635.

Bocquet-Appel, J.-P. 2009. The demographic impact of the agricultural system in human history. *Current Anthropology* 50:657–660.

Boehm, C. 1992. Segmentary 'warfare' and the management of conflict: comparison of East African chimpanzees and patrilineal-patrilocal humans. In A. H. Harcourt and F. B. M. de Waal, eds., *Coalitions and Alliances in Humans and Other Animals,* 137–173. Oxford: Oxford University Press.

———. 1994. Pacifying interventions at Arnhem Zoo and Gombe. In R. W. Wrangham, W. C. McGrew, F. B. M. de Waal, and P. G. Heltne, eds., *Chimpanzee Cultures,* 211–224. Cambridge, MA: Harvard University Press.

———. 1999. *Hierarchy in the Forest: The Evolution of Egalitarian Behavior.* Cambridge, MA: Harvard University Press.

Boesch, C. 1978. Nouvelles observations sur les chimpanzés de la Foret de Taï (Côte d'Ivoire). *La Terre et la Vie* 32:195–201.

———. 1991a. The effects of leopard predation on grouping patterns in forest chimpanzees. *Behaviour* 117:220–242.

———. 1991b. Handedness in wild chimpanzees. *International Journal of Primatology* 12:541–558.

———. 1993. Aspects of transmission of tool-use in wild chimpanzees. In K. R. Gibson and T. Ingold, eds., *Tools, Language and Cognition in Human Evolution,* 171–183. Cambridge: Cambridge University Press.

———. 1994a. Chimpanzees-red colobus monkeys: a predator-prey system. *Animal Behaviour* 47:1135–1148.

———. 1994b. Cooperative hunting in wild chimpanzees. *Animal Behaviour* 48:653–667.

———. 1994c. Hunting strategies of Gombe and Taï chimpanzees. In R. W. Wrangham, W. C. McGrew, F. B. M. de Waal and P. G. Heltne, eds., *Chimpanzee Cultures,* 77–91. Cambridge, MA: Harvard University Press.

———. 1995. Innovation in wild chimpanzees *(Pan troglodytes). International Journal of Primatology* 16:1–16.

———. 1996. Social grouping in Taï chimpanzees. In W. C. McGrew, L. F. Marchant, and T. Nishida, eds., *Great Ape Societies,* 101–113. Cambridge: Cambridge University Press.

———. 2002. Cooperative hunting roles among chimpanzees. *Human Nature* 13:27–46.

———. 2003. Is culture a golden barrier between human and chimpanzee? *Evolutionary Anthropology* 12:82–91.

———. 2009. *The Real Chimpanzee: Sex Strategies in the Forest.* Cambridge: Cambridge University Press.

———. 2013. Ecology and cognition of tool use in chimpanzees. In C. M. Sanz, J. Call, and C. Boesch, eds., *Tool Use in Animals. Cognition and Ecology,* 21–47. Cambridge: Cambridge University Press.

Boesch, C., Bi, Z. B. G., Anderson, D., and Stahl, D. 2006. Food choice in Taï chimpanzees: are cultural differences present? In G. Hohmann, M. M. Robbins, and C. Boesch, eds., *Feeding Ecology in Apes and Other Primates,* 83–201. Cambridge: Cambridge University Press.

Boesch, C., and Boesch, H. 1981. Sex differences in the use of natural hammers by wild chimpanzees: a preliminary report. *Journal of Human Evolution* 10:585–593.

———. 1982. Optimisation of nut-cracking with natural hammers by wild chimpanzees. *Behaviour* 83:265–286.

———. 1984a. The nut-cracking behavior and its nutritional importance in wild chimpanzees in the Taï National Park, Ivory Coast. *International Journal of Primatology* 5:323.

———. 1984b. Possible causes of sex differences in the use of natural hammers by wild chimpanzees. *Journal of Human Evolution* 13:415–440.

———. 1984c. Mental map in wild chimpanzees: an analysis of hammer transports for nut cracking. *Primates* 25:160–170.

———. 1989. Hunting behavior of wild chimpanzees in the Taï National Park. *American Journal of Physical Anthropology* 78:547–563.

———. 1990. Tool use and tool making in wild chimpanzees. *Folia Primatologica* 54:86–99.

———. 1993. Diversity of tool use and tool-making in wild chimpanzees. In A. Berthelet and J. Chavaillon, eds., *The Use of Tools by Human and Non-human Primates,* 158–168. New York: Oxford University Press.

Boesch, C., and Boesch-Achermann, H. 1991. Les chimpanzés et l'outil. *La Recherche* 22:724–731.

———. 2000. *The Chimpanzees of the Taï Forest.* Oxford: Oxford University Press.

Boesch, C., Crockford, C., Herbinger, I., Wittig, R., Moebius, Y., and Normand, E. 2008. Intergroup conflicts among chimpanzees in Taï National Park: lethal violence and female perspective. *American Journal of Primatology* 70:519–532.

Boesch, C., Head, J., and Robbons, M. M. 2009. Complex tool sets for honey extraction among chimpanzees in Loango National Park, Gabon. *Journal of Human Evolution* 56:560–569.

Boesch, C., Head, J., Tagg, N., Arandjelovic, M., Vigilant, L., and Robbons, M. M. 2009. Fatal chimpanzee attack in Loango National Park, Gabon. *International Journal of Primatology* 28:1025–1034.

Boesch, C., Kohou, G., Néné, H., and Vigilant, L. 2006. Male competition and paternity in wild chimpanzees of the Taï Forest. *American Journal of Physical Anthropology* 130:103–115.

Boesch, C., Marchesi, P., Marchesi, N., Fruth, B., and Joulian, F. 1994. Is nut cracking in wild chimpanzees a cultural behaviour? *Journal of Human Evolution* 26:325–338.

Boesch, C., and Tomasello, M. 1998. Chimpanzee and human cultures. *Current Anthropology* 39:591–614.

Bogart. S. L., and Pruetz, J. D. 2008. Ecological context of savanna chimpanzee *(Pan troglodytes verus)* termite fishing at Fongoli, Senegal. *American Journal of Primatology* 70:605–612.

———. 2011. Insectivory of savanna chimpanzees *(Pan troglodytes)* at Fongoli, Senegal. *American Journal of Physical Anthropology* 145:11–20.

Bogart. S. L., Pruetz, J. D., and Kante, D. 2008. Fongoli chimpanzee *(Pan troglodytes verus)* eats banded mongoose *(Mungos mungo)*. *Pan Africa News* 15:15–17.

Bogart. S. L., Pruetz, J. D., Ormiston, L. K., Russell, J. L., Meguerditchian, A., and Hopkins, W. D. 2012. Termite fishing laterality in the Fongoli savanna chimpanzees *(Pan troglodytes verus)*: further evidence of a left hand preference. *American Journal of Physical Anthropology* 149:591–598.

Boghi, A., Rasetti, R., Avidano, F., Manzone, C., Orsi, L., D'Agata, F., Caroppo, P., Bergui, M., Rocca, P., Puvlirenti, L., et al. 2006. The effect of gender on planning: An fMRI study using the Tower of London task. *NeuroImage* 33:999–1010.

Boinski, S. 1988. Use of a club by a wild white-faced capuchin *(Cebus capucinus)* to attack a venomous snake *(Bothrops asper)*. *American Journal of Primatology* 14:177–179.

Boinski, S., Quatrone, R. P., Sughrue, K., Selvaggi, L., Henry, M., Stickler, C. M., and Rose, L. M. 2003. Do brown capuchins socially learn foraging skills? In D. M. Fragaszy and S. Perry, eds., *The Biology of Traditions,* 365–390. Cambridge: Cambridge University Press.

Boinski, S., Quatrone, R. P., and Swartz, H. 2001. Substrate and tool use by brown capuchins in Suriname: ecological contexts and cognitive bases. *American Anthropologist* 102:741–761.

Bolster, R. B. 1978. Cross-modal matching in the monkey *(Macaca fascicularis)*. *Neuropsychologia* 16:407–416.

Bonnefille, R., and Riollet, G. 1987. Palynological spectra from the Upper Laetoli Beds. In M. D. Leakey and J. M. Harris, eds., *Laetoli: A Pliocene Site in Northern Tanzania,* 52–61. Oxford: Clarenden Press.

Bonnie, K. E., and de Waal, F. B. M. 2006. Affiliation promotes the transmission of a social custom: handclasp grooming among captive chimpanzees. *Primates* 47:27–34.

Bonvillian, J. D., and Patterson, F. G. P. 1999. Early sign-language acquisition: comparisons between children and gorillas. In S. T. Parker, R. W. Mitchell, and H. L. Miles, eds.,

The Mentalities of Gorillas and Orangutans, 240–264. Cambridge: Cambridge University Press.

Booth, F. W., and Neufer, P. D. 2005. Exercise controls gene expression. *American Scientist* 93:28–35.

Bordes, F., and Lafille, J. 1962. Découverte d'un squelette moustérien dans le gisement du Roc de Marsal, commune de Campagne-du-Bugue (Dordogne). *Comptes rendus des séances de l'Académie des sciences, Paris, série D* 254:714–715.

Borgerhoff Mulder, M., and Caro, T. M. 1983. Polygyny: definition and application to human data. *Animal Behaviour* 32:609–610.

Borod, J. C., Haywood, C. S., and Koff, E. 1997. Neuropsychological aspects of facial asymmetry during emotional expression: a review of the adult literature. *Neuropsychology Review* 7:41–60.

Borod, J. C., Koff, E., Yecker, S., Santschi, C., and Schmidt, M. 1998. Facial asymmetry during emotional expression: gender, valence and measurement technique. *Neuropsychology Review* 7:41–60.

Borofsky, R. 1994. *Assessing Cultural Anthropology.* New York: McGraw-Hill.

Borofsky, R., Barth, F., Shweder, R. A., Rodseth, L., and Stolzenberg, N. M. 2001. WHEN: A conversation about culture. *American Anthropologist* 103:432–446.

Borowik, O. A. 1995. Coding chromosomal data for phylogenetic analysis: phylogenetic resolution of the *Pan-Homo-Gorilla* trichotomy. *Systematic Biology* 44:563–570.

Boschetto, H. B., Brown, F. H., and McDougall, I. 1992. Stratigraphy of the Lothidok Range, northern Kenya, and K/Ar ages of its Miocene primates. *Journal of Human Evolution* 22:47–71.

Bosler, W. 1981. Species groupings of Early Miocene dryopithecine teeth from east Africa. *Journal of Human Evolution* 10:151–158.

Boubli, J. P. 1999. Feeding ecology of black-headed uacaris *(Cacajao melanocephalus melanocephalus)* in Pico da Neblina National Park, Brazil. *International Journal of Primatology* 20:719–749.

Boulle, P. 1963a. *La planète des singes.* Paris: R. Julliard.

———. 1963b. *Planet of the Apes.* New York: Vanguard.

Bourdieu, P. 1977. *Outline for a Theory of Practice.* Cambridge: Cambridge University Press.

Bourquin, L. D., Titgemeyer, E. C., and Fahey, G. C., Jr. 1993. Vegetable fiber fermentation by human fecal bacteria: cell wall polysaccharide disappearance and short-chain fatty acid production during in vitro fermentation and water-holding capacity of unfermented residues. *Journal of Nutrition* 123:860–869.

Bowdler, S. 2007. Liang Bua in the wider prehistoric context. In E. Indriati, ed., *Recent Advances in Southeast Asian Paleoanthropology and Archaeology,* 90–94. Yogyakarta, Indonesia: Gadjah Mada University.

Bowen-Jones, E., and Pendry, S. 1999. The threat to primates and other mammals from the bushmeat trade in Africa, and how this threat could be diminished. *Oryx* 33:233–246.

Bower, B. 2010. Ancient footprints yield oldest signs of upright gait. *ScienceNews,* March 22, www.sciencenews.org/view/generic/id/57513.

Bowles, S. 2011. Cultivation of cereals by the first farmers was not more productive than foraging. *Proceedings of the National Academy of Sciences of the United States of America* 108:4760–4765.

Bown, T. M., Kraus, M. J., Wing, S. L., Fleagle, J. G., Tiffney, B. H., Simons, E. L., Vondra, C. F. 1982. The Fayum forest revisited. *Journal of Human Evolution* 11:603–632.

Boysen, S. T. 1988. Kanting processes in the chimpanzee: what (and who) really counts? *Behavioral and Brain Sciences* 11:4.

———. 1996. "More is less": the elicitation of rule-governed resource distribution in chimpanzees. In A. E. Russon, K. A. Bard, and S. T. Parker, eds., *Reaching into Thought: The Minds of the Great Apes,* 177–189. Cambridge, MA: Cambridge University Press.

———. 1997. Representation of quantities by apes. *Advances in the Study of Behavior* 26:435–462.

Boysen, S. T., and Berntson, G. G. 1989. Numerical competence in a chimpanzee *(Pan troglodytes). Journal of Comparative Psychology* 103:23–31.

———. 1990. The development of numerical skills in the chimpanzee *(Pan troglodytes).* In S. T. Parker and K. R. Gibson, eds., *"Language" and intelligence in monkeys and apes,* 435–450. Cambridge: Cambridge University Press.

———. 1995. Responses to quantity: perceptual versus cognitive mechanisms in chimpanzees *(Pan troglodytes). Journal of Experimental Psychology* 21:82–86.

Boysen, S. T., Berntson, G. G., Hannan, M. B., and Cacioppo, J. T. 1996. Quantity-based interference and symbolic representations in chimpanzees *(Pan troglodytes). Journal of Experimental Psychology* 22:76–86.

Boysen, S. T., Berntson, and Mukobi, K. L. 2001. Size matters: impact of item size and quantity on array choice by chimpanzees *(Pan troglodytes). Journal of Comparative Psychology* 115:106–110.

Boysen, S. T., and Capaldi, E. J. 1993. *The Development of Numerical Competence: Animal and Human Models.* Hillsdale, NJ: Lawrence Erlbaum.

Boysen, S. T., and Kuhlmeier, V. A. 2002. Representational capacities for pretense with scale models and photographs in chimpanzees *(Pan troglodytes).* In R. W. Mitchell, ed., *Pretending and Imagination in Animals and Children,* 210–228. Cambridge: Cambridge University Press.

Boysen, S. T., Kuhlmeier, V. A., Halliday, P., and Halliday, Y. M. 1999. Tool use in captive gorillas. In S. T. Parker, R. W. Mitchell, and H. L. Miles, eds., *The Mentalities of Gorillas and Orangutans,* 179–187. Cambridge: Cambridge University Press.

Boysen, S. T., Mukobi, K. L., and Berntson, G. 1999. Overcoming response bias using symbolic representations of number by chimpanzees *(Pan troglodytes). Animal Learning and Behavior* 27:229–235.

Bozeat, S., Lambon Ralph, M. A., Patterson, K., Garrard, P., and Hodges, J. R. 2000. Nonverbal semantic impairment in semantic dementia. *Neuropsychologia* 38:1207–1215.

Brace, C. L. 1962. Cultural factors in the evolution of the human dentition. In M. F. A. Montagu, ed., *Culture and the Evolution of Man,* 343–354. New York: Oxford University Press.

———. 1963. Structural reduction in evolution. *American Naturalist* 97:39–49.

———. 1967. *The Stages of Human Evolution.* Englewood Cliffs, NJ: Prentice Hall.

———. 1982. The roots of the race concept in American physical anthropology. In F. Spencer, ed., *A History of American Physical Anthropology, 1930–1980,* 11–29. New York: Academic Press.

Brace, C. L., Rosenberg, K. R., and Hunt, K. D. 1967. Gradual change in human tooth size in the Late Pleistocene and post-Pleistocene. *Evolution* 41:705–720.

Bradley, B. J. 2008. Reconstructing phylogenies and phenotypes: a molecular view of human evolution. *Journal of Anatomy* 212:337–353.

Bradley, B. J., Doran-Sheehy, D. M., Lukas, D., Boesch, C., and Vigilant, L. 2004. Dispersed male networks in western gorillas. *Current Biology* 14:510–513.

Bradley, B. J., Robbins, M. M., Williamson, E. A., Steklis, H. D., Steklis, N. G., Eckhardt, N., Boesch, C., and Vigilant, L. 2005. Mountain gorilla tug-of-war: silverbacks have limited control over reproduction in multimale groups. *Proceedings of the National Academy of Sciences of the United States of America* 102:9418–9423.

Bradshaw, J. L. 1988. The evolution of human lateral asymmetries: new evidence and second thoughts. *Journal of Human Evolution* 17:615–637.

———. 1991. Animal asymmetry and human heredity: dextrality, tool use and language in evolution—10 years after Walker (1980). *British Journal of Psychology* 82:39–59.

Bradshaw, J. L., and Rogers, L. J. 1993. *The Evolution of Lateral Asymmetries, Language, Tool Use, and Intellect.* San Diego: Academic Press.

Brain, C. K. 1968a. New light on old bones. *Southern African Museums Association Bulletin* 9:22–27.

———. 1968b. Who killed the Swartkrans ape-men? *Southern African Museums Association Bulletin* 9:127–139.

———. 1969. The probable role of leopards as predators of the Swartkrans anstralopithecines. *South African Archaeological Bulletin* 24:170–171.

———. 1981. *The Hunters or the Hunted? An Introduction to African Cave Taphonomy.* Chicago: University of Chicago Press.

———. 1993a. The occurrence of burnt bones at Swartkrans and their implication for the control of fire by early hominids. In C. K. Brain, ed., *Swartkrans: A Cave's Chronicle of Early Man,* 229–242. Monograph No. 8. Pretoria: Transvaal Museum.

———. 1993b. A taphonomic overview of the Swartkrans fossil assemblages. In C. K. Brain, ed., *Swartkrans: A Cave's Chronicle of Early Man,* 257–264. Monograph No. 8. Pretoria: Transvaal Museum.

———. 1993c. *Swartkrans: A Cave's Chronicle of Early Man.* Monograph No. 8. Pretoria: Transvaal Museum.

Brain, C. K., and Shipman, P. 1993. The Swartkrans bone tools. In C. K. Brain, ed., *Swartkrans: A Cave's Chronicle of Early Man,* 195–215. Monograph No. 8. Pretoria: Transvaal Museum.

Brain, C. K., and Sillent, A. 1988. Evidence from the Swartkrans cave for the earliest use of fire. *Nature* 336:464–466.

Brakke, K. E., and Savage-Rumbaugh, E. S. 1995. The development of language skills in bonobo and chimpanzee—I. comprehension. *Language and Communication* 15:121–148.

———. 1996. The development of language skills in *Pan*—II. production. *Language and Communication* 16:361–380.

Bramble, D. M., and Carrier, D. R. 1983. Running and breathing in mammals. *Science* 250:251–256.

Bramble, D. M., and Lieberman, D. E. 2004. Endurance running and the evolution of *Homo*. *Nature* 432:345–352.

Brandon-Jones, D., Eudey, A. A., Geissmann, T., Groves, C. P., Melnick, D. J., Morales, J. C., Shekelle, M., and Stewart, C.-B. 2004. Asian primate classification. *International Journal of Primatology* 25:97–164.

Brannon, E. M. 2005. The independence of language and mathematical reasoning. *Proceedings of the National Academy of Sciences of the United States of America* 102:3177–3178.

Brass, M., and Rüschemeyer, S.-A. 2010. Mirrors in science: how mirror neurons changed neuroscience. *Cortex* I46:139–141.

Bräuer, G. 1982. Early anatomically modern man in Africa and the replacement of the Mediterranean and European Neandertals. In *L'Homo erectus et la place de l'Homme Tautavel parmi les hominidés fossiles,* H. de Lumley, ed., p. 112. Nice: Centre National de la Recherche Scientifique/Louis-Jean Scientific and Literary Publications.

———. 1984a. A craniological approach to the origin of anatomically modern *Homo sapiens* in Africa and implications for the appearance of modern Europeans. In F. H. Smith and F. Spencer, eds., *The Origin of Modern Humans,* 327–410. New York: Alan R. Liss.

———. 1984b. The 'Afro-European sapiens hypothesis' and hominid evolution in East Asia during the Late Middle and Upper Pleistocene. *Courier Forschungsinstitute Senkenberg* 69:145–165.

———. 1984c. Präsapiens-Hypothese oder Afro-europäische Sapiens-Hypothese? *Zeitschrift fuur Morphologie und Anthropologie* 75:1–25.

———. 1989. The Evolution of modern humans: a comparison of the African and non-African evidence. In P. Mellars and C. Stringer, eds., *The Human Revolution,* 123–154. Edinburgh: University of Edinburgh Press.

———. 1992. Africa's place in the evolution of *Homo sapiens.* In B. Bräuer and F. Smith, eds., *Continuity or Replacement Controversies in the Evolution of Homo sapiens,* 83–98. Rotterdam: Balkema.

———. 2008. The origin of modern anatomy: by speciation or interspecific evolution? *Evolutionary Anthropology* 17:22–37.

Bräuer, G., and Schultz, M. 1996. The morphological affinities of the Plio-Pleistocene mandible from Dmanisi, Georgia. *Journal of Human Evolution* 30:445–481.

Braun, D. R., Harrris, J. W. K., Levin, N. E., McCoy, J. T., Herries, A. I. R., Bamford, M. K., Bishop, L. C., Richmond, B. G., and Kibunjia, M. 2010. Early hominin diet included diverse terrestrial and aquatic animals 1.95 Ma in East Turkana, Kenya. *Proceedings of the National Academy of Sciences of the United States of America* 107:10002–10007.

Brent, L. 1995. Factors determining tool-using in two captive chimpanzee *(Pan troglodytes)* colonies. *Primates* 36:265–274.

Breuer, T. 2008. A window into the lives of wild gorillas. *Anthropology News* 49:49.

Breuer, T., Breuer-Ndoundou Hockemba, M., Olejniczak, C., Parnell, and Stokes, E. J. 2009. Physical maturation, life-history classes and age estimates of free-ranging western gorillas—insights from Mbeli Bai, Republic of Congo. *American Journal of Primatology* 71:106–119.

Breuer, T., Robbins, A. M., Olejniczak, C., Parnell, R. J., Stokes, E. J., and Robbins, M. M. 2010. Variance in the male reproductive success of western gorillas: acquiring females is just the beginning. *Behavioral Ecology and Sociobiology* 64:515–528.

Breuer, T., Robbins, M. M., and Boesch, C. 2007. Using photogrammetry and color scoring to assess sexual dimorphism in wild western gorillas *(Gorilla gorilla)*. *American Journal of Physical Anthropology* 134:369–382.

Brewer, S. M., and McGrew, W. C. 1990. Chimpanzee use of a tool-set to get honey. *Folia Primatologica* 54:100–104.

Bricknell, S. J. 1999. *Hybridisation and Behavioral Variation: a Socio-ecological Study of Hybrid Gibbons (Hylobates agilis albibarbis × H. muelleri) in Central Kalimantan Indonesia.* PhD diss., Australian National University, Canberra.

Briggs, A. W., Good, J. M., Green, R. E., Krause, J., Marcic, T., Stenzel, U., Lalueza-Fox, C., Rudan P., Brajkovic, D., Kucan, Z., et al. 2009. Targted retrieval and analysis of five Neandertal mtDNA genomes. *Science* 325:318–321.

Brightman, R. 1995. Forget culture: replacement, transcendence, reflexification. *Current Anthropology* 10:509–546.

Britten, R. J. 1986. Rates of DNA sequence evolution differ between taxonomic groups. *Science* 231:1393–1398.

———. 1989. Comment on a criticism of DNA hybridization measurements. *Journal of Human Evolution* 18:163–164.

———. 2002. Divergence between samples of chimpanzee and human DNA sequences is 5%, counting indels. *Proceedings of the National Academy of Sciences of the United States of America* 99:1363–1369.

Britten, R. J., Rowen, L., Williams, J., and Cameron, R. A. 2003. Majority of divergence between closely related DNA samples is due to indels. *Proceedings of the National Academy of Sciences of the United States of America* 100:4661–4665.

Brock, A., McFadden, P. L., and Partridge, T. C. 1977. Preliminary palaeomagnetic results from Makapansgat and Swartkrans. *Nature* 266:249–250.

Brockelman, W. Y. 1975. Gibbon populations and their conservation in Thailand. *Natural History Bulletin of the Siam Society* 26:133–157.

———. 1984. Social behaviour of gibbons: introduction. In H. Preuschoft, D. J. Chivers, W. Y. Brockelman, and N. Creel, eds., *The Lesser Apes: Evolutionary and Behavioural Biology,* 285–290. Edinburgh: Edinburgh University Press.

Brockelman, W. Y., and Gittins, S. P. 1984. Natural hybridization in the *Hylobates lar* species group: implications for speciation in gibbons. In H. Preuschoft, D. J. Chivers, W. Y. Brockelman, and N. Creel, eds., *The Lesser Apes: Evolutionary and Behavioural Biology,* 498–532. Edinburgh: Edinburgh University Press.

Brockelman W. Y., Reichard, U., Treesucon, U., and Raemaekers, J. J. 1998. Dispersal, pair formation, and social structure in gibbons *(Hylobates lar)*. *Behavioral Ecology and Sociobiology* 42:329–339.

Brockelman, W. Y., and Schilling, D. 1984. Inheritance of stereotyped gibbon calls. *Nature* 312:634–636.

Brockelman W. Y., and Srikosamatara, S. 1984. Maintenance and evolution of social structure in gibbons. In H. Preuschoft, D. J. Chivers, W. Y. Brockelman, and N. Creel, eds.,

The Lesser Apes: Evolutionary and Behavioural Biology, 298–323. Edinburgh: Edinburgh University Press.

Bromage, T. G. 1992. The ontogeny of *Pan troglodytes* craniofacial architectural relationships and implications for early hominids. *Journal of Human Evolution* 23:235–251.

Bromham, L., and Cardillo, M. 2007. Primates follow the "island rule": implications for interpreting *Homo floresiensis. Biology Letters* 3:390–400.

Bronikowski, A. M., Altmann, J., Brockman, D. K., Cords, M., Fedigan, L. M., Pusey, A., Stoinski, T., Morris, W. F., Strier, K. B., and Alberts, S. C. 2011. Aging in the natural world: comparative data reveal similar mortality patterns across primates. *Science* 331:1325–1328.

Bronowski, J., and Bellugi, U. 1970. Language, name and concept. *Science* 168:669–673.

Broom, D. M., Sena, H., and Moynihan, K. L. 2009. Pigs learn what a mirror image represents and use it to obtain information. *Animal Behaviour* 78:1037–1041.

Broom, R. 1936a. A new fossil anthropoid skull from South Africa. *Nature* 138:486–488.

———. 1936b. The dentition of Australopithecus. *Nature* 138:719.

———. 1938a. The Pleistocene anthropoid apes of South Africa. *Nature* 142:377–379.

———. 1938b. Further evidence on the structure of the South African Pleistocene anthropoids. *Nature* 142:897–899.

———. 1941. The origin of man. *Nature* 148:10–14.

———. 1949a. Another new type of fossil ape-man. *Nature* 163:57.

———. 1949b. The ape-men. *Scientific American* 181:20–24.

———. 1950. The genera and species of the South African fossil ape-men. *American Journal of Physical Anthropology* 8:1–13.

Broom, R., and Robinson, J. T. 1949a. A new type of fossil man. *Nature* 164:322–323.

———. 1949b. Thumb of the Swartkrans ape-man. *Nature* 164:841–842.

———. 1950. Man contemporaneous with the Swartkrans ape-man. *American Journal of Physical Anthropology* 8:151–155.

———. 1952. Swartkrans ape-man *Paranthropus crassidens. Memoirs of the Transvaal Museum,* no. 6, 1–123.

Broom, R., Robinson, J. T., and Schepers, G. W. H. 1950. Sterkfontein Ape-Man *Pleisianthropus. Memoirs of the Transvaal Museum,* No. 4, 1–117.

Broom, R., and Schepers, G. W. H. 1946. The South African fossil ape-men: the Australopithecinae. *Memoirs of the Transvaal Museum,* No. 2, 1–272.

Brosnan, S. J., Schiff, H. C., and de Waal, F. B. M. 2005. Tolerance for inequity may increase with social closeness in chimpanzees. *Proceedings of the Royal Society B* 272:253–258.

Brothers, L. 1990. The social brain: a project for integrating primate social behavior and neurophysiology in a new domain. *Concepts in Neuroscience* 1:27–51.

———. 2002. The social brain: a project for integrating primate social behavior and neurophysiology in a new domain. In J. T. Cacioppo, G. G. Berntson, R. Adolphs, C. S. Carter, R. J. Davidson, M. K. McClintock, B. S. McEwen, M. J. Meaney, D. L. Schacter, E. M. Sternberg, et al., eds., *Foundations in Social Neuroscience,* 367–385. Cambridge, MA: MIT Press.

Brown, B., Walker, A., Ward, C. V., and Leakey, R. E. 1993. New *Australopithecus boisei* calvaria from East Lake Turkana, Kenya. *American Journal of Physical Anthropology* 91:137–159.

Brown, B., and Ward, S. C. 1988. Basicranial and facial topography in *Pongo* and *Sivapithecus*. In J. H. Schwartz, ed., *Orang-utan Biology*, 247–260. New York: Oxford University Press.

Brown, D. P. F., Lenneberg, E. H., and Ettlinger, G. 1978. Ability of chimpanzees to respond to symbols of quantity in comparison with that of children and monkeys. *Journal of Comparative and Physiological Psychology* 92:815–820.

Brown, F. H. 1982. Tulu Bor Tuff at Koobi Fora correlated with the Sidi Hakoma Tuff at Hadar. *Nature* 300:631–633.

———. 1983. Brown replies. *Nature* 306:210.

———. 2000. Geochronometry. In E. Delson, I. Tattersall, J. A. Van Couvering, and A. S. Brooks, eds., *Encyclopedia of Human Evolution and Prehistory*, 285–286. New York: Garland.

Brown, F. H., and Feibel, C. S. 1988. "Robust" hominids and Plio-Pleistocene paleogeography of the Turkana Basin, Kenya and Ethiopia. In F. E. Grine, ed., *Evolutionary History of the "Robust" Australopithecines*, 325–341. Hawthorne, NY: Aldine de Gruyter.

Brown, F. H., Harris, J., Leakey, R., and Walker, A. 1985. Early *Homo erectus* from west Lake Turkana, Kenya. *Nature* 316:788–792.

Brown, J. H., and Maurer, B. A. 1989. Macroecology: the division of food and space among species on continents. *Science* 243:1145–1150.

Brown, P. 2012. LB1 and LB6 *Homo floresiensis* are not modern human *(Homo sapiens)* cretins. *Journal of Human Evolution* 62:201–224.

Brown, P. Sutikna, T., Morwood, M. J., Soejono, R. P., Jatmiko, Wayhu Saptomo, E., and Due, R. A. 2004. A new small-bodied hominin from the Late Pleistocene of Flores, Indonesia. *Nature* 431:1055–1061.

Brown, R. W. 1981. Symbolic and syntactic capacities. *Philosophical Transactions of the Royal Society, London B* 292:197–204.

Brown, S. The "musilanguage" model of music evolution. In N. L. Wallin, B. Merker and S. Brown, eds., *The Origins of Music*, 271–300. Cambridge, MA: MIT Press.

Brown, S., Merker, B., and Wallin, N. L. 2000. An introduction to evolutionary musicology. In N. L. Wallin, B. Merker, and S. Brown, eds., *The Origins of Music*, 3–24. Cambridge, MA: MIT Press.

Brown, W. M., George, M., Jr., Wilson, A. C. 1979. Rapid evolution of animal mitochondrial DNA. *Proceedings of the National Academy of Sciences of the United States of America* 76:1967–1971.

Brown, W. M., Prager, W. M., Wang, A., and Wilson, A. C. 1982. Mitochondrial DNA sequences of primates: tempo and mode of evolution. *Journal of Molecular Evolution* 18:225–239.

Brownlow, A. R., Plumptre, A. J., Reynolds, V., and Ward, R. 2001. Sources of variation in the nesting behavior of chimpanzees *(Pan troglodytes schweinfurthii)* in the Budongo Forest, Uganda. *American Journal of Primatology* 55:49–55.

Bruce, E. J., and Ayala, F. J. 1978. Humans and apes are genetically very similar. *Nature* 276:264–265.

———. 1979. Phylogenetic relationships between man and the apes: electrophoretic evidence. *Evolution* 33:1040–1056.

Brugiere, D., and Sakom, D. 2001. Population density and nesting behaviour of lowland gorillas *(Gorilla gorilla gorilla)* in the Ngotto forest, Central African Republic. *Journal of Zoology, London* 255:251–259.

Brumm, A., Aziz, F., Bergh, G. D. van den, Morwood, M. J., Moore, M. W., Kurniawan, I., Hobbs, D. R., and Fullagar, R. 2006. Early stone technology on Flores and its implications for Homo *floresiensis. Nature* 441:624–628.

Bruner, E., and Holloway, R. L. 2010. A bivariate approach to the widening of the frontal lobes in the genus *Homo. Journal of Human Evolution* 58:138–146.

Brunet, M., Beauvilain, A., Coppens, Y., Heintz, E., Moutaye, A. H. E., and Pilbeam, D. 1995. The first australopithecine 2,500 kilometers west of the Rift Valley (Chad). *Nature* 378:273–275.

———. 1996. *Australopithecus bahrelghazali,* une nouvelle espèce d'Hominidé ancien de la région de Koro Toro (Tchad). *Comptes Rendus de l'Académie des Sciences, Série II,* 322:907–913.

Brunet, M., Guy, F., Pilbeam, D., Lieberman, D. E., Likius, A., Mackaye, H. T., Ponce De León, M. S., Zollikofer, C. P. E., and Vignaud, P. 2005. New material of the earliest hominid from the Upper Miocene of Chad. *Nature* 434:752–755.

Brunet, M., Guy, F., Pilbeam, D., Mackaye, H. T., Likius, A., Ahounta, D., Beauvilain, A., Blondel, C., Bocherens, H., Boisserie, J.-R., et al. 2002. A new hominid from the Upper Miocene of Chad, Central Africa. *Nature* 418:145–151.

Brunet, M., Heintz, É., Jehenne, Y., and Sen, S. 1982. Der erste Primatenfund im Miozän von Afghanistan. *Zeitschrift für Geologie, Berlin* 10:891–897.

Bshary, R., and Noë, R. 1997. Anti-predation behaviour of red colobus monkeys in the presence of chimpanzees. *Behavioral Ecology and Sociobiology* 41:321–333.

Buchler, J. 1956. *The Philosophy of Peirce; Selected Writings.* London: Routledge & Kegan Paul.

Buettner-Janusch, J. 1973. *Physical Anthropology: A Perspective.* New York: Wiley.

———. 1974. Discussion. *Yearbook of Physical Anthropology* 17:140–146.

Bullock, J. M., and Hodder K. H. 1997. Reintroductions: challenges and lessons for basic ecology. *TREE* 12:68–69.

Bunn, H. T. 1981. Archaeological evidence for meat-eating by Plio-Pleistocene hominids from Koobi Fora and Olduvai Gorge. *Nature* 291:574–577.

———. 2007. Meat made us human. P. S. Ungar, ed., *Evolution of the Human Diet,* 191–211. Oxford: Oxford University Press.

Bunn, H. T., and Kroll, E. M. 1986. Systematic butchery by Plio/Pleistocene hominids at Olduvai Gorge, Tanzania. *Current Anthropology* 27:431–452.

Bunn, H. T., Kroll, E. M., Isaac, G. L., and Kaufulu, Z. M. 1997. FxJj 50. In G. L. Issac and B. Isaac, eds., *Koobi Fora Research Project,* Vol. 5: *Plio-Pleistocene Archaeology,* 192–211. Oxford: Oxford University Press.

Burbank, V. K. 1992. Sex, gender, and difference. Dimensions of aggression in an Australian aboriginal community. *Human Nature* 3:251–278.

Burbano, H. A., Hodges, E., Green, R. E., Briggs, A. W., Krause, J., Meyer, M., Good, J. M., Marcic, T., Johnson, P. L. F., Xuan, Z., et al. 2010. Targeted investigation of the Neandertal genome by array-based sequence capture. *Science* 328:723–725.

Burckhardt, von Haeseler, A., and Meyer, S. 1999. HvrBase: compilation of mtDNA control sequences from primates. *Nucleic Acids Research* 27:138–142.

Burger, J., Kirchner, M., Bramanti, B., Haak, W., and Thomas, M. G. 2007. Absence of lactase-persistence-associated allele in early Neolithic Europeans. *Proceedings of the National Academy of Sciences of the United States of America* 104:3736–3747.

Burley, N. The evolution of concealed ovulation. *American Naturalist* 114:835–858.

Burling, R. 1993. Primate calls, human language, and nonverbal communication. *Current Anthropology* 34:25–53.

———. 2005. *The Talking Ape: How Language Evolved.* New York: Oxford University Press.

Burr, D. B. 1976. Rhodesian man and the evolution of speech. *Current Anthropology* 17:762–763.

Bush, E. C., Simons, E. L., and Allman, J. M. 2004. High-resolution computed tomography study of the cranium of a fossil anthropoid primates, *Parapithecus grangeri:* new insights into the evolutionary history of primate sensory systems. In *Evolution of the Special Senses in Primates,* ed. T. D. Smith, C. F. Ross, N. J. Dominy, and J. T. Laitman, special issue, *Anatomical Record, Part A* 281A:1083–1087.

Bush, M. E., Lovejoy, C. O., Johanson, D. C., and Coppens, Y. 1982. Hominid carpal, metacarpal, and phalangeal bones recovered from the Hadar Formation: 1974–1977 collections. *American Journal of Physical Anthropology* 651–677.

Busse, C. D. 1977. Chimpanzee predation as a possible factor in the evolution of red colobus social organization. *Evolution* 31:907–911.

———. 1978. Do chimpanzees hunt cooperatively? *American Naturalist* 112:767–770.

Butterworth, B. 1999. A head for figures. *Science* 284:928–929.

Butynski, T. M., and Kalina, J. 1993. Gorilla tourism: a critical look. In E. J. Milner-Gullard and R. Mace, eds., *Conservation of Biological Resources,* 294–313. Malden MA: Blackwell Science.

Butynski, T. M., Werikhe, S. E., and Kalina, J. 1990. Status, distribution and conservation of the mountain gorilla in the Gorilla Game Reserve, Uganda. *Primate Conservation* 11:31–41.

Butzer, K. W. 1971. Another look at the australopithecine cave breccias of the Transvaal. *American Anthropologist* 73:1197–1201.

Bützler, W. 1980. Présence et répartition des gorillas, *Gorilla gorilla gorilla* (Savage & Wyman, 1847), au Cameroun. *Saugetierkundliche Mitteilungen,* no. 1, 69–79.

Buxhoeveden, D. P., Switala, A. E., Litaker, M., roy, E., and Casanova, M. F. 2001. Lateralization of minicolumns in human planum temporale is absent in nonhuman primate cortex. *Brain, Behavior and Evolution* 57:349–358.

Bygott, J. D. 1972. Cannibalism among wild chimpanzees. *Nature* 238:410–411.

————. 1974. Paleoecology of South African australopithecines: Taung revisited. *Current Anthropology* 15:367–382, 413–416, 421–426.

————. 1979. Agonistic behavior, dominance, and social structure in wild chimpanzees of the Gombe National Park. In D. A. Hamburg and E. R. McCown, eds., *The Great Apes,* 404–427. Menlo Park, CA: Benjamin/Cummings.

————. 1984. Archeogeology and Quaternary environment in the interior of southern Africa. In R. G. Klein, ed., *Southern African Prehistory and Paleoenvironments,* 1–64. Rotterdam: Balkema.

Byrne, R. W. 1994. Complex skills in wild mountain gorillas: techniques for gathering plant food. In J. R. Anderson, J. J. Roeder, B. Thierry, and N. Herrenschmidt, eds., *Current Primatology,* Vol. 3: *Behavioural Neuroscience, Physiology and Reproduction,* 51–59. Strasbourg: Université Louis Pasteur.

————. 1995. *The Thinking Ape: Evolutionary Origins of Intelligence.* Oxford: Oxford University Press.

————. 1996. The misunderstood ape: cognitive skills of the gorilla. In A. E. Russon, K. A. Bard, and S. T. Parker, eds., *Reaching into Thought: The Minds of the Great Apes,* 111–130. Cambridge: Cambridge University Press.

————. 1999. Object manipulation and skill organization in the complex food preparation of mountain gorillas. In S. T. Parker, R. W. Mitchell, and H. L. Miles, eds., *The Mentalities of Gorillas and Orangutans,* 147–159. Cambridge: Cambridge University Press.

————. 2001. Clever hands: the food-processing skills of mountain gorillas. In M. M. Robbins, P. Sicotte, and K. J. Steward, eds., *Mountain Gorillas. Three Decades of Research at Karisoke,* 291–313. Cambridge: Cambridge University Press.

————. 2007. Culture in great apes: using intricate complexity in feeding skills to trace the evolutionary origin of huan technical prowess. *Philosophical Transactions of the Royal Society, London B* 362:575–585.

Byrne, R. W., and Byrne J. M. E. 1991. Hand preferences in the skilled gathering tasks of mountain gorillas *(Gorilla g. berengei). Cortex* 27:521–536.

————. 1993. Complex leaf-gathering skills of mountain gorillas *(Gorilla g. beringei):* variability and standardization. *American Journal of Primatology* 31:241–261.

Byrne, R. W., and Corp, N. Acquisition of skilled gathering techiques in Mahale chimpanzees. *Pan Africa News* 10:4–7.

Byrne, R. W., Sanz, C. M., and Morgan, D. B. 2013. Chimpanzees plan their tool use. In C. M. Sanz, J. Call, C. Boesch, eds., *Tool Use in Animals. Cognition and Ecology,* 48–63. Cambridge: Cambridge University Press.

Byrne, R. W., and Whiten, A. 1988. *Machiavellian Intelligence: Social Expertise and the Evolution of Intellect in Monkeys, Apes, and Humans.* Oxford: Oxford University Press.

Caccone, A., and Powell, J. R. 1989. DNA divergence among hominoids. *Evolution* 43:925–942.

Cáceres, M., Lachuer, J., Zapala, M. A., Redmond, J. C., Kudo, L., Geschwind, D. H., Lockhart, D. J., Preuss, T. M., and Barlow, C. 2003. Elevated gene expression levels distinguish human from non-human primate brains. *Proceedings of the National Academy of Sciences of the United States of America* 100:13030–13035.

Cáceres, M., Suwyn, C., Maddox, M., Thomas, J. W., and Preuss, T. M. 2007. Increased cortical expression of two synaptogenic thrombospondins in human brain evolution. *Cerebral Cortex* 17:2312–2321.

Cachel, S. 1975. The beginnings of the Catarrhini. In R. H. Tuttle, ed., *Primate Functional Morphology and Evolution*, 23–36. The Hague: Mouton.

———. 1983. Diet of the Oligocene anthropoids *Aegyptopithecus* and *Apidium*. *Primates* 24:109–117.

Caeiro, C. C., Waller, B. M, Zimmermann, E., Burrows, A. M., and Davila-Ross, M. 2013. OrangFACS: a muscle-based facial movement coding system for orangutans (*Pongo* spp.). *International Journal of Primatology* 34:115–129.

Caillaud, D., Levréro, F., Gatte, S., Ménard, N., and Raymond, M. 2008. Influence of male morphology on male mating status and behavior during interunit encounters in western lowland gorillas. *American Journal of Physical Anthropology* 135:379–388.

Calarco, J. A., Xing, Y., Cáceres, M., Calarco, J. P., Xiao, X., Pan, Q., Lee, C., Preuss, T. M., and Blencowe, B. J. 2007. Global analysis of alternative splicing differences between humans and chimpanzees. *Genes and Development* 21:2963–2975.

Calcagno, J. M., and Gibson, K. R. 1991. Selective compromise: evolutionary trends and mechanisms of hominid tooth size. In M. A. Kelley and C. S. Larsen, eds., *Advances in Dental Anthropology*, 59–76. New York: Wiley-Liss.

Caldecott, J., and Kapos, V. 2005. Great ape habitats: tropical moist forests of the Old World. In J. Caldecott and L. Miles, eds., *World Atlas of Great Apes and Their Conservation*, 31–42. Berkeley: University of California Press.

Caldecott, J., and McConkey, K. 2005. Orangutan review. In J. Caldecott and L. Miles, eds., *World Atlas of Great Apes and Their Conservation*, 153–159 Berkeley: University of California Press.

Caldecott, J., and Miles, L. 2005. *World Atlas of Great Apes and Their Conservation*. Berkeley: University of California Press.

Caldecott, J. O. 1980. Habitat quality and populations of two sympatric gibbons (Hylobatidae) on a mountain in Malaya. *Folia Primatologica* 33:291–309.

Call, J. 2000. Estimating and operating on discrete quantities in orangutans *(Pongo pygmaeus)*. *Journal of Comparative Psychology* 114:136–147.

———. 2013. Three ingredients for becoming a creative tool user. In C. M. Sanz, J. Call, C. Boesch, eds., *Tool Use in Animals. Cognition and Ecology*, 3–20. Cambridge: Cambridge University Press.

Call, J., and Tomasello, M. 1994a. The social learning of tool use by orangutans *(Pongo pygmaeus)*. *Human Evolution* 9:297–313.

———. 1994b. Production and comprehension of referential pointing by orangutans *(Pongo pygmaeus)*. *Journal of Comparative Psychology* 108:307–317.

———. 2003. Social Cognition. In D. Maestripieri, ed., *Primate Psychology*, 234–253. Cambridge, MA: Harvard University Press.

———. 2007. *The Gestural Communication of Apes and Monkeys*. London: Lawrence Erlbaum.

Calvert, J. 1985. Food selection by western gorillas *(G. g. gorilla)* in relation to food chemistry. *Oecologia* 63:236–246.

Cameron, D. W. 1991. Sexual dimorphism in the Early Miocene species of *Proconsul* from the Kisingiri Formation of East Africa: a morphometric examination using multivariate statistics. *Primates* 32:329–343.

———. 1992. A morphometric analysis of extant and Early Miocene fossil hominoid maxillo-dental specimens. *Primates* 33:377–390.

———. 1997. A revised systematic scheme for the Eurasian Miocene fossil Hominoidea. *Journal of Human Evolution* 33:449–477.

Campbell, B. 1973. A new taxonomy of fossil man. *Yearbook of Physical Anthropology* 17:194–201.

———. 1985. *Human Evolution: An Introduction to Man's Adaptations.* 3rd ed. Hawthorne, NY: Aldine de Gruyter.

Campbell, M. C., and Tishkoff, S. A. 2008. African genetic diversity: implications for human demographic history, modern human origins, and complex disease mapping. *Annual Review of Genomics and Human Genetics* 9:403–433.

Campbell, M. W., Carter, J. D. Proctor, D., Eisenberg, M. L., and de Waal, F. B. M. 2009. Computer animations stimulate contagious yawning in chimpanzees. *Proceedings of the Royal Society B: Biological Sciences* 276:4255–4259.

Campbell, M. W., and de Waal, F. B. M. 2011. Ingroup-outgroup bias in contagious yawning by chimpanzees supports link with empathy. *PLoS One* 6:e18283.

Campbell, R. 1982. The lateralization of emotion: a critical review. *International Journal of Psychology* 17:211–229.

Canale, G. R., Guidorizzi, C. E., Kierulff, M. C. M., and Gatto, C. A. F. R. 2009. *American Journal of Primatology* 71:366–372.

Candland, D. K. 1987. Tool use. In G. Mitchell and J. Erwin, eds., *Comparative Primate Biology,* Vol. 2B: *Behavior, Cognition and Motivation,* 85–103. New York: Alan R. Liss.

Cann, R. L. 1988. DNA and human origins. *Annual Review of Anthropology* 17:127–143.

Cann, R. L., Brown, W. M., and Wilson, A. C. 1984. Polymorphic sites and the mechanism of evolution in human mitochondrial DNA. *Genetics* 106:479–499.

Cann, R. L., Stoneking, M., and Wilson, A. C. 1987. Mitochondrial DNA and human evolution. *Nature* 325:31–36.

Cant, J. G. H. 1987. Positional behavior of female Bornean orangutans *(Pongo pygmaeus). American Journal of Primatology* 12:71–90.

———. 1992. Positional behavior and body size of arboreal primates: a theoretical framework for field studies and an illustration of its application. *American Journal of Physical Anthropology* 88:273–283.

Cantalupo, C., and Hopkins, W. D. 2001. Asymmetric Broca's area in great apes. *Nature* 414:505.

———. 2010. The cerebellum and its contribution to complex tasks in higher primates: a comparative perspective. *Cortex* 46:821–830.

Cantalupo, C., Pilcher, D. L., and Hopkins, W. D. 2002. Are planum temporale and sylvian fissure asymmetries directly related? a MRI study in great apes. *Neuropsychologia* 41:1975–1981.

Cantlon, J. F. 2012. Math, monkeys, and the developing brain. *Proceedings of the National Academy of Sciences of the United States of America* 109:10725–10732.

Cantlon, J. F., and Brannon, E. M. 2006. Shared system for ordering small and large numbers in monkeys and humans. *Psychological Science* 17:401–407.

Capecchi, V. 1984. Reflections on the footprints of the hominids found at Laetoli. *Anthropologischer Anzeiger* 42:81–86.

Caplan, D. 2006. Why is Broca's area involved in syntax? *Cortex* 42:469–471.

Carbonell, E., Bermúdez de Castro, J. M., Arsuaga, J. L., Allue, E., Bastir, M., Benito, A., Cáceres, I., Canals, T., Dicz, J. C., van der Made, J., et al. 2005. An early Pleistocene hominin mandible from Atapuerca-TD6, Spain. *Proceedings of the National Academy of Sciences of the United States of America* 102:5674–5678.

Cardoso, F. L. 2004. *Male Sexual Behavior in Brazil, Turkey and Thailand among Middle and Working Social Classes.* PhD diss., Institute for Advanced Study of Human Sexuality, San Francisco.

Carey, S. 1998. Knowledge of number: its evolution and ontogeny. *Science* 282:641–642.

———. 2001a. Cognitive foundations of arithmetic: evolution and ontogenesis. *Mind and Language* 16:37–55.

———. 2001b. Bridging the gap between cognition and developmental neuroscience: the example of number representation. In C. A. Nelson and M. Luciana, eds., *The Handbook of Developmental Cognitive Neuroscience,* 415–431. Cambridge, MA: MIT Press.

Carini, L. 1970. On the origins of language. *Current Anthropology* 11:165–166.

Carlson, S. S., Wilson, A. C., and Maxson, R. D. 1978. Do albumin clocks run on time? *Science* 200:1183–1185.

Carlsöö, S. 1972. *How Man Moves.* London: Wm. Heinemann.

Caro, T. M. 1976. Observations on the ranging behaviour and daily activity of lone silverback mountain gorillas *(Gorilla gorilla beringei). Animal Behavior* 24:889–897.

Caro, T. M., Sellen, D. W., Parish, A., Frank, R., Brown, D. M., Voland, E., and Borgerhoff Mulder, M. 1995. Termination of reproduction in nonhuman female primates. *International Journal of Primatology* 16:205–220.

Carpenter, C. R. 1937. An observational study of two captive mountain gorillas *(Gorilla beringei). Human Biology* 9:175–196.

———. 1938. A survey of wild life conditions in Atjeh, North Sumatra, with special reference to the orangoutan. *Netherlands Committee for International Nature Protection,* Amsterdam, Communications no. 12, 1–34.

———. 1940. A field study in Siam of the behavior and social relations of the gibbon *(Hylobates lar). Comparative Psychology Monographs* 16:1–212.

Carretero, J. M., Arsuaga, J. L., and Lorenzo, C. 1997. Clavicles, scapulae and humeri from the Sima de los Huesos site (Sierra de Atapuerca, Spain). *Journal of Human Evolution* 33:357–408.

Carrier, D. R. 1984. The energetic paradox of human running and hominid evolution. *Current Anthropology* 25:483–495.

Carrier, D. R., Anders, C., and Schilling, N. 2011. The musculoskeletal system of humans is not tuned to maximize the economy of locomotion. *Proceedings of the National Academy of Sciences of the United States of America* 108:18631–18636.

Carrier, D. R., Heglund, N. C., and Earls, K. D. 1994. Variable gearing during locomotion in the human musculoskeletal system. *Science* 265:651–653.

Carroll, R. W. 1986. Status of the lowland gorilla and other wildlife in the Dzanga-Sangha region of southwestern Central African Republic. *Primate Conservation* 7:38–41.

———. 1988. Relative density, range extension, and conservation potential of the lowland gorilla *(Gorilla gorilla gorilla)* in the Dzanga-Sangha region of southwestern Central African Republic. *Mammalia* 52:309–323.

———. 1990. In the garden of the gorillas. *Wildlife Conservation* 93:50–63.

Carroll, S. B. 2003. Genetics and the making of *Homo sapiens. Nature* 422:849–857.

Carsten, J. 2000. *Cultures of Relatedness: New Approaches to the Study of Kinship.* Cambridge: Cambridge University Press.

Carter, M. L., Pontzer, H., Wrangham, R. W., and Peterhans, J. K. 2008. Skeletal pathology *Pan troglodytes schweinfurthii* in Kibale National Park. *American Journal of Physical Anthropology* 135:389–403.

Cartmill, E. A., and Byrne, R. W. 2007. Orangutans modify their gestural signaling according to their audience's comprehension. *Current Biology* 17:1345–1346.

———. 2010. Semantics of primate gestures: intentional meanings of orangutan gestures. *Animal Cognition* 13:793–804.

Cartmill, M. 1972. Arboreal adaptations and the origin of the order Primates. In R. H. Tuttle, ed., *The Functional and Evolutionary Biology of Primates,* 97–122. Chicago: Aldine.

———. 1974a. Rethinking primate origins. *Science* 184:436–443.

———. 1974b. Pads and claws in arboreal locomotion. In F. A. Jenkins Jr., ed., *Primate Locomotion,* 4–83. New York: Academic Press.

———. 1980. Morphology, function, and evolution of the anthropoid postorbital septum. In R. L. Ciochon and A. B. Chiarelli, eds., *Evolutionary Biology of the New World Monkeys and Continental Drift,* 243–274. New York: Plenum Press.

———. 1982a. Basic primatology and prosimian evolution. In F. Spencer, ed., *A History of American Physical Anthropology, 1930–1980,* 147–186. New York: Academic Press.

———. 1982b. Assessing tarsier affinities: is anatomical description phylogenetically neutral? *Geobios, mémoire spécial* 6:279–287.

———. 1983. "Four Legs Good, Two Legs Bad." *Natural History* 92:64–79.

———. 1993. *A View to a Death in the Morning.* Cambridge, MA: Harvard University Press.

———. 1997. Hunting hypothesis of human origins. In F. Spencer, ed., *History of Physical Anthropology: An Encyclopedia,* 508–512. New York: Garland.

Cartmill, M., and Kay, R. 1978. Cranio-dental morphology, tarsier affinities, and primate sub-orders. In D. J. Chivers and K. A. Joysey, ed., *Recent Advances in Primatology,* Vol. 3: *Evolution,* 205–214. London: Academic Press.

Cartmill, M., MacPhee, R. D. E., and Simons, E. L. 1981. Anatomy of the temporal bone in early anthropoids with remarks on the problem of anthropoid origins. *American Journal of Physical Anthropology* 56:3–21.

Cartmill, M., and Milton, K. 1977. The lorisiform wrist joint and the evolution of "brachiating" adaptations in the Hominoidea. *American Journal of Physical Anthropology* 47:249–272.

Cartmill, M., Pilbeam, D., and Isaac, G. 1986. One hundred years of paleoanthropology. *American Scientist* 74:410–420.

Caruana, M. V., d'Erico, F., and Blackwell, L. 2013. Early hominin social learning strategies underlying the use and production of bone and stone tools. In C. M. Sanz, J. Call, C. Boesch, eds., *Tool Use in Animals. Cognition and Ecology*, 242–385. Cambridge: Cambridge University Press.

Carvalho, S., Biro, D., McGrew, W. C., and Matsuzawa, T. 2009. Tool-composite reuse in wild chimpanzees *(Pan troglodytes):* archaeologically invisible steps in the technological evolution of early hominins? *Animal Cognition* 12 (Suppl. 1):S103–S114.

Cashdan, E. 1996. Women's mating strategies. *Evolutionary Anthropology* 5:134–143.

Casimir, M. J. 1975. Feeding ecology and nutrition of an eastern gorilla group in the Mt. Kahuzi Region (République du Zaïre). *Folia Primatologica* 24:81–136.

———. 1979. An analysis of gorilla nesting sites of the Mt. Kahuzi Region (Zaire). *Folia Primatologica* 32:290–308.

Casimir, M. J., and Burenandt, E. 1973. Migration and core area shifting in relation to some ecological factors in a mountain gorilla group *(Gorilla gorilla beringei)* in the Mt. Kahuzi region (Republique du Zaire). *Zeitschrift fur Tierpsychologie,* 33:514–522.

Cassidy, D., and Davy, K., dirs. 2003. *One Small Step: The Story of the Space Chimps.* Gainsville, FL: University of Florida Documentary Institute, DVD, 57 min., www.spacechimps.com.

Catani, M., Allen, P. G., Husain, M., Pugliese, L., Mesulam, M., Murray, R. M., and Jones, D. K. 2007. Symmetries in human brain language pathways correlate with verbal recall. *Proceedings of the National Academy of Sciences of the United States of America* 104:17163–17168.

Caton, J. M. 1999. A preliminary report on the digestive strategy of the western lowland gorilla. *Australasian Primatology* 13:2–7.

Caton, J. M., Hume, I. D., Hill, D. M., and Harper, P. 1999. Digesta retention in the gastrointestinal tract of the orang utan *(Pongo pygmaeus).* *Primates* 40:551–558.

Cavalieri, P. 1996. The Great Ape Project. *Etica & Animali* 8/96. Milan, Italy.

Cavalieri, P., and Singer, P. 1993. *The Great Ape Project: Equality beyond Humanity.* London: Fourth Estate.

Cavalli-Sforza, L. L., Kidd, J. R., Kidd, K. K., Bucci, C., Bowcock, A. M., Hewlett, B. S., and Friedlaender, J. S. 1986. DNA markers and genetic variation in the human species. *Cold Spring Harbor Symposia on Quantitative Biology* 51:41–417.

Cavalli-Sforza, L. L., Piazza, A., Menozzi, P., and Mountain, J. 1988. Reconstruction of human evolution: bringing together genetic, archaeological, and linguistic data. *Proceedings of the National Academy of Sciences of the United States of America* 85:6002–6006.

Cela-Conde, C. J. 1998. The problem of hominoid systematics, and some suggestions for solving it. *South African Journal of Science* 94:255–262.

Cela-Conde, C. J., and Ayala, F. J. 2003. Genera of the human ineage. *Proceedings of the National Academy of Sciences of the United States of America* 100:7684–7689.

Cela-Conde, C. J., Ayala, F. J., Munar, E., Maestú, F., Nadal, M., Capó, M. A., del Rio, D., López-Ibor, J. J., Ortiz, T., Mirasso, C., and Marty, G. 2009. Sex-related similarities and differences in the neural correlates of beauty. *Proceedings of the National Academy of Sciences of the United States of America* 106:3847–385.

Cerling, T. E., Mbua, E., Kirera, F. M., Manthi, F. K., Grine F. E., Leakey, M. G., Spon-heimer, M., and Uno, K. T. 2011. Diet of *Paranthropus boisei* in the early Pleistocene of East Africa. *Proceedings of the National Academy of Sciences of the United States of America* 108:9337–9341.

Chaimanee, Y. 2004. *Siamopithecus eocaenus,* anhropoid primate from the Late Eocene of Krabi, Thailand. In C. F. Ross and R. F. Kay, eds., *Anthropoid Origins,* 341–368. New York: Kluwer Academic/Plenum Press.

Chaimanee, Y., Chavasseau, O., Beard, K. C., Kyaw, A. A., Soe, A. N., Sein, C., Lazzari, V., Marivaux, L., Marandat, B., Swe, M., et al. 2012. Late Middle Eocene primate from Myanmar and the initial anthropoid colonization of Africa. *Proceedings of the National Academy of Sciences of the United States of America* 109:10293–10297.

Chaimanee, Y., Jolly, D., Benammi, M., Tafforeau, P., Duzer, D., Moussa, I., and Jaeger, J.-J. 2003. A Miocene hominoid from Thailand and orangutan origins. *Nature* 422:61–65.

Chaimanee, Y., Khansubha, S., and Jaeger, J.-J. 2000b. A new lower jaw of *Siamopithecus eocaenus* from the late Eocene of Thailand. *Comptes rendus de l'Académie des sciences, Paris* 323:235–241.

Chaimanee, Y., Suteethorn, V., Jaeger, J.-J., and Ducrocq, S. 1997. A new Late Eocene an-thropoid from Thailand. *Nature* 385:429–431.

Chaimanee, Y., Suteethorn, V., Jintasakul, P., Vidthayanon, C., Marandat, B., and Jaeger, J.-J. 2004. A new orang-utan relative from the Late Miocene of Thailand. *Nature* 427:439–441.

Chaimanee, Y., Thein, T., Ducrocq, S., Soe, A. N., Benammi, M., Tun, T., Lwin, T., Wai, S., and Jaeger, J.-J. 2000a. A lower jaw of *Pondaungia cotteri* from the late Middle Eo-cene Pondaung Formation (Myanmar) confirms its anthropoid status. *Proceedings of the National Academy of Sciences of the United States of America* 97:4102–4105.

Chaline, J., Durand, A., Marchand, D., Malassé, A. D., and Deshayes, M. J. 1996. Chromo-somes and the origins of apes and australopithecines. *Human Evolution* 11:43–69.

Chancellor, R. L., Rundus, A. S., and Nyandwi, S. 2012. The influence of seasonal variation on chimpanzee *(Pan troglodytes schweinfurthii)* fallback food consumption, nest group size, and habitat use in Gishwati, a montane rain forest fragment in Rwanda. *Interna-tional Journal of Primatology* 33:115–133.

Chang, L.-Y. E., and Slightom, J. L. 1984. Isolation and nucleotide sequence analysis of the beta-type globin pseudogene from human, gorilla and chimpanzee. *Journal of Molecu-lar Biology* 180:767–784.

Chang, S. W., Brent, L. J. N., Adams, G. K., Klein, J. T., Pearson, J. M., Watson, K. K., and Platt, M. L. 2013. Neuroethology of primate social behavior. *Proceedings of the National Academy of Sciences of the United States of America* 110 (Suppl. 2): 10387–10394.

Chapais, B., and Berman, C. M. 2004. Variation in nepotistic regimes and kin recognition: a major area for future research. In B. Chapais and C. M. Berman, eds., *Kinship and Behavior in Primates,* 477–489. Oxford: Oxford University Press.

Chapelain, A. S., Hogervorst, E., Mbonzo, P., and Hopkins, W. D. 2011. Hand preferences for bimanual coordination in 77 bonobos *(Pan paniscus):* replication and extension. *International Journal of Primatology* 32:491–510.

Chaplin, G., Jablonski, N. G., and Cable, N. T. 1994. Physiology, thermoregulation and bipedalism. *Journal of Human Evolution* 27:497–510.

Chapman, C. A., White, F. J., and Wrangham, R. W. 1993. Defining subgroup size in fission-fusion societies. *Folia Primatologica* 61:31–34.

———. 1994. Party size in chimpanzees and bonobos. In R. W. Wrangham, W. C. McGrew, F. B. M. de Waal, and P. G. Heltne, eds., *Chimpanzee Cultures,* 41–57. Cambridge, MA: Harvard University Press.

Chapman, C. A., and Wrangham, R. W. 1993. Range use of the forest chimpanzees of Kibale: implications for the understanding of chimpanzee social organization. *American Journal of Primatology* 31:263–273.

Chapman, C. A., Wrangham, R. W., and Chapman, L. J. 1995. Ecological constraints on group size: an analysis of spider monkey and chimpanzee subgroups. *Behavioral Ecology and Sociobiology* 36:59–70.

Charteris, J., Wall, J. C., and Nottrodt, J. W. 1981. Functional reconstruction of gait from Pliocene hominid footprints at Laetoli, northern Tanzania. *Nature* 290:496–498.

———. 1982. Pliocene hominid gait: new interpretations based on available footprint data from Laetoli. *American Journal of Physical Anthropology* 58:133–144.

Chase, P. G. 2006. *The Emergence of Culture: The Evolution of a Uniquely Human Way of Life.* New York: Springer.

Chasen, F. N., and Kloss, C. B. 1927. Spolia Mentaweiensia: Mammals. *Proceedings of the Zoological Society of London* 97:797–840.

Chatterjee, H. J. 2006. Phylogeny and biogeography of gibbons: a dispersal-vicariance analysis. *International Journal of Primatology* 27:699–712.

Chavaillon, J., Brahimi, C., and Coppens, Y. 1974. Première découverte d'hominidé dans l'un des sites acheuléens de Melka Kunturé (Ethiopie). *Comptes rendus hebdomadaires des séances de l'Académie des Sciences, Paris, série D* 278:3299–3302.

Chen, F.-C., and Li, W.-H. 2001. Genomic divergences between humans and other hominoids and the effective population size of the common ancestor of humans and chimpanzees. *American Journal of Human Genetics* 68:444–456.

Chen, N. 1995. Ecology of the Black-crested gibbon *(Hylobates concolor)* in the Ailao Mt. Reserve, Yunnan, China. MA Thesis. Maidol University, Bangkok, Thailand.

Cheney, D. L. 2011. Extent and limits of cooperation in animals. *Proceedings of the National Academy of Sciences of the United States of America* 108:10902–10909.

Cheney, D. L., and Seyfarth, R. M. 1980. Vocal recognition in free-ranging vervet monkeys. *Animal Behaviour* 28:362–367.

———. 1982a. How monkeys see the world: a review of recent research on East African vervet monkeys. In C. T. Snowdon, C. H. Brown, and M. R. Petersen, eds., *Primate Communication,* 239–252. Cambridge: Cambridge University Press.

———. 1982b. How vervet monkeys perceive their grunts: field playback experiments. *Animal Behaviour* 30:739–751.

———. 1990. *How Monkeys See the World: Inside the Mind of Another Species.* Chicago: University of Chicago Press.

———. 2004. The recognition of other individuals' kinship relationships. In B. Chapais and C. M. Berman, eds., *Kinship and Behavior in Primates,* 347–364. Oxford: Oxford University Press.

———. 2007. *Baboon Metaphysics: The Evolution of a Social Mind.* Chicago: University of Chicago Press.

Cheng, Z., Ventura, M., She, X., Khaitovich, P., Graves, T., Osoegawa, K., Church, D., DeJong, P., Wilson, R. K., Pääbo, S., et al. 2005. A genome-wide comparison of recent chimpanzee and human segmental duplications. *Nature* 437:88–93.

Chenn, A. 2002. Regulation of cerebral cortical size by control of cell cycle exit in neural precursors. *Science* 298:766–767.

Chenn, A., and Walsh, C. A. 2002. Regulation of cerebral cortical size by control of cell cycle exit in neural precursors. *Science* 297:365–369.

Cherry, L. M., Case, S. M., and Wilson, A. C. 1978. Frog perspective on the morphological difference between humans and chimpanzees. *Science* 200:209–211.

Chevalier-Skolnikoff, Suzanne 1973. Facial expression of emotion in nonhuman primates. In P. Ekman, ed., *Darwin and Facial Expression, a Century of Research in Review,* 11–89. New York: Academic Press.

Cheverud, J., Falk, D., Hildebolt, C., Moore, A. J., Helmkamp, R. C., and Vannier, M. 1990. Heritability and association of cortical petalias in rhesus macaques *(Macaca mulatta). Brain, Behavior and Evolution* 35:368–372.

Chiarelli, B. 1972. The karyotypes of the gibbons. In D. M. Rumbaugh, ed., *Gibbon and Siamang,* Vol. 1, 90–102. Basel: Karger.

Chivers, D. J. 1971a. The Malayan siamang. *Malayan Nature Journal* 24:78–86.

———. 1971b. Spatial relations within the siamang group. *Proceedings of the Third International Congress of Primatology, Zurich 1970,* Vol. 3, 14–21. Basel: Karger.

———. 1972. The gibbon and siamang in the Malay Peninsula. In D. M. Rumbaugh, ed., *Gibbon and Siamang,* Vol. 1, 103–135. Basel: Karger.

———. 1973. Introduction to the socio-ecology of Malayan forest primates. In R. P. Michael and J. H. Crook, eds., *Comparative Ecology and Behavior of Primates,* 101–146. London: Academic Press.

———. 1974. The siamang in Malaya. *Contributions to Primatology* 4:1–335. Basel: Karger.

———. 1975. Daily patterns of ranging and feeding in siamang. In S. Kondo, M. Kawai, and A. Ehara, eds., *Contemporary Primatology,* 362–372, Basel: Karger.

———. 1976. Communication within and between family groups of siamang *(Symphalangus syndactylus). Behaviour* 57:116–135.

———. 1977a. The lesser apes, In H. S. H. Prince Rainier and G. H. Bourne, eds., *Primate Conservation* 539–598. New York: Academic Press.

———. 1977b. The ecology of gibbons: some preliminary considerations based on observations in the Malay Peninsula. In M. R. N. Prasad and T. C. Anand Kumar, eds., *Use of Non-human Primates in Biomedical Research,* 85–105. New Delhi: Indian National Science Academy.

———. 1978. Sexual behaviour of wild siamang. In D. J. Chivers and J. Herbert, eds., *Recent Advances in Primatology,* Vol. 1: *Behaviour,* 609–610. London: Academic Press.

———. 1984. Feeding and ranging in gibbons: a summary. In H. Preuschoft, D. J. Chivers, W. Y. Brockelman, and N. Creel, eds., *The Lesser Apes: Evolutionary and Behavioural Biology,* 267–281. Edinburgh: Edinburgh University Press.

———. 1994. Functional anatomy of the gastrointestinal tract. In A. G. Davies and J. F. Oates, eds., *Colobine Monkeys: Their Ecology, Behaviour and Evolution,* 205–227. Cambridge: Cambridge University Press.

———. 2001. The swinging singing apes: fighting for food and family in far-east forests. In *The Apes: Challenges for the 21st Century, Conference Proceedings, May 10–12, 2000,* 1–28. Brookfield, IL: Brookfield Zoo.

———. 2005. Gibbons: the small apes. In J. Caldecott, and L. Miles, *World Atlas of Great Apes and Their Conservation,* eds., 205–214. Berkeley: University of California Press.

Chivers, D. J., and Chivers, S. T. 1975. Event preceding and following the birth of a wild siamang. *Primates* 16:227–230.

Chivers, D. J., and MacKinnon, J. 1977. On the behaviour of siamang after playback of their calls. *Primates* 18:943–948.

Chivers, D. J., and Raemaekers, J. J. 1980. Long-term changes in behaviour, In D. J. Chivers, ed., *Malayan Forest Primates: Ten Year's Study in Tropical Rain Forest,* 209–260. New York: Plenum Press.

Chivers, D. J., Raemaekers, J. J., and Aldrich-Blake, F. P. G. 1975. Long-term observations of siamang behaviour. *Folia Primatologica* 23:1–49.

Chomsky, N. 1965. *Aspects of a Theory of Syntax.* Cambridge, MA: MIT Press.

———. 1980. *Rules and Representations.* New York: Columbia University Press.

Chopra, S. R. K. 1983. Significance of recent hominoid discoveries from the Siwalik Hills of India. In R. L. Ciochon and R. S. Corruccini, eds., *New Interpretations of Ape and Human Ancestry,* 539–557. New York: Plenum Press.

Chopra, S. R. K., and Kaul, S. 1979. A new species of *Pliopithecus* from the Indian Siwaliks. *Journal of Human Evolution* 8:475–477.

Choudhury, A. 1990. Population dynamics of hoolock gibbons *(Hylobates hoolock)* in Assam, India. *Primates* 20:37–41.

———. 1991. Ecology of the hoolock gibbon *(Hylobates hoolock),* a lesser ape in the tropical forests of north-eastern India. *Journal of Tropical Ecology* 7:147–153.

Christel. M. 1993. Grasping techniques and hand preferences in Hominoidea. In H. Preuschoft and D. J. Chivers, eds., *Hands of Primates,* 910–108. Wien: Springer.

———. 1994. Catarrhine primates grasping small objects: techniques and hand preferences. In J. R. Anderson, J. J. Roeder, B. Thierry, and N. Herrenschmidt, eds., *Current Primatology,* Vol. 3: *Behavioural Neuroscience, Physiology and Reproduction,* 37–50. Strasbourg: Université Louis Pasteur.

Christel. M., Kitzel, S., and Niemitz, C. 1998. How precisely do bonobos *(Pan paniscus)* grasp small objects? *International Journal of Primatology* 19:165–194.

Christiansen, M. H., and Kirby, S. 2003. Language evolution: the hardest problem in science? In M. H., Christiansen, and S. Kirby, eds., *Language Evolution,* 1–15. Oxford: Oxford University Press.

Chua, H. F., Boland, J. E., and Nibett, R. E. 2005. Cultural variation in eye movements during scene perception. *Proceedings of the National Academy of Sciences of the United States of America* 102:12629–12633.

Churchill, S. E., Holliday, T. W., Carlson, K. J., Jashashvili, T., Macias, M. E., Mathews, S., Sparling, T. L., Schmid, P., de Ruiter, D. J., and Berger, L. R. 2013. The upper limb of *Australopithecus sediba. Science* 340, http://dx.doi.org/10.1126/science. 1233477.

Cianelli, S. N., and Fouts, R. S. 1998. Chimpanzee to chimpanzee American Sign Language. *Human Evolution* 13:147–159.

Ciochon, R. L. 1983. Hominid cladistics and the ancestry of modern apes and humans. A summary statement. In R. L. Ciochon and R. S. Corruccini, eds., *New Interpretations of Ape and Human Ancestry,* 783–843. New York: Plenum Press.

———. 1991. The ape that was. *Natural History* 100, no. 11: 54–63.

Ciochon, R. L., and Corruccini, R. S. 1977. The phenetic position of *Pliopithecus* and its phylogenetic relationship to the Hominoidea. *Systematic Zoology* 26:290–299.

Ciochon, R. L., Gingerich, P. D., Gunnell, G. F., and Simons, E. L. 2001. Primate postcrania from the late middle Eocene of Myanmar. *Proceedings of the National Academy of Sciences of the United States of America* 98:7672–7677.

Ciochon, R. L., and Gunnell, G. F. 2004. Eocene large-bodied primates of Myanmar and Thailand: morphological considerations and phylogenetic affinities. In C. F. Ross and R. F. Kay, eds., *Anthropoid Origins,* 249–282. New York: Kluwer Academic/Plenum Press.

Ciochon, R. L., and Olsen, J. W. 1986. Paleoanthropological and archaeological research in the Socialist Republic of Vietnam. *Journal of Human Evolution* 15:623–631.

Ciochon, R. L., Savage, D. E., Tint, T., and Ba Maw 1985. Anthropoid origins in Asia? New discovery of *Amphipithecus* from the Eocene of Burma. *Science* 229:756–759.

Cipolletta, C. 2003. Ranging patterns of a western gorilla group during habituation to humans in the Dzanga-Ndoki National Park, Central African Republic. *International Journal of Primatology* 24:1207–1226.

———. 2004. Effects of group dynamics and diet on the ranging patterns of a western gorilla group (Gorilla gorilla gorilla) at Bai Hokou, Central African Republic. *American Journal of Primatology* 24:193–205.

Cipolletta, C., Spagnoletti, N., Todd, A., Robbins, M. M., Cohen, H., and Pacyna, S. 2007. Termite feeding by *Gorilla gorilla gorilla* at Bai Hokou, Central African Republic. *International Journal of Primatology* 28:457–476.

Clark, A. G. 1999. Chips for chimps. *Nature Genetics* 22:119–120.

Clark, A. G., Glanowski, S., Nielsen, R., Thomas, P. D., Kejariwal, A., Todd, M. A., Tanenbaum, D. M., Civello, D., Lu, F., Murphy, B., et al. 2003. Inferring nonneutral evolution from human-chimp-mouse orthologous gene trios. *Science* 302:1960–1963.

Clark, A. P. 1993. Rank differences in the production of vocalizations by wild chimpanzees as a function of social context. *American Journal of Primatology* 31:159–179.

Clark, A. P., and Wrangham, R. W. 1993. Acoustic analysis of wild chimpanzee pant hoots: do Kibale Forest chimpanzees have an acoustially distinct food arrival pant hoot? *American Journal of Primatology* 31:99–109.

———. 1994. Chimpanzee arrival pant hoots: do they signify food or status? *International Journal of Primatology* 15:185–205.

Clark, C. B. 1977. A preliminary report on weaning among chimpanzees of the Gombe National Park, Tanzania. In S. Chevalier-Skolnikoff and F. E. Poirier, eds., *Primate Bio-social Development: Biological, social, and ecological determinants,* 235–260. New York: Garland.

Clark, G. A. 1988. Some thoughts on the black skull: an archeologist's assessment of WT-17000 *(A. boisei)* and systematics in human paleontology. *American Anthropologist* 90:357–371.

Clark, J. D. 1993. Stone artifact assemblages from Members 1–3, Swartkrans Cave. In C. K. Brain, ed., *Swartkrans: A Cave's Chronicle of Early Man,* 167–194. Monograph No. 8. Pretoria: Transvaal Museum.

Clark, J. D., Asfaw, B., Assefa, G., Harris, J. W. K., Kurashina, H., Walter, R. C., White, T. D., and Williams, M. A. J. 1984. Palaeoanthropological discoveries in the Middle Awash Valley, Ethiopia. *Nature* 307:423–428.

Clark, J. D., Beyene, Y., Wolde-Gabriel, G., Hart, W. K., Renne, P. R., Gilbert, H., Defleur, A., Suwa, G., Katoh, S., Ludwig, K. R., et al. 2003. Stratigraphic chronological and behavioural contexts of Pleistocene *Homo* from Middle Awash, Ethiopia. *Nature* 423:747–752.

Clark, J. D., and Harris, W. K. 1985. Fire and its roles in early hominid lifeways. *African Archaeological Review* 3:30–27.

Clarke, E., Reichard, U. H., and Zuberbühler, K. 2006. The syntax and meaning of wild gibbon songs. *PLoS One* 1, no. 1: e73.

Clarke, R. J. 1979. Early hominid footprints from Tanzania. *South African Journal of Science* 75:148–149.

———. 1985a. *Australopithecus* and early *Homo* in southern Africa. In E. Delson, ed., *Ancestors: The Hard Evidence,* 171–177. New York: Alan R. Liss.

———. 1985b. Comments on the Laetoli footprints. Taung Diamond Jubilee International Symposium, Mmabatho, South Africa, February 2, 1985 [taped commentary].

———. 1988. A new *Australopithecus* cranium from Sterkfontein and its bearing on the ancestry of *Paranthropus.* In F. E. Grine, ed., *Evolutionary History of the "Robust" Australopithecines,* 285–292. Hawthorne, NY: Aldine de Gruyter.

———. 1990. Observations on some restored hominid specimens in the Transvaal Museum, Pretoria. In G. H. Sperber, ed., *Apes to Angels: Essays in Anthropology in Honor of Phillip V. Tobias,* 135–151. New York: Wiley-Liss.

———. 1998. First ever discovery of a well-preserved skull and associated skeleton of *Australopithecus. South African Journal of Science* 94:460–463.

———. 1999. Discovery of complete arm and hand of the 3.3 million-year-old *Australopithecus* skeleton from Sterkfontein. *South African Journal of Science* 95:477–480.

———. 2002a. On the unrealistic 'revised age estimates' for Sterkfontein. *South African Journal of Science* 98:415–418.

———. 2002b. Newly revealed information on the Sterkfontein Member 2 *Australopithecus* skeleton. *South African Journal of Science* 98:523–526.

———. 2003. Bipedalism and arboreality in *Australopithecus. Courier Forschungsinstitut Senkenberg* 243:79–83.

Clarke, R. J., and Howell, F. C. 1972. Affinities of the Swartkrans 847 hominid cranium. *American Journal of Physical Anthropology* 37:319–335.

Clarke, R. J., and Tobias, P. V. 1995. Sterkfontein Member 2 foot bones of the oldest South African hominid. *Science* 269:521–524.

Claverie, J. M. 2001. What if there are only 30,000 human genes? *Science* 29:1255–1257.

Clay, Z., Smith, C. L., and Blumstein, D. T. 2012. Food-associated vocalizations in mammals and birds: what do these calls really mean? *Animal Behaviour* 83:323–330.

Clay, Z., and Zuberbühler, K. 2009. Food-associated calling sequences in bonobos. *Animal Behaviour* 77:1387–1396.

———. 2011. Bonobos extract meaning from call sequences. *PLoS One* 6:e18768.

Clemente, I. C., Ponsà, M., García, M., and Egozcue, J. 990. Evolution of the Simiiformes and the phylogeny of human chromosomes. *Human Genetics* 84:493–506.

Clifford, J. 1986. Introduction: Partial Truths. In J. Clifford and G. Marcus, eds., *Writing Culture: The Poetics and Politics of Ethnography,* 1–26. Berkeley, University of California Press.

———. 1988. *The Predicament of Culture.* Cambridge, MA: Harvard University Press.

Clifford, S. L., Abernethy, K. A., White, L. J. T., Tutin, C. E. G., Bruford, M. W., and Wickings, E. J. 2003. Genetic studies of western gorillas. In A. B. Taylor and M. L. Goldsmith, eds., *Gorilla Biology. A Multidisciplinary Perspective* 269–292. Cambridge: Cambridge University Press.

Clutton-Brock, T. H., and Harvey, P. 1977. Primate ecology and social organization. *Journal of Zoology, London* 183:1–39.

———. 1980. Primate brains and ecology. *Journal of Zoology (London)* 190:309–323.

Coffing, K., Feibel, C., Leakey, M., and Walker, A. 1994. Four-million-year-old hominids from East Lake Turkana, Kenya. *American Journal of Physical Anthropology* 93:55–65.

Coffman, K. A., Dum, R. P., and Strick, P. L. 2011. Cerebellar vermis is a target of projection from the motor areas in the cerebral cortex. *Proceedings of the National Academy of Science of the United States of America* 108:16068–16073.

Cohen, J. 1997. How many genes are there? *Science* 275:769.

———. 2007a. Relative differences: the myth of 1%. *Science* 316:1836.

———. 2007b. Venter's genome sheds new light on human variation. *Science* 317:1311.

———. 2010. Boxed about the ears, ape language research field is still standing. *Science* 328:38–39.

Cohen, J. E. 1975. The size and demographic composition of social groups of wild orangutans. *Animal Behaviour* 23:543–550.

Cole, M. 1996. *Cultural Psychology.* Cambridge, MA: Cambridge University Press.

Cole, S. 1975. *Leakey's Luck: The Life of Louis Leakey, 1903–1972.* London: Collins.

Colell, M., Segarra, D, and Sabater Pí, J. 1995a. Hand preferences in chimpanzees *(Pan troglodytes),* bonobos *(Pan paniscus),* and orangutans *(Pongo pygmaeus)* in food-reaching and other daily activities. *International Journal of Primatology* 16:413–434.

———. 1995b. Manual laterality in chimpanzees *(Pan troglodytes)* in complex tasks. *Journal of Comparative Psychology* 109:298–307.

Coleman, M. N. 2009. What do primates hear? A meta-analysis of all known nonhuman primate behavioral audiograms. *International Journal of Primatology* 30:55–91.

Coleman, M. N., and Ross, C. F. 2004. Primate auditory diversity and its influence on hearing performance. In *Evolution of the Special Senses in Primates,* ed. T. D. Smith, C. F. Ross, N. J. Dominy, and J. T. Laitman, special issue, *Anatomical Record Part A* 281A:1123–1137.

Collison, M. E., Andrews, P., and Bamford, M. K. 2009. Taphonomy of the early Miocene flora, Hiwegi Formation, Rusinga Island, Kenya. *Journal of Human Evolution* 57:149–162.

Condillac, E. B. de. 1746. *Essai sur l'origine des connaissance humaines.* Auvers-sur-Oise (Paris): Galilé.

Congdon, K. A. 2012. Interspecific and ontogenetic variation in proximal pedal phalangeal curvature of great apes (*Gorilla gorilla, Pan troglodytes,* and *Pan paniscus*). *International Journal of Primatology* 33:418–427.

Conklin, N. L., and Wrangham, R. W. 1994. The value of figs to hind-gut fermenting frugivore: a nutritional analysis. *Biochemical Systematics and Ecology* 22:137–151.

Conklin-Brittain, N. L., Knott, C. D., and Wrangham, R. W. 2006. Energy intake by wild chimpanzees and orangutans: methodological considerations and a preliminary comparison. In G. Hohmann, M. M. Robbins and C. Boesch, eds., *Feeding Ecology in Apes and Other Primates,* 445–471. Cambridge: Cambridge University Press.

Conroy, G. C. 1976. Primate postcranial remains from the Oligocene of Egypt. *Contributions to Primatology* 8:1–134.

———. 1980. New evidence of Middle Pleistocene hominids from the Afar Desert-Ethiopia. *Anthropos* 7:96–107.

———. 1987. Problems of body-weight estimation in fossil primates. *International Journal of Primatology* 8:115–137.

Conroy, G. C., and Bown, T. M. 1976. Anthropoid origins and differentiation: the Asian question. *Yearbook of Physical Anthropology* 18:1–6.

Conroy, G. C., Jolly, C. J., Cramer, D., and Kalb, J. E. 1978. Newly-discovered fossil hominid skull from the Afar Depression, Ethiopia. *Nature* 276:67–70.

Conroy, G. C., Pickford, M., Senut, B., and Mein, P. 1993a. Additional Miocene primates from the Otavi Mountains, Namibia, *Comptes rendus des séances de l'Academie des sciences, Paris, série II* 317:987–990.

———. 1993b. Diamonds in the desert: the discovery of *Otavipithecus namibiensis. Evolutionary Anthropology* 2:46–52.

Conroy, G. C., Pickford, M., Senut, B., van Couvering, J., and Mein, P. 1992a. *Otavipithecus namibiensis,* first Miocene hominoid from southern Africa (Berg Aukas, Namibia). *Nature* 356:144–148.

———. 1992b. The Otavi Mountain land of Namibia yields southern Africa's first Miocene hominoid. *National Geographic Research and Exploration* 8:492–494.

Conroy, G. C., and Pilbeam, D. 1975. *Ramapithecus:* a review of its hominid status. In R. H. Tuttle, ed., *Paleoanthropology: Morphology and Paleoecology,* 59–86. The Hague: Mouton.

Conroy, G. C., Senut, B., Gommery, D., Pickford, M., and Mein, P. 1996. Brief communication: new primate remains from the Miocene of Namibia, South Africa. *American Journal of Physical Anthropology* 99:487–492.

Conroy, G. C., and Smith, R. J. 2007. The size of scalable brain components in the human evolutionary lineage: with a comment on the paradox of *Homo floresiensis. HOMO—Journal of Comparative Human Biology* 58:1–12.

Conroy, G. C., Weber, G. W., Seidler, H., Tobias, P. V., Kane, A., and Brunsden, B. 1998. Endocranial capacity in an early hominid cranium from Sterkfontein, South Africa. *Science* 280:1730–1731.

Constable, J. L., Ashley, M. V., Goodall, J., and Pusey, A. E. 2001. Noninvasive paternity assignment in Gombe chimpanzees. *Molecular Ecology* 10:1279–1300.

Cook, D. C., Buikstra, J. B., DeRousseau, C. J., and Johanson, D. C. 1983. Vertebral pathology in the Afar australopithecines. *American Journal of Physical Anthropology* 60:83–101.

Cook, J., Stringer, C. B., Currant, A. P., Schwarcz, H. P., and Wintle, A. G. 1982. A review of the chronology of the European Middle Pleistocene hominid record. *Yearbook of Physical Anthropology* 25:9–65.

Cook, R. 1997. *Chromosome 6.* New York:. Putnam.

Cooke, H. B. S. 1978. Pliocene-Pleistocene Suidae from Hadar, Ethiopia. *Kirtlandia* 29:1–63.

———. 1983a. Horses, elephants and pigs as clues in the African later Cainozoic. In J. C. Vogel, ed., *Late Cainozoic Palaeoclimates of the Southern Hemisphere,* 473–482. Rotterdam: Balkema.

———. 1983b. Human evolution: the geological framework. *Canadian Journal of Anthropology* 3:143–161.

———. 1985. Plio-Pleistocene Suidae in relation to African hominid deposits. In M. M. Beden, A. K. Behrensmeyer, N. T. Boaz, R. Bonnefille, C. K. Brain, B. Cooke, Y. Coppens, R. Dechamps, V. Eisenmann, A. Gentry, et al., eds., *L'Environment des Hominidés au Plio-Pléistocène,* 101–117. Paris: Masson.

———. 1990. Taung fossils in the University of California collections. In G. H. Sperber, ed., *From Apes to Angels,* 119–134. New York: Wiley-Liss.

Coon, C. S. 1962. *The Origin of Races.* New York: Knopf.

Copeland, S. R., Sponheimer, M., de Ruiter, D. J., Lee-Thorp, J. A., Codron, D., le Roux, P. J., Grimes, V., and Richards, M. P. 2011. Strontium isotope evidence for landscape use by early hominins. *Nature* 474:76–78.

Coppens, Y. 1977. Evolution morphologique de la première prémolaire inférieure chez certains Primates supérieurs. *Comptes rendus hebdomadaires des séances de l'Académie des Sciences, Paris, série D* 285:1299–1302.

———. 1980. The differences between *Australopithecus* and *Homo:* preliminary conclusions from the Omo Research Expedition's studies. In L.-K. Konigsson, ed., *Current Argument on Early Man,* 207–225. Oxford: Pergamon Press.

———. 1981. Le cerveau des hommes fossiles. *Comptes rendus des séances de l'Académie des sciences, vie académique, supplément aux séries I-II-III* 292:3–24.

———. 1982a. Les origines de l'homme. *Histoire et Archéologie, Dijon,* no. 60: 8–17.

———. 1982b. Commencements de l'Homme. *Le Débat,* no. 20: 0–53.

———. 1983a. Systematique, phylogenie, environment et culture des australopitheques, hypotheses et synthese. *Bulletins et Mémoires de la Société d'Anthropologie de Paris, série XIII,* 10:273–284.

———. 1983b. Les hominidés du Pliocène d'Afrique orientale et leur environment. In M. Sakka, ed., *Morphologie Evolutive Morphogenèse du Crane et Origine de l'Homme,* 154–168. Paris: Éditions du Centre National de la Recherche Scientifique.

———. 1983c. Les plus anciens fossiles d'hominides. In C. Chagas, ed., *Working Group on Recent Advances in the Evolution of Primates, May 24–27, 1982* 1–9. *Pontificae Academiae Scientiarum Scripta Varia:* 50. Vatican: Pontificia Academia Scientiarum.

———. 1983d. Le singe, l'Afrique et l'homme. Paris: Fayard.

Coppens, Y., Gray, B. T., and Johanson, D. C. 1980. Biostratigraphie d'Hadar comparée à celle des autres gisements plio-pléistocènes est-africains. In R. E. Leakey and B. A. Ogot, eds., *Proceedings of the 8th Panafrican Congress of Prehistory and Quaternary Studies Nairobi, 5 to 10 September 1977,* 56–57. Nairobi: The International Louis Leakey Memorial Institute for African Prehistory.

Coppens, Y., Howell, F. C., Isaac, G. L., and Leakey, R. E. F. 1976. *Earliest Man and Environments in the Lake Rudolf Basin.* Chicago: University of Chicago Press.

Corballis, M. C. 1991. The Lopsided Ape. Evolution of the Generative Mind. New York: Oxford University Press.

———. 1998. Evolution of language and laterality: a gradual descent? *Communication, Politics and Culture* 17:1148–1155.

———. 1999. The gestural origins of culture. *American Scientist* 87:138–145.

———. 2002. *From Hand to Mouth: The Evolution of Language.* Princeton, NJ: Princeton University Press.

———. 2003a. From mouth to hand: gesture, speech, and the evolution of right-handedness. *Current Anthropology* 26:199–260.

———. 2003b. From mouth to hand: the gestural origins of language. In M. H. Christiansen and S. Kirby, eds., *Language Evolution,* 201–218. Oxford: Oxford University Press.

———. 2007. The uniqueness of human recursive thinking. *American Scientist* 95:240–248.

Corbey, R. 2005. *The Metaphysics of Apes: Negotiating the Animal-Human Boundary.* Cambridge: Cambridge University Press.

Cordoni, G., Palagi, E., and Tarli, S. B. 2006. Reconciliation and consolation in captive western gorillas. *International Journal of Primatology* 27:1365–1382.

Cords, M., and Aureli, F. 1996. Reasons for reconciling. *Evolutionary Anthropology* 5:42–45.

———. 2000. Reconciliation and relationship qualities. In F. Aureli and F. B. M. de Waal, eds., *Natural Conflict Resolution,* 177–198. Berkeley: University of California Press.

Corina, D. P., and McBurney, S. L. 2001. The neural representation of language in users of American Sign Language. *Journal of Communication Disorders* 34:455–471.

Corina, D. P., Poinzer, H., Bellugi, U., Feinberg, T., Dowd, D., and O'Grady-Batch, L. 1992. Dissociation between linguistic and nonlinguistic gestural systems: a case for compositionality. *Brain and Language* 43:414–447.

Corp, N., and Byrne, R. W. 2004. Sex differences in chimpanzee handedness. *American Journal of Physical Anthropology* 123:62–68.

Corruccini, R. S. 1975a. Multivariate analysis of *Gigantopithecus* mandibles. *American Journal of Physical Anthropology* 42:167–170.

———. 1975b. *Gigantopithecus* and hominids. *Anthropologischer Anzeiger* 35:55–57.

———. 1975c. Morphometric affinities in the forelimb of anthropoid primates. *Zeitschrift für Morphologie und Anthropologie* 67:19–31.

———. 1978. Comparative osteometrics of the hominoid wrist joint, with special reference to knuckle-walking. *Journal of Human Evolution* 7:307–321.

———. 1992. Bootstrap approaches to estimating confidence intervals for molecular dissimilarities and resultant trees. *Journal of Human Evolution* 23:481–493.

————. 1994. How certain are hominoid phylogenies? The role of confidence intervals in cladistics. In R. S. Corruccini and R. L. Ciochon, eds., *Integrative Paths to the Past,* 167–183, Englewood Cliffs, NJ: Prentice Hall.

Corruccini, R. S., Ciochon, R. L., and McHenry, H. M. 1975. Osteometric shape relationships in the wrist joint of some anthropoids. *Folia Primatologica* 24:250–274.

————. 1976. The postcranium of Miocene hominoids: were dryopithecines merely "dental apes"? *Primates* 17:205–223.

Corruccini, R. S., Cronin, J. E., and Ciochon, R. L. 1979. Scaling analysis and congruence among anthropoid primate macromolecules. *Human Biology* 51:167–185.

Corruccini, R. S., and McHenry, H. M. 1980. Cladometric analysis of Pliocene hominids. *Journal of Human Evolution* 9:209–221.

Cortright, G. W. 1983. *The Relative Mass of Hindlimb Muscles in Anthropoid Primates: Functional and Evolutionary Implications.* PhD diss., University of Chicago.

Corvinus, C., and Roche, H. 1980. Prehistoric exploration at Hadar in the Afar (Ethiopia) in 1973, 1974 and 1976. In R. E. Leakey and B. A. Ogot, eds., *Proceedings of the 8th Panafrican Congress of Prehistory and Quaternary Studies, Nairobi, 5 to 10 September 1977,* 186–188. Nairobi: International Louis Leakey Memorial Institute for African Prehistory.

Cousins, D., and Huffman, M. A. 2002. Medicinal properties in the diet of gorillas: an ethno-pharmacological evaluation. *African Study Monographs* 23:65–89.

Covert, H. H. 2004. Does overlap among the adaptive radiations of omomyoids, adapoids, and early anthropoids cloud out understanding of anthropoid origins? In C. F. Ross and R. F. Kay, eds., *Anthropoid Origins,* 139–155. New York: Kluwer Academic/Plenum Press.

Cowgill, G. 1993. Beyond criticizing new archaeology. *American Anthropologist* 95:551–573.

Cowley, A., and Weiskrantz, L. 1975. Demonstration of cross-modal matching in the rhesus monkeys, *Macaca mulatta. Neuropsychologia* 13:117–120.

Cramer, D. L. 1977. Craniofacial morphology of *Pan paniscus:* a morphometric and evolutionary appraisal. *Contributions to Primatology,* Vol. 10. Basel: Karger.

Crawford, G. E., Holt, I. E., Mullikin, J. C., Tai, D., National Institutes of Health Intramural Center, Green E. D., Wolfsberg, T. G., and Collins, F. S. 2004. Identifying gene regulatory elements by genome-wide recovery of DNase hypersensitive sites. *Proceedings of the National Academy of Sciences of the United States of America* 104:992–997.

Crelin, E. S. 1987. *The Human Vocal Tract: Anatomy, Function, Development, and Evolution.* New York: Vantage Press.

Crinion, J., Turner, R., Grogan, A., Hanakawa, T., Nppeney, U., Devlin, J. T., Aso, T., Urayama, S., Fukuyama, H., Stockton, K., Usui, K., Green, D. W., and Price, C. J. 2006. Language control in the bilingual brain. *Science* 312:1537–2540.

Crockford, C., and Boesch, C. 2003. Context-specific calls in wild chimpanzees, *Pan troglodytes verus:* an analysis of barks. *Animal Behaviour* 66:115–125.

————. 2005. Call combinations in wild chimpanzees. *Behaviour* 142:397–421.

Crockford, C., Herbinger, I., Vigilant, L., and Boesch, C. 2004. Wild chimpanzees produce group-specific calls: a case for vocal learning. *Ethology* 110:221–243.

Crompton, R. H., and Pataky, T. C. 2009. Stepping out. *Science* 323:1174–1175.

Crompton, R. H., Pataky, T. C., Savage, R., D'Août, K., Bennett, M. R., Day, M. H., Bates, K., Morse, S., and Sellers, W. I. 2011. Human-like function of the foot, and fully upright gait, confirmed in the 3.66 million year old Laetoli hominin footprints by topographic statistics, experimental footprint-formation and computer simulation. *Journal of the Royal Society Interface,* doi: 10.1098/rsif.2011.0258.

Crompton, R. H., Vereecke, E. E., and Thorpe, K. S. 2008. Locomotion and posture from the common hominoid ancestor to fully modern hominins, with special reference to the last common panin/hominin ancestor. *Journal of Anatomy* 212:501–543.

Cronin, J. E. 1983. Apes, humans, and molecular clocks, a reappraisal. In R. L. Ciochon and R. S. Corruccini, eds., *New Interpretations of Ape and Human Ancestry,* 115–150. New York: Plenum Press.

Cronin, J. E., Boaz, N. T., Stringer, C. B., and Rak, Y. 1981. Tempo and mode in hominid evolution. *Nature* 292:13–122.

Cronk, L. 1999. *That Complex Whole: Culture and the Evolution of Human Behavior.* Boulder CO: Westview Press.

Crook, J. H., and Gartlan, J. S. 1966. Evolution of primate societies. *Nature* 210:1200–1203.

Crook, J. H., and Goss-Custard, J. D. 1972. Social ethology. *Annual Review of Psychology* 23:277–312.

Cross, I., and Woodruff, G. E. 2009. Music as a communicative medium. In T. Botha and C. Knight, eds., *The Prehistory of Language,* 77–98. Oxford: Oxford University Press.

Crouau-Roy, B., Service, S., Slatkin, M., and Freimer, N. 1996. A fine-scale comparison of the human and chimpanzee genomes: linkage, linkage disequilibrium and sequence analysis. *Human Molecular Genetics* 5:1131–1137.

Crutzen, P. J. Geology of mankind. *Nature* 415:23.

Culotta, E. 2004. Spanish fossil sheds new light on the oldest great ape. *Science* 306:1273–1274.

———. 2005a. What genetic changes made us uniquely human? *Science* 309:91.

———. 2005b. Chimp genome catalogs differences with humans. *Science* 309:1468–1469.

———. 2009. Did humans learn from Hobbits? *Science* 324:447.

cummings, e.e. 1972. *Complete Poems.* New York: Harcourt Brace Jovanovich.

Cunha, Carniero da. 1986. *Antropologia do Brasil: Mito, História, Etnicidade.* Saõ Paulo: Editora Brasiliense.

Curtin, S. H., and Chivers, D. J. 1978. Leaf-eating primates of Peninsular Malaysia: the siamang and the dusky leaf-monkey. In G. G. Montgomery, ed., *The Ecology of Arboreal Folivores,* 441–464. Washington, DC: Smithsonian Institution Press.

Curtis, S. 1977. *Genie: A Psycholinguistic Study of a Modern-Day "Wild Child."* New York: Academic Press.

Custance, D., and Bard, K. A. 1994. The comparative and developmental study of self-recognition and imitation: the importance of social factors. In S. T. Parker, R. W. Mitchell, and M. L. Boccia, *Self-Awareness in Animals and Humans,* 207–226. Cambridge: Cambridge University Press.

Cybulski, J. S. 1981. *Homo erectus:* a synopsis, some new information, and a chronology. In B. A. Sigmon and J. S. Cybulski, eds., *Homo erectus: Papers in Honor of Davidson Black,* 227–236. Toronto: University of Toronto Press.

Czekala, N., and Sicotte, P. 2000. Reproductive monitoring of free-ranging female mountain gorillas by urinary hormone analysis. *American Journal of Primatology* 51:209–215.

Daegling, D. J. 2005. Functional morphology of the human chin. *Evolutionary Anthropology* 1:170–177.

Daegling, D. J., and Grine, F. E. 2007. Mandibular biomechanics and the paleontological evidence for the evolution of human diet. In P. S. Ungar, ed., *Evolution of the Human Diet,* 77–105. Oxford: Oxford University Press.

Dagosto, M. 2002. The origin and diversification of anthropoid primates: introduction. In W. C. Hartwig, ed., *The Primate Fossil Record,* 125–132. Cambridge: Cambridge University Press.

Dagosto, M., Marivaux, L., Gebo, D. L., Beard, K. C., Chaimanee, Y., Jaeger, J.-J., Marandat, B., Soe, A. N., and Kyaw, A. A. 2010. The phylogenetic affinities of the Pondaung tali. *American Journal of Physical Anthropology* 143:223–234.

Dahl, J. F. External genitalia. In J. H. Schwartz, ed., *Orang-utan Biology,* 133–144. New York: Oxford University Press.

Dahlberg, A. A. 1963. Dental evolution and culture. *Human Biology* 35:237–249.

Dahlberg, F. 1981. *Woman the Gatherer.* New Haven, CT: Yale University Press.

Daiger, S. P., Goode, M. E., and Trowbridge, B. D. 1987. Evolution of nuclear gene families in primates. Copy-number variation in the argininosuccinate synthetase (ASS) pseudogene family and the anonymous DNA sequence, D1S1. *Genetica* 73:91–98.

Dallmann, R., and Geissmann, T. 2009. Individual and geographical variability in the songs on wild silvery gibbons *(Hylobates Moloch)* on Java, Indonesia. In S. Lappan and D. J. Whittaker, eds., *The Gibbons: New Perspectives on Small Ape Socioecology and Population Biology,* 91–110. New York: Springer.

Dalrymple, G. B. 1979. Critical tables for conversion of K-Ar ages from old to new constants. *Geology* 7:558–560.

Dam, S. N. 2006. A short study on wild hoolock gibbons *(Hoolock hoolock)* in Assam and Bangladesh. *Gibbon Journal* 2:40–47.

D'Amour, D. E., Hohmann, G., and Fruth, B. 2006. Evidence of leopard predation on bonobos *(Pan paniscus). Folia Primatologica* 77:212–217.

D'Andrade, R., and Morin, P. A. 1996. Chimpanzee and human mitochondrial DNA: a principal components and individual-by-site analysis. *American Anthropologist* 98:352–370.

Dangel, A. W., Baker, B. J., Mendoza, A. R., and Yu, C. Y. 1995. Complement component C4 gene interon 9 as a phylogenetic marker for primates: long terminal repeats of the endogenous retrovirus ERV-K(C4) are a molecular clock of evolution. *Immunogenetics* 42:41–52.

Danish, L., Chapman, C. A., Hall, M. B., Rode, K. D., and Worman, C. O. 2006. The role of sugar in diet selection in redtail and red colobus monkeys. In G. Hohmann, M. M.

Robbins and C. Boesch, eds., *Feeding Ecology in Apes and Other Primates,* 473–487. Cambridge: Cambridge University Press.

Darian-Smith, I., Galea, M. P., and Darian-Smith, C. 1996. Manual dexterity: how does the cerebral cortex contribute? *Clinical and Experimental Pharmacology and Physiology* 23:948–956.

Dart, R. A. 1925. *Australopithecus africanus,* the man-ape of South Africa. *Nature* 115:195–199.

———. 1948. The Makapansgat proto-human *Australopithecus prometheus. American Journal of Physical Anthropology* 6:259–283.

———. 1949. The predatory implemental technique of *Australopithecus. American Journal of Physical Anthropology,* n.s., 6:1–38.

———. 1953. The predatory transition from ape to man. *International Anthropological and Linguistic Review* 1:201–218.

———. 1957. The osteodontokeratic culture of *Australopithecus prometheus.* Transvaal Museum Memoir No. 10. Pretoria: Transvaal Museum.

———. 1959. *Adventures with the Missing Link.* London: Hamish Hamilton.

———. 1960. The status of *Gigantopithecus. Anthropologischer Anzeiger* 24:139–145.

Darwin, C. 1859. *The Origin of Species by Means of Natural Selection or the Preservation of Favored Races in the Struggle for Life.* London: Murray.

———. 1871. *The Descent of Man and Selection in Relation to Sex.* London: Murray. Reprinted as New York: Modern Library, 1990.

———. 1872. *The Expressions of Emotions in Man and Animals.* London: Murray.

———. 1899. *The Expression of the Emotions in Man and Animals.* New York: D. Appleton.

DaSilva, J. M. 2011. A shift toward birthing relatively large infants early in human evolution. *Proceedings of the National Academy of Sciences of the United States of America* 108:1022–1027.

Davenport, Richard K. 1967. The orangoutan in Sabah. *Folia Primatologica* 5:247–263.

———. 1976. Cross-modal perception in apes. *Annals of the New York Academy of Sciences* 280:143–149.

———. 1977. Cross-modal perception: a basis for language? In D. M. Rumbaugh, ed., *Language Learning by a Chimpanzee: The Lana Project,* 73–83. New York: Academic Press.

Davenport, R. K., and Rogers, C. M. 1970. Intermodal equivalence of stimuli in apes. *Science* 168:279–280.

———. 1971. Perception of photographs by apes. *Behaviour* 39:318–320.

Davenport, R. K., Rogers, C. M., and Rumbaugh, D. M. 1973. Long-term cognitive deficits in chimpanzees associated with early impoverished rearing. *Developmental Psychology* 9:343–347.

Davenport, R. K., Rogers, C. M., and Russell, I. S. 1973. Cross-modal perception in apes. *Neuropsychologia* 11:21–28.

———. 1975. Cross-modal perception in apes: altered visual cues and delay. *Neuropsychologia* 13:229–235.

Davenport, T. R. B., Stanley, W. T., Sargis, E. J., De Luca, D. W., Mpunga, N. E., Machaga, S. J., and Olson, L. E. 2006. A new genus of African monkey, *Rungwacebus:* morphology, ecology and molecular phylogenetics. *Science* 312:1378–1381.

Davidson, D. W., Cook, S. C., Snelling, R. R., and Chua, T. H. 2003. Explaining the abundance of ants in lowland tropical rainforest canopies. *Science* 300:369–372.

Davidson, E. H., and Erwin, D. H. 2006. Gene regulatory networks and the evolution of animal body plans. *Science* 311:796–800.

Davidson, E. H., and Levine, M. 2005. Gene regulatory networks. *Proceedings of the National Academy of Sciences of the United States of America* 102:4935.

Davidson, R. K. 1992. Emotion and affective style: hemispheric substrates. *Psychological Science* 3:39–43.

Davies, A. G., Bennett, E. L., and Waterman, P. G. 1988. Food selection by two South-east Asian colobine monkeys (*Presbytis rubicunda* and *Presbytis melalophus*) in relation to plant chemistry. *Biological Journal of the Linnean Society* 34:33–56.

Davis, P. R., and Napier, J. 1963. A reconstruction of the skull of *Proconsul africanus (R.S. 51)*. *Folia Primatologica* 1:20–28.

Dawson, C., and Woodward, A. S. 1913. On the discovery of a Palaeolithic human skull and mandible in the flint-bearing gravel overlying the Wealden (Hastings Beds) at Piltdown, Fletching (Sussex). *Quarterly Journal of the Geological Society of London* 69:117–151.

Day, M. H. 1976a. Hominid postcranial remains for the East Rudolf Succession: a review. In Y. Coppens, F. C. Howell, G. L. Isaac, and R. E. F. Leakey, eds., *Earliest Man and Environments in the Lake Rudolf Basin,* 507–521. Chicago: University of Chicago Press.

———. 1976b. Hominid postcranial material from Bed I, Olduvai Gorge. In G. L. Isaac and E. R. McCown, *Human Origins,* eds., 363–374. Menlo Park, CA: W. A. Benjamin.

———. 1985. Hominid locomotion—from Taung to the Laetoli footprints. In P. V. Tobias, ed., *Hominid Evolution: Past, Present and Future,* 115–127. New York: Alan R. Liss.

———. 1986a. *Guide to Fossil Man.* 4th ed. Chicago: University of Chicago Press.

———. 1986b. Bipedalism: pressures, origins and modes. In B. Wood, L. Martin, and P. Andrews, eds., *Major Topics in Primate and Human Evolution,* 188–202. Cambridge: Cambridge University Press.

Day, M. H., Leakey, M. D., and Magori, C. 1980. A new hominid fossil skull (LH-18) from the Ngaloba Beds, Laetoli, northern Tanzania. *Nature* 284:55–56.

Day, M. H., and Wickens, E. H. 1980. Laetoli Pliocene hominid footprints and bipedalism. *Nature* 286:385–387.

Deacon, T. W. 1997. *The Symbolic Species.* New York: W. W. Norton.

Dean, C. D., and Smith, B. H. 2009. Growth and development of the Nariokotome youth KNM-WT-15000. In F. E. Grine, J. G. Fleagle, and R. E. Leakey, eds., *The First Humans: Origin and Early Evolution of the Genus Homo,* 101–120. New York: Springer Science + Business Media B. V.

Dean, D., and Delson, E. Features of the Dmanisi mandible. *Nature* 373:473.

Dean, M. C. 1986. *Homo* and *Paranthropus:* similarities in the cranial base and developing dentition. In B. Wood, L. Martin, and P. Andrews, eds., *Major Topics in Primate and Human Evolution,* 249–265. Cambridge: Cambridge University Press.

———. 1987. Growth layers and incremental markings in hard tissues, a review of the literature and some preliminary observations about enamel structure in *Paranthropus boisei. Journal of Human Evolution* 16:157–172.

————. 1988. Growth of teeth and development of the dentition in *Paranthropus*. In F. E. Grine, ed., *Evolutionary History of the "Robust" Australopithecines*, 43–53. Hawthorne, NY: Aldine de Gruyter.

Dean, M. C., and Benyon, A. D. 1991. Tooth crown heights, wear, sexual dimorphism and jaw growth in hominoids. *Zeitschrift für Morphologie und Anthropologie* 78:425–440.

Deblauwe, I. 2006. New evidence of honey-stick use by chimpanzees in southeast Cameroon. *Pan Africa News* 13:2–4.

Deblauwe, I., Dupain, J., Nguenang, G. M., Werdenich, D., and Van Elsacker, L. 2003. Insectivory by *Gorilla gorilla gorilla* in Southeast Cameroon. *International Journal of Primatology* 24:493–502.

Deblauwe, I., Guislain, J., Dupain, J., and Van Elsacker, L. 2006. Use of a tool-set by *Pan troglodytes* to obtain termites *(Macrotermes)* in the periphery of the Dja Biosphere Reserve, southeast Cameroon. *American Journal of Primatology* 68:1191–1196.

Deblauwe, I., and Janssens, P. J. 2008. New insights in insect prey choice by chimpanzees and gorillas in southeast Cameroon: the role of nutritional value. *American Journal of Physical Anthropology* 135:42–55.

de Bonis, L. 1986. *Homo erectus* et la transition vers *Homo sapiens* en Europe. In M. Sakka, ed., *Définition et Origines de l'Homme*, 253–261. Paris: Éditions du Centre National de la Recherche Scientifique.

————. 1987a. L'origine des hominidés. *L'Anthropologie (Paris)* 91:433–454.

————. 1987b. Les primates de l'ancien monde du Paléocène au Miocène. In M. Marois, ed., *L'évolution dans sa réalité et ses diverses modalités*, 93–130. Paris: Foundation Singer-Polignac Masson.

————. 1987c. Les racines des Hominidae: les recherches in Grèce. *La Vie des Sciences, Comptes rendus, série générale* 4:287–304.

de Bonis, L., Bouvrain, G., Geraads, D., and Koufos, G. 1990a. New remains of *Mesopithecus* (Primates, Cercopithecoidea) from the Late Miocene of Macedonia (Greece), with the description of a new species. *Journal of Vertebrate Paleontology* 10:473–483.

————. 1990b. New hominid skull material from the late Miocene of Macedonia in northern Greece. *Nature* 345:712–714.

————. 1991. New hominoid bearing locality of the late Miocene of Macedonia (Greece). *Bulletin of the Geological Society of Greece* 25:381–394.

de Bonis, L., Bouvrain, G., Geraads, D., and Koufos, G. D. 1992. diversity and paleoecology of Greek late Miocene mammalian faunas. *Palaeogeography, Palaeoclimatology, Palaeoecology* 91:99–121.

de Bonis, L., Bouvrain, G., Koufos, G., and Melentis, J. 1986. Succession and dating of the late Miocene primates of Macedonia. In J. G. Else and P. C. Lee, eds., *Primate Evolution*, 107–114. Cambridge: Cambridge University Press.

de Bonis, L., Jaeger, J.-J., Coiffait, B., and Coiffait, P.-E. 1988. Découverte du plus ancien primate Catarrhinien connu dans l'Éocène d'Afrique du Nord. *Comptes rendus hebdomadaires des séances de l'Académie des Sciences, Paris, série II* 306:929–934.

de Bonis, L., Johanson, D., Melentis, J., White, T., and Piveteau, J. 1981. Variations métriques de la denture chez les Hominidés primitifs: comparaison entre *Australopithecus afarensis* et *Ouranopithecus macedoniensis: Comptes rendus des séances de l'Academie des sciences, Paris, série II* 292:373–376.

de Bonis, L., and Koufos, G. 1993. The face and the mandible of *Ouranopithecus macedoniensis:* description of new specimens and comparisons. *Journal of Human Evolution* 24:469–491.

———. 1994. Our ancestor's ancestors: *Ouranopithecus* is a Greek link in human ancestry. *Evolutionary Anthropology* 3:75–83.

———. 2001. Phylogenetic relationships of *Ouranopithecus macedoniensis* (Mammalia, Primates, Hominoidea, Hominid of the late Miocene deposits of Central Macedonia (Greece). In L. de Bonis, G. D. Koufos, and P. Andrews, eds., *Hominoid Evolution and Climate Change in Europe,* Vol. 2: *Phylogeny of the Neogene Hominoid Primates of Eurasia,* 254–267. Cambridge: Cambridge University Press.

de Bonis, L., and Melentis, J. 1980. Nouvelles remarques sur l'anatomie d'un Primate hominoïde du Miocène: *Ouranopithecus macedoniensis.* Implications sur la phylogénie des Hominidés. *Comptes rendus hebdomadaires des séances de l'Académie des Sciences, Paris, série D* 290:755–758.

———. 1984. La position phylétique d'*Ouranopithecus. Courier Forschungsinstitut Senckenberg* 69:13–23.

———. 1987. Intérêt de l'anatomie naso-maxillaire pour la phylogénie des Hominidae. *Comptes rendus des séances de l'Académie des sciences, Paris, série D* 304:767–769.

———. 1991. Age et position phylétique du crane de Petralona (Grèce). In E. Bonifay and B. Vandermeersch, eds., *Les Premiers Européens: Actes du 114e Congrès National des Sociétés Savantes,* 285–289. Paris: Éditions du CTHS.

de Grouchy, J. 1987. Chromosome phylogenies of man, great apes, and old world monkeys. *Genetica* 73:37–52.

DeGusta, D., Gilbert, W. H., and Turner, P. 1999. Hypoglossal canal size and hominid speech. *Proceedings of the National Academy of Sciences of the United States of America* 96:1800–1804.

Dehaene, S. 2001. Précis of the number sense. *Mind and Language* 16:16–36.

Dehaene, S., Duhamel, J.-R., Hauser, M. D., and Rizzolatti, G. 2005. *From Monkey Brain to Human Brain.* Cambridge, MA: MIT Press.

Dehaene, S., Izard, V., Spelke, E., and Pica, P. 2008. Log or linear? distinct intuitions of the number scale in Western and Amazonian indigene cultures. *Science* 320:1217–1220.

Dehaene, S., Spelke, E., Pinel, P., Stanescu, R., and Tsivkin, S. 1999. Sources of mathematical thinking: behavioral and brain-imaging evidence. *Science* 284:970–974.

Deinard, A., and Kidd, K. 1999. Evolution of a HOXB6 intergenic region within the great apes and humans. *Journal of Human Evolution* 36:687–703.

Deinard, A., Sirugo, G., and Kidd, K. 1998. Hominiod phylogeny: inferences from a subterminal minisatellite analyzed by repeat expansion detection (RED). *Journal of Human Evolution* 35:313–317.

Deininger, P. L., and Daniels, G. R. 1986. The recent evolution of mammalian repetitive DNA elements. *Trends in Genetics* 2:76–80.

Deino, A. 2011. ^{40}Ar/^{39}Ar dating of Laetoli, Tanzania. In T. Harrison, ed., *Paleontology and Geology of Laetoli: Human Evolution in Context,* Vol. 1: *Geology, Geochronology,*

Paleoecology and Paleoenvironment, Vertebrate Paleobiology and Paleoanthropology, 77–97.

Delattre, A., and Fenart, R. 1956. Analyse morphologique du splanchnocrane chez les primates et ses rapports avec le prognathisme. *Mammalia* 20:168–214.

Delfour, F., and Marten, K. 2001. Mirror image processing in three marine mammal species: killer whales *(Orcinus orca),* false killer whales *(Pseudorca crassidens)* and California sea lions *(Zalophus californianus). Behavioural Processes* 53:181–190.

Delgado, R. A. 2007. Geographic variation in the long calls of male orangutans *(Pongo* spp.). *Ethology* 113:487–498.

Delgado, R. A., Lameira, A. R., Ross, M. D., Husson, S. J., Morrogh-Bernard, H. C., and Wich, S. A. 2009. Geographic variation in orangutan long calls. In S. A. Wich, S. S. Utami Atmoko, T. Mitra Setia, and C. P. van Schaik, eds., *Orangutans: Geographic Variation in Behavioral Ecology and Conservation,* 215–224. Oxford: Oxford University Press.

Delgago, R. A., and van Schaik, C. P. 2000. The behavioral ecology and conservation of the orangutan *(Pongo pygmaeus):* a tale of two islands. *Evolutionary Anthropology* 9:201–218.

Dellatore, D. F., Waitt, C. D., and Foitova, I. 2009. Two cases of mother-infant cannibalism in orangutans. *Primates* 50:277–281.

Deloison, Y. 1991. Les australopitheques marchaient-ils comme nous? In Y. Coppens & B. Senut, eds., *Origine(s) de la Bipédie chez les Hominidés,* 177–186. Cahiers de Paléoanthropologie. Paris: Éditions du Centre National de la Recherche Scientifique.

———. 1992. Empreintes de pas Laetoli (Tanzanie). Leur apport une meilleure connaissance de la locomotion des Hominids fossile. *Comptes rendus de l'Académie des sciences, Paris* 315 (II): 103–109.

———. 1997. The foot bones from Hadar, Ethiopia and the Laetoli, Tanzania footprints. Locomotion of *A. afarenis. American Journal of Physical Anthropology* 24 (Suppl.): 101.

———. 2004. Study of the Laetoli footprints compared with those of modern man and ape. In J. Y. Kim, K-S, Kim, S. I. Park, and M-K Shin, eds., *Proceeding of International Symposium on the Quaternary Footprints of Hominids and Other Vertebrates,* 106–110, Namjejugun, South Korea.

Delson, E. 1975a. Toward the origin of the Old World monkeys. In J.-P. Lehman, ed., *Colloque international du Centre National de la Recherche Scientifique,* No. 218, *Problèmes Actuels de Paléontologie-Évolution des Vertébrés,* 839–850. Paris: Éditions du CNRS.

———. 1975b. Evolutionary history of the Cercopithecidae. In *Approaches to Primate Paleobiology,* ed., F. S. Szalay, *Contributions to Primatology* 5:167–217. Basel: Karger.

———. 1979a. *Prohylobates* (Primates) from the early Miocene of Libya: a new species and its implications for cercopithecoid origins. *Geobios (Lyon),* 12:725–733.

———. 1979b. *Oreopithecus* is a cercopithecoid after all. *American Journal of Physical Anthropology* 50:431–432.

———. 1986a. An anthropoid enigma: historical introduction to the study of *Oreopithecus bambolii. Journal of Human Evolution* 15:523–531.

———. 1986b. Human phylogeny revised again. Nature 322:496–497.

————. 1987. Monkey. *McGraw-Hill Encyclopedia of Science and Technology,* 6th ed., 11:359–364.

————. 1988. Chronology of South African australopith site units. In F. E. Grine, ed., *Evolutionary History of the "Robust" Australopithecines,* 317–324. Hawthorne, NY: Aldine de Gruyter.

Delson, E., and Andrews, P. 1975. Evolution and interrelationships of the catarrhine primates. In W. P. Luckett and F. S. Szalay, eds., *Phylogeny of the Primates: A Multidisciplinary Approach,* 405–446. New York: Alan R. Liss.

Delson, E., and Rosenberger, A. 1984. Are there any anthropoid primate living fossils? In N. Eldredge and S. M. Stanley, eds., *Living Fossils,* 50–61. New York: Springer.

de Lumley, M.-A. 1981. Les anténéandertaliens en Europe. In B. A. Sigmon and J. S. Cybulski, eds., *Homo erectus: Papers in Honor of Davidson Black,* 115–132. Toronto: University of Toronto Press.

de Lumley, M.-A., Gabounia, L., Vekua, A., and Lordkipanidze, D. 2006. Les restes humains du Pliocène final et du début du Pléistocène inférieur de Dmanissi, Géorgie (1991–2000). 8—Les cranes, D 2280, D 2282, D 2700. *L'Anthropologie* 110:1–110.

Demars C., Berthomier, C., and Goustard, M. 1978. The ontogenesis of the 'great call' of gibbons *(Hylobates concolor).* In D. J. Chivers and J. Herbert, eds., *Recent Advances in Primatology,* Vol. 1: *Behaviour,* 827–830. London: Academic Press.

Demes, B., Larson, S. G., Stern, J. T., Jr., Jungers, W. L., Biknevicius, A. R., and Schmitt, D. 1994. The kinetics of primate quadrupedalism: "hindlimb drive" reconsidered. *Journal of Human Evolution* 26:353–374.

Demeter, F., Shackelford, L. L., Bacon, A.-M., Duringer, P., Westaway, K., Sayavongkhamdy, T., Braga, J., et al. 2012. Anatomically modern human skull in Southeat Asia (Laos) by 45 ka. *Proceedings of the National Academy of Sciences of the United States of America* 109:14375–14380.

de Moura, A. C., and Lee, P. C. 2004. Capuchin stone tool use in Caatinga dry forest. *Science* 306:1909.

Denham, T. P., Haberle, S. G., Lentifer, C., Fullagar, R., Field, J., Therin, M., Porch, N., and Winsborough. 2003. Origins of agriculture at Kuk Swamp in the highlands of New Guinea. *Science* 301:189–193.

Denham, T. P., and White, P. 2007. *The Emergence of Agriculture: A Global View.* London: Routledge.

DeNiro, M. J., and Epstein, S. 1978. Influence of diet on the distribution of carbon isotopes in animals. *Goechimica et Cosmochimica Acta* 42:495–506.

Dennis, C. 2005. Branching out. *Nature* 437:17–19.

Depue, B. E., Curran, T., and Banich, M. T. 2007. Prefrontal regions orchestrate suppression of emotional memories via a two-phase process. *Science* 317:215–219.

de Ruiter, D. J., DeWitt, T. J., Carlson, K. B., Bophy, J. K., Schroeder, L., Ackermann, R. R., Churchill, S. E., and Berger, L. R. 2013. Mandibular remains support taxonomic validity of *Australopithecus sediba. Science* 340, http://dx.doi.org/10.1126/science.1232997.

de Ruiter, D. J., Pickering, R., Steininger, C. M., Kramers, J. D., Hancox, P. J., Churchill, S. E., Berger, L. R., and Blackwell, L. 2009. New *Australopithecus robustus* fossils and associated U-Pb dates from Cooper's Cave (Gauteng, South Africa). *Journal of Human Evolution* 56:497–513.

Descartes, R. 1996. *Meditations on First Philosophy: With Selections from the Objections and Replies,* ed. J. Cottingham. New York: Cambridge University Press.

Deschner, T., Heistermann, M., Hodges, K., and Boesch, C. 2003. Timing and probability of ovulation in relation to sex skin swelling in wild West African chimpanzees, *Pan troglodytes verus. Animal Behaviour* 66:551–560.

———. 2004. Female sexual swelling size, timing of ovulation, and male behavior in wild West African chimpanzees. *Hormones and Behavior* 46:204–215.

De Silva, G. S. 1971. Notes on the orang-utan rehabilitation project in Sabah. *Malayan Nature Journal* 24:50–77.

———. 1972. The birth of an orang-utan *(Pongo pygmaeu)* at Sepilok Game Reserve. *International Zoo Yearbook* 12:104–105.

De Silva, J. M. 2009. Functional morphology of the ankle and the likelihood of climbing in early hominins. *Proceedings of the National Academy of Sciences of the United States of America* 106:6567–6572.

De Silva, J. M., Holt, K. G., Churchill, S. E., Carlson, K. J., Walker, C. S., Zipfel, B., and Berger, L. R. 2013. The lower limb and mechanics of walking in *Australopithecus sediba. Science* 340, http://ex.doi.org/10.1126/science.1232999.

de Veer, M. W., Gallup, G. G., Jr., Theall, L. A., van den Bos, R., and Povinelli, D. J. 2003. An 8-year longitudinal study of mirror self-recognition in chimpanzees *(Pan troglodytes). Neuropsychologia* 4:229–234.

de Veer, M. W., and van den Bos, R. 1999. A critical review of methodology and interpretation of mirror self-recognition research in nonhuman primates. *Animal Behaviour* 58:459–468.

De Vere, R. A., Warren, Y., Nicholas, A., Mackenzie, M. E., and Hiham, J. P. 2011. Nest site ecology of the Cross River gorilla at the Kagwene Gorilla Sanctuary, Cameroon, with special reference to anthropogenic influence. *American Journal of Primatology* 73:253–261.

de Villiers, H. 1973. Human skeletal remains from Border Cave, Ingwavuma District, Kwa-Zulu, South Africa. *Annals of the Transvaal Museum* 28:229–256.

———. 1976. A second adult human mandible from Border Cave, Ingwavuma District, KwaZulu, South Africa. *South African Journal of Science* 72:212–215.

De Vleeschouwer, K., Van Elsacker, L., and Verheyen, R. F. 1995. Effect of hand posture on hand preferences during experimental food reaching in bonobos *(Pan paniscus). Journal of Comparative Psychology* 109:203–207.

Devos, C., Sanz, C., Morgan, D. Onononga, J.-R., Laporte, N., and Huynin, M.-C. 2008. Comparing ape densities in habitats in Northern Congo: surveys of sympatric gorillas and chimpanzees in the Odzala and Ndoki regions, *American Journal of Primatology* 70:439–451.

de Vos, J. 1985. Faunal stratigraphy and correlation of the Indonesian hominid sites. In E. Delson, ed., *Ancestors: The Hard Evidence,* 215–220. New York: Alan R. Liss.

de Vos, J., and Sondaar, P. Y. 1994. Dating fossil sites in Indonesia *Science* 266:1726–1727.

de Waal, F. B. M. 1978. Exploitative and familiarity-dependent support strategies in a colony of semi-free living chimpanzees. *Behaviour* 66:268–312.

———. 1982. *Chimpanzee Politics: Power and Sex among Apes.* New York: Harper & Row.

———. 1984. Sex differences in the formation of coalitions among chimpanzees. *Ethology and Sociobiology* 5:239–255.

———. 1987. Tension regulation and nonreproductive functions of sex in captive bonobos *(Pan paniscus). National Geographic Research* 3:318–335.

———. 1988. The communicative repertoire of captive bonobos *(Pan paniscus),* compared to that of chimpanzees. *Behaviour* 106:183–251.

———. 1989a. Behavioral contrasts between bonobo and chimpanzee. In P. G. Heltne, L. A. Marquardt, eds., *Understanding Chimpanzees,* 154–175. Cambridge, MA: Harvard University Press.

———. 1989b. *Peacemaking among Primates.* Cambridge, MA: Harvard University Press.

———. 1990. Sociosexual behavior used for tension regulation in all age and sex combinations among bonobos. In F. R. Feierman, ed., *Pedophilia: Biological Dimensions,* 378–393. New York: Springer.

———. 1993. Reconciliation among Primates: a review of empirical evidence and unresolved issues. In W. A. Mason and S. P. Mendoza, eds., *Primate Social Conflict,* 111–144. New York: State University of New York Press.

———. 1995. Bonobo sex and society. *American Scientist* 272:82–88.

———. 1996a. Conflict as negotiation. In W. C. McGrew, L. F. Marchant, and T. Nishida, eds., *Great Ape Societies,* 59–172. Cambridge: Cambridge University Press.

———. 1996b. *Good Natured. The Origin of Right and Wrong in Humans and Other Animals.* Cambridge MA: Harvard University Press.

———. 1997. The chimpanzee's service economy: food for grooming. *Evolution and Human Behavior* 18:375–386.

———. 1999. Cultural primatology comes of age. *Nature* 399:635–636.

———. 2000. Primates—a natural heritage of conflict resolution. *Science* 289:586–590.

———. 2005. *Our Inner Ape.* New York: Riverhead Books.

———. 2007. *Chimpanzee Politics, Power and Sex among Apes.* Baltimore: Johns Hopkins University Press.

———. 2009. *The Age of Empathy: Nature's Lessons for a Kinder Society.* New York: Harmony Books.

———. 2013a. Animal conformists. *Science* 340:437–438.

———. 2013b. *The Bonobo and the Athiest: The Search for Humanism among the Primates.* New York: W. W. Norton.

de Waal, F. B. M., and Aureli, F. 1996. Consolation, reconciliation, and a possible cognitive difference between macaques and chimpanzees. In A. E. Russon, K. A. Bard, and S. T. Parker, eds., *Reaching into Thought: The Minds of the Great Apes,* 80–110. Cambridge, MA: Cambridge University Press.

de Waal, F. B. M., and Bonnie, K. S. 2009. In tune with others: the social side primate culture. In K. N. Laland and B. G. Galef, eds., *The Question of Animal Culture,* 19–40. Cambridge, MA: Harvard University Press.

de Waal, F. B. M., Dindo, M., Freeman, C. A., and Hall, M. J. 2005. The monkey in the mirror: hardly a stranger. *Proceedings of the National Academy of Sciences of the United States of America* 102:11140–11147.

de Waal, F. B. M., and Lanting, F. 1997. *Bonobo: The Forgotten Ape.* Berkeley: University of California Press.

de Waal, F. B. M., and Seres, M. 1997. Propagation of handclasp grooming among captive chimpanzees. *American Journal of Primatology* 43:339–346.

de Waal, F. B. M., and Tyack, P. L. 2003. *Animal Social Complexity.* Cambridge, MA: Harvard University Press.

de Waal, F. B. M., and Van Roosmalen, A. 1979. Reconciliation and consolation among chimpanzees. *Behavioral Ecology and Sociobiology* 5:55–66.

de Waal, F. B. M., and van Hooff, J. A. R. A M. 1981. Side-directed communication and agonistic interactions in chimpanzees. *Behaviour* 77:164–198.

Dewsbury, D. A. 1982. Ejaculate cost and male choice. *American Naturalist* 119:601–610.

———. 2006. *Monkey Farm.* Lewisburg, PA: Bucknell University Press.

Dewson, J. H., III. 1978. Some behavioural effects of removal of superior temporal cortex in the monkey. In D. J. Chivers and J. Herbert, eds., *Recent Advances in Primatology,* Vol. 1: *Behaviour,* 763–768. London: Academic Press.

Diamond, A. J. 1959. *The History and Origin of Language.* London: Methuen.

Diamond, J. M. 1984. DNA map of the human lineage. *Nature* 310:544.

———. 1988a. DNA-based phylogenies of the three chimpanzees. *Nature* 332:685–686.

———. 1988b. Relationships of humans to chimps and gorillas. *Nature* 334:656.

———. 1992. *The Third Chimpanzee: The Evolution and Future of the Human Animal.* New York: HarperCollins.

Dickemann, M. The ecology of mating systems in hypergynous dowry societies. *Social Science Information* 18:163–195.

Dickerson, R. E. 1971. The structure of cytochrome *c* and the rates of molecular evolution. *Journal of Molecular Evolution* 1:26–45.

Dietrich, C., Swingley, D., and Werker, J. F. 2007. Native language governs interpretation of salient speech differences at 18 months. *Proceedings of the National Academy of Sciences of the United States of America* 104:16027–16031.

Dijker, A. J. M. 2011. Physical constraints on the evolution of cooperation. *Evolutionary Biology* 38:124–143.

Dindo, M., Thierrry, B., and Whiten, A. 2008. Social diffusion of novel foraging methods in brown capuchin monkeys *(Cebus apella). Proceedings of the Royal Society B* 275:187–193.

Dirks, H. G. M., Kibii, J. M., Kuhn, B. F., Steininger, C., Churchill, S. E., Kramer, J. D., Pickering, R., Farber, D. L., Mériaux, A.-S., Herries, A. I. R., et al., 2010. Geological setting and age of *Australopithecus sediba* from Southern Africa. *Science* 328:205–208.

Dissanayake, E. 2000. Antecedents of the temporal arts in early mother-infant interaction. In C. Knight, M. Studdert-Kennedy, and J. R. Hurford, eds., *The Evolutionary Emergence of Language: Social Function and the Origins of Linguistic Form,* 389–410. Cambridge: Cambridge University Press.

Dixson, A. F. 1998. *Primate Sexuality: Comparative Studies of the Prosimians, Monkeys, and Apes, and Human Beings.* Oxford: Oxford University Press.

———. 2009. *Sexual Selection and the Origins of Human Mating Systems.* Oxford: Oxford University Press.

Djian, P., and Green, H. 1989. Vectorial expansion of the involucrin gene and the relatedness of the hominoids. *Proceedings of the National Academy of Sciences of the United States of America* 86:8447–8451.

Djojosudharmo, S., and van Schaik, C. P. 1992. Why are orangu utans so rare in the highlands? *Tropical Biodiverity* 1:11–22.

Do Amaral, L. Q. 1989. Early hominid physical evolution. *Human Evolution* 4:33–44.

———. 1996. Loss of body hair, bipedality and thermoregulation: comments on recent papers in the *Journal of Human Evolution. Journal of Human Evolution* 30:357–366.

Dobosi, V. T. 1990. Fireplaces of the settlement. In M. Kretzoi and V. T. Dobosi, ed., *Veeretesszölös. Site Man and Culture,* 519–521. Budapest: Akadémiai Kiadó.

Dobzhansky, T., Ayala, F. J., Stebbins, G. L., and Valentine, J. W. 1977. *Evolution.* San Francisco: W. H. Freeman.

Dolhinow, P. 1999. Review of *Demonic Males. American Anthropologist* 101:445–446.

Domínguez-Rodrigo, M., and Pickering, T. R. 2003. Early hominid hunting and scavanging: a zoolarcheological review. *Evolutionary Anthropology* 12:275–282.

Domínguez-Rodrigo, M., and Pickering, T. R., and Bunn, H. T. 2010. Configurational approach to identifying the earliest hominin butchers. *Proceedings of the National Academy of Sciences of the United States of America* 107:20929–20934.

———. 2011. Reply to McPherron et al.: doubting Dikika is about data, not paradigm. *Proceedings of the National Academy of Sciences of the United States of America* 108:E117.

Dominici, N., Ivanenko, Y. P., Cappellini, G., d'Avella, A., Mondi, V., Cicchese, M., Fabiano, A., et al. 2011. Locomotor primitives in newborn babies and their development. *Science* 334:997–999.

Dominy, N. J. 2004. Fruits, fingers, and fermentation: the sensory cues available to foraging primates. *Integrative and Comparative Biology* 44:295–303.

———. 2012. Hominins living on the sedge. *Proceedings of the National Academy of Sciences of the United States of America* 109:20171–20172.

Dominy, N. J., and Lucas, P. W. 2001. Ecological importance of trichromatic vision to primates. *Nature* 410:363–366.

———. 2004. Significance of color, calories, and climate to visual ecology of catarrhines. *American Journal of Primatology* 62:189–207.

Dominy, N. J., Lucas, P. W., and Noor, N. S. 2006. Primate sensory systems and foraging behavior. In G. Hohmann, M. M. Robbins and C. Boesch, eds., *Feeding Ecology in Apes and Other Primates,* 489–509. Cambridge: Cambridge University Press.

Dominy, N. J., Lucas, P. W., Osorio, D., and Yamashita, N. 2001. The sensory ecology of primate food selection. *Evolutionary Anthropology* 10:171–186.

Dominy, N. J., Ross, C. M., and Smith, T. D. 2004. Evolution of the special senses in primates: past, present, and future. In *Evolution of the Special Senses in Primates,* ed. T. D. Smith, C. F. Ross, N. J. Dominy, and J. T. Laitman, special issue, *Anatomical Record Part A* 281A:1078–1082.

Dominy, N. J., Svenning, J.-C., and Li, W.-H. 2003. Historical contingency in the evolution of primate color vision. *Journal of Human Evolution* 44:25–45.

Dominy, N. J., Vogel, E. R., Yeakel, J. D., Constantino, P., and Lucas, P. W. 2008. Mechanical properties of plant underground storage organs and implications for dietary models of early hominins. *Evolutionary Biology* 35:159–175.

Donald, M. 1993. Précis of *Origins of the modern mind: three stages in the evolution of culture and cognition. Behavioral and Brain Sciences* 16:737–791.

———. 1999. Preconditions for the evolution of protolanguages. In M. C. Corballis and S. E. G. Lea, eds., *The Descent of Mind*, 138–154. Oxford: Oxford University Press.

Donisthorpe, J. H. 1958. A pilot study of the mountain gorilla *(Gorilla gorilla beringei)* in South West Uganda, February to September, 1957. *South African Journal of Science*, 54:195–217.

Dooley, G. B., and Gill, T. 1977a. Mathematical capabilities of Lana chimpanzee. In G. H. Bourne, ed., *Progress in Ape Research*, 133–142. New York: Academic Press.

———. 1977b. Acquisition and use of mathematical skills by a linguistic chimpanzee. In D. M. Rumbaugh, ed., *The Lana Project*, 247–260. New York: Academic Press.

Doran, D. M. 1992. The ontogeny of chimpanzee and pygmy chimpanzee locomotor behavior: a case study of paedomorphism and its behavioral correlates. *Journal of Human Evolution* 23:139–157.

———. 1993. Comparative locomotor behavior of chimpanzees and bonobos: the influence of morphology on locomotion. *American Journal of Physical Anthropology* 91:83–98.

———. 1997a. Ontogeny of locomotion in mountain gorillas and chimpanzees. *Journal of Human Evolution* 32:323–344.

———. 1997b. Influence of seasonality on activity patterns, feeding behavior, ranging, and grouping patterns in Taï chimpanzees. *Inernational Journal of Primatology* 18:183–206.

Doran, D. M., and McNeilage, A. 1998. Gorilla ecology and behavior. *Evolutionary Anthropology* 6:120–131.

Doran, D. M., McNeilage, A. Greer, D., Bocian, C., Mehlman, P., and Shah, N. 2002. Western lowland gorilla diet and resource availability: new evidence, cross-site comparisons, and reflections on indirect sampling methods. *American Journal of Primatology* 58:91–116.

Doran-Sheehy, D. M., Ferández, D., and Borries, C. 2009. The strategic use of sex in wild female western gorillas. *American Journal of Primatology* 71:1011–1020.

Doran-Sheehy, D. M., Greer, D., Mongo, P., and Schwindt, D. 2004. Impact of ecological and social factors on ranging in western gorillas. *American Journal of Primatology* 64:207–222.

Douadi, M. I., Gatti, S., Levéro, Duhamel, F., Bermejo, M. A., Vallet, D., Ménard, N., and Petit, E. J. 2007. Sex-biased dispersal in western lowland gorillas *(Gorilla gorilla gorilla). Molecular Ecology* 16:2247–2259.

Dover, G. A. 1987. DNA turnover and the molecular clock. *Journal of Molecular Evolution* 26:47–58.

Drake, R. E., and Curtis, G. H. 1987. K-Ar geochronology of the Laetoli fossil localities. In M. D. Leakey and J. M. Harris, eds., *Laetoli, a Pliocene Site in Northern Tanzania*, 48–52. Oxford: Clarendon Press.

Drake, R. E., Van Couvering, J. A., Pickford, M. H., Curtis, G. H., and Harris, J. A. 1988. New chronology for the Early Miocene mammalian faunas of Kisingiri, western Kenya. *Journal of the Geological Society, London* 145:479–491.

Drapeau, M. S. M., and Ward, C. V. 2007. Forelimb segment length proportions in extant hominoids and *Australopithecus afarensis*. *American Journal of Physical Anthropology* 132:327–343.

DuBois, E. 1894. *Pithecanthropus erectus, eine menschenaehnliche Uebergangsform aus Java.* Batavia: Landesdruckerei.

Du Brul, E. L. 1958. *Evolution of the Speech Apparatus.* Springfield, IL: Charles C. Thomas.

———. 1962. The general phenomenon of bipedalism. *American Zoologist* 2:205–208.

———. 1976. Biomechanics of speech sounds. *Annals of the New York Academy of Sciences* 280:631–642.

———. 1977. Early hominid feeding mechanisms. *American Journal of Physical Anthropology* 47:305–320.

Du Chaillu, P. B. 1861. *Explorations and Adventures in Equatorial Africa.* New York: Harper & Brothers.

Duchin, L. E. 1990. The evolution of articulate speech: comparative anatomy of the oral cavity in *Pan* and *Homo*. *Journal of Human Evolution* 19:687–697.

Ducrocq, S. 1998. Eocene primates from Thailand: are Asian anthropoideans related to African ones? *Evolutionary Anthropology* 7:97–104.

———. 1999. *Siamopithecus eocaenus*, a late Eocene anthropoid primate from Thailand: its contribution to the evolution of anthropoids in Southeast Asia. *Journal of Human Evolution* 36:613–636.

Ducros, J., and Ducros, A. 1992. Le singe carnivore: la chasse chez les primates non humains. *Bulletins et Mémoires de la Société d'Anthropologie de Paris*, n.s., 4:243–264.

Dugoujon, J.-M., and Hazout, S. 1987. Evolution of immunoglobin allotypes and phylogeny of apes. *Folia Primatologica* 49:187–199.

Dugoujon, J.-M., and Hazout, S., Loirat, F., Mourrieras, B., Crouau-Roy, B., and Sanchez-Mazas, A. 2004. GM haplotype diversity of 82 populations over the world suggests centrifugal model of human migrations. *American Journal of Physical Anthropology* 125:175–192.

Dukelow, W. R. 1995. *The Alpha Males.* Lanham, MD: University Press of America.

Dunbar, R. I. M. 1993. Coevolution of neocortical size, group size and language in humans. *Behavioral and Brain Sciences* 16:681–735.

———. 1996. *Grooming, Gossip, and the Evolution of Language.* Cambridge, MA: Harvard University Press.

———. 1998. The social brain. *Evolutionary Anthropology* 6:178–190.

———. 2003a. Why are apes so smart? In P. M. Kappeler and M. E. Pereira, eds., *Primate Life Histories and Socioecology*, 285–298. Chicago: University of Chicago Press.

———. 2003b. The origin and subsequent evolution of language. In M. H. Christiansen and S. Kirby, eds., *Language Evolution*, 219–234. Oxford: Oxford University Press.

———. 2009. Why only humans have language. In T. Botha and C. Knight, eds., *The Prehistory of Language*, 12–35. Oxford: Oxford University Press.

Dunbar, R. I. M., and Shultz, S. 2007. Evolution in the social brain. *Science* 317:1344–1347.

Dunsworth, H. M., Warrener, A. G., Deacon, T., Ellison, P. T., and Pontzer, H. 2012. Metabolic hypothesis for human altriciality. *Proceedings of the National Academy of Sciences of the United States of America* 109:15212–15216.

Dupain, J., Guislain, P., Nguenang, G. M., De Vleeschouwer, K., and van Elsacker, L. 2004. High chimpanzee and gorilla densities in a non-protected area on the northern periphery of the Dja Faunal Reserve, Cameroon. *Oryx* 38:209–216.

Dupain, J., and van Elsacker, L. 2001a. The status of the bonobo *(Pan paniscus)* in the Democratic Republic of Congo. In B. M. F. Galdikas, N. E. Briggs, L. K. Sheeran, G. L. Shapiro, and J. Goodall, eds., *All Apes Great and Small,* Vol. 1: *African Apes,* 57–74. New York: Kluwer Academic/Plenum Press.

———. 2001b. Status of the proposed Lomako Forest Bonobo Reserve: a case study of the bushmeat trade. In B. M. F. Galdikas, N. E. Briggs, L. K. Sheeran, G. L. Shapiro, and J. Goodall, eds., *All Apes Great and Small,* Vol. 1: *African Apes,* 259–273. New York: Kluwer Academic/Plenum Press.

Dupain, J. van Krunkelsven, E., van Elsacker, L., and Verheyen, R. F. 2000. Current status of the bonobo *(Pan paniscus)* in the proposed Lomako Reserve (Democratic Republic of Congo). *Biological Conservation* 94:265–272.

Durkheim, E. 1938. *The Rules of Sociological Method.* Chicago: University of Chicago Press.

———. 1965. *The Elementary Forms of the Religious Life.* New York: Free Press.

Ebersberger, I., Metzler, D., Schwarz, C., and Pääbo, S. 2002. Genomewide comparison of DNA sequences between humans and chimpanzees. *American Journal of Human Genetics* 70:1490–1497.

Eckhardt, R. B. 1973. *Gigantopithecus* as a hominid ancestor. *Anthropologischer Anzeiger* 34:1–8.

———. 1974. The dating of *Gigantopithecus:* a critical reappraisal. *Anthropologischer Anzeiger* 34:129–139.

———. 1975. *Gigantopithecus* as a hominid. In R. H. Tuttle, ed., *Paleoanthropology, Morphology and Paleoecology,* 103–129. The Hague: Mouton.

———. 2007. Paleoanthropology, a science in need of a theoretical framework. In E. Indriati, ed., *Recent Advances in Southeast Asian Paleoanthropology and Archaeology,* 107–116. Yogyakarta, Indonesia: Gadjah Mada University.

Edgar, H. J. H., and Hunley, K. L. 2009. Race reconciled?: how biological anthropologists view human variation. *American Journal of Physical Anthropology* 139:1–4.

Edwards, S. D. 1982. Social potential expressed in captive, group-living orang utans. In L. E. M. De Boer, ed., *The Orang Utan: Its Biology and Conservation,* 249–255. The Hague: Dr W. Junk.

Edwards, S. D., and Snowdon, C. T. 1980. Social behavior of captive, group living orangeutans. *International Journal of Primatology* 1:39–62.

Eggan, F. 1975. *Essays in Social Anthropology and Ethnology.* Chicago: Department of Anthropology, University of Chicago.

Egi, N., Takai, M., Shigehara, N., and Tsubamoto, T. 2004. Body mass estimates for Eocene eosimiid and amphipithecid primates using prosimian and anthropoid scaling models. *International Journal of Primatology* 25:211–236.

Egi, N., Tun, S. T., Takai, M., Shigehara, N., and Tsubamoto, T. 2004. Geographical and body size distributions of the Pondaung primates with a comment on the taxonomic assignment of NMMP 20, postcranium of an amphipithecid. *Anthropological Science* 112:67–74.

Eichler, E. E., Johnson, M. E., Alkan, C., Tuzun, E., Sahinalp, C., Misceo, D., Archidi-acono, N. M., and Rocchi, M. 2001. Divergent origins and concerted expansion of two segmental duplications on chromosome 16. *Journal of Heredity* 92:462–468.

Eickhoff, R. 1988. Origin of bipedalism—when, why, how and where? *South African Journal of Science* 84:486–488.

Eisenberg, J. F. 1981. *The Mammalian Radiations: An Analysis of Trends in Evolution, Adaptation, and Behavior.* Chicago: University of Chicago Press.

Eisenberg, J. F., Muckenhirn, N. A., and Rudran, R. 1972. The relation between ecology and social structure in primates. *Science* 176:863–874.

Elder, A. A. 2009. Hylobatid diets revisited: the importance of body mass, fruit availability, and interspecific competition. In S. Lappan and D. J. Whittaker, eds., *The Gibbons: New Perspectives on Small Ape Socioecology and Population Biology,* 133–159. New York: Springer.

Eldredge, N., and Gould, S. J. 1972. Punctuated equilibria: an alternative to phyletic graduation. In T. J. M. Schopf, ed., *Models in Paleobiology,* 82–115. San Francisco: Freeman, Cooper.

Eldredge, N., and Tattersall, I. 1982. *The Myths of Human Evolution.* New York: Columbia University Press.

Ellefson, J. O. 1968. Territorial behavior in the common white-handed gibbon, *Hylobates lar* Linn. In P. Jay, ed., *Primates: Studies in Adaptation and Variability,* 180–199. New York: Holt, Rinehart and Winston.

———. 1974. A natural history of white-handed gibbons in the Malayan Peninsula. In ed. D. M. Rumbaugh, *Gibbon and Siamang,* 3:1–136. Basel: Karger.

Elliot, D. G. 1913. *A Review of the Primates,* Vol. 3: *Anthropoidea.* New York: American Museum of Natural History.

Elliot, D., and Roy, E. A. 1996. *Manual Asymmetries in Motor Performance.* Boca Raton, FL: CRC Press.

Elliott, R. C. 1976. Observations on a small group of mountain gorillas *(Gorilla gorilla beringei). Folia Primatologica,* 25:12–24.

———. 1977. Cross-modal recognition in three primates. *Neuropsychologia* 15:183–186.

Ellis, D. V. 1993. Wetlands or aquatic ape? Availability of food resources. *Nutrition and Health* 9:205–217.

Ellis, R. A., and Montagna, W. 1962. The skin of primates: VI: The skin of the gorilla *(Gorilla gorilla). American Journal of Physical Anthropology* 20:79–93.

Ely, J., Deka, R., Chakraborty, R., and Ferrell, R. E. 1992. Comparison of five tandem repeat loci between humans and chimpanzees. *Genomics* 14:692–698.

Ely, J. J., Frels, W. I., Howell, S., Izard, M. K., Keeling, M. E., and Lee, D. R. 2006. *American Journal of Physical Anthropology* 130:96–102.

Ely, J. J., Leland, M., Martino, M., Swett, W., and Moore, C. M. 1998. Technical Note: Chromosomal and mtDNA analysis of Oliver. *America Journal of Physical Anthropology* 105:395–403.

El Zaatari, S., Grine, F. E., Ungar, P. S., and Hublin, J.-J. 2011. Ecogeographic variation in Neandertal dietary habits: evidence from occlusal molar microwear texture analysis. *Journal of Human Evolution* 61:411–424.

Emery M. A., and Whitten, P. L. 2003. Size of sexual swellings reflects ovarian function in chimpanzees *(Pan troglodytes)*. *Behavioral Ecology and Sociobiology* 54:340–351.

Emery Thompson, M. 2005. Reproductive endocrinology of wild female chimpanzees *(Pan troglodytes schweinfurthii)*: methodological considerations in the role of hormones in sex and conception. *American Journal of Primatology* 67:137–158.

Emery Thompson, M., Kahlenberg, S. M., Gilby, I. C., and Wrangham, R. W. 2007. Core area quality is associated with variance in reproductive success among female chimpanzees at Kibale National Park. *Animal Behaviour* 73:501–512.

Emery Thompson, M., Muller, M. N., Kahlenberg, S. M., and Wrangham, R. W. 2010. Dynamics of social and energetic stress in wild female chimpanzees. *Hormones and Behavior* 58:440–449.

Emery Thompson, M., Wilson, M. L., Gobbo, G., Muller, M. N., and Pusey, A. E. 2008. Hyperpogestteronemia in response to *Vitex fischeri* consumption in wild chimpanzees *(Pan troglodytes schweinfurthii)*. *American Journal of Primatology* 70:1064–1071.

Emery Thompson, M., and Wrangham, R. W. 2008a. Diet and reproductive function in wild female chimpanzees *(Pan troglodytes schweinfurthii)* at Kibale National Park. *American Journal of Physical Anthropology* 135:171–181.

———. 2008b. Male mating interest varies with female fecundity in *Pan troglodytes schweinfurthii* of Kanyawara, Kibale National Park. *International Journal of Primatology* 29:885–905.

Emlen, J. T. 1962. The display of the gorilla. *Proceedings of the American Philosophical Society* 106:516–519.

Emlen, J. T., Jr., and Schaller, G. B. 1960a. Distribution and status of the mountain gorilla *(Gorilla gorilla beringei)*-1959. *Zoologica* 45:41–52.

———. 1960b. In the home of the mountain gorilla. *Animal Kingdom* 63:98–108.

Emlen, S. T., and Oring, L. W. 1977. Ecology, sexual selection, and the evolution of mating systems. *Science* 197:215–223.

Emmorey, K. 1999. Do signs gesture? In L. S. Messing and R. Campbell, eds., *Gesture, Speech, and Sign,* 133–159. Oxford: Oxford University Press.

———. 2002. *Language, Cognition, and the Brain: Insights from Sign Language Research.* Mahwah, NJ: Lawrence Erlbaum.

———. 2005. Sign languages are problematic for a gestural origins theory of language evolution. *Behavioral and Brain Sciences* 28:130–131.

Enard, W., Khaitovich, P., Klose, J. Zöllner, S., Heissig, F., Giavalisco, P., Nieselt-Struwe, K., Muchmore, E., Varki, A., Ravid, R., et al. 2002. Intra- and interspecific variation in primate gene expression patterns. *Science* 296:340–343.

Enard, W., Przewoski, M., Fisher, S. E., Lai, C. S. L., Wiebe, V., Kitano, T., Monaco, A. P., and Pääbo, S. 2002. Molecular evolution of *FOXP2,* a gene involved in speech and language. *Nature* 418:869–872.

Enfield, J. J. 2010. Without social context? *Science* 329:1600–1601.

Epstein, R., Lanza, R. P., and Skinner, B. F. 1980. Symbolic communication between two pigeons *(Columbia livia domestica)*. *Science* 207:543–545.

———. 1981. "Self-awareness" in the pigeon. *Science* 212:695–696.

Erikson, G. E. 1963. Brachiation in New World monkeys and in anthropoid apes. In J. Napier and N. A. Barnicot, eds., *Symposia of the Zoological Society of London*, no. 10, *The Primates*, 135–164.

Eriksson, A., and Manica, A. 2012. Effect of ancient population structure on the degree of polymorphism shared between modern human populations and ancient hominins. *Proceedings of the National Academy of Sciences of the United States of America* 109:13956–13960.

Ersoy, A., Kelley, J., Andrews, P., and Alpagut, B. 2008. Hominoid phalanges from the middle Miocene of Pasalar, Turkey. *Journal of Human Evolution* 54:518–529.

Essock, S. M., Gill, T. V., and Rumbaugh, D. M. 1977. Object- and color-naming skills of Lana chimpanzee. In G. H. Bourne, ed., *Progress in Ape Research*, 143–148. New York: Academic Press.

Essock, S. M., and Rumbaugh, D. M. 1978. Development and measurement of cognitive capabilities in captive nonhuman primates. In H. Markowitz and V. M. Stevern, eds., *Behaviour of Captive and Wild Animals*, 161–208. Chicago: Nelson-Hall.

Essock-Vitale, S. M. 1978. Comparison of ape and monkey modes of problem solving solution. *Journal of Comparative Physiological Psychology* 92:942–957.

Estep, D. Q., and Bruce, K. E. M. 1981. The concept of rape in non-humans: a critique. *Animal Behaviour* 29:1272–1273.

Estioko-Griffin, A. A., and Griffin, P. B. 1981. Woman the hunter: the Agta. In F. Dahlberg, ed., *Woman the Gatherer*, 121–151. New Haven, CT: Yale University Press.

———. 1985. Women hunters: the implications for Pleistocene prehistory and contemporary ethnography. In M. J. Goodman, ed., *Women in Asia and the Pacific: Towards an East-West Dialogue*, 61–81. Honolulu: University of Hawaii Press.

Etkin, W. 1954. Social behavior and the evolution of man's mental faculties. *American Naturalist* 88:129–142.

Etler, D. A. 1984. The fossil hominoids of Lufeng, Yunnan Province, The People's Republic of China: a Series of Translations. *Yearbook of Physical Anthropology* 27:1–55.

Ettinger, G., and Garcha, H. S. 1980. Cross-modal recognition by the monkey effects of cortical removals. *Neuropsychologia* 18:685–692.

Evans, E. M. N., van Couvering, J. A. H., and Andrews, P. 1981. Paleoecology of Miocene sites in western Kenya. *Journal of Human Evolution* 10:99–121.

Evans, P. D., Gilbert, S. L., Mekel-Bobrov, N., Vallender, E. J., Anderson, J. R., Vaez-Azizi, L. M., Tishkoff, S. A., Hudson, R. R., and Lahn, B. T. 2005. *Microcephalin*, a gene regulating brain size, continues to evolve adaptively in humans. *Science* 309:1717–1720.

Evans, T. A., and Beran, M. J. 2007. Chimpanzees use self-distraction to cope with impulsivity. *Biology Letters* 3:599–602.

Eyre-Walker, A. 2006. Size does not matter for mitochondrial DNA. *Science* 312:537–538.

Fabre, P.-H., Rodrigues, A., and Douzery, E. J. P. 2009. Patterns of macroevolution among Primates inferred from a supermatrix of mitochondrial and nuclear DNA. *Molecular Phylogenetics and Evolution* 53:808–825.

Fadiga, L, and Craighero, L. 2006. Hand actions and speech representation in Broca's area. *Cortex* 42:486–490.

Fadiga, L, Craighero, L., Buccino, G., and Rizzollati, G. 2002. Speech listening specifically modulates the excitability of tongue muscles: a TMS study. *European Journal of Neuroscience* 15:399–402.

Fadiga, L, Fogassi, L., Pavsi, G., and Rizzolatti, G. 1995. Motor facilitation during action observation: a magnetic stimulation study. *Journal of Neurophysiology* 73:2608–2611.

Fagot, J., and Vauclair, J. 1988. Handedness and bimanual coordination in the lowland gorilla. *Brain, Behavior and Evolution* 32:89–95.

———. 1991. Manual laterality in nonhuman primates: a distinction between handedness and manual specialization. *Psychological Bulletin* 109:76–89.

———. 1993. La latéralisation chez les singes. *La Recherche* 25:298–304.

Fagundes, N. J. R., Ray, N., Beaumont, M., Neuenschwander, S., alzano, F. M., Bonatto, S. L., and Excoffier, L. 2007. Statistical evaluation of alternative models of human evolution. *Proceedings of the National Academy of Sciences of the United States of America* 104:17614–17619.

Fahy, G. E., Richards, M., Riedel, J., Hublin, J.-J., and Boesch, C. 2013. Stable isotope evidence of meat eating and hunting specialization in adult male chimpanzees. *Proceedings of the National Academy of Sciences of the United States of America* 110:5829–5833.

Falk, D. 1975. Comparative anatomy of the larynx in man and the chimpanzee: implications for language in Neanderthal. *American Journal of Physical Anthropology* 43:123–132.

———. 1985. Hadar AL 162–28 endocast as evidence that brain enlargement preceded cortical reorganization in hominid evolution. *Nature* 313:45–47.

———. 1986. Evolution of cranial blood drainage in hominids: enlarged occipital/marginal sinuses and emissary foramina. *American Journal of Physical Anthropology* 70:311–324.

———. 1988. Enlarged occipital/marginal sinuses and emissary formaina: their significance in hominid evolution. In F. E. Grine, ed., *Evolutionary History of the "Robust" Australopithecines*, 85–96. Hawthorne, NY: Aldine de Gruyter.

———. 1990. Brain evolution of *Homo*: the radiator theory. *Behavioral and Brain Sciences* 13:333–381.

———. 1992. *Braindance*. New York: Henry Holt.

———. 2000. Hominid brain evolution and the origins of music. In N. L. Wallin, B. Merker and S. Brown, eds., *The Origins of Music*, 197–216. Cambridge, MA: MIT Press.

———. 2001. The evolution of sex differences in primate brains. In D. Falk and K. R. Gibson, eds., *Evolutionary Anatomy of the Primate Cerebral Cortex*, 98–112. Cambridge: University of Cambridge Press.

———. 2004. Prelinguistic evolution in early hominins: whence motherese? *Behavioral and Brain Sciences* 27:491–541.

Falk, D., and Clarke, R. 2007. Brief communication: new reconstruction of the Taung endocast. *American Journal of Physical Anthropology* 134:529–534.

Falk, D., and Conroy, G. C. 1983. The cranial venous sinus system in *Australopithecus afarensis*. *Nature* 306:779–781.

Falk, D., Froese, N., Sade, D. S., and Dudek, B. C. 1999. Sex differences in brain/body relationships of rhesus monkeys and humans. *Journal of Human Evolution* 36:233–238.

Falk, D., Hildebolt, C., Cheverud, J., Vannier, M., Helmkamp, R. C., and Konigsberg, L. 1990. Cortical asymmetries in frontal lobes of rhesus monkeys *(Macaca multta)*. *Brain Research* 512:40–45.

Falk, D., Hildebolt, C., Smith, K., Jungers, W., Larson, S., Morwood, M. J., Sutikna, T., Jatmiko, Saptomo, E. W., and Prior, F. 2009. The type specimen (LB1) of *Homo foresiensis* did not have Laron syndrome. *American Journal of Physical Anthropology* 140:52–63.

Falk, D., Hildebolt, C., Smith, K., Morwood, M. J., Sutikna, T., Brown, P., Jatmiko, Saptomo, E. W., Brunsden, B., and Prior, F. 2005. The brain of LB1, *Homo floresiensis*. *Science* 308:242–245.

Fan, P. F., and Jiang, X. L. 2008a. Sleeping sites, sleeping trees, and sleep-related behaviors of black crested gibbons *(Nomascus concolor jingdongensis)* at Mt. Wuliang, Central Yunnan, China. *American Journal of Primatology* 70:153–160.

———. 2008b. Effects of food and topography on ranging behavior of black crested gibbon *(Nomascus concolor jingdongensis)* in Wuliang Mountain, Yunnan, China. *American Journal of Primatology* 70:871–878.

———. 2009. Predation on giant flying squirrels *(Petaurista philippensis)* by black crested gibbons *(Nomascus concolor jingdongensis)* at Mt. Wuliang, Yunnan, China. *Primates* 50:45–49.

———. 2010. Maintenance of multifemale social organization in a group of *Nomascus concolor* at Wuliang Mountain, Yunnan, China. *International Journal of Primatology* 31:1–13.

Fan, P. F., Jiang, X. L., Liu, C. M., and Luo, W. S. 2006. Polygynous mating system and behavioural reason of black crested gibbon *(Hylobates concolor jingdongensis)* at Dazhaizi, Mt. Wuliang, Yunnan, China. *Zoological Research* 27:216–220. (Chinese)

Fan, P. F., Ni, Q. Y., Sun, G. Z., Huang, B., and Jiang, X. L. 2008. Seasonal variations in the activity budget of *Nomascus concolor jingdongensis*) at Mt. Wuliang, Central Yunnan, China: effects of diet and temperature. *International Journal of Primatology* 29:1047–1057.

———. 2009. Gibbons under seasonal stress: the diet of the black crested gibbon *(Nomascus concolor jingdongensis)* at Mt. Wuliang, Central Yunnan, China. *Primates* 50:37–44.

Fan, P. F., Xiao, W., Huo, S., and Jiang, X. L. 2009. Singing behavior and singing functions of black-crested gibbons *(Nomascus concolor jingdongensis)* at Mt. Wuliang, Central Yunnan, China. *American Journal of Primatology* 71:539–547.

Fawcett, K., and Muhumuza, G. 2000. Death of a wild chimpanzee community member: possible outcome of intense sexual competition. *American Journal of Primatology* 51:243–247.

Fay, J. M. 1989. Partial completion of a census of the western lowland gorilla *(Gorilla g. gorilla (Savage and Wyman))* in southwestern Central African Republic. *Mammalia* 53:203–215.

Fay, J. M., and Agnagna, M. 1992. Census of gorillas in northern Republic of Congo. *American Journal of Primatology* 27:275–284.

Fay, J. M., Agnagna, M., Moore, J., and Oko, R. 1989. Gorillas *(Gorilla gorilla gorilla)* in the Likouala swamp forests of north central Congo: preliminary data on populations and ecology. *International Journal of Primatology* 10:477–486.

Fay, J. M., and Carroll, R. 1994. Chimpanzee tool use for honey and termite extraction in Central Africa. *American Journal of Primatology* 34:309–317.

Fay, J. M., Carroll, R., Kerbis Peterhans, J. C., and Harris, D. 1995. Leopard attack on and consumption of gorillas in the Central African Republic. *Journal of Human Evolution* 19:93–99.

Fedigan, L. M. 1986. The changing role of women in models of human evolution. *Annual Review of Anthropology* 15:25–66.

Fehr, E., and Fischbacher, U. 2003. The nature of human altruism. *Nature* 425:785–791.

———. 2004. Social norms and human cooperation. *Trends in Cognitive Sciences* 8:185–190.

Fehr, E., and Schmitt, K. M. 1999. A theory of fairness, competition, and cooperation. *Quarterly Journal of Economics* 114:817–868.

Feibel, C. S., and Brown, F. H. 1991. Age of the primate-bearing deposits on Maboko Island, Kenya. *Journal of Human Evolution* 21:221–225.

Feibel, C. S., Brown, F. H., and McDougall, I. 1989. Stratigraphic context of fossil hominids from the Omo group deposits: northern Turkana Basin, Kenya and Ethiopia. *American Journal of Physical Anthropology* 78:595–622.

Feierman, J. R. *Pedophilia: Biosocial Dimensions.* New York: Springer.

Feistner, A. T. C., and McGrew, W. C. 1989. Food-sharing in primates: a critical review. In P. K. Seth and S. Seth, eds., *Perspectives in Primate Biology,* Vol. 3, 21–36. New Delhi: Today & Tomorrow's Printers and Publishers.

Feldesman, M. R. 1982. Morphometric analysis of the distal humerus of some Cenozoic catarrhines: the late divergence hypothesis revisited. *American Journal of Physical Anthropology* 59:73–95.

Feldman, M. W., Lewontin, R. C., and King, M.-C. 2003. Race: a genetic melting-pot. *Nature* 424:374.

Felsenstein, J. 1987. Estimation of hominoid phylogeny from a DNA hybridization data set. *Journal of Molecular Evolution* 26:123–131.

Felton, A. M., Engström, L. M., Felton, A., and Knott, C. D. 2003. Orangutan population density, forest structure and fruit availability in hand-logged and unlogged peat swamp forests in West Kalimantan, Indonesia. *Biological Conservation* 114:91–101.

Ferguson, W. W. 1983. An alternative interpretation of *Australopithecus afarensis* fossil material. *Primates* 24:397–409.

———. 1984. Revision of fossil hominid jaws from the Plio/Pleistocene of Hadar, in Ethiopia, including a new species of the genus *Homo* (Hominoidea: Hominiae). *Primates* 25:519–529.

———. 1986a. Taxonomic status of the hominid mandible KNM-ER TI 13150 from the Middle Pliocene of Tabarin, in Kenya. *Primates* 30:383–387.

———. 1986b. The taxonomic status of *Praeanthropus africanus* (Primates: Pongidae) from the Late Pliocene of eastern Africa. *Primates* 27:485–492.

———. 1992a. "*Australopithecus afarensis*": a composite species. *Primates* 33:273–279.

———. 1992b. Taxonomic status of the partial calvaria A. L. 333–45 from the Late Pliocene of Hadar, Ethiopia. *Palaeontologia Africana* 29:25–37.

Fernandes, M. E. B. 1991. Tool use and predation of oysters *(Crassostrea rhizophorae)* by tufted capuchin, *Cebus apella apella,* in brackish water mangrove swamp. *Primates* 32:529–531.

Fernández-Carriba, S., Loeches, Á., Morcillo, A., and Hopkins, W. D. 2002. Asymmetry in facial expression of emotions by chimpanzees. *Neuropsychologia* 40:1523–1533.

Ferrari, P. F., and Fogssi, L. 2012. The mirror neuron system in monkeys and its implications for social cognitive functions. In F. B. M. de Waal and P. F. Ferrari, eds., *The Primate Mind,* 13–31, Cambridge, MA: Harvard University Press.

Ferreira, R. G., Emikio, R. A., and Jerusalinsky, L. 2010. Three stones for three seeds: natural occurrence of selective tool use by capuchins *(Cebus libidinosus)* based on analysis of the weight of stones found at nutting sites. *American Journal of Primatology* 72:270–275.

Ferris, S. D., Brown, W. M., Davidson, W. S., and Wilson, A. C. 1981. Extensive polymorphism in the mitochondrial DNA of apes. *Proceedings of the National Academy of Sciences of the United States of America* 78:6319–6323.

Ferris, S. D., Wilson, A. C., and Brown, W. M. 1981. Evolutionary tree for apes and humans based on cleavage maps of mitochondrial DNA. *Proceedings of the National Academy of Sciences of the United States of America* 78:2432–2436.

Ferriss, S. 2005. Western gorilla *(Gorilla gorilla).* In J. Caldecott and L. Miles, eds., *World Atlas of Great Apes and Their Conservation,* 105–128. Berkeley: University of California Press.

Ferriss, S., Robbins, M. M., and Wiliamson, E. A. 2005. Eastern gorilla *(Gorilla beringei).* In J. Caldecott and L. Miles, eds., *World Atlas of Great Apes and Their Conservation,* 129–152. Berkeley: University of California Press.

Ferster, C. B., and Hammer, C. E., Jr. 1966. Synthesizing the components of arithmetic behavior. In W. K. Honig, ed., *Operant Behavior: Areas of Research and Application,* 634–676. Englewood Cliffs, NJ: Prentice-Hall.

Fields, C., Adams, M. D., White, O., and Vernier, J. C. 1994. How many genes in the human genome? *Nature Genetics* 7:345–346.

Fifer, F. C. 1987. The adoption of bipedalism by the hominids: a new hypothesis. *Human Evolution* 2:135–147.

Finch, C. 2010. Evolution of the human lifespan and diseases of aging: roles of infection, inflammation, and nutrition. *Proceedings of the National Academy of Sciences of the United States of America* 107 (Suppl. 1): 1718–1724.

Finch, G. 1941. Chimpanzee handedness. *Science* 94:117–118.

Fischer, E. 1906. Die Variationen an Radius und Ulna des Menschen. *Zeitschrift für Morphologie und Anthropologie* 9:147–247.

Fischman, J. 2005. The pathfinders. *National Geographic* 207:16–27.

Fish, J. L., and Lockwood, C. A. 2003. Dietary constraints on encephalization in primates. *American Journal of Physical Anthropology* 120:171–181.

Fitch, W. M. 1976. Molecular evolutionary clocks. In F. J. Ayala, ed., *Molecular Evolution,* 160–178. Sunderland, MA: Sinauer.

———. 1986. Commentary. *Molecular Biology and Evolution* 3:296–298.

Fitch, W. M., and Langley, C. H. 1976. Evolutionary rates in proteins: neutral mutations and the molecular clock. In M. Goodman and R. E. Tashian, eds., *Molecular Anthropology*, 197–219. New York: Plenum Press.

Fitch, W. T. 2000. The evolution of speech: a comparative review. *Trends in Cognition Science* 4:258–267.

———. 2002. Comparative vocal production and the evolution of speech: reinterpreting the descent of the larynx. In A. Wray, ed., *The Transition to Language*, 21–45. Oxford: Oxford University Press.

———. 2004. Kin selection and "mother tongues": a neglected component in language evolution. In D. K. Oller and U. Griebel, eds., *Evolution of Communication Systems: A Comparative Approach*, 275–296. Cambridge, MA: MIT Press.

———. 2005a. The evolution of music in comparative perspective. *Annals of the New York Academy of Science* 1060:1–21.

———. 2005b. The evolution of language: a comparative review. *Biology and Philosophy* 20:193–230.

———. 2005c. Protomusic and protolanguage as alternatives to protosign. *Behavioral and Brain Sciences* 28:132–133.

———. 2006. The biology and evolution of music: a comparative perspective. *Cognition* 100:173–215.

———. 2009a. Evolving meaning: the roles of kin selection, allomothering and paternal care in language evolution. In C. Lyon, C. L. Nehaniv, and A. Cangelosi, eds., *Emergence of Communication and Language*, 29–51. London: Springer.

———. 2009b. Fossil cues to the evolution of speech. In R. Botha and C. Knight, eds., *The Cradle of Language*, 112–134. Oxford: Oxford University Press.

———. 2010. *The Evolution of Language*. Cambridge: Cambridge University Press.

Fitch, W. T., and Hauser, M. D. 1995. Vocal production in nonhuman primates: acoustics, physiology, and functional constraints on "honest" advertisement. *American Journal of Primatology* 37:191–219.

Fitch, W. T., and Hauser, M. D., and Chomsky, N. 2005. The evolution of the language faculty: clarification and implications. *Cognition* 97:179–210.

Fitch, W. T., and Reby, D. 2001. The descended larynx is not uniquely human. *Proceedings of the Royal Society, London B* 268:1669–1675.

Fleagle, J. G. 1976. Locomotion and posture of the Malayan siamang and implications for hominid evolution. *Folia Primatologica* 26:245–269.

———. 1983. Locomotor adaptations of Oligocene and Miocene hominoids and their phyletic implications. In R. L. Ciochon and R. S. Corruccini, eds., *New Interpretations of Ape and Human Ancestry*, 301–324. New York: Plenum Press.

———. 1984. Are there any fossil gibbons? In H. Preuschoft, D. J. Chivers, W. Y. Brockelman and N. Creel, eds., *The Lesser Apes: Evolutionary and Behavioural Biology*, 431–447. Edinburgh: Edinburgh University Press.

———. 1986a. Early anthropoid evolution in Africa and South America. In J. G. Else and P. C. Lee, eds., *Primate Evolution*, 133–142. Cambridge: Cambridge University Press.

————. 1986b. The fossil record of early catarrhine evolution. In B. Wood, L. Martin and P. Andrews, eds., *Major Topics in Primate and Human Evolution,* 130–149. Cambridge: Cambridge University Press.

————. 1988, 1999. *Primate Adaptation and Evolution.* San Diego: Academic Press.

Fleagle, J. G., Bown, T. M., Obradovich, J. D., and Simons, E. L. 1986a. How old are the Fayum primates? In J. G. Else and P. C. Lee, eds., *Primate Evolution,* 3–17. Cambridge: Cambridge University Press.

————. 1986b. Age of the earliest African anthropoids. *Science* 234:1247–249.

Fleagle, J. G., and Kay, R. F. 1983. New interpretations of the phyletic position of Oligocene hominoids. In R. L. Ciochon and R. S. Corruccini, eds., *New Interpretations of Ape and Human Ancestry,* 181–210. New York: Plenum Press.

————. 1987. The phyletic position of the Parapithecidae. *Journal of Human Evolution* 16:483–532.

Fleagle, J. G., Kay, R. F., and Simons, E. L. 1980. Sexual dimorphism in early anthropoids. *Nature* 287:328–330.

Fleagle, J. G., and Simons, E. L. 1978. Humeral morphology of the earliest apes. *Nature* 276:705–707.

————. 1979. Anatomy of the bony pelvis in parapithecid primates. *Folia Primatologica* 31:176–186.

————. 1982a. Skeletal remains of *Propliopithecus chirobates* from the Egyptian Oligocene. *Folia Primatologica* 39:161–177.

————. 1982b. The humerus of *Aegyptopithecus zeuxis:* a primitive anthropoid. *American Journal of Physical Anthropology* 59:175–193.

————. 1995. Limb skeleton and locomotor adaptations of *Apidium phiomense,* an Oligocene anthropoid from Egypt. *American Journal of Physical Anthropology* 97:235–289.

Fleagle, J. G., Simons, E. L., and Conroy, G. C. 1975. Ape limb bone from the Oligocene of Egypt. *Science* 189:135–137.

————. 1983. The tibio-fibular articulation in *Apidium phiomense,* an Oligocene anthropoid. *Nature* 301:238–239.

Fletcher, A. W. 2006. Clapping in chimpanzees: evidence of exclusive hand preference in a spontaneous, bimanual gesture. *American Journal of Primatology* 68:1081–1088.

Flynn, L. J., Barry, J. C., Morgan, M. E., Pilbeam, D., Jacobs, L. L., and Lindsay, E. H. 1995. Neogene Siwalik mammalian lineages: species longevities, rates of change and modes of speciation. *Palaeogeography, Palaeoclimatology, Palaeoecology* 115:249–264.

Flynn, L. J., Pilbeam, D., Jacobs, L. L., Barry, J. C., Behrensmeyer, A. K., and Kappelman, J. W. 1990. The Siwaliks of Pakistan: time and faunas in a Miocene terrestrial setting. *Journal of Geology* 98:589–604.

Flynn, L. J., and Qi, G. 1982. Age of the Lufeng, China, hominoid locality. *Nature* 298:746–747.

Fodor, J. A. 1983. *The Modularity of Mind.* Cambridge, MA: MIT Press.

Fogassi, L., and Ferrari, P. F. 2007. Mirror neurons and the evolution of embodied language. *Current Directions in Psychological Science* 16:136–141.

Fogassi, L., Ferrari, P. F., Gesierich, B., Rozzi, S., Chersi, F., and Rizzolatti, G. 2005. Parietal lobe: from action organization to intention understanding. *Science* 38:662–667.

Foley, J. P., Jr. 1935. Judgement of facial expression of emotion in the chimpanzee. *Journal of Social Psychology* 6:31–67.

Foley, R. A. 1989. The evolution of hominid social behavior. In V. Standen and R. A. Foley, eds., *Comparative Socioecology,* 473–494. Oxford: Blackwell Scientific.

Foley, R. A., and Lee, P. C. 1989. Finite space, evolutionary pathways, and reconstructing hominid behavior. *Science* 243:901–906.

Foley, R., and Mirazón-Lahr, M. 2003. On stony ground: lithic technology, human evolution, and the emergence of culture. *Evolutionary Anthropology* 12:109–122.

Fooden, J. 1969. Taxonomy and evolution of the monkeys of Celebes. *Bibliotheca Primatologica* 10:1–148.

———. 1990. The bear macaque, *Macaca arctoides:* a systematic review. *Journal of Human Evolution* 19:607–686.

Fooden, J., Quan, G., and Luo, Y. 1987. Gibbon distribution in China. *Acta Theriologica Sinica* 7:161–167.

Fortuny, J. 2010. On the duality of patterning. In Jan-Wouter Zwart and Mark de Vries, eds., *Structure Preserved: Studies in Syntax for Jan Koster,* 131–140. Amsterdam/Philadelphia: John Benjamins.

Fossey, D. 1970. Making friends with mountain gorillas. *National Geographic* 137:48–67.

———. 1971. More years with mountain gorillas. *National Geographic* 140:574–585.

———. 1972. Vocalizations of the mountain gorilla *(Gorilla gorilla beriingei)*. *Animal Behaviour* 20:36–53.

———. 1974. Observations on the home range of one group of mountain gorillas *(Gorilla gorilla beringei)*. *Animal Behaviour* 22:568–581.

———. 1979. Development of the mountain gorilla *(Gorilla gorilla beringei):* the first thirty-six months. In D. A. Hamburg and E. R. McCown, eds., *The Great Apes,* 138–184. Menlo Park, CA: Benjamin/Cummings.

———. 1981. The imperiled mountain gorilla. *National Geographic* 159:501–523.

———. 1982. Reproduction among free-living mountain gorillas. *American Journal of Primatology* 1 (Suppl.): 97–104.

———. 1983. *Gorillas in the Mist.* New York: Houghton Mifflin.

———. 1984. Infanticide in mountain gorillas *(Gorilla gorilla beringei)* with comparative notes on chimpanzees. In G. Hausfater and S. B. Hrdy, eds., *Infanticide: Comparative and Evolutionary Perspectives,* 217–235. Hawthorne, NY: Aldine de Gruyter.

Fossey, D., and Harcourt, A. H. 1977. Feeding ecology of free-ranging mountain gorilla *(Gorilla gorilla beringei)*. In T. H. Clutton-Brock, ed., *Primate Ecology: Studies of Feeding and Ranging Behaviour in Lemurs, Monkeys and Apes.* 415–447. London: Academic Press.

Foster, M. L. 1990. Symbolic origins and transitions in the Paleolithic. In P. Mellars, ed., *The Emergence of Modern Humans: An Archaeological Perspective,* 517–539. Edinburgh: Edinburgh University Press.

Fouts, R. S. 1972. Use of guidance in teaching sign language to a chimpanzee *(Pan troglodytes)*. *Journal of Comparative and Physiological Psychology* 80:515–522.

———. 1973. Acquisition and testing of gestural sign in four young chimpanzees. *Science* 180:475–482.

————. 1974. Language: origins, definitions and chimpanzees. *Journal of Human Evolution* 3:475–482.

————. 1975a. Capacities for language in great apes. In R. H. Tuttle, ed., *Socioecology and Psychology of Primates,* 371–390. The Hague: Mouton.

————. 1975b. Communication with chimpanzees. In G. Kurth and I. Eibl-Eibesfeldt, eds., *Hominisation und Verhalten,* 137–158. Stuttgart: Gustav Fischer.

————. 1983. Chimpanzee language and elephant tails: a theoretical synthesis. In J. de Luce and H. T. Wildeer, eds., *Language in Primates: Perspectives and Implications,* 63–75. New York: Springer.

————. 1990. Sign language in chimpanzees. Implication of the visual mode and the comparative approach. In F. C. C. Peng, ed., *Sign Language and Language Acquisition in Man and Ape: New Dimensions in Comparative Pedolinguistics,* 121–136. Boulder, CO: Westview Press.

Fouts, R. S., Chown, B., and Goodin, L. 1976. Transfer of signed responses in American Sign Language from vocal English stimuli to physical object stimuli by a chimpanzee *(Pan). Learning and Motivation* 7:548–475.

Fouts, R. S., and Couch, J. B. 1976. Cultural evolution of learned language in chimpanzees. In M. E. Hahn and E. S. Simmel, eds., *Communicative Behavior and Evolution,* 141–161. New York: Academic Press.

Fouts, R. S., and Fouts, D. H. 1999. Chimpanzee sign language research. In P. Dolhinow and A. Fuentes, eds., *The Nonhuman Primates,* 252–256. Mountain View, CA: Mayfield.

Fouts, R. S., Fouts, D. H., and Van Cantfort, T. E. 1989. The infant Loulis learns signs from cross-fostered chimpanzees. In R. A. Gardner, B. T. Gardner and T. E. Van Cantfort, eds., *Teaching Sign Language to Chimpanzees,* 280–292. Albany: State University of New York Press.

Fouts, R., Hirsch, A. D., and Fouts, D. H. 1982. Cultural transmission of a human language in a chimpanzee mother-infant relationship. In H. E. Fitzgerald, J. A. Mullins and P. Gage, eds., *Child Nurturance,* Vol. 3: *Studies of Development in Nonhuman Primates,* 159–193. New York: Plenum Press.

Fouts, R., and Mills, S. T. 1997. *Next of Kin: What Chimpanzees Have Taught Me about Who We Are.* New York: William Morrow.

Fowler, A., and Hohmann, G. 2010. Cannibalism in wild bonobos *(Pan paniscus)* at Lui Kotale. *American Journal of Primatology* 72:509–514.

Fowler, A., Pascual-Garrido, A., Buba, U., Tranquilli, S., Akosim, C., Schöning, C., and Sommer, V. 2011. Pananthropology of the fourth chimpanzee: a contribution to cultural primatology. In V, Sommer and C. Ross, eds., *Primates of Gashaka: Socioecology and Conservation in Nigeria's Biodiversity Hotspot,* 451–492. New York: Springer.

Fowler, J. C. S., Skinner, J. D., Burgoyne, L. A., and Drinkwater, R. D. 1989. Satellite DNA and higher-primate phylogeny. *Molecular Biology and Evolution* 6:553–557.

Fowler, J. H., and Schreiber, D. 2008. Biology, politics and the emerging science of human nature. *Science* 322:912–914.

Fox, E. A. 2001. Homosexual behavior in wild Sumatran orangutans *(Pongo pygmaeus abelii). American Journal of Primatology* 55:177–181.

Fox, E. A., and bin'Muhammd, I. 2002a. Brief communication: new tool use by wild Sumatran orangutans *(Pongo pygmeus abelii). American Journal of Physical Anthropology* 119:186–188.

———. 2002b. Female tactics to reduce sexual harassment in the Sumatran orangutan *(Pongo pygmaeus abelii). Behavioral Ecology and Sociobiology* 52:93–101.

Fox, E. A., Sitompul, A., and van Schaik C. P. 1999. Intelligent tool use in wild Sumatran orangutans. In S. T. Parker, R. W. Mitchell, and H. L. Miles, eds., *The Mentalities of Gorillas and Orangutans,* 99–116, Cambridge: Cambridge University Press.

Fox, E. A., van Schaik, C. P., Sitompul, A., and Wright, D. N. 2004. Intra- and interpopulational differences in orangutan *(Pongo pygmaeus)* activity and diet: implications for the invention of tool use. *American Journal of Physical Anthropology* 125:162–174.

Fox, G. J. 1972. Some comparisons between siamang and gibbon behaviour. *Folia Primatologica* 18:122–139.

Fox, M. W. 1982. Are most animals "mindless automatons"?: a reply to Gordon G. Gallup Jr. *American Journal of Primatology* 3:341–343.

Fox, R. B. 1952. The Pinatubo Negritos, their useful plants and material culture. *Philippine Journal of Science* 81:173–414.

Fragaszy, D. M., and Adams-Curtis, L. E. 1989. Forelimb dimensions and goniometry of the wrist and fingers in tufted capuchin monkeys *(Cebus apella):* developmental and comparative aspects. *American Journal of Primatology* 17:133–146.

———. 1991. Generative aspects of manipulation in tufted capuchin monkeys *(Cebus apella). Journal of Comparative Psychology* 105:387–397.

Fragaszy, D. M., Izar, P., Visalberghi, E., Ottoni, E. B., and de Oliveira, M. G. 2004a. Wild capuchin monkeys (Cebus libidinosus) use anvils and stone pounding tools. *American Journal of Primatology* 64:359–366.

———. 2004b. Wild capuchin monkeys *(Cebus libidinosus)* use anvils and stone pounding tools. *American Journal of Primatology* 64:359–366.

Fragaszy, D. M., Johnson-Pynn, J., Hirsch, E., and Brakke, K. 2003. Strategic navigation of two-dimensional alley mazes: comparing capuchin monkeys and chimpanzees. *Animal Cognition* 6:149–160.

Fragaszy, D. M., and Perry, S. 2003. *The Biology of Traditions.* Cambridge: Cambridge University Press.

Fragaszy, D. M., and Visalberghi, E. 1989. Social influences on the acquisition of tool-using behaviors in tufted capuchin monkeys *(Cebus apella). Journal of Comparative Psychology* 103:159–170.

Fragaszy, D. M., Visalberghi, E., and Fedigan, L. M. 2004. *The Complete Capuchin.* Cambridge: Cambridge University Press.

Franciscus, R. G. 2009. When did the modern human pattern of childbirth arise? New insights from an old Neandertal pelvis. *Proceedings of the National Academy of Sciences of the United States of America* 106:9125–9126.

Franz, C. 1999. Allogrooming behavior and grooming site prefereces in captive bonobos *(Pan paniscus):* association with female dominance. *International Journal of Primatology* 20:525–546.

Fraser, O. N., and Aureli, F. 2008. Reconciliation, consolation and postconflict behavioral specificity in chimpanzees. *American Journal of Primatology* 70:1114–1123.

Fraser, O. N., Schino, G., and Aureli, F. 2008. Components of relationship quality in chimpanzees. *Ethology* 114:834–843.

Fraser, O. N., Stahl, D., and Aureli, F. 2008. Stress reduction through consolation in chimpanzees. *Proceedings of the National Academy of Sciences of the United States of America* 105:8557–8562.

Frayer, D. W. 1973. *Gigantopithecus* and its relationship to *Australopithecus*. *American Journal of Physical Anthropology* 39:413–426.

———. 1976. A reappraisal of *Ramapithecus*. *Yearbook of Physical Anthropology, 1974* 18:19–30.

———. 1978. The taxonomic status of *Ramapithecus*. In M. Malez, ed., *Krapinski pracovjek i evolucija hominida,* 255–268. ed. M. Malez, Zagreb: Jugoslavenska Akademija Znasosti I Umjetnosti.

———. 1993. The Kebara 2 Neanderthal hyoid only resembles humans. *American Journal of Physical Anthropology* 16 (Suppl.): 88.

Frayer, D. W., and Nicolay, C. 2000. Fossil evidence for the origin of speech sounds. In N. L. Wallin, B. Merker, and S. Brown, eds., *The Origins of Music,* 215–234. Cambridge, MA: MIT Press.

Freedman, L., and Lofgren, M. 1979a. The Cossack skull and a dihybrid origin of the Australian aborigines. *Nature* 282:298–300.

———. 1979b. Human skeletal remains from Cossack, Western Australia. *Journal of Human Evolution* 8:283–299.

Friedenthal, Hans. 1900. Über einen experimentellen Nachweis von Blutsverwandtschaft. *Archiv für Physiologie: Physiologische Abteilung des Archives für Anatomie und Physiologie,* 494–508. Leipzig: Veit & Comp.

Frisch, J. E. 1965. Trends in the evolution of the hominoid dentition. *Bibliotheca Primatologica,* Fascicle 3. Basel: Karger.

———. 1973. The hylobatid dentition. In D. M. Rumbaugh, ed., *Gibbon and Siamang,* Vol. 2, 55–95. Basel: Karger.

Frith, C. D. 2007. The social brain? *Philosophical Transactions of the Royal Society of London,* series B 362:671–678.

Fromkin, V. A. 1978. *Tone: A Linguistic Survey.* New York: Academic Press.

Fruth, B. 1995. *Nests and Nest Groups in Wild Bonobos (Pan paniscus): Ecological and Behavioural Correlates.* Aachen: Shaker.

Fruth, B., and Hohmann, G. 1993. Ecological and behavioural aspects of nest building in wild bonobos *(Pan paniscus). Ethology* 94:113–126.

———. 1994. Comparative analyses of nest building behavior in bonobos and chimpanzees. In R. W. Wrangham, W. C. McGrew, F. B. M. de Waal, and P. G. Heltne, eds., *Chimpanzee Cultures,* 109–128. Cambridge, MA: Harvard University Press.

———. 2000. Use and function of genital contacts among female bonobos. *Animal Behaviour* 60:107–120.

———. 2002. How bonobos handle hunts and harvests: why share food? In C. Boesch, G. Hohmann, and L. F. Marchant, eds., *Behavioural Diversity in Chimpanzees and Bonobos,* 231–243. Cambridge: Cambridge University Press.

———. 2006. Social grease for females? Same-sex genital contacts in wild bonobos. In V. Sommer and P. L. Vasey, eds., *Homosexual Behaviour in Animals: An Evolutionary Perspective,* 294–315. Cambridge: Cambridge University Press.

Fry, D. P. 2007. *Beyond War: The Human Potential for Peace.* New York: Oxford University Press.

Fuentes, A. 2000. Hylobatid communities: changing views on pair bonding and social organization in hominoids. 2000. Hylobatid communities: changing views on pair bonding and social organization in hominoids. *Yearbook of Physical Anthropology* 43:33–60.

———. 2007. Social organization. In C. J. Campbell, A. Fuentes, K. C. MacKinnon, M. Panger, and S. K. Bearder, eds., *Primates in Perspective,* 609–621. Oxford: Oxford University Press.

Fuentes, A., Malone, N., Sanz, C., Matheson, M., and Vaughan, L. 2002. Conflict and postconflict behavior in a small group of chimpanzees. *Primates* 43:222–235.

Fuiyama, A., Watanabe, H., Toyoda, A., Taylor, T. D., Itoh, T., Tsai, S.-F., Park, H.-S., Yaspo, M.-L., Lehrach, H., Chen, Z., et al. 2002. Construction and analysis of a human-chimpanzee comparative clone map. *Science* 295:131–134.

Fujimoto, M., and Shimada, M. 2008. Newly observed predation of wild birds by M-group chimpanzees *(Pan troglodytes schweinfurthii)* at Mahale, Tanzania. *Pan Africa News* 15:23–25.

Fukami-Kobayashi, K., Shiina, T., Anzai, T., Sano, K., Yamazaki, M., Inoko, H., and Tateno, Y. 2005. Genomic evolution of MHC class I region in primates. *Proceedings of the National Academy of Sciences of the United States of America* 102:9230–9234.

Furness, W. H. 1916. Observations on the mentality of chimpanzees and orang-utans. *Proceedings of the American Philosophical Society* 55:281–290.

Furuichi, T. 1987. Sexual swelling, receptivity, and grouping of wild pygmy chimpanzee females at Wamba, Zaïre. *Primates* 28:309–318.

———. 1989. Social interactions and the life history of female *Pan paniscus* in Wamba, Zaire. *International Journal of Primatology* 10:173–197.

———. 1997. Agonistic interactions and matrifocal dominance rank of wild bonobos *(Pan paniscus)* at Wamba. *International Journal of Primatology* 18:855–875.

———. 2011. Female contributions to the peaceful nature of bonobo society. *Evolutionary Anthropology* 20:131–142.

Furuichi, T., and Hashimoto, C. 2002. Why female bonobos have lower copulation rate during estrus than chimpanzees. In C. Boesch, G. Hohmann, and L. F. Marchant, eds., *Behavioural Diversity in Chimpanzees and Bonobos,* 156–167. Cambridge: Cambridge University Press.

———. 2004a. Botanical and topographical factors influencing nesting-site selection by chimpanzees in Kalinzu Forest, Uganda, *International Journal of Primatology* 25:755–765.

———. 2004b. Sex differences in copulation attempts in wild bonobos at Wamba. *Primates* 45:59–62.

Furuichi, T., Hashimoto, C., and Tashiro, Y. 2001a. Fruit availability and habitat use by chimpanzees in the Kalinzu forest, Uganda: examination of fallback foods. *International Journal of Primatology* 22:929–945.

———. 2001b. Extended application of a marked-nest census method to examine seasonal changes in habitat use by chimpanzees. *International Journal of Primatology* 22:913–928.

———. 2001c. What factors affect the size of chimpanzee parties in the Kalinzu Forest, Uganda? Examination of fruit abundance and number of estrous females. *International Journal of Primatology* 22:947–959.

Furuichi, T., Idani, G., Ihobe, H., Hashimoto, C., Tashiro, Y., Sakamaki, T., Mulavwa, M. N., Yangozene, K., and Kuroda, S. 2012. Long-term studies on wild bonobos at Wamba, Luo Scientific Reserve, D. R. Congo: towards an understanding of female life history in a male-philopatric species. In Kappeler, P. M., and Watts, D. P., eds., *Long-term Field Studies of Primates,* 413–433. Heidelberg: Springer.

Furuichi, T., Idani, G., Ihobe, H., Kuroda, S., Kiramura, K. Mori, A. Enomoto, T., Okayasu, N., Hashimoto, C., and Kano, T. 1998. Population dynamics of wild bonobos *(Pan paniscus)* at Wamba. *International Journal of Primatology* 19:1029–1043.

Furuichi, T., and Ihobe, H. 1994. Variation in male relationships in bonobos and chimpanzees. *Behaviour* 130:211–228.

Furuichi, T., Inagaki, H., and Angoue-Ovono, S. 1997. Population density of chimpanzees and gorillas in the Petit Loango Reserve, Gabon: employing a new method to distinguish between nests of the two species. *International Journal of Primatology* 18:1029–1046.

Furuichi, T., Mulavwa, M., Yangozene, K., Yamba-Yamba, M., Motema-Salo, B., Idani, G., Ihobe, H., Hashimoto, C., Tashiro, Y., and Mwanza, N. 2008. Relationships among fruit abundance, ranging rate, and party size and composition of bonobos at Wamba. In T. Furuichi and J. Thompson, eds., *The Bonobos: Behavior, Ecology and Conservation,* 135–149. New York: Springer.

Furuichi, T., and Thompson, J. 2008. Introduction. In T. Furuichi and J. Thompson, eds., *The Bonobos: Behavior, Ecology and Conseration,* 1–4. New York: Springer.

Futuyma, D. J. 1979. *Evolutionary Biology.* Sunderland, MA: Sinauer.

Gabounia, L., de Lumley, M.-A., Vekua, A., Lordkipanidze, D., and de Lumley, H. 2002. Découverte d'un nouvel hominidé à Dmanisi (Transcaucasie, Géorgie). *Comptes Rendus Palevol* 1:243–253.

Gabunia, L., Anton, S. Lordkipanidze, D., Vekua, A., Justus, A., and Swisher, C. C. III. 2001. Dmanisi and dispersal. *Evolutionary Anthropology* 10:158–170.

Gabunia, L., Gabashvili, E., Vekua, A., and Lordkipanidze, D. 1999. The taxonomic position of *Udapnopithecus garedziensis* Burth. et Gabash. (Udabno Georgia) and its geological age. *MOAMBE* 159:14.

Gabunia, L., and Vekua, A. 1995. A Plio-Pleistocene hominid from East Georgia, Caucasus. *Nature* 373:509–512.

Gabunia, L., Vekua, A., and Lordkipanidze, D. 2000. The environmental contexts of early human occupation of Georgia (Transcaucasia). *Journal of Human Evolution* 38:785–802.

Gabunia, L., Vekua, A., Lordkipanidze, D., Swisher, C. C. III, Ferring, R., Justus, A., Nioradze, M., Tvatchretidze, M., Anton, S. C., Bosinski, G., et al. 2000b. Earliest Pleistocene hominid cranial remains from Dmanisi, Republic of Georgia: taxonomy, geological setting, and age. *Science* 288:1019–1025.

Gadagkar, R. 1995. Can animals count? *Current Science* 68:1180–11182.

Gage, T. B., and DeWitte, S. 2009. What do we know about the agricultural demographic transition? *Current Anthropology* 50:649–655.

Gagneux, P. 2001. Retraction. *Nature* 414:508.

———. 2002. The genus *Pan:* population genetics of an endangered outgroup. *Trends in Genetics* 18:327–330.

Gagneux, P., Boesch, C., and Woodruff, D. S. 1999. Female reproductive strategies, paternity and community structure in wild West African chimpanzees. *Animal Behaviour* 57:19–32.

Gagneux, P., and Varki, A. 2001. Genetic differences between humans and great apes. *Molecular Phylogenetics and Evolution* 18:2–13.

Gagneux, P., Wills, C., Gerloff, U., Tautz, D., Morin, P. A., Boesch, C., Fruth, B., Hohmann, G., Ryder, O. A., and Woodruff, D. S. 1999. Mitochondrial sequences show diverse evolutionary histories of African hominoids. *Proceedings of the National Academy of Sciences of the United States of America* 96:5077–5082.

Gagneux, P., Woodruff, D. S., and Boesch, C. 1979. Furtive mating in female chimpanzees. *Nature* 387:358–359.

Galdikas, B. M. F. 1978. *Orangutan Adaptation in Tanjung Puting Reserve, Central Borneo.* PhD diss., University of California at Los Angeles.

———. 1979. Orangutan adaptation at Tanjung Puting Reserve: mating and ecology. In D. A. Hamburg and K. R. McCown, eds., *The Great Apes,* 194–233. Menlo Park, CA: Benjamin/Cummings.

———. 1981. Orangutan reproduction in the wild. In C. E. Graham, ed., *Reproductive Biology of the Great Apes,* 281–300, New York: Academic Press.

———. 1982a. Orang-utan tool-use at Tanjung Puting Reserve, Central Indonesian Borneo (Kalimantan Tengah). *Journal of Human Evolution* 10:19–33.

———. 1982b. An unusual instance of tool-use among wild orang-utans in Tanjung Puting Reserve, Indonesian Borneo. *Primates* 23:138–139.

———. 1982c. Wild orangutan birth at Tanjung Puting Reserve. *Primates* 23:500–510.

———. 1983. The orangutan long call and snag crashing at Tanjung Puting Reserve. *Primates* 24:371–384.

———. 1984. Adult female sociality among wild orangutans at Tanjung Puting Reserve. In M. F. Small, ed., *Female Primates: Studies by Women Primatologists,* 217–235, New York: Alan R. Liss.

———. 1985a. Adult male sociality and reproductive tactics among orangutans at Tanjung Puting. *American Journal of Primatology* 8:87–99.

———. 1985b. Orangutan sociality at Tanjung Puting. *American Journal of Primatology* 9:101–119.

———. 1985c. Subadult male orangutan sociality and reproductive behavior at Tanjung Puting. *American Journal of Primatology* 8:87–99.

———. 1988. Orangutan diet, range, and activity at Tanjung Puting, Central Borneo. *International Journal of Primatology* 9:1–35.

———. 1995. Social and reproductive behavior of wild adolescent female orangutans. In R. D. Nadler, B. F. M. Galdikas, L. S. Sheeran, and N. Rosen, eds., *The Neglected Ape,* 163–182. New York: Plenum Press.

Galdikas, B. M. F., and Insley, S. J. 1988. The fast call of the adult male orangutan. *Journal of Mammalogy* 69:371–375.

Galdikas, B. M. F., and Teleki, G. 1981. Variations in subsistence activities of female and male pongids: new perspectives on the origins of hominid labor division. *Current Anthropology* 22:241–2247.

Galdikas, B. M. F., and Wood, J. W. 1990. Birth spacing patterns in humans and apes. *American Journal of Physical Anthropology* 83:185–191.

Galef, B. G. 2009. Culture in Animals? In K. N. Laland and B. G. Galef, eds., *The Question of Animal Culture*, 222–246. Cambridge, MA: Harvard University Press.

Gallese, V. 2003. The roots of empathy: the shared manifold hypothesis and the neural basis of intersubjectivity. *Psychopathology* 36:171–180.

Gallese, V., Fadiga, L., Fogassi, L., and Rizzolatti, G. 1996. Action recognition in the premotor cortex. *Brain* 119:593–609.

Gallese, V., and Goldman, A. 1998. Mirror neurons and the simulation of theory of mind-reading. *Trends in Cognitive Sciences* 2:493–501.

Gallese, V., Keysers, C., and Rizzolatti, G. 1996. A unifying view of the basis of social cognition. *Trends in Cognitive Sciences* 8:396–403.

Gallup, G. G., Jr. 1966. Mirror-image reinforcement in monkeys. *Psychonomic Science* 5:39–40.

———. 1968. Mirror-image stimulation. *Psychological Bulletin* 70:782–793.

———. 1970. Chimpanzees: self-recognition. *Science* 167:86–87.

———. 1975. Towards an operational definition of self-awareness. In R. H. Tuttle, ed., *Socioecology and Psychology of Primates*, 309–341. The Hague: Mouton.

———. 1977a. Absence of self-recognition in a monkey *(Macaca fascicularis)* following prolonged exposure to a mirror. *Developmental Psychobiology* 10:281–284.

———. 1977b. Self-recognition in primates: a comparative approach to the bidirectional properties of consciousness. *American Psychologist* 32:329–338.

———. 1979. Self-awareness in primates. *American Scientist* 67:417–421.

———. 1982. Self-awareness and the emergence of mind in primates. *American Journal of Primatology* 2:237–248.

———. 1994. Self-recognition: research strategies and experimental design. In S. T. Parker, R. W. Mitchell, and M. L. Boccia, eds., *Self-Awareness in Animals and Humans*, 35–50. Cambridge: Cambridge University Press.

Gallup, G. G., Jr., Anderson, J. R., and Shillito, D. J., 2002. The mirror test. In M. Bekoff, C. Allen, and G. M. Burghardt, eds., *The Cognitive Animal*, 325–333. Cambridge, MA: MIT Press.

Gallup, G. G., Jr., Boren, J. L., Gagliardi, G. J., and Wallnau, L. B. 1977. A mirror for the mind of man, or will the chimpanzee create an identity crisis for *Homo sapiens? Journal of Human Evolution* 6:303–313.

Gallup, G. G., Jr., and McClure, M. K. 1971. Preference for mirror-image stimulation in differentially reared rhesus monkeys. *Journal of Comparative and Physiological Psychology* 5:5403–407.

Gallup, G. G., Jr., McClure, M. K., Hill, S. D., and Bundy, R. A. 1971. Capacity for self-recognition in differentially reared chimpanzees. *The Psychological Record* 21:69–71.

Gamble, C. 2008. Kinship and material culture. Archaeological implications of the human global diaspora. In N. J. Allen, H. Callan, R. Dunbar, and W. James, *Early Human Kinship: From Sex to Social Reproduction*, eds., 27–40. Malden, MA: Blackwell.

Ganas, J., Nkurunungi, J. B., and Robbins, M. M. 2009. A preliminary study of the temporal and spatial biomass patterns of herbaceous vegetation consumed by mountain gorillas in an afromontane rain forest. *Biotropica* 41:37–46.

Ganas, J., Ortmann, S., and Robbins, M. M. 2008. Food preferences of wild mountain gorillas. *American Journal of Primatology* 70:927–938.

Ganas, J., and Robbins, M. M. 2004. Intrapopulation differences in ant eating in the mountain gorillas of Bwindi Impenetrable National Park, Uganda. *Primates* 45:275–278.

———. 2005. Ranging behavior of the mountain gorillas *(Gorilla beringei beringei)* in Bwindi Impenetrable National Park, Uganda: a test of the ecological constraints model. *Behavioral Ecology and Sociobiology* 58:277–288.

Ganas, J., Robbins, M. M., Nkurunungi, J. B., Kaplin, B. A., and McNeilage, A. 2004. Dietary variability of mountain gorillas in Bwindi Impenetrable National Park, Uganda. *International Journal of Primatology* 25:1043–1072.

Gangestad, S. W., and Thornhill, R. 2004. Female multiple mating and genetic benefits in humans: investigations of design. In P. Kappeler and C. P. van Schaik, eds., *Sexual Selection in Primates: New and Comparative Perspectives*, 90–113. Cambridge: Cambridge University Press.

Gannon, P. J., Holloway, R. L., Broadfield, D. C., and Braun, A. R. 1998. Asymmetry of chimpanzee planum temporale: humanlike pattern of Wernicke's brain language area homolog. *Science* 279:220–222.

Gannon, P. J., Kheck, N. M., Braun, A. R., and Holloway, R. L. 2005. Planum parietale of chimpanzees and orangutans: a comparative resonance of human-like planum temporale asymmetry. *Anatomical Record, Part A* 287A:1128–1141.

Gantt, D. G. 1983. The enamel of Neogene hominoids. Structural and phyletic implications. In R. L. Ciochon and R. S. Corruccini, eds., *New Interpretations of Ape and Human Ancestry*, 249–298. New York: Plenum Press.

———. 1986. Enamel thickness and ultrastructure in hominoids: with reference to form, function and phylogeny. In D. R. Swindler and J. Erwin, eds., *Comparative Primate Biology*, Vol. 1: *Systematics, Evolution, and Anatomy*, 453–475. New York: Alan R. Liss.

Garcia, J. E., and Mba, J. 1997. Distribution, status and conservation of primates in Monte Alen National Park, Equatorial Guinea. *Oryx* 31:67–76.

Gardner, B. T., and Gardner, R. A. 1971. Two-way communication with an infant chimpanzee. In A. Schrier and F. Stollnitz, eds., *Behavior of Nonhuman Primates*, Vol. 4, 117–185. New York: Academic Press.

———. 1974. Comparing the early utterances of child and chimpanzee. In A. Pick, ed., *Minnesota Symposia on Child Psychology*, Vol. 8, 3–24. Minneapolis: University of Minnesota Press.

———. 1975. Evidence for sentence constituents in the early utterances of child and chimpanzee. *Journal of Experimental Psychology: General* 104:244–267.

———. 1980. Two comparative psychologists look at language acquisition. In K. E. Nelson, ed., *Children's Language*, Vol. 2, 331–369. New York: Gardner Press.

———. 1998. Development of phrases in the early utterances of children and cross-fostered chimpanzees. *Human Evolution* 13:161–188.

Gardner, H. 1983. *Frames of Mind: The Theory of Multiple Intelligences*. New York: Basic Books.

———. 2006. *Multiple Intelligences: New Horizons*. New York: Basic Books.

Gardner, R. A., and Gardner, B. T. 1969. Teaching sign language to a chimpanzee. *Science* 165:664–672.

———. 1972. Communication with a young chimpanzee: Washoe's vocabulary. *Colloques internationaux du Centre National de la Recherche Scientifique* 198:241–264.

———. 1978. Comparative psychology and language acquisition. *Annals of the New York Academy of Sciences* 309:37–76.

———. 1984. A vocabulary test for chimpanzees *(Pan troglodytes)*. *Journal of Comparative Psychology* 98:381–404.

———. 1988. The role of cross-fostering in sign language studies of chimpanzees. *Human Evolution* 3:65–79.

———. 1989. A cross-fostering laboratory. In R. A. Gardner and B. T. Gardner, eds., *Teaching Sign Language to Chimpanzees*, 1–28. Albany: State University of New York Press.

———. 1993. *Creating Minds: An Anatomy of Creativity Seen through the Lives of Freud, Einstein, Picasso, Stravinsky, Eliot, Graham, and Gandhi*. New York: Basic Books.

———. 1998. Ethological study of early language. *Human Evolution* 13:189–207.

Gardner, R. A., Van Cantfort, T. E., and Gardner, B. T. 1992. Categorical replies to categorical questions by cross-fostered chimpanzees. *American Journal of Psychology* 105:27–57.

Gargett, R. H. 1989. The evidence for Neandertal burial. *Current Anthropology* 30:157–190.

Garn, S. M., and Lewis, A. B. 1958. Tooth-size, body-size and "giant" fossil man. *American Anthropologist* 60:874–880.

Garner, K. J., and Ryder, O. A. 1996. Mitochondrial DNA diversity in gorillas. *Molecular Phylogenetics and Evolution* 6:39–48.

Garner, R. L. 1896. *Gorillas and Chimpanzees*. London: Osgood, McIlvaine.

Garrard, P., and Carroll, E. 2006. Lost in semantic space: a multi-modal, non-verbal assessment of feature knowledge in semantic dementia. *Brain* 129:1152–1163.

Gatti, S., Levéro, F., Ménard, Petit, E., and Gautier-Hion, A. 2003. Bachelor groups of western lowland gorillas *(Gorilla gorilla gorilla)* at Lokoue clearing, Odzala National Park, Republic of Congo. *Folia Primatologica* 74:196–196.

———. 2004. Population and group structure of western lowland gorillas *(Gorilla gorilla gorilla)* at Lokou, Odzala National Park, Republic of Congo. *American Journal of Primatology* 63:111–123.

Gautelli-Steinberg, D., Reid, D. J., Bishop, T. A., and Larsen, C. S. 2005. Anterior tooth growth periods in Neanderthals were comparable to those of modern humans. *Proceedings of the National Academy of Sciences of the United States of America* 102:14197–14202.

Gautier-Hion, A. 1990. Interactions among fruit and vertbrate-eaters in an African tropical rain forest. In K. S. Bawa and M. Hadley, eds., *Reproductive Ecology of Tropical Forest Plants*, 219–229. Carnforth Lanes, UK: Parthenon.

Gavrilets, S. 2012. Human origins and the transition from promiscuity to pair-bonding. *Proceedings of the National Academy of Sciences of the United States of America* 109:9923–9928.

Gazzola, V., Aziz-Zadeh, L., and Keysers, C. 2006. Empathy and the somatotopic auditory mirror system in humans. *Current Biology* 16:1924–1829.

Gebo, D. L. 1986. Anthropoid origins—the foot evidence. *Journal of Human Evolution* 15:421–430.

———. 1992. Plantigrady and foot adaptation in African apes: implications for hominid origins. *American Journal of Physical Anthropology* 89:29–58.

———. 1993a. Postcranial anatomy and locomotor adaptation in early African anthropoids. In D. L. Gebo, ed., *Postcranial Adaptation in Nonhuman Primates,* 220–234. DeKalb: Northern Illinois University Press.

———. 1993b. Reply to Meldrum. *American Journal of Physical Anthropology* 91:382–385.

———. 2002. Adapiformes: phylogeny and adaptation. In W. C. Hartwig, ed., *The Primate Fossil Record,* 21–43. Cambridge: Cambridge University Press.

Gebo, D. L., Beard, K. C., Teaford, M. F., Walker, A., Larson, S. G., Jungers, W. L., and Fleagle, J. G. 1988. A hominid proximal humerus from the Early Miocene of Rusinga Island, Kenya. *Journal of Human Evolution* 17:393–401.

Gebo, D. L., and Dagosto, M. 2004. Anthropoid origins: postcranial evidence from the Eocene of Asia. In C. F. Ross and R. F. Kay, eds., *Anthropoid Origins,* 369–380. New York: Kluwer Academic/Plenum Press.

Gebo, D. L., Dagosto, M., Beard, K. C., and Qi, T. 2000. The smallest primates. *Journal of Human Evolution* 38:585–594.

———. 2001. Middle Eocene tarsals from China: implications for haplorhine evolution. *American Journal of Physical Anthropology* 116:83–107.

Gebo, D. L., Dagosto, M., Beard, K. C., Qi, T., and Wang, J. 2000. The oldest known anthropoid postcranial fossils and the early evolution of higher primates. *Nature* 404:276–278.

Gebo, D. L., Dagosto, M., Beard, K. C., and Wang, J. 1999. A first metatarsal of *Hoanghonius stehlini* from the late Middle Eocene of Shanxi Province, China. *Journal of Human Evolution* 37:801–806.

Gebo, D. L., Gunnell, G. F., Ciochon, R. L., Takai, M. Tsubamoto, T., and Egi, N. 2002. New eosimiid primate from Myanmar. *Journal of Human Evolution* 43:549–553.

Gebo, D. L., MacLatchy, L., Kityo, R., Deino, A., Kingston, J., and Pilbeam, D. 1997. A hominoid genus from the Early Miocene of Uganda. *Science* 276:401–404.

Gebo, D. L., and Simons, E. L. 1987. Morphology and locomotor adaptations of the foot in Early Oligocene anthropoids. *American Journal of Physical Anthropology* 74:83–101.

Gebo, D. L., and Simons, E. L., Rasmussen, D. T., and Dagosto, M. 1994. Eocene anthropoid postcrania from the Fayum, Egypt. In J. G. Fleagle and R. F. Kay, eds., *Anthropoid Origins,* 203–233. New York: Plenum Press.

Geertz, C. 1973. *The Interpretation of Cultures.* New York: Basic Books.

Geissmann, T. 1984. Inheritance of song parameters in the gibbon song, analyzed in 2 hybrid gibbons (*Hylobates pileatus* X *H. lar*). *Folia Primatologica* 42:216–235.

———. 1994. Systematik der gibbons. *Zeitschrift des Kölner Zoo* 37:65–77.

————. 1995. Gibbon systematics and species identification. *International Zoo News* 42:467–501.

————. 2002. Duet-splitting and the evolution of gibbon songs. *Biological Reviews* 77:57–76.

————. 2007. Status reassessment of the gibbons: results of the Asian Primate Red List Workshop 2005. *Gibbon Journal* Nr. 3:5–16.

Geist, V. 1978. Life Strategies, Human Evolution, Environmental Design. *Toward a Biological Theory of Health*. New York: Springer-Verlag.

Gelman, R., and Gallistel, C. R. 2004. Language and the origin of numerical concepts. *Science* 306:441–443.

Gelvin, B. R. 1980. Morphometric affinities of *Gigantopithecus*. *American Journal of Physical Anthropology* 53:541–568.

Genet-Varcin, E. 1969. A la recherche du primate ancêtre de l'homme. Paris: Boubée.

Gentili, S., Mottura, A., and Rook, L. 1998. The Italian fossil primate record: recent finds and their geological context. *GEOBIOS* 31:675–686.

Gentilucci, M., Bernardis, P., Crisi, G., and Volta, R. D. 2006. Repetitive transcranial magnetic stimulation of Broca' area affects verbal responses to gesture observation. *Journal of Cognitive Neuroscience* 18:105–1074.

Gentilucci, M., and Corballis, M. C. 2006. From manual gesture to speech: a gradual transition. *Neuroscience and Biobehavioral Review* 30:949–960.

Gentry, A. W. 1981. Notes on Bovidae (Mammalia) from the Hadar Formation, and from Amado and Geraru, Ethiopia. *Kirtlandia* 33:1–30.

Gentry, E., Breuer, T., Hobaiter, C., and Byrne, R. W. 2009. Gestural communication of the gorilla *(Gorilla gorilla)*: repertoire, intentionality and possible origins. *Animal Cognition* 12:527–546.

Gerloff, U., Hartung, B., Fruth, B., Hohmann, G., and Tautz, D. 1999. Intracommunity relationships, dispersal pattern and paternity success in a wild living community of bonobos *(Pan paniscus)* determined from DNA analysis of faecal samples. *Proceedings of the Royal Society B* 266:1189–1195.

Gervais, P. 1872. Sur un singe fossile, d'espèce non encore décrite, qui a été découvert au Monte-Bamboli (Italie). *Comptes rendus hebdomadaires des séances de l'Académie des Sciences, Paris* 74:1217–1223.

Geschwind, N., and Galaburda, A. M. 1984. *Cerebral Dominance*. Cambridge, MA: Harvard University Press.

Geschwind, N., and Levitsky, W. 1968. Human brain: left-right asymmetries in temporal speech region. *Science* 161:186–187.

Gettler, L. T., McDade, T. W., Feranil, A. B., and Kuzawa, C. W. 2011. Longitudinal evidence that fatherhood decreases testosterone in human males. *Proceedings of the National Academy of Science of the United States of America* 108:16194–16199.

Ghazanfar, A. A., and Hauser, M. D. 1999. The neuroethology of primate vocal communication: substrates for the evolution of speech. *Trends in Cognitive Sciences* 3:377–384.

Gheerbrant, E., Thomas, H., Sen, S., and Al-Sulaimani, Z. 1995. Nouveau primate Oligopithecinae (Siimiformes) de l'Oligocène inférieur de Taqah, Sultanat d'Oman. *Comptes rendus de l'Académie des sciences, Paris, série II a* 321:425–432.

Ghiglieri, M. P. 1984a. *The Chimpanzees of Kibale Forest: A Field Study of Ecology and Social Structure*. New York: Columbia University Press.

———. 1984b. Feeding ecology and sociality of chimpanzees in Kibale Forest, Uganda. In *Adaptations for Foraging in Nonhuman Primates*, P. S. Rodman and J. G. H. Cant, eds., 161–194, New York: Columbia University Press.

———. 1987. Sociobiology of the great apes and the hominid ancestor. *Journal of Human Evolution* 16:319–357.

———. 1989. Hominoid sociobiology and hominid social evolution. In P. G. Heltne and L. A. Marquardt, eds., *Understanding Chimpanzees*, 370–379. Cambridge, MA: Harvard University Press.

Ghinassi, M., Libsekal, Y., Papini, M., and Rook, L. 2009. Palaeoenvironments of the Buia *Homo* site: high-resolution facies analysis and non-marine sequence stratigraphy in the Alat formation (Pleistocene Dandiero Basin, Danakil depression, Eritrea). *Palaeogeography, Palaeoclimatology, Palaeoecology* 280:415–431.

Gibbons, A. 1997. Archaeologists rediscover cannibals. *Science* 277:636–637.

———. 1998. Solving the brain's energy crisis. *Science* 280:1345–1347.

———. 2007a. A new body of evidence fleshes out *Homo erectus*. *Science* 317:1664.

———. 2007b. New fossils challenge line of descent in human family. *Science* 317:733.

———. 2008. Millennium ancestor gets its walking papers. *Science* 319:1599–1560.

———. 2009a. A new kind of ancestor: *Ardipithecus* unveiled. *Science* 326:36–40.

———. 2009b. Neandertals babies didn't do the twist. *ScienceNOW Daily News,* April 20.

———. 2011. African data bolster new view of modern human origins. *Science* 334:167.

Gibbs, S., Collard, M., and Wood, B. 2002. Soft-tissue anatomy of the extant hominoids: a review and phylogenetic analysis. *Journal of Anatomy* 200:3–49.

Gibert, J., Agustí, J., and Moyà-Solà, S. 1983. Presencia de *Homo* sp. en el Yacimiento del Pleistoceno Inferior de Venta Micena (Orce, Granada). *Paleontologia i Evolució* (special issue), 1–9. Sabadell: Institut de Paleontologia.

Gibson, K. 1991. Tools, language and intelligence evolutionary implications. *Man,* n.s., 26:255–164.

Gibson, K. R., and Ingold, T. 1993. *Tools, Language and Cognition in Human Evolution*. Cambridge: Cambridge University Press.

Gibson, K. R., Rumbaugh, D., and Beran, M. 2001. Bigger is better: primate brain size in relationship to cognition. In D. Falk and K. R. Gibson, eds., *Evolutionary Anatomy of the Primate Cerebral Cortex,* 79–97. Cambridge: University of Cambridge Press.

Gignoux, C. R., Henn B. M., and Mountain, J. L. 2011. Rapid, global demographic expansions after the origins of agriculture. *Proceedings of the National Academy of Sciences of the United States of America* 108:60447–60491.

Gilbert, W. H., and Asfaw, B. 2008. *Homo erectus: Pleistocene Evidence from the Middle Awash, Ethiopia*. Berkeley: University of California Press.

Gilbert, W. H., White, T. D., and Asfaw, B. 2003. *Homo erectus, Homo ergaster,* Homo *"cepanensis,"* and the Daka cranium. *Journal of Human Evolution* 45:255–259.

Gilby, I. C. 2001. Why do wild male chimpanzees share meat with females? One alternative to the "meat for sex" hypothesis. In *The Apes: Challenges for the 21st Century, Conference Proceedings, May 10–12, 2000,* 90–93. Brookfield, IL: Brookfield Zoo.

Gilby, I. C., Eberly, L. E., Pintea, L., and Pusey, A. E. 2006. Ecological and social influences on the hunting behaviour of wild chimpanzees, *Pan troglodytes schweinfurthii*. *Animal Behaviour* 72:169–180.

Gilissen, E. 2001. Structural symmetries and asymmetries in human chimpanzee brains. In D. Falk and K. R. Gibson, eds., *Evolutionary Anatomy of the Primate Cerebral Cortex*, 187–215. Cambridge: University of Cambridge Press.

Gill, T. V. 1977. Talking to Lana: the question of conversation. In *Progress in Ape Research*, G. H. Bourne, ed., 125–132. New York: Academic Press.

———. 1978. Ape language projects: a perspective. In D. Chivers and J. Herbert, eds., *Recent Advances in Primatology*, Vol. 1: *Behaviour*, 863–866. London: Academic Press.

Gill, T. V., and Rumbaugh, D. M. 1974a. Learning processes of bright and dull apes. *American Journal of Mental Deficiency* 78:683–687.

———. 1974b. Mastery of naming skills by a chimpanzee. *Journal of Human Evolution* 3:483–492.

Gillan, D. J. 1981. Reasoning in the chimpanzee: II. transitive inference. *Journal of Experimental Psychology: Animal Behavior Processes* 7:150–164.

Gillan, D. J., Premack, D., and Woodruff, G. 1981. Reasoning in the chimpanzee I. analogical reasoning. *Journal of Experimental Psychology: Animal Behavior Processes* 7:1–17.

Gillespie, D., Donehower, L., and Strayer, D. 1982. Evolution of primate DNA organization. In G. A. Dover and R. B. Flavell, eds., *Genome Evolution*, 113–133. London: Academic Press.

Gillespie, J. H. 1986. Natural selection and the molecular clock. *Molecular Biology and Evolution* 3:138–155.

Gilroy, P. 2000. *Against Race: Imagining Political Structure beyond the Color Line*. Cambridge, MA: Belknap Press.

Gil-White, J. 2001. Are ethnic groups biological "species" to the human brain? Essentialism in our cognition of some social categories. *Current Anthropology* 42:515–533.

Gingerich, P. D. 1980. Eocene Adapidae, paleobiogeography, and the origin of South American Platyrrhini. In R. L. Ciochon and A. B. Chiarelli, eds., *Evolutionary Biology of the New World Monkeys and Continental Drift*, 123–138. New York: Plenum Press.

———. 1981. Early Cenozoic Omomyidae and the evolutionary history of the tarsiiform primates. *Journal of Human Evolution* 10:345–374.

———. 1984a. Primate evolution. *University of Tennessee Studies in Geology* 8:167–184.

———. 1984b. Primate evolution: evidence from the fossil record, comparative morphology, and molecular biology. *Yearbook of Physical Anthropology* 27:57–72.

———. 1985. Nonlinear molecular clocks and ape-human divergence times. In P. V. Tobias, ed., *Hominid Evolution: Past, Present and Future*, 411–416. New York: Alan R. Liss.

———. 1986a. *Plesiadapis* and the delineation of the order Primates. In B. Wood, L. Martin, and P. Andrews, eds., *Major Topics in Primate and Human Evolution*, 32–46. Cambridge: Cambridge University Press.

———. 1986b. Temporal scaling of molecular evolution in primates and other mammals. *Molecular Biology and Evolution* 3:205–221.

Ginsburg, L. 1986. Chronology of the European pliopithecids. In J. G. Else and P. C. Lee, eds., *Primate Evolution,* 47–57. Cambridge: Cambridge University Press.

Ginsburg, L., and Mein, P. 1980. *Crouzelia rhodanica,* nouvelle espèce de Primate catarhinien, et essai sur la position systématique des Pliopithecidae. Bulletin du Muséum national d'Histoire naturelle, Paris, 4e série, 2:57–85.

Gittins, S. P. 1978. Hark! The beautiful song of the gibbon. *New Scientist* 80:832–834.

———. 1980. Territorial behavior in the agile gibbon. *International Journal of Primatology* 1:381–399.

———. 1982. Feeding and ranging in the agile gibbon. *Folia Primatologica* 38:39–71.

———. 1984. The vocal repertoire and song of the agile gibbon, In H. Preuschoft, D. J. Chivers, W. Y. Brockelman, and N. Creel, eds., *The Lesser Apes: Evolutionary and Behavioural Biology* 354–375. Edinburgh: Edinburgh University Press.

Gittins, S. P., and Raemaekers, J. J. 1980. Siamang, lar and agile gibbons. In D. J. Chivers, ed., *Malayan Forest Primates,* 63–105. New York: Plenum Press.

Givón, T. 1995. *Functionalism and Grammar.* Amsterdam: John Benjamins.

Glatston, A. R. 1996. Apes in zoos—prison or sanctuary? *International Zoo News* 43:228–231.

Glazko, G. V., and Nei, M. 2003. Estimation of divergence times for major lineages of primate species. *Molecular Biology and Evolution* 20:424–434.

Godefroit, P. 1990. Dental variation in the genus *Pan. Zeitschrift für Morphologie und Anthropologie* 78:175–195.

Godinot, M. 1994. Early North African Primates and their significance for the origin of Simiiformes (+anthropoidea). In J. G. Fleagle and R. F. Kay, eds., *Anthropoid Origins,* 235–295. New York: Plenum Press.

Godinot, M., and Mahboubi, M. 1992. Earliest known simian primate found in Algeria. *Nature* 357:324–326.

———. 1994. Les petits primates imiiformes de Glib Zegdou (Êocène inférieur à moyen d'Algérie). *Comptes rendus de l'Académie des sciences, Paris* 319:357–364.

Goebel, T., Waters, M. R., and O'Rourke, D. H. 2008. The Late Pleistocene dispersal of modern humans in the Americas. *Science* 319:1497–1502.

Gold, K. C. 2001. Group formation in captive bonobos: sex and a bonding strategy. In *The Apes: Challenges for the 21st Century, Conference Proceedings, May 10–12, 2000,* 90–93. Brookfield, IL: Brookfield Zoo.

———. 2002. Ladder use and clubbing by a bonobo *(Pan paniscus)* in Appenheul Primate Park. *Zoo Biology* 21:607–611.

Goldberg, P., Weiner, S., Bar-Yosef, O., Xu, Q., and Liu, J. 2001. Site formation processes at Zhoukoudian, China. *Journal of Human Evolution* 41:483–530.

Goldberg, T. L. 1998. Biogeographic predictors of genetic diversity in populations of eastern African chimpanzees *(Pan troglodytes schweinfurthii). International Journal of Primatology* 19:237–254.

Goldberg, T. L., and Ruvolo, M. 1997a. The geographic apportionment of mitochondrial genetic diversity in East African chimpanzees, *Pan troglodytes schweinfurthii. Molecular Biology and Evolution* 14:976–984.

———. 1997b. Molecular phylogenetics and historical biogeography of east African chimpanzees. *Biological Journal of the Linnean Society* 61:301–324.

Goldin-Meadow, S. 2005. Watching language grow. *Proceedings of the National Academy of Sciences of the United States of America* 102:2271–2272.

Goldin-Meadow, S., and McNeill, D. 1999. The role of gesture and mimetic representation in making language the province of speech. In M. C. Corballis and E. G. Lea, eds., *The Descent of Mind: Psychological Perspectives on Hominid Evolution*, 155–172. Oxford: Oxford University Press.

Goldman, D., Giri, P. R., and O'Brien, S. J. 1987. A molecular phylogeny of the hominoid primates as indicated by two-dimensional protein electrophoresis. *Proceedings of the National Academy of Sciences of the United States of America* 84:3307–3311.

Goldman, J. 1981. *The Lion in Winter*. New York: Random House. First published in 1966.

Goldsmith, M. L. 1999a. Ecological constraints on the foraging effort of western gorillas *(Gorilla gorilla gorilla)* at Bai Hoköu, Central African Republic. *International Journal of Primatology* 20:1–23.

———. 1999b. Gorilla socioecology. In P. Dohlinow and A. Fuentes, eds., *The Nonhuman Primates*, 58–63. Mountain View, CA: Mayfield.

———. 2003. Comparative behavioral ecology of a lowland and highland gorilla population: where to Bwindi gorillas fit? In A. B. Taylor and M. L. Goldsmith, eds., *Gorilla Biology: A Multidisciplinary Perspective*, 358–384. Cambridge: Cambridge University Press.

———. 2006. Gorillas living on the edge: literally and figuratively. In N. E. Newton-Fisher, H. Notman, J. D. Paterson, and V. Reynolds, eds., *Primates of Western Uganda*, 405–422. New York: Springer.

Goldstein, M. C. 1971. Stratification, polyandry, and family structure in Central Tibet. *Southwestern Journal of Anthropology* 27:64–74.

Gomes, C. M., and Boesch, C. 2009. Wild chimpanzees exchange meat for sex on a long-term basis. *PLoS One* 4:e5116.1–6.

Gómez, J. C. 1996. Ostensive behavior in great apes: the role of eye contact. In A. E. Russon, K. A. Bard, and S. T. Parker, eds., *Reaching into Thought: The Minds of the Great Apes*, 131–151. Cambridge: Cambridge University Press.

———. 2004. *Apes, Monkeys, Children, and the Growth of Mind*. Cambridge, MA: Harvard University Press.

———. 2005. Requesting gestures in captive monkeys and apes. *Gesture* 5:91–105.

———. 2007. Pointing behaviors in apes and human infants: a balanced interpretation. *Child Development* 78:729–734.

Gómez, J. C., Sarriá, E., and Tamarit, J. 1993. The comparative study of early communication and theories of mind: ontogeny, phylogeny and pathology. In S. Baron-Cohen, H. Tager-Flusberg, and D. Cohen, eds., *Understanding Other Minds: Perspectives from Autism*, 397–426. Oxford: Oxford University Press.

Gommery, D. 2005. Evolution of the vertebral column in Miocene Hominoids and Plio-Pleistocene hominids. In H. Ishida, R. H. Tuttle, M. Pickford, M. Nakatsukasa, and N. Ogihara, eds., *Human Origins and Environmental Backgrounds*, 31–43. New York: Springer.

Gonder, M. K., Locatelli, S., Ghobrial, L., Mitchell, M. W., Kujawski, J. T., Lankester, F. J., Stewart, C.-B., and Tishkoff, S. A. 2011. Evidence from Cameroon reveals differences in the genetic structure and histories of chimpanzee populations. *Proceedings of the National Academy of Science of the United States of America* 108:4766–4771.

Gonder, M. K., Oates, J. F., Disotell, T. R., Forstner, M. R., Morales, J. C., and Melnick, D. J. 1997. A new west African chimpanzee subspecies? Nature 388:337.

Gonzalez, I. L., Sylvester, J. E., Smith, T. F., Stambolian, D., and Schmickel, R. D. 1990. Ribosomal RNA gene sequences and hominoid phylogeny. *Molecular Biology and Evolution* 7:203–219.

Gonzalez-Kirchner, J. P. 1997. Census of western lowland gorilla population in Rio Muni Region, Equatorial Guinea. *Folia Zoologica* 46:15–22.

Goodall, A. G. 1977. Feeding and ranging behavior of a mountain gorilla group *(Gorilla gorilla beringei)* in the Tshibinda-Kahuzi region (Zaire). In T. H. Clutton-Brock, ed., *Primate Ecology: Studies of Feeding and Ranging Behaviour in Lemurs, Monkeys and Apes,* 449–479. London: Academic Press.

———. 1978. On habitat and home range in eastern gorillas in relation to conservation. In D. J. Chivers and W. Lant-Petter, eds., *Recent Advances in Primatology,* Vol. 2: *Conservation,* 81–83. London: Academic Press.

Goodall, J. 1963a. Feeding behaviour of wild chimpanzees—a preliminary report. *Symposia of the Zoological Society, London* 10:39–47.

———. 1963b. My life among wild chimpanzees. *National Geographic* 124:272–308.

———. 1964. Tool-using and aimed throwing in a community of free-living chimpanzees. *Nature* 201:1264–1266.

———. 1965. Chimpanzees of the Gombe Stream Reserve. In I. DeVore, ed., *Primate Behavior,* 425–473. New York: Holt, Rineart and Winston.

———. 1968. The behaviour of free-living chimpanzees in the Gombe Stream Reserve. *Animal Behaviour Monographs* 1, no. 3: 161–311.

———. 1977. Infant killing and cannibalism in free-living chimpanzees. *Folia Primatologica* 28:259–282.

———. 1979. Life and death at Gombe. National *Geographic Magazine* 155:592–621.

———. 1983. Population dynamics during a 15 year period in one community of free-living chimpanzees in the Gombe National Park, Tanzania. *Zeitschrift für Tierpsychologie* 61:1–60.

———. 1986. *The Chimpanzees of Gombe.* Cambridge, MA: Belknap Press.

Goodall J., Bandora A., Bergmann E., Busse C., Matama H., Mpongo E., Pierce A., and Riss D. 1979. Intercommunity interactions in the chimpanzee population of the Gombe National Park. In D. A. Hamburg and E. R. McCown, eds., *The Great Apes,* 13–53, Menlo Park, CA: Benjamin/Cummings.

Goodman, A. 2005. Three questions about race, human biological variation and racism. *Anthropology News,* September 2005, 18–19.

Goodman, M. 1962a. Evolution of the immunologic species specificity of human serum proteins. *Human Biology* 34:104–150.

———. 1962b. Immunochemistry of the Primates and primate evolution. *Annals of the New York Academy of Sciences* 102:219–234.

————. 1963a. Man's place in the phylogeny of the primates as reflected in serum proteins. In S. L. Washburn, ed., *Classification and Human Evolution,* 204–234. New York: Viking Fund Publications in Anthropology.

————. 1963b. Serological analysis of the systematics of recent hominoids. *Human Biology* 35:377–324.

————. 1981. Decoding the pattern of protein evolution. *Progress in Biophysics and Molecular Biology* 37:105–164.

————. 1982. Biomolecular evidence on human origins from the standpoint of Darwinian theory. *Human Biology* 54:247–264.

————. 1986. Molecular evidence of the ape subfamily Homininae. In H. Gershowitz, D. L. Rucknagel, and R. E. Tashian, eds., *Evolutionary Perspectives and the New Genetics,* 121–132. New York: Alan R. Liss.

————. 1989. Emerging alliance of phylogenetic systematics and molecular biology: a new age of exploration. In B. Fernholm, K. Bremer, and H. Jörnvall, eds., *The Hierarchy of Life,* 43–61. Amsterdam: Elsevier.

————. 1992. Hominoid evolution at the DNA level and the position of humans in a phylogenetic classification. In T. Nishida, W. C. McGrew, P. Marler, M. Pickford, and F. B. M. de Waal, eds., *Topics in Primatology,* Vol. 1: *Human Origins,* 331–346. Tokyo: University of Tokyo Press.

————. 1996. Epilogue: a personal account of the origins of a new paradigm. *Molecular Phylogenetics and Evolution* 5:269–285.

Goodman, M., Baba, M. L., and Darga, L. L. 1983. The bearing of molecular data on the cladogenesis and times of divergence of hominoid lineages. In R. L. Ciochon and R. S. Corruccini, eds., *New Interpretations of Ape and Human Ancestry,* 67–86. New York: Plenum Press.

Goodman, M., Bailey, W. J., Hayasaka, K., Stanhope, M. J., Slightom, J., and Czelusniak, J. 1994. Molecular evidence on primate phylogeny from DNA sequences. *American Journal of Physical Anthropology* 94:3–24.

Goodman, M., Barnabas, J., Matsuda, G., and Moore, G. W. 1971. Molecular evolution in the descent of man. *Nature* 233:604–613.

Goodman, M., Barnabas, J., and Moore, G. W. 1972. Man, the conservative and revolutionary mammal. Molecular findings on this paradox. *Journal of Human Evolution* 1:663–686.

Goodman, M., Braunitzer, G., Stangl, A., and Schrank, B. 1983. Evidence on human origins from haemoglobins of African apes. *Nature* 303:546–548.

Goodman, M., and Cronin, J. E. 1982. Molecular anthropology: its development and current directions. In F. Spencer, ed., *A History of American Physical Anthropology 1930–1980,* 105–146. New York: Academic Press.

Goodman, M., Czelusniak, J., and Beeber, J. E. 1985. Phylogeny of primates and other eutherian orders: a cladistic analysis using amino acid and nucleotide sequence data. *Cladistics* 1:171–185.

Goodman, M., Hewett-Emmett, D., and Beard, J. M. 1978. Molecular evidence on the phylogenetic relationships of *Tarsius.* In D. J. Chivers and K. A. Joysey, eds., *Recent Advances in Primatology,* Vol. 3: *Evolution,* 215–225. London: Academic Press.

Goodman, M., Koop, B. F., Czelusniak, J., Fitch, D. H. A., Tagle, D. A., and Slightom, J. L. 1989. Molecular phylogeny of the family of apes and humans. *Genome* 31:316–335.

Goodman, M., Olson, C. B., Beeber, J. E., and Czelusniak, J. 1982. New perspectives in the molecular biological analysis of mammalian phylogeny. *Acta Zoologica Fennica* 169:19–35.

Goodman, M., Porter, C. A., Czelusniak, J., Page, S. L., Schneider, H., Shoshani, J., Gunnell, G., and Groves, C. A. 1998. Toward a phylogenetic classification of primates based on DNA evidence complemented by fossil evidence. *Molecular Phylogenetics and Evolution* 9:585–598.

Goodman, M., Weiss, M. L., and Czelusniak, J. 1982. Molecular evolution above the species level: branching pattern, rates, and mechanisms. *Systematic Zoology* 31:376–399.

Goodman, M. J., Griffin, P. B., Estioko-Griffin, A. A., and Grove, J. S. 1985. The compatibility of hunting and mothering among the Agta hunter-gatherers of the Philippines. *Sex Roles* 12:1199–1209.

Goossens, B., Chikhi, L., Jalil, M. F., James, S., Ancrenaz, M., Lackman-Ancrenaz, I., and Bruford, M. W. 2009. Taxonomy, geographic variation and population genetics of Bornean and Sumatran orangutans. In S. A. Wich, S. S. Utami Atmoko, T. Mitra Setia, and C. P. van Schaik, eds., *Orangutans: Geographic Variation in Behavioral Ecology and Conservation,* 1–14. Oxford: Oxford University Press.

Goossens, B., Sechel, J. M., James, S. S., Funk, S. M., Chikhi, L., Abulani, A., Ancrenaz, M., Lackman-Ancrenaz, I., and Bruford, M. W. 2006. Philopatry and reproductive success in Bornean orang-utans *(Pongo pygmaeus). Molecular Ecology* 15:2577–2588.

Gordon, A. D., Green, D. J., and Richmond, B. G. 2008. Strong postcranial size dimorphism in *Australopithecus afarensis:* from two new resampling methods for multivariate data sets with missing data. *American Journal of Physical Anthropology* 135:311–328.

Gordon, P. Numerical cognition without words: evidence from Amazonia. *Science* 306:496–499.

Goren-Inbar, N., Alperson, N., Kislev, M. E., Simchoni, O., Melamed, Y., Ben-Nun, A., Werker, E. 2004. Evidence of hominin control of fire at Gesher Benot Ya-aqov, Isreal. *Science* 304:725–727.

Goren-Inbar, N., Feibel, C. S., Verosub, K. L., Melamed, Y., Kislev, M. E., Tchernov, E., Saragusti, I. 2000. Pleistocene milestones on the out-of-Africa corridor at Gesher Benot Ya-aqov, Isreal. *Science* 289:944–947.

Gould, J. J. 2004. Thinking about animals: how Donald R. Griffin (1915–2003) remade animal behavior. *Animal Cognition* 7:1–4.

Gould, K. G., Young, L. G., Smithwich, E. B., and Phythyon, S. R. 1993. Semen characteristics of the adult male chimpanzee *(Pan troglodytes). American Journal of Primatology* 29:221–232.

Gould, R. A. 1969. *Yiwara: Foragers of the Australian Desert.* New York: Charles Scribner's Sons.

Gould, S. J. 1996. *The Mismeasure of Man.* New York: W. W. Norton.

———. 2002. *The Structure of Evolutionary Theory.* Cambridge, MA: Belknap Press of Harvard University Press.

Goustard, M. 1965. Introduction á l'étude des émissions sonores des Hylobatinés. *Annales des Sciences Naturelles. Zoologie et Biologie Animale 12e série* 7:359–396.

———. 1976. The vocalizations of *Hylobates*. In D. M. Rumbaugh, ed., *Gibbon and Siamang*, Vol. 4, 135–166, Basel: Karger.

———. 1979. Les interactions acoustiques au cours des grandes émissions sonores des Gibbons *(Hylobates concolor leucogenys). Comptes Rendus Academie des Sciences Paris, Series* D, 288:1671–1673.

———. 1982. Les vocalisations des singes anthropomorphes. *Journal de Psychologie Normale et Pathologique* 79:141–166.

———. 1984. Patterns and functions of loud calls in the concolor gibbon In H. Preuschoft, D. J. Chivers, W. Y. Brockelman, and N. Creel, eds., *The Lesser Apes: Evolutionary and Behavioural Biology*, 404–415. Edinburgh: Edinburgh University Press.

———. 1986. L'utilisation d'instruments et la capacité d'adaptabilité chez le chimpanzé *(Pan troglodytes schweinfurthi)* observé en semi-liberté. *Journal de psychologie normale et pathologique* 81:395–412.

Gouzoules, S., Gouzoules, H., and Marler, P. 1984. Rhesus monkey *(Macaca mulatta)* screams: representational signaling in the recruitment of agonistic aid. *Animal Behaviour* 32:182–193.

Gowlett, J. A. J. 2008. Deep roots of kin: Developing the evolutionary perspective from prehistory. In N. J. Allen, H. Callan, R. Dunbar, and W. James, eds., *Early Human Kinship: From Sex to Social Reproduction*, 40–57. Malden, MA: Blackwell.

Gowlett, J. A. J., Harris, J. W. K., Walton, D., and Wood, B. A. 1981. Early archaeological sites, hominid remains and traces of fire from Chesowanja, Kenya. *Nature* 294:125–129.

Graczyk, T. K., DaSilva, A. J., Cranfield, M. R., Nizeyi, J. B., Kalema, G. R. N. N., and Pieniazek, N. J. 2001. *Cryptosporidium parvum* Genotype 2 infection is in free-ranging mountain gorillas *(Gorilla gorilla beringei)* of the Bwindi Impenetrable National Park, Uganda. *Parasitology Research* 87:368–360.

Graczyk, T. K., Mudakikwa, A. B., Cranfield, M. R., and Eilenberger, U. 2001. Hyperkerototic mange caused by *Sarcoptes scabiei* (Acariformes: Sarcoptidae) in juvenile mounain gorillas *(Gorilla gorilla beringei). Parasitology Research* 87:1024–1028.

Gradstein, F., Ogg, J., and Smith, A. 2004. *A Geological Time Scale 2004.* Cambridge: Cambridge University Press.

Grand, T. I. 1972. A mechanical interpretation of terminal branch feeding. *Journal of Mammalogy* 53:198–201.

Graubner, R. Buckwitz, R., Landmann, and Starke, A. 2009. Biomechanical analysis 12. IAAF World Championships in Athletics, Berlin. Available at: http://berlin.iaaf.org /records/biomechanics/.

Grausz, H. M., Leakey, R. E., Walker, A. C., and Ward, C. V. 1988. Associated cranial and postcranial bones of *Australopithecus boisei*. In *Evolutionary History of the "Robust" Australopithecines*, F. E. Grine, ed., 127–132. Hawthorne, NY: Aldine de Gruyter.

Graves, J. L. 2001. *The Emperor's New Clothes: Biological Theories of Race at the Millennium.* New Brunswick, NJ: Rutgers University Press.

Gravlee, C. C. 2009. How race becomes biology: embodiment of social inequality. *American Journal of Physical Anthropology* 139:47–57.

Gray, M., McNeilage, A., Fawcett, K., Robbins, M. M., Ssebide, B., Mbula, D., and Uwingeli, P. 2009. Censusing the mountain gorillas in the Virunga Volcanoes: complete sweep method versus monitoring. *African Journal of Ecology* 48:588–599.

Gray, P. B. 2011. The descent of a man's testosterone. *Proceedings of the National Academy of Science of the United States of America* 108:6141–16142.

Gray, P. M., Krause, B., Atema, J., Payne, R., Drumhansl, C., and Baptista, L. 2001. The music of nature and the nature of music. *Science* 291:52–54.

Green, D. J., and Alemseged, Z. 2012. *Australopithecus afarensis* scapular ontogeny, function, and the role of climbing in human evolution. *Science* 338:514–517.

Green, D. J., and Gordon, A. D. 2008. Metacarpal proportions in *Australopithecus africanus*. *Journal of Human Evolution* 54:705–719.

Green, R. E., Krause, J., Briggs, A. W., Marcic, T., Stenzl, U., Kircher, M., Patternson, N., Li, H., Zhai, W., Fritz, M. H-Y., et al. 2010. A draft sequence of the Neandertal genome. *Science* 328:710–722.

Greenberg, G., Partridge, T., and Ablah, E. 2007. The significance of the concept of emergence for comparative psychology. In D. A. Washburn, ed., *Primate Perspectives on Behaivor and Cognition*, 81–97. Washington, DC: American Psychological Association.

Greenberg, J. H. *Anthropological Linguistics: An Introduction*. New York: Random House.

Greenfield, L. O. 1974. Taxonomic reassessment of two *Ramapithecus* specimens. *Folia Primatologica* 22:97–115.

———. 1975. A comment on relative molar breadth in Ramapithecus. *Journal of Human Evolution* 4:267–273.

———. 1978. On the dental arcade reconstruction of Ramapithecus. *Journal of Human Evolution* 7:345–359.

———. 1979. On the adaptive pattern of "Ramapithecus." *American Journal of Physical Anthropology* 50:527–548.

———. 1980. A late divergence hypothesis. *American Journal of Physical Anthropology* 52:351–365.

———. 1983. Toward the resolution of discrepancies between phenetic and paleontological data bearing on the question of human origins. In R. L. Ciochon and R. S. Corruccini, eds., *New Interpretations of Ape and Human Ancestry*, 695–703. New York: Plenum Press.

Greenfield, P. M., Maynard, A. E., Boehm, C., and Schmidtling, E. Y. 2000. Cultural apprenticeship and cultural change. tool learning and imitation in chimpanzees and humans. In S. T. Parker, J. Langer, and M. L. McKinney, eds., *Biology, Brains and Behaviour*, 237–277. Santa Fe: SAR Press.

Greenfield, P. M., and Savage-Rumbaugh, E. S. 1984. Perceived variability and symbol use: a common language-cognition interface in children and chimpanzees *(Pan Troglodytes)*. *Journal of Comparative Psychology* 98:201–218.

———. 1990. Grammatical combination in *Pan paniscus:* processes of learning and invention in the evolution and development of language. In S. T. Parker and K. R. Gibson, eds., *"Language" and intelligence in monkeys and apes*, 540–578. Cambridge: Cambridge University Press.

———. 1993. Comparing communicative competence in child and chimp: *Journal of Child Language* the pragmatics of repetition. *Journal of Child Language* 20:1–26.

Gregory, W. K. 1916. Phylogeny of recent and extinct anthropoids, with special reference to the origin of man. *Bulletin American Museum of Natural History* 35:258–355.

———. 1927a. Dawn-man or ape? *Scientific American* 137:30–232.

———. 1927b. How near is the relationship of man to the chimpanzee-gorilla stock? *Quarterly Review of Biology* 2:549–560.

———. 1927c. The origin of man from the anthropoid stem—when and where? *Proceedings of the American Philosophical Society* 66:439–463.

———. 1927d. Two views on the origin of man. *Science* 65:601–605.

———. 1928. Were the ancestors of man primitive brachiators? *Proceedings of the American Philosophical Society* 67:129–150.

———. 1929. Is the pro-dawn man a myth? *Human Biology* 1:153–165.

———. 1930. A critique of Professor Osborn's theory of human origin. *American Journal of Physical Anthropology* 14:133–161.

———. 1934. *Man's Place among the Anthropoids.* Oxford: Clarendon Press.

———. 1950. *The Anatomy of the Gorilla.* New York: Columbia University Press.

Grehan, J. R. 2006. Mona Lisa smile: the morphological enigma of human and great ape evolution. *Anatomical Record* 289B:139–157.

Gremllion, K. J. 2011. *Ancestral Appetites Food in Prehistory.* Cambridge: Cambridge University Press.

Gresl, T. A., Baum, S. T., and Kemnitz, J. W. 2000. Glucose regulation in captive *Pongo pygmaeus abeli, P. p. pygmaeus,* and *P. p. abeli×P. p. pygmaeus* orangutans. *Zoo Biology* 19:193–208.

Grewe, T., Bornkessel, I., Zysset, S., Wiese, R., von Cramon, D. Y., and Schlesewsky, M. 2005. The emergence of the unmarked: a new perspective on the language-specific function of Broca's area. *Human Brain Mapping* 26:178–190.

Griffin, D. R. 1976. *The Question of Animal Awareness: Evolutionary Continuity and Mental Experience.* New York: Rockefeller University Press.

———. 1978. Prospects for a cognitive ethology. *Behavioral and Brain Sciences* 4:527–538, 555–629; 3:615–623.

———. 1981. *The Question of Animal Awareness: Evolutionary Continuity and Mental Experience.* New York: Rockefeller University Press.

———. 1983. Prospects for a cognitive ethology. In J. de Luce, J., and H. T. Wilder, eds., *Language in Primates: Perspectives and Implications,* 159–186. New York: Springer.

———. 1984a. *Animal Thinking.* Cambridge, MA: Harvard University Press.

———. 1984b. Animal thinking. *American Scientist* 72:456–464.

———. 1985. Animal consciousness. *Neuroscience and Biobehavioral Reviews* 9:615–622.

———. 2001a. *Animal Minds: Beyond Cognition to Consciousness.* Chicago: University of Chicago Press.

———. 2001b. Animals know more than we used to think. *Proceedings of the National Academy of Science of the United States of America* 98:4833–4834.

Griffin, D. R., and Speck, G. B. 2004. New evidence of animal consciousness. *Animal Cognition* 7:5–18.

Griffin, N. L., and Richmond, B. G. 2010. Joint orientation and function in great ape and human proximal phalanges. *American Journal of Physical Anthropology* 141:116–123.

Grillner, S. 2011. Human locomotor circuits conform. *Science* 334:912–913.

Grine, F. E. 1981. Trophic differences between 'gracile' and 'robust' australopithecines: a scanning electron microscopic analysis of occlusal events. *South African Journal of Science* 77:203–230.

———. 1982. A new juvenile hominid (Mammalia: Primates) from Member 3, Kromdraii Formation, Transvaal, South Africa. *Annals of the Transvaal Museum* 33:165–239.

———. 1984. Deciduous molar microwear of South African australopithecines. In D. J. Chivers, B. A. Wood and A. Bilsborough, eds., *Food Acquisition and Processing in Primates*, 525–534. New York: Plenum Press.

———. 1985a. Australopithecine evolution: the deciduous dental evidence. In E. Delson, ed., *Ancestors: The Hard Evidence*, 153–167. New York: Alan R. Liss.

———. 1985b. Dental morphology and the systematic affinities of the Taung fossil hominid. In ed. P. V. Tobias, *Hominid Evolution: Past, Present and Future*, 247–253. New York: Alan R. Liss.

———. 1986. Dental evidence for dietary differences in *Australopithecus africanus* and *Paranthropus*: a quantitative analysis of permanent molar microwear. *Journal of Human Evolution* 15:783–822.

———. 1988a. *Evolutionary History of the "Robust" Australopithecines.* Hawthorne, NY: Aldine de Gruyter.

———. 1988b. New craniodental fossils of *Paranthropus* from the Swartkrans Formation and their significance in "robust" australopithecine evolution. In F. E. Grine, ed., *Evolutionary History of the "Robust" Australopithecines*, 223–243. Hawthorne, NY: Aldine de Gruyter.

———. 1993a. Australopithecine taxonomy and phylogeny: historical background and recent interpretation. In R. L. Ciochon and J. G. Fleagle, eds., *The Human Evolution Source Book*, 198–210. Englewood Cliffs, NJ: Prentice Hall.

———. 1993b. Description and preliminary analysis of new hominid craniodental fossils from the Swartkrans Formation. In C. K. Brain, ed., *Swartkrans: A Cave's Chronicle of Early Man*, 75–116. Pretoria: Transvaal Museum.

Grine, F. E., Gwinnett, A. J., and Oaks, J. H. 1990. Early hominid dental pathology: interproximal caries in 1.5 million-year-old *Paranthropus robustus* from Swartkrans. *Archives of Oral Biology* 35:381–386.

Grine, F. E., and Kay, R. F. 1988. Early hominid diets from quantitative image analysis of dental microwear. *Nature* 333:765–768.

Grine, F. E., and Martin, L. B. 1988. Enamel thickness and development in *Australopithecus* and *Paranthropus*. In F. E. Grine, ed., *Evolutionary History of the "Robust" Australopithecines*, 3–42. Hawthorne, NY: Aldine de Gruyter.

Grodzinsky, Y., and Santi, A. 2008. The battle for Broca's region. *Cell* 12:474–480.

Gröning, F., Liu, J., Fagan, M. J., and O'Higgins, P. 2011. Why do humans have chins? Testing the mechanical significance of modern human symphyseal morphology with finite element analysis. *American Journal of Physical Anthropology* 144:593–606.

Grossi, G., Semenza, C., Corazza, S., and Volterra, V. 1996. Hemispheric specialization for sign language. *Neuropsychologia* 34:737–740.

Groves, C. P. 1970a. *Gigantopithecus* and the mountain gorilla. *Nature* 226:973–974.

———. 1970b. *Gorillas.* New York: Arco.

———. 1970c. Population systematics of the gorilla. *Journal of Zoology, London* 161:287–300.

———. 1986. Systematics of the great apes. In D. R. Swindler and J. Erwin, eds., *Comparative Primate Biology,* Vol. 1: *Systematics, Evolution, and Anatomy,* 187–217. New York: Alan R. Liss.

———. 1989. *A Theory of Human and Primate Evolution.* Oxford: Oxford University Press.

———. 2001. *Primate Taxonomy.* Washington, DC: Smithsonian Institution Press.

Groves, C. P., and Mazák, V. 1975. An approach to the taxonomy of the Hominidae: gracile Villafranchian hominids in Africa. *Casopis pro Mineralogii a Geologii* 20:225–247.

Grubb, P., Butynski, T. M., Oates, J. F., Bearder, S. K., Disotell, T. R., Groves, C. P., Struhsaker, T. T. 2003. Assessment of the Diversity of African Primates. *International Journal of Primatology* 24:1301–1357.

Gruber, T., Clay, A, and Zuberbühler, K. 2010. A comparison of bonobo and chimpanzee tool use: evidence for a female bias in the *Pan* lineage. *Animal Behaviour* 80:1023–1033.

Gruber, T., Muller, M. N., Strimling, P., Wranghm, R., and Zuberbühler, K. 2009. Wild chimpanzees rely on cultural knowledge to solve an experimental honey acquisition task. *Current Biology* 19:1806–1810.

Gu, Y. 1986. Preliminary research on the fossil gibbon of Pleistocene China. *Acta Anthropologica Sinica* 5:208–219.

———. 1989. Preliminary research on the fossil gibbons of the Chinese Pleistocene and recent. *Human Evolution* 4:509–514.

Gu, Y., and Lin, Y. 1983. First discovery of *Dryopithecus* in east China. *Acta Anthropologica Sinica* 2:305–314.

Güleç, E. S., Sevim, A., Pehevan, C., and Kaya, F. 2007. A new great ape from the late Miocene of Turkey. *Anthropological Science* 115:153–158.

Gumert, M. D., Hoong, L. K., and Malaivjitnond, S. 2011. Sex differences in the stone tool-use of a wild population of Burmese long-tailed macaques *(Macaca fascicularis aurea). American Journal of Primatology* 73:1239–1249.

Gunnell, G. F., Chiochon, R. L., Gingerich, P. D., and Holyroyd, P. A. 2002. New assessment of *Pondaungia* and *Amphipithecus* (Primates) from the late Middle Eocene of Myanmar, with a comment on 'Amphipithecidae.' *Contributions from the Museum of Paleontology, University of Michigan* 30:337–372.

Gunnell, G. F., and Miller, E. R. 2001. Origin of Anthropoidea: dental evidence and recognition of early anthropoids in the fossil record, with comments on the Asian anthropoid radiation. *American Journal of Physical Anthropology* 114:177–191.

Gunz, P. Booksein, F. L., Mitteroecker, P., Stadlmayr, A., Seidler, H., and Weber, G. W. 2009. Early modern human diversity suggests subdivided population structure and a complex out-of-Africa scenario. *Proceedings of the National Academy of Science of the United States of America* 106:6094–6098.

Gur, R. C., Turetsky, B. I., Matsui, M., Yan, M., Bilker, W., Hughett, P., and Gur, R. E. 1999. Sex differences in brain gray and white matter in healthy young adults: correlations with cognitive performance. *Journal of Neuroscience* 19:4065–4072.

Guschanski, K., Caillaud, D., Robbins, M. M., and Vigilant, L. 2008. Females shape the genetic structure of a gorilla population. *Current Biology* 18:1809–1814.

Guschanski, K., Vigilant, L., McNeilage, A., Gray, M., Kagoda, E., and Robbins, M. M. 2009. Counting elusive animals: comparing field and genetic census of the entire mountain gorilla population of Bwindi Impenetrable National Park, Uganda. *Biological Conservation* 142:290–300.

Guthrie, R. D. 1970. Evolution of human threat display organs. *Evolutionary Biology* 4:257–302.

Guy, F., Lieberman, D. E., Pilbeam, D., Ponce De León, M. S., Likius, A., Mackaye, H. T., Vignaud, P., Zollikofer, C. P. E., and Brunet, M., 2005. Morphological affinities of the *Sahelanthropus tschadensis* (Late Miocene hominid from Chad) cranium. *Proceedings of the National Academy of Science of the United States of America* 102:18836–18841.

Habib, M., Robichon, F., Lévroer. O., Khalil, R., and Salamon, G. 1995. Diverging asymmetries of temporo-parietal cortical areas: appraisal of Geschwind/Galaburda theory. *Brain and Language* 48:238–258.

Hacia, J. G. 2001. Genome of the apes. *Trends in Genetics* 17:637–645.

Hacia, J. G., Fan, J.-B., Ryder, O., Jin, L., Edgemon, K., Ghandour, G., Mayer, R. A., Sun, B., Hsie, L., Robbins, C. M., et al. 1999. Determination of ancestral alleles for human single-nucleotide polymorphisms using high-density oligonucleotide arrays. *Nature Genetics* 22:164–167.

Haeckel, E. 1966. *Generelle Morphologie.* Berlin: Reimer.

———. 1968. *Natürliche Schöpfungsgeschichte.* Berlin: Reimer.

———. 1974. *Anthropogenie oder Entwickelungsgeschichte des Menschen.* Leipzig: Engelmann.

Hagoort, P. 2005. On Broca, brain, and binding: a new framework. *Trends in Cognitive Sciences* 9:416–423.

Hagoort, P., Baggo, G., and Willems, R. M. 2009. Semantic unification. In M. S. Gazzaniga, ed., *The Cognitive Neurosciences,* 4th ed., 819–835. Cambridge, MA: MIT Press.

Haier, R. J., Jung, R. E., Yeo, R. A., Head, K., and Alkire, M. T. 2005 The neuroanatomy of general intelligence: sex matters. *NeuroImage* 25:320–327.

Haile-Selassie, Y., Latimer, B. M., Alene, M., Dieno, A. L., Gilbert, L., Melilo, S. M., Saylor, B. Z., Scott, G. R., and Lovejoy, C. O. 2010. An early *Australopithecus afarensis* postcranium from Woranso-Mille, Ethiopia. *Proceedings of the National Academy of Sciences of the United States of America* 107:12121–12126.

Haile-Selassie, Y., Saylor, B. Z., Dieno, A., Alene, M., and Latimer, B. M. 2010. New hominid fossils from Woranso-Mille (Central Afar, Ethiopia) and taxonomy of early *Australopithecus. American Journal of Physical Anthropology* 141:406–417.

Haile-Selassie, Y., Suwa, G., and White, T. D. 2003. Late Miocene teeth from Middle Awash, Ethiopia, and early hominid dental evolution. *Science* 203:1503–1505.

———. 2009. Hominidae. In Y. Haile-Selassie, and G. WoldeGabriel, eds., *Ardipithecus kadabba: Late Miocene Evidence from the Middle Awash Ethiopia,* 159–236. Berkeley: University of California Press.

Haimoff, E. H. 1981. Video analysis of siamang *(Hylobates syndactylus)* songs. *Behaviour,* 76:128–151.

———. 1983. Occurrence of anti-resonance in the song of the siamang *(Hylobates syndactylus). American Journal of Primatology* 5:249–256.

———. 1984. Acoustic and organizational features of gibbon songs, In H. Preuschoft, D. J. Chivers, W. Y. Brockelman, and H. Creel, eds., *The Lesser Apes: Evolutionary and Behavioural Biology,* 333–353. Edinburgh: Edinburgh University Press.

———. 1985. The organization of song in Müller's gibbon *(Hylobates muelleri). International Journal of Primatology* 6:173–192.

———. 1987. Preliminary observations of wild black-crested gibbons *(Hylobates concolor concolor)* in Yunnan Province, People's Republic of China. *Primates* 28:319–335.

Haimoff, E. H., Chivers, D. J., Gittins, S. P., and Whiten, A. J. 1982. A phylogeny of gibbons based on morphological and behavioral characters. *Folia Primatologica* 39:213–237.

Haimoff, E. H., and Gittins, S. P. 1985. Individuality in the songs of wild agile gibbons *(Hylobates agilis)* of Peninsular Malaysia. *American Journal of Primatology* 8:239–247.

Haimoff, E. H., Yang, X.-J., He, S.-J., and Chen, N. 1986. Census and Survey of wild black-crested gibbon *(Hylobates concolor concolor)* in Yunnan Province, People's Republic of China. *Folia Primatologica* 46:205–214.

Hall, C. M., Walter, R. C., Westgate, J. A., and York, D. 1984. Geochronology, stratigraphy and geochemistry of Cintery Tuff in Pliocene hominid-bearing sediments of the Middle Awash, Ethiopia. *Nature* 308:26–31.

Hall, J. S., Saltonstall, K., Inogwabini, B.-I., and Omari, I. 1998. Distribution, abundance and conservation status of Grauer's gorilla. *Oryx* 32:122–130.

Hall, J. S., White, L. J. T., Inogwabini, B.-I., Omari, I., Morland, H. S., Williamson, E. A., Saltonstall, K., Walsh, P., Sikubwabo, C., Bonny, D., et al. 1998. Survey of Grauer's gorillas *(Gorilla gorilla graueri)* and eastern chimpanzees *(Pan troglodytes schweinfurthii)* in the Kahuzi-Biega National Park lowland sector and adjacent forest of eastern Democratic Republic of Congo. *International Journal of Primatology* 19:207–235.

Hall, L. M., Jones, D., and Wood, B. 1996. Evolutionary relationships between gibbon subgenera inferred from DNA sequence data. *Biochemical Society Transactions* 24:416S.

Halperin, S. D. 1979. Temporary association patterns in free ranging chimpanzees: an assessment of individual grouping preferences. In D. A. Hamburg and E. R. McCown, eds., *The Great Apes,* 491–499. Menlo Park, CA: Benjamin/Cummings.

Hamai, M., Nishida, T., Takasaki, H., and Turner, L. A. 1992. New records of within-group infanticide and cannibalism in wild chimpanzees. *Primates* 33:151–162.

Hamard, M., Cheyne, S. M., and Nijman, V. 2010. Vegetation correlates of gibbon density in the peat-swamp forest of Sabangau Catchment, Central Kalimantan, Indonesia. *American Journal of Primatology* 72:607–616.

Hamburg, D. A. 1971. Aggressive behavior of chimpanzees and baboons in natural habitats. *Journal of Psychiatric Research* 8:385–398.

———. 2008. *Preventing Genocide: Practical Steps toward Early Detection and Effective Action.* Boulder, CO: Paradigm.

Hamdi, H., Nishio, H., Zielinski, R., and Dugaiczyk, A. 1999. Origin and phylogenetic distribution of Alu DNA repeats: irreversible events in the evolution of primates. *Journal of Molecular Biology* 289:861–871.

Hamer, D. 2002. Rethinking behavior genetics. *Science* 298:71–72.

Hamilton, M. J., Milne, B. T., Walker, R. S., and Brown, J. H. 2007. Nonlinear scaling of space use in human hunter-gatherers. *Proceedings of the National Academy of Sciences of the United States of America* 104:4765–4769.

Hamilton, R. A., and Galdikas, B. M. F. 1994. A preliminary study of food selection by the orangutan in relation to plant quality. *Primates* 35:255–263.

Hamilton, W. D. 1964a. The genetical evolution of social behavior. I. *Journal of Theoretical Biology* 7:1–16.

———. 1964b. The genetical evolution of social behavior. II. *Journal of Theoretical Biology* 7:17–52.

Hamilton, W. J., III, and Arrowood, P. C. 1978. Copulatory vocalizations of chacma baboons *(Papio ursinus)*, gibbons *(Hylobates hoolock)*, and humans. *Science* 200:1405–1409.

Hamilton, W. J., III, and Tilson, R. L. 1985. Fishing baboons at desert waterholes. *American Journal of Primatology* 8:255–257.

Hammer, M. F., Woerner, A. E., Mendez, F. L., Watkins, J. C., and Wall, J. D. 2011. Genetic evidence for archaic admixture in Africa. *Proceedings of the National Academy of Science of the United States of America* 108:15123–15128.

Handler, R. 2004. Afterword: mysteries of culture. *American Anthropologist* 106:488–494.

Hanna, J. B., Polk, J. D., and Schmitt, D. 2006. Forelimb and hindlimb forces in walking and galloping primates. *American Journal of Physical Anthropology* 130:529–535.

Haraway, D. 1989. *Primate Visions: Gender, Race, and Nature in the World of Modern Science.* New York: Routledge.

Haraway, M. M., and Maples E. G. 1998. Flexibility in the species-typical songs of gibbons. *Primates* 39:1–12.

Haraway, M. M., Maples E. G., and Tolson, S. 1981. Taped vocalization as a reinforcer of vocal behavior in a siamang gibbon *(Symphalangus syndactylus). Psychological Report* 49:995–999.

Harcourt, A. H. 1977. Virunga gorillas—the case against translocations. *Oryx* 13:469–472.

———. 1978a. Activity periods and patterns of social interaction: a neglected problem. *Behaviour* 66:121–135.

———. 1978b. Strategies of emigration and transfer by primates, with particular reference to gorillas. *Zeitschrift fur Tierpsychologie,* 48:401–420.

———. 1979a. Contrasts between male relationships in wild gorilla groups. *Behavioral Ecology and Sociobiology* 5:39–49.

———. 1979b. The social relations and group structure of wild mountain gorilla. In D. A. Hamburg and E. R. McCown, eds., *The Great Apes,* 186–192. Menlo Park, CA: Benjamin/Cummings.

———. 1979c. Social relationships among adult female mountain gorillas. *Animal Behaviour* 27:251–264.

———. 1979d. Social relationships between adult male and female mountain gorillas in the wild. *Animal Behaviour* 27:325–342.

———. 1980–1981. Can Uganda's gorillas survive? A survey of the Bwindi Forest Reserve. *Biological Conservation* 19:269–282.

———. 1981. Intermale competition and the reproductive behavior of the great apes. In C. E. Graham, ed., *Reproductive Biology of the Great Apes,* 301–318. New York: Academic Press.

———. 1986. Gorilla conservation: anatomy of a campaign. In K. Benirschke, ed., *Primates: The Road to Self-Sustaining Populations,* 32–46. New York: Springer.

———. 1996. Is the gorilla a threatened species? How should we judge? *Biological Conservation* 75:165–176.

———. 2003. An introductory perspective: gorilla conservation. In A. B. Taylor and M. L. Goldsmith, eds., *Gorilla Biology: A Multidisciplinary Perspective,* 407–413. Cambridge: Cambridge University Press.

Harcourt, A. H., and Curry-Lindahl, K. 1978. The FPS mountain gorilla project—a report from Rwanda. *Oryx* 14:316–324.

Harcourt, A. H., and Fossey, D. 1981. The Virunga gorillas: decline of an 'island' population. *African Journal of Ecology* 19:83–97.

Harcourt, A. H., Fossey, D., Sabater Pí, J. 1981. Demography of *Gorilla gorilla. Journal of Zoology, London* 195:215–233.

Harcourt, A. H., Fossey, D., Stewart, K. J., and Watts, D. P. 1980. Reproduction in wild gorillas and some comparisons with chimpanzees. *Journal of Reproduction and Fertility* 28 (Suppl.) :59–70.

Harcourt, A. H., and Greenberg, J. 2001. Do gorilla females join males to avoid infanticide? A quantitative model. *Animal Behaviour* 62:905–915.

Harcourt, A. H., and Groom, F. G. 1972. Gorilla census. *Oryx* 11:355–363.

Harcourt, A. H., Harvey, P. H., Larson, S. G., and Short, R. V. 1981. Testis weight, body weight and breeding system in primates. *Nature* 293:55–57.

Harcourt, A. H., Kineman, J., Campbell, G., Yamagiwa, J., Redmond, I., Aveling, C., and Condiotti, M. 1983. Conservation and the Virunga gorilla population. *African Journal of Ecology* 21:139–142.

Harcourt, A. H., and Stewart, K. J. 1977. Apes, sex, and societies. *New Scientist* 76:160–162.

———. 1978. Sexual behaviour of wild mountain gorillas. In D. J. Chivers and J. Herbert, eds., *Recent Advances in Primatology,* Vol. 1: *Behaviour,* 611–612. London: Academic Press.

———. 1981. Gorilla male relationships: can differences during immaturity lead to contrasting reproductive tactics in adulthood? *Animal Behaviour* 29:206–210.

———. 1983. Interactions relationships and social structure: the great apes. In *Primate Social Relationships: An Integrated Approach,* R. A. Hinde, ed., pp. 307–314. Sunderland, MA: Sinauer.

———. 1984. Gorillas' time feeding: aspects of methodology, body size, competition and diet. *African Journal of Ecology* 22:207–215.

———. 1987. The influence of help in contests on dominance rank in primates: hints from gorillas. *Animal Behaviour* 35:182–190.

———. 1989. Functions of alliances in contests within wild gorilla groups. *Behaviour* 109:176–190.

———. 1996. Function and meaning of wild gorilla 'close' calls. 2. Correlations with rank and relatedness. *Behaviour* 133:827–845.

———. 2001. Vocal relationships of wild mountain gorillas. In M. M. Robbins, P. Sicotte and K. J. Stewart, eds., *Mountain Gorillas,* 241–262. Cambridge: Cambridge University Press.

———. 2007. *Gorilla Society.* Chicago: University of Chicago Press.

Harcourt, A. H., Stewart, K. J., and Fossey, D. 1976. Male emigration and female transfer in wild mountain gorilla. *Nature* 263:226–227.

———. 1981. Gorilla reproduction in the wild. In C. E. Graham, ed., *Reproductive Biology of the Great Apes,* 265–279. New York: Academic Press.

Harcourt, A. H., Stewart, K. J., and Harcourt, D. E. 1986. Vocalizations and social relationships of wild gorillas: a preliminary analysis. In D. M. Taub and F. A. King, eds., *Current Perspectives in Primate Social Dynamics,* 346–356.

Harcourt, A. H., Stewart, K. J., and Hauser, M. 1993. Functions of wild gorilla 'close' calls. I. Repertoire, context, and interspecific comparison. *Behaviour* 124:89–122.

Harcourt, A. H., Stewart, K. J., and Inahoro, I. M. 1989. Gorilla quest in Nigeria. *Oryx* 23:7–13.

Harcourt-Smith, E. H., and Hilton, C. E. 2005. Did *Australopithecus afarensis* make the Laetoli footprint trails? New insights into an old problem. *American Journal of Physical Anthropology* 40 (Suppl.): 112.

Hardus, M. E., Lameira, A. R., Singleton, I., Morrogh-Bernard, H. C., Knott, C. D., Ancrenaz, M., Utami Atmoko, S. S., and Wich, S. A. 2009. A description of the orangutan's vocal and sound repertoire, with a focus on geographic variation. In S. A. Wich, S. S. Utami Atmoko, T. Mitra Setia, and C. P. van Schaik, eds., *Orangutans: Geographic Variation in Behavioral Ecology and Conservation,* 49–64. Oxford: Oxford University Press.

Hardy, A. C. 1960. Was man more aquatic in the past? New Scientist 7:642–645.

———. 1977. Was there a *Homo aquaticus? Zenith* 15:4–6.

Hare, B., Melis, A. P., Woods, V., Hastings, S., and Wrangham, R. 2007. Tolerance allows bonobos to outperform chimpanzees on a cooperative task. *Current Biology* 17:619–623.

Harlacker, L. 2006. Knowledge and know-how in the Oldowan: an experimental approach. In J. Apel and K. Knutsson, *Skilled Production and Social Reproduction,* eds. 219–243. Uppsala: Societas Archaeologica Upsaliensis.

Harlow, H. F. 1949. The formation of learning sets. *Psychological Review* 56:51–65.

Harris, D., Fay, J. M., and MacDonald, N. 1987. Report of gorillas from Nigeria. *Primate Conservation* 8:40.

Harris, D. R. 1980. Tropical savanna environments: definition, distribution, diversity and development. In D. R. Harris, ed., *Human Ecology in Savanna Environments,* 3–27. New York: Academic Press.

Harris, J. M. 1985. Age and paleoecology of the Upper Laetolil Beds, Laetoli, Tanzania. In E. Delson, ed., *Ancestors: The Hard Evidence,* 76–81. New York: Alan R. Liss.

———. 1987. Summary. In M. D. Leakey and J. M. Harris, eds., *Laetoli: A Pliocene Site in Northern Tanzania*, 524–531. Oxford: Clarenden Press.

Harris, J. W. K. 1983. Cultural beginnings: Plio-Pleistocene archaeological occurrences from the Afar, Ethiopia. *African Archaeological Review* 1:3–31.

Harris, M. 1999. *Theories of Culture in Postmodern Times*. Walnut Creek, CA: AltaMira Press.

Harris, R. A., Rogers, J., and Milosavljevic, A. 2007. Human-specific changes of genome structure detected by genomic triangulation. *Science* 316:235–237.

Harris, S., Thackeray, J. R., Jeffreys, A. J., and Weiss, M. L. 1986. Nucleotide sequence analysis of the lemur beta-globin gene family: evidence for major rate fluctuations in globin polypeptide evolution. *Molecular Biology and Evolution* 3:465–484.

Harrison, M. E., and Marshall, A. J. 2011. Strategies for the use of fallback foods in apes. *International Journal of Primatology* 32:531–565.

Harrison, T. 1981. New finds of small fossil apes from the Miocene locality at Koru in Kenya. *Journal of Human Evolution* 10:129–137.

———. 1986a. New fossil anthropoids from the Middle Miocene of East Africa and their bearing on the origin of the Oreopithecidae. *American Journal of Physical Anthropology* 71:265–284.

———. 1986b. A reassessment of the phylogenetic relationships of *Oreopithecus bambolii* Gervais. *Journal of Human Evolution* 5:541–583.

———. 1987. The phylogenetic relationships of the early catarrhine primates: a review of the current evidence. *Journal of Human Evolution* 16:41–80.

———. 1988. A taxonomic revision of the small catarrhine primates from the Early Miocene of East Africa. *Folia Primatologica* 50:59–108.

———. 1989a. New postcranial remains of *Victoriapithecus* from the middle Miocene of Kenya. *Journal of Human Evolution* 8:3–54.

———. 1989b. A new species of *Micropithecus* from the middle Miocene of Kenya. *Journal of Human Evolution* 18:537–557.

———. 1989c. New estimates of cranial capacity, body size and encephalization in *Oreopithecus bambolii*. *American Journal of Physical Anthropology* 78:237.

———. 1991a. Some observations on the Miocene hominoids of Spain. *Journal of Human Evolution* 20:515–520.

———. 1991b. The implications of *Oreopithecus bambolii* for the origins of bipedalism. In Y. Coppens and B. Senut, eds., *Origine(s) de la bipédie chez les hominidés*, 235–244. Paris: Éditions du Centre National de la Recherche Scientifique.

———. 1992. A reassessment of the taxonomic and phylogenetic affinities of the fossil catarrhines from Fort Ternan, Kenya. *Primates* 33:501–522.

———. 1993. Cladistic concepts and the species problem in hominoid evolution. In W. H. Kimbel and L. B. Martin, eds., *Species, Species Concepts, and Primate Evolution*, 345–371. New York: Plenum Press.

———. 1998. Evidence for a tail in *Proconsul heseloni*. *American Journal of Physical Anthropology* 26 (Suppl.): 93–94.

———. 2002. Late Oligocene to Middle Miocene catarrhines from Afro-Arabia. In W. C. Hartwig, ed., *The Primate Fossil Record*, 311–338. Cambridge: Cambridge University Press.

———. 2005. The zoogeographic and phylogenetic relationships of early catarrhine primates in Asia. *Anthropological Science* 13:43–51.

———. 2006. Taxonomy, phylogenetic relationships, and biogeography of Miocene hominoids from Yunnan, China. In Yang Decong, ed., *Collected Works for the 40th Anniversary of Yuanmou Man Discovery and the International Conference on Palaeoanthroploogical Studies,* 233–249. Kunming, China: Yunnan Science & Technology Press.

Harrison, T., Delson, E., and Guan, J. 1991. A new species of *Pliopithecus* from the middle Miocene of China and its implications for early catarrhine zoogeography. *Journal of Human Evolution* 21:329–361.

Harrison, T., and Gu, Y. 1999. Taxonomy and phylogenetic relationships of early Miocene catarrhines from Sihong, China. *Journal of Human Evolution* 37:225–277.

Harrison, T., Ji, X., and Su, D. 2002. On the systematic status of the late Neogene hominoids from Yunnan Province, China. *Journal of Human Evolution* 43:207–227.

Harrison, T., Made, J. van der, and Ribot, F. 2002. A new middle Miocene pliopithecid from Sant Quirze, northern Spain. *Journal of Human Evolution* 42:371–377.

Harrison, T., and Rook, L., Enigmatic anthropoid or misunderstood ape? The phylogenetic status of *Oreopithecus bambolii* reconsidered. 1997. In D. R. Begun, C. V. Ward, and M. D. Rose, eds., *Function, Phylogeny, and Fossils,* 327–362. New York: Plenum Press.

Harrison, T., and Yumin, G. 1999. Taxonomy and phylogenetic relationships of early Miocene catarrhines from Sihong, China. *Journal of Human Evolution* 37:225–277.

Harrison, T. S., and Harrison, T. 1989. Palynology of the late Miocene *Oreopithecus*-bearing lignite from Baccinello, Italy. *Palaeogeography, Palaeoclimatology, Palaeoecology* 76:45–65.

Harrisson, B. 1962. *Orang-utan,* London: Collins.

Hart D., and Karmel, M. P. 1996. Self-awareness and self-knowledge in humans, apes, and monkeys. In A. E. Russon, K. A. Bard, and S. Y. Parker, eds., *Reaching into Thought: The Minds of the Great Apes,* 325–347. Cambridge: Cambridge University Press.

Hart, D. L., and Sussman, R. W. 2009. *Man the Hunted.* Boulder CO: Westview Press.

Hart, J. A., Grossmann, F., Vosper, A., and Ilanga, J. 2008. Human hunting and its impact on bonobos in the Salonga National Park, Democratic Republic of Congo. In T. Furuichi and J. Thompson, eds., *The Bonobos: Behavior, Ecology, and Conservation,* 245–271. New York: Springer.

Hartman, S. E. 1988. A cladistic analysis of hominoid molars. *Journal of Human Evolution* 17:489–502.

———. 1989. Stereophotogrammetric analysis of occlusal morphology of extant hominoid molars: phenetics and function. *American Journal of Physical Anthropology* 80:145–166.

Hartung, J. 1982. Polygyny and inheritance of wealth. *Current Anthropology* 23:1–12.

Hartwig-Scherer, S., and Martin, R. D. 1991. Was "Lucy" more human than her "child?" Observations on early hominid postcranial skeletons. *Journal of Human Evolution* 21:439–449.

Harvey, P. H., Martin, R. D., and Clutton-Brock, T. H. 1987. Life histories in comparative perspective. In B. B. Smuts, D. L. Cheney, R. M. Seyfarth, R. W. Wrangham, and T. T. Struhsaker, eds., *Primate Societies,* 181–196. Chicago: University of Chicago Press.

Hasegawa, M. 1990. Phylogeny and molecular evolution in primates. *Japanese Journal of Genetics* 65:243–266.

———. 1991. Molecular phylogeny and man's place in Hominoidea. *Journal of the Anthropological Society of Nippon* 99:49–61.

———. 1992. Evolution of hominoids as inferred from DNA sequences. In T. Nishida, W. C. McGrew, P. Marler, M. Pickford, and F. B. M. de Waal, eds., *Topics in Primatology,* Vol. 1: *Human Origins,* 347–357. Tokyo: University of Tokyo Press.

Hasegawa, M., and Kishino, H. 1989. Confidence limits on the maximum-likelihood estimate of the hominoid tree from mitochondrial-DNA sequences. *Evolution* 43:672–677.

———. 1991. DNA sequence analysis and evolution of Hominoidea. In M. Kimura and N. Takahata, eds., *New Aspects of the Genetics of Molecular Evolution,* 303–317. Tokyo: Japan Science Society Press.

Hasegawa, M., Kishino, H., and Yano, T. 1985. Dating of the human-ape splitting by a molecular clock of mitochondrial DNA. *Journal of Molecular Evolution* 22:60–174.

———. 1987. Man's place in Hominoidea as inferred from molecular clocks of DNA. *Journal of Molecular Evolution* 26:132–147.

———. 1988. Phylogenetic inference from DNA sequence data. In K. Matusita, ed., *Statistical Theory and Data Analysis II,* 1–13. Amsterdam: North-Holland.

———. 1989. Estimation of branching dates among primates by molecular clocks of nuclear DNA which slowed down in Hominoidea. *Journal of Human Evolution* 18:461–476.

Hasegawa, M., and Yano, T. 1984. Phylogeny and classification of Hominoidea as inferred from DNA sequence data. *Proceedings of the Japan Academy, Series B* 60:389–392.

Hasegawa, T., Hiraiwa, M., Nishida, T., and Takasaki, H. 1983. New evidence on scavenging behavior in wild chimpanzees. *Current Anthropology* 24:231–232.

Hasegawa, T., and Hiraiwa-Hasegawa, M. 1983. Opportunistic and restrictive matings among wild chimpanzees in the Mahale Mountains, Tanzania. *Journal of Ethology* 1:75–85.

Hashimoto, C., and Furuichi, T. 2001. Current situation of bonobos in the Luo Reserve, Equateur, Democratic Republic of Congo. In B. M. F.,Galdikas, N. E. Briggs, L. K. Sheeran, G. L. Shapiro, and J. Goodall, eds., *All Apes Great and Small,* Vol. 1: *African Apes,* 83–93. New York: Kluwer Academic/Plenum Press.

———. 2006. Comparison of behavioral sequence of copulation between chimpanzees and bonobos. *Primates* 47:51–55.

Hashimoto, C., Furuichi, T., and Tashiro, Y. 2000. Ant dipping and meat eating by wild chimpanzees in the Kalinzu Forest, Uganda. *Primates* 41:103–108.

———. 2001. What factors affect the size of chimpanzee parties in the Kalinzu Forest, Uganda? Examination of fruit abundance and number of estrous females. *International Journal of Primatology* 22:947–959.

Hashimoto, C., Suzuki, S., Takenoshita, Y., Yamigawa, J., Basabose, A. K., and Furuichi, T. 2003. How fruit abundance affects the chimpanzee party size: a comparison between four study sites. *Primates* 44:77–81.

Hashimoto, C., Tashiro, Y., Hibino, E., Mulavwa, M., Yangozene, K., Furuchi, T., Idani, G., and Takenaka, O. 2008. Longitudinal structure of a unit-group of bonobos: male philopatry and possible fusion of unit-groups. In T. Furuichi and J. Thompson, eds., *The Bonobos: Behavior, Ecology, and Conservation,* 107–134. New York: Springer.

Hashimoto, C., Tashiro, Y., Kimura, D., Enomoto, T., Ingmanson, E. J., Idani, G., Furuchi, T. 1998. Habitat use and ranging of wild bonobos *(Pan paniscus)* in Wamba. *International Journal of Primatology* 19:1045–1060.

Hattori, M., Fujiyama, A., Taylor, T. D., Watanabe, H., Yada, T., Park, H.-S., Toyoda, A., Ishii, K., Totoki, Y., Choi, D.-K., et al. 2000. The DNA sequence of human chromosome 21. *Nature* 405:311–319.

Hauser, M. D. 1993a. Right hemisphere dominance for the production of facial expression in monkeys. *Science* 261:475–477.

———. 1993b. The role of articulation in the production of rhesus monkey, *Macaca mulatta,* vocalizations. *Animal Behaviour* 45:423=433.

———. 2000a. *Wild Minds: What Animals Really Think.* New York: Henry Holt.

———. 2000b. The sound and the fury: primate vocalizations as reflections of emotion and thought. In N. L. Wallin, B. Merker and S. Brown, eds., *The Origins of Music,* 77–102. Cambridge, MA: MIT Press.

———. 2005a. Beyond the chimpanzee genome: the threat of extinction. *Science* 309:1498–1499.

———. 2005b. Our chimpanzee mind. *Nature* 437:60–63.

Hauser, M. D., Chomsky, N., and Fitch, W. T. 2002. The faculty of language: what is it, who has it, and how did it evolve? *Science* 298:1569–1579.

Hauser, M. D., Evans, C. E., and Marler, P. 1993. The role of articulation in the production of rhesus monkey, *Macaca mulatta,* vocalizations. *Animal Behaviour* 45:423–433.

Hauser, M. D., and Fitch, W. T. 2003. What are the uniquely human components of the language faculty? In M. H. Christiansen and S. Kirby, eds., *Language Evolution,* 158–181. Oxford: Oxford University Press.

Hauser, M. D., MacNeilage, P., and Ware, M. 1996. Numerical representation in primates. *Proceedings of the National Academy of Science of the United States of America* 93:1514–1517.

Hauser, M. D., and McDermott, J. 2003. The evolution of the music faculty: a comparative perspective. *Nature Neuroscience* 6:663–668.

Häusler, M., and Schmid, P. 1995. Comparison of the pelves of Sts 14 and AL 288–1: implications for birth and sexual dimorphism in australopithecines. *Journal of Human Evolution* 29:363–383.

Hawkes, K. 1991. Showing off: tests of an hypothesis about men's foraging goals. *Ethology and Sociobiology* 12:29–54.

———. 1993. Why hunter-gatherers work: an ancient version of the problem of public goods. *Current Anthropology* 34:341–361.

Hawkes, K., O'Connell, J. F., Blurton Jones, N. G. 1989. In V. Standen and R. A. Foley, eds., *Comparative Socioecology: The Behavioural Ecology of Humans and Other Mammals,* 341–366. Oxford: Blackwell Scientific.

———. 1991. Hunting income patterns among Hadze: big game, common goods, foraging goals and the evolution of human diet. *Philosophical Transactions of the Royal Society of London, series B* 334:243–251.

———. 1997. Hadze women's time allocation, offspring, provisioning, and the evolution of long postmenopausal life spans. *Current Anthropology* 38:551–557.

Hawkes, K., O'Connell, J. F., Blurton Jones, N. G., Alvarez, H., and Charnov, E. L. 1998. Grandmothering, menopause, and the evolution of human life histories. *Proceedings of the National Academy of Science of the United States of America* 95:1336–1339.

Hawks, J., Wang, E. T., Cochran, G. M., Harpending, H. C., and Moyzis, R. K. 2007. Recent acceleration of human adaptive evolution. *Proceedings of the National Academy of Science of the United States of America* 104:20753–20758.

Haxton, H. A. 1947. Muscles of the pelvic limb: a study of the differences between bipeds and quadrupeds. *Anatomical Record* 98:337–345.

Hay, R. L. 1978. Melilitite-Carbonatite tuffs in the Laetolil Beds of Tanzania. *Contributions to Mineralogy and Petrology* 67:357–367.

———. 1981. Paleoenvironment of the Laetolil Beds, northern Tanzania. In G. Rapp Jr. and C. F. Vondra, eds., *Hominid Sites: their Geologic Settings,* 7–24. Boulder, CO: Westview Press.

———. 1987. Geology of the Laetoli area. In M. D. Leakey and J. M. Harris, eds., *Laetoli, a Pliocene Site in Northern Tanzania,* 23–47. Oxford: Clarendon Press.

Hay, R. L., and Leakey, M. D. 1982. The fossil footprints of Laetoli. *Scientific American* 246:50–57.

Hayakawa, T., Angata, T., Lewis, A. L., Mikkelsen, T. S., Varki, N. M., and Varki, A. 2005. A human-specific gene in microglia. *Science* 309:1693.

Hayasaka, K., Gojobori, T., and Horai, S. 1988. Molecular phylogeny and evolution of primate mitochondrial DNA. *Molecular Biology and Evolution* 5:626–644.

Hayashi, S., Hayasaka, K., Takenaka, O., and Horai, S. 1995. Molecular phylogeny of gibbons inferred from mitochondrial DNA sequences: preliminary report. *Journal of Molecular Evolution* 41:359–365.

Hayes, C. 1951. *The Ape in Our House.* New York: Harper.

Hayes, K. J., and Hayes, C. 1951. The intellectual development of a home-raised chimpanzee. *Proceedings of the American Philosophical Society* 95:105–109.

———. 1954. The cultural capacity of chimpanzee. *Human Biology* 26:288–303.

Hayes, K. J., and Nissen, C. H. 1971. Higher mental functions of a home-raised chimpanzee. In A. M. Schrier and F. Stollnitz, eds., *Behavior of Nonhuman Primates: Modern Research Trends,* Vol. 4, 59–115. New York: Academic Press.

Hazeltine, E., and Ivry, R. B. 2002. Can we teach the cerebellum new tricks? *Science* 296:1979–1980.

Head, J. S., Boesch, C., Makaga, L., and Robbins, M. M. 2011. Sympatric chimpanzees *(Pan troglodytes troglodytes)* and gorillas *(Gorilla gorilla gorilla)* in Loango National Park, Gabon: dietary composition, seasonality, and intersite comparisons. *International Journal of Primatology* 32:755–775.

Hebb, D. O. 1949. *The Organization of Behavior, a Neuropsychological Theory.* New York: Wiley.

Heberer, G. 1956. Die Fossilgeschichte der Hominoidea. In H. Hofer, A. H. Schultz, and D. Starck, eds., *Primatologia I: Systematik, Phylogenie, Ontogenie,* 379–560. Basel: Karger.

———. 1959. The descent of man and the present fossil record. *Cold Spring Harbor Symposia on Quantitative Biology* 24:235–244.

Heeney, J. L., Dalgleish, A. G., and Weiss, R. A. 2006. Origins of HIV and the evolution of resistance to AIDS. *Science* 313:462–466.

Heesy, C. P. 2004. On the relationship between orbit orientation and binocular visual field overlap in mammals. In *Evolution of the Special Senses in Primates,* ed. T. D. Smith, C. F. Ross, N. J. Dominy, and J. T. Laitman, special issue, *Anatomical Record Part A* 281A:1104–1110.

Heffner, H. E., and Heffner, R. S. 1984. Temporal lobe lesions and perception of species-specific vocalizations by macaques. *Science* 226:75–76.

Heffner, R. S. 2004. Primate hearing form a mammalian perspective. In *Evolution of the Special Senses in Primates,* ed. T. D. Smith, C. F. Ross, N. J. Dominy, and J. T. Laitman, special issue, *Anatomical Record Part A* 281A:1111–1122.

Heiervang, E., Hugdahl, K., Steinmetz, H., Smievoll, A. I., Stevenson, J., Lund, A., Ersland, L., and Lundervold, A. 2000. Planum temporale, planum parietale and dichotic listening in dyslexia. *Neuropsychologia* 38:1704–1713.

Heim, J.-L. 1968. Les restes Néandertaliens de La Ferrassie: nouvelles données sur la stratigraphie et inventaire des squelettes. *Comptes rendus de l'Académie des Sciences, Paris, série D* 266:576–578.

———. 1970. L'encéphale Néandertalien de L'Homme de La Ferrassie. *Anthropologie* 74:527–572.

———. 1974. Les hommes fossiles de La Ferrassie (Dordogne) et le problème de la définition des Néandertaliens classique. *Anthropologie* 78:81–112, 321–378.

———. 1976. *Les hommes fossiles de La Ferrassie,* Vol. 1: *Le Giesement: Les squelettes adultes (crâne et squelette du tronc).* Archives de l'Institut de Paléontologie Humaine, Mémoire 35. Paris: Masson.

———. 1982a. *Les enfants Néandertaliens de la Ferrassie.* Paris: Masson.

———. 1982b. *Les hommes fossiles de La Ferrassie,* Vol. 1: *Les squelettes d'adultes (squelette des membres).* Archives de l'Institut de Paléontologie Humaine, Mémoire 38. Paris: Masson.

———. 1984. Les squelettes moustériens de La Ferrassie. Principaux caractères et position anthropologique parmi les autres Néandertaliens, In E. Delporte, ed., *Le grande abri de La Ferrassie: fouilles 1968–1973,* 249–271. Etudes Quaternaires, Mémoire 7. Paris: Institut de Paléontologie Humaine.

Heintz, E., Brunet, M., and Battail, B. 1981. A cercopithecid primate from the Late Miocene of Molayan, Afghanistan, with remarks on *Mesopithecus. International Journal of Primatology* 2:273–284.

Heinzelin, J. de, Clark, J. D., White, T., Hart, W., Renne, P., Wolde-Gabriel, G., Beyene, Y., and Vrba, E. 1999. Environment and behavior of 2.5-million-year-old Bouri hominids. *Science* 284:625–629.

Heizmann, E. P. J., and Begun, D. R. 2001. The oldest Eurasian hominoid. *Journal of Human Evolution* 41:463–481.

Hellekant, G., Ninomiya, Y., and Danilova, V. 1998. Taste in chimpanzees III. Label-line coding in sweet taste. *Physiology and Behavior* 60:191–200.

Hellekant, G., Ninomiya, Y. DuBois, G. E., Danilova, V., and Roberts, T. W. 1996. Taste in chimpanzees I. The summated response to sweeteners and the effect of gymnemic acid. *Physiology and Behavior* 60:469–479.

———. 1997. Single chorda tympani fibers. *Physiology and Behavior* 60:829–841.

Heller, J. A., Knott, C. D., Conklin-Brittain, N. L., Rudel, L. L., Wison, M. D., and Froelich, J. W. 2002. *American Journal of Primatology* 57 (Suppl. 1): 44.

Heller, R., Sander, A. F., Wang, C. W., Usman, F., and Dabelsteen, T. 2010. Macrogeographical variability in the great call of *Hylobates agilis:* assessing the applicability of vocal analysis in studies of fine-scale taxonomy of gibbons. *American Journal of Primatology* 72:142–151.

Hellige, J. B. 1993. *Hemispheric Asymmetry: What's Right and What's Left.* Cambridge, MA: Harvard University Press.

Hemingway, C. A., and Bynum, N. 2005. The influence of seasonality in primate diet and ranging. In D. K. Brockman and C. P. van Schaik, eds., *Seasonality in Primates,* 57–104. Cambridge: Cambridge University Press.

Hemmer, H. 1972. Notes sur la position phylétique de l'homme de Petralona. *L'Anthropologie (Paris)* 76:155–161.

Henn, B. M., Bustamante, C. D., Mountain, J. L., and Feldman, M. W. 2011. Reply to Hublin and Klein: locating a geographic point of dispersion in Africa for contemporary humans. *Proceedings of the National Academy of Sciences of the United States of America* 108:E278.

Henn, B. M., Gignoux, C. R., Jobin, M., Granka, J. M., Macpherson, J. M., Kidd, J. M., Rodríguez-Botigue, L., Ramachandran, S., Hon, L., Brisbin, A., et al. 2011. Hunter-gatherer genomic diversity suggests a southern African origin for modern humans. *Proceedings of the National Academy of Sciences of the United States of America* 108:5154–5162.

Henneberg, M. The mode and rate of human evolution and the recent Liang Bua finds. In E. Indriati, ed., *Recent Advances in Southeast Asian Paleoanthropology and Archaeology,* 24–29. Yogyakarta, Indonesia: Gadjah Mada University.

Henneberg, M., and Thorne, A. 2004. Flores human may be pathological *Homo sapiens. Before Farming* 4:2–3.

Henry, A. G., Brooks, A. S., and Piperno, D. R. 2011. Microfossils in calculus demonstrate consumption of plants and cooked foods in Neanderthal diets (Shandar III, Iraq; Spy 1 and II, Belgium). *Proceedings of the National Academy of Sciences of the United States of America* 108:486–491.

Hens, S. M. 2005. Ontogeny of craniofacial sexual dimorphism in the orangutan *(Pongo pygmaeus).* I: face and palate. *American Journal of Primatology* 65:149–166.

Henshilwood, C. S., d'Errico, F., van Niekerk, K. L., Coquinot, Y., Jacobs, Z., Lauritzen, S.-E., Menu, M., and García-Moreno, R. 2011. A 100,000-year-old ochre-processing workshop at Blombos Cave, South Africa. *Science* 334:219–222.

Herbert, W. 1983. Lucy's family problems. *Science News* 124:8–11.

Herbinger, I., Papworth, S., Boesch, C., and Zuberbühler, K. 2009. Vocal, gestural and locomotor responses of wild chimpanzees to familiar and unfamiliar intruders: a playback study. *Animal Behaviour* 78:1389–1396.

Herculano-Houzel, S. 2012. The remarkable, yet not extraordinary, human brain is a scaled–up primate brain and its associated cost. *Proceedings of the National Academy of Science of the United States of America* 109:10661–10668.

Herculano-Houzel, S., Collins, C. E., Wong, P., Kaas, J. H., and Lent, R. 2008. The basic nonuniformity of the cerebral cortex. *Proceedings of the National Academy of Science of the United States of America* 105:12593–12598.

Hernandez-Aguilar, R. A., Moore, J., and Pickering, T. R. 2007. Savanna chimpanzees use tools to harvest the underground storage organs of plants. *Proceedings of the National Academy of Science of the United States of America* 104:19210–19213.

Herrmann, E., Call, J., Hernández-Lloreda, M. V., Hare, B., and Tomasello, M. 2007. Humans have evolved specialized skills of social cognition: the cultural intelligence hypothesis. *Science* 317:1360–1366.

Herrmann, E., Wobber, V., and Call, J. 2008. Great apes' *(Pan troglodytes, Pan paniscus, Gorilla gorilla, Pongo pygmaeus)* understanding of tool functional properties after limited experience. *Journal of Comparative Psychology* 122:220–230.

Hershkovitz, I., Kornreich, L., and Laron, Z. 2007. Comparative skeletal features between *Homo floresiensis* and patients with primary growth hormone insensitivity (Laron syndrome). *American Journal of Physical Anthropology* 134:198–208.

Hershkovitz, P. 1974. A new genus of Late Oligocene monkey (Cebidae, Platyrrhini) with notes on postorbital closure and platyrrhine evolution. *Folia Primatologica* 21:1–35.

Hess, E. 2008. *Nim Chimpsky: The Chimp Who Would Be Human.* New York: Bantam Books.

Hess, J. P. 1973. Some observations on the sexual behavior of captive lowland gorillas, *Gorilla gorilla.* In R. P. Michael and J. H. Crook, eds., *Comparative Ecology and Behaviour of Primates,* 507–581. London: Academic Press.

Hewes, G. W. 1961. Food transport and the origins of hominid bipedalism. *American Anthropologist* 63:687–710.

———. 1964. Hominid bipedalism: independent evidence for the food-carrying theory. *Science* 146:416–418.

———. 1973a. An explicit formulation of the relationship between tool-using, tool-making and the emergence of language. *Visible Language* 7:101–127.

———. 1973b. Primate communication and the gestural origin of language. *Current Anthropology* 14:5–24.

———. 1976. The current status of the gestural theory of language origin. *Annals of the New York Academy of Sciences* 280:482–504.

———. 1983. The evolution of phonemically-based language. In É. de Grolier, ed., *Glossogenetics: The Origin and Evolution of Language,* 143–162. New York: Harwood Academic.

Hewitt, G., MacLarnon, A., and Jones, K. E. 2002. The functions of laryngeal air sacs in Primates: a new hypothesis. *Folia Primatologica* 73:70–94.

Heyes, C. M. 1994. Reflections on self-recognition in primates. *Animal Behaviour* 47:909–919.

———. 1998. Theory of mind in nonhuman primates. *Behavioral and Brain Sciences* 21:101–148.

Hiatt, L. R. 1980. Polyandry in Sri Lanka: a test case for parental investment theory. *Man* 15:583–602.

Hicks. T. C., Fouts, R. S., and Fouts, D. H. 2005. Chimpanzee *(Pan troglodytes troglodytes)* tool use in the Ngotto Forest, Central African Republic. *American Journal of Primatology* 65:221–237.

Hill, A. 1985. Early hominid from Baringo, Kenya. *Nature* 315:222–224.

———. 2000. Taphonomy. In E. Delson, I. Tattersall, J. A. Van Couvering, and A. S. Brooks, eds., *Encyclopedia of Human Evolution and Prehistory* 686–689. New York: Garland.

Hill, A., Behrensmeyer, K., Brown, B., Deino, A., Rose, M., Saunders, J., Ward, S., and Winkler, A. 1991. Kipsaramon: a lower Miocene hominoid site in the Tugen Hills, Baringo District, Kenya. *Journal of Human Evolution* 20:67–75.

Hill, A., Curtis, G., and Drake, R. 1986. Sedimentary stratigraphy of the Tugen Hills, Baringo, Kenya. In L. E. Frostick, R. W. Renaut, I. Reid, and J. J. Tiercelin, eds., *Sedimentation in the African Rifts,* 285–295. Oxford: Blackwell Scientific.

Hill, A., Drake, R., Tauxe, L., Monaghan, M., Barry, J. C., Behrensmeyer, A. K., Curtis, G., Jacobs, B. F., Jacobs, L., Johnson, N., and Pilbeam, D. 1985. Neogene palaeontology and geochronology of the Baringo Basin, Kenya. *Journal of Human Evolution* 14:759–773.

Hill, A., and Ward, S. 1988. Origin of the Hominidae: the record of African large hominoid evolution between 14 my and 4 my. *Yearbook of Physical Anthropology* 31:49–83.

Hill, A., Ward, and Brown, B. 1992. Anatomy and age of the Lothagam mandible. *Journal of Human Evolution* 22:439–451.

Hill, A., Ward, S., Deino, A., Curtis, G., and Drake, R. 1992. Earliest *Homo. Nature* 355:719–722.

Hill, K. 1982. Hunting and human evolution. *Journal of Human Evolution* 11:521–544.

———. 2009. Animal "culture"? In K. N. Laland and B. G. Galef, eds., *The Question of Animal Culture,* 269–287. Cambridge, MA: Harvard University Press.

Hill, K. R., Walker, R. S., Bozicevic, M., Eder, J., Headland, T., Hewlett, B., Hurtado, A. M., Marlowe, F., Wiessner, P., and Wood, B. 2011. Co-residence patterns in hunter-gatherer societies show unique human social structure. *Science* 331:1286–1289.

Hill, S. D., Bundy, R. A., Gallup, G. G., Jr., and McClure, M. K. 1970. Responsiveness of young nursery reared chimpanzees to mirrors. *Proceedings of the Louisiana Academy of Sciences* 33:77–82.

Hill, W. C. O. 1955. *Primates: Comparative Anatomy and Taxonomy,* Vol. 2: *Haplorhini: Tarsioidea.* Edinburgh: Edinburgh University Press.

———. 1966. *Primates: Comparative Anatomy and Taxonomy,* Vol. 6: *Catarrhini: Cercopithecoidea: Cercopithecinae.* Edinburgh: Edinburgh University Press.

Hillix, W. A., and Rumbaugh, D. 2004. *Animal Bodies, Human Minds: Ape, Dolphin and Parrot Language Skills.* New York: Kluwer Academic/Plenum Press.

Hinrichsen, D. 1978. How old are our ancestors? *New Scientist* 78:571.

Hiraiwa-Hasegawa, M., Hasegawa, T., and Nishida, T. 1984. Demographic study of a large-sized unit-group of chimpanzees in the Mahale Mountains, Tanzania: a preliminary report. *Primates* 25:401–413.

Hirata, S. 2006. Tactical deception and understanding of others in chimpanzees. In T. Matsuzawa, M. Tomonaga, and M. Tanaka, eds., *Cognitive Development in Chimpanzees,* 265–276. Tokyo: Springer.

Hirata, S., Myowa, M., and Matsuzawa, T. 1998. Use of leaves as cushions to sit on wet ground by wild chimpanzees. *American Journal of Primatology* 44:215–220.

Hirata, S., Yamakoshi, G., Fujita, S., Ohashi, G., and Matsuzawa, T. 2001. Capturing and toying with hyraxes *(Dendrohyrax dorsalis)* by wild chimpanzees at Bossou, Guinea. *American Journal of Primatology* 53:93–97.

Hirata, S., Yamaoto, S., Takemoto, H., and Matsuzawa, T. 2010. A case report of meat and fruit sharing in a pair of wild bonobos. *Pan African News* 17:21–23.

Hirschfeld, L. A., and Gelman, S. A. 1994. *Mapping the Mind: Domain Specificity in Cognition and Culture.* Cambridge: Cambridge University Press.

Hirschler, P. 1942. Anthropoid and human endocranial casts. *Natuurkunde Verhandligen van de Noord-Hollandsche.* Amsterdam: Uitgevers Maatschappij.

Hixson, M. D. 1998. Ape language research: a review and behavioral perspective. *Analysis of Verbal Behavior* 15:17–39.

Hladik, C. M. 1973. Alimentation et activité d'un groupe de chimpanzés réintroduits en foret gabonaise. *Terre et la vie, revue decalogue appliquée* 27:343–413.

———. 1977. Chimpanzees of Gabon and chimpanzees of Gombe: some comparative data on diet. In T. H. Clutton-Brock, ed., *Primate Ecology: Studies of Feeding and Ranging Behaviour in Lemurs, Monkeys and Apes,* 481–501. London: Academic Press.

Hladik, C. M., Pasquet, P., and Simmen, B. 2002. New perspectives on taste and primate evolution: the dichotomy in gustatory coding for perception of beneficent versus noxious substances as supported by correlations among human thresholds. *American Journal of Physical Anthropology* 117:342–348.

Hladik, C. M., and Simmen, B. 1996. Taste perception and feeding behavior in nonhuman primates and human populations. *Evolutionary Anthropology* 5:58–71.

Ho, C. K. 1988. Human origins in Asia? *Human Evolution* 3:357–365.

———. 1990. A new Pliocene hominoid skull from Yuanmou southwest China. *Human Evolution* 5:309–318.

Ho, K.-C., Roessmann, U., Straumfjord, J. V., and Monroe, G. 1980. Analysis of brain weight. II. Adult brain weight in relation to body height, weight, and surface area. *Archives of Pathology and Laboratory Medicine* 104:640–645.

Hoang, X. C., Nguyen, L. C., and Vu, T. L. 1979. First discoveries on pleistocenian man, culture and fossilized fauna in Vietnam. In Committee for Social Sciences of Vietnam, eds. *Recent Discoveries and New Views on Some Archaeological Problems in Vietnam,* 14–20. Hanoi: Institute of Archaeology.

Hockett, C. F. 1960. Logical considerations in the study of animal communication. In W. E. Lanyon and W. N. Tavolga, eds., *Animal Sounds and Communication,* 392–430. Washington, DC: American Institute of Biological Sciences.

————. 1966. The problem of universals in language. In J. H. Greenberg, ed., *Universals of Language,* 2nd ed., 1–29. Cambridge, MA: MIT Press.

————. 1978. In search of Jove's brow. *American Speech* 53:243–313.

Hockett, C. F., and Ascher R. 1964. The human revolution. *Current Anthropology* 5:135–168.

Hockings, K. J., Anderson, J. R., and Matsuzawa, T. 2009. Use of wild and cultivated foods by chimpanzees at Bossou, Republic of Guinea: feeding dynamics and a human-influenced environment. *American Journal of Primatology* 71:636–646.

————. 2010. Flexible feeding on cultivated underground storage organs by rainforest-dwelling chimpanzees at Bossou, West Africa. *Journal of Human Evolution* 58:227–233.

Hodder, I. 1982. Symbols in action: ethnoarchaeological studies of material culture. Cambridge: Cambridge University Press.

Hofer, H. 1952. Der Gestaltwandel des Schädels der Säugetiere und Vögel, mit besonderer Berücksichtigung der Knickungstypen und der Schädelbasis. *Verhandlungen der Anatomischen Gesellschaft,* Jena 50:102–113.

————. 1972. Uber den Gesang des Orang-Utan *(Pongo pygmaeus). Der* Zoologisch Garten, *N. F., Leipzig* 41:299–302.

Hoffecker, J. F. 2009. The spread of modern humans in Europe. *Proceedings of the National Academy of Sciences of the United States of America* 106:16040–16045.

Hoffmann, J. N., Montag, A. G., and Dominy, N. J. 2004. Meissner corpuscles and somatosensory acuity: the prehensile appendages of primates and elephants. In *Evolution of the Special Senses in Primates,* ed. T. D. Smith, C. F. Ross, N. J. Dominy, and J. T. Laitman, special issue, *Anatomical Record Part A* 281A:1138–1147.

Hoffstetter, R. 1972. Relationships, origins, and history of the ceboid monkeys and caviomorph rodents: a modern reinterpretation. In T. Dobzhansky, M. K. Hecht and W. C. Steere, eds., *Evolutionary Biology,* Vol. 6, 323–347. New York: Appleton-Century-Crofts.

————. 1980. Origin and deployment of New World monkeys emphasizing the southern continents route. In R. L. Ciochon and A. B. Chiarelli, ed., *Evolutionary Biology of the New World Monkeys and Continental Drift,* 103–122. New York: Plenum Press.

————. 1988. Relations phylogéniques et position systématique de *Tarsius:* nouvelles controverses. *Comptes rendus hebdomadaires des séances de l'Académie des Sciences, Paris,* série II 307:1837–1840.

Hofreiter, M., Siedel, H., Van Neer, W., and Vigilant, L. 2003. Mitochondrial DNA sequence from an enigmatic gorilla population *(Gorilla gorilla uellensis). American Journal of Physical Anthropology* 121:361–368.

Hohenegger, J., and Zapfe, H. 1990. Craniometric investigations on *Mesopithecus* in comparison with two recent colobines. *Beiträge zur Paläontologie von Österreich* 16:111–143, Wien.

Hohmann, G. 1988. A case of simple tool use in wild liontailed macaques *(Macaca silenus). Primates* 29:565–567.

Hohmann, G., and Fruth, B. 1993. Field observations on meat sharing among bonobos *(Pan paniscus). Folia Primatologica* 60:225–229.

———. 1994. Structure and use of distance calls in wild bonobos *(Pan paniscus)*. *International Journal of Primatology* 15:767–782.

———. 1995. Loud calls of great apes: sex differences and social correlates. In *Current Topics in Primate Vocal Communication,* E. Zimmermann, J. D. Newman, and U. Jürgens, eds., pp, 161–184. New York: Plenum Press.

———. 1996. Food sharing and status in unprovisioned bonobos. In P. Wiessner and W. Schiefenhövel, eds., *Food and the Status Quest: An Interdisciplinary Perspective,* 47–67. Providence, RI: Berghahn Books.

———. 2000. Use and function of genital contacts among female bonobos. *Animal Behaviour* 60:107–120.

———. 2008. New records on prey capture and meat eating by bonobos at Lui Kotale, Salonga National Park, Democratic Republic of Congo. *Folia Primatologica* 79:103–110.

Hohmann, G., Gerloff, U., Tautz, D., and Fruth, B. 1999. Social bonds and genetic ties: kinship, association and affiliation in a community of bonobos *(Pan paniscus)*. *Behaviour* 136:1219–1235.

Hohmann, G., Potts, K., N'Guessan, A., Fowler, A. Mundry, R., Ganzhorn, J. U., and Ortmann, S. 2010. Plant foods consumed by *Pan:* exploring the variation of nutritional ecology across Africa. *American Journal of Physical Anthropology* 141:476–485.

Holekamp, K. 2007. Questioning the social intelligence hypothesis. *Trends in Cognitive Sciences* 11:65–69.

Holliday, T. W. 2000. Evolution at the crossroads: modern human emergence in Western Asia. *American Anthropologist* 102:54–68.

Hollihn, U., and Jungers, W. L. 1984. Kinesiologische Untersuchungen zür Brachiation bei Weißhandgibbons *(Hylobates lar)*. *Zeitschrift für Morphologie und Anthropologie* 74:275–293.

Holloway, R. L. 1967. The evolution of the primate brain: some aspects of quantitative relations. *Brain Research* 7:121–172.

———. 1976. The fossil record and neural organization. *Annals of the New York Academy of Sciences* 280:330–348.

———. 1979. Brain size, allometry, and reorganization: toward a synthesis. In M. E. Hahn, C. Jensen, and B. C. Dukek, eds., *Development and Evolution of Brain Size,* 59–88. New York: Academic Press.

———. 1983a. Cerebral brain endocast pattern of *Australopithecus afarensis* hominid. *Nature* 303:420–422.

———. 1983b. Human paleontological evidence relevant to language behavior. *Human Neurobiology* 2:105–114.

———. 2002. Brief communication: how much larger is the relative volume of area 10 of the prefrontal cortex in humans? *American Journal of Physical Anthropology* 118:399–401.

———. 2009. Brain fossils: endocasts. In L. R. Squire, ed., *Encyclopedia of Neuroscience,* Vol. 2, 353–361. Oxford: Academic Press.

Holloway, R. L., Broadfield, D. C., and Yuan, M. S. 2004. *The Human Fossil Record,* Vol. 3: *Brain Endocasts: The Paleoneurological Evidence.* Hoboken, NJ: Wiley-Liss.

Holloway, R. L., and de la Coste-Lareymondie, M. C. 1982. Brain endocast asymmetry in pongids and hominids: some preliminary findings on the paleontology of cerebral dominance. American *Journal of Physical Anthropology* 58:101–110.

Holmes, E. C., Pesole, G., and Saccone, C. 1989. Stochastic models of molecular evolution and the estimation of phylogeny and rates of nucleotide substitution in the hominoid primates. *Journal of Human Evolution* 18:775–794.

Holmquist, R., Miyamoto, M. M., and Goodman, M. 1988a. Higher-primate phylogeny— why can't we decide? *Molecular Biology and Evolution* 5:201–216.

———. 1988b. Analysis of higher-primate phylogeny from transversion differences in nuclear and mitochondrial DNA by Lake's methods of evolutionary parsimony and operator metrics. *Molecular Biology and Evolution* 5:217–236.

Hooijer, D. A. 1948. Prehistoric teeth of man and the orang-utan from central Sumatra, with notes on the fossil orang-utan from Java and southern China. *Zoologische Mededlingen Museum (Leiden)* 29:175–301.

———. 1960. Quaternary gibbons from the Malay archipelago. Zoologische Verhandlingen, Leiden 46:1–41.

Hook, M. A. 2004. The evolution of lateralized motor function. In L. J. Rogers and G. Kaplan, eds., *Comparative Vertebrate Cognition,* 325–370. New York: Kluwer Academic/ Plenum Press.

Hooker, J. J., Russell, D. E., and Phélizon, A. 1999. A new family of Pleiadapiformes (Mammalia) from the Old World Lower Paleogene. *Palaeontology* 42:377–407.

Hopkins, W. D. 1993. Posture and reaching in chimpanzees *(Pan troglodytes)* and orangutans *(Pongo pygmaeus). Journal of Comparative Psychology* 107:162–168.

———. 1994. Hand preferences for bimanual feeding in 140 captive chimpanzees *(Pan troglodytes):* rearing and ontogenetic determinants. *Developmental Psychobiology* 27:395–407.

———. 1995. Hand preferences for a coordinated bimanual task in 110 chimpanzees *(Pan troglodytes):* cross-sectional analysis. *Journal of Comparative Psychology* 109:291–297.

———. 1996. Chimpanzee handedness revisited: 55 years since Finch (1941). *Psychonomic Bulletin and Review* 3:449–457.

———. 1998. The evolution and genetic basis (possibly?) of hemispheric specialization and language: what can the great apes tell us? *CPC* 17:1167–1175.

———. 2008. Brief communication: locomotor limb preferences in captive chimpanzees *(Pan troglodytes):* implications for morphological asymmetries in limb bones. *American Journal of Physical Anthropology* 137:113–118.

Hopkins, W. D., and Bard, K. A. 1993. The ontogeny of lateralized behavior in nonhuman primates with special reference to chimpanzees *(Pan troglodytes).* In J. P. Ward and W. D. Hopkins, eds., *Primate Laterality: Current Behavioral Evidence of Primate Asymmetries,* 251–265. New York: Springer.

Hopkins, W. D., Bard, K. A., Jones, A., and Bales, S. L. 1993. Chimpanzee hand preference in throwing and infant cradling: implications for the origin of human handedness. *Current Anthropology* 34:786–790.

Hopkins, W. D., and Bennett, A. J. 1994. Handedness and approach—avoidance behavior in chimpanzees *(Pan). Journal of Experimental Psychology: Animal Behavior Processes* 20:413–418.

Hopkins, W. D., Bennett, A. J., Bales, S. L., Lee, J., and Ward, J. P. 1993. Behavioral laterality in captive bonobos *(Pan paniscus)*. *Journal of Comparative Psychology* 107:403–410.

Hopkins, W. D., and Cantalupo, C. 2004. Handedness in chimpanzees *(Pan troglodytes)* is associated with asymmetries of the primary cortex but not with homologous language areas. *Behavioral Neuroscience* 118:1176–1183.

———. 2005. Individual and setting differences in the hand preferences of chimpanzees *(Pan troglodytes):* a critical analysis of some alternative explanations. *Laterality* 10:65–80.

Hopkins, W. D., Cantalupo, C., Wesley, M., Hostetter, A. B., and Pilcher, D. 2002. Grip morphology and hand use in chimpanzees *(Pan troglodytes):* evidence of a left hemisphere specialization in motor skill. *Journal of Experimental Psychology* 131:412–423.

Hopkins, W. D., and Cantero, M. 2003. From hand to mouth in the evolution of language: the influence of vocal behavior on lateralized hand use in manual gestures by chimpanzees *(Pan troglodytes)*. *Developmental Science* 6:55–61.

Hopkins, W. D., and de Waal, F. B. M. 1995. Behavioral laterality in captive bonobos *(Pan paniscus):* replication and extension. *International Journal of Primatology* 16:261–276.

Hopkins, W. D., and Fernández-Carriba, S. 2000. The effect of situational factors on hand preferences for feeding in 177 captive chimpanzees *(Pan troglodytes)*. *Neuropsychologia* 38:403–409.

———. 2002. Laterality of communicative behaviours in non-humn primates: a critical analysis. In L. G. Rogers and R. J. Andrew, eds., *Comparative Vertebrate Lateralization*, 445–479. New York: Cambridge University Press.

Hopkins, W. D., and Fowler, L. A. 1998. Lateralized changes in tympanic membrane temperature in relation to different cognitive tasks in chimpanzees *(Pan troglodytes)*. *Behavioral Neuroscience* 112:83–88.

Hopkins, W. D., and Leavens, D. A. 1998. Hand use and gestural communication in chimpanzees *(Pan troglodytes)*. *Journal of Comparative Psychology* 112:95–99.

Hopkins, W. D., Lyn, H., and Cantalupo, C. 2009. Volumetric and lateralized differences in selected brain regions of chimpanzees *(Pan troglodytes)* and bonobos *(Pan paniscus)*. *American Journal of Primatology* 71:988–997.

Hopkins, W. D., and Marino, L. 2000. Asymmetries in cerebral width in nonhuman primate brains as revealed by magnetic resonance imaging (MRI). *Neuropsychologia* 38:493–499.

Hopkins, W. D., Marino, L., Rilling, J. K., and MacGregor, L. A. 1998. Planum temporale asymmetries in great apes as revealed by magnetic resonance imaging (MRI). *NeuroReport* 9:2913–1918.

Hopkins, W. D., and Morris, R. D. 1993. Handedness in great apes: a review of findings. *International Journal of Primatology* 14:1–25.

Hopkins, W. D., and Pearson, K. 2000. Chimpanzee *(Pan troglodytes)* handedness: variability across multiple measures of hand use. *Journal of Comparative Psychology* 114:126–135.

Hopkins, W. D., Phillips, K. A., Bania, A., Calcutt, S. E., Gardner, M., Russell, J., Schaeffer, J., Lonsdorf, E. V., Ross, S. R., and Schapiro, S. J. 2011. Hand preferences for coordinated bimanual actions in 777 great apes: implications for the evolution of handedness in hominins. *Journal of Human Evolution* 60:605–611.

Hopkins, W. D., and Pilcher, D. L. 2001. Neuroanatomical localization of the motor hand area with magnetic resonance imaging: the left hemisphere is larger in great apes. *Behavioral Neuroscience* 115:1159–1164.

Hopkins, W. D., Pilcher, D. L., and Cantalupo, C. 2003. Brain structures for communication, cognition, and handedness. In D. Maestripieri, ed., *Primate Psychology*, 424–450. Cambridge, MA: Harvard University Press.

Hopkins, W. D., Pilcher, D. L., and MacGregor, L. 2000. Sylvian fissure asymmetries in nonhuman primates revisited: a comparative MRI study. *Brain, Behavior and Evolution* 56:293–299.

Hopkins, W. D., and Russell, J. L. 1997. Manual specialisation and tool use in captive chimpanzees *(Pan troglodytes):* the effect of unimanual and bimanual strategies on hand preference. *Laterality* 2:267–277.

———. 2004. Further evidence of a right hand advantage in motor skill by chimpanzees *(Pan troglodytes)*. *Neuropsychologia* 42:990–996.

Hopkins, W. D., Russell, J., Cantalupo, C., Freeman, H., and Shapiro, S. J. 2005. Factors influencing the prevalence and handedness for throwing in captive chimpanzees *(Pan troglodytes)*. *Journal of Comparative Psychology* 119:363–370.

Hopkins, W. D., Russell, J., Freeman, H., Buehler, N., Reynolds, E., and Shapiro, S. J. 2005. The distribution and development of handedness for manual gestures in captive chimpanzees *(Pan troglodytes)*. *Psychological Science* 16:487–493.

Hopkins, W. D., Russell, J., Hook, M., Braccini, S., and Shapiro, S. J. 2005. Simple reaching is not so simple: association between hand use and grip preferences in captive chimpanzees. *International Journal of Primatology* 26:259–277.

Hopkins, W. D., Russell, J., Hostetter, A., Pilcher, D., and Dahl, J. F. 2005. Grip preference, dermatoglyphics, and hand use in captive chimpanzees *(Pan troglodytes)*. *American Journal of Physical Anthropology* 128:57–62.

Hopkins, W. D., Russell, J., Remkus, M., Freeman, H., and Shapiro, S. J. 2007. Handedness and grooming in *Pan troglodytes:* comparative analysis between findings in captive and wild individuals. *International Journal of Primatology* 28:1315–1326.

Hopkins, W. D., and Savage-Rumbaugh, E. S. 1991. Vocal communication as a function of differential rearing experiences in *Pan paniscus:* a preliminary report. *International Journal of Primatology* 12:559–583.

Hopkins, W. D., Stoinski, T. S., Lukas, K. E., Ross, S. R., and Wesley, M. J. 2003. Comparative assessment of handedness for a coordinated bimanual task in chimpanzees *(Pan troglodytes)*, gorillas *(Gorilla gorilla)* and orangutans *(Pongo pygmaeus)*. *Journal of Comparative Psychology* 117:302–308.

Hopkins, W. D., Taglialatela, J. P., and Leavens, D. A. 2007. Chimpanzees differentially produce novel vocalizations to capture the attention of a human. *Animal Behaviour* 73:281–286.

Hopkins, W. D., Washburn, D. A., and Rumbaugh, D. M. 1989. Note on hand use in the manipulation of joysticks by rhesus monkeys *(Macaca mulatta)* and chimpanzees *(Pan troglodytes)*. *Journal of Comparative Psychology* 103:91–94.

Hopkins, W. D., and Wesley, M. J. 2002. Gestural communication in chimpanzees *(Pan troglodytes):* the influence of experimenter position on gesture type and hand preference. *Laterality* 7:19–30.

Hopkins, W. D., Wesley, M. J., Izard, M. K., Hook, M., and Shapiro, S. 2004. Chimpanzees *(Pan troglodytes)* are predominantly right-handed: replication in three populations of apes. *Behavioral Neuroscience* 118:659–663.

Horai, S., Gojobori, T., and Matsunaga, E. 1986. Distinct clustering of mitochondrial DNA types among Japanese, Caucasians and Negroes. *Japanese Journal of Genetics* 61:271–275.

Horai, S., Hayasaka, K., Kondo, R., Tsugane, K., and Takahata, N. 1995. Recent African origin of modern humans revealed by complete sequences of hominoid mitochondrial DNAs. *Proceedings of the National Academy of Science of the United States of America* 92:532–536.

Horai, S., Satta, Y., Hayasaka, K., Kondo, R., Inoue, T., Ishida, T., Hayashi, S., Takahata, N. 1992. Man's place in Hominoidea revealed by mitochondrial DNA genealogy. *Journal of Molecular Evolution* 35:32–43.

———. 1993. Erratum. *Journal of Molecular Evolution* 37:89.

Horn, A. D. 1980. Some observations on the ecology of the bonobo chimpanzee *(Panpaniscus,* Schwartz 1929) near Lake Tumba, Zaire. *Folia Primatologica* 34:145–169.

Horr, D. A. 1972. The Borneo orang-utan. *Borneo Research Bulletin* 4:46–50.

———. 1975. The Borneo Orang-utan: population structure and dynamics in relationship to ecology and reproductive strategy, In L. A. Rosenblum, ed., *Primate Behavior,* Vol. 4, 307–323. New York: Academic Press.

———. 1977. Orang-utan maturation: growing up in a female world, In S. Chevalier-Skolnikoff and F. E. Poirier, eds., *Primate Bio-Social Development,* 289–321. New York: Garland Press.

Horváth, E. 1990. Charcoal remains of the upper loess section. In M. Kretzoi and V. T. Dobosi, eds., *Veeretesszölös: Site, Man and Culture,* 137–143. Budapest: Akadémiai Kiadó.

Hosaka, K., Nishida, T., Hamai, M., Matusumoto-Oda, A., and Uehara, S. 2001. In B. M. F. Galdikas, N. E. Briggs, L. K. Sheeran, G. L. Shapiro, and J. Goodall, eds., *All Apes Great and Small,* Vol. 1: *African Apes,* 107–130. New York: Kluwer Academic/Plenum Press.

Höss, M., Jaruga, P., Zastawny, T. H., Dizdaroglu, M., and Pääbo, S. 1996. DNA damage and DNA sequence retrieval from ancient tissues. *Nucleic Acids Research* 24:1304–1307.

Hostetter, A. B., Canero, M., and Hopkins, W. D. 2001. Differential use of vocal and gestural communication by chimpanzees *(Pan troglodytes)* in response to the attentional status of a human *(Homo sapiens). Journal of Comparative Psychology* 115:337–343.

Houghton, P. Neandertal supralaryngeal vocal tract. *American Journal of Physical Anthropology* 90:139–146.

Houle, A., Chapman, C. A., and Vickery, W. L. 2007. Variation in fruit production and implications for primate foraging. *International Journal of Primatology* 28:1197–1217.

Howard, H. E. 1920. *Territory in Bird Life.* New York: Dutton.

Howell, F. C. 1959. The Villafranchian and human origins. *Science* 130:831–844.

———. 1960. European and northwest African Middle Pleistocene hominids. *Current Anthropology* 1:195–232.

———. 1965. Comments. *Current Anthropology* 6:399–401.

———. 1978. Hominidae. In V. J. Maglio and H. B. S. Cooke, eds., *Evolution of African Mammals,* 154–248. Cambridge, MA: Harvard University Press.

———. 1981. Some views of *Homo erectus* with special reference to its occurrence in Europe. In B. A. Sigmon and J. S. Cybulski, *Homo erectus: Papers in Honor of Davidson Black,* 153–157. Toronto: University of Toronto Press.

Howell, F. C., and Coppens, C. 1976. An overview of Hominidae from the Omo Succession, Ethiopia. In Y. Coppens, F. C. Howell, G. L., and R. E. F. Leakey, eds., *Earliest Man and Environments in the Lake Rudolf Basin,* 522–532. Chicago: University of Chicago Press.

Howell, F. C., Haesaerts, P., and de Heinzelin, J. 1987. Depositional environments, archeological occurrences and hominids from Members E and F of the Shungura Formation (Omo basin, Ethiopia). *Journal of Human Evolution* 16:665–700.

Howells, W. W. 1973. *Evolution of the Genus Homo.* Reading, Massachusetts: Addison-Wesley.

———. 1977. The importance of being human. In J. M. Tanur, F. Mosteller, W. H. Kruskal, R. F. Link, R. S. Pieters, G. R. Rising, and E. L. Lehmann, eds., *Statistics: a Guide to the Study of the Biological and Health Sciences,* 65–73. San Francisco: Holden-Day.

———. 1980. *Homo erectus*—who, when and where: a survey. *Yearbook of Physical Anthropology* 23:1–23.

———. 1981. *Homo erectus* in human descent: ideas and problems. In B. A. Sigmon and J. S. Cybulski, eds., *Homo erectus: Papers in Honor of Davidson Black,* 63–85. Toronto: University of Toronto Press.

———. 1993. *Getting Here. The Story of Human Evolution.* Washington, DC: Compass.

Hoyer, B. H., van de Velde, N. W., Goodman, M., and Roberts, R. B. 1972. Examination of hominid evolution by DNA sequence homology. *Journal of Human Evolution* 1:645–649.

Hrdy, S. B. 1977. *The Langurs of Abu: Female and Male Strategies of Reproduction.* Cambridge, MA: Harvard University Press.

———. 1979. Infanticide among animals: a review, classification, and examination of the implications for the reproductive strategies of females. *Ethology and Sociobiology* 1:13–40.

———. 1981. *The Woman That Never Evolved.* Cambridge, MA: Harvard University Press.

———. 1999a. *Mother Nature: Natural Selection and the Female of the Species.* London: Chatto & Windus.

———. 1999b. *Mother Nature: Maternal Instincts and How They Shape the Human Species.* New York: Ballantine Books.

———. 2009. *Mothers and Others: The Evolutionary Origins of Mutual Understanding.* Cambridge, MA: Harvard University Press.

Hrdy, S. B., and Bennett, W. 1981. Lucy's husband: what did he stand for? *Harvard Magazine* July–August, 7–9, 46.

Hu, Y., Xu, H., and Yang, D. 1989. The studies on ecology of *Hylobates concolor leucogenys.* [Article in Chinese.] *Zoology Research* 10 (Suppl.) :61–67.

———. 1990. Feeding ecology of the white-cheeked gibbon *(Hylobates concolor leucogenys).* [Article in Chinese.] *Acta Ecologica Sinica* 10:155–159.

Huan, D. B. M., Rapold, C. J., Call, J., Janzen, G., and Levinson, S. C. 2006. Cognitive cladistics and cultural override in hominid spatial cognition. *Proceedings of the National Academy of Sciences of the United States of America* 103:17568–17573.

Hublin, J. J. 1985. Human fossils from the North African Middle Pleistocene and the origin of *Homo sapiens*. In E. Delson, ed., *Ancestors: The Hard Evidence*, 283–288. New York: Alan R. Liss.

———. 2009. The origin of Neanderthals. *Proceedings of the National Academy of Sciences of the United States of America* 106:16022–16027.

———. 2012. The earlist modern human colonization of Europe. *Proceedings of the National Academy of Sciences of the United States of America* 109:13471–13472.

Hublin, J. J., and Klein, R. G. 2011. Northern Africa could also have housed the source population for living humans. *Proceedings of the National Academy of Sciences of the United States of America* 108:E277.

Hublin, J.-J., Talamo, S., Julien, M., David, F., Conet, N., Bodu, P., Vandermeersch, B., and Richards, M. P. 2012. Radiocarbon dates from the Grotte du Renne and Saint-Césaire support a Neandertal origin for the Châtelperronian. *Proceedings of the National Academy of Sciences of the United States of America* 109:18743–18748.

Hudjashov, G., Kivisild, T., Underhill, P. A., Endicott, P., Sanchez, J. J., Lin, A. A., Shen, P., Oefner, P., Renfrew, C., Villems, R., and Forster, P. 2007. Revealing the prehistoric settlement of Australia by Y chromosome and mtDNA analysis. *Proceedings of the National Academy of Sciences of the United States of America* 104:8726–8730.

Huffman, M. A. 1995. La pharmacopée des chimpanzés. *Le Recherche* 280:66–71.

———. 1997. Current evidence for self-medication in primates: a multidisciplinary perspective. *Yearbook of Physical Anthropology* 40:171–200.

———. 2001. Self-medicative behavior in the African great apes: an evolutionary perspective into the origins of human traditional medicine. *BioScience* 51:651–661.

———. 2007. Primate self-medication. In C. J. Campbell, A. Fuentes, K. C. MacKinnon, M. Panger, and S. K. Bearder, eds., *Primates in Perspective*, 677–690. Oxford: Oxford University Press.

Huffman, M. A., and Chapman, C. A. 2009. *Primate Parasite Ecology*. Cambridge: Cambridge University Press.

Huffman, M. A., Gotoh, S., Izutsu, D., Koshimizu, K., and Kalunde, M. S. 1993. Further observations on the use of the medicinal plant, *Vernonia amygdalina* (Del) by a wild chimpanzee, its possible effect on parasite load, and its phytochemistry. *African Study Monongraphs* 14:227–240.

Huffman, M. A., Gotoh, Turner, L. A., Hamai, M., and Yoshida, K. 1997. Seasonal trends in intestinal nematode infection and medicinal plant use among chimpanzees in the Mahale Montains, Tanzania. *Primates* 38:111–125.

Huffman, M. A., and Kalunde, M. S. 1993. Tool-assisted predation on a squirrel by a female chimpanzee in the Mahale Mountains, Tanzania. *Primates* 34:93–98.

Huffman, M. A., Koshimizu, K., and Ohigahi, H. 1996. Ethnobotany and zoopharmacognosy of *Vernonia amygdalina*, a medicinal plant used by humans and chimpanzees. In P. Caligari and D. J. N. Hind, eds., *Proceedings of the International Compositae*

Conference, Kew, 1994, Vol. 2: *Biology and Utilization,* 351–360. Kew, UK: Royal Botanical Gardens.

Huffman, M. A., Ohigashi, H., Kawanaka, M., Page, J. E., Kirby, G. C., Gasquet, M. Murakami, A., and Koshimizu, K. 1998. African great ape self-medication: a new paradigm for treating parasite disease with natural medicines? In H Ageta, N. Aimi, Y. Ebiuka, T. Fujita, and G. Honda, eds., *Towards Natural Medicine Research in the 21st Century,* 113–123. Philadelphia: Elsevier.

Huffman, M. A., Page, J. E., Sukhdeo, V. K., J. E. Gotoh, S., Kalunde, M. S., Chandrasiri, T., and Towers, G. H. N. 1996. Leaf-swallowing by chimpanzees: a behavioral adaptation for the control of strongyle nematode infections. *International Journal of Primatology* 17:475–503.

Huffman, M. A., and Seifu, M. 1989. Observations on the illness and consumption of a possibly medicinal plant *Vernonia amygdalina* (DEL.), by a wild chimpanzee in the Mahale Mountains National Park, Tanzania *Primates* 30:51–63.

Huffman, M. A., and Wrangham, R. W. 1994. Diversity of medicinal plant use by chimpanzees in the wild. In R. W. Wrangham, W. C. McGrew, F. B. M. de Waal, and P. G. Heltne, *Chimpanzee Cultures,* 129–148. Cambridge, MA: Harvard University Press.

Hughes, A. L. 1988. *Evolution and Human Kinship.* Oxford: Oxford University Press.

Hughes, A. R., and Tobias, P. V. 1977. Fossil skull probably of the genus *Homo* from Sterkfontein, Transvaal. *Nature* 265:310–312.

Huijbregts, B., De Wachter, P., Obiang, L. S. N., Akou, M. E. 2003. Ebola and the decline of gorilla *Gorilla gorilla* and chimpanzee *Pan troglodytes* populations in Minkebe Forest, north-eastern Gabon. *Oryx* 37:437–443.

Humle, T., and Matsuzawa, T. 2002. Ant-dipping among the chimpanzees of Bossou, Guinea, and some comparisons with other sites. *American Journal of Primatology* 58:133–148.

———. 2009. Laterality in hand use across four tool-use behaviors among the wild chimpanzees of Bossou, Guinea, West Africa. *American Journal of Primatology* 71:40–48.

Humphrey, N. K. 1976. The social function of intellect. In P. P. G. Bateson and R. A. Hinde, eds., *Growing Points in Ethology,* 303–317. Cambridge: Cambridge University Press.

Hunley, K. L., Healy, M. E., and Long, J. C. 2009. The global pattern of gene identity variation reveals a history of long-range migrations, bottlenecks, and local mate exchanges: implications for biological race. *American Journal of Physical Anthropology* 139:35–46.

Hunt, K. D. 1994. The evolution of human bipedality: ecology and functional morphology. *Journal of Human Evolution* 26:183–202.

Hunt, K. D., and McGrew, W. C. 2002. Chimpanzees in the dry habitats of Assirik, Senegal and Semliki Wildlife Reserve, Uganda. In C. Boesch, G. Hohmann, and L. F. Marchant, eds., *Behavioural Diversity in Chimpanzees and Bonobos,* 35–51. Cambridge: Cambridge University Press.

Hunter, A. G. W. 2006. Brain. In R. E. Stevenson and J. G. Hall, eds., *Human Malformations and Related Anomalies,* 2nd ed., 469–511. Oxford: Oxford University Press.

Hunter, M. 2007. *Honor Betrayed: Sexual Abuse in the Military.* Fort Lee, NJ: Barricade.

Hurford, J. R. 2003. The language mosaic and its evolution, In M. H. Christiansen and S. Kirby, eds., *Language Evolution*, 38–57. Oxford: Oxford University Press.

———. 2004. Language beyond our grasp: what mirror neurons can, and cannot, do for the evolution of language. In D. K. Oller and U. Griebel, eds., *Evolution of Communication Systems: A Comparative Approach*, 297–313. Cambridge, MA: MIT Press.

———. 2007. *The Origins of Meaning: Language and Its Evolution*. Oxford: Oxford University Press.

———. 2012. *The Origins of Grammar: Language in the Light of Evolution*. Oxford: Oxford University Press.

Hürzeler, J. 1954. Zur systematischen Stellung von *Oreopithecus*. *Verhandlungen der Naturforschenden Gesselschaft in Basel* 65:88–95.

———. 1956. *Oreopithecus*, un point de repère pour l'histoire de l'humanité a l'ère Tertiare. *Problèmes Actuels de Paléontologie. Colloques Internationaux du Centre National de la Recherche Scientifique, Paris* 60:115–123.

———. 1958. *Oreopithecus bambolii* Gervais, a preliminary report. *Verhandlungen der Naturforschenden Gesselschaft in Basel* 69:1–48.

———. 1960. The significance of *Oreopithecus* in the genealogy of man. *Triangle* 4:164–174.

———. 1968. Questions et réflexions dur l'histoire des anthropomorphes. *Annales de Paléontologie (Vertébrés)* 54:95–233.

———. 1978. L'origine de l'homme vue par un paléontologiste. In E. Boné, Y. Coppens, E. Genet-Varcin, P.-P. Grassé, J.-L., W. W. Howells, J. Hürzeler, S. Kurkoff, H. de Lumley, M.-A. de Lumley, J. Piveteau, et al., eds., *Les origines humaines et les époques de l'intelligence*, 5–12. Paris: Masson.

Huxley, T. H. 1963. *Evidence as to Man's Place in Nature*. London: Williams and Norgate.

———. 1968. Mr. Darwin's critics. In *Collected Essays*, Vol. 2: *Darwinia*. New York: Greenwood Press.

Hyatt, C. W. 1988. Responses of gibbons *(Hylobates lar)* to their mirror images. *American Journal of Primatology* 45:307–311.

Hyatt, C. W., and Hopkins, W. D. 1994. Self-awareness in bonobos and chimpanzees: a comparative perspective. In S. T. Parker, R. W. Mitchell, and M. L. Boccia, eds., *Self-Awareness in Animals and Humans*, 248–253. Cambridge: Cambridge University Press.

———. 1998. Interspecies object exchange: bartering in apes? *Behavioural Processes* 42:177–187.

Hylander, W. L. 1988. Implications of in vivo experiments for interpreting the functional significance of "robust" australopithecine jaws. In Grine, F. E., ed., *Evolutionary History of the "Robust" Australopithecines*, 55–83. Hawthorne, NY: Aldine de Gruyter.

———. 2013. Functional links between canine height and jaw gape in catarrhines with special refrence to early hominins. *American Journal of Physical Anthropology* 150:247–259.

Iacoboni, M. 2012. The human mirror neuron system and its role in imitations and empathy. In F. B. M. de Waal and P. F. Ferrari, eds., *The Primate Mind*, 23–47, Cambridge, MA: Harvard University Press.

Iacoboni, M., Molnar-Szakacs, I., Gallese, V., Buccino, G., Mazziotta, J. C., and Rizzolatti, G. 2005. Grasping the intentions of others with one's own mirror neuron system. *PLoS Biology* 3:e79.

Iacoboni, M., Woods, R. P., Brass, M., Bekkering, H., Mazziotta, J. C., and Rizzolatti, G. 2005. Cortical mechanisms of human imitation. *Science* 286:2526–2528.

Idani, G. 1990. Relations between unit-groups of bonobos at Wamba, Zaire: encounters and temporary fusions. *African Study Monographs* 11:153–186.

―――. 1991. Social relationships between immigrant and resident bonobo *(Pan paniscus)* females at Wamba. *Folia Primatologica* 57:83–95.

―――. 1995. Function of peering behavior among bonobos *(Pan paniscus)* at Wamba, Zaire. *Primates* 36:377–383.

Idani, G., Kuroda, S., Kano, T., and Asato, R. 1994. Flora and vegetation of Wamba Forest, Central Zaire with reference to bonobo 332. *(Pan paniscus)* foods. *Tropics* 3:309–332.

Idani, G., Mwanza, N., Ihobe, H., Hashimoto, C., Tashiro, Y., and Furuichi, T. 2008. Changes in the status of bonobos, their habitat, and the situation of humans at Wamba in the Luo Scientific Reserve, Democratic Republic of Congo. In T. Furuichi and J. Thompson, eds., *The Bonobos: Behavior, Ecology, and Conservation*, 291–302. New York: Springer.

Ihobe, H. 1990. Interspecific interactions between wild pygmy chimpanzees *(Pan paniscus)* and red colobus *(Colobus badius)*. *Primates* 31:109–112.

―――. 1992a Male-male relationships among wild bonobos *(Pan paniscus)* at Wamba, Republic of Zaire. *Primates* 33:163–179.

―――. 1992b Observations on the meat-eating behavior of wild bonobos *(Pan paniscus)* at Wamba, Republic of Zaire. *Primates* 33:247–250.

―――. 1997a. Evolution of hunting and meat-eating behavior in Hominoidea. [Article in Japanese.] *Primate Research* 13:203–213.

―――. 1997b. Non-antagonistic relations between wild bonobos and two species of guenons. *Primates* 38:351–357.

Inaba, A. Power takeover occurred in M group of the Mahale Mountains, Tanzania, in 2007. *Pan Africa News* 16:13–15.

Indriati, E., Swisher III, C. C., Lepre, C., Quinn, R. L., Suriyanto, R. A., Hascaryo, A. T., Feibel, C. S., Pobiner, B. L., and Antón, S. C. 2010. The age of the 20 meter Solo River Terrace, Ngandong (Java, Indonesia) reconsidered. *American Journal of Physical Anthropology* 50 (Suppl.): 132–133.

Ingman, M., Kaessmann, H., Pääbo, S., and Gyllensten, U. 2000. Mitochondrial genome variation and the origin of modern humans. *Nature* 408:708–713.

Ingmanson, E. J. 1996. Tool-using behavior in wild *Pan paniscus*: social and ecological considerations. In A. E. Russon, K. A. Bard, and S. T. Parker, eds., *Reaching into Thought: The Minds of the Great Apes,* 190–210. Cambridge: Cambridge University Press.

Ingold, T. 2001. The use and abuse of ethnography. *Behavioral and Brain Sciences* 124:337.

Inman, V. T., Ralston, H. J., and Todd, F. 1981. *Human Walking.* Baltimore: Williams & Wilkins.

Inogwabini, B.-I., Bewa, M., Longwango, M., Abokome, M., and Vuvu, M. 2008. The bonobos in the Lake Tumba hinterland: threats and opportunities for population conservation. In T. Furuichi and J. Thompson, eds., *The Bonobos: Behavior, Ecology, and Conservation,* 273–290. New York: Springer.

Inogwabini, B.-I., Hall, J. S., Vedder, A., Curran, B., Yamagiwa, J., and Basabose, K. 2000. Status of large mammals in the mountain sector of Kahuzi-Biega National Park, Democratic Republic of Congo, in 1996 *African Journal of Ecology* 38:269–276.

Inoue, E., Inoue-Murayama, M., Vigilant, L., Takenaka, O., and Nishida, T. 2008. Relatedness in wild chimpanzees: influence of paternity, male philopatry, and demographic factors. *American Journal of Physical Anthropology* 137:256–262.

Inoue-Nakamura, N. 2001. Mirror self-recognition in Primates: an ontogenetic and a phylogenetic approach. In T. Matzuzawa, ed., *Primate Origins of Human Cognition and Behavior,* 297–312. Tokyo: Springer.

Inoue-Nakamura, N., and Matzuzawa, T. 1997. Development of stone tool use by wild chimpanzees *(Pan troglodytes). Journal of Comparative Psychology* 111:159–173.

Inouye, S. E. 1990. Variation in the presence and development of the dorsal ridge of the metacarpal head in African apes. *American Journal of Physical Anthropology* 81:243.

———. 1991. Ontogeny and allometry in African ape fingers. In A. Ehara, T. Kimura, O. Takenaka, and M. Iwamoto, eds., *Primatology Today,* 537–538. Amsterdam: Elsevier Science Publishers.

———. 1992. Ontogeny and allometry of African ape manual rays. *Journal of Human Evolution* 23:107–138.

———. 1994. Ontogeny of knuckle-walking hand postures in African apes. *Journal of Human Evolution* 26:459–485.

Inskipp, T. 2005. Chimpanzee *(Pan troglodytes).* In J. Caldecott and L. Miles, eds., *World Atlas of Great Apes and Their Conservation,* 53–82. Berkeley: University of California Press.

International Gorilla Conservation Programme (IGCP). 2010. Census confirms increase in population of the critically endangered Virunga mountain gorillas. IGCP Blog, 7 December, www.igcp.org/2010-mountain-gorilla-census/.

Irish, J. D., Guitalli-Steinberg, D., Legge, S. S., de Ruiter, D. J., and Berger, L. R. 2013. Dental morphology and the phylogenetic "place" of *Australopithecus sediba. Science* 340, http://dx.doi.org/10.1126/science.1233062.

Isaac, G. 1971. The diet of early man: aspects of archaeological evidence from Lower and Middle Pleistocene sites in Africa. *World Archaeology* 2:278–299.

———. 1976. The activities of early African hominids: a review of the archaeological evidence from the time span two and a half to one million years ago. In G. L. Isaac and E. R. McCown, eds., *Human Origins: Louis Leakey and the East African Evidence,* 483–514. Menlo Park, CA: W. A. Benjamin.

———. 1977. *Olorgesailie: Archeological Studies of a Middle Pleistocene Lake Basin in Kenya.* Chicago: University of Chicago Press.

———. 1978. The food-sharing behavior of protohuman hominids. *Scientific American* 238:90–108.

———. 1981. Emergence of human behaviour patterns. *Philosophical Transactions of the Royal Society of London, Series B* 292:177–188.

―――. 1983a. Aspects of human evolution. In D. S. Bendall, ed., *Evolution from Molecules to Men*, 509–543. Cambridge: Cambridge University Press.

―――. 1983b. Bones in contention: competing explanations for the juxtaposition of Early Pleistocene artefacts and faunal remains. In J. Clutton-Brock and C. Grigson, eds., *Animals and Archaeology: Hunters and Their Prey*, 3–19. International Series 163. Oxford: British Archaeological Reports.

Isaac, G. L., and Crader, D. C. 1981. To what extent were early hominids carnivorous? In R. S. O. Harding and G. Teleki, eds., *Omnivorous Primates: Gathering and Hunting in Human Evolution*, 37–103. New York: Columbia University Press.

Isaac, G. L., and Isaac, B. 1997. *Koobi Fora Research Project*, Vol. 5: *Plio-Pleistocene Archaeology*. Oxford: Oxford University Press.

Isabirye-Basuta, C. M., and Lwanga, J. S. 2008. Primate populations and their interactions with changing habitats. *International Journal of Primatology* 29:35–48.

Isbell, L. A. 2006. Snakes as agents of evolutionary change in primate brains. *Journal of Human Evolution* 51:1–15.

―――. 2009. *The Fruit, the Tree, and the Serpent*. Cambridge, MA: Harvard University Press.

Ishida, H. 1984. Outline of 1982 survey in Samburu Hills and Nachola area, Northern Kenya. *African Study Monographs* 2 (Suppl.): 1–13.

Ishida, H., Kimura, T. Okada, M., and Yamazaki, N. 1976. Kinesiological aspects of bipedal walking in gibbons. In H. Preuschoft, D. J. Chivers, W. Y. Brockelman, and N. Creel, eds., *The Lesser Apes: Evolutionary and Behavioural Biology*, 135–145. Edinburgh: Edinburgh University Press.

Ishida, H., Kumakura, H., and Kondo, S. 1985. Primate bipedalism and quadrupedalism: comparative electromyography. In S. Kondo, H. Ishida, T. Kimura, M. Okada, N. Yamazaki, and J. Prost, eds., *Primate Morphophysiology, Locomotor Analyses and Human Bipedalism*, 59–79. Tokyo: University of Tokyo Press.

Ishida, H., Kunimatsu, Y., Nakutsukasa, M., and Nakano, Y. 1999. New hominoid genus from the middle Miocene of Nachola, Kenya. *Anthropological Science* 107:189–191.

Ishida, H., Kunimatsu, Y., Takano, T., Nakano, Y., and Nakutsukasa, M. 2004. *Nacholapithecus* skeleton from the Middle Miocene of Kenya. *Journal of Human Evolution* 46:69–103.

Ishida, H., Okada, M., Tuttle, R. H., and Kimura, T. 1978. Activities of hindlimb muscles in a bipedal gibbon. In D. J. Chivers and K. A. Joysey, eds., *Recent Advances in Primatology*, Vol. 3: *Evolution*, 459–462. London: Academic Press.

Ishida, H., and Pickford, M. 1997. A new Late Miocene hominoid from Kenya: *Samburupithecus kiptalami* gen. et sp. nov. *Comptes rendus des séances de l'Académie des Sciences Paris, Sciences de la terre et des planètes* 325:823–829.

Ishida, H., Pickford, M., Nakaya, H., and Nakano, Y. 1984. Fossil anthropoids from Nachola and Samburu Hills, Samburu District, Kenya. *African Study Monographs* 2 (Suppl.): 73–85.

Islam, M. A., and Feeroz, M. M. 1992. Ecology of hoolock gibbon of Bangladesh. *Primates* 33:451–464.

Itani, J. 1980. Social structure of African great apes. *Journal of Reproduction and Fertility* 28 (Suppl.): 33–41.

———. 1982. Intraspecific killing among non-human primates. *Journal of Social and Biological Structures* 5:361–368.

Itani, J., and Suzuki, A. 1967. The social unit of wild chimpanzees. *Primates* 8:355–381.

Iwata, Y., and Ando, C. 2007. Bed and bed-site reuse by western lowland gorillas *(Gorilla g. gorilla)* in Moukalaba-Doudou National Park, Gabon. *Primates* 48:77–80.

Iwamoto, M., Hasegawa, Y., and Koizumi, A. 2005. A Pliocene colobine from the Nakatsu Group, Kanagawa, Japan. *Anthropological Science* 113:123–127.

Izawa, K. 1970. Unit-groups of chimpanzees and their nomadism in the savannah woodland. *Primates* 11:1–46.

Izawa, K., and Itani, J. 1966. Chimpanzees in Kasakati Basin, Tanza nia, (I) ecological study in the rainy season 1963–1964. *Kyoto University African Studies* 1:73–156.

Jablonski, N. G. 1990. A brief review of the fossil record of nonhuman primates in China. *Primate Report* 26:29–44.

———. 1992. The evolution of primates in Asia. In P. K. Seth and S. Seth, eds. *Perspectives in Primate Biology,* Vol. 4, 35–41. New Delhi: Today & Tomorrow's Printers and Publishers.

———. 1993. Quaternary environments and the evolution of primates in East Asia, with notes on two new specimens of fossil Cercopithecidae from China. *Folia Primatologica* 60:118–132.

———. 2002. Fossil Old World monkeys: the late Neogene radiation. In W. C. Hartwig, ed., *The Primate Fossil Record,* 255–299. Cambridge: Cambridge University Press.

Jablonski, N. G., and Chaplin, G. 1992. The origin of hominid bipedalism re-examined. *Archaeology in Oceania* 27:113–119.

———. 1993. Origin of habitual terrestrial bipedalism in the ancestor of the Hominidae. *Journal of Human Evolution* 24:259–280.

Jablonski, N. G., and Gu, Y. 1991. A reassessment of *Megamacaca lantianensis,* a large monkey from the Pleistocene of north-central China. *Journal of Human Evolution* 20:51–66.

Jablonski, N. G., and Pan, Y. 1988. The evolution and palaeobiogeography of monkeys in China. In P. Whyte, J. Aigner, N. G. Jablonski, G. Taylor, D. Walker, and P. X. Wang, eds., *The Palaeoenvironment of East Asia from the Mid-Tertiary,* 849–867. Hong Kong: University of Hong Kong Centre of Asian Studies.

Jack, R. E., Garrod, O. G. B., Yu, H., Caldara, R., and Schyns, P. G. 2012. Facial expressions of emotion are not culturally universal. *Proceedings of the National Academy of Sciences of the United States of America* 109:7241–7244.

Jackendoff, R. 1999. Possible stages in the evolution of language capacity. *Trends in Cognitive Science* 3:272–279.

———. 2002. *Foundations of Language: Brain, Meaning, Grammar, Evolution.* New York: Oxford University Press.

———. 2007. *Language, Consciousness, Culture: Essays on Mental Structure.* Cambridge, MA: MIT Press.

Jackendoff, R., and Lerdahl, F. 2006. The capacity for music: what is it, and what's special about it? *Cognition* 100:33–72.

Jackendoff, R., and Pinker, S. 2005. The nature of language faculty and its implications for evolution of language (reply to Fitch, Hauser, and Chompsky). *Cognition* 97:211–225.

Jacob, T. 1975. Morphology and paleoecology of early man in Java. In R. H. Tuttle, ed., *Paleoanthropology: Morphology and Paleoecology*, 311–325. The Hague: Mouton.

———. 1981. Solo man and Peking man. In B. A. Sigmon and J. S. Cybulski, eds., *Homo erectus: Papers in Honor of Davidson Black*, 87–104. Toronto: University of Toronto Press.

Jacob, T., Indriati, E., Soejono, R. P., Hsü, K., Frayer, D. W., Eckhardt, R. B., Kupervage, A. J., Thorne, A., and Henneberg, M. 2006. Pygmoid Australomelenesian *Homo sapiens* skeletal remains from Liang Bua, Flores: population affinities and pathological abnormalities. *Proceedings of the National Academy of Sciences of the United States of America* 103:13421–13426.

Jacob-Friesen, K. H. 1959. *Einfürung in Niedersachsens Urgeschichte*, Vol. 1: *Steinzeit*. Hildesheim, Germany: August Lax.

Jacobs, G. H. 1993. The distribution and nature of colour vision among the mammals. *Biological Reviews* 68:413–471.

———. 2002. Progress toward understanding the evolution of primate color vision. *Evolutionary Anthropology* 11 (Suppl. 1): 132–135.

———. 2005. Variations in primate color vision: mechanisms and utility. *Evolutionary Anthropology* 3:196–205.

Jacobs, L. L., and Pilbeam, D. 1980. Of mice and men: fossil-based divergence dates and molecular "clocks". *Journal of Human Evolution* 9:551–555.

Jaeger, J.-J. 1975. The mammalian faunas and hominid fossils of the Middle Pleistocene of the Maghreb. In K. W. Butzer and G. L. Isaac, eds., *After the Australopithecines*, 399–418. The Hague: Mouton.

———. 1981. Les hommes fossiles du Pléistocène moyen du Maghreb dans leur cadre géologique, chronologique, et paléoécologique. In B. A. Sigmon and J. S. Cybulski, *Homo erectus: Papers in Honor of Davidson Black*, 159–187. Toronto: University of Toronto Press.

Jaeger, J.-J., Beard, K. C., Chaimanee, Y., Salem, M., Benammi, M., Hlal, O., Coster, P., Bilal, A. A., Duringer, P., Schuster, M., et al. 2010. Late middle Eocene epoch of Libya yields earliest known radiation of African anthropoids. *Nature* 467:1095–1098.

Jaeger, J.-J., and Marivaux, L. 2005. Shaking the earliest branches of anthropoid primate evolution. *Science* 310:244–245.

Jaeger, J.-J., Thein, T., Benammi, M., Chaimanee, Y., Soe, A. N., Lwin, T., Tun, T., Wai, S., and Ducrocq, S. 1999. A new primate from the middle Eocene of Myanmar and the Asian early origin of anthropoids. *Science* 286:528–530.

Jaeggi, A. V., Dunkel, L. P., van Noordwijk, M. A., Wich, S. A., Sura, A. A. L., and Van Schaik, C. P. 2010. Social learning of diet and foraging skills by wild immature Bornean orangutans: implications for culture. *American Journal of Primatology* 72:62–71.

Jaeggi, A. V., Stevens, J. M. G., and Van Schaik, C. P. 2010. Tolerant food sharing and reciprocity is precluded by despotism among bonobos but not chimpanzees. *American Journal of Physical Anthropology* 143:41–51.

Jaeggi, A. V., van Noordwijk, M. A., and Van Schaik, C. P. 2008. Begging for information: mother-offspring food sharing among wild Bornean orangutans. *American Journal of Primatology* 70:533–541.

Jaeggi, A. V., and Van Schaik, C. P. 2011. The evolution of food sharing in primates. *Behavioral Ecology and Sociobiology* 65:2125–2140.

Jalles-Filho, E. 1995. Manipulative propensity and tool use in capuchin monkeys. *Current Anthropology* 36:664–667.

James, S. R. 1989. Hominid use of fire in the Lower and Middle Pleistocene. *Current Anthropology* 30:1–26.

———. 1996. Early hominid use of fire: recent approaches and methods for evaluation of the evidence. In O. Bar-Yosef, L. L. Cavalli-Sforza, R. J. March, and M. Piperno, eds., *The Lower and Middle Paleolithic*, 65–75. Forlì, Italy: Abaco.

James, W. Why 'kinship'? New questions on an old topic. In N. J. Allen, H. Callan, R. Dunbar, and W. James, eds., *Early Human Kinship: From Sex to Social Reproduction*, 3–20. Malden, MA: Blackwell.

Jäncke, L., Schlaug, G., Huang, Y., and Steinmetz, H. 1994. Asymmetry of the planum parietale. *NeuroReport* 5:1161–1163.

Janczewski, D. N., Goldman, D., and O'Brien, S. J. 1990. Molecular genetic divergence of orang utan *(Pongo pygmaeus)* subspecies based on isozyme and two-dimensional gel electrophoresis. *Journal of Heredity* 81:375–387.

Janecka, J. E., Miller, W., Pringle, T. H., Wiens, F., Zitzmann, A., Helgin, K. M., Springer, M. S., and Murphy, W. J. 2007. Molecular and genomic data identify the closest living relative of Primates. *Science* 318:792–794.

Janik, V. M., and Slater, P. J. B. 1997. Vocal learning in mammals. *Advances in the Study of Behavior* 26:59–99.

Janke, A., Feldmaier-Fuchs G., Thomas. W. K., von Haeseler, A., and Pääbo, S. 1994. The marsupial mitochondrial genome and the evolution of placental mammals. *Genetics* 137:243–256.

Januszkiewicz, K. 2000. Speed of the fastest human, running. In G. Elert, ed., *The Physics Factbook,* http://hypertextbook.com/facts.

Janzen, D. H. 1997. Wildland biodiversity management in the tropics. In M. L. Reaka-Kudla, D. E. Wilson, and E. O. Wilson, eds., *Biodiversity II,* 411–431. Washington, DC: National Academy of Sciences.

Jarvis, M. J., and Ettinger, G. 1977. Cross-modal recognition in chimpanzees and monkeys. *Neuropsychologia* 15:499–506.

———. 1978. Jarvis, M. J., and Ettinger, G. 1977. Cross-modal recognition in monkeys and apes. In D. J. Chivers and J. H. Herbert, eds., *Recent Advances in Primatology,* Vol. 1: *Behaviour,* 953–956. London: Academic Press.

Jasienska, G., Thune, I., and Ellison, P. T. 2006. Fatness at birth predicts adult susceptibility to ovarian suppression: an empirical test of the predictive adaptive response hypothesis.

Proceedings of the National Academy of Sciences of the United States of America 103:12759–12762.

Jefferies, E., and Lambon Ralph, M. A. 2006. Semantic impairment in stroke aphasia versus semantic dementia: a case-series comparison. *Brain* 129:2132–2147.

Jelinek, J. 1978. *Homo erectus* or *Homo sapiens?* In D. J. Chivers and K. A. Joysey, eds., *Recent Advances in Primatology,* Vol. 3: *Evolution,* 419–429. London: Academic Press.

———. 1980. European *Homo erectus* and the origin of *Homo sapiens.* In L.-K. Königsson, ed., *Current Argument on Early Man,* 137–144. Oxford: Pergamon Press.

Jenkins, F. A., Jr. 1981. Wrist rotation in primates: a critical adaptation for brachiators. In M. H. Day, ed., *Vertebrate Locomotion, Symposia of the Zoological Society of London,* no. 48, 429–451.

Jenkins, F. A., Jr., and Fleagle, J. G. 1975. Knuckle-walking and the functional anatomy of the wrists in living apes. In R. H. Tuttle, ed., *Primate Functional Morphology and Evolution,* 213–231. The Hague: Mouton.

Jenny, D., and Zuberbühler, K. 2005. Hunting behaviour in West African forest leopards. *African Journal of Ecology* 43:197–200.

Jensen, K., Call, J., and Tomasello, M. 2007. Chimpanzees are vengeful but not spiteful. *Proceedings of the National Academy of Sciences of the United States of America* 104:13046–13050.

Jensen, K., Hare, B., Call, J., and Tomasello, M. 2006. What's in it for me? Self-regard precludes altruism and spite in chimpanzees. *Proceedings of the Royal Society B* 273:1013–1021.

Jensen-Seaman, M. I., Deinard, A. S., and Kidd, K. K. 2003. Mitochondrial and nuclear DNA estimates of divergence between western and eastern gorillas. In A. B. Taylor and M. L. Goldsmith, eds., *Gorilla Biology: A Multidisciplinary Perspective,* 247–268. Cambridge: Cambridge University Press.

Jerison, H. J. 1973. *Evolution of the Brain and Intelligence.* New York: Academic Press.

———. 2000. Paleoneurology and the biology of music. In N. L. Wallin, B. Merker and S. Brown, eds., *The Origins of Music,* 177–196. Cambridge, MA: MIT Press.

Jespersen, O. 1922. *Language: Its Nature, Development and Origin.* London: Allen and Unwin.

Jiang, X., Luo, A., Zhao, S., Li, R., and Liu, C. 2006. Status and distribution pattern of black crested gibbons *(Nomascus concolor jingdongensis)* in Wuliang Mountains, Yunnan, China: implications for conservation. *Primates* 47:264–271.

Jiang, X., Wan, Y., and Wang, Q. 1999. Coexistence of monogamy and polygyny in black-crested gibbon *(Hylobates concolor). Primates* 40:607–611.

Jisaka, M., Ohigashi, H., Takagaki, T., Nozaki, H., Tada, T., Hirota, M., Irie, R., Huffman, M. A., Nishida, T., Kaji, M., and Koshimizu, K. 1992. Bitter steroid glucosides, vernoniosides A_1, A_2, and A_3, and related B_1 from a possible medicinal plant, *Vernonia amygdalina,* used by wild chimpanzees. *Tetrahedron* 48:625–632.

Jóhannesson, A. 1949. *Origin of Language: Four Essays.* Reykjavik: Leiftur.

———. 1950. The gestural origin of language; evidence from six 'unrelated' languages. *Nature* 166:60–61.

Johanson, D. C. 1974. Some metric aspects of the permanent and deciduous dentition of the pygmy chimpanzee *(Pan paniscus). American Journal of Physical Anthropology* 41:39–48.

———. 1976. Ethiopia yields first "family" of early man. *National Geographic* 150:790–811.

———. 1977. Rethinking the origins of the genus *Homo*. In R. Duncan and M. Weston-Smith, eds., *The Encyclopaedia of Ignorance*, 243–250. Oxford: Pergamon Press.

———. 1980. Early African hominid phylogenesis: a re-evaluation. In L.-K. Königsson, ed., *Current Argument on Early Man*, 31–69. Oxford: Pergamon Press.

———. 1985. The most primitive *Australopithecus*. In P. V. Tobias, ed., *Hominid Evolution: Past, Present and Future*, 203–212. New York: Alan R. Liss.

———. 1989. The current status of Australopithecus. In G. Giacobini, ed., *Hominidae: Proceedings of the 2nd International Congress of Human Paleontology*, 77–96. Milan: Jaca Book.

Johanson, D. C., and Coppens, Y. 1976. A preliminary anatomical diagnosis of the first Plio/Pleistocene hominid discoveries in the central Afar, Ethiopia. *American Journal of Physical Anthropology* 45:217–234.

———. 1980. A provisional interpretation of the Hadar hominids. In R. E. Leakey and B. A. Ogot, eds., *Proceedings of the 8th Panafrican Congress of Prehistory and Quaternary Studies Nairobi, 5 to 10 September 1977*, 157–160. Nairobi: International Louis Leakey Memorial Institute for African Prehistory.

Johanson, D., and Edey, M. 1981. *Lucy: The Beginnings of Humankind*. New York: Simon & Schuster.

Johanson, D., and Edgar, B. 1996. *From Lucy to Language*. New York: Simon & Schuster.

Johanson, D. C., Lovejoy, C. O., Kimbel, W. H., White, T. D., Ward, S. C., Bush, M. E., Latimer, B. M., and Coppens, Y. 1982. Morphology of the Pliocene hominid skeleton (A. L. 288–1) from the Hadar Formation, Ethiopia. *American Journal of Physical nthropology* 57:403–451.

Johanson, D. C., and Taieb, M. 1976. Plio-Pleistocene hominid discoveries in Hadar, Ethiopia. *Nature* 260:293–297.

———. 1978. Plio-Pleistocene hominid discoveries in Hadar, central Afar, Ethiopia. In C. Jolly, ed., *Early Hominids of Africa*, 29–44. New York: St. Martin's Press.

Johanson, D. C., Taieb, M., and Coppens, Y. 1982. Pliocene hominids from the Hadar Formation, Ethiopia (1973–1977): stratigraphic, chronologic, and paleoenvironmental contexts, with notes on hominid morphology and systematics. *American Journal of Physical Anthropology* 57:373–402.

Johanson, D. C., Taieb, M., Coppens, Y., and Roche, H. 1978. Nouvelles découvertes d'Hominidés et découvertes d'industries lithiques pliocènes à Hadar. *Comptes rendus hebdomadaires des séances de l'Académie des Sciences, Paris*, série D 287:237–240.

Johanson, D. C., and White, T. D. 1979. A systematic assessment of early African hominids. *Science* 203:321–30.

Johanson, D. C., White, T. D., and Coppens, Y. 1978. A new species of the genus *Australopithecus* (Primates: Hominidae) from the Pliocene of eastern Africa. *Kirtlandia* 28:1–14.

Johnson, C. M., Frank, R. E., Flynn, D. 1999. Peering in mature, captive bonobos *(Pan paniscus)*. *Primates* 40:397–407.

Johnson, G. D., Opdyke, N. D., Tandon, S. K., Nanda, A. C. 1983. The magnetic polarity stratigraphy of the Siwalik Group at Haritalyangar (India) and a new last appearance datum for *Ramapithecus* and *Sivipithecus* in Asia. *Palaeogeography, Palaeoclimatology, Palaeoecology* 44:223–249.

Johnson, M. E., Viggiano, L., Bailey, J. A., Abdul-Rauf, M., Goodwin, G., Rocchi, M., and Eichler, E. E. 2001. Positive selection of a gene family during the emergence of humans and African apes. *Nature* 413:514–519.

Johnston, F. E. 1982. Pliocene hominid fossils from Hadar, Ethiopia. *American Journal of Physical Anthropology* 57:373–719.

Jolly, A. 1966. Lemur behavior and primate intelligence. *Science* 153:501–506.

Jolly, C. J. 1967. The evolution of baboons. In H. Vagtborg, ed., *The Baboon in Medical Research*, Vol. 2, 23–50. Austin: University of Texas Press.

———. 1970. The seed eaters: a new model of hominid differentiation based on baboon analogy. *Man* 5:5–27.

———. 1972. The classification and natural history of *Theropithecus* (Simopithecus) (Andrews, 1916), Baboons of the African Plio-Pleistocene. *Bulletin of the British Museum (Natural History) Geology* 22:1–123.

Jolly, C. J., Oates, J. F., and Disotell, T. R. 1995. Chimpanzee kinship. *Science* 268:185–186.

Jolly, C. J., and Plog, F. 1986. *Physical Anthropology and Archeology*. New York: Knopf.

Jones, C., and Sabater Pí. 1971. Comparative ecology of *Gorilla gorilla* (Savage and Wyman) and *Pan troglodytes* (Blumenbach) in Reio Muni, West Africa. *Bibliotheca Primatologica*, No. 13:1–96. Basel: S. Karger.

Jones, D. 2003. Kinship and deep history: exploring connections between culture areas, genes, and languages. *American Anthropologist* 103:501–514.

———. 2010. Human kinship from conceptual structure to grammar. *Behavioral and Brain Sciences* 33:367–416.

Jones, F. W. 1940. Attainment of upright posture of man. *Nature* 146:26–27.

Jones, P. R. 1987. Recording the hominid footprints. In M. D. Leakey and J. M. Harris, eds., *Laetoli: A Pliocene Site in Northern Tanzania*, 551–558. Oxford: Clarendon Press.

Jones-Engel, L. E., and Bard, K. A. 1996. Precision grips in young chimpanzees. *American Journal of Primatology* 39:1–15.

Jouffroy, F. K., Godinot, M., and Nakano, Y. 1991. Biometrical characteristics of primate hands. *Human Evolution* 6:269–306.

Joulian, F. 1994. Culture and material culture I chimpanzees and early hominids. In J. J. Roeder, B. Thierry, J. R. Anderson, and N. Herrenschmidt, eds., *Current Primatology*, Vol. 2: *Social Development, Learning and Behaviour*, 397–404. Strasbourg: Université Louis Pasteur.

———. 1995. Mis en évidence de différences traditionnelles dans le cassage des noix chez les chimpanzés *(Pan troglodytes)* de la Côte d'Ivoire, implications paleoanthropologiques. *Journal des africanistes* 65:57–77.

———. 1996. Comparing chimpanzee and early hominid techniques: some contributions to cultural and cognitive questions. In P. Mellars and K. Gibson, eds., *Modelling the Early Human Mind*, 173–189.

Jukes, T. H. 1980. Silent nucleotide substitutions and the molecular evolutionary clock. *Science* 210:973–978.

Jungers, W. L. 1977. Hindlimb and pelvic adaptations to vertical climbing and clinging in *Megaladapis,* a giant subfossil prosimian from Madagascar. *Yearbook of Physical Anthropology* 20:508–524.

———. 1982. Lucy's limbs: skeletal allometry and locomotion in *Australopithecus afarensis. Nature* 297:676–677.

———. 1984. Aspects of size and scaling in primate biology with special reference to the locomotor skeleton. *Yearbook of Physical Anthropology* 27:73–97.

———. 1985. Body size and scaling of limb proportions in primates. In W. L. Jungers, ed., *Size and Scaling in Primate Biology,* 345–381. New York: Plenum Press.

———. 1987. Body size and morphometric affinities of the appendicular skeleton in *Oreopithecus bambolii* (IGF 11778). *Journal of Human Evolution* 16:445–456.

———. 1988. New estimates of body size in australopithecines. In F. E. Grine, ed., *Evolutionary History of the "Robust" Australopithecines,* 115–125. Hawthorne, NY: Aldine de Gruyter.

Jungers, W. L., and Cole, M. S. 1992. Relative growth and shape of the locomotor skeleton in lesser apes. *Journal of Human Evolution* 3:93–105.

Jungers, W. L., and Grine, F. E. 1986. Dental trends in the australopithecines: the allometry of mandibular molar dimensions. In eds. B. Wood, L. Martin, and P. Andrews, *Major Topics in Primate and Human Evolution,* 203–219. Cambridge: Cambridge University Press.

Jungers, W. L., Harcourt-Smith, W. E. H., Wunderlich, R. E., Tocheri, M. W., Larson, S. G., Sutikna, T., Due, R. A., and Morwood, M. J. 2009. The foot of *Homo floresiensis. Nature* 459:81–84.

Jungers. W. L., Pokempner, A. A., Kay, R. F., and Cartmill, M. 2003. Hypoglossal canal size in living hominoids and the evolution of human speech. *Human Biology* 75:473–484.

Jungers, W. L., and Stern, J. L., Jr. 1980. Telemetered electromyography of forelimb muscle chains in gibbons *(Hylobates lar). Science* 208:617–619.

———. 1983. Body proportions, skeletal allometry and locomotion in the Hadar hominids: a reply to Wolpoff. *Journal of Human Evolution* 12:673–684.

Jungers, W. L., and Susman, R. L. 1984. Body size and skeletal allometry in African apes. In R. L. Susman, ed., *The Pygmy Chimpanzee,* 131–177. New York: Plenum Press.

Jurmain, R. 1997. Skeletal evidence of trauma in African apes, with special reference to the Gombe chimpanzees. *Primates* 38:1–14.

Jusczyk, P. W., and Hohne, E. A. 1997. Infant's memory for spoken words. *Science* 277:1984–1986.

Kaas, J. H. 1987. The organization and evolution of neocortex. In S. P. Wise, ed., *Higher Brain Functions: Recent Explorations of the Brain's Emergent Properties,* 347–378. New York: Wiley.

Kaessmann, H., Wiebe, V., Pääbo, S. 1999. Extensive nuclear DNA sequence diversity among chimpanzees. *Science* 286:1159–1162.

Kaessmann, H., Wiebe, V., Weiss, G., and Pääbo, S. 2001. Great ape DNA sequences reveal a reduced diversity and an expansion of humans. *Nature Genetics* 27:155–156.

Kahlenberg, S. M., Emery Thompson, M., Muller, M. N., and Wrangham, R. W. 2008. Immigration costs for female chimpanzees and male protection as an immigrant counter-strategy to intrasexual aggression. *Animal Behaviour* 76:1497–1509.

Kahlenberg, S. M., Emery Thompson, M., and Wrangham, R. W. 2008. Female competition over core areas in *Pan troglodytes schweinfurthii,* Kibale National Park, Uganda. *International Journal of Primatology* 29:931–947.

Kaifu, Y., Baba, H., Kurniawan, I., Sutikna, T., Saptomo, E. W., Jatmiko, Awe, R. D., Kaneko, T., Aziz, F., and Djubiantono, T. 2009. Brief communication: "pathological" deformation in the skull of LB1, the type specimen of *Homo floresiensis. American Journal of Physical Anthropology* 140:177–185.

Kajikawa, S., and Hasegawa, T. 2000. Acoustic variation of pant hoot calls by male chimpanzees: a playback experiment. *Journal of Ethology* 18:133–139.

Kako, 1999a. Elements of syntax in the systems of three language-trained animals. *Animal Learning and Behavior* 27:1–14.

———. 1999b. Response to Pepperberg; Herman and Uyeyama; and Shanker, Savage-Rumbaugh, and Taylor. *Animal Learning and Behavior* 27:26–27.

Kalb, J. 2001. *Adventures in the Bone Trade.* New York: Copernicus Books.

Kalema-Zikusoka, G., Kock, R. A., and Macfie, E. J. 2002. Scabies in free-ranging mountain gorillas *(Gorilla beringei beringei)* in Bwindi Impenetrable National Park, Uganda. *Veterinary Record* 150:12–15.

Kalpers, J., Williamson, E. A., Robbins, M. M., McNeilage, A., Nzamurambaho, A., Lola, N., and Mugiri, G. 2003. Gorillas in the crossfire: population dynamics of the Virunga Mountain gorillas over the past three decades. *Oryx* 37:326–337.

Kamenya, S. 2002. Human baby killed by Gombe chimpanzee. *Pan Africa News* 9:26.

Kanamori, T., Kuze, N., Bernard, H., Malim, T., Koshima. S. 2010. Feeding ecology of Bornean Orangutans *(Pongo pygmaeus morio)* in Danum Valley, Sabah, Malaysia: a 3-Year record including two mast fruitings. *American Journal of Primatology* 72:820–840.

Kanngiesser, P., Sueur, C., Riedl, K., Grossmann, J., and Call, J. 2011. Grooming network cohesion and the role of individuals in a captive chimpanzee group. *American Journal of Primatology* 73:758–767.

Kano, T. 1971. The chimpanzee of Filabanga, western Tanzania. *Primates* 12:229–246.

———. 1979. A pilot study on the ecology of pygmy chimpanzees, *Pan paniscus.* In D. A. Hamburg and E. R. McCown, eds., *The Great Apes,* 122–135. Menlo Park, CA: Benjamin/Cummings.

———. 1980. Social behavior of wild pygmy chimpanzees *(Pan paniscus)* of Wamba: a preliminary report. *Journal of Human Evolution* 9:243–260.

———. 1982a. The social group of pygmy chimpanzees *(Pan paniscus)* of Wamba. *Primates* 23:171–188.

———. 1982b. The use of leafy twigs for rain cover by the pygmy chimpanzee of Wamba. *Primates* 23:453–457.

———. 1983. An ecological study of the pygmy chimpanzee *(Pan paniscus)* of Yalisidi, Republic of Zaire. *International Journal of Primatology* 4:1–31.

———. 1984. Distribution of pygmy chimpanzees *(Pan paniscus)* in the Central Zaire Basin. *Folia Primatologica* 43:36–52.

———. 1989. The sexual behavior of pygmy chimpanzees. In P. G. Heltne ad L. A. Marquardt, eds. *Understanding Chimpanzees,* 176–183. Cambridge, MA: Harvard University Press.

————. 1992. *The Last Ape: Pygmy Chimpanzee Behavior and Ecology.* Stanford, CA: Stanford University Press.

————. 1996. Male rank order and copulation rate in a unit-group of bonobos at Wamba, Zaïre. In W. C. McGrew, L. F. Marchant, and T. Nishida, eds., *Great Ape Societies,* 135–145. Cambridge: Cambridge University Press.

Kano, T., and Asato, R. 1994. Hunting pressure on chimpanzees and gorillas in the Motoba River Area, Northeastern Congo. *African Study Monographs* 15:143–162.

Kano, T., Lingomo-Bongoli, Idani, G., and Hashimoto, C. 1996. The challenge of Wamba. In P. Cavaliere, and P. Singer, eds. *Etica e Animali* 8/96:68–74. London: Fourth Estate.

Kano, T., and Mulavwa, M. 1984. Feeding ecology of the pygmy chimpanzees *(Pan paniscus)* of Wamba. In R. L. Susman, ed. *The Pygmy Chimpanzee,* 233–274. New York: Plenum Press.

Kaplan, H., Hill, K., Lancaster, J., and Hurtado, A. M. 2000. A theory of human life history evolution: diet, intelligence, and longevity. *Evolutionary Anthropology* 9:156–185.

Kappeler, P. 1984a. Diet and feeding behaviour of the moloch gibbon. In H. Preuschoft, D. J. Chivers, W. Y. Brockelman, and N. Creel, eds., *The Lesser Apes: Evolutionary and Behavioural Biology,* 228–241. Edinburgh: Edinburgh University Press.

————. 1984b. Vocal bouts and territorial maintenance in the moloch gibbon. In H. Preuschoft, D. J. Chivers, W. Y. Brockelman, and N. Creel, eds., *The Lesser Apes: Evolutionary and Behavioural Biology,* 376–389. Edinburgh: Edinburgh University Press.

Kappelman, J. 1991. The paleoenvironment of Kenyapithecus at Fort Ternan. *Journal of Human Evolution* 20:95–129.

————. 1992. The age of the Fayum primates as determined by paleomagnetic reversal stratigraphy. *Journal of Human Evolution* 22:495–503.

Kappelman, J., Kelly, J., Pilbeam, D., Sheikh, K. A., Ward, S., Anwar, M., Barry, J. C., Brown, B., Hake, P., Johnson, N. M., et al. 1991. The earliest occurrence of Sivapithecus from the middle Miocene Chinji Formation of Pakistan. *Journal of Human Evolution* 21:61–73.

Kass, J. H. 2012. Evolution of columns, modules, and domains in the neocortex of primates. *Proceedings of the National Academy of Sciences of the United States of America* 109:10655–10660.

Katz, J. 1976. A hypothesis about the uniqueness of natural language. *Annals of the New York Academy of Sciences* 280:33–41.

Kawabe, M. 1966. One observed case of hunting behavior among wild chimpanzees living in the savanna woodland of western Tanzania. *Primates* 6:393–396.

————.1970. A preliminary study of the wild siamang gibbon *(Hylobates syndactylus)* at Fraser's Hill, Malaysia. *Primates* 11:285–291.

Kawai, M., and Mizuhara, H. 1959–1960. An ecological study on the wild mountain gorilla *(Gorilla gorilla beringei). Primates* 2:1–42.

Kawamura, S., Saitou, N., and Ueda, S. 1992. Concerted evolution of the primate immunoglobulin alpha-gene through gene conversion. *Journal of Biological Chemistry* 267:7359–7367.

Kawamura, S., Tanabe, H., Watanabe, Y., Kurosaki, K., Saitou, N., and Ueda, S. 1991. Evolutionary rate of immunoglobin alpha noncoding region is greater in hominoids than in Old World monkeys. *Molecular Biology and Evolution* 8:743–752.

Kawamura, S., and Ueda, S. 1992. Immunoglobin CH gene family in hominoids and its evolutionary history. *Genomics* 13:194–200.

Kawanaka, K. 1981. Infanticide and cannibalism in chimpanzees, with special reference to the newly observed case in the Mahale Mountains. *African Study Monographs* 1:69–99.

———. 1982a. A case of inter-unit-group encounter in chimpanzees of the Mahale Mountains. *Primates* 23:558–562.

———. 1982b. Further studies on predation by chimpanzees in the Mahale Mountains. *Primates* 23:364–384.

———. 1984. Association, ranging, and the social unit in chimpanzees of the Mahale Mountains, Tanzania. *International Journal of Primatology* 5:411–434.

Kawanaka K., and Nishida, T. 1975. Recent advances in the study of inter-unit-group relationships and social structure of wild chimpanzees of the Mahali Mountains. In S. Kondo, M. Kawai, A. Ehara, and S. Kawamura, eds., *Proceedings from the Symposia of the Fifth Congress of the International Primatological* Society, 173–186. Tokyo: Japan Science Press.

Kay, R. F. 1975. The functional adaptations of primate molar teeth. *American Journal of Physical Anthropology* 43:195–216.

———. 1977. The evolution of molar occlusion in the Cercopithecidae and early catarrhines. *American Journal of Physical Anthropology* 46:327–352.

———. 1978. Molar structure and diet in extant Cercopithecidae. In P. M. Butler and K. A. Joysey, eds., *Development, Function, and Evolution of Teeth,* 309–339. London: Academic Press.

———. 1981. The nut-crackers—a new theory of the adaptations of the Ramapithecinae. *American Journal of Physical Anthropology* 55:141–151.

———. 1982a. Sexual dimorphism in Ramapithecinae. *Proceedings of the National Academy of Sciences of the United States of America* 79:209–212.

———. 1982b. *Sivapithecus simonsi,* a new species of Miocene hominoid, with comments on the phylogenetic status of the Ramapithecinae. *International Journal of Primatology* 3:113–173.

———. 1985. Dental evidence for the diet of *Australopithecus. Annual Review of Anthropology* 14:315–341.

———. 1988. Parapithecidae. In I. Tattersall, E. Delson and J. Van Couvering, eds., *Encyclopaedia of Human Evolution and Prehistory* 440–443. New York: Garland.

Kay, R. F., Cartmill, M., and Balow, M. 1998. The hypoglossal canal and the origin of human vocal behavior. *Proceedings of the National Academy of Sciences of the United States of America* 98:5417–5419.

Kay, R. F., and Frine, F. E. 1988. Tooth morphology, wear and diet in *Australopithecus* and *Paranthropus* from southern Africa. In F. E. Grine, ed., *Evolutionary History of the Robust" Australopithecines,* 427–447. Hawthorne, NY: Aldine de Gruyter.

Kay, R. F., and Hylander, W. L. 1978. The dental structure of mammalian folivores with special reference to Primates and Phalangeroidea (Marsupialia). In G. G. Montgomery, ed., *The Ecology of Arboreal Folivores*, 173–191. Washington, DC: Smithsonian Institution Press.

Kay, R. F., Ross, C. F., and Williams, B. A. 1997. Anthropoid origins. *Science* 275:797–804.

Kay, R. F., and Simons, E. L. 1980. The ecology of Oligocene African Anthropoidea. *International Journal of Primatology* 1:21–37.

———. 1983a. Dental formulae and dental eruption patterns in Parapithecidae (Primates, Anthropoidea). *American Journal of Physical Anthropology* 62:363–375.

———. 1983b. A reassessment of the relationship between later Miocene and subsequent Hominoidea. In R. L. Ciochon and R. S. Corruccini, eds., *New Interpretations of Ape and Human Ancestry*, 577–624. New York: Plenum Press.

———. 2005. A synopsis of the phylogeny and paleobiology of Amphiphithecidae, South Asian middle and late Eocene primates. *Anthropological Science* 113:33–42.

Kay, R. F., and Van Couvering, J. A. 1988. Fayum. In I. Tattersall, E. Delson, and J. Van Couvering, eds., *Encyclopaedia of Human Evolution and Prehistory*, 206–206. New York: Garland.

Kay, R. N. B., and Davies, A. G. 1994. Digestive physiology. In A. G. Davies and J. F. Oates, eds., *Colobine Monkeys: Their Ecology, Behaviour and Evolution*, 229–249. Cambridge: Cambridge University Press.

Keeley, L. H. 1977. The functions of Paleolithic flint tools. *Scientific American* 237:103–109.

Keeley, L. H., and Newcomer, M. H. 1977. Analysis of experimental flint tools: a test case. *Journal of Archaeological Science* 4:29–62.

Keeley, L. H., and Toth, N. 1981. Microwear polishes on early stone tools from Koobi Fora, Kenya. *Nature* 293:464–465.

Keith, A. 1903. The extent to which the posterior segments of the body have been transmuted and suppressed in the evolution of man and allied primates. *Journal of Anatomy and Physiology* 37:18–40.

———. 1912a. Certain phases in the evolution of man. *British Medical Journal* 1:734–737, 788–790.

———. 1912b. Important phases in the evolution of man. *Medical Press* 93:271–273.

———. 1923. Man's posture: its evolution and disorders. *British Medical Journal* 1:451–454, 499–502, 545–548, 587–590, 624–626, 669–672.

———. 1926. *The Engines of the Human Body*. Philadelphia: Lippincott.

———. 1940. Fifty years ago. *American Journal of Physical Anthropology* 26:251–267.

Keith, S. A., Waller, M. S., and Geissmann, T. 2009. Vocal diversity of Kloss's gibbons *(Hylobates Klossii)* in the Mentawai Islands, Indonesia. In S. Lappan and D. J. Whittaker, eds., *The Gibbons: New Perspectives on Small Ape Socioecology and Population Biology*, 51–71. New York: Springer.

Kelley, J. 1986. Species recognition and sexual dimorphism in *Proconsul* and *Rangwapithecus*. *Journal of Human Evolution* 15:461–495.

———. 1988. A new large species of *Sivapithecus* from the Siwaliks of Pakistan. *Journal of Human Evolution* 17:305–324.

———. 1993. Taxonomic implications of sexual dimorphism in *Lufengpithecus*. In W. H. Kimbel and L. B. Martin, eds., *Species, Species Concepts, and Primate Evolution*, 429–558, New York: Plenum Press.

———. 2001. Phylogeny and sexually dimorphic characters: canine reduction in *Ouranopithecus*. In L. de Bonis, G. D. Koufos, and P. Andrews, eds., *Hominoid Evolution and Climatic Change in Europe*, Vol. 2: *Phylogeny of the Neogene Hominoid Primates of Eurasia*, 269–283. Cambridge: Cambridge University Press.

Kelley, J., and Etler, D. 1989. Hominoid dental variability and species number at the Late Miocene site of Lufeng, China. *American Journal of Primatology* 18:5–34.

Kelley, J., and Gao, F. 2012. Juvenile hominoid cranium from the late Miocene of southern China and hominoid diversity in Asia. *Proceedings of the National Academy of Sciences of the United States of America* 109:6882–6885.

Kelley, J., and Pilbeam, D. 1986a. Kenyan finds not early Miocene *Sivapithecus*. *Nature* 321:475–476.

———. 1986b. The dryopithecines: taxonomy, comparative anatomy, and phylogeny of Miocene large hominoids. In J. Erwin, ed., *Comparative Primate Biology*, Vol. 1: *Systematics, Evolution, and Anatomy*, 361–411. New York: Alan R. Liss.

Kelley, J., and Xu, Q. 1991. Extreme sexual dimorphism in a Miocene hominoid. *Nature* 352:151–153.

Kellner, C. M., and Schoeninger, M. J. 2007. A simple carbon isotope model for reconstructing prehistoric human diet. *American Journal of Physical Anthropology* 133:1112–1127.

Kellogg, W. N. 1969. Research on the home raised chimpanzee. In G. H. Bourne, ed., *The Chimpanzee*, Vol. 1: *Anatomy, Behavior, and Diseases of Chimpanzees*, 369–392. Basel: Karger.

Kellogg, W. N., and Kellogg, L. A. 1933. *The Ape and the Child: A Study of Environmental Influence upon Early Behavior*. New York: McGraw-Hill.

Kelly, R. C. 2000. *Warless Societies and the Origin of War*. Ann Arbor: University of Michigan Press.

———. 2005. The evolution of lethal intergroup violence. *Proceedings of the National Academy of Sciences of the United States of America* 102:15294–15298.

Kelly, R. L. 1995. *The Foraging Spectrum: Diversity in Hunter-Gatherer Lifeways*. Washington, DC: Smithsonian Institution Press.

Keltner, D., Marsh, J., and Smith, J. A. 2010. *The Compassionate Instinct: The Science of Human Goodness*. New York: W. W. Norton.

Keltner, R. L. 2009. *Born to Be Good*. New York: W. W. Norton.

Kemp, C., and Regier, T. 2012. Kinship categories across languages reflect general communicative principles. *Science* 336:1049–1054.

Kennedy, G. E. 1991. On the autapomorphic traits of Homo erectus. *Journal of Human Evolution* 20:375–412.

Kennedy, K. A. R. 1960. The phylogenetic tree: an analysis of its development in studies of human evolution. *Kroeber Anthropological Society Papers*, no. 23: 7–53.

Kennedy, K. A. R., Sonakia, A., Chiment, J., and Verma, K. K. 1991. Is the Narmada hominid an Indian *Homo erectus*? *American Journal of Physical Anthropology* 86:475–496.

Kennett, D. J., and Winterhalder, B. 2006. *Behavioral Ecology and the Transition to Agriculture*. Berkeley: University of California Press.

Kevles, D. J. 1995. In the Name of Eugenics. *Genetics and the Uses of Human Heredity*. Cambridge, MA: Harvard University Press.

Keyser, A. W. 2000. The Drimolen skull: the most complete australopithecine cranium and mandible to date. *South African Journal of Science* 96:189–197.

Keysers, C., Kohler, E., Umiltà, M. A., Nanetti, L., Fogassi, L., and Gallese, V. 2003. Audiovisual mirror neurons and action recognition. *Experimental Brain Research* 153:628–636.

Khatri, A. P. 1975. The early fossil hominids and related apes of the Siwalik foothills of the Himalayas: recent discoveries and new interpretations. In R. H. Tuttle, ed., *Paleoanthropology, Morphology and Paleoecology*, 31–58. The Hague: Mouton.

Khudr, G., Benirschke, K., and Sedgwick, C. J. 1973. Man and *Pan paniscus:* a karyologic comparison. *Journal of Human Evolution* 2:323–331.

Kim, H.-S., and Takenaka, O. 1996. A comparison of TSPY genes from Y-chromosomal DNA of the great apes and humans: sequence, evolution and phylogeny. *American Journal of Physical Anthropology* 100:301–309.

Kim, S., Lappan, S., and Choe, J. C. 2011. Diet and ranging behavior of the endangered Javan gibbon *(Hylobates moloch)* in a submontane tropical rainforest. *American Journal of Primatology* 73:270–280.

Kimbel, W. H. 1984. Variation in the pattern of cranial venous sinuses and hominid phylogeny. *American Journal of Physical Anthropology* 63:243–263.

Kimbel, W. H., Johanson, D. C., and Copppens, Y. 1982. Pliocene hominid cranial remains from the Hadar Formation, Ethiopia. *American Journal of Physical Anthropology* 57:453–499.

Kimbel, W. H., Lockwood, C. A., Ward, C. V., Leakey, M. G., Rak, Y., Johanson, D. C. 2006. Was *Australopithecus anamensis* ancestral to *A. afarensis?* A case of anagenesis in the hominin fossil record. *Journal of Human Evolution* 51:134–152.

Kimbel, W. H., and Rak, Y. 1985. Functional morphology of the asterionic region in extant hominoids and fossil hominids. *American Journal of Physical Anthropology* 66:31–54.

Kimbel, W. H., and White, T. C. 1988. Variation, sexual dimorphism and the taxonomy of *Australopithecus*. In F. E. Grine, ed., *Evolutionary History of the "Robust" Australopithecines*, 175–192. Hawthorne, NY: Aldine de Gruyter.

Kimbel, W. H., White, T. D., and Johanson, D. C. 1984. Cranial morphology of *Australopithecus afarensis*: a comparative study based on a composite reconstruction of the adult skull. *American Journal of Physical Anthropology* 64:337–388.

———. 1985. Craniodental morphology of the hominids from Hadar and Laetoli: evidence of *"Paranthropus"* and *Homo* in the mid-Pliocene of eastern Africa? In E. Delson, ed., *Ancestors: The Hard Evidence*, 120–137. New York: Alan R. Liss.

———. 1986. On the phylogenetic analysis of early hominids. *Current Anthropology* 27:361–362.

———. 1988. Implications of KMN-WT 17000 for the evolution of "robust" *Australopithecus*. In F. E. Grine, ed., *Evolutionary History of the "Robust" Australopithecines*, 259–268. Hawthorne, NY: Aldine de Gruyter.

Kimura, D. 1979. Neuromotor mechanisms in the evolution of human communication. In H. D. Steklis and M. J. Raleigh, eds., *Neurobiology of Social Communication in Primates,* 197–219. New York: Academic Press.

Kimura, M. 1983. *The Neutral Theory of Molecular Evolution.* Cambridge: Cambridge University Press.

———. 1987. Molecular evolutionary clock and the neutral theory. *Journal of Molecular Evolution* 26:24–33.

Kimura, T. 1992. Hindlimb dominance during primate high-speed locomotion. *Primates* 33:465–476.

King, B. J. 1986. Extractive foraging and the evolution of primate intelligence. *Human Evolution* 1:361–372.

———. 2003. Alternative pathways for the evolution of gesture. *Sign Language Studies* 4:68–82.

———. 2004. *The Dynamic Dance: Nonvocal Communication in African Great Apes.* Cambridge, MA: Harvard University Press.

———. 2007. *Evolving God: A Provocative View of the Origins of Religion.* New York: Doubleday.

King, B. J., and Shanker, S. G. 2003. How can we know the dancer from the dance? The dynamic nature of African great ape social communication. *Anthropological Theory* 31:5–15.

King, J. E., and Fobes, J. J. 1982. Complex learning by primates. In *Primate Behavior,* J. L. Fobes and J. E. King, eds., p. 327–360. New York: Academic Press.

King, M.-C., and Wilson, A. C. 1975. Evolution at two levels in humans and chimpanzees. *Science* 188:107–116.

King, T. 2001. Dental microwear and diet in Eurasian Miocene catarrhines. In L. de Bonis, G. D. Koufos, and P. Andrews, eds., *Hominoid Evolution and Climatic Change in Europe,* Vol. 2: *Phylogeny of the Neogene Hominoid Primates of Eurasia,* 102–117. Cambridge: Cambridge University Press.

King, T., Aiello, L. C., and Andrews, P. 1999a. Dental microwear *of Griphopithecus alpani. Journal of Human Evolution* 36:3–31.

King, T., Andrews, P., and Boz, B. 1999b. Effect of taphonomic processes on dental wear. *American Journal of Physical Anthropology* 108:359–373.

Kingdon, J. 2003. *Lowly Origin: Where, When and Why Our Ancestors First Stood Up.* Princeton, NJ: Princeton University Press.

Kingsley, S. 1982. Causes of non-breeding and the development of the secondary sexual characteristics in the male orangutan: a hormonal study. In L. E. M. de Boer, ed., *The Orang Utan: Its Biology and Conservation,* 215–229. The Hague: Dr W. Junk.

———. 1988. Physiological development in orang-utans and gorillas. In J. H. Schwartz, ed., *Orang-Utan Biology,* 123–131. Oxford: Oxford University Press.

Kingston, J. D. 2007. Shifting adaptive landscapes: progress and challenges in reconstructing early hominid environments. *Yearbook of Physical Anthropology* 50:20–58.

Kingston, J. D., and Hill, A. Isotopic evidence for Neogene hominid paleoenvironments in the Kenya Rift Valley. *Science* 264:955–959.

Kinsella, A. R. 2009. *Language Evolution and Syntactic Theory.* Camridge, MA: Cambridge University Press.

Kinsey, A. C., Pomeroy, W. B., and Martin, C. E. 1948. *Sexual Behavior of the Human Male.* Philadelphia: W. B. Saunders.

Kinzey, W. G. 1970. Basic rectangle of the mandible. *Nature* 288:289–290.

———. 1984. The dentition of the pygmy chimpanzee, *Pan paniscus.* In R. L. Susman, ed., *The Pygmy Chimpanzee,* 65–88. New York: Plenum Press.

Kirby, S. 1999. *Function, Selection and Innateness: The Emergence of Language Universals.* Oxford: Oxford University Press.

———. 2000. Syntax without natural selection: how compositionality emerges from vocabulary in a population of learners. In C. Knight, M. Studdert-Kennedy, and J. R. Hurford, eds., *The Evolutionary Emergence of Language: Social Function and the Origins of Linguistic Form,* 303–323. Cambridge: Cambridge University Press.

Kirk, E. C. 2004. Comparative morphology of the eye in Primates. In *Evolution of the Special Senses in Primates,* ed. T. D. Smith, C. F. Ross, N. J. Dominy, and J. T. Laitman, special issue, *Anatomical Record Part A* 281A:1095–1103.

Kirk, E. C., and Simons, E. L. 2001. Diets of fossil primates from the Fayum Depression of Egypt: a quantitative analysis of molar shearing. *Journal of Human Evolution* 40:203–229.

Kishino, H., and Hasegawa, M. 1989. Evaluation of the maximum likelihood estimate of the evolutionary tree typologies from DNA sequence data, and the branching order in Hominoidea. *Journal of Molecular Evolution* 29:170–179.

Kitamura, K. 1983. Pygmy association patterns in ranging. *Primates* 24:1–12.

Kittles, R. A., and Weiss, K. M. 2003. Race, ancestry, and genes: implications for defining disease risk. *Annual Review of Genomics and Human Genetics* 4:33–67.

Kivell, T. L., and Begun, D. R. 2009. New primate carpal bones from Rudabànya (late Miocene, Hungary): taxonomic and functional implications. *Journal of Human Evolution* 57:697–709.

Kivell, T. L., and Schmitt, D. 2009. Independent evolution of knuckle-walking in African apes shows that humans did not evolve from a knuckle-walking ancestor. *Proceedings of the National Academy of Sciences of the United States of America* 106: 14241–14246.

Klailova, M., Hodgkinson, C., and Lee, P. C. 2010. Behavioral responses of one western lowland gorilla *(Gorilla gorilla gorilla)* group at Bai Hokou, Central African Republic, to tourists, researchers and trackers. *American Journal of Primatology* 72:897–906.

Kleiman, D. G. 1977. Monogamy in mammals. *Quarterly Review of Biology* 52:39–69.

———. 1981. Correlations among life history characteristics of mammalian species exhibiting two extreme forms of monogamy. In R. D. Alexander and D. W. Tinkle, eds., *Natural Selection and Social Behavior: Recent Research and Theory,* 332–344. New York: Chiron Press.

Klein, R. G. 1995. Anatomy, behavior, and modern human origins. *Journal of World Prehistory* 9:167–198.

Kluge, A. G. 1983. Cladistics and the classification of the great apes. In R. L. Ciochon and R. S. Corruccini, *New Interpretations of Ape and Human Ancestry,* 151–177. New York: Plenum Press.

Knauft, B. M. 1991. Violence and sociality in human evolution. *Current Anthropology* 32:391–428.

Knight, C. Play as precursor of phonology and syntax. In C. Knight, M. Studdert-Kennedy, and J. R. Hurford, eds., *The Evolutionary Emergence of Language: Social Function and the Origins of Linguistic Form,* 99–119. Cambridge: Cambridge University Press.

Knott, C. D. Field collection and preservation of urine in orangutans and chimpanzees. *Tropical Biodiversity* 4:95–102.

———. 1998. Changes in orangutan caloric intake, energy balance, and ketones in response to fluctuating fruit availability. *International Journal of Primatology* 19:1061–1079.

———. 1999. Orangutan behavior and ecology. In P. Dohlinow and A. Fuentes, eds., *The Nonhuman Primates,* 50–57. Mountain View, CA: Mayfield.

———. 2001. Female reproductive ecology of the apes: implications for human evolution. In P. T. Ellison, ed., *Reproductive Ecology and Human Evolution,* 429–463. New Brunswick, NJ: Aldine Transaction.

———. 2005. Energetic responses to food availability in the great apes: implications for hominin evolution. In D. K. Brockman and C. P. van Schaik, eds., *Seasonality in Primates,* 351–378. Cambridge: Cambridge University Press.

———. 2009. Orangutans: sexual coercion without sexual violence. In M. N. Muller and R. W. Wrangham, eds., *Sexual Coercion in Primates and Humans,* 81–111. Cambridge, MA: Harvard University Press.

———. 2011. Female reproductive ecology of the apes. Implications for human evolution. In P. T. Ellison, ed., *Reproductive Ecology and Human Evolution,* 429–463. New Brunswick, NJ: Aldine Transaction.

Knott, C. D., Beaudrot, L., Snaith, T., White, S., Tschauner, H., and Planasky, G. 2008. Female-female competition in Bornean orangutans. *International Journal of Primatology* 29:975–997.

Knott, C. D., Emery Thompson, M., and Stumpf, R. M. 2007. Sexual coercion and mating strategies of wild Bornean orangutans. *American Journal of Physical Anthropology* 44 (Suppl.): 145.

Knott, C. D., Emery Thompson, M., and Wich, S. A. 2009. The ecology of female reproduction in wild orangutans. In S. A. Wich, S. S. Utami Atmoko, T. Mitra Setia, and C. P. van Schaik, eds., *Orangutans: Geographic Variation in Behavioral Ecology and Conservation,* 171–188. Oxford: Oxford University Press.

Knott, C. D., and Kahlenberg, S. M. 2007. Orangutans in perspective. Forced copulations and female resistance. In C. J. Campbell, A. Fuentes, K. C., MacKinnon, M. Panger, and S. Bearder, eds., *Primates in Perspective* 290–305. Oxford: Oxford University Press.

Knüsel, C. J. 1992. The throwing hypothesis and hominid origins. *Human Evolution* 7:1–7.

Knussmann, R. 1967. Humerus, Ulna und Radius der Simiae. *Bibliotheca Primatologia,* Fascicle 5. Basel: Karger.

Koechlin, E., and Hyafil, A. 2007. Anterior prefrontal function and the limits of human decision-making. *Science* 318:594–598.

Koenig, B. A., Lee, S. S-J., and Richardson, S. S. 2008. *Revisiting Race in a Genomic Age.* New Brunswick, NJ: Rutgers University Press.

Koenigswald, G. H. R. von. 1935. Eine fossile Säugetierfauna mit Simia aus Südchina. *Proceedings of the Koninklijke Akademie van Wetenschappen te Amsterdam* 38:872–879.

———. 1952. *Gigantopithecus blacki* von Koenigswald, a giant fossil hominoid from the Pleistocene of southern China. *Anthropological Papers of the American Museum of Natural History* 43:295–325.

———. 1957. Remarks on *Gigantopithecus* and other hominoid remains from southern China. *Proceedings of the Koninklijke Nederlandse Akademie van Wetenschappen, Amsterdam, series B* 60:872–879.

———. 1958. *Gigantopithecus* and *Australopithecus. Leech* 27:101–105.

———. 1975. Early man in Java: catalogue and problems. In R. H. Tuttle, ed., *Paleoanthropology: Morphology and Paleoecology,* 303–309. The Hague: Mouton.

———. 1981. A possible ancestral form of *Gigantopithecus* (Mammalia, Hominoidea) from the Chinji layers of Pakistan. *Journal of Human Evolution* 10:511–515.

Kohler, E., Keysers, C., Umiltà, M. A., Fogassi, L., Gallese, V., and Rizzolatti, G. 2002. Hearing sounds, understanding actions: action representation in mirror neurons. *Science* 297:846–848.

Köhler, M., Alba, D. M., Moyà-Solà, S., and MacLatchy, L. 2002. Taxonomic affinities of the Eppelsheim femur. *American Journal of Physical Anthropology* 119:297–304.

Köhler, M., Moyà-Solà, S. 1997. Ape-like or hominid-like? The positional behavior of *Oreopithecus bambolii* reconsidered. *Proceedings of the National Academy of Sciences of the United States of America* 94:11741–11750.

Köhler, M., Moyà-Solà, S., and Alba, D. M. 2001. Eurasian hominoid evolution in the light of recent *Dryopithecus* findings. L. de Bonis, G. D. Koufos, and P. Andrews, In eds., *Hominoid Evolution and Climatic Change in Europe,* Vol. 2: *Phylogeny of the Neogene Hominoid Primates of Eurasia,* 192–212. Cambridge: Cambridge University Press.

Köhler, W. 1925. *The Mentality of Apes.* New York: Humanities Press.

———. 1959. *The Mentality of Apes.* New York: Vintage Books.

Kohne, D. E. 1970. Evolution of higher-organism DNA. *Quarterly Reviews of Biophysics* 33:327–375.

———. 1975. DNA evolution data and its relevance to mammalian phylogeny. In W. P. Luckett and F. S. Szalay, eds., *Phylogeny of the Primates. A Multidisciplinary Approach,* 249–261. New York: Plenum Press.

Kohne, D. E., Chiscon, J. A., and Hoyer, B. H. 1972. Evolution of primate DNA sequences. *Journal of Human Evolution* 1:627–644.

Kolbert, E. 2011. Enter the age of man. *National Geographic* 219:60–85.

Köndgen, S., Kühl, H., N'Goran, P. K., Walsh, P. D., Schenk, S., Ernst, N., Biek, R., Formenty, P., Mätz-Rensing, K., Schweiger, B., et al. 2008. Pandemic human viruses cause decline of endangered great apes. *Current Biology* 18:260–264.

Konopka, G., Bomar, J. M., Winden, K., Coppola, G., Jonssen, Z. O., Gao, F., Peng, S., Preuss, T. M., Wohlschlegel, J. A., and Geschwind, D. H. 2009. Human-specific transcriptional regulation of CNS development by FOXP2. *Nature* 462:213–217.

Koop, B. F., Goodman, M., Xu, P., Chan, K., and Slightom, J. L. 1986. Primate beta-globin DNA sequences and man's place among the great apes. *Nature* 319:234–238.

Koop, B. F., Siemieniak, D., Slightom, J. L., Goodman, M., Dunbar, J., Wright, P. C., and Simons, E. L. 1989. *Tarsius* delta- and beta- globin genes: conversions, evolution, and systematic implications. *Journal of Biological Chemistry* 264:68–79.

Koops, K., Humle, T., Sterck, H. M., and Matsuzawa, T. 2007. Ground-nesting by the chimpanzees of the Nimba Mountains, Guinea: environmentally or socially determined? *American Journal of Primatology* 69:407–419.

Kordos, L. 1987. Description and reconstruction of the skull of *Rudapithecus hungaricus* Kretzoi (Mammalia). *Annales Historico-Naturales Musei Nationalis Hungarici* 79:77–88.

———. 1988. Comparison of early primate skulls from Rudabánya (Hungary) and Lufeng (China). *Anthropologia Hungarica* 20:9–22.

———. 1991. Le Rudapithecus hungaricus de Rudabànya (Hongrie). *L'Anthropologie, (Paris)* 95:343–362.

———. 2000. New results of hominoid research in the Carpathian Basin. *Acta Biologica Szegediensis* 44:71–74.

Kordos, L., and Begun, D. R. 1998. Encephalization and endocranial morphology in *Dryopithecus brancoi:* implications for brain evolution in early hominids. *American Journal of Physical Anthropology* 26 (Suppl.): 141–142.

———. 2001. A new cranium of *Dryopithecus* from Rudagánya, Hungary. *Journal of Human Evolution* 41:689–700.

Korstjens, A. H. 2008. The importance of kinship in monkey society. In N. J. Allen, H. Callan, R. Dunbar, and W. James, eds., *Early Human Kinship: From Sex to Social Reproduction,* 151–159. Malden, MA: Blackwell.

Kortlandt, A. 1962. Chimpanzees in the wild. *Scientific American* 206:128–138.

———. 1968. Handgebrauch bei freilebenden Schimpansen. In B. Rensch, ed., *Handgebrauch und Verstandigung bei Affen und Fruhmenschen,* 59–102. Bern: Hans Huber.

———. 1972. *New Perspectives on Ape and Human Evolution.* Amsterdam: Stichting voor Psychobiologie.

———. 1980. How might early hominids have defended themselves against large predators and food competitors? *Journal of Human Evolution* 9:79–112.

———. 1995. A survey of the geographic range, habitats and conservation of the pygmy chimpanzee *(Pan paniscus):* an ecological perspective. *Primate Conservation* 16:21–36.

Kortlandt, A., and Holzhaus, E. 1987. New data on the use of stone tools by chimpanzees in Guinea and Liberia. *Primates* 28:473–496.

Koshimizu, K., Ohigashi, H., and Huffman, M. A. 1994. Use of *Veronia amygdalina* by wild chimpanzee: possible roles of its bitter and related constituents. *Physiology and Behavior* 56:1209–1216.

Koski, S. E., and Sterck, E. H. M. 2007. Triadic postconflict affiliation in captive chimpanzees: does consolation console? *Animal Behaviour* 73:133–142.

Kostopoulos, D., Koliadimou, K., and Koufos, G. D. 1996. Giraffids at Nikiti. *Palaeontographica* 239:61–68.

Kostopoulos, D., and Koufos, G. D. 1996. Late Miocene bovids (Mammalia, Artiodactyla) from the locality of Nikiti 1, Macedonia, Greece. *Annales de Paléontologie* 87:251–300.

Kouakou, C. Y., Boesch, C., and Kuehl, H. 2009. Estimating chimpanzee population size with nest counts: validating methods in Taï National Park. *American Journal of Primatology* 71:447–457.

Koufos, G. D. 1993. Mandible of *Ouranopithecus macedoniensis* (Hominidae, Primates) from a new Late Miocene locality of Macedonia (Greece). *American Journal of Physical Anthropology* 91:225–234.

———. 1995. The first female maxilla of the hominoid *Ouranopithecus macedoniensis* from the late Miocene locality of Macedonia (Greece). *Journal of Human Evolution* 29:385–389.

———. 2002. The hominid radiation in Asia. In W. C. Hartwig, ed., *The Primate Fossil Record*, 369–384. Cambridge: Cambridge University Press.

———. 2008. Identification of a single birth cohort in *Kenyapithecus kizili* and the nature of sympatry between *K. kizili* and *Griphopithecus alpani* at Paşalar. *Journal of Human Evolution* 54:530–537.

Koufos, G. D., Syrides, G. E., Koliadimou, K. K., and Kostopoulos, D. S. 1991. Un nouveau gisement de vertébrés avec hominoïde dans le Miocène supérieur de Macédoine (Grèce). *Comptes rendus hebdomadaires des séances de l'Académie des Sciences, Paris, Série II* 313:691–696.

Koyama, N. 1971. Observations on mating behavior of wild siamang gibbons at Fraser's Hill, Malaysia. *Primates* 2:183–189.

Kraatz, R. 1985. A review of recent research on Heidelberg Man, *Homo erectus heidelbergensis*. In E. Delson, ed., *Ancestors: The Hard Evidence*, 268–271. New York: Alan R. Liss.

Kramer, A. 1986. Hominid-pongid distinctiveness in the Miocene-Pliocene fossil record: the Lothagam mandible. *American Journal of Physical Anthropology* 70:457–473.

———. 1993. Human taxonomic diversity in the Pleistocene: does *Homo erectus* represent multiple hominid species? *American Journal of Physical Anthropology* 91:161–171.

Kramer, P. A. Brief communication: could Kadanuumuu (KDS-VP-A/A) and Lucy (AL 288–1) have walked together comfortably? *American Journal of Physical Anthropology* 149:616–621.

Krantz, G. S. 1987. A reconstruction of the skull of *Gigantopithecus blacki* and its comparison with a living form. *Cryptozoology* 6:24–39.

Krause, J., Fu, Q., Good, J. M., Viola, B., Shunkov, M., Derevianko, A. P., and Pääbo, S. 2010. The complete mitochondrial DNA of an unknown hominin from southern Siberia. *Nature* 464:894–897.

Krause, J., Lalueza-Fox, C., Orlando, L., Enard, W., Green, R. E., Burbano, H. A., Hublin, J-J., Hänni, C., Fortea, J., de la Rasilla, et al. 2007. The derived FOXP2 variant of modern humans was shared with Neandertals. *Current Biology* 17:1–5.

Krause, J., Orlando, L., Seere, D., Viola, B., Prüfer, K., Richards, M. P., Hublin, J-J., Hänni, C., Derevianko, A. P., and Pääbo, S. 2007. Neandertals in central Asia and Siberia. *Nature* 449:902–904.

Krause, M. A., and Fouts, R. S. 1997. Chimpanzees *(Pan troglodytes)* pointing: hand shapes, accuracy, and the role of eye gaze. *Journal of Comparative Psychology* 111:330–336.

Kretzoi, M. 1975. New ramapithecines and *Pliopithecus* from the lower Pliocene of Rudabánya in north-eastern Hungary. *Nature* 257:578–581.

Krings, M., Stone, A., Schmitz, R. W., Krainitzki, H., Stoneking, M., and Pääbo, S. 1997. Neandertal DNA sequences and the origin of modern humans. *Cell* 90:19–30.

Kroeber, A. L., and Kluckhohn, C. 1952. *Culture: A Critical Review of Concepts and Definitions.* New York: Random House.

Krubitzer, L., and Kahn, D. M. 2004. The evolution of human neocortex: is the human brain fundamentally different than that of other mammals? In N. Kanwisher and J. Duncan, eds., *Functional Neuroimaging of Visual Cognition: Attention and Performance XX,* 57–81. Oxford: Oxford University Press.

Krüger, O., Affeldt, E., Brackmann, M., and Milhahn, K. 1998. Group size and composition of *Colobus guereza* in Kyambura Gorge, Southwest Uganda, in relation to chimpanzee activity. *International Journal of Primatology* 19:287–297.

Krupinski, T., and Rajchel, Z. Odtworzenie czaszki *Gigantopithecus blacki* III—przedstawiciela kopalnych Hominoidea [Reconstruction of the skull of *Gigantopithecus blacki* III—fossil hominoid]. *Materialy i Prace Antropologiczne,* no. 106: 59–65.

Kudo, H., and Mitani, M. 1985. New record of predatory behavior by the mandrill in Cameroon. *Primates* 26:161–167.

Kuhlmeier, V. A., and Boysen, S. T. 2001. The effect of response contingencies on scale model task performance by chimpanzees *(Pan troglodytes)*. *Journal of Comparative Psychology* 115:300–306.

———. 2002. Chimpanzees *(Pan troglodytes)* recognize spatial and object correspondences between a scale model and its referent. *Psychological Science* 13:60–63.

Kuhlmeier, V. A., Boysen, S. T., and Mukobi, K. L. 1999. Scale-model comprehension by chimpanzees *(Pan troglodytes)*. *Journal of Comparative Psychology* 113:396–402.

Kullmer, O., Schrenk, F., and Dörrhöffer. B. 2003. High resolution 3D-image analysis of ape, hominid and human footprints. *Courier Forschungsinstitut Senkenberg* 243:85–91.

Kumar, S., Filipski, A., Swarna, V., Walker, A., and Hedges, S. B. 2005. Placing confidence limits on the molecular age of the human-chimpanzee divergence. *Proceedings of the National Academy of Sciences of the United States of America* 102:18842–18847.

Kummer, H. 1982. Social knowledge in free-ranging primates. In D. R. Griffen, ed., *Animal Mind-Human Mind,* 113–130. Berlin: Springer.

Kunimatsu, Y. 1992a. New finds of a small anthropoid primate from Nachola, northern Kenya. *African Study Monographs* 13:237–249.

———. 1992b. A revision of the hypodigm of *Nyanzapithecus vancouveringi*. *African Study Monographs* 13:231–235.

———. 1997. New species of Nyanzapithecus from Nachola, northern Kenya. *Anthropological Science* 105:117–141.

Kunimatsu, Y., Nakatsukasa, M., Nakano, Y., and Ishida, H. 1999. The jaws and dentition of the large Miocene hominoid from Nachola, northern Kenya. In H. Ishida, ed., *Abstracts of the International Symposium "Evolution of Middle-to-Late Miocene Hominoids in Africa," July 11–13.* Abstract presented at the meeting, Takaragaike, Kyoto, Japan.

Kunimatsu, Y., Nakatsukasa, M., Sawada, Y., Sakai, T., Hyodo, M., Hyodo, H., Itaya, T., Nakaya, H., Saegusa, H., Mazurier, A., et al. 2007. A new Late Miocene great ape from Kenya and its implications for the origins of African great apes and humans. *Proceedings of the American Academy of Sciences of the United States of America* 104:19220–19225.

Kunimatsu, Y., Ratanasthien, B., Nakaya, H., Saegusa, H., and Nagaoka, S. 2005. Hominoid fossils discovered from Chiang Muan, northern Thailand: the first step towards understanding hominoid evolution in Neogene Southeast Asia. *Anthropological* Science 113:85–93.

Kuper, A. 1999. *Culture: The Anthropologists' Account.* Cambridge: Harvard University Press.

Kuroda, S. 1979. Grouping of the pygmy chimpanzees. *Primates* 20:161–183.

———. 1980. Social behavior of the pygmy chimpanzees. *Primates* 21:181–197.

———. 1984. Interactions over food among pygmy chimpanzees. In R. L. Susman, ed., *The Pygmy Chimpanzee: Evolutionary Biology and Behavior,* 301–324. New York: Plenum Press.

———. 1992. Ecological interspecies relationships between gorillas and chimpanzees in the Ndoki-Nouabale Reserve, Northern Congo. In eds. N. Itoigawa, Y., G. P. Sackett, and R. K. R. Thompson, *Topics in Primatology,* Vol. 2: *Behavior, Ecology, Conservation,* 385–394. Tokyo: University of Tokyo Press.

Kuroda, S., Nishihara, T., Suzuki, S., and Oko, R. A. 1996. Sympatric chimpanzees in the Ndoki Forest, Congo. In W. C. McGrew, L. F. Marchant, and T. Nishida, eds., *Great Ape Societies,* 71–81. Cambridge: Cambridge University Press.

Kuroda, S., Suzuki, S., and Nishihara, T. 1996. Preliminary report on predatory behavior and meat sharing in Tschego chimpanzees *(Pan troglodytes troglodytes)* in the Ndoki Forest, Northern Congo. *Primates* 37:353–359.

Kurth, G. 1956. *Oreopithecus bambolii* Gervais, ein Hominide von der Wende Miozän/ Pliozän. *Naturwissenschaftliche Rundschau* 9:57–61.

———. 1958. Neue Befunde zu Oreopithecus bambolii Gervais. *Naturwissenschaftliche Rundschau* 11:420–426.

Kutsukake, S., and Matsusaka, T. 2002. Incident of intense aggression by chimpanzees against an infant from another group by chimpanzees in Mahale Mountains National Park, Tanzania. *American Journal of Primatology* 58:175–180.

Kutsukake, Y., and Castles, D. I. 2004. Reconciliation and post-conflict third-party affiliation among wild chimpanzees in the Mahale Mountains, Tanzania. *Primates* 45:157–165.

Kuze, N., Malim, T. P., and Koshima, S. 2005. Developmental changes in the facial morphology of the Borneo orangutan *(Pongo pygmaeus):* possible signals in visual communication. *American Journal of Primatology* 65:353–376.

Lacambra, C., Thompson, J., Furuchi, T., Vervaecke, H., and Stevens, J. 2005. Chimpanzee *(Pan troglodytes).* In J. Caldecott and L. Miles, eds., *World Atlas of Great Apes and Their Conservation,* 83–96. Berkeley: University of California Press.

Laden, G., and Wrangham, R. 2005. The rise of the hominids as an adaptive shift in fallback foods: plant underground storage organs (USOs) and australopith origins. *Journal of Human Evolution* 49:482–498.

Ladygina-Kohts, N. N. 2002. *Infant Chimpanzee and Human Child.* Oxford: Oxford University Press. Originally published in Russian in 1935.

Ladygina-Kots, N. N. 1935. Ditia schimpanze i ditia cheloveka. In *Scientific Memoirs of the Museum Darwinianum in Moscow,* Vol. 3. Moscow: Museum Darwinianum.

Lahr, M. M., and Foley, R. 2004. Human evolution writ small. *Nature* 431:1043–1044.

Lai, C. S. L., Fisher, S. E., Hurst, J. A., Vargha-Khadem, F., and Monaco, A. P. 2001. A forkhead-domain gene is mutated in a severe speech and language disorder. *Nature* 413:519–523.

Laidler, K. 1978. Language in the orang-utan. In A. Lock, ed., *Action, Gesture and Symbol: The Emergence of Language,* 133–155. London: Academic Press.

———. 1980. *The Talking Ape.* New York: Stein & Day.

Laird, C. D., McConaughy, B. L., and McCarthy, B. J. 1969. Rate of fixation of nucleotide substitutions in evolution. *Nature* 224:149–154.

Laitman, J. T. 1983. The evolution of the hominid upper respiratory system and implications for the origins of speech. In E. de Grolier, ed., *Glossogenetics: The Origin and Evolution of Language,* 63–90. Paris: Harwood Academic.

———. 1985a. Later Middle Pleistocene hominids. In E. Delson, ed., *Ancestors: The Hard Evidence,* 265–267. New York: Alan R. Liss.

———. 1985b. Evolution of the hominid upper respiratory tract: the fossil evidence. In P. V. Tobias, ed., *Hominid Evolution: Past, Present and Future,* 281–286. New York: Alan R. Liss.

Laitman, J. T., and Heimbuch, R. C. 1982. The basicranium of Plio-Pleistocene hominids as an indicator of the upper respiratory systems. *American Journal of Physical Anthropology* 59:323–343.

Laitman, J. T., Heimbuch, R. C., and Crelin, E. S. 1978. Developmental change in a basicranial line and its relationship to the upper respiratory system in living primates. *American Journal of Anatomy* 152:467–482.

———. 1979. The basicranium of fossil hominids as an indicator of their upper respiratory systems. *American Journal of Physical Anthropology* 51:15–33.

Laitman, J. T., Reidenberg, J. S., and Gannon, P. J. 1992. Fossil skulls and hominid vocal tracts: new approaches to charting the evolution of human speech. In J. Wind, B. Chiarelli, B. Bichakjian, A. Nocentini, and A. Jonker, eds., *Language Origin: A Multidisciplinary Approach,* 385–397. Dordrecht: Kluwer Academic.

Laitman, J. T., Reidenberg, J. S., Gannon, P. J., and Johansson, B. 1990. The Kebara hyoid: what can it tell us about the evolution of the hominid vocal tract? *American Journal of Physical Anthropology* 81:254.

Laland, K. N., and Galef, B. G. 2009. Introduction. In K. N Laland and B. G. Galef, eds., *The Question of Animal Culture,* 1–18. Cambridge, MA: Harvard University Press.

Laland, K. N., and Hoppit, W. 2003. Do animals have culture? *Evolutionary Anthropology* 12:150–159.

Laland, K. N., Kendal J. R., and Kendal, R. L. 2009. Animal culture: problems and solutions. In K. N. Laland and B. G. Galef, eds., *The Question of Animal Culture,* 174–19. Cambridge, MA: Harvard University Press.

La Lumière, L. P. 1981. Evolution of human bipedalism: a hypothesis about where it happened. *Philosophical Transactions of the Royal Society of London, series B* 292:103–107.

Lambert, J. E. 1998. Primate digestion: interactions among anatomy, physiology, and feeding ecology. *Evolutionary Anthropology* 7:8–20.

———. 1999. Seed handling in chimpanzees *(Pan troglodytes)* and redtail monkeys *(Cercopithecus ascanius):* implications for understanding hominoid and cercopithecine fruit-processing strategies and seed dispersal. *American Journal of Physical Anthropology* 109:365–386.

———. 2007a. Book Review *of Feeding Ecology in Apes and Other Primates. International Journal of Primates* 28:1181–1183.

———. 2007b. Primate nutritional ecology. In C. J. Campbell, A. Fuentes, K. C. MacKinnon, M. Panger, and S. K. Bearder, eds., *Primates in Perspective,* 482–495. Oxford: Oxford University Press.

———. 2007c. Seasonality, fallback strategies, and natural selection: a chimpanzee and cercopithecoid model for interpreting the evolution of hominin diet. In P. S. Ungar, ed., *Evolution of the Human Diet,* 324–343. Oxford: Oxford University Press.

Lambert, P. M. 2009. Health versus fitness. *Current Anthropology* 50:649–655.

Lameira, A. R., Hardus, M. E., and Wich, S. A. 2012. Orangutan instrumental gesture-calls: reconciling acoustic and gestural speech evolution models. *Evolutionary Biology* 39:415–418.

Lameira, A. R., and Wich, S. A. 2008. Orangutan long call degradation and individuality over distance: a playback approach. *International Journal of Primatology* 29:615–625.

Lamprecht, J. 1970. Duettgesang beim Siamang, *Symphalangus syndactylus* (Hominoidea, Hylobatinae). *Zeitschrift für Tierpsychologie* 27:186–204.

Lan, D. Y. 1993. Feeding and vocal behavior of black gibbons *(Hylobates concolor)* in Yunnan: a preliminary study. *Folia Primatologica* 60:94–105.

Lanave, C., Tommasi, S., Preparata, G., and Saccone, C. 1986. Transition and transversion rate in the evolution of animal mitochondrial DNA. *Biosystems* 19:273–283.

Lancaster, J. B., and Lancaster, C. S. 1983. Parental investment: the hominid adaptation. In D. Ortner, ed., *How Humans Adapt: A Biocultural Odyssey,* 33–66. Washington, DC: Smithsonian Institution Press.

Landsmeer, J. M. F. 1955. Anatomical and functional investigations of the articulation of the human fingers. *Acta Anatomica* 55 (Suppl. 24): 1–69.

———. 1984. The human hand in phylogenetic perspective. *Bulletin of the Hospital for Joint Diseases Orthopaedic Institute* 44:276–287.

———. 1986. A comparison of the fingers and hand in varanus, opossum and primates. *Acta morphologica neerlando-scandinavica* 24:193–221.

———. 1987. The hand and hominisation. *Acta morphologica neerlando-scandinavica* 25:83–93.

Landsmeer, J. M. F., and Long, C. 1965. The mechanism of finger control, based on electromyograms and location analysis. *Acta Anatomica* 60:330–347.

Landsoud-Soukate, J., Tutin, C. E. G., and Fernandez, M. 1995. Intestinal parasites of sympatric gorillas and chimpanzees in Lopé Reserve, Gabon. *Annals of Tropical Medicine and Parasitology* 89:73–79.

Langbroek, M., and Roebroeks, W. 2000. Extraterrestrial evidence on the age of the hominids from Java. *Journal of Human Evolution* 38:595–600.

Langdon, J. H. 1986. *Functional Morphology of the Miocene Hominoid Foot.* Contributions to Primatology, Vol. 22. Basel: Karger.

Langergraber, K. E., Prüfer, K., Rowney, C., Boesch, C., Crockford, C., Fawcett, K., Inoue, E., Inoue-Muruyama, M., Mitani, J. C., Muller, M. N., et al. 2012. Generation times in wild chimpanzees and gorillas suggest earlier divergence times in great ape and human evolution. *Proceedings of the National Academy of Sciences of the United States of America* 109:15716–15721.

Langguth A., and Alonso C. 1997. Capuchin monkeys in the Caatinga: tool use and food habits during drought. *Neotropical Primates* 5:77–78.

Lanjouw, A. 2002. Behavioural adaptations to water scarcity in Tongo chimpanzees. In C. Boesch, G. Hohmann, and L. F. Marchant, eds., *Behavioural Diversity in Chimpanzees and Bonobos,* 52–60. Cambridge: Cambridge University Press.

Lappan, S. 2007. Patterns of dispersal in Sumatran siamangs *(Symphalangus syndactylus):* preliminary mtDNA evidence suggests more frequent male that female dispersal to adjacent groups. *American Journal of Primatology* 69:692–698.

———. 2008. Male care of infants in a siamang *(Symphalangus syndactylus)* population including socially monogamous and polyandrous groups. *Behavioral Ecology and Sociobiology* 62:137–1317.

———. 2009a. Flowers are an important food for small apes in Southern Sumatra. *American Journal of Primatology* 71:624–635.

———. 2009b. Patterns of infant care in wild siamangs *(Symphalangus syndactylus)* in Southern Sumatra. In S. Lappan and D. J. Whittaker, eds., *The Gibbons: New Perspectives on Small Ape Socioecology and Population Biology,* 327–345. New York: Springer.

Larsen, C. S. 2003. Equality for the sexes in human evolution? Early hominid sexual dimorphism and implications for mating systems and social behavior. *Proceedings of the National Academy of Sciences of the United States of America* 100:9103–9104.

Larson, S. G. 2012. Did australopiths climb trees? *Science* 338:478–479.

Larson, S. G., Jungers, W. L., Morwood, M. J., Sutikna, T., Jatmiko, Saptomo, E. W., Due, R. A., and Djubiantono, T. 2007. *Homo floresiensis* and the evolution of the hominin shoulder. *Journal of Human Evolution* 53:718–731.

Latimer, B. 1991. Locomotor adaptations in *Australopithecus afarensis:* the issue of arboreality. In Y. Coppens and B. Senut, eds., *Origine(s) de la bipédie chez les Hominidés,* 169–176. Cahiers de Paléoanthropologie. Paris: Éditions du Centre National de la Recherche Scientifique.

Latimer, B. Ohman, J. C., Lovejoy, C. O. 1987. Talocrural joint in African hominoids: implications for *Australopithecus afarensis. American Journal of Physical Anthropology* 74:155–175.

Lau, C. 2001. Speed of the fastest human, walking. In G. Elert, ed., *The Physics Factbook,* http://hypertextbook.com/facts.

Laursen, H. B., Jørgensen, A. L., Jones, C., and Bak, A. L. 1992. Higher rate of evolution of X chromosome alpha-repeat DNA in human than in the great apes. *EMBO Journal* 11:2367–2372.

Lazenby, R. A., Skinner, M. M., J.-J. Hublin, and Boesch, C. 2011. Metacarpal trabecular architecture variation in the chimpanzee *(Pan troglodytes):* evidence for locomotion and tool-use? *American Journal of Physical Anthropology* 144:215–2011.

Leach, E. R. 1955. Polyandry, inheritance and the definition of marriage. *Man* 55:182–186.

Leakey, L. S. B. 1937. *White African.* London: Hodder & Stoughton.

———. 1952. The Olorgesailie prehistoric site. In L. S. B. Leakey and S. Cole, eds., *Proceedings of the Pan-African Congress on Prehistory, 1947,* 209. New York: Philosophical Library.

———. 1959. A new fossil skull from Olduvai. *Nature* 184:491–493.

———. 1960a. *Adam's Ancestors.* 4th ed. London: Methuen.

———. 1960b. Finding the world's earliest man. *National Geographic* 118:420–435.

———. 1961. New finds at Olduvai Gorge. *Nature* 189:649–650.

———. 1962. A new lower Pliocene fossil primate from Kenya. *Annals and Magazine of Natural History,* series 13, 4:689–696.

———. 1963. East African Hominoidea and the classification within this super-family. In S. L. Washburn, ed., *Classification and Human Evolution,* 32–49. New York: Wenner-Gren Foundation for Anthropological Research.

———. 1965. *Olduvai Gorge 1951–1961, Fauna and Background.* Cambridge: Cambridge University Press.

———. 1967. An early Miocene member of Hominidae. *Nature* 213:155–163.

———. 1974. *By the Evidence.* New York: Harcourt Brace Jovanovich.

Leakey, L. S. B., Tobias, P. V., and Napier, J. R. 1964. A new species of genus *Homo* from Olduvai Gorge. *Nature* 202:7–9.

Leakey, M. D. 1971. *Olduvai Gorge,* Vol. 3: *Excavations in Beds I and II, 1960–1963.* Cambridge: Cambridge University Press.

———. 1975. *Olduvai Gorge: My Search for Early Man.* London: Collins.

———. 1979. Footprints in the ashes of time. *National Geographic* 155:446–457.

———. 1980. Early man, environment and tools. In L.-K. Königsson, ed., *Current Argument on Early Man,* 114–133. Oxford: Pergamon Press.

———. 1981. Tracks and tools. *Philosophical Transactions of the Royal Society of London, series B* 292:95–102.

———. 1984. *Disclosing the Past.* New York: Doubleday.

———. 1987a. The hominid footprints. Introduction. In M. D. Leakey and J. M. Harris, eds., *Laetoli: A Pliocene Site in Northern Tanzania,* 490–496. Oxford: Clarendon Press.

———. 1987b. The Laetoli hominid remains. In M. D. Leakey and J. M. Harris, eds., *Laetoli: A Pliocene Site in Northern Tanzania,* 108–117. Oxford: Clarendon Press.

Leakey, M. D., and Harris, J. M. 1987. *Laetoli: A Pliocene Site in Northern Tanzania.* Oxford: Clarendon Press.

Leakey, M. D., and Hay, R. L. 1979. Pliocene footprints in the Laetoli Beds at Laetoli, northern Tanzania. *Nature* 278:317–323.

Leakey, M. D., Hay, R. L., Curtis, G. H., Drake, R. E., Jackes, M. K., and White, T. D. 1976. Fossil hominids from the Laetoli Beds. *Nature* 262:460–466.

Leakey, M. G. 1977. Skeletal remains of *Theropithecus (Simopithecus) oswaldi* from the Site DE/89, Horizon A. In G. L. Isaac, ed., *Olorgesailie. Archeological Studies of a Middle Pleistocene Lake Basin in Kenya,* 229–232. Chicago: University of Chicago Press.

———. 1985. Early Miocene cercopithecids from Buluk, northern Kenya. *Folia Primatologica* 44:1–14.

Leakey, M. G., Feibel, C., McDougall, I., and Walker, A. 1995. New four million-year-old species from Kanapoi and Allia Bay, Kenya. *Nature* 376:565–571.

Leakey, M. G., and Leakey, R. E. 1973. Further evidence of *Simopithecus* (Mammalia, Primates) from Olduvai and Olorgesailie. In L. S. B. Leakey, R. J. G. Savage, and S. C. Coryndon, eds., *Fossil Vertebrates of Africa,* Vol. 3, 191–120. New York: Academic Press.

———. 1978. *Koobi Fora Research Project,* Vol. 1: *The Fossil Hominids and an Introduction to Their Context, 1968–1974.* Oxford: Clarendon Press.

Leakey, M. G., Leakey, R. E., Richtsmeier, J. T., Simons, E. L., and Walker, A. C. 1991. Similarities in *Aegyptopithecus* and *Afropithecus* facial morphology. *Folia Primatologica* 56:65–85.

Leakey, M. G., Spoor, F., Brown, F. H., Gathogo, P. N., Klarie, C., Leakey, L. N., and McDougall, I. 2001. New hominin genus from eastern Africa shows diverse middle Pliocene lineages. *Nature* 410:433–440.

Leakey, M. G., Unger, P. S., and Walker, A. 1995. A new genus of large primate from the Late Oligocene of Lothidok, Turkana District, Kenya. *Journal of Human Evolution* 28:519–531.

Leakey, R. E., and Leakey, M. G. 1986a. A new Miocene hominoid from Kenya. *Nature* 324:143–146.

———. 1986b. A second new Miocene hominoid from Kenya. *Nature* 324:146–148.

———. 1987. A new Miocene small-bodied ape from Kenya. *Journal of Human Evolution* 16:369–387.

Leakey, R. E., Leakey, M. G., and Behrensmeyer, A. K. 1978. The hominid catalogue, In M. G. Leakey and R. E. Leakey, eds., *Koobi Fora Research Project,* Vol. 1: *The Fossil Hominids and an Introduction to Their Context, 1968–1974,* 86–182. Oxford: Clarendon Press.

Leakey, R. E., Leakey, M. G., and Walker, A. 1988a. Morphology of *Turkanapithecus kalakolensis* from Kenya. *American Journal of Physical Anthropology* 76:277–288.

———. 1988b. Morphology of *Afropithecus turkanensis* from Kenya. *American Journal of Physical Anthropology* 76:289–307.

Leakey, R. E., and Lewin, R. 1977. *Origins.* New York: E. P. Dutton.

———. 1978. *People of the Lake: Mankind and Its Beginnings.* Garden City, NY: Anchor Press.

———. 1992. *Origins Reconsidered.* New York: Doubleday.

Leakey, R. E. F. 1970. In search of man's past at Lake Rudolf. *National Geographic* 137:711–732.

———. 1973a. Skull 1470. *National Geographic* 143:818–829.

———. 1973b. Evidence for an advanced Plio-Pleistocene hominid from East Rudolf, Kenya. *Nature* 242:447–450.

————. 1980. How many species of hominids at Lake Turkana? In L.-K. Königsson, ed., *Current Argument on Early Man,* 29–30. Oxford: Pergamon Press.

————. 1981. *The Making of Mankind.* New York: E. P. Dutton.

Leakey, R. E. F., and Walker, A. C. 1976. *Australopithecus, Homo erectus* and the single species hypothesis. *Nature* 261:572–574.

————. 1985a. *Homo erectus* unearthed. *National Geographic Magazine* 1 68:624–629.

————. 1985b. New higher primates from the early Miocene of Buluk, Kenya. *Nature* 318:173–175.

————. 1988. New *Australopithecus boisei* specimens from East and West Lake Turkana, Kenya. *American Journal of Physical Anthropology* 76:1–24.

Leavens, D. A. 2004. Manual deixis in apes and humans. *Interaction Studies* 5:387–408.

Leavens, D. A., and Hopkins, W. D. 1998. Intentional communication by chimpanzees: a cross-sectional study of the use of referential gestures. *Developmental Psychology* 34:813–822.

————. 1999. The whole-hand point: the structure and function of pointing from a comparative perspective. *Journal of Comparative Psychology* 113:417–425.

————. 2005. Multimodal concomitants of manual gesture by chimpanzees *(Pan troglodytes). Gesture* 5:75–90.

————. 2007. Multimodal concomitants of manual gesture by chimpanzees *(Pan troglodytes).* In K. Liebal, C. Müller, and S. Pika, eds., *Gestural Communication in Nonhuman and Human Primates,* 68–82. Amsterdam: John Benjamins.

Leavens, D. A., Hopkins, W. D., and Bard, K. A. 1996. Indexical and referential pointing in chimpanzees *(Pan troglodytes). Journal of Comparative Psychology* 110:346–353.

————. 2005. Understanding the point of chimpanzee pointing. Epigenesis and ecological validity. *Current Directions in Psychological Science* 14:185–189.

Leavens, D. A., Hopkins, W. D., and Thomas, R. K. 2004. Referential communication by chimpanzees *(Pan troglodytes). Journal of Comparative Psychology* 118:48–57.

Leavens, D. A., Racine, T. P., and Hopkins, W. D. 2009. The ontogeny and phylogeny of non-verbal deixis. In R. Rotha and C. Knight, eds., *The Prehistory of Language,* 142–166. Oxford: Oxford University Press.

Leavens, D. A., Russell, J. L., and Hopkins, W. D. 2005. Intentionality as measured in the persistence and elaboration of communication by chimpanzees *(Pan troglodytes). Child Development* 76:291–306.

Lebatard, A.-E., Bourlès, D. L., Duringer, P., Jolivet M., Braucher, R., Carcaillet, J., Schuster, M., Arnaud, N., Monié, P., Lihoreau, F., et al. 2008. Cosmogenic nuclide dating of *Sahelanthropus tchadensis* and *Australopithecus bahrelghazali:* Mio-Plliocene hominids from Chad. *Proceedings of the National Academy of Sciences of the United States of America* 105:3226–3231.

Lee, R. B. 1968. What hunters do for a living, or, how to make out on scarce resources. In R. B. Lee and I. DeVore, eds., *Man the Hunter,* 30–48. Chicago: Aldine.

————. 1969. !Kung Bushman subsistence: an input-output analysis. In A. P. Vaydam, ed., *Environment and Cultural Behavior. Ecological Studies in Cultural Anthropology,* 47–79. New York: Natural History Press.

————. 1979. *The !Kung San. Men, Women and Work in a Foraging Society.* Cambridge: Cambridge University Press.

Lee, R. B., and Daly, R. 1999. *The Cambridge Encyclopedia of Hunters and Gatherers.* Cambridge: Cambridge University Press.

Lee, R. B., and DeVore, I. 1968. *Man the Hunter.* Chicago: Aldine.

———, eds. 1976. *Kalahari Hunter-Gatherers: Studies of the !Kung San and Their Neighbors.* Cambridge, MA: Harvard University Press.

Leeflang, E. P., Liu, W.-M., Hashimoto, C., Choudary, P. V., Schmid, C. W. 1992. Phylogenetic evidence for multiple Alu source genes. *Journal of Molecular Evolution* 35:7–16.

Lee-Thorp, J. 2002. Hominid dietary niches from proxy chemical indicators in fossils: the Swartkrans example. In P. S. Ungar and M. F. Teaford, eds., *Human Diet: Its Origin and Evolution,* 123–141. Westport, CT: Bergin & Garvey.

———. 2011. The demise of "Nutcracker Man." *Proceedings of the National Academy of Sciences of the United States of America* 108:9319–9320.

Lee-Thorp, J., Likius, A., Mackaye, H. T., Vignaud, P., Sponheimer, M., and Bunet, M. 2012. Isotopic evidence for an early shift to C_4 resources by Pliocene hominins in Chad. *Proceedings of the National Academy of Sciences of the United States of America* 109:20369–20372.

Lee-Thorp, J., and Sponheimer, M. 2006. Contributions of biogeochemistry to understanding hominin dietary ecology. *Yearbook of Physical Anthropology* 49:131–148.

Lee-Thorp, J., van der Merwe, N. J., and Brain, C. K. 1994. Diet of *Australopithecus robustus* at Swartkrans from stable carbon isotopic analysis. *Journal of Human Evolution* 27:361–372.

Le Gros Clark, W. E. 1940. Palaeontological evidence bearing on human evolution. *Biological Review* 5:202–230.

———. 1950. New palaeontological evidence bearing on the evolution of the Hominoidea. *Quarterly Journal of the Geological Society of London* 105:225–259.

———. 1955. *The Fossil Evidence for Human Evolution.* Chicago: University of Chicago Press.

———. 1967. *Man-apes or Ape-men.* New York: Holt, Rinehart and Winston.

———. 1971. *The Antecedents of Man.* 3rd ed. Chicago: Quadrangle Books.

———. 1978. *The Fossil Evidence for Human Evolution.* 3rd ed. Chicago: University of Chicago Press.

Le Gros Clark, W. E., and Leakey, L. S. B. 1950. Diagnoses of East African Miocene Hominoidea. *Quarterly Journal of the Geological Society of London* 105:260–262.

———. 1951. The Miocene Hominoidea of East Africa. *Fossil Mammals of Africa,* no. 1. London: British Museum (Natural History).

Lehmann, J. 2008. Meaning and relevance of kinship in great apes. In N. J. Allen, H. Callan, R. Dunbar, and W. James, eds., *Early Human Kinship: From Sex to Social Reproduction,* 160–167. Malden, MA: Blackwell.

Lehmann, J., and Boesch, C. 2003. Social influences on ranging patterns among chimpanzees *(Pan troglodytes verus)* in the Taï National Park, Côte d'Ivoire. *Behavioral Ecology* 14:642–649.

Lehmann, J., Fickenshcer, G., and Boesch, C. 2006. Kin biased investment in wild chimpanzees. *Behaviour* 143:931–955.

Lei, C. 1985. Study on the mid-Miocene apes discovered in Jiangsu, China. *Acta Geologica Sinica* 59:17–24.

Leibowitz, L. 1983. Origins of the sexual division of labor. In eds. M. Lowe and R. Hubbard, *Woman's Nature: Rationalizations of Inequality,* 123–147. New York: Pergamon.

Leigh, S. R. 1992. Patterns of variation in the ontogeny of primate body size dimorphism. *Journal of Human Evolution* 23:27–50.

Leighton, D. R. 1987. Gibbons: territoriality and monogamy. In B. B. Smuts, D. L. Cheney, R. M. Seyfarth, R. W. Wrangham, and T. T. Struhsaker, eds., *Primate Societies,* 135–145. Chicago: University of Chicago Press.

Leighton, M. 1993. Modeling dietary selectivity by Bornean orangutans: evidence for integration of multiple criteria in fruit selection. *International Journal of Primatology* 14:257–313.

Leighton, M., and Leighton, D. R. 1983. Vertebrate responses to fruiting seasonality within a Bornean forest. In S. L. Sutton, T. C. Whitmore, and A. C. Chadwick, eds., *Tropical Rain Forest: Ecology and Management,* 181–196. Oxford: Blackwell Scientific Publications.

Leinders J. J. M., Aziz, F., Sondaar, P. Y., and de Vos, J. 1985. The age of the hominid-bearing deposits of Java: state of the art. *Geologie en Mijnbouw* 64:167–173.

LeMay, M. 1976. Morphological cerebral asymmetries of modern man, fossil man, and non-human primate. *Annals of the New York Academy of Sciences* 280:349–366.

LeMay, M., and Culebras, A. 1972. *Human brain: morphologic differences in the hemispheres demonstrable by carotid arteriography.* New England Journal of Medicine 287:168–170.

Leonard, W. R. 1991. *Australopithecus afarensis* and the single species hypothesis. *Primates* 32:125–130.

Leonard, W. R., and Hegmon, M. 1987. Evolution of P3 morphology in *Australopithecus afarensis. American Journal of Physical Anthropology* 73:41–63.

Leonard, W. R., and Robertson, M. L. 1995. Energetic efficiency of human bipedality. *American Journal of Physical Anthropology* 97:335–338.

Leonard, W. R., Robertson, M. L., and Snodgrass, J. J. 2007. Energetic models of human nutritional evolution. In C. B. Stanford and H. T. Bunn, eds., *Meat-Eating and Human Evolution,* 344–359. New York: Oxford University Press.

Lerdahl, F., and Jackendoff, R. 1983. *A Generative Grammar of Tonal Music.* Cambridge, MA: MIT Press.

Leridon, H. 2004. Can assisted reproduction technology compensate for the natural decline in fertility with age: a model assessment. *Human Reproduction* 19:1548–1553.

Leroy, E. M., Rouquet, P., Formenty, P., Souquière, S., Kilbourne, A., Forment, J.-M., Bermejo, M., Smit, S., Karesh, W., Swanepoel, R., et al. 2004. Multiple Ebola virus transmission events and rapid decline of Central African wildlife. *Science* 303:387–390.

Lethmate, J. 1977. Instrumentelles Verhalten zoolebender Orang-Utans. *Zeitschrift für Morphologie und Anthropologie* 68:57–87.

———. 1979. Instrumental behaviour of zoo orang-utans. *Journal of Human Evolution* 8:741–744.

———. 1982. Tool-using skills of orang-utans. *Journal of Human Evolution* 11:49–64.

Leutenegger, W. 1984. Encephalization in *Proconsul africanus. Nature* 309:287.

————. 1987. Origin of hominid bipedalism. *Nature* 325:305.

Leutenegger, W., and Shell, B. 1987. Variability and sexual dimorphism in canine size of *Australopithecus* and extant hominoids. *Journal of Human Evolution* 16:359–367.

Lévêque, B., and Vandermeersch, B. 1980. Découverte de restes humains dans un niveau castelperronien à Saint-Césaire (Charente-Maritime). *Comptes rendus de l'Académie des Sciences Paris* série D 291:187–189.

Levinson, S. C. Kinship and human thought. *Science* 336:988–989.

Lewin, R. 1987. *Bones of Contention.* New York: Simon & Schuster.

Lewis, G. E. 1934. Preliminary notice of new man-like apes from India. *American Journal of Science* 227:161–79.

Lewis, O. J. 1977. Joint remodelling and the evolution of the human hand. *Journal of Anatomy* 123:157–201.

————. 1989. *Functional Morphology of the Evolving Hand and Foot.* Oxford: Clarendon Press.

Lewontin, R. C. 1972. The apportionment of human diversity. T. Dobzhansky, M. K. Hecht and W. C. Steere, eds., *Evolutionary Biology,* Vol. 6, 381–398. New York: Appleton-Century-Crofts.

Li, C. 1978. A Miocene gibbon-like primate from Shihhung, Kiangsu Province. *Vertebrata PalAsiatica* 16:187–192.

Li, W.-H., and Saunders, M. A. 2005. The chimpanzee and us. *Nature* 437:50–51.

Li, W.-H., and Tanimura, M. 1987. The molecular clock runs more slowly in man than in apes and monkeys. *Nature* 326:93–96.

Li, W.-H., Wolfe, K. H., Sourdis, J., and Sharp, P. M. 1987. Reconstruction of phylogenetic trees and estimation of divergence times under nonconstant rates of evolution. *Cold Spring Harbor Symposia on Quantitative Biology* 52:847–856.

Liberman, A. M., and Whalen, D. H. 2002. On the relation of speech to language. *Trends in Cognitive Sciences* 4:187–196.

Liebal, K., Call, J., and Tomasello, M. 2004. Use of gesture sequences in chimpanzees. *American Journal of Primatology* 64:377–396.

Liebal, K., Pika, S., and Tomasello, M. 2004. Social communication in siamangs *(Symphalangus syndactylus):* use of gestures and facial expressions. *Primates* 45:41–57.

Lieberman, D. E. 2001. Another face in our family tree. *Nature* 410:419–420.

————. 2003. Motor control, speech and the evolution of human language. In M. H. Christiansen and S. Kirby, eds., *Language Evolution,* 255–271. Oxford: Oxford University Press.

————. 2007. Homing in on early *Homo. Nature* 449:291–292.

————. 2009. *Homo floresiensis* from head to toe. *Nature* 459:41–42.

Lieberman, D. E., and McCarthy, R. C. 1999. The ontogeny of cranial base angulation in humans and chimpanzees and its implications for reconstructing pharyngeal dimensions. *Journal of Human Evolution* 36:487–517.

Lieberman, D. E., Raichlen, D. A., Pontzer, H., Bramble, D. M., and Cutright-Smith, E. 2006. The human gluteus maximus and its role in running. *Journal of Experimental Biology* 209:2143–2155.

Lieberman, L., and Jackson, F. L. C. 1995. Race and three models of human origin. *American Anthropologist* 97:231–242.

Lieberman, L., Kirk, R. C., and Littlefield, A. 2003. Perishing paradigm: race—1931–99. *American Anthropologist* 105:110–113.

Lieberman, P. 1975. *On the Origins of Language.* New York: Macmillan.

———. 1976. Interactive models for evolution: neural mechanisms, anatomy, and behavior. *Annals of the New York Academy of Sciences* 280:660–672.

———. 1984. *The Biology and Evolution of Language.* Cambridge, MA: Harvard University Press.

———. 1991. *Uniquely Human: The Evolution of Speech, Thought, and Selfless Behavior.* Cambridge, MA: Harvard University Press.

———. 1993. On the Kebara KMH 2 hyoid and Neanderthal speech. *Current Anthropology* 34:172–175.

———. 1994a. Human language and human uniqueness. *Language Communication* 14:87–95.

———. 1994b. Hyoid bone position and speech: reply to Dr. Arensburg (1990). *American Journal of Physical Anthropology* 94:275–278.

———. 2007. The evolution of human speech. Its anatomical and neural bases. *Current Anthropology* 48:39–66.

Lieberman, P., and Crelin, E. S. 1971. On the speech of Neanderthal man. *Linguistic Inquiry* 2:203–222.

Lieberman, P., Klatt, D. H., and Wilson, W. H. 1969. Vocal tract limitations on the vowel repertoire of rhesus monkeys and other nonhuman primates. *Science* 164:1185–1187.

Liebhaber, S. A., and Begley, K. A. 1983. Structural and evolutionary analysis of the two chimpanzee alpha-globin mRNAs. *Nucleic Acid Research* 11:8915–8929.

Lilly, A. A., Mehlman, P. T., and Doran, D. 2002. Intestinal parasites in gorillas, chimpanzees, and humans at Mondika Research Site, Dzanga-Ndoki National Park, Central African Republic. *International Journal of Primatology* 23:555–573.

Limber, J. 1977. Language in child and chimp? *American Psychologist* 32:280–295.

Limongelli, L., Boysen S. T., and Visalberghi, E. 1995. Comprehension of cause-effect relations in a tool-using task by chimpanzees *(Pan troglodytes). Journal of Comparative Psychology* 109:18–26.

Lin, A. C., Bard, K. A., and Anderson, J. R. 1992. Development of self-recognition in chimpanzees *(Pan troglodytes). Journal of Comparative Psychology* 106:120–127.

Linden, E. 1974. *Apes, Men and Language.* New York: Penguin Books.

———. 1981. *Apes, Men and Language.* 2nd ed. New York: Penguin Books.

Lingnau, A., Gesierich, B., and Caramazza, A. 2009. Asymmetric fMRI reveals no evidence for mirror neurons in humans. *Proceedings of the National Academy of Sciences of the United States of America* 106:9925–9930.

Linton, S. 1971. Woman the gatherer: male bias in anthropology. In S. E. Jacobs, ed., *Women in Cross-cultural Perspective: A Preliminary Sourcebook,* 9–21. Urbana: University of Illinois Press.

Lipson, S., and Pilbeam, D. 1982. *Ramapithecus* and hominoid evolution. *Journal of Human Evolution* 11:545–548.

Littlefield, A., Lieberman, L., and Reynolds, L. 1982. Redefining race: the potential demise of a concept in physical anthropology. *Current Anthropology* 23:641–655.

Littleton, J. 2005. Fifty years of chimpanzee demography at Taronga Park Zoo. *American Journal of Primatology* 67:281–298.

Liu, L., Bestel, S., Shi, J., Song, Y., and Chen, X. 2013. Paleolithic human exploitation of plant foods during the last glacial maximum in North China. *Proceedings of the National Academy of Sciences of the United States of America* 110:5380–5385.

Liu, Q., Simpson, K., Izar, P., Ottoni, E., and Fragaszy, D. 2009. Kinematics and energetics of nut-cracking in wild capuchin monkeys *(Cebus libidinosus)* in Piauí, Brazil. *American Journal of Physical Anthropology* 138:210–220.

Liu., R. Shi, L., and Chen, Y. 1987. A study on the chromosomes of white-browed gibbon *(Hylobates hoolock leuconedys)*. *Acta Theriologica Sinica* 7:1–7.

Liu, W., Gao, F., and Zheng, L. 2002. The diet analysis from tooth size and morphology for Yuanmou hominoids, Yunnan Province, China. *Anthropological Science* 110:149–163.

Liu, W., Jin, C.-Z., Zhang, Y.-Q., Cai, Y.-J., Xing, S., Wu, X.-J., Cheng, H., Edwards, R. L., Pan, W.-S., Qin, D.-G., et al. 2010. Human remains from Zhirendong, South China, and modern human emergence in East Asia. *Proceedings of the National Academy of Sciences of the United States of America* 107:19201–19206.

Liu, W., Miller, B. L., Kramer, J. H., Rankin, K., Wyss-Coray, C., Gearhart, R., Phengrasamy, L., Weiner, M., and Rosen, H. J. 2004. Behavioral disorders in the frontal and temproal variants of frontotemporal dementia. *Neurology* 62:742–748.

Liu, W., and Zheng, L. 2005a. Comparisons of tooth size and morphology between the late Miocene hominoids from Lufeng and Yuanmou, China, and their implications. *Anthropological Science* 113:73–77.

———. 2005b. Tooth wear differences between the Yuanmou and *Lufengpithecus*. *International Journal of Primatology* 26:491–506.

Liu, Z., Zhang, Y., Jiang, H., and Southwick, C. 1989. Population structure of *Hylobates concolor* in Bawanglin Nature Reserve, Hainan, China. *American Journal of Primatology* 19:247–254.

Livak, K. J., Rogers, J., and Lichter, J. B. 1995. Variability of dopamine D4 receptor (DRD4) gene sequence within and among nonhuman primate species. *Proceedings of the National Academy of Sciences of the United States of America* 92:427–431.

Livingstone, F. B. 1962a. On the non-existence of human races. *Current Anthropology* 3:279–281.

———. 1962b. Reconstructing man's Pliocene pongid ancestor. *American Anthropologist* 64:301–305.

———. 1973. Did the Australopithecines sing? *Current Anthropology* 14:25–29.

Livingstone, F. B., Cowgill, G., and Howell, F. C. 1961. More on Middle Pleistocene hominids. *Current Anthropology* 2:117–120.

Llorente, M., Mosquera, M., and Fabré, M. 2009. Manual laterality for simple reaching and bimanual coordinated task in naturalistic housed *Pan troglodytes*. *International Journal of Primatology* 30:183–197.

Llorente, M., Riba, D., Palou, L., Carrasco, L., Mosquera, M., Colell, M., and Feliu, O. 2011. Population-level right-handedness for a coordinated bimanual task in naturalistic housed chimpanzees: replication and extension in 114 animals from Zambia and Spain. *American Journal of Primatology* 73:281–290.

Locke, D. P., Hillier, L. W., Warren, W. C., Worley, K. C., Nazareth, L. V., Muzny, D. M., Yang, S. P.., Wang Z., Chinwalla A. T., Minx P., et al. 2011. Comparative and demographic analysis of orang-utan genomes. *Nature* 469:529–533.

Locke, J. L., Bekken, K. E., McMinn-Larson, L., and Wein, D. 1995. Emergent control of manual and vocal-motor activity in relation to the development of speech. *Brain and Language* 51:498–508.

Lockley, M., Roberts, G., and Kim, J. Y. 2008. In the footprints of our ancestors: an overview of the hominid track record. *Ichnos* 15:106–125.

Lockwood, C. A., Kimbel, W. H., and Lynch, J. M. 2004. Morphometrics and hominoid phylogeny: support for a chimpanzee-human clade and differentiation among great ape subspecies. *Proceedings of the National Academy of Sciences of the United States of America* 101:4356–4360.

Lockwood, C. A., Menter, C. G., Moggi-Cecchi, J., and Keyser, A. W. 2007. Extended male growth in a fossil hominin species. *Science* 318:1443–1446.

Lodwick, J. L., Borries, C., Pusey, A. E., Goodall, J., and McGrew, W. C. 2004. From nest to nest—influences and reproduction on the active period of adult Gombe chimpanzees. *American Journal of Primatology* 64:249–260.

Logan, T. R., Lucas, S. G., and Sobus, J. C. 1983. The taxonomic status of *Australopithecus afarensis* Johanson in Hinrichsen 1978 (Mammalia, Primates). *Haliksa'i: UNM Contributions to Anthropology* 2:16–27.

Long, C., Conrad, P. W., Hall, E. A., and Furler, S. L. 1970. Intrinsic-extrinsic muscle control of the hand in power grip and precision handling. *Journal of Bone and Joint Surgery* 52-A:853–867.

Long, J. C., Li, J., and Healy, M. E. 2009. Human DNA sequences: more variation less race. *American Journal of Physical Anthropology* 139:23–34.

Lonsdorf, E. V. 2005. Sex differences in the development of termite-fishing skills in the wild chimpanzees, *Pan troglodytes schweinfurthii,* of Gombe National Park, Tanzania. *Animal Behaviour* 70:673–683.

Lonsdorf, E. V., Eberly, L. E., and Pusey, A. E. 2004. Sex differences in learning in chimpanzees. *Nature* 428:715–716.

Lonsdorf, E. V., and Hopkins, W. D. 2005. Wild chimpanzees show population-level handedness for tool use. *Proceedings of the National Academy of Sciences of the United States of America* 102:12634–12638.

Lonsdorf, E. V., Travis, D., Pusey, A. E., and Goodall, J. 2006. Using retrospective health data from the Gombe chimpanzee study to inform future monitoring efforts. *American Journal of Primatology* 68:987–908.

Lordkipanidze, D., Jashashvili, T., Vekua, A., Ponce de León, M. S., Zollikofer, C. P. E., Rightmire, G. P., Pontzer, H., Ferring, R., Oms, O., Tappen, M., et al. 2007. Postcranial evidence from early *Homo* from Dmanisi, Georgia. *Nature* 449:305–310.

Lordkipanidze, D., Vekua, A., Ferring, R., Rightmire, G. P., Augsti, J., Kiladze, G., Mouskhelishvili, A., Nioradze, M., Ponce de León, M., Tappen, M., and Zollikofer, C. 2005. The earliest toothless hominin skull. *Nature* 434:717–718.

Lorenz, K. 1963. *On Aggression.* New York: Harcourt, Brace & World.

Lovejoy, C. O. 1980. Evolution of human walking. *Scientific American* 259:118–125.

————. 1981. The origin of man. *Science* 211:341–350.

————. 1988. Evolution of human walking. *Scientific American* 259, no. 5: 118–125.

————. 2007. An early ape shows its hand. *Proceedings of the Royal Society B* 274:2373–2374.

————. 2009a. Reexamining human origins in light of *Ardipithecus ramidus*. *Science* 326:74e1–74e8.

————. 2009b. Reexamining human origins in light of *Ardipithecus ramidus*. *Science* 326:74.

Lovejoy, C. O., Burstein, A. H., and Heiple, K. G. 1972. Primate phylogeny and immunological distance. *Science* 176:803–805.

Lovejoy, C. O., Heiple, K. G., and Burstein, A. H. 1973. The gait of *Australopithecus*. *American Journal of Physical Anthropology* 38:757–780.

Lovejoy, C. O., Latimer, Suwa, G., Asfaw, B., and White, T. D., 2009a. Combining prehension and propulsion: the foot of *Ardipithecus ramidus*. *Science* 326:72.

————. 2013b. Combining prehension and propulsion: the foot of *Ardipithecus ramidus*. *Science* 326:72e1-e8.

Lovejoy, C. O., Meindl, R. S., Ohman, J. C., Heiple, K. G., and White, T. D. 2002. The Maka femur and its bearing on the antiquity of human walking: applying contemporary concepts of morphogenesis to the human fossil record. *American Journal of Physical Anthropology* 119:97–133.

Lovejoy, C. O., Simpson, S. W., White, T. D., Asfaw, B., and Suwa, G. 2009a. Careful climbing in the Miocene: the forelimbs of *Ardipithecus ramidus* and humans are primitive. *Science* 326:70e1–70e8.

———— 2009b. Careful climbing in the Miocene: the forelimbs of *Ardipithecus ramidus* and humans are primitive. *Science* 326:70.

Lovejoy, C. O., Suwa, G., Simpson, S. W., Matternes, J. H., and White, T. D. 2009a. The great divides: *Ardipithecus ramidus* reveals the postcrania of our last common ancestors with African apes. *Science* 326:73.

———— 2009b. The great divides: *Ardipithecus ramidus* reveals the postcrania of our last common ancestors with African apes. *Science* 326:100–106.

Lovejoy, C. O., Suwa, G., Spurlock, L., Asfaw, B., and White, T. D. 2009a. The pelvis and femur of *Ardipithecus ramidus:* the emergence of upright walking. *Science* 326:71e1–71e6.

———— 2009b. The pelvis and femur of *Ardipithecus ramidus:* the emergence of upright walking. *Science* 326:71.

Low, B. S. 1988. Measures of polygyny in humans. *Current Anthropology* 29:189–194.

Lowe, J., Barton, N., Blockley, S., Ramsey, C. B., Cullen, V. L., Davies, W., Gamble, C., et al. 2012. Volcanic ash layers illuminate the resilience of Neanderthals and early modern humans to natural hazards. *Proceedings of the National Academy of Sciences of the United States of America* 109:13532–13537.

Lu, Q., Xu, Q., and Zhao, Z. 1988. The reconstruction of the head of the female Lufeng ape. *Acta Anthropologica Sinica* 7:9–16.

Lu, Q., Xu, Q., and Zheng, L. 1981. Preliminary research on the cranium of *Sivapithecus yunnanensis*. *Vertebrata PalAsiatica* 19:101–106.

Lucas, P. W. 2007. The evolution of the hominin diet from a dental functional perspective. In P. S. Ungar, ed., *Evolution of the Human Diet*, 31–55. Oxford: Oxford University Press.

Lucas, P. W., Darvell, B. W., Lee, P. K. D., Yuan, T. D. B., and Choong, M. F. 1998. Colour cues for leaf food selection by long-tailed macaques *(Macaca fascicularis)* with a new suggestion for the evolution of trichromatic colour vision. *Folia Primatologia* 69:139–154.

Lucas, P. W., Peters, C. R., and Arrandale, S. R. 1994. Seed-breaking forces exerted by orang-utans with their teeth in captivity and a new technique of estimating forces produced in the wild. *American Journal of Physical Anthropology* 94:365–378.

Luckett, W. P. 1975. Ontogeny of the fetal membranes and placenta. Their bearing on primate phylogeny. In W. P. Luckett and F. S. Szalay, eds., *Phylogeny of the Primates: A Multidisciplinary Approach*, 157–182. New York: Plenum Press.

———. 1980. Monophyletic or diphyletic origins of Anthropoidea and Hystricognathi: evidence of the fetal membranes. In R. L. Ciochon and A. B. Chiarelli, eds., *Evolutionary Biology of the New World Monkeys and Continental Drift*, 347–368. New York: Plenum Press.

Lucotte, G. 1988. African pygmies have the more ancestral gene pool when studied for Y-chromosome DNA haplotypes. In B. Bräuer and F. Smith, eds., *Continuity or Replacement Controversies in the Evolution of Homo sapiens*, 75–81. Rotterdam: Balkema.

Lucotte, G., Barriel, V., Guerin, P., Abbas, N., and Ruffie, J. 1990. Rétro-transposition de la séquence humaine homologue à la p49f sur le chromosome Y au cours de l'évolution des singes anthropoïdes. *Biochemical Systematics and Ecology* 18:199–204.

Lucotte, G., and Hazout, S. 1986. Distances éntre l'homme et les singes anthropoïdes basées sur la mobilité électrophorétique des protéines et enzymes. *Biochemical Systematics and Ecology* 14:135–140.

Lukas, D., Bradley, B. J., Nsubuga, A. M., Doran-Sheehy, D., Robbins, M. M., and Vigilant, L. 2004. Major histocompatibility complex and microsatellite variation in two populations of wild gorillas. *Molecular Ecology* 13:3389–3402.

Lukas, D., Reynolds, V., Boesch, C., and Vigilant, L. 2005. To what extent does living in a group mean living with kin? *Molecular Ecology* 14:2181–2196.

Lumier, A. R., and Wich, S. A. 2008. Orangutan long call degradation and individuality over distance: a playback approach. *International Journal of Primatology* 29:615–625.

Lyn, H., Greenfield, P. M., Savage-Rumbaugh, E. S., Gillespie-Lybch, K., and Hopkins, W. D. 2011. Nonhuman primates do declare! A comparison of declarative symbol and gesture use in two children, two bonobos, and a chimpanzee. *Language and Communication* 31:63–74.

Lyn, H., and Savage-Rumbaugh, E. S. 2000. Observational word learning in two bonobos *(Pan paniscus):* ostensive and non-ostensive contexts. *Language & Communication* 20:255–273.

Ma, S., Wang, Y., Poirier, F. E. 1988. Taxomony, distribution and status of gibbons *(Hylobates)* in southern China and adjacent areas. *Primates* 29:277–286.

Macaulay, V., Hill, C., Achilli, A., Rengo, C., Clarke, D., Meehan, W., Blackurn, J., Semino, O., Scozzari, R., Cruciani, F., et al. 2005. Single, rapid coastal settlement of Asia revealed by analysis of complete mitochondrial genomes. *Science* 308:1034–1036.

MacConaill, M. A., and Basmajian, J. V. 1969. *Muscles and Movements: A Basis for Human Kinesiology.* Baltimore: Williams & Wilkins.

Macdonald, S. J., and Long, A. D. 2005. Prospects for identifying functional variation across the genome. *Proceedings of the National Academy of Sciences of the United States of America* 102:6614–6621.

Machado, C. A., Robbins, N., Gilbert, T. P., and Herre, E. A. 2005. Critical review of host specificity and its coevolutionary implications in the fig/fig-wasp mutualism. *Proceedings of the National Academy of Sciences of the United States of America* 102:6558–6565.

Mâche, F.-B. 2000. The necessity of and problems with a universal musicality. In N. L. Wallin, B. Merker and S. Brown, eds., *The Origins of Music,* 473–479. Cambridge, MA: MIT Press.

Macho, G. A., Shimizu, D., Jiang, Y., and Spears, I. R. 2005. *Australopithecus anamensis:* a finite-element approach to studying the functional adaptations of extinct hominins. *Anatomical Record Part A* 283A:310–318.

MacKinnon, J. R. 1971. The orang-utan in Sabah today. *Oryx* 11:141–191.

———. 1974a. The behavior and ecology of wild orang-utans *(Pongo pygmaeus). Animal Behaviour* 22:3–74.

———. 1974b. *In Search of the Red Ape,* London: Collins.

———. 1976. Mountain gorillas and bonobos. *Oryx* 13:372–382.

———. 1977. A comparative ecology of Asian apes. *Primates* 18:747–772.

———. 1978. *The Ape within Us.* London: William Collins Sons.

———. 1979. Reproductive behavior in wild orangutan populations. In D. A. Hamburg and E. R. McCown, eds., *The Great Apes,* 256–273. Menlo Park, CA: Benjamin/Cummings.

MacKinnon, J. R., and MacKinnon, K. S. 1977. The formation of a new gibbon group. *Primates* 18:701–708.

———. 1978. Comparative feeding ecology of six sympatric primates in West Malaysia. In D. J. Chivers and J. Herbert, *Recent Advances in Primatology,* Vol. 1: *Behaviour,* 305–321. London: Academic Press.

———. 1980. Niche differentiation in a primate community. In D. J. Chivers, ed., *Malayan Forest Primates,* 167–190. New York: Plenum Press.

MacKinnon, K., Hatta, G., Halim, H., Mangalik, A. 1996. *The Ecology of Kalimantan.* Singapore: Periplus Editions.

MacLarnon, A., and Hewitt, G. 2004. Increased breathing control: another factor in the evolution of human language. *Evolutionary Anthropology* 13:181–197.

MacLatchy, L. M., and Bossert, W. H. 1996. An analysis of the articular surface distribution of the femoral head and acetabulum in anthropoids, with implications for hip function in Miocene hominoids. *Journal of Human Evolution* 31:425–453.

MacLeod, C. E., Zilles, K., Schleicher, A., Rilling, J. K., and Gibson, K. R. 2003. Expansion of the neocerebellum in Hominoidea. *Journal of Human Evolution* 44:401–429.

MacNeilage, P. F. 1992. Evolution and lateralization of the two great primate action systems. In J. Wind, B. H. Bichakjian, A. Nocentini, and B, Chiarelli, eds., *Language Origin: A Multidisciplinary Approach,* 281–300. Boston: Kluwer Academic.

———. 1998a. Evolution of the mechanism of language output: comparative neurobiology of vocal and manual communication. In J. R. Hurfod, M. G. Studdert-Kennedy, and C. Knight, *Approaches to the Evolution of Language: Social and Cognitive Bases,* 222–241. Cambridge: Cambridge University Press.

———. 1998b. The frame/content theory of evolution of speech production. *Behavioral and Brain Sciences* 21:499–546.

———. 1998c. Towards a unified view of cerebral hemispheric specializations in vertebrates. In A. D. Milner, ed., *Comparative Neuropsychology,* 167–183. Oxford: Oxford University Press.

———. 2008. *The Origin of Speech.* Oxford: Oxford University Press.

MacNeilage, P. F., and Davis, B. L. 2005. The frame/content theory of evolution of speech. *Interaction Studies* 6:173–199.

MacPhee, R. D. E., Beard, K. C., and Qi, T. 1995. Significance of a primate petrosal from Middle Eocene fissure-fillings at Shanghuang, Jiangsu Province, People's Republic of China. *Journal of Human Evolution* 29:501–514.

Madden, C. T. 1980. New *Proconsul (Xenopithecus)* from the Miocene of Kenya. *Primates* 21:241–252.

Maeda, N., Wu, C.-I, Bliska, J., and Reneke, J. 1988. Primates: pattern of DNA changes, molecular clock, and evolution of repetitive sequences. *Molecular Biology and Evolution* 5:1–20.

Maestripieri, D. 2003. *Primate Psychology,* Cambridge, MA: Harvard University Press.

———. 2007. *Machiavellian Intelligence: How Rhesus Monkeys and Human Have Conquered the World.* Chicago: University of Chicago Press.

Maggioncalda, A. N., Czekala, N. M., and Sapolsky, R. M. 2000. Growth hormone and thyroid stimulating hormone concentrations in captive male orangutans: implications for understanding developmental arrest. *American Journal of Primatology* 50:67–76.

———. 2002. Male orangutan subadulthood: a new twist on the relationship between chronic stress and developmental arrest. *American Journal of Physical Anthropology* 118:25–32.

Maggioncalda, A. N., Sapolsky, R. M., and Czekala, N. M. 1999. Reproductive hormone profiles in captive male orangutans: implications for understanding developmental arrest. *American Journal of Physical Anthropology* 109:19–32.

Magill, C. R., Ashley, G. M., and Freeman, K. H. 2013a. Ecosystem variability and early human habitats in eastern Africa. *Proceedings of the National Academy of Sciences of the United States of America* 110:1167–1174.

———. 2013b. Water, plants, and early human habitats in eastern Africa. *Proceedings of the National Academy of Sciences of the United States of America* 110:1175–1180.

Magliocca, F., and Gautier-Hion, A. 2002. Mineral content as a basis for food selection by western lowland gorillas in a forest clearing. *American Journal of Primatology* 57:67–77.

Magliocca, F., Querouil, S., and Gautier-Hion, A. 1999. Population structure and group composition of western lowland gorillas in north-western Republic of Congo. *American Journal of Primatology* 48:1–14.

Magori, C., and Day, M. H. 1983a. Laetoli Hominid 18: an early *Homo sapiens* skull. *Journal of Human Evolution* 12:747–753.

———. 1983b. An early *Homo sapiens* skull from the Ngaloba Beds, Laetoli, northern Tanzania. *Anthropos* 10:143–183.

Maguire, J. M. 1985. Recent geological, stratigraphic and palaeontological studies at Makapansgat Limeworks. In P. V. Tobias, ed., *Hominid Evolution: Past, Present and Future,* 151–164. New York: Alan R. Liss.

Mahaney, W. C. 1993. Scanning electron microscopy of earth mined and eaten by mountain gorillas in the Virunga Mountains, Rwanda. *Primates* 34:311–319.

Mahaney, W. C., Aufreiter, S., and Hancock, R. G. V. 1995. Mountain gorilla geophagy: a possible seasonal behavior for dealing with the effects of dietary changes. *International Journal of Primatology* 16:475–488.

Mahaney, W. C., Hancock, R. G. V. Aufreiter, S., and Huffman, M. A. 1996. Geochemistry and clay mineralogy of termite mound soil and the role of geophagy in chimpanzees of the Mahale Mountains, Tanzania. *Primates* 37:121–134.

Mahaney, W. C., Milner, M. W., Aufreiter, S., Hancock, R. G. V. Wrangham, R. W., and Campbell, S. 2005. Soils consumed by chimpanzees of the Kanyawara community in the Kibale Forest, Uganda. *International Journal of Primatology* 26:1375–1398.

Mahaney, W. C., Milner, M. W., Sanmugadas, K., Hancock, R. G. V. Aufreiter, S., Wrangham, R. W., and Pier, H. W. 1997. Analysis of geophagy soils in Kibale Forest, Uganda. *Primates* 38:159–176.

Mahaney, W. C., Stambolic, A., Milner, M. W., Russon, A., Hancock, R. G. V., and Aufreiter, S. 1996. *Geochemistry of soils eaten by orangutans in Indonesia.* SLOWPOKE Reactor Facility, Annual Report.

Mahaney, W. C., Watts, D. P., and Hancock, R. G. V. 1990. Geophagia by mountain gorillas *(Gorilla gorilla beringei)* in the Virunga Mountains, Rwanda. *Primates* 31:113–121.

Mahaney, W. C., Zippin, J., Milner, M. W., Sanmugadas, K., Hancock, R. G. V., Aufreiter, S., Campbell, S., Huffman, M. A., Wink, M., Malloch, D, and Kalm, V. 1999. Chemistry, mineralogy and microbiology of termite mound soil eaten by chimpanzees of the Mahale Mountains, Western Tanzania. *Journal of Tropical Ecology* 15:565–588.

Mahmoudzadeh, M., Dehaene-Lambertz, G., Fournier, M., Kongolo, G., Goudjil, S., Dubois, J., Grebe, R., and Wallois, F. 2013. Syllabic discrimination in premature human infants prior to complete formation of cortical layers. *Proceedings of the National Academy of Sciences of the United States of America* 110:4846–4851.

Mai, L. L. 1983. A model of chromosome evolution and its bearing on cladogenesis in the Hominoidea. In R. L. Ciochon and R. S. Corruccini, ed., *New Interpretations of Ape and Human Ancestry,* 87–114. New York: Plenum Press.

Maier, W. O., and Nkini, A. T. 1984. Olduvai hominid 9: new results of investigation. *Courier Forschungsinstitut Senckenberg* 69:123–130.

———. 1985. The phylogenetic position of Olduvai hominid 9, especially as determined from basicranial evidence. In E. Delson, ed., *Ancestors: The Hard Evidence,* 249–254. New York: Alan R. Liss.

Makova, K. D., and Li, W.-H. 2002. Strong male-driven evolution of DNA sequences in humans and apes. *Nature* 416:624–626.

Malaivijitnond, S., Lekprayoon, C., Tandavanittj, N., Panha, S., Cheewatham, C., and Hamada, Y. 2007. Stone-tool usage by Thai long-tailed macaques *(Macaca fasicularis)*. *American Journal of Primatology* 69:227–233.

Mallavarapu, S., Stoinski, T. S., Bloomsmith, M. A., and Maple, T. L. 2006. Postconflict behavior in captive western lowland gorillas *(Gorilla gorilla gorilla)*. *American Journal of Primatology* 68:789–801.

Mallengi, F. *Homo cepranensis sp. nov.* and the evolution of African-European Middle Pleistocene hominids *Comptes Rendus Palevol* 2:153–159.

Malone, N., and Fuentes, A. 2009. The ecology and evolution of hylobatid communities: causal and contextual factors underlying inter- and intraspecific variation. In S. Lappan and D. J. Whittaker, *The Gibbons: New Perspectives on Small Ape Socioecology and Population Biology,* 241–264. New York: Springer.

Manduell, K. L., Morrogh-Bernard, H. C., and Thorpe, S. K. S. 2011. Locomotor behavior of wild orangutans *(Pongo pygmaeus wurmbii)* in disturbed peat swamp forest, Sabangau, Central Kalimantan, Indonesia. *American Journal of Physical Anthropology* 145:348–349.

Manega, P. 1993 *Geochronology, Geochemistry and Isotopic Study of the Plio-Pleistocene Hominid Sites and the Ngorongoro Volcanic Highland in Northern Tanzania.* Ann Arbor, MI: University Microfilms.

Mania, D., and Vlcek, E. 1981. *Homo erectus* in middle Europe: the discovery from Bilzingsleben. In B. A. Sigmon and J. S. Cybulski, eds., *Homo erectus: Papers in Honor of Davidson Black,* 133–151. Toronto: University of Toronto Press.

Maniacky, J. *Pan paniscus,* sometimes a linguistic issue. *Pan Africa News* 13:4–6.

Mankoto, M. O., Yamagiwa, J., Steinhauer-Burkart, B., Mwanza, N., Maruhashi, T., and Yumoto, T. 1994. Conservation of eastern lowland gorillas in the Kahuzi-Biega National Park, Zaire. In B. Thierry, J. R. Anderson, J. J. Roeder, and N. Herrenschmidt, eds., *Current Primatology,* Vol. 1: *Ecology and Evolution,* 113–122. Strasbourg: Université Louis Pasteur.

Manni, E., and Petrosini, L. 2004. A century of cerebellar somatotopy: a debated presentation. *Nature Reviews Neuroscience* 5:241–249.

Mannu, M., and Ottoni, E. B. 2009. The enhanced tool-kit of two groups of wild bearded capuchin monkeys in Caatinga: tool making, associative use, and secondary tools. *American Journal of Primatology* 71:242–251.

Manson, J. H., Perry, S., and Parish, A. R. 1997. Nonconceptive sexual behavior in bonbos and capuchins. *International Journal of Primatology* 18:767–786.

Manson, J. H., and Wrangham, R. W. 1991. Intergroup aggression in chimpanzees and humans. *Current Anthropology* 32:369–390.

Manzi, G., Bruner, E., Passarello, P. 2003. The one-million-year-old *Homo* cranium from Bouri (Ethiopia): a reconsideration of its *H. erectus* affinities. *Journal of Human Evolution* 44:731–736.

Maple, T. L. 1980. *Orang-utan Behavior.* New York: Van Nostrand Reinhold.

March, E. W. 1957. Gorillas of eastern Nigeria. *Oryx* 4:30–34.

Marchant, L. F., and McGrew, W. C. 1991. Laterality of function in apes: a meta-analysis of methods. *Journal of Human Evolution* 21:425–438.

———. 1996. Laterality of limb function in wild chimpanzees of Gombe National Park: comprehensive study of spontaneous activities. *Journal of Human Evolution* 30:427–443.

Marchi, D. Articular to diaphyseal proportions of human and great ape metatarsals. *American Journal of Physical Anthropology* 143:198–207.

Marcus, G. E., and Fischer, M. M. J. 1986, 1999. *Anthropology as Cultural Critique: An Experimental Moment in the Human Sciences.* Chicago: University of Chicago Press.

Margoliash, E. 1963. Primary structure and evolution of cytochrome c. *Proceedings of the National Academy of Sciences of the United States of America* 50:672–679.

Margoliash, E., and Fitch, W. M. 1968. Evolutionary variability of cytochrome *c* primary structure. *Annals of the New York Academy of Sciences* 151:359–381.

Marino, L., Reiss, D., and Gallup, G. G., Jr. 1994. Mirror self-recognition in bottlenose dolphins: implication for comparative investigations of highly dissimilar species. In S. T. Parker, R. W. Mitchell, and M. L. Boccia, eds., *Self-Awareness in Animals and Humans,* 380–391. Cambridge: Cambridge University Press.

Marivaux, L., Antoine, P.-O., Baqri, S. R. H., Benammi, M., Chaimanee, Y., Crochet, J.-Y., de Franceschi, D., Iqbal, N., Jaeger, J.-J., Métais, G., et al. 2005. Anthropoid primates from the Oligocene of Pakistan (Bugti Hills): data on early anthropoid evolution and biogeography. *Proceedings of the National Academy of Sciences of the United States of America* 102:8436–8441.

Marivaux, L., Beard, K. C., Chaimanee, Y., Dagosto, M., Gebo, D. L., Guy, F., Marandat, B., Kyaw, K., Kyaw, A. A., et al. 2010. Talar morphology, phylogenetic affinities and locomotor adaptation of a large-bodied aphipithecid primate from the Late Middle Eocene of Myanmar. *American Journal of Physical Anthropology* 143:208–222.

Marivaux. L., Chaimanee, Y., Ducrocq, S., Marandat, B., Sudre, J., Soe, A. N., Tun, S. T., Htoon, W., and Jaeger, J.-J. 2003. The anthropoid status of a primate from the late middle Eocene Pondaung Formation (Central Myanmar): tarsal evidence. *Proceedings of the National Academy of Sciences of the United States of America* 100:13173–13178.

Markham, R., and Groves, C. P. 1990. Brief communication: weights of wild orang utans. *American Journal of Physical Anthropology* 81:1–3.

Marks, J. 1983a. Hominoid cytogenetics and evolution. *Yearbook of Physical Anthropology* 26:131–159.

———. 1983b. Rates of karyotype evolution. *Systematic Zoology* 32:207–209.

———. 1984. On the classification of *Homo. Current Anthropology* 25:131–132.

———. 1986. Evolutionary epicycles. In K. M. Flanagan, J. A. Lillegraven, and G. G. Simpson, eds., *Vertebrates, Phylogeny, and Philosophy,* 339–350. Contributions to Geology, University of Wyoming, Special Paper 3. Laramie: Department of Geology and Geophysics, University of Wyoming.

———. 1988. Relationships of humans to chimps and gorillas. *Nature* 334:656.

———. 1992. Genetic relationships among the apes and humans. *Current Biology* 2:883–889.

———. 1993. Hominoid heterochromatin: terminal C-bands as a complex genetic trait linking chimpanzee and gorilla. *American Journal of Physical Anthropology* 90:237–246.

———. 1995a. *Human Diversity: Genes, Race, and History.* Hawthorne, NY: Aldine de Gruyter.

———. 1995b. Learning to live with the trichotomy. *American Journal of Physical Anthropology* 98:211–232.

———. 2002. *What It Means to Be 98% Chimpanzee: Apes, People and Their Genes.* Berkeley: University of California Press.

———. 2005. Phylogenetic trees and evolutionary forests. *Evolutionary Anthropology* 14:49–53.

Marks, J., Schmid, C. W., and Sarich, V. M. 1988. DNA hybridization as a guide to phylogeny: relations of the Hominoidea. *Journal of Human Evolution* 17:769–786.

Marler, P. 1965. Communication in monkeys and apes. In I. DeVore, ed., *Primate Behavior,* 544–584, New York: Holt, Rinehart & Winston.

———. 1969. Vocalizations of wild chimpanzees: an introduction. C. R. Carpenter, ed., *Proceedings of the Second International Congress of Primatology, Atlanta, GA 1968,* Vol. 1, 94–100. Basel: Karger.

———. 1976. Social organization, communication and graded signals: the chimpanzee and the gorilla. In P. P. G. Bateson and RA. Hinde, eds., *Growing Points in Ethology,* 239–280. Cambridge: Cambridge University Press.

———. 1977. The predator alarm calls of free-ranging vervet monkeys. *National Geographic Society Research Reports* 18:505–516.

———. 1985. Representational vocal signals of primates. *Fortschritte der Zoologie* 31:211–221.

———. 1991. The instinct to learn. In S. Carey and R. Gelman, eds., *Epigenesis of Mind: Essays on Biology and Cognition,* 37–66. Hillsdale, NJ: Lawrence Erlbaum.

———. 2000. Origins of music and speech: insights from animals. In N. L. Wallin, B. Merker, and S. Brown, eds., *The Origins of Music,* 31–48. Cambridge, MA: MIT Press.

Marler, P., and Hobbett, L. 1975. Individuality in a long-range vocalization of wild chimpanzees. *Zeitschrift für Tierpsychologie* 38:97–109.

Marler, P., and Tenaza, R. 1977. Signaling behavior of apes with special reference to vocalization. In T. A. Sebeok, ed., *How Animals Communicate,* 965–1033. Bloomington: Indiana University Press.

Marlowe, F. W. 2005. Hunter-gatherers and human evolution. *Evolutionary Anthropology* 14:54–67.

———. 2010. *The Hadza Hunter-Gatherers of Tanzania.* Berkeley: University of California Press.

Marr, M. J. 2007. The emergence of emergents: one behaviorist's perspective. In D. A. Washburn, ed., *Primate Perspectives on Behavior and Cognition,* 99–108. Washington, DC: American Psychological Association.

Marsh, J., dir. 2011. *Project Nim.* Santa Monica, CA: Lionsgate, DVD, 93 min.

Marsh, L. K. 2003. *Primates in Fragments: Ecology and Conservation.* New York: Kluwer Academic/Plenum Press.

Marshack, A. 1972. *The Roots of Civilization.* New York: McGraw-Hill.

————. 1990. Early hominid symbol and evolution of the human capacity. In P. Mellars, ed., *The Emergence of Modern Humans: An Archaeological Perspective*, 457–498. Edinburgh: Edinburgh University Press.

Marshall, A. J., Ancrenez, M., Brearley, F. Q., Fredriksson, G. M., Ghaffar, N., Heydon, M., Husson, S. J., Leighton, M., McConkey, K. R., Morrogh-Bernard, H. C., et al. 2009. The effects of forest phenology and floristics on populations of Bornean and Sumatran orangutans. In S. A. Wich, S. S. Utami Atmoko, T. Mitra Setia, and C. P. van Schaik, eds., *Orangutans: Geographic Variation in Behavioral Ecology and Conservation*, 97–117. Oxford: Oxford University Press.

Marshall, A. J., Cannon, C. H., and Leighton, M. 2009. Competition and niche overlap between gibbons *(Hylobates albibarbis)* and other frugivorous vertebrates in Gunung Palung National Park, West Kalimantan, Indonesia. In S. Lappan and D. J. Whittaker, *The Gibbons: New Perspectives on Small Ape Socioecology and Population Biology*, 161–188. New York: Springer.

Marshall, A. J., and Hohmann, G. 2005. Urinary testosterone levels in wild male bonobos *(Pan paniscus)* in the Lomako Forest, Democratic Republic of Congo. *American Journal of Primatology* 65:87–92.

Marshall, A. J., and Leighton, M. 2006. How does food availability limit the population density of white bearded gibbons? In G. Hohmann, M. M. Robbins and C. Boesch, eds., *Feeding Ecology in Apes and Other Primates*, 313–335. Cambridge: Cambridge University Press.

Marshall, A. J., Salas, L. A., Stephens, S., Nardiyono, Engsrom, L., Meijaard, E., and Stanley, S. A. 2007. Use of limestone karst forests by Bornean orangutans *(Pongo pygmaeus morio)* in the Sangkulirang Peninsula, East Kalimantan, Indonesia. *American Journal of Primatology* 69:212–219.

Marshall, A. J., and Wrangham, R. W. 2006. Evolutionary consequences of fallback foods. *International Journal of Primatology* 28:1219–1235.

Marshall, A. J., Wrangham, R. W., and Arcadi, A. C. 1999. Does learning affect the structure of vocalizations in chimpanzees? *Animal Behaviour* 58:825–830.

Marshall, C. R. 1991. Statistical tests and bootstrapping: assessing the reliability of phylogenies based on distance data. *Molecular Biology and Evolution* 8:386–391.

Marshall, J. T., Jr., and Marshall, E. R. 1976. Gibbons and their territorial songs. *Science* 193:235–237.

Marshall, J. T., Jr., Ross, B. A., and Chantharojvong, S. 1972. The species of gibbons in Thailand. *Journal of Mammalogy* 53:479–486.

Marshall, J. T., Jr., and Sugardjito, J. 1986. Gibbon systematics. In D. R. Swindler and J. Irwin, eds., *Comparative Primate Biology*, Vol. 1: *Systematics, Evolution, and Anatomy*, 137–185. New York: Alan R. Liss.

Marshall, L. 1976. *The !Kung of Nyae Nyae*. Cambridge, MA: Harvard University Press.

Marshall-Pescini, S., and Whiten, A. 2008. Chimpanzees *(Pan troglodytes)* and the question of cumulative culture: an experimental approach. *Animal Cognition* 11:449–456.

Marson, J, Meuris, S., Moysan, F., Gervais, D., Cooper, R. W., and Jouannet, P. 1988. Cellular and biochemical characteristics of semen obtained from pubertal chimpanzees by masturbation. *Journal of Reproduction and Fertility* 82:199–207.

———. 1989. Influence of ejaculation frequency on semen characteristics in chimpanzees *(Pan troglodytes)*. *Journal of Reproduction and Fertility* 82:43–50.

Marten, K., and Marler, P. 1977. Sound transmission and its significance for animal vocalization. I. Temperate habitats. *Behavioral Ecology and Sociobiology* 2:271–290.

Marten, K., Quine, D., Marler, P. 1977. Sound transmission and its significance for animal vocalization. I. Tropical forest habitats. *Behavioral Ecology and Sociobiology* 2:291–302.

Martin, C. P. 1934. A comparison of the joints of the arm and leg and the significance of the structural differences between them. *Journal of Anatomy* 68:511–520.

Martin, K., and Psarakos, S. Evidence of self-awareness in the bottlenose dolphin *(Tursiops trundatus)*. In S. T. Parker, R. W. Mitchell, and M. L. Boccia, eds., *Self-Awareness in Animals and Humans,* 361–379. Cambridge: Cambridge University Press.

Martin, L. 1986. Relationships among extant and extinct great apes and humans. In B. Wood, L. Martin, and P. Andrews, eds., *Major Topics in Primate and Human Evolution,* 161–187. Cambridge: Cambridge University Press.

———. 1991. Teeth, sex and species. *Nature* 352:111–112.

Martin, L. B., and Andrews, P. 1993. Species recognition in Middle Miocene hominoids. In W. H. Kimbel and L. B. Martin, eds., *Species, Species Concepts, and Primate Evolution,* 393–427. New York: Plenum Press.

Martin, M. M., Rayner, J. C., Gagneux, P., Barnwell, J. W., Varki, A. 2005. Evolution of human-chimpanzee differences in malaria susceptibility: relationship to human genetic loss of *N*-glycolylneuraminic acid. *Proceedings of the National Academy of Sciences of the United States of America* 102:12819–12824.

Martin, P., and Caro, T. M. 1985. On the function of play and its role in behavioral development. *Advances in the Study of Behavior* 15:59–103.

Martin, R. D. 1981. Relative brain size and basal metabolic rate in terrestrial vertebrates. *Nature* 293:57–60.

———. 1982. *Human Brain Evolution in an Ecological Context.* New York: American Museum of Natural History.

———. 1986. Primates: a definition. In B. Wood, L. Martin, and P. Andrews, eds., *Major Topics in Primate and Human Evolution,* 1–31. Cambridge: Cambridge University Press.

———. 1990. *Primate Origins and Evolution: A Phylogenetic Reconstruction.* Princeton, NJ: Princeton University Press.

———. 1993. Primate origins: plugging the gaps. *Nature* 363:223–224.

———. 2006. *New Light on Primate Evolution: Ernst Mayr Lecture 2003.* Berlin: Akademie Verlag.

———. 2007. Problems with the tiny brain of the Flores hominid. In E. Indriati, ed., *Recent Advances in Southeast Asian Paleoanthropology and Archaeology,* 9–23. Yogyakarta, Indonesia: Gadjah Mada University.

———. 2012. Primates. *Current Biology* 22:R785–R790.

Martin, R. D., MacLarnon, A. M., Phillips, J. L., and Dobyns, W. B. 2006a. Comment on "The brain of LB1, *Homo floresiensis.*" *Science* 312:999b.

———. 2006b. Flores hominid: new species of microcephalic dwarf? *Anatomical Record Part A* 288A:1123–1145.

Martinet, A. 1960. *Éléments de linguistique générale.* Paris: Armand Colin.

Martinón-Torres, M., Bermúdez de Castro, J. M., Gómez-Robles, A., Arsuaga, J. L., Carbonell, E., Lordkipanidze, D., Manzi, G., and Margvelashvili, A. 2007. Dental evidence on the hominin dispersals during the Pleistocene. *Proceedings of the National Academy of Sciences of the United States of America* 104:13279–13282.

Marvan, R., Stevens, J. M. G. Roeder, A. D., Mazura, I., Bruford, M. W., and de Ruiter, J. R. 2006. Male dominance rank, mating and reproductive success in captive bonobos *(Pan paniscus). Folia Primalogica* 77:364–376.

Marzke, M. W. 1971. Origin of the human hand. *American Journal of Physical Anthropology* 34:61–84.

———. 1983. Joint functions and grips of the *Australopithecus afarensis* hand, with special reference to the region of the capitate. *Journal of Human Evolution* 12:197–211.

———. 1986. Tool use and the evolution of hominid hands and bipedality. In J. G. Else and P. C. Lee, eds., *Primate Evolution. Selected Proceedings of the Tenth Congess of the International Primatological Society, held in Nairobi, Kenya, in July 1984,* Vol. 1, 203–209. Cambridge: Cambridge University Press.

———. 1992. Evolutionary development of the human thumb. *Hand Clinics* 8:1–8.

———. 1997. Precision grips, hand morphology, and tools. *American Journal of Physical Anthropology* 102:91–110.

Marzke, M. W., and Marzke, R. F. 1987. The third metacarpal styloid process in humans: origin and functions. *American Journal of Physical Anthropology* 73:415–431.

———. 1999. Chimpanzee thumb muscle cross sections, moment arms and potential torques, and comparisons with humans. *American Journal of Physical Anthropology* 110:163–178.

Marzke, M. W., and Shackley, M. S. 1986. Hominid hand use in the Pliocene and Pleistocene: evidence from experimental archaeology and comparative morphology. *Journal of Human Evolution* 15:439–460.

Marzke, M. W., Toth, N., Schick, K., Reece, S., Steinberg, B., Hunt, K., Linscheid, R. L., and An, K.-N. 1998. EMG study of hand muscle recruitment during hard hammer percussion manufacture of Oldowan tools. *American Journal of Physical Anthropology* 105:315–332.

Marzke, M. W., and Wullstein, K. L. 1996. Chimpanzee and human grips: a new classification with a focus on evolutionary morphology. *International Journal of Primatology* 17:117–139.

Marzke, M. W., Wullstein, K. L., and Viegas, S. F. 1992. Evolution of the power ("squeeze") grip and its morphological correlates in hominids. *American Journal of Physical Anthropology* 89:283–298.

———. 1994. Variability at the carpometacarpal and midcarpal joints involving the fourth metacarpal, hamate, and lunate in Catarrhini, *American Journal of Physical Anthropology* 93:229–240.

Mascaro, O., and Csibra, G. 2012. Representation of stable social dominance relations by human infants. *Proceedings of the National Academy of Sciences of the United States of America* 109:6862–6867.

Maschenko, E. N. 2005. Cenozoic primates of eastern Eurasia (Russia and adjacent areas). *Anthropological Science* 113:103–115.

Masi, S., Cipolletta, C., and Robbins, M. M. 2009. Western lowland gorillas *(Gorilla gorilla gorilla)* change their activity patterns in response to frugivory. *American Journal of Primatology* 71:91–100.

Masquelet, A. C., Salama, J., Outrequin, G., Serrault, M, and Chevrel, J. P. 1986. Morphology and functional anatomy of the first dorsal interosseous muscle of the hand. *Surgical and Radiologic Anatomy* 8:19–28.

Masterson, T. J., and Leutenegger, W. 1990. The ontogeny of sexual dimorphism in the cranium of Bornean orang-utans *(Pongo pygmaeus pygmaeus)* as detected by principal-components analysis. *International Journal of Primatology* 11:517–539.

———. 1992. Ontogenetic patterns of sexual dimorphism in the cranium of Bornean orang-utans *(Pongo pygmaeus pygmaeus)*. *Journal of Human Evolution* 23:3–26.

Matano, S. 2001. Brief communication: proportions of the ventral half of the cerebellar dentate nucleus in humans and great apes. *American Journal of Physical Anthropology* 114:163–165.

Mathew, S., and Boyd, R. 2011. Punishment sustains large-scale cooperation in prestate warfare. *Proceedings of the National Academy of Sciences of the United States of America* 108:11375–11380.

Mathur, V. A., Harada, T. Lipke, T., and Chiao, J. Y. 2010. Neural basis of extraordinary empathy and altruistic motivation. *NeuroImage* 51:1468–1475.

Mathur, Y. K. 1984. Cenozoic palynofossils, vegetation, ecology and climate of the north and northwestern Subhimalayan Region, India. In R. O. Whyte, ed., *The Evolution of the East Asian Environment,* Vol. 2, 504–551. Hong Kong: University of Hong Kong Centre of Asian Studies.

Matsuda, T., Torii, M., Koyaguchi, T., Makinouchi, T., Mitsushio, H., and Ishida, S. 1986. Geochronology of Miocene hominoids east of the Kenya Rift Valley. In J. G. Else and P. C. Lee, eds., *Primate Evolution,* 35–45. Cambridge: Cambridge University Press.

Matsumoto-Oda, A. 1998. Injuries to the sexual skin of female chimpanzees at Mahale and their effect on behavior. *Folia Primatologica* 69:400–404.

———. 1999. Female choice in the opportunistic mating of wild chimpanzees *(Pan troglodytes schweinfurthii)* at Mahale. *Behavioral Ecology and Sociobiology* 46:258–266.

Matsumoto-Oda, A., and Hayashi, Y. 1999. Nutritional aspects of fruit choice by chimpanzees. *Folia Primatologica* 70:154–162.

Matsumoto-Oda, A., and Kasuya, E. Proximity and estrous synchrony in Mahale chimpanzees. *American Journal of Primatology* 66:159–166.

Matsumura, H., Nakatsukasa, H., and Ishida, H. 1992. Comparative study of crown cusp areas in the upper and lower molars of African apes. *Bulletin of the National Museum, Tokyo, series D* 18:3–15.

Matsuno, T., Kawai, N., and Matsuzawa, T. 2006. Color recognition in chimpanzees *(Pan troglodytes)*. In T. Matsuzawa, M. Tomonaga, and M. Tanaka, eds., *Cognitive Development in Chimpanzees.,* 317–329. Tokyo: Springer.

Matsu'ura, S. 1986. Age of the early Javanese hominids: a review. J. G. Else and P. C. Lee, eds., In *Primate Evolution,* 115–121. Cambridge: Cambridge University Press.

Matsuzawa, T. 1985. Use of numbers by a chimpanzee. *Nature* 315:57–59.

———. 1994. Field experiments on use of stone tools by chimpanzees in the wild. In R. W. Wrangham, W. C. McGrew, F. B. M. de Waal, and P. G. Heltne, eds., *Chimpanzee Cultures,* 351–370. Cambridge, MA: Harvard University Press.

———. 2001. *Primate Origins of Human Cognition and Behavior.* Tokyo: Springer.

Matsuzawa, T., Asano, T., Kuboda, K., and Murofushi, K. 1986. Acquisition and generalization of numerical labeling by a chimpanzee. In D. M. Taub and F. A. King, eds., *Current Perspectives in Primate Social Dynamics,* 416–430. New York: Van Nostrand Reinhold.

Matsuzawa, T., Tomonaga, M., and Tanaka, M. 2006 *Cognitive Development in Chimpanzees.* Tokyo: Springer.

Matsuzawa, T., and Yamakoshi, G. 1996. Comparison of chimpanzee material culture between Bossou and Nimba, West Africa. In A. E. Russon, K. A. Bard, and S. T. Parker, eds., *Reaching into Thought: The Minds of the Great Apes,* 211–232. Cambridge: Cambridge University Press.

Matthews, A., and Matthews, A. 2004. Survey of gorillas *(Gorilla gorilla gorilla)* and chimpanzees *(Pan troglodytes troglodytes)* in southwestern Cameroon. *Primates* 45:15–24.

Maung, M., Htike, T., Tsubamoto, T., Suzuki, H., Sein, C., Egi, N. Win, Z., Thein, Z. M. M., and Aung, A. K. 2005. Stratigraphy of the primate-bearing beds of the Eocene Pondung Formation at Paukkaung area, Myanmar. *Anthropological Science* 113:1–15.

Mautz, B. S., Wong, B. B. M., Peters, R. A., and Jennions, M. D. 2012. Penis size interacts with body shape and height to influence male attractiveness. *Proceedings of the National Academy of Sciences of the United States of America* 110:6925–1930.

Mayer, W. E., Jonker, M., Klein, D., Ivanyi, P., van Seventer, G., and Klein, J. 1988. Nucleotide sequences of chimpanzee MHC class I alleles: evidence for *trans*-species mode of evolution. *EMBO Journal* 7:2765–2774.

Mayes, R. W. 2006. The possible application of novel marker methods for estimating dietary and nutritive value in primates. In G. Hohmann, M. M. Robbins and C. Boesch, eds., *Feeding Ecology in Apes and Other Primates,* 421–444. Cambridge: Cambridge University Press.

Maynard Smith, J., and Price, G. R. 1973. The logic of animal conflict. *Nature* 246:15–18.

Mayr, E. 1963. *Animal Species and Evolution.* Cambridge: Harvard University Press.

McBrearty, S., Bishop, L., and Kingston, J. 1996. Variability in traces of Middle Pleistocene hominid behavior in the Kapthurin Formation, Baringo, Kenya. *Journal of Human Evolution* 30:563–580.

McBrearty, S., and Brooks, A. S. 2000. The revolution that wasn't: a new interpretation of the origin of modern human behavior. *Journal of Human Evolution* 39:453–563.

McBrearty, S., and Jablonski, N. G. 2005. First fossil chimpanzee. *Nature* 437:105–108.

McBrearty, S., and Moniz, M. 1991. Prostitutes or providers? Hunting, tool use, and sex roles in earliest *Homo.* In D. Walde and D. Willows, eds., *The Archaeology of Gender,* 71–82. Calgary: University of Calgary Archaeological Association.

McCann, C. 1933. Notes on the colouration and habits of the white-browed gibbon or hoolock (*Hylobates hoolock* Harl.). *Journal of the Bombay Natural History Society* 36:395–405.

McClelland, J. L., and Rogers, T. T. 2003. The parallel distributed processing approach to semantic cognition. *Nature Reviews Neuroscience* 4:310–322.

McClure, H. E. 1964. Some observations of primates in climax dipterocarp forest near Kuala Lumpur, Malaya. *Primates* 5:39–58.

McCollum, M. A., Grine, F. E., Ward, S. C., and Kimbel, W. H. 1993. Subnasal morphological variation in extant hominoids and fossil hominids. *Journal of Human Evolution* 24:87–111.

McConkey, K. 2005a. Bornean orangutan *(Pongo pygmaeus)*. In J. Caldecott and K. Miles, eds., *World Atlas of Great Apes and Their Conservation,* 161–184. Berkeley: University of California Press.

———. 2005b. Sumatran orangutan *(Pongo abelii)*. In J. Caldecott and K. Miles, eds., *World Atlas of Great Apes and Their Conservation,* 185–204. Berkeley: University of California Press.

McConkey, K. R., Aldy, F., Ario, A., and Chivers, D. J. 2002. Selection of fruits by gibbons (*Hylotates muelleri × agilis*) in the rain forests of central Borneo. *International Journal of Primatology* 23:123–145.

McCrossin, M. L. 1992. An oreopithecid proximal humerus from the Middle Miocene of Maboko Island, Kenya. *International Journal of Primatology* 13:659–677.

———. 1999. New postcranial remains of *Kenyapithecus* and their implications for understanding the origins of hominoid terrestriality. In H. Ishida, ed., *Abstracts of the International Symposium "Evolution of Middle-to-Late Miocene Hominoids in Africa," July 11–13, 1999.* Abstract presented at the meeting, Takaragaike, Kyoto, Japan.

McCrossin, M. L., and Benefit, B. R. 1992. Comparative assessment of the ischial morphology of *Victoriapithecus macinnesi*. *American Journal of Physical Anthropology* 87:277–290.

———. 1993. Recently recovered *Kenyapithecus* mandible and its implications for great ape and human origins. *Proceedings of the National Academy of Sciences of the United States of America* 90:1962–1966.

———. 1994. Maboko Island and the evolutionary history of Old World monkeys and apes. In R. S. Corruccini and R. L. Ciochon, eds., *Integrative Paths to the Past,* 95–122. Englewood Cliffs, NJ: Prentice Hall.

———. 1997. On the relationships and adaptations of *Kenyapithecus,* a large-bodied hominoid from the Middle Miocene of eastern Africa. In D. R. Begun, C. V. Ward, and M. D. Rose, eds., *Function, Phylogeny, and Fossils,* 241–267. New York: Plenum Press.

McCrossin, M. L., Benefit, B. R., Gitau, S. N., Palmer, A. K., and Blue, K. T. 1998. Fossil evidence for the origins of terrestriality among Old World higher primates. In E. Strasser, J. Fleagle, A. Rosenberger, and H. McHenry, eds., *Primate Locomotion: Recent Advances,* 353–396. New York: Plenum Press.

McCrossin, M. L., and Reyes, L. D. 2010. *Australopithecus afarensis* exhibited a chimpanzee-like pattern of female transfer. *American Journal of Physical Anthropology* 50 (Suppl.): 166–167.

McDougall, I., and Watkins, R. T. 1985. Age of hominoid-bearing sequence at Buluk, northern Kenya. *Nature* 318:175–178.

McElreath, R., Boyd, R., and Richerson, P. J. 2003. Shared norms and evolution of ethnic markers. *Current Anthropology* 44:122–129.

McFadden, P. L. 1980. An overview of palaeomagnetic chronology with special reference to the South African hominid sites. *Palaeontologia Africana* 23:35–40.

McFadden, P. L., Brock, A., and Partridge, T. C. 1979. Palaeomagnetism and the age of the Makapansgat hominid site. *Earth and Planetary Science Letters* 44:373–382.

McGinnis, Patrick R. 1979. Sexual behavior in free-living chimpanzees: consort relationships In: D. A. Hamburg and E. R. McCown, eds., *The Great Apes,* 429–439. Menlo Park, CA: Benjamin/Cummings.

McGraw, W. S., Cooke, C., and Shultz, S. 206. Primate remains from African crowned eagle *(Stephanoaetus coronatus)* nests in Ivory Coast's Taï forest: implications for primate predation and early hominid taphonomy in South Africa. *American Journal of Physical Anthropology* 131:151–165.

McGrew, W. C. 1974. Tool use by wild chimpanzees feeding upon driver ants. *Journal of Human Evolution* 3:501–505.

———. 1977. Socialization and object manipulation of wild chimpanzees. In S. Chevalier-Skolnikoff and F. E. Poitier, eds., *Primate Bio-Social Development,* 261–288. New York: Garland Publishing.

———. 1979. Evolutionary implications of sex differences in chimpanzee predation and tool-using. In D. A. Hamburg and E. R. McCown, eds., *The Great Apes,* 441–464. Menlo Park, CA: Benjamin/Cummings.

———. 1983. Animal foods in the diets of wild chimpanzees *(Pan troglodytes):* why cross-cultural variation? *Journal of Ethology* 1:46–61.

———. 1992a. *Chimpanzee Material Culture: Implications for Human Evolution.* Cambridge: Cambridge University Press.

———. 1992b. Tool-use by free-ranging chimpanzees: the extent of diversity. *Journal of Zoology, London* 228:689–694.

———. 1993. The intelligent use of tools. twenty propositions. In R. K. Gibson and T. Ingold, eds., *Tools, Language and Cognition in Human Evolution,* 151–170. Cambridge: Cambridge University Press.

———. 1996. Dominance status, food sharing, and reproductive success in chimpanzees. In P. Wiessner and W. Schiefenhövel, eds., *Food and the Status Quest: An Interdisciplinary Perspective,* 39–45. Providence, RI: Berghahn Books.

———. 1998a. Behavioral diversity in populations of free-ranging chimpanzees in Africa: is it culture? *Human Evolution* 13:209–220.

———. 1998b. Comments. *Current Anthropology* 39:607–608.

———. 1998c. Culture in nonhuman primates? *Annual Review of Anthropology* 27:301–328.

———. 2001. The other faunivory. Primate insectivory and early human diet. In C. B. Stanford and H. T. Bunn, eds., *Meat-Eating and Human Evolution,* 160–178. Oxford: Oxford University Press.

———. 2004. *The Cultured Chimpanzee: Reflections on Cultural Primatology.* Cambridge: Cambridge University Press.

————. 2007. Savanna chimpanzees dig for food. *Proceedings of the National Academy of Sciences of the United States of America* 104:19167–19168.

————. 2009. Ten dispatches from the chimpanzee culture wars, plus (revisiting the battlefronts). In K. N Laland and B. G. Galef, eds., *The Question of Animal Culture,* 41–69. Cambridge, MA: Harvard University Press.

McGrew, W. C., Baldwin, P. J., Marchant, L. F., Pruetz, J. D., Scott, S. E., and Tutin, C. E. G. 2003. Ethnoarchaeology and elementary technology of unhabituated wild chimpanzees at Assirik, Senegal, West Africa. *PaleoAnthropology* 1:1–20.

McGrew, W. C., Baldwin, P. J., and Tutin, C. E. G. 1982. Recherches scientifiques dans les pares nationaux du Senegal. XXV. Observations préliminaires sur les chimpanzés *(Pan troglodytes verus)* du Parc National du Niokolo Koba. *Memoires de l'Institut Fondamental d'Afrique Noire,* Ifan-Dakar 92:335–340.

————. 1988. Diet of wild chimpanzees *(Pan troglodytes verus)* at Mt. Assirik, Senegal: I. composition. *American Journal of Primatology* 16:213–226.

McGrew, W. C., and Collins, D. A. 1985. Tool use by wild chimpanzees *(Pan troglodytes)* to obtain termites *(Macrotermes herus)* in the Mahale Mountains, Tanzania. *American Journal of Primatology* 9:47–62.

McGrew, W. C., Ham, R. M., White, L. J. T., Tutin, C. E. G., and Fernandez, M. 1997. Why don't chimpanzees in Gabon crack nuts?. *International Journal of Primatology* 18:353–374.

McGrew, W. C., and Marchant, L. F. 1973. Are gorillas right-handed or not? *Human Evolution* 8:17–23.

————. 1992. Chimpanzees, tools, and termites: hand preference or handedness? *Current Anthropology* 33:114–119.

————. 1996. On which side of the apes? Ethological study of laterality of hand use. In W. C. McGrew, L. F. Marchant, and T. Nishida, eds., *Great Ape Societies,* 255–272. Cambridge: Cambridge University Press.

————. 1997a. Using the tools at hand: manual laterality and elementary technology in *Cebus* spp. and *Pan* spp. *International Journal of Primatology* 18:787–810.

————. 1997b. On the other hand: current issues in and meta-analysis of the behavioral laterality of hand function in nonhuman primates. *Yearbook of Physical Anthropology* 40:201–232.

McGrew, W. C., Marchant, L. F., Beuerlein, M. M., Vrancken, D., Fruth, B., and Hohmann, G. 2007. Prospects for bonobo insectivory at Lui Kotal, Democratic Republic of Congo. *International Journal of Primatology* 28:1237–1252.

McGrew, W. C., Marchant, L. F., Wrangham, R. W., and Klein, H. 1999. Manual laterality in anvil use: wild chimpanzees cracking *Strychnos* fruits. *Laterality* 4:79–87.

McGrew, W. C., Tutin, C. E. G., Collins, D. A., and File, S. K. 1989. Intestinal parasites of sympatric *Pan troglodytes* and *Papio* spp. at two sites: Gombe (Tanzania) and Mt. Assirik (Senegal). *American Journal of Primatology* 17:147–155.

McHenry, H. M. 1975. Fossils and the mosaic nature of human evolution. *Science* 190:425–431.

————. 1976. Multivariate analysis of early hominid humeri. In E. Giles and J. S. Friedlaender, eds., *The Measures of Man,* 338–371. Cambridge, MA: Peabody Museum Press.

————. 1983. The capitate of *Australopithecus afarensis* and *A. africanus*. *American Journal of Physical Anthropology* 62:445–454.

————. 1984. Relative cheek-tooth size in *Australopithecus*. *American Journal of Physical Anthropology* 64:187–198.

————. 1985. Implications of postcanine megadontia for the origin of *Homo*. In E. Delson, ed., *Ancestors: The Hard Evidence*, 178–183. New York: Alan R. Liss.

————. 1986. The first bipeds: a comparison of the *A. afarensis* and *A. africanus* postcranium and implications for the evolution of bipedalism. *Journal of Human Evolution* 15:177–191.

————. 1988. New estimates of body weight in early hominids and their significance to encephalization and megadontia in "robust" australopithecines. In F. E. Grine, ed., *Evolutionary History of the "Robust" Australopithecines*, 133–148. Hawthorne NY: Aldine de Gruyter.

————. 1991a. Petite bodies of "robust" australopithecines. *American Journal of Physical Anthropology* 86:445–454.

————. 1991b. Sexual dimorphism in *Australopithecus afarensis*. *Journal of Human Evolution* 20:21–32.

————. 1994a. Behavioral ecological implications of early hominid body size. *Journal of Human Evolution* 27:77–87.

————. 1994b. Tempo and mode in human evolution. *Proceedings of the National Academy of Sciences of the United States of America* 91:6780–6786.

McHenry, H. M., Andrews, P., and Corruccini, R. S. 1980. Miocene hominoid palatofacial morphology. *Folia Primatologica* 33:241–252.

McHenry, H. M., and Berger, L. R. 1998a. Body proportions in *Australopithecus afarensis* and *A. africanus* and the origin of the genus *Homo*. *Journal of Human Evolution* 35:1–22.

————. 1998b. Limb lengths in *Australopithecus* and the origin of the genus *Homo*. *South African Journal of Science* 94:447–450.

McHenry, H. M., and Coffing, K. 2000. Australopithecus to Homo: transformations in body and mind. *Annual Reviews in Anthropology* 29:125–146.

McHenry, H. M., and Corruccini, R. S. 1975. Distal humerus in hominoid evolution. *Folia Primatologica* 23:227–244.

————. 1976. Affinities of Tertiary hominoid femora. *Folia Primatologica* 26:139–150.

————. 1980. Late Tertiary hominoids and human origins. *Nature* 285:397–398.

————. 1983. The wrist of *Proconsul africanus* and the origin of hominoid postcranial adaptations. In R. L. Ciochon and R. S. Corruccini, eds. *New Interpretations of Ape and Human Ancestry*, 353–367. New York: Plenum Press.

————. 1991. First steps? Analysis of the postcranium of early hominids. In *Origine(s) de la Bipédie chez les Hominidés*, Y. Coppens and B. Senut, eds., R. L. Ciochon and R. S. Corruccini, eds., 133–141. Paris: Éditions du Centre National de la Recherche Scientifique.

McHenry, H. M., and Skelton, R. R. 1985. *Australopithecus africanus* ancestral to *Homo?* In P. V. Tobias, ed., *Hominid Evolution: Past, Present and Future*, 221–226. New York: Alan R. Liss.

McKee, J. K. 1993. Faunal dating of the Taung hominid fossil deposits. *Journal of Human Evolution* 25:363–376.

McKee, J. K., Thackery, J. F., and Berger, L. R. 1995. Faunal assemblage seriation of Southern African Pliocene and Pleistocene fossil deposits. *American Journal of Physical Anthropology* 96:235–250.

McKey, D. B., Gartlan, S. J., Waterman, P. G., and Choo, G. M. 1981. Food selection by black colobus monkeys *(Colobus satanus)* in relation to plant chemistry. *Biological Journal of the Linnaean Society* 16:115–146.

McLeod, J., and Gold, R. Z. 1952. The male factor in fertility and infertility. V. Effect of continence on semen quality. *Fertility and Sterility* 3:297–315.

McMurray, B. Defusing the childhood vocabulary explosion. 2007. *Science* 317:631.

McNamara, P., Barton, R. A., and Nunn, C. L. 2009. *Evolution of Sleep.* Cambridge: Cambridge University Press.

McNeilage, A. 1996. Ecotourism and mountain gorillas in the Virunga Volcanoes. In V. J. Taylor and N. Dunstone, eds., *The Exploitation of Mammal Populations,* 334–150. London: Chapman & Hall.

———. 2001. Diet and habitat use of two mountain gorilla groups in contrasting habitats in the Virungas. In M. M. Robbins, P. Sicotte and K. J. Stewart, eds., *Mountain Gorillas,* 265–292. Cambridge: Cambridge University Press.

McNeilage, A., Plumptre, A., Brock-Doyle, A., and Vedder, A. 2001. Bwindi Impenetrable National Park, Uganda Gorilla and Large Mammal Census, 1997. *Oryx* 35:39–47.

McNeilage, A., Robbins, M. M., Gray, M., Olupot, W., Babaasa, D., Bitariho, R., Kasangaki, A, Rainer, H., Asuma, S., Mugiri, G., and Baker, J. 2006a. Census of the mountain gorilla *Gorilla beringei beringei* population in Bwindi Impenetrable National Park, Uganda. *Oryx* 40:419–427.

McNeilage, A., Robbins, M. M., Guschanski, K., Gray, M., and Kagoda, E. 2006b. *Mountain Gorilla Census—Bwindi Impenetrable National Park Summary Report,* www.igcp.org/wp-content/themes/igcp/docs/pdf/Bwindicensus2006resultssummary .pdf.

McNeill, D. 1992. *Hand and Mind.* Chicago: University of Chicago Press.

McNeill, D., Bertenthal, B., Cole, J., and Galagher, S. 2005. Gesture-first, but no gestures? *Behavioral and Brain Sciences* 28:138–139.

McPherron, S. P. 2000. Handaxes as a measure of the mental capabilities of early hominids. *Journal of Archaeological Science* 27:655–663.

———. 2013. Perspectives on stone tools and cognition in the early Paleolithic record. In C. M. Sanz, J. Call, and C. Boesch, eds., *Tool Use in Animals. Cognition and Ecology,* 286–309. Cambridge: Cambridge University Press.

McPherron, S. P., Alemseged, Z., Marean, C. W., Wynn, J. G., Reed, D., Geraads, D., Bobe, R., and Béarat, H. A. 2010. Evidence for stone-tool-assisted consumption of animal tissues before 3.39 million year ago at Dikika, Ethiopia. *Nature* 466:857–860.

———. 2011. Tool-marked bones from before the Oldowan change the paradigm. *Proceedings of the National Academy of Sciences of the United States of America* 108:E116.

McRae, M. 1994. Creature cures. *Equinox* 75:47–55.

Meador, D. M., Rumbaugh, D. M., Pate, J. L., and Bard, K. A. 1987. Learning, problem solving, cognition, and intelligence. In G. Mitchell, J. Erwin, eds., *Comparative Biology*, Vol. 2B: *Behavior, Cognition, and Motivation*, 17–83. New York: Alan R. Liss.

Medina, M. 2005. Genomes, phylogeny, and evolutionary systems biology. *Proceedings of the National Academy of Sciences of the United States of America* 102:6630–6635.

Meggitt, M. J. 1962. *Desert People: A Study of the Walbiri Aborigines of Central Australia.* Sydney: Angus & Robertson.

Meguerditchian, A., Vauclair, J., and Hopkins, W. D. 2010. Captive chimpanzees use their right hand to communicate with each other: implications for the origin of the cerebral substrate for language. *Cortex* 46:40–48.

Mehlman, P. T., and Doran, D. M. 2002. Influencing western gorilla nest construction at Mondika Research Center. *International Journal of Primatology* 23:1257–1285.

Meijaard, E. 2004. Biogeographic history of the Javan leopard *PANTHERA PARDUS* based on craniometric analysis. *Journal of Mammology* 85:302–310.

Mein, P. 1986. Chronological succession of hominoids in the European Neogene. In J. G. Else and P. C. Lee, eds., *Primate Evolution,* 59–70. Cambridge: Cambridge University Press.

Mekel-Bobrov, N., Gilbert, S. L., Evans, P. D., Vallender, E. J., Anderson, J. R., Hudson, R. R., Tishkoff, S. A., and Lahn, B. T. 2005. Ongoing adaptive evolution of *ASPM,* a brain size determinant in *Homo sapiens. Science* 309:1720–1722.

Meldrum, D. J. Fossilized Hawaiian footprints compared with Laetoli hominid footprints. In D. J. Meldrum and C. E. Hilton, eds., *From Biped to Strider: The Emergence of Modern Human Walking, Running, and Resource Transport,* 63–83, New York: Kluwer Academic/Plenum Press.

Meldrum, D. J., and Pan, Y. 1988. Manual phalanx of *Laccopithecus robustus* from the latest Miocene site of Lufeng. *Journal of Human Evolution* 17:719–731.

Melenky, R. K., Kuroda, S., Vineberg, E. O., and Wrangham, R. W. 1994. The significance of terrestrial herbaceous foods for bonobos, chimpanzees, and gorillas. In R. W. Wrangham, W. C. McGrew, F. B. M. de Waal, and P. G. Heltne, eds., *Chimpanzee Cultures,* 59–75. Cambridge, MA: Harvard University Press.

Melenky, R. K., and Stiles, E. W. 1991. Distribution of terrestrial herbaceous food consumption by *Pan paniscus* in the Lomako Forest, Zaire. *American Journal of Primatology* 23:153–169.

Melenky, R. K., and Wrangham, R. W. 1994. A quantitative comparison of terrestrial herbaceous vegetation and its consumption by *Pan paniscus* in the Lomako Forest, Zaire, and *Pan troglodytes* in the Kibale Forest, Uganda. *American Journal of Primatology* 32:1–12.

Melis, A. P., Hare, B., and Tomasello, M. 2006. Chimpanzees recruit the best collaborators. *Science* 311:1297–1300.

Mellars, P. 1996. *The Neanderthal Legacy.* Princeton, NJ: Princeton University Press.

Meltzoff, A. N., Kuhl, P. K., Movellan, J., and Sejnowski, T. J. 2009. Foundations for a new science of learning. *Science* 325:284–288.

Menzel, C., Fowler, A., Tennie, C., and Call, J. 2013. Leaf surface roughness elicits leaf swallowing behavior in captive chimpanzees *(Pan troglodytes)* and bonobos *(P. panis-*

cus), but not in gorillas *(Gorilla gorilla)* or orangutans *(Pongo abelii). International Journal of Primatology* 34:533–553.

Menzel, C. R. 1997. Primates' knowledge of their natural habitat: As indicated in foraging. In A. Whiten and R. W. Byrne, eds., *Machiavellian Intelligence II: Extensions and Evaluations,* 207–239. Cambridge: Cambridge University Press.

———. 1999. Unprompted recall and reporting of hidden objects by a chimpanzee *(Pan troglodytes)* after extended delays. *Journal of Comparative Psychology* 113:426–434.

———. 2005. Progress in the study of chimpanzee recall and episodic memory. In H. S. Terrace and J. Metcalf, eds., *The Missing Link in Cognition: Origins of Self-Reflective Consciousness,* 188–224. New York: Oxford University Press.

Menzel, C. R., Savage-Rumbaugh, E. S., and Menzel, E. W., Jr. 2002. Bonobo *(Pan paniscus)* spatial memory and communication in a 20-hectare forest. *International Journal of Primatology* 23:601–619.

Menzel, E. W., Jr. 1971a. Group behavior in young chimpanzees: responsiveness to cumulative novel changes in a large outdoor enclosure. *Journal of Comparative and Physiological Psychology* 74:46–51.

———. 1971b. Communication about the environment in a group of young chimpanzees. *Folia Primatologica* 15:220–232.

———. 1973a. Further observations on the use of ladders in a group of young chimpanzees. *Folia Primatologica* 19:450–457.

———. 1973b. Leadership and communication in young chimpanzees. In E. W. Menzel, Jr., ed., *Symposia of the Fourth international Congress of Primatology, Portland, Ore. 1972,* Vol. 1: *Precultural Primate Behavior,* 192–225. Basel: Karger.

———. 1974. A group of young chimpanzees in a one-acre field. In A. M. Schrier and F. Stollnitz, ed., *Behavior of Nonhuman Primates: Modern Research Trends,* 83–153. New York: Academic Press.

———. 1978. Cognitive mapping in chimpanzees. In S. A. Hulse, H. Fowler, and W. K. Honig, eds., *Cognitive Processes in Animal Behavior,* 375–422. Hillsdale, NJ: Lawrence Erlbaum.

———. 1979. Communication of object-locations in a group of young chimpanzees. In D. A. Hamburg and E. R. McCown, eds., *The Great Apes,* 359–371. Menlo Park, CA: Benjamin/Cummings.

Menzel, E. W., Jr., and Halperin, S. 1975. Purposive behavior as a basis for objective communication between chimpanzees. *Science* 189:652–654.

Menzel, E. W., Jr., Premack, D., and. Woodruff, G. 1978. Map reading in chimpanzees. *Folia Primatologica* 29:241–249.

Menzel, E. W., Jr., Savage-Rumbaugh, E. S., and Lawson, J. 1985. Chimpanzee *(Pan troglodytes)* spatial problem solving with the use of mirrors and televised equivalents of mirrors. *Journal of Comparative Psychology* 99:211–217.

Mercader, J., Barton, H., Gillespie, J., Harris, J., Kuhn, S., Tyler, R., and Boesch, C. 2007. 4,300-year-old chimpanzee sites and the origins of percussive stone technology. *Proceedings of the National Academy of Sciences of the United States of America* 104:3043–3048.

Mercader, J., Panger, M., and Boesch, C. 2002. Excavation of a chimpanzee stone tool site in the African rainforest. *Science* 296:1452–1455.

Meredith, R. W., Janecka, J. E., Gatesy, J., Ryder, O. A., Fisher, C. A., Teeling, E. C., Goodbla, A., et al. 2011. Impacts of the Cretaceous terrestrial revolution and KPg extinction on mammalian diversification. *Science* 334:521–524.

Merker, B. 1984. A note on hunting and hominid origins. *American Anthropologist* 86:112–114.

———. 2000. Synchronous chorusing and human origins. In N. L. Wallin, B. Merker and S. Brown, eds., *The Origins of Music*, 315–327. Cambridge, MA: MIT Press.

Meulman, E. J. M., and van Schaik, C. P. 2013. Orangutan tool use and the evolution of technology. In C. M. Sanz, J. Call, C. Boesch, eds., *Tool Use in Animals. Cognition and Ecology*, 176–202. Cambridge: Cambridge University Press.

Meyer, E., Wiegand, P., Rand, S. P., Kuhlmann, D., Brack, M., and Brinkmann, B. 1995. Microsatellite polymorphisms reveal phylogenetic relationships in Primates. *Journal of Molecular Evolution* 41:10–14.

Michael, J. 1984. Verbal behavior. *Journal of the Experimental Analysis of Behavior* 42:363–376.

Michilsens, F., D'Août, K., and Aerts, P. 2011. How pendulum-like are siamangs? energy exchange during brachiation. *American Journal of Physical Anthropology* 145:581–591.

Midlo, C. 1934. Form of hand and foot in primates. *American Journal of Physical Anthropology* 19:337–389.

Mikkelsen, T. S., Hillier, L. W., Eichler, E. E., Zody, M. C., Jaffe, D. B., Yang, S-P., Enard, W., Hellmann, I., Lindblad-Toh, K., Altheide, T. K., et al., for the Chimpanzee Sequencing and Analysis Consortium. 2005. Initial sequence of the chimpanzee genome and comparison with the human genome. *Nature* 437:69–87.

Miles, H. L. 1978. Language acquisition in apes and children. In F. C. Peng, ed., *Sign Language and Language Acquisition in Man and Ape: New Dimensions in Comparative Psycholinguistics*, 103–120. Boulder, CO: Westview Press.

———. 1983. Apes and language: the search for communicative competence. In J. de Luce and H. T. Wilder, eds., *Language in Primates: Perspectives and Implications*, 43–61. New York: Springer.

———. 1990 The cognitive foundations for reference in a signing orangutan. In S. T. Parker and K. R. Gibson, eds., *"Language" and Intelligence in Monkeys and Apes*, 511–539. Cambridge: Cambridge University Press.

———. 1993. Language and the Orang-utan: the old "person" of the forest. In P. Cavalieri and P. Singer, eds., *The Great Ape Project: Equality Beyond Humanity*, 42–57. London: Fourth Estate.

———. 1999. Symbolic communication with and by great apes. In S. T. Parker, R. W. Mitchell, and H. L. Miles, eds., *The Mentalities of Gorillas and Orangutans*, 197–219. Cambridge: Cambridge University Press.

Miller, D. J., Duka, T., Stimpson, C. D., Shapiro, S. J., Baze, W. B., McArthur, M. J. Fobbs, A. J., Sousa, A. M. M., Sestan, N., Wildman, D. E., et al. 2012. Prolonged myelination in human neocortical evolution. *Proceedings of the National Academy of Sciences of the United States of America* 109:16480–16485.

Miller, E. R. 1999. Faunal correlations of Wadi Moghara, Egypt: implications for the age of *Prohylobates tandyi. Journal of Human Evolution* 36:519–533.

Miller, E. R., and Simons, E. L. 1997. Dentition of *Proteopithecus sylviae,* an archaic anthropoid from the Fayum, Egypt. *Proceedings of the National Academy of Sciences of the United States of America* 94:13760–13764.

Miller, G. 2006. The thick and thin of brainpower: developmental timing linked to IQ. *Science* 311:1851.

———. 2009. Fingerprints enhance the sense of touch. *Science* 323:572

———. 2011. The brain's social network. *Science* 334:578–579.

Miller, G. F. 2000. Evolution of human music through sexual selection. In N. L. Wallin, B. Merker and S. Brown, eds., *The Origins of Music,* 329–360. Cambridge, MA: MIT Press.

Miller, J. D., Scott, E. C., and Okamoto, S. 2006. Public acceptance of evolution. *Science* 313:765–766.

Miller, R. A. 1945. The ischial callosities of primates. *Journal of Anatomy* 76:67–91.

Miller, S. F., White, J. L., and Ciochon, R. L. 2008. Assessing mandibular shape variation within *Gigantopithecus* using a geometric morphometric approach. *American Journal of Physical Anthropology* 137:201–212.

Milton, K. 1982. Distribution patterns of tropical plant foods as an evolutionary stimulus to primate mental development. *American Anthropologist* 83:534–548.

———. 1984. The role of food-processing factors in primate food choice. In P. S. Rodman and J. G. H. Cant, eds., *Adaptations for Foraging in Nonhuman Primates,* 249–279. New York: Columbia University Press.

———. 1987. Primate diets and gut morphology: implications for hominid evolution. M. Harris and E. B. Ross, eds., *Food and Evolution: Toward a Theory of Human Food Habits,* 93–115. Philadelphia: Temple University Press.

———. 1993. Diet and primate evolution. Scientific American 269:86–93.

———. 1999a. Back to basics: why food of wild primates have relevance for modern human health. *Nutrition* 16:480–483.

———. 1999b. Nutritional characteristics of wild primate foods: do the diets of our closest living relatives have lessons for us? *Nutrition* 15:448–498.

———. 2000. A hypothesis to explain the role of meat-eating in human evolution. *Evolutionary Anthropology* 8:11–21.

Milton, K., and Demment, M. W. 1988. Digestion and passage kinetics of chimpanzees fed high and low fiber diets and comparison with human data. *Journal of Nutrition* 118:1082–1088.

Minghetti, P. O., and Dugaiczyk, A. 1993. The emergence of new DNA repeats and the divergence of primates. *Proceedings of the National Academy of Sciences of the United States of America* 90:1872–1876.

Mischel, W., Shoda, Y., Rodriguez, M. L. 1989. Delay of gratification in children. *Science* 244:933–938.

Mistler-Lachman, J. L., and Lachman, R. 1974. Language in man, monkeys, and machines. *Science* 185:871–872.

Mitani, J. C. 1984. The behavioral regulation of monogamy in gibbons *(Hylobates muelleri). Behavioral Ecology and Sociobiology* 15:225–229.

———. 1985a. Gibbon song duets and intergroup spacing. *Behaviour* 92:59–96.

———. 1985b. Mating behaviour of male orangutans in the Kutai Game Reserve, Indonesia. *Animal Behaviour* 33:392–402.

———. 1985c. Responses of gibbons *(Hylobates muelleri)* to self, neighbor, and stranger song duets. *International Journal of Primatology* 6:193–200.

———. 1985d. Sexual selection and adult male orangutan long calls. *Animal Behaviour* 33:272–283.

———. 1987. Species discrimination of male song in gibbons. *American Journal of Primatology* 13:413–423.

———. 1988. Male gibbon *(Hylobates agilis)* singing behavior: natural history, song variations and function. *Ethology* 79:177–194.

———. 1992. Preliminary results of the studies on wild western lowland gorillas and other sympatric diurnal primates in the Ndoki forest, Northern Congo. In N. Itoigawa, Y. Sugiyama, G. P. Sackett, and R. K. R. Thompson, eds., *Topics in Primatology,* Vol. 2: *Behavior, Ecology, Conservation,* 215–3224. Tokyo: University of Tokyo Press.

———. 1994. Ethological studies of chimpanzee vocal behavior. In R. W. Wrangham, W. C. McGrew, F. B. M. de Waal, and P. G. Heltne, eds., *Chimpanzee Cultures,* 195–210. Cambridge, MA: Harvard University Press.

———. 1996. Comparative studies of African ape vocal behavior. In W. C. McGrew, L. F. Marchant, and T. Nishida, eds., *Great Ape Societies,* 241—254. Cambridge: Cambridge University Press.

———. 2006. Demographic influences on the behavior of chimpanzees. *Primates* 47:6–13.

Mitani, J. C., and Brandt, K. L. 1994. Social factors influence the acoustic variability in the long distance calls of male chimpanzees. *Ethology* 96:233–252.

Mitani, J. C., Grether, G. F., Rodman, P. S., and Priatna, D. 1991. Associations among wild orang-utans: sociality, passive aggregations or chance? *Animal Behavior* 42:33–46.

Mitani, J. C., and Gros-Louis, J. 1995. Species and sex differences in the screams of chimpanzees and bonobos. *International Journal of Primatology* 16:393–411.

Mitani, J. C., Gros-Louis, J., and Macedonia, J. M. 1996. Selection for acoustic individuality within the vocal repertoire of wild chimpanzees. *International Journal of Primatology* 17:569–583.

———. 1998. Chorusing and call convergence in chimpanzees: tests of three hypotheses. *Bahviour* 135:1041–1064.

Mitani, J. C., Hasegawa, T., Gros-Louis, J. Marler, P., and Byrne, R. 1992. Dialects in chimpanzees? *American Journal of Primatology* 27:233–243.

Mitani, J. C., Hunley, K., and Murdoch, M. E. 1999. Geographic variation in the calls of wild chimpanzees: a reassessment. *American Journal of Primatology* 47:133–151.

Mitani, J. C., McGrew, W. C., and Wrangham, R. 2006. Toshidada Nichida's contributions to primatology. *Primates* 47:2–5.

Mitani, J. C., Merriwether, D. A., and Zhang, C. 2000. Male affiliation, cooperation and kinship in wild chimpanzees. *Behaviour* 59:885–893.

Mitani, J. C., and Nishida, T. 1993. Contexts and social correlates of long-distance calling by male chimpanzees. *Animal Behaviour* 45:735–746.

Mitani, J. C., and Stuht, J. 1998. The evolution of nonhuman primate loud calls: acoustic adaptation for long-distance transmission. *Primates* 39:171–182.

Mitani, J. C., and Watts, D. P. 1999. Demographic influences on the hunting behavior of chimpanzees. *American Journal of Physical Anthropology* 109:439–454.

———. 2001. Why do chimpanzees hunt and share meat? *Animal Behaviour* 61:915–924.

Mitani,. J. C., Watts, D. P., and Lwanga, J. S. 2002. Ecological and social correlates of chimpanzee party size and composition. In C., Boesch, G. Hohmann, and L. F. Marchant, eds., *Behavioural Diversity in Chimpanzes and Bonobos.* 102–11. Cambridge: Cambridge University Press.

Mitani, J. C., Watts, D. P., and Muller, M. N. 2002. Recent developments in the study of wild chimpanzee behavior. *Evolutionary Anthropology* 11:2–25.

Mitani, M., Yamagiwa, J., Oko, R. A., Moutsamboté, J.-M., Yumoto, T., and Maruhashi, T. 1993. Approaches in density estimates and reconstruction of social groups in western lowland gorilla population in the Ndoki forest northern Congo. *Tropics* 2:219–229.

Mitchell, J. P., and Heatherton, T. F. Components of a social brain. In M. S. Gazzaniga, ed., *The Cognitive Neurosciences,* 953–960. Cambridge, MA: MIT Press.

Mitchell, R. W. 1991. Deception and hiding in captive lowland gorillas *(Gorilla gorilla gorilla). Primates* 32:523–527.

———. 1992. Developing concepts in infancy: animals, self-perception, and two theories of mirror self-recognition. *Psychological Inquiry* 3:127–130.

———. 1993a. Mental models of mirror-self-recognition: two theories. *New Ideas in Psychology* 11:295–325.

———. 1993b. Recognizing one's self in a mirror? A reply to Gallup and Povinelli, de Lannoy, Anderson, and Byrne. *New Ideas in Psychology* 11:351–377.

———. 1997. Kinesthetic-visual matching and the self-concept as explanations of mirror-self-recognition. *Journal for the Theory of Social Behavior* 27:17–39.

———. 1999. Scientific and popular conceptions of the psychology of great apes from the 1790s to the 1970s: déjà vu all over again. *Primate Report* 53:3–118.

———. 2002. Kinesthetic-visual matching, imitation, and self-recognition. In M. Bekoff, C. Allen, and G. M. Burghardt, eds., *The Cognitive Animal,* 345–351. Cambridge, MA: MIT Press.

Mithen, S. J. 1996. *The Prehistory of the Mind: The Cognitive Origins of Art, Religion and Science.* London: Thames and Hudson.

———. 2005. *The Singing Neanderthals: The Origins of Music, Language, Mind, and Body.* Cambridge, MA: Harvard University Press.

———. 2009. Holistic communication and the co-evolution of language and music: resurrecting an old idea. In T. Botha and C. Knight, eds., *The Prehistory of Language,* 58–76. Oxford: Oxford University Press.

Mitra Setia, T., Delgado, A., Atmoko, S. S. U., Singleton, I., and van Schaik, C. P. 2009. Social organization and male-female relationships. In S. A. Wich, S. S. U. Atmoko, and C. P. van Schaik, eds., *Orangutans: Geographic Variations in Behavioral Ecology and Conservation,* 245–253. Oxford: Oxford University Press.

Mitra Setia, T., and van Schaik, C. P. 2007. The response of adult orangu-utans to flanged male long calls: inferences about their function. *Folia Primatologica* 78:215–226.

Mittermeier, R. A. 2002. Primate conservation in the first decade of the 21st century. *Abstracts of the XIXth Congress of the International Primatological Society*, p. 3. Beijing: Mammalogical Society of China.

Mittermeier, R. A., Wallis, J., and Rylands, A. B. 2011. The World's Top 25 Most Endangered Primates 2010–2012. wwwsoc.nii.ac.jp/psj2/ips/public/ips_synopsis/IPS10-597 -W.pdf.

Miyamoto, M. M., and Goodman, M. 1990. DNA systematics and evolution of primates. *Annual Review of Ecology and Systematics* 21:197–220.

Miyamoto, M. M., Koop, B. F., Slightom, J. L., Goodman, M., and Tennant, M. R. 1988. Molecular systematics of higher primates: genealogical relations and classification. *Proceedings of the National Academy of Sciences o the United States of America* 85:7627–7631.

Miyamoto, M. M., Slightom, J. L., and Goodman, M. 1987. Phylogenetic relations of human and African apes from DNA sequences in the psi-eta-globin region. *Science* 238:369–373.

Moens, H. M. B. 1908. *Truth: Experimental Researches about the Descent of Man.* London: A. Owen.

Mohammad-Ali, K., Eladari, M.-E., and Galibert, F. 1995. Gorilla and orangutan c-myc nucleotide sequences: inference on hominoid phylogeny. *Journal of Molecular Evolution* 41:262–276.

Mollon J. D. 1989. "Tho' she neel'd in that place where they grew . . ." The uses and origins of primate colour vision. *Journal of Experimental Biology* 146:21–38.

Montagna, W. 1965. The skin. *Scientific American* 212:56–66.

———. 1982. The evolution of human skin. In A. B. Chiarelli and R. S. Corruccini, eds., *Advanced Views in Primate Biology*, 35–41. Berlin: Springer-Verlag.

———. 1985. The evolution of human skin (?). *Journal of Human Evolution* 14:3–32.

Montgomery, G. D. 1988. Rhythmical man: an alternative hypothesis to ape-human evolution. *Speculations in Science and Technology* 11:153–159.

Moore, J. 1994. Plant of the Tongwe East Forest Reserve (Ugalla), Tanzania. *Tropics* 3:333–340.

Moore, M. W., Sutikna, T., Jatmiko, Morwood, M. J., and Brumm, A. 2009 Continuities in stone flaking technology at Liang Bua, Flores, Indonesia. *Journal of Human Evolution* 57:503–526.

Moorman, S., Gobes, M. H., Kuijpers, M., Kerkhofs, A., Zandbergen, M. A., and Bolhuis, J. 2012. Human-like brain hemispheric dominance in birdsong learning. *Proceedings of the National Academy of Sciences of the United States of America* 109:12782–12787.

Mootnick, A., and Fan, P.-P. 2011. A comparative study of crested gibbons *(Nomascus)*. *American Journal of Primatology* 26:983–988.

Mootnick, A., and Groves, C. P. 2005. New generic name for hoolock gibbons (Hylobatidae). *International Journal of Primatology* 26:983–988.

Morange, F. 1994. Handedness in two chimpanzees in captivity. In J. R. Anderson, J. J., Roeder, B. Thierry, and N. Herrenschmidt, eds., *Current Primatology*, Vol. 3: *Behavioral Neuroscience, Physiology, and Reproduction*, 61–67. Strasbourg: Université Louis Pasteur.

Morbeck, M. E. 1975. *Dryopithecus africanus* forelimb. *Journal of Human Evolution* 4:39–46.

———. 1976. Problems in reconstruction of fossil anatomy and locomotor behaviour: the *Dryopithecus* elbow complex. *Journal of Human Evolution* 5:223–233.

———. 1983. Miocene hominoid discoveries from Rudabánya: implications from the postcranial skeleton. In R. L. Ciochon and R. S. Corruccini, eds., *New Interpretations of Ape and Human Ancestry,* 369–404. New York: Plenum Press.

Morbeck, M. E., and Zihlman, A. L. 1989. Body size and proportions in chimpanzees, with special reference to *Pan troglodytes schweinfurthii* from Gombe National Park, Tanzania. *Primates* 30:369–382.

Morell, V. 1995. *Ancestral Passions.* New York: Simon & Schuster.

———. 2011. Why do parrots talk? Venzuelan site offers clue. *Science* 333:398–400.

Morgan, E. 1972. *The Descent of Woman.* New York: Stein and Day.

———. 1990. *The Scars of Evolution.* London: Souvenir Press.

Morgan, L. H. 1871. *Systems of Consanguinity and Affinity of the Human Family.* Washington, DC: Smithsonian Institution.

———. 1877. *Ancient Society or Researches in the Lines of Human Progress from Savagery through Barbarism to Civilization.* London: MacMillan.

Mori, Akio 1982. An ethological study on chimpanzees at the artificial feeding place in the Mahale Mountains, Tanzania—with special reference to the booming situation. *Primates* 23:45–65.

———. 1983. Comparison of the communicative vocalizations and behaviors of group ranging in eastern gorillas, chimpanzees and pygmy chimpanzees. *Primates* 24:486–500.

———.1984. An ethological study of pygmy chimpanzees in Wamba, Zaire: a comparison with chimpanzees. *Primates* 25:255–278.

Morin, P. A., Chambers, K. E., Boesch, C., and Vigilant, L. 2001. Quantitative polymerase chain reaction analysis of DNA from noninvasive samples for accurate microsatellite genotyping of wild chimpanzees *(Pan troglodytes verus). Molecular Ecology* 10:1835–1844.

Morin, P. A., and Goldberg, T. L. 2004. Determination of genealogical relationships from genetic data: a review of methods and applications. In B. Chapais and C. M. Berman, eds., *Kinship and Behavior in Primates,* 15–45. Oxford: Oxford University Press.

Morin, P. A., Moore, J. J., Chakraborty, R., Jin, L., Goodall, J., and Woodruff, D. S. 1994. Kin selection, social structure, gene flow, and the evolution of chimpanzees. *Science* 265:1193–1201.

Morino, L. 2011. Left-hand preference for a complex manual task in a population of wild siamangs *(Symphalangus syndactylus). International Journal of Primatology* 32:793–800.

Morris, D. 1967. *The Naked Ape: A Zoologist's Study of the Human Animal.* New York: McGraw-Hill.

Morris, R. D., and Hopkins, W. D. 1993. Perception of human chimeric faces by chimpanzees: evidence for a right hemisphere advantage. *Brain and Cognition* 21:111–122.

Morris, R. D., Hopkins, W. D., and Bolser-Gilmore, L. 1993. Assessment of hand preference in two language-trained chimpanzees *(Pan troglodytes):* a multimethod analysis. *Journal of Clinical and Experimental Neuropsychology* 15:487–502.

Morris, R. D., Hopkins, W. D., Bolser-Gilmore, L., and Washburn, D. A. 1993. Behavioral lateralization in language-trained chimpanzees. In J. P. Ward and W. D. Hopkins, eds., *Primate Laterality: Current Behavioral Evidence of Primate Asymmetries,* 207–233. New · York: Springer.

Morrogh-Bernard, H. C., Husson, S. J., Knott, C. B., Wich, S. A., van Schaik, C. P., van Noordwijk, M. A., Lackman-Ancrenez, J., Marshall, A. J., at al. 2009. Orangutan activity budgets and diet. In S. A. Wich, S. S. Utami Atmoko, T. Mitra Setia, and C. P. van Schaik, eds., *Orangutans: Geographic Variation in Behavioral Ecology and Conservation,* 119–123. Oxford: Oxford University Press.

Morrogh-Bernard, H. C., Morf, N. V., Chivers, D. J., and Krützen, M. 2011. Dispersal patterns of orang-utans (*Pongo* spp.) in a Bornean peat-swamp forest. *International Journal of Primatology* 32:362–376.

Morton, D. J. 1922. Evolution of the human foot. I. *American Journal of Physical Anthropology* 5:305–336.

———. 1924a. Evolution of the human foot. II. *American Journal of Physical Anthropology* 7:1–52.

———. 1924b. Evolution of the longitudinal arch of the human foot. *Journal of Bone and Joint Surgery* 6:56–90.

———. 1926. Evolution of man's erect posture. *Journal of Morphology and Physiology* 43:147–179.

———. 1935. *The Human Foot.* New York: Columbia University Press.

Morwood, M., Sutikna, T., and Roberts, R. 2005. World of the little people. *National Geographic* 207:2–15.

Morwood, M. J., O'Sullivan, P. B., Aziz, F., and Raza, A. 1998. Fission-track ages of stone tools and fossils on the east Indonesian island of flores. *Nature* 392:173–176.

Morwood, M. J., Soejono, R. P., Roberts, R. G., Sutikna, T., Turney, C. S. M., Westaway, K. E., Rink, W. J., Zhao, J.-X., Van den Bergh, G. D., Rokus Awe Due, et al. 2004. Archaeology and age of a new hominin from Flores in eastern Indonesia. *Nature* 431:1087–1091.

Moscovice, L. R., Issa, M. H., Petrzelkova, K. J., Keuler, N. S., Snowdon, C. T., and Huffman, M. A. 2007. Fruit availability, chimpanzee diet, and group patterns on Rubondo Island, Tanzania. *American Journal of Primatology* 69:487–502.

Mottl, M. 1957. Bericht über die neuen Menschenaffenfunde aus Österreich, von St. Stefen im Levanttal, Kärnten. *Carinthia II, Naturwissenschaftliche Beiträge zur Heimatkunde Kärntens, Mitteilungen des Naturwissenschaftlichen Vereins für Kärnten* 67:39–84.

Mounin, G. 1976. Language, communication, chimpanzees. *Current Anthropology* 17:1–21.

Moutsamboté, J.-M., Yumoto, T., Mitani, M., Nishihara, T., Suzuki, S., and Kuroda, S. 1994. Vegetation and list of plant species identified in the Nouabalé-Ndoki Forest, Congo. *Tropics* 3:277–293.

Moyà-Solà, S., Alba, D. M., Almécija, S., Casanovas-Vilar, I., Köhler, M., De Esteban-Trivigno, S., Robles, J. M., Galindo, J., and Fortuny, J. 2009. A unique Middle Miocene European hominoid and the origins of the great ape and human clade. *Proceedings of the National Academy of Sciences of the United States of America* 106:9601–9606.

Moyà-Solà, S., and Köhler, M. 1993. Recent discoveries of *Dryopithecus* shed new light on evolution of great apes. *Nature* 365:543–545.

———. 1995. New partial cranium of *Dryopithecus* Lartet, 1863 (Hominoidea, Primates) from the upper Miocene of Can Llobateres, Barcelona, Spain. *Journal of Human Evolution* 29:101–139.

———. 1996. A *Dryopithecus* skeleton and the origins of great-ape locomotion. *Nature* 379:156–159.

———. 1997. The phylogenetic relationships of *Oreopithecus bambolii* Gervais, 1872. *Comptes rendus des séances de l'Academie des sciences, Paris, série IIa* 306:141–148.

Moyà-Solà, S., Köhler, M., and Alba, D. M. 2001. *Egarapithecus narcisoi,* a new genus of Pliopithecidae (Primates, Catarrhini) from the Late Miocene of Spain. *American Journal of Physical Anthropology* 114:312–324.

Moyà-Solà, S., Köhler, M., Alba, D. M., Almécija, S., Casanovas-Vilar, I., Galindo, J. Robles, J. M., Cabrera, L., Garcéw, M., Almécija, S., and Beamud, E. 2009. First partial face and upper dentition of the Middle Miocene *Dryopithecus fontani* from Abocador de Can Mata (Vallès-Penedès Basin, Catalonia, NE Spain): taxonomic and phylogenetic implications. *American Journal of Physical Anthropology* 139:126–145.

Moyà-Solà, S., Köhler, M., Alba, D. M., Casanovas-Vilar, I., and Galindo, J. 2004. *Pierolapithecus catalaunicus,* a new Middle Miocene great ape from Spain. *Science* 306:1339–1344.

———. 2005. Response to comment on "*Pierolapithecus catalaunicus,* a new Middle Miocene great ape from Spain." *Science* 308:203.

Moyà-Solà, S., Köhler, M., and Rook, L. 1999. Evidence of hominid-like precision grip capability n the hand of the Miocene ape *Oreopithecus. Proceedings of the National Academy of Sciences of the United States of America* 96:313–317.

———. 2005. The *Oreopithecus bambolii* thumb: a strange case in hominoid evolution. *Journal of Human Evolution* 49:395–404.

Mubalamata, K. 1984. Will the pygmy chimpanzee be threatened with extinction as are the elephant and white rhinoceros in Zaire? In R. L. Susman, ed., *The Pygmy Chimpanzee: Evolutionary Biology and Behavior,* 415–419. New York: Plenum Press.

Mudakikwa, A. B., Cranfield, M. R., Sleeman, J. M., and Eilenberger, U. 2001. Clinical medicine, preventive health care and research on mountain gorillas in the Virunga Volcanoes region. In M. M. Robbins, P. Sicotte, and K. J. Steward, eds., *Mountain Gorillas: Three Decades of Research at Karisoke,* 342–360. Cambridge: Cambridge University Press.

Muehlenbein, M. P. 2005. Parasitological analyses of the male chimpanzees *(Pan troglydytes schweinfurthii)* at Ngogo, Kibale National Park, Uganda. *American Journal of Primatology* 65:167–179.

Muir, C. C., Galdikas, B. M. F., and Beckenbach, A. T. 1998. Is there sufficient evidence to elevate the orangutan of Borneo and Sumatra to separate species? *Journal of Molecular Evolution* 46:378–379.

Mukherjee, R. P. Caudhuri, S., and Murmu, A. 1988 Hoolock gibbons in Arunachal Pradesh, Northeastern India. *Primate Conservation* 9:121–123.

————. 1991–92. 1988 Hoolock gibbons *(Hylobates hoolock)* in Arunachal Pradesh, Northeastern India: the Lohit District. *Primate Conservation* 12–13:31–33.

Mulavwa, M., Furuichi, T., Yangozene, K., Yamba-Yamba, M., Motema-Salo, B., Idani, G., Ihobe, H., Hashimoto, C., Tashiro, Y., and Mwanz, N. 2008. Seasonal changes in fruit production and party size of bonobos at Wamba. In T. Furuich and J. Thompson, eds., *The Bonobos: Behavior, Ecology, and Conservation*, 121–134. New York: Spinger.

Mulcahy, N. J., and Call J. 2006. Apes save tools for future use. *Science* 312:1038–1040.

Muller, M. N. 2002. Agonistic relations among Kanyawara chimpanzees. In C. Boesch, G. Hohmann, and L. Marchant, eds., *Behavioural Diversity in Chimpanzees and Bonobos*, 112–124. Cambridge: Cambridge University Press.

Muller, M. N., Emery Thompson, M., Kahlenberg, S. M., and Wrangham, R. W. 2011. Sexual coercion by male chimpanzees shows that female choice may be more apparent than real. *Behavioral Ecology and Sociobiology* 65:921–933.

Muller, M. N., Emery Thompson, M., and Wrangham, R. W. 2006. Male chimpanzees prefer mating with old females. *Current Biology* 16:2234–2238.

Muller, M. N., Kahlenberg, S. M., Emery Thompson, M., and Wrangham, R. W. 2007. Male coercion and the costs of promiscuous mating for female chimpanzees. *Proceedings of the Royal Society B* 274:1009–1014.

Muller, M. N., and Mitani, J. C. 2005. Conflict and cooperation in chimpanzees. *Advances in the Study of Behavior* 35:275–331.

Muller, M. N., Mpongo, E., Stanford, C. B., and Boehm, C. 1995. A note on scavenging by wild chimpanzees. *Folia Primatologica* 65:43–47.

Muller, M. N., and Wrangham, R. W. 2001. The reproductive ecology of male hominoids. In P. T. Ellison, ed., *Reproductive Ecology and Human Evolution*, 397–427. New Brunswick, NJ: Aldine Transaction.

————. 2004a. Dominance, aggression and testosterone in wild chimpanzees: a test of the 'challenge hypothesis.' *Animal Behaviour* 67:113–123.

————. 2004b. Dominance, cortisol and stress in wild chimpanzees *(Pan troglodytes schweinfurthii)*. *Behavioral Ecology and Sociobiology* 55:332–340.

————. 2005. Testosterone and energetics in wild chimpanzees *(Pan troglodytes schweinfurthii)*. *American Journal of Primatology* 66:119–130.

Muncer, S. J. 1983a. "Conversations" with a chimpanzee. *Developmental Psychobiology* 16:1–11.

————. 1983b. Is Nim, the chimpanzee, problem-solving? *Perceptual and Motor Skills* 57:132–134.

Munthe, J., Dongol, B., Hutchison, J. H., Kean, W. F., Munthe, K., and West, R. M. 1983. New fossil discoveries from the Miocene of Nepal include a hominoid. *Nature* 303:331–333.

Murdoch, B. E. 2010. The cerebellum and language: historical perspective and review. *Cortex* 46:858–868.

Murdock, G. P. 1960. *Social Structure*. New York: Macmillan.

Murnyak, D. F. 1981. Censusing the gorillas in Kahuzi-Biega National Park. *Biological Conservation* 21:163–176.

Muroyama, Y. 1991. Chimpanzees' choices of prey between two sympatric species of *Macrotermes* in the Campo Animal Reserve, Cameroon. *Human Evolution* 6:143–151.

Murray, C. M., Eberly, L. E., and Pusey, A. E. 2006. Foraging strategies as a function of season and rank among wild female chimpanzees *(Pan troglodytes)*. *Behavioral Ecology* 17:1020–1028.

Murray, C. M., Mane, S. V., and Pusey, A. E. 2007. Dominance rank influences female space use in wild chimpanzees, *Pan troglodytes:* towards an ideal despotic society. *Animal Behaviour* 74:1795–1804.

Murray, C. M., Wroblewski, E., and Pusey, A. E. 2007. New case of intragroup infanticide in the chimpanzees of Gombe National Park. *International Journal of Primatology* 28:23–37.

Murray, M. P., Gore, D. E., and Clarkson, B. H. 1971. Walking patterns of patients with unilateral hip pain due to osteo-arthritis and avascular necrosis. *Journal of Bone and Joint Surgery* 53-A: 259–274.

Murray, P. F. 1975. The role of cheek pouches in cercopithecine monkey adaptive strategy. In R. H. Tuttle, ed., *Primate Functional Morphology and Evolution,* 151–194. The Hague: Mouton.

Murrill, R., and Wallace, D. T. 1971. A method for making an endocranial cast through the foramen magnum of an intact skull. *American Journal of Physical Anthropology* 34:441–446.

Musgrave, J. H. 1969. A comparative study of the hand bones of the Neanderthal man. *Human Biology* 41:587–588.

———. 1971. How dextrous was Neanderthal man? Nature 233:538–541.

———. 1973. The phalanges of Neanderthal and Upper Paleolithic hands. In M. H. Day, ed., *Human Evolution,* 59–85. London: Taylor & Francis.

———. 1977. The Neanderthals from Krapina, northern Yugoslavia: an inventory of the hand bones. *Zeitschrift für Morphologie und Anthropologie* 68:150–171.

Musiba, C., Tuttle, R. H., Hallgrimsson, B., and Webb, D. M. 1997. Swift and sure-footed on the savanna: a study of Hadza gaits and feet in northern Tanzania. *American Journal of Human Biology* 9:303–321.

Muzaffar, S. B., Islam, M. A., Feeroz, M. M., Kabir, M., Begum, S., Mahmud, M. S., Chakma, S., and Hasan, M. K. 2007. Habitat characteristics of the endangered hoolock gibbons of Bangladesh: the role of plant species richness. *Biotropica* 39:539–545.

Mwanza, N., Maruhashi, T. Yumoto, T., and Yamagiwa, J. 1988. Conservation of eastern lowland gorillas in the Masisi Region, Zaire. *Primate Conservation* 9:111–113.

Mwanza, N., Yamagiwa, J., Yumoto, T., and Maruhashi, T. 1992. Distribution and range utilization of eastern lowland gorillas. In N. Itoigawa, Y. Sugiyama, G. P. Sackett, and R. K. R. Thompson, eds., *Topics in Primatology,* Vol. 2: *Behavior, Ecology, Conservation,* 283–300. Tokyo: University of Tokyo Press.

Myatt, J. P., and Thorpe, S. K. S. 2011. Postural strategies employed by orangutans *(Pongo abelii)* during feeding in the terminal branch niche. *American Journal of Physical Anthropology* 146:73–82.

Nadler, R. D. 1977. Sexual behavior of captive orangutans. *Archives of Sexual Behavior* 6:457–475.

————. 1989. Sexual initiation in wild mountain gorillas. *International Journal of Primatology* 10:81–92.

Nagatoshi, K. R. 1984. *Systematics and Paleoecology of European Tertiary Apes.* PhD diss., University of Chicago.

————. 1987. Miocene hominoid environments of Europe and Turkey. *Palaeogeography, Palaeoclimatology, Palaeoeololgy* 61:145–154.

————. 1990. Molar enamel thickness in European Miocene and extant Hominoidea. *International Journal of Primatology* 11:283–295.

Nakamichi, M. 1998. Stick throwing by gorillas *(Gorilla gorilla gorilla)* at the San Diego Wild Animal Park. *Folia Primatologica* 69:291–295.

————. 1999. Spontaneous use of sticks as tools by captive gorillas *(Gorilla gorilla gorilla).* *Primates* 40:487–498.

Nakamura, M. 1997. First observed case of chimpanzee predation on yellow baboons *(Papio cynocephalus)* at the Mahale Mountains National Park. *Pan African News* 4:9–11.

Nakamura, M., and Nishida, T. 2006. Subtle behavioral variation in wild chimpanzees, with special reference to Imanishi's concept of *kaluchua. Primates* 47:35–42.

————. 2012a. Long-term studies of chimpanzees at Mahali Mountains National Park, Tanzania. In Kappeler, P. M., and Watts, D. P., eds., *Long-term Field Studies of Primates,* 339–356. Heidelberg: Springer.

————. 2012b. Ontogeny of a social custom in wild chimpanzees: age changes in grooming hand-clasp at Mahale. *American Journal of Primatology* 75:186–196.

Nakaya, H. 1989. Late Miocene mammalian interchange of subsaharan Africa palaeoenvironments of large hominoids evolution. In G. Giacobini, ed., *Hominidae,* 67–70. Milan: Jaca Book.

Nakatsukasa, M., Kunimatsu, Y., Nakano, Y., and Ishida, H. 1999. Postcranial features of the large Miocene hominoid from Nachola in comparison with *Kenyapithecus africanus* and *K. wickeri.* In H. Ishida, ed., *Abstracts of the International Symposium "Evolution of Middle-to-Late Miocene Hominoids in Africa," July 11–13, 1999.* Abstract presented at the meeting, Takaragaike, Kyoto, Japan.

————. 2007. Postcranial bones of infant *Nacholopithecus:* ontogeny and positional behavioral adaptation. *Anthropological Science* 115:201–213.

Nakatsukasa, M., Kunimatsu, Y., Nakano, Y., Takano, T., and Ishida, H. 2003. Comparative and functional anatomy of phalanges in *Natholapithecus kerioi,* a Middle Miocene hominoid from northern Kenya. *Primates* 44:371–412.

Nakatsukasa, M., Nakano, Y, Kunimatsu, Y., Ogihara, N., and Tuttle, R. H. 2005. Hidemi Ishida: 40 years of footprints in Japanese primatology and paleoanthropology. In H. Ishida, R. H. Tuttle, M. Pickford, M. Nakatsukasa, and N. Ogihara, eds., *Human Origins and Environmental Backgrounds,* 1–14. New York: Springer.

Nakatsukasa, M., Shimizu, D., Nakano, Y., and Ishida, H. 1996. Three-dimensional morphology of the sigmoid notch of the ulna in *Kenyapithecus* and *Proconsul. African Study Monographs* 24 (Suppl.): 57–71.

Nakatsukasa, M., Tsujikawa, H., Shimizu, D., Takano, T., Kunimatsu, Y., Nakano, Y., and Ishida, H. 2003b. Definitive evidence for tail loss in *Nacholapithecus,* an East African Miocene hominoid. *Journal of Human Evolution* 45:179–186.

Nakatsukasa, M., Yamanaka, A., Kunimatsu, Y., Shimizu, D., and Ishida, H. 1998. A newly discovered *Kenyapithecus* skeleton and its implications for the evolution of positional behavior in Miocene East African hominoids. *Journal of Human Evolution* 34:657–664.

Nakatsukasa, M., Ward, C. V., Walker, A., Teaford, M. F., Kunimatsu, Y., and Ogihara, N. 2004. Tail loss in *Proconsul heseloni*. *Journal of Human Evolution* 46:777–784.

Nakurunungi, J. B., and Stanford, C. B. 2006. Preliminary GIS analysis of range use by sympatric mountain gorillas and chimpanzees in Bwindi Impenetrable National Park, Uganda. In N. E. Newton-Fisher, H. Notman, J. D. Paterson, and V. Reynolds, eds., *Primates of Western Uganda*, 193–205. New York: Springer.

Naour, P. 2009. E. O. Wilson and B. F. Skinner. *A Dialogue between Sociobiology and Radical Behaviorism*. New York: Springer.

Napier, J. R. 1952. The attachments and function of the abductor pollicis brevis. *Journal of Anatomy* 86:335–341.

———. 1955. The form and function of the carpo-metacarpal joint of the thumb. *Journal of Anatomy* 89:362–369.

———. 1956. The prehensile movements of the human hand. *Journal of Bone and Joint Surgery* 38B:902–913.

———. 1959. Fossil metacarpals from Swartkrans. *Fossil Mammals of Africa*, no. 17: 1–18. London: British Museum (Natural History).

———. 1960. Studies on the hands of living primates. *Proceedings of the Zoological Society of London* 134:647–657.

———. 1961a. Hands and handles. *New Scientist* 9:797–799.

———. 1961b. Prehensility and opposability in the hands of primates. *Symposia of the Zoological Society of London*, no. 5: 115–132.

———. 1962a. The evolution of the hand. *Scientific American* 207:56–62.

———. 1962b. Fossil hand bones from Olduvai Gorge. *Nature* 196:409–411.

———. 1965. Evolution of the human hand. *Proceedings of the Royal Institution of Great Britain* 40:544–557.

———. 1966. Functional aspects of the anatomy of the hand. In C. Rob and R. Smith, eds., *Clinical Surgery*, 1–31. London: Butterworths.

———. 1980. *Hands*. New York: Pantheon Books.

———. 1993. *Hands* (revised by R. H. Tuttle). Princeton, NJ: Princeton University Press.

Napier, J. R., and Davis, P. R. 1959. The forelimb skeleton and associated remains of *Proconsul africanus*. *Fossil Mammals of Africa*, no. 16. London: British Museum (Natural History).

Naylor, G. J. P., and Brown, W. M. 1998. Amphioxus mitochondrial DNA, chordate phylogeny, and the limits of inference based on comparison of sequences. *Systematic Biology* 47:61–76.

Nei, M. 1978. The theory of genetic distance and evolution of human races. *Japanese Journal of Human Genetics* 23:341–369.

———. 1986. Stochastic errors in DNA evolution and molecular phylogeny. In H. Gershowitz, D. L. Rucknagel, and R. E. Tashian, eds., *Evolutionary Perspectives and the New Genetics*, 133–147. New York: Alan R. Liss.

Nei, M., and Roychoudhury, A. K. 1982. Genetic relationship and evolution of human races. M. K. Hecht, B. Wallace and G. T. Prance, eds., *Evolutionary Biology*, Vol. 14, 1–59. New York: Plenum Press.

Nei, M., Stephens, J. C., and Saitou, N. 1985. Methods for computing the standard errors of branching points in an evolutionary tree and their application to molecular data from humans and apes. *Molecular Biology and Evolution* 2:66–85.

Nei, M., and Tajima, F. 1985. Evolutionary change of restriction cleavage sites and phylogenetic inference for man and apes. *Molecular Biology and Evolution* 2:189–205.

Nei, M., Zhang, J., and Yokoyama, S. 1997. Color vision of ancestral organisms and higher primates. *Molecular Biology and Evolution* 14:611–618.

Nelissen, K., Luppino, G., Vanduffel, W., Rizzolatti, G., and Orban, G. A. 2005. Observing others: multiple action representation in the frontal lobe. *Science* 310:332–336.

Nelson, S. V. 2003. *The Extinction of Sivapithecus.* Boston: Brill Academic.

New, J., Cosmides, L., and Tooby, J. 2007. Category-specific attention for animals reflects ancestral priorities, not expertise. *Proceedings of the National Academy of Sciences of the United States of America* 104:16598–16603.

Newman, R. W. 2002. Why man is such a sweaty and thirsty naked animal: a speculative review. *Human Biology* 42:12–27.

Newmeyer, F. J. 1991. Functional explanation in linguistics and the origin of language. *Language and Communication* 11:3–28.

Newton, Y., Hasegawa, T., and Nishida, T. 1984. Chimpanzee predation in the Mahale Mountains from August 1979 to May 1982. *International Journal of Primatology* 5:213–233.

Newton-Fisher, N. E. 1999a. The diet of chimpanzees in the Budongo Forest Reserve, Uganda. *African Journal of Ecology* 37:344–354.

———. 1999b. Termite eating and food sharing by male chimpanzees in the Budongo Forest, Uganda. *African Journal of Ecology* 37:369–371.

———. 1999c. Infant killers of Budongo. *Folia Primatologica* 70:167–169.

———. 1999d. Association by male chimpanzees: a social tactic? *Behaviour* 136:705–730.

———. 2002. Relationships of male chimpanzees in the Budongo Forest, Uganda. In C. Boesch, G. Hohmann, and L. F. Marchant, eds., *Behavioural Diversity in Chimpanzees and Bonobos,* 125–137. Cambridge: Cambridge University Press.

Newton-Fisher, N. E., Notman, H., and Reynolds, V. 2002. Hunting of mammalian prey by Budongo Forest chimpanzees. *Folia Primatolgica* 73:281–283.

Newton-Fisher, N. E., Reynolds, V., and Plumptre, A. J. 2002. Food supply and chimpanzee *(Pan troglodytes schweinfurthii)* party size in the Budongo Forest Reserve, Uganda. *International Journal of Primatology* 21:613–628.

N'guessan, A. K., Ortmann, S., and Boesch, C. 2009. Daily energy balance and protein gain among *Pan troglodytes verus* in the Taï National Park, Côte d'Ivoire. *International Journal of Primatology* 30:481–496.

Ni, X., Hu, Y., Wang, Y., and Li, C. 2005. A clue to the Asian origins of euprimates. *Anthropological Science* 113:3–9.

Nicholson, A. J. 1954. An outline of the dynamics of animal populations. *Australian Journal of Zoology* 2:9–65.

Nicol, A. C., and Paul, J. P. 1988. Biomechanics. In B. Helal and D. Wilson, eds., *The Foot*, Vol. 1, 75–86. Edinburgh: Churchill Livingstone.

Nicolson, N. A. 1977. A comparison of early behavioral development in wild and captive chimpanzees. In S. Chevalier-Skolnikoff and F. E. Poirier, eds., *Primate Bio-social Development: Biological, Social, and Ecological Determinants*, 529–560. New York: Garland Publishing.

Niemitz, C., and Kok, D. 1976. Observations on the vocalization of a captive infant Orang utan *(Pongo pygmaeus). Sarawak Museum Journal* 24:237–250.

Niewoehner, W. A., Weaver, A. H., and Trinkaus, E. 1997. Neandertal capitate-metacarpal articular morphology. *American Journal of Physical Anthropology* 103:219–233.

Nimchinsky, E. A., Gilissen, E., Allman, J. M., Perl, D. P. Erwin, J. M., and Hof, P. R. 1999. A neuronal morphologic type unique to humans and great apes. *Proceedings of the National Academy of Sciences of the United States of America* 96:5268–5273.

Nishida, T. 1968. The social group of wild chimpanzees in the Mahali Mountains. *Primates* 9:167–224.

———. 1970. Social behavior and relationship among wild chimpanzees of the Mahali Mountains. *Primates* 11:47–87.

———. 1972. Preliminary information on the pygmy chimpanzees *(Pan paniscus)* of the Congo Basin. *Primates* 13:415–425.

———. 1973. The ant-gathering behaviour by the use of tools among wild chimpanzees of the Mahali Mountains. *Journal of Human Evolution* 2:357–370.

———. 1979. The social structure of chimpanzees of the Mahale Mountains. In D. A. Hamburg and E. R. McCown, eds., *The Great Apes*, 72–121. Menlo Park, CA: Benjamin/Cummings.

———. 1983a. Alpha status and agonistic alliance in wild chimpanzees *(Pan troglodytes). Primates* 24:318–336.

———. 1983b. Alloparental behavior in wild chimpanzees of the Mahale Mountains, Tanzania. *Folia Primatologica* 41:1–33.

———. 1990. A quarter century of research in the Mahale mountains: an overview. In T. Nishida, ed., *The Chimpanzees of the Mahale Mountains*, 3–35. Tokyo: University of Tokyo Press.

———. 1994. Review of recent findings on Mahale chimpanzees. Implications and future directions. In R. W. Wrangham, W. C. McGrew, F. B. M. de Waal, and P. G. Heltne, eds. *Chimpanzee Cultures*, 373–396. Cambridge, MA: Harvard University Press.

———. 1996. The death of Ntologi, the unparalleled leader of M group. *Pan Africa News* 3:1–4.

———. 2008. Forty years of chimpanzee research at Mahale: traditions, changes and future. Plenary Lecture, *The International Primatological Society, XXII Congress, Edinburgh, UK,* August 7.

———. 2012. *Chimpanzees of the Lakeshore: Natural History and Culture at Mahale.* Cambridge: Cambridge University Press.

Nishida, T., Corp, N., Hamai, M., Hasegawa, T., Hiraiwa- Hasegawa, M., Hosaka, K., Hunt, K. D., Itoh, M., Kawanaka, K., Matsumoto-Oda, A., et al. 2003. Demography, female life history, and reproductive profiles among the chimpanzees of Mahale. *American Journal of Primatology* 59:99–121.

Nishida, T., Hasegawa, T., Hayaki, H., Takahata, Y., and Uehara, S. 1992. Meat-sharing as a coalition strategy by a alpha male chimpanzee. In T. Nishida, W. C. McGrew, P. Marler, M. Pickford, and F. B. M. de Waal, eds., *Topics in Primatology,* Vol. 1: *Human Origins,* 159–174. Tokyo: University of Tokyo Press.

Nishida, T., and Hiraiwa, M. 1982. Natural history of a tool-using behavior by wild chimpanzees in feeding upon wood-boring ants. *Journal of Human Evolution* 11:73–99.

Nishida, T., and Hiraiwa-Hasegawa, M. 1984. Behavior of an adult male in one-male unit-group of chimpanzees in the Mahale Mountains, Tanzania. *International Journal of Primatology* 5:367.

———. 1985. Responses to a stranger mother-son pair in the wild chimpanzees: a case report. *Primates* 26:1–13.

———. 1987. Chimpanzees and bonobos. Cooperative relationsips among males. In B. B. Smuts, D. L. Cheney, R. M. Seyfarth, R. W. Wrangham, and T. T. Struhsaker, eds., *Primate Societies,* 165–177. Chicago: University of Chicago Press.

Nishida, T., Hiraiwa-Hasegawa, M., Hasegawa, T., Takahata, Y. 1985. Group extinction and female transfer in wild chimpanzees in the Mahale National Park, Tanzania. *Zeitschrift für Tierpsychologie* 67:284–301.

Nishida, T., Kano, T., Goodall, J., McGrew, W. C., and Nakamura, M. 1990. Ethogram and ethnography of Mahale chimpanzees. *Anthropological Science* 107:141–188.

Nishida, T., and Kawanaka, K. 1972. Inter-unit-group relationships among wild chimpanzees of the Mahale Mountains. *Kyoto University African Studies* 7:131–169.

Nishida, T., and Nakamura, M. 1993. Chimpanzee tool use to clear a blocked nasal passage. *Folia Primatologica* 61:218–220.

Nishida, T., Takasaki, H., and Takahata, Y. 1990. Demography and reproductive profiles. In T. Nishida, ed., *The Chimpanzees of the Mahale Mountains,* 63–97. Tokyo: Univesity of Tokyo Press.

Nishida, T., and Turner, L. A. 1996. Food transfer between mother and infant chimpanzees of the Mahale Mountains National Park, Tanzania. *International Journal of Primatology* 17:947–968.

Nishida, T., and Uehara, S. 1983. Natural diet of chimpanzees *(Pan troglodytes schweinfurthii):* long-term record from the Malhale Mountains, Tanzania. *African Study Monographs* 3:109–130.

Nishida, T., Uehara, S., and Nyundo, R. 1979. Predatory behavior among wild chimpanzees of the Mahale Mountains. *Primates* 20:1–10.

Nishida, T., Uehara, S., and Ramadhani, N. 1983. Predatory behavior among wild chimpanzees of the Mahale Mountains. *Primates* 20:1–10.

Nishida, T., Wrangham, W., Goodall, J., and Uehara, S. 1983. Local differences in plant-feeding habits of chimpanzees between the Mahale Mountains and Gombe National Park, Tanzania. *Journal of Human Evolution* 12:467–480.

Nishihara, T. 1992. A preliminary report on the feeding habits of western lowland gorillas *(Gorilla gorilla gorilla)* in the Ndoki Forest, Northern Congo. In N. Itoigawa, Y. Sugiyama, G. P. Sackett, and R. K. R. Thompson, eds. *Topics in Primatology,* Vol. 2: *Behavior, Ecology and Conservation,* 225–240. Tokyo: University of Tokyo Press.

———. 1994. Population density and group organization of gorillas *(Gorilla gorilla gorilla)* in the Nouabalé-Ndoki National park, congo. *African Studies* 44:29–45.

———. 1995. Feeding ecology of western lowland gorillas in the Nouabalé-Ndoki National Park, Congo. *Primates* 36:151–168.

Nishimura, T. 2005. Developmental changes in the shape of the supralaryngeal vocal tract in chimpanzees. *American Journal of Physical Anthropology* 126:193–204.

———. 2006. Descent of the larynx in chimpanzees: mosaic and multiple-step evolution of the foundations for human speech. In T. Matsuzawa, M. Tomonaga, and M. Tanaka, eds., *Cognitive Development in Chimpanzees,* 75–95. Tokyo: Springer.

Nishimura, T., Mikami, A., Suzuki, J., and Matsuzawa, T. 2003. Descent of the larynx in chimpanzee infants. *Proceedings of the National Academy of Sciences of the United States of America* 100:6930–6933.

Nishimura, T., Oishi, T., Suzuki, J., Matsuda, K., and Takahashi, T. 2008. Development of the supralaryngeal vocal tract in Japanese macaques: implications for the evolution of the descent of the larynx. *American Journal of Physical Anthropology* 135:182–194.

Nishimura, T., Okayasu, N., Hamada, Y., and Yamagiwa, J. 2003. A case report of a novel type of stick use by wild chimpanzees. *Primates* 44:199–201.

Nissen, H. W. 1931. A field study of the chimpanzee. *Comparative Psychology Monographs* 8:1–122.

Nkurunguni, J. B., Ganas, J., Robbins, M. M., and Stanford, C. B. 2004. A comparison of two mountain gorilla habitats in Bwindi Impenetrable National Park, Uganda. *African Journal of Ecology* 42:289–297.

Norconk, M. A., and Conklin-Brittain, N. L. 2004. Variation on frugivory: the diet of Venezuelan white-faced sakis. *International Journal of Primatology* 25:1–26.

Norikoshi, K. 1982. One observed case of cannibalism among wild chimpanzees of the Mahale Mountain. *Primates* 23:66–74.

———. 1983. Prevalent phenomenon of predation observed among wild chimpanzees of the Mahale Mountains. *Journal of the Anthropological Society, Nippon,* 91:475–480.

Noss, A. J., and Hewlett, B. S. 2001. The contexts of female hunting in Central Africa. *American Anthropologist* 103:1024–1040.

Notman, H., and Rendall, D. 2005. Contextual variation in chimpanzee pant hoots and its implication for referential communication. *Animal Behaviour* 70:177–190.

Novacek, M. J., and Norell, M. A. 1982. Fossils, phylogeny, and taxonomic rates of evolution. *Systematic Zoology* 31:366–375.

Nozawa, M. Kawahara, Y., and Nei, M. 2007. Genomic drift and copy number variation of sensory receptor genes in humans. *Proceedings of the National Academy of Sciences of the United States of America* 104:20421–20426.

Nsubuga, A. M., Robbins, M. M., Boesch, C., and Vigilant, L. 2008. Patterns of paternity and group fission in wild multimale mountain gorilla groups. *American Journal of Physical Anthropology* 135:263–274.

Nunn, C. L., and Altizer, S. 2006. *Infectious Diseases in Primates: Behavior, Ecology and Evolution.* New York: Oxford University Press.

Nuttal, G. H. 1904. *Blood Immunity and Blood Relationship.* Cambridge: Cambridge University Press.

Oakley, K. P. 1959. *Man the Tool-maker.* Chicago: University of Chicago Press.

Oakley, K. P., Andrews, P., Keeley, L. H., and Clark, J. D. 1977. A reappraisal of the Clacton spear point. *Proceedings of the Prehistoric Society* 43:13–30.

Oates, J. F. African primate conservation: general needs and specific priorities. In K. Benirschke, ed., *Primates: The Road to Self-Sustaining Populations,* 21–29. New York: Springer-Verlag.

Oates, J. F., McFarland, K. L., Groves, J. L., Bergl, R. A., Linder, J. M., and Disotell, T. R. 2003. The Cross River gorilla: natural history and status of a neglected and critically endangered subspecies. In A. B. Taylor and M. L. Goldsmith, eds., *Gorilla Biology: A Multidisciplinary Perspective,* 472–497. Cambridge: Cambridge University Press.

Oates, J. F., Sunderland-Groves, J., Bergl, R., Dunn, A. Nicholas, A., Takang, E., Omeni, F., Imong, I., Fotso, R., Nkembi, L., and Williamson, L. 2007. Regional action plan for the conservation of the Cross River gorilla *(Gorilla gorilla diehli).* Arlington VA: IUCN/SSC Primate Specialist Group and Conservation International.

Obendorf, P. J., Oxnard, C. E., and Kefford, B. J. 2008. Are the small human-like fossils found on Flores human endemic cretins? *Proceedings of the Royal Society B* 275:1287–1296.

O'Brien, S. J., Menotti-Raymond, M., Murphy, W. J., Nash, W. G., Wienberg, J., Stanyon, R., Copeland, N. G., Jenkins, N. A., Womack, J. E., and Graves, J. A. M. 1999. The promise of comparative genomics in mammals. *Science* 286:458–481.

O'Brien, T. G., and Kinnaird, M. F. 2011. Demography of agile gibbons *(Hylobates agilis)* in a lowland tropical rain forest of southern Sumatra, Indonesia: problems in paradise. *International Journal of Primatology* 32:1203–1217.

O'Brien, T. G., Kinnaird, M. F., Dierenfeld, E. S., Conklin-Brittain, N. L., Wrangham, R. W., and Silver, S. C. 1998. What's so special about figs? *Nature* 392:668.

Ochman, H., and Wilson, A. C. 1987. Evolution in bacteria: evidence for a universal substitution rate in cellular genomes. *Journal of Molecular Evolution* 26:74–86.

O'Connell, J. F., Hawkes, K., and Blurton Jones, N. G. 1999. Grandmothering and the evolution of *Homo erectus. Journal of Human Evolution* 36:461–485.

Odling-Smee, J., and Laland, K. N. 2009. Cultural niche construction: evolution's cradle of language. In T. Botha and C. Knight, eds., *The Prehistory of Language,* 99–121. Oxford: Oxford University Press.

Oelze, V. M., Fuller, B. T., Richards, M. P., Fruth, B., Surbeck, M., J.-J. Hublin, and G. Hohmann. 2011. Exploring the contribution and significance of animal protein in the diet of bonobos by stable isotope ration analysis of hair. *Proceedings of the National Academy of Sciences of the United States of America* 108:9792–9797.

Ogawa, H., Idani, G., Moore, J., Piltea, L., and Hernandez-Aguilar, A. 2007. Sleeping parties and nest distribution of chimpanzees in the savanna woodland, Ugalla, Tanzania. *International Journal of Primatology* 28:1397–1412.

O'Hara, S. J., and Lee, P. C. 2006. High frequency of postcoital penis cleaning in Budongo chimpanzees. *Folia Primatologica* 77:353–358.

Ohigashi, H., Huffman, M. A., Izutsu, D., Kishimizu, K., Kawanaka, M., Sugiyama, Y., Kirby, G. C., Warhurst, D. C., Allen, D., Wright, C. W., et al. 1994. Toward the chemical ecology of medicinal plant use in chimpanzees: the case of *Vernonia aygda-lina.*, a plant used by wild chimpanzees possibly for parasite-related diseases. *Journal of Chemical Ecology* 20:541–553.

O'Higgins, P., and Dryden, I. L. 1993. Sexual dimorphism in hominoids: further studies of craniofacial shape differences in *Pan, Gorilla* and *Pongo. Journal of Human Evolution* 24:183–205.

O'Higgins, P., and Elton, S. 2007. Walking on trees. *Science* 316:1292–1294.

O'Higgins, P., Moore, W. J., Johnson, D. R., and McAndrew, T. J. 1990. Patterns of cranial sexual dimorphism in certain groups of extant hominoids. *Journal of Zoology, London* 222:399–420.

Öhman, A. 2007. Has evolution primed humans to "beware the beast"? *Proceedings of the National Academy of Sciences of the United States of America* 104:16396–16397.

Oko, R. A. 1992. The present situation of conservation for wild gorillas in the Congo. In N. Itoigawa, Y. Sugiyama, G. P. Sackett, and R. K. R. Thompson, eds., *Topics in Primatology*, Vol. 2: *Behavior, Ecology, Conservation*, 241–243. Tokyo: University of Tokyo Press.

Oldham, M. C. Horvath, S., and Geschwind, D. H. 2006. Conservation and evolution of gene expression networks in human and chimpanzee brains. *Proceedings of the National Academy of Sciences of the United States of America* 103:17973–17978.

O'Leary, M. A., Bloch, J. I., Flynn, J. J., Gaudin, T. J., Giallombardo, A., Giannini, N. P., Goldberg, S. L., et al. 2013. The placental mammal ancestor and the post-K-Pg radiation of placentals. *Science* 339:662–667.

Olejniczak, C. 2001. The 21st century gorilla: progress or perish? In *The Apes: Challenges for the 21st Century, Conference Proceedings, May 10–12, 2000*, 36–42. Brookfield IL: Brookfield Zoo.

Oller, D. K., Buder, E. H., Ramsdell, H. L., Warlaumont, A. S., Chorna, L., and Bakeman, R. 2013. Functional flexibility of infant vocalization and the emergence of language. *Proceedings of the National Academy of Sciences of the United States of America* 110:6318–6323.

Olson, D. A., Ellis, J. A., and Nadler, R. D. 1990. Hand preferences in captive gorillas, orang-Utans and gibbons. *American Journal of Primatology* 20:83–94.

Olson, M. V., and Varki, A. 2003. Sequencing the chimpanzee genome: insights into human evolution and disease. *Nature Reviews Genetics* 4:20–28.

———. 2004. The chimpanzee genome—a bittersweet celebration. *Science* 305:191–192.

Olson, S. L., and Rasmussen, D. T. 1986. Paleoenvironment of the earliest hominoids: new evidence from the Oligocene avifauna of Egypt. *Science* 233:1202–1204.

Olson, T. R. 1981. Basicranial morphology of the extant hominoids and Pliocene hominids: the new material from the Hadar Formation, Ethiopia, and its significance in early human evolution and taxonomy. In C. B. Stringer, ed., *Aspects of Human Evolution*, 99–128. London: Taylor & Francis.

———. 1985a. Cranial morphology and systematics of the Hadar Formation hominids and *"Australopithecus" africanus*. In E. Delson, ed., *Ancestors: The Hard Evidence*, 102–119. New York: Alan R. Liss.

———. 1985b. Taxonomic affinities of the immature hominid crania from Hadar and Taung. *Nature* 316:539–540.

O'Malley, R. C. Two observations of galago predation by the Kasakela chimpanzees of Gombe Stream National Park, Tanzania. *Pan African News* 17:17–19.

O'Malley, R. C., and Power, M. L. 2012. Nutritional composition of actual and potential insect prey for the Kasekela chimpanzees of Gombe National Park, Tanzania. *American Journal of Physical Anthropology* 149:493–503.

Omari, I., Hart, J. A., Butynski, M., Birhashirwa, N. R., Upoki, A., M'Keyo, Y., Bengana, F., Bashonga, M., and Bagurubumwe, N. 1999. The Itombwe Massif, Democratic Republic of Congo: biological surveys and conservation, with an emphasis on Grauer's gorilla and birds endemic to the Albertine Rift. *Oryx* 33:301–322.

Omasombo, V., Bokelo, D., and Dupain, J. 2005. Current status of bonobos and other large mammals in the proposed forest reserve of Lomako-Yokokala, Equateur Province, Democratic Republic of Congo. *Pan Africa News* 12:14–17.

O'Neil, P. B., and Doolittle, R. F. 1973. Mammalian phylogeny based on fibrinopeptide amino acid sequences. *Systematic Zoology* 22:590–595.

Ono, A. Man the tool-maker and Pan the tool-maker. *Primate Research* 11:239–246.

O'Rourke, D. H. 2003. Anthropological genetics in the genomic era: a look back and ahead. *American Anthropologist* 105:101–109.

Ortmann, S., Bradley, B. J., Stolter, C., and Ganzhorn, J. U. 2006. Estimating the quality and composition of wild animal diets—a critical survey of methods. In G. Hohmann, M. M. Robbins and C. Boesch, eds., *Feeding Ecology in Apes and Other Primates*, 397–420. Cambridge: Cambridge University Press.

Ortner, S. B. 1984. Theory in anthropology since the sixties. *Comparative Studies in Society and History* 26:126–166.

Osborn, H. F. 1921. The Dawn Man of Piltdown, Sussex. *Natural History* 21:577–590.

———. 1927. Recent discoveries relating to the origin and antiquity of man. *Science* 65:481–488.

———. 1929. Is the ape-man a myth? *Human Biology* 1:4–9.

Osborn, R. M. 1963. Observations on the behaviour of the mountain gorilla. *Symposia of the Zoological Society, London* 10:29–37.

Osmond, J. K. 1990. Th^{230}/U^{234} dating of *Veeretesszölös*. In *Veeretesszölös: Site Man and Culture*, M. Kretzoi and V. T. Dobosi, eds., p. 545. Budapest: Akadémiai Kiadó.

Osorio, D., and Vorobyev, M. 1996. Colour vision as an adaptation to frugivory in primates. *Proceedings of the Royal Society, London B* 263:593–599.

Otari, E., and Gilchrist, J. S. 2006. Why chimpanzee *(Pan troglodytes schweinfurthii)* mothers are less gregarious than nonmothers and males: the infant safety hypothesis. *Behavioral Ecology and Sociobiology* 59:561–570.

Ottoni, E. B., and Mannu, M. 2001. Semifree-ranging tufted capuchins *(Cebus apella)* spontaneously use tools to crack open nuts. *International Journal of Primatology* 22:347–358.

Owen, R. 1859. *On the Classification and Geographic Distribution of the Mammalia*. London: Parker.

Owren, M. J., Amos, R. T., and Rendall, D. 2010. Two organizing principles of vocal production: implications for nonhuman and human primates. *American Journal of Primatology* 73:530–544.

Owren, M. J., and Rendall, D. 2001. Sound on the rebound: bringing form and function back to the forefront in understanding nonhuman primate vocal signaling. *Evolutionary Anthropology* 10:58–71.

Oxnard, C. E. 1968. A note on the fragmentary Sterkfontein scapula. *American Journal of Physical Anthropology* 28:213–217.

———. 1975a. The place of australopithecines in human evolution: grounds for doubt? *Nature* 258:389–395.

———. 1975b. *Uniqueness and Diversity in Human Evolution: Morphometric Studies of Australopithecines.* Chicago: University of Chicago Press.

———. 1983a. Evolutionary radiations in humans and great apes. In eds. P. K. Seth and S. Seth, *Perspectives in Primate Biology,* Vol. 2, 1–31. New Delhi: Today and Tomorrow's Printers & Publishers.

———. 1983b. Sexual dimorphisms in the overall proportions of primates. *American Journal of Primatology* 4:1–22.

Oyen, O. J. 1979. Tool-use in free-ranging baboons of Nairobi National Park. *Primates* 20:595–597.

Ozansoy, F. 1957. Faunes de mammifères du Tertiare de Turquie et leur révision stratigraphiques. *Bulletin of the Mineral Research and Exploration Institute of Turkey* (Foreign Edition) 49:29–48.

———. 1965. Étude des gisements continentaux et des mammifères du Cénozoïque de Turquie. *Mémoires de la Société géologique de France (nouvelle série)* 44:1–92.

Packer, C., and Ruttan, L. 1988. The evolution of cooperative hunting. *American Naturalist* 132:159–161.

Page, S. L., and Goodman, M. 2001. Catarrhine phylogeny: noncoding DNA evidence for a diphyletic origin of the mangabeys and for a human-chimpanzee clade. *Molecular Phylogenetics and Evolution* 18:14–25.

Paget, R. A. 1930. *Human Speech: Some Observation, Experiments, and Conclusions as to the Nature, Origin, Purpose and Possible Improvement of Human Speech.* London: Kegan Paul, Trence, Trubner.

———. 1944. The origin of language. *Science* 99:14–15.

Palagi, E. 2006. Social play in bonobos *(Pan paniscus)* and chimpanzees *(Pan troglodytes):* implications for natural social systems and interindividual relationships. *American Journal of Physical Anthropology* 129:418–426.

Palagi, E., Cordoni, G., and Borgognini Tarli, S. 2006. Possible roles of consolation in captive chimpanzees *(Pan troglodytes). American Journal of Physical Anthropology* 129:106–111.

Palmer, A. K., Benefit, B. R., and McCrossin, M. L. 1999. Was *Kenyapithecus* a sclerocarp feeder? An exploration of the dietary adaptations of a middle Miocene hominoid through anterior dental microwear analysis. *American Journal of Physical Anthropology* 28 (Suppl.): 217.

Palmer, A. K., Benefit, B. R., McCrossin, M. L., and Gitau, S. N. 1998. Paleoecological implications of dental microwear analysis for the middle Miocene primate fauna from Maboko Island, Kenya. *American Journal of Physical Anthropology* 26 (Suppl.): 175.

Palmer, A. R. 2002. Chimpanzee right-handedness reconsidered: evaluating the evidence with funnel plots. *American Journal of Physical Anthropology* 118:191–199.

Palmié, S. 2008. Genomics, Divination, "Racecraft." *American Ethnologist* 34:205–222.

Palombit, R. A. 1993. Lethal territorial aggression in a white-handed gibbon. *American Journal of Primatology* 31:311–318.

———. 1994a. Dynamic pair bonds in hylobatids: implications regarding monogamous social systems. *Behaviour* 128:65–101.

———. 1994b. Extra-pair copulations in a monogamous ape. *Animal Behaviour* 47:721–723.

———. 1995. Longitudinal patterns of reproduction in wild female siamang *(Hylobates syndactylus)* and white-handed gibbons *(Hylobates lar)*. *International Journal of Primatology* 16:739–760.

———. 1996. Pair bonds in monogamous apes: a comparison of the siamang *Hylobates syndactylus* and the *white-handed gibbon Hylobates lar*. *Behaviour* 133:321–356.

———. 1997. Inter- and intraspecific variation in the diets of sympatric siamang *(Symphalangus syndactylus)* and lar gibbons *(Hylobates lar)*. *Folia Primatologica* 68:337.

Pan, R., Groves, C., and Oxnard, C. 2004. Relationships between the fossil colobine *Mesopithecus pentelicus* and extant cercopithecoids, based on dental metrics. *American Journal of Primatology* 62:287–299.

Pan, Y. 1988. Small fossil primates from Lufeng, a latest Miocene site in Yunnan Province, China. *Journal of Human Evolution* 17:359–366.

Pan, Y-R., and Jablonski, N. G. 1987. The age and geographical distribution of fossil cercopithecids in China. *Human Evolution* 2:59–69.

Pan, Y.-R., Waddle, D. M., and Fleagle, J. G. 1989. Sexual dimorphism in *Laccopithecus robustus*, a Late Miocene hominoid from China. *American Journal of Physical Anthropology* 79:137–158.

Panger, M. A. 1998. Object-use in free-ranging white-faced capuchins *(Cebus capucinus)* in Costa Rica. *American Journal of Physical Anthropology* 106:311–321.

Panger, M. A., Brooks, A. S., Richmond, B. G., and Wood, B. 2002. Older than the Oldowan? Rethinking the emergence of hominin tool use. *Evolutionary Anthropology* 11:235–245.

Panksepp, J. 2011. Toward a cross-species neuroscientific understanding of the affective mind: do animals have emotional feelings? *American Journal of Primatology* 73:545–561.

Paoli, T., Palagi, E., and Borgognini Tarli, S. M. 2006. Reevaluation of dominance hierarchy in bonobos *(Pan paniscus)*. *American Journal of Physical Anthropology* 130:116–122.

Papadopoulos, N., Lykaki-Anastopoulou, G., and Alvanidou, E. 1989. The shape and size of the human hyoid bone and a proposal for an alternative classification. *Journal of Anatomy* 163:249–260.

Papaioannou, J. 1973. Observations on locomotor and general behavior of the siamang. *Malayan Nature Journal* 26:46–52.

Pappu, S., Gunnell, Y., Akhilesh, K., Braucher, R., Taieb, M., Demory, F., and Touveny, N. 2011. Early Pleistocene presence of Acheulian hominids in South India. *Science* 331:1596–1599.

Paquette, D. 1992. Discovering and learning tool-use for fishing honey by captive chimpanzees. *Human Evolution* 7:17–30.

———. 1994. Can chimpanzees use tools by observational learning? In R. A. Gardner, B. Chiarelli and F. C. Plooij, eds., *The Ethological Roots of Culture*, 155–172. Dordrecht: Kluwer Academic.

Parish, A. R. 1994. Sex and food control in the "uncommon chimpanzee": how bonobo females overcome a phylogenetic legacy of male dominance. *Ethology and Sociobiology* 15:157–179.

———. 1996. Female relationships in bonobos *(Pan paniscus)*. Evidence for bonding, cooperation, and female dominance in a male-philopatric species. *Human Nature* 7:61–96.

Parish, A. R., and de Waal, F. B. M. 2000. The other "closest living relative" How bonobos *(Pan paniscus)* challenge traditional assumptions about females, dominance, intra- and intersexual interactions, and hominid evolution. *Annals of the New York Academy of Sciences* 907:97–113.

Parker, S. T. 1987. A sexual selection model for hominid evolution. *Human Evolution* 2:235–253.

———. 1990. Why big brains are so rare: energy costs of intelligence and brain size in anthropoid primates. In S. T. Parker and K. R. Gibson, eds., *"Language" and intelligence in monkeys and apes,* 129–154. Cambridge: Cambridge University Press.

Parker, S. T., and Gibson, K. R. 1977. Object manipulation, tool use and sensorimotor intelligence as feeding adaptations in cebus monkeys and great apes. *Journal of Human Evolution* 6:623–641.

———. 1979. A developmental model for the evolution of language and intelligence in early hominids. *Behavioral and Brain Sciences* 2:367–407.

———. 1990. *"Language" and intelligence in monkeys and apes.* Cambridge: Cambridge University Press.

Parker, S. T., Kerr, M., Markwitz, H., and Gould, J. 1999. A survey of tool use in zoo gorillas. In S. T. Parker, R. W. Mitchell, and H. L. Miles, eds., *The Mentalities of Gorillas and Orangutans* 188–193. Cambridge: Cambridge University Press.

Parker, S. T., Langer, J., and McKinney, M. L. 2000. *Biology, Brains, and Behavior: The Evolution of Human Development.* Santa Fe, NM: SAR Pess.

Parnell, R. J. 2002. Group size and structure in western lowland gorillas *(Gorilla gorilla gorilla)* at Mbeli Bai, Republic of Congo. *American Journal of Primatology* 56:193–206.

Parnell, R. J., and Buchanan-Smith, H. M. 2001. An unusual social display by gorillas. *Nature* 412:294.

Parr, L. A., and de Waal, F. B. M. 1999. Visual kin recognition in chimpanzees. *Nature* 399:647–648.

Parra, E. J. 2007. Human pigmentation variation: evolution, genetic basis, and implications for public health. *Yearbook of Physical Anthropology* 50:85–105.

Partridge, T. C. 1973. Geomorphological dating of cave openings at Makapansgat, Sterkfontein, Swartkrans and Taung. *Nature* 246:75–79.

———. 1978. Re-appraisal of lithostratigraphy of Sterkfontein hominid site. *Nature* 275:282–287.

———. 1982. Dating of South African hominid sites. *South African Journal of Science* 78:300–301.

———. 1985. Spring flow and tufa accretion at Taung. In P. V. Tobias, ed., *Hominid Evolution: Past, Present and Future,* 171–187. New York: Alan R. Liss.

———. 1986. Palaeoecology of the Pliocene and Lower Pleistocene hominids of southern Africa: how good is the chronological and palaeoenvironmental evidence? *South African Journal of Science* 82:80–83.

Passingham, R. 1979. Specialization and the language areas. In H. D. Steklis and M. J. Raleigh, eds., *Neurobiology of Social Communication in Primates,* 221–256. New York: Academic Press.

———. 1982. *The Human Primate.* Oxford: W. H. Freeman.

———. 2006. Brain development and IQ. *Nature* 440:619–620.

Pate, J. L. 2007. The transfer index as a precursor of nonhuman language research and emergents. In D. A. Washburn, ed., *Primate Perspectives on Behavior and Cognition,* 137–142. Washington, DC: American Psychological Association.

Pate, J. L., and Rumbaugh, D. M. 1983. The language-like behavior of Lana chimpanzee: is it merely discrimination and paired-associate learning? *Animal Learning and Behavior* 11:134–138.

Patel, A. D. 2008. *Music, Language, and the Brain.* Oxford: Oxford University Press.

Patel, B. A., Susman, R. L., Rosie, J. B., and Hill, A. 2009. Terrestrial adaptations in the hands of *Equatorius africanus* revisited. *Journal of Human Evolution* 57:763–772.

Patel, T. 1995. Burnt stones and rhino bones hint at earliest fire. *New Scientist* 46:5.

Patterson, B., and Howells, W. W. 1967. Hominid humeral fragment from Early Pleistocene of northwestern Kenya. *Science* 156:64–66.

Patterson, C., Williams, D. M., and Humphries, C. J. 1993. Congruence between molecular and morphological phylogenies. *Annual Review of Ecology and Systematics* 24:153–188.

Patterson, F. G., and Gordon, W. 1993. The case for personhood of gorillas. In P. Cavalieri and P. Singer, eds., *The Great Ape Project,* 58–77. New York: St. Martin's Press.

———. 2001. Twenty-seven years of project Koko and Michael. In B. M. F. Galdikas, N. E. Briggs, L. K. Sheeran, G. L. Shapiro, and J. Goodall, eds., *All Apes Great and Small,* Vol. 1: *African Apes,* 165–176. New York: Kluwer Academic/Plenum Press.

Patterson, F. G., and Linden, E. 1981. *The Education of Koko.* New York: Holt, Rinehart and Winston.

Patterson, F. G. 1978a. Conversations with a gorilla. *National Geographic* 154:438–465.

———. 1978b. The gestures of a gorilla: language acquisition in another pongid. *Brain and Language* 5:72–97.

———. 1978c. Linguistic capabilities of a lowland gorilla. In F. C. C. Peng, ed., *Sign Language and Language Acquisition in Man and Ape: New Dimensions in Comparative Pedolinguistics,* 161–201. Boulder, CO: Westview Press.

———. 1980. Innovative uses of language by a gorilla: a case study. In K. E. Nelson, ed., *Children's Language,* Vol. 2, 497–561. New York: Gardner Press.

Patterson, F. G. P., and Cohn, R. H. 1980. The mind of the gorilla: conversation and conservation. In K. Benirschke, ed., *Primates: The Road to Self-Sustaining Populations,* 933–947. New York: Springer-Verlag.

————. 1990. Language acquisition by a lowland gorilla: Koko's first ten years of vocabulary development. *Word* 41:97–143.

————. 1994. Self-recognition and self-awareness in lowland gorillas. In S. T. Parker, R. W. Mitchell, and M. L. Boccia, eds., *Self-Awareness in Animals and Humans,* 273–290. Cambridge: Cambridge University Press.

Patterson, N., Richter, D. J., Gnerre, S., Lander, E. S., and Reich, D. 2006. Genetic evidence for complex speciation of humans and chimpanzees. *Nature* 1103–1108.

Patterson, T. 1979. The behavior of a group of captive pygmy chimpanzees *(Pan paniscus). Primates* 20:341–354.

Patterson, T. L., and Tzeng, O. J. L. 1979. Long-term memory for abstract concepts in the lowland gorilla *(Gorilla g. Gorilla). Bulletin of the Psychonomic Society* 13:279–282.

Payne, K. 2000. The progressively changing songs of humpback whales: a window on the creative process in a wild animal. In N. L. Wallin, B. Merker and S. Brown, eds., *The Origins of Music,* 135–150. Cambridge, MA: MIT Press.

Payne, M. M. 1965. Family in search of prehistoric man. *National Geographic* 127:194–321.

————. 1966. Preserving the treasures of Olduvai Gorge. *National Geographic* 130: 700–709.

Pearson, O. M. 2004. Has the combination of genetic and fossil evidence solved the riddle of modern human origins? *Evolutionary Anthropology* 13:145–159.

Peccei, J. S. 1995. The origin and evolution of menopause: the altriciality-lifespan hypothesis. *Ethology and Sociobiology* 16:425–449.

————. 2001. Menopause: adaptation or epiphenomenon? *Evolutionary Anthropology* 10:43–57.

Pécsi, M. Geomorphological position and absolute age of the *Veeretesszölös Lower Paleolithic.* In M. Kretzoi and V. T. Dobosi, eds., *Veeretesszölös: Site, Man and Culture,* 27–41. Budapest: Akadémiai Kiadó.

Pedersen, J., and Fields, W. M. 2009. Aspects of repetition in bonobo-human conversation: creating cohesion in a conversation between species. *Integrative Psychological and Behavioral Science* 43:22–41.

Peeters, M. 2004. Cross-species transmission of simian retroviruses in Africa and risk for human health. *Lancet* 363:911–912.

Pei, W. C. 1957. Giant ape's jaw bone discovered in China. *American Anthropologist* 59:834–838.

Pellegrini, A. D., and Smith, P. K. 2005. *The Nature of Play: Great Apes and Humans.* New York: Guillford Press.

Pellicciari, C., Formenti, D., Redi, D. A., and Manfredi Romanini, M. G. 1982. DNA content variability in primates. *Journal of Human Evolution* 11:131–141.

Pellicciari, C., Ronchetti, E., Formenti, D., Stanyon, R., and Manfredi Romanini, M. G. 1990. Genome size and "C-heterochromatic-DNA" in man and the African apes. *Human Evolution* 5:261–267.

Pennisi, E. 1999. Did cooked tubers spur the evolution of big brains? *Science* 283:2004–2005.

————. 2003. Cannibalism and prion disease may have been rampant in ancient humans. *Science* 300:227–229.

————. 2005. Why do humans have so few genes? *Science* 309:80.

————. 2006a. Mining the molecules that made our mind. *Science* 313:1908–1911.

————. 2006b. Social animals prove their smarts. *Science* 312:1734–1738.

————. 2007a. Geomicists tackle the primate tree. *Science* 316:218–221.

————. 2007b. Human genetic variation. *Science* 318:1842–1843.

————. 2007c. No sex please, we're Neandertals. *Science* 316:967.

————. 2007d. Working the (gene count) numbers: finally, a form answer? *Science* 316:1113.

————. 2013. More genomes from Denisova Cave show mixing of early human groups. *Science* 34:799.

Pepper, J. W., Mitani, J. C., and Watts, D. P. 1999. General gregariousness and specific social preferences among wild chimpanzees. *International Journal of Primatology* 20:613–632.

Pepperberg, I. M. 1999. Rethinking syntax: commentary on E. Kako's "elements of syntax in the systems of three language-trained animals." *Animal Learning and Behavior* 27:15–17.

Perica, S. 2001. Seasonal fluctuation and intracanopy variation in leaf nitrogen level in olive. *Journal of Plant Nutrition* 24:779–787.

Peritz, E., and Rust, P. G. 1972. On the estimation of the nonpaternity rate using more than one blood-group system. *American Journal of Human Genetics* 24:46–53.

Perkins, W. H., and Kent, R. D. 1986. *Functional Anatomy of Speech, Language, and Hearing: A Primer.* San Diego: College-Hill Press.

Perry, G. H., Dominy, N. J., Stone, A. C., Claw, K. G., Lee, A. S. Fiegler, H., Redon, R., Werner, J., Villanea, F. A., Mountain, J. L., et al. 2007. Diet and the evolution of human amylase gene copy number variation. *Nature Genetics* 39:1256–2660.

Perry, G. H., Tshinda, J., McGrath, S. D., Zhang, J., Picker, S. R., Cáceres, A. M., Iafrate, A. J., Tyler-Smith, C., Scherer, S. W., Eichler, E. E., et al. 2006. Hotspots for copy number variation in chimpanzees and humans. *Proceedings of the National Academy of Sciences of the United States of America* 103:8006–8011.

Perry, S. 2006. What cultural anthropology can tell anthropologists about the evolution of culture. *Annual Review of Anthropology* 35:171–190.

————. 2009. Are nonhuman primates likely to exhibit cultural capacities like those of humans? In K. N. Laland and B. G. Galef, eds., *The Question of Animal Culture,* 247–268. Cambridge, MA: Harvard University Press.

Perry, S., and Manson, J. H. 2003. Tradition in monkeys. *Evolutionary Anthropology* 12:71–81.

Peters, C. R. 1981. The early hominid plant-food niche: insights from an analysis of plant exploitation by *Homo, Pan,* and *Papio* in Eastern and Southern Africa. *Current Anthropology* 22:127–140.

————. 2007. Theoretical and actualistic ecobotanical perspectives on early hominin diets and paleoecology. In P. S. Ungar, ed., *Evolution of the Human Diet,* 233–261. Oxford: Oxford University Press.

Peters, H. G. 2001, Tool use to modify calls by wild orang-utans. *Folia Primatologica* 72:242–244. Peters, H. H., and Rogers, L. J. 2008. Limb use and preferences in wild orang-utans during feeding and locomotor behavior. *American Journal of Primatology* 70:261–270.

Peterson, D., and Goodall, J. 1993. *Visions of Caliban.* Boston: Houghton Mifflin.

Peterson, J. T. 1978. *The Ecology of Social Boundaries: Agta Foragers of the Philippines.* Urbana: University of Illinois Press.

Petrinovich, L. F. 2000. *The Cannibal Within.* Hawthorne, NY: Aldine de Gruyter.

Pfungst, O. 1911. *Clever Hans (The Horse of Mr. Von Osten): A Contribution to Experimental Animal and Human Psychology.* Bristol, UK: Thoemmes Press.

Phillips, K. A. 1998. Tool use in wild capuchin monkeys *(Cebus albifrons trinitatis). American Journal of Primatology* 46:259–261.

Phillips, K. A., and Hopkins, W. D. 2007. Exploring the relationship between cerebellar asymmetry and handedness in chimpanzees *(Pan troglodytes)* and capuchins *(Cebus apella). Neuropsychologia* 45:2333–2339.

Phoonjampa, R., Koenig, A., Borries, C., Gale, G. A., and Savini, T. 2010. Selection of sleeping trees in pileated gibbons *(Hylobates pileatus). American Journal of Primatology* 72:617–625.

Piaget, J. 1952. *The Origins of Intelligence in Children.* New York: International Universities Press.

———. 1973. *The Psychology of Intelligence.* Totowa, NJ: Littlefield.

Pica, P., Lemer, C., Izard, V., and Dehaene, S. 2004. Exact and approximate arithmetic in an Amazonian indigene group. *Science* 306:499–507.

Pickering, T. R. 2001. Taphonomy of the Swartkrans hominid postcrania and its bearing on issues of meat-eating and fire management. In C. B. Stanford and H. T. Bunn, eds., *Meat-Eating and Human Evolution,* 33–51. Oxford: Oxford University Press.

Pickering, T. R., Clarke, R. J., and Heaton, J. L. 2004. The context of Stw 573, an early hominid skull and skeleton from Sterkfontein Member 2: taphonomy and paleoenvironment. *Journal of Human Evolution* 46:279–297.

Pickering, T. R., and Domínguez-Rodrigo, M. 2006. The acquisition and use of large mammal carcasses by Oldowan hominins in eastern and southern Africa: a selected review and assessment. In N. Toth and K. Schick, eds., *The Oldowan: Case Studies into the Earliest Stone Age,* 113–128. Gosport, IN: Stone Age Institute Press.

Pickford, M. 1975. Late Miocene sediments and fossils from the Northern Kenya Rift Valley. *Nature* 256:279–284.

———. 1982. New higher primate fossils from the Middle Miocene deposits at Majiwa and Kaloma, western Kenya. *American Journal of Physical Anthropology* 58:1–19.

———. 1983. Sequence and environments of the Lower and Middle Miocene hominoids of western Kenya. In R. L. Ciochon and R. S. Corruccini, eds., *New Interpretations of Ape and Human Ancestry,* 421–439. New York: Plenum Press.

———. 1985a. *Kenyapithecus:* a review of its status based on newly discovered fossils from Kenya. In P. V. Tobias, ed., *Hominid Evolution: Past, Present and Future,* 107–112. New York: Alan R. Liss.

———. 1985b. A new look at *Kenyapithecus* based on recent discoveries in western Kenya. *Journal of Human Evolution* 14:113–143.

————. 1986a. Cainozoic paleontological sites of western Kenya. *Münchner Geowissenschaftliche Abhandlungen,* Reihe A, 8:1–151.

————. 1986b. The geochronology of Miocene higher primate faunas of East Africa. In J. G. Else and P. C. Lee, eds., *Primate Evolution,* 19–33. Cambridge: Cambridge University Press.

————. 1986c. Geochronology of the Hominoidea: a summary. In J. G. Else and P. C. Lee, eds., *Primate Evolution,* 123–128. Cambridge: Cambridge University Press.

————. 1986d. Hominoids from the Miocene of East Africa and the phyletic position of *Kenyapithecus. Zeitschrift für Morphologie und Anthropologie* 76:117–130.

————. 1986e. A reappraisal of *Kenyapithecus.* In J. G. Else and P. C. Lee, eds., *Primate Evolution,* 163–171. Cambridge: Cambridge University Press.

————. 1986f. Sexual dimorphism in *Proconsul. Human Evolution* 1:111–148.

————. 1998. Dater les anthropoïdes néogènes de l'Ancien Monde: une base essentielle pour l'analyse phylogénétique, la biogéographie et la paléoécologie. *Primatologie* 1:27–92.

Pickford, M., and Ishida, H. 1998. Interpretation of *Samburupithecus,* an Upper Miocene homoinoid from Kenya. *Comptes rendus des séances de l'Académie des Sciences Paris, Sciences de la terre et des planètes* 326:299–306.

Pickford, M., Ishida, H., Nakano, Y, and Nakaya, H. 1984. Fossiliferous localities of the Nachola-Samburu Hills area, northern Kenya. *African Study Monographs* 2 (Suppl.): 45–56.

Pickford, M., Johanson, D. C., Lovejoy, C. O., White, T. D., and Aronson, J. L. 1983. A hominoid humeral fragment from the Pliocene of Kenya. *American Journal of Physical Anthropology* 60:337–346.

Pickford, M., Moyà-Solà, S., and Köhler, M. 1997. Implications phylogénétiques du premier os frontal d'un Hominoïdea du Miocène moyen d'Otavi, Namibie. *Comptes rendus des séances de l'Académie des Sciences Paris, Sciences de la terre et des planètes* 325:459–466.

Pickford, M., Senut, B., and Gommery, D. 1999. Sexual dimorphism in *Morotopithecus bishopi,* and early Middle Miocene hominoid from Uganda, and a reassessment of its geological and biological contexts. In P. Andrews and P. Banham, eds., *Late Cenozoic Environments and Hominid Evolution: a Tribute to Bill Bishop,* 27–38. London: Geological Society of London.

Pickford, M., Senut, B., Gommery, D., and Musiime, E. 2003. New catarrhine fossils from Moroto II, early Middle Miocene (ca 17.5 Ma) Uganda. *Comptes Rendus Palevol* 2:649–662.

Pickford, M., Senut, B., Gommery, D., and Treil, J. 2002. Bipedalism in *Orrorin tugenensis* revealed by its femora. *Comptes Rendus Palevol* 1:191–203.

Pickford, M. L., and Senut, B. 1988. Habitat and locomotion in Miocene cercopithecids. In A. Gautier-Hion, F. Boulière, J.-H. Gautier, and J. Kingdon, eds., *A Primate Radiation: Evolutionary Biology of the African Guenons,* 35–53. Cambridge: Cambridge University Press.

————. 2001. The geological and faunal context of Late Miocene hominid remains from Lukeino, Kenya. *Comptes Rendus de l'Académie des Sciences, Série II. Fascicule a, Sciences de la Terre et des planètes* 332:145–1452.

————. 2005. Hominoid teeth with chimpanzee- and gorilla-like features from the Miocene of Kenya: implications for the chronology of ape-human divergence and biogeography of Miocene hominoids. *Anthropological Science* 113:95–102.

Pieta, K. 2008. Female mate preferences among *Pan troglodytes schweinfurthii* of Kanyawara, Kibale National Park, Uganda. *International Journal of Primatology* 29:845–864.

Pika, S., Liebal, K., Call, J., and Tomasello, M. 2003. Gestural communication in young gorillas *(Gorilla gorilla):* gestural repertoire, learning, and use. *American Journal of Primatology* 60:95–111.

————. 2005. The gestural communication of apes. *Gesture* 5:41–56.

————. 2007. The gestural communication of apes. In K. Liebal, C. Müller, and S. Pika, eds., *Gestural Communication in Nonhuman and Human Primates,* 37–51. Amsterdam: John Benjamins Publishing Company.

Pika, S., Liebal, K., and Tomasello, M. 2005. Gestural communication in subadult bonobos *(Pan paniscus):* repertoire and use. *American Journal of Primatology* 65:39–61.

Pika, S., and Mitani, J. C. 2006. Referential gesture communication in wild chimpanzees *(Pan troglodytes). Current Biology* 16:R191–192.

————. 2009. The directed scratch: evidence for a referential gesture in chimpanzees? In T. Botha and C. Knight, eds., *The Prehistory of Language,* 166–180. Oxford: Oxford University Press.

Pilbeam, D. R. 1966. Notes on *Ramapithecus,* the earliest known hominid, and *Dryopithecus. American Journal of Physical Anthropology* 25:1–6.

————. 1967. Man's earliest ancestors. *Science Journal* 3:47–53.

————. 1968. The earliest hominids. *Nature* 219:1335–1338.

————. 1969a. Newly recognized mandible of *Ramapithecus. Nature* 222:1093–1094.

————. 1969b. Tertiary Pongidae of East Africa: evolutionary relationships and taxonomy. *Peabody Museum Bulletin* 31:1–185.

————. 1970a. *The Evolution of Man.* London: Thames and Hudson.

————. 1970b. *Gigantopithecus* and the origins of Hominidae. *Nature* 225:516–519.

————. 1972. *The Ascent of Man.* New York: Macmillan.

————. 1978. Rethinking human origins. *Discovery* 13:2–9.

————. 1979. Recent finds and interpretations of Miocene hominoids. *Annual Review of Anthropology* 8:333–352.

————. 1980. Major trends in human evolution. In L.-K. Konigsson, ed., *Current Argument on Early Man,* 261–285. Oxford: Pergamon Press.

————. 1982. New hominoid skull material from the Miocene of Pakistan. *Nature* 295:232–234.

————. 1983. Hominoid evolution and hominid origins. In C. Chagas, ed., *Working Group on Recent Advances in the Evolution of Primates, May 24–27, 1982,* 43–61. Pontificae Academiae Scientiarum Scripta Varia, 50. Città del Vaticano: Pontifica Academia Scientiarum.

————. 1984. The descent of hominoids and hominids. *Scientific American* 250:84–96.

————. 1985. Patterns of hominoid evolution. In E. Delson, ed., *Ancestors: The Hard Evidence,* 51–59. New York: Alan R. Liss.

———. 1986. Distinguished lecture: hominoid evolution and hominoid origins. *American Anthropologist* 88:295–312.

———. 1996. Genetic and morphological records of the Hominoidea and hominid origins: a synthesis. *Molecular Phylogenetics and Evolution* 5:155–168.

Pilbeam, D. R., Behrensmeyer, A. K., Barry, J. C., and Shah, S. M. I. 1979. Miocene sediments and faunas of Pakistan. *Postilla,* no. 179: 1–45.

Pilbeam, D., and Jacobs, L. L. 1978. Changing views of human origins. *Plateau* 51:18–30.

Pilbeam, D., Meyer, G. E., Badgley, C., Rose, M. D., Pickford, M. H. L., Behrensmeyer, A. K., Shah, S. M. I. 1977. New hominoid primates from the Siwaliks of Pakistan and their bearing on hominoid evolution. *Nature* 270:689–695.

Pilbeam, D. R., Rose, M. D., Badgley, C., and Lipschutz, B. 1980. Miocene hominoids from Pakistan. *Postilla,* no. 181: 1–94.

Pilbeam, D. R., Rose, M. D., Barry, J. C., and Shah, S. M. I. 1990. New *Sivapithecus* humeri from Pakistan and the relationship of *Sivapithecus* and *Pongo. Nature* 229:406–407.

Pilbeam, D. R., and Simons, E. L. 1965. Some problems of hominid classification. *American Scientist* 53:237–259.

———. 1971. On the humerus of *Dryopithecus* from Saint-Gaudens, France. *Nature* 229:406–407.

Pilbeam, D., and Walker, A. 1968. Fossil monkeys from the Miocene of Napak, North-East Uganda. *Nature* 220:657–660.

Pilbeam, D. R., and Young, N. M. 2001. *Sivapithecus* and hominoid evolution: some brief comments. In L. de Bonis, G. D. Koufos, and P. Andrews, eds., *Hominoid Evolution and Climatic Change in Europe,* Vol. 2: *Phylogeny of the Neogene Hominoid Primates of Eurasia,* 349–364. Cambridge: Cambridge University Press.

Pilcher, D. L., Hammock, E. A. D., and Hopkins, W. D. 2001. Cerebral volumetric asymmetries in non-human primates: a magnetic resonance imaging study. *Laterality* 6:165–179.

Pinhasi, R., Higham, T. F. G., Golovanova, L. V., and Boronichev, V. B. 2011. Revised age of late Neanderthal occupation and the end of the Middle Paleolithic in the Northern Caucasus. *Proceedings of the National Academy of Sciences of the United States of America* 108:8611–8616.

Pinker, S. 1994. *The Language Instinct: How the Mind Creates Language.* New York: William Morrow.

———. 1997. *How the Mind Works.* New York: W. W. Norton.

———. 2001. Talk of genetics and vice versa. *Nature* 413:465–466.

———. 2003. Language as an adaptation to the cognitive niche. In M. H. Christiansen and S. Kirby, eds., *Language Evolution,* 16–37. Oxford: Oxford University Press.

Pinker, S., and Bloom, P. 1990. Natural language and natural selection. *Behavioral and Brain Sciences* 13:707–784.

Pinker, S., and Jackendoff, R. 2005. The faculty of language: what's special about it? *Cognition* 95:201–236.

Platek, S. M., Criton, S. R., Myers, T., and Gallup, G. G., Jr. 2003. Contagious yawning: the role of self-awareness and mental state attribution. *Cognitive Brain Research* 17:223–224.

Platek, S. M., Mohamed, F. B., and Gallup, G. G., Jr. 2002. Contagious yawning: the role of self-awareness and mental state attribution. *Cognitive Brain Research* 17:223–224.

Plavcan, J. M. 2004a. Evidence for early anthropoid social behavior. In C. F. Ross and R. F. Kay, eds., *Anthropoid Origins*, 383–412. New York: Kluwer Academic/Plenum Press.

———. 2004b. Sexual selection, measures of sexual selection, and sexual dimorphism in primates. In P. Kappeler and C. P. van Schaik, eds., *Sexual Selection in Primates: New and Comparative Perspectives*, 230–252. Cambridge: Cambridge University Press.

Plavcan, J. M., van Schaik, C. P., and McGraw, W. S. 2005. In D. K. Brockman and C. P. van Schaik, eds., *Seasonality in Primates*, 401–441. Cambridge: Cambridge University Press.

Plooij, F. X. 1978a. Some basic traits of language in wild chimpanzees? In A. Lock, ed., *Action, Gesture and Symbol, the Emergence of Language*, 111–113. London: Academic Press.

———. 1978b. Tool use during chimpanzees' bushpig hunt. *Carnivore* 1:103–106.

———. 1979. How wild chimpanzee babies trigger the onset of mother infant play and what the mother makes of it. In M. Bullowa, ed., *Before Speech, the Beginnings of Human Communication*, 223–243. Cambridge, England: Cambridge University Press.

Plotnik, J. M., de Waal, F. B. M., and Reiss, D. 2006. Self-recognition in an Asian elephant. *Proceedings of the National Academy of Sciences of the United States of America* 103:17053–17057.

Plumptre, A. J. 1996. Modelling the impact of large herbivores on the food supply of mountain gorillas and implications for management. *Biological Conservation* 75:147–155.

Plumptre, A. J., McNeilage, A., Hall, J. S., and Williamson, E. A. 2003. The current status of gorillas and threats to their existence at the beginning of a new millennium. In A. B. Taylor and M. L. Goldsmith, eds., *Gorilla Biology. A Multidisciplinary Perspective*, 414–431. Cambridge: Cambridge University Press.

Pobinar, B. L., Rogers, M. J. Monahan, C. M., and Harris, J. W. K. 2008. New evidence for hominin carcass processing strategies at 1.5 Ma, Koobi Fora, Kenya. *Journal of Human Evolution* 55:103–130.

Poe, E. A. 1975. The murders in the Rue Morgue. In *The Compete Tales and Poems of Edgar Allan Poe*, 141–168. New York: Random House.

Pohlig, H. 1895. *Paedopithex rhenanus*, n.g.n.s., le singe anthropomorphe du Pliocene rhenan. *Bulletin de la Société belge de géologie, de paléontologie et d'hydrologie* 9:149–151.

Pollard, K. S., Salama, S. R., Lambert, N., Lambot, M.-A., Coppens, S., Pedersen, J. S., Katzman, S., King, B., Onodera, C., Siepel, et al. 2006. An RNA gene expressed during cortical development evolved rapidly in humans. *Nature* 443:167–172.

Pollick, A. S., and de Waal, F. B. M. 2007. Ape gestures and language evolution. *Proceedings of the National Academy of Sciences of the United States of America* 104:8184–8189.

Pollick, A. S., Jeneson, A., and de Waal, F. B. M. 2008. Gestures and multimodal signaling in bonobos. In T. Furuich and J. Thompson, eds., *The Bonobos: Behavior, Ecology, and Conservation,* 75–94. New York: Spinger.

Pond, C. 1987a. Fat and figures. *New Scientist,* No. 1563, 62–66.

———. 1987b. The great ape debate. *New Scientist* 12 November, 39–42.

Pontzer, H., and Wrangham, R. W. 2004. Climbing and the daily energy cost of locomotion in wild chimpanzees: implications for hominoid locomotor adaptations. *Journal of Human Evolution* 46:317–335.

Pope, G. G. 1983. Evidence on the age of the Asian Hominidae. *Proceedings of the National Academy of Sciences of the United States of America* 80:4988–4992.

———. 1988. Recent advances in Far Eastern paleoanthropology. *Annual Review of Anthropology* 17:43–75.

Popesco, M. C., MacLaren, E. J., Hopkins, J., Dumas, L., Cox, M., Meltesen, L., McGavran, L., Wykoff, G. J., and Sikela, J. M. 2006. Human lineage-specific amplification, selection and neuronal expression of DUF1220 domains. *Science* 313:1304–1307.

Popovich, D. G., Jenkins, D. J. A., Kendall, C. W. C., Dierenfeld, E. S., Carroll, R. W., Tariq, N., Vidgen, E. 1997. The western lowland gorilla diet has implications for the health of humans and other hominoids. *Journal of Nutrition* 127:2000–20005.

Posada, S., and Colell, M. 2007. Another gorilla *(Gorilla gorilla gorilla)* recognizes himself in a mirror. *American Journal of Primatology* 69:567–583.

Poss, S. R., Kuhar, C., Stoinski, T. S., and Hopkins, W. D. 2006. Differential use of attentional and visual communicative signaling by orangutans *(Pongo pygmaeus)* and gorillas *(Gorilla gorilla)* in response to the attentional status of a human. *American Journal of Primatology* 68:978–992.

Posthuma, D., de Geus, E. J. C., and Deary, I. J. 2009. The genetics of intelligence. In T. E. Goldberg and D. R. Weinberger, eds., *The Genetics of Cognitive Neuroscience,* 97–121. Cambridge, MA: MIT Press.

Potts, K. B., Watts, D. P., and Wrangham, R. W. 2011. Comparative feeding ecology of two communities of chimpanzees *(Pan troglodytes)* in Kibale National Park, Uganda. *International Journal of Primatology* 32:669–690.

Potts, R. 1989. Olorgesailie: new excavations and findings in Early and Middle Pleistocene contexts, southern Kenya rift valley. *Journal of Human Evolution* 18:477–484.

Potts, R., Behrensmeyer, A. K., Deino, A., Ditchfield, P., and Clark, J. 2004. Small Mid-Pleistocene hominin associated with East African Acheulean technology. *Science* 305:75–78.

Potts, R., Behrensmeyer, A. K., and Ditchfield, P. 1999. Paleolandscape variation and Early Pleistocene hominid activities: Members 1 and 7, Olorgesailie Formation, Kenya. *Journal of Human Evolution* 37:747–788.

Potts, R., and Shipman, P. 1981. Cutmarks made by stone tools on bones from Olduvai Gorge, Tanzania. *Nature* 291:577–580.

Poulsen, J. R., and Clark, C. J. 2004. Densities, distributions, and seasonal movements of gorillas and chimpanzees in swamp forest in Northern Congo. *International Journal of Primatology* 25:285–306.

Poulsen, J. R., Clark, C. J., and Smith, T. B. 2001. Seed dispersal by a diurnal primate community in the Dja Reserve, Cameroon. *Journal of Tropical Ecology* 17:787–808.

Povinelli, D. J. 1994. How to create self-recognizing gorillas (but don't try it on macaques). Self-recognition and self-awareness in lowland gorillas. In S. T. Parker, R. W. Mitchell, and M. L. Boccia, eds., *Self-Awareness in Animals and Humans,* 291–300. Cambridge: Cambridge University Press.

Povinelli, D. J., Bering, J. M., and Giambrone, S. 2003. Chimpanzees' "pointing": another error of the argument by analogy? In S. Kita, ed., *Pointing: Where Language, Culture and Cognition Meet,* 35–68. New York: Psychology Press.

Povinelli, D. J., and Eddy, T. J. 1996a. Factors influencing young chimpanzees' *(Pan troglodytes)* recognition and attention. *Journal of Comparative Psychology* 110:336–345.

Povinelli, D. J., Eddy, T. J., Hobson, R. P., and Tomasello, M. 1996. What young chimpanzees know about seeing. *Monographs of the Society for Research in Child Development* 61, no. 3: i–191.

Povinelli, D. J., Nelson, K. E., and Boysen, S. T. 1992. Comprehension of social role reversal by chimpanzees: evidence for empathy? *Animal Behaviour* 43:633–640.

Povinelli, D. J., Reaux, J. E., Theall, L. A., and Giambrone, S. 2000. *Folk Physics for Apes.* OxfordUK: Oxford University Press.

Povinelli, D. J., Rulf, A. B., Landau, K. R., Bierschwale, D. T. 1993. Self-recognition in chimpanzees *(Pan troglodytes):* distribution, ontogeny, and patterns of emergence. *Journal of Comparative Psychology* 107:347–372.

Power, M. 1991. *The Egalitarians—Human and Chimpanzee: An Anthropological View of Social Organization.* Cambridge: Cambridge University Press.

Pradhan, G. R., Engelhardt, A., van Schaik, C. P., and Maestripieri, D. 2006. The evolution of female copulation calls in primates: a review and a new model. *Behavioral Ecology and Sociobiology* 59:333–343.

Pradhan, G. R., van Noordwijk, M. A., and van Schaik, C. P. 2012. A model for the evolution of developmental arrest in male orangutans. *American Journal of Physical Anthropology* 149–125.

Prasad, K. N. 1969. Fossil anthropoids from the Siwalik system of India. *Proceedings of the Second International Congress of Primatology Atlanta, GA 1968* 2:131–134. Basel: Karger.

———. 1975. Observations on the paleoecology of South Asian Tertiary primates. In R. H. Tuttle, ed., *Paleoanthropology: Morphology and Paleoecology,* 21–30. The Hague: Mouton.

———. 1982. Was *Ramapithecus* a tool-user? *Journal of Human Evolution* 11:101–104.

Prasetyo, D., Ancrenaz, M. Morrogh-Bernard, H. C., Atmoko, S. S. U., Wich, S. A., and van Schaik, C. P. 2009. Nest building in orangutans. In S. A. Wich, S. S. Utami Atmoko, T. Mitra Setia, and C. P. van Schaik, eds., *Orangutans: Geographic Variation in Behavioral Ecology and Conservation,* 269–277. Oxford: Oxford University Press.

Premack, A. 1976. *Why Chimps Can Read.* New York: Harper and Rowe.

———. 1984. Pedagogy and aesthetics as sources of culture. In M. S. Gazziniga, ed., *Handbook of Cognitive Neuroscience,* 15–35. New York: Plenum Press.

Premack, A. J., and Premack, D. 1972. Teaching language to an ape. *Scientific American* 227:92–99.

———. 1975. Le pouvoir du mot chez les chimpanzés. *La Recherche* 6:918–925.

Premack, D. 1970. A functional analysis of language. *Journal of the Experimental Analysis of Behavior* 14:107–125.

———. 1971a. Language in chimpanzees? *Science* 172:808–822.

———. 1971b. On the assessment of language competence in the chimpanzee. In A. M. Schrier and F. Stollnitz, *Behavior of Nonhuman Primates: Modern Research Trends,* Vol. 4, 185–228. New York: Academic Press.

———. 1971c. Some general characteristics of a method for teaching language to organisms that do not ordinarily acquire it. In L. E. Jarrard, ed., *Cognitive Processes of Nonhuman Primates,* 47–82. New York: Academic Press.

———. 1975. Putting a face together. *Science* 188:288–236.

———. 1976a. *Intelligence in Ape and Man.* Hillsdale, NJ: Lawrence Erlbaum.

———. 1976b. Language and intelligence in ape and man. *American Scientist* 64:674–683.

———. 1978. Comparison of language-related factors in ape and man. In D. J. Chivers and J. Herbert, eds., *Recent Advances in Primatology,* Vol. 1: *Behaviour,* 867–881. London: Academic Press.

———. 2007. Human and animal cognition: continuity and discontinuity. *Proceedings of the National Academy of Science of the United States of America* 104:13861–13867.

———. 2003. *Original Intelligence: Unlocking the Mystery of Who We Are.* New York: McGraw-Hill.

Premack, D., and Hines, P. 1976. On the study of intelligence in chimpanzees. *Current Anthropology* 17:516–521.

Premack, D., and Premack, A. J. 1983. *The Mind of an Ape.* New York: W. W. Norton.

Premack, D., and Schwartz. A. 1966. Preparations for discussing behaviorism with chimpanzee. In F. Smith and G. A., Miller, eds., *The Genesis of Language: A Psycholinguistic Approach,* 295–335. Cambridge, MA: MIT Press.

Premack, D., and Woodruff, G. 1978a. Chimpanzee problem-solving: a test for comprehension. *Science* 202:532–535.

———. 1978b. Does the chimpanzee have a theory of mind? *Behavioral and Brain Sciences* 4:515–526.

Preparata, G., and Saccone, C. 1987. A simple quantitative model of the molecular clock. *Journal of Molecular Evolution* 26:7–15.

Preston, S. D., and de Waal, F. B. M., 2002. Empathy: its ultimate and proximate bases. *Behavioral and Brain Sciences* 25:1–72.

Preuschoft, H. 1973a. Body posture and locomotion in some East African Miocene Dryopithecinae. In M. H. Day, ed., *Human Evolution,* 13–46. London: Taylor and Francis.

———. 1973b. Functional anatomy of the upper extremity. In G. H. Bourne, ed., *The Chimpanzee,* Vol. 6, 34–120. Basel: Karger.

Preuschoft, S., Wang, X., Aureli, F., and de Waal, F. B. M. 2002. Reconciliation in captive chimpanzees: a reevaluation with controlled method. *International Journal of Primatology* 23:29050.

Preuss, T. M. 1982. The face of *Sivapithecus indicus:* description of a new, relatively complete specimen from the Siwaliks of Pakistan. *Folia Primatologica* 38:141–157.

———. 2000a. Taking the measure of diversity: comparative alternatives to the model-animal paradigm in cortical neuroscience. *Brain, Behavior and Evolution* 55:287–299.

———. 2000b. What's human about the human brain? In M. S. Gassaniga, ed., *The New Cognitive Neurosciences,* 2nd ed., 1219–1234. Cambridge, MA: MIT Press.

———. 2001. The discovery of cerebral diversity: an unwelcome scientific revolution. In D. Falk and K. R. Gibson, eds., *Evolutionary Anatomy of the Primate Cerebral Cortex,* 138–164. Cambridge: University of Cambridge Press.

———. 2004. What is it like to be human? In M. S. Gazzaniga, ed., *The Cognitive Neurosciences,* 3rd ed., 5–22. Cambridge, MA: MIT Press.

———. 2007a. Evolutionary specializations of primate brain systems. In M. J. Ravosa and M. Dagasto, eds., *Primate Origins: Adaptations and Evolution,* 625–675. New York: Springer.

———. 2007b. Primate brain evolution in phylogenetic context. In J. L. Kaas and T. M. Preuss, eds., *Evolution of Nervous Systems,* Vol. 4: *Primates,* 1–34. Amsterdam: Elsevier.

———. 2008. Human brain evolution. In L. R. Squire, D. Berg, F. E. Bloom, S. du Lac, A. Ghosh, and Spitzer, N. C., eds., *Fundamental Neuroscience,* 3rd ed., 1019–1037. San Diego: Academic Press.

———. 2009. The cognitive neuroscience of human uniqueness. In M. S. Gazzaniga, ed., *The New Cognitive Neurosciences,* 4th ed., 49–66. Cambridge, MA: MIT Press.

Preuss, T. M., Cáceres, M., Oldham, M. C., and Geschwind, D. H. 2004. Human brain evolution: insights from microarrays. *Nature Reviews Genetics* 5:850–860.

Preuss, T. M., and Kaas, J. H. 1999. Human brain evolution. In M. J. Zigmond, F. E. Bloom, S. C. Landis, J. L. Robert, and L. R. Squire, eds., *Fundamental Neuroscience,* 1283–1311. San Diego: Academic Press.

Pribrow, V. 2006. Lingual incisor traits in modern hominoids and an assessment of their utility for fossil hominoid taxonomy. *American Journal of Physical Anthropology* 129:323–338.

Prince Peter of Greece and Denmark. 1955. Polyandry and the kinship group. *Man* 55:179–181.

Prior, H., Schwarz, A., and Güntürkün, O. 2008. Mirror-induced behavior in the magpie *(Pica pica):* evidence for self-recognition. *PLoS Biology* 6:1642–1650.

Proctor, D. J. 2010. Brief communication. Shape analysis of the MT 1 proximal articular surface in fossil hominins and shod and unshod humans. *American Journal of Physical Anthropology* 143:631–637.

Proctor, D. J., Broadfield, D., and Proctor, K. 2008. Quantitative three-dimensional shape analysis of the proximal hallucal metatarsal articular surface in *Homo, Pan, Gorilla,* and *Hylobates. American Journal of Physical Anthropology* 135:216–224.

Prost, J. H. 1967. Bipedalism of man and gibbons compared using estimates of joint motion. *American Journal of Physical Anthropology* 26:135–148.

Protsch, R. 1981. The Kohl-Larsen Eyasi and Garusi hominid finds in Tanzania and their relation to *Homo erectus.* In B. A. Sigmon and J. S. Cybulski, eds., *Homo erectus: Papers in Honor of Davidson Black,* 217–226. Toronto: University of Toronto Press.

Prouty, L. A., Buchanan, P. D., Pollitzer, W. S., and Mootnick, A. R. 1983. A presumptive new hylobatid subgenus with 38 chromosomes. *Cytogenetics and Cell Genetics* 35:141–142.

Pruetz, J. D. 2001. Use of caves by savanna chimpanzees *(Pan troglodytes verus)* in the Tomboronkoto Region of Southeastern Senegal. *Pan Africa News* 8:26–28.

———. 2006. Feeding ecology of savanna chimpanzees *(Pan troglodytes verus)* at Fongoli, Senegal. In G. Hohmann, M. M. Robbins and C. Boesch, eds., *Feeding Ecology in Apes and Other Primates,* 161–182. Cambridge: Cambridge University Press.

———. 2007. Evidence of cave use by savanna chimpanzees *(Pan troglodytes verus)* at Fongoli, Senegal: implications for thermoregulatory behavior. *Primates* 48:316–319.

Pruetz, J. D., and Bertolani, P. 2007. Savanna chimpanzees, *Pan troglodytes verus,* hunt with tools. *Current Biology* 17:412–417.

Pruetz, J. D., Fulton, S. J., Marchant, L. F., McGrew, W. C., Schiel, M., and Walter, M. 2008. Arboreal nesting as anti-predator adaptation by savanna chimpanzees *(Pan troglodytes verus)* in Southeastern Senegal. *American Journal of Primatology* 70:393–401.

Pruetz, D., and Marshack, J. L. 2009. Savanna chimpanzees *(Pan troglodytes verus)* prey on patas monkeys *(Erythrocebus patas)* at Fongoli, Senegal. *Pan Africa News* 16:15–17.

Puech, P.-F. 1984. Acidic-food choice in *Homo habilis* at Olduvai Gorge. *Current Anthropology* 25:349–350.

Puech, P.-F., and Albertini, H. 1984. Dental microwear and mechanisms of early hominds from Laetoli and Hadar. *American Journal of Physical Anthropology* 65:87–91.

Puech, P., Cianfarani, F., and Albertini, H. 1986. Dental microwear features as an indicator for plant food in early hominids: a preliminary study of enamel. *Human Evolution* 1:507–515.

Pusey, A. E. 1979. Intercommunity transfer of chimpanzees in Gombe National Park. In D. A. Hamburg and E. R. McCown, eds., *The Great Apes,* 465–479, Menlo Park, CA: Benjamin/Cummings.

———. 1980. Inbreeding avoidance in chimpanzees. *Animal Behaviour* 28:543–552.

Pusey, A. E., Murray, C., Wallauer, W., Wilson, M., Wroblewski, E., and Goodall, J. 2008. Severe aggression among female *Pan troglodytes schweinfurthii* at Gombe National Park, Tanzania. *International Journal of Primatology* 29:949–973.

Pusey, A. E., and Packer, C. 1987. Dispersal and philopatry. In B. B. Smuts, D. L. Cheney, R. M. Seyfarth, R. W. Wrangham, and T. T. Struhsaker, eds., *Primate Societies,* 250–266. Chicago: University of Chicago Press.

Pusey, A. E., Williams, J., and Goodall, J. 1997. The influence of dominance rank on the reproductive success of female chimpanzees. *Science* 277:828–831.

Qi, G. 1985. Stratigraphic summarization of *Ramapithecus* fossil locality Lufeng, Yunnan. *Acta Anthropologica Sinica* 4:67–69.

———. 1993. The environment and ecology of the Lufeng hominoids. *Journal of Human Evolution* 24:13–11.

Qi, T., and Beard, K. C. 1998. Late Eocene sivaladapid primate from Guangxi Zhuang Atuonomous Region, People's Republic of China. *Journal of Human Evolution* 35:211–220.

Qiu, Z., and Guan, J. 1986. A lower molar of *Pliopithecus* from Tongxin, Nigxia Hui Autonomous Region. *Acta Anthropologica Sinica* 5:201–207.

Qiu, Z., Han, D., Qi, G., and Lin, Y. 1985. A preliminary report on a micromammalian assemblage from the hominoid locality of Lufeng, Yunnan. *Acta Anthropologica Sinica* 4:29–32.

Quade, J., Cerling, T. E., Andrews, P., Alpagut, B. 1995. Paleodietary reconstruction of Miocene faunas from Pasalar, Turkey using stable carbon and oxygen isotopes of fossil tooth enamel. *Journal of Human Evolution* 28:373–384.

Quallo, M. M., Price, C. J., Ueno, K., Asamizuya, T., Cheng, K., Lemon, R. M., and Iriki, A. 2009. Gray and white matter changes associated with tool-use learning in macaque monkeys. *Proceedings of the National Academy of Science of the United States of America* 106:18379–18384.

Rabanal, L. I., Kuehl, H. S., Mundry, R., Robbins, M. M., and Boesch, C. 2010. Oil prospecting and its impact on large rainforest mammals in Loango National Park, Gabon. *Biological Conservation* 143:1017–1034.

Radacille, D. 2000. *The Scalpel and the Butterfly: The War between Animal Research and Animal Protection.* New York: Farrar, Straus and Giroux.

Radcliffe-Brown, A. R. 1922. *The Adaman Islanders.* Cambridge: Cambridge University Press.

Radick, G. 2008. *The Simian Tongue: The Long Debate about Animal Language.* Chicago: University of Chicago Press.

Radinsky, L. 1972. Endocasts and studies of primate brain evolution. In R. H. Tuttle, ed., *Primate Functional Morphology and Evolution,* 175–184. Chicago: Aldine Atherton.

———. 1978. Do albumin clocks run on time? *Science* 200:182–183.

Rae T. C. 1999. Mosaic evolution in the origin of the Hominoidea. *Folia Primatologica* 70:125–135.

Raemaekers, J. J. 1978a. Changes through the day in the food choice of wild gibbons. *Folia Primatologia* 30:194–205.

———. 1978b. The sharing of food sources between two gibbon species in the wild. *Malayan Nature Journal* 31:181–188.

———. 1979. Ecology of sympatric gibbons. *Folia Primatologica* 31:227–45.

———. 1984. Large versus small gibbons: relative roles of bioenergetics and competition in their ecological segregation in sympatry. In H. Preuschoft, D. J. Chivers, W. Y. Brockelman, and N. Creed., eds., *The Lesser Apes,* 109–218. Edinburgh: Edinburgh Univesity Press.

Raemaekers, J. J., and Raemaekers, P. M. 1985. Field playback of loud calls to gibbons *(Hylobates lar):* territorial, sex-specific and species-specific responses. *Animal Behaviour* 33:481–493.

Raffaele, P. 2006. The smart and swinging bonobo. *Smithsonian Magazine,* November 2006, 66–75.

Raghanti, M. A., Stimpson, C. D., Marcinkiewicz, J. L., Erwin, J. M., Hof, P. R., and Sherwood, C. C. 2008a. Differences in cortical serotonergic innervation among humans, chimpanzees, and macaque monkeys: a comparative study. *Cerebral Cortex* 18:584–597.

———. 2008b. Cholinergic innervation of the frontal cortex: differences among humans, chimpanzees, and macaque monkeys. *Journal of Comparative Neurology* 506:409–424.

Ragir, S. 1985. Retarded development: the evolutionary mechanism underlying the emergence of the human capacity for language. *Journal of Mind and Behavior* 6:451–468.

———. 2000. Diet and food preparation: rethinking early hominid behavior. *Evolutionary Anthropology* 9:153–155.

Ragir, S., Rosenberg, M., and Tierno, P. 2000. Gut morphology and the avoidance of carrion among chimpanzees, baboons and early hominids. *Journal of Anthropological Research* 56:477–512.

Rahm, U. 1967. Observations during chimpanzee captures in the Congo. In D. Starck, R. Schneider, and H.-J-Kuhn, eds., *Neue Ergebninsse der Primatologie,* 195–207. Stuttgart: Gustav Fischer Verlag.

Raichlen, D. A., Gordon, A. D., Harcourt-Smith, E. H., Foster, A. D., and Hass, Jr., W. 2010. Laetoli footprints preserve earliest direct evidence of human-like bipedal biomechanics. *PLos One* 5:e0769.

Rainger, R. 1991. *An Agenda for Antiquity: Henry Fairfield Osborn and Vertebrate paleontology at the American Museum of Natural History, 1890–1935.* Tuscaloosa: University of Alabama Press.

Rak, Y. 1983. *The Australopithecine Face.* New York: Academic Press.

———. 1985. Sexual dimorphism, ontogeny and the beginning of differentiation of the robust australopithecine clade. In P. V. Tobias, ed., *Hominid Evolution: Past, Present and Future,* 233–237. New York: Alan R. Liss.

———. 1988. On variation in the masticatory system of *Australopithecus boisei.* In F. E. Grine, ed., *Evolutionary History of the "Robust" Australopithecines,* 193–198. Hawthorne, NY: Aldine de Gruyter.

Rak, Y., and Arensburg, B. 1987. Kebara 2 pelvis: first look at a complete inlet. *American Journal of Physical Anthropology* 73:227–231.

Rak, Y., Ginzburg, A., and Geffen, E. 2002. Does *Homo neanderthalensis* play a role in modern human ancestry? the mandibular evidence. *American Journal of Physical Anthropology* 119:199–204.

———. 2007. Gorilla-like anatomy on *Australopithecus afarensis* mandibles suggests *Au. Afarensis* link to robust australopiths. *Proceedings of the National Academy of Science of the United States of America* 104:6568–6572.

Rakic, P. 2008. Confusing cortical columns. *Proceedings of the National Academy of Science of the United States of America* 105:12099–12100.

Ramharack, R., and Deeley, R. G. 1987. Structure and evolution of primate cytochrome *c* oxidase subunit II gene. *Journal of Biological Chemistry* 262:14014–14021.

Ramnani. M., Behrens. T. E. J., Johansen-Berg, H., Richter, M. C., Pinsk, M. A., Anderson, J. L. R., Rudebeck, P., Ciccarelli, O., Richter, W., Thompson, A. J., et al. 2006. The evolution of prefrontal inputs to the cortico-pontine system: diffusion imaging evidence from macaque monkeys and humans. *Cerebral Cortex* 16:811–818.

Rapacz, J., Chen, L., Butler-Brunner, E., Wu, M.-J., Hasler-Rapacz, J. O., Butler, R., and Schumaker, V. N. 1991. Identification of the ancestral haplotype for apolipoprotein B suggests an African origin of *Homo sapiens* and traces their subsequent migration to Europe and the Pacific. *Proceedings of the National Academy of Sciences of the United States of America* 88:1403–1406.

Rapchan, E. S. 2011. Culture: what can anthropologists and chimpanzees teach us? *Anthropologie* 49:1–12.

Rapp, G., Jr., and Vondra, C. F. 1981. *Hominid Sites: Their Geologic Settings.* Boulder, CO: Westview Press.

Rappaport, R. A. 1999. *Ritual and Religion in the Making of Humanity.* Cambridge: Cambridge University Press.

Rasheed, B. K. A., Whisenant, E. C., Fernandez, R., Ostrer, H., and Bhatnagar, Y. M. 1991. A Y-chromosomal DNA fragment is conserved in human and chimpanzee. *Molecular Biology and Evolution* 8:416–432.

Rasmussen, D. T. 1986. Anthropoid origins: a possible solution to the Adapidae-Omomyidae paradox. *Journal of Human Evolution* 15:1–12.

———. 2002. Early catarrhines of the African Eocene and Oligocene. In W. C. Hartwig, ed., *The Primate Fossil Record,* 203–220. Cambridge: Cambridge University Press.

Rasmussen, D. T., Bown, T. M., and Simons, E. L. 1992. The Eocene-Oligocene transition in continental Africa. In D. R. Prothero and W. A. Berggren, eds., *Eocene-Oligocene Climatic and Biotic Evolution,* 548–566. Princeton: Princeton University Press.

Rasmussen, D. T., and Simons, E. L. 1988. New specimens of *Oligopithecus savagei,* Early Oligocene primate from the Fayum, Egypt. *Folia Primatologica* 51:182–208.

———. 1992. Paleobiology of the Oligopithecines, the earliest known anthropoid primates. *International Journal of Primatology* 13:477–508.

Ravey, M. 1978. Bipedalism: an early warning system for Miocene hominoids. *Science* 199:372.

Ravosa, M. J., and Dagosto, M. 2007. *Primate Origins: Adaptations and Evolution.* New York: Springer.

Rayadin, Y., and Saitoh, T. 2009. Individual variation in nest size and nest site features of the Bornean orangutans *(Pongo pygmaeus). American Journal of Primatology* 71:393–399.

Raz, N., Dupuis, J. H., Briggs, S. D., McGavran, C., and Acker, J. D. 1998. Differential effects of age and sex on the cerebellar hemispheres and the vermis: a prospective MR study. *AJNR American Journal of Neuroradiology* 19:65–71.

Read, D. W., and Lestrel, P. E. 1970. Hominid phylogeny and immunology: a critical appraisal. *Science* 168:578–580.

Reader, J. 1988. *Missing Links: The Hunt for Earliest Man.* London: Penguin.

Read-Martin, C. E., and Read, D. W. 1975. Australopithecine scavenging and human evolution: an approach from faunal analysis. *Current Anthropology* 16:359–368.

Reaux, J. E., Theall, L. A., and Povinelli, D. J. 1999. A longitudinal investigation of chimpanzees' understanding of visual perception. *Child Development* 70:275–290.

Reeves, R. H. 2000. Recounting a genetic story. *Nature* 405:283–284.

Reed, K. E., and Fleagle, J. G. 1995. Geographic and climatic control of primate diversity. *Proceedings of the National Academy of Sciences of the United States of America* 92:7874–7876.

Reed, K. E., and Rector, A. L. 2007. African Pliocene paleoecology: hominin habitats, resources, and diets. In P. S. Ungar, ed., *Evolution of the Human Diet,* 262–288. Oxford: Oxford University Press.

Rees, A. *The Infanticide Controversy: Primatology and the Art of Field Science.* Chicago: University of Chicago Press.

Regan, B. C., Julliot, C., Simmen, B., Viénot, F., Charles-Dominique, P., and Mollon, J. D. 2001. Fruits, foliage and the evolution of primate colour vision. *Philosophical Transactions of the Royal Society, London, B* 356:229–283.

Reichard, U. H. 1998. Sleeping sites, sleeping places, and presleep behavior of gibbons *(Hylotates lar). American Journal of Primatology* 46:35–62.

———. 2003a. Monogamy: past and present. In U. H. Reichard and C. Boesch, eds., *Monogamy: Mating Strategies and Partnerships in Birds, Humans and Other Mammals,* 3–20. Cambridge: Cambridge University Press.

———. 2003b. Social monogamy in gibbons: the male perspective. In U. H. Reichard and C. Boesch, eds., *Monogamy: Mating Strategies and Partnerships in Birds, Humans and Other Mammals,* 190–213. Cambridge: Cambridge University Press.

Reichard, U. H., and Barelli, C. 2008. Life history and reproductive strategies of Khao Yai *Hylobates lar:* implications for social evolution in apes. *International Journal of Primatology* 29:823–844.

Reichard, U. H., Ganpanakngan, M., and Barelli, C. 2012. White-handed gibbons of Khao Yai: reproductive strategies, and a slow life history. In Kappeler, P. M., and Watts, D. P., eds., *Long-term Field Studies of Primates,* 237–258. Heidelberg: Springer.

Reinartz, G., and Inogwabini, B. I. 2001. Bonobo survival and a wartime conservation mandate. In *The Apes: Challenges for the 21st Century,* 52–56. Brookfield, IL: Chicago Zoological Society.

Reiss, A. L., Abrams, M. T., Singer, H. S., Ross, J. L., and Denckla, M. B. 1996. Brain development, gender and IQ in children: a volumetric imaging study. *Brain* 119:1763–1774.

Reiss, D., and Marino, L. 2001. Mirror self-recognition in the bottlenose dolphin: a case of cognitive convergence. *Proceedings of the National Academy of Sciences of the United States of America* 98:5937–5942.

Relethford, J. H. 2009. Race and global patterns of phenotypic variation. *American Journal of Physical Anthropology* 139:16–22.

Remedios, R., Logothetis, N. K., and Kayser, C. 2009. Monkey drumming reveals common networks for perceiving vocal and nonvocal communication sounds. *Proceedings of the National Academy of Sciences of the United States of America* 106:18010–18015.

Remington, C. L. 1971. An experimental study of man's genetic relationship to great apes, by means of interspecific hybridization. In J. Katz, ed., *Experimentation with Human Beings,* 461–464. New York: Russel Sage.

Remis, M. J. 1993. Nesting behavior of lowland gorillas in the Dzanga-Sangha Reserve, Central African Republic: implications for population estimates and understandings of group dynamics. *Tropics* 2:245–255.

———. 1997a. Western lowland gorillas *(Gorilla gorilla gorilla)* as seasonal frugivores: use of variable resources. *American Journal of Primatology* 43:87–109.

———. 1997b. Ranging and grouping patterns of a western lowland gorilla group at Bai Hokou, Central African Republic. *American Journal of Primatology* 43:110–133.

———. 2000. Initial studies on the contributions of body size and gastrointestinal passage rates to dietary flexibility among gorillas. *American Journal of Physical Anthropology* 112:171–180.

———. 2002. Food preferences among captive western gorillas *(Gorilla gorilla gorilla)* and chimpanzees *(Pan troglodytes)*. *International Journal of Primatology* 23:231–249.

———. 2003. Are gorillas vacuum cleaners of the forest floor? The roles of body size, habitat, and food preferences on dietary flexibility and nutrition. In A. B. Taylor and M. L. Goldsmith, eds., *Gorilla Biology: A Multidisciplinary Perspective,* 385–404. Cambridge: Cambridge University Press.

———. 2006. The role of taste in food selection by African apes: implications for niche separation and overlap in tropical forests. *Primates* 47:56–64.

Remis, M. J., and Dierenfeld, E. S. 2004. Digesta passage, digestibility and behavior in two captive gorillas under two dietary regimens. *International Journal of Primatology* 25:825–845.

Remis, M. J., Dierenfeld, E. S., Mowry, C. B., and Carroll, R. W. 2001. Nutritional aspects of western lowland gorilla *(Gorilla gorilla gorilla)* diet during seasons of fruit scarcity at Bai Hokou, Central African Republic. *International Journal of Primatology* 22:807–836.

Rendall, D. 2004. "Recognizing" kin: mechanisms, media, minds, modules, and muddles. In B. Chapais and C. M. Berman, eds., *Kinship and Behavior in Primates,* 295–316. Oxford: Oxford University Press.

Rendell, L., and Whitehead, H. 2001. Culture in whales and dolphins. *Behavioral and Brain Sciences* 24:309–382.

Renne, P. R., WoldeGabriel, G., Hart, W. K., Heiken, G., and White, T. D. 1999. Chronostratigraphy of the Miocene-Pliocene Sagantole Formation, Middle Awash Valley, Afar Rift, Ethiopia. *Geological Bulletin of America Bulletin* 111:869–885.

Reno, P. L., Meindl, R. S., McCollum, M. A., and Lovejoy, C. O. 2003. Sexual dimorphism in *Australopithecus afarensis* was similar to that of modern humans. *Proceedings of the National Academy of Sciences of the United States of America* 100:9404–9409.

Rensberger, B. 1984. Bones of our ancestors. *Science 84* 5:28–39.

Revedin, A., Aranguren, B., Becattini, R., Longo, L., Marconi, E., Lippi, M. M., Skakun, N., Sinitsyn, A., Spiridonova, E., and Svoboda, J. 2010. Thirty thousand-year-old evidence of plant food processing. *Proceedings of the National Academy of Sciences of the United States of America* 107:18815–18819.

Reynolds, P. C. 1968. Evolution of primate vocal-auditory communication systems. *American Anthropologist* 70:300–308.

Reynolds, V. 1964. Écologie et comportement social des chimpanzés de la foret de Budongo, Ouganga. *La Terre et la vie* 2:155–166.

———. 1965. *Budongo, a Forest and Its Chimpanzees.* London: Metheun.

———. 1975. Reynolds, V. 1975. How wild are the Gombe chimpanzees? *Man* 10:123–125.

———. 2005. *The Chimpanzees of the Budongo Forest.* Oxford: Oxford University Press.

Reynolds, V., and Luscombe, G. P. 1969a. Chimpanzee rank order and the function of displays. In *Social Behavior of Chimpanzees in an Open Environment, IV.* Holloman Air Force Base, New Mexico: 6571st Aeromedical Research Laboratory.

———. 1969b. Chimpanzee social behavior. In *Social Behavior of Chimpanzees in an Open Environment, IV.* Holloman Air Force Base, New Mexico: 6571st Aeromedical Research Laboratory.

———. 1969c. Chimpanzee rank order and the function of displays. In C. R. Carpenter, ed., *Proceedings of the Second International Congress of Primatology, Atlanta, Ga, 1968,* Vol. 1: *Behavior,* 81–86. Basel: Karger.

———. 1976. Greeting behaviour, displays and rank order in a group of free-ranging chimpanzees. In M. R. A. Chance and R. R. Larsen, eds., *The Social Structure of Attention,* 105–115, London: Wiley.

Reynolds, V., Plumptre, A. J., Greenham, J., and Harborne, J. 1998. Condensed tannins and sugars in the diet of chimpanzees *(Pan troglodytes schweinfurthii)* in the Budongo forest, Uganda. *Oecologia* 11:331–336.

Reynolds, V., and Reynolds, F. 1965. Chimpanzees of the Budongo Forest. In I. DeVore, ed., *Primate Behavior,* 368–424. New York: Holt, Rinehart & Winston.

Richards, G. D. 2006. Genetic, physiologic and ecogeographic factor contributing to variation in *Homo sapiens: Homo floresiensis* reconsidered. *Evolutionary Biology* 19:1744–1767.

Richards, M. P., and Trinkaus, E. 2009. Isotopic evidence for the diets of European Neanderthals and early modern humans. *Proceedings of the National Academy of Sciences of the United States of America* 106:16034–16039.

Richards, P. W. 1952. *The Tropical Rain Forest.* Cambridge: Cambridge University Press.

Richman, B. 1993. On the evolution of speech: singing as the middle term. *Current Anthropology* 34:721–722.

Richmond, B. G., Aiello, L. C., Wood, B. A. 2002. Early hominin limb proportions. *Journal of Human Evolution* 43:529–548.

Richmond, B. G., and Jungers, W. L. 2008. Orrorin tugenensis femoral morphology and the evolution of hominin bipedalism. *Science* 319:1662–1665.

Ricklan, D. E. 1987. Functional anatomy of the hand of *Australopithecus africanus. Journal of Human Evolution* 16:643–664.

———. 1990. The precision grip in *Australopithecus africanus:* anatomical and behavioral correlates. In G. H. Sperber, ed., *From Apes to Angels: Essays in Honor of Phillip V. Tobias,* 171–183. New York: Wiley-Liss.

Riedel, J., Franz, M., and Boesch, C. 2011. How feeding competition determines female chimpanzee gregariousness and ranging in the Taï National Park, Côte d'Ivoire. *American Journal of Primatology* 73:305–313.

Rifkin, J. 2009. *The Empathetic Civilization: The Race to Global Consciousness in a World in Crisis.* New York: J. P. Tarcher/Penguin.

Rightmire, G. P. 1972. Multivariate analysis of an early hominid metacarpal from Swartkrans. *Science* 176:149–161.

———. 1979. Cranial remains of *Homo erectus* from Beds II and IV, Olduvai Gorge, Tanzania. *American Journal of Physical Anthropology* 51:99–116.

———. 1980. *Homo erectus* and human evolution in the African Middle Pleistocene. In L.-K. Königsson, ed., *Current Argument on Early Man,* 70–85. Oxford: Pergamon Press.

———. 1981. *Homo erectus* at Olduvai Gorge, Tanzania. In B. A. Sigmon and J. S. Cybulski, eds., *Homo erectus: Papers in Honor of Davidson Black,* 189–192. Toronto: University of Toronto Press.

———. 1983. The Lake Ndutu cranium and early *Homo Sapiens* in Africa. *American Journal of Physical Anthropology* 61:245–254.

———. 1984a. The fossil evidence for hominid evolution in southern Africa. In R. G. Klein, ed., *Southern African Prehistory and Paleoenvironment,* 147–168. Rotterdam: A. A. Balkema.

———. 1984b. Comparisons of *Homo erectus* from Africa and Southeast Asia. *Courier Forschungsinstitut Senckenberg* 69:83–98.

———. 1984c. *Homo sapiens* in sub-Saharan Africa. In F. H. Smith and F. Spencer, eds., *The Origin of Modern Humans,* 295–325. New York: Alan R. Liss.

———. 1985. The tempo of change in the evolution of Mid-Pleistocene *Homo.* In E. Delson, ed., *Ancestors: The Hard Evidence,* 255–264. New York: Alan R. Liss.

———. 1986. Body size and encephalization in *Homo erectus. Anthropos (Brno)* 23:139–150.

———. 1988. *Homo erectus* and later Middle Pleistocene humans. *Annual Review of Anthropology* 17:239–259.

———. 1990. *The Evolution of Homo erectus.* Cambridge: Cambridge University Press.

———. 2008. Homo in the Middle Pleistocene: hypodigms, variation, and species recognition. *Evolutionary Anthropology* 17:8–21.

———. 2009. Middle and later Pleistocene hominins in Africa and Southwest Asia. *Proceedings of the National Academy of Sciences of the United States of America* 106:16040–16045.

Rightmire, G. P., and Deacon, H. J. 1991. Comparative studies of Late Pleistocene human remains from Klasies River Mouth, South Africa. *Journal of Human Evolution* 20:131–156.

Rijksen, H. D. 1975. Social structure in a wild orangutan population in Sumatra. In S. Kondo, M. Kawai, and A. Ehara, eds., *Contemporary Primatology,* 373–379, Basel: Karger.

———. 1978. *A Field Study on Sumatran Orang Utans (Pongo pygmaeus abelii* Lesson 1827). Wageningen, Nederland: H. Veenman & Zonen.

Rijksen, H. D., and Meijaard, E. 1999. *Our Vanishing Relative: The Status of Wild Orangutans at the Close of the Twentieth Century.* Dordrecht: Kluwer.

Rilling, J. K. 2006. Human and nonhuman primate brains: are they allometrically scaled versions of the same design? *Evolutionary Anthropology* 15:65–77.

Rilling, J. K., Barks, S. K., Parr, L. A., Preuss, T. M., Faber, T. L., Pagnoni, G., Bremner, J. D., and Votaw, J. R. 2007. A comparison of resting-state brain activity in humans and chimpanzees. *Proceedings of the National Academy of Sciences of the United States of America* 104:17146–17151.

Rilling, J. K., Glasser, M. F., Preuss, T. M., Ma, X., Zhao, T., Hu, X., and Behrens, T. I. J. 2008. The evolution of the arcuate fasciculus revealed with comparative DTI. *Nature Neuroscience* 11:426–428.

Rilling, J. K., and Insel, T. R. 1999a. Differential expansion of neural projection systems in primate brain evolution. *NeuroReport* 10:1453–1459.

———. 1999b. The primate neocortex in comparative perspective using magnetic resonance imaging. *Journal of Human Evolution* 37:191–223.

Rilling, J. K., Scholz, J., Preuss, T. M., Glasser, M. F., Errangi, B. K., and Behrens, T. E. 2012. Differences between chimpanzees and bonobos in neural systems supporting social cognition. *Social Cognitive and Affective Neuroscience* 7:369–379.

Rilling, J. K., and Seligman R. A. 2002. A quantitative morphometric comparative analysis of the primate temporal lobe. *Journal of Human Evolution* 42:505–533.

Ringo, J. L., Doty, R. W., Demeter, S., and Simard, P. Y. 1994. Time is of the essence: a conjecture that hemispheric specialization arises from interhemispheric conduction delay. *Cerebral Cortex* 4:331–343.

Riss, D., and Goodall, J. 1977. The recent rise to the alpha-rank in a population of free-living chimpanzees. *Folia Primatologica* 27:134–151.

Riss, D. C., and Busse, C. D. 1977. Fifty-day observation of a free-ranging adult male chimpanzee. *Folia Primatologica* 28:283–297.

Ristau, C. A. 1983. Language, cognition, and awareness in animals? *Annals of the New York Academy of Sciences* 406:170–186.

———. 1996. Animal language and cognition projects. In A. Lock and C. R. Peters, eds., *Handbook of Symbolic Evolution*, 644–685. Oxford: Clarendon Press.

Ristau, C. A., and Robbins, D. 1982. Language in the great apes: a critical review. *Advances in the Study of Behavior* 12:141–255.

Rizzolatti, G., and Arbib, M. A. 1998. Language within our grasp. *Trends in Neuroscience* 21:188–194.

Rizzolatti, G., and Buccino, G. 2005. The mirror neuron system and its role in imitation and language. In S. Dehaene, J.-R. Duhamel, M. D. Hauser, and G. Rizzolatti eds., *From Monkey Brain to Human Brain*, 213–233. Cambridge, MA: MIT Press.

Rizzolatti, G., and Craighero, L. 2004. The mirror-neuron system. *Annual Review of Neurosciences* 27:169–192.

Rizzolatti, G. Fadiga, L., Gallese, V., and Fogassi, L. 1996a. Premotor cortex and the recognition of motor actions. *Cognitive Brain Research* 3:131–141.

Rizzolatti, G. Fadiga, L., Matelli, M., Bettinardi, V., Paulesu, E., Perani, D., and Fazio, F. 1996b. Localization of grasp representations in humans by PET: 1. Observation versus execution. *Experimental Brain Research* 111:247–252.

Rizzolatti, G. Fogassi, L., and L., Gallese, V. 2001. Neurophysiological mechanisms underlying the understanding of imitation of action. *Nature Reviews Neuroscience* 2:661–667.

Rizzolatti, G., and Sinigaglia, C. 2006. *Mirrors in the Brain: How Our Minds Share Actions and Emotions*. New York: Oxford University Press.

Roach, M. 2008. Almost human. *National Geographic* 214:124–144.

Robbins, A. M., and Robbins, M. M. 2005. Fitness consequences of dispersal decisions for male mountain gorilla *(Gorilla beringei beringei)*. *Behavioral Ecology and Sociobiology* 58:295–309.

Robbins, A. M., Robbins, M. M., and Fawcett, K. A. 2007. Maternal investment of the Virunga mountain gorillas. *Ethology* 113:235–245.

Robbins, A. M., Robbins, M. M., Gerald-Steklis, N., and Steklis, H. D. 2006. Age-related patterns of reproductive success among female mountain gorillas. *American Journal of Physical Anthropology* 131:511–521.

Robbins, A. M., Stoinski, T. S., Fawcett, K. A., and Robbins, M. M. 2009a. Does dispersal cause reproductive delays in female mountain gorillas? *Behaviour* 146:525–549.

———. 2009b. Evidence of a second folivore paradox. *Behaviour* 146:525–549.

———. 2011. Lifetime reproductive success of female mountain gorillas. *American Journal of Physical Anthropology* 145:582–593.

Robbins, L. M. 1987. Hominid footprints from Site G. In M. D. Leakey and J. M. Harris, eds., *Laetoli: A Pliocene Site in Northern Tanzania,* 497–502. Oxford: Clarendon Press.

Robbins, M. M. 1995. A demographic analysis of male life history and social structure of mountain gorillas. *Behaviour* 132:21–47.

———. 1996. Male-male interactions in heterosexual and all-male wild mountain gorilla group. *Ethology* 102:942–965.

———. 1997. A demographic analysis of male life history and social structure of mountain gorillas. *Behaviour* 132:21–47.

———. 1999. Male mating patterns in wild multimale mountain gorilla groups. *Behaviour* 57:1013–1020.

———. 2003. Behavioral aspects of sexual selection in mountain gorillas. In C. B. Jones, ed., *Sexual Selection and Reproductive Competition in Primates: New Directions and Perspectives,* 476–501. Norman, OK: American Society of Primatologists.

———. 2007. Gorillas. Diversity in ecology and behavior. In C. J. Campbell, A. Fuentes, K. C. MacKinnon, M. Panger, and S. K. Bearder, eds., *Primates in Perspective,* 305–321. Oxford: Oxford University Press.

———. 2008. Feeding competition and agonistic relationships among Bwindi *Gorilla beringei. International Journal of Primatology* 29:999–1018.

———. 2009. Male aggression against females in mountain gorillas: courtship or coercion? In M. N. Muller and R. W. Wrangham, eds., *Sexual Coercion in Primates and Humans,* 112–127.

Robbins, M. M., Bermejo, M., Cipolletta, C., Magliocca, F., Parnell, R. J., and Stokes, E. 2004. Social structure and life-history patterns in western gorillas *(Gorilla gorilla gorilla). American Journal of Primatology* 64:145–159.

Robbins, M. M., Gray, M., Kagoda, E., and Robbins, A. M. 2009. Population dynamics of the Bwindi mountain gorillas. *Biological Conservation* 142:2886–2895.

Robbins, M. M., and McNeilage, A. 2003. Home range and frugivory patterns of mountain gorillas in Bwindi Impenetrable National Park, Uganda. *International Journal of Primatology* 24:467–491.

Robbins, M. M., and Robbins, A. M. 2004. Simulation of the population dynamics and social structure of the Virunga Mountain gorillas. *American Journal of Primatology* 63:201–223.

Robbins, M. M., and Sawyer, S. C. 2007. Intergroup encounters in mountain gorillas of Bwindi Impenetrable National Park, Uganda. *Behaviour* 144:1497–1519.

Robinson, G. E., Fernald, R. D., and Clayton, D. F. 2008. Genes and social behavior. *Science* 322:896–899.

Robinson, J. T. 1954. The genera and species of the Australopithecinae. *American Journal of Physical Anthropology* 12:181–200.

———. 1956. The dentition of the Australopithecinae. *Memoirs of the Transvaal Museum*, no. 9: 1–179.

———. 1958. Cranial cresting patterns and their significance in the Hominoidea. *American Journal of Physical Anthropology* 16:397–428.

———. 1961. The australopithecines and their bearing on the origin of man and of stone tool-making. *South African Journal of Science* 57:3–13.

———. 1963. Adaptive radiation in the australopithecines and the origin of man. In F. C. Howell and F. Bourlière, eds., *African Ecology and Human Evolution*, 385–416. New York: Wenner-Gren Foundation for Anthropological Research.

———. 1965. *Homo 'habilis'* and the australopithecines. *Nature* 205:121–124.

———. 1966. The distinctiveness of *Homo habilis*. *Nature* 209:957–960.

———. 1967. Variation and taxonomy of the early hominids. In T. Dobzhansky, M. K. Hecht, and W. C. Steere, eds., *Evolutionary Biology*, Vol. 1, 69–100. New York: Appleton-Century-Crofts.

———. 1968a. The origin and adaptive radiation of the australopithecines. In G. Kurth, ed., *Evolution und Hominisation*, 2nd ed., 150–175. Stuttgart: Gustav Fischer.

———. 1968b. Dentition and adaptation in early hominids. *Proceedings VIIIth International Congress of Anthropological and Ethnological Sciences*, Vol. 1, 302–305. Tokyo: Science Council of Japan.

———. 1972a. *Early Hominid Posture and Locomotion*. Chicago: University of Chicago Press.

———. 1972b. The bearing of East Rudolf fossils on early hominid systematics. *Nature* 240:239–240.

Robinson, J. T., and Steudel, K. 1973. Multivariate discrimination analysis of dental data bearing on early hominid affinities. *Journal of Human Evolution* 2:509–527.

Roche, H., and Tiercelin, J.-J. 1980. Industries lithiques de la formation plio-pléistocène d'Hadar Ethiopie (campagne 1976). In R. E. Leakey and B. A. Ogot, eds., *Proceedings of the 8th Panafrican Congress of Prehistory and Quaternary Studies, Nairobi, 5 to 10 September 1977*, 194–199. Nairobi: International Louis Leakey Memorial Institute for African Prehistory.

Rockel, A. J., Hiorns, R. W., and Powell, P. S. 1980. The basic uniformity in structure of the neocortex. *Brain* 103:221–244.

Rodman, P. S. 1973. Population composition and adaptive organization among orang-utans of the Kutai Reserve, In R. P. Michael and J. H. Crook, eds., *Comparative Ecology and Behaviour of Primates*, 171–209. London: Academic Press.

———. 1977. Feeding behaviour of orang-utans of the Kutai Nature Reserve, East Kalimantan. In T. H. Clutton-Brock, ed., *Primate Ecology: Studies of feeding and ranging behaviour in lemurs, monkeys and apes*, 383–413. London: Academic Press.

———. 1978. Diets, densities, and distributions of Bornean primates. In G. G. Montgomery, ed., *The Ecology of Arboreal Folivores,* 465–478. Washington, DC: Smithsonian Institution Press.

———. 1979. Individual activity pattern and the solitary nature of orangutans. In D. A. Hamburg and E. R. McCown, eds., *The Great Apes,* 234–255. Menlo Park, CA: Benjamin/Cummings.

———. 1984. Foraging and social systems of orangutans and chimpanzees. In P. R. S. Rodman and J. G. H. Cant, eds., *Adaptations for Foraging in Nonhuman Primates,* 134–160. New York: Columbia University Press.

———. 1988. Diversity and consistency in ecology and behavior. In J. H. Schwartz, ed., *Orang-utan Biology,* 31–51. Oxford: Oxford University Press.

Rodman, P. S., and McHenry, H. M. 1980. Bioenergetics and the origin of hominid bipedalism. *American Journal of Physical Anthropology* 52:103–106.

Rodman, P. S., and Mitani, J. C. 1986. Orangutans: sexual dimorphism in a solitary species. In B. B. Smuts, D. L. Cheney, R. M. Seyfarth, R. W. Wrangham, and T. T. Struhsaker, eds., *Primate Societies,* 146–154. Chicago: University of Chicago Press.

Rodseth, L., Wrangham, R. W., Harrigan, A. M., and Smuts, B. B. 1991. The human community as a primate society. *Current Anthropology* 32:221–254.

Roebroeks, W. 2008. Time for the Middle to Upper Paleolithic transition in Europe. *Journal of Human Evolution* 55:918–926.

Roebroeks, W., and Villa, P. 2011. On the earliest evidence for habitual use of fire in Europe. *Proceedings of the National Academy of Sciences of the United States of America* 108:5209–5212.

Roffman, I., Savage-Rumbaugh, S., Rubert-Pugh, E., Ronen, A., and Nevo, E. 2012. Stone tool production and utilization by bonobo-chimpanzees *(Pan paniscus). Proceedings of the National Academy of Sciences of the United States of America* 109:14500–14503.

Rogers, C. M., and Davenport, R. K. 1975. Capacities of nonhuman primates for perceptual integration across sensory modalities. In R. H. Tuttle, ed., *Socioecology and Psychology of Primates,* 343–352. The Hague: Mouton.

Rogers, J. 1993. The phylogenetic relationships among *Homo, Pan,* and *Gorilla:* a population genetics perspective. *Journal of Human Evolution* 25:201–215.

———. 1994. Levels of genealogical hierarchy and the problem of hominoid phylogeny. *American Journal of Physical Anthropology* 94:81–88.

Rogers, J., and Comuzzie, A. G. 1995. When is ancient polymorphism a potential problem for molecular phylogenetics? *American Journal of Physical Anthropology* 98:216–218.

Rogers, L. J. 1995. Evolution and development of brain asymmetry, and its relevance to language, tool use and consciousness. *International Journal of Comparative Psychology* 8:1–15.

Rogers, L. J., and Andrew, R. J. 2002. *Comparative Vertebrate Lateralization.* New York: Cambridge University Press.

Rogers, L. J., and Kaplan, G. 1994. A new form of tool use by orang-utans in Sabah, East Malaysia. *Folia Primatologica* 63:50–52.

———. 1996. Hand preferences and other lateral biases in rehabilitated orang-utans, *Pongo pygmaeus pygmaeus. Animal Behaviour* 51:13–25.

Rogers, M. E., Abernethy, K., Bermejo, M., Cipolletta, C., Doran, D., McFarland, K., Nishihara, T., Remis, M., and Tuttin, D. E. G. 2004. Western gorilla diet: a synthesis from six sites. *American Journal of Primatology* 64:173–192.

Rogers, M. E., Maisels, F., Williamson, E. A., Fernandez, M., and Tutin, D. E. G. 1990. Gorilla diet in the Lopé Reserve, Gabon: a nutritional analysis. *Oecologia* 84:326–339.

Rogers, M. E., Maisels, F., Williamson, E. A., Tutin, D. E. G., and Fernandez, M. 1992. Nutritional aspects of gorilla food choice in the Lopé Reserve, Gabon. In N. Itoigawa, Y. Sugiyama, G. P. Sackett, and R. K. R. Thompson, eds., *Topics in Primatology*, Vol. 2: *Behavior, Ecology and Conservation*, 255–266. Tokyo: University of Tokyo Press.

Rogers, M. E., Tutin, D. E. G., Williamson, E. A., Parnell, R. J., Voysey, B. C., and Fernandez, M. 1994. Seasonal feeding on bark by gorillas: an unexpected keystone food? In B. Thierry, J. R. Anderson, J. J. Roeder, and N. Herrenschmidt, eds., *Current Primatology*, Vol. 1: *Ecology and Evolution*, 37–43. Strasbourg: Université Louis Pasteur.

Rogers, M. E., Voysey, B. C., McDonald, K. E., Parnell, R. J., and Tutin, D. E. G. 1998. Lowland gorillas as seed dispersers: the importance of nest sites. *American Journal of Primatology* 45:45–68.

Rogers, M. E., Williamson, E. A., Tutin, D. E. G., and Fernandez, M. 1988. Effects of the dry season on gorilla diet in Gabon. *Primate Report* 22:25–33.

Rogers, R., and Hammerstein, O., II, 1949. *South Pacific.* New York: Random House.

Rogers, T. T., Lambon Ralph, M. A., Garrard, P., Bozeat, S., McClelland, J. L., Hodges, J. R., and Patterson, K. 2004. Structure and deterioration of semantic memory: a neuropsychoogical and computational investigation. *Psychological Review* 111:205–235.

Rögl, F. 1999. Mediterranean and Paratethys paleogeography during the Oligocene and Miocene. In J. Agusti, L. Rook, and P. Andrews, eds., *Hominoid Evolution and Climatic Change in Europe*, Vol. 1: *The Evolution of Neogene Terrestrial Ecosystems in Europe*, 8–22. Cambridge: Cambridge University Press.

Rögl, F., and Daxner-Höck, G. 1996. Late Miocene Paratethys correlations. In R. L. Bernor, V. Fahlbusch, and H.-W. Mittmann, eds., *The Evolution of Western Eurasian Neogene Mammal Faunas,* 47–55. New York: Columbia University Press.

Rohleder, H. O. 1918. Künstliche Zeugung und Anthropogenie. *Monographien über Zeugung beim Menschen* 6:1–243.

Röhrer-Ertl, O. 1988. Research history, nomenclature, and taxonomy of the orang-utan. In J. H. Schwartz, ed., *Orang-utan Biology* 7–18. New York: Oxford University Press.

Roma, P. G., Silberberg, A., Huntsberry, M. E., Christensen, C. J., Ruggiero A. M., and Soumi, S. J. 2007. Mark tests for mirror self-recognition in capuchin monkeys *(Cebus apella)* trained to touch marks. *American Journal of Primatology* 69:989–1000.

Romanes, G. J. 1989. *Mental Evolution in Man: Origin of Human Faculty.* New York: Appelton.

Romanski, E. M. 2013. Integration of faces and vocalizations in ventral prefrontal cortex: implications for the evolution of audiovisal speech. *Proceedings of the National Academy of Sciences of the United States of America* 109:10717–10724.

Romero, T., and de Waal, F. B. M. 2011. Third-party postconflict affiliation of aggressors in chimpanzees. *American Journal of Primatology* 73:397–404.

Romero-Herrera, A. E., Lieska, N., Goodman, M., and Simons, E. L. 1979. The use of amino acid sequence analysis in assessing evolution. *Biochimie* 61:767–779.

Roof, K. A., Hopkins, W. D., Izard, M. K., Hook, M., and Schapiro, S. J. 2005. Maternal age, parity, and reproductive outcome in captive chimpanzees *(Pan troglodytes)*. *American Journal of Primatology* 67:199–207.

Rook, L. 1993. A new find of *Oreopithecus* (Mammalia, Primates) in the Baccinello basin (Grosseto, Southern Tuscany). *Rivista Italiana di Paleontologia e Stratigrafica* 99:255–262.

Rook, L., Bondioli, L., Köhler, M., Moyà-Solà, S., Macchiarelli, R. 1999. *Oreopithecus* was a bipedal ape after all: evidence from the iliac cancellous architecture. *Proceedings of the National Academy of Sciences of the United States of America* 96:8795–8799.

Rook, L., Renne, P., Benvenuti, M., and Papini, M. 2000. Geochronology of *Oreopithecus*-bearing succession at Baccinello (Italy) and the extinction pattern of European Miocene hominoids. *Journal of Human Evolution* 39:577–558.

Rosaldo, R. 1989. *Culture and Truth: The Remaking of Social Analysis.* Boston: Beacon Press.

Rosas, A., and Bermúdez de Castro, J. M. 1998. On the taxonomic affinities of the Dmanisi mandible (Georgia). *American Journal of Physical Anthropology* 107:145–162.

Rose, L. M. 1997. Vertebrate predation and food-sharing in *Cebus* and *Pan. International Journal of Primatology* 18:727–765.

Rose, L. M., Perry, S., Panger, M. A., Jack, K., Manson, J. H., Gros-Louis, J., Mackinnon, K. C., and Vogel, E. 2003. Interspecific interactions between *Cebus capucinus* and other species: data from three Costa Rican sites. *International Journal of Primatology* 24:759–796.

Rose, M. D. 1973. Quadrupedalism in primates. *Primates* 14:337–357.

———. 1974a. Ischial tuberosities and ischial callosities. *American Journal of Physical Anthropology* 40:375–383.

———. 1974b. Postural adaptations in New and Old World monkeys. In F. A. Jenkins, ed., *Primate Locomotion,* 201–222. New York: Academic Press.

———. 1976. Bipedal behavior of olive baboons *(Papio anubis)* and its relevance to an understanding of the evolution of human bipedalism. *American Journal of Physical Anthropology* 44:247–261.

———. 1983. Miocene hominoid postcranial morphology: monkey-like, ape-like, neither, or both? In R. L. Ciochon and R. S. Corruccini, ed., *New Interpretations of Ape and Human Ancestry,* 405–417. New York: Plenum Press.

———. 1984a. Hominoid postcranial specimens from the Middle Miocene Chinji Formation, Pakistan. *Journal of Human Evolution* 13:503–516.

———. 1984b. Food acquisition and the evolution of positional behaviour: the case of bipedalism. In D. J. Chivers, B. A. Wood, and A. Bilsborough, eds., *Food Acquisition and Processing in Primates,* 509–524. New York: Plenum.

———. 1986. Further hominoid postcranial specimens from the Late Miocene Nagri Formation of Pakistan. *Journal of Human Evolution* 15:333–367.

———. 1988. Another look at the anthropoid elbow. *Journal of Human Evolution* 17:193–224.

―――. 1989. New postcranial specimens of catarrhines from the Middle Miocene Chinji Formation, Pakistan: descriptions and a discussion of proximal humeral functional morphology in anthropoids. *Journal of Human Evolution* 18:131–162.

―――. 1991. The process of bipedalization in hominids. In Y. Coppens and B. Senut, *Origine(s) de la bipédie chez les Hominidés,* eds., 37–48. Paris: Éditions du Centre National de la Recherche Scientifique.

―――. 1992. Kinematics of the trapezium-1st metacarpal joint in extant anthopoids and Miocene hominoids. *Journal of Human Evolution* 22:255–266.

―――. 1993. Locomotor anatomy of Miocene hoinoids. In D. L. Gebo, ed., *Postcranial Adaptation in Nonhuman Primates,* 252–272 DeKalb: Northern Illinois University Press.

―――. 1994. Quadrupedalism in some Miocene catarrhines. *Journal of Human Evolution* 26:387–411.

Rose, M. D., and Fleagle, J. G. 1981. The fossil history of nonhuman primates in the Americas. In A. F. Coimbra-Filho, R. A. Mittermeier, eds., *Ecology and Behavior of Neotropical Primates,* Vol. 1: 111–167. Rio de Janeiro: Academia Brasileira de Ciências.

Rose, M. D., Leakey, M. G., Leakey, R. E. F., and Walker, A. C. 1992. Postcranial specimens of *Simiolus enjiessi* and other primitive catarrhines from the early Miocene of Lake Turkana, Kenya. *Journal of Human Evolution* 22:171–237.

Rose, M. D., Nakano, Y., and Ishida, H. 1996. *Kenyapithecus* postcranial specimens from Nachola, Kenya. *African Study Monographs* 24 (Suppl.): 3–56.

Rosenbaum, S., Silk, J. B., and Stoinski, T. S. 2011. Male-immature relationships in multimale groups of mountain gorillas *(Gorilla beringei beringei). American Journal of Primatology* 73:356–365.

Rosenberg, K., and Trevathan, W. 1995. Bipedalism and human birth: the obstetrical dilemma revisited. *Evolutionary Anthropology* 4:161–168.

Rosenberger, A. L. 1986. Platyrrhines, catarrhines and the anthropoid transition. In B. Wood, L. Martin, and P. Andrews, eds., *Major Topics in Primate and Human Evolution,* 66–88. Cambridge: Cambridge University Press.

Ross, C. F. 2000. Into the light: the origin of Anthropoidea. *Annual Review Anthropology* 29:147–194.

Ross, C. F., and Kay, R. F. 2004. *Anthropoid Origins: New Visions.* New York: Plenum Press.

Ross, M. D., and Geissmann, T. 2007. Call diversity of wild male orangutans: a phylogenetic approach. *American Journal of Primatology* 69:305–324.

Rossianov, K. 2002. Beyond species: Il'ya Ivanov and his experiments on cross-breeding humans with anthropoid apes. *Science in Context* 15:277–316.

Rossie, J. B. 2005. Anatomy of the nasal cavity and paranasal sinuses in *Aegyptopithecus* and Early Miocene African catarrhines. *American Journal of Physical Anthropology* 126:250–267.

Rossie, J. B., and MacLatchy, L. 2006. A new pliopithecoid genus from the early Miocene of Uganda. *Journal of Human Evolution* 50:568–586.

Rothman, J. M., Bowman, D. D., Kalema-Zikusoka, G., and Nkurunungi, J. B. 2006. The parasites of the gorillas in Bwindi Impenetrable National Park Uganda. In N. E.

Newton-Fisher, H. Notman, J. D. Paterson, and V. Reynolds, eds., *Primates of Western Uganda,* 171–192. New York: Springer.

Rothman, J. M., Chapman, C. A., and Pell, A. N. 2008. Fiber-bound nitrogen in gorilla diets: implications for estimating dietary protein intake of primates. *American Journal of Primatology* 70:690–694.

Rothman, J. M., Dierenfeld, E. S., Hintz, H. F., and Pell, A. N. 2008. Nutritional quality of gorilla diets: consequences of age, sex and season. *Oecologia* 155:111–122.

Rothman, J. M., Dierenfeld, E. S., Molina, D. O., Shaw, A. V., Hintz, H. F., and Pell, A. N. 2006. Nutritional chemistry of foods eaten by gorillas in Bwindi Impenetrable National Park, Uganda. *American Journal of Primatology* 68:675–691.

Rothman, J. M., Pell, A. N., Dierenfeld, E. S., and McCann, C. M. 2006. Plant choice in the construction of night nests by gorillas in the Bwindi Impenetrable National Park, Uganda. *American Journal of Primatology* 68:361–368.

Rothman, J. M., Pell, A. N., Nkurunungi, J. B., and Dierenfeld, E. S. 2006. Nutritional aspects of the diet of wild gorillas. In N. E. Newton-Fisher, H. Notman, J. D. Paterson, and V. Reynolds, eds., *Primates of Western Uganda,* 152–169. New York: Springer.

Rothman, J. M., Van Soest, P. J., and Pell, A. N. 2006. Decaying wood is a sodium source for mountain gorillas. *Biology Letters* 2:321–324.

Rougier, N. P., Noelle, D. C., Braver, T. S., Cohen, J. D., and O'Reilly, R. C. 2005. Prefrontal cortex and flexible cognitive control: rules without symbols. *Proceedings of the National Academy of Sciences of the United States of America* 102:7338–7343.

Rouquet, P., Froment. J.-M., Bermejo, M., Kilbourn, A., Karesh, W., Reed, P., Kumulungui, B., Yaba, P., Délicat, A., Rollin, P. E., and Leroy, E. M. 2005. Wild animal mortality monitoring and human Ebola outbreaks, Gabon and Republic of Congo, 2001–2003. *Emerging Infectious Diseases* 11:2883–290.

Roush, W. 1997. Chimp retirement plan proposed. *Science* 277:471.

Rousseau, J.-J. 1755. *Essai sur l'origine des langues ou il est parlé de la mélodie et de l'imitation musicale.* Reprint edition, introduction and notes by Charles Porset (1970). Bordeaux: Ducros.

Rowe, N. 1996. *The Pictorial Guide to the Living Primates.* East Hampton, NY: Pogonias Press.

Rowlett, R. M. 1990. Burning issues in fire taphonomy. In F. Lopez, ed., *Communicaciones de la Reunion de Tafonomia y Fosilizacion,* 320–336. Madrid: Universidad Complutense de Madrid.

Ruano, G., Rogers, J., Ferguson-Smith, A. C., and Kidd, K. K. 1992. DNA sequence polymorphism within hominoid species exceeds the number of phylogenetically informative characters for a HOX2 locus. *Molecular Biology and Evolution* 9:575–586.

Ruff, C. 1990. Body mass and hindlimb bone cross-sectional and articular dimensions in anthropoid primates. In J. Damuth and B. J. MacFadden, eds., *Body Size in Mammalian Paleobiology: Estimation and Biological Implications,* 119–149, 367–369. Cambridge: Cambridge University Press.

———. 2007. Body size prediction from juvenile skeletal remains. *American Journal of Physical Anthropology* 133:698–716.

————. 2010. Body size and body shape in early hominins—implications of the Gona pelvis. *Journal of Human Evolution* 58:166–178.

Ruff, C. B., Trinkaus, E., Walker, A., and Larsen, C. S. 1993. Postcranial robusticity in *Homo*. I. Temporal trends and mechanical interpretation. *American Journal of Physical Anthropology* 91:21–53.

Ruff, C. B., and Walker, A. 1993. Body size and body shape. In A. Walker and R. Leakey, eds., *The Nariokotome Homo erectus Skeleton,* 234–265. Cambridge, MA: Cambridge University Press.

Ruff, C. B., Walker, A., and Teaford, M. F. 1989. Body mass, sexual dimorphism and femoral proportions of *Proconsul* from Rusinga and Mfangano Islands, Kenya. *Journal of Human Evolution* 18:515–536.

Ruff, C. B., Walker, A., and Trinkaus, E. 1994. Postcranial robusticity in *Homo*. III. Ontogeny. *American Journal of Physical Anthropology* 93:35–54.

Rui, H., Gerhardt, P., Mevag, B., Thomassen, Y., and Parvis, K. 1984. Seminal plasma characteristics during frequent ejaculation. *International Journal of Andrology* 7:119–128.

Rumbaugh, D. M. 1965. Maternal care in relation to infant behavior in the squirrel monkey. *Psychological Reports* 16:171–176.

————. 1969. The transfer index: an alternative measure of learning set. In C. R. Carpenter, ed., *Proceedings of the Second International Congress of Primatology, Atlanta, GA 1968,* Vol. 1, 267–173. Basel: Karger.

————. 1970. Learning skills of anthropoids. In L. A. Rosenblum, ed., *Primate Behavior: Developments in Field and Laboratory,* Vol. 1, 1–70. New York: Academic Press.

————. 1971a. Chimpanzee intelligence. In G. H. Bourne, ed., *The Chimpanzee,* Vol. 4, 19–45. Basel: Karger.

————. 1971b. Evidence of qualitative differences in learning processes among primates. *Journal of Comparative and Physiological Psychology* 76:250–255.

————. 1974. Comparative primate learning and its contributions to understanding development, play, intelligence, and language. In B. Chiarelli, ed., *Perspectives in Primate Biology,* 253–281. New York: Plenum Press.

————. 1975. The learning and symbolizing capacities of apes and monkeys. In R. H. Tuttle, ed., *Socioecology and Psychology of Primates,* 353–365. The Hague: Mouton.

————. 1977a. *Language Learning by a Chimpanzee: The Lana Project.* New York: Academic Pres.

————. 1977b. Language behavior of apes. In A. M. Schrier, ed., *Behavioral Primatology: Advances in Research and Theory,* Vol. 1, 105–138. Hillsdale, NJ: Lawrence Erlbaum.

————. 1977c. The emergence and state of ape language research. In G. H. Bourne, ed., *Progress in Ape Research,* 75–83. New York: Academic Press.

————. 1978. Ape language projects: a perspective. In D. Chivers and J. Herbert, eds., *Recent Advances in Primatology.* Vol. 1: *Behaviour,* 855–859. London: Academic Press.

————. 1985. Comparative psychology: patterns in adaptation. In A. M. Rogers and C. J. Scheirer, eds., *The G. Stanley Hall Lecture Series* 5:7–53, Washington, DC: American Psychological Association.

———. 1990. Comparative psychology and the great apes: their competence in learning, language, and numbers. *Psychological Record* 40:15–39.

———. 1993. Learning about primates' learning, language, and cognition. In G. Brannigan and M. Merrens, eds., *The Undaunted Psychologist: Adventures in Research,* 90–109. Philadelphia: Temple University Press.

———. 1997. The psychology of Harry F. Harlow: a bridge from radical to rational behaviorism. *Philosophical Psychology* 10:197–210.

———. 2002. Emergents and rational behaviorism. *Eye on Psi Chi* 6:8–14.

Rumbaugh, D. M., and Beran, M. J. 2005. Language acquisition by animals. In L. Nadel, ed., *Encyclopedia of Cognitive Science,* Vol. 1, 700–707. New York: Wiley.

Rumbaugh, D. M., Beran, M. J., and Savage-Rumbaugh, E. S. 2003. Language. In D. Maestripieri, ed., *Primate Psychology,* 395–423. Cambridge, MA: Harvard University Press.

Rumbaugh, D. M., and Gill, T. V. 1973. The learning skills of great apes. *Journal of Human Evolution* 2:171–179.

———. 1974. Language in man, monkey and machine. *Science* 185:872–873.

———. 1975. Language, apes, and the apple which—is orange, please. In S. Kondo, M. Kawai, E. Ehara, and S. Kawamura, *Symposia of the Fifth International Primatological Society,* eds., 247–257. Tokyo: Japan Science Press.

———. 1976a. Language and acquisition of language-type skills by a chimpanzee *(Pan). Annals of the New York Academy of Sciences* 270:90–123.

———. 1976b. The mastery of language-type skills by a chimpanzee *(Pan). Annals of the New York Academy of Sciences* 280:562—578.

———. 1977. Language and language-type communication: studies with a chimpanzee. In M. Lewis and L. A. Rosenblum, eds., *Interaction, Conversation, and the Development of Language,* 115–131. New York: Wiley.

Rumbaugh, D. M., Gill, T. V., Brown, J. V., von Glasersfeld, E., Pisani, P., Warner, H., and Bell, C. L. 1973. A computer-controlled language training system for investigating the language skills of young apes. *Behavioral Research Methods and Instrumentation* 5:385–392.

Rumbaugh, D. M., Gill, T. V., and von Glasersfeld, E. 1973. Reading and sentence completion by a chimpanzee *(Pan). Science* 182:731–733.

———. 1974. Language in man, monkeys and machines. *Science* 185:871–872.

Rumbaugh, D. M., Gill, T. V., von Glasersfeld, E., Warner, H., and Pisani, P. 1975. Conversations with a chimpanzee in a computer-controlled environment. *Biological Psychiatry* 10:627–641.

Rumbaugh, D. M., Hopkins, W. D., Washburn, D. A., and Savage-Rumbaugh, E. S. 1989. Lana chimpanzee learns to count by "NUMATH": a summary of a videotaped experimental report. *Psychological Record* 39:459–470.

———. 1991. Comparative perspectives of brain, cognition, and language. In N. A. Krasnegor, D. M. Rumbaugh, R. L. Schiefelbusch and M. Studdert-Kennedy, eds., *Biological and Behavioral Determinants of Language Development,* 145–164. Hillsdale, NJ: Lawrence Erlbaum.

Rumbaugh, D. M., King, J. E., Beran, M. J., Washburn, D. A., and Gould, K. L. 2007. A salience theory of language and behavior: with perspectives on neurobiology and cognition. *International Journal of Primatology* 28:973–996.

Rumbaugh, D. M., and Pate, J. L. 1984a. The evolution of cognition in primates: a comparative perspective. In H. L. Roitblat, T. G. Bever, and H. S. Terrace, eds., *Animal Cognition,* 569–587. Hillsdale, NJ: Lawrence Erlbaum.

———. 1984b. Primates' learning by levels. In G. Greenberg and E. Tobach, eds., *Behavioral Evolution and Integrative Levels,* 221–240. Hillsdale, NJ: Lawrence Erlbaum.

Rumbaugh, D. M, Riesen, A. H., and Wright, S. C. 1972. Creative responsiveness to objects: a report of a pilot study with young apes. *Folia Primatologica* 17:397–403.

Rumbaugh, D. M., and Savage-Rumbaugh, E. S. 1990. Chimpanzees: competencies for language and numbers. In W. C. Stebbin and M. A. Berkley, eds., *Comparative Perception,* Vol. 2: *Complex Signals,* 409–441. New York: Wiley.

Rumbaugh, D. M., Savage-Rumbaugh, E. S., and Gill, T. V. 1978. Language skills, cognition and the chimpanzee. In F. C. C. Peng, ed., *Sign Language and Language Acquisition in Man and Ape: New Dimensions in Comparative Pedolinguistics,* 137–159. Boulder, CO: Westview Press.

Rumbaugh, D. M., Savage-Rumbaugh, E. S., and Hagel, M. T. 1987. Summation in the chimpanzee *(Pan troglodytes). Journal of Experimental Psychology: Animal Behavior Processes* 13:107–115.

Rumbaugh, D. M, Savage-Rumbaugh, E. S., and Pate, J. L. 1988. Addendum to "Summation in the chimpanzee *(Pan troglodytes)." Journal of Experimental Psychology: Animal Behavior Processes* 14:118–120.

Rumbaugh, D. M., Savage-Rumbaugh, E. S., and Scanlon, J. L. 1982. The relationship between language in apes and human beings. In *Primate Behavior,* J. L. Fobes and J. E. King, eds., p. 361–385. New York: Academic Press.

Rumbaugh, D. M., Savage-Rumbaugh, E. S., and Sevcik, R. A. 1994. Biobehavioral roots of language: a comparative perspective of chimpanzee, child, and culture. In R. W. Wrangham, W. C. McGrew, F. B. M. de Waal, and P. G. Heltne, eds., *Chimpanzee Cultures,* 319–334. Cambridge, MA: Harvard University Press.

Rumbaugh, D. M., Savage-Rumbaugh, E. S., and Washburn, D. A. 1996. Toward a new outlook on primate learning and behavior: complex learning and emergent processes in comparative perspective. *Japanese Psychological Research* 38:113–125.

Rumbaugh, D. M., and Steinmetz, G. T. 1971. Discrimination reversal skills of the lowland gorilla *(Gorilla gorilla). Folia primatologica* 16:144–152.

Rumbaugh, D. M., von Glasersfeld, E., Gill, T. V., Warner, H., and Pisani, P., Brown, J. V., and Bell, C. L. 1975. The language skills of a young chimpanzee in a computer-controlled training situation. In R. H. Tuttle, ed., *Socioecology and Psychology of Primates,* 391–401. The Hague: Mouton.

Rumbaugh, D. M., von Glasersfeld, E., Warner, H., and Pisani, P., and Gill, T. V. 1974. Lana (chimpanzee) learning language: a progress report. *Brain and Language* 1:205–212.

Rumbaugh, D. M., von Glasersfeld, E., Warner, H., Pisani, P., Gill, T. V., Brown, J. V., and Bell, C. L. 1973. Exploring the language skills of Lana chimpanzee. *International Journal of Symbology* 4:1–9.

Rumbaugh, D. M., and Washburn, D. A. 1993. Counting by chimpanzees and ordinality judgements by macaques in video-formatted tasks. In S. T. Boysen, and E. J. Capaldi, eds., *The Development of Numerical Competence: Animal and Human Models,* 87–106. Hillsdale, NJ: Lawrence Erlbaum.

———. 2003. *Intelligence of Apes and Other Rational Beings.* New Haven, CT: Yale University Press.

Rumbaugh, D. M., Washburn, D. A., and Hillix, W. A. 1996. Respondents, operants and *Emergents:* toward an integrated perspective on behavior. In K. Pribram and J. King, eds., *Learning as a Self-Organizing Process,* 57–73. Hillsdale, NJ: Lawrence Erlbaum.

Rumbaugh, D. M., Washburn, D. A., King, J. E., Beran, M. J., Gould, K., and Savage-Rumbaugh, S. 2008. Why some apes imitate and/or emulate observed behavior and others do not: fact, theory and implications for our kind. *Journal of Cognitive Education and Psychology* 7:101–110.

Rumpler, Y., and Dutrillaux, B. 1990. Chromosomal evolution and speciation in primates. *Cell Biology Reviews* 23:1–136.

Russak, S. M., and McGrew, W. C. 2008. Chimpanzees as fauna: comparisons of sympatric large mammals across long-term study sites. *American Journal of Primatology* 70:402–409.

Russon, A. E. 1999. Orangutans' imitation of tool use: a cognitive interpretation. In S. T. Parker, R. W. Mitchell, and H. L. Miles, eds., *The Mentalities of Gorillas and Orangutans,* 117–146. Cambridge: Cambridge University Press.

Russon, A. E., Bard, K. A., and Parker, S. T. 1996. *Reaching into Thought: The Minds of the Great Apes.* Cambridge: Cambridge University Press.

Russon, A. E., and Galdikas, B. M. F. 1995. Imitation and tool use in rehabilitant orangutans. In R. D. Nadler, B. M. F. Galdikas, L. K. Sheernan, and N. Rosen, eds., *The Neglected Ape,* 191–197. New York: Plenum Press.

Russon, A. E., Wich, S. A., Ancrenaz, M., Kanamori, T., Knott, C. D., Kuze, N., Morrogh-Bernard, H. C., Pratje, P., Ramlee, H., Rodman, P., et al. 2009. Geographic variation in orangutan diets. In S. A. Wich, S. S. Utami Atmoko, T. Mitra Setia, and C. P. van Schaik, eds., *Orangutans: Geographic Variation in Behavioral Ecology and Conservation,* 135–156. Oxford: Oxford University Press.

Rutberg, A. T. 1983. The evolution of monogamy in primates. *Journal of Theoretical Biology* 104:93–112.

Ruvolo, M. 1994. Molecular evolutionary processes and conflicting gene trees: the hominoid case. *American Journal of Physical Anthropology* 94:89–113.

———. 1996. A new approach to studying modern human origins: hypothesis testing with coalescence time distributions. *Molecular Phylogenetics and Evolution* 5:202–219.

Ruvolo, M., Disotell, T. R., Allard, M. W., Brown, W. M., and Honeycutt, R. L. 1991. Resolution of the African hominoid trichotomy by use of a mitochondrial gene sequence. *Proceedings of the National Academy of Sciences of the United States of America* 88:1570–1574.

Ruvolo, M., Pan, D., Zehr, S., Goldberg, T., Disotell, T. R., and von Dornum, M. 1994. Gene trees and hominoid phylogeny. *Proceedings of the National Academy of Sciences of the United States of America* 91:8900–8904.

Ruvolo, M., and Pilbeam, D. 1986. Hominoid evolution: molecular and palaeontological patterns. In B. Wood, L. Martin, and P. Andrews, eds., *Major Topics in Primate and Human Evolution,* 157–160. Cambridge: Cambridge University Press.

Ruvolo, M., and Smith, T. F. 1986. Phylogeny and DNA-DNA hybridization. *Molecular Biology and Evolution* 3:285–289.

Ruvolo, M., Zehr. S., von Dornum, M., Pan, D., Chang, B., and Lin, J. 1993. Mitochondrial COII sequences and modern human origins. *Molecular Biology and Evolution* 10:1115–1135.

Ryan, A. S., and Johanson, D. C. 1989. Anterior dental microwear in *Australopithecus afarensis:* comparisons with human and nonhuman primates. *Journal of Human Evolution* 18:235–268.

Ryder, O. A. 2003. An introductory perspective: *Gorilla* systematics, taxonomy, and conservation in the era of genomics. In A. B. Taylor and M. L. Goldsmith, eds. *Gorilla Biology: A Multidisciplinary Perspective,* 239–246. Cambridge: Cambridge University Press.

Ryder, O. A., and Chemnick, L. G. 1993. Chromosomal and mitochondrial DNA variation in orang utans. *Journal of Heredity* 84:405–409.

Rymer, R. 1994. *Genie: A Scientific Tragedy.* New York: HarperPerennial.

Sabater Pí, J. 1977. Contribution to the study of alimentation of lowland gorillas in the natural state, in Río Muni, Republic of Equatorial Guinea (West Africa). *Primates* 18:183–204.

———. 1979. Feeding behaviour and diet of chimpanzees *(Pan troglodytes troglodytes)* in the Okorobikó Mountains of Rio Muni (West Africa). *Zeitschrift für Tierpsychologie* 50:265–281.

———. 1980–81. Exploitation of gorillas *Gorilla gorilla gorilla* (Savage & Wyman 1847) in Rio Muni, Republic of Equatorial Guinea, West Africa. *Biological Conservation* 19:131–140.

Sabater Pí, J., Bermejo, M., Illera, G., and Veà, J. J. 1993. Behavior of bonobos *(Pan paniscus)* following their capture of monkeys in Zaire. *International Journal of Primatology* 14:797–804.

Sabater Pí, J., and Veà, J. J. 1994. Comparative inventory of foods commonly consumed by the wild pygmy chimpanzees *(Pan paniscus)* in the Lilungu-Lokofe region of the Republic of Zaire. *African Journal of Zoology* 108:381–396.

Sabater Pí, J., Veà, J. J., and Serrallonga, J. 1997. Did the first hominids build nests? *Current Anthropology* 38:914–916.

Sadier, B., Delanoy, J.-J., Benedetti, L., Bourlès, D. L., Jaillet, S., Geneste, J. M., Lebatard, A.-E., and Arnold, M. 2012. Further constraints on Chauvet cave artwork elaboration. *Proceedings of the National Academy of Sciences of the United States of America* 109:802–8006.

Sahlins, M. 1976. *Culture and Practical Reason.* Chicago: University of Chicago Press.

———. 1981. *Historical Metaphors and Mythical Realities.* Ann Arbor: University of Michigan Press.

———. 2013. *What Kinship Is—And Is Not.* Chicago: University of Chicago Press.

Saitou, N. 1991. Reconstruction of molecular phylogeny of extant hominoids from DNA sequence data. *American Journal of Physical Anthropology* 84:75–85.

Saitou, N., and Nei, M. 1986. The number of nucleotides required to determine the branching order of three species, with special reference to the human-chimpanzee-gorilla divergence. *Journal of Molecular Evolution* 24:189–204.

Sakai, K. L. 2005. Language acquisition and brain development. *Science* 310:815–819.

Sakamaki, T., Itoh, N., and Nishida, T. 2001. An attempted within-group infanticide in wild chimpanzees. *Primates* 42:359–366.

Sakoyama, Y., Hong, K.-J., Byun, S. M., Hisajima, H., Ueda, S., Yaoita, Y., Hayashida, H., Miyata, T., and Honjo, T. 1987. Nucleotide sequences of immunoglobin epsilon genes of chimpanzee and orangutan: DNA molecular clock and hominoid evolution. *Proceedings of the National Academy of Sciences of the United States of America* 84:1080–1084.

Sakura, O. Factors affecting party size and composition of chimpanzees *(Pan troglodytes verus)* at Bossou, Guinea. *International Journal of Primatology* 15:167–183.

Salem, A.-H., Ray, D. A., Xing, J., Callinan, P. A., Myers, J. S., Hedges, D. J., Garber, R. K., Witherspoon, D. J., Jorde, L. B., and Batzer, M. A. 2003. Alu elements and hominid phylogenetics. *Proceedings of the National Academy of Sciences of the United States of America* 100:12787–12791.

Sallet, J., Mars, R. B., Noonan, M. P., Andersson, J. L., O'Reilly, J. X., Jbabdi, S., Croxson, P. L., Jenkinson, M., Mille, K. S., and Rushworth, M. F. S. 2011. Social network size affects neural circuits in macaques. *Science* 334:697–700.

Sally, A., Dutheil, J. Y., Hillier, L. W., Jordan, G. E., Goodhead, I., Herrero, J., Hobolth, A., et al. 2012. Insights into hominid evolution from the gorilla genome sequence. *Nature* 483:169–175.

Saltonstall, K., Amato, G., and Powell, J. 1998. Mitochondrial DNA variability in Grauer's gorillas of Kahuzi-Biega National Park. *Journal of Heredity* 89:129–135.

Samollow, P. B., Cherry, L. M., White, S. M., and J. Rogers. 1996. Interspecific variation at the Y-linked *RPS4Y* locus in hominoids: implications for phylogeny. *American Journal of Physical Anthropology* 101:333–343.

Sanders, R. J. 1985. Teaching apes to ape language: explaining the imitative and nonimitative signing of a chimpanzee *(Pan troglodytes)*. *Journal of Comparative Psychology* 99:197–210.

Sanders, W. J., and Bodenbender, B. E. 1994. Morphometric analysis of lumbar vertebra UMP 67–28: implications for spinal function and phylogeny of the Miocene Moroto hominoid. *Journal of Human Evolution* 26:203–237.

Sanders, W. J., Trapani, J., and Mitani, J. C. 2003. Taphonomic aspects of crowned hawk-eagle predation on monkeys. *Journal of Human Evolution* 44:87–105.

Sandgathe, D. M., Dibble, H. L., Godlberg, P., and McPherron, S. P. 2011. The Roc de Marsal Neandertal child: a reassessment of its status as a deliberate burial. *Journal of Human Evolution* 61:243–253.

Sandler, W., Meir, I., Padden, C., and Aronoff, M. 2005. The emergence of grammar: systematic structure in a new language. *Proceedings of the National Academy of Sciences of the United States of America* 102:2661–2665.

Santa Luca, A. P. 1980. The Ngandong fossil hominids: a comparative study of a Far Eastern *Homo erectus* group. *Yale University Publications in Anthropology* 78:1–175.

Sanz, C. M., and Morgan, D. B. 2007. Chimpanzee tool technology in the Goualougo Triangle, Republic of Congo. *Journal of Human Evolution* 52:420–433.

———. 2009. Flexible and persistent tool-using strategies in honey-gathering by wild chimpanzees. *International Journal of Primatology* 30:411–427.

Sanz, C. M., Morgan, D. B., and Gulick, S. 2007. New insights into chimpanzees, tools, and termites from the Congo Basin. *American Naturalist* 164:567–581.

Sanz, C. M., Schöning, C., and Morgan, D. B. 2010. Chimpanzees prey on army ants with specialized tool set. *American Journal of Primatology* 72:17–24.

Sargeant, B. L., and Mann, J. 2009. From social learning to culture: intrapopulation variation in bottlenose dolphins. In K. N. Laland and B. G. Galef, eds., *The Question of Animal Culture,* 152–173. Cambridge, MA: Harvard University Press.

Sargis, E. J., Boyer, D. M., Bloch, J. I., and Silox, M. T. 2007. Evolution of pedal grasping in Primates. *Journal of Human Evolution* 53:103–107.

Sarich, V. M. 1968. The origin of the hominids: an immunological approach. In S. L. Washburn and P. C. Jay, eds., *Perspectives on Human Evolution 1,* 94–121. New York: Holt, Rinehart and Winston.

———. 1971. A molecular approach to the question of human origins. In P. Dolhinow and V. M. Sarich, eds., *Background for Man,* 60–81, Boston: Little, Brown.

Sarich, V. M., and Cronin, J. E. 1976. Molecular systematics of the Primates. In M. Goodman and R. E. Tashian, eds., *Molecular Anthropology,* 141–170. New York: Plenum Press.

Sarich, V. M., Schmid, C. W., and Marks, J. 1989. DNA hybridization as a guide to phylogenies: a critical analysis. *Cladistics* 5:3–32.

Sarich, V. M., and Wilson, A. C. 1967a. Immunological time scale for hominid evolution. *Science* 158:1200–1203.

———. 1967b. Rates of albumin evolution in primates. *Proceedings of the National Academy of Sciences of the United States of America* 58:142–148.

———. 1973. Generation time and genomic evolution in primates. *Science* 179:1144–1147.

Sarmiento, E. E. 1987. The phylogenetic position of *Oreopithecus* and its significance in the origin of the Hominoidea. *American Museum Novitates,* no. 2881:1–44.

———. 1988. Anatomy of the hominoid wrist joint: its evolutionary and functional implications. *International Journal of Primatology* 9:281–345.

———. 1996. Gorillas of the Bwindi Impenetrable Forest and the Virunga volcanoes: taxonomic implications of morphological and ecological differences. *American Journal of Primatology* 40:1–21.

———. 2003. Distribution, taxonomy, genetics, ecology, and casual links of gorilla survival: the need to develop practical knowledge for gorilla conservation. In A. B. Taylor and M. L. Goldsmith, eds., *Gorilla Biology: A Multidisciplinary Perspective,* 432–471. Cambridge: Cambridge University Press.

Sarmiento, E. E., and Oates, J. F. 2000. The Cross River gorillas: a distinct subspecies, *Gorilla gorilla diehli* (Matchie 1904). *American Museum Novitates,* no. 3304: 55.

Sarrazin F, and Barbault R. 1996. Reintroductions: challenges and lessons for basic ecology. *TREE* 11:474–478.

Sarringhaus, L. A., McGrew, W. C., and Marchant, L. F. 2005. Misuse of anecdotes in primatology: lessons from citation analysis. *American Journal of Primatology* 65:283–288.

Sarringhaus, L. A., Stock, J. T. Marchant, L. F., and McGrew, W. C. 2005. Bilateral asymmetry of the limb bones of the chimpanzee *(Pan troglodytes). American Journal of Physical Anthropology* 128:840–845.

Sartono, S. 1975. Implications arising from *Pithecanthropus* VIII. In R. H. Tuttle, ed., *Paleoanthropology: Morphology and Paleoecology,* 327–360. The Hague: Mouton.

Sarukhán, J. 1997. Global issues. *Science* 275:175.

Sati, J. P., and Alfred, J. R. B. 2002. Locomotion and posture in hoolock gibbon. *Annals of Forestry* 10:298–306.

Saussure, Ferdinand de. 1972. *Cours de linguistique générale.* Paris: Payot.

———. 2011. *Course in General Linguistics.* Trans. by W. Baskin. New York: Columbia University Press.

Savage-Rumbaugh, E. S. 1979. Symbolic communication—its origins and early development in the chimpanzee. *New Directions for Child Development* 3:1–15.

———. 1981. Can apes use symbols to represent their world? *Annals of the New York Academy of Sciences* 364:35–59.

———. 1984a. Verbal behavior at a procedural level in the chimpanzee. *Journal of the Experimental Analysis of Behavior* 41:223–250.

———. 1984b. *Pan paniscus* and *Pan troglodytes:* contrasts in preverbal communicative competence. In R. L. Susman, ed., *The Pygmy Chimpanzee: Evolutionary Biology and Behavior,* 395–413. New York: Plenum Press.

———. 1986. *Ape Language: From Conditioned Response to Symbol.* New York: Columbia University Press.

———. 1987. Communication, symbolic communication, and language: reply to Seidenberg and Petitto. *Journal of Experimental Psychology: General* 116:288–292.

———. 1988. A new look at ape language: comprehension of vocal speech and syntax. In D. W. Leger, ed., *Comparative Perspective in Modern Psychology, Nebraska Symposium on Motivation,* Vol. 35, 201–255. Lincoln: University of Nebraska Press.

———. 1990. Language acquisition in a nonhuman species: implications for the innateness debate. *Developmental Psychobiology* 23:599–620.

———. 1991. Language learning in the bonobo: how and why they learn. In N. A. Krasnegor, D. M. Rumbaugh, R. L. Schiefelbusch, and M. Studdert-Kennedy, eds. *Biological and Behavioral Determinants of Language Development,* 209–233. Hillsdale, NJ: Lawrence Erlbaum.

———. 1994. Hominid evolution: looking to modern apes for clues. In D. Quiatt and J. Itani, eds., *Hominid Culture in Primate Perspective,* 7–49. Niwot: University Press of Colorado.

Savage-Rumbaugh, E. S., Brakke, K. E., and Hutchins, S. S. 1992. Linguistic development: contrasts between co-reared *Pan troglodytes* and *Pan paniscus.* In T. Nishida, W. C. McGrew P. Marler, M. Pickford, and F. B. M. de Waal, eds., *Topics in Primatology,* Vol. 1: *Human Origins,* 51–66. Tokyo: University of Tokyo Press.

Savage-Rumbaugh, E. S., and Fields, W. M. 2000. Linguistic, cultural and cognitive capacities of bonobos *(Pan paniscus). Culture and Psychology* 6:131–153.

Savage-Rumbaugh, E. S., Fields, W. M., and Taglialatela, J. P. 2000. Language, speech, tools and writing. *Journal of Consciousness Studies* 8:273–292.

———. 2006. Rules and tools: beyond anthropomorphism. In N. Toth and K. Schick, eds., *The Oldowan: Case Studies into the Earliest Stone Age,* 223–241. Gosport, IN: Stone Age Institute Press.

Savage-Rumbaugh, E. S., and Lewin, R. 2004. *Kanzi: The Ape at the Brink of the Human Mind*. New York: Wiley.

Savage-Rumbaugh, E. S., and McDonald, K. 1988. Deception and social manipulation in symbol-using apes. In R. Byrne and A. Whiten, eds., *Machiavellian Intelligence: Social Expertise in the Evolution of Intellect in Monkeys, Apes, and Humans*, 224–237.

Savage-Rumbaugh, E. S., McDonald, K., Sevcik, R. A., Hopkins, W. D., and Rubert, E. 1986. Spontaneous symbol acquisition and communicative use by pygmy chimpanzees *(Pan paniscus)*. *Journal of Experimental Psychology: General* 115:211–235.

Savage-Rumbaugh, E. S., Murphy, J., Sevcik, R. A., Brakke, K. E., Williams, S. L., Rumbaugh, D. M., and Bates, E. 1993. Language comprehension in ape and child. *Monographs of the Society for Research in Child Development* 233, vol. 58, nos. 3–4.

Savage-Rumbaugh, E. S., Pate, J. L., Lawson, J., Smith, S. T., and Rosenbaum, S. 1983. Can a chimpanzee make a statement? *Journal of Experimental Psychology: General* 112:457–492.

Savage-Rumbaugh, E. S., Romski, M. A., Hopkins, W. D., and Sevcik, R. A. 1989. Symbol acquisition and use by *Pan troglodytes, Pan paniscus,* and *Homo sapiens*. In P. G. Heltne and L. A. Marquardt, eds., *Understanding Chimpanzees*, 266–295. Cambridge, MA: Harvard University Press.

Savage-Rumbaugh, E. S., Romski, M. A., Sevcik, R. A., and Pate, J. L. 1983. Assessing symbol usage versus symbol competency. *Journal of Experimental Psychology: General* 112:508–512.

Savage-Rumbaugh, E. S., and Rumbaugh, D. M. 1978. Symbolization, language, and chimpanzees: a theoretical reevaluation based on initial language acquisition processes in four young *Pan troglodytes*. *Brain and Language* 6:265–300.

———. 1979. Chimpanzee problem comprehension: insufficient evidence. *Science* 206:1201–1202.

———. 1980. Language analogue project, phase II: theory and tactics. In K. E. Nelson, ed. *Children's Language*, Vol. 2, 267–307. New York: Gardner Press.

———. 1982. Ape language research is alive and well: a reply. *Anthropos* 77:568–573.

———. 1993. The emergence of language. In K. R. Gibson and T. Ingold, eds., *Tools, Language and Cognition in Human Evolution*, 86–108. Cambridge: Cambridge University Press.

Savage-Rumbaugh, E. S., Rumbaugh, D. M., and Boysen, S. 1978a. Symbolic communication between two chimpanzees *(Pan troglodytes)*. *Science* 201:641–644.

———. 1978b. Linguistically mediated tool use and exchange by chimpanzees *(Pan troglodytes)*. *Behavioral and Brain Sciences* 4:539–554.

———. 1980. Do apes use language? One research group considers the evidence for representational abilities in apes. *American Scientist* 68:49–61.

Savage-Rumbaugh, E. S., Rumbaugh, D. M., and Fields, W. M. 2006. Language as a window on rationality. In S. Hurley and M. Nudds, eds., *Rational Animals?* 513–552. Oxford: Oxford University Press.

———. 2009. Empirical Kanzi: the ape language controversy revisited. *Skeptic* 15:25–33.

Savage-Rumbaugh, E. S., Rumbaugh, D. M., and McDonald, K. 1985. Language learning in two species of apes. *Neuroscience and Biobehavioral Reviews* 9:653–665.

Savage-Rumbaugh, E. S., Rumbaugh, D. M., Smith, S. T., and Lawson, J. 1980. Reference: the linguistic essential. *Science* 210:922–925.

Savage-Rumbaugh, E. S., Segerdahl, P., and Fields, W. M. 2005. Individual differences in language competencies in apes resulting from unique rearing conditions imposed by different first epistemologies. In L. L. Namy, ed., *Symbol Use and Symbolic Representation: Developmental and Comparative Perspectives,* 199–219. Mahwah, NJ: Lawrence Erlbaum.

Savage-Rumbaugh, E. S., and Sevcik, R. A. 1984. Levels of communicative competency in the chimpanzee: Pre-representational and representational. In G. Greenberg and E. Tobach, eds., *Behavioral Evolution and Integrative Levels,* 197–219. Hillsdale, NJ: Lawrence Erlbaum.

Savage-Rumbaugh, E. S., Sevcik, R. A., Brakke, K. E., Rumbaugh, D. M., and Greenfield, P. M. 1990. Symbols: their communicative use, comprehension, and combination by bonobos *(Pan paniscus).* In C. Rovee-Collier and L. P. Lipsitt, eds., *Advances in Infancy Research,* Vol. 6, 221–271. Norwood, NJ: ABLEX.

Savage-Rumbaugh, E. S., Sevcik, R. A., and Hopkins, W. D. 1988. Symbolic cross-modal transfer in two species of chimpanzees. *Child Development* 59:617–625.

Savage-Rumbaugh, E. S., Sevcik, R. A., Rumbaugh, D. M., and Rubert, E. 1985. The capacity of animals to acquire language: do species differences have anything to say to us? *Philosophical Transactions of the Royal Society of London, B* 308:177–185.

Savage-Rumbaugh, E. S., Shanker, S. G., and Taylor, T. J. 1998. *Apes, Language, and the Human Mind.* Oxford: Oxford University Press.

Savage-Rumbaugh, E. S., Toth, N., and Schick, K. 2007. Kanzi learns to knap stone tools. In D. A. Washburn, ed., *Primate Perspectives on Behavior and Cognition,* 279–291. Washington, DC: American Psychological Association.

Savage-Rumbaugh, E. S., Wamba, K., and Wamba, P. 2007. Welfare of apes in captive environments: comments on, and by, a specific group of apes. *Journal of Applied Animal Welfare Science* 10:7–19.

Savage-Rumbaugh, E. Sue, and Wilkerson, Beverly J. 1978. Socio-sexual behavior in *Pan paniscus* and *Pan troglodytes:* a comparative study. *Journal of Human Evolution* 7:327–344.

Savini, T., Boesch, C., and Reichard, U. H. 2009. Varying ecological quality influences the probability of polyandry in white-handed gibbons *(Hylobates lar)* in Thailand. *Biotropica* 41:503–513.

Sawada, Y., Pickford, M., Itaya, T., Mikinouchi, T., Tateishi, M., Kabeto, K., Ishida, S., and Ishida, H. 1998. K-Ar ages of Miocene Hominoidea (*Kenyapithecus* and *Samburupithecus*) from Samburu Hills, northern Kenya. *Comptes rendus des séances de l'Académie des Sciences Paris, Sciences de la Terre et des planètes* 326:445–451.

Sawada, Y., Saneyoshi, M., Nakayama, K., Sakai, T., Itaya, T., Hyodo, M., Mukokya, Y., Pickford, M., Senut, B., Tanaka, S., et al. 2006. The ages and geological backgrounds of Miocene hominoids *Nacholapithecus, Samburupithecus,* and *Orrorin* from Kenya. In Ishida, H., Tuttle, R., Pickford, M., Ogihara, N., and Masato, N., eds., *Hominid Origins and Environmental Backgrounds,* 71–96. New York: Springer.

Sawyer, G. J., Deak, V., Sarmiento, E., Milner, R., Johanson, D. C., Leakey, M., and Tattersall, I. 2007. *The Last Human: A Guide to Twenty-Two Species of Extinct Humans.* New Haven, CT: Yale University Press.

Sawyer, S. C., and Robbins, M. M. 2009. A novel food processing technique by wild mountain gorilla *(Gorilla beringei beringei). Folia Primatologica* 80:83–88.

Scacco, A. M., Jr. 1975. *Rape in Prison.* Springfield, IL: Charles C. Thomas.

————. 1982. *Male Rape: A Casebook of Sexual Aggressions.* New York: AMS Press.

Schaller, G. B. 1960. The preservation of the gorillas in the Parc National Albert is threatened. *Current Anthropology* 1:331.

————. 1961. The orang-utan in Sarawak. *Zoologica* 46:73–82.

————. 1963. *The Mountain Gorilla: Ecology and Behavior.* Chicago: University of Chicago Press.

————. 1965. The behavior of the mountain gorilla. In I. DeVore, ed., *Primate Behavior, Field Studies of Monkeys and Apes,* 324–367. New York: Holt, Rinehart and Winston.

————. 1970. Mountain gorilla displays. In M. Bates, ed., *Field Studies in Natural History,* 193–201. New York: Van Nostrand Reinhold.

Schaller, G. B., and Emlen, J. T., Jr. 1963. Observations on the ecology and social behavior of the mountain gorilla. In F. C. Howell and F. Bourliere, eds., *African Ecology and Human Evolution,* 368–384. Viking Fund Publications in Anthropology, no. 36. New York: Wenner-Gren Foundation for Anthropological Research.

Schaller, G. B., and Lowther, G. R. 1969. The relevance of carnivore behavior to the study of early hominids. *Southwestern Journal of Anthropology* 25:307–341.

Schaumburg, F., Mugisha, L., Peck, B., Becker, K., Gillespie, T. R., Peters, G., and Leendertz, F. H. 2012. Drug-resistant human *Staphylococcus aureus* in sanctuary apes pose threat to endangered wild ape populations. *American Journal of Primatology* 74:1071–1075.

Scheibert, J., Leurent, S., Prevost, A., and Debrégeas, G. 2009. The role of fingerprints in the coding of tactile information probed with a biomimetic sensor. *Science* 323:1503–1506.

Schepartz, L. A. 1993. Language and modern human origins. *Yearbook of Physical Anthropology* 36:91–126.

Schick, K. D., and Toth, N. 1993. *Making Silent Stones Speak: Human Evolution and the Dawn of Technology.* New York: Simon & Schuster.

Schick, K. D., Toth, N., Garufi, G., Savage-Rumbaugh, E. S., Rumbaugh, D., and Sevick, R. 1999. Continuing investigations into the stone tool-making and tool-using capabilities of a bonobo *(Pan paniscus). Journal of Archaeological Science* 26:821–832.

Schilhab, T. S. S. 2004. What mirror self-recognition in nonhumans can tell about aspects of self. *Biology and Philosophy* 19:111–126.

Schilling, D. 1984. Song bouts and duetting in the concolor gibbon In H. Preuschoft, D. J. Chivers, W. Y. Brockelman, and N. Creel, eds., *The Lesser Apes: Evolutionary and Behavioural Biology,* 390–403, Edinburgh: Edinburgh University Press.

Schlaug, G., Jäncke, L., Schlaug, G., Huang, Y., and Steinmetz, H. 1995. In vivo evidence of structural brain asymmetry in musicians. *Science* 267:699–701.

Schlebusch, C. M., Skoglund, P., Sjödin, P., Gattepaille, L. M., Hernandez, D., Jay, F., Li, S., De Johgh, M., Singleton, A., Blum, M. G. B., Soodjall, H., and Jakobsson, M. 2012. Genomic variation in seven Khoe-San groups reveals adaptation and complex African history. *Science* 338:374–379.

Schlosser, M. 1924. Fossil primates from China. *Palaeontologica Sinica*, n.s., series D, 1, no. 2: 1–16.

Schmahmann, J. D., and Pandya, D. N. 1997. The cerebrocerebellar system. *International Review of Neurobiology* 41:31–60.

Schmelz, M., Call, J., and Tomasello, M. 2011. Chimpanzees know that others make inferences. *Proceedings of the National Academy of Sciences of the United States of America* 108:3077–3079.

Schmid, P. 1983. Eine Rekonstruktion des Skelettes von A. L. 288–1 (Hadar) und deren Konsequenzen. *Folia Primatologica* 40:283–306.

———. 2004. Functional interpretation of the Laetoli footprints. In D. J. Meldrum and C. E. Hilton, eds., *From Biped to Strider: The Emergence of Modern Human Walking, Running, and Resource Transport*, 49–62. New York: Kluwer Academic/Plenum Press.

Schmid, P., Churchill, S. E., Nallah, S., Weissen, E., Carlson, K. J., de Ruiter, D. J., and Berger, L. R. 2013. Mosaic morphology in the thorax of *Australopithecus sediba*. *Science* 340, http://dx.doi.org/10.1126/science1234598.

Schmid, P., and Piaget, A. 1994. Three-dimensional kinematics of bipedal locomotion. *Zeitschrift für Morphologie und Anthropologie* 80:79–87.

Schmidt, T. R., Wildman, D. E., Uddin, M., Opazo, J. C., Goodman, M., and Grossman, L. I. 2005. Rapid electrostatic evolution at the binding site for cytochrome c on cytochrome c oxidase in anthropoid primates. *Proceedings of the National Academy of Sciences of the United States of America* 102:6379–6384.

Schmitt, D., and Larson, S. G. 1995. Heel contact as a function of substrate type and speed in primates. *American Journal of Physical Anthropology* 96:39–50.

Schmitt, T. J., and Nairn, A. E. M. 1984. Interpretations of the magnetostratigraphy of the Hadar hominid site, Ethiopia. *Nature* 309:704–706.

Schmitt, T. J., Walter, R. C., Taieb, M., Tiercelin, J.-J., and Page, N. 1980. Magnetostratigraphy of the Hadar Formation of Ethiopia. In R. E. Leakey and B. A. Ogot, eds., *Proceedings of the 8th Panafrican Congress of Prehistory and Quaternary Studies Nairobi, 5 to 10 September 1977*, 53–55. Nairobi: International Louis Leakey Memorial Institute for African Prehistory.

Schmitz, R. W., Serre, D., Bonani, G., Feine, S., Hillgruber, F., Krainitzki, H., Pääbo, S., and Smith, F. H. 2002. The Neandertal type site revisited: interdisciplinary investigations of skeletal remains from the Neander Valley, Germany. *Proceedings of the National Academy of Sciences of the United States of America* 99:13342–13347.

Schneider, D. M. 1968. *American Kinship: A Cultural Account*. Englewood Cliffs, NJ: Prentice-Hall.

———. 1984. *A Critique of the Study of Kinship*. Ann Arbor: University of Michigan Press.

Schoeninger, M. J. 2007. Reconstructing early hominin diets. Evaluating tooth chemistry and macronutrient composition. In P. S. Ungar, ed., *Evolution of the Human Diet*, 150–164. Oxford: Oxford University Press.

————. 2011. In search of the australopithecines. *Nature* 474:43–45.

Schoeninger, M. J., Bunn, H. T., Murray, S., Pickering, T., and Moore, J. 2007. Meat-eating in the fourth African ape. In C. B. Stanford and H. T. Bunn, eds., *Meat-Eating and Human Evolution*, 179–195. Oxford: Oxford University Press.

Scholz, C. A., Johnson, T. C., Cohen, A. S., King, J. W., Peck, J. A., Overpeck, J. T., Talbot, M. R., Brown, E. T., Kalindekafe, L., Amoako, P. Y. O., et al. 2007. East Africa mega-droughts between 135–75 thousand years ago and bearing on early-modern human origins. *Proceedings of the National Academy of Sciences of the United States of America* 104:16416–16421.

Schöning, C., Humle, T., Möbius, Y., and McGrew, W. C. 2008. The nature of culture: technological variation in chimpanzee predation on army ants. *Journal of Human Evolution* 55:48–59.

Schön Ybarra, M. A., and Conroy, G. C. 1978. Non-metric features in the ulna of *Aegyptopithecus, Alouatta, Ateles*, and *Lagothrix*. *Folia Primatologica* 29:178–195.

Schrauf, C., Huber, L., and Visalberghi, E. 2008. Do capuchin monkeys use weight to select hammer tools? *Animal Cognition* 11:413–422.

Schroepfer, K. K., Rosati, A. G., Chartrand, T., and Hare, B. 2011. Use of "entertainment" chimpanzees in commercials distorts public perception regarding their conservation status. *PLoS One* 6:e26048.

Schultz, A. H. 1926. 1926. Fetal growth in man and other primates. *Quarterly Review of Biology* 1:465–521.

————. 1927. Studies on the growth of gorilla and of other higher primates with special reference to a fetus of gorilla, preserved in the Carnegie Museum. *Memoirs of the Carnegie Museum* 11:1–87.

————. 1933. Observations on the growth, classification and evolutionary specializations of gibbons and siamangs. *Human Biology* 5:385–428.

————. 1936. Characters common to higher primates and characters specific for man. *Quarterly Review of Biology* 11:259–283, 425–455.

————. 1937. Fetal growth and development of the rhesus monkey. *Contributions to Embryology* 26:71–99.

————. 1938. The relative weight of the testes in primates. *Anatomical Record* 72:387–394.

————. 1950. The specializations of man and his place among the catarrhine primates. *Cold Spring Harbor Symposia on Quantitative Biology* 15:37–52.

————. 1955. Primatology in its relation to anthropology. *Yearbook of Anthropology* 1:47–60.

————. 1956. Postembryonic age changes. In *Primatologia* 1:887–964, H. Hofer, A. H. Schultz, and D. Starck, eds. Basel: Karger.

————. 1960. Einige Beobachtungen und Masse am Skelett von *Oreopithecus Zeitschrift für Morphologie und Anthropologie* 50:136–149.

————. 1961. Vertebral column and thorax. In H. Hofer, A. H. Schultz, and D. Starck eds., *Primatologia*, Vol. 4,1–66. Basel: Karger.

————. 1962. Die Schädelkapazität männlicher Gorillas und ihr Höchstwert. *Anthropologischer Anzeiger* 25:197–203.

————. 1963a. Age changes, sex differences, and variability as factors in the classification of primates. In S. L. Washburn, ed., *Classification and Human Evolution,* 85–115. New York: Wenner-Gren Foundation for Anthropological Research.

————. 1963b. Relations between the lengths of the main parts of the foot skeleton in primates. *Folia Primatologica* 1:150–171.

————. 1969a. The skeleton of the chimpanzee. In G. H. Bourne, ed., *The Chimpanzee,* Vol. 1, 50–103. Basel: Karger.

————. 1969b. *The Life of Primates.* New York: Universe Books.

————. 1973. The skeleton of the Hylobatidae and other observations on their morphology. In D. M. Rumbaugh, ed., *Gibbon and Siamang,* Vol. 2, 1–54. Basel: Karger.

Schürmann, C. L. 1981. Courtship and mating behavior of wild orangutans in Sumatra. In A. B. Chiarelli and RS. Corruccini, eds., *Primate Behavior and Sociobiology,* 130–135. Berlin: Springer-Verlag.

————. 1982. Mating behaviour of wild orang utans. In L. E. M. De Boer, ed., *The Orang Utan: Its Biology and Conservation,* 269–294. The Hague: Dr W. Junk.

Schürmann, C. L., and van Hooff, J. A. R. A. M. 1981. Reproductive strategies of the orang-utan: new data and the reconsideration of existing sociosexual models. *International Journal of Primatology* 7:265–287.

Schusterman, R. J. 1962. Transfer effects of successive discrimination-reversed training in chimpanzees. *Science* 137:422–423.

Schwartz, G. T. 2000. Taxonomic and functional aspects of the patterning of enamel thickness distribution in extant large-bodied hominoids. *American Journal of Physical Anthropology* 111:221–244.

Schwartz, J. H. 1984a. On the evolutionary relationships of humans and orang-utans. *Nature* 308:501–505.

————. 1984b. Hominid evolution: a review and a reassessment. *Current Anthropology* 25:655–672.

————. 1988. History, morphology, paleontology, and evolution. In J. H. Schwartz, ed., *Orang-utan Biology,* 69–85. Oxford: Oxford University Press.

————. 1990. *Lufengpithecus* and its potential relationship to an orang-utan clade. *Journal of Human Evolution* 19:591–605.

————. 2005. *The Red Ape: Orangutans and Human Origins.* Boulder, CO: Westview Press.

————. 2007. Defining Hominidae. In W. Henke and I. Tattersall, eds., *Handbook of Paleoanthropology: Phylogeny of Hominids,* Vol. 3, 1379–1408. Berlin: Springer.

Schwartz, J. H., Vu The Long, N. L. C., Le Trung Kha, and Tattersall, I. 1994. A diverse hominoid fauna from the late Middle Pleistocene breccia cave of Tham Khuuyen, Socialist Republish of Vietnam. *Anthropological Papers of the American Museum of Natural History,* no. 73, 1–11.

Scotland, R. W., Olmstead, R. G., and Bennett, R. 2003. Phylogeny reconstruction: The role of morphology. *Systematic Biology* 52:539–548.

Scott, J. E., and Lockwood, C. A. 2004. Patterns of tooth crown size and shape variation in great apes and humans and species recognition in the hominid fossil record. *American Journal of Physical Anthropology* 125:303–319.

Scott, R. S., Kappelman, J., and Kelley, J. 1999. The paleoenvironment of *Sivapithecus parvada*. *Journal of Human Evolution* 36:245–274.

Scott, R. S., Ungar, P. S., Bergstrom, T. S., Brown, C. A., Grine, F. E., Teaford, M. F., and Walker, A. 2005. Dental microwear texture analysis shows within-species diet variability in fossil hominins. *Nature* 436:693–695.

Sebeok, T. A., and Rosenthal, R. 1981. The Clever Hans phenomenon: communication with horses, whales, apes and people. *Annals of the New York Academy of Sciences* 364:1–309.

Sebeok, T. A., and Umiker-Sebeok, J. 1980. *Speaking of Apes: A Critical Anthology of Two-Way Communication with Man.* New York: Plenum Press.

Segerdahl, P., Fields, W., and Savage-Rumbaugh, E. S. 2005. *Kanzi's Primal Language: The Cultural Initiation of Primates into Language.* New York: Palgrave Macmillan.

Seidenberg, M. S., and Petitto, L. A. 1979. Signing behavior in apes: a critical review. *Cognition* 7:177–215.

———. 1987. Communication, symbolic communication, and language: comment on Savage-Rumbaugh, McDonald, Sevcik, Hopkins, and Rupert (1986). *Journal of Experimental Psychology: General* 116:279–287.

Seidler, R. D., Purushotham, A., Kim, S.-G., Ugurbil, K., Willingham, D., and Ashe, J. 2002. Cerebellum activation associated with performance change but not motor learning. *Science* 296:1979–1980.

Seiffert, E. R. 2006. Revised age estimates for the later Paleogene mammal faunas of Egypt and Oman. *Proceedings of the National Academy of Sciences of the United States of America* 103:5000–5005.

Seiffert, E. R., Simons, E. L., and, Attia, Y. 2003. Fossil evidence for an ancient divergence of lorises and galagos. *Nature* 422:421–424.

Seiffert, E. R., Simons, E. L., Clyde, W. C., Rossie, J. B., Attia, Y., Bown, T. M., Chatrath, P., and Mathison, M. E. 2005. Basal anthropoids from Egypt and the antiquity of Africa's higher primate radiation. *Science* 310:300–304.

Seiffert, E. R., Simons, E. L., and Fleagle, J. G. 2000. Anthropoid humeri from the late Eocene of Egypt. *Proceedings of the National Academy of Sciences of the United States of America* 97:10062–10067.

Seiffert, E. R., Simons, E. L., and Simons, C. V. M. 2004. Phylogenetic, biogeographic, and adaptive implications of new fossil evidence bearing on crown anthropoid origins and early stem catarrhine evolution. In C. F. Ross and R. F. Kay, eds., *Anthropoid Origins: New Visions,* 157–181. New York: Kluwer Academic/Plenum Press.

Sémah, F. 1984. The Sangiran Dome in the Javanese Plio-Pleistocene chronology. *Courier Forschungsinstitut Senckenberg* 69:245–252.

Sémah, F., Falgueres, C., Yokoyama, Y., Féraud, G., Saleki, H., and Djubiantono, T. 1977. Arrivée et disparition des *Homo erectus* à Java, les données actuelles. *Abstracts of the Third Meeting of the European Association of Archaeologists,* 11–12.

Semaw, S. 2006. The oldest stone artifacts from Gona (2.6–2.5 MA), Afar, Ethiopia: implications for understanding the earliest stages of stone knapping. In N. Toth and K. Schick, eds., *The Oldowan: Case Studies into the Earliest Stone Age,* 43–75. Gosport, IN: Stone Age Institute Press.

Semaw, S., Renne, P., Harris, J. W. K., Feibel, C. S., Bernor, R. L., Fesseha, N., and Mowbray, K. 1997. 2.5-million-year-old stone tools from Gona, Ethiopia. *Nature* 385:333–336.

Semaw, S. Rogers, M. J., Quade, J., Renne, P. R., Butler, R. F., Domínguez-Rodrigo, M., Stout, D., Hart, W. S., Pickering, T., and Simpson, S. W. 2003. 2.6-million-year-old stone tools associated with bones from OGS-6 and OGS-7, Gona, Afar, Ethiopia. *Journal of Human Evolution* 45:169–177.

Semaw, S., Simpson, S. W., Quade, J., Renne, P. R., Butler, R. F., McIntosh, W. C., Levin, N., Dominguez-Rodrigo, M., and Rogers, M. J. 2005. Early Pliocene hominids from Gona, Ethiopia. *Nature* 433:301–305.

Semenderferi, K., Armstrong, E., Schleicher, A., Zilles, K., and Van Hoesen, G. W. 2001. Prefrontal cortex in humans and apes: a comparative study of area 10. *American Journal of Physical Anthropology* 114:224–241.

Semendeferi, K., and Damasio, H. 2000. The brain and its main anatomical subdivisions in living hominoids using magnetic resonance imaging. *Journal of Human Evolution* 38:317–332.

Sen, S., Koufos, G. D., Kondopoulou, D., and de Bonis, L. 2000. Magnetostratigraphy of the late Miocene continental deposits of the lower Axios valley, Macedonia, Greece. *Bulletin of the Geological Society of Greece* 9:197–206.

Senut, B. 1979. Comparaison des hominidés de Gomboré et de Kanapoi: deux pièces du genre *Homo? Bulletins et Mémoires de la Société d'Anthropologie de Paris, série XIII,* 6:111–117.

———. 1980. New data on the humerus and its joints in Plio-Pleistocene hominids. *Collegium Anthropologium* 4:87–94.

———. 1983a. Quelques remarques a propos d'un humérus d'hominoide pliocene provenant de Chemeron (bassin du lac Baringo, Kenya). *Folia Primatologica* 41:267–276.

———. 1983b. Les hominides Plio-Pleistocenes: essai taxinomique et phylogenetique a partir de certains os longs. *Bulletins et Mémoires de la Société d'Anthropologie de Paris, série XIII* 10:325–334.

———. 1986. Long bones of the primate upper limb: monomorphic or dimorphic? *Human Evolution* 1:7–22.

———. 1987. Upper limb skeletal elements of Miocene cercopithecoids from East Africa: implications for function and taxonomy. *Human Evolution* 2:97–106.

———. 1988a. Du nouveau sur les Primates paléogènes du continent arabo-africain. *Bulletins et Mémoires de la Société d'Anthropologie de Paris, série XIV,* 5:123–126.

———. 1988b. Taxonomie et fonction chez les Hominoidea Miocènes Africains: exemple de l'articulation du coude. *Annales de Paléontologie (Vert.-Invert.)* 74:129–154.

———. 1989a. *Le Coude des Primates Hominoïdes: Anatomie, Fonction, Taxonomie, Évolution.* Cahiers de Pléoanthropologie. Paris: Éditions du Centre National de la Recherche Scientifique.

———. 1989b. La locomotion des pré-hominidés. In G. Giacobini, ed., *Hominidae: Proceedings of the 2nd International Congress of Human Paleontology,* 53–60. Milan: Jaca Book.

————. 1991. Forme et mouvement chez des primates Néogènes de l'ancien monde. *Geobios* 24:193–199.

————. 2003. Paleontological approach to the evolution of hominid bipedalism: the evidence revisited. *Courier Forschungsinstitut Senkenberg* 243:125–134.

————. 2005. Arboreal origin of bipedalism. In H. Ishida, M. Pickford, M. Nakatsukasa and N. Ogihara and R. H. Tuttle, eds., *Human Origins and Environmental Backgrounds,* 195–204. New York: Kluwer Academic/Plenum Press.

————. 2012. From hominoid arboreality to hominid bipedalism. In S. C. Reynolds and A. N. Gallagher, eds., *African Genesis: Perspectives on Hominin Evolution,* 77–98. Cambridge: Cambridge University Press.

Senut, B., Nakatsukasa, M., and Ishida, H. 1999. Preliminary analysis of *Kenyapithecus* shoulder joint. In: *Abstracts of the International Symposium "Evolution of Middle-to-Late Miocene Hominoids in Africa," July 11–13,* Takaragaike, Kyoto, Japan.

Senut, B., Pickford, M., Gommery, D., and Kunimatsu, Y. 2000. Un nouveau genre d'hominoïde du Miocène inférieur d/Afrique orientale: *Ugandapthecus major* (Le Gros Clark & Leakey, 1950). *Comptes Rendus de l'Académie des Sciences, Paris* 331 IIa:227–233.

Senut, B., Pickford, M., Gommery, D., Mein, P., Cheboi, K., and Coppens, Y. 2001. First hominid from the Miocene (Lukeino Formation, Kenya). *Comptes Rendus de l'Académie des Sciences, Paris, Série II: Fascicule a, Sciences de la Terre et des planètes* 332:137–144.

Senut, B., and Tardieu, C. 1985. Functional aspects of Plio-Pleistocene hominid limb bones: implications for taxonomy and phylogeny. In E. Delson, ed., *Ancestors: The Hard Evidence,* 193–201. New York: Alan R. Liss.

Sept, J. M. 1990. Vegetation studies in the Semliki Valley, Zaire as a guide to paleoanthropological research. In N. T. Boaz, ed., *Evolution of Environments and Hominidae in the African Western Rift Valley,* 95–121. Virginia Museum of Natural History Memoir 1. Martinsville: Virginia Museum of Natural History.

————. 1992a. Archaeological evidence and ecological perspectives for reconstructing early hominid behavior. *Archaeological Method and Theory* 4:1–56.

————. 1992b. Was there no place like home? *Current Anthropology* 33:187–207.

————. 1994. Beyond bones: archaeological sites, early hominid subsistence, and the costs and benefits of exploiting wild plant foods in east African riverine landscapes. *Journal of Human Evolution* 27:295–320.

————. 2001. Modeling the edible landscape. In C. B. Stanford and H. T. Bunn, eds., *Meat-Eating and Human Evolution,* 73–98. Oxford: Oxford University Press.

————. 2007. Modeling the significance of paleoenvironmental context for early hominin diets. In P. S. Ungar, ed., *Evolution of the Human Diet,* 289–307. Oxford: Oxford University Press.

Sept, J. M., and Brooks, G. E. 1994. Reports of chimpanzee natural history, including tool use, in 16th- and 17th-century Sierra Leone. *International Journal of Primatology* 15:867–878.

Sepulchre, P., Ramstein, G., Fluteau, F., Schuster, M., Tiercelin, J.-J., and Brunet, M. 2006. Tectonic uplift and eastern Africa aridification. *Science* 313:1419–1423.

Seringhaus, M., and Gerstein, M. 2008. Genomics confounds gene classification. *American Scientist* 96:466–473.

Serre, D., Langaney, A., Chech, M., Teschler-Nicola, M., Paunovic, M., Mennecier, P., Hofreiter, M., Possnert, G., and Pääbo, S. 2004. No evidence of Neandertal mtDNA contribution to early modern humans. *PLoS Biology* 2:0313–0317.

Seth, S., and Seth, P. K. 1986. A review of evolutionary and genetic differentiation in primates. In J. G. Else and P. C. Lee, ed., 291–306. Cambridge: Cambridge University Press.

Setiawan, E., Knott, C. D., and Budhi, S. 1996. Preliminary assessment of vigilance and predator avoidance behavior of orangutans in Gunung Palung National Park Indonesia. *Tropical Biodiversity* 3:269–279.

Sevcik, R. A., and Savage-Rumbaugh, E. S. 1994. Language comprehension and use by great apes. *Language and Communication* 14:37–58.

Seyfarth, R. M., Cheney, D. L. 2013. Affiliation, empathy, and the origins of theory of mind. *Proceedings of the National Academy of Sciences of the United States of America* 110 (Suppl. 2): 10349–10356.

Seyfarth, R. M., Cheney, D. L., Harcourt, A. H., and Stewart, K. J. 1994. The acoustic features of gorilla double grunts and their relation to behavior. *American Journal of Primatology* 33:31–50.

Shafer, D. A., Myers, R. H., and Saltzman, D. 1984. Biogenetics of the siabon (gibbon-siamang hybrids). In H. Preuschoft, D. J. Chivers, W. Y. Brockelman, and N. Creel, eds., *The Lesser Apes: Evolutionary and Behavioural Biology,* 486–497. Edinburgh: Edinburgh University Press.

Shafer, D. D. 1993. Patterns of handedness: comparative study of nursery school children and captive gorillas. In J. P. Ward and W. D. Hopkins, eds., *Primate Laterality: Current Behavioral Evidence of Primate Asymmetries,* 267–283. New York: Springer.

———. 1997. Hand preference behaviors shared by two groups of captive bonobos. *Primates* 38:303–313.

Shankar, S. G., Savage-Rumbaugh, E. S., and Taylor, T. J. 1999. Kanzi: a new beginning. *Animal Learning and Behavior* 27:24–25.

Shapiro, G. L. 1982. Sign acquisition in a home-reared/free-ranging orangutan, comparisons with other signing apes. *American Journal of Primatology* 3:121–129.

Shapiro, G. L., and Galdikas 1999. Early sign performance in a free-ranging adult orangutan. In S. T. Parker, R. W. Mitchell, and H. L. Miles, eds., *The Mentalities of Gorillas and Orangutans,* 265–279. Cambridge: Cambridge University Press.

Shariff, G. A. 1953. Cell counts in the primate cerebral cortex. *Journal of Comparative Neurology* 98:381–400.

Sharma, K. K. 1984. The sequence of phased uplift of the Himalaya. In R. O. Whyte, ed., *The Evolution of the East Asian Environment,* volume 1, 56–70. Hong Kong: University of Hong Kong Centre of Asian Studies.

Sharp, D. J., Scott, S. K., and Wise, R. J. S. 2004. Retrieving meaning after temporal lobe infarction: the role of the basal language area. *Annals of Neurology* 56:836–846.

Sharp, P. M., Shaw, G. M., and Hahn, B. H. 2004. Simian immunodeficiency virus infection of chimpanzees. *Journal of Virology* 79:3891–3902.

Shaw, P., Greenstein, D., Lerch, J., Clasen, L., Lenroot, R., Gogtay, N., Evans, A., Rapoport, J., and Giedd, J. Intellectual ability and cortical development in children and adolescents. 2006. *Nature* 440:676–679.

Shea, B. T. 1981. Relative growth of the limbs and trunk in the African apes. *American Journal of Physical Anthropology* 56:179–201.

———. 1983. Paedomorphosis and neoteny in the pygmy chimpanzee. *Science* 222:521–522.

———. 1984. An allometric perspective on the morphological and evolutionary relationships between pygmy *(Pan paniscus)* and common *(Pan troglodytes)* chimpanzees. In R. L. Susman, ed., *The Pygmy Chimpanzee: Evolutionary Biology and Behavior,* 89–130. New York: Plenum Press.

———. 1985. On aspects of skull form in African apes and orangutans, with implications for hominoid evolution. *American Journal of Physical Anthropology* 68:329–342.

———. 1986. Scapula form and locomotion in chimpanzee evolution. *American Journal of Physical Anthropology* 70:475–488.

———. 1988. Phylogeny and skull form in the hominoid primates. In J. H. Schwartz, *Orang-utan Biology,* ed., 233–245. New York: Oxford University Press.

Shea, B. T., and Coolidge, H. J., Jr. 1988. Craniometric differentiation and systematics in the genus *Pan. Journal of Human Evolution* 17:671–685.

Shea, B. T., and Inouye, S. E. 1993. Knuckle-walking ancestors. *Science* 259:293–294.

Shea, B. T., Leigh, S. R., and Groves, C. P. 1993. Multivariate craniometric variation in chimpanzees. Implications for species identification in paleoanthropology. In W. H. Kimbel and L. B. Martin, eds., *Species, Species Concepts, and Primate Evolution,* 265–296. New York: Plenum Press.

Shea, J. J. 2007. Lithic archaeology, or what stone tools can (and can't) tell us about early hominin diets. In P. S. Ungar, ed., *Evolution of the Human Diet,* 212–229. Oxford: Oxford University Press.

Sheets-Johnstone, M. 1989. Hominid bipedality and sexual-selection theory. *Evolutionary Theory* 9:57–70.

Shellis, R. P., Benyon, A. D., Reid, D. J., and Hiiemae, K. M. 1998. Variations in molar enamel thickness among primates. *Journal of Human Evolution* 35:507–522.

Shen, S., Lin, L., Cai, J. J., Jiang, P., Kenkel, E. J., Stroik, M. R., Sato, S., Davidson, B. L., and Ying, Y. 2011. Widespread establishment and regulatory impact of Alu exons in human genes. *Proceedings of the National Academy of Sciences of the United States of America* 108:2837–2842.

Sherrow, H. M. 2005. Tool use in insect foraging by the chimpanzees of Ngogo, Kibale National Forest, Uganda. *American Journal of Primatology* 65:377–383.

Sherrow, H. M., and Amsler, A. J. 2007. New intercommunity infanticides by the chimpanzees of Ngogo, Kibale National Park, Uganda. *International Journal of Primatology* 28:9–22.

Sherwood, C. C., Broadfield, D. C., Holloway, R. L., Gannon, P. J., and Hof, P. R. 2003. Variability of Broca's area homologue in African great apes: implications for language evolution. *Anatomical Record Part A* 271A:26–285.

Sherwood, C. C., Holloway, R. L., Erwin, J. M., Schleicher, A., Zilles, K., and Hof, P. R. 2004. Cortical orofacial motor representation in Old World monkeys, great apes and humans. *Brain, Behavior and Evolution* 63:61–81.

Shillito, D. J., Gallup, G. G., Jr., and Beck, B. B. 1999. Factors affecting mirror behaviour in western lowland gorillas, *Gorilla gorilla. Animal Behaviour* 57:999–1004.

Shipman, P. 1986a. Paleoecology of Fort Ternan reconsidered. *Journal of Human Evolution* 15:193–204.

———. 1986b. Baffling limb on the family tree. *Discover* 7:86–93.

———. 2000. Doubting Dmanisi. *American Scientist* 88:491–494.

———. 2002. A worm's view of human evolution. *American Scientist* 90:508–510.

Shipman, P., Bosler, W., and Davis, K. L. 1981. Butchering of giant geladas at an Acheulian site. *Current Anthropology* 22:257–268.

Shipman, P., and Walker, A. 1989. The costs of becoming a predator. *Journal of Human Evolution* 18:373–392.

Shipman, P., Walker, A., van Couvering, J. A., Hooker, P. J., and Miller, J. A. 1981. The Fort Ternan hominoid site, Kenya: geology, age, taphonomy and paleoecology. *Journal of Human Evolution* 10:49–72.

Shir-Vertesh, D. 2012. "Flexible personhood": loving animals as family members in Israel. *American Anthropologist* 114:420–432.

Sholley, C. R. 1991. Conserving gorillas in the midst of guerillas. *American Association of Zoological Parks and Aquariums, Annual Proceedings,* 30–37. Oslo: A. Wilhelmsen Foundation.

Shore, B. 1996. *Culture in Mind.* Oxford: Oxford University Press.

Short, R. V. 1979. Sexual selection and its component parts, somatic and genital selection, as illustrated by man and the great apes. *Advances in the Study of Behavior* 9:131–158.

———. 1980. The great apes of Africa. *Journal of Reproduction and Fertility* 28:3–11.

———. 1981. Sexual selection in man and the great apes. In C. E. Graham, ed., *Reproductive Biology of the Great Apes,* 319–341. New York: Academic Press.

Shoshani, J., Groves, C. P., Simons, E. L., and Gunnell, G. F. 1996. Primate phylogeny: morphological vs molecular results. *Molecular Phylogenetics and Evolution* 5:102–154.

Shreeve, J. 1995. *The Neandertal Enigma.* New York: Avon Books.

Shrewsbury, M. M., and Johnson, R. K. 1983. Form, function and evolution of the distal phalanx. *Journal of Hand Surgery* 8:475–479.

Shrewsbury, M. M., Marzke, M. W., Linscheid, R. L., and Reece, S. P. 2003. Comparative morphology of the pollical distal phalanx. *American Journal of Physical Anthropology* 121:30–47.

Shriver, M. D., and Kittles, R. A. 2004. Genetic ancestry and the search for personalized genetic histories. *Nature Reviews Genetics* 5:611–718.

Shumaker, R. W., Palkovich, A. M., Beck, B. B., and Guagnano, G. A. 2001. Spontaneous use of magnitude discrimination and ordination by the orangutan *(Pongo pygmaeus). Journal of Comparative Psychology* 115:385–391.

Shumaker, R. W., Walkup, K. R., and Beck, B. B. 2011. *Animal Tool Behavior: The Use and Manufacture of Tools by Animals.* Baltimore: Johns Hopkins University Press.

Sibley, C. H., and Ahlquist, J. E. 1984. The phylogeny of the hominoid primates, as indicated by DNA-DNA hybridization. *Journal of Molecular Evolution* 20:2–15.

———. 1987. DNA hybridization evidence of hominoid phylogeny: results from an expanded data set. *Journal of Molecular Evolution* 26:99–121.

Sibley, C. H., Comstock, J. A., and Ahlquist, J. E. 1990. DNA hybridization evidence of hominoid phylogeny: a reanalysis of the data. *Journal of Molecular Evolution* 30:202–236.

Sicotte, P. 1993. Inter-group encounters and female transfer in mountain gorillas: influence of group composition on male behavior. *American Journal of Primatology* 30:21–36.

———. 1994. Effect of male competition on male-female relationships in bi-male groups of mountain gorillas *Ethology* 97:47–64.

———. 1995. Interpositions in conflicts between males in bimale groups of mountain gorillas. *Folia Primatologica* 65:14–24.

———. 2000. A case study of mother-son transfer in mountain groups. *Primates* 41:93–101.

———. 2001. Female mate choice in mountain gorillas. In M. M. Robbins, P. Sicotte, and K. J. Stewart, eds., *Mountain Gorillas: Three Decades of Research at Karisoke,* 59–87. Cambridge: Cambridge University Press.

———. 2002. The function of male aggressive displays towards females in mountain gorillas. *Primates* 4:277–289.

Siddiqi, N. A. 1986. Gibbons *(Hylobates hoolock)* in the West Banugach Reserved Forest of Sylhet District, Bangladesh. *Tigerpaper* 13:29–31.

Sigé, B., Jaeger, J.-J., Sudre, J., and Vianey-Liaud, M. 1990. *Altiatlasius koulchii* n. gen. et. sp., Primate Omomyidé du Paléocène supérieur de Maroc, et les origines des Euprimates. *Palaeontographica Abteilung A* 214:31–56.

Sigmon, B. A. 1971. Bipedal behavior and the emergence of erect posture in man. *American Journal of Physical Anthropology* 34:55–60.

———. 1975. Functions and evolution of hominid hip and thigh musculature. In R. H. Tuttle, ed., *Primate Functional Morphology and Evolution,* 235–252. The Hague: Mouton.

Sikela, J. M. 2006. The jewels of our genome: the search for the genomic changes underlying the evolutionarily unique capacities of the human brain. *PLoS Genetics* 2:646–655.

Sikes, N. E., Potts, R., and Behrensmeyer, A. K. 1999. Early Pleistocene habitat in Member 1 Ororgesailie based on paleosol stable isotopes. *Journal of Human Evolution* 37:721–746.

Silberbauer, G. B. 1981. *Hunter & Habitat in the Central Kalahari Desert.* Cambridge: Cambridge University Press.

Silk, J. B. 1996. Why do primate reconcile? *Evolutionary Anthropology* 5:39–42.

———. 1998. Making amends: adaptive perspectives on conflict remediation in monkeys, apes and humans. *Human Nature* 9:341–368.

———. 2002. The form and function of reconciliation in primates. *Annual Review of Anthropology* 31:21–44.

———. 2007. Chimps don't just get mad, they get even. *Proceedings of the National Academy of Sciences of he United States of America* 104:13537–13538.

Silk, J. B., Brosnan, S. F., Vonk, J., Henrich, J., Povinelli, D. J., Richardson, A. S., Lambeth, S. P., Mascaro, J., and Shapiro, S. J. 2005. Chimpanzees are indifferent to the welfare of unrelated group members. *Nature* 437:1357–1359.

Silk, J. B., and House, B. R. 2011. Evolutionary foundations of human prosocial sentiment. *Proceedings of the National Academy of Sciences of the United States of America* 108:10910–10917.

Sillen, A. 1992. Strontium calcium ratios (Sr/Ca) of *Australopithecus robustus* anassociated fauna from Swartkrans. *Journal of Human Evolution* 23:495–516.

Sillen, A., Hall, G., and Armstrong, R. 1995. Strontium calcium ratios (Sr/Ca) and strontium isotopic ratios (^{87}Sr/^{86}Sr) of *Australopithecus robustus* and *Homo* sp. from Swartkrans. *Journal of Human Evolution* 28:277–285.

Silox, M. T., Boyer, D. M., Bloch, J. I., and Sargis, E. J. 2007. Evolution of pedal grasping in Primates. *Journal of Human Evolution* 53:321–324.

Simmen, B., and Hladik, C. M. 1998. Sweet and bitter taste discrimination in Primates: scaling effect across species. *Folia Primatologia* 69:129–138.

Simmons, L. W., Firman, R. C., Rhodes, G., and Peters, M. 2004. Human sperm competition: testis size, sperm production and rates of extrapair copulations. *Animal Behaviour* 68:297–302.

Simons, E. L. 1961. The phyletic position of *Ramapithecus. Postilla* 57:1–9.

———. 1963. Some fallacies in the study of hominid phylogeny. *Science* 141:879–889.

———. 1964a. On the mandible of *Ramapithecus. Proceedings of the National Academy of Sciences of the United States of America* 51:528–535.

———. 1964b. The early relatives of man. *Scientific American* 211:51–62.

———. 1965. New fossil apes from Egypt and the initial differentiation of Hominoidea. *Nature* 205:135–139.

———. 1967. The significance of primate paleontology for anthropological studies. *American Journal of Physical Anthropology* 27:307–332.

———. 1968a. A source for dental comparison of *Ramapithecus* with *Australopithecus* and *Homo. South African Journal of Science* 64:92–112.

———. 1968b. On the mandible of *Ramapithecus.* In G. Kurth, ed., *Evolution und Hominisation,* 139–149. Stuttgart: G. Fischer.

———. 1969. Late Miocene hominid from Fort Ternan, Kenya. *Nature* 221:448–451.

———. 1970. The deployment and history of Old orld monkeys (Cercopithecidae, Primates). In J. R. Napier, and P. H. Napier, eds., *Old World Monkeys,* 97–113. New York: Academic Press.

———. 1971. Relationships of *Amphipithecus* and *Oligopithecus. Nature* 232:489–491.

———. 1972. *Primate Evolution.* New York: Macmillan.

———. 1977. *Ramapithecus. Scientific American* 236:28–35.

———. 1978. Diversity among the early hominids: a vertebrate paleontologist's viewpoint. In C. Jolly, ed., *Early Hominids of Africa,* 543–566. New York: St. Martin's Press.

———. 1981. Man's immediate forerunners. *Philosophical Transactions of the Royal Society, London* B 292:21–41.

———. 1987. New faces of *Aegyptopithecus* from the Oligocene of Egypt. *Journal of Human Evolution* 16:273–289.

————. 1989a. Description of two genera and species of Late Eocene Anthropoidea from Egypt. *Proceedings of the National Academy of Sciences of the United States of America* 86:9956–9960.

————. 1989b. Human origins. *Science* 245:1343–1350.

————. 1990. Discovery of the oldest known anthropoidean skull from the Paleogene of Egypt. *Science* 247:1567–1569.

————. 1992. Diversity in the early Tertiary anthropoidean radiation in Africa. *Proceedings of the National Academy of Sciences of the United States of America* 89:10743–10747.

————. 1997. Discovery of the smallest Fayum Egyptian primates (Anchomomyini, Adapidae). *Proceedings of the National Academy of Sciences of the United States of America* 94:180–184.

————. 2004. The cranium and adaptations of a stem anthropoid from the Fayum Oligocene of Egypt. In C. F. Ross and R. F. Kay, eds., *Anthropoid Origins,* 183–204. New York: Kluwer Academic/Plenum Press.

Simons, E. L., and Bown, T. M. 1985. *Afrotarsius chatrathi,* first tarsiiform primate (?Tarsiidae) from Africa. *Nature* 313:475–477.

Simons, E. L., Bown, T. M., and Rasmussen, D. T. 1987. Discovery of two additional prosimian primate families (Omomyidae, Lorisidae) in the African Oligocene. *Journal of Human Evolution* 15:431–437.

Simons, E. L., and Chopra, S. R. K. 1969. *Gigantopithecus* (Pongidae, Hominoidea) a new species from North India. *Postilla* 138:1–18.

Simons, E. L., and Ettel, P. C. 1970. *Gigantopithecus. Scientific American* 222:76–85.

Simons, E. L., and Fleagle, J. G. 1973. The history of extinct gibbon-like primates. In D. M. Rumbaugh, ed., *Gibbon and Siamang,* Vol. 2, 121–148. Basel: Karger.

Simons, E. L., and Kay, R. F. 1983. *Qatrania,* new basal anthropoid primate from the Fayum, Oligocene of Egypt. *Nature* 304:624–626.

————. 1988. New material of *Qatrania* from Egypt with comments on the phylogenetic position of the Parapithecidae (Primates, Anthropoidea). *American Journal of Primatology* 15:337–347.

Simons, E. L., and Meinel, W. 1983. Mandibular ontogeny in the Miocene great ape *Dryopithecus. International Journal of Primatology* 4:331–337.

Simons, E. L., and Pilbeam, D. R. 1965. Preliminary revision of the Dryopithecinae (Pongidae, Anthropoidea). *Folia Primatologica* 3:81–152.

————. 1972. Hominoid paleoprimatology. In R. Tuttle, ed., *The Functional and Evolutionary Biology of Primates,* 36–62. Chicago: Aldine.

————. 1978. *Ramapithecus* (Hominidae, Hominoidea). In V. J. Maglio and H. B. S. Cooke, ed., *Evolution of African Mammals,* 147–153. Cambridge, MA: Harvard University Press.

Simons, E. L., Plavcan, J. M., and Fleagle, J. G. 1999. Canine sexual dimorphism in Egyptian Eocene anthropoid primates: *Catopithecus* and *Proteopithecus. Proceedings of the National Academy of Sciences of the United States of America* 96:2559–2562.

Simons, E. L., and Rasmussen, D. T. 1989. Cranial morphology of *Aegyptopithecus* and *Tarsius* and the question of the tarsier-anthropoidean clade. *American Journal of Physical Anthropology* 79:1–23.

————. 1991. The generic classification of Fayum Anthropoidea. *International Journal of Primatology* 12:163–178.

Simons, E. L., Rasmussen, D. T., and Gebo, D. L. 1987. A new species of *Propliopithecus* from the Fayum, Egypt. *American Journal of Physical Anthropology* 73:139–147.

Simons, E. L., Seiffert, E. R. 1999. A partial skeleton of *Proteopithecus sylviae* (Primates, Anthropoidea): first associated dental and postcranial remains of an Eocene anthropoidean. *Comptes rendus hebdomadaires des séances de l'Académie des Sciences, Paris, Série IIA* 329:921–927.

Simons, E. L., Seiffert, E. R., Chatrath, P. S., and Attia, Y. 2001. Earliest record of a parapithecid anthropoid from the Jebel Qatrani Formation, Northern Egypt. *Folia Primatologica* 72:316–331.

Simons, E. L., Seiffert, E. R., Ryan T. M., and Attia, Y. 2007. A remarkable female cranium of the early Oligocene anthropoid *Aegyptopithecus zeuxis* (Catarrhini, Propliopithecidae). *Proceedings of the National Academy of Sciences of the United States of America* 104:8731–8736.

Simpson, G. G. 1945. The principles of classification and a classification of mammals. *Bulletin of the American Museum of Natural History* 85:1–350.

————. 1961. *Principles of Animal Taxonomy.* New York: Columbia University Press.

————. 1963. The meaning of taxonomic statements. In S. L. Washburn, ed., *Classification and Human Evolution,* 1–31. New York: Wenner-Gren Foundation for Anthropological Research.

————. 1975. Recent advances in methods of phylogenetic inference. In W. P. Luckett and F. S. Szalay, eds., *Phylogeny of the Primates: A Multidisciplinary Approach,* 3–19. New York: Plenum Press.

Simpson, M. J. A. 1973. The social grooming of male chimpanzees. In R. P. Michael and J. H. Crook, eds., *Comparative Ecology and Behaviour of Primates.* London: Academic Press.

Simpson, S. W., Quade, J., Levin, N. E., Butler, R., Dupont-Nivet, G., Everett, M., and Semaw, S. 2008. A female *Homo erectus* pelvis from Gona, Ethiopia. *Science* 322:1089–1091.

Sinclair, A. R. E., Leakey, M. D., and Norton-Griffiths, M. 1986. Migration and hominid bipedalism. *Nature* 324:307–308.

Singer, M. 1980. Signs of the self: an exploration in semiotic anthropology. *American Anthropologist* 82:485–507.

Singer, R., and Wymer, J. 1982. *The Middle Stone Age at Klasies River Mouth in South Africa.* Chicago: University of Chicago Press.

Singer, T., and Hein, G. 2012. Human empathy through the lens of psychology and social neuroscience. In F. B. M. de Waal and P. F. Ferrari, eds., *The Primate Mind,* 158–174, Cambridge, MA: Harvard University Press.

Singleton, I., Knott, C. D., Morrogh-Bernard, H. C., Wich, S. A., and van Schaik, C. P. 2009. Ranging behavior of orangutan females and social organization. In S. A. Wich, S. S. Utami Atmoko, T. M Setia, and C. P. van Schaik, eds, *Orangutans: Geographic Variation in Behavioral Ecology and Conservation,* 205–213. Oxford: Oxford University Press.

Singleton, I., and van Schaik, C. P. 2002. The social organisation of a population of Sumatran orang-utans. *Folia Primatologica* 73:1–20.

Skelton, R. R., McHenry, H. M., and Drawhorn, G. M. 1986. Phylogenetic analysis of early homininds. *Current Anthropology* 27:21–43.

Skipper, J. I., Goldin-Meadow, S., Nusbaum, H. C., and Small, S. L. 2007. Speech-associated gestures, Broca's area, and the human mirror system. *Brain and Language* 101:260–277.

Slater, J. B. 2000 Birdsong repertoires: their origins and use. In N. L. Wallin, B. Merker and S. Brown, eds., *The Origins of Music,* 49–63. Cambridge, MA: MIT Press.

Slightom, J. L., Chang, L.-Y. E., Koop, B. F., and Goodman, M. 1985. Chimpanzee G-gamma and A-gamma globin gene nucleotide sequences provide further evidence of gene conversions in hominine evolution. *Molecular Biology and Evolution* 2:370–389.

Slimak, L., Svendsen, J. I., Mangerud, J., Plisson, H., Heggen, H. P., Brugère, A., and Pavlov, P. Y. 2011. Late Mousterian persistence near the Arctic Circle. *Science* 332:841–845.

Slocombe, K. E., Kaller, T., Call, J., and Zuberbühler, K. 2010. Chimpanzees extract social information from agonistic screams. *PLoS One* 5:e11473.

Slocombe, K. E., and Newton-Fisher, N. E. 2005. Fruit sharing between wild adult chimpanzees *(Pan troglodytes):* a socially significant event? *American Journal of Primatology* 65:385–391.

Slocombe, K. E., Townsend, S. W., and Zuberbühler, K. 2009. Wild chimpanzees *(Pan troglodytes schweinfurthii)* distinguish between different scream types: evidence from a playback study. *Animal Cognition* 12:441–449.

Slocombe, K. E., and Zuberbühler, K. 2005. Functionally referential communication in a chimpanzee. *Current Biology* 15:1779–1784.

———. 2007. Chimpanzees modify recruitment screams as a function of audience composition. *Proceedings of the National Academy of Sciences of the United States of America* 104:17228–172.

———. 2011. Vocal communication in chimpanees. In E. V. Lonsdorf, S. R. Ross, and T. Matsuzawa, eds., *The Mind of the Chimpanzee,* 190–207. Chicago: University of Chicago Press.

Slocum, S. 1975. *Woman the Gatherer: Male Bias in Anthropology.* In R. R. Reiter, ed., *Toward an Anthropology of Women,* 36–50. New York: Monthly Review Press.

Smith, B. D. 1998. *The Emergence of Agriculture.* New York: Scientific American Library.

Smith, B. F. 1994. Sequence of emergence of the permanent teeth in *Macaca, Pan, Homo* and *Australopithcus:* its evolutionary significance. *American Journal of Human Biology* 6:61–76.

Smith, F. H. 1982. Upper Pleistocene evolution in south-central Europe: a review of evidence and analysis of trends. *Current Anthropology* 23:667–703.

———. 1984. Fossil hominids from the Upper Pleistocene of Central Europe and the origins of modern Europeans. In F. H. Smith and F. Spencer, eds., *The Origin of Modern Humans,* 137–209. New York: Alan R. Liss.

———. 1985. Continuity and change in the origin of modern *Homo sapiens. Zeitschrift für Morphologie und Anthropologie* 75:197–222.

———. 1992. The role of continuity in modern human origins. In B. Bräuer and F. Smith, eds., *Continuity or Replacement Controversies in the Evolution of Homo sapiens,* 145–156. Rotterdam: Balkema.

Smith, F. H., Falsetti, A. B., and Donelly, S. M. 1989. Modern Human Origins. *Yearbook of Physical Anthropology* 32:35–68.

Smith, F. H., and Paquette. S. P. 1989. The adaptive basis of Neandertal facial form, with some thoughts on the nature of modern human origins. In E. Trinkaus, ed., *The Emergence of Modern Humans,* 181–210. Cambridge: Cambridge University Press.

Smith, F. H., and Trinkaus, E. 1991. Modern human origins in Central Europe: a case of continuity. In J.-J. Hublin and A.-M. Tillier, eds., *Aux origines de la diversité humaine,* 251–290. *Nouvelle Encyclopédie des Sciences et des Techniques Fondation Diderot.* Paris: Presses Universitaires de France.

Smith, G. E. 1927. *The Evolution of Man.* London: Oxford University Press.

Smith, J. Z. 1990. The necessary lie: duplicity in the disciplines. In D. M. Enerson, ed., *Teaching at Chicago,* 60–64. Chicago: The University of Chicago.

Smith, K. D., Young, K. E., Talbot, C. C., Jr., and Schmeckpeper, B. J. 1987. Repeated DNA of the human Y chromosome. *Development* 101 (Suppl.): 77–92.

Smith, R., and Walker, A. 1984. Smith and Walker reply. *Nature* 309:287–288.

Smith, R. J. 1984. Comparative functional morphology of maximum mandibular openness (gape) in Primates. In D. J. Chivers, B. A. Wood, and A. Bilsborough, eds., *Food Acquisition and Processing,* 231–255. New York: Plenum Press.

Smith, R. J., and Jungers, W. L. 1997. Body mass in comparative primatology. *Journal of Human Evolution* 32:523–559.

Smith, S. L. 1995. Pattern profile analysis of hominid and chimpanzee hand bones. *American Journal of Physical Anthropology* 113:329–348.

———. 2000. Shape variation of the human pollical distal phalanx and metacarpal. *American Journal of Physical Anthropology* 96:283–300.

Smith, T., Rose, K. D., and Gingerich, P. D. 2006. Rapid Asia-Europe-North America geographic dispersal of earliest Eocene primate *Teilhardina* during the Paleocene-Eocene thermal maximum. *Proceedings of the National Academy of Sciences of the United States of America* 103:11223–11227.

Smith, T. M., Machanda, Z., Bernard, A. B., Donovan, R. M., Papakyrikos, A. M., Muller, M. N. and Wrangham, R. 2013. First molar eruption, weaning, and life history in living wild chimpanzees. *Proceedings of the National Academy of Sciences of the United States of America* 110: 2787–2791.

Smith, T. M., Tafforeau, P., Reid, D. J., Grün, R., Eggins, S., Boutakiout, M., and Hublin, J.-J. 2007. Earliest evidence of modern human life history in North African early *Homo sapiens. Proceedings of the National Academy of Sciences of the United States of America* 107:1628–6133.

Smith, T. M., Tafforeau, P., Reid, D. J., Pouech, J., Lazzari, V., Zermeno, J. P., Guatelli-Steinberg, D., Olejniczak, A. J., Hoffman, A., Radovic, J., et al. 2010. Dental evidence for ontogenetic differences between modern humans and Neanderthals. *Proceedings of the National Academy of Sciences of the United States of America* 107:20923–20928.

Smith, T. M., Toussaint, M., Reid, Olejniczak, A. J., and Hublin, J.-J. 2007. Rapid dental development in a Middle Paleolithic Belgian Neanderthal. *Proceedings of the National Academy of Sciences of the United States of America* 104:20220–20225.

Smouse, P. E., and Li, W.-H. 1987. Likelihood analysis of mitochondrial restriction-cleavage patterns for the human-chimpanzee-gorilla trichotomy. *Evolution* 41:1162–1176.

Smuts, B. B. 1999. *Sex and Friendship in Baboons: With a New Preface.* Cambridge, MA: Harvard University Press.

Smuts, B. B., and Smuts, R. W. 1993. Male aggression and sexual coercion of females non-human primates and other mammals: evidence and theoretical implications. *Advances in the Study of Behavior* 22:1–63.

Snyder, P. J., Bilder, R. M., Wu, H., Bogerts, B., and Lieverman, J. A. 1995. Cerebellar volume asymmetries are related to handedness: a quantitative MRI study. *Neuropshchologia* 4:407–419.

Sockol, M. D., Raichlen, D. A., and Pontzer, H. 2007. Chimpanzee locomotor energetics and the origin of human bipedalism. *Proceedings of the National Academy of Sciences of the United States of America* 104:12265–12269.

Solà, S. M., and Köhler, M. 1993. Recent discoveries of *Dryopithecus* shed new light on evolution of great apes. *Nature* 365:543–545.

Solecki, R. S. 1971. *Shanidar: The First Flower People.* New York: Knopf.

Soligo, C., Will, O. A., Tavaré, S., Marshall, C. R., and Martin, R. D. 2007. New light on the dates of primate origins and divergence. In M. J. Ravosa and M. Dagosto, eds., *Primate Origins,* 29–49. New York: Springer.

Sommer, I. E. C., Aleman, A., Bouma, A., and Kahn, R. S. 2004. Do women really have more bilateral language representation that men? A meta-analysis of functional imaging studies. *Brain* 127:1845–1852.

Sommer, V., and Reichard, U. 2000. Rethinking monogamy: the gibbon case. In P. M. Kappeler, ed., *Primate Males: Causes and Consequences of Variation in Group Composition,* 159–168. Cambridge: Cambridge University Press.

Sommer, V., and Ross, C. 2011. *Primates of Gashaka: Socioecology and Conservation in Nigeria's Biodiversity Hotspot.* New York: Springer.

Sonakia, A. 1985. Early *Homo* from Narmada Valley, India. In E. Delson, ed., *Ancestors: The Hard Evidence,* 334–38. New York: Alan R. Liss.

Spelke, E. 2004. Core knowledge. In N. Kanwisher and J. Duncan, eds., *Functional Neuroimaging of Visual Cognition: Attention and Performance,* Vol. 20, 29–55. Oxford: Oxford University Press.

Spencer, F. 1990a. *Piltdown: A Scientific Forgery.* New York: Oxford University Press.

———. 1990b. *The Piltdown Papers 1908–1955.* New York: Oxford University Press.

Sperber, D. 1994. The modularity of thought and the epidemiology of representations. In L. Hirschfeld and S. Gelman, eds., *Mapping the Mind: Domain Specificity in Cognition and Culture,* 39–67. New York: Cambridge University Press.

Speth, J. D., and Tchernov, E. 2001. Neandertal hunting and meat-processing in the Near East. Evidence from Kababa Cave (Israel). In C. B. Stanford and H. T. Bunn, eds., *Meat-Eating and Human Evolution,* 52–72. New York: Oxford University Press.

Spillmann, B., Dunkel, L. P., van Noordwijk, M. A., Amda, R. N. A., Lameira, A. R., Wich, S. A., and van Meijaard, C. P. 2010. Acoustic properties of long calls given by flanged male orang-utans *(Pongo pygmaeus)* reflect both individual identity and context. *Ethology* 116:385–395.

Spitsnya, G., Warren, J. E., Scott, S. K., Turkheimer, F. E., and Wise, R. J. S. 2006. Converging language streams in the human temporal lobe. *Journal of Neuroscience* 26:7328–7336.

Sponheimer, M., and Lee-Thorp, J. 1999. Isotopic evidence for the diet of an early hominid, *Australopithecus africanus*. *Science* 283:368–370.

———. 2007. Hominin paleodiets: the contribution of stable isotopes. In W. Henke and I. Tattersall, eds., *Handbook of Paleoanthropology*, Vol. 1: *Principles, Methods, Approaches*, 555–585. Berlin: Springer.

Sponheimer, M., Lee-Thorp, J., and de Ruiter, D. 2007. Icarus, isotopes, and australopith diets. In P. S. Ungar, ed., *Evolution of the Human Diet*, 132–149. Oxford: Oxford University Press.

Sponheimer, M., Lee-Thorp, J., de Ruiter, D., Cordon, D. Codron, J., Baugh, A. T., and Thackeray, F. 2005. Hominins, sedges and termites: new carbon isotope data from the Sterkfontein valley and Kruger National Park. *Journal of Human Evolution* 48:301–312.

Sponheimer, M., Passey, B. H., de Ruiter, D. J., Guatelli-Steinberg, D., Cerling, T. E., and Lee-Thorp, J. 2006. Isotopic evidence for dietary variability in the early hominid *Paranthropus robustus*. *Science* 314:98–982.

Sponsel, L. E. 1997. The human niche in Amazonia: explorations in ethnoprimatology. In W. G. Kinzey, ed., *New World Primates*, 143–165. Hawthorne, NY: Aldine de Gruyter.

Spoor, C. F., Sondaar, P. Y., and Hussain, S. T. 1991. A new hominoid hamate and first metacarpal from the Late Miocene Nagri Formation of Pakistan. *Journal of Human Evolution* 21:413–424.

Spoor, F., Leakey, M. F., Gathogo, P. N., Brown, F. H., Antón, S. C., McDougall, I., Kiarie, C., Manthi, F. K., and Leakey, L. N. 2007. Implications of new early *Homo* fossils from Ileret, east of Lake Turkana, Kenya. *Nature* 448:688–691.

Spruijt, B. M., Van Hoof, J. A. R. A. M., and Gipsen, W. H. 1992. Ethology and neurobiology of grooming behavior. *Physiological Reviews* 72:825–852.

Spuhler, J. N. 1988. Evolution of mitochondrial DNA in monkeys, apes, and humans. *Yearbook of Physical Anthropology* 31:15–48.

Srikosamatara, S. 1984. Ecology of pileated gibbons in south-east Thailand. In H. Preuschoft, D. J. Chivers, W. Y. Brockelman, and N. Creel, eds., *The Lesser Apes: Evolutionary and Behavioural Biology*, 242–257. Edinburgh: Edinburgh University Press.

Srikosamatara, S., and Brockelmann, W. Y. 1987. Polygyny in a group of pileated gibbons via a familial route. *International Journal of Primatology* 8:389–393.

Srivastava, S. K., and Binda, P. L. 1991. Depositional history of the Early Eocene Shumaysi Formation, Saudi Arabia. *Palynology* 15:47–61.

Stafford, D. K., Milliken, G. W., and Ward, J. P. 1990. Lateral bias in feeding and brachiation in *Hylobates*. *Primates* 31:407–414.

Stamenov, M. I. 2002. Some features that make mirror neurons and human language faculty unique. In M. L. Stamenov and V. Gallese, eds., *Mirror Neurons and the Evolution of Brain and Language*, 249–271. Amsterdam: John Benjamins.

Stanford, C. B. 1995a. The influence of chimpanzee predation on group size and anti-predator behaviour in red colobus monkeys. *Animal Behaviour* 49:577–587.

———. 1995b. Chimpanzee hunting behavior and human evolution. *American Scientist* 83:256–261.

———. 1996. The hunting ecology of wild chimpanzees: implications for the evolutionary ecology of Pliocene hominids. *American Anthropologist* 98:96–113.

———. 1998a. *Chimpanzee and Red Colobus.* Cambridge, MA: Harvard University Press.

———. 1998b. The social behavior of chimpanzees and bonobos. *Current Anthropology* 39:399–420.

———. 1999. *The Hunting Apes: Meat Eating and the Origins of Human Behavior.* Princeton, NJ: Princeton University Press.

———. 2006. The behavioral ecology of sympatric African apes: implications for understanding fossil hominid ecology. *Primates* 47:91–101.

———. 2008. *Apes of the Impenetrable Forest.* Upper Saddle River, NJ: Pearson Prentice Hall.

Stanford, C. B., and Allen, J. S. 1991. On strategic storytelling: current models of human behavioral evolution. *Current Anthropology* 32:58–61.

Stanford, C. B., and Nkurunguni, J. B. 2003. Behavioral ecology of sympatric chimpanzees and gorillas in Bwindi Impenetrable National Park, Uganda: diet. *International Journal of Primatology* 24:901–918.

Stanford, C. B., and O'Malley, R. C. 2008. Sleeping tree choice by Bwindi chimpanzees. *American Journal of Primatology* 70:642–649.

Stanford, C. B., Wallis, J., Matama, H., and Goodall, J. 1994a. Patterns of predation by chimpanzees on red colobus monkeys in Gombe National Park, 1982–1991. *American Journal of Physical Anthropology* 94:213–228.

Stanford, C. B., Wallis, J., Mpongo, E., and Goodall, J. 1994b. Hunting decisions in wild chimpanzees. *Behaviour* 131:1–18.

Stanyon, R. 1989. Implications of biomolecular data for human origins with particular reference to chromosomes. In G. Giacobini, ed., *Hominidae,* 35–44. Milan: Jaca Book.

Stanyon, R., and Chiarelli, B. 1982. Phylogeny of the Hominoidea: the chromosome evidence. *Journal of Human Evolution* 11:493–504.

Steele, T. E., 2010. A unique hominin menu dated to 1.95 million years ago. *Proceedings of the National Academy of Sciences of the United States of America* 107:10771–10772.

Steels, L. 2009. Is sociality a crucial prerequisite for the emergence of language? In T. Botha and C. Knight, eds., *The Prehistory of Language,* 36–57. Oxford: Oxford University Press.

Steiner, S. M. 1990. Handedness in chimpanzees. *Friends of Washoe* 9:9–19.

Steinmetz, H., and Galaburda, A. M. 1991. Planum temporale asymmetry: *in-vivo* morphometry affords a new perspective for neuro-behavioral research. *Reading and Writing: An Interdisciplinary Journal* 3:331–343.

Steinmetz, H., Rademacher, J., Jäncke, L., Huang, Y., Thron, A., and Zilles, K. 1990. Total surface of temporoparietal intrasylvian cortex, diverging left-right asymmetries. *Brain and Language* 39:357–372.

Steiper, M. E. 2006. Population history, biogeography, and taxonomy of orangutans (Genus: *Pongo*) based on a population genetic meta-analysis of multiple loci. *Journal of Human Evolution* 50:509–522.

Steiper, M. E., and Young, N. M. 2008. Timing primate evolution: lessons from the discordance between molecular and paleontological estimates. *Evolutionary Anthropology* 17:179–188.

Steklis, H. D. 1985. Primate communication, comparative neurology, and the origin of language re-examined. *Journal of Human Evolution* 14:157–173.

Steklis, H. D., and Gerald-Steklis, N. 2001. Status of Virunga mountain gorilla population. In M. M. Robbins, P. Sicotte, and K. J. Stewart, eds., *Mountain Gorillas: Three Decades of Research at Karisoke,* 391–412. Cambridge: Cambridge University Press.

Steklis, H. D., and Raleigh, M. J. 1979. Requisites for language: interspecific and evolutionary aspects. In H. D. Steklis and M. J. Raleigh, eds., *Neurobiology of Social Communication: An Evolutionary Perspective,* 283–314. New York: Academic Press.

Sterelny, K. Peacekeeping in the culture wars. In K. N. Laland and B. G. Galef, eds., *The Question of Animal Culture,* 288–304. Cambridge, MA: Harvard University Press.

Stern, J. T. Jr., and Susman, R. L. 1983. The locomotory anatomy of *Australopithecus afarensis. American Journal of Physical Anthropology* 60:279–317.

Stevens, J. M. G., Vervaecke, H., de Vries, H, and van Elsacker, L. 2005. Peering is not a formal indicator of subordination in bonobos *(Pan paniscus). American Journal of Primatology* 65:255–267.

———. 2007. Sex differences in the steepness of dominance hierarchies in captive bonobo groups. *International Journal of Primatology* 28:1417–1430.

Stevens, N. J., Seiffert, E. R., O'Connor, P. M., Roberts, E. M., Schmitz, M. D., Krause, C., Gorscak, E., Ngasala, S., Hieronymus, T. L., and Temu, J. 2013. Paleontological evidence for an Oligocene divergence between Old World monkeys and apes. *Nature* 947:611–614.

Stewart, F. A., Piel, A. K., and O'Malley, R. C. 2012. Responses of chimpanzees to a recently dead community member at Gombe National Park, Tanzania. *American Journal of Primatology* 74:1–7.

Stewart, K. J. 1977. The birth of a wild mountain gorilla *(Gorilla gorilla). Primates* 18:965–976.

———. 1988. Suckling an lactational anoestrus in wild gorillas *(Gorilla gorilla). Journal of Reproduction and Fertility* 83:627–634.

———. 2001. Social relationships of immature gorillas and silverbacks. In M. M. Robbins, P. Sicotte, and K. J. Stewart, eds., *Mountain Gorillas: Three Decades of Research at Karisoke,* 183–213. Cambridge: Cambridge University Press.

Stewart, K. J., and Harcourt, A. H. 1994. Gorillas' vocalizations during rest periods: signals of impending departure? *Behaviour* 133:827–845.

Stewart, K. J., Sicotte, P., and Robbins, M. M. 2001. Mountain gorillas of the Virungas: a short history. In M. M. Robbins, P. Sicotte, and K. J. Stewart, eds., *Mountain Gorillas: Three Decades of Research at Karisoke,* 1–26. Cambridge: Cambridge University Press.

Stiner, M. C. 1990. The use of mortality patterns in archaeological studies of hominid predatory adaptations. *Journal of Anthropological Archaeology* 9:305–351.

———. 1991a. An interspecific perspective on the emergence of the modern human predatory niche. In M. C. Stiner, ed., *Human Predators and Prey Mortality,* 149–185. Boulder CO: Westview Press.

———. 1991b. Food procurement and transport by human and non-human predators. *Journal of Archaeological Science* 18:455–482.

———. 1993. Modern human origins—faunal perspectives. *Annual Review of Anthropology* 22:55–82.

———. 1994. *Honor among Thieves: A Zooarchaeological Study of Neandertal Ecology.* Princeton, NJ: Princeton University Press.

Stiner, M. C., Barkai, R., and Gopher, A. 2009. Cooperative hunting and meat sharing 400–200 kya at Qesem Cave, Israel. *Proceedings of the National Academy of Sciences of the United States of America* 106:13207–13212.

Stokes, E. J. 2004. Within-group social relationships among female and adult males in wild western lowland gorillas *(Gorilla gorilla gorilla). American Journal of Primatology* 64:233–246.

Stokes, E. J., Parnell, R. J., and Olejniczak, C. 2003. Female dispersal and reproductive success in wild western lowland gorillas *(Gorilla gorilla gorilla). Behavioral Ecology and Sociobiology* 54:329–339.

Stokoe, W. C., Jr. 1960. Sign language structure: an outline of the visual communication systems of the American deaf. *Studies in Linguistics,* Occasional Papers 8. Buffalo, NY: University of Buffalo.

———. 1970. *The Study of Sign Language.* Washington, DC: Center for Applied Linguistics.

———. 1980. Sign language structure. *Annual Review of Anthropology* 9:365–390.

Stone, A. C., Griffiths, R. C., Zegura, S., and Hammer, M. F. 2002. High levels of Y-chromosome nucleotide diversity in the genus *Pan. Proceedings of the National Academy of Sciences of the United States of America* 99:43–48.

Stone, W. H. 1987. Genetic research with nonhuman primates: serving the needs of mankind. Symposium summary and future prospects. *Genetica* 73:169–177.

Stoneking, M. 1994. In defence of "Eve"—a response to Templeton's critique. *American Anthropologist* 96:131–155.

Stoneking, M., Bhatia, K., and Wilson, A. C. 1986. Rate of sequence divergence estimated from restriction maps of mitochondrial DNAs from Papua New Guinea. *Cold Spring Harbor Symposia on Quantitative Biology* 51:433–439.

Stoneking, M., and Cann, R. L. 1989. African origin of human mitochondrial DNA. In P. Mellars and C. Stringer, *The Human Revolution,* eds., 17–30. Edinburgh: University of Edinburgh Press.

Stoner, K. E. 1995. Dental pathology in *Pongo satyrus borneensis. American Journal of Physical Anthropology* 98:307–321.

Stoodley, C. J., and Schmahmann, J. D. 2009. Functional topography in the human cerebellum: a meta-analysis of neuroimaging studies. *NeuroImage* 44:489–501.

Stout, D., and Chaminade, T. 2012. Stone tools, language and the brain in human evolution. *Philosophical Transactions of the Royal Society, London B* 367:75–87.

Stout, D., Toth, N., Schick, K., Stout, J., and Hutchins, G. 2000. Stone tool-making and brain activation: position emission tomography (PET) studies. *Journal of Archaeological Science* 27:1215–1223.

Strait, D. A., Grine, F. E., and Moniz, M. A. 1997. A reappraisal of early hominid phylogeny. *Journal of Human Evolution* 32:17–62.

Strait, D. A., Weber, G. W., Neubauer, S., Chalk J., Richmond, B. G., Lucas, P. W., Spencer, M. A., Schrein, C., Dechow, P. C., Ross, C. F., et al. 2009. The feeding biomechanics and dietary ecology of *Australopithecus africanus*. *Proceedings of the National Academy of Sciences of the United States of America* 106:2124–2129.

Strassmann, B. J. 2011. Coopertion and competition in a cliff-dwelling people. *Proceedings of the National Academy of Sciences of the United States of America* 108:10894–10901.

Straus, W. L., Jr. 1940. The posture of the great ape hand in locomotion, and its phylogenetic implications. *American Journal of Physical Anthropology* 27:199–207.

——. 1942. Rudimentary digits in primates. *Quarterly Review of Biology* 17:228–243.

——. 1949. The riddle of man's ancestry. *Quarterly Review of Biology* 24:200–223.

——. 1962. Fossil evidence of the evolution of the erect, bipedal posture. *Clinical Orthopaedics* 25:9–19.

——. 1963. The classification of *Oreopithecus*. In S. L. Washburn, ed., *Classification and Human Evolution*, 146–177. New York: Wenner-Gren Foundation for Anthropological Research.

Straus, W. L., Jr., and Schön, M. A. 1960. Cranial capacity of *Oreopithecus bambolii*. *Science* 132:670–672.

Strier, K. B. 1997. Behavioral ecology and conservation biology of primates and other animals. *Advances in the Study of Behavior* 26:101–158.

——. 2000. *Primate Behavioral Ecology*. Boston: Allyn and Bacon.

——. 2003. Primate behavioral ecology: from ethnography to ethology and back. *American Anthropologist* 105:16–27.

Stringer, C. B. 1984a. The definition of *Homo erectus* and the existence of the species in Africa and Europe. *Courier Forschungsinstitut Senckenberg* 69:131–143.

——. 1984b. Human evolution and biological adaptation in the Pleistocene. In R. Foley, *Hominid Evolution and Community Ecology*, ed., 55–83. London: Academic Press.

——. 1985. Middle Pleistocene hominid variability and the origin of Late Pleistocene humans. In E. Delson, ed., *Ancestors: The Hard Evidence*, 289–295. New York: Alan R. Liss.

——. 1986. The credibility of *Homo habilis*. In B. Wood, L. Martin, and P. Andrews, eds., *Major Topics in Primate and Human Evolution*, 266–294. Cambridge: Cambridge University Press.

——. 1989a. Documenting the origin of modern humans. In E. Trinkaus, ed., *The Emergence of Modern Humans*, 67–96. Cambridge: Cambridge University Press.

——. 1989b. The origin of early modern humans: a comparison of the European and Non-European evidence. In P. Mellars and C. Stringer, eds., *The Human Revolution*, 232–244. Edinburgh: University of Edinburgh Press.

——. 1990a. The Asian connection. *New Scientist* 128:33–37.

——. 1990b. The emergence of modern humans. *Scientific American* 263:98–105.

——. 1992. Replacement, continuity and the origin of *Homo sapiens*. In B. Bräuer and F. Smith, eds., *Continuity or Replacement Controversies in the Evolution of Homo sapiens*, 9–24. Rotterdam: Balkema.

Stringer, C. B., and Andrews, P. 1988. Genetic and fossil evidence for the origin of modern humans. *Science* 239:1263–1268.

Stringer, C. B., and Gamble, C. 1993. *In Search of the Neanderthals.* New York: Thames and Hudson.

Stringer, C. B., Howell, F. C., and Melentis, J. K. 1979. The significance of the fossil hominid skull from Petralona, Greece. *Journal of Archaeological Science* 6:235–253.

Struhsaker, T. T. 1967. Auditory communication in vervet monkeys *(Cercopithecus aethiops)*. In S. A. Altmann, ed., *Social Communication among Primates,* 281–364. Chicago: University of Chicago Press.

———. 1997. *Ecology of an African Rain Forest: Logging in Kibale and the Conflict between Conservation and Exploitation.* Gainesville: University Press of Florida.

Strum, S. C. 2012. Darwin's monkey: why baboons can't become human. *Yearbook of Physical Anthropology* 55:3–23.

Strum, S. C., and Fedigan, L. M. 2000. *Primate Encounters: Models of Science, Gender, and Society.* Chicago: University of Chicago Press.

Studdert-Kennedy, M., and Goldstein, L. 2003. Launching language: the gestural origin of discrete infinity. In M. H. Christiansen and S. Kirby, eds., *Language Evolution,* 235–254. Oxford: Oxford University Press.

Stumpf, R. 2007. Chimpanzees and bonobos. Diversity within and between species. In C. J. Campbell, A. Fuentes, K. C. MacKinnon, M. Panger, and S. K. Bearder, eds., *Primates in Perspective,* 321–344. Oxford: Oxford University Press.

Stumpf, R., Emery Thompson, M., and Knott, C. D. 2008. A comparison of female mating strategies in *Pan troglodytes* and *Pongo* sp. *International Journal of Primatology* 29:865–884.

Stumpf, R., Emery Thompson, M., Muller, M. N., and Wrangham, R. W. 2009. The context of female dispersal in Kanyawara chimpanzees. *Behaviour* 146:629–656.

Suarez, S. D., and Gallup, G. G., Jr. 1981. Self-recognition in chimpanzees and orangutans, but not gorillas. *Journal of Human Evolution* 10:175–188.

Sugardjito, J. 1983. Selecting nest-sites of Sumatran orang-utans, *Pongo pygmaeus abelii* in the Gunung Leuser National Park, Indonesia. *Primates* 24:467–474.

———. 1988. Use of forest strata by the Sumatran orang-utans: a consideration of functional aspects. *Treubia* 29:255–265.

Sugardjito, J., Boekhorst, I. J. A. te, and van Hooff, J. A. R. A. M. 1987. Ecological constraints on the grouping of wild orang-utans *(Pongo pygmaeus)* in the Gunung Leuser National Park, Sumatra, Indonesia. *International Journal of Primatology* 8:17–41.

Sugardjito, J., and Cant, J. G. H. 1994. Geographic and sex differences in positional behavior of orang-utans. *Treubia* 31:31–41.

Sugardjito, J., and Nurhuda, N. 1981. Meat-eating behavior in wild orang utans, *Pongo pygmaeus. Primates* 22:414–416.

Sugardjito, J., and van Hooff, J. A. R. A. M. 1986. Age-sex class differences in the positional behaviour of the Sumatran orang-utan *(Pongo pygmaeus abelii)* in the Gunung Leuser National Park, Indonesia. *Folia Primatologica* 47:14–25.

Sugiyama, Y. 1968. Social organization of chimpanzees in the Budongo Forest, Uganda. *Primates* 9:225–258.

————. 1969. Social behavior of chimpanzees in the Budongo Forest, Uganda. *Primates* 10:197–225.

————. 1973. The social structure of wild chimpanzees -a review of field studies. In R. P. Michael and J. H. Crook, eds., *Comparative Ecology and Behaviour of Primates,* 376–410, London: Academic Press.

————. 1981. Observations on the population dynamics and behavior of wild chimpanzees at Bossou, Guinea, between 1979–1980. *Primates* 22:435–444.

————. 1984. Population dynamics of wild chimpanzees at Bossou, Guinea, between 1976 and 1983. *Primates* 25:391–400.

————. 1989. Description of some characteristic behaviors and discussion on their propagation process among chimpanzees of Bossou, Guinea, In Y. Sugiyama, ed., *Behavioural Studies of Wild Chimpanzees at Bossou, Guinea,* 43–76. Inuyama, Japan: Kyoto University Primate Research Institute.

————. 1993. Local variation of tools and tool use among wild chimpanzee populations. In A. Berthelet and J. Chavaillon, eds., *The Use of Tools by Human and Non-human Primates,* 175–187. New York: Oxford University Press.

————. 1994a. Tool use by chimpanzees. *Nature* 367:327.

————. 1994b. Age-specific birth rate and lifetime reproductive success of chimpanzees at Bossou, Guinea. *American Journal of Primatology* 32:311–318.

————. 1995. Tool-use for catching ants by chimpanzees at Bossou and Monts Nimba, West Africa. *Primates* 36:193–205.

————. 1999. Socioecological factors of male chimpanzee migration at Bossou, Guinea. *Primates* 4:61–68.

Sugiyama, Y., and Koman, J. 1979. Social structure and dynamics of wild chimpanzees at Bossou, Guinea. *Primates* 20:323–339.

————. 1987. A preliminary list of chimpanzees' alimentation at Bossou, Guinea. *Primates* 28:133–147.

————. 1992. The flora of Bossou: its utilization by chimpanzees and humans. *African Study Monographs* 13:127–169.

————. 1997. Social tradition and the use of tool-composites by wild chimpanzees. *Evolutionary Anthropology* 6:23–27.

Sumita, K., Kitahara-Frisch, J., and Norikoshi, K. 1985. The acquisition of stone-tool use in captive chimpanzees. *Primates* 26:168–181.

Sumner, P., and Mollon, J. D. 2000a. Catarrhine photopigments are optimized for detecting targets against a foliage background. *Journal of Experimental Biology* 203:1963–1986.

————. 2000b. Chromaticity as a signal of ripeness in fruits taken by primates. *Journal of Experimental Biology* 203:1987–2000.

Sun, G.-Z., Huang, B., Guan, Z.-H., Geissmann, T., and Jiang, X.-L. 2011. Individuality in male songs of wild black crested gibbons *(Nomascus concolor). American Journal of Primatology* 73:431–438.

Sun, T., Patoine, C., Abu-Khalil, A., Visvader, J., Sum, E., Cherry, T. J., Orkin, S. H., Geschwind, D. H., and Walsh, C. A. 2005. Early asymmetry of gene transcription in embryonic human left and right cerebral cortex. *Science* 308:1794–1798.

Sunderland-Groves, J. L., Ekinde, A., and Mboh, H. 2009. Nesting behavior of *Gorilla gorilla diehli* at Kagwene Mountain, Cameroon: implications for assessing group size and density. *International Journal of Primatology* 30:253–266.

Sunstein, C. R., and Nussbaum, M. C. 2004. *Animal Rites: Current Debates and New Directions.* Oxford: Oxford University Press.

Surbeck, M., and Hohmann, G. 2008. Primate hunting by bonobos at LuiKotale, Salonga National Park. *Currrent Biology* 18:R906–R907.

Surbeck, M., Mundry, R., and Hohmann, G. 2011. Mothers matter! Maternal support dominance status and mating success in male bonobos *(Pan paniscus). Proceedings of the Royal Society B* 278:590–598.

Surridge, A. K., Osorio, D., and Mundy, N. I. 2003. Evolution and selection of trichromatic vision in primates. *Trends in Ecology and Evolution* 8:198–205.

Susman, R. L. 1974. Facultative terrestrial hand postures in an orangutan *(Pongo pygmaeus)* and pongid evolution. *American Journal of Physical Anthropology* 40:27–38.

———. 1979a. Comparative and functional morphology of hominoid fingers. *American Journal of Physical Anthropology* 50:215–236.

———. 1979b. Functional and morphological affinities of the subadult hand (O. H. 7) from Olduvai Gorge. *American Journal of Physical Anthropology* 51:311–331.

———. 1984. The locomotor behavior of *Pan paniscus* in the Lomako forest. In R. L. Susman ed., *The Pygmy Chimpanzee,* 369–394. New York: Plenum Press.

———. 1988a. Hand of *Paranthropus robustus* from Member 1, Swartkrans: fossil evidence for tool behavior. *Science* 240:781–784.

———. 1988b. New postcranial remains from Swartkrans and their bearing on the functional morphology and behavior of *Paranthropus robustus.* In F. E. Grine, ed., *Evolutionary History of the "Robust" Australopithecines,* 149–172. Hawthorne, NY: Aldine de Gruyter.

———. 1989. New hominid fossils from the Swarkrans Formation (1979–1986 excavations): postcranial specimens. *American Journal of Physical Anthropology* 79:451–474.

———. 1991. Species attribution of the Swartkrans thumb metacarpals: reply to Drs. Trinkaus and Long. *American Journal of Physical Anthropology* 86:549–552.

———. 1993. Homonid postcranial remains from Swartkrans. In C. K. Brain, ed., *Swartkrans: A Cave's Chronicle of Early Man,* 117–136. Pretoria, South Africa: Transvaal Museum.

———. 1994. Fossil evidence for early hominid tool use. *Science* 2651570–1573.

———. 1998. Hand function and tool behavior in early hominids. *Journal of Human Evolution* 35:23–46.

———. 2004. *Oreopithecus bambolii:* an unlikely case of hominidlike grip capability in a Miocene ape. *Journal of Human Evolution* 46:105–117.

———. 2005. *Oreopithecus:* still apelike after all these years. *Journal of Human Evolution* 49:405–411.

Susman, R. L., Badrian, N. L., and Badrian, A. J. 1980. Locomotor behavior of *Pan paniscus* in Zaire. *American Journal of Physical Anthropology* 53:69–80.

Susman, R. L., and Creel, N. 1979. Functional and morphological affinities of the subadult hand (O. H. 7) from Olduvai Gorge. *American Journal of Physical Anthropology* 51:311–332.

Susman, R. L., and Stern, J. T., Jr. 1979. Telemetered electromyography of flexor digitorum profundus and flexor digitorum superficialis in *Pan troglodytes* and implications for interpretation of the O. H. 7 hand. *American Journal of Physical Anthropology* 50:565–574.

———. 1980. EMG of the interosseous and lumbrical muscles in the chimpanzee *(Pan troglodytes)* hand during locomotion. *American Journal of Physical Anthropology* 157:389–397.

———. 1982. Functional morphology of *Homo habilis*. *Science* 217:931–933.

Susman, R. L., Stern, J. T., Jr., and Jungers, W. L. 1984. Arboreality and bipedality in the Hadar hominids. *Folia Primatologica* 43:113–156.

———. 1985. Locomotor adaptations in the Hadar hominids. In E. Delson, ed., *Ancestors: The Hard Evidence*, 184–192. New York: Alan R. Liss.

Susman, R. L., and Tuttle, R. H. 1976. Knuckling behavior in captive orangutans and a wounded baboon. *American Journal of Physical Anthropology* 45:123–124.

Sussman, R. W. 1997. *The Biological Basis of Human Behavior*. Needam Heights, MA: Simon & Schuster.

———. 1999. The myth of man the hunter, man the killer and the evolution of human morality. *Zygon* 34:453–471.

———. 2000. Piltdown Man: the father of American field primatology. In S. C. Strum and L. M. Fedigan, eds., *Primate Encounters*, 85–103. Chicago: University of Chicago Press.

———. 2004. Are humans inherently violent? In R. Selig, M. R. London, and P. A. Kaupp, eds., *Anthropology Explored: Revised and Expanded*. 30–45. Washington, DC: Smithsonian Institution Press.

Sussman, R. W., Cheverud, J. M., and Bartlett, T. Q. 1994. Infant killing as an evolutionary strategy: reality or myth? *Evolutionary Anthropology* 3:149–151.

Sussman, R. W., and Cloninger, C. R. 2011. *Origins of Altruism and Cooperation*. New York: Springer.

Sussman, R. W., Garber, P. A., and Cheverud, J. M. 2005. Importance of cooperation and affiliation in the evolution of primate sociality. *American Journal of Physical Anthropology* 128:84–97.

Sussman, R. W., and Kinzey, W. G. 1984. The ecological role of the Callitrichidae: a review. *American Journal of Physical Anthropology* 64:419–449.

Suteethorn, V., Buffetaut, E., Buffetaut-Tong, H., Ducrocq, S., Helmcke-Ingavat, R., Jaeger, J.-J., Jongkanjanasoontorn, Y. 1990. A hominoid locality in the Middle Miocene of Thailand. *Comptes rendus hebdomadaires des séances de l'Académie des Sciences, Paris, Série II* 311:1449–1454.

Suwa, G. 1988. Evolution of the "robust" australopithecines in the Omo Succession: evidence from mandibular promolar morphology. In F. E. Grine, ed., *Evolutionary History of the "Robust" Australopithecines*, 199–222. Hawthorne, NY: Aldine de Gruyter.

Suwa, G., Asfaw, B., Beyene, Y., White, T. D., Katoh, S., Nagaoka, S., Nakaya, H., Uzawa, K., Renne, P., WoldeGabriel, G. 1997. The first skull of *Australopithecus boisei*. *Nature* 389:489–492.

Suwa, G., Asfaw, B., Kono, R. T., Kubo, D., Lovejoy, C. O., and White, T. D. 2009a. The *Ardipithecus ramidus* skull and its implications for hominid origins. *Science* 326:68e1–7.

———. 2009b. The *Ardipithecus ramidus* skull and its implications for hominid origins. *Science* 326:68.

Suwa, G., Kono, R. T., Katoh, S., Asfaw, B., and Beyene, Y. 2007. A new species of great ape from the late Miocene epoch in Ethiopia. *Nature* 448:921–924.

Suwanvecho, U. 2003. *Ecology and Interspecific Relationships of Two Sympatric Hylobates species (H. lar and H. pileatus) in Khao Yai National Park, Thailand.* PhD diss., Mahidot University, Bangkok.

Suzuki, A. 1969. An ecological study of chimpanzees in a savanna woodland. *Primates* 10:103–148.

———. 1971. Carnivory and cannibalism observed among forest-living chimpanzees. *Journal of the Anthropological Society, Nippon* 79:30–48.

———. 1975. The origin of hominid hunting: a primatological perspective. In R. H. Tuttle, ed., *Socioecology and Psychology of Primates,* 259–278. The Hague: Mouton.

Suzuki, H., Kawamoto, Y., Takenaka, O., Munechika, I., Hori, H., and Sakurai, S. 1994. Phylogenetic relationships among *Homo sapiens* and related species based on restriction site variations in rDNA spacers. *Biochemical Genetics* 32:257–269.

Suzuki, H., and Takai, F. 1970. *The Amud Man and His Cave Site.* Tokyo: University of Tokyo.

Suzuki, S., Kuroda, S., and Nishihara, T. 1995. Tool-set for termite-fishing by chimpanzees in the Ndoki Forest, Congo. *Behaviour* 132:219–235.

Suzuki, S., and Nishihara, T. 1992. Feeding strategies of sympatric gorillas and chimpanzees in the Ndoki-Nouabalé Forest, with special reference to co-feeding behavior by both species. *Abstracts of the XVIth Congress of the International Primatological Society,* p. 86.

Swartz, K. B., and Evans, S. 1991. Not all chimpanzees *(Pan troglodytes)* show self-recognition. *Primates* 32:483–496.

Swartz, S. M. 1989. Pendular mechanics and the kinematics and energetics of brachiating locomotion. *International Journal of Primatology* 10:387–418.

———. 1993. Biomechanics of primate limbs. In D. L. Gebo, ed., *Postcranial Adaptation in Nonhuman Primates,* 5–42. DeKalb: Northern Illinois University Press.

Swartz, S. M., Bertram, J. E. A., and Biewener, A. A. 1989. Telemetered in vivo strain analysis of locomotor mechanics of brachiating gibbons. *Nature* 342:270–272.

Swindler, D. R. 1976. *Dentition of Living Primates.* New York: Academic Press.

———. 2002. *Primate Dentition: An Introduction to the Teeth of Non-human Primates.* Cambridge: Cambridge University Press.

Swindler, D. R., and Olshan, A. F. 1988. Comparative and evolutionary aspects of the permanent dentition. In J. H. Schwartz, ed., *Orang-utan Biology,* 271–284. New York: Oxford University Press.

Swindler, D. R., and Wood, C. D. 1973. *An Atlas of Primate Gross Anatomy: Baboon, Chimpanzee, and Man.* Seattle: Washington University Press.

Swisher, C. C., Curtis, G. H., Jacob, T., Getty, A. G., Suprojo, A., and Widiasmoro. 1994. Age of the earliest known hominids in Java, Indonesia. *Science* 263:1118–1121.

Swisher, C. C., W. J. Rink, S. C. Anton, Schwarcz, H. P., Curtis, G. H., Juprijo, A., Widiasmoro. 1996. Latest *Homo erectus* of Java: potential contemporaneity with *Homo sapiens* in Southeast Asia. *Science* 274:1870–1871.

Symington, M. M. 1990. Fission-fusion social organization in *Ateles* and *Pan*. *International Journal of Primatology* 11:47–61.

Syvanen, M. 1984. Conserved regions in mammalian beta-globins: could they arise by cross-species gene exchange? *Journal of Theoretical Biology* 107:685–696.

———. 1987. Molecular clocks and evolutionary relationships: possible distortions due to horizontal gene flow. *Journal of Molecular Evolution* 26:16–23.

Szabó, C. Á., Lancaster, J. L., Xiong, J., Cook, C., and Fox, P. 2003. MR imaging volumetry of subcortical structures and cerebellar hemispheres in normal persons. *AJNR American Journal of Neuroradiology* 24:644–647.

Szalay, F. S. 1970. Late Eocene *Amphipithecus* and the origins of catarrhine primates. *Nature* 227:355–357.

———. 1975a. Phylogeny of primate higher taxa. The basicranial evidence. In W. P. Luckett and F. S. Szalay, eds., *Phylogeny of the Primates: A Multidisciplinary Approach*, 91–125. New York: Plenum Press.

———. 1975b. Hunting-scavenging protohominids: a model for hominid origins. *Man* 10:420–429.

Szalay, F. S., and Berzi, A. 1973. Cranial anatomy of *Oreopithecus*. *Science* 180:183–185.

Szalay, F. S., and Delson, E. 1979. *Evolutionary History of the Primates*. New York: Academic Press.

Szalay, F. S., and Langdon, J. H. 1986. The foot of *Oreopithecus:* an evolutionary assessment. *Journal of Human Evolution* 15:585–621.

Szathmáry, E. 2001. The origin of the human language faculty: the language amoeba hypothesis. In J. Trabant and S. Ward, eds., *New Essays on the Origin of Language*, 41–51. Berlin: Mouton de Gruyter.

Tabuce, R., Marivaux, L., Lebrun, R., Adaci, M., Bensalah, M., Fabre, P.-H., Fara, E., Rodrigues, H. G., Hautier, L., Jaeger, J.-J., et al. 2009. Anthropoid versus strepsirrhine status of the African Eocene primates *Algeripithecus* and *Azibius:* craniodental evidence. *Proceedings of the Royal Society B* 276:4087–4094.

Taglialatela, J. P., Russell, J. L., Schaeffer, J. A., and Hopkins, W. D. 2008. Communicative signaling activates 'Broca's' homolog in chimpanzees. *Current Biology* 18:343–348.

Taglialatela, J. P., Savage-Rumbaugh, S., and Baker, L. A. 2003. Vocal production by a language-competent *Pan paniscus*. *International Journal of Primatology* 24:1–17.

Taglialatela, J. P., Savage-Rumbaugh, S., Rumbaugh, D. M., Benson, J., and Greaves, W. 2004. Language, apes and meaning-making. In G. Williams and A. Lukin, eds., *The Development of Language: Functional Perspectives on Species and Individuals*, 91–111. London: Continuum.

Taieb, M. 1982. Le Rift est-africain: la plus riche réserve d'hominidés fossiles. *Histoire et Archeologie* 60:18–37.

Taieb, M., Johanson, D. C., and Coppens, Y. 1975. Expédition internationale de l'Afar, Ethiopie (3e campagne 1974), découverte dj'Hominidés plio-pléistocènes à Hadar. *Comptes rendus hebdomadaires des séances de l'Académie des Sciences, Paris, série D* 281:1297–1300.

Taieb, M., Johanson, D. C., Coppens, Y., and Aronson, J. L. 1976. Geological and palaeontological background of Hadar hominid site, Afar, Ethiopia. *Nature* 260:289–293.

Taieb, M., Johanson, D. C., Coppens, Y., Bonnefille, R., and Kalb, J. 1974. Découverte d'Hominidés dans les séries plio-pléistocènes d'Hadar (Gassin de l'Awash, Afar, Ethiopie). *Comptes rendus hebdomadaires des séances de l'Académie des Sciences, Paris, série D* 279:735–738.

Taieb, M., Johanson, D. C., Coppens, Y., and Tiercelin, J.-J. 1978. Chronostratigraphie des gisements à Hominidés pliocènes d'Hadar et corrélations avec les sites préhistoriques du Kada Gona. *Comptes rendus hebdomadaires des séances de l'Académie des Sciences, Paris, série D* 287:459–461.

Taieb, M., and Tiercelin, J.-J. 1980. La stratigraphie et paléoenvironnements sédimentaires de la formation d'Hadar, dépression de l'Afar, Ethiopie. In R. E. Leakey and B. A. Ogot, eds., *Proceedings of the 8th Panafrican Congress of Prehistory and Quaternary Studies Nairobi, 5 to 10 September 1977,* 109–114. Nairobi: International Louis Leakey Memorial Institute for African Prehistory.

Takahashi, L. K. 1990. Morphological basis of arm-swinging: multivariate analyses of the forelimbs of *Hylobates* and *Ateles*. *Folia Primatologica* 54:70–85.

Takahata, N., and Satta, Y. 1997. Evolution of the primate lineage leading to modern humans: phylogenetic and demographic inferences from DNA sequences. *Proceedings of the National Academy of Sciences of the United States of America* 94:4811–4815.

Takahata, N., Satta, Y., and Klein, J. 1995. Divergence time and population size in the lineage leading to modern humans. *Theoretical Population Biology* 48:198–221.

Takahata, Y. 1985. Adult male chimpanzees kill and eat a male newborn infant: newly observed intragroup infanticide and cannibalism in Mahale National Park, Tanzania. *Primates* 44:161–170.

Takahata, Y., Ihobe, H., and Idani, G. 1996. Comparing copulations of chimpanzees and bonobos: do females exhibit proceptivity and receptivity? In W. C. McGrew, L. F. Marchant, and T. Nishida, eds., *Great Ape Societies,* 135–145. Cambridge: Cambridge University Press.

Takai, M., and Shigehara, N. 2004. The morphology of two maxillae of Pondong primates (*Pondaungia cotteri* and *Amphipithecus mogaungensis*) (middle Eocene, Myanmar) In C. F. Ross and R. F. Kay, eds., *Anthropoid Origins,* 283–321. New York: Kluwer Academic/Plenum Press.

Takai, M., Shigehara, N., Aung, A. K., Tun, S. T., Soe, A. N., Tsubamoto, T., and Thein, T. 2001. A new anthropoid from the latest middle Eocene of Pondaung, central Myanmar. *Journal of Human Evolution* 40:393–409.

Takai, M., Shigehara, N., Tsubamoto, T., Egi, N., Aye, K. A., Soe, T., Tin, T., and Maung, M. 2002. Morphology of the frontal bones of *Amphipithecus* discovered from the Pondaung formation (latest middle Eocene, central Myanmar). *Anthropological Sciences* 110:92.

Takai, M., Sien, C., Tsubamoto, T., Egi, N., Maung, M., and Shigehara, N. 2005. A new eosimiid from the latest middle Eocene in Pondaung, central Myanmar. *Anthropological Science* 113:17–25.

Takenoshita, Y., Ando, C., Iwata, Y., and Yamagiwa, J. 2008. Fruit phenology of the great ape habitat in the Moukalaba-Doudou National Park, Gabon. *African Study Monographs* 39:23–39.

Takenoshita, Y., and Yamagiwa, J. 2008. Estimating gorilla abundance by dung count in the northern part of Moukalaba-Doudou National Park, Gabon. *African Study Monographs* 39:41–54.

Takeshita, H., and van Hooff, J. A. R. A. M. 1996. Tool use by chimpanzees *(Pan troglodytes)* of the Arnhem Zoo community. *Japanese Psychological Research* 38:163–173.

———. 2001. Tool use by chimpanzees *(Pan troglodytes)* of the Arnhem Zoo community. In T. Matsuzawa, ed., *Primate Origins of Human Cognition and Behavior,* 519–536. Tokyo: Springer.

Takeshita, H., and Walraven, V. 1996. A comparative study of the variety and complexity of object manipulation in captive chimpanzees *(Pan troglodytes)* and bonobos *(Pan paniscus).* *Primates* 37:423–441.

Tanaka, J. 1980. *The San: Hunter-Gatherers of the Kalahari.* Tokyo: University of Tokyo Press.

Tanner, J. A., Patterson, F. G., and Byrne, R. W. 2006. The development of spontaneous gestures in zoo-living gorillas and sign-taught gorillas: from action and location to object representation. *Journal of Developmental Processes* 1:69–103.

Tanner, N. M. 1981. *On Becoming Human.* Cambridge: Cambridge University Press.

———. 1987. The chimpanzee model revisited and the gathering hypothesis. In W. G. Kinzey, ed., *The Evolution of Human Behavior: Primate Models,* 3–27. Albany: State University of New York Press.

Tanner, N., and Zihlman, A. 1976. Women in evolution. part I: innovation and selection in human origins. *Signs: Journal of Women in Culture and Society* 1:585–608.

Tappen, M. 2001. Deconstructing the Serengeti. In C. B. Stanford and H. T. Bunn, eds., *Meat-Eating and Human Evolution,* 13–32.

Tardieu, C. 1982. Caractères plésiomorphes et apomorphes de l'articulation du genou chez les primates hominoïdes. *Geobios, mémoire spécial* 6:321–334.

———. 1983a. *L'Articulation du Genou, analyse morpho fonctionnelle chez les primates et les hominidés fossiles.* Cahiers de Pléoanthropologie. Paris: Éditions du Centre National de la Recherche Scientifique.

———. 1983b. L'articulation du genou des primates catarhiniens et hominids fossiles. Implications phylogenetique et taxinomique. *Bulletins et Mémoires de la Société d'Anthropologie de Paris, série XIII* 10:355–372.

———. 1986. The knee joint in three hominoid primates: application to Plio-Pleistocene hominids and evolutionary implications. In D. M. Taub and F. A. King, eds., *Current Perspectives in Primate Biology,* 182–192. New York: Van Nostrand Reinhold.

———. 1991. Étude comparative des déplacements du centre de gravité du corps pendant la marche par une nouvelle méthode d'analyse tridimensionnelle. Mis à l'épreuve d'une hypothèse évolutive. In Y. Coppens and B. Senut, eds., *Origine(s) de la Bipédie chez les Hominidés*, 49–58. Paris: Éditions du Centre National de la Recherche Scientifique.

———. 1992a. *Le centre de gravité du corps et sa trajectoire pendant la marche: évolution de la locomotion des hommes fossiles.* Cahiers de Pléoanthropologie. Paris: Éditions du Centre National de la Recherche Scientifique.

———. 1992b. Location of the body center of gravity in primates and other mammals: implications for the evolution of hominid body shape and bipedalism. In S. Matano, R. H. Tuttle, H. Ishida, and M. Goodman, eds., *Topics in Primatology*, Vol. 3: *Evolutionary Biology, Reproductive Endocrinology and Virology*, 191–208. Tokyo: University of Tokyo Press.

———. 1999. Ontogeny and phylogeny of femoro-tibial characters in humans and hominid fossils: functional influence and genetic determinism. *American Journal of Physical Anthropology* 110:365–377.

Tardieu, C., Aurengo, A., and Tardieu, B. 1993. New method of three-dimensional analysis of bipedal locomotion for the study of displacements of the body and body-parts centers of mass in man and non-human primates: evolutionary framework. *American Journal of Physical Anthropology* 90:455–476.

Tattersall, I. 1969a. Ecology of North Indian *Ramapithecus. Nature* 221:451–452.

———. 1969b. More on the ecology of North Indian *Ramapithecus. Nature* 224:821–822.

———. 1986. Species recognition in human paleontology. *Journal of Human Evolution* 15:165–175.

———. 2000. Once we were not alone. *Scientific American* 282:56–62.

———. 2001. Evolution, genes, and behavior. *Zygon* 36:657–665.

Tattersall, I., Delson, E., and Van Couvering, J. 1988. Victoriapithecae. *Encyclopedia of Human Evolution and Prehistory*, p. 594. New York: Garland.

Tattersall, I., and Schwartz, J. H. 2000. *Extinct Humans.* Boulder CO: Westview Press.

———. 2000. The morphological distinctiveness of Homo sapiens and its recognition in the fossil record: clarifying the problem. *Evolutionary Anthropology* 17:49–54.

Tauxe, L., Monaghan, M., Drake, R., Curtis, G., and Staudigel, H. 1985. Paleomagnetism of Miocene East African Rift sediments and the calibration of the geomagnetic reversal time scale. *Journal of Geophysical Research* 90:4639–4646.

Tay, S.-K., Blythe, J., and Lipovich, L. 2009. Global discovery of primate-specific genes in the human genome. *Proceedings of the National Academy of Sciences of the United States of America* 106:12019–12024.

Taylor, A. B. 1997. Relative growth, ontogeny, and sexual dimorphism in *Gorilla* (*Gorilla gorilla gorilla* and *G. g. beringei*): evolutionary and ecological considerations. *American Journal of Primatology* 43:1–31.

———. 2006. Diet and mandibular morphology in African apes. *International Journal of Primatology* 27:181–201.

———. 2009. The functional significance of variation in jaw form in orangutans. In S. A. Wich, S. S. Utami Atmoko, T. M. Setia, and C. P. van Schaik, *Orangutans: Geographic Variation in Behavioral Ecology and Conservation*, 15–31. Oxford: Oxford University Press.

Taylor, W. W. 1948. *A Study of Archeology*. Memoir 69, American Anthropological Association. Carbondale: Southern Illinois University Press.

Teaford, M. F. 2007. What do we know and not know about diet and enamel structure? In P. S. Ungar, ed., *Evolution of the Human Diet,* 56–76. Oxford: Oxford University Press.

Teaford, M. F., Beard, K. C., Leakey, R. E., and Walker, A. 1988. New hominoid facial skeleton from the Early Miocene of Rusinga Island, Kenya, and its bearing on the relationship between *Proconsul nyanzae* and *Proconsul africanus*. *Journal of Human Evolution* 17:461–477.

Teaford, M. F., Maas, M. C., and Simons, E. L. 1996. Dental microwear and microstructure in Early Oligocene primates from the Fayum, Egypt: implications for diet. *American Journal of Physical Anthropology* 101:527–543.

Teaford, M. F., and Ungar, P. S. 2000. Diet and the evolution of the earliest human ancestors. *Proceedings of the National Academy of Sciences of the United States of America* 97:13506–13511.

Teaford, M. F., Ungar, P. S., and Grine, F. E. 2002. Paleontological evidence for the diets of African Plio-Pleistocene hominins with special reference to early *Homo*. In P. S. Ungar and M. F. Teaford, eds., *Human Diet: Its Origin and Evolution,* 143–166. Westport CT: Bergin & Garvey.

Teaford, M. F., and Walker, A. 1984. Quantitative differences in dental microwear between primate species with different diets and a comment on the presumed diet of *Sivapithecus*. *American Journal of Physical Anthropology* 64:191–200.

Teaford, M. F., Walker, A., and Mugaisi, G. S. 1993. Species discrimination in *Proconsul* from Rusinga and Mfangano Islands, Kenya. In W. H. Kimbel and L. B. Martin, eds., *Species, Species Concepts, and Primate Evolution,* 373–392. New York: Plenum Press.

te Boekhorst, I. J. A., Schürmann, and C. L., Sugardjito, J. 1990. Residential status and seasonal movements of wild orangutans in the Gunung Leuser Reserve (Sumatra, Indonesia). *Animal Behavior* 39:1098–1109.

Teelen, S. Primate abundance along five transect lines at Ngogo, Kibale National Forest, Uganda. *American Journal of Primatology* 69:1030–1044.

Teleki, G. 1973a. *The Predatory Behavior of Wild Chimpanzees*. Lewisburg, PA: Bucknell University Press.

———. 1973b. The omnivorous chimpanzee. *Scientific American* 228:32–42.

———. 1973c. Group response to the accidental death of a chimpanzee in Gombe National Park, Tanzania. *Folia Primatologica* 20:81–94.

Teleki, G., Hunt, E. E., Jr., and Pfifferling, J. H. 1976. Demographic observations (1963–1973) in chimpanzees of Gombe National Park, Tanzania. *Journal of Human Evolution* 5:559–598.

Tembrock, G. 1974. Sound production of *Hylobates* and *Symphalangus*. In D. M. Rumbaugh, ed., *Gibbon and Siamang,* Vol. 3, 176–205. Basel: Karger.

Temerlin, M. K. 1975. *Lucy: Growing Up Human*. Palo Alto, CA: Science and Behavior.

Templeton, A. R. 1993. The "Eve" hypothesis: a genetic critique and reanalysis. *American Anthropologist* 95:51–72.

Tenaza, R. R. 1975. Territory and monogamy among Kloss' gibbons *(Hylobates klossii)* in Siberut Island, Indonesia. *Folia Primatologica* 24:60–80.

———. 1976. Songs, choruses and countersinging of Kloss' gibbons *(Hylobates klossii)* in Siberut Island, Indonesia. *Zeitschrift für Tierpsychologie* 40:37–52.

———. 1985. Songs of hybrid gibbons (*Hylobates lar*× *H. muelleri*). *American Journal of Primatology* 8:249–253.

Tenaza, R. R., and Hamilton, W. J. III. 1971. Preliminary observations of the Mentawai Islands gibbons, *Hylobates klossii. Folia Primatologica* 15:201–211.

Tenaza, R. R., and Tilson, R. L. 1977. Evolution of long-distance alarm calls in Kloss's gibbons. *Nature* 268:233–235.

Tennie, C., Hedwig, d., Call, J., and Tomasello, M. 2008. An experimental study of nettle feeding in captive gorillas. *American Journal of Primatology* 70:584–593.

Terrace, H. S. 1979a. *Nim: A Chimpanzee Who Learned Sign Language.* New York: Knopf.

———. 1979b. Is problem-solving language? *Journal of the Experimental Analysis of Behavior* 31:161–175.

———. 1981. A report to an academy, 1980. *Annals of the New York Academy of Sciences* 384:94–114.

———. 1982. Why Koko can't talk. *The Sciences* 22:8–10.

Terrace, H. S., and Bever, T. G. 1976. What might be learned from studying language in the chimpanzee? The importance of symbolizing oneself. *Annals of the New York Academy of Sciences* 280:579–588.

Terrace, H. S., Petitto, L. A., Sanders, R. J., and Bever, T. G. 1979. Can an ape create a sentence? *Science* 206:891–902.

Terrace, H. S., Straub, R., Bever, T. G., and Seidenberg, M. 1977. Representation of a sequence by a pigeon. *Bulletin of the Psychonomic Society* 10:269.

te Vilde, E. R., Scheffer, G. J., Dorland, M., Broekmans, F. J., and Fauser, B. C. J. M. 1998. Developmental and endocrine aspects of normal ovarian aging. *Molecular and Cellular Endocrinology* 145:67–73.

Thanaraji, K., Chaubey, G., Kivisild, T., Reddy, A. G., Singh, V. K., Rasalkar, A. A., and Singh, L. 2005. Reconstructing the origin of Adaman Islanders. 2005. *Science* 308:996.

Theall, L. A., and Povinelli, D. J. 1999. Do chimpanzees tailor their gestural signals to fit the attentional states of others? *Animal Cognition* 2:207–214.

Thein, T. 2004. A review of the large-bodied Pondaung primates of Myanmar. In C. F. Ross and R. F. Kay, eds., *Anthropoid Origins,* 219–47. New York: Kluwer Academic/Plenum Press.

Thenius, E. 1982. Ein Menschenaffenfund (Primates: Pongidae) aus dem Pannon (Jung-Miozän) von Niederösterreich. *Folia Primatologica* 39:187–200.

Thieme, H. 1997. Lower Paleolithic hunting spears from Germany. *Nature* 385:807–810.

Thieme, H., and Viel, S. 1985. Neue Untersuchungen zum eemzeitlichen Elefanten-Jagdplatz Lehrigen, Ldkr. Verden. *Die Kunde N. F.* 36:11–58.

Thinh, V. N., Rawson, B., Hallam, C., Kenyon, M., Nadler, T., Lutz, W., and Roos, C. 2010. Phylogeny and distribution of crested gibbons (genus Hylobates) based on mitochondrial cytochrome b gene sequential data. *American Journal of Primatology* 72:1047–1054.

Thoma, A. 1981. The position of the Vértesszöllös find in relation to *Homo erectus.* In B. A. Sigmon and J. S. Cybulski, eds., *Homo erectus: Papers in Honor of Davidson Black,* 105–114. Toronto: University of Toronto Press.

Thomas, D. H. 2000. Pollen analysis. In E. Delson, I. Tattersall, J. A. Van Couvering, and A. S. Brooks, eds., *Encyclopedia of Human Evolution and Prehistory,* 578–579. New York: Garland.

Thomas, H. 1985. The Early and Middle Miocene land connection of the Afro-Arabian plate and Asia: a major event for hominoid dispersal? In E. Delson, ed., *Ancestors: The Hard Evidence,* 42–50. New York: Alan R. Liss.

Thomas, H., Roger, J., Sen, S., and Al-Sulaimani, Z. 1988. Découverte des plus anciens "Anthropoïdes" du continent arabo-africain et d'un Primate tarsiiforme dans l'Oligocène du Sultanat d'Oman. *Comptes rendus des séances de l'Académie des sciences, Paris, série II* 306:823–829.

Thomas, H., Roger, J., Sen, S., Bourdillon-de-Grissac, C., and Al-Sulaimani, Z. 1989. Découverte de vertébrés fossiles dans l'Oligocène inférieur du Dhofar (Sultantat d'Oman). *Geobios,* no. 22: 101–120.

Thomas, H., Roger, J., Sen, S., Dejax, J., Schuler, M., Al-Sulaimani, Z., Bourdillon de Grissac, C., Breton, G., de Broin, F., Camoin, G., et al. 1991. Essai de reconstitution des milieux de sédimentation et de vie des primates anthropoïdes de l'Oligocène de Taqah (Dhofar, Sultanat d'Oman). *Bulletin de Société Géologique de France* 162:713–724.

Thomas, H., Sen, S., Khan, M., Battail, B., and Ligabue, G. 1982. The Lower Miocene fauna of Al-Sarrar (eastern province, Saudi Arabia). *ATLAL: Journal of Saudi Arabian Archaeology* 5:109–136.

Thomas, H., Sen, S., Roger, J., and Al-Sulaimani, Z. 1991. The discovery of *Moeripithecus markgrafi* Schlosser (Propliopithecidae, Anthropoidea, Primates), in the Ashawq Formation (Early Oligocene of Dhofar Province, Sultantate of Oman). *Journal of Human Evolution* 20:33–49.

Thompson, C. R., and Church R. M. 1980. An explanation of the language of a chimpanzee. *Science* 208:313–314.

Thompson, J. A. M. 2001. The status of bonobos in their southernmost geographic range. In B. M. F. Galdikas, N. E. Briggs, L. K. Sheeran, G. L. Shapiro, and J. Goodall, eds., *All Apes Great and Small,* Vol. 1: *African Apes,* 75–81. New York: Kluwer Academic/Plenum Press.

Thompson, J. M., Nestor, L. M., and Kabanda, R. B. 2008. Traditional land-use practices for bonobo conservation. In T. Furuichi and J. Thompson, eds., *The Bonobos: Behavior, Ecology, and Conservation,* 227–244. New York: Springer.

Thompson, R. K. R., and Contie, C. L. 1994. Further reflections on mirror usage by pigeons: lessons from Winnie-the-Pooh and Pinocchio too. In S. T. Parker, R. W. Mitchell, and M. L. Boccia, eds., *Self-Awareness in Animals and Humans,* 392–409. Cambridge: Cambridge University Press.

Thompson-Handler, N., Malenky, R. K., and Badrian, N. 1984. Sexual behavior of *Pan paniscus* under natural conditions in the Lomako Forest, Equateur, Zaire. In R. L. Susman, ed., *The Pygmy Chimpanzee: Evolutionary Biology and Behavior,* 347–368. New York: Plenum Press.

Thorén, S., Lindenfors, P., and Kappeler, P. M. 2006. Phylogenetic analyses of dimorphism in Primates: evidence for stronger selection on canine size than body size. *American Journal of Physical Anthropology 130:50–59.*

Thorndike, E. L. 1898. Animal intelligence: an experimental study of the associative processes in animals. *Psychological Review Monograph Supplements 2,* no. 8.

Thorne, A. G. 1971a. Mungo and Kow Swamp: morphological variation in Pleistocene Australians. *Mankind* 8:85–89.

———. 1971b. The racial affinities and origins of the Australian aborigines. In J. D. Mulvaney and J. Golson, eds., *Aboriginal Man and Environment in Australia,* 316–325. Canberra: Australian National University.

———. 1976. Morphological contrasts in Pleistocene Australians. In R. L. Kirk and A. G. Thorne, eds., *The Origin of the Australians,* 95–112. Canberra: Australian National University.

———. 1980. The longest link: human evolution in Southeast Asia and the settlement of Australia. In J. J. Fox, R. G. Garnaut, P. T. McCawley, and J. A. C. Maukie, eds., *Indonesia: Australian Perspectives,* 35–43. Canberra: Research School of Pacific Studies.

Thorne, A. G., and Henneberg, A. 2007. Argue and colleagues: testing the Howells's data. In E. Indriati, ed., *Recent Advances in Southeast Asian Paleoanthropology and Archaeology,* 30–36. Yogyakarta, Indonesia: Gadjah Mada University.

Thorne, A. G., and Macumber, P. G. 1972. Discoveries of Late Pleistocene man at Kow Swamp, Australia. *Nature* 328:316–319.

Thorne, A. G., and Wolpoff, M. H. 1992. The multiregional evolution of humans. *Scientific American* 266:76–79.

Thornhill R., and Alcock J. 1983. *The Evolution of Insect Mating Systems.* Cambridge, MA: Harvard University Press.

Thorpe, S. K. S., and Crompton, R. H. 2005. Locomotor ecology of wild orangutans *(Pongo pygmaeus abelii)* in the Gunung Leuser ecosystem, Sumatra, Indonesia: a multivariate analysis using log-linear modelling. *American Journal of Physical Anthropology* 127:58–78.

———. 2006. Orangutan positional behavior and the nature of arboreal locomotion in Hominoidea. *American Journal of Physical Anthropology* 131:384–401.

———. 2009. Orangutan positional behavior. In eds. S. A. Wich, S. S. Utami Atmoko, T. Mira Setia, and C. P. van Schaik, *Orangutans: Geographic Variation in Behavioral Ecology and Conservation,* 33–47. Oxford: Oxford University Press.

Thorpe, S. K. S., Holder, R. L., and Crompton, R. H. 2007. Origin of human bipedalism as an adaptation for locomotion on flexible branches. *Science* 316:1328–1331.

Tiercelin, J. J. 1986. The Pliocene Hadar Formation, Afar depression of Ethiopia. In L. E. Frostick, R. W. Renaut, I. Reid, and J. J. Tiercelin, eds., *Sedimentation in the African Rifts,* 221–240. Oxford: Blackwell Scientific.

Tiger, L. 1969. *Men in Groups.* New York: Random House.

Tiger, L., and Fox, R. 1971. *The Imperial Animal.* New York: Holt, Rinehart and Winston.

Tilson, R. L. 1979. Behaviour of hoolock gibbon *(Hylobates hoolock)* during different seasons in Assam, India. *Journal of the Bombay Natural History Society* 76:1–16.

―――. 1981. Family formation strategies of Kloss' gibbons. *Folia Primatologica* 35:259–287.

Tilson, R. L., and Tenaza, R. R. 1982. Interspecific spacing between gibbons *(Hylobates klossii)* and langurs *(Pesbytis potenziani)* on Siberut Island, Indonesia. *American Journal of Primatology* 2:355–361.

Tipple, B. J. 2013. Capturing climate variability during our ancestors' earliet days. *Proceedings of the National Academy of Sciences of the United States of America* 110:1144–1145.

Tishkoff, S. A., Reed, F. A., Friedlaender, F. R., Ehret, C., Ranciaro, A., Froment, A., Hirbo, J. B., Awomoyi, A. A., Bodo, J.-M., Doumbo, O., et al. 2009. The genetic struture and history of Africans and African Americans. *Science* 324:1035–1044.

Tobias, P. V. 1965. New discoveries in Tanganyika: their bearing on hominid evolution. *Current Anthropology* 6:391–411.

―――. 1966. Fossil hominid remains from Ubeidiya, Israel. *Nature* 211:30–133.

―――. 1967a. Cultural hominization among the earliest African Pleistocene hominids. *Proceedings of the Prehistoric Society for 1967* 33:367–376.

―――. 1967b. *The Cranium and Maxillary Dentition of Australopithecus (Zinjanthropus) boisei.* Vol. 2 of *Olduvai Gorge,* ed. L. S. B. Leakey, Cambridge: Cambridge University Press.

―――. 1971. *The Brain in Hominid Evolution.* New York: Columbia University Press.

―――. 1973. Implications of the new age estimates of the early South African hominids. *Nature* 246:79–83.

―――. 1974. Taxonomy of the Taung skull. *Nature* 252:85–86.

―――. 1978a. The earliest Transvaal members of the genus *Homo* with another look at some problems of hominid taxonomy and systematics. *Zeitschrift für Morphologie und Anthropologie* 69:225–265.

―――. 1978b. Position et rôle des australopithécinés dans la phylogenèse humaine, avec étude particulière de *Homo habilis* et des théories controversées avancées a propos des premiers hominidés fossiles de Hadar et de Laetoli. In E. Bone, Y. Coppens, E. Genet-Varcin, P.-P. Grasse, J.-L. Heim, W. W. Howells, J. Hurzeler, S. Krukoff, H. de Lumley, M.-A. de Lumley, et al., eds., *Les Origines humaines et les époques de l'intelligence,* 38–77. Paris: Masson.

―――. 1978c. The place of *Australopithecus africanus* in hominid evolution. In D. J. Chivers and K. A. Joysey, eds., *Recent Advances in Primatology,* Vol. 3: *Evolution,* 373–394. London: Academic Press.

―――. 1978d. The South African australopithecines in time and hominid phylogeny, with special reference to the dating and affinities of the Taung skull. In C. Jolly, ed., *Early Hominids of Africa,* 45–84. New York: St. Martin's Press.

―――. 1980a. "*Australopithecus afarensis*" and *A. africanus:* critique and an alternative hypothesis. *Palaeontologia Africana* 23:1–17.

―――. 1980b. A survey and synthesis of the African hominids of the late Tertiary and early Quaternary periods. In L.-K. Königsson, ed., *Current Argument on Early Man,* 86–113. Oxford: Pergamon Press.

―――. 1981a. The emergence of man in Africa and beyond. *Philosophical Transactions of the Royal Society, London B* 292:43–56.

———. 1981b. The anatomy of hominization. In E. A. Vidrio and M. A. Galina, eds., *Eleventh International Congress of Anatomy: Advances in the Morphology of Cells and Tissues,* 101–110. New York: Alan R. Liss.

———. 1987. The brain of *Homo habilis:* a new level of organization in cerebral evolution. *Journal of Human Evolution* 16:741–761.

———. 1988. Numerous apparently synapomorphic features in *Australopithecus robustus, Australopithecus boisei* and *Homo habilis:* support for the Skelton-McHenry-Drawhorn hypothesis. In F. E. Grine, ed., *Evolutionary History of the "Robust" Australopithecines,* 293–308. Hawthorne, NY: Aldine de Gruyter.

———. 1991. *Olduvai Gorge,* Vol. 4: *Homo habilis: Skulls, Endocasts and Teeth.* Cambridge: Cambridge University Press.

———. 1996. The dating of linguistic beginnings. *Behavioral and Brain Sciences* 19:789–793.

———. 2003. Encore Olduvai. *Science* 299:1193–1194.

Tocheri, M. W., Marzke, M. W., Liu, D., Bae, M., Jones, G. P., Williams, R. C., and Razdan, A. 2003. Functional capabilities of modern and fossil hominid hands: three-dimensional analysis of trapezia. *American Journal of Physical Anthropology* 122:101–112.

Tocheri, M. W., Orr, C. M., Larson, S. G., Sutikna, T., Jatmiko, Saptomo, E. W., Due, R. A., Djubiantono, T., Morwood, M. J., and Jungers, W. L. 2007. The primitive wrist of *Homo floresiensis* and its implications for hominin evolution. *Science* 317:1743–1745.

Tolman, E. C. 1978. Cognitive maps in rats and men. *Psychological Review* 55:189–208.

Tomasello, M. 1994. The question of culture. In R. W. Wrangham, W. C. Mc Grew, F. B. M. de Waal, and P. G. Heltne, eds., *Chimpanzee Cultures,* 301–317. Cambridge, MA: Harvard University Press.

———. 1996. Do apes ape? In C. M. Heyes and B. G. Galef Jr., *Social Learning in Animals: The Roots of Culture,* 319–346. San Diego: Academic Press.

———. 1999a. *The Cultural Origins of Human Cognition.* Cambridge, MA: Harvard University Press.

———. 1999b. Having intentions, understanding intentions, and understanding communicative intentions. In P. D. Zelazo, J. W. Astington, and D. R. Olson, eds., *Developing Theories of Intention: Social Understanding and Self-control,* 63–75. Mahwah, NJ: Lawrence Erlbaum.

———. 2002. Not waving but speaking: how important were gestures in the evolution of language? *Nature* 417:971–972.

———. 2003. On the different origins of symbols and grammar. In M. H., Christiansen, and S. Kirby, eds., *Language Evolution,* 94–110. Oxford: Oxford University Press.

———. 2008. *Origins of Human Communication.* Cambridge, MA: MIT Press.

———. 2009. The question of chimpanzee culture, plus postscript (chimpanzee culture, 2009). In K. N Laland and B. G. Galef, eds., *The Question of Animal Culture,* 198–221. Cambridge, MA: Harvard University Press.

Tomasello, M., and Call, J. 1997. *Primate Cognition.* New York: Oxford University Press.

———. 2007. Ape gestures and the origins of language. In J. Call and M. Tomasello, eds., *The Gestural Communication of Apes and Monkeys,* 221–239. London: Lawrence Erlbaum.

Tomasello, M., Carpenter, M., Call, J., Behne, T., and Moll, H. 2005. Understanding and sharing intentions: the origins of cultural cognition. *Behavioral and Brain Sciences* 28:675–735.

Tomasello, M., George, B. L., Kruger, A. C., Farrar, M. J., and Evans, A. 1985. The development of gestural communication in young chimpanzees. *Journal of Human Evolution* 14:175–186.

Tomilin, M. I., and Yerkes, R. M. 1935. Mother-infant relationships in chimpanzees. *Journal of Comparative Psychology* 20:321–358.

Tonooka, R., Inoue, N., and Matsuzawa, T. 1994. Leaf-folding behavior for drinking water by wild chimpanzees at Bossou, Guinea: a field experiment and leaf selectivity. [Article in Japanese.] *Primate Research* 10:307–313.

Tonooka, R., and Matsuzawa, T. 1995. Hand preferences of captive chimpanzees *(Pan troglodytes)* in simple reaching for food. *International Journal of Primatology* 16:17–35.

Tooby, J., and DeVore, I. 1987. The reconstruction of hominid behavioral evolution through strategic modeling, In W. G. Kinzey, *The Evolution of Human Behavior: Primate Models,* ed., 183–237. Albany: State University of New York Press.

Topping, D. L., and Clifton, P. M. 2001. Short-chain fatty acids and human colonic function: roles of resistant starch and nonstarch polysaccharides. *Physiological Reviews* 81:1031–1064.

Toth, N. 1985. Archaeological evidence for preferential right handedness in the Lower and Middle Pleistocene and its possible implications. *Journal of Human Evolution* 14:607–614.

Toth, N., Schick, K. D., Savage-Rumbaugh E. S., Sevick, R. A., and Rumbaugh, D. M. 1993. *Pan* the tool-maker: investigations into the stone tool-making and tool-using capabilities of a bonobo *(Pan paniscus). Journal of Archaeological Science* 20:81–91.

Toth, N., Schick, K. D., and Semaw, S. 2006. A comparative study of the stone tool-making skills of *Pan, Australopithecus,* and *Homo sapiens.* In N. Toth and K. Schick, eds., *The Oldowan: Case Studies into the Earliest Stone Age,* 155–222. Gosport, IN: Stone Age Institute Press.

Tóth, T. 1965. The variability of the weight of the brain of *Homo.* In *Homenaje a Juan Comas en su 65 aniversario,* Vol. 2, 391–402. México DF: Editorial Libros de Mexico.

Townsend, S. W., Deschner, T., and Zuberbühler, K. 2008. Female chimpanzees use copulation calls flexibly to prevent social competition. *PLoS One* 3, no. 6: e2431.

———. 2011. Copulation calls in female chimpanzees *(Pan troglodytes schweinfurthii)* convey identity but do not accurately reflect fertility. *International Journal of Primatology* 32:914–923.

Townsend, S. W., Siocombe, K. E., Emery Thompson, M., and Zuberbühler, K. 2007. Female-led infanticide in wild chimpanzees. *Current Biology* 17:R355–R356.

Townsend, S. W., and Zuberbühler, K. 2009. Audience effect in chimpanzee copulation calls. *Communicative and Integrative Biology* 2:282–284.

Trainor, L. J., and Trehub, S. E. 1992. A comparison of infants' and adults' sensitivity to Western musical structure. *Journal of Experimental Psychology: Human Perception and Performance* 18:394–402.

Trainor, L. J., Tsang, C. D., and Cheung, V. H. W. 2002. Preference for sensory consonance in 2- and 4-month-old infants. *Music Perception* 20:187–194.

Tramo, M. J. 2001. Music of the hemispheres. *Science* 291:54–56.

Travis-Henikoff, C. A. 2008. *Dinner with a Cannibal: The Complete History of Mankind's Oldest Taboo.* Santa Monica, CA: Santa Monica Press.

Treesucon, U., and Raemaekers, J. J. 1984. Group formation in gibbon through displacement of an adult. *International Journal of Primatology* 5:387.

Trehub, S. 2000. Human processing predispositions and musical universals. In N. L. Wallin, B. Merker and S. Brown, eds., *The Origins of Music,* 427–448. Cambridge, MA: MIT Press.

————. 2001. Musical predispositions in infancy. *Annals of the New York Academy of Music* 930:1–16.

————. 2003. The developmental origins of musicality. *Nature Neuroscience* 6:669–673.

Trehub, S. E., and Hannon, E. E. 2006. Infant music perception: domain-general or domain-specific mechanisms? *Cognition* 100:73–99.

Trinkaus, E. 1983a. *The Shanidar Neandertals.* New York: Academic Press.

————. 1983b. Neandertal postcrania and the adaptive shift to modern humans. In *The Mousterian Legacy: Human Biocultural Change in the Upper Pleistocene,* ed. E. Trinkaus, special issue, *British Archaeological Reports* 164 (Suppl.): 165–200.

————. 1984. Neandertal pubic morphology and gestation length. *Current Anthropology* 25:509–514.

————. 1986. The Neandertals and modern human origins. *Annual Review of Anthropology* 15:193–218.

————. 1989a. Olduvai hominid 7 trapezial metacarpal 1 articular morphology: contrasts with recent humans. *American Journal of Physical Anthropology* 80:411–416.

————. 1989b. Neandertal upper limb morphology and manipulation. In G. Giacobini, ed., *Hominidae: Proceedings of the 2nd International Congress of Human Paleontology,* 331–338. Milan: Jaca Book.

————. 2003. Early modern human cranial remains from the Pestera cu Oase, Romania. *Journal of Human Evolution* 45:245–253.

————. 2007. European early modern humans and the fate of the Neandertals. *Proceedings of the National Academy of Sciences of the United States of America* 104:7367–7372.

————. 2011. Late Pleistocene adult mortality patterns and modern human establishment. *Proceedings of the National Academy of Sciences of the United States of America* 108:1267–1271.

Trinkaus, E., Churchill, S. E., and Ruff, C. B. 1994. Postcranial robusticity in *Homo,* II: Humeral bilateral asymmetry and bone plasticity. *American Journal of Physical Anthropology* 93:1–34.

Trinkaus, E., Churchill, S. E., Vilemeur, I., Riley, K. G., Heller, J. A., and Ruff, C. B. 1991. Robusticity *versus* shape: the functional interpretation of Neandertal appendicular morphology. *Journal of the Anthropological Society of Tokyo* 99:257–278.

Trinkaus, E., and Long, J. C. 1990. Species attribution of the Swartkrans Member 1 first metacarpals: SK 84 and SKX 5020. *American Journal of Physical Anthropology* 83:419–424.

Trinkaus, E., and Shipman, P. 1993. *The Neandertals: Changing the Image of Mankind.* New York: Knopf.

Trinkaus, E., and Tompkins, R. L. 1990. The neandertal life cycle: the possibility, probability, and perceptibility of contrasts with recent humans. In C. J. DeRousseau, ed., *Primate Life History and Evolution,* 153–180. New York: Wiley-Liss.

Trivers, R. L. 1971. The evolution of reciprocal altruism. *Quarterly Review of Biology* 46:35–57.

———. 1972. Parental investment and sexual selection. In B. Campbell, ed., *Sexual Selection and the Descent of Man, 1871–1971,* 136–179. Chicago: Aldine.

Trivers, R. L., and Willard, D. E. 1973. Natural selection of parental ability to vary the sex ratio of offspring. *Science* 179:90–92.

Tsubamoto, T., Takai, M., Shigehara, N., Egi. N., Tun, S. T., Aung, A. K., Maung, M., Danhara, T., Suzuki, H. 2002. Fission-track zircon age of the Eocene Pondaung Formation, Myanmar. *Journal of Human Evolution* 42:361–369.

Tsukahara, T. 1993. Lions eat chimpanzees: the first evidence of predation by lions on wild chimpanzees. *American Journal of Primatology* 29:1–11.

Turnbull, C. M. 1965. *Wayward Servants: The Two Worlds of the African Pygmies.* Garden City, NY: Natural History Press.

Turner, C. G II and Turner, J. A. 1999. *Man Corn: Cannibalism and Violence in the Prehistoric American Southwest.* Salt Lake City: University of Utah Press.

Turner, J. H. 2000. *On the Origins of Human Emotions.* Stanford, CA: Stanford University Press.

Turner, V. 1967. *The Forest of Symbols.* New York: Cornell University Press.

———. 1969. *The Ritual Process.* Chicago: Aldine.

Tutin, C. E. G. 1975. Exceptions to promiscuity in a feral chimpanzee community. In S. Kondo, M. Kawai, and A. Ehara, eds., *Contemporary Primatology,* 445–449, Basel: Karger.

———. 1979a. Mating patterns and reproductive strategies in a community of wild chimpanzees *(Pan troglodytes schweinfurthii). Behavioral Ecology and Sociobiology* 6:29–38.

———. 1979b. Responses of chimpanzees to copulation, with special reference to interference by immature individuals. *Animal Behaviour* 27:845–854.

———. 1980. Reproductive behaviour of wild chimpanzees in the Gombe National Park, Tanzania. *Journal of Reproduction and Fertility* 28 (Suppl.): 43–57.

———. 1996. Ranging and social structure of lowland gorillas in the Lopé Reserve, Gabon. In W. C. McGrew, L. F. Marchant, and T. Nishida, eds., *Great Ape Societies,* 58–70. Cambridge: Cambridge University Press.

———. 2001. Saving the gorillas *(Gorilla g. gorilla)* and chimpanzees *(Pan t. troglodytes)* of the Congo Basin. *Reproduction, Fertility and Development* 13:469–476.

Tutin, C. E. G., and Fernandez, M. 1983. Gorillas feeding on termites in Gabon, West Africa. *Journal of Mammalogy* 64:530–531.

———. 1984. Nationwide census of gorilla *(Gorilla g. gorilla)* and chimpanzee *(Pan t. troglodytes)* populations in Gabon. *American Journal of Primatology* 6:313–336.

———. 1985. Foods consumed by sympatric populations of *Gorilla g. gorilla* and *Pan t. troglodytes* in Gabon: some preliminary data. *International Journal of Primatology* 6:27–43.

———. 1992. Insect-eating by sympatric lowland gorillas *(Gorilla g. gorilla)* and chimpanzees *(Pan t. troglodytes)* in the Lopé Reserve, Gabon. *American Journal of Primatology* 28:29–40.

———. 1993a. Composition of the diet of chimpanzees and comparisons with that of sympatric lowland gorillas in the Lopé Reserve, Gabon. *American Journal of Primatology* 30:195–211.

———. 1993b. Faecal analysis as a method of describing diets of apes: examples from sympatric gorillas and chimpanzees at Lopé, Gabon. *Tropics* 2:187–197.

———. 1994. Comparison of food processing by sympatric apes in the Lopé Reserve, Gabon. In B. Thierry, J. R. Anderson, J. J. Roeder, and N. Herrenschmidt, eds., *Current Primatology,* Vol. I: *Ecology and Evolution,* 29–36, Strasbourg: Université Louis Pasteur.

Tutin, C. E. G., Fernandez, M., Rogers, and M. E., Williamson, E. A. 1992. A preliminary analysis of the social structure of lowland gorillas in the Lopé Reserve, Gabon. In N. Itoigawa, Y. Sugiyama, G. P. Sackett, and R. K. R. Thompson, eds., 245–253. Tokyo: University of Tokyo Press.

Tutin, C. E. G., Fernandez, M., Rogers, M. E., Williamson, E. A., and McGrew, M. C. 1991. Foraging profiles of sympatric lowland gorillas and chimpanzees in the Lopé Reserve, Gabon. *Philosophical Transactions of the Royal Society, London* B-334:179–186.

Tutin, C. E. G., Ham, R. M., White, L. J. T., and Harrison, M. J. S. 1997. The primate community of the Lopé Reserve, Gabon: diets, responses to fruit scarcity, and effects on biomass. *American Journal of Primatology* 42:1–24.

Tutin, C. E. G., and McGinnis, P. R. 1981. Chimpanzee reproduction in the wild. In C. E. Graham, ed., *Reproductive Biology of the Great Apes,* 230–264. New York: Academic Press.

Tutin, C. E. G., and McGrew, W. C. 1973. Chimpanzee copulatory behaviour. *Folia Primatologica.* 19:237–256.

Tutin, C. E. G.; McGrew, W. C.; and Baldwin, P. J. 1983. Social organization of savanna-dwelling chimpanzees, *Pan troglodytes verus,* at Mt. Assirik, Senegal. *Primates* 24:154–173.

Tutin, C. E. G., Parnell, R. J., White, L. J. T., and Fernandez, M. 1995. Nest building by lowland gorillas in the Lopé Reserve, Gabon: environmental influences and implications for censusing. *International Journal of Primatology* 16:53–76.

Tutin, C. E. G., White, L. J. T., Williamson, E. A., Fernandez, M., and McPhearson, G. 1994. List of plant species identified in the northern part of the Lopé Reserve, Gabon. *Tropics* 3:249–276.

Tutin, C. E. G., Williamson, E. A., Fernandez, M., Rogers, M. E., and Fernandez. 1991. A case study of a plant-animal relationship: *Cola lizae* and lowland gorillas in the Lopé Reserve, Gabon. *Journal of Tropical Ecology* 7:181–199.

Tuttle, R. H. 1965. The functional morphology and evolution of hominoid hands. *American Journal of Physical Anthropology* 23:333.

———. 1967. Knuckle-walking and the evolution of hominoid hands. *American Journal of Physical Anthropology* 26:171–206.

———. 1969a. Knuckle-walking and the problem of human origins. *Science* 166:953–961.

——. 1969b. Quantitative and functional studies on the hands of the Anthropoidea. I. the Hominoidea. *Journal of Morphology* 128:309–364.

——. 1969c. Terrestrial trends in the hands of the Anthropoidea: a preliminary report. In H. Hofer, ed., *Proceedings of the 2nd International Congress of Primatology, Atlanta, GA,* Vol. 2, 192–200, Basel: Karger.

——. 1970. Postural, propulsive, and prehensile capabilities in the cheiridia of chimpanzees and other great apes. G. H. Bourne, ed., *The Chimpanzee,* Vol. 2, 167–253. Basel: Karger.

——. 1972. Functional and evolutionary biology of hylobatid hands and feet. In D. M. Rumbaugh, ed., *Gibbon and Siamang,* Vol. 1, 136–206. Basel: Karger.

——. 1974. Darwin's apes, dental apes and the descent of man. *Current Anthropology* 15:389–426.

——. 1975a. Parallelism, brachiation and hominoid phylogeny. In W. P. Luckett and F. S. Szalay, eds., *Phylogeny of the Primates: A Multidisciplinary Approach,* 447–480. New York: Plenum Press.

——. 1975b. Knuckle-walking and knuckle-walkers: a commentary on some recent perspectives on hominoid evolution. In R. H. Tuttle, ed., *Primate Functional Morphology and Evolution,* 203–212. The Hague: Mouton.

——. 1975c. *Socioecology and Psychology of Primates.* The Hague: Mouton.

——. 1977. Naturalistic positional behavior of apes and models of hominid evolution, 1929–1976. In G. H. Bourne, ed., *Progress in Ape Research,* 277–296. New York: Academic Press.

——. 1981a. Paleoanthropology without inhibitions. *Science* 212:798.

——. 1981b. Evolution of hominid bipedalism and prehensile capabilities. *Philosophical Transactions of the Royal Society, London* B-292:89–94.

——. 1985. Ape footprints and Laetoli impressions. A response to the SUNY claims. In P. V. Tobias, ed., *Hominid Evolution: Past, Present, and Future,* 129–133. New York: Alan R. Liss.

——. 1986. *Apes of the World: Their Social Behavior, Communication, Mentality and Ecology.* Park Ridge, NJ: Noyes.

——. 1987. Kinesiological inferences and evolutionary implications from Laetoli bipedal trails G-1, G-2/3, and A. In M. D. Leakey and J. M. Harris, eds., *Laetoli: A Pliocene Site in Northern Tanzania,* 503–523. Oxford: Clarendon Press.

——. 1988a. What's new in African paleoanthropology? *Annual Review of Anthropology* 17:391–426.

——. 1988b. Comparative anatomy and evolution. In B. Helal and D. Wilson, eds., *The Foot,* Vol. 1, 12–24. Edinburgh: Churchill Livingstone.

——. 1990a. The pitted pattern of Laetoli feet. *Natural History* 99:60–65.

——. 1990b. Apes of the world. *American Scientist* 78:115–123.

——. 1992. Hands from newt to Napier. In S. Matano, R. H. Tuttle, H. Ishida and M. Goodman, eds., *Topics in Primatology,* Vol. 3: *Evolutionary Biology, Reproductive Endocrinology and Virology,* 3–20. Tokyo: University of Tokyo Press.

——. 1994a. Up from electromyography: primate energetics and the evolution of human bipedalism. In R. S. Corruccini and R. L. Ciochon, eds., *Integrative Paths to the Past:*

Paleoanthropological Advances in Honor of F. Clark Howell, 269–284. Englewood Cliffs, NJ: Prentice Hall.

———. 1994b. A trans-specific agenda. *Science* 264:602–603.

———. 1996. The Laetoli hominid G footprints: where do they stand today? *Kaupia* 6:97–102.

———. 1998. Global primatology in a new millennium. *International Journal of Primatology* 19:1–12.

———. 2001. Phylogenies, Fossils, and Feelings. In B. B. Beck, T. S. Stoinski, M. Hutchins, T. L. Maple, B. Norton, A. Rowan, E. F. Stevens, and A. Arluke, eds., *Great Apes and Humans: The Ethics of Coexistence*, 178–190. Washington, DC: Smithsonian Institution Press.

———. 2002. Paleoanthropology read in tooth and nail. *Reviews in Anthropology* 31:103–128.

———. 2003. An introductory perspective: gorillas—how important, how many, how long? In A. B. Taylor and M. L. Goldsmith, eds., *Gorilla Biology: A Multidisciplinary Perspective*, 11–14. Cambridge: Cambridge University Press.

———. 2006a. Are human beings apes, or are apes people too? In H. Ishida, R. H. Tuttle, M. Pickford, N. Ogihara, and M. Nakatsukasa, eds., *Human Origins and Environmental Backgrounds*, 249–258. New York: Springer.

———. 2006b. Animalia, *Homo*, and the Kingdom of God. *Zygon* 41:139–168.

———. 2006c. Seven decades of East African Miocene anthropoid studies. In H. Ishida, R. H. Tuttle, M. Pickford, N. Ogihara, and M. Nakatsukasa, eds., *Human Origins and Environmental Backgrounds*, 15–29. New York: Springer.

———. 2006d. Neontological perspectives on East African Middle and Late Miocene Anthropoidea. In H. Ishida, R. H. Tuttle, N. Ogihara, and M. Nakatsukasa, eds., *Human Origins and Environmental Backgrounds*, 209–223. New York: Springer.

———. 2007. Apes, intelligent science, and conservation. In D. A. Washburn, ed., *Primate Perspectives of Behavior and Cognition*, 17–28. Washington, DC: American Psychological Association.

———. 2008. Footprint clues in hominid evolution and forensics: lessons and limitations. *Ichnos* 15:158–165.

———. 2009. Darwin descended. Unpublished sermon delivered at the Rockefeller Memorial Chapel of the University of Chicago, Chicago, IL, February 1, 2009.

———. 2011. Forward. K. D'Août and E. E. Vereecke, eds., *Primate Locomotion: Linking Field and Laboratory Research*, v–ix. New York: Springer.

Tuttle, R. H., and Basmajian, J. V. 1974. Electromyography of brachial muscles in *Pan gorilla* and hominoid evolution. *American Journal of Physical Anthropology* 41:71–90.

———. 1975. Electromyography of *Pan gorilla* an experimental approach to hominization. S. Kondo, M. Kawai, A. Ehara, and S. Kawamura, eds., *Proceedings from the Symposia of the Fifth Congress of the International Primatological Society*, 303–314. Tokyo: Japan Science Press.

———. 1977. Electromyography of pongid shoulder muscles and hominoid evolution. I. retractors of the humerus and rotators of the scapula. *Yearbook of Physical Anthropology* 20:491–497.

Tuttle, R. H., Basmajian, J. V., and Ishida, H. 1975. Electromyography of the gluteus maximus muscle in gorilla and the evolution of hominid bipedalism. In R. H. Tuttle, ed., *Primate Functional Morphology and Evolution*, 253–269. The Hague: Mouton.

———. 1978. Electromyography of pongid gluteal muscles and hominid evolution. In D. J. Chivers and K. A. Joysey, eds., *Recent Advances in Primatology*. Vol. 3: *Evolution*, 463–468. London: Academic Press.

———. 1979. Activities of pongid thigh muscles during bipedal behavior. *American Journal of Physical Anthropology* 50:123–136.

Tuttle, R. H., and Beck, B. B. 1972. Knuckle walking hand posture in an orangutan *(Pongo pygmaeus). Nature* 236:33–34.

Tuttle, R. H., Buxhoeveden, D. P., and Cortright, G. W. 1979. Anthropology on the move, progress in studies of non-human primate positional behavior. *Yearbook of Physical Anthropology—1979* 22:187–214.

Tuttle, R. H., and Cortright, G. W. 1988. Positional behavior, adaptive complexes, and evolution. In J. H. Schwartz, ed., *Orang-utan Biology*, 311–330. Oxford: Oxford University Press.

Tuttle, R. H., Hallgrimsson, B., and Basmajian, J. V. 1994. Electromyography and elastic mechanisms in knuckle-walking *Pan gorilla* and *Pan troglodytes*. In B. Thierry, J. R., Anderson, J. J. Roeder, and N. Herrenschmidt, eds., *Current Primatology*, Vol. 1: *Ecology and Evolution*, 215–222. Strasbourg: Université Louis Pasteur.

———. 1999. Electromyography, elastic energy and knuckle-walking: a lesson in experimental anthropology. In S. Strum, D. G. Lindburg, and D. Hamburg, ed., *The New Physical Anthropology*, 32–41. Englewood Cliffs, NJ: Prentice-Hall.

Tuttle, R. H., Hallgrimsson, B., and Stein, T. 1998. Heel, squat, stand, stride: function and evolution of hominoid feet. In E. Strasser, J. Fleagle, A. Rosenberger, and H. McHenry, eds., *Primate Locomotion*, 435–448. New York: Plenum Press.

Tuttle, R. H., and Mirsky, D. E. 2007. Beware of Hobbits: the occurrence of pathology in famous fossils. In E. Indriati, ed., *Recent Advances in Southeast Asian Paleoanthropology and Archaeology*, 3–8. Yogyakarta, Indonesia: Gadjah Mada University.

Tuttle, R. H., and Rogers, C. M. 1966. Genetic and selective factors in reduction of the hallux in *Pongo pygmaeus. American Journal of Physical Anthropology* 24:191–198.

Tuttle, R. H., Velte, M. J., and Basmajian, J. V. 1983. Electromyography of brachial muscles in *Pan troglodytes* and *Pongo pygmaeus. American Journal of Physical Anthropology* 61:75–83.

Tuttle, R. H., and Watts D. P. 1985. The positional behavior and adaptive complexes of *Pan gorilla*. In S. Kondo, ed., *Primate Morphophysiology, Locomotor Analyses and Human Bipedalism*, 261–288. Tokyo: University of Tokyo Press.

Tuttle, R. H., Webb, D. M., and Baksh, M. 1991. Laetoli toes and *Australopithecus afarensis. Human Evolution* 6:193–222.

Tuttle, R. H., Webb, D. M., and Tuttle, N. I. 1991. Laetoli footprint trails and the evolution of hominid bipedalism. In Y. Coppens and B. Senut, eds., *Origine(s) de la Bipédie chez les Hominidés*, 203–218. Paris: Éditions du Centre National de la Recherche Scientifique.

Tuttle, R. H., Webb, D. M., and Tuttle, N. I., and Baksh, M. 1992. Footprints and gaits of bipedal apes, bears, and barefoot people: perspective on Pliocene tracks. In S. Matano,

R. H. Tuttle, H. Ishida, and M. Goodman, eds., *Topics in Primatology,* Vol. 3: *Evolutionary Biology, Reproductive Endocrinology and Virology,* 221–242. Tokyo: University of Tokyo Press.

Tuttle, R., Webb, D., Weidl, E., and Baksh, M. 1990. Further progress on the Laetoli trails. *Journal of Archaeological Science* 17:347–362.

Tweheyo, M., and Babweteera, F. 2007. Production, seasonality and management of chimpanzee food trees in Budongo Forest, Uganda. *African Journal of Ecology* 45:535–544.

Tweheyo, M., and Lye, K. A. 2003. Phenology of figs in Budongo Forest Uganda and its importance for the chimpanzee diet. *African Journal of Ecology* 41:306–316.

Tweheyo, M., and Obua, J. 2001. Feeding habits of chimpanzees *(Pan troglodytes),* red-tail monkeys *(Cercopithecus ascanius schmidt)* and blue monkeys *(Cercopithecus mitis stuhlmanii)* on figs in Budongo Forest Reserve, Uganda. *African Journal of Ecology* 39:133–139.

Tweheyo, M., Reynolds, V., Huffman, M. A., Pebsworth, P., Goto, S., Mahaney, W. C., Milner, M. W., Waddell, A., Dirszowsky, R., and Hancock, R. G. V. 2006. Geophagy in chimpanzees *(Pan troglodytes schweinfurthii)* of the Budongo Forest Reserve, Uganda. In N. E. Newton-Fisher, H. Notman, J. D. Paterson, and V. Reynolds, eds., *Primates of Western Uganda,* 135–152. New York: Springer.

Tyler, D. E. 1991. The problems of the Pliopithecidae as a hylobatid ancestor. *Human Evolution* 6:73–80.

Tylor, E. B. 1868. On the origin of language. *Fortnightly Review* 4:544–549.

———. 1874. *Primitive Culture: Researches into the Development of Mythology, Philosophy, Religion, Language, Art and Custom.* Boston: Estes & Lauriat.

———. 1877. *Ancient Society.* Tucson: University of Arizona Press. Reprinted in 1985.

Tyron, C. A., and McBrearty, S. 2002. Tephrostratigraphy and the Acheulian to Middle Stone Age transition in the Kapthurin Formation, Kenya. *Journal of Human Evolution* 42:211–235.

Uchida, A. 1996. What we don't know about great ape variation. *TREE* 11:163–168.

———. 1998a. Design of the mandibular molar in the extant great apes and Miocene fossil Hominoida. *Anthropological Science* 106 (Suppl.): 119–126.

———. 1998b. Variation in tooth morphology of *Gorilla gorilla. Journal of Human Evolution* 34:55–70.

———. 1998c. Variation in the tooth morphology of *Pongo pygmaeus. Journal of Human Evolution* 34:71–79.

Uddin, M., Goodman, M., Erez, O., Romero, R., Liu, G., Islam, M., Opazo, J. C., Sherwood, C. C., Grossman, L. I., and Willdman, D. E. 2008. Distinct genomic signatures of adaptation in pre- and postnatal environments during human evolution. *Proceedings of the National Academy of Sciences of the United States of America* 105:3215–3220.

Uddin, M., Wildman, D. E., Liu, G., X., W., Johnson, R. M., Hof, P. R., Kapatos, G., Grossman, L. I., and Goodman, M. 2004. Sister grouping of chimpanzees and humans as revealed by genome-wide phylogenetic analysis of brain gene expression profiles. *Proceedings of the National Academy of Sciences of the United States of America* 101:2957–2962.

Ueda, S., Watanabe, Y., Hayashida, H., Miyata, T., Matsuda, F., and Honjo, T. 1986. Hominoid evolution based on the structures of immunoglobin epsilon and alpha genes. *Cold Spring Harbor Symposia on Quantitative Biology* 51:429–432.

Ueda, S., Watanabe, Y., Saitou, N., Omoto, K., Hayashida, H., Miyata, T., Hisajima, H., and Honjo, T. 1989. Nucleotide sequences of immunoglobulin-epsilon pseudogenes in man and apes and their phylogenetic relationships. *Journal of Molecular Biology* 205:85–90.

Uehara, S. 1981. The social unit of wild chimpanzees: a reconsideration based on the diachronic data accumulated at Kasoje in the Mahale Mountains, Tanzania. *Africa-Kenkyu (Journal of African Studies)* 20:15–32.

———. 1982. Seasonal changes in the techniques employed by wild chimpanzees in the Mahale Mountains, Tanzania, to feed on termites *(Pseudocanthotermes spiniger)*. *Folia Priamtologica* 37:44–76.

———. 1984. Sex differences in feeding on *Camponotus* ants among wild chimpanzees in the Mahale Mountains, Tanzania. *International Journal of Primatology* 5:389.

—— 1990. Utilization patterns f a marsh grassland within the tropical rain forest by the bonobos *(Pan paniscus)* of Yalosidi, Republic of Zaire. *Primates* 31:311–322.

———. 1997. Predation on mammals by the chimpanzee *(Pan troglodytes)*. *Primates* 38:193–214.

Uehara, S., and Ihobe, H. 1998. Distribution and abundance of diurnal mammals, especially monkeys, at Kasoje, Mahale Mountains, Tanzania. *Anthropological Science* 106:349–369.

Uehara, S., and Nishida, T. 1987. Body weights of wild chimpanzees *(Pan troglodytes schweinfurthii)* of the Mahale Mountains National Park, Tanzania. *American Journal of Physical Anthropology* 72:315–321.

Uehara, S., Nishida, T., Hamai, M., Hasegawa, T., Hayaki, H., Huffman, M. A., Kawanaka, K., Kobayashi, S., Mitani, J. C., Takahata, Y., et al. 1992. Characteristics of predation by chimpanzees in the Mahale Mountains National Park, Tanzania. In T. Nishida, W. C. McGrew, P. Marler, M. Pickford, and F. B. M. de Waal, eds., *Topics in Primatology,* Vol. 1: *Human Origins,* 143–158. Tokyo: University of Tokyo Press.

Uehara, S., and Nyundo, R 1983. One observed case of temporary adoption of an infant by unrelated nulliparous females among wild chimpanzees in the Mahale Mountains, Tanzania. *Primates* 24:456–466.

Uhlenhuth, P. 1904. Ein neuer biologischer Beweis für die Blutsverwandtschaft zwischen Menschen- und Affengeschlecht. *Archiv für Rassen- und Gesellschafts-Biologie* 1:682–688.

Ujhelyi, M. 2000. Social organization as a factor in the origin of language and music. In N. L. Wallin, B. Merker and S. Brown, eds., *The Origins of Music,* 125–134. Cambridge, MA: MIT Press.

Ulijaszek, S., Mann, N., and Elton, S. 2012. *Evolving Human Nutrition: Implications for Public Health.* Cambridge: Cambridge University Press.

Umiker-Sebeok, J., and Sebeok T. A. 1981. Clever Hans and smart simians: the self-fulfilling prophecy and kindred methodological pitfalls. *Anthropos* 76:89–165.

———. 1982. Rejoinder to the Rumbaughs. *Anthropos* 77:574–578.

Ünay, E., de Bruijn, H., and Suata-Alpaslan, F. 2006. Rodents from the Upper Miocene Locality Çorakyerler (Anatolia). *Beitrage zur Palaontologie* 30:453–467.

Ungar, P. S. 1994. Patterns of ingestive behavior and anterior tooth use differences in sympatric anthropoid primates. *American Journal of Physical Anthropology* 95:197–219.

———. 1995. Fruit preferences of four sympatric primate species at Ketambe, Northern Sumatra, Indonesia. *International Journal of Primatology* 16:221–245.

———. 1996a. Dental microwear of European Miocene catarrhines: evidence for diets and tooth use. *Journal of Human Evolution* 31:335–366.

———. 1996b. Feeding height and niche separation in sympatric Sumatran monkeys and apes. *Primates* 67:163–168.

———. 1998. Dental allometry, morphology, and wear as evidence or diet in fossil primates. *Evolutionary Anthropology* 6:205–217.

———. 2004. Dental topography and diets of *Australopithecus afarensis* and early *Homo*. *Journal of Human Evolution* 46:605–622.

———. 2007a. *Evolution of the Human Diet*. Oxford: Oxford University Press.

———. 2007b. Limits to knowledge on the evolution of hominin diet. In P. S. Ungar, ed., *Evolution of the Human Diet*, 395–5407. Oxford: Oxford University Press.

———. 2007c. Dental functional morphology. The known, the unknown, and the unknowable. In P. S. Ungar, ed., *Evolution of the Human Diet*, 39–55. Oxford: Oxford University Press.

Ungar, P. S., and Grine, F. E. 1991. Incisor size and wear in *Australopithecus africanus* and *Paranthropus robustus*. *Journal of Human Evolution* 20:313–340.

Ungar, P. S., and Kay, R. F. 1995. The dietary adaptations of European Miocene catarrhines. *Proceedings of the National Academy of Sciences of the United States of America* 92:5479–5481.

Ungar, P. S., and Scott, R. S. 2009. Dental evidence for diets of *Homo*. In F. E. Grine, J. G. Fleagle, and R. E. Leakey, eds., *The First Humans: Origin and Early Evolution of the Genus Homo*, 121–134. New York: Springer.

Ungar, P. S., and Teaford, M. F. 2002. *Human Diet: Its Origins and Evolution*. Westport CT: Bergin & Garvey.

Ungar, P. S., Walker, A., and Coffing, K. 1994. Reanalysis of the Lukeino molar (KNM-LU 335). *American Journal of Physical Anthropology* 94:165–173.

Uno, K. T., Cerling, T. E., Harris, J. M., Kunimatsu, Y., Leakey, M. G., Nakatsukasa, M., Nakaya, H. 2011. Late Miocene to Pliocene carbon isotope record of differential diet change among East African herbivores. *Proceedings of the National Academy of Sciences of the United States of America* 108:6509–6514.

Urbani, B. 1999. Spontaneous use of tools by wedge-capped capuchin monkeys *(Cebus olivaceus)*. *Folia Primatologica* 70:172–174.

Usherwood, J. R., Larson, S. G., and Bertram, J. E. A. 2003. Mechanisms of force and power production in unsteady ricochetal brachiation. *American Journal of Physical Anthropology* 120:384–372.

Utami, S. S., Goossens, B., Bruford, M. W., de Ruiter, J., and van Hooff, J. A. R. A. M. 2002. Male bimaturism and reproductive success in Sumatran orang-utans. *Behavioral Ecology* 13:643–652.

Utami, S. S., and Mitra Setia, T. 1995. Behavioral changes in wild male and female Sumatran orangutans *(Pongo pygmaeus abelii)* during and following a resident male takeover. In R. D. Nadler, B. F. M. Galdikas, L. S. Sheeran, and N. Rosen, eds., *The Neglected Ape,* 183–190. New York: Plenum Press.

Utami Atmoko, S. S., Mitra Setia, T., Goossens, B., James, S. S. Knott, C. D., Borrogh-Bernard, H. C., van Schaik, C. P., and van Noordwjik, M. A. 2009. Orangutan mating behaviour and strategies. In S. A. Wich, S. S. Utami Atmoko, T. Mitra Setia, and C. P. van Schaik, eds., *Orangutans: Geographic Variation in Behavioral Ecology and Conservation,* 235–244. Oxford: Oxford University Press.

Utami Atmoko, S. S., Singleton, I., van Noordwjik, M. A., van Schaik, C. P., and Mitra Setia, T. 2009. Male-male relationships in orangutans. In S. A. Wich, S. S. Utami Atmoko, T. Mitra Setia, and C. P. van Schaik, eds., *Orangutans: Geographic Variation in Behavioral Ecology and Conservation,* 225–233. Oxford: Oxford University Press.

Utami Atmoto, S. S., and Van Hooff, J. A. R. A. M. 1997. Meat-eating by adult female Sumatran orangutans *(Pongo pygmaeus abelii). American Journal of Primatology* 43:159–165.

———. 2004. Alternative male reproductive strategies: male bimaturism in orangutans. In P. Kappeler and C. P. van Schaik, eds., *Sexual Selection in Primates: New and Comparative Perspectives,* 196–207. Cambridge: Cambridge University Press.

Utami, S. S., Wich, S. A., Sterck, H. M., and Van Hooff, J. A. R. A. M. 1997. Food competition between wild orangutans in large fig trees. *International Journal of Primatology* 18:909–927.

Uzzell, T., and Pilbeam, D. 1971. Phyletic divergence dates of hominoid primates: a comparison of fossil and molecular data. *Evolution* 25:615–635.

Van Cantfort, T. E., Gardner, B. T., and Gardner, R. A. 1989. Developmental trends in replies to Wh-questions by children and chimpanzees. In R. A. Gardner and B. T. Gardner, eds., *Teaching Sign Language to Chimpanzees,* 198–239. Albany NY: SUNY Press.

Van Cantfort, T. E., and Rimpau, J. B. 1982. Sign language studies with children and chimpanzees. *Sign Language Studies* 34:15–72.

van Casteren, A., Sellers, W. I., Thorpe, K. S., Coward, S. Crompton, R. H., Myatt, J. P., and Ennos, R. 2012. Nest-building orangutans demonstrate engineering know-how to produce safe, comfortable beds. *Proceedings of the National Academy of Sciences of the United States of America* 109:6873–6877.

Van Couvering, J. A., and Harris, J. A. 1991. Late Eocene age of Fayum mammal faunas. *Journal of Human Evolution* 21:241–260.

Vandenbergh, B. The dog-eat-dog world of carnivores. A review of past and present carnivore community dynamics. In C. B. Stanford and H. T. Bunn, eds., *Meat-Eating and Human Evolution* 101–121. Oxford: Oxford University Press.

Vandenberghe, R., Nobre, A. C., and Price, C. J. 2002. The response of left temporal cortex to sentences. *Journal of Cognitive Neuroscience* 14:550–560.

van den Bos, R. 1999. Reflections on self-recognition in nonhuman primates. *Animal Behaviour* 58:F1–F9.

van der Merwe, N. J., Masao, F. T., and Bramford, M. K. 2008. Isotopic evidence for contrasting diets of early hominins *Homo habilis* and *Australopithecus boisei* of Tanzania. *South African Journal of Science* 104:153–155.

van der Merwe, N. J., Thackeray, J. F., Lee-Thorp, J. A., and Luyt, J. 2003. The carbon isotope ecology and diet of *Australopithecus africanus* at Sterkfontein, South Africa. *Journal of Human Evolution* 44:581–597.

van de Waal, E., Borgeaud, C., and Witten, A. 2013. Potent social learning and conformity shape a wild primate's foraging decisions. *Science* 340:483–485.

van Elsacker, L., Vervaecke, H., and Verheyen, R. F. 1995. A review of terminology on aggregation patterns in bonobos *(Pan paniscus)*. *International Journal of Primatology* 16:37–52.

van Elsacker, L., and Walraven, V. 1994. The spontaneous use of a pineapple as a recipient by a captive bonobo *(Pan paniscus)*. *Mammalia* 58:159–162.

van Hooff, J. A. R. A. M. 1962. Facial expressions in higher primates *Symposia of the Zoological Society, London* 8:97–125.

———. 1963. Facial expressions in higher primates. *Symposia of the Zoological Society, London* 10:103–104.

———. 1967. The facial displays of the catarrhine monkeys and apes. In D. Morris, ed., *Primate Ethology*, 7–68. London: Weidenfeld and Nicolson.

———. 1971. *Aspecten van het Sociale Gedrag en de Communicatie bij Humane en Hogere Niet-humane Primaten*. Rotterdam: Bronder-Offset.

———. 1972. A comparative approach to the phylogeny of laughter and smiling. In R. A. Hinde, ed., *Non-verbal Communication*, 209–241. Cambridge: Cambridge University Press.

———. 1973. A structural analysis of the social behaviour of a semi-captive group of chimpanzees. In Mario von Cranach and I. Vine, eds., *Social Communication and Movement, Studies in Interaction and Expression in Man and Chimpanzee*, 75–162. London: Academic Press.

———. 1976. The comparison of facial expression in man and higher primates. In M. von Cranach, *Methods of Inference from Animal to Human Behaviour*, 165–196. Chicago: Aldine.

———. 1981. Facial expressions. In D. McFarland, ed., *The Oxford Companion to Animal Behaviour*, 165–176, Oxford: Oxford University Press.

———. 1995. The orangutan: a social outsider. In R. D. Nadler, B. F. M. Galdikas, L. S. Sheeran, and N. Rosen, eds., *The Neglected Ape*, 153–162. New York: Plenum Press.

Van Krunkelsven, E. 2001. Density estimation of bonobos *(Pan paniscus)* in Salonga National Park, Congo. *Biological Conservation* 99:387–391.

Van Krunkelsven, E., Inogwabini, B.-I., and Draulans, D. 2000. A survey of bonobos and other large mammals in the Salonga National Park, Democratic Republic of Congo. *Oryx* 34:180–187.

van Lawick-Goodall, J. 1965. New discoveries among Africa's chimpanzees. *National Geographic* 128:802–831.

———. 1967. *My Friends the Wild Chimpanzees*. Washington, DC: National Geographic Society.

———. 1968a. The behaviour of free-living chimpanzees in the Gombe Stream area. *Animal Behaviour Monographs* 1:161–311.

———. 1968b. A preliminary report on expressive movements and communication in the Gombe Stream chimpanzees, In P. C. Jay, ed., *Primates: Studies in Adaptation and Variability*, 313–374. New York Holt, Rinehart & Winston.

———. 1970. Tool-using in primates and other vertebrates. In D. S. Lehrman, R. A. Hinde, and E. Shaw, eds., *Advances in the Study of Behaviour*, Vol. 3, 195–249. New York: Academic Press.

———. 1971. *In the Shadow of Man*. Boston: Houghton Mifflin.

———. 1973a. Cultural elements in a chimpanzee community. In E. W. Menzel Jr., ed., *Precultural Primate Behavior, Symposia of the Fourth International Congress of Primatology*, Vol. 1, 144–184. Basel: S. Karger.

———. 1973b. The behavior of chimpanzees in their natural habitat. *American Journal of Psychiatry* 130:1–12.

———. 1975. The behaviour of the chimpanzee. In G. Kurth and I. Eibl-Eibesfeldt, eds., *Hominisation und Verhalten*, 74–136. Stuttgart: Gustav Fischer.

van Noordwijk, M. A., Sauren, S. E. B., Nuzuar, Abulani, A., Morrogh-Bernard, H. C., Utami Atmoko, S. S., and van Schaik, C. P. 2009. Development of independence. Sumatran and Bornean orangutans compared. In S. A. Wich, S. S. Utami Atmoko, T. Mitra Setia, and C. P. van Schaik, eds., *Orangutans: Geographic Variation in Behavioral Ecology and Conservation*, 189–203. Oxford: Oxford University Press.

van Noordwijk, M. A., and van Schaik, C. P. 2000. Reproductive patterns in eutherian mammals: adaptations against infanticide? In C. P. van Schaik and C. H. Janson, eds., *Infanticide by Males and Its Implications*, 322–3660. Cambridge: Cambridge University Press.

———. 2004. Sexual selection and the careers of primate males: paternity concentration, dominance-acquisition tactics and transfer decisions. In P. Kappeler and C. P van Schaik, eds., *Sexual Selection in Primates: New and Comparative Perspectives*, 208–229. Cambridge: Cambridge University Press.

———. 2005. Development of ecological competence in Sumatran orangutans. *American Journal of Physical Anthropology* 127:79–94.

Vannucci, R. C., Barrron, T. F., and Holloway, R. L. 2011. Craniometric ratios of microcephaly and LB1, *Homo floresiensis*, using MRI and endocasts. *Proceedings of the National Academy of Sciences of the United States of America* 108:14043–14048.

van Roosmalen, M. G. M., Mittermeier, R. A., and Fleagle, J. G. 1988. Diet of the northern bearded saki *(Chiropotes satanas chiropotes)*: a Neotropical seed predator. *American Journal of Primatology* 14:11–25.

van Schaik, C. P. 1999. The socioecology of fission-fusion sociality in orangutans. *Primates* 40:69–86.

———. 2000. Social counterstrategies against male infanticide in primates and other mammals. In P. M. Kappeler, ed., *Primate Males*, 34–52. Cambridge: Cambridge University Press.

———. 2004. *Among Orangutans*. Cambridge, MA: Harvard University Press.

———. 2009. Geographic variation in the behavior of wild great apes: is it really culture? In K. N Laland and B. G. Galef, eds., *The Question of Animal Culture*, 70–98. Cambridge, MA: Harvard University Press.

van Schaik, C. P., Ancrenaz, M., Borgen, G., Galdikas, B., Knott, C. D., Singleton, I., Suzuki, A., Utami, S. S., and Merrill, M. 2003. Orangutan cultures and the evolution of material culture. *Science* 299:102–105.

van Schaik, C. P., and Aureli, F. 2000. Reconciliation and relationship qualities. In F. Aureli and F. B. M. de Waal, eds., *Natural Conflict Resolution,* 307–333. Berkeley: University of California Press.

van Schaik, C. P., and Brockman, D. K. 2005. Seasonality in primate ecology, reproduction and life history: an overview. In D. K. Brockman and C. P. van Schaik, eds., *Seasonality in Primates: Studies of Living and Extinct Human and Non-human Primates,* 3–20. Cambridge: Cambridge University Press.

van Schaik, C., and Deaner, R. O. 2003. Life history and cognitive evolution in Primates. In F. B. M. de Waal, and P. L. Tyack, eds., *Animal Social Complexity,* 5–25. Cambridge, MA: Harvard University Press.

van Schaik, C. P., Deaner, R. O., and Merrill, Y. 1999. The conditions for tool use in primates: implications for the evolution of material culture. *Journal of Human Evolution* 36:719–741.

van Schaik, C. P., and Dunbar, R. I. M. 1990. The evolution of monogamy in large primates: a new hypothesis and some crucial tests. *Behaviour* 115:30–62.

van Schaik, C. P., Fox, E. A., and Fechtman, L. T. 2003. Individual variation in the rate of use of tree-hole tools among wild orang-utans: implications for hominin evolution. *Journal of Human Evolution* 44:11–23.

van Schaik, C. P., Fox, E. A., and Sitompul, A. F. 1996. Manufacture and use of tools in wild Sumatran orangutans. *Naturwissenschaften* 83:186–188.

van Schaik, C., and Griffiths, M. 1996. Activity periods of Indonesian rain forest mammals. *Biotropica* 28:105–112.

van Schaik, C. P., and Kappeler, P. M. 1997. Infanticide risk and the evolution of male-female association in primates. *Proceedings of the Royal Society B* 264:1687–1694.

van Schaik, C. P., and Knott, C. D. 2001. Geographic variation in tool use on *Neesia* fruits in orangutans. *American Journal of Physical Anthropology* 114:331–342.

van Schaik, C. P., Madden, R., and Ganzhorn, J. U. 2005. Seasonality ad primate communities. In D. K. Brockman and C. P. van Schaik, eds., *Seasonality in Primates,* 445–463. Cambridge: University Press.

van Schaik, C. P., and Pfannes, K. R. 2005. Tropical climates and phenology: a primate perspective. In D. K. Brockman and C. P. van Schaik, eds., *Seasonality in Primates,* 23–54. Cambridge: Cambridge University Press.

van Schaik, C. P., Pradhan, G. R., and van Noordwijk, M. A. 2004. Mating conflict in primates: infanticide, sexual harassment and female sexuality. In P. Kappeler and C. P. van Schaik, eds., *Sexual Selection in Primates: New and Comparative Perspectives,* 131–150. Cambridge: Cambridge University Press.

van Schaik, C. P., Preuschoft, S., and Watts, D. P. 2004. Great ape social systems. In A. E. Russon and D. R. Begun, eds., *The Evolution of Thought: Evolutionary Origins of Great Ape Intelligence,* 190–209.

van Schaik, C. P., van Noordwijk, M. A., and Vogel, E. R. 2009. Ecological sex differences in wild orangutans. In S. A. Wich, S. S. Utami Atmoko, T. Mitra Setia, and C. P. van

Schaik, eds., *Orangutans: Geographic Variation in Behavioral Ecology and Conservation,* 255–268. Oxford: Oxford University Press.

van Schaik, C. P., van Noordwijk, M. A., and Wich, S. A. 2006. Innovation in wild Bornean orangutans *(Pongo pygmaeus wurmbii). Behaviour* 143:839–876.

van Tuinen, P., and Ledbetter, D. H. 1983. Cytogenetic comparison and phylogeny of three species of Hylobatidae. *American Journal of Physical Anthropology* 61:53–466.

Varki, A., and Altheide, T. K., 2005. Comparing the human and chimpanzee genomes: searching for needles in a haystack. *Genome Research* 15:1746–1758.

Varley, R. A., Klessinger, J. C., Romanowski, A. J., and Siegal, M. 2005. Agrammatic but numerate. *Proceedings of the National Academy of Sciences of the United States of America* 102:3519–3524.

Vasey, P. L. Homosexual behavior in primates: a review of evidence and theory. *International Journal of Primatology* 16:173–204.

Vauclair, J., and Anderson, J. R. 1994. Object manipulation and tool use, and the social context in human and non-human primates. *Techniques and Culture* 23–24:121–136.

Vauclair, J., and Faot, J. 1993. Manual specializations in gorillas and baboons. In J. P. Ward and W. D. Hopkins, eds., *Primate Laterality: Current Behavioral Evidence of Primate Asymmetries,* 193–205. New York: Springer.

Vawater, L., and Brown, W. M. 1986. Nuclear and mitochondrial DNA comparisons reveal extreme rate variation in the molecular clock. *Science* 234:194–196.

Veà, J. J., and Sabater Pí, J. 1998. Spontaneous pointing behaviour in the wild pygmy chimpanzee *(Pan paniscus). Folia Primatologica* 69:289–290.

Vecellio, V. 2007. Female mountain gorillas continue to try out new groups. *Dian Gorilla Fund International Field News,* June 26, http://gorillafund.org/page.aspx?pid=517.

Vedder, A. L. 1984. Movement patterns of a group of free-ranging mountain gorillas *(Gorilla gorilla beringei)* and their relation to food availability. *American Journal of Primatology* 7:73–88.

Vekua, A., Lordkipanidze, D., Rightmire, G. P., Augsti, J., Ferring, R., Maisuradze, G., Mouskhelishvili, A., Nioradze, M., Ponce de Leon, M., Tappen, M., et al. 2002. A new skull of early *Homo* from Dmanisi, Georgia. *Science* 297:85–89.

Venter, J. C., Adams, M. D., Myers, E. W., Li, P. W., Mural, R. J., Sutton, G. G., Smith, H. O., Yandell, M., Evans, C. A., Holt, R. A., et al. 2001. The sequence of the human genome. *Science* 291:1304–1351.

Vereecke, E., D'Août, K., Van Elsacker, L., De Clercq, D., and Aerts, P. 2005. Functional analysis of the gibbon foot during terrestrial bipedal walking: plantar pressure distributions and three-dimensional ground reaction forces. *American Journal of Physical Anthropology* 128:659–669.

Verhaegen, M. J. B. 1990. African ape ancestry. *Human Evolution* 5:295–297.

Vermiere, B. A., Hamilton, C. R., and Erdmann, A. L. 1998. Right-hemispheric superiority in split-brain monkeys for learning and remembering facial discriminations. *Behavioral Neuroscience* 112:1048–1061.

Vervaecke, H., Stevens, J., and Van Elsacker, L. 2004. Pan continuity: bonobos-chimpanzee hybrids. *Folia Primatologica* 75:59.

Vervaecke, H., de Vries, H., and Van Elsacker, L. 2000a. Dominance and its behavioral measures in a captive group of bonobos *(Pan paniscus)*. *International Journal of Primatology* 21:47–68.

———. 2000b. Function and distribution of coalitions in captive bonobos *(Pan paniscus)*. *Primates* 41:249–265.

Vervaecke, H., and Van Elsacker, L. 1992. Hybrids between common chimpanzees *(Pan troglodytes)* and pygmy chimpanzees *(Pan paniscus)* in captivity. *Mammalia* 56:667–669.

Vick, S.-J., and Paukner, A. 2010. Variation and context of yawns in captive chimpanzees *(Pan troglodytes)*. *American Journal of Primatology* 72:262–269.

Videan, E. N. 2006. Bed-building in captive chimpanzees *(Pan troglodytes):* the importance of early rearing. *American Journal of Primatology* 68:745–751.

Vigilant, L., and Bradley, B. J. 2004. Genetic variation in gorillas. *American Journal of Primatology* 64:161–172.

Vigilant, L., Hofreiter, M., Siedel, H., and Boesch, C. 2001. Paternity and relatedness in wild chimpanzee communities. *Proceedings of the National Academy of Sciences of the United States of America* 98:12890–12895.

Vigilant, L., Stoneking, M., Harpending, H., Hawkes, K., and Wilson, A. C. 1991. African populations and the evolution of human mitochondrial DNA. *Science* 253:1503–1507.

Vignaud, P., Duringer, P., Mackaye, H. T., Likius, A., Blondel, C., Boisserie, J.-R., de Bonis, L., Eisenmann, V., Etienne, M.-E., Geraads, D., et al. 2002. Geology and palaeontology of the Upper Miocene Toros-Menalla hominid locality, Chad. *Nature* 418:152–155.

Viranta, S., and Andrews, P. 1995. Carnivore guild structure in the Pasalar Miocene fauna. *Journal of Human Evolution* 28:359–372.

Viret, J. 1955. A propos de l'Oréopithèque. *Mammalia* 19:320–324.

Visalberghi, E. 1990. Tool use in *Cebus*. *Folia Primatologica* 54:146–154.

———. 1993a. Capuchin monkeys. a window into tool use in apes and humans. In K. R. Gibson and T. Ingold, eds., *Tools, Language and Cognition in Human Evolution,* 138–150. Cambridge: Cambridge University Press.

———. 1993b. Tool use in a South American monkey species: an overview of the characteristics and limits of tool use in *Cebus apella*. In A. Berthelet and J. Chavaillon, eds., *The Use of Tools by Humans and Non-human Primates,* 118–131. New York: Oxford University Press.

Visalberghi, E., Addessi, E., Trppa, V., Spagnoletti, N., Ottoni, E., Izar, P., and Fragaszy, D. 2009. Selection of effective stone tools by wild bearded capuchin monkeys. *Current Biology* 19:213–217.

Visalberghi, E., and Fragaszy, D. M. 1990. Food-washing behaviour in tufted capuchin monkeys, *Cebus apella,* and crab-eating macaques, *Macaca fascicularis. Animal Behaviour* 40:829–836.

———. 2000. Tool use by monkeys and apes. In *Frontiers of Life,* Vol. 4: *The Living World,* 221–230. San Diego: Academic Press.

———. 2013. The Etho-*Cebus* Project: Stone-tool use by wild capuchin monkeys. In C. M. Sanz, J. Call, C. Boesch, eds., *Tool Use in Animals. Cognition and Ecology,* 203–222. Cambridge: Cambridge University Press.

Visalberghi, E., Fragaszy, D. M., Ottoni, E., Izar, P., de Oliveira, M. G., and Andrade, F. R. D. 2007. Characteristics of hammer stones and anvils used by wild bearded capuchin monkeys *(Cebus libidinosus)* to crack open palm nuts. *American Journal of Physical Anthropology* 132:426–444.

Visalberghi, E., Fragaszy, D. M., and Savage-Rumbaugh, S. 1995. Performance in a tool-using task by common chimpanzees *(Pan troglodytes)*, bonobos *(Pan paniscus)*, an orangutan *(Pongo pygmaeus)*, and capuchin monkeys *(Cebus apella)*. *Journal of Comparative Psychology* 109:52–60.

Visalberghi, E., and Limongelli, L. 1994. Lack of comprehension of cause-effect relations in tool-using capuchin monkeys *(Cebus apella)*. *Journal of Comparative Psychology* 108:15–22.

———. 1996. Acting and understanding: tool use revisited through the minds of capuchin monkeys. In A. E. Russon, K. A. Bard, and S. T. Parker, eds., *Reaching into Thought: The Minds of the Great Apes*, 57–79. Cambridge, MA: Cambridge University Press.

Visalberghi, E., and Néel, C. 2003. Tufted capuchins *(Cebus apella)* use weight and sound to choose between full and empty nuts. *Ecological Psychology* 15:215–228.

Visalberghi, E., Sabratini, G., Spagnoletti, N., Andrade, F. R. D., Ottoni, E., Izar, P., and Fragaszy, D. M. 2008. Physical properties of palm fruits processed with tools by wild bearded capuchin monkeys *(Cebus libidinosus)*. *American Journal of Primatology* 70:884–891.

Visalberghi, E., and Tomasello, M. 1998. Primate causal understanding in the physical and psychological domains. *Behavioural Processes* 42:189–203.

Visalberghi, E., and Trinca, L. 1989. Tool use in capuchin monkeys: distinguishing between performing and understanding. *Primates* 30:511–521.

Vishnu-Mittre. 1984. Floristic change in the Himalaya (southern slopes) and Siwaliks from the mid-Tertiary to Recent times. In R. O. Whyte, ed., *The Evolution of the East Asian Environment*, Vol. 2, 483–503. Hong Kong: University of Hong Kong Centre of Asian Studies.

Vitousek, P. M., Ehrlich, P. R., Ehrlich, A. H., and Matson, P. A. 1986. Human appropriation of the products of photosynthesis. *BioScience* 36:368–373.

Vlcek, E. 1978. A new discovery of *Homo erectus* in Central Europe. *Journal of Human Evolution* 7:239–251.

———. 1980. Die mittelpleistozänen Hominidenreste von der Steinrinne bei Bilzingsleben. *Veröffentlichungen des Landesmuseums für Vorgeschichte in Halle* 32:91–130, 165–175.

———. 1983. Die Neufunde vom *Homo erectus* aus dem mittelpleistozänen Travertinkomplex bei Bilzingsleben aus den Jahren 1977 bis 1979. *Veröffentlichungen des Landesmuseums für Vorgeschichte in Halle* 36:189–199, 253.

Vogel, C. 1975. Remarks on the reconstruction of the dental arcade of *Ramapithecus*. In R. H. Tuttle, ed., *Paleoanthropology: Morphology and Paleoecology*, 87–98. The Hague: Mouton.

Vogel, E. R., Knott, C. D., Crowley, B. E., Blakely, M. D., Larsen, M. D., and Dominy, N. J. 2012. Bornean orangutans on the brink of protein deficiency. *Biology Letters* 8:333–336.

Vogel, E. R., van Woerden, J. T., Lucas, P. W., Atmoko, S. S. U., van Schaik, C. P., and Dominy, N. J. 2008. Functional ecology and evolution of hominoid molar enamel thickness: *Pan troglodytes schweinfurthii* and *Pongo pygmaeus wurmbii*. *Journal of Human Evolution* 55:60–74.

Vogel, G. 2003. Can great apes be saved from Ebola? *Science* 300:1645.

———. 2006. Tracking Ebola's deadly march among wild apes. *Science* 314:1522–1523.

———. 2007. Scientists say Ebola has pushed western gorillas to the brink. *Science* 317:1484.

Vogel, J. C. 1980. Dating possibilities for the South African hominid sites. *Palaeontologia Africana* 23:41–44.

———. 1985. Further attempts at dating the Taung tufas. In P. V. Tobias, ed., *Hominid Evolution: Past, Present and Future,* 189–194. New York: Alan R. Liss.

von Glasersfeld, E. 1976. The development of language as purposive behavior. *Annals of the New York Academy of Sciences* 280:212–226.

von Glasersfeld, E., Warner, H., and Pisani, P., Rumbaugh, D. M., Gill, T. V., and Bell, C. 1973. A computer mediates communication with a chimpanzee. *Computers and Automation* 22:9–11.

von Wright, J. M. 1970. Cross-modal transfer and sensory equivalence—a review. *Scandinavian Journal of Psychology* 11:21–30.

Voracek, M., Haubner, T., and Fisher, M. L. 2008. Recent decline in nonpaternity rates: a cross-temporal meta-analysis. *Psychological Reports* 103:799–811.

Voysey, B. C., McDonald, K. E., Rogers, M. E., Tutin, C. E. G., and Parnell, R. J. 1999a. Gorillas and seed dispersal in the Lopé Reserve, Gabon I: Gorilla acquisition by trees. *Journal of Tropical Ecology* 15:23–38.

———. 1999b. Gorillas and seed dispersal in the Lopé Reserve, Gabon, II: Survival and growth of seedlings. *Journal of Tropical Ecology* 15:39–60.

Vrba, E. S. 1975. Some evidence of chronology and palaeoecology of Sterkfontein, Swartkrans and Kromdraai from the fossil Bovidae. *Nature* 254:301–304.

———. 1985. Early hominids in southern Africa: updated observation on chronological and ecological background. In P. V. Tobias, ed., *Hominid Evolution: Past, Present and Future,* 195–200. New York: Alan R. Liss.

Vrba, E. S., Denton, G. H., Partridge, T. C., and Burckle, L. H. 1995. *Paleoclimate and Evolution with Emphasis on Human Origins.* New Haven, CT: Yale University Press.

Waga, I. C., Dacier, A. K., Pinha, P. S., and Tavares, M. C. H. 2006. Spontaneous tool use by wild capuchin monkeys *(Cebus libinosus)* in the Cerrado. *Folia Primatologia* 77:337–344.

Wagner, G. A. Krbetschek, M., Degering, D. Bahain, J.-J., Shao, Q., Falgères, C., Voinchet, P., Dolo, J.-M., Garcia, T., and Rightmire, G. P. 2010. Radiometric dating of the type-site for *Homo heidelbergensis* at Mauer, Germany. *Proceedings of the National Academy of Sciences of the United States of America* 107:19726–19730.

Wagner, R. 1981. *The Invention of Culture.* Englewood Cliffs, NJ: Prentice-Hall.

Wainscoat, J. S., Hill, A. V. S., Boyce, A. L., Flint, J., Hernandez, M., Thein, S. L., Old, J. M., Lynch, J. R., Falusi, A. G., Weatherall, D. J., and Clegg, J. B. 1986. Evolutionary relationships of human populations from an analysis of nuclear DNA polymorphisms. *Nature* 319:491–493.

Wakefield, M. L. 2008. Grouping patterns and competition among female *Pan troglodytes schweinfurthii* at Ngogo, Kibale National Park, Uganda. *International Journal of Primatology* 29:907–929.

Walker, A. 1981a. The Koobi Fora hominids and their bearing on the origins of the genus *Homo*. In B. A. Sigmon and J. S. Cybulski, eds., *Homo erectus: Papers in Honor of Davidson Black,* 193–215. Toronto: University of Toronto Press.

———. 1981b. Dietary hypotheses and human evolution. *Philosophical Transactions of the Royal Society, London* B-292:57–64.

———. 1992. Louis Leakey, John Napier and the history of *Proconsul. Journal of Human Evolution* 22:245–254.

———. 2007. Early hominin diets. Overview and historical perspectives. In P. S. Ungar, ed., *Evolution of the Human Diet,* 3–10. Oxford: Oxford University Press.

Walker, A., and Andrews, P. 1973. Reconstruction of the dental arcades of *Ramapithecus wickeri. Nature* 244:313–314.

Walker, A., Falk, D., Smith, R., and Pickford, M. 1983. The skull of *Proconsul africanus: reconstruction and cranial capacity. Nature* 305:525–527.

Walker, A., and Leakey, R. E. 1978. The hominids of East Turkana. *Scientific American* 239:54–66.

———. 1984. New fossil primates from the lower Miocene site of Buluk, N. Kenya. *American Journal of Physical Anthropology* 63:232.

———. 1988. The evolution of *Australopithecus boisei.* In F. E. Grine, ed., *Evolutionary History of the "Robust" Australopithecines,* 247–258. Hawthorne, NY: Aldine de Gruyter.

———. 1993. *The Nariokotome Homo erectus.* Cambridge, MA: Harvard University Press.

Walker, A., Leakey, R. E., Harris, J. M., and Brown, F. H. 1986. 2.5-myr *Australopithecus boisei* from west of Lake Turkana, Kenya. *Nature* 322:517–522.

Walker, A., and Pickford, M. 1983. New postcranial fossils of *Proconsul africanus* and *Proconsul nyanza.* In R. L. Ciochon and R. S. Corruccini, ed., *New Interpretations of Ape and Human Ancestry,* 325–351. New York: Plenum Press.

Walker, A., and Rose, M. D. 1968. Fossil hominoid vertebra from the Miocene of Uganda. *Nature* 217:980–981.

Walker, A., and Shipman, P. 2005. *The Ape in the Tree.: An Intellectual and Natural History of Proconsul.* Cambridge, MA: Harvard University Press.

Walker, A., and Teaford, M. 1988. The Kaswanga primate site: an Early Miocene hominoid site on Rusinga Island, Kenya. *Journal of Human Evolution* 17:539–544.

———. 1989. The hunt for *Proconsul. Scientific American* 260:76–82.

Walker, A., Teaford, M. F., and Leakey, R. E. 1986. New information concerning the R114 *Proconsul* site, Rusinga Island, Kenya. In J. G. Else and P. C. Lee, eds., *Primate Evolution, Selected Proceedings of the Tenth Congress of the International Primatological Society, held in Nairobi, Kenya, in July 1984,* Vol. 1, 143–149. Cambridge: Cambridge University Press.

Walker, A., Teaford, M. F., Martin, L., and Andrews, P. 1993. A new species of *Proconsul* from the early Miocene of Rusinga/Mfangano Islands, Kenya. *Journal of Human Evolution* 25:43–53.

Walker, J., Cliff, R. A., and Latham, A. G. 2006. U-Pb isotopic age of the StW 573 hominid from Sterkfontein, South Africa. *Science* 314:1592–1594.

Walker, M. J., Ortega, J., López, M. V., Parmová, K., and Trinkaus, E. 2011. Neanderthal postcranial remains from the Sima de las Palomas del Cabezo Gordo, Murcia, southeastern Spain. *American Journal of Physical Anthropology* 144:505–515.

Walker, R., Hill, K., Burger, O., and Hurtado, M. 2006. Life in the slow lane revisited: ontogenetic separation between chimpanzees and humans. *American Journal of Physical Anthropology* 129:577–583.

Wall, C. E., Larson, S. G., and Stern, J. T., Jr. 1994. EMG of the digastric muscle in gibbon and orangutan: functional consequences of the loss of the anterior digastric in orangutans. *American Journal of Physical Anthropology* 94:549–567.

Wallace, A. R. 1881. Review of *Anthropology: An Introduction to the Study of Man and Civilization* by Edward B. Tylor. *Nature* 24:242–245.

———. 1895. The expressiveness of speech, or, mouth-gesture as a factor in the origin of language. *Fortnightly Review*, n.s., 64:528–543.

Wallace, John A. 1975. Did La Ferrassie I use his teeth as a tool? *Current Anthropology* 16:393–401.

Waller, B. M., Lembeck, M., Kuchenbuch, P., Burrows, A., and Liebal, K. 2012. Gibbon-FACS: a muscle-based facial movement coding sytem for hylobatids. *International Journal of Primatology* 33:809–821.

Wallis, J. 1997. *Primate Conservation: The Role of Zoological Parks*. San Antonio, TX: American Society of Primatologists.

Wallis, J., and Lee, D. R. 1999. Primate conservation: the prevention of disease transmission. *International Journal of Primatology* 20:803–826.

Walsh, P. D., Abernethy, K. A., Bermejo, M., Beyers, R., De Wachter P., Akou, M. E., Huijbregts, B., Mambounga, D. I., Toham, A. K., Kilbourn, A. M., et al. 2003. Catastrophic ape decline in western equatorial Africa. *Nature* 42:611–614.

Walter, R. C., and Aronson. J. L. 1982. Revisions of K/Ar ages for the Hadar hominid site, Ethiopia. *Nature* 296:122–127.

Wang, W., Tian, F., and Mo, J.-y. 2007. Recovery of *Gigantopithecus blacki* fossils from Mohui Cave in the Bubing Basin, Guangxi, South China. *Acta Anthropological Sinica* 26:329–343.

Wang, X. 1984. The palaeoenvironment of China from the Tertiary. In R. O. Whyte, ed., *The Evolution of the East Asian Environment*, Vol. 2: *Palaeobotany, Palaeozoology and Palaeoanthropology*, 472–482. Hong Kong: University of Hong Kong Centre of Asian Studies.

Ward, C. V. 1993. Torso morphology and locomotion in *Proconsul nyanzae*. *American Journal of Physical Anthropology* 92:291–328.

Ward, C. V., Kimbel, W. H., and Johanson, D. C. 2011. Complete fourth metatarsal and arches in the foot of *Australopithecus afarensis*. *Science* 331:750–751.

Ward, C. V., Leakey, M. G., and Walker, A. 2001. Morphology of *Australopithecus anamensis* from Kanapoi and Allia Bay, Kenya. *Journal of Human Evolution* 41:255–368.

Ward, C. V., Ruff, C. B., Walker, A, Teaford, M. F., Rose, M. L., and Nengo, I. O. 1995. Functional morphology of *Proconsul* patellas from Rusinga Island, Kenya, with implications for other Miocene-Pliocene catarrhines. *Journal of Human Evolution* 29:1–19.

Ward, C. V., Walker, A, Teaford, M. F., and Odhiambo, I. 1993a. Partial skeleton of *Proconsul nyanzae* from Mfangano Island, Kenya. *American Journal of Physical Anthropology* 90:77–111.

———. 1993b. *Proconsul* did not have a tail. *Journal of Human Evolution* 21:215–220.

Ward, J. P., and Hopkins, W. D. 1993. *Primate Laterality: Current Behavioral Evidence of Primate Asymmetries.* New York: Springer.

Ward, S. C., and Brown, B. 1986. The facial skeleton of *Sivapithecus indicus.* In D. R. Swindler and J. Erwin, eds., *Comparative Primate Biology,* Vol. 1: *Systematics, Evolution, and Anatomy,* 413–452. New York: Alan R. Liss.

Ward, S. C., Brown, B., Hill, A., Kelley, J., and Downs, W. 1999. *Equatorius:* a new hominoid genus from the Middle Miocene of Kenya. *Science* 285:1382–1386.

Ward, S. C., and Hill, A. 1987. Pliocene hominid partial mandible from Tabarin, Baringo, Kenya. *American Journal of Physical Anthropology* 72:21–37.

Ward, S. C., and Kimbel, W. H. 1983. Subnasal alveolar morphology and the systematic position of *Sivapithecus. American Journal of Physical Anthropology* 61:957–161.

Ward, S. C., and Pilbeam, D. R. 1983. Maxillofacial morphology of Miocene hominoids from Africa and Indo-Pakistan. In R. L. Ciochon and R. S. Corruccini, eds., *New Interpretations of Ape and Human Ancestry,* 211–238. New York: Plenum Press.

Waser, P. M. 1982. Primate polyspecific associations: do they occur by chance? *Animal Behaviour* 30:1–8.

Washburn, D. A. 1991. Ordinal judgements of numerical symbols by macaques *(Macaca mulatta). Psychological Science* 2:190–193.

———. 2007a. *Primate Perspectives on Behavior and Cognition.* Washington, DC: American Psychological Association.

———. 2007b. The perception of emergents. In D. A. Washburn, ed., *Primate Perspectives on Behavior and Cognition,* 109–123. Washington, DC: American Psychological Association.

Washburn, S. L. 1950. The analysis of primate evolution with particular reference to the origin of man. *Cold Spring Harbor Symposia on Quantitative Biology* 15:67–78.

———. 1951. The new physical anthropology. *Transactions of the New York Academy of Sciences* 13:298–304.

———. 1957. *Australopithecus:* the hunters or the hunted? *American Anthropologist* 59:612–614.

———. 1959. Speculations on the interrelations of the history of tools and biological evolution. *Human Biology* 31:21–31.

———. 1960. Tools and human evolution. *Scientific American* 203:63–75.

———. 1963a. The study of race. *American Anthropologist* P65:521–531.

———. 1963b. Behavior and human evolution. In S. L. Washburn, ed., *Classification and Human Evolution,* 190–203. New York: Wenner-Gren Foundation for Anthropological Research.

———. 1967. Behaviour and the origin of man. *Proceedings of the Royal Anthropological Institute of Great Britain and Ireland 1967,* 21–27.

————. 1968a. The study of human evolution. *Condon Lectures.* Eugene, Oregon: University of Oregon Books.

————. 1968b. Speculations on the problem of man's coming to the ground. In B. Rothblatt, ed., *Changing Perspectives on Man,* 193–206. Chicago: University of Chicago Press.

————. 1972. Human evolution. T. Dobzhansky, M. K. Hecht, and W. C. Steere, eds., *Evolutionary Biology,* Vol. 6, 349–361. New York: Appleton-Century-Crofts.

————. 1973a. Primate studies and human evolution. In G. H. Bourne, ed., *Nonhuman Primates and Medical Research,* 467–485. New York: Academic Press.

————. 1973b. Human evolution: science or game? *Yearbook of Physical Anthropology* 17:67–70.

Washburn, S. L., and Avis, V. 1958. Evolution of human behavior. In G. G. Simpson and A. Roe, eds., *Behavior and Evolution,* 421–436. New Haven, CT: Yale University Press.

Washburn, S. L., and Lancaster, C. S. 1968. The evolution of hunting. In. R. B. Lee and I. DeVore, eds., *Man the Hunter,* 293–303. Chicago: Aldine.

Washburn, S. L., and Patterson, B. 1951. Evolutionary importance of the South African "man-apes." *Nature* 167:651–652.

Watanabe, H., Fujiyama, A., Hattori, M., Taylor, T. D., Toyoda, A., Kuroki, Y., Noguchi, H., BenKahla, A., Lehrach, H., Sudbrak, R., et al. 2004. DNA sequence and comparative analysis of chimpanzee chromosome 22. *Nature* 429:382–388.

Waterman, P. G., Choo, G. M., Vedder, A. L., and Watts, D. P. 1983. Digestibility, digestion-inhibitors, and green stems from an African montane flora and comparison with other tropical flora. *Oecologia (Berlin)* 60:244–249.

Waterman, P. G., and Kool, K. M. 1994. Colobine food selection and plant chemistry. In A. G. Davies and J. F. Oates, eds., *Colobine Monkeys: Their Ecology, Behaviour and Evolution,* 251–284, Cambridge: Cambridge University Press.

Waterman, P. G., Mbi, C. N., McKey, D. B., and Gartlan, J. S. 1980. African rainforest vegetation and rumen microbes: phenolic compounds and nutrients as correlates of digestibility. *Oecologia (Berlin)* 47:22–33.

Watson, L. 1999–2000. Leopard's pursuit of a lone lowland gorilla *Gorilla gorilla gorilla* within the Danga-Sangha Reserve, Central African Republic. *African Primates* 41:74–75.

Watson, P. J. 1995. Archaeology, anthropology, and the culture concept. *American Anthropologist* 97:683–694.

Watts, D. P. 1984. Composition and variability of mountain gorilla diet in the Central Virungas. *American Journal of Primatology* 7:323–356.

————. 1985a. Observations on the ontogeny of feeding behavior in mountain gorillas *(Gorilla gorilla beringei). American Journal of Primatology* 8:1–10.

————. 1985b. Relations between group size and composition and feeding competition in mountain gorilla groups. *Animal Behaviour* 33:72–85.

————. 1987. Effects of mountain gorilla foraging activities on the productivity of their food plant species. *African Journal of Ecology* 25:155–163.

————. 1988. Environmental influences on mountain gorilla time budgets. *American Journal of Primatology* 15:1995–211.

————. 1989a. Ant eating behavior of mountain gorillas. *Primates* 30:121–125.

————. 1989b. Infanticide in mountain gorillas: new cases and a reconsideration of the evidence. *Ethology* 81:1–18.

————. 1990a. Ecology of gorillas and its relation to female transfer in mountain gorillas. *International Journal of Primatology* 11:2145.

————. 1990b. Mountain gorilla life histories, reproductive competition and sociosexual behavior and some implications for captive husbandry. *Zoo Biology* 9:185–200.

————. 1991. Mountain gorilla reproduction and sexual behavior. *American Journal of Primatology* 24:211–25.

————. 1992. Social relationships of immigrant and resident female mountain gorillas, I: Male-female relationships. *American Journal of Primatology* 28:159–181.

————. 1994a. The influence of male mating tactics on habitat use in mountain gorillas *(Gorilla gorilla beringei). Primates* 35:35–47.

————. 1994b. Agonistic relationships between female mountain gorillas *(Gorilla gorilla beringei). Behavioral Ecology and Sociobiology* 34:347–358.

————. 1994c. Social relationships of immigrant and resident female mountain gorillas, I: Relatedness, residence, and relationships between females. *American Journal of Primatology* 32:13–10.

————. 1995a. Post-conflict social events in wild mountain gorillas (Mammalia, Hominoidea), I: social interactions between opponents. *Ethology* 100:139–157.

————. 1995b. Post-conflict social events in wild mountain gorillas, II: Redirection, side direction and consolidation. *Ethology* 100:158–174.

————. 1996. Comparative socio-ecology of gorillas. In W. C. cGrew, L. F. Marchant, and T. Nishida, eds., *Great Ape Societies,* 16–28. Cambridge: Cambridge University Press.

————. 1997. Agonistic interventions in wild mountain gorilla groups. *Behaviour* 134:23–57.

————. 1998a. Long-term habitat use by mountain gorillas *(Gorilla gorilla beringei):* 1. Consistency, variation, and home range size and stability. *International Journal of Primatology* 19:651–680.

————. 1998b. Long-term habitat use by mountain gorillas *(Gorilla gorilla beringei):* 2. Reuse of foraging areas in relation to resource abundance, quality, and depletion. *International Journal of Primatology* 19:681–702.

————. 1998c. Seasonality in the ecology and life histories of mountain gorillas *(Gorilla gorilla beringei). International Journal of Primatology* 19:929–948.

————. 1998d. Coalitionary mate guarding by male chimpanzees at Ngogo, Kibale National Park, Uganda. *Behavioral Ecology and Sociobiology* 44:43–55.

————. 2000a. Causes and consequences of variation in male mountain gorilla life histories and group membership. In P. Kappeler, ed., *Primate Males: Causes and Consequences of Variation in Group Composition,* 169–179. Cambridge: Cambridge University Press.

————. 2000b. Mountain gorilla habitat use strategies and group movement. In S. Boinski and P. A. Garber, eds., *On the Move: How and Why Animals Travel in Groups,* 351–374. Chicago: University of Chicago Press.

———. 2001. Social relationships of female mountain gorillas. In M. M. Robbins, P. Sicotte, and K. J. Stewart, eds., *Mountain Gorillas: Three Decades of Research at Karisoke,* 215–240. Cambridge: Cambridge University Press.

———. 2002. Reciprocity and interchange in the social relationships of wild male chimpanzees. *Behaviour* 139:343–370.

———. 2003. Gorilla social relationships: a comparative overview. In A. B. Taylor and M. L. Goldsmith, eds., *Gorilla Biology: Multidisciplinary Perspective,* 302–327. Cambridge: Cambridge University Press.

———. 2006. Conflict resolution in chimpanzees and the valuable-relationships hypothesis. *International Journal of Primatology* 27:1337–1364.

———. 2008. Scavenging by chimpanzees at Ngogo and the relevance of chimpanzee scavenging to early hominin behavioral ecology. *Journal of Human Evolution* 54:125–133.

———. 2012. Long-term Research on Chimpanzee Behavioral Ecology in Kibale National Park, Uganda. In Kappeler, P. M., and Watts, D. P., eds., *Long-term Field Studies of Primates,* 313–338. Heidelberg: Springer.

Watts, D. P., Colmenares, F., and Arnold, K. 2000. Redirection, consolation, and male policing. In F. Aurelli and F. B. M. de Waal, eds., *Natural Conflict Resolution,* 281–301. Berkeley: University of California Press.

Watts, D. P., and Mitani, J. C. 2000. Infanticide and cannibalism by male chimpanzees at Ngogo, Kibale National Park, Uganda. *Primates* 41:35–365.

———. 2001. Boundary patrols and intergroup encounters in wild chimpanzees. *Behaviour* 138:299–327.

———. 2002. Hunting behavior of chimpanzees at Ngogo, Kibale National Park, Uganda. *International Journal of Primatology* 23:1–28.

Watts, D. P., Mitani, J. C., and Sherrow, H. M. 2002. New cases of inter-community infanticide by male chimpanzees at Ngogo, Kibale National Park, Uganda. *Primates* 43:263–270.

Watts, D. P., Muller, M., Amsler, S. J., Mbabazi, G., and Mitani, J. 2006. Lethal intergroup aggression by chimpanzees in Kibale National Park, Uganda. *American Journal of Primatology* 68:161–180.

Watts, D. P., and Pusey A. E. 1993. Behavior of juvenile and adolescent great apes. In M. E. Pereira and L. A. Fairbanks, eds., *Juvenile Primates: Life History, Development and Behavior,* 148–167. New York: Oxford University Press.

Weaver, K. F. 1985. The search for our ancestors. *National Geographic* 168:560–623.

Weaver, T. D., and Hublin, J.-J. 2009. Neandertal birth canal shape and the evolution of human childbirth. *Proceedings of the National Academy of Sciences of the United States of America* 106:8151–8156.

Weaver, T. D., and Roseman, CC. 2008. New developments in the genetic evidence for modern human origins. *Evolutionary Anthropology* 17:69–80.

Webb, D., and Fabiny, S. 2009. Cheiridial genetics and tool use among later homininids. *Anthropologie* 47:47–56.

Webb, D., Tuttle, R. H., and Baksh, M. 1994. Pendular activity of human upper limbs during slow and normal walking. *American Journal of Physical Anthropology* 93:477–489.

Weber, A. W., and Vedder, A. 1983. Population dynamics of the Virunga gorillas: 1959–1978. *Biological Conservation* 26:341–366.

Weber, B., Schempp, W., and Wiesner, H. 1986. An evolutionarily conserved early replicating segment on the sex chromosomes of man and the great apes. *Cytogenetics and Cell Genetics* 43:72–78.

Weber, J., Czarnetzki, A., and Pusch, C. M. 2005. Comment on "The brain of LB1, *Homo floresiensis*." *Science* 310:236b.

Weber, W. 1993. Primate conservation and ecotourism in Africa. C. S. Potter, J. I. Cohen, and D. Janczewski, eds., *Perspectives on Biodiversity*, 129–159. Washington, DC: American Association for Advancement of Science Press.

Weidenreich, F. 1935. The *Sinanthropus* population of Choukoutien (locality 1) with a preliminary report on new discoveries. *Bulletin of the Geological Society of China* 14:427–468.

———. 1936. The mandibles of *Sinanthropus pekinensis:* a comparative study. *Palaeontologia Sinica*, n.s., series D, 7:1–162.

———. 1937a. The new discoveries of *Sinanthropus pekinensis* and their bearing on the *Sinanthropus* and *Pithecanthropus* problems. *Bulletin of the Geological Society of China* 16:439–471.

———. 1937b. The dentition of *Sinanthropus pekinensis:* a comparative odontography of the hominids. *Palaeontologia Sinica*, n.s., series D, 1:1–180.

———. 1943. The skull of *Sinanthropus pekinensis,* a comparative study on a primitive hominid skull. *Palaeontologia Sinica*, n.s., series D, 10:1–485.

———. 1945. Giant early man from Java and South China. *Anthropological Papers of the American Museum of Natural History* 40:1–134.

———. 1946. *Apes, Giants and Man.* Chicago: University of Chicago Press.

Weiner, J. S., Oakley, K. P., and Clark, W. E. Le Gros. 1953. The solution of the Piltdown problem. *Bulletin of the British Museum of Natural History (Geology)* 2:141–146.

Weiner, S., Bar-Yosef, O. Xu, Q., Goldberg, P., Xu, Q.-q., and Liu, J.-Y., 2000. Evidence for the use of fire at Zhoukoudian. *Acta Anthropologica Sinica* 19 (Suppl.): 218–223.

Weiner, S., Xu, Q., Goldberg, P., Liu, J., Bar-Yosef, O. 1998. Evidence for the use of fire at Zhoukoudian, China. *Science* 281:251–252.

Weiss, A., King, J. E., and Murray, L. 2011. *Personality and Temperament in Nonhuman Primates.* New York: Springer.

Weiss, M. L. 1987. Nucleic acid evidence bearing on hominoid relationships. *Yearbook of Physical Anthropology* 30:41–73.

Welcomme, J.-L., Aguilar, J.-P., and Ginsburg, L. 1991. Découverte d'un nouveau Pliopithèthique (Primates, Mammalia) associé à des rongeurs dans les sables du Miocène supérieur de Priay (Ain, France) et remarques sur la paléogéographie de la Brese au Vallésien. *Comptes rendus des séances de l'Académie des sciences, Paris, série II* 313:723–729.

Wells, J. C. K., and Stock, J. T. 2007. The biology of the colonizing ape. *Yearbook of Physical Anthropology* 50:191–222.

Werdelin, L., and Peigné, S. 2010. Carnivora. In L. Werdelin and W. J. Sanders, eds., *Cenozoic Mammals of Africa*, 603–657. Berkeley: University of California Press.

Werdelin, L., and Sanders, W. J. 2010. *Cenozoic Mammals of Africa*. Berkeley: University of California Press.

Werner, D. 2006. The evolution of male homosexuality and its implications for human psychological and cultural variations. In V. Sommer and P. L. Vasey, eds. *Homosexual Behaviour in Animals: An Evolutionary Perspective*, 316–346. Cambridge: Cambridge University Press.

Wescott, R. W. 1967. Hominid uprightness and primate display. *American Anthropologist* 69:738.

West, K. 1981. *The Behavior and Ecology of the Siamang in Sumatra*. MA thesis, University of California, Davis.

West, R. M. 1984. Siwalik faunas from Nepal: paleoecologic and paleoclimatic implications. In R. O. Whyte, ed., *The Evolution of the East Asian Environment*, Vol. 2, 724–744. Hong Kong: University of Hong Kong Centre of Asian Studies.

Westaway, K. E., Sutikna, T., Saptomo, W. E., Jatmiko, Morwood, M. J., Roberts, R. G., and Hobbs, D. R. 2009. Reconstructing the geomorphic history of Liang Bua, Flores, Indonesia: a stratigraphic interpretation of the occupational environment. *Journal of Human Evolution* 57:465–483.

West-Eberhard, M. J. 2005. Developmental plasticity and the origin of species differences. *Proceedings of the National Academy of Sciences of the United States of America* 102:6543–6549.

Westenhöfer, M. 1942. *Der Eigenweg des Menschen*. Mannstaedt: Verlag Die Medizinische Welt.

Westergaard, G. C. 1994. The subsistence technology of capuchins. *International Journal of Primatology* 15:899–906.

———. 1995. The stone-tool technology of capuchin monkeys: possible implications for the evolution of symbolic communication in hominids. *World Archaeology* 27:1–9.

Westergaard, G. C., and Fragaszy, D. M. 1985. Effects of manipulatable objects on the activity of captive capuchin monkeys *(Cebus apella). Zoo Biology* 4:317–327.

———. 1987. The manufacture and use of tools by capuchin monkeys. *Journal of Comparative Psychology* 101:159–168.

Westergaard, G. C., Greene, J. A., Babitz, M. A., and Soumi, S. J. 1995. Pestle use and modification by tufted capuchins *(Cebus apella). International Journal of Primatology* 16:643–651.

Westergaard, G. C., Greene, J. A., Munuhin-Hauser, C., and Soumi, S. J. 1996. The use of naturally-occurring copper and iron tools by monkeys: possible implications for the emergence of metal-tool technology in hominids. *Human Evolution* 11:17–25.

Westergaard, G. C., and Hyatt, C. W. 1994. The responses of bonobos *(Pan paniscus)* to their mirror images: evidence of self-recognition. *Human Evolution* 9:273–279.

Westergaard, G. C., Kuhn, H. E., and Soumi, S. J. 1998. Bipedal posture and hand preference in humans and other primates. *Journal of Comparative Psychology* 112:55–64.

Westergaard, G. C., Lundquist, A. L., Haynie, M. K., Kuhn, H. E., and Soumi, S. J. 1998. Why some capuchin monkeys *(Cebus apella)* use probing tools (and others do not). *Journal of Comparative Psychology* 112:207–211.

Westergaard, G. C., Lundquist, A. L., Kuhn, H. E., and Soumi, S. J. 1997. Ant-gathering by captive tufted capuchins *(Cebus apella). International Journal of Primatology* 18:95–103.

Westergaard, G. C., and Soumi, S. J. 1993. Use of a tool-set by capuchin monkeys *(Cebus apella). Primates* 34:459–462.

———. 1994a. The use and modification of bone tools by capuchin monkeys. *Current Anthropology* 35:75–77.

———. 1994b. Stone-tool bone-surface modification by monkeys. *Current Anthropology* 35:468–470.

———. 1994c. A simple stone-tool technology in monkeys. *Journal of Human Evolution* 27:399–404.

———. 1994d. Hierarchical complexity of combinatorial manipulation in capuchin monkeys *(Cebus apella). American Journal of Primatology* 32:171–176.

———. 1994e. Aimed throwing by tufted capuchin monkeys *(Cebus apella). Human Evolution* 9:323–329.

———. 1995a. The manufacture and use of bamboo tools by monkeys: possible implications for the development of material culture among East Asian hominids. *Journal of Archaeological Science* 22:677–681.

———. 1995b. Mirror inspection varies with age and tool-using ability in tufted capuchin monkeys *(Cebus apella). Human Evolution* 10:217–223.

———. 1995c. The production and use of digging tools by monkeys: a nonhuman primate model of a hominid subsistence activity. *Journal of Anthropological Research* 51:1–8.

———. 1995d. The stone tools of capuchins *(Cebus apella). International Journal of Primatology* 16:1017–1024.

———. 1995e. Stone-throwing by capuchins *(Cebus apella):* a model of throwing capabilities in *Homo habilis. Folia Primatologica* 65:234–238.

Weston, E. M., and Lister, A. M. 2009. Insular dwarfism in hippos and a model for brain size reduction in *Homo floresiensis. Nature* 459:85–88.

Weyland, P. G., Sternlight, D. B., Bellizzi, M. J., and Wright, S. 2000. Faster top running speeds are achieved with greater ground forces not more rapid leg movements. *Journal of Applied Physiology* 89:1991–1999.

Whaling, C. 2000. What's behind a song? The neural basis of song learning in birds. In N. L. Wallin, B. Merker and S. Brown, eds., *The Origins of Music,* 65–76. Cambridge, MA: MIT Press.

Wheatley, B. P. 1982. Energetics of foraging in *Macaca fascicuaris* and *Pongo pygmaeus* and a selective advantage of large body size in the orang-utan. *Primates* 23:348–363.

Wheeler, P. E. 1984. The evolution of bipedality and loss of functional body hair in humans. *Journal of Human Evolution* 13:91–98.

———. 1991a. The thermoregulatory advantages of hominid bipedalism in open equatorial environments: the contribution of increased convective heat loss and cutaneous evaporative cooling. *Journal of Human Evolution* 21:107–115.

———. 1991b. The influence of bipedalism on the energy and water budgets of early hominids. *Journal of Human Evolution* 21:117–136.

————. 1992. The thermoregulatory advantages of large body size for hominids foraging in savannah environments. *Journal of Human Evolution* 23:351–362.

————. 1993. The influence of stature and body on the hominid energy and water budgets, a comparison of *Australopithecus* and *Homo* physiques. *Journal of Human Evolution* 24:13–28.

————. 1994a. The thermoregulatory advantages of heat storage and shade-seeking behaviour to hominids foraging in equatorial savannah environments. *Journal of Human Evolution* 26:339–350.

————. 1994b. The foraging times of bipedal and quadrupedal hominids in open equatorial environments (a reply to Chaplin, Jablonski & Cable 1994). *Journal of Human Evolution* 27:511–517.

Whitam, F. L. 1983. Culturally invariable properties of male homosexuality: tentative conclusions from cross-cultural research. *Archives of Sexual Behavior* 12:207–226.

Whitcome, K. K., Shapiro, L. J., and Lieberman, D. E. 2007. Fetal load and the evolution of lumbar lordosis in bipedal evolution. *Nature* 405:1075–1080.

White, F. J. 1988. Party composition an dynamics in *Pan paniscus*. *International Journal of Primatology* 9:179–193.

————. 1989a. Ecological correlates of pygmy chimpanzee social structure. In V. Standen and R. A. Foley, eds., *Comparative Socioecology*, 151–164. Oxford: Blackwell Scientific.

————. 1989b. Social organization of pygmy chimpanzees. In P. G. Heltne and L. A. Marquardt, eds., *Understanding Chimpanzees*, 194–207. Cambridge, MA: Harvard University Press.

————. 1992a. Eros of the apes. *BBC Wildlife Magazine* 10:38–47.

————. 1992b. Activity budgets, feeding behavior, and habitat use of pygmy chimpanzees at Lomako, Zaire. *American Journal of Primatology* 26:215–223.

————. 1992c. Pygmy chimpanzee social organization: variation with party size and between study sites. *American Journal of Primatology* 26:203–214.

————. 1994. Food sharing in wild pygmy chimpanzees, *Pan paniscus*. In J. J. Roeder, B. Thierry, J. R. Anderson, and N. Herrenschmidt, eds., *Current Primatology*, Vol. 2: *Social Development, Learning and Behavior*, 1–10. Strasbourg: Université Louis Pasteur.

————. 1996a. *Pan paniscus* 1973–1996: twenty-three years of field research. *Evolutionary Anthropology* 5:11–17.

————. 1996b. Comparative socio-ecology of *Pan paniscus*. In W. C. McGrew, L. F. Marchant, and T. Nishida, eds., *Great Ape Societies*, 29–41. Cambridge: Cambridge University Press.

————. 1998. Seasonality and socioecology: the importance of variation in fruit abundance to bonobo sociality. *International Journal of Primatology* 19:1013–1027.

White, F. J., and Burgman, M. A. 1990. Social organization of pygmy chimpanzees *(Pan paniscus)*: multivariate analysis of intracommunity associations. *American Journal of Primatology* 83:193–201.

————. 1992. Feeding competition in Lomako bonobos: variation in social cohesion. In T. Nishida, W. C. McGrew, P. Marler, M. Pickford, and F. B. M. de Waal, eds., *Topics in Primatology*, Vol. 1: *Human Origins*, 67–79. Tokyo: University of Tokyo Press.

White, F. J., and Chapman, C. A. 1994. Contrasting chimpanzees and bonobos: nearest neighbor distances and choices. *Folia Primatologica* 63:181–191.

White, F. J., and Lanjouw, M. A. 1992. Feeding competition in Lomako bonobos: variation in social cohesion. In T. Nishida, W. C. McGrew, P. Marler, M. Pickford, and F. B. M. de Waal, eds., *Topics in Primatology,* Vol. 1: *Human Origins,* 67–79. Tokyo: University of Tokyo Press.

White, F. J., and Wood, K. D. 2007. Female feeding priority in bonobos, *Pan paniscus,* and the question of female dominance. *American Journal of Primatology* 69:837–850.

White, F. J., and Wrangham, R. W. 1988. Feeding competition and patch size in the chimpanzee species *Pan paniscus* and *Pan troglodytes. Behaviour* 105:148–164.

White, L. 1959. *The Evolution of Culture.* New York: McGraw-Hill.

White, L. A., and B. Dillingham. 1973. *The Concept of Culture.* Minneapolis: Burgess.

White, L. J. T., and Tutin, C. E. G. 2001. Why chimpanzees and gorillas respond differently to logging. In W. Weber, L. J. T., White, A. Vedder, and L. Naughton-Treves, eds., *African Rain Forest Ecology and Conservation,* 449–462. New Haven, CT: Yale University Press.

White, R., Mensan, R., Bourrillon, R., Cretin, C., Higham, F. G., Clark, A. E., Sisk, M. L., Tartar, E., Gardère P., Goldberg, P., et al. 2012. Context and dating of Aurignacian vulvar representations from Abri Castanet, France. *Proceedings of the National Academy of Sciences of the United States of America* 109:8450–8455.

White, T. D. 1975. Geomorphology to paleoecology: *Gigantopithecus* reappraised. *Journal of Human Evolution* 4:219–233.

———. 1977. New fossil hominids from Laetoli, Tanzania. *American Journal of Physical Anthropology* 46:197–230.

———. 1980a. Additional fossil hominids from Laetoli, Tanzania: 1976–1979 specimens. *American Journal of Physical Anthropology* 53:487–504.

———. 1980b. Evolutionary implications of Pliocene hominid footprints. *Science* 208:175–176.

———. 1981. Primitive hominid canine from Tanzania. *Science* 213:348–349.

———. 1982. Les Australopithèques. *La Recherche* 13:1258–1270.

———. 1984. Pliocene hominids from the Middle Awash, Ethiopia. *Courier Forschungsinstitut Senckenberg* 69:57–68.

———. 1985. The hominids of Hadar and Laetoli: an element-by-element comparison of the dental samples. In E. Delson, ed., *Ancestors: The Hard Evidence,* 138–152. New York: Alan R. Liss.

———. 1986. *Australopithecus afarensis* and the Lothagam mandible. *Anthropos (Brno)* 23:79–90.

———. 1992. *Prehistoric Cannibalism at Mancos 5MTUMR-2346.* Princeton, NJ: Princeton University Press.

———. 2002. Earliest hominids. In W. C. Hartwig, ed., *The Primate Fossil Record,* 407–417. Cambridge: Cambridge University Press.

———. 2003. Early hominids—diversity or distortion? *Science* 299:1994–1997.

White, T. D., Asfaw, B., Beyene, Y., Haile-Selassie, Y., Lovejoy, C. O., Suwa, G., and WoldeGabriel, G. 2009. *Ardipithecus ramidus* and the paleobiology of early hominids. *Science* 326:75–86.

White, T. D., Asfaw, B., DeGusta, D., Gilbert, H., Richards, G. D., Suwa, G., and Howell, F. C. 2003. Pleistocene *Homo sapiens* from Middle Awash, Ethiopia. *Nature* 423:742–747.

White, T. D., and Johanson, D. C. 1989. The hominid composition of Afar Locality 333: some preliminary observations. In G. Giacobini, ed., *Hominidae: Proceedings of the 2nd International Congress of Human Paleontology,* 97–101. Milan: Jaca Book.

White, T. D., Johanson, D. C., and Kimbel, W. H. 1981. *Australopithecus afarensis:* its phyletic position reconsidered. *South African Journal of Science* 77:445–470.

White, T. D., Moore, R. V., and Suwa, G. 1984. Hadar biostratigraphy and hominid evolution. *Journal of Vertebrate Paleontology* 4:575–583.

White, T. D., and Suwa, G. 1987. Hominid footprints at Laetoli: facts and interpretations. *American Journal of Physical Anthropology* 72:485–514.

White, T. D., Suwa, G., and Asfaw, B. 1994. *Australopithecus ramidus,* a new species of early hominid from Aramis, Ethiopia. *Nature* 371:306–312.

———. 1995. Corrigendum. *Australopithecus ramidus,* a new species of early hominid from Aramis, Ethiopia. *Nature* 375:88.

White, T. D., Suwa, G., Simpson S., and Asfaw, B. 2000. Jaws and teeth of *Australopithecus afarensis* from Maka, Middle Awash, Ethiopia. *American Journal of Physical Anthropology* 111:45–68.

White, T. D., WoldeGabriel, G., Asfaw, B., Ambrose, S., Beyene, Y., Bernor, R. L., Boisserie, J.-R., Currie, B., Gilbert, H., Haile-Selassie, Y., et al. 2006. Asa Issie, Aramis and the origin of *Australopithecus. Nature* 40:883–889.

Whitehead, H. 2009. How might we study culture? A perspective from the ocean. In K. N Laland and B. G. Galef, eds., *The Question of Animal Culture,* 125–151. Cambridge, MA: Harvard University Press.

Whitehead, P. 1993. Aspects of the anthropoid wrist and hand. In D. L., Gebo, ed., *Postcranial Adaptation in Nonhuman Primates,* 96–120. DeKalb: Northern Illinois University Press.

Whiten, A. 1998. Evolutionary and developmental origins of the mindreading system. In J. Langer and M. Killen, eds., *Piaget, Evolution and Development,* 73–99. Hillsdale, NJ: Lawrence Erlbaum.

———. 2009. The identification and differentiation of culture in chimpanzees and other animals: from natural history to diffusion experiments. In K. N. Laland and B. G. Galef, eds., *The Question of Animal Culture,* 99–124. Cambridge, MA: Harvard University Press.

Whiten, A., and Byrne, R. W. 1988. Tactical deception in primates. *Behavioral and Brain Sciences* 11:233–273.

———. 1997. *Machiavellian Intelligence II: Extensions and Evaluations.* Cambridge: Cambridge University Press.

Whiten, A., Goodall, J., McGrew, W. C., Nishida, T., Reynolds, V., Sugiyama, Y., Tutin, C. E. G., Wrangham, R. W., and Boesch, C. 1999. Cultures in chimpanzees. *Nature* 399:682–685.

———. 2001. Charting cultural variation in chimpanzees. *Behaviour* 138:1481–1516.

———. 2005. Conformity to cultural norms of tool use in chimpanzees. *Nature* 437:737–740.

Whiten, A., Horner, V., and Marshall-Pescini, S. 2003. Cultural panthropology. *Evolutionary Anthropology* 12:92–105.

Whiten, A., Spiteri, A., Horner, V., Bonnie, K. E., Lambeth, S. P., Schapiro, S. J., and de Waal, F. B. M. 2007. Transmission of multiple traditions within and between chimpanzee groups. *Current Biology* 17:1038–1943.

Whiten, A., and van Schaik, C. P. 2007. The evolution of animal "cultures" and social intelligence. *Philosophical Transactions of the Royal Society, London B* 362:603–620.

Whitesides, G. H. 1985. Nut cracking by wild chimpanzees in Sierra Leone, West Africa. *Primates* 26:91–94.

Whitington, C. L. 1992. Interactions between lar gibbons and pig-tailed macaques at fruit sources. *American Journal of Primatology* 26:61–64.

Whitman, W. 1982. *Complete Poetry and Collected Prose.* New York: Library of America.

Whitmore, T. C. 1990. *An Introduction to Tropical Rain Forests.* Oxford: Clarendon Press.

Whitney, E. N., and Rolfes, S. R. 2002. *Understanding Nutrition.* 9th ed. Belmont, CA: Wadsworth.

Whitten, A. J. 1982. Home range use by Kloss gibbons *(Hylobates klossii)* on Siberut Island, Indonesia. *Animal Behaviour* 30:182–198.

Whitten, P. L. 1987. Infants and adult males. In B. B. Smuts, D. L. Cheney, R. M. Seyfarth, R. W. Wrangham, and T. T. Struhsaker, eds., *Primate Societies,* 343–357. Chicago: University of Chicago Press.

Whitten, T. 1982. *The Gibbons of Siberut,* London: J. M. Dent & Sons.

———. 1984. The trilling handicap in Kloss gibbons. In H. Preuschoft, D. J. Chivers, W. Y. Brockelman, and N. Creel, eds., *The Lesser Apes: Evolutionary and Behavioural Biology,* 416–419. Edinburgh: Edinburgh University Press.

Whybrow, P. J. 1984. Geological and faunal evidence from Arabia for mammal "migrations" between Asia and Africa during the Miocene. *Courier Forschungsinstitut Senckenberg* 69:89–198.

———. 1987. Miocene geology and palaeontology of Ad Dabtiyah, Saudi Arabia. *Bulletin of the British Museum of Natural History (Geology)* 41:367–457.

Whybrow, P. J., and Bassiouni, M. A. 1986. The Arabian Miocene: rocks, fossils, primates and problems. In J. G. Else and P. C. Lee, eds., *Primate Evolution,* 85–91. Cambridge: Cambridge University Press.

Whybrow, P. J., Hill, A., al-Tikriti, W. Y., and Hailwood, E. A. 1990. Late Miocene primate fauna, flora and initial palaeomagnetic data from the Emirate of Abu Dhabi, United Arab Emirates. *Journal of Human Evolution* 19:583–588.

Wich, S. A., de Vries, H., Ancrenaz, M., Perkins, L., Shumaker, R. W., Suzuki, A., and van Schaik, C. P. 2009. Orangutan life history variation. In In S. A. Wich, S. S. Utami Atmoko, T. Mitra Setia, and C. P. van Schaik, eds., *Orangutans: Geographic Variation in Behavioral Ecology and Conservation,* 65–75. Oxford: Oxford University Press.

Wich, S. A., Geurts, M. L., Mitra Setia, T., and Utami-Atmoko, S. S. 2006a. Influence of fruit availability on Sumatran orangutan sociality and reproduction. In G. Hohmann,

M. M. Robbins and C. Boesch, eds., *Feeding Ecology in Apes and Other Primates,* 337–358. Cambridge: Cambridge University Press.

Wich, S. A., and Nunn, C. L. 2002. Do male "long-distance calls" function in mate defense? A comparative study of long-distance calls in primates. *Behavioral Ecology and Sociobiology* 54:474–484.

Wich, S. A., Sterck, E. H. M., and Utami, S. S. 1999. Are orang-utan females as solitary as chimpanzee females? *Folia Primatologica* 70:23–28.

Wich, S. A., Swartz, K. B., Hardus, M. E., Lameira, A. R., Stromberg, E., and Shumaker, R. W. 2009. A case of spontaneous acquisition of a human sound by an orangutan. *Primates* 50:56–64.

Wich, S. A., Utami-Atmoko, S. S., Mitra Setia, T., Djoyosudharmo, S., and Geurts, M. L. 2006b. Dietary and energetic responses of *Pongo abelii* to fruit availability fluctuations. *International Journal of Primatology* 27:1535–1550.

Wich, S. A., Utami-Atmoko, S. S., Mitra Setia, T., Rijksen, H. D., Schürmann, C., van Hooff, J. A. R. A. M., and van Schaik, C. P. 2004. Life history of wild Sumatran orangutans *(Pongo abelii). Journal of Human Evolution* 47:385–398.

Wickler, W., and Seibt, U. 1983. Monogamy: an ambiguous concept. In P. Bateson, ed. *Mate Choice,* 33–50. Cambridge: Cambridge University Press.

Wildman, D. E., Grossman, L. I., and Goodman, M. 2002. Functional DNA in humans and chimpanzees shows they are more similar to each other than either is to other apes. In M. Goodman and A. S. Simon, eds., *Probing Human Origins,* 1–10. Cambridge, MA: American Academy of Arts and Sciences.

Wildman, D. E., Uddin, M., Liu, G., Grossman, L. I., and Goodman, M. 2003. Implications of natural selection in shaping 99.4% nonsynonymous DNA identity between humans and chimpanzees: enlarging genus *Homo. Proceedings of the National Academy of Sciences of the United States of America* 100:7181–7188.

Wiley, E. O. 1981. *Phylogenetics: The Theory and Practice of Phylogenetic Systematics.* New York: Wiley.

Wilkerson, B. J., and Rumbaugh, D. M. 1979. Learning and intelligence in prosimians. In G. A. Doyle and R. D. Martin, eds., *The Study of Prosimian Behavior,* 207–246. New York: Academic Press.

Wilkie, D. S. 2001. Bushmeat trade in the Congo Basin. In B. B. Beck, T. S. Stoinski, M. Hutchins, T. L. Maple, B. Norton, A. Rowan, E. F. Stevens, and A. Arluke, eds., *Great Apes and Humans: The Ethics of Coexistence,* 86–109. Washington, DC: Smithsonian Institution Press.

Wilkins, W. K., and Wakefield, J. 1995. Brain evolution and neurolinguistic preconditions. *Behavioral and Brain Sciences* 18:161–226.

Wilkinson, M. J. 1985. Lower-lying and possibly older fossiliferous deposits at Sterkfontein. In P. V. Tobias, ed., *Hominid Evolution: Past, Present and Future,* 165–170. New York: Alan R. Liss.

Willems, R. M., and Hagoort, P. 2007. Neural evidence for the interplay between language, gesture, and action: a review. *Brain and Language* 101:278–289.

———. 2009. Broca's region: battles are not won by ignoring half the facts. *Cell* 13:101.

Williams, G. C. 1957. Pleitropy, natural selection, and the evolution of senescence. *Evolution* 11:398–411.

———. 1975. *Sex and Evolution*. Princeton, NJ: Princeton University Press.

Williams, J. M., Oehlert, G. W., Carlis, J. V., and Pusey, A. E. 2004. Why do male chimpanzees defend a group range? *Animal Behaviour* 68:523–532.

Williams, J. M., Pusey, A. E., Carlis, J. V., Farm, B. P., and Goodall, J. 2002. Female competition and male territorial behaviour influence female chimpanzees' ranging patterns. *Animal Behaviour* 63:347–360.

Williams, M. A. J., Assefa, G., and Adamson, D. A. 1986. Depositional context of Plio-Pleistocene hominid-bearing formations in the Middle Awash valley, southern Afar Rift, Ethiopia. In L. E. Frostick, R. W. Renaut, I. Reid, and J. J. Tiercelin, eds., *Sedimentation in the African Rift*, 241–251. Oxford: Blackwell Scientific.

Williams, S. A., and Goodman, M. A statistical test that supports a human/chimpanzee clade based on noncoding DNA sequence data. *Molecular Biology and Evolution* 6:325–330.

Williamson, E. A., and Feistner, A. T. C. 2003. Habituating primates: processes, techniques, variables and ethics. In J. M. Setchell and D. J. Curtis eds., *Field and Laboratory Methods in Primatology: A Practical Guide*, 25–39, Cambridge: Cambridge University Press.

Williamson, E. A., Tutin, C. E. G., and Fernandez. 1988. Western lowland gorillas feeding in streams and on savannas. *Primate Report* 19:29–34.

Williamson, E. A., Tutin, C. E. G., Rogers, M. E., and Fernandez. 1990. Composition of the diet of lowland gorillas at Lopé in Gabon. *American Journal of Primatology* 21:265–277.

Williamson, L., and Usongo, L. 1996. Survey of gorillas *Gorilla gorilla* and chimpanzees *Pan troglodytes* in the Réserve de Faune du Dja, Cameroon. *African Primates* 2:67–72.

Willis, D. 1989. *The Hominid Gang*. New York: Viking.

Wilson, A. C., Bush, G. L., Case, S. M., and King, M.-K. 1975. Social structuring of mammalian populations and rate of chromosomal evolution. *Proceedings of the National Academy of Sciences of the United States of America* 72:5061–5065.

Wilson, A. C., and Cann, R. 1992. The recent African genesis of humans. *Scientific American* 266:2–27.

Wilson, A. C., Carlson, S. S., and White, T. J. 1977. Biochemical evolution. *Annual Review of Biochemistry* 46:573–639.

Wilson, A. C., Maxson, L. R., and Sarich, V. M. 1974a. Two types of molecular evolution: evidence from studies of interspecific hybridization. *Proceedings of the National Academy of Sciences of the United States of America* 71:2843–2847.

———. 1974b. The importance of gene rearrangement in evolution: evidence from studies on rates of chromosomal, protein, and anatomical evolution. *Proceedings of the National Academy of Sciences of the United States of America* 71:3028–3030.

Wilson, A. C., and Sarich, V. M. 1969. A molecular time scale for human evolution. *Proceedings of the National Academy of Sciences of the United States of America* 63:1088–1093.

Wilson, D. R. 1972. Tail reduction in *Macaca*. In R. Tuttle, ed., *The Functional and Evolutionary Biology of Primates*, 241–261. Chicago: Aldine.

Wilson, E. O. 1975. *Sociobiology*. Cambridge, MA: Harvard University Press.

———. 2005. Systematics and the future of biology. *Proceedings of the National Academy of Sciences of the United States of America* 102:6520–6521.

Wilson, G. N., Knoller, M., Szura, L. L., and Schmickel, R. D. 1984. Individual and evolutionary variation of primate ribosomal DNA transcription initiation regions. *Molecular Biology and Evolution* 2:221–237.

Wilson, M. L. 2012. Long-term studies of the chimpanzees of Gome National Park, Tanzania. In Kappeler, P. M., and Watts, D. P., eds., *Long-term Field Studies of Primates*, 357–384. Heidelberg: Springer.

Wilson, M. L., Britton, N. F., and Franks, N. R. 2002. Chimpanzees and the mathematics of battle. *Proceedings of the Royal Society B* 269:1107–1112.

Wilson, M. L., Hauser, M. D., and Wrangham, R. 2001. Does participation in intergroup conflict depend on numerical assessment, range location, or rank for wild chimpanzees? *Animal Behaviour* 61:1203–1216.

Wilson, M. L., Wallauer, W. R., and Pusey, A. E. 2004. New cases of intergroup violence among chimpanzees in Gombe National Park, Tanzania. *International Journal of Primatology* 25:523–549.

Wilson, M. L., and Wrangham, R. W. 2003. Intergroup relations in chimpanzees. *Annual Review of Anthropology* 32:363–392.

Winckler, W., Myers, S. R., Richter, D. J., Onofrio, R. C., McDonald, G. J., Bontrop, R. E., McVean, G. A. T., Gabriel, S. B., Reich, D., Donnelly, P., and Altshuler, D. 2005. Comparison of fine-scale recombination rates in humans and chimpanzees. *Science* 308:107–111.

Wind, J. Phylogeny of the human vocal tract. *Annals of the New York Academy of Sciences* 280:612–630.

Wingfield, J. C., Ball, G. F., Dufty, A. M., Jr., Hegner, R. E., and Ramenofsky, M. 1987. Testosterone and aggression in birds. *American Scientist* 75:602–608.

Wingfield, J. C., Hegner, R. E., Dufty, A. M., Jr., and Ball, G. F. 1990. The "challenge hypothesis": theoretical implications for patterns of testosterone secretion, mating systems, and breeding strategies. *American Scientist* 136:829–846.

Wingfield, J. C., Jacobs, J., and Hillgarth, N. 1997. Ecological constraints and the evolution of hormone-behavior interrelationships. *Annals of the New York Academy of Sciences* 807:22–41.

Wingfield, J. C., Jacobs, J. D., Tramontin, A. D., Perfito, N., Meddle, S., Maney, D. L., and Soma, K. 2000. Toward an ecological basis of hormone-behavior interactions in reproduction of birds. In K. Wallen and J. E. Schneider, eds., *Reproduction in Context: Social and Environmental Influences on Reproductive Physiology and Behavior*, 85–128. Cambridge, MA: MIT Press.

Winkler, L. A., Conroy, G. C., and Vannier, M. W. 1988. Sexual dimorphism in exocranial and endocranial dimensions. In J. H. Schwartz, eds., *Orang-utan Biology*, 225–232. New York: Oxford University Press.

Winterhalder, B. 1996. Social foraging and the behavioral ecology of intragroup resource transfers. *Evolutionary Anthropology* 5:46–57.

Wise, C. A., Sraml, M., Rubinsztein, D. C., and Easteal, S. 1997. Comparative nuclear and mitochondrial genome diversity in humans and chimpanzees. *Molecular Biology and Evolution* 14:707–716.

Wittenberger, J., and Tilson, R. L. 1980. The evolution of monogamy: hypotheses and evidence. *Annual Review of Ecology and Systematic* 11:197–232.

Wittenberger, J. F. 1979. The evolution of mating systems in birds and mammals. In P. Marler and J. G. Vandenbergh, eds., *Handbook of Behavioral Neurobiology*, Vol. 3: *Social Behavior and Communication, 271–349.* New York: Plenum Press.

Wittig, R. M., and Boesch, C. 2003a. Food competition and linear dominance hierarchy among female chimpanzees of the Taï National Park. *International Journal of Primatology* 24:847–867.

———. 2003b. The choice of post-conflict interactions in wild chimpanzees *(Pan troglodytes). Behaviour* 140:1527–1559.

———. 2005. How to repair relationships—reconciliation in wild chimpanzees *(Pan troglodytes). Ethology* 111:736–763.

Wittiger, L., and Sunderland-Groves, J. L. 2007. Tool use during display behavior in wild Cross River gorillas. *American Journal of Primatology* 69:1307–1311.

Wobber, V., Hare, B., Maboto, J., Lipson, S., Wrangham, R., and Ellison, P. T. 2010. Differential changes in steroid hormones before competition in bonobos and chimpanzees. *Proceedings of the National Academy of Sciences of the United States of America* 107:12457–12462.

Wolf, E. R. 1984. Culture: panacea or problem? *American Antiquity* 49:393–400.

Wolfe, C. 2003. *Animal Rites: American Culture, the Discourse of Species, and Posthumanist Theory.* Chicago: University of Chicago Press.

Wolfe, N. D., Switzer, W. M., Carr, J. K., Bhullar, V. B., Shanmugam, V. Tamoufe, U., Prosser, A. T., Torimiro, J. N., Wright, A., Mpoudi-Ngole, E., et al. 2004. Naturally acquired simian retrovirus infections in central African hunters. *Lancet* 363:932–937.

Wolfe, T. 1979. *The Right Stuff.* New York: Picador.

Wolff, J. O., and Macdonald, D. W. 2004. Promiscuous females protect their offspring. *Trends in Ecology and Evolution* 19:127–134.

Wolkin, J. R., and Myers, R. H. 1980. Characteristics of a gibbon-siamang hybrid ape. *International Journal of Primatology* 1:203–221.

Woll, B. 2009. Do mouths sign? Do hand speak?: Echo phonology as a window on language genesis. In R. Botha and H. de Swart, eds., *Language Evolution: The View From Restricted Linguistic Systems, 203–224.* Utrecht: LOT.

Wolpoff, M. H. 1977. Some notes on the Vértesszöllös occipital. *American Journal of Physical Anthropology* 47:357–363.

———. 1980a. *Paleoanthropology.* New York: Knopf.

———. 1980b. Cranial remains of Middle Pleistocene European hominids. *Journal of Human Evolution* 9:339–358.

———. 1982. *Ramapithecus* and hominid origins. *Current Anthropology* 23:501–522.

————. 1983a. Australopithecines: the unwanted ancestors. In K. J. Reichs, ed., *Human Origins*, 109–126. Washington, DC: University Press of America.

————. 1983b. Lucy's lower limbs: long enough for Lucy to be fully bipedal? *Nature* 304:59–61.

————. 1983c. Lucy's little legs. *Journal of Human Evolution* 12:443–453.

————. 1985. Human evolution at the peripheries: the pattern at the eastern edge. In P. V. Tobias, ed., *Hominid Evolution: Past, Present and Future*, 355–365. New York: Alan R. Liss.

————. 1989a. The place of Neandertals in human evolution. In E. Trinkaus, ed., *The Emergence of Modern Humans*, 97–141. Cambridge: Cambridge University Press.

————. 1989b. Multiregional revolution: the fossil alternative to Eden. In P. Mellars and C. Stringer, eds., *The Human Revolution*, 62–108. Edinburgh: University of Edinburgh Press.

————. 1992. Theories of modern human origins. In G. Bräuer and F. H. Smith, eds., *Continuity or Replacement*, 25–63. Rotterdam: Balkema.

Wolpoff, M. H., Hawks, J., Senut, B., Pickford, M., and Ahern, J. 2006. An ape or *the* ape: is the Toumaï cranium TM 266 a hominid? *PaleoAnthropology* 2006:36–60.

Wolpoff, M. H., and Nkini, A. 1985. Early and early Middle Pleistocene hominids from Asia and Africa. In E. Delson, ed., *Ancestors: The Hard Evidence*, 202–205. New York: Alan R. Liss.

Wolpoff, M. H., Spuhler, J. N., Smith, F. H., Radovcic, J., Pope, G., Frayer, D. W., Eckhardt, R., and Clark, G. 1988. Modern human origins. *Science* 241:772–774.

Wolpoff, M. H., Wu, X. Z., and Thorne, A. G. 1984. Modern *Homo sapiens* origins: a general theory of hominid evolution involving the fossil evidence from East Asia. In F. H. Smith and F. Spencer, eds., *The Origin of Modern Humans*, 411–483. New York: Alan R. Liss.

Wong, K. 2005. Footprints to fill. *Scientific American* 293:18–19.

Woo, J. 1957. *Dryopithecus* teeth from Keiyuan, Yunnan Province, *Vertebrata PalAsiatica* 1:25–32.

————.1958. New materials of *Dryopithecus* from Keiyuan, Yunnan. *Vertebrata PalAsiatica* 2:38–43.

————. 1962. The mandibles and dentition of *Gigantopithecus*. *Palaeontologia Sinica*, n.s., series D, 146:1–94.

————. 1980. Palaeoanthropology in the new China. In L.-K. Königsson, ed., *Current Argument on Early Man*, 182–206. Oxford: Pergamon Press.

Woo, J., and Wu, R. 1982. Paleoanthropology in china, 1949–1979. *Current Anthropology* 23:473–477.

Wood, B., and Lieberman, D. E. 2001. Craniodental variation in *Paranthropus boisei:* a developmental and functional perspective. *American Journal of Physical Anthropology* 116:13–35.

Wood, B., and Richmond, B. G. 2000. Human evolution: taxonomy and paleobiology. *Journal of Anatomy* 196:19–60.

Wood, B. A. 1984. The origin of *Homo erectus*. *Courier Forschungsinstitut Senckenberg* 69:99–111.

———. 1985a. A review of the definition, distribution and relationships of *Australopithecus africanus*. In P. V. Tobias, ed., *Hominid Evolution: Past, Present and Future*, 227–232. New York: Alan R. Liss.

———. 1985b. Early *Homo* in Kenya, and its systematic relationships. In E. Delson, ed., *Ancestors: The Hard Evidence*, 206–214. New York: Alan R. Liss.

———. 1988. Are "robust" australopithecines a monophyletic group? In F. E. Grine, ed., *Evolutionary History of the "Robust" Australopithecines*, 269–284. Hawthorne, NY: Aldine de Gruyter.

———. 1991. *Koobi Fora Research Project*, Vol. 4: *Hominid Cranial Remains*. Oxford: Clarendon Press.

———. 1992a. Old bones match old stones. *Nature* 355:678–679.

———. 1992b. Origin and evolution of the genus *Homo*. *Nature* 355:783–690.

Wood, B. A., and Chamberlain, A. T. 1986. *Australopithecus*: grade or clade? In B. Wood, L. Martin, and P. Andrews, eds., *Major Topics in Primate and Human Evolution*, 220–248. Cambridge: Cambridge University Press.

Wood, B. A., and Collard, M. 1999. The human genus. *Science* 284:65–71.

Wood, B. A., and Constantino, P. 2007. *Paranthropus boisei*: fifty years of evidence and analysis. *Yearbook of Physical Anthropology* 50:106–132.

Wood, B. A., and Harrison, T. 2011. The evolutionary context of the first hominins. *Nature* 470:347–352.

Wood, B. A., and Stack, C. G. 1980. Does allometry explain the differences between "gracile" and "robust" australopithecines? *American Journal of Physical Anthropology* 52:55–62.

Wood, J. W., Johnson, P. L., and Campbell, K. L. 1985. Demographic and endocrinological aspects of low natural fertility in Highland New Guinea. *Journal of Biosocial Science* 17:57–79.

Woodford, M. H., Butynski, M., and Karesh, W. B. 2002. Habituating the great apes: the disease risks. *Oryx* 36:153–160.

Wooding, S., Bufe, B., Grassi, C., Howard, M. T., Stone, A. C., Vazquez, M., Dunn, D. M., Meyerhof, W., Weiss, R. B., and Bamshad, M. J. 2006. Independent evolution of bitter-taste sensitivity in humans and chimpanzees. *Nature* 440:930–934.

Woodruff, G., and Premack, D. 1981. Primitive mathematical concepts in the chimpanzee: proportionality and numerosity. *Nature* 293:568–570.

Woodruff, G., Premack, D., and Kennel, K. 1978. Conservation of liquid and solid quantity by the chimpanzee. *Science* 202:991–994.

———. 1979. Intentional communication in the chimpanzee: the development of deception. *Cognition* 7:333–362.

Wooton, W. S., and Parker, J. 1982. *Men behind Bars: Sexual Exploitation in Prison*. New York: Plenum Press.

Wrangham, R. W. 1974a. Predation by chimpanzees in the Gombe National Park, Tanzania. *Primate Eye* 2:6.

———. 1974b. Artificial feeding of chimpanzees and baboons in their natural habitat. *Animal Behaviour* 22:83–93.

———. 1977. Feeding behaviour of chimpanzees in Gombe National Park, Tanzania. In T. H. Clutton-Brock, ed., *Primate Ecology: Studies of Feeding and Ranging Behaviour in Lemurs, Monkeys and Apes,* 503–538, London: Academic Press.

———. 1979a. On the evolution of ape social systems. *Social Science Information* 18:335–368.

———. 1979b. Sex differences in chimpanzee dispersion. In D. A. Hamburg and E. R. McCown, eds., *The Great Apes,* 481–489. Menlo Park, CA: Benjamin/Cummings.

———. 1980. Bipedal locomotion as a feeding adaptation in gelada baboons, and its implications for hominid evolution. *Journal of Human Evolution* 9:329–331.

———. 1986. Ecology and social relationships in two species of chimpanzees. In D. L. Rubenstein and R. W. Wrangham, eds., *Ecological Aspects of Social Evolution,* 352–378. Princeton, NJ: Princeton University Press.

———. 1995a. Relationship of chimpanzee leaf-swallowing to a tapeworm infection. *American Journal of Primatology* 37:297–303.

———. 1995b. Ape cultures and missing links. *Symbols,* Spring, 2–20.

———. 1999. Evolution of coalitionary killing. *Yearbook of Physical Anthropology* 42:1–30.

———. 2000. Why are male chimpanzees more gregarious than mothers? A scramble competition hypothesis. In P. Kappeler, ed., *Primate Males: Causes and Consequences of Variation in Group Composition,* 248–258. Cambridge: Cambridge University Press.

———. 2002. The cost of sexual attraction: is there a trade-off in female *Pan* between sex appeal and received coercion? In C. Boesch, G. Hohmann, and L. F. Marchant, eds., *Behavioural Diversity in Chimpanzees and Bonobos,* 204–215. Cambridge: Cambridge University Press.

———. 2007. The cooking enigma. In P. S. Ungar, ed., *Evolution of the Human Diet,* 308–323. Oxford: Oxford University Press.

———. 2009. *Catching Fire: How Cooking Made Us Human.* New York: Basic Books.

Wrangham, R. W., Chapman, C. A., and Chapman, L. J. 1994. Seed dispersal by forest chimpanzees in Uganda. *Journal of Tropical Ecology* 10:355–368.

Wrangham, R. W., Clark, A. P., and Isabirye-Basuta, G. 1992. Female social relationships and social organization of Kibale Forest chimpanzees. In T. Nishida, W. C. McGrew, P. Marler, M. Pickford, and F. B. M. de Waal, eds., *Topics in Primatology,* Vol. 1: *Human Origins,* 81–98. Tokyo: University of Tokyo Press.

Wrangham, R. W., and Conklin-Brittain, N. L. 2003. Cooking as a biological trait. *Comparative Biochemistry and Physiology* Part A 136:35–46.

Wrangham, R. W., Conklin-Brittain, N. L., and Hunt, K. D. 1998. Dietary response of chimpanzees and cercopithecines to seasonal variation in fruit abundance, I: Antifeedants. *International Journal of Primatology* 19:949–970.

Wrangham, R. W., Conklin, N. L., Chapman, C. A., and Hunt, K. D. 1991. The significance of fibrous foods for Kibale Forest chimpanzees. *Philosophical Transactions of the Royal Society, London B* 334:171–178.

Wrangham, R. W., de Waal, F. M. B., and McGrew, W. C. 1994. The challenge of behavioral diversity. In R. W. Wrangham, W. C. McGrew, F. B. M. de Waal and P. G. Heltne, eds., *Chimpanzee Cultures,* 1–18. Cambridge, MA: Harvard University Press.

Wrangham, R. W., Jones, J. H., Laden, G., Pilbeam, D., and Conklin-Brittain, NL. 1999. The raw and stolen: cooking and the ecology of human origins. *Current Anthropology* 40:567–594.

Wrangham, R. W., and Peterson, D. 1996. *Demonic Males: Apes and the Origins of Human Violence.* Boston: Houghton Mifflin.

Wrangham, R. W., and Smuts, B. B. 1980. Sex differences in the behavioural ecology of chimpanzees in the Gombe National Park, Tanzania. *Journal of Reproduction and Fertility* 28 (Suppl.): 13–31.

Wrangham, R. W., and Wilson, M. L. 2004. Collective violence. comparisons between youths and chimpanzees. *Annals of the New York Academy of Sciences* 1036:233–256.

Wrangham, R. W., Wilson, M. L., and Muller, M. N. 2006. Comparative rates of violence in chimpanzees and humans. *Primates* 47:14–26.

Wrangham, R. W., and Zinnicq Bergmann Riss, E. 1990. Rates of predation on mammals by Gombe chimpanzees, 1972–1975. *Primates* 31:157–170.

Wray, A. 1998. Protolanguage as a holistic system for social interaction. *Language and Communication* 18:47–67.

———. 2000. Holistic utterances in protolanguage: the link from primates to humans. In C. Knight, M. Studdert-Kennedy, and J. R. Hurford, eds., *The Evolutionary Emergence of Language: Social Function and the Origins of Linguistic Form,* 285–302. Cambridge: Cambridge University Press.

———. 2002. Dual processing in protolanguage: performance without competence. In A. Wray, ed., *The Transition to Language,* 113–137. Oxford: Oxford University Press.

Wright, C. 1997. *Black Zodiak.* New York: Farrar, Straus, and Giroux.

Wright, P. C. 1990. Patterns of paternal care in primates. *International Journal of Primatology* 11:89–102.

Wright, R. V. S. 1972. Imitative learning of a flaked stone technology—the case of an orangutan. *Mankind* 8:296–306.

———. 1978. Imitative learning of a flaked stone technology—the case of an orangutan. In S. L. Washburn and L. R. McCown, eds., *Human Evolution: Biosocial Perspectives,* 215–236. Menlo Park, CA: Benjamin/Cummings.

Wroblewski, E. E. 2008. An unusual incident of adoption in a wild chimpanzee *(Pan troglodytes)* population at Gombe National Park. *American Journal of Primatology* 70:995–998.

Wu, C.-I., and Li, W.-H. 1985. Evidence for higher rates of nucleotide substitution in rodents than in man. *Proceedings of the National Academy of Sciences of the United States of America* 82:1741–1745.

Wu, C.-I., and Maeda, N. 1987. Inequality in mutation rates of the two strands of DNA. *Nature* 327:69–170.

Wu, R. 1984. New Chinese *Homo erectus* and recent work at Zhoukoudian. In E. Delson, ed., *Ancestors: the Hard Evidence,* 245–248. New York: Alan R. Liss.

———. 1987. A revision of the classification of the Lufeng great apes. *Acta Anthropologica Sinica* 6:265–271.

Wu, R., and Dong, X. 1985. *Homo erectus* in China. In R. Wu and J. W. Olsen, eds., *Palaeoanthropology and Palaeolithic Archaeology in the People's Republic of China* 79–89. Orlando, FL: Academic Press.

Wu, R., and Lin, S. 1983. Peking man. *Scientific American* 248:86–94.

Wu, R., Lu, Q., and Xu, Q. 1984. Morphological features of *Ramapithecus* and *Sivapithecus* and their phylogenetic relationships—morphology and comparison of the mandibles. *Acta Anthropologica Sinica* 3:1–10.

Wu, R., and Oxnard, C. E. 1983a. *Ramapithecus* and *Sivapithecus* from China: some implications for higher primate evolution. *American Journal of Primatology* 5:303–344.

———. 1983b. Ramapithecines from China: evidence from tooth dimensions. *Nature* 306:258–260.

Wu, R., and Pan, Y. 1985. Preliminary observations on the cranium of *Laccopithecus robustus* from Lufeng, Yunnan with reference to its phylogenetic relationship. *Acta Anthropologica Sinica* 4:7–12.

Wu, R., and Wang, L. 1987. Sexual dimorphism of fossil apes in Lufeng. *Acta Anthropologica Sinica* 6:169–174.

Wu, R., and Xu, Q. 1985. *Ramapithecus* and *Sivapithecus* from Lufeng, China. In R. Wu and J. W. Olsen, eds., *Palaeoanthropology and Palaeolithic Archaeology in the People's Republic of China*, 53–68. Orlando, FL: Academic Press.

Wu, R., Xu, Q., and Lu, Q. 1983. Morphological features of *Ramapithecus* and *Sivapithecus* and their phylogenetic relationships—morphology and comparisons of the crania. *Acta Anthropologica Sinica* 2:1–10.

———. 1985. Morphological features of *Ramapithecus* and *Sivapithecus* and their phylogenetic relationships—morphology and comparisons of the teeth. *Acta Anthropologica Sinica* 4:197–204.

———. 1986. Relationship between Lufeng *Sivapithecus* and *Ramapithecus* and their phylogenetic position. *Acta Anthropologica Sinica* 5:1–30.

Wu, X. 1981. A well preserved cranium of an archaic type of early *Homo sapiens* from Dali, China. *Scientia Sinica* 24:530–541.

Wu, X., Holloway, R. L., Schepartz, L. A., and Xing, S. 2011. A new brain endocast of *Homo erectus* from Hulu Cave, Nanjing, China. *American Journal of Physical Anthropology* 145:452–460.

Wu, X., and Wang, L. 1985. Chronology in Chinese palaeoanthropology. In R. Wu and J. W. Olsen, eds., *Palaeoanthropology and Palaeolithic Archaeology in the People's Republic of China*, 29–51. Orlando, FL: Academic Press.

Wuethrich, B. 1998. Geological analysis damps ancient Chinese fires. *Science* 281:165–166.

Wundt, W. M. 1911. *Völkerpsychologie: Ein Untersuchung der Entwicklungsgesetze von Sprache, Mythus und Sitte*. Leipzig: W. Engelmann.

Wüst, K. 1951. Über den Unterkiefer von Mauer (Heidelberg) im Vergleich zu anderen fossilen und rezenten Unterkiefern von Anthropoiden und Hominiden, mit besonderer Berücksichtigung der phyletischen Stellung des Heidelberger Fossils. *Zeitschrift für Morphologie und Anthropologie* 42:1–112.

Wynn, J. G., Alemseged, Z, Bobe, R., Geraads, D., Reed, D., and Roman, D. C. 2006. Geological and palaeontological context of a Pliocene juvenile hominin at Dikika, Ethiopia. *Nature* 443:332–336.

Wynn, T. 1979. The intelligence of later Acheulian hominids. *Man* 14:371–391.

————. 1981. The intelligence of Oldowan hominids. *Journal of Human Evolution* 10:529–541.

Wynn, T., and McGrew, W. C. 1989. An ape's view of the Oldowan. *Man* 24:383–398.

Wynne, C. 2008a. Rosalià Abreu and the apes of Havana. *International Journal of Primatology*. 29:289–302.

————. 2008b. Aping language: a skeptical analysis of the evidence for nonhuman primate language. *Skeptic* 13:9–14.

Xing, J., Wang, H., Belancio, V. P., Cordaux, R., Deininger, P. L., and Batzer, M. A. 2006. Emergence of primate genes by retro transposon-mediated sequence transduction. *Proceedings of the National Academy of Sciences of the United States of America* 103:17608–17613.

Xing, J., Witherspoon, D. J., Ray, D. A., Batzer, M. A., and Jorde, L. B. 2007. Mobile DNA elements in primate and human evolution. *Yearbook of Physical Anthropology* 50:2–19.

Xu, J., Kemeny, S., Park, G., Frattali, C., and Braun, A. 2005. Language in context: emergent features of word, sentence, and narrative comprehension. *NeuroImage* 25:1002–1015.

Xu, Q., and Lu, Q. 1979. The mandibles of *Ramapithecus* and *Sivapithecus* from Lufeng, Yunnan. *Vertebrata Palasiatica* 17:1–13.

————. 2008. *Lufengpithecus lufengensis: An Early Member of the Hominidae*. Beijing: Science Press.

Xu, X., and Arnason, U. 1996. The mitochondrial DNA molecule of Sumatran orangutan and a molecular proposal for two (Bornean and Sumatran) species of orangutan. *Journal of Molecular Evolution* 43:431–437.

Xu, X., Zou, X., Wang, X., and Han, S. 2009. Do you feel my pain? Racial group membership modulates empathetic neural responses. *Journal of Neuroscience* 29:8525–8529.

Xue, X.-X., and Delson, E. 1988. A new species of *Dryopithecus* from Gansu, China. *Kexue Tongbao* 33:449–452. (in Chinese)

————. 1989. A new species of *Dryopithecus* from Gansu, China. *Chinese Science Bulletin* 34:223–229.

Yamagiwa, J. 1983. Diachronic changes in two eastern lowland gorilla groups *(Gorilla gorilla graueri)* in the Mt. Kahuzi region, Zaire. *Primates* 24:174–183.

————. 1986. Activity rhythm and the ranging of a solitary male mountain gorilla *(Gorilla gorilla beringei)*. *Primates* 27:273–282.

————. 1987a. Intra- and inter-group interactions of an all-male group of Virunga mountain gorillas *(Gorilla gorilla beringei)*. *Primates* 28:1–30.

————. 1987b. Male life history and social structure of wild mountain gorillas *(Gorilla gorilla beringei)*. In S. Kawano, J. H. Connell, and T. Hidaka, eds., *Evolution and Coadaptation in Biotic Communities,* 31–51. Tokyo: University of Tokyo Press.

————. 1992. Functional analysis of social staring behavior in an all-male group of mountain gorillas. *Primates* 33:523–544.

————. 1999. Socioecological factors influencing population structure of gorillas and chimpanzees. *Primates* 40:87–104.

————. 2001. Factors influencing the formation of ground nests by eastern lowland gorillas in Kahuzi-Biega National Park: some evolutionary implications of nesting behavior. *Journal of Human Evolution* 40:99–109.

————. 2003. Bushmeat poaching and the conservation crisis in Kahuzi-Biega National Park, Democratic Republic of the Congo. *Journal of Sustainable Forestry* 16:111–130.

————. 2004. Diet and foraging of the great apes: ecological constraints on their social organizations and implications for their divergence. In A. E. Russon and D. R. Begun, eds., *The Evolution of Thought: Evolutionary Origins of Great Ape Intelligence,* 210–133. Cambridge: Cambridge University Press.

————. 2006. Playful encounters: the development of homosexual behaviour in male mountain gorillas. In V. Sommer and P. L. Vasey, eds., *Homosexual Behaviour in Animals: An Evolutionary Perspective,* 273–293. Cambridge: Cambridge University Press.

Yamagiwa, J., Angoue-Ovono, S., and Kassi, R. 1995. Densities of apes' food trees and primates in the Petit Loango Rese[a]rve, Gabon. *African Study Monographs* 16:181–193.

Yamagiwa, J., and Basabose, A. K. 2006a. Diet and seasonal changes in sympatric gorillas and chimpanzees at Kahuzi-Biega National Park. *Primates* 47:74–90.

————. 2006b. Effects of fruit scarcity on foraging strategies of sympatric gorillas and chimpanzees. In M. M. Robbins and C. Boesch, eds., *Feeding Ecology in Apes and Other Primates,* G. Hohmann, 73–96. Cambridge: Cambridge University Press.

————. 2009. Fallback foods and dietary partitioning among *Pan* and *Gorilla. American Journal of Physical Anthropology* 140:739–750.

Yamagiwa, J., Basabose, A. K., Kahekwa, J., Bikaba, D., Ando, C., Matsubara, M., Iwasaki, N., and Sprague, D. S. 2012. Long-term research on Grauer's gorillas in Kahuzi-Biega National Park, DRC: life history, foraging strategies, and ecological differentiation from sympatric chimpanzees. In Kappeler, P. M., and Watts, D. P., eds., *Long-term Field Studies of Primates,* 385–412. Heidelberg: Springer.

Yamagiwa, J., Basabose, K., Kaleme, K., and Yumoto, T. 2003. Within-group feeding competition and socioecological factors influencing social organization of gorillas in the Kahuzi-Biega National Park, Democratic Republic of Congo. In A. B. Taylor and M. L. Goldsmith, eds., *Gorilla Biology: Multidisciplinary Perspective,* 328–357. Cambridge: Cambridge University Press.

————. 2005. Diet of Grauer's gorillas in the montane forest of Kahuzi, Democratic Republic of Congo. *International Journal of Primatology* 26:1345–1373.

Yamagiwa, J., and Goodall, A. G. 1992. Comparative socio-ecology and conservation of gorillas. In N. Itoigawa, Y. Sugiyama, G. P. Sackett, and R. K. R. Thompson, eds., *Topics in Primatology,* Vol. 2: *Behavior, Ecology, Conservation,* 209–213. Tokyo: University of Tokyo Press.

Yamagiwa, J., and Kahekwa, J. 2001. Dispersal patterns, group structure, and reproductive parameters of eastern lowland gorillas at Kahuzi in the absence of infanticide. In M. M. Robbins, P. Sicotte, and K. J. Stewart, eds., *Mountain Gorillas: Three Decades of Research at Karisoke,* 89–122. Cambridge: Cambridge University Press.

————. 2004. First observations of infanticides by a silverback in Kahuzi-Biega. *Gorilla Journal* 29:6–9.

Yamagiwa, J., Kahekwa, J., and Basabose, K. 2003. Intra-specific variation in social organization of gorillas: implications for their social evolution. *Primates* 44:359–369.

————. 2009. Infanticide and social flexibility in the genus *Gorilla. Primates* 50:293–202.

Yamagiwa, J., Kaleme, K., Milinganyo, M., and Basabose, K. 1996. Food density and ranging patterns of gorillas and chimpanzees in the Kahuzi-Biega National Park, Zaire. *Tropics* 6:65–77.

Yamagiwa, J., Kaleme, K., and Yumoto, T. 2008. Phenology of fruits consumed by a sympatric population of gorillas and chimpanzees in Kahuzi-Biega National Park, Democratic Republic of Congo. *African Study Monographs* 39 (Suppl.): 3–22.

Yamagiwa, J., Maruhashi, T., Yumoto, T., and Mwanza, N. 1996. Dietary and ranging overlap in sympatric gorillas and chimpanzees in Kahuzi-Biega National Park, Zaïre. In W. C. McGrew, L. F. Marchant, and T. Nishida, eds., *Great Ape Societies*, 82–98. Cambridge: Cambridge University Press.

Yamagiwa, J., and Mwanza, N. 1994. Day-journey length and daily diet of solitary male gorillas in lowland and highland habitats. *International Journal of Primatology* 15:207–224.

Yamagiwa, J., Mwanza, N., Spangenberg, A., Maruhashi, T., Yumoto, T., Fischer, A., and Steinhauer-Burkart, B. 1993. A census of the eastern lowland gorillas *Gorilla gorilla graueri* in Kahuzi-Biega National Park with reference to mountain gorillas *G. g. beringei* in the Virunga Region, Zaire. *Biological Conservation* 64:83–89.

Yamagiwa, J., Mwanza, N., Spangenberg, A., Maruhashi, T., Yumoto, T., Fischer, A., Steinhauer-Burkart, B, and Refisch, J. 1992. Population density and ranging pattern of chimpanzees in Kahuzi-Biega National Park, Zaire: a comparison with a sympatric population of gorillas. *African Study Monographs* 13:217–230.

Yamagiwa, J., Mwanza, N., Yumoto, T., and Maruhashi, T. 1991. Ant eating by eastern lowland gorillas. *Primates* 32:247–253.

———. 1992. Travel distances and food habits of eastern lowland gorillas: a comparative analysis. In N. Itoigawa, Y. Sugiyama, G. P. Sackett, and R. K. R. Thompson, eds., *Topics in Primatology*, Vol. 2: *Behavior, Ecology, Conservation*, 267–281. Tokyo: University of Tokyo Press.

———. 1994. Seasonal change in the composition of the diet of eastern lowland gorillas. *Primates* 35:1–14.

Yamagiwa, J., Yumoto, T., Maruhashi, T., and Mwanza, N. 1993. Field methodology for analyzing diets of eastern lowland gorillas in Kahuzi-Biega National Park, Zaire. *Tropics* 2:209–218.

Yamagiwa, J., Yumoto, T., Mwanza, N., and Maruhashi, T. 1988. Evidence of tool-use by chimpanzees *(Pan troglodytes schweinfurthii)* for digging out a bee-nest in the Kahuzi-Biega National Park, Zaire. *Primates* 29:405–411.

Yamakoshi, G. 1998. Dietary responses to fruit scarcity of wild chimpanzees at Bossou, Guinea: possible implications for ecological importance of tool use. *American Journal of Physical Anthropology* 106:283–295.

Yamakoshi, G., and Sugiyama, Y. 1995. Pestle-pounding behavior of wild chimpanzees at Bossou, Guinea: a newly observed tool-using behavior. *Primates* 36:489–500.

Yamamoto, S., Yakoshi, G., Hule, T., and Matsuzawa, T. 2008. Invention and modification of a new tool use behavior: ant-fishing in trees by a wild chimpanzee *(Pan troglodytes verus)* at Bossou, Guinea. *American Journal of Primatology* 70:699–702.

Yamazaki, N. 1985. Primate bipedal walking: computer simulation. In S. Kondo, ed., *Primate Morphophysiology, Locomotor Analyses and Human Bipedalism,* 105–130. Tokyo: University of Tokyo Press.

———. 1990. The effects of gravity on the interrelationship between body proportions and brachiation in the gibbon. In F. K. Jouffroy, M. S. Stack, and C. Niemitz, eds., *Gravity, Posture and Locomotion in Primates,* 157–172. Firenze: Editrice "Il Sedicesimo."

Yamazaki, N., and Ishida, H. 1984. A biomechanical study of vertical climbing and bipedal walking in gibbons. *Journal of Human Evolution* 13:563–571.

Yang, C. 2013. Ontogeny and phylogeny of language. *Proceedings of the National Academy of Sciences of the United States of America* 110:6324–6327.

Yang, Z. 1996. Maximum-likelihood models for combined analyses of multiple sequence data. *Journal of Molecular Evolution* 42:587–596.

———. 2002. Likelihood and Bayes estimation of ancestral population sizes in hominoids using data from multiple loci. *Genetics* 162:1811–1823.

Yaroch, L. A., and Vitzthum, V. J. 1984. Was *Australopithecus africanus* ancestral to the genus *Homo? American Journal of Physical Anthropology* 63:237.

Yengoyan, A. A. 1986. Theory in anthropology: on the demise of the concept of culture. *Comparative Studies in Society and History* 28:368–374.

Yeni-Komshian, G. H., and Bensin, D. A. 1976. Anatomical study of cerebral asymmetry in the temporal lobe of humans, chimpanzees, and rhesus monkeys. *Science* 192:387–389.

Yerkes, R. M. 1925. *Almost Human.* New York: Century.

———. 1939. The life history and personality of the chimpanzee. *American Naturalist* 73:97–112.

———. 1943. *Chimpanzees: A Laboratory Colony.* New Haven, CT: Yale University Press.

Yerkes, R. M., and Learned, B. W. 1925. *Chimpanzee Intelligence and Its Vocal Expressions,* Baltimore: Williams & Wilkins.

Yerkes, R. M., and Yerkes, A. W. 1929. *The Great Apes.* New Haven, CT: Yale University Press.

Yollin, P. 2005a. Gorilla Foundation rocked by breast display lawsuit. *San Francisco Chronicle,* February 18, 2005.

———. 2005b. Ex-worker is third to sue over gorilla. *San Francisco Chronicle,* February 26, 2005.

———. 2005c. Lips are sealed in settlement of Koko case. *San Francisco Chronicle,* December 2, 2005.

Yoshiba, K. 1964. Report of the preliminary survey on the orangutan in North Borneo. *Primates* 5:11–26.

Yu, N., Chen, F.-C., Ota, S., Jorde, L. B., Pamilo, P., Ratthy, L., Ramsay, M., Jenkins, T., Shyue, S.-K., and Li, W.-H. 2002. Larger genetic differences within Africans than between Africans and Europeans. *Genetics* 161:269–274.

Yu, N., Jensen-Seaman, M. I., Chemnick, L., Kidd, J. R., Deinard, A. S., Ryder, O., Kidd, K. K., and Li, W.-H. 2003. Low nucleotide diversity in chimpanzees and bonobos. *Genetics* 164:1511–1518.

Yulish, S. M. 1970. Anterior tooth reductions in *Ramapithecus. Primates* 11:255–263.

Yumoto, T., Yamagiwa, J., Mwanza, N., and Maruhashi, T. 1994. List of plant species identified in Kahuzi-Biega National Park, Zaire. *Tropics* 3:295–308.

Yunis, J. J., and rakash, O. 1982. The origin of man: a chromosomal pictorial legacy. *Science* 215:1525–1530.

Zahn, R., Moll, J., Krueger, F., Huey, E. D., Garrido, G., and Grafman, J. 2007. Social concepts are represented in the superior anterior temporal cortex. *Proceedings of the National Academy of Sciences of the United States of America* 104:6430–6435.

Zalasiewicz, J., Williams, M., Smith, A., Barry, T. L., Coe, A. L., Brown, P. R., Brenchley, P., Cantrill, D., Gale, A., Gibbard, P., et al. 2008. Are we now living in the Anthropocene? *GSA Today* 18:2–8.

Zalmout, I. S., Sanders, W. J., MacLatchy, L. M., Gunnell, G. F., Al-Mufarreh, Y. A., Ali, M. A., Nasser, A.-A. H., Al-Masari, A. M., Al-Sobhi, S. A., Nadhra, A. O., et al. 2010. New Oligocene primate from Saudi Arabia and the divergence of apes and Old World monkeys. *Nature* 466:360–364.

Zambon, S. N., McCrossin, M. L., and Benefit, B. R. 1999. Estimated body weight and degree of sexual dimorphism for *Victoriapithecus macinnesi*, a Miocene cercopithecoid. *American Journal of Physical Anthropology* 28 (Suppl.): 284.

Zapfe, H. 1960. Die Primatenfunde aus der miozänen Spaltenfüllung von Neudorf an der March (Devínská Nová Ves) Tschechoslowakei. Mit Anhang: Der Primatenfund aus dem Miozän von Klein Hadersdorf in Niederösterreich. *Schweizeriche Palaeontologische Abhandlungen* 78:1–293.

Zatorre, R. J., and Salimpoor, V. N. 2013. From perception to pleasure: music and its neural substrates. *Proceedings of the National Academy of Sciences of the United States of America* 110 (Suppl. 2): 10430–10437.

Zeitoun, T. V., Widianto, H., and Djubiantono, T. 2007. The phylogeny of the Flores Man: the cladisticladistic answer. In E. Indriati, ed., *Recent Advances in Southeast Asian Paleoanthropology and Archaeology,* 54–60. Yogyakarta, Indonesia: Gadjah Mada University.

Zhang, J. 2007. The drifting human genome. *Proceedings of the National Academy of Sciences of the United States of America* 104:20147–20148.

Zhang, R., Peng, Y., Wang, W., and Su, B. 2007. Rapid evolution of an X-linked microRNA cluster in primates. *Genome Research* 17:612–617.

Zhang, X. Y., Jiang, Z., and Lin, I. P. 1987. Further discussions on the Central Yunnan Plateau and human origin in terms of the new discoveries of *Homo orientalis* and *Ramapithecus hudiensis. Yunnan Social Science Journal* 3:48–50.

Zhang, X. Y., Lin, I. P., Jang, Z., and Xiao, L. 1987. New species of *Ramapithecus* from Yuanmou, Yunnan. *Sixiang Zhanxian* 3:54–56.

Zhang, Y. 1982. Variability and evolutionary trends in tooth size of *Gigantopithecus blacki. American Journal of Physical Anthropology* 59:21–32.

———. 1985a. *Gigantopithecus* and *"Australopithecus"* in China. In R. Wu and J. W. Olsen, eds., *Palaeoanthropology and Palaeolithic Archaeology in the People's Republic of China,* 69–78. Orlando, FL: Academic Press.

———. 1985b. Occlusal relationships between canine and premolar of *Gigantopithecus blacki. Acta Anthropologica Sinica* 4:97–104.

Zhang, Z., and Harrison, T. 2008. A new middle Miocene pliopithecid from Inner Mongolia. *Journal of Human Evolution* 54:444–447.

Zhou, J., Wei, F., Li, M., Bosco, C., Lok, P., and Wang, D. 2008. Reproductive characters and mating behaviour of wild *Nomascus hainanus*. *International Journal of Primatology* 29:1037–1046.

Zhou, L.-X. 2004. Linear enamel hypoplasia of *Lufengpithecus lufengensis*. *Acta Anthropologica Sinica* 23:118.

Zihlman, A. 1978. Women in evolution, part II: subsistence and social organization among early hominids. *Signs: Journal of Women in Culture and Society* 4:4–20.

———. 1985. *Australopithecus afarensis*: two sexes or two species? In P. V. Tobias, ed., *Hominid Evolution: Past, Present and Future*, 213–220. New York: Alan R. Liss.

———. 1987. American Association of Physical Anthropologists Annual Luncheon Address, April 1985: sex, sexes, and sexism in human origins. *Yearbook of Physical Anthropology* 30:11–19.

Zihlman, A., and Lowenstein, J. 1983. A few words with Ruby. *New Scientist* 98:81–83.

Zihlman, A., McFarland, R. K., and Underwood, C. E. 2011. Functional anatomy and adaptation of male gorillas *(Gorilla gorilla gorilla)* with comparison to male orangutans *(Pongo pygmeus)*. *Anatomical Record* 294:1842–1855.

Zihlman A. L., Mootnick A. R., Underwood C. E. 2011. Anatomical contribution to hylobatid taxonomy and adaptation. *International Journal of Primatology* 32:865–877.

Zihlman, A., Stahl, D., and Boesch, C. 2008. Morphological variation in adult chimpanzees *(Pan troglodytes verus)* of the Taï National Park, Côte d'Ivoire. *American Journal of Physical Anthropology* 135:34–41.

Zihlman, A., and Tanner, N. 1978. Gathering and the hominid adaptation. In L. Tiger and H. T. Fowler, eds., *Female Hierarchies*, 163–194. Chicago: Beresford Book Service.

Zimmer, C. 2004. Faster than a hyena? Running may make humans special. *Science* 306:1283.

Zimmermann, F., Zemke, F., Call, J., and Gómez, J. C. 2009. Orangutans *(Pongo pygmaeus)* and bonobos *(Pan paniscus)* point to inform a human about the location of a tool. *Animal Cognition* 12:347–358.

Zinner, D. P., Nunn, C. L., van Schaik, C. P., and Kappeler, P. M. 2004. Sexual selection and exaggerated sexual swellings of female primates. In P. Kappeler and C. P. van Schaik, eds., *Sexual Selection in Primates: New and Comparative Perspectives*, 71–89. Cambridge: Cambridge University Press.

Zollikofer, C. P. E., and Ponce De León, M. S. 2007. Early *Homo* from Dmanisi and its relationship to African and Asian *Homo erectus*. In E. Indriati, ed., *Recent Advances in Southeast Asian Paleoanthropology and Archaeology*, 61–69. Yogyakarta, Indonesia: Gadjah Mada University.

Zollikofer, C. P. E., Ponce De León, M. S., Lieberman, D. E., Guy, F., Pilbeam, D., Likius, A., Mackaye, H. T., Vignaud, P., and Brunet, M., 2005. Virtual cranial reconstruction of *Sahelanthropus tchadensis*. *Nature* 434:755–759.

Zuberbühler, K. 2007. Predation and primate cognitive evolution. In S. Gursky and K. A. I. Nekaris, eds., *Primate Anti-predator Strategies*, 3–26. New York: Springer.

Zuberbühler, K., Cheney, D. L., and Seyfarth, R. M. 1999. Conceptual semantics in a non-human primate. *Journal of Comparative Psychology* 113:33–42.

Zuberbühler, K., and Jenny, D. 2002. Leopard predation and primate evolution. *Journal of Human Evolution* 43:873–886.

Zuckerkandl, E. 1978. Molecular evolution as a pathway to man. *Zeitschrift für Morphologie und Anthropologie* 69:117–142.

———. 1987. On the molecular evolutionary clock. *Journal of Molecular Evolution* 26:34–46.

Zuckerkandl, E., and Pauling, L. 1962. Molecular disease, evolution, and genic heterogeneity. In M. Kasha and B. Pullman, eds., *Horizons in Biochemistry*, 189–225. New York: Academic Press.

———. 1965. Evolutionary divergence and convergence in proteins. In V. Bryson and H. J. Vogel, eds., *Evolving Genes and Proteins*, 97–166. New York: Academic Press.

Illustration Credits

Figures 2.2, 2.3, 2.4, 2.5C and D, 2.8, 2.11, 2.17, 2.18, 3.1, 3.5A, 4.21E, 5.1, 5.5, 5.9, 5.10, 5.11, 5.14, 6.2, 6.3, 6.5B, 6.6 (right), 6.12, 6.16, 6.17, 10.1, 11.1, 11.2, 12.2, and 12.4 were drawn by Ms. Alba Tomasula y Garcia © 2014.

Figure 2.1. Courtesy of Robert D. Martin. Reprinted by permission from Macmillan Publishers Ltd: 1993. Martin, R. D. Primate origins: plugging the gaps. *Nature* 363:223–224. © 1993 Nature Publishing Group.

Figure 2.3. Based on Michael D. Rose and John G. Fleagle, 1981, The fossil history of non-human Primates in the Americas, A. F Coimbra-Filho et al., eds. *Ecology and Behavior of Neotropical Pirmates,* volume 1, 111–167. © 1981 Academia Brasileira de Ciencias.

Figure 2.5 A and B. Based on Peter F. Murray, 1975, The role of cheek pouches in cercopithecine monkey adaptive strategy, Figure 1 in *Primate Functional Morphology and Evolution*, Russell H. Tuttle, ed., 151–194. © 1975 by Mouton & Co. By permission of de Gruyter Mouton.

Figure 2.6. Based on Adolph H. Schultz, 1969, *The Life of Primates,* Figure 33, p. 81, Universe Books. © 1969 by Adolph H. Schultz. Courtesy of Carel van Schaik and Anthropological Institute & Museum, University of Zürich.

Figure 2.7. Based on Adolph H. Schultz, 1969, *The Life of Primates,* Figure 27, p. 66, Universe Books. © 1969 by Adolph H. Schultz. Courtesy of Carel van Schaik and Anthropological Institute & Museum, University of Zürich.

Figure 2.9. Elliot 1913, Plate XX.

Figure 2.10. Elliot 1913, Plate XXI.

Figure 2.12. Elliot 1913, Plate XXVII.

Figure 2.13. Elliot 1913, Plate XXX.

Figure 2.14. Elliot 1913, Plate XXXVIII.

Figure 2.15. Based on Adolph H. Schultz, 1926, Fetal growth of man and other primates. *The Quarterly Review of Biology,* p. 466. Courtesy of Carel van Schaik and Anthropological Institute & Museum, University of Zürich.

Figure 2.16. Based on Adolph H. Schultz, 1969, *The Life of Primates,* Figure 30, p. 77, Universe Books. © 1969 by Adolph H. Schultz. Courtesy of Carel van Schaik and Anthropological Institute & Museum, University of Zürich.

Figure 3.2. From Simons, E. L., Seiffert, E. R., Ryan T. M., and Attia, Y. 2007. A remarkable female cranium of the early Oligocene anthropoid *Aegyptopithecus zeuxis* (Catarrhini, Propliopithecidae). *Proceedings of the National Academy of Sciences of the United States of America* 104:8731–8736. © 2007 Elwyn L. Simons et al. Courtesy of Elwyn L. Simons and Erik Seiffert.

Figure 3.3. Courtesy of David R. Pilbeam.

Figure 3.4. From Langergraber et al. 2012. Generation times in wild chimpanzees and gorillas suggest earlier divergence times in great ape and human evolution. *Proceedings of the National Academy of Sciences of the United States of America* 109:15716–15721. © 2012 Kevin E. Langergraber et al. Courtesy of Linda Vigilant.

Figure 3.5B. Courtesy of Masato Nakatsukasa.

Figure 3.6A. Courtesy of Naturhistorisches Museum, Wien.

Figure 3.6B. This figure was published in *Primate Evolution, an introduction to man's place in nature*, Elwyn L. Simons, Page 223. © 1972 Elwyn L. Simons. Reprinted by permission of Elsevier.

Figure 3.7. From Moyà-Solà, S., Köhler, M., Alba, D. M., Casanovas-Vilar, I., and Galindo, J. 2004. *Pierolapithecus catalaunicus,* a new Middle Miocene great ape from Spain. *Science* 306:1339–1344. Reprinted with permission from AAAS. Courtesy of Salvador Moyà-Solà and David Alba.

Figure 3.8. Courtesy of Salvador Moyà-Solà and David Alba.

Figure 3.9. © 1985 David L. Brill. Paleontological Museum, University of Thessaloniki.

Figure 3.10. Courtesy of David R. Pilbeam. Photos by D. Sacco.

Figure 3.11. Courtesy of Lu Qingwu.

Figure 3.12. Courtesy of Eric Delson.

Figure 3.13A. Courtesy of Louis de Bonis.

Figure 3.13B. This figure was published in *Primate Adaptation and Evolution*, John G. Fleagle, Page 481. © 1999, 1988 Academic Press. Reprinted by permission of Elsevier. Courtesy of John G. Fleagle.

Figure 3.13C. *Function, Phylogeny and Fossils. Miocene Hominoid Evolution and Adaptations,* 1997, page 51, Chapter 3, Interrelationships between functional morphology and paleoenvironments in Miocene hominoids by Peter Andrews, David R. Begun, and Myriam Zylstra, Figure 5. © 1997 Plenum Press with kind permission from Springer Science+Business Media B. V. Courtesy of Peter Andrews.

Figure 3.14A. *New Interpretations of Ape and Human Ancestry,* p. 823, Chapter 30, Hominoid cladistics and the ancestry of modern apes and humans. A summary statement by R. L. Ciochon, Figure 7. © 1983 Plenum Press with kind permission from Springer Science+Business Media B. V.

Figure 3.14B. Courtesy of Salvador Moyà-Solà.

Figure 3.14C. Courtesy of Lu Qingwu.

Figure 3.14D. *Function, Phylogeny and Fossils. Miocene Hominoid Evolution and Adaptations,* 1997, p. 356, Chapter 16, Enigmatic anthropoid or misunderstood ape? The phylogenetic status of *Oreopithecus bambolii* reconsidered by Terry Harrison and Lorenzo Rook, Figure 2. © 1997 Plenum Press with kind permission from Springer Science+Business Media B. V.

Figure 3.14E. Andrews, P., and Bernor, R. L. 1999. Vicariance biogeography and paleoecology of Eurasian Miocene hominoid primates. In J. Agusti, L. Rook, and P. Andrews, eds., *Hominoid Evolution and Climatic Change in Europe,* Vol. 1: *The Evolution of Neogene Terrestrial Ecosystems in Europe,* 454–487. © 1999 Cambridge University Press. Reprinted with permission of Cambridge University Press.

Figure 3.14F. Harrison, T. 2002. Late Oligocene to Middle Miocene catarrhines from Afro-Arabia. In W. C. Hartwig, ed., *The Primate Fossil Record,* 311–338. © 2002 Cambridge University Press. Reprinted with permission of Cambridge University Press.

Figure 3.14G. Begun, D. R. 2001. African and Eurasian Miocene hominoids and the origins of the Hominidae. In L. de Bonis, G. D. Koufos, and P. Andrews, eds., *Hominoid Evolution and Climate Change in Europe,* Vol. 2: *Phylogeny of the Neogene Hominoid Primates of Eurasia,* 231–253. © 2001 Cambridge University Press. Reprinted with permission of Cambridge University Press.

Figure 4.1A and B. Reprinted Figure 22.5 from W. C. Hartwig, ed., *The Primate Fossil Record,* p. 394. © 2002 Cambridge University Press. Reprinted with permission of Cambridge University Press.

Figure 4.1C. Courtesy of Michel Brunet.

Figure 4.1D and F. © David L. Brill. National Museum of Ethiopia, Addis Ababa.

Figure 4.1E. Courtesy of Martin Pickford and Brigitte Senut.

Figure 4.2. Courtesy of Elisabetta Cioppi. With permission of Museo di Storia Naturale, Sezione di Geologia e Paleontolgia, Università di Firenze, Italy.

Figure 4.3. © Russell H. Tuttle.

Figure 4.4. Courtesy of Alan Walker. Reprinted Figure 22.3D from W. C. Hartwig, ed., *The Primate Fossil Record,* p. 392. © 2002 Cambridge University Press. Reprinted with permission of Cambridge University Press.

Figure 4.5A. Courtesy of Daniel Lieberman. Reprinted by permission from Macmillan Publishers Ltd: 2001. Lieberman, Daniel, Another face in our family tree. *Nature* 410:419–420. © 2001 Nature Publishing Group.

Figure 4.5B. Courtesy of Daniel Lieberman. Reprinted by permission from Macmillan Publishers Ltd: 2009. Lieberman, D. *Homo floresiensis* from head to toe. *Nature* 459:41–42. © 2009 Nature Publishing Group.

Figure 4.6. Courtesy of Cleveland Museum of Natural History.

Figure 4.7. Courtesy of Charles Musiba.

Figure 4.8. Courtesy of Mary D. Leakey. Reprinted by permission of Taylor & Francis Ltd. http://www.informaworld.com: 2008. Russell H. Tuttle, Footprint clues in hominid evolution and forensics: lessons and limitations. *Ichnos* 15:158–165.

Figure 4.10A–C. © Russell H. Tuttle.

Figure 4.10D. © Tim D. White. Transvaal Museum, Pretoria.

Figure 4.10E. © 1985 David L. Brill. Transvaal Museum, Pretoria.

Figure 4.11. © David L. Brill. National Museum of Ethiopia, Addis Ababa.

Figure 4.12. Courtesy of Michel Brunet. Reprinted by permission from Macmillan Publishers Ltd: 1995. Michel Brunet, The first australopithecine 2,500 kilometers west of the Rift Valley (Chad). *Nature* 378:273–275. © 1995 Nature Publishing Group.

Figure 4.13A. © David L. Brill. National Museums of Kenya, Nairobi.

Figure 5.5. Based on Plate 3 in Erik Erikson 1963, Brachiation in New World monkeys and in anthropoid apes, John Napier and Nigel A. Barnicot, eds., *Symposia of the Zoological Society of London*, no. 10, *The Primates*.

Figure 5.6. Based on Farish A. Jenkins, Jr. and John G. Feagle, 1975, Knuckle-walking and the functional anatomy of the wrists of living apes, Figure 1, p. 217, *Primate Functional Morphology and Evolution*, Russell H. Tuttle, ed., © Mouton & Co.

Figure 5.7. © Russell H. Tuttle.

Figure 5.8. © Russell H. Tuttle.

Figure 5.12. © Russell H. Tuttle.

Figure 5.13. Based on Becky A. Sigmon, 1975, Functions and evolution of hominid hip and thigh musculature, Figures 1–4, pp. 238–241, *Primate Functional Morphology and Evolution*, Russell H. Tuttle, ed., © Mouton & Co.

Figure 5.15. © Russell H. Tuttle.

Figure 5.16. © Russell H. Tuttle.

Figure 5.17. © Russell H. Tuttle.

Figure 5.18. © Russell H. Tuttle.

Figure 5.19. © Russell H. Tuttle.

Figure 5.20. Based on Adolph H. Schultz, 1969, *The Life of Primates,* Figure 23, p. 58, Universe Books. © 1969 by Adolph H. Schultz. Courtesy of Carel van Schaik and Anthropological Institute & Museum, University of Zürich.

Figure 6.1. Based on Adolph H. Schultz, 1969, *The Life of Primates,* Figure 29, p. 71, Universe Books. © 1969 by Adolph H. Schultz. Courtesy of Carel van Schaik and Anthropological Institute & Museum, University of Zürich.

Figure 6.4. © Russell H. Tuttle.

Figure 6.6 (left). © Russell H. Tuttle.

Figure 6.7. © Russell H. Tuttle.

Figure 6.8. © Russell H. Tuttle.

Figure 6.9A. This figure is from *How Man Moves. Kinesiological Methods and Studies* by Sven Carlsöö, published by William Heinemann. Reprinted by permission of The Random House Group Limited.

Figure 6.9B. This figure was published in *Recent Advances in Primatology, Volume 3 Evolution*, D. J. Chivers and K. A. Joysey, eds., Page 460 of Activities of hindlimb muscles in bipedal gibbons, H. Ishida et al. © 1978 Academic Press Inc. Reprinted by permission of Elsevier.

Figure 6.10. © Russell H. Tuttle.

Figure 6.11. Reprinted by permission from Macmillan Publishers Ltd: Dennis M. Bramble and Daniel E. Lieberman 2004. Endurance running and the evolution of *Homo*. *Nature* 432(7015), page 346, Figure 1. © 2004 Nature Publishing Group.

Figure 6.12. Based on Bramble and Lieberman 2004 and Zimmer 2005.

Figure 6.13. © Russell H. Tuttle.

Figure 6.14. © Russell H. Tuttle.

Figure 6.15. Courtesy of Bruce Latimer.

Figure 6.16. Based on photos of fresh dissections by Russell H. Tuttle.

Figure 6.17. Based on undated slide by Donald C. Johanson.

Figure 6.18. Courtesy of John Gurche.

Figure 6.19. © Russell H. Tuttle.

Figure 7.1. Photo by David Watts.

Figure 8.1. Courtesy of Kenji Kawanaka.

Figure 8.2. Courtesy of Kenji Kawanaka.

Figure 9.1. Courtesy of Caroline E. G. Tutin.

Figure 9.2. Courtesy of Caroline E. G. Tutin.

Figure 9.3A–C. Courtesy of Christophe Boesch.

Figure 9.4. Based on Adolph H. Schultz, 1969, *The Life of Primates,* Figure 24, p. 59, Universe Books. © 1969 by Adolph H. Schultz. Courtesy of Carel van Schaik and Anthropological Institute & Museum, University of Zürich.

Figure 9.5. © Russell H. Tuttle.

Figure 9.6A. By permission of University of Chicago Press. *Guide to Fossil Man,* Fourth Edition by Michael H. Day, Figure 56, Page 166 © 1986 by Michael H. Day.

Figure 9.6B. © Russell H. Tuttle.

Figure 9.6C. This figure was published in Orang-like manual adaptations in the fossil hominoid *Hispanopithecus laietanus:* first steps towards great ape suspensory behaviours, S. Almécija et al., *Proceedings of the Royal Society B* 274(1614), Page 2377, Figure 1. © 2007 The Royal Society.

Figure 9.6D. By permission of the Natural History Museum Picture Library of the British Museum of Natural History, London UK. Based on Figure 26 in Plate 9 from *Fossil Mammals of Africa,* Number 16 (1959).

Figure 10.2A–B. © Russell H. Tuttle.

Figure 10.2C. Courtesy of Duane M. Rumbaugh.

Figure 10.3. © Russell H. Tuttle.

Figure 11.1. Based on photo by Takayoshi Kano.

Figure 11.2. Based on photo by Takayoshi Kano.

Figure 12.1. © Russell H. Tuttle.

Figure 12.3. Courtesy of Richard W. Wrangham.

Figure 12.4. Based on photos by R. Allen and Beatrice T. Gardner.

Figure 12.5. Courtesy of Duane M. Rumbaugh.

Figure 12.6. Courtesy of E. Sue Savage-Rumbaugh.

Figure 13.1. Tobias, P. V. 1991. *Olduvai Gorge,* Vol. 4: *Homo habilis: Skulls, Endocasts and Teeth,* Plate 2. © 1991 Cambridge University Press. Reprinted with permission of Cambridge University Press.

Figure 13.2. Courtesy of Ray Jackendoff. *Foundations of Language* by Jackendoff (2002) fig. 8, p. 238. By permission of Oxford University Press.

Figure 13.3. Courtesy of John Mitani.

Figure 13.4 (left) Photo from www.DaveLiggett.com. Courtesy of David Liggett.

Figure 13.4 (right) Photo by James Bellucci.

Plate 1 (top left). Photo by Jim Schultz © Chicago Zoological Society.

Plate 1 (top right). Courtesy of Shutterstock.

Plate 1 (bottom). Elliot 1913, Plate XXI.

Plate 2 (top). Courtesy of Durrell Wildlife Conservation Trust.

Plate 2 (bottom). © Russell H. Tuttle.

Plate 3. Courtesy of Carel Van Schaik, photos by Perry van Duijnhoven.

Plate 4 (top and above left). Courtesy of David J. Chivers.

Plate 4 (right). Courtesy of Carel Van Schaik, photo by Perry van Duijnhoven.

Plate 5 (top left, middle and bottom). Photos by and courtesy of Nathaniel Dominy.

Plate 5 (top right). Photo by and courtesy of Alain Houle.

Plate 6. Photos by and courtesy of Alain Houle.

Plate 7 (top, middle and bottom left). Photos by and courtesy of John Mitani.

Plate 7 (bottom right). Photo by and courtesy of Laura Watilo Blake.

Plate 8. Photos by and courtesy of David Watts.

Plate 9. Photos by and courtesy of David Watts.

Plate 10. Photos by and courtesy of John Mitani.

Plate 11 (top). Courtesy of Sue Savage-Rumbaugh.

Plate 11 (middle and bottom). Photos by and courtesy of Melissa Gatter.

Plate 12. Photos by and courtesy of David Watts.

Plate 13 (top and bottom left). Photos by and courtesy of David Watts.

Plate 13 (bottom right). © Frans Lanting/www.lanting.com.

Plate 14. Photos by and courtesy of John Mitani.

Plate 15. Photos by and courtesy of Melissa Gatter.

Plate 16 (top left). Courtesy of Tom Bourdon.

Plate 16 (top right). Courtesy of Elizabeth Garland.

Plate 16 (above left). Courtesy of Michael Gurven.

Plate 16 (above right). Courtesy of Michele Borzoni.

Plate 16 (bottom). © Russell H. Tuttle; Photo by Matthew Tuttle.

Index

Abang, 333
abortion, 441–442
Abrue, R., 33
abstraction, 381–382
Abuqatrania basiodontos, 72
Acacia xanthophloea, 94
Adam, 41
adaptive complex: development of hominoid, 26, 216; development of human, 189; dietary, 266; evolution of, 213; hand of *Pierolapithecus catalaunicus,* 221; hylobatid forelimb, 190–194; knuckle-walking, 206–212; ricochetal arm-swinging, 190–194
adaptive radiation: Anthropoidea, 20; *Australopithecus* spp., 149, 156; Catarrhini, 115–125; Ceboidea, 20; Cercopithecoidea, 20, 82, 89; *Dryopithecus* spp., 102; eastern African Miocene apes, 88–92, 116; Hominidae, 115, 149, 156; Hominoidea, 20; *Homo* spp., 159–185; *Paranthropus* spp., 156–158
adaptive zone: Hominidae, 115; *Homo sapiens,* 115
Aegyptopithecus zeuxis, 67, 115; brain, 71–72; craniodental features, 70–71, 95; diet, 73; habitat, 73; in hominoid phylogeny, 115–116; hypodigm, 70; positional behavior, 215; postcranial features, 71; sexual dimorphism, 70–71; social structure, 70
Affe(n), defined, 15
Aframomum, 287–288
Africa, 2; catarrhine fossil-bearing area, 74, 125; colonized by Asian anthropoid clades, 70, 124; colonized by Eurasian ancestors of extant Hominoidea, 125; cradle of human evolution, 29; homeland of the Catarrhini, 69; homeland of earliest Anthropoidea, 74; homeland of the Hominoidea, 119, 124–125; homeland of the

Homininae, 129; homeland of *Homo sapiens,* 174–185; homeland of Platyrrhini, 72; interchange with Eurasian mammals, 95; lowland rain forest, 74; Neogene aridification and expansion of grasslands, 89, 129; seasonal deciduous woodlands, 100; separate from Eurasia, 63
African apes, 16; adaptive compromise, 204–205; altruism, 590; arithmetic and number, 256; bipedalism, 33, 199, 212, 235–237, 239, 335–338, 343, 371, 531, 533, 540–541, 543; craniodental features, 47–52; footprints, 243; fossil record 86–88; genetic diversity, 42; genitalia, 52–54; parasite, 282, 294–296, 302; phylogeny, 117–137; positional behavior, 203; postcranial features, 52, 193, 203; propinquity with humans, 26–29, 57, 115, 137–138; sexual dimorphism, 40, 52, 266, 362–363; squatting, 242; use of medicinal plants, 282; vertical climbing, 203–204
African hunting (= wild) dog. See *Lycaon pictus*
African leaders of precolonial heritage, 600
Afro-Arabian plate, 63
Afropithecus turkanensis, 84, 85, 95; body mass, 217; craniodental features, 91; diet, 91; geochronological age, 83; phylogeny, 86, 116–123; postcranial features, 216–217; type KNM-WK 16999, 91
Afrotarsius chatrathi, 72
Afrotarsius libycus, 69
Aggression: in human evolution, 7, 10, 305, 595–596; interspecific, 285–286, 290. *See also individual spp.*
Agha, A., 572–573
agile gibbon. See Hylobatidae spp.
Aguirre, E., 136

airorhynchy: defined, 39; gibbon, 39; orangutan, 44; *Sivapithecus sivalensis,* 110

Alemseged, Z., 254, 574

Alexeev, V. P., 162

Algeripithecus minutus, 64; body mass, 69; dentition, 69; stem Strepsirrhini, not Anthropoidea, 69

Alia, 564

Allbrook, D., 84

Allen, J. S., 360

allogrooming. *See under individual taxa*

Ally, 552, 555

Alouatta, 21, 215, 220

Alpagut, B., 106

Altiatlasius koulchii, 73–74

altruism, 312, 408; defined, 432, 590; in humans, 501, 589

Amazonian people, 247, 383

American Museum of Natural History, 4

American Sign Language (ASL), 385, 386, 549, 561; chimpanzee competence, 549–553, 556; defined, 549; gorilla competence, 553–554; orangutan competence, 554–556

Ameslan. *See* American Sign Language (ASL)

amino acid: FOXP2, 366; geochemical dating, 179; in figs, 262; in insects, 262

Amphipithecidae, 64, 214; diet, 66; positional behavior, 214; postcranial features, 214

Amphipithecus mogaungensis, 66–68

Amunts, K., 359

amylase. *See* digestion

Anapithecus hernyaki, 97; diet, 98; habitat, 96; positional behavior, 220, 222

Andrews, P., 82, 94, 95, 98, 102, 103, 105, 106, 107, 116, 118, 119

Animalia, 15

Ankarapithecus meteai, 104, 113; body mass, 106; craniodental features, 106–107; diet, 17; geochronological age, 106; hypodigm, 196; phylogeny 119–123

Ankel, F., 220

Anoiapithecus brevirostris, 125

ant dipping: Gombe chimpanzee, 335–336; handedness, 371; learning, 337, 340

antelope: as competitors for fallen fruit, 284; in Dmanisi fauna, 172; Pliocene proliferation of, 149; postorbital bar in, 17; prey of mandrills, 93

Anthropoidea, 15, 34, 69, 111; basal Paleogene families in, 64–65; craniodental features, 19, 64; Eastern African Miocene, 215–219; Eurasian Miocene, 219–224; Fayum, 70–73; identification of fossil, 63, 74; infraorder, 15; members, 19;

mosaic evolution of traits, 63; stem, 69, 74; suborder, 15, 19; superfamilies, 20

Anthropoid Research Station, Tenerife, 378

anthropology(-ist), 8, 21, 32, 129, 137, 189, 442, 586–587, 599; cultural, 60, 500; evolutionary, 1, 7, 11, 331, 345, 573, 576; molecular, 138, 179; paleo-, 2–4, 60–61, 115, 117, 137–138, 156, 160, 162, 170, 226, 292, 294, 317, 356, 374, 500, 574, 591, 593; physical, 5, 356; sociocultural, 356, 500, 587–588

antler, 357

anvil, 333, 337, 339–340, 344–345

Aotus spp., 19, 21

ape: as amoral being, 597; art, 382, 550, 554; brain specializations, 358–359; drawing, 382, 550, 554; drink, 265, 332, 336–337, 548, 552, 555–556, 560–561; classification, 15, 19–20, 22, 26–33, 35–54, 156; competitors, 284–290; dangers to, 261, 294–296, 596–598; dental, 70, 72, 74, 82, 94, 117, 215–216, 244, 349; disease, 55, 297, 302; ecological restrictions, 32; ethical policy on, 597–601; flagship species, 600; fossil, 60, 70, 72, 74, 82–91, 94–125, 129, 132, 220, 222, 224, 236; life span, 54, 504; in models of human phylogeny, 2–6, 10, 129, 137, 189, 226–230, 305, 349, 393, 407, 498–499, 502, 505, 517, 524, 535–536, 541–542, 547–548, 562–563, 583–585, 592–593, 596; people as, xi; in popular culture, xi; rights, xi; theory of mind, 389, 392; umbrella species, 600

ape language study: bonobo, 385, 563–565, 588; chimpanzee, 381–382, 385, 390–391, 534–535, 547–553, 556–563, 565, 588; controversy, 556–557, 562–565; gorilla, 385, 533–534; orangutan, 385, 548, 554–556; results, 392, 507, 562–565

Apenheul Primate Park, 446, 458, 471

aphasia, 580

Apidium spp., 215

Apidium bowni, 72

Apidium moustafi, 72

Apidium phiomense, 72; adapted for saltation, 215

appeasement, 229, 431, 471, 532, 544

aquatic activity: bipedalism advantageous for, 229; in bonobo foraging, 282; in gorilla foraging, 279, 281, 289

arboreal highway, 199, 213, 407, 490

arboreality: all New World monkeys, 21; bonobo, 205; chimpanzee, 93, 204–205, 213, 339; curved phalanges indicators of, 243, 253–254, 303, 351, 353; gorilla, 204–205; as major selective factor for hominid bipedalism, 228, 237; orangutan, 45, 199, 228; primate characteristic, 16

archeological traces of technology, 331
archeologist, 325, 333, 356, 500
Ardipithecus kadabba, 87, 126–127, 142
Ardipithecus ramidus: as earliest Homindae,
126–127, 136; geochronological age, 246;
habitat, 246; as last common ancestor of
chimpanzees and hominids, 246; phylogeny,
142; positional behavior, 246; postcranial
features, 246, 349–351; skeleton
(ARA-VP-6/500), 246
Ardrey, R., 5, 6
Arensburg, B., 574
arm-swinging: hylobatid, 219; richochetal,
190–194, 219, 223. *See also* brachiation
Aronson, J. L., 147
Arsinoea kallimos, 64, 72, 214
art, 589; abstract, 550; graphic, 3, 354, 590; as
indication of cognitive skills, 382, 590; as
indication of culture, 590; plastic, 3, 354, 574,
590; representational, 550; Upper Paleolithic,
254, 574
arthropodan prey of chimpanzees: *Dorylus*
(Anomma) nigricans, 335, 338; *Macrotermes,*
288; *Macrotermes lilljegorgi,* 288; *Macrotermes*
vitrialatus, 288; stingless bee, 338, 339, 342;
termites, alate, 335
arthropodan prey of gorillas, 265; *Cubitermes* sp.,
279, 288; *Toracotermes,* 288
arthropodan prey of orangutans, 265, 268–274,
342
artifact, 321; Acheulean, 326, 354; bead, 254;
bonobo, 333–334, 580; chimpanzee, 332, 577,
580; demonstrate symbolism in Middle
Pleistocene, 590; flint, 326; Herto, 179; Kada
Gona, 332; language, 507, 548–585; Liang Bua,
168–170; Oldowan, 171–172, 334, 352–353;
orangutan, 333, 580; pendant, 590; symbols,
365, 385, 588, 590
Asia: anthropoid homeland, 67; cradle of humanity,
4; earliest Primates in, 17; emigration of *Homo*
sapiens to, 178–179; forests, 284; homeland of
Homo erectus, 165, 171; homeland of Hylobatidae, *Pongo,* and *Gigantopithecus,* 125
Asian great ape, 42, 265, 284, 421; clade, 124;
origin and deployment, 124–125
ASL. *See* American Sign Language
Aspilia sp., 282
assault: chimpanzee, 528; environmental, 32, 527,
540, 541; human, 10; orangutan, 408
assertiveness: chimpanzee, 420, 529; primate
female, 8
asymmetry: cerebral, 368–374; chimpanzee
cerebellum, 358; human cingulate gyrus, 360;

human metacarpal heads, 349; human motor
cortex, 359; human neuropil volume, 360
Ateles spp., 193, 215, 220, 381; torso and scapulae
resemble those of apes, 21; societies resemble
those of *Pan* spp., 21
Austin, 363, 388, 560, 561, 562, 563, 564, 565;
brain mass, 363; competence communicating,
363; share food with Sherman, 561
Australians of precolonial heritage, role of women, 7
Australopithecinae: adjectival form of, 34; confined
to Africa, 153; language ability, 583; members,
140, 162; musical ability, 583
Australopithecus spp.: body mass, 259; brain size,
149, 374; carnivory, 291; cranial capacity, 576;
dentition, 149, 291; euryphagous diet, 291;
maturation, 173; phylogeny, 138, 155; positional
behavior, 259, 303
Australopithecus afarensis (= *Praeanthropus*
africanus), 86, 137–138, 149, 158; A.L. 162-28,
575; A.L. 288-1 (Lucy), 254–258; arboreal
climbing abilities, 254; body size, 9–10;
challenges to taxonomic validity, 147–148;
craniodental features, 147–148, 159, 291–292,
357; diet, 149, 291–292; DIK-1, 254; endocasts,
575; geochronological age, 136, 140; habitat,
149, 254; Hadar Formation, 138, 143–145,
147–148, 169, 226, 248, 251, 253–254, 256, 292,
332, 351, 498; hands, 351; hyoid bone, 574;
hypodigm, 140, 351; knuckle-walking, 351;
model of social structure, 498; as part of the
hypodigm of *Praeanthropus africanus,* 86;
phylogenic history, 142, 149–156, 184;
postcranial features, 148, 159, 253–258;
questionable makers of Laetoli G footprints, 148,
246–254; sexual dimorphism, 9–10, 498; Sidi
Hakoma Member, 254, 332; as tool user, 226,
332, 353; torso shape, 256–257
Australopithecus africanus: body mass, 158; brachial
index, 258; cannibal, 5; craniodental features,
155, 158–159; diet, 293; geochronological age,
140, 293; headhunter, 5; holotype, 5; as killer
ape, 5; osteodontokeratic culture, 5; phylogeny,
142, 149–151; positional behavior, 352;
postcranial features, 148, 253, 257–258, 351, 353;
prey of carnivores, 7; as proximate ancestor to
Homo, 150, 153; social structure, 498; stature,
158; Sts 14, 258; StW 573, 351; Taung child, 7,
575
Australopithecus anamensis (= *Praeanthropus*
africanus), 86; diet, 129; geochronological age,
137; habitat, 129; hypodigm, 129–130;
phylogeny, 142, 149; thick molar enamel, 129
Australopithecus bahrelghazali, 142, 149, 292

Australopithecus boisei. See *Paranthropus boisei*

Australopithecus crassidens. See *Paranthropus robustus*

Australopithecus garhi: arboreal activity, 149; bipedalism, 149; as maker of earliest lithic artifacts, 332; phylogeny, 142, 149; postcranial features, 258

Australopithecus habilis (= *Homo habilis*), 149, 162

Australopithecus prometheus (= *Australopithecus africanus*), 156

Australopithecus robustus. See *Paranthropus robustus*

Australopithecus sediba: arboreality, 259; bipedalism, 259; body mass, 259; brain size, 148; craniodental features, 155; geochronological date, 154; hypodigm, 154; phylogeny, 155; postcranial features, 259; sexual dimorphism, 155; similarity to *Australopithecus africanus*, 155

Austriacopithecus weinfurteri, 220; positional behavior, 221

awareness: death-, 355, 554, 589, 592; self-, 355, 386–389, 592

axon, 359–360

aye-aye. See *Daubentonia madagascariensis*

Azzaroli, A., 131

babble, 359, 548–549, 552

Babette, 33

baboon. *See Papio* spp.

Badgely, C., 108

Badrian, A., 449, 453, 542

Badrian, N., 449, 453, 542

Bahinia pondaungensis, 68; arboreal, 67; body mass, 67

Bai Hokou, 264, 289, 301, 461

Bakewell, M. A., 29

Baldwin, L. A., 509

bamboo, 134, 281

banana, 379, 543; plantation, 416; provisioning, 305, 310–313, 341, 419, 421, 434–435

Bard, K. A., 387

Barelli, C., 487

barter, 384

Bartlett, T. Q., 267

Bat. *See* Chiroptera

bear, 172, 243, 326

Beard, K. C., 64, 67, 68

bearded saki. See *Chiropotes satanas*

beastiality, 502

begging: bonobo, 322–323, 446, 544–545; chimpanzee, 307, 311–314, 317, 340, 431; estrous effects, 313; success in bonobos, 323; success in chimpanzees, 313

Begun, D. R., 102, 103, 117, 119, 124, 125, 221

belief, 115, 570; afterlife, 592; angel, 1; anthropocentric, 416; entrenched, 386; evil spirit, 1; extraterrestrial being, 1; folk, 11; ghost, 1; God, 596; harmful, 11, 24, 596; as human collective cognitive emergent, 588; intangible, 596; kin, 501–502; moral, 597; race, 596–597; ritual, 500; spiritual, 355, 389, 396, 500; symbolically mediated, 585

Belohdelie, 143

Benson, J., 565

Berger, L. R., 153, 154, 155, 259

Bermejo, M., 544, 545

Bermúdez de Castro, J. M., 171

Bernor, R. L., 98, 99, 117, 119

Bertolani, P., 307, 341

Berzi, A., 131

bias, 8, 556; data, 439; Laetoli hominid feet, 251; male, 8; old ungulates in Mousterian caves, 326; popular and scientific writing, 11

Bickerton, D., 579, 580

Binford, L. R., 9

biome: with caves and rock shelters, 302; Late Miocene East Africa, 129; shared, 284; tropical rain forest, 284

bipedal complex, 345

bipedalism, 356; adaptive advantages, 225, 229–330, 345; bonobo, 199, 212, 371, 543; capuchin, 344–345; chimpanzee, 199, 212, 235–237, 335–337, 431, 531, 533; as defining hominid characteristic, 29, 115, 224; evolution of, 224, 227–228, 246–260; facilitates carrying objects, 9; gorilla, 199, 212, 235–237, 343, 371, 540–541; human, 2, 199, 225–226, 230–232, 236, 239, 242–245, 248, 256, 259, 304, 356; hylobatid, 190, 198–199, 227, 237; Laetoli evidence, 143, 226–253; linked to tool use, 4, 6, 135, 225, 304–305; mechanisms of, 230–246; orangutan, 203, 237, 371, 522; pelvic tilt, 236, 239; rhesus monkey, 371; roots of obligate human, 2, 4, 6, 9, 29, 87, 115, 132, 135–136, 138, 143, 149, 151, 159, 189, 213, 224–230, 246–249, 253–260, 304, 352–353, 356; stance, 228–234; stride, 6, 225, 239, 248–251; terrestrial, 229; threat displays, 229

Birdsell, J. B., 8

Biretia piveteaui, 69, 72

birth canal, 259; of Gona pelvis, 159–160; model of *Australopithecus*, 258

Bishop, W. W., 84

black gibbon. See *Nomascus concolor*

black-headed uacari. See *Cacajao melanocephalus*

Bloch, J. I., 17

Blom, A., 281

Blue, K. T., 219
Boaz, N. T., 147, 148
bodily ornamentation, 3
Bodvapithecus altipalatus (= *Dryopithecus brancoi*),
   102
Boehm, C., 593
Boekhorst, T., 404
Boesch, C., 306, 307, 316, 426, 586, 587
Boesch-Achermann, H., 426
Bolt, U., 225
bond: bonobo male-female, 453–454; bonobo
   mother-son, 445; captive bonobo, 451;
   chimpanzee, 423, 426–427, 525; father-daughter,
   473; female, 10, 450, 453, 505; gorilla
   male-female, 463, 473; human, 506; intermale,
   313, 315, 432, 505; lifetime, 506; mother-infant,
   7; mother-offspring, 421; pair, 7, 9, 487, 493,
   496, 498, 505–506, 595; social-emotional, 8,
   458, 517, 594
bone: breccia, 5; coccyx, 230; collar bone
   (= clavicle), 23–24, 100, 111, 160, 193, 221,
   222, 227, 254; cut-marked, 8–9, 11, 172, 179,
   325–326, 332; frontal, 19, 64, 91, 94, 159, 173;
   humerus(-i), 10, 24, 66, 73, 97, 100, 128,
   192–193, 212, 214–215, 218–219, 221–223, 246,
   254, 258, 260, 372; ilium(-a), 203, 230, 239,
   354; inform re muscle action, 214; inform re
   positional behavior, 214; ischial tuberosity, 22,
   239, 259; mandible, 5, 39, 44, 47, 51, 56–57, 66,
   68–69, 71, 82, 87, 95, 97, 102, 104–108,
   110–111, 113–114, 129, 131, 133, 136, 147, 154,
   171, 173–174, 178–179, 264–265, 279, 531;
   mastoid process, 231; mental symphysis, 19, 51,
   68, 90; patella(-ae=kneecap), 100, 254; pollical
   metacarpal bone, 349, 352; scapula (= shoulder
   blade), 5, 21, 23–24, 85, 111, 193, 195, 216, 218,
   221–223, 226–227, 254, 257; tooth-marked, 172,
   325, 499; trapezium, 349–350, 353
bonobo. See *Pan paniscus*
Bonobo Hope Sanctuary, 563
booboisie, 34
Booee, 551
Bornean agile gibbon. See *Hylobates agilis*
Bornean orangutan. See *Pongo pygmaeus*
Bornean white-bearded gibbon. See *Hylobates
   albibarbis*
Bossou, 275, 307, 320, 339, 421, 455–456, 595
Boubli, J. P., 92
Boulle, P., 600
Boysen, S. T., 384
brachial index, 258
brachiation: defined, 190; gibbon, 190–191, 227;
   overarm in orangutan, 200; rare in great apes,

190; role in human evolution, 226–228. *See also*
   arm-swinging
Brachyteles, 21, 193
Bradley, B. J., 475
brain: areas, 365, 385, 563, 575–576; bonobo, 362,
   393, 565; cellular composition, 363–364, 394;
   chimpanzee, 362–363, 393; dendritic connections,
   364; elephant, 355; endocranial estimates of size,
   274, 374; enlargement, 10, 16, 226, 374; evolution
   of human, 2, 6, 8, 345, 374–375, 504; gibbon,
   362; gorilla, 362–363; hemispheric dominance,
   368–374; hominoid, 355, 358; human, 356;
   human cooling mechanism, 245; human qualities
   of, 379, 392, 394; human sexual dimorphism, 357,
   359–361; human specializations, 359, 569;
   interhemispheric transfer speed, 363–364; limbic
   system, 360; marmoset, 363; metabolically
   expensive, 375; monkey, 363, 372, 392; neuron
   (= nerve cell), 363–364; ontogenetic development
   in human, 359, 361; orangutan, 362–363; problems
   in estimating size from endocranial volume, 374;
   reorganization, 9, 364; size, 9, 58, 355–356, 359,
   500, 574, 576, 590; social, 419, 425; tarsier, 363;
   whale, 355. *See also* cerebellum; cerebrum
Brain, C. K., 6
Bramble, D. M., 245, 259
Brannon, E. M., 383
Bräuer, G., 171
Brazil nut *(Bertholletia excelsa)*, 92
breccia: Makapansgat cave, 5; Sterkfontein cave,
   156, 352
Breuer, T., 474, 475
Briggs, A. W., 182
Britten, R. J., 29
Broca's area: function, 368; homologue in great
   apes, 373; imagined on endocasts, 575;
   lateralization, 360, 368, 372
Brockelman, W. Y., 485, 486
Brodmann's area 4, 44, 45, 360, 364, 372–373, 576
Broom, R., 132, 133, 156
Bruno, 551
Budongo Forest, 266, 299, 312–313, 316–317, 337,
   417, 424–425, 432, 437, 441, 526–529, 531
buffalo, 303, 538
buff-cheeked gibbon (= *Nomascus gabriellae*). See
   Nomascus spp.
Bugtipithecus inexpectans, 68–69, 117
Burgman, M. A., 450
Busidima Formation, Gona, habitat, 259
Busse, C. D., 306
Bwindi Impenetrable National Park, 286, 290,
   299, 463
Bygott, J. D., 317

^{13}C and ^{15}N isotopes, 322
^{13}C/^{12}C ratio, 292–293
C$_3$ photosythesis, 292
C$_4$ photosynthesis, 292
Cacajao melanocephalus, 219–220; diet, 92–93
Cachel, S., 72
Caesalpiniaceae, 289
Cain, 10
calcium, 262, 293
call system, 580
cannibalism: australopithecine, 5; bonobo, 319–320, 595; chimpanzee,10, 317–319, 442–443; endocannibalism defined, 317; exocannibalism defined, 317; gorilla, 320; human, 10, 317; orangutan, 320; stem hominids, 593; symbolic meaning, 317
Cantalupo, C., 358
Cantlon, J. F., 383
Capuchin. See *Cebus* spp.
cardinal number, 383–384
Carnivora, 306, 316, 319, 325–326, 573, 593
Carpenter, C. R., 397, 508, 509
Carpenter's white-handed gibbon. See *Hylobates lar*
carrying: adaptive advantage of, 6, 8, 32, 229–230, 346, 504; bonobo, 320; chimpanzee, 339, 345; devices, 8, 246, 543; gorilla, 471; siamang, 386
Cartesian dualism, 386, 392
cash economy, 384
Catarrhini: Africa homeland to, 69; canial features, 71; earliest species in Eurasia, 99; members, 20; phylogeny, 115; Propliopithecoidea as basal to, 64
Catopithecus browni, 69, 72; body mass, 69, 214; geochronological age, 73; positional behavior, 215
Ceboidea (= Platyrrhini and New World monkeys), 20–22, 71–72, 219–220
Cebus spp., 89; cerebellar asymmetry, 370; cognitive skill, 381; handedness, 370; Herschl's gyrus, 372; manual prehension, 346; as predators, 21, 93, 305; tool use, 344
Cebus apella, 344
Cebus capucinus, 21
Cebus libidinosus, 344–345
cecum: as digestive organ, 262, 288. See digestion
cemengang *(Neesia malayana),* 342, 371
census, Virunga Massif 2010, 466
central chimpanzee (= *Pan troglodytes troglodytes*). See *Pan troglodytes*
central nervous system (CNS), 359
Cercopithecidae. See Cercopithecoidea
Cercopithecinae, 219; charateristics, 22–23; diet, 22, 263, 284; diversity, 22; geographic distribution, 22; in Late Miocene Eurasia, 114; learning processes, 381; prey of bonobos, 320

Cercopithecoidea, 20, 27; adaptive radiation, 89; characteristics, 22, 25, 35, 131, 216, 227, 265, 295, 346, 363, 380, 487; diet, 22, 93, 266; distribution, 21–22; habitat, 219; Parapithecidae as ancestors of, 72; phylogeny, 82, 95, 114, 116, 120, 123; positional behavior, 116, 216, 218; Propliopithecidae as ancestors of, 72; terrestrial activity, 22, 219
Cercopithecus aethiops: alarm call as indexical symbol, 573, 580; baboon prey, 94; body mass, 219; chimpanzee prey, 307; kinship ties, 572; lack a symbolic concept of kinship, 572; matrilines indexical of kinship ties, 572
Cercopithecus ascanius, feeding ecology, 263
cerebellum(-a): chimpanzee, 358; dentate nucleus, 358; FOXP2 expression in, 366; function, 357, 368; great ape, 357–358; human, 357–358, 360; lateral asymmetry, 358; lateralized functions, 358; neocerebellum, 358
cerebral cortex: association, 357–359, 364, 366–369; cell assembly, 391; cell columns, 364, 372; cingulate gyrus, 360; frontal, 358, 368; grey matter, 359, 363, 392; medial frontal, 374; motor, 359–360, 364, 578; neocortex, 355, 357, 359, 364, 366; neural module, 355–356; neuron, 359, 364–365; neuropil, 360, 365; organization, 364; planum parietale, 373; planum temporale, 369, 372–373; precentral gyrus, 360, 367, 369, 373; prefrontal, 359, 366, 374; thickness, 363, 365; visual, 16, 364; white matter, 359
cerebrospinal fluid, 359
cerebrum(-a): amygdala, 374; anterior temporal lobe, 374; connections to cerebellum, 359; frontal lobe, 357–358, 360, 369, 372, 375; function, 358; gyrus(-i), 363, 372; hemispheric dominance, 368, 373; inferior parietal lobes, 365; interhemispheric transfer, 363; neural connectivity, 355; occipital lobe, 36, 369; right temporal lobe, 369; sulcus(-i), 16, 359, 363, 369, 374, 576
Cerling, T. E., 292
cetacean behavioral capacities, 387, 573, 585
chacma baboon. See *Papio ursinus*
Chaimanee, Y., 114
chance similarity, 593
Chantek, 555, 556
Chapman, C. A., 453
Chase, P., 589
cheek flange (= cheek pad), 43, 484
cheek pouch, 22–23, 25
Chemeron proximal humerus (KNM-BC 1745), 128
chereme, 549

Chester Zoo, 430
chicken, 387
chimpanzee. See *Pan troglodytes*
Chimpanzee and Human Communication Institute, 550
chimpanzee predation, 93; arboreal, 306–308; capture, 311; consumption, 311–312; cooperation, 306; distribution of carcass, 307, 311–315; effect of provisioning, 305, 311; effect on monkey populations, 308, 311, 315; female, 307; function 312, 315–316; on humans, 310–311; killing methods, 305, 308; male, 305–306, 313; prey size, 305; prey species, 305, 307–311, 315; pursuit of prey, 311; reaction to beggars (*see* begging); sharing prey, 312–313; success rate, 308; tactics, 306, 308; tool assisted, 307–308, 341
Chiropotes satanas, 92
Chiroptera, 171; bonobo prey, 94, 320–321; relation to the Primates, 17
Chivers, D. J., 490–491, 509, 515
Chordata, 15
Chororapithecus abyssinicus, 87–88
chrysalides, 343
cingulum(a), 41, 45, 52, 62, 91, 97, 104, 106–108, 113
Ciochon, R. L., 66, 117
civet, 17, 284, 310
clade, 70, 119, 124, 131; chimpanzee-bonobo-human, 26, 115; chimpanzee-human, 28; defined, 115; from DNA sequences, 28; Hominidae, 117; hominine, 117; *Homo/Gorilla*, 28; from karyology, 28; *Pan/Australopithecus/Homo*, 117; *Pan/Gorilla*, 28; *Sivapithecus/Pongo*, 113, 116–118
cladistic analysis, 82, 119
cladistics, 26, 584; cladist (= phylogenetic systematist) defined, 115; cladogram, 115; sister group, 115; synapomorphic (= shared derived) feature, 112, 115, 117
Clark, J. D., 61, 179
Clarke, R. J., 156, 351, 510
classism(-ist), 4, 34
Clay, Z., 543
Clever Hans, 535, 556
clothing, 246, 302
clouded leopard, 199, 296–297, 406. See *Neofelis diardi*
Cody, 548
cognition: advance in, 390; asymmetric hemispheric, 373; brain size and, 355, 576; CNS underpinnings of, 346; cultural, 586; development, 57, 183, 375, 391, 595; inclusion in rational behaviorism, 390; for nut-cracking, 339; primate,

379; processing speed, 363; role in meat acquisition, 10; role of cortical cell columns, 364; role of human cerebellum, 357–359; sex-related differences, 360–362; social, 374, 392–394; spatial, 358; specific differences, 392
cognitive mapping, 384–385
cognitive skills: archaeological indications of, 3, 180; belief, 596; demands on, 8; distinction among primate species, 381, 507, 528, 555, 598; emergent, 588; prehuman, 579; selective factors for human, 394
Colobinae, 19, 22, 88, 284; Angola black-and-white colobus, 94, 320; *Colobus guereza*, 89, 311; diet, 22, 263; distribution, 22; features of, 22–23, 263; in Late Miocene Eurasia, 114; *Piliocolobus rufomitratus tephrosceles*, 263; red colobus, 305–311, 315–316, 320
color vision: advantage, 265, 285; catarrhine, 265
colugo. See Dermoptera
communication: bird, 573, 582; cetacean, 387, 573; effective, 525, 568–570; human (*see* facial expression; gesture; speech); nonhuman primate, 573; shared semiotic properties of, 571. See also *individual taxa*
community. See unit-group
comparative method, 2–3, 32, 60, 137, 213, 287, 302, 584, 593; chief function of, 379
comparative psychology(-ist), 356, 378–380, 390, 548, 553
competition: contest, 402, 423; female, 8, 429, 450, 453, 454, 470, 474, 481; grooming, 445; interspecific, 284, 286–287, 290, 411, 429, 476, 486, 491; intraspecific, 313, 323–324, 337, 384, 394, 407–408, 426, 429, 436, 446, 453, 462, 484, 510, 594, 596; male, 10, 403, 411, 429, 434, 437, 439, 456, 476, 480, 483, 504; mother/child, 279; scramble, 402, 423; sperm, 433, 456, 499
concolor gibbon. See *Nomascus concolor*
conditional, 557
Congo River, 46
Conochaetes spp., 262
Conroy, G., 134
conservation, xii, 32, 599–601
consolation: African ape, 431, 590; bonobo, 431; chimpanzee, 431–432; defined, 431; function, 431; gorilla, 471; in Hominidae, 591
consonant, 581
contraception, 436, 503
Cooke, B., 158, 291
cooking: archaeological evidence, 293, 303; benefits of, 291, 316, 375; tooth size reduction, 293; weaning food, 291

cooperation, 394; African ape, 561, 590; bonobo, 453; chimpanzee, 306, 315, 322–323, 423, 435, 562; early hominid, 591, 593, 597; human, 500–501; intermale, 6–7; nonhuman primate, 597

coordination: chimpanzee, 339; motor, 357; sexual, 547

Copeland, S. R., 499

Coppens, Y., 61, 128, 147, 148

coprophagy: chimpanzee, 265; defined, 265; gorilla, 265

copulation: bonobo, 323, 445–447, 450–451, 453–459, 545, 547; chimpanzee, 313–315, 428, 433–440, 456, 484, 528, 530, 532–533; early hominid, 9; gibbon, 492, 494, 497; gorilla, 475–484, 538; human, 239, 502–503, 547; orangutan, 201, 274, 407–415, 519, 521; siamang, 491–492

Corballis, M. C., 389

Cordoni, G., 471

coregulation, defined, 571

Corruccini, R. S., 128, 208

cortex, visual, 364

Coula edulis, 339–341

Cours de linguistique générale, 572

courtship: bonobo, 343, 456; chimpanzee, 435; gibbon, 492, 494; gorilla, 469, 483; human, 517

Cramer, D. L., 363

Creel, N., 353

crested serpent eagle, 510

Crocuta crocuta, 244

Crompton, R. H., 199

cross-modal skills, 385–386, 577; defined, 385; facilitate human language, 386

Cross River gorilla (= *Gorilla gorilla diehli*). See *Gorilla* spp.

cross-sectional data, 399, 465

Crouzelia auscitanensis, 97

Crouzelia rhodanica, 97

Crouzeliinae, 97, 99, 117; members, 97, 120

crowned hawk-eagle. See *Stephanoaetus coronatus*

Cuban missile crisis, 5

cultural anthropology(-ist), 60, 317, 356, 500, 585–588, 599

cultural emergent: altruism, 589; death-awareness, 589; economic ideology, 589; God, 596; love, 505–506, 588, 596; morality, 589; political ideology, 589; religious ideology, 589; scientific ideology, 589; spirituality, 589

cultural primatology, 585, 588

cultural process, defined, 586

cultural psychology(-ist), 586

cultural transmission, 533, 583

culture, adaptive advantages, 54; animal, 585; archeological indications, 3; chimpanzee, 585; collective cognitive emergent, 588; development, 500; definitions, 585–589; derived from killing, 7; human, 355, 385, 392, 582, 587; nonhuman primate, 587–588; non-material, 356; osteodontokeratic, 5–6; popular, xi; problem of, 585; symbol-based, 115, 584

cummings, e.e., 556

dance, 584, 589

Dar, 550, 552, 553

Dart, R. A., 5, 6, 133, 156, 577

Darwin, C. R., 2, 4, 5, 29, 34, 135, 225, 226, 407, 514, 517, 582

Darwinian sexual selection: defined, 407–408; in great apes, 408, 484

Daubentonia madagascariensis, 19

Davenport, R. K., 381, 385

David Greybeard, 585

Dean, C. D., 173

Dean, D., 171

death, 111, 173, 443, 464, 467, 470, 472–474, 477, 553

de Bonis, L., 69, 86, 105, 106, 116, 117, 118

deception, 355, 389, 392, 556

deer, 172, 284, 326

deforestation, 497, 520

deictics, 573

Deloison, Y., 253

Delson, E., 104, 171

Demonic Males, 10, 593; definition of demonic, 593; scenario, 10

Dendrohyrax dorsalis, 320

Dendropithecoidea, 223

Dendropithecus macinnesi, 83, 86, 116, 120, 123, 223; as ancestor to *Victoriapithecus,* 82; body mass, 215; positional behavior, 82, 215; postcranial features, 82, 215

Dendropithecus orientalis, 98

Dermoptera, 284; relation to primates, 17

de Saussure, F., 571, 572

Descent of Man and Selection in Relation to Sex, The, 407

determiners, 598

detoxification, 270–271, 288

de Veer, M. W., 387

developmental psychologist, 586

de Waal, F. B. M., 431, 440, 505, 545

Diamond, J. M., 27

diaphragm, 25, 227, 569

diarrhea, 423

diastema(ta), 91, 131, 133, 135, 148; defined, 40; rare in *Homo sapiens,* 57

dichromatism: asexual, 38; sexual, 38

dig: bonobo, 334; chimpanzee, 288, 332, 335; human, 504; tool, 325

digestion, 22, 262–263, 266, 271, 288, 375; amylase, 266; cecum, 262, 288; colon, digestive function, 262–264, 288; hemicellulose, 263; lignin, 263; lipid, 262–263, 271, 287, 359

Dionysopithecidae, 117

Dionysopithecus sp., 86, 98, 119–120, 124

Dionysopithecus shuangouensis, 98–99, 117; body mass, 99

diploid chromosome number (2n): great apes, 26; *Homo sapiens,* 26; *Hoolock* spp., 35; *Hylobates* spp., 35; indicator of genomic relationships, 3; *Nomascus* spp., 35; *Symphalangus syndactylus,* 35

Dipterocarpaceae, 267, 271, 284

Dirks, H. G. M., 154

discrimination learning, "win-stay; lose-shift" strategy, 379, 381

discrimination-reversal tests, 373, 379–381, 558

disease, 54–55, 58–59, 302, 316, 366, 443, 473, 598; cancer, 54; diabetes, 271; Ebola hemorrhagic fever, 460; epidemic, 54, 262, 443; ischemia, 359; Mendelian, 29; Neolithic, 294; neurodegeneration, 54; pandemic, 32, 59, 473; pleurisy, 539; pneumonia, 539

display, 38, 43, 190, 200, 229, 286, 306, 335, 341, 343, 383, 387, 409, 421–422, 438, 448–449, 458, 467–470, 472, 476, 480, 486–487, 489–490, 492, 494–495, 502, 509, 511–512, 514–515, 520–522, 524, 531, 536, 539, 540–545, 558, 598, 600; phallic, 229, 456, 553

division of labor, 6, 9, 589, 594

Dmanisi hominids. See *Homo georgicus*

DNA (deoxyribonucleic acid), 26, 28–29, 262, 298, 415, 423, 462, 475, 501; hybridization, 28; mitochondrial (mtDNA), 3, 29, 33, 138, 179, 182–183, 185, 436, 440–441, 453, 491; nuclear, 3, 29, 179, 183, 421, 453, 525; recombinant, 58

Dobzhansky, T., 591, 592

Dolhinow, P. J., 10

Domínguez-Rodrigo, M., 332

Donisthorpe, J. H., 462, 463

Doran, D. M., 208

Doran-Sheehy, D. M., 481

Douadi, M. I., 462

double articulation (= duality of patterning), 570, 579, 581; defined, 570

Dryopithecidae, 95; geochronological age, 100; members, 100–104

Dryopithecine, 117–119

Dryopithecus spp., 82, 84–87, 96, 117, 119, 126; adaptive radiation, 102, 125; craniodental features, 41, 125; diet, 102; habitat, 96, 102; persistence in Europe, 102, 104; positional behavior, 124, 221; postcranial features, 220–221, 223; sexual dimorphism, 101–102

Dryopithecus brancoi, 96, 101, 103, 112, 221–222

Dryopithecus carinthiacus, 102–103

Dryopithecus crusafonti, 102

Dryopithecus fontani, 101–103, 221

Dryopithecus laietanus (= *Hispanopithecus laietanus*), 101–103, 117, 220–221, 350

Dryopithecus wuduensis, 104

Du Chaillu, P. B., 540

Ducrocq, S., 67

Ducros, A., 93

Ducros, J., 93

Dur At-Talah anthropoids, 69–70

Earth, 1, 11, 231, 261, 294, 301, 585, 598, 600

eastern chimpanzee, (= *Pan troglodytes schweinfurthi*). See *Pan troglodytes*

Egarapithecus narcisoi, 97

ejaculate, 433

elastic recoil, 210–211

electromyography (EMG), 193

elephant (= *Elephas maximus*), 17, 149, 281, 287, 293, 303, 326, 355

Ellefson, J. O., 190, 509

emergent behavior: acquisition, 390–391; ape, 390–391; collective cognitive, 588; defined, 390–391; human, 390, 505–506, 588–589, 596; insight, 333, 355, 378, 390–391, 555

empathy, 554, 590; in Hominidae, 355, 365

emulation, 392, 570

endocrine shift, 323

energetics, 172, 210–211, 213, 229–230, 237, 243–244, 259, 261–264, 266–267, 271, 274–275, 284, 291, 294, 297, 314–315, 339, 375, 384, 407, 422, 428, 438, 485, 504

English, 15, 34, 548–549, 552–554, 562, 564, 572–573

Enos, 379

environment, 28, 32, 58, 111, 114, 118, 129, 149, 189, 230, 284, 343, 366, 381, 386, 394, 500–501, 548, 556, 561, 564–565, 585, 589–590, 596, 599

Eoanthropus dawsoni (= Piltdown forgery), 4

Eosimias centennicus, 68

Eosimias dawsonae, 65, 68

Eosimias paukkaungensis, 62, 65, 67

Eosimias sinensis, 62, 65, 67–68

Eosimiidae spp., 64–65, 68–69, 214

epigamic feature, 42–43, 55, 398, 407, 483–484, 522

epoch, 61; Anthropocene, 61; arbitrary boundaries, 61; Eocene, 16–17, 62–70, 73, 214; Miocene, 35, 42, 61, 63, 70, 72, 74–80, 82–83, 85–89, 91–92, 94–95, 98–99, 101, 104–108, 110–111, 113–119, 124–129, 132, 134–135, 137–138, 144, 156, 167, 215–224, 292–293, 349; Oligocene, 61, 63–64, 66, 68, 70, 74–75, 79–80, 82, 115, 137, 214, 223–224; Paleocene, 16, 63, 74, 129; Pleistocene, 6, 8–9, 11, 35, 42, 46, 54, 61, 77, 114, 116–118, 132–133, 137, 148, 159, 162, 165, 170–174, 176, 182, 259, 291, 293, 302, 319, 325, 331, 351–353, 372, 375, 498–500, 576–577, 590, 594; Pliocene, 61, 63, 87–88, 114, 126–128, 136, 138, 140, 143, 147–148, 172, 174, 249, 253, 291, 293, 303, 319, 332, 349, 351–352, 499, 576, 590, 592–594; Recent (Holocene), 54, 61, 140, 144, 167, 293, 325–326, 354, 497, 500, 589

Equatorius africanus (= *Kenyapithecus africanus*), 85, 104, 218

Equus spp., 4, 17, 172, 244, 326

estrous swelling. *See* perineal swell

estrus, 6

ethogram, 530

ethology, 599

Etler, D. A., 112

eugenics, 4, 11, 58

Eurasia: deployment of *Homo sapiens* to, 29, 179, 181; earliest Catarrhini, 99; Eocene cooling, 69; fossils, 2; joined by African plate, 63; links to African Hominoidea, 116; Middle Miocee hominoid decline,114; Middle Miocene hotbed of hominoid evolution, 119, 125; Neogene aridification, 89; Neogene faunal exchange with Africa, 94–95, 124–225

European, 1, 10, 61, 83, 97, 100, 117, 183, 303, 416, 445

Evans, S., 387

Eve, 41

evolution: acceptance, 1; human, 1–2, 4–11, 15–16, 34, 54–55, 58–60, 125; primate, 15–16, 63–125

evolutionary anthropology(-ist), 1, 7, 11, 331, 345, 442, 573

evolutionary biology(-ist), 3, 8, 115, 442, 504, 599

evolutionary rates, 138

evolutionary stable strategy (ESS), 442–443

evolutionary theory, 2, 60, 226, 304, 370

experiment, 33, 193, 213, 253, 290–291, 312, 319, 325, 332–334, 337, 355, 372–373, 378–379, 385–388, 481, 485, 534–535, 558, 560, 563–564, 590, 596–597; ablation, 365–366; playback, 507–510, 512, 517, 519, 526

eye, 16, 19, 68, 382, 391, 503, 524, 528, 530, 544–545, 547

facial expression, 365, 373, 509, 515, 517, 522, 524, 529–530, 534, 536, 541, 545, 570, 574

fallback food, 45, 111, 213, 265–266, 275, 292, 403; fig, 274, 281, 285; filler, 263, 289–290; flower, 267; seasonal, 263, 394; staple, 263, 289–290

false killer whale *(Pseudorca crassidens). See* cetacean behavioral capacities

family, 7, 296, 390, 398, 405, 409, 416, 475, 485, 489, 492–495, 503, 506, 511, 515–516, 553, 556

Fan, P. F., 324

farming, 264, 293–294

fat depots, 55, 263

Father Knows Best, 9

fatty acids, 263–264, 270, 280

faunal analysis, 2, 7, 11, 61, 66, 68, 73–74, 88–89, 95, 100, 104–105, 107–108, 110–111, 113, 115, 131, 136, 143, 147, 149, 158, 171–172, 254, 281, 285, 325–327

Fayum dental pattern, 72

Feldesman, M. R., 212

Felidae, 295, 405, 409, 510, 591

female repression, 11

Fernandez, M., 287

fetus, 33, 258–259, 261, 375, 499

fever tree. See *Acacia xanthophloea*

fibrofatty pad, 47

Ficus spp. (= figs), 266, 286, 402–403, 488; ape reliance on, 263, 290; calcium-to-phosphorus ratio, 262; as fallback food, 274, 285, 289; nutrient, 262, 271; predilection for, 271, 281, 285, 289, 299; as staple, 267, 275; strangler, 274, 286; wasp, 262

Fields, W. M., 565

finger spelling, 549

finite-element stress analysis, 129

fire, 6, 32; control of, 3, 246, 302–303, 375, 499, 591; earliest use, 302

Fitch, W. T., 579, 582, 583

fitness, 58, 390, 427, 432, 434, 442–443, 446, 454, 490, 500, 505, 514, 589

Fleagle, J. G., 70, 72, 73, 86, 99, 115, 116, 117, 128, 219

flying lemur. *See* Dermoptera

Foley, J. P., 541

folivory, 267, 284; dental adaptation, 41, 92–93, 98, 132, 288; intestinal adaptation, 288

food sharing, 324; early hominids, 8–9, 294, 536; early male hunters, 7, 9, 304; human, 294, 504, 590 foot (= pes), 16–17, 243; African ape, 204, 212–213, 235; axis of balance, 242; bonobo, 333; chimpanzee, 213, 236–237, 247, 423, 532–533; fossil, 100, 132, 160, 216, 222–223, 246–254, 256, 259–260, 351, 353; gibbon, 38, 194, 197, 199, 241; gorilla, 213, 540; great ape, 232; hallux (first toe), 16, 216, 220, 223, 243, 246–248, 253; human, 212, 232, 239, 242–243, 247–248, 249, 253–254, 303; medial longitudinal arch, 242, 248; orangutan, 19, 45, 199–200, 201–203, 342, 522; pedal digits, 201, 213, 243

forced copulation: gorilla, 480; human, 10, 408; orangutan, 407–413, 519

forest zones: montane rain, 281; transition, 281; tropical rain, 281

Fossey, D., 279, 320, 466, 467, 468, 471, 472, 479, 536, 538, 539

fossil endocasts, 374, 377, 590; A.L. 162-28, 575; KNM-ER 1470, 575; KNM-ER 3732, 575; KNM-ER 3733, 575; KNM-ER 3883, 575; KNM-WT 15000, 575; OH 24, 975; SK 1585, 370; Sts 60, 370; Taung 1, 151, 175

Fouts, R. S., 550, 552, 553, 563

Franz, C., 445

frontal sinus: *Ankarapithecus meteai*, 106; *Australopithecus turkanensis*, 91; bonobo, 196; defined, 39; *Dryopithecus brancoi*, 104; *Dryopithecus laietanus*, 102; gibbon, 39; gorilla, 47, 106; orangutan, 44; *Saadanius hijazensis*, 95; *Sivapithecus sivalensis*, 108; *Turkanapithecus kalakolensis*, 91

frottage, 445; bonobo, 322, 444, 446–450, 452, 457–459, 483, 543–545, 547; functional hypotheses, 448, 458–459; gorilla, 468, 482–483

frugivory, 73

fruit, 100, 262, 265, 284, 391, 552

fruiting season, 111, 270, 282, 284, 290, 299, 315, 402, 408, 414, 417, 425, 461, 475, 514

Fruth, B., 322, 323, 459

functional magnetic resonance imaging (fMRI), 374

funeral, 500

Furness, W. H., 547

Furuichi, T., 446, 448, 453, 455, 456, 545

Gabunia, L. (= Gabounia), 171, 172

Gagneux, P., 441

galago, 19, 94, 275, 307–308, 320, 341

Galdikas, B. M. F., 401, 410, 411, 414, 518, 520, 521

Gallup, G. G., 386, 387, 388

gardening, 266

Gardner, B. T., 548, 549, 550, 563, 584, 589, 590

Gardner, R. A., 548, 549, 550, 563, 584, 589, 590

Garner, R. L., 507

Gardnerian intelligences, 358, 373, 584, 589–590

gas model equation, 472

gatherer, 54, 60, 292, 294, 500–501, 504–505, 589; sex of, 6–8, 341

Gebo, D. L., 507

Geismann, T., 38

Gelada. See *Theropithecus gelada*

gene: aerobic energy metabolism, 375; combination, 391; and intelligence, 361–362; neuronal function, 375; recombination, 58

gene flow, 180–181, 183; interspecific, 138

genome: chimpanzee, 28–29; gibbon, 28; human, 28–29, 361–362, 393; macaque, 28; Neandertal, 182–183; orangutan, 28

genomics, 291, 366; comparative, 46, 54, 182, 392–393; genetic load, 58; metaphor, 505–506; study of, 3, 33, 35, 42, 46, 183, 393, 413, 497, 584

Gentry, E., 541

Geochelone sp. (= giant tortoise), 171

geochronology, 60, 115, 375; $^{40}Ar/^{39}Ar$, 165, 171, 179; defined, 61; faunal, 2, 61, 66, 74, 88, 99, 105, 111, 131, 143–147, 171, 325; fission track, 61, 66; geochemical, 2, 61, 143; potassium-argon (K/AR), 61, 74, 131, 143, 171; problems with indirect methods, 61; radiocarbon, 61; radiometric, 61, 70, 143, 147; sedimentology, 61; uranium-disequilibrium, 61

geology, 60

geophagy, 265

Gerloff, U., 453

Gervais, P., 129

gesticulations, 507, 517, 524, 546–547, 554, 571

gesture, 312, 337, 371, 373, 391, 432, 507, 509, 515, 523, 528–529, 532–534, 541–543, 545, 547, 560, 570–571, 574–575, 577–578, 584, 590; protolanguage, 575, 577–579

Gheerbrant, E., 73

Ghiglieri, M. P., 424, 426

giant flying squirrel, 324

giant gelada (*Theropithecus oswaldi*), 326

gibbon. *See* Hylobatidae spp.

Gigantopithecus spp., 82, 108, 118–119, 125, 129, 132

Gigantopithecus blacki, 113, 117; as ancestor of *Homo erectus*, 132; body mass, 107, 134; dental features, 118, 132–133; diet, 134; hypodigm, 132; phylogeny, 120, 132; sexual dimorphism, 133

Gigantopithecus giganteus (= *G. bilaspurensis* and *Indopithecus giganteus*), 117; as ancestor to *Gigantopithecus blacki*, 120–133; body mass, 134; dental features, 108, 118, 133; sexual dimorphism, 133; stature, 134

Gilby, I. C., 314–315

Ginsburg, L., 97–98

Gittins, S. P., 489–490, 511

glans penis, 54

global climatic change, 149

gloss, 548–550, 552–553, 555, 557, 559–560, 562, 572

glottogony, 576

Godinot, M., 69

Goldman, J., 389

Goliath (Gombe alpha chimpanzee), 421

Gomes, C. M., 316

Gona pelvis, 259–260

Gondwanaland, 17

Goodall, J., 298, 305, 310, 316, 318, 335, 417–419, 422, 524. *See also* Lawick-Goodall, J. van

Goodman, M., 26, 137–138

Gordon, A. D., 10, 351, 553

Gorilla spp., 54, 89, 110, 134, 284, 416, 460; aggression, 286, 460–461, 464, 467–474, 469, 476, 479–480, 482–483, 536–537; allogrooming, 465, 468–469, 479, 539; body mass, 47, 52; brain size, 362–363, 377; captive, 332, 380, 389, 504, 523, 541; cognitive abilities, 355, 371–373, 380–381, 385, 388–389, 397, 553–554; cohesiveness, 289, 300, 463–465, 467, 473, 483, 537; craniodental features, 25, 34, 46–47, 50–53, 82, 84–85, 87, 107, 264, 364–365, 369; Cross River, 30, 264, 279–280, 460–461, 473; diet, 264–266, 279–282, 286–290, 299, 301, 320, 342, 460–461, 466, 469, 475–476; distribution, 281, 460; dominance, 465, 468–469, 475–476, 478–479; enamel thickness, 52, 88; epigamic features, 47, 50; estrus, 468, 470, 472, 477, 479–484; female choice, 478–479; female transfer, 466–467, 469, 472–479, 483, 497, 499; foraging, 264, 286–287; forced copulation, 480; Grauer's, 30, 264, 279–281, 290, 299, 460–461, 466, 472–473, 543; group formation, 461, 466–467, 471–472, 474–478, 482; group size, 460–464, 467, 471, 474; habitat, 46, 48–49, 204, 264, 281, 286, 460, 462–463, 473; harassment, 479; hierarchy, 465, 468–469, 478, 481; infanticide, 466, 472–475, 478–479; interbirth interval, 474, 477; life-history pattern, 461, 466, 476–477, 504; lodging, 294, 299–302, 320, 461, 463; male emigration, 473–477, 497, 499; mountain, 30, 88, 204, 264, 279–281, 286, 290, 299–300, 324, 397, 459–460, 462–484, 518, 536–541; nonvocal communication, 536, 541; pigmentation, 47; positional behavior, 86, 199, 204, 206–209, 211–212, 221, 223–224, 228, 240, 242–243, 346, 536, 541; postcranial features, 46, 52–54, 195, 208, 210–213, 242, 346; range, 279–280, 289–290, 460, 475; reconciliation, 469, 471, 490; self-recognition, 388; sexual behavior, 473–474, 477, 479, 480–484, 538 (*see also* copulation); sexual dimorphism, 10, 47, 50, 52, 106–107, 113, 251, 362, 483–484; sharing, 324–325; silverback, 47, 264, 282, 286, 300–301, 320, 460–484, 536, 541; social relations and organization (= society), 397–398, 459, 459–461, 463, 544, 547; taxonomy, 26–30, 32–33, 45–46, 86–88, 115–117, 119–125, 137–139, 158, 212; tool use, 264, 332, 335, 343; vocalizations, 480, 482, 508, 518, 536–544; weaning, 465, 467, 470, 474; western, 30, 50, 53, 264–265, 282, 286–290, 300–302, 324, 460–462, 464, 471–474, 481–482, 536, 538, 541, 553

Gorilla beringei beringei (= mountain gorilla). See *Gorilla* spp.

Gorilla beringei graueri (= Grauer's). See *Gorilla* spp.

Gorilla gorilla diehli (= Cross River). See *Gorilla* spp.

Gorilla gorilla gorilla (= western gorilla). See *Gorilla* spp.

Gorillinae, 26

gorp, 264

Graecopithecus sp., 104, 119

Graecopithecus freybergi (= *Ouranopithecus macedoniensis*), 105

grammar, 366, 558, 580

Grauer's gorilla. See *Gorilla* spp.

Great Ape Project, 600

Great Ape Survival Plan, 600

Great Ape Trust of Iowa, 563

Great Ape World Heritage Species Project, 600

Greaves, W., 565

Green, R. E., 183, 351

Gregory, W. K., 137, 227

Grehen, J. R., 28

Griffin, D. R., 386

Griphopithecidae, 77, 95, 100

Griphopithecus spp., 96, 102, 104, 119, 124–126, 220

Griphopithecus alpani, 100

Griphopithecus darwini (= *Sivapithecus darwini*), 100, 117, 124

Groves, C. P., 46, 162

Gu, Y., 98, 99

Gua, 548

Gunnell, G. F., 73

Guschanski, K., 475

habitat, 4, 55, 69, 88, 95, 102, 114, 171, 189, 213, 230, 266, 274–275, 284–285, 287, 294, 298–302, 384, 403, 416, 422, 424–425, 429,

452, 459, 461, 464, 466, 468, 491–492, 496, 505, 517, 536, 573, 598–601; caatinga forest, 92; chavascal (= floodplain) forest, 92; closed-canopy rain forest, 46; deciduous woodland, 46, 74, 99–100, 110, 281, 315; deltaic, 254; dolomitic landscapes, 499; dryland forest, 403; ecotone, 46; forest-savannah, 46, 113; heterogeneic mosaic, 275, 403; Late Miocene habitat, 110; lost, 473, 485, 596, 598; lowland dipterocarp forest, 267, 271; lowland swamp forest, 342; in model of ancient food sources, 291; montane forest, 46, 204, 216, 264, 281, 287, 298, 300, 465, 470; mosaic, 88, 149; Neogene, 63, 89, 129; Oligocene, 63; open, 105–106, 118, 134, 188, 246, 315, 344; savanna, 5, 74, 107, 134, 228–229, 259, 264, 284, 293; savanna woodland, 46, 264, 275; secondary forest, 46, 282, 299–300, 448; subtropical forest, 96, 221; swamp forest, 46, 70, 96, 113, 132, 222, 279, 282, 289, 343, 403, 415, 462, 536; terra firma forest, 92; woodland, 219, 292, 344; woodland-grassland mosaic, 104

habituation, 279, 286, 295, 418, 423, 456, 463, 466, 473, 497, 535, 537, 539

Hadzabe, 248–249

Haimoff, E. H., 496, 512

Hainan black-crested gibbon. See *Nomascus hainanus*

Hainan Island, 35, 495

hair, 3, 38, 43, 47, 55–56, 244–245, 262, 271, 298, 322, 342–343, 398, 408, 462, 464, 522, 531, 550

Halperin, S. D., 424

Ham, 379

Hamilton, M. J., 500

hand (= manus), 2, 6, 16, 19, 57, 100, 160, 190–191, 194, 196–197, 199–201, 203–212, 215–216, 221–222, 226, 228–229, 232, 246, 266, 311, 317, 331, 333, 335–337, 342–343, 345–354, 388, 391, 412, 423, 431, 456, 518, 522–523, 531–532, 540, 547, 549, 553, 556, 570–571, 577–578

handedness, 6, 333, 358–360, 368–374, 577

Haplorhini: ancestral traits, 63; members, 19

Haraway, D. J., 11

Harcourt, A. H., 279, 463, 468, 471, 472, 475, 476, 478, 482, 537, 538

Hardus, M. E., 518

Harrison, T., 73, 83, 97, 98, 99, 116, 117, 119, 129, 131, 222

Hartmann, S. E., 87

Hasegawa, T., 138, 439, 440

Hashimoto, C., 425, 456

hate, 596

Hayes, C., 547, 548

Hayes, K. J., 547, 548

Hebb, D. O., 391

Heberer, G., 129, 132, 133

Heliopithecus leakeyi: affinitiy with *Afropithecus*, 99; diet, 95; habitat, 95; hypodigm, 94–95

Herculano-Houzel, S., 364

heritability, 561–562

Herrmann, E., 392

Hess, J. P., 538

Heterohominidae, 129

Hewes, G. W., 577–578

high-resolution magnetic resonance imaging (MRI), 360, 369, 391, 578

Hill, A., 129, 147

Hill, K. R., 9, 500

hippopotamus(-i), 179, 293, 326

Hiraiwa-Hasegawa, M., 439, 440

Hispanopithecus laietanus, 101–103, 221, 350

history, 143, 209, 214, 291, 302, 317, 325, 379, 390, 394, 398, 412, 452, 461, 466, 469, 482, 497, 504, 573, 592, 600–601

Ho, C. K., 136

Hoanghonius stehlini, 68

Hoffstetter, R., 72

Hohmann, G., 321, 322, 323, 343, 453, 454, 459

Holloway, R. L., 575

home base, 8–9, 229

Hominidae: basal, 87, 117, 125, 133; evolution, 9, 11, 137, 162, 305, 547, 593–594. *See also individual taxa*

Hominina, 26

Homininae, 26, 117–119, 583

Hominoidea, 20, 29, 55, 95, 119, 159; members, 15, 22, 27, 30–31, 70, 86, 108, 114–116, 124, 127–128

Homo antecessor, 141, 159, 162, 169, 174–175, 184–185, 354, 357

Homo erectus, 6, 132, 141, 150, 153, 159–160, 162, 164–169, 171–174, 245, 257, 259–260, 325–326, 352–354, 357, 374–375, 411, 499, 575–576, 583

Homo ergaster, 141–142, 159, 162–164, 169, 171–175, 184, 257, 353–354, 357

Homo floresiensis (Hobbit), 141–142, 165, 168–170, 184, 260

Homo georgicus, 141, 162, 169–172, 184, 353, 357, 576

Homo habilis (= *Australopithecus habilis*), 128, 140, 142, 149–150, 160–162, 169, 171–174, 254, 257, 259, 292, 350, 352–353, 357, 375, 574–576

Homo heidelbergensis, 141–142, 159, 162, 169, 173–176, 185, 357

homology, 3, 117, 365, 393, 505, 562–563, 571, 585, 588, 593

Homo neanderthalensis, 141–142, 159, 162, 174–175, 177–178, 182–185, 257, 325–326, 331, 357, 375, 574

Homo orientalis, 136

Homo paniscus, 27

Homo rhodesiensis, 159

Homo rudolfensis, 141–142, 159, 162–163, 169, 353, 357, 575–576

Homo sapien, 34

Homo sapiens, 2, 26, 165, 170, 309, 384, 411, 574, 580; adoption, 58, 502; aggression, 7, 305, 589, 592–596; altruism (*see* altruism); as animal, 15; as apes, 4–5, 26–29, 33–34, 584–585, 600–601; archaic, 179, 181–182, 352, 354; artifact (*see under Homo sapiens:* technology); behavioral flexibility, 390, 502; bilocality, 594; bipedalism (*see* bipedalism); bonding, 7, 505–506; brain size, 8–9, 58, 185, 355–358, 359, 363, 377, 379, 574; breathing, 245, 544; center of mass, 231–235, 238, 244, 259; ceremony, 594; cheating, 501; child care, 502–505, 581, 583, 589–590, 594; cloning, 59; cognitive development, 57, 183, 346, 358–361, 375, 377, 385, 387, 391–394, 528, 569, 573, 588–589, 595–596; cooking, 291, 293, 303, 375; cooperation, 4, 6–7, 326, 394, 500–501, 591, 593–594, 596–597; concealed ovulation, 54, 414; craniodental features, 45, 56–57, 110, 158, 160, 178–180, 245, 291, 293, 369, 375, 575–577; culture (*see* culture); curiosity, 2; dad (= father), 9, 501–502, 505–506; deployment, 29, 178, 181–182, 359; designer baby, 58; diet, 58, 262, 291, 293, 303, 326, 375; distribution, 21, 29, 165, 178, 180–182; division of labor, 6, 9, 589, 594; earliest anatomically modern, 174, 179, 181, 574; emotions (include jealousy, etc.), 8, 345, 359–361, 365, 373–374, 528, 571, 583; endurance runner, 243–245; epigamic features (= secondary sexual characteristics), 55–57; ethics, 32–33, 365, 597, 600–601; euphenic recombinant DNA interventions, 58; fairness, 501; fecundity, 291, 436, 503; female agency, 594; female genitalia, 52, 55; foraging, 7, 229–230, 243, 261, 293–294, 303, 325, 375, 500, 505, 593–594; future, 11, 596; genomic diversity, 54, 59, 182–183; geochronological age, 29, 141, 169, 180, 248; grandfather, 503; grandmother, 503–504; grave goods, 500; hair, 55–56, 244–245; hunting, 4–8, 10–11, 54, 60, 292, 294–296, 304–305, 326; ideology, 115, 594; incisiform canines, 57; infant, 7, 9, 178–179, 310–311, 356, 388; jealousy, 505; juvenile, 54,

294; kin (*see* belief); kin selection, 500, 582; language, 115, 183, 355–356, 358–361, 366, 368, 381, 383–386, 390–391, 392, 500, 534, 541, 557, 570–571, 573–575, 581–582, 585, 589–590, 594, 599; lethal behavior, 593–595; life history, 54, 173, 361, 379, 394, 592; male genitalia, 4, 33, 54, 499; manual dexterity, 6, 456; marriage, 436, 506; matrilocality, 594; menopause, 54, 504; mental protuberance (= mental eminence), 57, 293; monogamy, 7, 9–10, 498–499, 503, 506; morality, 115, 589, 597–598, 600–601; mother, 7, 180, 501, 503–505; ontogeny, 294, 359, 361, 379; origins of society, 294, 356, 448, 501–502; pair bond, 7, 9, 505–506; partnership, 502, 505; paternity, 7, 501–504; patrilocality, 594; pelvic tilt, 236, 239; phylogyny, 2, 16, 26–30, 46, 116–124, 128–129, 133, 136, 138, 142–143, 149–150, 153, 159, 162, 182–185, 228, 230, 503, 585, 593–596; play, 379; politics, 355, 375, 442–443, 459, 466, 531, 588–589, 596, 600–601; polyandry, 503; polygamy, 506; polygyny, 503; population bottleneck, 59; population growth, 54, 294; postcranial features, 16, 24, 55, 151, 160, 224, 228, 233, 243–246, 248–250, 253–260, 331, 345, 347, 351–354, 574, 591; predation, 8–9, 281, 308, 341, 452, 466, 497, 514, 517; public policy, 598; race, 29–30, 596–597; recursion, 389, 570; referential pointing, 523, 528; reproductive behavior, 54, 58, 182, 291, 436, 499, 503–504, 593; residence patterns, 21, 54, 181, 594–595; ritual, 500, 541, 583, 594; sanctions, 501, 503, 594; self-awareness, 355, 386, 388, 592; selfishness, 11, 501; sexual behavior, 7, 497–498, 502–503, 516, 583, 594; sexual dimorphism, 9–10, 47, 55, 57, 251, 258–260, 498, 542, 591; shame, 505; sharing, 7–11, 294, 304, 536, 585, 588, 590, 598–599; sire, 502; skin pigmentation, 55–56; social code, 115, 501, 570, 589, 596–597; social cognition, 374, 392–394; social learning, 392, 500; social relationships and organization, 7, 9, 425, 499–502, 504–505; stewardship, 598; sweat glands, 245; technology, 3, 32, 293, 326, 331, 354–355, 358, 359, 375, 574, 577; truncal erectness, 230; trustworthiness, 374, 501, 591; tuition, 392–393, 501, 589–590, 594; unity, 25, 29; Upper Paleolithic and Holocene, 54, 140, 144, 167, 293, 303, 325–327, 353–354, 497, 500, 574, 589; writing, 4–5, 11, 59, 368, 505–506

Homo troglodytes, 27

honey, 104, 270, 274, 282, 288–289, 290–291, 303, 335, 338–339, 342

Hooff, J.A.R.A.M. van, 199, 412

Hoolock hoolock (western hoolock gibbon), 31–32, 35–36, 38, 99, 191, 268, 495, 515–516

Hoolock leuconedys (eastern hoolock gibbon), 31–32, 35, 268

Hopkins, W. D., 358, 369, 370, 371

Horn, A. D., 443

Horr, D. A., 399, 401, 407, 409, 411

Howell, F. C., 61, 128, 156

howler. See *Alouatta*

howling monkey. See *Alouatta*

Hrdy, S. B., 8

Huffman, M. A., 282

human biological engineering, 11, 33–34

human gatherer-hunters, 7–9, 54, 60, 292, 294, 341, 355, 500–501, 504–506, 589

human nature, 8, 185

human rights, 585, 600–601

Humle, T., 370

humor-response, 360–361

Hunting Apes, The, 10

Hurford, J. R., 578

Hürzeler, J., 129, 131–132

Huxley, T. H., 137, 226

Hyatt, C. W., 388

hybrid, 180–181; ape-human, 33–34; bonobo-chimpanzee, 33; hylobatid, 267, 285, 485, 511; protohuman-protochimpanzee, 28, 138

Hylobates agilis, 30, 32, 38, 42, 268–269, 285, 296, 489–490, 492, 511, 515

Hylobates albibarbis, 30, 32, 269

Hylobates klossii, 31–32, 35–36, 38, 267–268, 295–296, 376, 486, 492–494, 513–514

Hylobates lar, 31–32, 38–39, 41–42, 192, 193, 194, 197–198, 243, 265, 267, 285–287, 295–296, 324–325, 376, 397, 486–492, 495, 508–511, 515

Hylobates moloch, 31–32, 36, 38, 268, 513

Hylobates muelleri, 31–32, 38, 268–269, 286, 512–513

Hylobates pileatus, 31–32, 36, 38, 267, 269, 295–296, 510–512, 516

hylobatid apes. See Hylobatidae spp.

Hylobatidae spp., 55, 70, 82, 93, 98–99, 116, 121–123, 125, 131, 284–285, 335, 380–381, 397, 416; allogrooming, 196, 486–489, 496; asexual dichromatism, 38; basic social unit, 485, 492; birth interval, 487; body mass, 26, 35, 42, 63, 66, 97, 227, 498, 517; brain size, 362; captive, 38, 371, 380, 508, 516–517, 523; craniodental features, 38–42, 56–57, 82, 99, 356–357, 362, 369, 372, 374; diet, 41, 191, 265–268, 285, 324, 485–486; display (*see* display); distribution, 35, 37, 42, 114; diversity, 30–31, 35; extra-pair mating, 485; foraging, 191–194, 226, 285,

486–487, 489, 496; great-call, 486, 488, 508–517; home range, 485–486; insectivory, 265, 267; intermembral index, 35, 42; lodging, 295–296, 486–487, 514; meat eating, 265, 324; monochromatism, 38; nonvocal communication, 517; obligate monogamy, 485–486; origin and deployment, 35, 124–125, 139; pair-bonding, 487, 496, 498; phylogeny, 25–32, 35, 118–120, 122–123, 125, 324; polychromatism, 38; polygamy, 485, 495–496, 498; polygyny, 495; positional behavior, 190–194, 196, 198, 200, 203, 207, 216, 219, 223–224, 226–229, 237, 241, 371, 524 (*see also* brachiation); postcranial features, 22, 24, 35, 38, 42, 190–193, 196, 198–199, 202, 210, 212, 220, 224, 228–229, 237, 243, 257, 346–348, 499; ranging, 268, 486, 489, 492–493, 517; self-recognition, 388; sexual behavior, 54, 487; sexual dichromatism, 38; sexual dimorphism, 35, 38, 57, 362, 498; sharing, 324–325; social behavior, 285–286, 296, 485–486, 488–489, 491, 493–498, 511; sociality, 398, 485, 487, 491, 497, 517, 536; species of, 26, 30–31, 35, 509; territoriality, 485–486, 488–490, 492–494, 510–511, 514–515; vocalization, 38, 486, 488, 492, 495–496, 508–517. *See also individual taxa*

hypoglossal canal, 574–575

hypothermia, 297

hypothesis: alternative tactics, 415; central-place foraging, 9; challenge, 428; cooking influenced size of posterior teeth, 291, 293, 375; evolution of adaptive complexes, 189–190, 193, 206, 213, 266; evolution of bipedalism, 2, 4, 6, 9, 87, 115, 135, 189, 213, 223–230, 246–260, 304, 356; false, 356; food attraction, 404; food sharing, 6–7, 9, 320, 594; food sharing occurs in species with partner choice, 324–325; food sharing traded for mating and coalition support, 8–9, 294, 323, 325; gene acceleration of human origins, 29; grandmother, defined, 503–505; high frequencies of copulation in chimpanzees, bonobos, etc., 456, 491; hit 'em-where-it-hurts, 225; hunting, 4–7, 9–10–11, 229, 304–305, 325–327, 341; imbalance-of-power, defined, 593, 595; learning, 525; male-bonding, 313; man the hunter, disputed, 6–8, 305, 341; meat-for-sex, 9, 315–316; Newman emergent human habitat, 245; pathology, 249–250, 253; range expansion, 442; reconciliation, 431–432, 458; risk-aversion, 475; sex-for-social bonding, 458; sexual attraction, 404; sexual selection, 442; sharing-people, 341; sharing under pressure, 314; showoff hypothesis, defined, 504; social brain, defined, 425; social

hypothesis *(continued)*
  life, defined, 394; social status, 459; stopping
    early, 504; tension-relief, 458; test with fossil
    record, 213–214; tool use had major influence on
    anterior teeth, 293; woman-the-gatherer, 6, 8, 341

Ichnanthropus bipes, 248
iconic sign, 547, 549, 552, 570–573
iconic symbol, 363, 385, 552, 558
Idani, G., 447, 545, 546
idea, 6, 9, 26, 28, 129, 133, 138, 157, 182, 189, 218,
    291, 295, 304, 326, 356, 408, 413, 415–417, 428,
    437, 453, 457, 570, 575, 582, 585, 587–588, 592,
    595
ideology, 115, 594
Ihobe, H., 321, 446
imitation, 356, 365, 387, 392, 407, 556, 570,
    578–579, 582
immunological diffusion analysis, 137
incest, 6, 320, 455, 478, 502
inclusive fitness, 454, 500
index finger *(digitus indicis),* 42, 52, 202, 322, 335,
    346–347, 349, 352, 523, 546, 559, 572
indexical sign, 546–547, 570–573
Indochinese gibbon. See *Nomascus concolor*
infanticide, 319, 409, 415, 434, 442, 455, 496;
    chimpanzee, 317–319, 427, 435, 437, 442–443;
    gorilla, 466, 472–474, 478–479; human, 441,
    592; intragroup, 317–318, 429, 436, 443;
    sexually selected, 442
insectivory, 21, 41, 63, 67–68, 92, 264–265, 267,
    270, 274, 288–289, 291, 313, 337, 341–342, 386,
    458; nutritional value, 262, 267, 289
insemination, 33, 408, 411
insight, 333, 355, 378, 390–391, 419, 422, 555,
    578, 580
instinct, 296, 390
Institute of Primate Studies (IPS), Norman,
    Oklahoma, 556
instrumental intelligence, 11, 331–333, 360–362,
    375, 380, 390–392, 394, 425, 524, 554,
    584–585, 589–590
intentionality, 386
interdigital webbing, 35, 213
Isbell, L. A., 266
ischial callosity, 22, 35, 72, 219, 227, 295
Ishida, H., 86, 128
Itani, J., 418, 424
Ivanov, I. I., 33
ivory, 254, 327

Jackendoff, R., 580
Jaeger, J. J., 69, 70

Jaeggi, A. V., 324, 325, 451
Janecka, J. E., 17
Japanese macaque. See *Macaca fuscata*
Javan gibbon. See *Hylobates moloch*
Jespersen, O., 582
Jiang, X. L., 324
Johanson, D. C., 147, 148, 153
Johnson, C. M., 546
joint: ankle, 203, 222, 231–232, 234, 243, 253;
    ball-and-socket, 193–194, 203, 346; carpal
    (= wrist), 193, 196, 207–209, 211, 219, 222–223;
    carpometacarpal, 196, 216, 345–349, 351, 353;
    elbow, 190, 192–193, 212, 214, 216–218, 220,
    222–223, 344; hip, 203, 216, 221–222,
    230–235, 237–239, 256, 259; humeral, 193,
    216, 223; interpalangeal, 196, 211, 213, 349;
    knee, 196, 198, 202–203, 216, 222, 230–235,
    237–238, 256; metacarpophalangeal, 209–211,
    221, 349; pollical carpometacarpal, 194, 196,
    216, 262, 271, 282, 345–349, 353; shoulder, 23,
    52, 190, 193, 203, 214, 219–220, 222–223, 227,
    231, 236–237, 260, 344, 445, 472, 480, 522,
    531; wrist, 193–194, 203
Jones, C., 287
joystick, 384
just-so story, 576

Kada Gona, Hadar, Ethiopia, 226, 332
Kalepithecus songhorensis, 79, 81, 83, 86, 116, 215
Kama Sutra, 547
Kamoyapithecus hamiltoni, 74, 79, 86, 123
Kanapoi distal humerus (KNM-KP 271), 128
Kano, T., 418, 443, 445–447, 455, 542, 545
Kanzi, 333–334, 373, 542–543, 563–564, 565
Kaplan, H., 504
Kappleman, J., 110, 219
Karanisia arenula, 62, 70
Kawai, M., 463
Kawanaka, K., 422, 423
Kay, R. F., 41, 63, 72, 73, 98, 102, 107, 108, 111,
    574
Keith, A., 137, 226, 227
Kelley, J., 89, 100, 103, 105, 106, 112
Kellogg, D., 548
Kellogg, L. A., 547, 548
Kellogg, W. N., 547, 548
Kenyapithecus spp., 83, 85–86, 92, 94, 116–119,
    124, 128, 221
Kenyapithecus africanus, 79, 81, 83, 85, 104; body
    size, 94, 219; dentition, 92, 116; diet, 92; as
    predator on *Simiolus* and *Mabokopithecus,* 93;
    sexual dimorphism, 219
Kenyapithecus kizili, 100

Kenyapithecus wickeri (= *Ramapithecus wickeri*), 135–136, 218–219
Kenyathropus platyops (= *Australopithecus afarensis*), 149
ketone, 271, 274
Khoratpithecus chiangmuanensis, 77, 81, 114
Khoratpithecus piriyai, 77, 81, 114
Khosian, 385
Kimbel, W. H., 110
Kimeu, 143
King Henry II, 389
Kingston, J. D., 129
kin selection, 500, 582
kinship system, 294, 312, 430, 440–442, 447, 454, 469, 497, 501, 503–504, 506, 510, 537, 572, 589, 591, 594, 596
kiss, 319, 431, 532
Kitamura, K., 447
Kluckhohn, C., 587
Knott, C. D., 271, 402, 408, 414
knuckle-walking, 86, 204, 206, 226, 348–349, 351; adaptive complex, 209, 211, 221, 228–229; advantage of, 205; bonobo, 46, 208–209, 224; chimpanzee, 46, 208–209, 211–212, 224, 372; features, 205–206, 208–209, 212, 218, 223, 246, 349, 353; gorilla, 46, 208–209, 211–212, 224; mechanism, 207, 210, 211; metacarpal tori, 209, 211, 228; obligate, 158, 210, 216; pads, 206–211
Koenigswald, G. H. R. von, 132, 133
Kogolepithecus morotoensis, 78, 85
Köhler, M., 117, 124, 132
Köhler, W., 332, 333, 378, 379
Koko, 553, 554, 556
Komodo dragon. See *Varanus komodoensis*
Konopka, G., 366
koolokamba, 33
Kordos, L., 102, 103
Kortlandt, A., 416, 417
Koufos, G. D., 105, 106, 117
Krause, J., 183
Kretzoi, M., 102
Krings, M., 182
Krishnapithecus posthumus, 79, 98
Kroeber, A., 587
Kunimatsu, Y., 83, 85, 88, 93, 98
Kuroda, S., 444, 445
Kurth, G., 129

Laccopithecus robustus, 76, 81, 117; basal to gibbons, 125; body mass, 99, 111, 223; craniodental features, 99; habitat, 99, 113; positional behavior, 223; sexual dimorphism, 99
lactation, 261, 274, 428

lactational amenorrhea, 411, 436, 477
Ladygina-Kohts, N. N., 378
Laetoli G trackway, 143, 146–148, 226, 246–251, 253–254
Lagothrix, 21, 193
Laidler, K., 548
Lambert, J. E., 263
Lana, 558, 560, 561, 565
landbridge, 95
Landsmeer, J. M. F., 346
Langdon, J. H., 132
Langergraber, K. E., 138
language, 6, 304, 373–375, 501, 524, 547, 565, 569, 576–580, 583–584; ape artifactual (*see* ape language study); bilaterality of, 360; child, 570, 580; defined, 585; lateralization, 360; as mediator of cross-modal skills, 385; origin and evolution of, 6, 304, 356, 373–375, 524, 547, 576–580, 582–584; pidgin, 549, 580; second acquisition, 580; sign, 368, 385, 548–549, 551, 570, 577–578, 582; symbolically mediated, 500, 565, 569–570, 572–573, 589; tonal, 571; verbal, 570
Language Research Center, Georgia State University (LRCGSU), Atlanta, 563
langur. See *Semnopithecus entellus*
Larsen, C. S., 498
laryngeal air sac, 35, 43, 398, 515, 522, 540
La Société de Linguistique de Paris, 576
lateral meniscus, 256
Latimer, B., 253
laughter, 527, 529, 537, 544, 571
Lawick-Goodall, J. van, 316. *See also* Goodall, J.
Leakey, L. S. B., 60, 116, 135, 136, 143, 159, 352
Leakey, M. D., 60, 89, 143, 253
Leakey, M. G., 116, 128, 143, 147
Leakey, R. E. F., 61, 83, 143
learning, 271, 333–334, 337, 340, 358, 360, 365–367, 373, 378–380, 384–385, 390–394, 500–501, 504, 507, 525, 549–551, 577, 579, 581–582, 585, 588–590; analytic, 583; catarrhine, 381, 392, 399, 529, 541, 546, 552–555, 557–558, 560–561, 564–565; comparative primate, 380–381, 384; imitation, 356; interspecific, 271, 337, 340; modular, 355–356
learning set, 379, 573
leaves, food item, 21–22, 41, 44, 73, 92, 98, 113, 191, 213, 262–263, 265–270, 272, 274–275, 281–283, 287–292, 297, 312, 317, 335, 343, 403
Lee, R. B., 7
Lehmann, J., 441

Leighton, M., 267, 271
lemur, 17, 19, 20, 380
leopard. See *Panthera pardus*
lesser apes. *See* Hylobatidae spp.
Lewis, G. E., 134, 136
lexical protolanguage, 575, 579–580
lexicon, 579–580
lexigram, 373, 385, 391, 557–565
Li, C., 99
Lieberman, D. E., 245
Lieutenant Cable, 596
ligament, 259; extensor expansion, 209, 349;
    function, 194, 228, 230, 244, 348–349;
    iliofemoral, 259; metacarpophalangeal, 349;
    pollical carpometacarpal ligament, 348
Limnopithecus spp., diet, 92
Limnopithecus evansi, 83
Limnopithecus legetet, 79, 83
Lin, A. C., 387
line of gravity, 231
Lingnau, A., 365
linguist, 562, 571, 573–574, 576, 578
linguistic functions, 358, 368
linguistic productivity, 558
Linton (= Slocum) S., 7, 9
lion. See *Panthera leo*
Lion in Winter, The, 389
lion-tailed macaque. See *Macaca silenus*
lip-smacking, 530–531
Liu, Z., 496
Lockwood, C. A., 151
Logan, T. R., 147
logging, 32, 266, 401, 460, 492
Lomorupithecus harrisoni, 77, 82
Long, J. C., 352
longevity: captive ape, 54; fossil hominids, 173;
    human, 54–55, 504; primate, 17, 284, 504
long-tailed macaque. See *Macaca fascicularis*
Lonsdorf, E. V., 370, 371
lordosis, 230
Lorenz, K. Z., 5
loris, 19, 93, 214
Lothagam mandibular fragment (KNM-LT 329),
    128
Loulis, 552, 553
love, 355, 459, 502, 505, 600; asexual, 506; defined,
    506; as emergent quality, 505, 596
Lovejoy, C. O., 9, 246
Lu, Q., 111, 118, 136
Lucas, P. W., 293
Lucy, 551, 552
Lufengpithecus hudienensis, 77, 81, 111, 114
Lufengpithecus keiyuanensis, 77, 81, 111, 113

Lufengpithecus lufengensis, 77, 81, 118–119, 124,
    220; body mass, 111; craniodental features,
    111–114; diet, 113; habitat, 111–113; hypodigm,
    111; positional behavior, 136, 223; postcranial
    features, 111, 223; as proximate ancestor of
    Australopithecus, 126; sexual dimorphism, 111,
    113
Lukeino molar (KNM-LU 335), 128
Lycaon pictus, 425

Maboko Island, Kenya, 72, 74, 81–83, 85, 92–93,
    116, 218–219
Mabokopithecus clarki, 75, 81, 83, 93
Mabokopithecus harrisoni, 218–219
Mabokopithecus pickfordi, 93
Macaca spp., 23–24, 28, 110, 116, 218, 285,
    364–365, 372, 552, 578; caudal reduction,
    216
Macaca fascicularis, 343, 372, 381, 385
Macaca fuscata, 21–22, 372, 392
Macaca mulatta, 366, 367, 371–373, 383, 385
Macaca silenus, 343
Macho, G. A., 129
MacKinnon, J. R., 401, 405, 407, 409–411, 488,
    515, 517, 518, 520, 522, 542
MacKinnon, K. S., 488
MacLeod, C. E., 358
Mahboubi, M., 69
Malenky, R. K., 282
mammal. *See* Mammalia
Mammalia, 15–17, 45, 68, 83, 85, 93–95, 110,
    134, 146, 179, 192, 200, 210, 214, 225, 229,
    264–265, 282, 284, 290–291, 305, 307–311,
    316–317, 320–322, 325–326, 355, 364, 366,
    368, 374, 378, 387, 392–393, 398, 430, 434,
    458, 485, 498, 501, 514, 526, 535, 592–593,
    595
mandrill. See *Mandrillus sphinx*
Mandrillus sphinx, 93, 110
Manega, P., 179
mangabeys, 372, 542
Manson, J. H., 593–595
Maple, T. L., 518
Marchant, L. F., 370
Marchi, D., 254
Marivaux, L., 68, 69, 117
Marks, J. M., 28
Marler, P., 512, 524, 530, 536, 537
marmoset, 19, 67, 69, 363
marrow, 9, 291, 325–326, 332
Mars, 7
Marshall, A. J., 267, 454
Marshall, E. R., 512, 514

Marshall, J. T., 512, 514
Marshall, L. J., 7
Martin, L., 95, 116
Martin, R. D., 16, 17
Marvan, R., 456
mast: definition, 284; occurrence, 270
masturbation (= onanism), 502; bonobo, 459;
    chimpanzee, 433; human, 502; orangutan, 412
Matata, 373, 564
mating, 485; bonobo, 451, 453–457, 547;
    chimpanzees, 315–316, 319, 424, 430, 433–437,
    439–440, 485; fossil species, 152, 497–498, 584;
    gorilla, 469, 479–483; human, 501, 504, 547;
    Hylobatidae, 485, 491, 496–498, 517; orangutan,
    403, 405, 412–415, 521; primate, 325, 434
Matsumura, H., 82
Matsuzawa, T., 370
maximum parsimony analysis, 124
Mazàk, V., 162
McCrossin, M. L., 83, 93, 116, 218, 219
McGrew, W. C., 370, 585
McHenry, H., 128, 148
McKee, J. K., 158
McPherron, S. P., 332
meat, 4, 10, 149, 172, 290–293, 305, 312–313,
    315–316, 320–327, 332, 450, 458, 460; control
    system, 10–11, 314, 316; providers of, 6, 8–9,
    304, 313, 504; sharing, 8–10, 307, 311–313, 426
megafaunal analysis, 147
Mein, P., 97–98, 102
Meissner's corpuscles, 346
memory, 360, 381–382, 392, 556, 573
menstrual cycle, 411
Mentawai gibbon. See *Hylobates klossii*
Menzel, C., 390–391
Menzel, E. W., 534–535
Mercury astronauts, 379
metalinguistic phenomenon(-a), 558
metaphor, 393, 592; mitochondrial Eve, 505; not an
    evolutionary mechanism, 505; selfish gene, 505
Michael, 553, 554
Micropithecus clarki, 75, 83; as ancestor to
    Victoriapithecus, 82; body mass, 215; body size,
    82, 89; craniodental features, 82–83, 117;
    positional behavior, 215; postcranial features, 215
Micropithecus songhorensis, 83
Mike, 421
Miles, H. L., 555, 556, 563
Milk, H., 553
Miller, G. S., 73, 394
mind, 7, 214, 355–356, 379, 386, 388–389,
    392–393, 498, 507, 550, 571, 573, 585–586
mineral, 61, 143, 165, 262, 270

mirror, 368, 388; social facilitator, 387
mirror-image recognition, 386–388, 592
mirror-image stimulation (MIS), 386–388
mirror neuron: audiovisual, 578; critique, 578;
    human, 365, 578–579; macaque, 365, 578;
    mechanism, 365; multimodal, 578; single-cell
    recording, 365, 579
MIS. *See* mirror-image stimulation
Mitani, J. C., 281, 405, 413–414, 461, 512–513,
    519–520, 525
Mitchell, R. W., 389
Mitteilungsbedürfis (defined), 579
Mittermeier, R., 599
Mizuhara, H., 463
models, theories, and hypotheses: African
    hybridization and replacement, 181; African
    replacement (= little out-of-Africa), 180–181,
    185; androcentric chimpanzee society, 424, 426,
    450; assimilation, 181; behavioral, 2, 4, 226;
    caudal reduction in *Proconsul*, 216–217, 220;
    chimpanzee, 427; Darwin's musical or prosodic
    protolanguage, 517, 582; ecological, 4; on
    emergence of bipedalism, 226; evolution of
    human language, 373, 517, 571, 576–580, 582;
    evolution of speech, 365, 573–574, 582; gestural,
    576–579; Horr-Rodman's orangutan reproduc-
    tive strategies, 411; Jackendoff's, 580;
    Johanson-White, 147–148; Keith's troglodytian,
    226–227; knuckle-walking, 206, 216, 223–224,
    228–229, 348–349, 351, 353; Lucas's cooking,
    293; MacKinnon's orangutan male sexual
    strategy, 409–410; Morton's hylobatian,
    226–227; multiregional evolution (= regional
    continuity), 180–182; nontroglodytian, 246;
    orangutan society, 404, 421; pongid versatile
    climbing, 228, 348; scientifically informed, 2, 4,
    577; speculative, 4; Tuttle's hylobatian, 226,
    228–229; unreliable replication of evolutionary
    history, 370, 499; use of extant creatures in, 3–4,
    17, 69, 82, 86–89, 93, 111, 125, 136–137, 148,
    176, 219–220, 223–224, 227, 291, 294, 356,
    574, 584; vocal mechanisms of fossil hominids,
    574, 577; Washburn's knuckle-walker, 226–228;
    Wrangham's demonic male, 10, 593–595
module, 355–356
Moeripithecus markgrafi, 66–67, 70, 73, 81;
    hypodigm, 70
Moja, 550, 552, 553
molding, 548–551, 553–556
molecular biology, 28, 60, 137
molecular evolutionary clock, 17; accuracy
    perturbed by different generation times, 138;
    calibrated with arguably dated fossils, 138; of

molecular evolutionary clock *(continued)*
Hominoidea, 137–138; mitochondrial DNA
(mDNA), 138; modelers sustain markability
of, 138
monochromatism, 38
monogamy, 506; Aves, 485; defined, 485; female as
key, 485; hylobatid, 485, 487, 489, 492,
495–496, 498; Mammalia, 485; obligate, 485;
primate, 7, 9–10, 498–499, 503
monomorphy, 498
morality, 115, 589, 597–598, 600–601
Morgan, E., 8
Mori, A., 543–544
Moroto, Uganda, 81, 84–86, 217
Morotopithecus bishop, 85–86, 120, 122–123, 217
morphemes, 570, 579, 581, 583
morphology, 60, 63, 68–69, 82, 87–89, 93, 99,
110–113, 117, 119, 149, 208, 222–223, 227, 249,
264, 266, 292–293, 331, 334, 345, 354, 570, 574,
584
mortality, 403, 422, 443, 466, 500, 591; infant,
435, 442, 445, 462, 474, 477
Morton, D. J., 137, 226–227
mortuary practice, 3, 179
Moscone, G., 553
motivation, 380, 408, 526, 579, 589
Mottl, M., 102
mountain gorilla. See *Gorilla beringei beringei*
mouth, 134, 262–263, 311, 317, 336–337, 339, 342,
382, 387, 399, 407, 410, 412, 431, 518, 522,
530–531, 541, 545, 553, 578
movie, 558–559
Moyà Solà, S., 97, 124–125, 132
Mulika, 563, 564, 565
Müller's gibbon. See *Hylobates muelleri*
multiple microsatellite loci analysis, 464
murder, 10, 429
muriquis. See *Brachyteles*
muscle: abdominal, 584; abductor pollicis brevis,
347–348; digital flexor, 196, 201–203, 254, 348,
353; extensor pollicis brevis, 349; first dorsal
interosseous, 349; flexor carpi ulnaris, 254; flexor
digitorum profundus, 228, 347–349; flexor
digitorum superficialis, 210–211, 228, 259, 349,
353; flexor pollicis longus, 346–348, 351–352;
gluteus maximus, 204, 230–231, 239–241, 254;
gluteus medius, 204, 230–231, 236, 239–240,
254; gluteus minimus, 204, 230, 231–232, 236,
239, 254; ischiofemoralis, 239, 241; latissimus
dorsi, 203, 205; lesser gluteal, 236, 239; manual
terminal phalanx, 348; one-joint, 234–235;
peroneus longus, 253; quadriceps femoris, 230,
232; soleus, 231, 234, 240–241; thumb

(= pollical), 346, 348; triceps surae (= calf), 196,
199, 213, 230, 232; two-joint, 234–235
Museum Darwinianum of Moscow, 378
music, 356, 514, 517, 558, 576–577, 581–582, 584,
589
musical protolanguage, 576–577, 583
musician, 369
mutualism, 8, 402, 419, 421–422, 432, 453, 472,
584, 591
Myanmarpithecus yarshensis, 62, 65, 67

Nacholapithecus kerioi, 75, 81, 90, 92, 220, 224;
body mass, 218; craniodental features, 128;
positional behavior, 86, 218; postcranial features,
86, 92, 218
Nadler, R. D., 480
Nagatoshi, K. R., 41
Nairn, A. E. M., 147
Nairobi National Park, Kenya, 343
Nakalipithecus nakayamai, 78, 81, 87–88
Nakatsukasa, M. Y., 216
Napier, J. R., 159–160, 345, 353
Natural History, 4
natural selection, 29, 183, 190, 203, 226–227, 366,
407, 425, 429, 442, 595
nature and nurture, 356
negative sign, 550
Nelson, S. V., 110–111
Neofelis diardi, 199, 296–297, 406
Neogene, 89, 95–96, 129, 331; defined, 63
neoteny, 9, 51
nest (= tree platform), 26, 35, 229, 287, 294–295,
296–302, 307, 320, 399, 409–410, 434, 449,
454, 460, 465, 470, 520–521, 535, 541; indicate
group composition, 295, 300, 443, 461, 539,
544; indicate population distribution and
density, 295, 461, 463
neuroimaging, 365–366
New, J., 265
Newman, R. W., 245
Newmeyer, F. J., 578
Newton-Fisher, N. E., 425
Ngorora molar (KNM-BN 378), 128
Nigeria chimpanzee (= *Pan troglodytes vellerosus*).
See *Pan troglodytes*
Nim Chimpsky, 556, 561
Nishida, T., 282, 418–419, 422, 427, 439, 443
Nishihara, T., 286, 461
Nissen, H. W., 416, 524
Nomascus spp., 35, 38, 495, 516
Nomascus concolor, 26, 32, 267, 324, 495–496, ·
516
Nomascus gabriellae, 32, 516

Nomascus hainanus (= Hainan black crested gibbon), 32, 36, 495–496, 516
Nomascus leucogenys, 32, 267, 516–517
Nomascus nasutus, 32
Nomascus siki, 32
nomina, 34, 156
nominal, 550, 554, 588
nonhuman, 7, 15, 19, 26, 137, 231, 305, 324, 326, 331, 364, 370, 379, 383, 385, 389, 392–393, 442, 498, 502, 505, 530, 570–571, 573–574, 580, 582, 585, 587–589, 590; as amoral being, 597–598
Northern white-cheeked gibbon. See *Nomascus leucogenys*
Notman, H., 526
Nova, 554
Nozawa, M., 29
Ntologi, 319
nuclear Armaggedon, 9
nuclear DNA analysis, 3, 29, 179, 183, 421, 525
nucleotide, 29
nut-cracking: capuchin, 344–345; chimpanzee, 337, 339–341, 546
nutrient, 262–263, 281, 288, 291, 315
Nyanzapithecus harrisoni, 85
Nyanzapithecus pickfordi, 83, 218–219
Nyanzapithecus vancouveringorum (= *Proconsul [Rangwapithecus] vancouveringi*), 75, 81, 83, 85
Nycticebus coucang, 93
Nyota, 563

observational learning, 390, 392–393, 565, 570, 589
Old World monkey. See Cercopithecoidea
Oligopithecus rogeri, 66, 73, 81
Oligopithecus savagei, 62, 66, 72–73, 81
olive baboon. See *Papio anubis*
Oliver, 33
Omedes, A., 544–545
operant conditioning, 379, 573
orangutan. See *Pongo abelii*; *Pongo pygmaeus*
orca (= *Orcinus orca*). See cetacean behavioral capacities
order (Linnaean), 15, 27, 34
Oreopithecidae, 86, 95, 116
Oreopithecus bambolii, 78, 83, 117–119, 131–132, 220, 222–224
organism, 1, 11, 15, 28, 290, 390, 598–599
organismal biology(-ist), 66, 599
Orrorin tugenensis, 87; classification, 162; geochronological age, 79, 126, 141; hypodigm, 127–128; phylogeny, 149–150, 184; positional behavior, 87; postcranial features, 87, 260
orthographic rules of classification, 34

Osborn, H. F., 4
Osborn, R. M., 462
Otavipithecus namibiensis, 78, 86, 94
Ouranopithecus spp., 104
Ouranopithecus macedoniensis (= *Graecopithecus freybergi*), 88; body mass, 105; craniodental features, 105–107, 112; diet, 106–107; geochronological age, 78, 105; habitat, 105; phylogeny, 117–121, 126, 132; as proximate ancestor of *Australopithecus*, 105, 117–119; sexual dimorphism, 106
Ouranopithecus turkae, 78, 106–107
ovulation, 54, 407, 414–415, 437, 439, 442, 456, 487

Paidopithex rhenanus, 77, 220
pair-bond, 7, 9, 284, 487, 496, 498, 505–506
paleoanthropology (-ist), 3, 61, 117, 138, 156, 160, 162, 170, 226, 292, 294, 356, 374, 500, 574, 591, 593; multidisciplinary, historical science, 2, 60; overrun by molecular biology, 137; status 4; ultimate goal, 115
paleoenvironment, 2, 63, 219
Paleogene, 63–64
paleontology(-ist), 4, 17, 60–61, 129, 131–132, 137, 143, 180, 577, 599
paleoprimatology(-ist), 60
Palmer, A. K., 92–93
palpate, 266
palynology, 60, 95, 113, 292
Pan, Y., 223
Pan spp.: distribution, 46, 48–49, 460; ecological versatility, 46; enamel thickness, 45; furcation of *P. paniscus* and *P. troglodytes*, 46, 138; taxonomy, 26–27, 30, 46
Panbanisha, 334, 563, 565
Panidae, 29–30, 115, 125
Paninae, 26, 30
Pan paniscus, 26, 32, 46, 48–49, 355, 397, 416, 443, 482–484, 595, 600; adult band, 448; affinity between females, 445, 448, 450, 454; affinity between males, 445, 448; affinity between males and females, 445, 448, 452–454; aggression, 444, 446, 454; all-male band, 450; allogrooming, 444–445, 448–451, 454, 545–546; birth interval, 455–456; bisexual dyads, 443, 448; body mass, 26, 47; brain asymmetry, 369, 372; brain size, 46, 362, 376; close-calling, 544, 547; cognition, 381, 384–385; cohesiveness, 398, 46, 448, 452, 487; consortship, 457; craniodental features, 50–52; cross-modal ability, 385; diet, 263–265, 275, 282–283, 298, 322–323, 448–449, 453, 455, 458; distribution, 46; dominance, 445–446, 450–456; estrus, 454, 457;

Pan paniscus (continued)

female choice, 456; female coalition, 446, 450; female genitalia, 52; female influence on males, 450; food sharing, 305, 319–324, 446, 450–451; foraging, 282, 452–453; foraging party, 426, 444, 453, 542; greeting, 459, 544, 547; habitat, 46, 452–453; interfemale bonds, 448; intragroup relations, 448; knapping skill, 333–334; Lake Tumba, 443, 449; Lilungu, 94, 320, 543–544; linear hierarchy, 451; lodging, 204, 297–298; Lomako, 93, 278, 282–283, 297–298, 321–323, 449–450, 452–453, 456–459, 542; lone male, 448; Lui Kotale, 94, 321; Lukuru, 278; male-female dyad, 454; male genitalia, 54, 484; male philopatry, 453; male rump touching, 459; meat-eating, 305, 320–322; mixed band, 444–445, 448; model of prehuman nature and society, 10, 448; mother-son bond, 446–448; neoteny, 51; nonvocal communication, 542–543, 545–547 (*see also* ape language study); party size, 443, 449, 452–453; peering, 545–546; perianal tuft, 47; perineal swell, 52, 454–458; pigmentation, 47, 55; positional behavior, 46, 199, 204, 208; postcranial features, 21, 46, 52, 206, 208–210, 212–213, 243; predation, 265, 305, 320–322; range size, 278; reconciliation, 431, 458, 590; referential signaling, 523, 546–547; reunion behavior, 444, 447, 449, 457; self-recognition, 388; sexual behavior, 434, 443–445, 449, 453–459, 483, 485, 547; sexual dimorphism, 47, 52, 362; social cognition, 393–394; social relations and organization (= society), 443–445, 448–450, 452–458, 505; tool behavior, 332–335, 341, 343–344; transfer of nulliparae, 447; unit-group (= community) size, 426, 444, 449; unit-group encounters, 449; unit-group structure, 447–448; vocal flexibility, 542–543; vocalization, 449, 518, 542–545; Wamba, 93, 278, 282, 297, 320–323, 444–449, 452–458, 543, 545–546; Yalosidi (Ikela, Iyoko), 282–283, 297, 448

Pan sapiens, 27

Panthera leo, 296, 298, 425

Panthera pardus, 296, 298, 301, 316, 406, 425–426, 573

Panthera tigris sumatrae, 199

Pan troglodytes, 416, 578, 585; activity pattern, 287, 311, 430, 530; adolescent sterility, 438; age of first reproduction, 433; aggression, 279, 306, 316–319, 341, 343, 394, 417, 420, 422–432, 435–439, 441–444, 446, 448, 454, 528–533, 591–595; alliance, 314, 420, 431–432, 441, 526; allogrooming, 312, 317–318, 419, 421–424, 426,

430–432, 436, 438, 440–441, 529, 531, 585–586; alpha male, 311, 314, 319, 421–422, 437–441, 446, 485, 525, 528; analogical reasoning, 381–383; androcentric unit-group, 424–426, 450; attempt to humanize, 34, 547–548; bilocal residency, 595; birth interval, 427, 429, 433, 455; bisexual adult band, 416; body mass, 9, 46–47, 89, 110, 542; boundary patrol, 422, 426; captive, 33, 264, 306, 312, 322, 332, 337, 341, 369, 371–372, 378–394, 427–428, 431–434, 440, 451, 523–530, 534–535, 541, 545–546, 551, 559, 588, 590; central, 30, 33, 287–289, 338–339; cognition, 10, 339, 355, 363, 373, 379–392, 393–394, 416, 419, 425, 552, 595, 598; cohesiveness, 298, 421–422, 452, 487; competition, 10, 275, 279, 281, 284, 286–290, 296, 299, 308, 313, 323–324, 337, 379, 383, 423, 425–426, 429–430, 434, 436–437, 439, 441, 453, 484, 531, 535, 594, 596; consortship, 298, 313, 434–439, 457, 484–485; conspecific killing, 309, 317–319, 422–423, 429, 442, 479, 595; cooperation (*see* cooperation); core area, 424, 427–429; cranial capacity, 362–363, 376–377; craniodental features, 5, 47, 50–52, 127–128, 148, 288; cultural being, 57, 553, 584–588; development, 57, 173; diet, 82, 111, 213, 263–265, 274–279, 281–282, 286–290, 292, 298–299, 308–310, 315, 323, 332, 336, 371, 417, 425–428, 430, 453, 526, 531, 534; dominance, 419, 429, 441, 454; drumming, 417, 527, 529; eastern, 30, 266, 279, 282, 286, 290, 298–299, 307, 339, 418, 422, 425–427, 441, 450, 525–528, 533, 544; estrus, 313, 315, 345–436, 444; fecundity, 433, 435–436; female choice, 433–440, 485; female genitalia, 54; female transfer, 422–424, 430, 438; foraging, 282, 286–290, 298–299, 425, 427, 453; greeting, 428; habitat, 46, 264, 275, 453, 596; hunting, 10, 93, 274, 305–316, 595; intention, 306, 312, 386, 389; kin, 430, 440–441; lodging, 297–299, 461; male band, 424; male genitalia, 54, 484, 499; male mating behavior, 435, 484–485; mate guarding, 437; mother/offspring ties, 279, 417, 419; neural circuitry for social cognition, 393–394; Nigeria, 30, 339; nonvocal communication, 312, 371–373, 385, 431–432, 522–524, 528–534, 536–537, 543, 545–565, 577; nursery group, 416, 424; open mouth-to-mouth contact (kiss), 319, 431, 532; perianal tuft, 47; perineal swell, 54, 437; phylogeny, 86, 88, 138; pigmentation, 47; play, 343, 378, 421, 531, 541, 551; positional behavior, 46, 199, 209, 204–213,

212, 227–228, 235–238; postcranial features, 21, 24, 26, 52, 202, 204–214, 212, 234, 243, 247, 254–257, 345–346, 348; predation (*see* chimpanzee predation); presenting, 532; prototype for human biological engineering, 379; range, 453; reassurance, 483, 533; reconciliation, 431–432, 532, 590; restrictive association, 319, 435–436, 440, 455; reunion behavior, 431, 434–444, 529; scavenging, 316; self-concept, 387–388; self-recognition, 386–388, 592; sexual behavior, 433–440, 444, 455–456, 481–482, 498, 528, 530, 532–533; sexual dimorphism, 47, 50, 251, 362, 376–377, 438, 451, 484, 560–561; sharing, 279, 305, 311–315, 323–324, 451; sleeping associations, 298–299; social cognition, 393–394; socialization, 420; social relations and organization (= society), 416–432, 450, 452–453, 461, 475, 498–499, 505, 536, 594; territorial behavior, 422, 424; tool behavior, 332–342, 335–345; triadic awareness, 528; unit-group (= community), 418–420, 423–425, 444, 525; vocal control, 437, 525, 528; vocalization, 312, 423, 431, 508, 518, 524–537, 541–545, 547, 582; weaning, 455, 528; western, 30, 46, 286, 306–307, 337, 340, 344, 397, 423–424, 437–438, 441, 527, 529

Panzee [a.k.a. Panpanzee], 390, 391

Papio spp.: hunting, 22, 305; lack Herschl's gyrus, 372; ventral digitigrade hand posture, 205; vocalization, 516

Papio anubis, 309, 343

Papio cynocephalus, 94

Papio ursinus, 343

Paranthropinae, 133, 140, 162, 583

Paranthropus spp., 133, 156, 158, 173, 184, 351, 375; adaptive radiation, 149; ancestor, 151, 153–154, 156; brain size, 151; craniodental features,151, 156, 158–159, 291, 369, 370, 576; geochronological age, 155, 157; postcranial features, 128, 151, 158, 353

Paranthropus aethiopicus (= *Australopithecus aethiopicus*), 140, 142, 151, 154, 156, 162, 169, 357; KNM-WT 17000 (black skull), 155–156, 159

Paranthropus boisei, 142, 153, 156, 162, 357; craniodental features, 154, 158, 292; diet, 292; geochronological age, 140, 153, 169, 352; OH 5 (type), 157; with Oldowan artifacts, 352; positional behavior, 353; postcranial features, 353

Paranthropus crassidens (= *Paranthropus robustus*), 156

Paranthropus robustus, 140, 142, 153, 156, 162; craniodental features, 151, 292, 352, 357, 498; with Developed Oldowan artifacts, 352; diet,

293; geochronological span, 140, 151, 169; hand bones from Olduvai Gorge might belong to, 353; ichium, 353; positional behavior, 352; postcranial features, 253, 260, 352–353

paraphyletic taxon: defined, 115; Panidae, 115

Parapithecidae, 64–65, 67, 69, 72–73; as ancestors of Cercopithecoidea, 72; dentition, 72; platyrrhine characteristics, 72

Parapithecoidea, 69, 72

Parapithecus (= *Simonsius*) *grangeri,* 72, 81

Parapithecus fraasi, 72, 81

parasites, 261–262, 294–295, 302

Paraustralopithecus aethiopicus. See Paranthropus aethiopicus

parental care, 9, 435

parental investment, 407, 479, 558

Parr, L. A., 440

particles, 290, 293

parturition (= birth): chimpanzee, 443, 552; gorilla, 467, 469, 474, 477; orangutan, 399, 521; siamang, 490

Passingham, R. E., 357, 366

Patel, B. A., 218

paternity, 7, 413, 415, 434, 436–437, 441–442, 453, 476, 481, 501–502; testing, 413, 415, 434, 441–442

paternity test, 415, 441–442

Patterson, B., 156

Patterson, F. C., 553

Patterson, F. G. P., 553–554, 563

Patterson, N., 138

peaceful postconflict interactions (PPCI), 432

pectin, 263

pedophilia, 502

Peirce, C. S., 571–572

Peking man. See *Homo erectus*

pelvis, 231, 256, 259–260, 533; bodily position, 230, 232, 236, 239; defined, 230; pubic symphysis, 230; pubis, 230

penis: bonobo, 52, 54, 484; chimpanzee, 54, 484, 533; gorilla, 54, 483; human, 33, 54; orangutan, 412–413, 484

perception, 324, 359, 365, 380, 382, 416, 528

perianal tuft: bonobo, 47, 51; chimpanzee, 47; gibbon, 487; gorilla, 47

perineal swell (= estrous swelling, sexual swelling): bonobo, 52, 54, 451, 455–458; chimpanzee, 54, 433–434, 437, 440, 455–456; gorilla, 477, 483; orangutan, 407, 414; white-handed gibbon, 487

perineal tumescence. *See* perineal swell

perineum(-a), 47, 317, 410, 413, 428, 456

peripatetic, 199, 418, 492

peripheralization, 6, 491–495

personality, 506, 591–592
pest, 151, 294, 582
petalia, 369
Peterson, D. A., 10, 595
Phenacopithecus krishtalkai, 62, 65, 68
Phenacopithecus xueshii, 62, 65; body mass, 68;
    hypodigm, 68
phenol, 271, 281, 288
phenology, 284–285
Phileosimias brahuiorum, 62, 65, 69
Phileosimias kamali, 62–65, 69
philopatric, 403, 415, 453, 476, 498–499, 593
phloem, 270
phoneme, 549, 569–570, 580–581
phonology, 383, 562, 570–571, 579–583
photograph, 382, 440, 536, 560, 572
phylogenetic classification, 115
phylogenic scheme, 3
phylogenic tree, 115
physical anthropology(-ist), 356
physiology, 26, 245, 281–282, 416, 428, 430, 438,
    589–599
Piaget, J., 382
Pica pica (= magpie), 387
Pickford, M., 82–83, 85–86, 89, 95, 128
pidgin/creole transition, 580
pidgin sign language, 549
Pierolapithecus catalaunicus, 77, 80, 100–101,
    220–221
pig. See *Sus scrofa*
pig Latin, 554
Pika, S., 541
Pilbeam, D. R., 84–87, 89, 102–103, 105, 128,
    134–135, 137
pileated gibbon. See *Hylobates pileatus*
Pili, 550
Piliocolobus rufomitratus tephrosceles. See Colobinae
piloerection, 43, 522, 572
pinniped, 205–206
pitch, 369
Pithecanthropus (= Javanese *Homo erectus*), 132
Pithecia, 92
placenta, 355, 375
Plackendael Wild Animal Park, 451
Planet of the Apes, 600
plantation, 264, 287, 416
plantigrade, 199, 203–204, 212, 216, 218, 226–227,
    246, 253, 421
plants, 6–7, 9, 61, 93, 100, 113, 129, 246, 262,
    264–265, 275, 279, 281, 284, 292–293,
    300–301, 322–323, 325–326–337, 403, 504;
    availability of nutriments, 262–263, 282, 289;
    self-protective toxins, 262, 266, 270–271, 288;

variation in parts, 22, 98, 113, 262–263, 266,
    274–275, 279, 281, 290–291, 313, 453
Platodontopithecus jianghuaiensis, 76, 81, 98–99
Platyrrhini, 20–21, 64, 67, 72–73, 97, 215, 369
play, 286, 343, 378–379, 394, 398–400, 406, 421,
    441, 449, 457, 465, 486–490, 492, 496, 502,
    522–523, 530–531, 539, 541, 544, 548–549, 551,
    554, 556
Plesiadapiformes, 16–17
Plesianthropus transvaalensis (= *Australopithecus
    africanus*), 156
Plesiopliopithecus lockeri, 76, 97
Pliopithecidae, 69, 72–74, 115, 215; members, 64,
    66, 70, 76, 95–99, 117
Pliopithecinae, 117; members, 76, 97, 99
Pliopithecus spp., 80, 95–100, 102, 116, 217,
    219–220
Pliopithecus (= *Plesiopliopithecus*) *priensis*, 76, 80, 97
Pliopithecus antiquus, 76, 96–98
Pliopithecus krishnaii, 62, 65, 98
Pliopithecus piveteaui, 97
Pliopithecus platyodon, 76, 97–98
Pliopithecus vindobonensis, 76, 80, 96–98, 219–220
Pliopithecus zhanxiangi, 76, 99
plus-4 dental pattern, 52; defined, 42
poaching, 32, 443, 462, 466, 473, 538
poison, 327, 591
polar bear (*Ursus maritimus*), 134
politics, 10–11, 58, 214, 314, 355, 375, 432,
    442–443, 459, 466, 531, 588–589, 596,
    600–601; *Homo habilis*, 353; *Homo sapiens*,
    345–346, 348–349; Machiavellian, 11; *Proconsul*
    spp., 216
Pollick, A. S., 545
polyandry, 487–488, 503
polychromatism. See Hylobatidae spp.
polygamy, 152, 463, 485, 495–496, 498, 506
polygyny, 63, 70, 324, 487, 495–496, 498–499,
    503
polysaccharide cellulose, 263
Pondaung Formation, 62, 66–67
Pondaungia spp., 66–68
Pondaungia cotteri, 62, 64, 66–67
Pondaungia savagei, 62, 66–67, 214
Pongidae, 30
pongid apes. See Pongidae
Pongo spp., 26–27, 46, 284–286, 407–408, 598;
    aggression, 401–403, 409; allogrooming,
    399–400, 405–406, 409–410, 421, 523, 555;
    bimaturism of male, 398; birth interval, 399,
    407; brain size, 362–364, 369, 372–373;
    cannibalism, 319–320; captive, 33, 43, 208, 333,
    371, 380, 398, 406, 517–518, 521, 523–524,

554–555; cheek flange, 42–43, 270, 297, 398, 407, 410, 414–415, 518; cognitive abilities, 333, 355, 380–381, 383–386, 388, 392, 555; cohesiveness, 402, 405, 487; competition, 402–403, 407–408, 410–411; concealed ovulation, 54, 407, 414; consortship, 297, 405, 408–414, 484; craniodental features, 43–45, 47, 52, 108, 408; diet, 44–45, 93, 200, 265, 267–268, 270–272, 274–275, 285, 292, 324, 342, 398, 400, 402–404, 407–408, 412, 414; display, 43, 200, 409, 520–522; distribution, 42; dominance, 402–403, 519, 521; emigration, 404, 415; endangered, 42; epigamic features, 42–43, 398, 522; estrus, 407–408, 410, 484; fat metabolism, 271; fecundity, 414; female choice, 409, 411, 414–415; female genitalia, 54, 413, 484; flanged male, 297, 398, 403, 414–415; foraging, 200, 270, 285, 402, 405, 407; genetic diversity, 42; habitat, 36–37, 42; learning, 271, 399–400; life history, 398–399; lodging, 297–298, 399; long call, 43, 398, 401, 405–406, 410–411, 413, 518–522; longevity, 399; male genitalia, 54, 484; memory, 402; nonvocal communication, 413, 517, 519–524, 536, 553–555; ontogeny, 399, 405; paternity, 413, 415; phylogeny, 26, 28, 30, 42, 119, 139; pigmentation, 43; play, 398–400, 406, 556; positional behavior, 199–201, 203, 206–207, 207–208, 228, 237–238, 346–347, 371, 521, 524; postcranial features, 45, 52, 94, 193, 201–203, 207–208, 210, 212, 222–223, 228, 257, 265, 346, 348; prey, 297; ranging, 114, 400–404, 406; rehabilitant, 411–412, 518, 521, 523, 554; sexual behavior, 201, 406–407, 409, 413–415; sexual dimorphism, 10, 42, 45, 47, 50, 113, 251, 362, 407, 484; sharing, 399–400; social behavior, 297, 397–402, 404, 406, 416, 421, 423, 524; social organization, 297, 398, 400–401, 403–404, 406; territoriality, 401; tool behavior, 297, 332–333, 335, 342; versatile climbing, 199–201, 203, 212, 221, 228, 285, 348, 416; vocalization, 43, 517–523, 528, 537, 547–548
Pongo abelii (Sumatran orangutan). See *Pongo* spp.
Pongo pygmaeus (Bornean orangutan). See *Pongo* spp.
Pontzer, H., 213
population, 29, 32, 42, 46, 54, 57–59, 98, 102, 152–153, 165, 173, 176, 179–183, 185, 266–267, 271, 274–275, 284–285, 287, 292, 294–295, 308, 311–312, 315, 318, 335, 337, 342–343, 362, 369–371, 387, 397, 401, 403, 406, 415–416, 424, 442–443, 460, 462–467, 469, 473, 475, 477, 482, 485, 496–497, 500, 508–509, 511, 519–520, 525, 535, 585, 588–589, 599

positional behavior. See *individual taxa*
positron emission tomography (PET), 358
posterior superior temporal sulcus, 374
Posthuma, D., 361
postorbital bar, 16–17
postorbital constriction: chimpanzee, 50; gorilla, 50; *Homo erectus*, 173; human, 56; orangutan, 44; *Sivapithecus*, 109
posture. See *individual taxa*
potassium, 61, 131
Povinelli, D. J., 387
Praeanthropus africanus, 86
Prasad, K. N., 136
predation, 9, 93, 199, 229, 261, 265, 281, 297–298, 302, 306, 308, 316, 341, 375, 409, 411, 425, 452, 466, 472, 497, 514, 517, 524, 591
predator, 6, 108, 151, 189, 213, 218, 261, 265–266, 284, 294–296, 298–300, 303, 305–306, 308, 311, 326, 335, 341, 394, 405, 407, 425, 427, 486, 492, 510, 514, 521, 524, 529, 535, 537, 573, 591, 593
Premack, A. J., 385, 556–558, 563
Premack, D., 389, 392, 556–558, 563
preposition, 557–558
presenting, 391, 411–413, 532
primate culturology, 588
Primates, 2, 7–8, 35; definition, 15–16; earliest Eocene, 16–17; emergence of order, 15, 18–19; features, 16–17; handedness, 370; members, 2, 15, 17, 20–21, 25, 27, 30; revision, 16; suborders, 19
Primate Visions, 11
primatology, 60, 72, 295, 356, 394, 418, 587, 599
Prince Geoffrey, 390
Prince John, 389
Prince Richard, 389
Princess, 554
probe, 2, 298, 307–308, 332, 335–338, 341–343
problem, 61, 87, 108, 125, 143, 162, 172, 183, 290, 302, 325–326, 333, 356, 375, 380, 382, 389, 391, 394, 417, 517, 560, 562, 580, 582–583, 585, 599
problem-solving, 335, 378–380, 382, 386, 389, 391, 394, 561–562
Proconsul spp., 82–86, 91, 92, 110, 116, 118–123, 216–218, 220–221, 223–224, 227
Proconsul africanus, 83, 86, 89, 101
Proconsul gordoni, 75, 83, 92. See also *Rangwapithecus gordoni*
Proconsul heseloni, 75, 81; body mass, 89; craniodental features, 89–91; KNM-RU 2036, 89; KNM-RU 7290, 89; positional behavior, 216; postcranial features, 216–217, 220, 350
Proconsul major (= *Ugandapithecus major*), 81; as ancestor to *Gorilla*, 86; craniodental features, 83–84, 91; diet, 86; hypodigm, 84; as

Proconsul major (continued)
knuckle-walker, 75, 81, 84–85; positional behavior, 216; taxonomic affiliation, 217
Proconsul nyanzae, 75; body mass, 89, 101, 216; positional behavior, 216–217
Proconsuloidea, 349; arguably linked to extant African apes, 82; members, 75, 116, 223
productivity, 294, 504, 556, 558
Prohylobates spp., 72
Prohylobates macinnesi, 85
Prohylobates tandyi, 72
Propliopithecoidea, 64, 66, 70, 73, 82
Propliopithecus ankeli, 66, 70
Propliopithecus chirobates, 66, 70
Propliopithecus haeckeli, 66, 70
propulsive lever, 234–235
propulsive strut, 198, 234–235
proscriptions, 502
Prosimii, 19–21, 60, 63, 66, 68–71, 73, 97, 226
prosody, 571
protein, 3, 26, 137, 172, 179, 262, 265, 267, 271, 281–282, 288–289, 322, 325, 453
Proteopithecidae, 65, 72
Proteopithecus sylviae, 69, 72–73, 214–215
protohominid, 227–229
protolanguage, 576–580, 582–583
protopongid, 229, 277
provisioner, 7, 9; monogamous, 9
provisioning, 462, 504, 548; bonobo, 320, 322–323, 445–446, 448–449, 452; chimpanzee, 305, 310–312, 315, 332, 417–419, 421, 423, 524, 535, 537, 543; effect, 305, 310–312, 341, 418; purpose, 305, 418
proximity-matrix analysis, 423
Pruetz, J. D., 275, 307, 341
Pseudorca crassidens (= false killer whale), 387
psycholinguist, 356
public policy, 598
pudenda, 446, 482
pulp-to-pulp opposition of the thumb and index finger: capuchin, 346; human, 345–346, 349; Old World monkey, 346
punctuated equilibrium: defined, 185; human career, 185
Pusey, A. E., 424, 429
pygmy chimpanzee. See *Pan paniscus*

Qatrania wingi, 62, 65, 69, 72–73
quadrilaterals of support, humans vs. quadrupeds, 232–234
quadruped, 46, 63, 86, 100, 199, 201, 204–205, 211–212, 214–221, 223–224, 227, 229–232, 234–235, 239, 244, 249, 351, 522, 532

Quallo, M. M., 391
Quatrania fleaglei, 72

race, unsupportable genomically, behaviorally, and phenotypically, 31–32, 596–597
racism(-ist), 4, 34
Rae, T. C., 116
rainfall, 111, 300–301; predictor of folivory and frugivory, 267
Ramapithecus spp., as Miocene contender for earliest known hominid, 129, 133–138
Ramapithecus hudienensis, 136, 166; affinities with *Lufengpithecus lufengensis,* 136
Ramapithecus punjabicus (= *Sivapithecus sivalensis*), 108, 160; as biped, 135; cranial features, 134–136; credibility weakened by biomolecular evidence, 137; as first hominid, 135–136; hypodigm, 135; as tool user, 135
Rangwapithecus sp., 75, 83, 86, 116, 120, 123
Rangwapithecus gordoni, 83, 92
rape. See forced copulation
Rasmussen, D. T., 73
rational behaviorism, defined, 390
reading, 5, 391, 442, 554, 558, 560, 577, 591
Reaux, J. E., 528
reciprocal altruism, 432
reconciliation, 431–432, 471
reconstruction: misleading term, 3; phylogeny, 2–3
recursion, 389, 393, 557, 564, 570–571, 576
redundancy, 15, 59, 556, 587
referent, 383, 523, 528, 535, 546–549, 552, 555, 561, 563, 570, 572–573, 580
rehabilitant, 411–412, 518, 521, 523, 554
Reichard, U. H., 487
religion, 596–597
Rendall, D., 526
Reno, P. L., 9
resin, 21, 73, 264
restoration, 3, 575
reticulated python, 510
Reyes, L. D., 498
Reynolds, F., 417, 444
Reynolds, V., 417, 444, 524
rhesus macaque, *Macaca mulatta,* 23–24, 364–366, 372, 383, 552, 578
rhinoceros, 95, 172, 293, 303, 326
Ricklan, D. E., 351
Rijksen, H. D., 286, 406, 411, 518, 521–522
Rilling, J. K., 358, 393, 514
Rinnie, 554, 555
Ristau, C. A., 388
Rizzolatti, G., 578
Robbins, A. M., 479

Robbins, C., 251
Robbins, D., 388
Robbins, M. M., 461, 480
Robinson, J. T., 129, 132, 133, 156, 158, 353
Rockel, A. J., 364, 485
Rodman, P. S., 401, 407, 408, 409, 411, 519
Rogers, C. M., 203, 279, 385
role model, 400
Romero, T., 431
Rook, L., 119, 131, 132
Rosas, A., 171
Rose, M. D., 84, 217, 219, 223
Rudapithecus hungaricus (= *Dryopithecus brancoi*), 77, 80, 96, 101–104, 112, 221–222
Ruff, C. B., 89, 173, 260
rules, 8, 34, 158, 365, 379, 393, 477, 556, 558, 564, 572, 581, 598
Rumbaugh, D. M., 380, 390–391, 558, 560, 563–564
running, 116, 190, 198, 210, 215, 225–226, 228–236, 239, 242–245, 248, 256, 259, 532, 540–541
Rusinga Island, Kenya, 74, 81, 83, 89, 227

Saadanioidea, 66, 95
Saadanius hijazensis, 66, 81, 95
Sabangau peat-swamp, 36, 415
Sabater Pí, J., 94, 287
sacrum, 216, 220, 222, 227, 230, 233
Sahelanthropus tchadensis, 81, 87, 126, 141–142, 162, 169, 357
Saimiri sp., 215, 372
Samburupithecus kiptalami, 78, 81, 87, 126–128
San, 7
San Diego Zoo, 380, 543, 546
San Francisco Zoo, 533
Sarah, 381–383, 389, 556–557, 561
Sarich, V., 137–138
Sarmiento, E. E., 223
Savage-Rumbaugh, E. S., 547, 557, 560–561, 563, 565
Scandentia, 16, 17
scar, 325, 490
scenario, 2, 4; aggressive male, 5–7, 10, 305, 592–595; androcentric, 6; bisexual sharing, 8–9, 294, 305; Dawn Man, 4–5; evolution of human bipedalism, 2, 6, 9, 189, 213, 225–228, 230, 237, 242–243, 248–249, 253, 304, 352–353, 356; evolution of speech, 183, 304, 365–366, 547, 569, 573–578, 582, 584; killer ape, 4–5, 10, 595; man-the-hunter, 4–8, 10–11, 304–305; meat as money, 9–10, 304; pair-bonded, nuclear family, 6–7, 9; Plio-Pleistocene hominid mating patterns, 152, 501; polygamy, 152, 498, 503; polygyny, 63,

70, 498–499, 503; scavenging, 4, 149, 325–326; scientifically informed, 2, 4; structure of past human societies, 6–7, 9; versus model, 4; woman the gatherer, 6–8
Schaller, G. B., 463–468, 471–472, 521, 536–537, 540–541
Schick, K., 334
Schleichermacher, 95
Schmid, P., 256
Schmitt, T. J., 147
Schön, M. A., 131
Schultz, A. H., 363
Schultz, M., 171
Schürmann, C. L., 412–413
Schwartz, J. H., 28, 119
science, 2, 11, 58, 61, 95, 571, 596, 598–599
science fiction, 4, 59, 590
science journalism(-ist), 591, 599
scientist, 4–6, 8, 33, 58, 61, 99, 153, 170, 317, 356, 379, 385–386, 392, 397, 463, 473, 508, 569, 570–571, 573, 584–585, 590, 596–597, 599, 600
scrotum, 17, 532
Seiffert, E. R., 215
Seifu, M., 288
Sein, C., 67
Selatan, 36, 371
self-concept, 387–388, 556
self-consciousness. *See* awareness
self-distraction: children, 384; chimpanzee, 384; self-identity (*see* awareness: self-)
self-recognition, 386–388, 592; age of emergence, 387; dye test, 386–388; relation to self-awareness, 387–388
semantics, 360, 517, 558, 562, 576–577, 579
semiology, 571
semiotics, 571
Semnopithecus entellus, 442
sense: chromatic, 265; haptic, 346, 385, 577; olfactory, 16, 64, 385, 482, 577; taste, 264, 266, 296; vision, 16, 19, 334, 358, 364, 372, 385, 386, 417, 515, 528, 530, 534, 541, 544, 577, 589–590
sentence, 34, 359, 391, 549, 557, 564, 570–571
sentential modifier, 558
Senut, B., 85, 128, 217
septuagenarian, 505
Sepulchre, P., 129
Serapia eocaena, 65, 72–73
sesamoid bone, 242
sex, 239, 304, 315, 379, 387, 394, 398, 400, 404–405, 407, 409–413, 427, 433–435, 437, 439, 443–444, 449, 453–459, 479–485, 487, 491, 502–503, 505–506, 520, 554, 582, 589; play, 413; solitary, 406

sexual coercion, 408, 411–412, 414, 428–429, 436, 456

sexual experimentation, 438, 481

sexual swelling. *See* perineal swell

Shapiro, G., 554, 555

Shellis, R. P., 41, 45

shelter, 353; archeological trace, 302; cave, 302, 499; fabricated, 3; from inclement weather, 294; rock, 302, 499

Sherman, 388, 560–565

Sherman-Austin study, 561, 563; results, 563

Sherwood, C. C., 5, 137, 373

Shipman, P., 326

short-chain fatty acid, 263–264, 288

siamang. *See Symphalangus syndactylus*

Sianopithecus socaenus, 62, 65, 67

Sianthropus (= Peking Man, Chinese *Homo erectus*). *See Homo erectus*

Siberut, 36, 268, 492

sibling species, defined, 585

Sicotte, P., 470–471, 480

sign, 5, 368, 385, 548–555, 557, 561–562, 570–572, 577–578, 583–584, 592; arbitrary, 549, 572–573; comprehension of meaning, 549; defined, 562; iconic, 549, 572–573; indexical, 572–573; sequences, 555; symbolic, 557, 562, 572

signal, 5, 119, 312, 359, 372, 386, 432, 471, 505, 507, 509, 511, 517, 521, 524, 527–530, 532, 535, 540, 542, 545–547, 558, 569, 571–572, 577–579, 581–582, 585, 590

sign-vehicle, 572

Silberbauer, G. B., 7

Silk, J. B., 432

silvery gibbon. *See Hylobates moloch*

simian, 15, 51, 136

Simiolus spp., 83, 86, 92–93, 120, 123

Simiolus enjiessi, 76, 81, 84

Simiolus leakeyorum, 81, 83, 93, 315

Simons, E. L., 64, 70, 72–73, 97, 102, 111, 115, 128, 134–137, 156

Simpson, G. G., 156, 422

singe(s), defined, 15

singing, 486, 488, 492, 494, 511–517, 584

Singleton, I., 403

sitting pad. *See* ischial callosity

Sivapithecus spp., 78, 81–82, 85, 105, 113, 122–125, 136; body size, 110; craniodental features, 102, 104, 106–108, 112, 117, 158; diet, 110–111; geochronological span, 104, 107, 117; habitat, 104, 107, 110, 114, 118, 124; members, 104, 107, 108; nominally confined to South Asia, 104; positional behavior, 223; postcranial anatomy, 220, 223; sexual dimorphism, 107

Sivapithecus africanus (= *Kenyapithecus africanus*), 104

Sivapithecus darwini, 117

Sivapithecus indicus, 103; dental sexual dimorphism, 107; thick enamel, 108

Sivapithecus parvada, 223; body mass, 110; habitat, 108; thick enamel, 108

Sivapithecus simonsi (= *Sivapithecus sivalensis*), 108

Sivapithecus sivalensis, 78, 81, 87, 108; airorhynchy, 110; as ancestral *Pongo* spp., 117–121; cranial features, 108–109, 112, 134; dental features, 108, 110; diet, 110–111; GSP-15000, 108; habitat, 107–108, 110; no frontal sinus, 108; positional behavior, 223; postcranial features, 110, 134; sexual dimorphism, 107; thick enamel, 111

skin pigmentation: bonobo, 47; chimpanzee, 47; gorilla, 47; human, 55, 56; orangutan, 43

sleep, 35, 192, 229; deprivation of, 294; diurnally, 294, 298

Slocombe, K. E., 528, 535

slow loris, 93, 214, 270

small intestine, 375; function, 262–263

smell, 266, 387, 552

Smith, H., 173, 508

Smithsonian Institution, 508

Smuts, B. B., 436

snake, 171, 265–266, 295, 529, 573

social facilitation, 589

social humanist, 356

sociality, 394, 403, 445, 449–450, 453, 497, 576

socialization: of humans, 393, 589; of nonhuman animals, 393

social scientist, 356

sociobiological theory, 407

sociobiology(-ist), 8, 356

sociocultural anthropology(-ist), 356, 587–588

sociologist, 587

sodium, 290, 462

soil, 265, 270, 274, 343

solicitation, 279, 312–313, 446, 456, 459, 468, 478, 480–481, 531, 544, 560

Sommer, I. E. C., 360

song tree, 513

sonogram, 516

sound spectrograph, 524, 543

South Pacific, 596

spear, 326, 341

spectrographic analysis, 524, 543

speech, 372, 379, 383, 392, 549, 562–564, 575, 583–584; analogue nature of, 571, 577; breath control, 584; evolution of, 365, 573–574, 576–578, 582; FOXP2 gene, 183, 366; function of abdominal muscles, 584; human capability,

365, 368, 570–571, 574; physical production, 359, 368, 569, 574; vocal apparatus, 569, 574
speed, 4, 242, 244, 249, 270, 288, 363, 485, 516, 558, 584; Olympian four-minute miler, 225; pedestrian walking, 225
sperm count, 414, 433, 483–484
spider monkey. See *Ateles* spp.
spinal cord, 359
Spoor, F., 149, 173
sports team, 525
spotted hyena. See *Crocuta crocuta*
sprint, 242, 248
squat, 205, 213, 229, 231, 239, 242
squirrel, 17, 93, 215, 267, 284, 307, 310, 321–322, 324, 372
squirrel monkey. See *Saimiri* sp.
Srikosamatara, S., 485–486
stable isotope analysis, 88, 100, 129, 262, 292–293
Stamenov, M. I., 578
Stanford, C. B., 10, 286, 308, 313–315, 553–554
Stanford-Binet Intelligence Test, 554
Stanford University, 553
statistical tests, 450
Stegadon, 171
Stephanoaetus coronatus, 316
Stevens, J. M. G., 451
Stewart, K. J., 537, 598
Stiner, M. C., 326
Stokes, E. J., 473–474
stomach contents, 262; function, 262–263, 375; sacculated, 22–23, 266; simple, 23, 263, 266
Stout, D., 358
straight-tusked elephant *(Elephas antiquus)*, 326
Straus, W. L. Jr., 131, 201
Strepsirhini, 69; members, 19
Stringer, C. B., 159, 162
Strix woodfordii, 307
strontium calcium ratios (Sr/Ca), 293
strontium isotopic ratios (^{87}Sr/^{86}Sr), 293, 498
student, 159, 162, 253, 379, 383, 387, 517, 537, 541, 557, 585
studies on prelinguistic human children, 383
Stumpf, R., 275, 430, 434
Suaq Balimbing, Gunung Leuser National Park, 36, 270, 274, 342, 403–404, 407, 414
subsistence pattern, 9
subspecies, 32–35, 38, 46–47, 179, 265, 275–276, 280, 520, 596, 599
substrate, 199, 205, 214, 218, 246, 248, 250, 256, 261, 263, 344, 347, 394, 499, 527, 577, 585
sugar, 263–266, 270–271, 282, 428; disaccharides (= sucrose), 262; monosaccharide (fructose and glucose), 262; quick source of metabolic energy,

263; simple, 263; soluble, 288–289; stored as glucose in liver, 263
Sugardjito, J., 199, 402
Sugiyama, Y., 424, 426
Sukhumi, Georgia, 33
sulcus (-i), 16, 104, 106, 108, 111, 359–360, 363, 369, 374, 576; impressions on endocasts, 374
Sumatran orangutan. See *Pongo abelii*
Sumatran tiger. See *Panthera tigris sumatrae*
Sumatran white-handed gibbon. See *Hylobates lar*
sunda clouded leopard. See *Neofelis diardi*
superior parietal lobe, 358
Surbeck, M., 321
surgical ablation study, 366
Susman, R. L., 352–353
suspended feeding, 190, 194, 196, 200–201, 215–216, 358; advantage of, 192–193
suspensory behavior. *See individual taxa*
Sus scrofa, 17, 148, 284, 305, 310, 341, 387
Suteethorn, V., 98
Suzuki, A., 286, 317, 418
Swartz, K. B., 387
Swisher, C. C., 165
Sylvette, 33
Sylvian fissure, 369
symbol, 26, 115, 375, 381–383, 385–386, 393, 556–558, 560–561, 563–564, 570–571, 573, 580, 586, 588–589, 590; communication with artifactual, 365; defined, 572–573; key to concept of culture, 585–587; neuroendorinological capacity for, 585, 589; phonological, 580
symbolically encoded meaning: neurological substrates, 585; nonhuman homology, 585
symbolic behavior, 317; honeybee, 386; signing chimpanzee, 386
symbolic capabilities, 562; communication, 8, 507, 563; language, 500, 565, 573; maps, 385; mediation, 365, 588–589, 597; niche, 590; process, 584; sign, 570, 572
symbolism: emergence of, 571; lexical, 570; reliance on, 587; unique to human, 393, 571
sympathy, 590
Symphalangus syndactylus, 27, 31, 35–36, 38, 285, 371, 380, 488, 509, 514–516; aggression, 285–286, 405; allogrooming, 286, 490–491; brain size, 362; cohesiveness, 296, 405, 487, 490, 515; craniodental features, 39–41; daily routine, 490; diet, 41, 267, 269, 285; females dispersal, 491; females leading, 491; food sharing, 324–325; group composition, 490–491; intergroup conflicts, 286, 492; intragroup aggression, 491–492; juvenile sleeps with male, 490; laryngeal sac, 35, 514; paternal care,

Symphalangus syndactylus (continued)
490–491; peripheralization, 491, 495; positional behavior, 237, 491; postcranial features, 39, 42, 223; roosting, 298; sexual behavior, 491; sexual dimorphism, 35, 362; sleeping associations, 296, 490; territoriality, 492; territory size, 285, 486
synapse, 391
syntax, 391, 547, 557–558, 561–562, 570–571, 573, 576–577, 579–580, 582–583
systematics, 60, 599
Szalay, F., 131–132

Tabarin mandible, 143
Tabuce, R., 69
tactical deception, 556; chimpanzee, 389; humans, nonhominoid primates, 355, 389, 392; gorilla, 389
Taieb, M., 147
tail, 5, 219–220, 324, 344; absence, 25, 86, 131, 216, 218, 222, 227; dermatoglyphics, 16, 21; prehensile, 21–22, 220
Takahata, Y., 456
Takai, M., 67
Talahpithecus parvus, 62, 66, 69
Talebia hammadae, 62, 64–65, 69
tamarin, 19
Tanaka, J., 7
Tanjung Puting, Kalimantan Tengah, 26, 399, 401, 403, 410–411, 518–522, 554
Tanner, N. M., 8
tannin, 264, 266, 271, 288, 289
Tantrum: chimpanzee, 528, 533; gorilla, 538; human, 528; orangutan, 399, 521, 528
taphonomic analysis, 6, 74
taphonomic processes, simulate artifactual cutmarks, 291, 332
taphonomy, 2, 60, 89, 111, 149, 183, 291, 332
Tardieu, C., 236
Tarsiidae, 19, 68, 74, 363
Tarzan, 33, 533
Tatu, 550, 552, 553
taxonomy, 72, 82–83, 98, 102, 107, 115, 136, 156, 160, 170, 185, 260, 370, 403; Linnaean system, 15, 27, 34; subspecific classification, 32
Teaford, M., 111
Te Boekhorst, I. J. A., 404
tectonic, 129, 343, 401
Telanthropus capensis (= Homo erectus), 156
Teleki, G., 305
television, 384–385, 388
temporal lobe, 360, 369, 374
Tenaza, R. R., 295, 492, 494–495, 512, 514, 530

tendon: Achilles (= tendo calcaneus), 199, 234; flexor digitorum profundus, 228, 348–349; tendo calcaneus (= Achilles tendon), 199, 234
Tenerife, Canary Islands, 378
termite fishing, 265, 335–337, 339–340, 370
Terrace, H. S., 165, 556–557, 563
terrestrial herbaceous vegetation (THV), 275–277, 279, 281–283, 287–290, 301, 453, 544
territoriality, 190, 319, 422, 424–425, 429, 464, 468, 485–489, 492–494, 509, 511, 513–515, 517, 542
testes (= testicle), 17, 499; bonobo, 54, 484; chimpanzee, 54, 484; gibbon, 499; gorilla, 483; human, 54, 499; orangutan, 484
testosterone, 324, 428, 454, 484
theory, 7, 60, 179–180, 226–227, 407, 479, 587; cost-benefit, 593; hominid language origin, 577–578; of mind, 355–356, 389, 392; regional-transition (= African mother hypothesis), 179–180; single origin, 179
thermoregulation, 229, 230
Theropithecus gelada, 22
think, 385–386, 389–390
Thomas, H., 73, 137
Thompson-Handler, N., 457
Thorpe, S. K. S., 199
throw, 22, 351, 353, 358, 371; bonobo, 333–334, 343, 371; chimpanzee, 332, 341, 371, 541; orangutan, 23, 371
tibial plateau, 256
tickle, 529, 552
Tilson, R. L., 295, 493–495
Tobias, P., 6, 147–148, 156–157, 159–160, 575
Tocheri, M. W., 353
Tomasello, M., 306, 389, 586, 587
tool (= artifact), 6, 32, 135, 325–326, 331; Acheulean cleaver, 165, 179, 325, 354; bifacial and polyfacial cores, 334; bone, 5, 7, 327, 577; chimpanzee communication about, 561; container, 8, 264; daily importance of, 331, 375; debitage, 334; definition, 331; hammer, 333, 337–339, 341–342, 344–345, 351; handaxe, 7, 165; Herto, 178–179; important cultural invention, 375; Lehringen yew wood lance, 326; lever, 234–235, 242, 332, 335, 558, 573; manufacture, 3, 342, 345, 348, 353, 577; Mousterian lithics, 183, 326; Oldowan assemblage, 171–172, 334, 352, 354; Oldowan Dmanisi, 171–172, 334, 352, 354; polish and wear patterns, 325; sling, 6; stick, 7, 231, 288, 307–308, 332, 335, 337–339, 342–343, 353, 384, 562; stone, 6, 9, 168, 170–171, 226, 292–293, 325, 332–334, 337, 339, 340,

343–345, 351–354, 358, 375; stone, produced by Early and Middle Pleistocene hominids, 372; stone, right hand preferred for knapping flakes, 372; wipe, 332, 336, 342

tool behavior, 115, 331; adaptive advantage of, 229, 335; agonistic versus non-agonistic, 335; apes compared, 335, 342–343, 585; chimpanzee, 288–289, 307–308, 332, 335–341, 387, 550, 560–561; as culture, 356; gorilla, 264, 335, 343; hylobatid, 335; imitation, 387; as indicator of intelligence, 331–332, 365, 375, 378, 394, 500, 574; monkey, 343–344; ontogeny of, 293, 345, 375, 394, 500, 574, 577–578; orangutan, 270, 333, 342

tool-making, 345, 348, 351–352, 358, 390, 577; bonobo, 333–334; chimpanzee, 332, 335–339, 577; in Late Pliocene, 293; orangutan, 333, 342; sex of maker, 8, 11, 341

tool use, 225–226, 304, 345, 352–353, 358, 369, 390, 577; bonobo, 333–335, 343; chimpanzee, 288–289, 307–308, 332, 335–341, 387, 561, 577; earliest stages of stone, 292, 325; orangutan, 270, 333, 342; sex of user, 8, 11, 341

tooth enamel, 41, 45, 57, 63, 66–67, 69, 87–88, 92–93, 97, 100, 102, 104, 106–108, 111, 113–114, 117, 129, 133, 136, 149, 158, 291–292, 298

torturer, 379

Tóth, H., 377

Toth, N., 333–334, 372

Townsend, S. W., 437

toxin, 262, 266, 271, 275

toy, 550, 558, 563

tradition: in chimpanzees, 433, 585; in gorillas, 541; in humans, 481, 487

transfer index, 367, 380–381; defined, 380

transitive inference, 382, 392

trap, 327

tree hyrax *(Dendrohyrax dorsalis)*, 320

tree shrew (Tupaiidae), 16–17

trek, 229–230

Trinkaus, E., 352

Trivers, R. L., 407

tuition, 548, 589–590, 594

Tupaiidae (= tree shrew), 16–17, 34

Turciops truncates, 387

Turkanapithecus kalakolensis, 75, 81, 84, 86; brow ridge, 91; compared with *Pliopithecus,* 217; craniodental feartures, 91–92; elbow complex, 217; positional behavior, 217; postcranial features, 216–218, 221; similar to *Alouatta,* 217; skull KNM-WK 16950, 91

tutelage, 392

Tutin, C. E. G., 287, 301, 439, 460

Tuttle, R. H., 203, 249

twins, 429, 434, 487

Twycross Zoo, Leicestershire, United Kingdom, 456

Tyler, D. E., 99

Tylor, E. B., 588

Uchida, A., 82

Udabnopithecus garedziensis, 79–80, 107

Uehara, S., 265, 423

Ugandapithecus major (= Proconsul major), 85, 217

Ulu Segama Reserve, 36, 273, 400–401

underground storage organs, 275, 291, 293; tubers, 106, 149, 337

Ungar, P., 41, 44, 98, 128

United States, 1, 33–34, 305

unit-group: basic social unit of bonobo, 444–449, 452, 455; basic social unit of chimpanzee, 418–420, 423–425, 525, 545

unit-group (= community), 418. See also *Pan* spp.

University of Tennessee, Chattanooga, 555

Ursus maritimus, 134; body mass, 134

Utami, S. S., 402, 415

Varanus komodoensis (= Komodo dragon), 171

Varley, R. A., 383

Vedder, A. L., 279, 281

Vekua, A., 171

velum palatinum, 569

verbalization, 385–386

Verhaegen, M. J., 158

vermiform appendix, 25

versatile climbing, 132, 194, 196, 199, 203, 212–213, 215, 217–218, 220–224, 228, 285, 348

vertical climbing, 190, 194, 198–199, 201, 203, 204, 212, 214, 218, 221–222, 226, 228, 235, 237, 239, 246, 254, 303, 348

vervet. See *Cercopithecus aethiops*

Victoriapithecus spp., 72, 85, 92–94, 116

Victoriapithecus macinnesi, 83, 218–219

videotape, 382, 389, 543, 556

Vietnam, 35, 42, 125, 132, 305, 599; conflagration, 5

vigilance behavior, 229; male bonobo, 449

Viki, 529, 548

vines, 194, 196, 201, 254, 268, 270, 279, 281, 295, 300, 335, 337, 347, 351

Viret, J., 129

visceral attachment in abdominal cavity, 27, 227

visual perception, 528

vitamins, 262

vocabulary, 27, 189
vocal tract, 542, 570, 577; cords, 569; refinement, 584
Vogel, E. R., 45
voxel-based morphometry, 391
Vrba, E. S., 7

wadge, 266, 312, 317, 337, 387
wading, 32, 229; gorilla, 343
waggle dance of bees as indexical icons, 572
Walker, A., 84, 89, 111, 143, 173
Wallace, A. R., 170, 577
Waller, B. M., 509
Walter, R. C., 147
Wang, J., 68
war, 4, 10, 33, 54, 59, 137, 320, 536, 593–595
Ward, C. V., 110, 217, 254
Ward, S., 85, 147
Washburn, S. L., 5–6, 137, 156, 190, 226–228, 390
Washoe, 548–561; understand spoken English, 553
water, 7, 32, 58, 94–95, 104, 245–246, 254, 261–262, 282, 336–337, 342–343, 371, 379, 549, 552, 559
Watts, D. P., 279, 281, 432, 437, 468–469, 471–472, 481
Way Canguk Research Station, 491
weaning, 291, 294, 504, 506; bonobo, 544; chimpanzee, 420, 443, 528; gorilla, 460–461, 465, 470, 472, 474, 538; orangutan, 399–400; siamang, 490–491
weapon, 4–6, 134, 225, 304, 332; hafting, 327; long-range, 326; manual, 591; projectile, 229, 327, 517, 591
weaver ants, as food species, 288
Weidenreich, F., 132
Wernicke's area, 369, 575; connection with Broca's area, 360–361; function, 368, 372
Western chimpanzee. See *Pan troglodytes*
Western gorilla. See *Gorilla gorilla gorilla*
Western hoolock gibbon. See *Hoolock hoolock*
Western Javan silvery gibbon. See *Hylobates moloch*
Westminster Abbey, 5
Whitcome, K. K., 258
White, D., 553
White, F. J., 450, 452, 453
White, T. D., 128, 147–149, 152–153, 323
white-faced saki *(Pithecia pithecia)*, 92
Whitten, A. J., 295–296, 493, 514
Wich, S. A., 399, 518
Williams, J. M., 279, 429, 504
Williamson, E. A., 279
Wilson, A. C., 137–138

Wilson, E. O., 407
wipe, 332, 336, 342
wolf (= *Canis lupus*), 172, 244, 595
Woman That Never Evolved, The, 8
Woo, 129, 132–133
Wood, B. A., 156, 159, 436
Woodruff, G., 389
Woodside, California, 553
woolly monkey. See *Lagothrix*
Woranso-Mille craniodental fragments, 149
word, 137, 359–360, 368, 383, 391, 548–550, 552, 554–555, 557, 560, 562, 564–565, 570–571, 573, 576, 579–581, 583
World Heritage Site, 600
World's Top 25 Most Endangered Primates 2010–12, 600
wound, 319, 341, 408, 422, 469; bite, 431, 471–472
Wrangham, R. W., 10, 213, 282, 314, 407, 423–424, 427, 453, 594–595

Xenopithecus, 83
Xu, Q., 111, 113, 118, 136
Xue, X. X., 104

Y-5 *(Dryopithecus)* pattern, 25, 41–42, 52, 91, 133
Yale Laboratories of Comparative Psychobiology 368
Yale Laboratories of Primate Biology, Orange Park, Florida, 378
Yale University Primate Facility, New Haven, Connecticut, 378
Yamagiwa, J., 281, 472
Yankee, 548; diet, 305; flag, 572
yawn, 522, 530
Y-chromosome analysis, 462
Yerkes, A. W., 379, 517, 541
Yerkes, R. M., 33–34, 378–379, 397, 517, 541, 548
Yerkes National Primate Research Center of Emory University, Atlanta, Georgia, 363, 380, 431, 543, 555, 558
Yerkish, 558, 560
Yunnan black (crested) gibbon. See *Nomascus concolor*

Zalmout, I. S., 95
Zambia, 172, 259
Zhang, X. Y., 136
Zhang, Y., 133
Zihlman, A. L., 8, 363
Zinjanthropus boisei (= *Paranthropus boisei*), 156
Zuberbühler, K., 535